Fruit Oils: Chemistry and Functionality

Mohamed Fawzy Ramadan

Editor

Fruit Oils: Chemistry and Functionality

 Springer

Editor
Mohamed Fawzy Ramadan
Agricultural Biochemistry Department
Faculty of Agriculture
Zagazig University
Zagazig, Egypt

ISBN 978-3-030-12475-5   ISBN 978-3-030-12473-1   (eBook)
https://doi.org/10.1007/978-3-030-12473-1

This Springer imprint is published by the registered company Springer Nature Switzerland AG
The registered company address is: Gewerbestrasse 11, 6330 Cham, Switzerland

Preface

Over the last years, oils and fats production from fruits (fruit seeds, fruit pulp, fruit peel, and fruit pomace) has increased significantly to become one of the most important global sources of specialty oils/fats. New and superior nontraditional fruit varieties have been developed and released in main oil-bearing plants in several countries.

It is acknowledged that the contribution of fruit oils and fats to our health and well-being is recognized by their chemical composition. Fatty acid profile (i.e., the relative levels of saturated, monounsaturated, and polyunsaturated fatty acids) and a wide range of specific bioactive lipids (i.e., polar lipids, tocols, sterols, phenolics) have been shown to affect the biological functions of our bodies.

Upon invitation by the Springer Nature, this book was edited, which contains several chapters that describe different fruit oils and fats. With the aim at providing major reference work for those involved with the oils and fats industry as well as undergraduate and graduate students, this volume presents a comprehensive review of the results that have led to the advancements in fruit oils production, processing, functionality, and applications. As possible, the chapters followed a similar outline describing properties and processing of fruit oils and fats, with a focus on the extraction, chemical composition, quality of different fats and oils, and applications of fruit oils and fats in food and nonfood applications as well as nutraceutical products. I hope that the book will be a valuable source for people involved in oils and fats applications.

I sincerely thank all authors for their contributions and for their cooperation during the book preparation. The help and support given to me by the Springer Nature staff, especially *Daniel Falatko* and *Sofia Priya Dharshini*, was essential for the completion of my task and is appreciated.

Zagazig, Egypt   Mohamed Fawzy Ramadan

Fruit Oils: Chemistry and Functionality

Description

Fruit Oils: Chemistry and Functionality reported on the fruit oils and fats currently in use in food and nonfood applications, as well as those with significant commercial potential. Fruit oils and fats have an increasing number of applications in the food and pharmaceutical industry, due to the increase interest in "clean label" novel foods and the emerging markets in "free-from" and specialist foods.

Fruit Oils: Chemistry and Functionality covers several topics with a focus on lipid chemistry, technology, and functionality. The book covers specific topics including properties and processing of fruit oils and fats, extraction, quality of different fats and oils, and applications of fruit oils and fats in food and nonfood applications as well as nutraceutical products.

Fruit Oils: Chemistry and Functionality is a key text for functional food developers as well as research and development (R&D) managers working in all sectors using fats and oils. It is a useful reference work for companies reformulating their products or developing new products, as well as academics and students with a research interest in the area, such as lipid scientists, food scientists, and horticulturists.

Key Features

- Broad coverage encompasses traditional and nontraditional fruit oils and fats.
- Authored by international academics and industry experts.
- Addresses growing application areas including functional foods, pharmaceuticals, nutraceuticals, and cosmetics.

Readership

- Academics and students with a research interest in the area (food chemists, lipid scientists, food scientists, horticulturists, and agronomists)
- Functional food developers and R&D managers working in all sectors using fats and oils

Contents

Contributors

Ali Abbas Department of Chemistry, Government Postgraduate Taleem-ul-Islam College, Chenab Nagar, Chiniot, Pakistan

Anjana Adhikari-Devkota Graduate School of Pharmaceutical Sciences, Kumamoto University, Kumamoto, Japan

Ahmad Adnan Department of Chemistry, Government College University, Lahore, Pakistan

Naveed Ahmad Department of Chemistry, University of Education, Faisalabad, Pakistan

Sumia Akram Department of Chemistry, Minhaj University, Lahore, Pakistan
Department of Chemistry, Kinnaird College for Women University, Lahore, Pakistan

Farooq Anwar Department of Chemistry, University of Sargodha, Sargodha, Pakistan

Anam Arain National Centre of Excellence in Analytical Chemistry, University of Sindh, Jamshoro, Pakistan

Rizwan Ashraf Department of Chemistry, University of Agriculture Faisalabad, Faisalabad, Pakistan

Hamide Filiz Ayyildiz Faculty of Pharmacy, Department of Basic Pharmaceutical Sciences, Selcuk University, Konya, Turkey

Tarun Belwal G. B. Pant National Institute of Himalayan Environment and Sustainable Development, Almora, Uttarakhand, India

Dhaka Ram Bhandari Institute of Inorganic and Analytical Chemistry, Giessen, Germany

Ronnie Böck University of Namibia, Windhoek, Namibia

Zoubida Charrouf Université Mohammed V-Agdal, Laboratoire de Chimie des Plantes, Synthèse Organique et Bioorganique, Faculté des Sciences, Rabat, Morocco

Ahmad Cheikhyoussef University of Namibia, Windhoek, Namibia

Natascha Cheikhyoussef Ministry of Higher Education, Training and Innovation, Windhoek, Namibia

Sook Chin Chew Department of Food Science and Nutrition, Faculty of Applied Sciences, UCSI University, Kuala Lumpur, Malaysia

Sofien Chniti Ecole des Métiers de l'Environnement, Bruz, France

Université de Rennes 1, ENSCR, CNRS, Rennes, France

Monika Choudhary Punjab Agricultural University, Ludhiana, India

Slavica Čolić Institute for Science Application in Agriculture, Belgrade, Republic of Serbia

Caroline Mariana de Aguiar Federal Technological University of Parana, Toledo, Paraná, Brazil

Frédéric Debaste Transfers, Interfaces and Processes- Chemical Engineering Unit, Ecole polytechnique de Bruxelles, Université libre de Bruxelles, Brussels, Belgium

Amélia M. Delgado MeditBio – Centre for Mediterranean Bioresources and Food, University of Algarve, Faro, Portugal

Hari Prasad Devkota Graduate School of Pharmaceutical Sciences, Kumamoto University, Kumamoto, Japan

Alessandra Durazzo CREA Research Centre for Food and Nutrition, Rome, Italy

Milica Fotirić-Akšić University of Belgrade, Faculty of Agriculture, Belgrade, Republic of Serbia

Giuseppe Fregapane Faculty of Chemistry, Castilla-La Mancha University, Ciudad Real, Spain

Ines Gharbi Laboratory LR12ES05 Lab-NAFS 'Nutrition – Functional Food & Vascular Health', Faculty of Medicine, University of Monastir, Monastir, Tunisia

Hasanah Mohd Ghazali Department of Food Science, Faculty of Food Science and Technology, Universiti Putra Malaysia, Serdang, Selangor, Malaysia

Annalisa Giovannini CREA Research Centre for Vegetable and Ornamental Crops, Sanremo, Italy

Kiran Grover Department of Food and Nutrition, College of Home Science, Punjab Agricultural University, Ludhiana, India

Dominique Guillaume Faculté de Médecine/Pharmacie, URCA, Reims, France

Mohamed Hammami Laboratory LR12ES05 Lab-NAFS 'Nutrition – Functional Food & Vascular Health', Faculty of Medicine, University of Monastir, Monastir, Tunisia

Syeda Mariam Hasany Department of Chemistry, Kinnaird College for Women University, Lahore, Pakistan

Muhammed Imran Institute of Home and Food Sciences, Faculty of Life Sciences, Government College University Faisalabad, Faisalabad, Pakistan

Saira Ishaq Department of Chemistry, Government College University, Lahore, Pakistan

Manel Issaoui Lab-NAFS 'Nutrition – Functional Food & Vascular Health', Faculty of Medicine, University of Monastir, Monastir, Tunisia

Faculty of Science and Technology of Sidi Bouzid, University of Kairouan, Sidi Bouzid, Tunisia

Monia Jemni Laboratoire Technologies de dattes, Centre Régional de Recherches en Agriculture Oasienne Degueche, Tozeur, Tunisie

Martha Kandawa-Schulz University of Namibia, Windhoek, Namibia

Aftab Ahmed Kandhro Dr. M.A. Kazi Institute of Chemistry, University of Sindh, Jamshoro, Pakistan

Erkan Karacabey Faculty of Engineering, Department of Food Engineering, Suleyman Demirel University, Isparta, Turkey

Huseyin Kara Faculty of Science, Department of Chemistry, Selcuk University, Konya, Turkey

Johannes Kiefer Technische Thermodynamik, Universität Bremen, Bremen, Germany

Mustafa Kiralan Faculty of Engineering, Department of Food Engineering, Balıkesir University, Balıkesir, Turkey

S. Sezer Kiralan Faculty of Engineering, Department of Food Engineering, Balıkesir University, Balıkesir, Turkey

L. Klavina University of Latvia, Riga, Latvia

M. Klavins University of Latvia, Riga, Latvia

Abdul Hameed Kori National Centre of Excellence in Analytical Chemistry, University of Sindh, Jamshoro, Pakistan

Erdogan Kucukoner Faculty of Engineering, Department of Food Engineering, Suleyman Demirel University, Isparta, Turkey

Zahid Husain Laghari National Centre of Excellence in Analytical Chemistry, University of Sindh, Jamshoro, Pakistan

Massimo Lucarini CREA Research Centre for Food and Nutrition, Rome, Italy

Sarfaraz Ahmed Mahesar National Centre of Excellence in Analytical Chemistry, University of Sindh, Jamshoro, Pakistan

Huynh Cang Mai Department of Chemical Engineering, Nong Lam University, Ho Chi Minh City, Vietnam

Clayton Antunes Martin Federal Technological University of Parana, Toledo, Paraná, Brazil

Q. Maxhuni Kosovo Institute for Nature Protection, Prishtine, Kosovo

Mohamed Elwathig Saeed Mirghani Department of Biotechnology Engineering, Kulliyyah of Engineering, International Islamic University Malaysia (IIUM), Kuala Lumpur, Malaysia

Adel Abdel Razek Abdel Azim Mohdaly Food Science and Technology Department, Faculty of Agriculture, Fayoum University, Fayoum, Egypt

Jörg-Thomas Mörsel UBF-Untersuchungs-, Beratungs-, Forschungslaboratorium GmbH, Altlandsberg, Berlin, Germany

Muhammad Mushtaq Department of Chemistry, Government College University, Lahore, Pakistan

Muhammad Nadeem Faculty of Animal Production and Technology, University of Veterinary and Animal Sciences, Lahore, Pakistan

Maja Natić University of Belgrade, Faculty of Chemistry, Belgrade, Republic of Serbia

Omprakash H. Nautiyal Department of Chemistry, Lovely Professional University, Chaheru, Punjab, India

Kar Lin Nyam Department of Food Science and Nutrition, Faculty of Applied Sciences, UCSI University, Kuala Lumpur, Malaysia

Rosa M. Ojeda-Amador Faculty of Chemistry, Castilla-La Mancha University, Ciudad Real, Spain

Ali Osman Agricultural Biochemistry Department, Faculty of Agriculture, Zagazig University, Zagazig, Egypt

M. Mustafa Ozcelik Faculty of Engineering, Department of Food Engineering, Suleyman Demirel University, Isparta, Turkey

Gülcan Özkan Faculty of Engineering, Department of Food Engineering, Suleyman Demirel University, Isparta, Turkey

Daniel Pioch CIRAD, UR BioWooEB Biorefinery Team, Montpellier, France

Antonio Raffo CREA Research Centre for Food and Nutrition, Rome, Italy

Mohamed Fawzy Ramadan Agricultural Biochemistry Department, Faculty of Agriculture, Zagazig University, Zagazig, Egypt

Rr. Ferizi University of Prishtina "Hasan Prishtina", Faculty of Medicine, Paramedical Department, Prishtine, Kosovo

Maria Desamparados Salvador Faculty of Chemistry, Castilla-La Mancha University, Ciudad Real, Spain

Sílvio César Sampaio State University of Western Paraná, Cascavel, Paraná, Brazil

Kátia Andressa Santos State University of Western Paraná, Toledo, Paraná, Brazil

Syed Tufail Hussain Sherazi National Centre of Excellence in Analytical Chemistry, University of Sindh, Jamshoro, Pakistan

Sirajuddin National Centre of Excellence in Analytical Chemistry, University of Sindh, Jamshoro, Pakistan

Said Saad Soliman Department of Horticultural Crops Technology, National Research Centre, Cairo, Dokki, Egypt

Bushra Sultana Department of Chemistry, University of Agriculture Faisalabad, Faisalabad, Pakistan

Chin Xuan Tan Department of Food Science, Faculty of Food Science and Technology, Universiti Putra Malaysia, Serdang, Selangor, Malaysia

Seok Shin Tan Department of Nutrition and Dietetics, School of Health Sciences, International Medical University, Bukit Jalil, Kuala Lumpur, Malaysia

K. Thirugnanasambandham Department of Chemistry, ECET, Coimbatore, Tamil Nadu, India

Mustafa Topkafa Vocational School of Technical Sciences, Department of Chemistry and Chemical Technologies, Konya Technical University, Konya, Turkey

Gurcan Yildirim Department of Mechanical Engineering, Abant Izzet Baysal University, Bolu, Turkey

Gordan Zec University of Belgrade, Faculty of Agriculture, Belgrade, Republic of Serbia

About the Editor

Mohamed Fawzy Ramadan is a Professor in the Agricultural Biochemistry Department, Faculty of Agriculture, at Zagazig University in Zagazig, Egypt. Professor Ramadan obtained his Ph.D. (*Dr. rer. nat.*) in Food Chemistry from Berlin University of Technology (Germany, 2004). Professor Ramadan continued his postdoctoral research in ranked universities in different countries such as the University of Helsinki (Finland), the Max Rubner Institute (Germany), Berlin University of Technology (Germany), and the University of Maryland (USA). In 2010, he was appointed to be Visiting Professor (100% research) at King Saud University in Saudi Arabia. In 2012, he was appointed to be Visiting Professor (100% teaching) in School of Biomedicine, Far Eastern Federal University, Vladivostok, Russian Federation. Since 2013, he is a Research Consultant and Professor of Biochemistry in the Deanship of Scientific Research at Umm Al-Qura University in Saudi Arabia.

Professor Ramadan published more than 180 research papers and reviews in international peer-reviewed journals as well as several books and book chapters (Scopus *h*-index is 32 and more than 3000 citations). He was an invited speaker at several international conferences. Since 2003, Prof. Ramadan is a Reviewer and Editor in several highly cited international journals such as the *Journal of Medicinal Food*.

Professor Ramadan received the Abdul Hamid Shoman Prize for Young Arab Researcher in Agricultural Sciences (2006); Egyptian State Prize for Encouragement in Agricultural Sciences (2009); European Young Lipid Scientist Award (2009); AU-TWAS Young Scientist National Awards (Egypt) in Basic Sciences, Technology, and Innovation (2012); TWAS-ARO Young Arab Scientist (YAS) Prize in Scientific and Technological Achievement (2013); and Atta-ur-Rahman Prize in Chemistry (2014).

Part I
General Aspects

Chapter 1
Chemistry and Functionality of Fruit Oils: An Introduction

Mohamed Fawzy Ramadan

Abstract Oil-bearing fruits are of great economic importance and nutritional value. The contribution of lipid-soluble phytochemicals to nutrition and health is acknowledged by their composition. Fatty acids and minor bioactive lipids (i.e., glycolipids, phospholipids, tocols, phytosterols, aroma compounds and phenolic compounds) exhibited health-promoting and functional properties in several products. Oils (fixed or essential) and fats extracted from fruit parts (fruit seeds, fruit pulp, fruit peels, fruit pomace, and nuts) generally differ in terms of their major and minor constituents. The methods used to extract oils and fats as well as the processing techniques such as refining, bleaching and deodorization affect their major and minor constituents. In addition, different post-processing treatments (i.e. heating and storage) of fruit oils may alert or degrade important bioactive constituents. This book aims at creating a multidisciplinary forum of discussion on recent advances in chemistry and functionality of fruit oils/fats mainly focusing on their chemical composition, physicochemical characteristics, organoleptic attributes, nutritional quality, oxidative stability, food/non-food applications as well as functional and health-promoting properties.

Keywords Fruits · Lipids · Oils · Fats · Antioxidants · Fatty acids

1 Challenge of Fruit Oils Production

In 2050, it is expected that the world population will reach 9.1 billion (United Nations 2017). More than 30% of food will be necessary compared to the current days, while several challenges are coming, specially those related to food security (Antónia Nunes 2018). Over the last years, the demand for higher amounts of food production has been increased, which has been raising a challenge to agriculture. The challenge of the future is the adequate food production for an increasing

M. F. Ramadan (✉)
Agricultural Biochemistry Department, Faculty of Agriculture, Zagazig University, Zagazig, Egypt
e-mail: mframadan@zu.edu.eg

© Springer Nature Switzerland AG 2019   3
M. F. Ramadan (ed.), *Fruit Oils: Chemistry and Functionality*,
https://doi.org/10.1007/978-3-030-12473-1_1

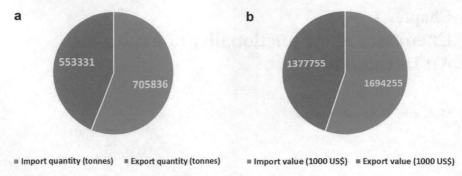

Fig. 1.1 International records for edible oils trades (**a**) quantity and (**b**) value in 2016. (http://www.fao.org/faostat/en/#data)

population using a decreasing area of arable land and the keeping of food production in harmony with the environment.

Production of oils/fats from different sources is important to provide edible oil or to be applied in technical applications (Matthäus 2012). According to FAOSTAT (http://www.fao.org/faostat/en/#data) in 2016, the international market recorded a high increase in the edible oils market (Fig. 1.1). The oil-bearing fruits become more important because an increasing number of people need more oils/fats for nutrition, pharmaceuticals and technical applications. Many fruit plants are highly seasonal so for maximum utilization of a processing plant, it would be important to be able to process multiple varieties. It is essential that the cost of fruit oils production remain reasonable once a successful recovery operation established. Oils recovered from fruits or fruit processing by-products have characteristics that make them desirable in cosmetics and medical formulations. Such specialized application may command a premium price (Weihrauch and Teter 1994; Matthäus 2012).

Modern fruit processing techniques produce significant amounts of by-products that are now used as fertilizer or just dumped creating pollution problems. As the cost of handling waste increases, a point might be reached where it becomes economically feasible to recover oils from the fruit wastes. The recovery of edible oil from fruit by-products (pomace) increases the source of the edible commodity and turns potential pollutants, which might be costly to control, into a useful value-added materials. Several things are important to ensure profitability with the processing of fruit by-products for oils/fats and other edible products. The by-product should be available in a highly localized area and the by-products should be stable so that it could be stored for short time without degradation (Weihrauch and Teter 1994).

2 Extraction of Fruit Oils

Different extraction methods could be applied for fruit oils recovery including solvent extraction, pressing, and a combination of both techniques. The efficiency of these methods could be improved with the assistance of microwave, enzymes or carbon dioxide (Ramadan and Moersel 2009). The aim of the extraction is to optimize the oil yield while maintaining the oil quality. Not only the removal of contaminants or substances affecting the sensory quality or storage stability of the oil are important, but also harmful compounds formed during the processing such as *trans* fatty acids (Matthäus 2012).

The handling of the raw material after harvest strongly depends on the type of raw material. Under appropriate conditions, fruits such as palm, olives, or avocado have to be processed rapidly after harvest because degradation processes start directly, impairing the oil quality with regard to sensory quality and storage stability. Mechanical extraction, a very old method for oil processing is still used for fruit oils or specialty oils. This results in a crude oil that has to run through different steps of a refining process to make the oil usable (Matthäus 2012).

Most of oil-bearing fruits consist of fruit flesh (mesocarp) and seed kernel, which contain oils or fats. In the international market, the most commonly processed oil-bearing fruits the olive (*Olea europaea*), and palm tree (*Elaeis guineensis*). The production of olive oil is a very old technique and the basic steps have not really changed over hundreds of years. The use of a solvent is not allowed for the production of extra virgin olive oil (EVOO) or other virgin oils and it is not useful for the production of fruit oils. From this high susceptibility of the oil-bearing raw material, it could be concluded that it is necessary to maintain the good quality of fruits after harvest until processing, which means for oil-bearing fruits such as palm fruits or olives the processing of the matured fruits has to be done in a very short time. For example, palm fruits are harvested manually, whereas pickers test the fruit ripeness before they cut them from the tree. Very important is how fast the fruits come to the process, because the fruits contain high amounts of water that favor the condition for metabolic processes if the involucre is destroyed. In the case of palm oil extraction, lipases have to be inactivated directly after harvest, because with progressing maturity of the fruits enzyme and substrate come together, leading to the degradation of triacylglycerols. In addition, microorganisms cause oil degradation through an attack on bruised and overripe fruits, thus causing an increase in the free fatty acid (FFA). To avoid this degradation, fruits have to be sterilized or cooked before processing (Matthäus 2012).

Many fruit oils retain their flavor and taste because they are produced at low quantities and are not processed by traditional refining processes usually used to remove impurities and extend the oil shelf life (McKevith 2005). Some fruit oils are produced by cold-pressing, but other extraction methods are also applied. The unique flavors of some fruit seed oils, such as pumpkin seed oil, are generated either by toasting the seeds or by treating the oils at high temperatures. In this book, some chapters describe the different extraction techniques used for fruit oils processing and the different requirements necessary for the production of high-quality oils. The book also discusses the different aspects of fruit oil processing from the pretreatment of the raw material to the purification steps.

3 Chemistry and Functionality of Fruit Oils

Fruit oils originate from diverse botanical families and are considerably different in their chemical compositions. Fruit oils are recovered from various fruit parts that include seeds, nuts, and mesocarps. Most of the fruit oils are extracted from fruit seeds where they serve as a storage form of energy and carbon, which are used when the seeds germinate. However, the most commercially important fruit oils (i.e. olive oil and avocado oil) are produced from fruits mesocarp that surround the seeds.

Most of the fruit oils are characterized by their aroma and taste, mainly resulting from the fact that those oils are not refined. According to FAOSTAT (http://www.fao.org/faostat/en/#data) the most internationally popular fruit oils are palm and olive oils (Fig. 1.2). Other commercially produced fruit oils/fats that found in the international market include avocado oil and some tree nut oils. In this volume conventional fruit oils/fats (i.e., olive oil, and palm oil) and non-conventional fruit oils/fats are included because each of them has a definite chemical composition, functionality,

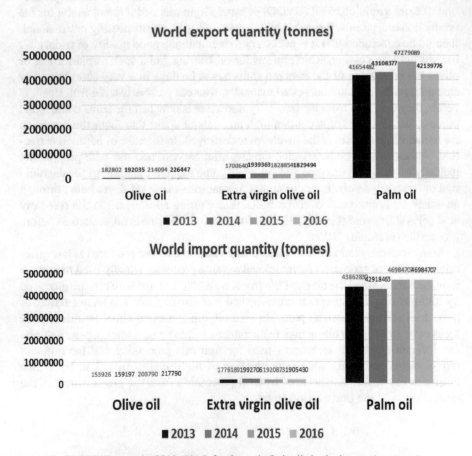

Fig. 1.2 FAOSTAT records (2013–2016) for the main fruit oils in the international market

health-promoting traits, and applications that will be described in detail in the respective chapters.

Although the fatty acid composition of oils/fats is the property that receives much attention, a number of minor lipid-soluble components are known to have a significant contribution to the oil nutritional properties. Fruit oils are convenient sources of the essential fatty acids and the other fatty acids derived from them. It is acknowledged that an adequate balance of essential fatty acids is needed to serve as precursors to produce prostaglandins and leukotrienes, in order to provide optimal pro-inflammatory and anti-inflammatory status in the human body. On the other hand, crude and unrefined oils/fats retain their non-acyl lipid constituents (unsaponifiables). These include lipid-soluble vitamins, polar lipids, tocols, phytosterols, phenolics and aroma compounds. For example, the health benefits of EVOO are attributed to the synergistic effects of oleic acid, squalene, and phenolic compounds (Moreau and Kamal-Eldin 2009). It is well known that lipid-soluble bioactive compounds have numerous health-promoting impacts. Therefore, fruit oils/fats rich in lipid-soluble bioactive compounds are considered novel foods. In addition, the oxidative stability of oils/fats is mainly affected by the presence of the anti- and pro-oxidant compounds (Gunstone 2006). These components are especially important in the case of unrefined fruit oils. Most of the chapters in this book reviewed the profiles of fatty acid and bioactive lipids in each fruit oil or fat as well as the influence of this specific composition on the stability, functionality, and applications of each oil.

Most of the fruit oils are characterized by (1) gentle processing (i.e., cold pressing and no refining), (2) unique aroma and health-promoting traits, and (3) low production yield and high price. Because of the small quantities that produced and their high prices, adulteration of more expensive fruit oils with cheaper commercial oils is sometimes encountered. International Olive Council (IOC) is involved in preventing frauds and protecting olive oil consumers (IOC 2006). The result is that olive oil is the strictest regulated oil in the world. This book included also an important topic regarding grading, labeling and standardization of edible oils which highlight the standardization of oils to define grades for the final products, specify purity and quality parameters, requirements for labeling, and methods of analysis.

The following chapters in this volume describe several fruit oils and fats. Most of the fruit oils/fats have unique composition and properties that make them valuable source for functional foods, pharmaceuticals, nutraceuticals and technical applications.

References

Antónia Nunes, M., Costa, A. S. G., Bessada, S., Santos, J., Puga, H., Alves, R. C., Freitas, V., & Oliveira, M. B. P. P. (2018). Olive pomace as a valuable source of bioactive compounds: A study regarding its lipid- and water-soluble components. *Science of the Total Environment, 644*, 229–236.

Gunstone, F. D. (2006). Minor specialty oils. In F. Shahidi (Ed.), *Nutraceutical and specialty lipids and their co-products* (Nutraceutical science and technology series) (Vol. 5, pp. 91–125). Boca Raton: Taylor & Francis.

IOC, International Olive Council. (2006). *Trade standard applying to olive oil and olive-pomace oil*. COI/T.15/NC n°3/Rev. 2. Madrid.

Matthäus, B. (2012). Oil technology. In S. K. Gupta (Ed.), *Technological innovations in major world oil crops* (Volume 2: Perspectives) (pp. 23–82). New York: Springer-Verlag. https://doi. org/10.1007/978-1-4614-0827-7_2.

McKevith, B. (2005). Nutritional aspects of oilseeds, British Nutrition Foundation. *Nutrition Bulletin, 30*, 13–26.

Moreau, R. A., & Kamal-Eldin, A. (2009). Introduction. In R. A. Moreau & A. Kamal-Eldin (Eds.), *Gourmet and health-promoting specialty oils* (pp. 1–13). Urbana: AOCS Press.

Ramadan, M. F., & Moersel, J. (2009). Oil extractability from enzymatically treated goldenberry (*Physalis peruviana* L.) pomace: Range of operational variables. *International Journal of Food Science & Technology, 44*, 435–444.

United Nations. (2017). *World population prospects: The 2017 revision, key findings and advance tables*. Working Paper No. ESA/P/WP/248. Department of Economic and Social Affairs, Population Division.

Weihrauch, J. L., & Teter, B. B. (1994). Fruit and vegetable by-products as sources of oil. In B. S. Kamel & Y. Kakuda (Eds.), *Technological advances in improved and alternative sources of lipids* (pp. 177–208). Boston: Springer US.

Chapter 2
Grading, Labeling and Standardization of Edible Oils

Manel Issaoui and Amélia M. Delgado

Abstract Fats and oils are omnipresent in several dishes around the world despite the variability of use and culinary arts, which place them among staple foods in many countries. Edible oils have different origins, forms, aspects and flavors. A multitude of technical processes exists for the extraction of oils of the same origin, making possible the elaboration of final products that differ on physic-chemical proprieties and nutritional values, hence, grading oils accordingly seems to be primordial. Although categorization gives consumer more freedom of choice in selecting their wanted product, it also makes them confused and the question is how to guide the consumer. Cooking oil is a strategic product that forces governments to take regulatory measures with respect to controlling the marketing of edible oils. The key word is standardization. It is the aim of standardization of edible oils to define grades for the final products, specify purity and quality parameters, establish requirements for labeling, and list adequate methods of analysis. International policies in the edible oil sector have evolved, and the levels of standardization at the international level are found. Codex Alimentarius standards are voluntary but serve as a reference in the settlement of international trade disputes. In Europe and USA, the regulation on the marketing and analysis of oils is harmonized. Apart from the aforementioned levels, the national regulations have certain texts specific to them.

Keywords Halal, kosher · Vegetable oil · Refined oil · Virgin oil

M. Issaoui (✉)
Lab-NAFS 'Nutrition – Functional Food & Vascular Health', Faculty of Medicine, University of Monastir, Monastir, Tunisia

Faculty of Science and Technology of Sidi Bouzid, University of Kairouan, Sidi Bouzid, Tunisia

A. M. Delgado
MeditBio – Centre for Mediterranean Bioresources and Food, University of Algarve, Faro, Portugal
e-mail: ameliad@netcabo.pt

© Springer Nature Switzerland AG 2019
M. F. Ramadan (ed.), *Fruit Oils: Chemistry and Functionality*,
https://doi.org/10.1007/978-3-030-12473-1_2

Abbreviations

2- or 3-MCPD ester	2- or 3-monochloropropane-1,2-diol
ARSO	African Organization for Standardization
CAC	Codex Alimentarius Commission
CCFO	Codex Committee on Fats and Oils
COI	International Olive Oil Council
EC	European Commission
EFSA	European Food Safety Authority
EU	European Union
EVOO	Extra virgin olive oil
FAO	Food and Agriculture Organization of the United Nations
FDA	US Food and Drug Administration
FFA	Free fatty acid
ISO	International Organization for Standardization
i-TFA	*Trans*-fat of industrial origin
MOAH	Mineral oil aromatic hydrocarbons
MOSH	Mineral oils saturated hydrocarbons
MUFA	Monounsaturated fatty acid
OO	Olive oil
PDO	Protected designation of origin
PGI	Protected geographical indication
PUFA	Polyunsaturated fatty acid
ROO	Refined olive oil
SFA	Saturated fatty acid
TFA	*Trans*-fatty acid or trans-fat
UN	United Nations
VOO	Virgin olive oil

1 Introduction

Fats and oils are omnipresent in several dishes around the world as lipids are important nutrients and strongly influence the structure and taste of many foods. Besides their well-known uses in culinary arts, oils and fats play important roles in industrial processes. They contribute to produce and maintain aeration in ice cream, moisture retention in cakes, and glossy appearance in pastries. In respect to the source, oils and fats can be of animal origin (e.g. lard, butter, and fish oils) or obtained from vegetables (fruits or seeds). In respect to physical state, lipids that are solid at room temperature are known as fats (as those originated from the mesocarp of coconut or palm fruit), while those that are liquid under the same conditions are known as oils (e.g. extracted from soy or sunflower seeds). Mainly due to the convenience and consumer preferences, edible oils and fats are mostly from vegetable origin.

Table 2.1 Designations for edible pork fat fractions and other animal fats, according to CODEX STAN 211-1999 (last revision 2015)

Named animal fat		Definition
Lard	Pure rendered lard	Fat rendered from fresh, clean, sound fatty tissues from swine (*Sus scrofa*) in good health, at the time of slaughter, and fit for human consumption. The tissues do not include bones, detached skin, head skin, ears, tails, organs, windpipes, large blood vessels, scrap fat, skimmings, settlings, pressings, and the like, and are reasonably free from muscle tissues and blood
	Lard subject to processing	May contain refined lard, lard stearin and hydrogenated lard, or be subject to processes of modification provided that it is clearly labeled
Rendered pork fat	Rendered pork fat	Fat rendered from the tissues and bones of swine (*Sus scrofa*) in good health, at the time of slaughter, and fit for human consumption. It may contain fat from bones (properly cleaned), from the detached skin, from the head skin, from ears, from tails and from other issues fit for human consumption
	Rendered pork fat subject to processing	May contain refined lard, refined rendered pork fat, hydrogenated lard, hydrogenated rendered pork fat, lard stearin and rendered pork fat stearin provided that it is clearly labeled
Premier jus (oleo stock)		Obtained by rendering at low heat the fresh fat (killing fat) of heart, caul, kidney and mesentery collected at the time of slaughter of bovine animals in good health at the time of slaughter and fit for human consumption, as well as cutting fats
Edible tallow	Edible tallow (dripping)	Obtained by rendering the clean, sound, fatty tissues (including trimming and cutting fats), attendant muscles and bones of bovine animals and/or sheep (*Ovis aries*) in good health at the time of slaughter and fit for human consumption
	Edible tallow subject to processing	May contain refined edible tallow, provided that it is clearly labeled

Adapted from Codex Alimentarius

According to data from FAO's food balance sheets in 2013, the world average availability of vegetable oils and fats and those of animal origin were respectively 271 and 61 Kcal/capita/day (FAO 2017). In other words, the global average intake[1] of vegetable oils is more than fourfold higher than that of fats of animal origin, and the trend for the preference for vegetable fats maintains. Pork fat fractions are herein presented as examples of usable ingredients for food and feed, and are thus discriminated in Table 2.1.

Only some fruits, like olive, coconut and palm may provide oil, while a large variety of seeds are used for oil extraction. An overview of sources of edible vegetable oils is provided in Table 2.2, with the notable exception of olive oil. Olive oil is highly valued for their sensorial properties and health claims due to their balanced

---

[1] Average daily intake of fat is herein considered as an approximation, based on the food availability of aggregated items (animal or vegetable origin), expressed in Kcal/capita/day and complying to the "food availability" definition provided in http://www.fao.org/faostat/en/#definitions.

Table 2.2 Designations and origin of edible vegetable oils according to CODEX STAN 210–1999

Vegetable oil	Origin (derived from)
Arachis oil (peanut oil; groundnut oil)	Groundnuts (seeds of *Arachis hypogaea* L.)
Babassu oil	The kernel of the fruit of several varieties of the palm *Orbignya* spp.
Coconut oil	The kernel of the coconut (*Cocos nucifera* L.)
Cottonseed oil	Seeds of various cultivated species of *Gossypium* spp.
Grapeseed oil	Seeds of the grape (*Vitis vinifera* L.)
Maize oil (corn oil)	Maize germ (the embryos of *Zea mays* L.)
Mustardseed oil	The seeds of white mustard (*Sinapis alba* L. or *Brassica hirta* Moench), brown and yellow mustard (*Brassica juncea* (L.) Czernajew and Cossen) and of black mustard (*Brassica nigra* (L.) Koch)
Palm kernel oil	The kernel of the fruit of the oil palm (*Elaeis guineensis*)
Palm kernel olein (liquid fraction)	Fractionation of palm kernel oil
Palm kernel stearin (solid fraction)	Fractionation of palm kernel oil
Palm oil	Fleshy mesocarp of the fruit of the oil palm (*Elaeis guineensis*)
Palm olein (liquid fraction)	Fractionation of palm oil
Palm stearin (high-melting fraction)	Fractionation of palm oil
Palm superolein (liquid fraction)	Palm oil produced through a specially controlled crystallization process to achieve an iodine value of 60 or higher
Rapeseed oil (turnip rape oil; colza oil; ravison oil; sarson oil: toria oil)	Seeds of *Brassica napus* L., *Brassica rapa* L., *Brassica juncea* L. and *Brassica tournefortii* Gouan species
Rapeseed oil – low erucic acid (low erucic acid turnip rape oil; low erucic acid colza oil; canola oil)	Low erucic acid oil-bearing seeds of varieties derived from the *Brassica napus* L., *Brassica rapa* L. and *Brassica juncea* L., species
Rice bran oil (rice oil)	Bran of rice (*Oryza sativa* L)
Safflower seed oil (safflower oil; carthamus oil; kurdee oil)	Safflower seeds (seeds of *Carthamus tinctorious* L.)
Safflower seed oil – high oleic acid (high oleic acid safflower oil; high oleic acid carthamus oil; high oleic acid kurdee oil)	High oleic acid oil-bearing seeds of varieties derived from *Carthamus tinctorious* L.
Sesame seed oil (sesame oil; gingelly oil; benne oil; ben oil; till oil; tillie oil)	Sesame seeds (seeds of *Sesamum indicum* L.)
Soya bean oil (soybean oil)	Soya beans (seeds of *Glycine max* (L.) Merr.)
Sunflower seed oil (sunflower oil)	Sunflower seeds (seeds of *Helianthus annuus* L.)
Sunflower seed oil – high oleic acid (high oleic acid sunflower oil)	High oleic acid oil-bearing seeds of varieties derived from sunflower seeds (seeds of *Helianthus annuus* L.)
Sunflower seed oil – mid-oleic acid (mid-oleic acid sunflower oil)	Mid-oleic acid oil-bearing sunflower seeds (seeds of *Helianthus annuus* L.)

Adapted from Codex Alimentarius

fatty acid composition (containing the eicosanoids precursors, linoleic and alfa-linolenic acids, in an equilibrated n-6:n-3 ratio). In addition, olive oil generally contains significant levels of a wide range of health beneficial compounds. Therefore, food safety authorities granted health claims to olive oil (Delgado et al. 2017; EFSA 2011, 2012; Reg (EC) 432/2012 n.d.). According to EU standards and following Reg (EC) 432/2012 some of these health claims (e.g. ALA, DHA, low SFA, oleic acid) may also apply to other oils and fats if compliance to specified concentration levels in the oil/fat and daily intakes are ensured.[2]

The most valued type, extra virgin olive oil (EVOO), is directly obtained in the liquid form, after a few physical unit operations at ambient or mild temperatures (T <30 °C), and without any subsequent refining or chemical change. Care is taken during pickling, transportation and storage of fruits to avoid deterioration of olive oil' sensorial qualities. Yet, other types of lower quality olive oil exist, mainly due to the separation of fractions along the extraction process. A designation corresponds to each fraction of different quality attributes, as explained below (please see Sect. 5, on grading).

The lipid fraction of the other above-mentioned fruits (palm, and coconut) can also be extracted by squeezing the fruit pulp but, at room temperature, a solid paste (fat) is obtained instead. However, subsequent processing steps allow the separation of liquid (olein) from solid (stearin) fractions, which are used for different purposes by food industries.

2 The Relevance of Oils and Fats in the Global Market

With the increased global circulation of information, people and goods, a "global consumer" has been idealized by industries, facilitating the standardization of products and processes, which can be decentralized. Food industries can obtain raw materials from worldwide taking advantage of the lowest production costs, almost independently of natural seasons. The increased relevance of processed foods in human nutrition, and particularly the global dissemination of fast food chains, have been influencing the trade market for oils and fats, given their high-required amounts of lipids.

Gunstone (2013) compiled the usage of lipids for food, feed and other uses over a period of 100 years, starting before World War I. This author found that oil production grew accordingly to population and economic growth, increasing modestly in the first half of the twentieth century. However, in the last 50 years, world population roughly doubled, while production of oils and fats has increased fourfold. This observation is confirmed by data shown in Fig. 2.1, which exhibits the growth of world food supply from 1961 to 2013 and the evolution of the ratio of fat to protein supply (world average in the same period). It can be observed, on one side that, in

---

[2] As explained later on in this chapter, issues regarding analytical methodologies (e.g. detection levels and standardization) may impair the practical use of such allegations in labels.

Food supply   Fat/protein
(Kg/capita/day)   ratio

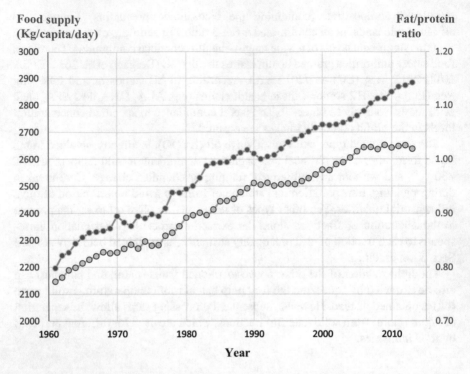

Year

Fig. 2.1 Evolution of Global Food Supply, from 1961 to 2013 (in Kcal/capita/day) – vertical axis on the left; blue (darker) marks; and the evolution of the ratio fat/protein supply (world average) in the same period -vertical axis on the right; pink (lighter) marks. Values were calculated from statistical data from FAO's Food Balance Sheets (FAO 2017)

average, the world food supply has been continuously improving towards the UN's target "zero hunger". In fact, the proportion of the undernourished decreased to 11% in 2014–2016 (UN 2017). On the other hand, a close exam of the parcels (data for individual countries) shows asymmetries: the persistence of hunger in certain regions of the world (low daily supply of energy) and a rising trend of obesity in developed countries and emerging economies (many average values above 3000 Kcal/capita/day, when the reference value is 2000 Kcal/capita/day). Of particular concern is the rising trend of obesity in childhood, with overweight and obesity affecting 6% of children under 5 years of age worldwide, in 2016 (UN 2017).

As it can also be note from Fig. 2.1, in spite of the growth in the demand for sugar, protein and fats (and total food supply), the ratio of fat to protein supply has been changing along time. Until 2004, fat/protein ratio was lower than one, meaning that average protein supply was higher than the average supply for edible oils and fats. After 2004, the situation reverted with more and more lipids being consumed. This dramatic raising demand of oils and fats was first noticed after World War II (Gunstone 2013) and can be attributed to the rise of mass production of foods, the establishment and growth of multinational food companies, offering affordable, safe and convenient foodstuffs for the fast-growing world population and emerging economies.

The advancements in food processing, focused on product standardization, have been based in formulating a growing number of food-containing lipids, thus generating a mounting demand in fats that has been mainly addressed by palm oil and derivatives. Consequently, palm oil's offer increased more than 1000% in the period 1975–2005, and it is nowadays one of the most used edible fat, in line with the consumer's trend for the replacement of animal fats by vegetable oils in the diet (Nunes 2007).

The rise of obesity has been worrying health services, policymakers and international organizations. In fact, dietary shifts have been noticed, towards a global non-sustainable model of low-nutrient/energy-dense foods, with negative consequences on human health and the environment (Delgado et al. 2017; FAO 2018). In one hand, joint efforts of governments, food industries and not-for-profit organizations have been raising awareness of consumers about health; on the other hand they have been improving the overall nutritious value of foods, translated into regulations, as is the case of the nutritional label (more details in Sect. 7). On the other hand, it was noticed that the increased availability of palm oil in the global market was due to the increase of new plantations, rather than from the increase in yield of the existing ones (Gunstone 2013). These intensive monoculture agricultural systems and deforestation are a non-sustainable model of exploiting natural resources, and awareness on their serious consequences has been raised by the scientific community (Voigt et al. 2018), food activists, ecologists and consumers. As a consequence, many foods manufactures and retail companies worldwide responded by including allegations in their products' label as "sustainable palm oil" or "palm oil-free", leading to discussions and even to legal battles on regulations (Smith 2017; EPOA 2017; WWF 2018). Still, some brands publicly commit to excluding palm oil from their ingredient's list due to the difficulty of tracing the sustainability of the production (Aware Environmental 2018; Business Green 2018). The replacement of palm by other oils in foods presents technological challenges that many brands have been successfully solving, aiming at better performing nutritionally (better balance of fatty acids) and environmentally (avoiding the association of the brand to malpractices).

The development of products of reduced fat content and the replacement of some lipids for healthier and more sustainable ones is a trend anticipated to continue influencing the sector in the mid and long terms. In this regard, olive oil is the main lipid source in the Mediterranean diet, referred to as an example of a healthier and sustainable diet by FAO (2010). The history of olive oil's use goes back to 8000 BC. Olive groves have been expanding to new areas and soils, so do the interest in olive oil and their well-established health benefits. Based on the recent information from IOC, the world harvest of olive oil is expected to increase by 14% (2894 thousand tonnes) while consumption is expected to increase by 5% (2954 thousand tonnes). According to IOC estimates, foreign trades in olive oil (outside its traditional Mediterranean area) have been increasing in Brazil, Australia, China, Canada, Japan and in the United States. It is noteworthy that 95% of Chinese imports come from the EU (80% from Spain, 13% from Italy, and 2% from Greece). This market has an interest in quality olive oil. Thus, the olive oils imported by China are for 77% virgin (VOO) and EVOO. South European and Maghreb countries are the largest world producers and the largest consumers. For the EU and during the season from October 2016 to March 2017, imports increased by 11% at the intra-EU level, and decreased by 12% at the extra-EU level. Prices of VOO have increased thanks to the international trade of this product around the world, in comparison to other vegetable oils. Italian olive oils were the most valued in 2016–2018, ranging from 4.4 €/kg (Oct 2016) to 6.14 €/kg (March 2017), followed by Tunisian olive oils -the second preferred, although traded at lower prices (3.9 €/kg in average, for the same period). In third and fourth positions for consumer preferences (and hence prices), are Spanish and Greek olive oils.

In what respects to quality, EVOO and VOO are unique, as olive oil does not need any refining process, keeping their authenticity and their typical flavor, which are much appreciated. As a consequence, in EU, about 12 Protected Geographical Indication (PGI) and around 90 Protected Designation of Origin (PDO) are registered in the DOOR database, in 2018 (EC 2018). Some of them are listed in Table 2.3. Most of these registries refer to EVOO produced from different cultivars and regions of Greece, Italy, Portugal, Spain, Croatia, and Slovenia. In second place but with a much lower number of registries are butter, which flavor reflects the regional character and the way cattle are raised and fed. Only a few registries correspond to non-refined seed oils obtained from traditional extraction processes and following specificities identical to EVOO. Table 2.3 shows an example of such designations, randomly omitting many EVOO registries, while exhibiting practically all other fats' to which TSG, PGI or PDO was granted by EC.

One of the reasons why there are not so many seed oils in conditions to request this quality label is because most seed oils may present toxicity issues (e.g. risk of accumulating mycotoxins during seed storage) as well as considerable amounts of strong bitter and pungent flavors. Some so-called virgin seed oils, and cold-pressed seed oils can be found in the global market. However, the designations "virgin" and "cold-pressed" are not consensual. Even for olive oil, a well-established oil in the global market, divergences in designations are found, particularly in countries that do not follow IOC standards, as is the case of USA (for more details on the grading of olives oils, please see Fig. 2.2 and Sect. 5). Labeling issues, also concerning other oils and fats, are more in-depth discussed below.

Squeezing oil from seeds demands high shear forces, and even if heat is not applied prior to pressing (in the case of cold-pressed oils), heat is generated in the process. Temperature generally raises to 80 °C or higher (Sect. 3). A previous seed conditioning operation with "dry steam[3]" is common, exposing the seeds to high temperatures and pressures aiming at inactivating enzymes and microorganisms, as well as to soften husks and cell walls. If the oil extraction process only includes such mechanical unit operations, the oil thus obtained can exhibit the label "virgin". Olive oil, is generally extracted right after harvest, an operation which is conducted

---

[3] Dry steam is a designation for steam with virtually no water phase, and with a higher enthalpy hence ensuring a better heating.

Table 2.3 Examples of quality labels, linked to a region and to intellectual property, for food products belonging to the class of oils and fats (butter, margarine, and oil), from the DOOR database, for EU countries (until 2017)

Country	Product	Specific designation	Date of registration
	TSG – Traditional Specialty Guaranteed		
Poland	Camelina's oil	Olej rydzowy tradycyjny	16\06\2009
	PGI – Protected Geographical Indication		
Italy	EVOO	Marche	20\04\2017
Slovenia	Pumpkin seed oil	Štajersko prekmursko bučno olje	03\10\2012
Greece	EVOO	Ζάκυνθος	18\07\1998
Italy	EVOO	Toscano	21\03\1998
Germany	Linseed oil	Lausitzer Leinöl	25 11\1997
Austria	Pumpkin seed oil	Steirisches Kürbiskernöl	02\071996
Greece	EVOO	Χανιά Κρήτης	21\06\1996
	PDO – Protected Designation of Origin		
Croatia	EVOO	Šoltansko maslinovo ulje	21\10\2016
France	Butter	Beurre de Bresse	15\04\2014
Spain	EVOO	Aceite Sierra del Moncayo	04\12\2013
Greece	EVOO	Μεσσαρά	12\10\2013
Italy	EVOO	Vulture	13\01\2012
Spain	Butter	Mantequilla de Soria	16\02\2007
France	EVOO	Huile d'olive de Corse	16\02\2007
France	EVOO	Huile d'olive de Nîmes	16\02\2007
Portugal	EVOO	Azeite do Alentejo Interior	16\02\2007
Slovenia	EVOO	Ekstra deviško oljčno olje Slovenske Istre	16\02\2007
Spain	Butter	Mantequilla de l'Alt Urgell y la Cerdanya	22\11\2003
Luxembourg	Butter	Beurre rose – Marque Nationale du Grand-Duché de Luxembourg	13\07\2000
Belgium	Butter	Beurre d'Ardenne	13\11\1996
Portugal	EVOO	Azeites do Ribatejo	13\11\1996
Greece	EVOO	Κρανίδι Αργολίδας	21\06\1996
France	Butter	Beurre Charentes-Poitou	21\06\1996
France	Butter	Beurre d'Isigny	21\06\1996
France	EVOO	Huile d'olive de Nyons	21\06\1996
Portugal	EVOO	Azeite de Moura	21\06\1996

EVOO is Extra Virgin Olive Oil, which is obtained from the fruits of *Olea Europeaea* by mechanical cold extraction only (no further processing or blends)

in such a way as to preserve the physical integrity of fruits, aiming at preserving flavor and simultaneously minimizing the chances for microbial degradation. Considerable yields of oil can be extracted from olives at room temperature, and EVOO is highly-valued economically. Moreover, while olive oil has a balanced healthy composition, in other oils the concentrations of elaidic acid, lauric and, and most particularly of erucic acid, are a matter of concern due to the risk of exceeding tolerable daily intake values, mainly in children (EFSA 2016).

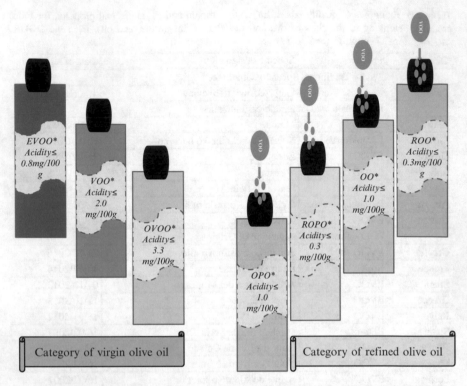

Fig. 2.2 Grading of olive oil according to CODEX STAN 33-1981- Standard for Olive Oils and Olive Pomace Oils –modified in 2017. *EVOO* Extra virgin olive oil, *VOO* Virgin olive oil, *OVOO* Ordinary virgin olive oil, *ROO* Refined olive oil, *OO* Olive oil, *ROPO* Refined olive pomace oil, *OPO** Olive pomace oil, *and whose other characteristics correspond to those laid down for this category

3 Unit Operations in Oil Extraction and Their Impact on Quality

The unit operations in edible oil extraction and further processing determine physic-chemical properties and nutritional quality of the final products, and are briefly described herein, in a generic manner. The fruits and seeds are first harvested, transported and stored before processing. With the exception of olive oil (mills are generally located nearby the orchards and extraction takes place shortly after harvesting), for other oils harvesting site may be distant from the processing plant, and the storage time may be quite long.

Due to their physical characteristics, palm oil's paste and kernel, as well as copra (dried meat or kernel of the coconut) are stored in warehouses, while seeds are stored in silos.

Table 2.4 Quality and composition features for edible oils and fats covered by CODEX STAN 19-1981

Quality characteristic	
Color	Characteristic of the designated product
Odor and taste	Characteristic of the designated product and free from foreign and rancid odor and taste
Maximum level	
Matter volatile at 105 °C	0.2% w/w
Insoluble impurities	0.05% w/w
Soap content	0.005% w/w
Iron (Fe)	
Refined fats and oils	2.5 mg/kg
Virgin fats and oils	5.0 mg/kg
Cold pressed fats and oils	5.0 mg/kg
Copper (Cu)	
Refined fats and oils	0.1 mg/kg
Virgin fats and oils	0.4 mg/kg
Cold pressed fats and oils	0.4 mg/kg
Acidity value	
Refined fats and oils	0.6 mg KOH/g fat or oil
Virgin fats and oils	4.0 mg KOH/g fat or oil
Cold pressed fats and oils	4.0 mg KOH/g fat or oil
Peroxide value	
Virgin oils and cold pressed fats and oils	Up to 15 milliequivalents of active oxygen/kg oil
Other fats and oils	Up to 10 milliequivalents of active oxygen/kg oil

Adapted from Codex Alimentarius

Storage conditions are important for the yield and the quality of the oil. Cell respiration and endogenous lipase activity, pests, microbial growth and moisture may have adverse effects in oil quality and safety, namely by increasing free fatty acids (FFA) and hence the acid value (Table 2.4). Then, raw materials are prepared for oil extraction steps, trough husking, cleaning, crushing, and conditioning. Cleaning usually involves the removal of metal (by the use of a rotary magnet separator) and of sand and other particulate materials (using sieves). After a first mechanical crushing, the paste is heated to temperatures higher than 80 °C (conditioning), a step that may consist of cooking the paste with steam at variable pressure, in an autoclave-type conditioner. This heating step aims to sterilizing the paste, inactivate lipases and other deleterious enzymes, soften cell walls facilitating oil extraction and obtain the proper elasticity for efficient pressing (Keller 2018). However, conditioning also eliminates vitamins and many nutraceutical compounds that might be present. Flaking (deforming seeds into flakes) may take place before or after the heat treatment and always prior to mechanical oil extraction in expeller presses. The resulting cake is submitted to solvent extraction (usually hexane) for further oil withdrawal. The crude oil is separated, washed/clarified, and the solvent is evaporated and recovered. Residues are conditioned (for example, dried) and reprocessed to yield by-products such as animal feed (Kemper 2018). During solvent extraction, some chemical reactions may occur with the accumulation of minor quantities of by-products.

Further processing of crude oil includes clarification and fractionation prior to refining. Refining aims at eliminating unwanted compounds, standardizing the oil characteristics and leveraging the quality (Zeldenrust 2018). Refining starts with the degumming operation (the elimination of gums: mainly phosphatides but also some glycolipids and proteins). Gums are a major problem when solvent extraction is used (Dijkstra 2018) and they have to be eliminated due to their emulsifying properties and thermal instability. Degumming can be performed by chemical or enzymatic methods. Chemical degumming is more common and consists in a treatment with phosphoric or citric acids at temperatures of about 90 °C, aiming at forming a precipitate separated upon neutralization with NaOH/KOH or by adsorption on bleaching earth (primarily hydrous aluminum silicates, and other clay minerals). After degumming, food or feed grade lecithin (phosphatidylcholine) may be obtained as a by-product, and the oil will then undergo further physical or chemical (alkali) refining. Such designations (physical or chemical) come from the process used to remove the FFA that are responsible for the oil acidity. The process' choice will depend on the features of the crude oil. Thus, physical refining makes use of the lower boiling point of the FFA when compared to the boiling point of the triglycerides (bulk oil fraction), while in chemical refining (the most common and traditional method) alkali is used to neutralize FFA. Both processes occurring at high temperatures (using steam under pressure), increasing the probability of occurrence of side reactions and the accumulation of a variety of impurities (Zeldenrust 2018).

Chemical refining mainly consists in an acid-base reaction between the OH- (from the alkali) and the H+ from the carboxyl group (COOH) of the FFA, with the obtainment of soap, which is diluted in the water phase and removed by separators. The neutral oils are subsequently bleached and deodorized (viz by adsorption). Other non-glyceride materials are also removed, as remaining gums, metal ions, pigments and insoluble impurities. The steps of chemical refining can be optimized in accordance with the crude oil and by balancing final quality with acceptable oil losses (Gunstone et al. 2007; Zeldenrust 2018).

The unit operations involved in oil extraction, refining and subsequent processes aim at addressing or surpassing the quality parameters listed in Table 2.4. However, some side reactions may occur during the refining process. Some maybe innocuous, others may affect sensorial characteristics, and the risk of accumulation of harmful compounds exists. For example, in the presence of chlorinated substances (even at very small levels) 3-MCPD esters can be formed and in some cases critical levels may be attained. For these compounds, the tolerable daily intake (TDI) is 2.0 µg/kg body weight per day (benchmark) and EFSA noted a risk of exceeding the TDI, in the case of children that have a high-fat diet (EFSA 2018). In this respect, EFSA recommended that the EU member states implement vigilance programs for it and for the presence of glycidic esters and other genotoxic compounds in edible oils and fats, as well as in fat-containing foods, and additives (recommendations 2014/661/UE and 2017/84/UE).

Whatever the individual reasons are, the fact is that most consumers prefer vegetable fats to those of animal origin, and this is a consistent trend given the abundance of plant origin oils and fats. For these reasons and the commercial benefits of launching new products, food industries have developed margarine and other spreads from vegetable oils. This goal (of addressing consumers' preferences) has been achieved by partially converting unsaturated fatty acids from plant origin to saturated fatty acids, by a relatively simple hydrogenation reaction (e.g. $H_2C=CH_2 + H_2 \rightarrow H_3C-CH_3$). Vegetable oils mainly contain polyunsaturated fatty acids (PUFA) that once partially saturated have higher melting points. The degree of hydrogenation of unsaturated oils determines the consistency of the product (from a creamy paste to a hard block).

Margarine, spreads and foodstuffs containing these *trans*-fats became popular since the 1950s. During partial hydrogenation, some of the unsaturated fatty acids, which are normally found as *cis*-isomers, are changed to *trans*-isomers remaining unsaturated. These *trans* fatty acids (i-TFA) are of the same length and mass as the original *cis* fatty acids but they do not occur in nature and are deleterious to human health. According to WHO (2018), diets high in *trans*-fat increase heart disease risk by 21% and deaths by 28%. To alleviate the burden on public health, the recommended intake of *trans*-fat (including those from dairy and beef) is 1% of the total energy intake (WHO 2018). Moreover the so-called "REPLACE" action plan is on-going since May 2018, aiming at eliminating industrially-produced *trans*-fatty acids (i-TFA) from the global food supply, as part of UNSDG (WHO 2018).

Conversely, virgin olive shows health-promoting features and it is valued for their minor components and sensorial proprieties, which contribute to endogenous typicity. All these privileges make this flagship product a vulnerable target to fraudulent practices, and it ranks as number one in a list of tampered foods (above wine and honey for example) according to the report from European Parliament 2013/2091(INI) on food fraud and on how to control it (European Parliament 2013). A consensual and comprehensive definition of food fraud does not exist. Thus, as the main reason for this malpractice is essentially economic, a widely accepted definition is that from Moore et al. (2012). These authors defined food fraud as 'a collective term used to encompass the deliberate and intentional substitution, addition, tampering, or misrepresentation of food, food ingredients, or food packaging; or false or misleading statements made about a product for economic gain'.

A common way of tampering olive oil is by blending refined seed and palm oils, which are much cheaper. Changes in physical properties and sensorial characteristics may be more or less evident, and depend on the expertise of the consumer. However, at least in EU, random inspections and the specific chemical analysis are used to detect and to prove the fraud. These analyses consist on the detection/quantification of markers for the presence of refined factions blended in products labeled as "virgin", or to track the region of origin of olives, for those products holding PDO/TGI/TSG.

4 Standardization

As defined by Collins online Dictionary, "to standardize things means to change them so that they all have the same features". The verb "to standardize" is synonymous with "to bring into line", "to regularize', "to stereotype", "to regiment" (Collins 2018). Alternatively, the definition given by the Business Dictionary online is "formulation, publication, and implementation of guidelines, rules, and specifications for common and repeated use, aimed at achieving the optimum degree of order or uniformity in a given context, discipline, or field"(Business Dictionary 2018). The African Organization for Standardization (ARSO) defines it as "(…) a set of rules for ensuring quality; they define how most products, processes, and people interact with each other and their environments. They enhance competitiveness by offering proof that products and services adhere to requirements of governments or the marketplace" (ARSO 2018).

This concept, as is known, has appeared before the World War I, and thanks to the multitude of the advantages given by the standardization, a world standards day is celebrated each 14th October. In fact, standardization and international trade are evolving in parallel. Whenever an increase in international trade is noticeable, a requirement to standardize the rules appears. ARSO notes that "(…) in today's multipolar world, where commerce is increasingly international and the growing economic weight of emerging economies is shifting the balance of power, international standards are essential to safeguarding consumer health and safety and stimulating national and international diversification". Standardization is a way for all actors involved in the trade of edible fats and oils to speak the same language. A common language translates into a harmonization of means of control, analysis, classification, labeling, sale, purchase of these products, etc.

Standardization of the method of analysis and the standardization of product's features are closely linked. Hence, the control procedure and analytical methods should be set and standardized at the international level, to prevent fraud and ensure that the fats and oils marketed under a given designation comply with the corresponding specifications. Standardization is a pertinent and crucial step to make international transactions and trade more reliable, to restore trust between vendors and clients, and to guarantee the credibility of official and private control services.

Common international standards for edible fats and oils would help in resolving conflicts, and in preventing arbitration disputes, which occur whenever different standards are used, thus leading to dissimilar results. Normally, standardization simplifies and clarifies the relationship between companies: on the one hand, to develop markets by harmonizing practices and reducing technical barriers to trade and, on the other hand, to clarify transactions, especially by a better definition of the needs.

It is noteworthy that there are many kinds of the standard: specification standards, standard analytical methods, standard substances used in performing the analysis, and standards of management. In what concerns the basic terminology:

- The specification standards set the characteristics of the products and the performance thresholds to be achieved
- Standards for testing and analytical methods indicate how to measure the characteristics of the products

– Specification standards define the characteristics required for the product
– Management standards describe the rules of organization and operation of companies (Canard 2009)

Nowadays, there are so many standardization bodies; each one has a specific function and its own rules. In what concerns edible oils and fats, the most relevant organizations involved in standardization worldwide are the International Organisation for Standardization ISO (which issues standards covering a wide range of activities); Codex Alimentarius (a Commission dependent on UN issuing recommendations and rules concerning food quality and safety); still at an international level, the role of International sectorial associations is noteworthy. Relevant examples are: FOSFA, the Federation of Oils, Seeds and Fats Associations Ltd. (regulating seed oils trades), The American Oil Chemists' Society, AOCS (supplying different types of reference materials) and International Olive Oil Council, IOC (specific standards for olive oil and olive products).

European Commission, EC and European Food Safety Authority, EFSA are the most prominent organizations in Europe, dealing with trade, quality and safety. US Food and Drug Administration (FDA) and US Department of Agriculture (USDA) play equivalent roles in the USA. At National levels, ministries, food safety agencies and sectorial organizations act at different levels in the sector.

4.1 Types of Lipids Found in the Global Market

Most edible oils and fats are obtained from the seeds or fruits of oleaginous plants. In addition, some fats and oils of animal origin are commercialized in pure form (e.g. butter) or mixtures (spreads). To facilitate comparisons and trade, the corresponding designations are standardized and frequently reviewed and updated by Codex Alimentarius (Fig. 2.3).

Seed oils, and derivatives from palm and coconut are named after Codex Stan 210-1999 (Table 2.2), while the different fractions obtained from olives are named after a different standard (Codex Stan 33-1981), given the larger number of oils of different qualities and prices that can be obtained from olives (Fig. 2.2). Due to the economic relevance, increased demand, and price, IOC issues standards that are specific for olive oils and pomace oils. Special attention is paid to VOO, given its high market price, which IOC defines as *"Virgin olive oils are the oils obtained from the fruit of the olive tree (Olea europaea L.) solely by mechanical or other physical means under conditions, particularly thermal conditions, that do not lead to alterations in the oil, and which have not undergone any treatment other than washing, decantation, centrifugation and filtration"* (IOC 2018).

Fats and oils of animal origin, such as butter (produced from the extraction of the milk fat of ruminants, by mechanical methods only) or lard (obtained from the lipid tissues of pork) are suitable for human consumption. Their pure forms and their lipid fractions are named after Codex Stan 211-1999. Fish oils are named in a specific

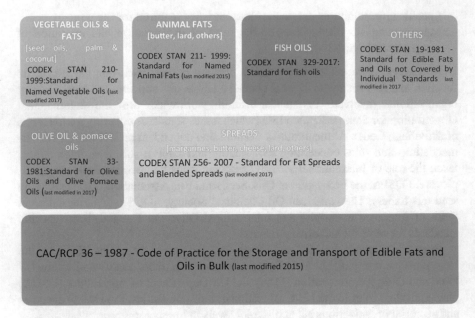

Fig. 2.3 The seven most important standards, referred to edible fats and oils, issued by the Codex Committee on Fats and Oils (CCFO) from Codex Alimentarius. These standards are regularly updated and are often in the basis of International regulations and of National laws

category after Codex Stan 329-2017 (Table 2.1). Fish oils are valued for their high content in n-3 fatty acids and increasingly used in the manufacturing of food supplements. In order to anticipate novel forms of obtaining edible oils, Codex Alimentarius issued Codex Stan 19-1981, covering those edible lipid fractions not covered by the other categories.

In what concerns their use, oils and fats are mainly used to fry foods, incorporated as ingredients during food processing or consumed directly (e.g. salad dressings and spreads). Spreads generally undergo some modifications, mainly to adjust their consistency or to lower their fat content (light spreads). These type of industrial products include from "light butter" (obtained by the emulsification of water in butter), cheese-based products, vegetable margarines produced from partially hydrogenised oils (with adjusted consistencies and fatty acid composition), as well as products that result from mixtures of fats from animal and plant origin (blended spreads). Designations in this category are standardized by Codex Stan 256-2007. The standards for labeling these commercial products in the international market, as issued by Codex Alimentarius, are presented in Table 2.5.

Following the WHO plan to eliminate *trans*-fats from foodstuffs (WHO 2018) changes to the composition of current fats in the market are expected, including the introduction of new products by food industries. These measures will take place stepwise as part of UNSDG and are expected to have an impact on many regulations, namely on food labeling. In short, the Codex Committee on Fats and Oils

Table 2.5 Codex Alimentarius' Standards for grading and labeling of edible oils and fats

Category of fats and oils	Appropriate standard
Edible fats and oils not covered by individual standards	General Standard for the labeling of prepackaged foods – CDEX STAN 1- 1985- last modification 2010 and the recommendation of Chap. 6 of the CODEX STAN 19-1981-v 2017
Olive oils and olive pomace oils	General Standard for the labeling of prepackaged foods – CDEX STAN 1- 1985- last modification 2010 and the recommendation of Chap. 7 of CODEX STAN 33-1981- v 2017
Named vegetable oils	General Standard for the labeling of prepackaged foods – CDEX STAN 1- 1985- last modification 2010 and the recommendation of Chap. 7 of the CODEX STAN 210-1999-v 2017
Named animal fats	General Standard for the labeling of prepackaged foods – CDEX STAN 1- 1985- last modification 2010 and the recommendation of Chap. 7 of the CODEX STAN 211- 1999 v 2015
Fat spreads and blended spreads	General Standard for the labeling of prepackaged foods – CDEX STAN 1- 1985- last modification 2010 and the recommendation of Chap. 7 of the CODEX STAN 256- 2007 v 2017
Fish oils	General Standard for the labeling of prepackaged foods – CDEX STAN 1- 1985- last modification 2010 and the recommendation of Chap. 7 of the CODEX STAN 329- 2017

(CCFO) has elaborated seven important references covering a wide range of edible fats and oils. These references are generally transposed into national laws in order to reach an effective regulation (Fig.2.3, Table 2.5).

4.2   Standardization Bodies and Respective Standards for to Oils and Fats

Nowadays, there are many standardization bodies, each one with a specific function and their own rules. These organizations issue guidelines, product specifications and/ or manage specific labels, thus ensuring certain attributes or qualities of the products. Some of these trade standards reflect the concerns of certain groups of consumers (halal, kosher, and vegan requisites), while others may be related to overall food safety and quality. Main organizations that make standards related to food safety and quality are globally known and widely accepted. Nevertheless, the landscape of standardization appears both dense and varied: alongside regulations and "traditional" collective norms, there are multiple tools from various circuits: referential formalized by the public authorities, private specifications, and contracts.

The criteria for characterizing standardization devices are four in number: 1st, the methods of elaboration, 2nd, the functions, 3th, the methods of implementation and 4th, the methods of demonstrating compliance. Standardization schemes are also distinguished by the way they are used and implemented by the actors. Their application may be mandatory, voluntary or contractually required by other actors. Their field of application can be very broad (the whole food chain) or on the contrary, be limited to a chain of production, or to a particular link of production.

4.2.1 International Organization for Standardization (ISO)

The acronym ISO is derived from the Greek *isos*, meaning equal, and spells for the International Organization for Standardization. This well-known organization was officially created on the 23/02/1947 with the objective "to facilitate the international coordination and unification of industrial standards". The ISO is an independent, non-governmental body and the world's largest elaborator of voluntary international standards. The ISO has 780 technical committees and subcommittees that have published so far 22088 international standards, covering a wide range of technology and manufacturing aspects.

The SC11 and the SC2 are the two subcommittees concerning mainly edible fats and oils and belong to the 34 technical committees responsible for the elaboration of the international standard of analytical methods. The subcommittee ISO/TC 34/SC 11 is entitled "Animal and vegetable fats and oils", and hence elaborates standard analytical methods on that topic (fats and oils of animal and of plant origin). Until now, this subcommittee published 85 standards and 9 are under development. On its turn, the subcommittee ISO/TC 34/SC2 elaborates standards on analytical methods on "oleaginous seeds and fruits and oilseed meals". The SC2 has published 27 standards and 4 are currently under development (www.iso.org).

4.2.2 Codex Alimentarius

The Codex Alimentarius is a Commission operating under the auspices of FAO and WHO. It was created with the purpose to protect the consumer, ensure fair business practices and harmonize global food standards. This triple objective guarantees the free movement of food products between nations. The Codex Alimentarius, or "Food Code" is thus a collection of standards, guidelines and codes of practice adopted by the Codex Alimentarius Commission (also known as CAC).

In this regard, Codex Alimentarius has adopted some measures to implement international standards for the sector of edible oils and fats, often proposing open discussions with stakeholders, which include regulatory bodies and not-for-profit organizations (the most prominent of them are listed below) (www.fao.org/fao-who-codexalimentarius).

The Codex Committee on Fats and Oils (CCFO) has elaborated seven important references covering a wide range of edible fats and oils from different origins, as listed in Fig. 2.3.

4.2.3 American Oil Chemists' Society (AOCS)

AOCS, American Oil Chemists' Society, has been created in 1909 in the USA as a regional professional organization and started an active international expansion in the 1970s. Nowadays, AOCS is a prominent global organization with members from all around the globe, and it has a relevant role in the development of methods of analysis, proficiency testing, and in providing reference materials. AOCS coordinates actions with other standards' developers, including ISO and Codex Alimentarius. To achieve these goals, AOCS is led by a Governing Board that works with strategic working groups. Within each of the working groups are valued centers that function as a "think tank". Worldwide acceptance has made the AOCS Methods a requirement wherever fats and oils are analyzed, e.g. accreditation of laboratories by ISO 17025, and the adoption of several AOCS' methods by Codex Alimentarius (www.aocs.org).

4.2.4   Fédération de l'industrie de l'huilerie de la CE (FEDIOL)

FEDIOL is the federation representing the European Vegetable Oil and Protein meal Industry in Europe, representing the interests of the European oilseed crushers, vegetable oil refiners and bottlers, currently from five EU countries. FEDIOL's mission is to represent the interests of associates towards public and private organizations, notably European Institutions, contributing to a constructive regulatory framework within the EU. FEDIOL also coordinates activities with international bodies and stakeholders, such as Codex Alimentarius. FEDIOL develops and disseminates principles of good manufacturing practices, ensuring the quality and safety of products. In this regard, FEDIOL releases position papers, codes of practice and specifications (www.fediol.be).

4.2.5   International Olive Council (IOC)

The International Olive Council (IOC) is the largest association of olive oil producing countries. IOC was created in 1956 following a first agreement on the trade of olive oil, which was signed in Geneva the year before. In December 2017, the IOC counted among its member's EU producing countries and 18 extra-EU countries, in addition to observers participating in IOC activities. Palestine is the newest member of this intergovernmental association (2017). These states account for more than 98% of world olive production. The body of the IOC is very simple, and mainly based on the advice of the members and five committees. The IOC publishes international trade standards exclusively concerning olive oil, olive pomace oil and table olives. The major standards have been issued in 1987 and since then, amendments have been made to it. Later amendments concern mainly food protection (food safety and detection of food fraud), as olive oil is the food product more susceptible to fraud (www.internationaloliveoil.org).

Although of wide acceptance, IOC standards are not universal. Different standards around the world show some differences in defining olive oil quality grades, particularly in non-member countries of IOC. The major differences refer to the designation of EVOO, a category not separated from VOO under some standards, and mixtures of refined and non-refined fractions may not be well distinguished either. It is noteworthy that in 2010, following a petition of the California Olive Oil

Council, a trade association of American olive oil producers, the USA olive oil standards were revised towards the harmonization with Codex and IOC standards (USDA 2010).

Prominent organizations that issue standards for olive oil and olive pomace oil are listed in Fig. 2.4, along with their site's address. These include organizations of local ambit that cooperate with IOC (e.g. Afidol), international organizations generally compliant with IOC (as EC), as well as organizations that emit their own standards valid in their countries and territories of influence (as are the cases of American and Australian organizations).

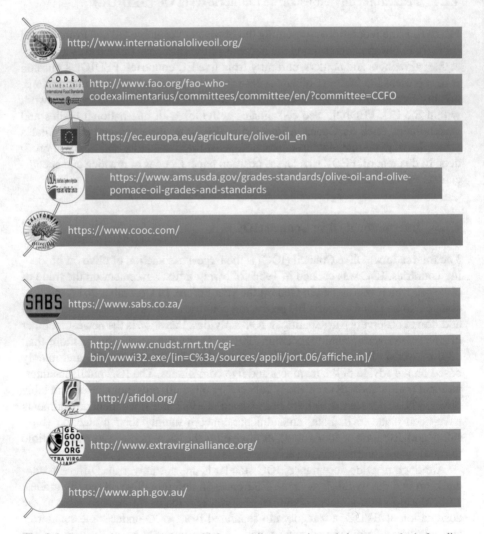

Fig. 2.4 Organizations around the world that contribute to or issue their own standards for olive oils and olive pomace oils, logos and hyperlinks to their sites

The divergence in standards is notorious with olive oil, given its economic relevance. Such a matter is comprehensively discussed by García-González et al. (2017) in their paper "A study of the differences between trade standards inside and outside Europe". These authors refer to the methods used to analyze olive oil (OO) established by the different standards. Different methodologies lead to different results and conclusions. Olive oil is a product of great natural diversity, expensive, risking food fraud, and therefore it is highly regulated.

In short, the most widely accepted international standards for olive oils and olive-pomace oils are:

- International Olive Council Trade Standard Applying to Olive oils and Olive – Pomace Oils – COI/T. 15/NC No 3/Rev. 8 February 2015
- Codex Standards for Olive Oils and Olive Pomace Oils – CODEX STAN 33 – 1981 Adopted in 1981. Revision: 1989, 2003, 2015. Amendment: 2009, 2013

To close this section, we can deduce that OO is marketed in accordance with some designations. A general consensus of the majority of standard classifies OO as:

- VOO: fit for consumption as they are, including EVOO and VOO (VOO is not fit for consumption as it is, is designated lampante VOO and can be refined to be suitable for consumption).
- OO: is the oil consisting of a blend of refined OO and VOO fit for consumption as they are.
- OPO: is the oil obtained by treating olive pomace with solvents or other physical treatments (it can be refined to be suitable for consumption).

5 Grading

According to Oxford online Dictionary (2018), the origin of the word "grade" comes back to sixteenth century. Etymologically speaking, the word "grade" comes from the French, or from Latin *gradus* 'step'. "Originally used as a unit of measurement …, the term later referred to degrees of merit or quality". As defined by Collins online Dictionary, the grade is "synonym of classifying, rate, order, class". The grade is a verb, which has the meaning of "if somethings is graded, its quality is judged, and it is often given a number or a name that indicates how good or bad it is". The grade is also a countable noun "the grade of a product is its quality, especially when this has been officially judged".

It follows from these definitions that "grading a product" has two steps:

1. Making a quality judgment
2. Giving a quality indicator (number/name)

In order to judge the quality of a food product, an official document is needed. Such a document containing scientific criteria to compare the required characteristics to the appropriate standard. The result of this step is the compliance or non-compliance of the product with the standard. If the product complies with the standard, then it can be classified in the appropriate grade.

Again, when grading oils, there is a marked distinction between grading olive oil and grading any other oil, the first one being more complex, as it may attend to sensorial parameters in addition to composition parameters. In other words, generally two orders of parameters are used to classify olive oil in the grade corresponding to the appropriate standard: purity and quality parameters.

COI releases the official methods for the organoleptic evaluation of olive oils, as well as corresponding standards, panel methodologies and data analysis. These methods are adopted by COI member countries. In EU, they are especially relevant for PGI[4] and PDO[5] characterization (examples of European PGI and PDO olive oils are shown in Table 2.3).

IOC standards (2018) for olive oil are mainly based in free acidity and sensorial evaluation. IOC grade edible olive oils in:

> **Extra virgin olive oil (EVOO)**: VOO which has a free acidity, expressed as oleic acid, of not more than 0.8 grams per 100 grams, and the other characteristics of which correspond to those fixed for this category in this standard.
> **Virgin olive oil (VOO)**: VOO which has a free acidity, expressed as oleic acid, of not more than 2 grams per 100 grams and the other characteristics of which correspond to those fixed for this category in this standard.
> **Ordinary virgin olive oil**: VOO which has a free acidity, expressed as oleic acid, of not more than 3.3 grams per 100 grams and the other characteristics of which correspond to those fixed for this category in this standard.
> **Refined olive oil** is the olive oil obtained from VOO by refining methods, which do not lead to alterations in the initial glyceridic structure. It has a free acidity, expressed as oleic acid, of not more than 0.3 grams per 100 grams and its other characteristics correspond to those fixed for this category in this standard.
> **Olive oil** is the oil consisting of a blend of refined olive oil and VOO fit for consumption as they are. It has a free acidity, expressed as oleic acid, of not more than 1 gram per 100 grams and its other characteristics correspond to those fixed for this category in this standard.

Besides olive oil, the demand for olive pomace oil has been increasing in some markets, thus paving the way of the olive sector in the circular economy, by valorizing pomace (the by-product of virgin olive oil production). IOC defines olive-pomace oil as *"the oil obtained by treating olive pomace with solvents or other physical treatments, to the exclusion of oils obtained by re-esterification processes and of any mixture with oils of other kinds"*.

Edible olive pomace oil should be labeled according to the following designations and definitions by IOC (2018):

> <u>Refined olive pomace oil</u> is the oil obtained from crude olive pomace oil by refining methods, which do not lead to alterations in the initial glyceridic structure. It has a free acidity, expressed as oleic acid, of not more than 0.3 grams per 100 grams and its other characteristics correspond to those fixed for this category in this standard.

---

[4] Protected geographical indication.
[5] Protected designation of origin.

Olive pomace oil is the oil comprising the blend of refined olive-pomace oil and VOO fit for consumption as they are. It has a free acidity of not more than 1 gram per 100 grams and its other characteristics correspond to those fixed for this category in this standard. In no case shall this blend be called olive oil.

There is some harmony between the Codex Alimentarius designations and those of IOC, listed above, as the reader can deduce by comparing the above descriptions (IOC designations) to those below and summarized in Fig. 2.2 (corresponding to Codex standards). Harmonization is important because such regulations are used by the World Trade Organization in the framework of agreements on the application of sanitary and phytosanitary measures and on technical barriers to trade.

Thus, according to Codex Alimentarius, olive oil is graded in seven classes based on the kind of extraction process (physical or chemical), physical-chemical and sensorial parameters as adopted by CODEX STAN 33-1981 Standard for Olive Oils and Olive Pomace Oils -modified in 2017. Firstly, it is stated that olive oil extracted by a chemical process is excluded from the class of VOO. Secondly, olive oil extracted from pomace (residue of olive fruit and stone) is also excluded from the list of VOO. Thirdly, only olive oil produced by mechanical processes and complying with the criteria of edibility can be considered as a VOO. Inside the category of VOO, three grades are presented: Extra, Virgin, and Ordinary Virgin Oil. The criteria used to discriminate between these three grades of VOO are the organoleptic properties, mainly the median of fruity attributes and the median of the defects, acidity, and absorbance in the ultra-violet (Fig. 2.2). It is noticeable that IOC and Codex Alimentarius are the only international standards that define ordinary VOO, which is an intermediate grade between virgin and lampante oils. Figure 2.2 shows an overview of olive oil fractions obtained by the grading methods of CODEX STAN 33-1981.

It is noticeable that pomace olive oil (obtained after the extraction of EVOO and VOO, and refined) is preferred to VOO by certain consumers, due to the mildness of their characteristics (viscosity, color, and flavor). Pomace olive oil is thus a category of "olive oil" comparable to most oils found in the market. Edible virgin fats and oils must be obtained only by mechanical processes (e.g. expelling or pressing) without altering the proprieties of the lipids. If necessary they may be clarified with water (generally hot), filtered and centrifuged.

Cold-pressed edible vegetable fats and oils are obtained by the same process but without the application of heat to the paste, that is, without the conditioning/cooking step (please see Sect. 3 for details on oil extraction and refining processes). Table 2.4 lists the general characteristics as well as specific quality features of each category, according to CODEX STAN 19-1981. Table 2.5 displays the regulations, labeling criteria and requirements within each grade. The use of food additives is not authorized in products covered by this standard (mainly virgin and cold-pressed oils).

Colorings are generally not allowed in vegetable oils, with the exception of curcumin, beta-carotenes (of vegetable origin, synthetic or microbial origin- *Blakeslea trispora*), beta-apo-8'-Carotenal, beta-apo-8'-carotenoic acid, methyl or ethyl ester and annatto extracts; bixin-based colorings are authorized only below a fixed maximum level.

Flavorings used to aromatize edible fats and oils shall comply with the guidelines for the use of flavorings (CAC/GL 66-2008) as indicated by the present standard. Antioxidants, antioxidant synergists and antifoaming agents authorized by the new version of the CODEX STAN 19-1981 are listed in the positive list inside the standard with their maximum use level.

In addition to these recommendations, some supplementary obligations have been added to the CODEX STAN 19-1981/2017.

The standard for olive oils and olive-pomace oils – codex stan 33 – 1981 was adopted in 1981, revised four times (1989, 2003, 2015 and 2017) and amended in 2009 and 2013. As entitled, the scope of the standard covers olive oils and olive-pomace oils, very similarly to IOC. However, when referring to olive-pomace oils and other refined fractions, the present standard gives the industrials the opportunity to compensate for the losses in vitamins during refining. Thus, vitamin E can be added in such a way to do not exceed 200 mg/kg in the final product of alpha-tocopherols (d-alpha tocopherol -INS 307a), mixed tocopherol concentrate (INS 307b) and dl-alpha-tocopherol (INS 307c).

As in IOC definition, in order to be mentioned as "Virgin", olive oil must be produced only by mechanical/physical processes under controlled thermal conditions and without the addition of any kind of additives. Operations of washing, decanting, centrifuging and filtration can be used according to the recommendation of the present standard.

Edible vegetable oils (EVO) covered by the present standard must meet the following definition: EVO "are foodstuffs which are composed primarily of glycerides of fatty acids being obtained only from vegetable sources. They may contain small amounts of other lipids such as phosphatides, unsaponifiable constituents and of FFA naturally present in the fat or oil". Besides the common proprieties of EVO exhibited later, each vegetable oil has specific characteristics, represented mainly by the fatty acid profile. Essential composition and quality factors of edible vegetable oils are presented in detail by the CODEX STAN 210 -1999 (last revision in 2017) (Tables 2.4 and 2.6).

The standard for named animal fats, CODEX STAN 211-1999, was amended three times, firstly in 2009, secondly in 2013 and recently in 2015. Authorized food additives, such as color and antioxidants, are exhibited in the present standard with the corresponding maximum use level. Essential composition and quality factors of named animal fats covered by the standard are mainly the percentages of the fatty acid of each kind of fat. The standards that cover animal fats (Tables 2.1 and 2.5) are as follows:

Spreads and blended spreads are emulsions (plastic or fluid) mainly of water and edible fats and oils, and regulated by Codex Stan 256-2007, which has been adopted for the first time in 1999, revised twice and amended in 2017. Codex stan 256-2007 covers products intended mainly for spreading and containing a percentage of fat ranging from 10% to 90%, that is, only margarine and products having the same purposes. However, this standard excludes products with a fat content

Table 2.6 Fatty acid profile (% w/w) of some edible fats and oils (adopted by Codex Alimentarius and IOC)

Fatty acid	Virgin oil	Refined oil				Animal fat	
	VOO	Arachis/peanut	Palm	Soybean	Sunflower	Lard fat	Cod liver
C6:0	NA	ND	ND	ND	ND	≤0.5	NA
C8:0	NA	ND	ND	ND	ND	≤0.5	NA
C10:0	NA	ND	ND	ND	ND	≤0.5	NA
C12:0	NA	ND - 0.1	0.05–0.5	0.05–0.1	0.05–0.1	≤0.5	NA
C14:0	≤0.03	ND - 0.1	0.5–2.0	0.05–0.2	0.05–0.2	1.0–2.5	2.0–6.0
C16:0	7.5–20.0	5.0–14.0	39.3–47.5	8.0–13.5	5.0–7.6	20–30	7.0–14.0
C16:1	0.30–3.50	ND - 0.1	0.05–0.6	0.05–0.2	0.05–0.3	2.0–4.0	4.5–11.5
C17:0	≤0.40	ND - 01	0.05–0.2	0.05–0.1	0.05–0.2	≤1	NA
C17:1	≤0.60	ND - 01	ND	0.05–0.1	0.05–0.1	≤1	NA
C18:0	0.50–5.00	1.0 - 4.5	3.5–6.0	2.0–5.4	2.7–6.5	8–22	1.0–4.0
C18:1	55.0–83.0	35.0–80.0	36.0–44.0	17–30	14.0–39.4	35–55	(n – 7) 2–7 (n – 9) 12–21
C18:2	2.50–21.00	4.0–43.0	9.0–12.0	48.0–59.0	48.3–74.0	4–12	0.5–3.0
C18:3	≤1.00	ND - 0.5	0.05–0.4	4.5–11.0	0.05–0.3	≤1.5	0.05–2.0
C20:0	≤0.60	0.7–2.0	ND	0.1–0.6	0.1–0.5	≤1.0	NA
C20:1	≤0.50	0.7–3.2	0.05–0.2	0.05–0.5	0.05–0.3	≤1.5	5.0–17.0
C20:2	ND	ND	ND	0.05–0.1	ND	≤1.0	NA
C22:0	≤0.20	1.5–4.5	ND	0.05–0.7	0.3–1.5	<0.1	NA
C22:1	ND	ND - 0.6	ND	0.05–0.3	0.05–0.3	<0.5	5.0–12.0
C22:2	ND	ND	ND	ND	0.05–0.3	NA	NA
C24:0	≤0.20	0.5–2.5	ND	0.05–0.5	0.05–0.5	NA	NA
C24:1	ND	ND - 0.3	ND	ND	0.05	NA	NA

ND not-detectable or defined as ≤0.05%, *NA* not applicable or not available

lower than two-third of the dry matter, and fat spreads obtained solely from milk and/or milk products.

Codex Stan 329-2017 describes fish oils as products derived from the raw material as defined in Sect. 2 of the code of practice for fish and fishery products (CAC/RCP 52-2003). This standard has recently been adopted and covers exclusively fish oils.

5.1 Regulations on Quality Parameters

The CCFO has adopted (since 1987) a standard that recommends a particular practice for the storage and the transportation of edible fats and oils: "code of practice for the storage and transport of edible fats and oils in bulk CAC/RCP 36-1987". Such matters are correlated to food safety and to food fraud with economic impact and potential threats to the human health. Given the serious consequences of the non-observance of such code of practice, it is translated or adapted nationally into regulations with the force of law. Most countries issue their own laws to regulate such operations and thus avoid problems related to human health and food fraud. Examples of such regulations of national ambit are the Tunisian law (Arrêté du ministre de l'industrie, de l'energie et des petites et moyennes enterprises du 11 février 2005, portant approbation du cahier des charges relative à l'organisation de l'activité de conditionnement des huiles alimentaires et à la creation d'une commission de contrôle technique, J Officiel de la Republique Tunisiene n°14/2055), the Brasilian law (BRASIL.Portaria SVS/MS n°. 326-1997, Seção I) or the transposition to the European national laws of Directive n ° 93/43/CEE. In this last case, a special emphasis goes to the maritime transportation in bulk.

Even when human health is not at stake, fraud can ruin the reputation of a brand name or even an entire class of products, afflicting many actors along the value chain. Mass media and social media amplify the issue, and usually do not distinguish between perpetrators from isolated non-intentional flaws in quality control, or even stakeholders simply affected by the fraud. Following the broadcast of a fraud episode, the entire foodstuff market is afflicted by the doubts of the consumers on the safety and authenticity, with the risk that many opt out to do not consume that product anymore.

Within each country, local food safety authorities and other agents ensure the application of the regulations, and the conformity to their contents (including composition, contaminant levels, and procedures during frying operations), namely through regular analysis and random inspection actions. Quality may thus be viewed as depending on regulations, which include a set of definitions of product designations, a set of analytical methods to assess product composition and other relevant attributes, as well as a list of maximum and minimum values for the relevant parameters. Standard methods and limits for parameters are the basis for assessing purity, authenticity, safety and other quality features. It means that analytical methods on which legislation is based on have to be validated and their accuracy and reproducibility need to be assessed.

In what concerns the safety aspects of edible oils and fats, nutritional features are monitored in addition to the control of contaminants. The toxic compounds that may appear in oils and fats can be:

(a) Toxins of biological origin, which is the case of substances naturally occurring in certain seeds (erucic acid, mostly from rapeseed, and glucosinolates from *Brassicaceae*) or mycotoxins that may be produced by fungi during seed storage (aflatoxins A and B)
(b) Chemicals assimilated by animals from feed or by plants during cultivation, and dependent on environment and farming practices: e.g. pesticide residues, heavy metals, and genotoxic compounds
(c) Toxic compounds accumulated during oil processing steps: as by-products -cyclic hydrocarbons, glycidic esters, i-TFA; or contaminants- MOSH, MOAH, volatiles from fumes
(d) Toxic compounds produced during frying and other cooking uses: acrylamide, breakdown compounds (short chain bound compounds)

The next section deals with specific analytical methods for relevant parameters in oils and fats, and labeling will be the object of Sect. 7, but let's first discuss the relationship between analytical control, labeling, safety and authenticity of products.

To limit the undesirable compounds mentioned in (a) firstly, deterioration during storage and transportation of the seeds and fruits should be prevented thus minimizing the risk of mycotoxin accumulation. Secondly, only authorized plant species and or animal fats shall be used in blended oils and spreads; a good traceability system can be an important tool in this control. Although safety is of utmost importance, authenticity aspects cannot be disregarded. Some quality labels (IP-related to a region; halal, kosher) require the authenticity and purity of oils and fats to be analyzed, to ensure the trust of consumers. Risk management is thus an important tool in these label-associated certification processes.

In the case of the chemicals mentioned in (b), epidemiological and mechanistic studies have been demonstrating the deleterious health impact of pesticide residues, pollutants, antibiotics and other substances used to increase production yield. Food safety authorities have been adjusting TDI (tolerable daily intake) accordingly to new scientific evidence. As many consumers are worried about the toxic compounds that can be found in foods, a specific quality label –the so-called "organic label"– intends to address food safety and environmental concerns of consumers. Certified organic foods are intended to comply with a framework of rules, which however are not globally consensual –e.g. differences between EU and USA definitions of "organic foods" (see Sect. 7 for details).

In what concerns contaminants referred to in (c), it should be recalled that most edible oils and fats require a stepwise processing (Sect. 3). Refining, on one hand, eliminates unwanted components (gums, natural toxins) and, on the other hand, may cause the accumulation of toxic by-products, such as 2- and 3-MCPD esters and glycidyl esters. Contamination with MOSH and MOAH may be due to unsuitable equipment and/or lack of control during some unit operations. The so-called i-TFA, are formed during some processing steps of oils and fats, as transesterifications and partial hydrogenations. i-TFA are *trans* isomers which double bounds (C=C) are located in positions different from the TFA of natural origin (e.g. spreads from hydrogenated seed oil *vs* butter). Evidence point towards the particularly harmful effects of these industrially produced TFA in human health. Since several years ago, some governments and large food industries have been working towards the stepwise reduction of i-TFA, namely by reformulating products and avoiding partially hydrogenated fats. More recently WHO (2018) called for total elimination of i-TFA from foods by 2023, in order to mitigate the burden of obesity and non-communicable diseases on health-care systems, and to improve the well-being of populations. Worldwide, task forces involving different stakeholders have been working for years with policymakers and compliant legislation is expected to be released from 2018 onwards.

However, the implementation of the REPLACE's plan of action, recommended by WHO (2018), follows different paces. Many food industries are committed to replacing palm and other hydrogenated fats with healthier options, reformulating products and developing new healthy foods. However, some other food industries and governments prefer to buy some more time, in order to develop strategies that please several stake-holders. This different pace in eliminating i-TFA may create obstacles to trade. The label will be the expression of the extent of reduction in i-TFA, and if standards and limits are not harmonized, the same wording may have different meanings and that will be a labeling issue. It is necessary to note that labeling is linked to aspects of traceability, analytical control and transparency, and it is of utmost importance to communicate with the consumer, ensuring trust or allowing mistrust.

In last, concerning the toxic compounds produced during food processing (as mentioned above in d) these are related to the global changes in food habits that have been noticed by international organizations. A state of "nutrient transition" of high-caloric/low nutrient diet has been registered worldwide. Highly processed foods are on the rise and almost all are highly rich in lipids. Only recently, the avail-ability of fat for food uses seems to be decelerating (Fig. 2.1). UN organizations have been making efforts to inform and motivate consumers and food industries to shift current mainstream direction (non-sustainable and a burden to health) towards more healthy and sustainable diets, namely by setting goals and plans of action (UN 2017). Independently of the dietary patterns, the use of edible oils by food indus-tries, hotels, restaurants and catering are more and more strictly regulated and inspected (despite the differences, stressed above).

Labeling claims and logos are major mechanisms to communicate "quality attri-butes" to consumers. Examples include claims as "contains-no xxx", "halal", "kosher", "organic", "vegan", or the logos corresponding to the designations in Table 2.3. These claims and logos are displayed on the product's label, and are asso-ciated with a certain number of attributes required by market niches or larger groups of consumers. In most cases they correspond to strict requirements, which compli-ance is ensured by renowned organizations and their mechanisms of action. In the global market, it would be desirable to have a consensus on standards, on analytical methodologies and on surveillance programs, which is unfortunately still an utopia. Not all situations are clear and the same label designation may translate differently according to the region and to the organization responsible to verify compliance. Viz. definitions of "organic" in EU and in the USA, as well as the differences in the requirements for i-TFA. Divergences in the meaning and in the limits of relevant parameters may create barriers to trade and confuse consumers.

6   Analytical Methodologies in Support of Consumers and Regulatory Bodies

Food industries use several mechanisms to guarantee the authenticity of their prod-ucts to the consumer. Several of these mechanisms are connected to the labeling process. Generally, inspection services verify the accuracy of the information con-tained on the label, and different degrees of penalties are applied in case of non-compliance. For this purpose, and for each parameter on the label, inspection services rely on rules and specifications to compare the required characteristics to the appropriate standard. For example, when considering nutritional labeling, inspection services rely on standard analysis methods and appropriate apparatus. As referred to before, the harmonization of analytical methods is of upmost importance for the product categorization, and for the verification of the authenticity of edible fats and oils. This aspect is of primordial importance in the case of EVOO, because of the high probability of fraud involving olive oil (European Parliament 2013). Thus, details on analytical methods to detect fraud will be presented and further discussed herein, using olive oil as an example.

A first broad distinction among oils and/or fats can be achieved by comparing fatty acid composition, because each oil or fat has a characteristic fatty acid profile, as illustrated by the examples presented in Table 2.6. Separation, detection and quantification of fatty acids are achieved by transforming FA C12 to C24 to their methyl esters, which can then be quantified by gas chromatography (GC). The methylation method consists of a *trans* esterification with methanol in the presence of KOH, at room temperature. IOC developed and validated their own method for the determination of the fatty acid composition from C12 to C24 as well as *trans* fatty acid content, in olive oil: COI/T.20/Doc. No 33 "Determination of fatty acid methyl esters by gas-chromatography". The recent version of the IOC's trade standard uses the FA profile as a purity criterion.

Codex Alimentarius recommends, as trade standards, the COI regulation (COI\T.20\Doc. no 24), as well as methods from ISO (ISO 5508:1990) and AOCS (AOCS Ch 2-91 (02) or AOCS Ce 1f-96 (02)). Like IOC, EU regulates the analysis of fatty acid composition from C12 to C24 (including saturated, *cis* and *trans* mono-unsaturated, and *cis* and *trans* PUFA) although not only for olive oil but also for a variety of edible oils and fats.

The fatty acid profile also gives an indication about the authenticity of a given oil or fat. Olive oil has a high content of monounsaturated fatty acids (MUFA), a fraction dominated by C18:1 (oleic acid) (Table 2.6). The remaining fractions are a combination of saturated (SFA) and PUFA, including the essential fatty acid and eicosanoid precursor C18:3, *n3*-ALA (alfa-linolenic acid), which is thought to play an important role in the health-related effects of the Mediterranean dietary pattern (Delgado et al. 2017).

In the case of arachis (or peanut), soybean and sunflower oils the dominant fraction is that of PUFA, while palmitic acid (SFA) is predominant in palm fat. In the case of lard (fat from pork) a variety of SFA constitutes the predominant fraction. The high SFA proportion (in relation to MUFA and PUFA) found in lard and palm explain why they are normally solid at room temperature. However, fractionation steps allow separation of SFA from MUFA and PUFA, a common practice in the industrial processing of palm oil and animal fats (see Sect. 3, and Tables 2.1 and 2.2, for more details).

In addition, to the content in each FA, their distribution in the glycerol molecule indicates the origin of the triacylglyceride (TAG). Even if two botanical species would produce lipids with identical FA profiles, the esterification of these FA units to the glycerol molecule would occur by a different order, thus forming different TAGs. After analysis of these TAGs, with an equivalent carbon number equal to 42

(ECN42), experimental and theoretical profiles can be compared, by computing ΔECN42, which expresses the difference between the theoretical and experimental model of the profile of certain TAGs. The method is approved by COI and coded as COI/T.20/Doc No 20/rev.3 (2010). According to García-González et al. (2017) it is equivalent to the method of AOCS coded 5b-89. It is noteworthy that this parameter (ΔECN42) is particularly sensitive to the presence of seed oil (detectable limit ≥1%). Moreover, the characterization of fatty acids in position 2 (i.e., in the central position of the "lyre") and more particularly the search for palmitic acid in position 2 informs on the possibility of an inter-esterified oil, re-esterified or the presence of palm oil (Pouyet and Ollivier 2014).

The peroxide value is the main parameter to characterize the quality of an oil. Peroxides are intermediates of autoxidation reactions underwent by oils, giving thus a measure of the extent of primary oxidation of a given oil sample (oxidative rancidity). Peroxide value is used to grade oil samples by quality. The method involves dissolving the sample in acetic acid/chloroform before being treated with potassium iodide; the released iodine is finally titrated with sodium thiosulphate. Despite being laborious and dependent on the analyst' skills and practice, this method is adopted by COI, ISO and AOCS, with no major differences among procedures and limits (García-González et al. 2017).

The absorbance in the ultraviolet allows the determination of conjugated dienes (-C=C-C=C-) or trienes (-C=C-C=C-C=C-) resulting from oxidation processes and/or refining providing information on the quality of a fat, its state of preservation and the effect of technological processes. Iso-octane is the solvent used to measure the absorbance at 232 and 268 nm, while cyclohexane is used at 232 and 270 nm. IOC, USDA and others recommend this analysis to investigate the adulteration of VOO with refined seed oils. On the other hand, AOCS and AOAC support the use of vibrational spectroscopy (mid-infra-red) for the rapid determination of total SFA, *trans*, MUFA, and PUFA in oils, namely by the standard methods AOCS Cd 14-95 and AOAC 965.3, as well as AOCS Cd 14d-99 and AOAC 2000.1. According to Mossoba et al. (2012), the discrimination between functional groups, and subsequent identification of individual fatty acids can be achieved by analyzing the IR spectra trough quantitative chemometrics tools and multivariate analysis. As GC is the reference method, the use of GC's results to calibrate IR spectroscopic analysis would allow for routine measurements, probably with sufficient accuracy to address nutritional labeling requirements.

Regarding alkyl esters, these are formed by esterification of FFA with low molecular alcohols and it is an analysis particularly relevant, again, in detecting food fraud in olive oil. It is a quality parameter for EVOO allowing detecting mixtures with deodorized oils and lower quality olive oils. Gas chromatography is the reference analytical technique, but liquid or thin layer chromatographic methods may be required for sample preparation. This method is coded by COI as COI/T.20/Doc No 31 (2012).

Moisture and volatile matter in oils are determined by gravimetric methods; the sample is heated at 105 °C (sand-bath) to evaporate volatiles and water, quantifying the absence, in accordance to IOC and ISO methods.

The unsaponifiable matter of an oil includes lipids of natural origin, such as sterols, hydrocarbons and alcohols, aliphatic and terpenic alcohols, as well as contaminants extracted by the solvent and not volatile at 103 °C (e.g. mineral oils). Similarly to the methods for the determination of the peroxide value, also the quantification of the unsaponifiable fraction (expressed per Kg of oil) lack accuracy and precision. Procedures are described by IOC, ISO (namely ISO18609:2000) and AOCS (AOCS ca 6b-53). The unsaponifiable fraction includes beneficial and harmful compounds that can be further discriminated in other analysis. Some constituents of the unsaponifiable fraction can be used as indicators of the authenticity of olive oils from certain regions/fruit cultivar, as well as to detect adulteration by blending olive-pomace oils and others in EVOO. To detect such a fraud, IOC presents an analytical method for total phytosterols, or 4-desmethylsterols, erythrodiol, and uvaol, coded COI\T.20\Doc. No 30, "Determination of the composition and content of sterols and triterpene dialcohols by capillary column gas chromatography". This method is somewhat different from the EU's analytical procedure for determining the individual and the total sterols and triterpendialcohols content of olive oils and olive-pomace oils. García-González et al. (2017) have confirmed that the limit of each sterol determined by both analytical procedures (IOC and EU) result in the same limit values for individual and total sterols and the sum of erythrodiol and uvaol.

The refining of the oil triggers a dehydration of phytosterols, which leads to their transformation into sterenes. Virgin oils contain little or no stigmastadienes, and therefore their presences above the limits is an indicator of the presence of blended refined oils. In order to investigate that type of fraud, the IOC recommends the analysis of stigmastadienes according to COI/T.20/Doc. No 11, "Determination of stigmastadienes in vegetable oils", or COI/T.20/Doc. no. 16, "Determination of sterenes in refined vegetable oils". Alternatively, ISO 15788-1 or AOCS Cd 26-96 can be used. The IOC and EU recently lowered the limit for stigmastadiene from 0.1 to 0.05 mg\kg in EVOO.

Other compounds may be used to distinguish refined and non-refined oils, from other sources than olive. It is the case of phytol, the tetramethyl-branched, monoenoic alcohol, which is widespread in nature as a part of chlorophyll. In chlorophyll, only *trans*-phytol is found and thus the absence of *cis*-phytol can be used as a marker for non-refined (e.g. cold-pressed) edible oils (Vetter et al. 2012).

Finally, the organoleptic profile is an important parameter in the grading of olive oils, distinguishing regions, cultivars and levels of quality. Organoleptic features may determine the classification of an oil in a superior or inferior category.

The organoleptic profile is complementary to physicochemical parameters and only applies to virgin oils, particularly to EVOO and VOO. This analysis makes possible to judge both, the condition of the olives before trituration and the state of the oil. The IOC created a strong framework for the sensorial aspects of VOO, after many norms were created, validated and revised. An organoleptic analysis relies on the opinion of at least eight panelists or tasters that are members of a panel, and are rigorously selected (for their discriminatory abilities on smell and taste). Panels members are trained to recognize the five defects of olive oil, as well as to distinguish

Table 2.7 Norms established by the International Olive oil Council (IOC) covering the sensorial analysis of VOO

Standard	Objective
COI/T.20/Doc. No 4/Rev. 1, 2007	General basic vocabulary for the sensory analysis
COI/T.20/Doc. No 5/Rev. 1, 2007	Kind of glasses for virgin olive oil tasting
COI/T.20/Doc. No 6/Rev.1, 2007	Guide for the installation of a test room
COI/T.20/Doc. No 14/Rev. 4, 2013	Guide for the selection, training and monitoring of skilled VOO tasters
COI/T.20/Doc. No 15/Rev. 8, 2015	Method for the organoleptic assessment of VOO
COI/T.28/Doc. No 1, 2007	Guidelines for the accreditation of laboratories undertaking the sensory analysis of VOO
Annex I of COI/T.20/Doc. No15/Rev. 8	Method for calculating the median and the confidence intervals
COI/T.20/Doc. No 22, 2005	Organoleptic assessment particularly focused on designations of origin

the fruity, bitter and spicy grades, as defined in Appendix XII of RCE 2568/91 and in the COI/T.20/Doc. No 4/Rev. 1, 2007. This panel of experts is led by a panel leader, whom should be recognized by the IOC. Such a panel has to decide on the category of a virgin oil (extra, virgin or lampante), as defined by the trade standard COI/T.15/NC No 3/Rev. 11 July 2016. Thus, the panel of tasters must be able to grade with expertise such as the variation of the notes of the jury members around the median score of the panel should not exceed 20% (Pouyet and Ollivier 2014). The framework of norms that regulate organoleptic tests on olive oil is listed in Table 2.7.

When applying analytical methods many sources of error and variation should be considered. The first set of variation comes from the way a procedure is written and interpreted. In the case of research papers, summarizing methods and omitting details is a common procedure. Even the interpretation of a detailed norm may depend on the expertise and mindset of the analyst. Finally, the implementation of a method varies with the laboratory conditions, from ambient temperature to the equipment's (which performance may vary with a great number of factors) and even with the brand of certain reagents (even of the same analytical grade). In addition, different regulatory bodies propose a diversity of analytical methods for the same purpose, as happens with the limit values for each parameter, as well as established standards, as explained in-depth by García-González et al. (2017).

7 Aspects on Labeling

"Label", as defined by Oxford Online Dictionary, is "a small piece of paper, fabric, plastic, or similar material attached to an object and giving information about it". In practice, the label is the major communication system between the stakeholders in a food value chain and the consumer. The label shows the image of a brand, transmits trust through a certified allegation or logo, informs the consumer of relevant nutritional and food safety aspects, and facilitates traceability. Food labeling may not be limited to a paper stamped on a food package. In fact, labels and food packaging is expected to greatly evolve in a near future, by means of ICT applications. Food labeling needs to comply with legal frameworks and is crucial in trade, like it is related to analysis, limits for parameters, grading, standards and designations. In fact, the label contains and condenses many pieces of information and is a way to guarantee the authenticity, quality, and safety of a product.

In what concerns oils and fats, the majority of standards establishes labeling requirements in relation to the origin of the oil or fat (animal and/or vegetable), information on processing (refined or virgin), nutritional and quality details, handling and shelf-life information, and country of origin. It is important to note that according to Reg (EU) No 1169/2011 (also called INCO regulation), oils produced from vegetable origin should be mentioned as "vegetable oils", along with an indication for the specific origin. In addition, the American Nutrition Labeling and Education Act (NLEA), which amended the FFDCA, requires some more specific indication to the edible fats and oils. Table 2.8 discriminates the main requirements on origin for the labeling oils and fats, according to a variety of regulatory bodies.

7.1 Labelling in Accordance to Codex Alimentarius

The Codex Committee on Food Labelling (CCFL) has elaborated eight standards related to the labeling procedure:

- General Guidelines on claims -CA GL 1-1979- last modification on 2009
- Guidelines on Nutrition Labelling CAC/GL 2- 1985 last modification on 2017
- Guidelines for use of Nutrition and Health Claims CAC/GL 23-1997 last modification on 2013
- General Guidelines for use of the Term "Halal" – CAC/GL 24-1997- last modifications on 1997
- Guidelines for the Production, Processing, Labelling and Marketing of Organically Produced Foods -CAC/GL 32-1999- last modification on 2013
- Compilation of Codex texts relevant to the labeling of foods derived from modern biotechnology -CAC/GL 76-2011- last modification 2011
- General Standard for the labeling of prepackaged foods -CDEX STAN 1-1985- Last modification 2010
- General Standard for the labeling of and claims for prepackaged foods for special dietary uses -CODEX STAN 146-1985- last modification on 2009.

In the case of edible fats and oils, the general standard for the labeling of prepackaged foods may be used jointly with a specific recommendation given by the appropriate standard.

Table 2.8 Ingredient categories for which the category indication may replace that of the specific name -based on specific requirement of some international standards

Standard	Name of classes	Class name
General standard for labeling of prepackaged foods (CODEX STAN 1-1985)	Refined oils other than olive	"'Oil' together with either the term 'vegetable' or 'animal', qualified by the term 'hydrogenated' or' partially-hydrogenated', as appropriate"
	Refined fats	"'Fat' together with either, the term 'vegetable' or 'animal', as appropriate"
Regulation (EU) No 1169/2011	Refined oils of vegetable origin	"May be grouped together in the list of ingredients under the designation 'vegetable oils' followed immediately by a list of indications of specific vegetable origin,
		The expression 'fully hydrogenated' or 'partly hydrogenated', as appropriate, must accompany the indication of a hydrogenated oil"
	Refined fats of vegetable origin	"May be grouped together in the list of ingredients under the designation 'vegetable fats' followed immediately by a list of indications of specific vegetable origin,…
		The expression 'fully hydrogenated' or 'partly hydrogenated', as appropriate, must accompany the indication of a hydrogenated fat"
	Refined oils of animal origin	"'Oil', together with either the adjective 'animal', or the indication of specific animal origin. The expression 'fully hydrogenated' or 'partly hydrogenated', as appropriate, must accompany the indication of a hydrogenated oil"
	Refined fats of animal origin	"'Fat', together with either the adjective 'animal' or the indication of specific animal origin. The expression 'fully hydrogenated' or 'partly hydrogenated', as appropriate, must accompany the indication of a hydrogenated fat"

U.S. FDA Food, beverage, and supplement labeling requirements	"…(14) Each individual fat and/or oil ingredient of a food intended for human consumption shall be declared by its specific common or usual name (e.g., "beef fat", "cottonseed oil") in its order of predominance in the food except that blends of fats and/or oils may be designated in their order of predominance in the foods as "Ill shortening" or "blend of Ill oils", the blank to be filled in with the word "vegetable", "animal", "marine", with or without the terms "fat" or "oils", or combination of these, whichever is applicable if, immediately following the term, the common or usual name of each individual vegetable, animal, or marine fat or oil is given in parentheses, e.g., "vegetable oil shortening (soybean and cottonseed oil)". For products that are blends of fats and/or oils and for foods in which fats and/or oils constitute the predominant ingredient, i.e., in which the combined weight of all fat and/or oil ingredients equals or exceeds the weight of the most predominant ingredient that is not a fat or oil, the listing of the common or usual names of such fats and/or oils in parentheses shall be in descending order of predominance. In all other foods in which a blend of fats and/or oils is used as an ingredient, the listing of the common or usual names in parentheses need not be in descending order of predominance if the manufacturer, because of the use of varying mixtures, is unable to adhere to a constant pattern of fats and/or oils in the product. If the fat or oil is completely hydrogenated, the name shall include the term partially hydrogenated. If each fat and/or oil in a blend or the blend is completely hydrogenated, the term "hydrogenated" may precede the term(s) describing the blend, e.g., "hydrogenated vegetable oil (soybean, cottonseed, and palm oils)", rather than preceding the name of each individual fat and/or oil; if the blend of fats and/or oils is partially hydrogenated, the term "partially hydrogenated" may be used in the same manner. Fat and/or oil ingredients not present in the product may be listed if they may sometimes be used in the product. Such ingredients shall be identified by words indicating that they may not be present, such as "or", "and/or", "contains one or more of the following:", e.g., "vegetable oil shortening (contains one or more of the following: cottonseed oil, palm oil, soybean oil)". No fat or oil ingredient shall be listed unless actually present if the fats and/or oils constitute the predominant ingredient of the product, as defined in this paragraph (b)(14)"		
Tunisian Order of the Ministers for Trade and Handicrafts, Public Health, Industry, Energy and Small and Medium Enterprises of 3 September 2008 on the labeling and presentation of prepackaged foods		Refined oils other than olive oil	"Oil", qualified as "vegetable" or "animal" and "hydrogenated" or "partially hydrogenated" as appropriate
		Refined fat	"Fat", qualified as "vegetable" or "animal", as the case may be

7.2 Labelling in Accordance to IOC

The Chap. 10 of the IOC's trade standard discuss the labeling procedure of olive oil. In addition to the requirements of the Codex General Standard for the Labelling of Prepackaged Foods (CODEX STAN 1-1985), it recommends the use of some more specific information. A label must indicate a specific designation of the olive oil (as previously explained). The net content shall be mentioned by volume in the metric system ("SI" units). Name and address of the manufacturer, packer, distributor, importer, exporter or seller shall be declared. The label of the virgin of olive oil (EVOO or VOO) should indicate their country of origin, geographical indications and designation of origin. To recall/withdrawal efficiently non-conform products, an identification of the lot must be stamped on the container. The date of minimum durability should be indicated by the month and year. The durability date may be preceded by the word "best before end". The consumer must be informed about the adequate conditions of storage.

7.3 Quality Labels Managed by Other Organizations and Regulatory Bodies

Both official and private organizations manage product certification procedures aiming at adding value to noble edible fats and oils (when compared to cheaper and less nutritious ones), and to guarantee their authenticity, and specific attributes. Since the last century, there has been a significant wave of labeling of processed and unprocessed foods. Besides, the basic information given by the label, consumer selects their food product on the basis of some new criteria such as the origin of the product and the link to a particular territory and to the traditional methods of production.

The regional character of food products is linked to the protection of biodiversity and often to local traditions. These attributes are protected and displayed in labels that link IP to regions of which those of EU are noteworthy: Traditional specialty guaranteed (TSG), Protected Geographical Indication (PGI) and Protected Designation of Origin (PDO). These designations correspond to specific logos, and were created by the European Commission, Department of Agricultural and Rural Development. Such quality designations correspond to a strict list of conditions and attributes and the management and inspection are frequently of the competence of not-for-profit organizations (e.g. association of producers).

PGI cover agricultural products and foods linked to the geographical area where they are produced, processed, or prepared. PDO is a designation used for agricultural products and foodstuffs produced, processed and prepared in an area, and whose quality or properties are significantly or exclusively determined by the geographical environment, including natural and human factors. However, the TSG does not certify that the food product has a link to a specific geographical zone. Based on article 17 of the regulation (EU) n° 1151\2012 on quality schemes for agricultural products and foodstuffs, the objective of the TSG is mainly to safeguard traditional methods of production and recipes. In the sector of edible oils and fats only Poland has a TSG. The PDO covers 116 products, in which fats and oils produced in Italy and in Greece dominate the list (Table 2.3).

Concerning labels related to regional character, the landscape of certification in the Maghreb area is not as dense and rich as in Europe. For example, in Tunisia only one of such labels has been registered "IP – huile d'olive Monastir". Morocco has more quality label than Tunisia, e.g. 'IG huiles d'olive la Vierge Extra Ouezzane' and the 'AO l'huile d'olive Tyout Chiadma and l'huile d'olive Extra Vierge Aghmat Aylane'.

Returning to EU labels, the same department of EC (Department of Agricultural and Rural Development) also manages the certification and attribution of the logo for organic foods. In EU, organic foods must comply to Reg (CE) N° 834/2007 meaning that a given organic oil or fat has to be controlled along manufacturing processing in order to ensure that at least 95% w/w of their ingredients have been certified as originated from organic farming (organic farming systems and practices are also comprehensively defined). The global demand for food products labeled as "organic" is on the rise, but different regulations for the same designation co-exist in the global market.

In the USA, regulation of organic food production is of the competence of USDA, under the Organic Foods Production Act and CFR, Title 7: Agriculture, PART 205-National Organic Program, last amended in 2017, and under discussion. In the EU, organic farming (including aquaculture) and organic food production are regulated by Reg. (EC) n° 834/2007 and 889/2008, currently under discussion with amendments expected to 2020. European organic certification corresponds to a label linked to a geographical indication of raw materials and a comprehensive code of practice and rules. Due to non-compliance to these premises, cannot be labeled as "organic products": products in conversion, processed food and feed containing less than 95% organic ingredients, and products produced according to private or national standards. On the other hand, in USA the claims, "100 percent organic," "organic," or "made with organic (specified ingredients or food group(s))," are allowed, and in the case of agricultural products containing less than 70% organic ingredients, the term, "organic," is allowed in the ingredients panel. A wide variety of operations and practices are allowed, according to the specificities of each farm and given basic principles of the Organic Foods Production Act are respected. In short, an organic food from EU must comply to a stricter regulation than USA, with the consequent impacts on quality and safety aspects.

Another relevant trend among consumers from many countries concerns environmental and sustainability claims. The first thing is to distinguish between certifying a process as environmentally friendly, or attributing such claims to a given product. In respect to Eco-conception (of a product), the underlying processes must comply with Reg (CE) No 2009/125 or ISO 14062, or another equivalent. The allegations exhibited in the label should clearly point out the savings or other benefits, as savings in energy and/or water and/or packaging material, as well as information about the reduction in GHG emissions.

In respect to allegations directly linked to a product, an emblematic example is the recent massive rejection of palm oil, by many consumers, an action supported by several governments and large food companies, which voluntarily claim on the label "contains no palm oil". The reason for this palm oil rejection is mainly a response to the appeals of ecologists and scientists that showed as palm oil industries have been responsible for massive deforestations, destruction of rain-forest habitats, and increased risk of extinction of threatened species. Negotiations and compromises lead to the implementation of the claim "sustainable palm oil" based on a list of requisites and corresponding to a logo managed by the well-known not-for-profit global organization World Wildlife Fund (WWF).

Nutritional labeling is of utmost importance and is currently under changes worldwide, following the WHO recommendation to reduce fats in the diet, and now the WHO call to eliminate industrially produced i-TFA from foodstuffs. Many food companies have been making efforts to develop new products and to adapt existing foodstuffs, by reducing and replacing SFA and i-TFA for healthier and sustainable options. In some cases, such replacements are technically puzzling and in practice, some may result in the increased content of SFA. Such challenges posed to industries, to analysts, and to regulatory bodies, maybe somewhat explain the different pace of elimination of i-TFA that is observed, even in developed countries.

The European Food Safety Authority (EFSA) states that 'TFA intakes should be as low as is possible within the context of a nutritionally adequate diet', and surveillance programs are ongoing, as some subgroups may be exposed to high intake levels. EU strategy is aligned to that of WHO and the matter is under discussion in EC during 2018, concerning labeling and legal limits (Mouratidou et al. 2014). In Denmark TFAs are limited by law to 2% of total fat, since 2003 (FEDIOL 2017). A similar situation is reported in Brazil (According to RDC Resolution 360/2003 of ANVISA, TFA must be declared if >0.2 g per food portion; foods may be labeled as "zero *trans*." if containing less than 0.2 g/per portion). In Australia and New Zealand, no obligation exists to declare TFA in food labels, in 2018 (Food Standards Australia New Zealand 2018). Nutritional labels in EU and Canada, in 2018, call only for total fat and SFA but both, EU and Canada, are expected to release stricter legislation (including limits for i-TFA) during 2018. On the other hand, in the USA, FDA initiated the i-TFA ban in 2015, stating the obligation to declare and express *trans*-fat content of foods as grams per serving to the nearest 0.5-g increment; "if a serving contains less than 0.5 g, the content, when declared, must be expressed as "0 g" (FDA 2018). Recently, the FDA extended the compliance date to this rule (at least) until June 2019, for a wide category of food industries. These divergences between different world regions create asymmetries, potential mistrust of consumers and probable barriers to trade.

Other types of concerns of consumers are related to beliefs and lifestyles, and are normally reflected on labels too, although sometimes not so well regulated. Examples are claims of "fair trade", suitable for "vegans", as well as Halal and Kosher labels. In certain countries, food producers may blend vegetable oils with lard in order to reduce the production costs. In some food formulations, animal fats can replace vegetable oils to obtain specific texture and taste. Some examples are certain margarine, shortenings, and some specialty commodity oils. Lard fat (pig fat) and beef tallow can be blended in different proportions with plant oils such olive, palm, palm or coconut kernel, and canola.

Current consumer trends are in favor of fats and oils of vegetable origins, and even butter is experiencing a decline in demand (see Sect. 2). Nevertheless, blends of animal and vegetable fats are feasible, and hence not fraudulent, if every ingredient is detailed in the label. The blend of animal and vegetable fats may be more important than a simple dietary choice. In some communities, it can become an issue, as is the case of Islam, Judaism and Hinduism due to the religious prohibition of these commodities (lard and beef tallow). Thus, the detection of animal fat adulterants in vegetable oils is very relevant to these communities. The investigation of this kind of fraud is based on the analysis of fatty acid profile. As previously mentioned, each fat and oil has a typical composition in fatty acids and triacylglycerides. However, when animal fats (lar fat or beef tallow) are blended at low levels in vegetable oils, they become difficult to detect, due to limitations in sensitivity and reproducibility of methods and instruments, as explained by Al-Kahtani et al. (2014). Until now, no standard method was adopted to detect this kind of adulteration, although many analytical procedures based on the TAG profile can be found in the literature (Al-Kahtani et al. 2014; Man and Rohman 2011; Marikkar et al. 2005) with potential for method development and optimization. Marrikar et al. (2005) stress that pork fat has larger amounts of TAG, containing SFA at the Sn-2 position, than fats of another origin. Thus, it means that if a certain fat has been adulterated with lard, the ratio of TAG-containing SFA at position 2 (SSU) over TAG-containing MUFA/PUFA at the same (Sn-2) position (SUS) tends to increase. Analytical techniques proposed to detect and to quantify lard in vegetable oils include High performance liquid Chromatography (HPLC), Fourier transform infrared (FTIR) spectroscopy, differential scanning calorimetry (DSC), chromatographic-based techniques, and electronic nose (EN).

A typical consumer, concerned with the presence of lard in foodstuffs will have to rely on a specific label, as are the cases of the ritual labels: halal and Kosher. For food, halal certification -that is to say, in accordance with Islamic law- gradually extends to food products even outside the classical Islamic territories. To be labeled as Halal, foodstuffs must be manufactured without alcohol and swine fat, bovine fat, as prescribed in Islam and, more broadly, without any animal fat.

In such a booming market niche, the tendency to claim to be halal generates profits, and fraud is becoming more and more frequent. The imposition of a halal standard of worldwide application (as evidenced by the numerous initiatives to build a "halal standard") is the only guarantee that these products are genuine and will not mislead the consumer. Hence, detection of animal fat adulterants in vegetable oils is of great importance from economic, health and religion points of view. Some Islamic countries require halal certification for food imports. In addition, halal and kosher labels satisfy not only religious groups but also vegans, and persons suffering from some food allergies, as explained by Regenstein et al. (2003).

8 Concluding Remarks

Oils and fats are vital nutrients of the human diet, and may contain precursors of important human metabolites (e.g. ALA), oil-soluble vitamins, carotenoids, antioxidants and anti-carcinogenic compounds. The type, quantities and regularity of the intake of oils and fats in a diet determine how healthy and sustainable it is. In the Mediterranean diet, a major example of healthy and sustainable diet (Delgado et al. 2017), olive oil plays a central role, as a base for cooking, and as salad dressing, and may, at least in part, explain health benefits of this dietary pattern. On the opposite side, lays the so-called western diet, the global expression of the fast-food culture, led by a few multinational food industries. This low-nutrient, highly caloric western diet mostly consists on highly processed foods generally containing large quantities of fat, mainly SFA and i-TFA. To address the demand for such foodstuffs, industries have been widening the applications of oils and fats to a multitude of products (aiming from desirable textures, to longer shelf life), simultaneously increasing total fat in foods, until the use of oils and fats reached unsustainable levels, threatening the health of the populations and the environment. The recent polemic with palm oil and derivatives is an example of how the option of consumers can make a difference: after announcements of massive deforestation, habitat destruction, and threat to wildlife by palm growers, a large number of brands, consumer associations and even governments raised awareness on those aspects and reduced or eliminated palm oil from foodstuffs. In addition to sustainability issues, industrially used palm oil fractions raise concerns on health, as they are sources of many well-known deleterious components, from SFA to i-TFA, not to mention non-negligible risk for the presence of contaminants (e.g. mycotoxins, hydrocarbons from fuels). A category of so-called "sustainable palm oil" is arising in the global market, aiming at mitigating some of these negative aspects.

Olive oil is on the opposite side of nutritional and sustainability "scores". Olive is a xerophytic tree, thus able to withstand draughts and to cope with some climate change issues. Cold mechanical extraction of olive oil is an easy process, not requiring so much water or energy to obtain quite high yields, i.e. with an expected lower C and Water footprint than in most of the other oil processing. The olive oil industry can easily be profitable and sustainable, while valuing biodiversity and its ancestral identity. Olive is native from the Mediterranean region but olive oil is nowadays produced in several regions of the world. As it is highly valued for its aroma and health-promoting properties, it is also at the top of products that are the object of economic food fraud. To combat food fraud, a multitude of norms, analytical procedures, laws and actions have been developed. Unfortunately, the globalized market is not subjected to the same standards, procedures and norms. Despite some agreements on food trades, different countries and regions make use of different food control measures, their institutional and regulatory frameworks diverge widely, as do the definitions and labeling, as explained by Chammem et al. (2018). The lack of harmonization of analytical methods, inspection services and labeling cause barriers to trade and may also cause confusion and the mistrust of consumers. The availability and ease of use of information technologies may constitute an opportunity to evolve towards the utopic vision of global transparency and harmonization of standards.

Although the market is global, no global homogenous consumer exists but rather different groups or niches of consumers that require quality, safety, sustainability and a set of other conditions related to beliefs and way of life. Trade in the sector of edible fats and oils is very dynamic and a dense standardization has been developed, particularly by ISO and Codex Alimentarius, which are widely well accepted. A consensus regarding the definition of oil and fat categories, its grading and labeling requirements have been adopted, although with some changes by most international standards. Still, disagreements on the limit values of parameters and criterions have to be noted (e.g. allowed levels of i-TFA in USA foods may be higher than elsewhere; the definition of organic foods in EU is far more restrictive than in the USA). In addition, each organization has created and validated its own methods of analysis, and thus many results are not directly comparable. These issues pose a crucial impairment for exporters and importers, creating barriers to trade. Even with transparency and good communication, it is difficult to avoid the rise of consumers mistrust in certain products. The key is on the open dialogue, the acceptance of common standards (e.g. from widely accepted international organizations), dissemination of good practices and harmonization of methods and standards.

References

African Organization for Standardization, ARSO. (2018). Standards and global trade. http://www.arso-oran.org/standards-and-global-trade/. Accessed 28 Apr 2018.

Al-Kahtani, H. A., Abou Arab, A. A., & Asif, M. (2014). Detection of lard in binary animal fats and vegetable oils mixtures and in some commercial processed foods. *IJNFE, 8*(11), 1244–1252.

Aware Environmental. (2018). Why we choose to be palm oil free. http://awareenvironmental.com.au/whats-new/choose-palm-oil-free/. Accessed 28 Apr 2018.

BG. (2018). Iceland pledges to remove palm oil from own-brand food. Supermarket chain claims it is not possible to source mass market palm oil in a verifiably sustainable way. In J. Murray (Ed.), *Business green*. Available via https://www.businessgreen.com/bg/news/3029843/iceland-pledges-to-remove-palm-oil-from-own-brand-food. Accessed 28 Apr 2018.

Business Dictionary. (2018). Web Finance Inc. Austin. http://www.businessdictionary.com/. Accessed 28 Mar 2018.

Canard, F. (2009). *Management de la qualité*. Paris: Gualino Editeur. Lextenso Éditions.

Chammem, N., Issaoui, M., De Almeida, A. I. D., & Delgado, A. M. (2018). Food crises and food safety incidents in European Union, United States, and Maghreb area: Current risk communication strategies and new approaches. *Journal of AOAC International, 101*, 1–16. https://doi.org/10.5740/jaoacint.17-0446.

Collins English Dictionary. (2018). https://www.collinsdictionary.com/us/dictionary/english. Accessed 28 Mar 2018.

Delgado, A. M., Parisi, S., & Almeida, M. D. V. (2017). *Chemistry of the Mediterranean diet*. Cham: Springer International Publishing.

Dijkstra, A. J. (2018). Edible oil processing: Introduction to degumming. In *AOCS lipid library, oils and fats*. Available via http://lipidlibrary.aocs.org/OilsFats/content.cfm?ItemNumber=40325. Accessed 12 Mar 2018.

EFSA CONTAM Panel (EFSA Panel on Contaminants in the Food Chain). (2018). Scientific opinion on the update of the risk assessment on 3-monochloropropane diol and its fatty acid esters. *EFSA Journal, 16*(1), 5083, 48 pp. https://doi.org/10.2903/j.efsa.2018.5083.

EFSA Panel on Contaminants in the Food Chain (CONTAM). (2016). Scientific Opinion on erucic acid in feed and food. *EFSA Journal, 14*(11), 4593., 173 pp. https://doi.org/10.2903/j.efsa.2016.4593.

EFSA Panel on Dietetic Products, Nutrition and Allergies (NDA). (2011). Scientific Opinion on the substantiation of health claims related to olive oil and maintenance of normal blood LDL-cholesterol concentrations (ID 1316, 1332), maintenance of normal (fasting) blood concentrations of triglycerides (ID 1316, 1332), maintenance of normal blood HDL-cholesterol concentrations (ID 1316, 1332) and maintenance of normal blood glucose concentrations (ID 4244) pursuant to Article 13(1) of Regulation (EC) No 1924/2006. *EFSA Journal, 9*(4), 2044. https://doi.org/10.2903/j.efsa.2011.2044.

EFSA Panel on Dietetic Products, Nutrition and Allergies (NDA). (2012). Scientific Opinion on the substantiation of a health claim related to polyphenols in olive and maintenance of normal blood HDL-cholesterol concentrations (ID 1639, further assessment) pursuant to Article 13(1) of Regulation (EC) No 1924/2006. *EFSA Journal, 10*(8), 2848. https://doi.org/10.2903/j.efsa.2012.2848.

European Commission, EC. (2018). *Agriculture and rural development*. Brussels: DOOR. http://ec.europa.eu/agriculture/quality/database/index_en.htm. Accessed 28 Mar 2018.

European Palm Oil Alliance, EPOA. (2017). EUFIC study: Free from labels potentially misleading. https://www.palmoilandfood.eu/en/news/eufic-study-free-labels-potentially-misleading. Accessed 28 Mar 2018.

European Parliament. (2013). Procedure: 2013/2091(INI). Report 4 Dec 2013, on the food crisis, fraud in the food chain and the control thereof. http://www.europarl.europa.eu/sides/getDoc.do?pubRef=-//EP//TEXT+REPORT+A7-2013-0434+0+DOC+XML+V0//EN. Accessed 23 May 2018.

FEDIOL Aisbl- The EU Vegetable Oil And Protein Meal Industry, FEDIOL. (2017). FEDIOL view on an EU TFA limit vs. a ban on partially hydrogenated oil – why applying the US approach does not fit the EU context, 17NUT054. In http://www.fediol.be/data/17NUT054. Accessed 2 June 2018.

Food and Agriculture Organization of the United Nations (FAO). (2010). Nutrition and consumer protection division. Sustainable diets and biodiversity, directions and solutions for policy, research and action In B. Burlingame, & S. Denini (Eds.), *Proceedings of the international scientific symposium: Biodiversity and sustainable diets united against hunger*. Rome, 3–5 2010.

Food and Agriculture Organization of the United Nations (FAO). (2017). *Food balance sheets*. FAOStat. http://www.fao.org/faostat/en/#data/FBS. Accessed 12 Mar 2018.

Food and Agriculture Organization of the United Nations, FAO. (2018). *Shifting the balance: Getting the private sector to favour nutritious, affordable and accessible diets. Urgent action needed to tackle undernutrition, mineral deficiencies and escalating obesity*. Joint Press Release. 18 April 2018. Rome. Available via http://www.fao.org/news/story/en/item/1118441/icode/. Accessed 30 Apr 2018.

Food Standards Australia New Zealand. (2018). Nutrition and fortification. Trans-fatty acids. In: http://www.foodstandards.gov.au/consumer/nutrition/transfat/Pages/default.aspx. Accessed 14 June 2018.

García-González, D. L., Tena, N., Romero, I., Aparicio-Ruiz, R., Morales, M. T., & Aparicio, R. (2017). A study of the differences between trade standards inside and outside Europe. *Grasas y Aceites, 68*(3), e210. https://doi.org/10.3989/gya.0446171.

Gunstone, F. D. (2013). Oils and fats in the market place. In *AOCS lipid library, oils and fats*. Available via http://lipidlibrary.aocs.org/OilsFats/content.cfm?ItemNumber=39453. Accessed 12 Mar 2018.

Gunstone, F. D., Harwood, J. L., & Dijkstra, A. J. (Eds.). (2007). *The lipid. Handbook*. Boca Raton: CRC Press.

International Olive Oil Council, IOC. (2018). Olive world. http://www.internationaloliveoil.org/web/aa-ingles/oliveWorld/aceite1.html. Accessed 24 May 2018.

Keller, U. V. (2018). Seed preparation. In *AOCS lipid library, oils and fats*. Available via http://lipidlibrary.aocs.org/OilsFats/content.cfm?ItemNumber=40335. Accessed 22 Mar 2018.

Kemper, T. G. (2018). Meal desolventizing, toasting, drying and cooling. In *AOCS lipid library, oils and fats*. Available via http://lipidlibrary.aocs.org/OilsFats/content.cfm?ItemNumber=40327. Accessed 22 Mar 2018.

Man, Y. B. C., & Rohman, A. (2011). Detection of lard in vegetable oils. *Lipid Technology, 23*, 180–182. https://doi.org/10.1002/lite.201100128.

Marikkar, J. M. N., Ghazali, H. M., Man, Y. B. C., Peiris, T. S. G., & Lai, O. M. (2005). Distinguishing lard from other animal fats in admixtures of some vegetable oils using liquid chromatographic data coupled with multivariate data analysis. *Food Chemistry, 91*, 5–14. https://doi.org/10.1016/j.foodchem.2004.01.080.

Moore, J. C., Spink, J., & Lipp, M. (2012). Development and application of a database of food ingredient fraud and economically motivated adulteration from 1980 to 2010. *Journal of Food Science, 77*, R118–R126. https://doi.org/10.1111/j.1750-3841.2012.02657.x.

Mossoba, M. M., Azizian, H., Kramer, J. K. G. (2012). Application of infrared spectroscopy to the rapid determination of total saturated, trans, monounsaturated, and polyunsaturated fatty acids In *AOCS lipid library, oils and fats*. Available via http://lipidlibrary.aocs.org/Analysis/content.cfm?ItemNumber=40380. Accessed 2 June 2018.

Mouratidou, T., Livaniou, A., Saborido, C. M., Wollgast, J., Caldeira, S. (2014). Trans fatty acids in Europe: Where do we stand? A synthesis of the evidence: 2003–2013 In *JRC science and policy reports*. European Commission. Ispra: Joint Research Center, Institute for Health and Consumer Protection.

Nunes, S. P. (2007). Produção e consumo de óleos vegetais no Brasil. In *Deser. Boletim electrónico. Conjuntura Agrícola. Departamento de Estudos Sócio-Econômicos Rurais.* N° 159-June 07. Available via http://www.deser.org.br/documentos/boletim_completo/Boletim_159.pdf. Accessed 13 Mar 2018.

Pouyet, B., & Ollivier, V. (2014). Réglementations sur l'étiquetage et la présentation des huiles d'olive. *OCL, 21*(5), D508. https://doi.org/10.1051/ocl/2014005.

Reg (EC) 432/2012. (n.d.). Commission Regulation (EU) No 432/2012, of 16 May 2012, establishing a list of permitted health claims made on foods, other than those referring to the reduction of disease risk and to children's development and health. OJL 404, 30.12.2006.

Regenstein, J. M., Chaudry, M. M., & Regenstein, C. E. (2003). The kosher and halal food laws. *Comprehensive Reviews in Food Science and Food Safety, 2*, 111–127.

Smith, G. (2017). *Displaced orangutan becomes the face of new ethical label. The world's biggest plant oil industry has become infamous for devastating habitats. Now a charity has created a logo to help consumers cut ties with the palm oil market.* New Food, 27 Sept 2017. Russell Publishing Ltd. https://www.newfoodmagazine.com/news/44610/palm-oil-label/. Accessed 12 Mar 2018.

UN Sustainable Development Knowledge Platform. (2017). Food security and nutrition and sustainable agriculture. Goal 2. Progress and Info. https://sustainabledevelopment.un.org/sdg2. Accessed 6 June 2018.

United States Department of Agriculture, USDA. (2010). Agricultural marketing service. USDA revises the grade standards for olive oil. AMS No 039-10. https://www.ams.usda.gov/press-release/usda-revises-grade-standards-olive-oil. Accessed 6 Jun 2018.

US Food and Drug Administration, FDA. (2018). Labelling & nutrition. Guidance for industry: Trans fatty acids in nutrition labeling, nutrient content claims, health claims; small entity compliance guide. In: https://www.fda.gov/Food/GuidanceRegulation/GuidanceDocuments RegulatoryInformation/LabelingNutrition/ucm053479.htm. Accessed 15 June 2018.

Vetter, W., Schröder, M., & Lehnert, K. (2012). Differentiation of refined and virgin edible oils by means of the trans- and cis-phytol isomer distribution. *Journal of Agricultural and Food Chemistry, 60*, 6103–6107. https://doi.org/10.1021/jf301373k.

Voigt, M., Wich, S. A., Ancrenaz, M., et al. (2018). Global demand for natural resources elimi-
    nated more than 100,000 Bornean orangutans. *Current Biology, 28*, 761–769.e5. https://doi.
    org/10.1016/j.cub.2018.01.053.
WHO, World Health Organization. (2018). News. WHO plan to eliminate industrially-produced trans-
    fatty acids from global food supply. http://www.who.int/news-room/detail/14-05-2018-who-
    plan-to-eliminate-industrially-produced-trans-fatty-acids-from-global-food-supply.   Accessed
    6 June 2018.
World Wildlife Fund, WWF. (2018). Palm oil. https://www.worldwildlife.org/pages/which-every-
    day-products-contain-palm-oil. Accessed 12 Mar 2018.
Zeldenrust, R. S. (2018). Alkali refining. In *AOCS lipid library, oils and fats*. Available via http://
    lipidlibrary.aocs.org/OilsFats/content.cfm?ItemNumber=40319. Accessed 22 Mar 2018.

RETRACTED CHAPTER: Fruit Oils in Kosovo: Chemistry and Functionality

Rr. Ferizi and Q. Maxhuni

This chapter [3] has been retracted by the publisher. The chapter contains sections that substantially overlap with the following articles (among others) [2–10].

Rr. Ferizi and Q. Maxhuni do not agree to this retraction.

[1] Ferizi R, Maxhuni Q. Fruit Oils in Kosovo: Chemistry and Functionality. InFruit Oils: Chemistry and Functionality 2019 (pp. 53–83). Springer, Cham.

[2] Pulaj B, Mustafa B, Nelson K, Quave CL, Hajdari A. Chemical composition and in vitro antibacterial activity of Pistacia terebinthus essential oils derived from wild populations in Kosovo. BMC complementary and alternative medicine. 2016 Dec;16(1):147.

[3] Hajdari A, Mustafa B, Nebija D, Miftari E, Quave CL, Novak J. Chemical composition of Juniperus communis L. Cone essential oil and its variability among wild populations in Kosovo. Chemistry & biodiversity. 2015 Nov;12(11):1706–17.

[4] Hajdari A, Mustafa B, Gashi V, Nebija D, Ibraliu A, Novak J. Chemical composition of the essential oils of ripe berries of Juniperus oxycedrus L., growing wild in Kosovo. Biochemical Systematics and Ecology. 2014 Dec 1;57:90–4.

[5] Hajdari A, Novak J, Mustafa B, Franz C. Essential oil composition and antioxidant activity of Stachys sylvatica L.(Lamiaceae) from different wild populations in Kosovo. Natural product research. 2012 Sep 1;26(18):1676–81.

[6] Hajdari A, Mustafa B, Ahmeti G, Pulaj B, Lukas B, Ibraliu A, Stefkov G, Quave CL, Novak J. Essential oil composition variability among natural populations of Pinus mugo Turra in Kosovo. SpringerPlus. 2015 Dec 1;4(1):828.

Retraction note: Several conference proceedings have been infiltrated by fake submissions generated by the SCIgen computer program. Due to the fictional content the chapter {title} by {author} has been retracted by the publisher. Measures are being taken to avoid similar breaches in the future.

[7] Franz CH, Novak J, Hajdari A, Mustafa B. Total flavonoids, total phenolics and antioxidant activity of Betonica officinalis L. from Kosovo. InIV International Symposium on Breeding Research on Medicinal and Aromatic Plants-ISBMAP2009 860 2009 Jun 17 (pp. 75–80).

[8] Moghaddam M, Mehdizadeh L. Chemistry of Essential Oils and Factors Influencing Their Constituents. InSoft chemistry and food fermentation 2017 Jan 1 (pp. 379–419). Academic Press.

[9] Pieroni A, Soukand R, Quave CL, Hajdari A, Mustafa B. Traditional food uses of wild plants among the Gorani of South Kosovo. Appetite. 2017 Jan 1;108:83–92.

[10] Hajdari A, Mustafa B. Essential oils from Kosovar aromatic plants. Your hosts Macedonian Pharmaceutical Association and Faculty of Pharmacy, Ss Cyril and Methodius University in Skopje. 2016:467.

RETRACTED CHAPTER

Chapter 4
Olive Oil Properties from Technological Aspects to Dietary and Health Claims

Manel Issaoui and Amélia M. Delgado

Abstract The Mediterranean is tightly bond with *Olea europaea* var. sativa, the olive tree, endemic in the region. The edible products of the olive tree are the table olives (fermented and processed in many styles) and olive oil. Olive oil can be described as the pure olive juice obtained after a mechanical extraction process under controlled environmental conditions. The popularity of olive oil has been steadily increasing worldwide as its market demand. Since the last century, olive oil production and consumption have increased sharply. Olive oil demands are expanding conquering new markets and reaching new consumers, thanks to the numerous scientific discoveries that have been highlighting their nutritional and therapeutic properties, as well as its role as a key component of a balanced diet. Previously, the nutritional quality of olive oil was only attributed to its high level of oleic acid. However, that is a feature shared with many other oils, which do not display so many beneficial health effects. Recently, key roles have been attributed to the minor components that are thought to make extra virgin olive oil, a fatty substance with exclusive chemical characteristics. Although representing only 2% of the weight of the oil, the minor components' fraction includes more than 230 substances, belonging to varied chemical categories: tocopherols, polyphenols, squalene, flavors, pigments, sterols, and aroma. Evidence-based scientific outcomes, namely cause-effect relationships, have led competent authorities in many countries to formally recognize the nutritional and health properties of this Mediterranean Diet component.

M. Issaoui (✉)
Lab-NAFS 'Nutrition – Functional Food & Vascular Health', Faculty of Medicine, University of Monastir, Monastir, Tunisia

Faculty of Science and Technology of Sidi Bouzid, University of Kairouan, Sidi Bouzid, Tunisia

A. M. Delgado
MeditBio – Centre for Mediterranean Bioresources and Food, University of Algarve, Faro, Portugal
e-mail: amdelgado@ualg.pt

© Springer Nature Switzerland AG 2019
M. F. Ramadan (ed.), *Fruit Oils: Chemistry and Functionality*,
https://doi.org/10.1007/978-3-030-12473-1_4

Keywords Minor bioactive compounds · Mediterranean diet · Nutritional claim · Health claim

1 Introduction

Olive oil and the Mediterranean are tightly bonds, given the symbolism of the tree, its origin, the economic importance to the region and the fact olive oil in one of the pillars of the Mediterranean Diet. Olive oil has been a long times and should continue to be the main dietary source of fats in the Mediterranean Dietary pattern, which can be followed worldwide (Bach-Faig et al. 2011, Keys and Fidanza 1960; Keys 1995).

Most edible oils and fats undergo a harsh extraction process typically involving high temperatures and pressure and/or solvents. A refining process is generally required to eliminate deleterious compounds. Only a few edible virgin oils can be found. Generally, refining steps include degumming, decoloring, deodorization and other operations that lead to a homogenous mild product, to be released in the global market. Thus, refined oils are practically colorless, odorless and tasteless. For more information on other oils and refining processes, please see other chapters in this book.

Conversely, olive oil can easily be obtained from the juice of a fruit, the olive, at ambient temperature, by simple physical processes (Picture 4.1), with no need for any further processing, refining or additives. In fact, traditional consumers of olive oil make educated choices between the variety of products with distinctive flavors, viscosity and colors that depend on the region, the fruit variety and a multitude of other factors, mainly of environmental nature. Most valued olive oils are those types graded as an extra-virgin olive oil (EVOO) and virgin olive oil (VOO), which obtainment process uses much milder conditions than the other few types of virgin vegetable oils that can be found in the market.

VOO can be found in a large variety of aromas, tastes, and colors as well as from an intense bitterness to sweet and pungent flavors. According to the variety and degree of fruit maturation, the color can vary from deep green, to light green and from pale yellow to golded yellow. In respect to flavor, one can find notes of artichoke, tomatoes, grass, fruits such as bananas or apples grouped together under the name of aromas resulting from the presence of volatile compounds in VOO. The organoleptic properties of olive oil are the mirror of its complex, varied and unique chemical composition. By adding market value to such differences and peculiarities, dependent on the fruit cultivar and on the region, biodiversity and local economies are indirectly supported. EVOO and VOO are also acknowledged by the unique set of benefits they bring to human nutrition and health. In a simplistic approach, a couple of tablespoons of this golden liquid help control weight, prevent heart and coronary diseases, and probably some other ailments. In addition, olive oil is a flavor enhancer.

Olive oil is the most regulated edible fat regarding quality control parameters, analytical methodologies, grading and labeling, namely by EC, IOOC, Codex

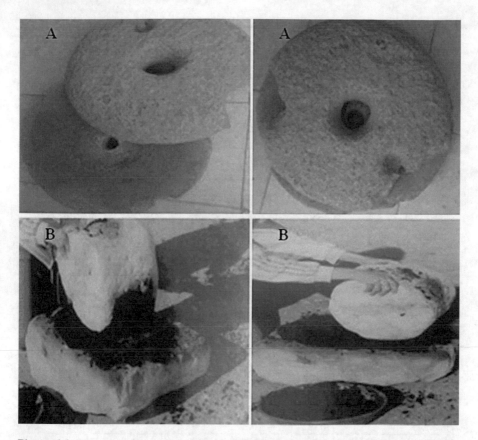

Picture 4.1 The artisanal processes of crushing olive fruit to produce olive oil in Tunisia. (Photos by Issaoui M). (**a**) (Top row, left and right): Granite stone wheels that rotate around an axis to grind the olives. (**b**) (Bottom row, left and right): Manualartisanal method for crushing full ripe olives to obtain olive oil. The paste is repeatedly pressed between stones, while oil drains into a container, held at a lower level. (Photos by Issaoui M. and taken in Tunisia; Issaoui et al. 2009)

Alimentarius, ISO, some national laws and regulations. The International Olive Council (IOC) and some other standards around the world have established very strict criteria for defining the different categories of olive oils available on the market, such as Extra virgin, Virgin Ordinary, and virgin lampante, as well as refined olive oil and refined pomace oil (Table 4.1 for details).

Olive oil is the only fat for which a specific sensorial profile has been implemented (García-Gonzáleza et al. 2017). It is among the foods with the higher number of entries in the DOOR database,[1] in the category of oils and fats (EC 2018a), and it is also represented at the level of WIPO database (world intellectual property

---

[1] DOOR database is the portal of the Department of Agriculture and Rural Development of the European Commission that manages the designations (and corresponding logos) related with intellectual property linked to the regional character of foods: Protected Designation of Origin (PDO), Protected Geographical Indication, and Traditional Specialty Guaranteed (TSG).

Table 4.1 Quality parameters of olive oil produced from some international standards

Standard\Parameters	Extra	Virgin	Ordinary	Lampante
Free fatty acid (%)				
Codex Alimentarius	<0.8	<2.0	<3.3	
IOC	≤0.8	≤2.0	≤3.3	>3.3
EC	≤0.8	≤2.0		>2.0
USA	≤0.8	≤2.0		>2.0
Australia	≤0.8	≤2.0		>2.0
Peroxide value (meqO$_2$\kg)				
Codex Alimentarius	≤20.0	≤20.0	≤20.0	
IOC	≤20.0	≤20.0	≤20.0	No limit
EC	≤20.0	≤20.0		—
USA	≤20.0	≤20.0		No limit
Australia	≤20.0	≤20.0		>20
Absorbance K232				
Codex Alimentarius	≤2.50	≤2.604		
IOC	≤2.50	≤2.60		
EC	≤2.50	≤2.60		
USA	≤2.50	≤2.60		N\A
Australia	≤2.50	≤2.60		>2.60
UV absorbance K270				
Codex Alimentarius	≤0.22	≤0.25	≤0.30	
IOC	≤0.22	≤0.25	≤0.30	
EC	≤0.22	≤0.25		
USA	≤0.22	≤0.25		N\A
Australia	≤0.22	≤0.25		>2.60
Median of defect (Md) and median of fruity attribute (Mf)				
Codex Alimentarius	Md = 0	0 < Md ≤ 2.5	2.5 < Md ≤ 6	
	Mf > 0	Mf > 0	Mf > 0	
IOC	Md = 0	0 < Md ≤ 3.5	3.5 < Md ≤ 6	Md > 6
	Mf > 0	Mf > 0	Mf > 0	
EC	Md = 0	Md ≤ 2.5		Md > 2.5
	Mf > 0	Mf > 0		
USA	Md = 0	0 < Md ≤ 2.5		Md > 2.5
	Mf > 0	Mf > 0		Mf N\A
Australia	Md = 0	0 < Md ≤ 2.5		>2.5
	Mf > 0	Mf > 0		N\A

organization) in the category of Appellations of Origin[2] (WIPO 2018). Olive oil is also the first fat officially recognized as healthy (in addition to its nutritional favorable acknowledgment) by several authorities, particularly by the United States of America, Department of Agriculture (USDA) and in Europe by European Food

[2] Appellation of Origin is a form of protecting the intellectual property of a geographical name or a traditional designation used on products which have a specific quality or characteristics that are essentially due to the geographical environment in which they are produced (e.g. Oporto wine, Parmigiano-Regianocheese).

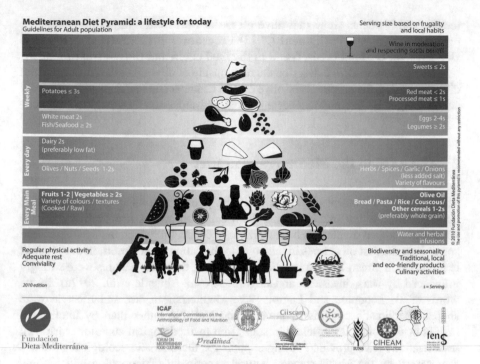

Fig. 4.1 Mediterranean diet pyramid updated to nowadays lifestyle. (Reproduced from the site of Fundación Dieta Mediterrânea – https://dietamediterranea.com/)

Safety Authority (EFSA) as well as by the European Council (EC). It is in fact the only fat that comes under the umbrella of the Mediterranean diet (MD).

Ancel Keys, an American Medical doctor, was the first to note the health benefits of such food pattern and lifestyle, introducing the concept of Mediterranean Diet (Keys and Fidanza 1960; Keys 1970; Keys et al. 1980; Keys and Keys, 1975). Several years later, the United Nations Educational, Scientific and Cultural Organization (UNESCO) recognized the Mediterranean Diet, in its broad sense, as an intangible heritage of humanity. Although UNESCO definition of Mediterranean Diet encompasses many other dimensions (viz. historical, cultural, sociological, and environmental) but the agricultural aspects and the associated food pattern are indubitably the cement of the concept. Mediterranean dietary (MD) pattern is an almost vegetarian dietary regimen where olive oil, wheat and wine/tea play central roles. The UNESCO distinction has been helping to raise awareness on this healthy and sustainable diet, and hence to olive oil. According to the Mediterranean pyramid by Bach-Faig et al. (2011) presented in Fig. 4.1, olive oil should be consumed daily and included in all main meals. Unlike other dietary fats, olive oil has never ceased to steer debates towards its countless health benefits for the consumer. A growing number of scientific evidence points in that direction (Arvaniti et al. 2011; Bonaccio et al. 2013, 2014, 2018; Huijbregts et al. 1997; Kastorini et al. 2011; Lassale et al. 2018; Yang et al. 2014). On the other hand, recent investigations, recalls, and incidents show that olive oil is the world's first target for fraudulent practices **(European Parliament (2013) Procedure: 2013/2091(INI), Report 4 Dec 2013, on the food crisis, fraud in the food chain and the control thereof)**.

In the present work, we describe the relationship of olive oil to the Mediterranean diet, the market, the properties of olive oil and its impact on human health, officially acknowledged by international organizations and supported by scientific evidence.

2   Olive Oil in the Mediterranean Diet

The UNESCO's concept of MD is multidisciplinary and regarded as a common heritage linked to the area bordering the Mediterranean sea, irrespectively of religious and national divides, as in the Braudel's concept in which the Mediterranean spans from the first olive trees in the north to the palms of the desert, in the south. It is grounded in Greco-Roman and Islamic cultures, and their features were strongly influenced by wars, invasions and migrations (Vaz-Almeida et al. 2017a). In the words of Trichopoulou and Lagiou "The Mediterranean diet and lifestyle were shaped by climatic conditions, poverty and hardship rather than by intellectual insight or wisdom. Nevertheless, results from methodological superior nutritional investigations have provided strong support for the dramatic ecological evidence represented by the Mediterranean natural experiment (Tricopoulou and Lagiou, 1997). In fact, the MD food pattern is undoubtedly correlated with social habits and cultural heritage, and is rooted in respect for the territory and biodiversity. Traditionally Mediterranean dietary pattern was based on three pillars: olive oil, wheat and wine.

The so-called "good or ancient" MD, as described by Keys (Keys and Keys 1975) is a quasi-vegetarian dietary regimen including daily consumption of olive oil, accounting for most of the energy intake. Nuts and table olives were frequent snacks. Large quantities of varied vegetables and fruits, of the season, were important sources of vitamins, fibers and phytonutrients. Pulses and cheeses were the main contributors of protein to the diet. Limited amounts of smoked pork preserves were also relevant in southern European countries, particularly in rural isolated communities. Meat and fish were consumed with parsimony, and even rarely. Wheat, rice and potatoes were the main carbohydrate sources. Liquid milk was not commonly consumed by adults, as milk was preferentially preserved in the form of cheese or yogurt. In southern European countries, adults often take a glass of wine during meals, a habitude that has been referred to as beneficial to health. Tea and coffee can be placed in a similar category. These daily meals drastically contrasted with meals associated with celebrations, mostly of religious nature. While in daily meals, the quantities of animal-origin protein were small and fruit was the "desert", in celebrations like Christmas and Easter with Christians, and the Eid Al-Adha (or the Festival of Sacrifice) and the Eid Al Fitr (that marks the end of Ramadan) with Muslins, long lasting convivial meals include large amounts of meats, with lamb playing a central role. Olive oil is

Table 4.2 Main nutritional features of the Mediterranean dietary pattern

Nutrient	% Total energy intake	Particularities
Carbohydrates	60–70	Of which 50% starch, associated with oligosaccharides and fibers, mainly from bread, pasta, rice and potato
Protein	Approx. 10	Of high biological value: From pulses and other vegetables, complemented with dairy products and some meat and fish
Lipids	20–32	MUFA and PUFA; oleic acid and α-linolenic acid, mainly from olive oil; nuts are also relevant; modest SFA intake
Fiber	NA	Soluble and insoluble forms, from fresh fruit, vegetables cereals and nuts
Vitamins	NA	Mainly from vegetable sources; olive oil, fish and nuts as preferential sources of oil-soluble vitamins
Phytonutrients	NA	Include phenols, tannins, and other compounds nowadays recognized as very relevant for the prevention of non-communicable diseases; supplied by vegetables, grains, pulses and olive oil

Adapted from Vaz-Almeida et al. (2017b)
MUFA Monounsaturated fatty acid(s), which carries a single double bond between carbons of the backbone chain, *PUFA* Polyunsaturated fatty acid(s), which carries (normally) one to three double bonds between carbons of the backbone chain, *SFA* Saturated fatty acid(s), which carries only single bonds between carbons of the backbone chain, *NA* not applicable (vestigial amounts involved)

omnipresent in the confection of dishes and even in some desserts. Sweet deserts, rich in eggs and nuts, are reserved to these occasions.

Scarcity and parsimony was the rule. Abundance was the exception, reserved to festivities, when peoples indulged themselves to eat and drink abundantly (Vaz-Almeida et al. 2017b). This particular dietary pattern has been the object of many studies, since the Seven countries' study by Keys mounting evidence has been collected on health benefits, their association to specific components of the diet and the underlying mechanisms of action. The update of this healthy dietary pattern is presented in Fig. 4.1- the MD Pyramid, a lifestyle for today.

As can be observed in Fig. 4.1 and discriminated in Table 4.2, in the Mediterranean dietary pattern, olive oil accounts for a relevant portion of the daily total energy intake, essential lipids, oil-soluble vitamins and phytonutrients. It is noteworthy that the FAO and WHO defines energy requirement as "the amount of food energy needed to balance energy expenditure in order to maintain body size, body composition and a level of necessary and desirable physical activity consistent with long-term health". This includes the energy needed for the optimal growth and development of children, for the deposition of tissues during pregnancy, and for the secretion of milk during lactation consistent with the good health of mother and child (FAO 2001).

Although the energy requirement is highly variable with numerous factors, like age, ambient temperature, sex, body size, and level of physical activity, just to name a few, a reference indicative value of 2000 Kcal/day has been set and is widely accepted for most purposes. As noted by Keys et al. (1980), a westernization of food habits was already occurring in the Mediterranean area. These new food habits of the Mediterranean peoples have been favored by policies that have been promoting intensive farming aiming at feeding everyone, food industries, including fast-food chains, supplying ready-to-eat and frozen processed meals as ideal solutions today's busy life. Marketing messages are conveyed to consumers in ways more and more effective and present these highly processed, highly caloric, and nutrient-poor foods, as fashionable, convenient, affordable, and globally available. The dark side of such western globalized diets is evident from the rising trend of obesity and non-communicable diseases worldwide. An action plan implemented by the World Health Organization (WHO) aims at reverting this trend, particularly by combating obesity and diabetes encouraging changes in the diet and physical activity (WHO 2018a, b).

Changing food habits towards a nutrition transition state, loss of cooking skills and non-valorisation of traditional foods have been observed in countries from both shores of the Mediterranean and noted by several authors (da Silva et al. 2009; Vareiro et al. 2009; Vaz-Almeida et al. 2017c, d). These authors refer to average energy intakes, above 3500 kcal/day/person, in such countries, which largely surpasses the reference value of 2000 Kcal/day/person. Moreover, as noted by Vaz-Almeida et al. (2017c), main changes are the rising consumption of meat, milk, other fats rather than olive oil (including saturated and trans-fats), simple sugars and salt, along with a decreased consumption of cereal grains, pulses, greens and other vegetables, and the rise of fast-food. As many processing steps in food industries use other fats, often palm-derived, these dietary shifts will certainly be reflected in the decreased consumption in olive oil in the Mediterranean region.

3   The Market of Olive Oil

Globalization has affected the olive oil economy in recent years, which has led to a growing demand. At the same time, supply has evolved based on new models of agricultural production. The world production was estimated at 3,133,50 thousand tons in 2017–2018 campaign (IOC 2018), surpassing consumption, as can be shown in Fig. 4.2. In addition to the offer of the IOC member countries, new actors emerge, alongside the traditional ones from the Mediterranean area. Australia and USA are now established producers.

EC previsions for European production in 2017/2018 are of 15% above 5-year average, of previous campaigns, mainly due to the increased production share of Italy, Greece and Portugal (EC 2018b). Despite the recent tendency for stabilizing or even decreasing production, Spain is still the largest producer in the world (more than the 1/3 of world production), followed by Italy, Greece and Portugal, in EU, and Tunisia, outside EU (Table 4.3). As can be verified in Table 4.3, Italy is the first consumer of olive oil in the world representing the 1/4 of global consumption. USA consumption of olive oil has tripled over the past two decades, and it is now the third world consumer, with 312,00 thousand tons/year/capita in 2017/2018.

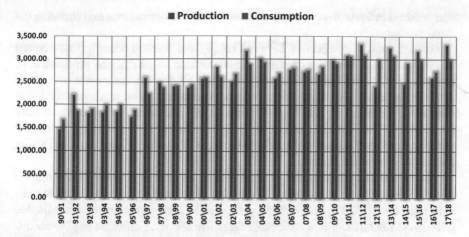

Fig. 4.2 Global production and consumption levels of olive oil during a time-span of 27 years from the 90\91 to 2017\2018 campaign (values for global production and consumption of olive oil are expressed in thousand tonnes/year). (Source: IOOC since 90/91 campaign to 2017/2018 campaign http://www.internationaloliveoil.org/estaticos/view/131-world-olive-oil-figures)

Table 4.3 Statistics related to commercial trades of olive oil for the campaign of 2017–2018 (production, consumption, import and export quantities in thousand tonnes)

Country	Production	Rank	Consumption	Rank	Import	Rank	Export	Rank
World	3,133,50		2,978,0		880,00		971,0	
Spain	1,251,30	1	470,00	2	61,70	4	304,2	1
Turkey	263,00	5	160,00	4	0	NR	90,0	4
Greece	346,00	3	130,00	5	0	NR	9,8	8
Italy	428,90	2	560,70	1	84,60	2	217,7	2
Morocco	140,00	6	120,00	6	6	8	15,0	9
Syria	100,00	8	100,00	7	NC	NR	0	NR
Tunisia	280,00	4	33,00	13	NC	NR	200,0	3
Portugal	134,80	7	80,00	9	NC	NR	39,7	5
Algeria	80,00	9	85,00	8	NC	NR	NC	NR
Japan	NC	NR	55,00	11	55,00	5	NC	NR
USA	16,00	11	312,00	3	305,00	1	12,0	7
Australia	21,00	10	45,00	12	29,00	6	4,5	10
Brazil	NC	NR	70,00	10	70,00	3	0	NR
Argentina	NC	NR	8,00	15	0	NR	36,0	6
Russia	NC	NR	20,00	14	20,00	7	0	NR

NR not ranked

In some cases, Tunisia herein presented as an example, a disproportionality exists between the level of production and the level of consumption. In fact, Tunisia is the fourth producer country in the world, however occupying the 13th place in the category of olive oil consumption (Table 4.3). In contrast, a perfect proportionality was deduced in Syria, the country produced 100,000 thousand tons and consume the same average.

Other cases of disproportion between supply and demand occur in USA, Brazil and Russia. These countries have an internal demand larger than the production. In the case of USA, despite the increase in internal production, most olive oil continues to be imported and other countries do not produce olive oil at all, mostly due to climacteric constraints. The imports from Brazil dramatically increased recently (EC 2018b). The historical and cultural linkage to Portugal and Italy, as well as recent migratory fluxes to EU (mainly Portugal) may have contributed to raising awareness of new consumers.

The global rising in demand of olive oil can be attributed to the dissemination of scientific evidence about olive oil's health benefits, its adequacy for most culinary uses, and its linkage to the MD, presented by FAO as healthy and sustainable (CIHEAM/FAO 2015). It is noticeable that olive oils have different physic-chemical and sensorial properties that depend on many factors, in particular the origin, the olive variety and the extraction process. If not accounting for sensorial preferences, EVOO can eventually replace one another but the same set of properties cannot be found in any other vegetable oil. This constitutes a relevant product' strength. The olive oil market is very dynamic and boosted by two kinds of coupled drivers. The first one is the pair {Production, Consumption} and the other driver's pair is the {Import Export}.

3.1 Production-Consumption Trends

The statistical evolution of the pair {Production/Consumption} during the time lapse from the 90/91 harvesting/producing season to the 2017/2018 olive campaign is presented in Fig. 4.2. It is noteworthy that the evolution of the pair has doubled during that time span of 27 years, from about 1,500.00 thousand tons to more than 3000,000.00 thousand tons.

Three events characterize the equilibrium production /consumption:

1. The overall production was higher than the consumption 12 times during the time span analyzed in Fig. 4.2.
2. The overall consumption was higher than the production, 14 times during the period from 90/91 to 17/18.
3. Finally, the situation when the production was equal to the consumption never happened, although in 98/99 the difference between both factors was not significant. As can be deduced from Fig. 4.2 and statistically proved, olive oil is a highly demanded product, in the global market.

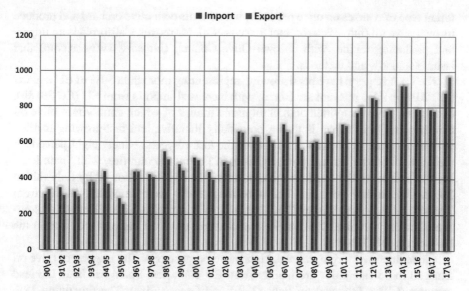

Fig. 4.3 Global market trends for imports and exports of olive oil from 1990\1991 campaign until 2017\2018 (in thousands of tonnes). (Source: IOOC since 90/91 campaign to 2017/2018 campaign http://www.internationaloliveoil.org/estaticos/view/131-world-olive-oil-figures)

3.2 Import-Export Trends

Figure 4.3 exhibits the behavior for the pair {Import, Export} during the period of 1990/1991 to 2017/2018. A regular increase is observed. Moreover, the values of the pair (in thousand tons) tripled, during that time lapse. This increase in the global trade for olive oil is due to the rise of new consumers mainly in USA, Australia, Japan, Russia, Brazil and China.

The evolution of olive oil consumption has been described as spectacular over the past 25 years in the United States, as noticed by IOC (2018). On the other hand, EC (2018b) describes the global olive oil market in a quasi-steady state for the recent years, with an increase of 1% in world consumption, and a decrease of 2% in EU consumption (5-year average).

The US, the world's largest importer of olive oil, imported a total of 316,758 thousand tons in the 2016/2017 season. Imports of olive oil by the US accounted for nearly 30% of world imports (IOC 2018a). The largest suppliers of bottled olive oil are Italy, followed by Spain, Greece, Tunisia and Turkey, while imports of bulk olive oil are mainly from Italy, followed by Spain, Tunisia, Argentina, Turkey and Chile.

According to IOC (2018), this organization launched a US "Olive Oil Promise" at the University of California, UC Davis, to help to raise awareness on olive oil, and probably to cooperate with American researchers and associations of producers. In this respect, it is noteworthy that UC Davis has an olive center that offers a different type of courses on olive products, manages its own olive orchard and produce its own olive oil (https://olivecenter.ucdavis.edu). Moreover, California is an important producing region, with its own Olive Council (https://www.cooc.com) that issues standards and methods.

The USA imports have been growing representing more than 30% of EU exports (EC 2018b). Also relevant are Brazil, with increased imports from EU (EC 2018b), and Japan, which imported 56,853 thousand tonnes of which extra virgin olive oil accounts for 71%. Japan is thus a potentially lucrative and fast-growing market, with an increase in imports of 51% over the last 10 years. Ninety eight percent of the olive oil consumed in Japan originates in European countries, with Spain leading that market in the last four seasons (more than 50%), followed by Italy (about 30%), Greece and Portugal. The remaining 3–4% were imported mostly from Tunisia and also from Turkey (IOC 2015). Russia imported olive oil and olive pomace oil mainly from EU. About 97% of Russian olive oil imports are from in the last-season (**2017/2018**).

China imported virgin, refined, blends and pomace oils. The imports of olive oil from all these categories are mainly from EU countries: Spain has been the lead exporter (61%), followed by Italy (23%) and Greece (4%). The remaining 12% were imported from Tunisia (4%), Turkey (3%), Morocco (2%), Australia (2%), and 1% for other countries (IOC 2015). From 2012 onwards, Spain and Tunisia dramatically increased their production, due, at least partly, to the strategies and political orientations aiming at intensifying the cultivation of the olive tree and the rising of intensive and hyper-intensive orchards. In addition, many producers have invested in the genetic selection of more productive varieties, even if they are not autochthonous. Other non-sustainable practices have been followed by many, at expenses of losses on product quality, biodiversity, issues with pests and water management. Another important argument to explain the enlargement of the olive market is the rise of new producers in Australia, USA and South America (i.e., Chile).

Commitment to the action plans for the implementation of the UN sustainable development goals (particularly UN-SDG 2, 12, 13 and 15) calls for changes in policies in favour of the adoption of healthier and sustainable diets, the circular economy, more sustainable agricultural practices, mitigation of climate change, and protection of biodiversity (UN 2018). Moreover, disruptive innovations in technology and business models continue its fast-growing trend, and impacts to the sector are to be expected. New producing countries, the valuation of organic production, the introduction of IT technologies in agriculture and industry, and the valuation of former by-products are challenges that may lead to a changing paradigm of the olive sector.

3.3   Trends on Olive Oil Prices

According to IOC (2018) and EC (2018b), the price of EVOO, at the consumer, in EU markets may vary substantially. In a period of 5–6 years the price in the main internal EU markets (Spain, Italy, Greece and Portugal) ranged from about 2.5 €/Kg to more than 6.0 €/Kg, depending on the campaign. A clear tendency for a higher valorization of EVOO in the Italian domestic market (taking Bari as a reference) is evident, in opposition to the tendency for the depreciation of olive oil grades in the Spanish market.

In what concerns non-EU Mediterranean countries, Tunisia stands out as an important EU partner and relevant actor in the global olive oil market. Tunisian olive oils have been awarding important world prizes. The valorization of olive oil in Tunisian markets is similar to that in EU countries. A reference value of 3.34€/Kg for VOO (4 year average value) was registered. It is noteworthy that non-negligible amounts of Tunisian olive oil are traded for EU inward processing (meaning that some Tunisian olive oil is processed in EU to be exported to third countries or distributed within EC markets) (EC 2018b).

Refined olive oil prices may approach those of VOO, in certain markets. Hence, in 2018 the difference between the two categories was about 0.16€/Kg in the case of the Spanish market (IOC 2018). Conversely, in Italy, the difference in price between the two categories was about 3.28€/Kg in 2015, and 2.28€/Kg in 2017 (IOC 2018). One probable explanation lays on the distinct perception of quality by Italian consumers and Spanish consumers; with the Italian consumer probably acting as "connoisseur" in search of specific sensory attributes and geographical origins, while the choices of an average Spanish consumer seems to be less educated. In fact, as shown by Salazar-Ordóñez et al. (2018), the average Spanish consumer generally buys olive oil in 5-L package, from Spanish cooperatives, putting brand loyalty over all. One third of inquiries (from a representative sample of 700 regular buyers) responded yes to the closed question: "do you know what an oil refining process is?" and then explained (in a subsequent open question) that the refining process always improve the quality of olive oil, which is obviously wrong. These authors noted that the self-perceived knowledge of the average Spanish consumer about olive oil is far from reality.

On its turn, Caporaso et al. (2015) sampled 32 EVOO bottles from the Italian retail market and registered differences in price from 2.89€/kg to 20.00€ per kilo.[3] Italian consumers seem to value differences in sensorial properties, origin and composition of olive oils, which are related to total polyphenols content (between 406 and 594.2 ppm) of the VOO, their high antioxidant activity and tocopherols content (more than 300 ppm for α-tocopherols and around 5 ppm for the sum of β and γ-tocopherols). In such a market, sensorial panel scores and quality parameters have a strong impact on the price of olive oil. Those VOO with higher scores and lower mereological parameters have the highest price. In addition, as the same authors noticed, in general, mono-varietal olive oils have higher prices than blend oils. In short, consumers do not perceive the balance of price vs quality in the same way. Differences are observed between different markets, including in traditional producing and consumer countries in EU, as explained above.

Thus, improving the literacy of the average consumer about olive oil will enable valuing endogenous parameters, with consequent impacts on the price, allowing a

---

[3] The corresponding official market price in this period was 2.95€ per kilo.

more clear discrimination of types and grades. Other parameters that may affect consumers preferences, and hence the market price, are the presence of reliable quality and origin logos. Many consumers search for organic produces and the display of the EU eco-label (or equivalent certification), increase the perceived value of the product by a wide range of consumers, thus affecting its price. In addition, in EU, organic agriculture is encouraged and subsidized. Outside EU, Tunisia stands out for the efforts toward the adoption of similar strategy (ISOFAR 2016). Official and private quality and origin logos play an important role in product-related differentiation and on the growth of SME, also providing a way of guaranteeing the quality, origin and traceability.

4 Extraction of Olive Oil and the Subsequent Impact on the Composition

4.1 The Regulatory Aspects and Definitions of Olive Oil

The regulations that define olive oil have always been conditioned by its extraction techniques. Some insights are herein included for a better clarification of the reader. The term "Virgin" guarantees that the oil has not been mixed (or blended) with other factions of different grade or with other oils, or have been subjected to the refining process. Conversely, "Olive oil" and "Pomace oil" are respectively examples of a blend of different fractions (olive oil), and of a by-product (pomace oil). In both cases extraction with solvents and refining processes are used.

The designation "Virgin" also indicates that no other inputs, besides water, are used for oil extraction, and any chemical, enzyme or other substance is added to facilitate the release of the oil from the fruit. VOO is extracted by crushing only, as explained above. The designation "virgin olive oil" also indicates that the oil has been exposed to heat as little as possible during all process. "Virgin olive oil" can then be viewed as a synonymous of authenticity and purity, obtained from the juice of the fruit of *Olea europaea* L, which is a drupe. In fact, unlike most vegetable oils (mainly seed oil obtained by solvent extraction and further submitted to diverse refining processes), extraction of virgin olive oil is based on a series of mechanical operations. Such operations begin by a cold extraction of the oil droplets from the cells of the pulp of olives, which soften during maturation as happens with other drupes. Next operations eventually include washing with water, filtering, and centrifugation. The environmental conditions must also be controlled: the room temperature must not exceed 28 °C, the humidity must be adjusted, and the unpleasant odors in the air must be avoided and/or removed, because olive oil has the ability to capture volatiles, meaning all pleasant or unpleasant airborne smells, which may ruin its flavor.

The different designations and grading of olive oil accepted in different countries are given in Table 4.4. There is some general consensus around the definition of

Table 4.4 Legal definitions of olive oil, around the world, and standards in which they are based on

Standard: IOC: International Olive Council Trade Standard Applying to Olive Oils and Olive-Pomace Oils – COI/T.15/NC N° 3/Rev. 3 November 2008
Definition
Virgin olive oils are the oils obtained from the fruit of the olive tree solely by mechanical or other physical means under conditions, particularly thermal conditions, that do not lead to alterations in the oil, and which have not undergone any treatment other than washing, decantation, centrifugation and filtration"
Standard: Codex Alimentarius (FAO\WOH): Codex Standard for Olive Oils and Olive Pomace Oils - Codex Stan 33–- 1981 (Rev. 22003)
Definition
Virgin olive oil are the oils obtained from the fruit of the olive tree solely by mechanical or other physical means under conditions, particularly thermal conditions, that do not lead to alterations in the oil, and which have not undergone any treatment other than washing, decanting, centrifuging and filtration
Standard: European Commission Regulation (EEC) N° 2568/91 of 11 July 1991 on the characteristics of olive oil and olive-residue oil and on the relevant methods of analysis and subsequent amendments
Definition
Olive oil: Directly obtained from olives solely by mechanical or other physical means under conditions that do not lead to alteration in the oil and which have not undergone any treatment other than washing, decantation, centrifugation or filtration. Excluded are: – Oils obtained by the use of solvents, adjuvants having a chemical or biochemical action, or re-esterification methods; – Oils mixed with oils from other sources"
Country: USA –California State. **Standard**: State of California. Department of Food and Agriculture. Grade and Labeling Standards for Olive Oil, Refined Olive Oil and Olive-Pomace Oil. Effective September 26, 2014
Definition
Olive oil is the oil obtained solely from the fruit of the olive tree (*Olea europaea* L.), solely by mechanical or other physical means under conditions, including thermal conditions, that do not lead to alterations in the oil, and which has not undergone any treatment other than washing, crushing, malaxing, decantation, pressing, centrifugation, and filtration and to the exclusion of oils obtained using solvents or re-esterification processes and of any mixture with oils of other kinds
Country: United States. **Standard**: Standards for Grades of Olive Oil – Effective date March 22, 1948 together with their current Proposed United States Standards for Grades of Olive Oil and Olive-Pomace Oil – Release date March 28, 2008
Definition
Virgin olive oil: s are the oils obtained from the fruit of the olive tree solely by mechanical or other physical means under conditions, including thermal conditions, that do not lead to alterations in the oil, and which have not undergone any treatment other than washing, decantation, centrifugation, and filtration and shall meet the minimum requirements of Table I, found in §52.1539 of these grade standards. No additives of any kind are permitted
Country: Australia. **Standard**: AS 5264–2011- "Olive oils and olive-pomace oils"

(continued)

Table 4.4 (continued)

Definition
Olive oil is the oil obtained solely from the fruit of the olive tree (*Olea europaea* L.), excluding oils obtained using solvents or re-esterification processes and any mixture with other kinds of oils
Natural olive oils are olive oils obtained solely by mechanical or other physical means under conditions, including thermal conditions, that do not lead to alterations in the oil, and which have not undergone any treatment other than washing, crushing, malaxing, decantation, pressing, centrifugation, and filtration
Country: China. **Standard**: General Administration of Quality Supervision, Inspection and Quarantine of the People's Republic of China (AQSIQ) National Standard of the People's Republic of China ICS 67.200.10
Definition
Olive oil The fruit of the olive tree (*Olea europaea* L) is the raw material for the extraction of oils and fats, with the exception of oils and fats acquired from solvent extraction or heavy esterification technology
Virgin olive oil adopts the physical method of mechanical extraction to extract oil products directly from the olive fruit
Note: In the process of oil extraction, external factors such as temperature, etc., should not cause any changes to the oil and fat ingredients. Oil products may only be handled by means of cleaning, decantation, centrifugation or filtering technology
Country: Brazil. **Standard**: Ministério da agricultura, pecuária e abastecimento. Gabinete do ministro. Instrução normativa n° 1, de 30 de Janeiro de 2012
Definition
Olive oil: the product obtained only from the olive tree (*Olea europaea* L.) excluding any oil obtained by the use of a solvent, by re-esterification process or by mixing with other oils, irrespective of their proportions;
Virgin olive oil: the product extracted from the fruit of the olive tree solely by mechanical means or other physical means, under appropriate temperature control, while retaining the original nature of the product; the oil thus obtained may also be subjected to the washing, decantation, centrifugation and filtration treatments, observing the values of the quality parameters provided in Annex I of this Normative Instruction;
Country: India. **Standard**: India Standard: Draft Indian Standard olive oil - specification ICS No. 67.200 Doc No.: FAD 13 (2505)
Definition
Virgin Olive Oil: Virgin olive oil means the oil obtained from the fruit of the olive tree by mechanical or other physical means under conditions, particularly thermal, which do not lead to alteration of the oil
Country: Argentina. **Standard**: Article 535 (Joint Resolution SPReI No. 64/2012 and SAGyP No. 165/2012)
Definition
Olive oil: "It is understood by Olive oil, the one obtained from the fruits of *Olea europaea* L
Virgin olive oils are those obtained from the fruit of the olive tree exclusively by appropriate mechanical and technical procedures and purified only by washing, sedimentation, filtration and/or centrifugation (excluding extraction by solvents)
Virgin olive oil: is the oil obtained from the fruit of the olive only by mechanical methods or by other physical means in conditions, especially thermal, that do not produce the alteration of the oil, and that has not had more treatments than washing, decanting, centrifugation and filtering

(continued)

Table 4.4 (continued)

Country: Tunisia. **Standard**: Arrêté du ministre de l'industrie, de l'énergie et des petites et moyennes entreprises, du ministre du commerce et de l'artisanat, du ministre de l'agriculture et des ressources hydrauliques et du ministre de la santé publique du 26 mai 2008, fixant les catégories, caractéristiques et les conditions de conditionnement, d'emballage et d'étiquetage des huiles d'olive et des huiles de grignons d'olive

Definition

Virgin olive oil: the designation 'virgin olive oil' is attributed to the oil obtained from the olive fruit solely by mechanical processes or other physical processes under conditions, in particular thermal conditions, which do not lead to no alteration of the oil, and no treatment other than washing, decanting, centrifugation and filtration"

olive oil (and virgin olive oil) as noted by Garcia-Gonzaleza et al. (2017). VOO definition is more or less identical for all the international regulatory bodies. However, in some standards, the values for some chemical parameters, which define the different categories, are not the same. This feature can be deduced from Table 4.1, especially for lower grades. That is, a difference in the percentage of free fatty acids, hydroperoxides, dienes and trienes' levels impact the intensity of positive attributes, and the presence or absence of defects. Thus, the same designation may correspond to oils of the different quality level, according to the standard that was followed for the classification or grading. All these quality parameters are strongly impacted by the conditions and the type of extraction system (Di Giovacchino et al. 2002; Issaoui et al. 2009; Issaoui et al. 2015).

The olive oil comes from the pressure of the drupes of the olive tree, in which oils represent 17–30% of the weight of the fresh olive. Most of these oils are localized in the mesocarp and in the epicarp (Balatsouras 1997). While the term "olive oil" refers to all oils derived from olive extraction, it is important to distinguish the different categories and to fully understand the qualitative differences. In order to achieve that, an overview of the extraction methods will be useful, framing up the evolution from ancient to modern extraction methods and the impacts on quality and yield.

Over the years, the technology has designed olive oil extraction equipment to guarantee its utmost authenticity and preserve its endogenous biochemical composition. Advances in technology have enabled to crush and press fruits, just like the old way, as well as maintain and ensure the highest of quality of the final product. The majority of recent research has been conducted to improve these processes, to reduce waste and to enhance resource use.

4.2 Equipment Used to Produce Olive Oil

There are several methods for extracting olive oil. The main extraction methods include traditional press, selective filtration and decantation method. Pressing is the oldest method used in the extraction of olive oil. Grinding was traditionally done by the means of granite stone wheels (Picture 4.1a). An alternative method is shown in Picture 4.1b. In this ancient process, after collecting the olives those are placed, in specific quantities, in an area made of marble and having the size of one square meter. After that, a huge rock made of pebbles and gypsum is used to squeeze the olives manually by passing the rock on all the olives on the area. About an hour after, a dough made of olives and their kernels is obtained. This technique dates back to the Roman Era. A small vessel is used to collect the dough to knead it again manually until it becomes more solid meaning that the olive oil is almost completely extracted. This step takes about a half an hour to 2 h, depending on the number of olives. In the third and last step, the dough is transferred to a larger vessel, where about two pails of water are poured on the dough, which will be rubbed manually until it becomes disintegrated in the water. This step allows the olive oil to float over the water, while the solids and impurities will remain at the bottom. As the olive oil rises over the water, it becomes easy to extract it manually and carefully poured into a filtering tool to get rid of the impurities. This step is repeated as many times as necessary, until the obtainment of almost all the oil the dough may contain.

Picture 4.2 Steps of olive oil production by discontinuous extraction system. (**a**) Stone olive mills for grinding; (**b**, **e**, **g**) preparation of the needle; (**c**) mats; (**h**) hydraulic press piston; (**i**) natural decantation; (**f**) clarification. (Photos taken by authors)

Finally, this obtained pure olive oil is poured inside a vessel. This traditional manual method is still used in many Mediterranean countries, including Tunisia, in every olive oil harvest season. It permits the production of olive oil, at a family scale, with minimal costs.

The traditional press (Picture 4.2) also used stone olive mills for grinding. The stone cylinders turn around in a round tray whose floor is also of stone. The wheels used for grinding are slightly off-center with respect to the axis of rotation, which accentuates the possibility of crushing olives (Pictures 4.2a, b). The grinding mill (between 2 and 6 rotating cylinders) has a low working capacity because it takes a long time (between 15 and 25 min) to grind each batch of olives, and a low retention speed (12–15 rpm). However, thanks to the slow and non-violent crushing action, a good preparation of the dough is achieved, and the obtained oils are generally harmonious and balanced, from the sensory point of view.

The kneading is done by scrapers constantly bringing the dough under the grinding wheels, which then play the role of kneaders. The paste is then placed in a layer on 2 cm thick nylon fiber discs (the mats), themselves stacked on top of each other, around a central pivot (called needle) mounted on a small carriage (Picture 4.2b, d, e, g). The assembly is placed on a hydraulic press piston, which allows the pulp to undergo a pressure of the order of 100 kg/cm^2. The liquid phase flows into a tank, while the pomace remains on the mats (Picture 4.2e). Then each mat is stripped of its pomace by tapping it like a carpet. Finally, a natural decantation process separates oil and water by the difference of density. The pressing process is discontinuous and filters can easily be contaminated. Hygienic conditions and good manufacturing practices must be applied for the final product to have a good quality.

Picture 4.2 summarizes the steps required to produce olive oil through a discontinuous traditional artisanal system. Fruits are first crushed by a pair of stones, as explained above. The resulting solid paste is spread in thin layers on disks of filter material (canvas or, more recently, plastic fiber) called mats (Picture 4.2, left photo in the middle row). Mats are piled on top of one another in a wagon, and they are guided by a central needle. The assembly formed by the wagon, the needle and the crushed mats coated with dough is called the load (Picture 4.2e). This load is subjected to pressing with a hydraulic press (Picture 4.2h). The pressure is generated by a group of hydraulic pumps, located in the pump box (Picture 4.2h). The application of the pressure makes it possible to obtain a liquid, which flows on the wagon. At first, the liquid obtained is a must rich in oil; with the increase of the extraction pressure, its quality decreases. When the pressing is completed, the liquid phase is transposed into tanks (Picture 4.2i), his favors the natural decantation (separation of the aqueous phase and the oily phase).

Current VOO mechanical extraction systems defined as "continuous-type" generally include a continuous mechanical mill, a batch mixer, and a continuous horizontal axis centrifugal separator (decanter) (Picture 4.3). Grinding is carried out by mechanical grinders with discs or hammers. These grinders can work continuously. Metal grinders have a higher yield than discontinuous processes because they grind

Picture 4.3 Steps of olive oil production by continuous extraction system. (**a, b, e, i**) triage, selection, leaf removing and washing; (**c, d, g, h**) decanter and separation of phases; (**k–p**) clarification of olive oil. (Photos taken by authors)

olives in a very fast way. Their action is more or less violent depending on the type of grinder used. Metal hammers with stationary hammers and those with discs are the most violent, whereas metal mills with blades, with moving hammers, with cones or with rollers, act less violently. The oils obtained from the olive dough prepared with the metal grinders are generally more bitter and pungent when compared to those obtained from olive paste prepared with the stone grinders (Allogio et al. 1996; Angerosa and Giacinto 1995). The impact of the type of grinder on the purity and quality parameters have been raising many discussions, with arguments in favor and against (Di Giovacchino et al. 2002; Issaoui et al. 2009, 2015). The paste is then poured into a stainless steel tank, in which a spiral or an endless screw rotates. That endless screw is also of stainless steel: It is the malaxer, a stainless steel key piece in the continuous process. The aim of malaxation is to facilitate the coalescence of oil droplets into larger drops, which are more easily separable by a centrifugal field, reducing the viscosity of the olive paste and optimizing the separation of the oil inside the decanter. Malaxer also works as a heat exchanger and it is the bottleneck of the continuous extraction process. In short, malaxation is a slow and continuous stirring of the dough, generated by the rotation of helicoidal bands or pallets welded on a horizontal axis, which rotates slowly, at 20–30 rpm. The technological parameters of mixing (time and temperature) are important for the yield, as well as for the quality of the final product (Di Giovacchino et al. 2002; Stefanoudaki et al. 2011). The malaxation can last from 2 to 30 min, when using the pressure system, and from 45 to 60 min when using the two- or three-phase centrifugation system.

After the malaxation step, the kneaded dough is injected (by a pump) into a centrifuge whose axis is horizontal. This unit is called horizontal decanter. In the decanter manufacturing process (two- or three-phase process), an olive paste containing both water and oil may remain in containers until the oil is separated from the remaining water and the solid material. This natural separation takes its time. Modern decanters are large horizontal centrifuges that separate oil from solids and water by a similar way as in a settling tank, but more quickly. Time savings increase the efficiency of the system and reduce the contact time with the vegetation or wastewater, closely related to oil quality (chemical composition and sensorial parameters). Some of the disadvantages of the three phases' centrifugal decanter are the high volume of vegetation water and oil, which results in a lower total phenol content. This inconvenience was solved by the introduction of two-stage decanters in the market. The two-stage decanter produces only oil and pomace (Picture 4.3). Also a two-and-a-half-stage decanter produces oil, pomace and small amounts of vegetation water. In these decanters, the separation of the oil from the olive paste can be carried out without the addition of dilution water, or, if water is added, it will be in reduced quantities, which makes possible to minimize the production of wastewater. The two-decanter scale produces only a limited volume of wastewater (10–30 L/100 kg of olives) and obtains oils with a high content of total phenols (Di Giovacchino et al. 1994).

As a conclusion, like the traditional method (Picture 4.2), the continuous extraction method (Picture 4.3) requires prior grinding, in this case performed in hammer mills or disc mills. A comparison between continuous and discontinuous grinding steps is presented in Table 4.5. In the continuous process, when the grinding is finished, and with the aid of a variable speed metering pump, the pulp is sent to a horizontal centrifuge, where occurs the separation in three phases (solid phase, oil, and vegetable water) or in two phases (oil and humid pomace). The solid phase, called pomace contains most of the solids present in the olive: the skin, the pulp, the nucleus and a small portion of oil. The aqueous residue called vegetable water is originally a dark liquid. The quantity and quality of the generated vegetable water are variable, and depends on the system, the type of fruit, and the water used.

The two-scale decanter has been used for the first time in Spain at the beginning of the 1990s of the last decade. It is known by the "ecological decanter". This is not only due to water savings and more than the substantial removal of wastewaters, but factors such as lower operating costs, and the quality of the oil produced with the

Table 4.5 Advantages and disadvantages of stone and hammer mills (respectively discontinuous and continuous process) for olive oil extraction

	Stone mills	Hammer mills
Advantages	Gentle system for olives, without no risks of overheating	Allows continuous milling operation, taking better advantage of equipment, facilities and labor
	Limited formation of emulsions due to reduced mechanical stress; and negligible risk of metal contamination	High working hourly capacity less expensive and cumbersome system
	Flexibility to adapt the pressing according to the characteristics of the olives and the size of the fragments of the stones, favoring the formation of larger droplets of oil and replacing in part to the subsequent grinding	Cleaning is easier
	Oil is less bitter	
Disadvantages	Time-consuming process: 20–50 min	Non-optimal preparation of the paste with the formation of stable emulsions, whose breakage requires higher times and temperatures of grinding
	Bulky and expensive installations	
	Slow and discontinuous work	
	Long storage times detrimental to the quality of the oil	
		The paste may heat up, with negative consequences in oil quality

two-phase system. The oil is slightly higher or different in terms of oxidative stability, bitterness and more intense aromatic bouquet. All these factors have reinforced this conviction among the players in the olive sector. A comparison of the three most commonly used types of extraction systems should consider the following three criteria:

- The energy used and the cost (mainly water): discontinuous system requires the lowest energy in comparison to continuous system independently to the kind of decanter used. The three scale decanter use a huge amount of water.
- The quality and quantity of oil produced: in general, both the quantity and the quality of the oil obtained by the continuous system are superior to that produced by the traditional process. The two scale decanter produce oil with the highest level of polyphenols and oxidative stability.
- The type and quantity of waste re-generated: the three scale decanter produce a huge amount of wastewater.

5 Composition of the Olive Oil

The composition of olive oil is largely influenced by the process and pieces of equipment used for harvesting, transport and conditioning of fruits, and oil extraction. Most olive oil producers aim at maximizing the obtainment of VOO, the most valued fraction, thus using care in handling the fruits, short transportation and storage periods, and especially mild conditions for the extraction of the first valuable oil fractions. Thus, VOO's composition can be viewed as the juice of olive fruit with its unique composition and flavor. As VOO is consumed directly without any further refining treatment, it retains much of the original fruit components. In other words, while most fats and oils are almost exclusively a mixture of triacylglycerols (TAG) (eventually with very small amounts of other components), VOO contains a wide variety (in non-negligible amounts) of natural components of the olive fruit, besides TAG. The TAG fraction is known as a saponifiable fraction, because it can react with an aqueous solution of a strong base (NaOH or KOH) to form soap and glycerol. The other fraction, which remains mostly associated with the water phase and/or can be recovered by solvents, is known as unsaponifiable fraction and is formed by many compounds in minor concentrations. Saponifiable fraction, or the lipid phase, represents around 98.5–99.5% of the total composition and is mainly formed by TAG, diacylglycerols, monoacylglycerols, free fatty acids, and phospholipids. An unsaponifiable fraction consists of phenolic compounds, tocopherols, hydrocarbons, pigments, sterols, triterpenes and other compounds, which represent 0.5–1.5% of the olive oil composition.

5.1 Saponifiable Fraction of Olive Oil

Mono-, di- and tri-glycerides, are glycerol esters of fatty acids (FA) formed by the union of a molecule of glycerol (trifunctional alcohol) with three FA. Glycerol (also called glycerine) is a small 3-carbon molecule, each C carrying a hydroxyl group (OH). On its turn, a FA is a molecule formed of a chain of carbons bound to hydrogens (also called hydrocarbon) terminated by an acidic carboxyl group (COOH) to form an acylglycerol, each fatty acid undergoes an esterification reaction to a specific position in the glycerol molecule, in which the COOH, of the FA reacts with a OH of the glycerol. In monoglyceride, only one position of the glycerol molecule is esterified with FA, a diglyceride corresponds to the esterification of two positions and in a TAG, all the positions of the glycerol backbone are occupied with FA chains. Olive oil is mostly formed by TAG, in which contribute to 94–96% of the total weight.

The number of double bonds in the carbon chain, also called unsaturations, affects the physical and chemical properties of fats, and the presence and balance between different types is relevant to human health. Olive oil contains saturated fatty acids (SFA), monounsaturated (MUFA), and polyunsaturated fatty acids (PUFA). It is considered an equilibrated oil as it contains larger amounts of PUFA than of SFA, and MUFA dominates over PUFA (Table 4.6). As can be deduced from Table 4.6, MUFA profile is a significant feature through which olive oil is distinguished from other fats. The most common MUFA in olive oil is oleic acid (18:1 n-9[4]), present in a

---

[4] n-9 or omega 9, means that the first (and this case single) double bound in the FA chain is located

Table 4.6 Fatty acid composition of olive oil, saturated, mono-unsaturated and poly-unsaturated fatty acids, and their discriminated composition

Composition data from USDA food composition database	
Lipids – olive oil	Value per 100 g
Fatty acids, total saturated	13.808
Fatty acids, total monounsaturated	72.961
Fatty acids, total polyunsaturated	10.523
Composition data from Ciqual food composition nutritional table	
FA Saturated	11.9
FA monounsaturated	75.2
FA polyunsaturated	7.39
FA4: 4.0	< 0.05
FA6:0	<0.05
FA 8:0	< 0.05
FA 10: 0	< 0.05
FA 12:0	<0.05
FA14:0	<0.05
FA16:0	8.26
FA18:0	2.99
FA18:1 n -9 cis	71.0
FA 18: 2 9c, c12, c15 (n − 3)	6.75
FA 18: 3 c9, c12, c15 (n − 3)	0.64

Data was collected from USDA food composition database (https://ndb.nal.usda.gov/ndb/search/list?) and CIQUAL table (https://ciqual.anses.fr/#/aliments/17270/olive-oil-extra-virgin)

range from 55% to 83% of the total FA's composition (IOC 2015; CODEX STAN 33-1981–rev 2017).

PUFA of olive oil is mostly linoleic acid (C18:2) and linolenic acid (C18:3), which have, respectively, two and three conjugated double bonds in the carbon chain. Linoleic acid (n-6) is an omega-6 fatty acid[5] that makes up about 3.5–21% of total FA in olive oil composition (IOC 2015; CODEX STAN 33-1981–rev 2017). It is an essential component of structural membrane lipids and it is involved in cell signaling (Delgado et al. 2017). Linolenic acid is a polyunsaturated omega-3 fatty acid that makes up 0–1.5% of olive oil. It is an essential FA, as the n-3 α-linolenic acid (ALA) is involved in neurological development and growth and is a precursor of eicosanoids (Delgado et al. 2017). Olive oil has the appropriate balance between n-3 and n-6 fatty acids that should be maintained to limit lipid peroxidation, which is thought to be a component in the development of atherosclerotic plaques (Delgado et al. 2017).

in the 9th carbon, counting from the opposite side to the functional group (alfa extremity), which in the case of TAG is the unattachedend of the FA chain (omega extremity).

[5] Meaning that the first double bound of the carbon chain is located in the n-6 position, or omega-6 carbon.

Saturated fatty acids that can be found in olive oil are lauric, myristic, palmitic, stearic, arachidic, behenic and lignoceric acids. No essential role, other than as an energy source, has been found for SFA and TFA or cholesterol in the diet, as humans can synthesize these compounds according to their needs.

TAGs contain FA with 16–24 carbon atoms such as palmitic, palmitoleic, stearic, oleic, linoleic, and linolenic acids. The position 2 is reserved to an unsaturated fatty acid and it is dominated by linolenic acid. The distribution of FAs in the glycerol moiety allows 20 different combinations of TAG, however only four are frequent, in significant proportions: oleic-oleic- oleic (OOO: 27.53–59.34%), palmitic-oleic-oleic and stearic-oleic-linoleic (POO + SOL: 12.42–30.57%), oleic-oleic-linoleic and linolenic-palmitic-palmitic (OOL + LnPP, 4.14-17.46%), palmitic-oleic-linoleic and stearic-linoleic-linoleic (POL+SLL, 2.69–12.31%).

Admittedly, TAGs dominate the saponifiable fraction but a set of so-called minor components are of paramount importance for the distinctive composition of VOO. These liposoluble compounds, as vitamins and vitamin precursors, as well as other minor components, some of which are water soluble (such as phenols, pigments, sterols and volatile materials) and are thus integrated in the unsaponifiable fraction, defined as the fraction extracted with solvents after saponification of the oil, and the soluble fraction. Non-saponifiable compounds give olive oil its unique flavor.

5.2 Unsaponifiable Fraction of Olive Oil

This fraction generally varies from 1% to 2%, and this variability depends on the species, the cultivar, the stage of maturation, the state of health of the raw material before extraction, and the techniques of extraction and preservation.

Unsaponifiables are natural or accidental constituents that do not react with strong bases (sodium hydroxide and potassium hydroxide) to produce soaps but remain soluble in hexane or petroleum ether after saponification. These constituents of olive oil (saponifiable and unsaponifiables) can be used as biomarkers of its authenticity, and are useful in various analyses and food fraud investigations. Unsaponifiables are mostly removed during refining and, their presence and concentration ranges are used in characterization, grading and discrimination of VOO. Minor components of VOO have then different degrees of polarity or lipophilic character.

Tocopherols are known as lipophilic phenols that include eight occurring forms: four forms of tocopherols and four forms of tocotrienols (α, β, γ and δ). In EVOO, the most predominant is the α-tocopherol (up to 90% of total tocols) that has vitamin E activity (Delgado et al. 2017; Kalogeropoulos and Tsimidou 2014). Tunisian VOO were found to have the highest α-tocopherol content, when compared to VOO from other origins (e.g. >459 mg\kg in Oueslati VOO inform the region of Kairouan) (Dabbou et al. 2009). According to Sarolić et al. (2014), the recommended daily intake of tocopherols is the amount equivalent to 1 mg of α-tocopherol (Vit. E),

Table 4.7 Indicative levels of tocopherols in olive oil, according to values from the USDA food composition database and CIQUAL, food composition nutritional table

USDA – food composition database	
Tocopherol – olive oil	Value per 100 g
Alpha tocopherol (mg\100 g)	14.35
Ciqual food composition nutritional table	
Vitamin E (mg\100 g)	21.7

Data was collected from USDA, food composition database (https://ndb.nal.usda.gov/ndb/search/list?) and CIQUAL, food composition nutritional table (https://ciqual.anses.fr/#/aliments/17270/olive-oil-extra-virgin)

although each form plays its own role in the organism. Table 4.7 shows indicative values for tocopherols content in VOO, according to data from two different food composition databases (USDA in the USA, and Ciqual in France).

Simple phenolics and polyphenols are important soluble minor components of olive oil. Phenolics are a broad range of compounds that have in common the presence of one or more 6C aromatic rings, attached to hydroxyl groups. Simple phenolics are those with a single ring, like hydroxytyrosol, tyrosol, which are colorless and odorless, thus not transmitting any particular taste to VOO. However, both compounds are known for their strong anti-oxidant properties (Delgado et al. 2017). On the other hand, secoridoids are modified polyphenols (e.g. with sugar moieties), as is the case of oleuropein, the aglycone of ligstroside, which is typical from olives and confers a characteristic bitter stringent note to VOO. Another relevant class for the flavor of VOO is lignans (i.e., pinoresinol).

As explain by Boskou et al. (2005), polyphenols are "polar phenolic compounds", designation preferred by these authors to distinguish them from the non-polar phenols, as the tocopherols. Qualitatively, tyrosol, hydroxytyrosol, and their secoiridoid derivatives dominated the polar phenolic fraction, which makes up around 90% of the total phenolic content of virgin olive oil as reported by De la Torre et al. (2005). The polar phenolic fraction can be distinguished as simple or complex. The first class covered 3,4-dihydroxyphenyl-ethanol, or hydroxytyrosol, and p-hydroxyphenyl-ethanol, or tyrosol, and phenolic acids including caffeic, vanillic, syringic, protocatechuic, p-coumaric and o-coumaric, gallic, 4-hydroxybenzoic acids (Mannino et al. 1993; Montedoro 1972; Tsimidou et al. 1996). Secondary glycosidic compounds are included in the class of complex phenols: secoiridoidsoleuropein, demethyloleuropein, ligstroside and nüzhenide are the main representative of this category (Bianco and Uccella 2000). Regarding sensorial properties, polyphenols are responsible for the bitterness and astringency taste of the oil. Quantitatively and based on the research published by Perez-Jimenez et al. (2010), virgin olive oil ranked 61st position of a list of the 100 richest foods in polyphenols and antioxidant content with a concentration of 62 mg/100g.

Table 4.8 lists the phenolic compounds that can be found in olive oil, discriminated within each class of compounds, as well as indicative values for their concentration. Phenolic concentration in VOO is not constant, depending on the olive

Table 4.8 Phenolic compounds' composition of olive oil – indicative values based on the phenol explorer database http://phenol-explorer.eu/contents/food/822

Flavonolds		
Flavones	Apigenin	1.17 mg/100 g
	Luteolin	0.36 mg/100 g
Phenolic acids		
Hydroxybenzoicacids	4-Hydroxybenzoic acid	3.13e-0.3 mg/100 g
	Synergic acid	0.02 mg/100 g
	Vanillic acid	0.06 mg/100 g
Hydroxycinnamicacids	Caffeic acid	0.02 mg/100 g
	Cinnamic acid	0.02 mg/100 g
	Ferulic acid	0.02 mg/100 g
	Hydroxycaffeic acid	5.00e-0.3 mg/100 g
	m-Coumaric acid	4.5oe-03 mg/100 g
	p-Coumaric acid	0.09 mg/100 g
Hydroxyphenylaceticacids	4-Hydroxyphenylacetic acid	0.05 mg/100 g
	Homovanillic acid	9.00e-0.3 mg/100
Lignans		
Lignans	1-Acetoxypinoresinol	0.66 mg/100 g
	Pinoresinol	0.42 mg/100 g
Other phenols		
Tyrosols	3,4-DHPEA-EA	7.22 mg/100 g
	3,4-DHPEA-EDA	25.16 mg/100 g
	Hydroxytyrosol	0.77 mg/100 g
	Ligstroside	1.56 mg/100 g
	Ligstroside-aglycone	1.74 mg/100 g
	Oleuropein	0.17 mg/100 g
	Oleuropein-aglycone	3.66 mg/100 g
	p-HPEA-EA	3.80 mg/100 g
	p-HPEA-EDA	14.28 mg/100 g
	Tyrosol	1.13 mg/100 g

cultivar, the maturation index of fruit, pedoclimatic factors, extraction and separation process of the oil from fruit, as well as storage conditions and time. Some fruit varieties are known for their high level of polyphenols. This is the case of Coratina, Coroneiki, Rekhami, Jarboui (>600 mg/kg), others have a medium content of polyphenols, such as Arbosana, Leccino, Chétoui, Sayalie (between 600 and 200 mg/kg), and a third group shows a polyphenol content lower than 200 mg/kg. This is the case of Arbequina, Chemlali, and Zarrazi Zarzis varieties (Issaoui et al. 2009; Rigacci and Stefani 2016). It is well established that the content of polyphenols decreases during maturation, until ripeness (Amiot et al. 1986). Ranalli et al. (2009), have shown that green fruits have the highest content on the iridoid oleuropein, which decreased during the ripening process. The storage conditions and the period of time and prior to milling of the fruits have a considerable effect on polyphenols' content.

Salvador et al. (2003) showed that the lowest concentration of polyphenols found in the oil produced by the three-phase centrifuges if compared with any other types of extraction (pressure and dual phase decanter). The three-phase decanter use a higher quantity of water, which reduces the concentration of the polar phenolic compounds (Cert et al. 1996; Di Giovacchino 2001; Issaoui et al. 2009). On the other hand, Owen et al. (2000) reported that about 70% of total phenols are lost during the refining process of olive oil (registering a reduction from 232 ± 15 mg/kg to 62 ± 12 mg/kg).

EVOO is the richest grade, simultaneously higher in polyphenols simple phenols (such as secoiridoids and lignans) than others. Repassed olive oil, classified as ordinary olive oil is produced from the second extraction of the dual phase paste by the introduction of hot water into the centrifugation phase. This oil fraction exhibits a lower concentration of polyphenols in comparison to those produced from the dual and three-phase decanter, as reported by Ammar et al. (2014).

Based on our previous work (Issaoui et al. 2009), the content of polyphenols may relate to environmental factors, and most probably increase with the altitude. Hence the polyphenols content of VOO obtained from cvs Chétoui and Chemlali cultivated at high altitude showed higher concentrations of polyphenols (551.1 mg/kg and 572.5 mg/kg, respectively) in comparison with oil produced and obtained in the same way from fruit from the same cvs cultivated at low altitude (274.0 mg/kg and 172.5 mg/kg, respectively). An important protective measure to preserve antioxidants and flavor is the choice of packaging materials. In this respect, the report "how packaging influences olive oil quality (produced by UC Davis Olive center)" summarizes and compares the advantages and disadvantages of each material such as glass, aluminum, tinplate cans, stainless steel, plastic, coated paperboard, and bag-in-box containers. Based on such report, the ideal packaging material should be impermeable to the light and to the air. Moreover, to better preserve its properties EVOO should ideally be stored in the dark at 16 °C.

The desirable levels of bitterness are due to the presence of high amount of phenolics. However, the odor and the flavor of olive oil is also due to the presence of aroma compounds that develop mainly during and after oil extraction from the olive fruit and from the degradation of PUFA as a result of lipoxygenase (LOX) enzyme activity. More than 180 volatile compounds are at the base of the many aromatic notes of virgin olive oil including herbaceous, green apple, fruity green intense, almond, artichoke, and tomato have been noted (Fedeli 1977; Montedoro et al. 1978). These authors have attributed the aroma bouquet occurring in the olive oil matrix to the presence of aliphatic and aromatic hydrocarbons, aliphatic and triterpenic alcohols, aldehydes, ketones, ethers, esters, and furan and thiophene derivatives. Kiritsakis et al. (1998) have listed dozens of odor descriptors that were attributed to some volatile compounds identified in VOO samples. Based on this list, aromatic compounds produced through the LOX exhibited usually a green note (e.g. hexanal, E-3-hexenal, Z-3- hexenal, 3-hexyl acetate and E-2-hexen-1-ol).

Pigments such as chlorophylls and pheophytins are responsible for the green color of the oil while the carotenes impart the yellow color to the oil. The USA standard for Grades of Olive Oil and Olive-Pomace Oil is the only standard that required color as a parameter to classify olive oil. Thus, according to this standard, VOO (of any of the four classes) must be yellow to green (US EVOO, USVOO and lampante olive oil), whereas, US refined olive oil must be light yellow. US olive oil and US pomace oil are included in the light yellow to the green color palette. Light Yellow to Brownish Yellow and Dark Green, Brown (or Black) are allowed colors for the US Refined Olive-pomace Oil, and US Crude Olive-pomace Oil respectively.

The composition of olive oil is richer and complicated. Recently, a database has been created by a number of experts in the field of olive and olive oil. OliveNet™, is a free, available and comprehensive database, detailing a total of 676 compounds derived from the various olive matrices (Bonvino et al. 2018).

6   Olive Oil and Health

6.1   Beneficial Effects of Olive Oil

Olive oil can be ingested at all ages, from toddlers to centenarians. According to the consensual report of International Conference on the healthy effect of VOO "The protective effect of VOO can be most important in the first decades of life, which suggests that the dietetic benefit of VOO intake should be initiated before puberty, and maintained through life" (Perez-Jimenez et al. 2011).

Many manufacturers take advantage of the fact that nutritional properties of olive oil are similar to the composition of breast milk and have it adapted to infants over 6 months (http://www.internationaloliveoil.org/web/aa-ingles/oliveWorld/salud10. html). In the Middle East, during the weaning period, infants are fed with olive oil as an alternative to breast milk. Olive oil is a relevant source of vitamins and phyto-nutrients, polyphenols, squalene, tocopherols and phylloquinone (Vit. K1), as well as a source of macronutrients (Fig. 4.4).

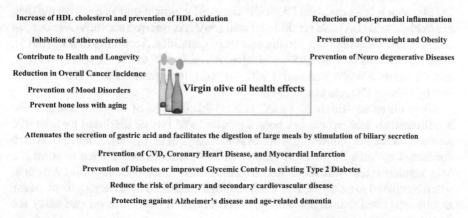

Increase of HDL cholesterol and prevention of HDL oxidation   Reduction of post-prandial inflammation

Inhibition of atherosclerosis   Prevention of Overweight and Obesity

Contribute to Health and Longevity   Prevention of Neuro degenerative Diseases

Reduction in Overall Cancer Incidence

Prevention of Mood Disorders   **Virgin olive oil health effects**

Prevent bone loss with aging

Attenuates the secretion of gastric acid and facilitates the digestion of large meals by stimulation of biliary secretion

Prevention of CVD, Coronary Heart Disease, and Myocardial Infarction

Prevention of Diabetes or improved Glycemic Control in existing Type 2 Diabetes

Reduce the risk of primary and secondary cardiovascular disease

Protecting against Alzheimer's disease and age-related dementia

Fig. 4.4 Nutrient values and weights are for edible portion:100g of olive oil (Based on National Nutrient Database for Standard Reference Release 1 April, 2018, https://ndb.nal.usda.gov/ndb/search/list?)

As a lipid source, VOO has a high content of essential fatty acids (as ALA[6]) that support the nutritional needs during the growth of infants and children. Essential FA are those that the body cannot synthesize and must be provided by food, as demonstrated in 1963 by Hansen et al. (Gómez Candela et al. 2011). Thus, besides the supply of specific essential FA, the balance between n-3 and n-6 FA is also important for human health, and is often referred to as omega 3/omega 6 or W6/W3 ratio. n-3 and n-6 FA have important but opposite physiological effects in terms of inflammatory activities and cardiovascular health. Thus, omega 3 FA reduce the synthesis and transport of fats, and play an antagonistic role to omega 6 in the development of the number of fat cells (Gomez Gandela et al 2011). Numerous epidemiological studies and clinical trials have shown the relationship between the consumption of W3 and the beneficial effects in different diseases such as cardiovascular disorders, various cancers (breast, colorectal, and prostate), asthma, inflammatory disease, and osteoporosis (Simopoulos 1991; Mozaffarian and Wu 2011; Mori 2014). n-6 FA tends to have a pro-inflammatory action and when consumed in excess they may contribute to persistent inflammation, which is known to cause many chronic diseases.

Eating habits have been disrupted worldwide, and the W6/W3 ratio in the diet dramatically changed. Nowadays, in Europe, the average intake of n-6 FA is about 10–20 times higher than the intake of n-3 FA. In the USA, the scenario is even worse, and the W6/W3 ratio can reach values of 50 to 1 (the intake of n-6 FA is 40–50 times higher than the intake of n-3 FA). Simopoulos (2006) is among the countless authors that refer an adequate balance between n-6 FA and n-3 FA to be important for the prevention and treatment of cardiovascular disease. For secondary prevention of cardiovascular disease, a ratio of 4:1 was found to be associated with a 70% reduction in total mortality. Dietary needs of an average adult male are about 2 g of omega-3 FA per day, and 10 g of omega 6, while an average adult female, needs 1.6 g and 8 g, respectively. These values are based on a balanced diet with a daily reference caloric import of 2200 kcal/day for men and of 1800 kcal/day, for women. The most common omega 3 fatty acid, in the human diet is α-linolenic acid (ALA), which is an essential FA, while the most common omega 6 FA, in the diet, are linoleic acid (LA) and arachidonic acid (AA). As referred above, VOO contains all those FA, including ALA, in the necessary quantities. According to IOC (2015), olive oil contains between 2.5% and 21% of LA, 0.60% to 1% (or more) of ALA, and the balance W6/W3 complies with the ratio recommended by WHO as exhibited by Gómez Candela et al. (2011).

Olive oil as an "Elixir of Youth" is a mythical belief of the inhabitants of the Mediterranean area, which has been supported and further disclosed by scientific research, since the pioneering work of Ancel Keys et al. (1980). A recent research conducted by Valls-Pedret et al. (2015) on 344 participants, intended to attest if a MD supplemented with foods rich in antioxidants influences cognitive function, when compared to a control diet. Their findings confirmed the later judgment. Many results suggested that in the elderly, a MD enriched with olive oil can delay the

---

[6]Alfa-linolenic acid.

cognitive decline associated with age (REF, you may add for example one recent study published by Harvard T Chan school of public health or Harvard medical school, 2018).

Based on the epidemiological study by Valls-Pedret et al. (2015), a multivariate analysis adjusted for confounders, allowed to demonstrate that the participants in a MD supplemented with olive oil had better results in the Rey Auditory Verbal Learning Test (to rate immediate and delayed episodic verbal memory). They also performed better in the Color Trail Test part 2 (to measure attention, visuomotor speed, and cognitive flexibility) than participants from the control cohort.

Several studies pointed toward the beneficial effects of VOO to be mainly due to the richness in antioxidant compounds inclduing phenolics, in vitamin E, and in anti-inflammatory factors (W6/W3 ratio). However, other compounds have proven to have bioactive activity. This is the case of oleocanthal, an important phenol from VOO, which has potent antioxidant and anti-inflammatory effects similar to the non-steroidal anti-inflammatory drug ibuprofen (Abuznait et al. 2013; Qosa et al. 2015). Another example is squalene, an alkane triterpene that is an intermediate in the synthesis of cholesterol, thus conditioning the lipid metabolism (Delgado et al. 2017). In addition, several recent studies showed that oleuropein may possess important neuroprotective activities against Alzheimer's disease (Abuznait et al. 2013, 2017; Batarseh et al. 2017; Beauchamp et al. 2005; Qosa et al. 2015).

Figure 4.4 summarizes many benefits of olive oil in human health, as mentioned in research papers reviewed by experts and published in recognized international scientific journals. Some of these benefits have been substantiated into health claims by authorities, in other cases the mechanisms of action and the dose-response levels are yet to be elucidated.

In this respect, the International Olive Council (IOC) organized a symposium to analyze the large amount of research that continues to be produced reflecting the benefits of olive oil on human health, in order to avoid confusion between science

Calories (kcal): 100g = 884
    Total fat (g): 100g

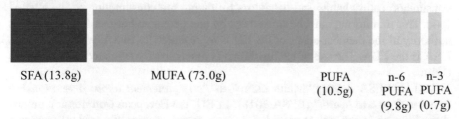

| SFA (13.8g) | MUFA (73.0g) | PUFA (10.5g) | n-6 PUFA (9.8g) | n-3 PUFA (0.7g) |

Minerals. Calcium (Ca) 1mg; Iron, Fe 0,56mg; Potassium (K) 1mg; Sodium (Na) 2mg

Vitamins. Vitamin E (alpha tocopherol) 14,35mg; Vitamin K (phylloquinone) 60,2μg

Fig. 4.5 Some beneficial health effects of olive oil as reported in literature and scientific research

facts and marketing messages. Thus, in January 2018, IOC invited renowned international experts to the Robert Mondavi Institute (at UC-Davis, USA), to consult, analyze and summarize the available data on the topic. Visioli et al. (2018) summarized the main findings, conclusions, and proposed areas of research, on the relationship between olive oil and human health.

7   Nutritional Claims, Health Claims and Recommended Daily Intake of Olive Oil

At this step, it is important to clarify the difference between a health claim and nutritional claim. The definitions by Codex Alimentarius, EFSA, and the US - FDA is herein present, as these organizations are global references in these aspects.

As defined by Codex Alimentarius (CAC/GL 23-1997-2013) **nutrition claim** is *"any representation which states, suggests or implies that a food has particular nutritional properties including but not limited to the energy value and to the content of protein, fat and carbohydrates, as well as the content of vitamins and minerals"* (CAC 2013).

The EFSA definition is similar, limiting such claims to those listed in the Annex of Regulation (EC) No 1924/2006, lastly amended by Regulation (EU) No 1047/2012. On its turn, US-FDA mentions "nutrient content claims" as *label claims that characterize the level of a nutrient in a food.* Terms such as *"free"*, *"healthy"*, *"high"*, and *"low"* are authorized if in accordance with the guidelines that ensure coherence and meaning to the consumers. Claims can also compare a food to another food. In addition, in the USA, food labels may also contain Structure/function claims for conventional foods (which focus on effects derived from nutritive value), as well as disclaimers.

Again, starting by the definition of Codex Alimentarius (CAC 2013), **health claim** *"means any representation that states, suggests, or implies that a relationship exists between a food or a constituent of that food and health"*. The Codex Alimentarius stresses the importance of monitoring the effects of health claims on consumers' eating habits and behaviors following the consumption of the alleged food. This monitoring must be carried out by competent authorities. The competent authority of the member-states of the EU is represented by EFSA, while that of the USA is the FDA. The use of health claims is much strictly regulated than nutrition claims.

In EU, EFSA defines "health claim" as *"any statement about a relationship between food and health"* (EFSA 2018). In EU, the European Commission authorizes different health claims provided they are based on scientific evidence and can be easily understood by consumers. Therefore, health claims are regulated by legislation and monitored, at European and at the national level. EFSA distinguishes 3 types of health claims: The so-called 'Function Health Claims' (or Article 13 claims), e.g. on slimming or weight-control, "'Risk Reduction Claims' (or Article

14(1) (a) claims) on reducing a risk factor in the development of a disease, and 'Claims referring to children's development' (Article 14(1)(b) claims). Permitted Function-health claims are listed in the Regulation (EU) No 432/2012, which is regularly updated. For other claims, which must be based on substantial scientific evidence, applications are analyzed case by case (EFSA 2018).

In USA, FDA defines "health claim" as *"Health claims describe a relationship between a food substance (a food, food component, or dietary supplement ingredient), and reduced risk of a disease or health-related condition"* (FDA 2018a); such a health claim statement has two essential components: the substance (whether a food, food component, or dietary ingredient) and the disease or health-related condition. Statements that address a role of a specific substance in maintaining normal healthy structures or functions of the body are considered to be structure/function claims, and do not require review and authorization by FDA (FDA 2018a).

According to the recommendations of the Codex Alimentarius, any claim (e.g. nutrition, nutrition function, and health claims) can only be used if they comply with national requirements for nutrition and health, respectively. One of the key objectives of these regulations is to ensure that any claim on the label of a food being sold is clear and supported by scientific evidence (CAC 2013). As can be easily deduced from what is above explained, differences in definitions and regulations can be found, namely between USA and EU, at present.

7.1 Nutritional Claims for Olive Oil

Saillard (2014) summarized the nutritional claims that olive oil can exhibit in the label as follows:

(a) Nutritional claims related to FA composition (the conditions of use as given by the Annex to Regulation (EC) No 1924/2006):

- **Rich in unsaturated fats**: are "The product must contain at least 70% of fatty acids derived from unsaturated fats, and the energy provided by unsaturated fats accounts for more than 20% of the energy intake of the product". As can be seen in Table 4.7 olive oil is mostly formed by MUFA and PUFA.
- **Rich in monounsaturated fats**: "The product must contain at least 45% of fatty acids derived from monounsaturated fats, and the energy provided by the unsaturated fats represents more than 20% of the energy intake of the product". As can be seen in Table 4.7 olive oil contains at least 70% of MUFA, doubling the required value.

(b) Nutritional claims related to vitamin content (the conditions of use as given by the Annex to Regulation (EC) No 1924/2006):

- **Source of vitamin E**: "The product contains at least 15% of the Recommended Daily Allowances, 1.8 mg/100 g. As can be seen in Table 4.8, olive oil contains Vit. E (or α-tocopherol) in a concentration much above RDA, according to either database (USDA or CIQUAL), as shown in Table 4.9.

Table 4.9 Health claims authorized for olive oil in accordance to EC regulation (EU) No 432/2012 of 16 May 2012 establishing a list of permitted health claims made on foods, other than those referring to the reduction of disease risk and to children's development and health

Nutriment, substance, food or food category	Health claim	Conditions of use	Olive oil composition
Oleic acid	Replacing saturated fats in the diet with unsaturated fats contributes to the maintenance of normal blood cholesterol levels. Oleic acid is an unsaturated fat	The claim may be used only for food, which is high in unsaturated fatty acids, as referred to in the claim HIGH UNSATURATED FAT as listed in the Annex to Regulation (EC) No 1924/200	71g/100 g[a]
Monounsaturated and/ or polyunsaturated fatty acids	Replacing saturated fats with unsaturated fats in the diet contributes to the maintenance of normal blood cholesterol levels [MUFA and PUFA are unsaturated fats]	The claim may be used only for food which is high in unsaturated fatty acids, as referred to in the claim HIGH UNSATURATED FAT as listed in the Annex to Regulation (EC) No 1924/2006	72.961 g/100 g and 10.523 g/100g, respectively
Olive oil polyphenols	Olive oil polyphenols contribute to the protection of blood lipids from oxidative stress	The claim may be used only for olive oil which contains at least 5 mg of hydroxytyrosol and its derivatives (e.g. oleuropein complex and tyrosol) per 20 g of olive oil. In order to bear the claim information shall be given to the consumer that the beneficial effect is obtained with a daily intake of 20 g of olive oil	
Vitamin E	Vitamin E contributes to the protection of cells from oxidative stress	The claim may be used only for food which is at least a source of vitamin E as referred to in the claim SOURCE OF [NAME OF VITAMIN/S] AND/OR [NAME OF MINERAL/S] as listed in the Annex to Regulation (EC) No 1924/2006	14.35 mg/100 g (α tocopherols) 21.7 mg/100 g
Vitamin k	Vitamin K contributes to normal blood clotting	The claim may be used only for food which is at least a source of vitamin K as referred to in the claim SOURCE OF [NAME OF VITAMIN/S] AND/OR [NAME OF MINERAL/S] as listed in the Annex to Regulation (EC) No 1924/2006	60.2 μg/100 g
Vitamin k	Vitamin K contributes to the maintenance of normal bones	The claim may be used only for food which is at least a source of vitamin K as referred to in the claim SOURCE OF [NAME OF VITAMIN/S] AND/OR [NAME OF MINERAL/S] as listed in the Annex to Regulation (EC) No 1924/2006	60.2 μg/100 g

[a]Reference values presented above are based in data from CIQUAL, food composition nutrition table (https://ciqual.anses.fr/#/aliments/17270/olive-oil-extra-virgin) and from USDA food composition database https://ndb.nal.usda.gov/ndb/search/list?

- **Rich in vitamin E**: "The product contains at least 2 times the content required for the claim "source vitamin E of [name of vitamins] that according to the Annex to Regulation (EC) No 1924/2006, should be at least 3.6 mg/100 g" Olive oil contains Vit. E (or α-tocopherol) in a concentration much above the required, according to either database (USDA or CIQUAL)
- **A source of Vitamin K**: "The product contains at least 15% of the Recommended Daily Allowances, 11.25 µg/100 g". As can be seen from Table 4.9 (values from two distinct databases), olive oil concentration in phylloquinone, or Vit. K1, largely surpasses RDA
- **Rich in vitamin K**: "The product contains at least 2 times the content required for the claim "source of vitamin K that is 22.5 µg/100 g". As can be seen from Table 4.9 (values from two distinct databases), olive oil concentration in phylloquinone, or Vi. K1, surpasses 2 x RDA, as required to hold the claim

7.2   Health Claims for Olive Oil

The EU published a list of nutrients, foods or food components for which an evidence-based relationship between daily intake and claimed an effect on health was scientifically proved, upon analysis by EFSA. Olive Oil generally complies with several claims of the Annex, list of permitted health claims, of EC Reg. (EU) No 432/2012. The health claims that may apply to olive oil, in compliance to this EU regulation are listed in Table 4.9. These olive oil constituents fulfill the necessary concentrations in VOO to provide the daily intake associated with the health benefit, to a consumer with a regular and equilibrated intake of VOO. Thus, the constituents of olive oil that comply with such requirements are oleic acid, MUFA and PUFA, Vit. K (phylloquinone) and Vit. E (tocopherols).

In addition to those constituents, olive oil contains other relevant compounds that are also listed in Annex, list of permitted health claims, of EC Reg. (EU) No 432/2012. Although the regular consumption of olive oil may not provide enough quantities to fulfill daily requirements, it can contribute to a better health namely by the presence of ALA, which contributes to the maintenance of normal blood cholesterol levels (the beneficial effect is obtained with a daily intake of 2 g of ALA). Also noteworthy is the presence of linoleic acid (LA), which contributes to the maintenance of normal blood cholesterol levels (the beneficial effect is obtained with a daily intake of 10 g of LA).

Currently, the health effect claimed for olive oil is "Protection of blood lipids against oxidative stress" which is correlated to the intake of an adequate amount of hydroxytyrosol. However, The Panel of EFSA (EFSA 2012), recommended that only olive oil that contain a minimum of 5 mg of hydroxytyrosol per 20 g might make use of this claim: "may be used only for olive oil which contains at least 5 mg of hydroxytyrosol and its derivatives (e.g. oleuropein complex and tyrosol) per 20 mg of olive oil. In order to bear the claim, information shall be given to the consumer that the beneficial effect is obtained with a daily intake of 20 mg of olive oil". In fact, most of the beneficial effects can be obtained from a daily dose of 20 g, which is roughly equivalent to two tablespoons of olive oil (the differences in the composition of different grades of olive oil should be taken into account by the consumer).

In the USA, health claims were also granted to olive oil. The FDA (2018b) reports that "*Limited and not conclusive scientific evidence suggests that eating about 2 tablespoons (23 grams) of olive oil daily may reduce the risk of coronary heart disease due to the monounsaturated fat in olive oil. To achieve this possible benefit, olive oil is to replace a similar amount of saturated fat and not increase the total number of calories you eat in a day. One serving of this product [Name of food] contains [x] grams of olive oil.*"(Docket No. 2003Q-0559, 11/01/2004).

According to their findings, the FDA has determined that a minimum daily intake of 17.5 g of MUFA from olive oil was needed to exert a positive effect on the reduction of coronary heart disease. Taking a monounsaturated oil content of up to 73%, only 23 g of olive oil is needed to provide the required 17.5 g of MUFA, which is equivalent to 1.7 tablespoons of olive oil. Consequently, the daily intake of two tablespoons of olive oil was included in the US-FDA approved qualified health claim.

It is widely accepted that "Replacing saturated fats in the diet with unsaturated fats contributes to the maintenance of normal blood cholesterol levels." Oleic acid is the most common MUFS and it is one of the fundamental components of olive oil. Olive oil contains between 55% and 83% oleic acid (Šarolić 2014) and according to CIQUAL data (2018), it contains more than 70 g/100 g of oleic acid. Oleic acid decreases the total cholesterol level in the blood and increases that of "good" cholesterol. As a consequence, olive oil contributes to the maintenance of normal blood's cholesterol levels.

There is as many as 5 mg of polyphenols in every 10 grams of olive oil, as reported by Psaltopoulou et al. (2004). The study investigated the relationship between olive oil consumption and blood pressure in the Greek population. The authors found that Greeks consume, on average, about five to six tablespoons of olive oil each day, which means that the daily intake of polyphenols by a Greek consumer is about 33.75–40.5 mg. Polyphenols, as referred before, are actually a complex mosaic of compounds with variable chemical structure, which have in common the fact of being natural antioxidants that contribute to the bitter taste, astringency, and resistance to oxidation of the VOO. Phenols from olive oil, including polyphenols, are listed in Table 4.8, with indicative concentration values. They are becoming increasingly important, particularly because of their beneficial effects on health. Indeed, their role as natural antioxidants is attracting more and more interest in the prevention and treatment of cancer, inflammatory diseases, cardiovascular and neurodegenerative diseases.

The consensus is found between different organizations and countries, in what concerns daily intakes of the different types of fatty acids. The consensus is also found in respect to the health benefits of polyphenols. In what concerns health claims and recommended intakes of vitamins, some divergences were found, namely in the Recommended Daily Allowances (RDA) for tocopherols. Thus, Japanese authorities recommend 6.5 mg Vit. E/day, while in EU, RDA established by EFSA is of 12 mg. Finally, the Codex Committee on Nutrition and Foods for Special Dietary Uses (CCNFSDU) of Codex Alimentarius validated a value of 9 mg following numerous consultations with EFSA (EC), NHMRC/MOH (Australia & New Zealand), NIHN (Japan) and WOH/FAO (Codex Alimentarius). The RDA recommended by CCNFSDU was calculated as explained below:

Means EFSA, NHMRC\MOH, NIHN, WOH\FAO	
	$(12+8.5+6.8+8.8)/4 = 9.025 \approx 9$

Source: CAC (2016)

In short, Codex Alimentarius established a consensus value for Vit. E RDA = 9 mg (CAC 2016).

It is noteworthy that Vit. E activity in olive oil is due to tocopherols, which fraction consist in: α-tocopherol (trimethylocotol), β-tocopherol, γ-tocopherol (dimethylocotol) and δ-tocopherol (monomethylocotol). The tocopherols of olive oil are mostly (90–95%) of α form (Blekas et al. 1995). Health claims listed in Table 4.9 concern the ability of tocopherols to protect cells against oxidative stress. Vitamin E acts as a "chain break" compound, preventing the propagation of lipid peroxidation. EFSA's scientific experts have concluded that vitamin E helps protect DNA, proteins, and lipids from oxidative damage. This health claim can only be used for foods that provide a significant source of vitamin E (at least 15% of the RDA). The recommended daily intake of vitamin E is 12 mg. According to USDA and CIQUAL databases (USDA 2018; Ciqual 2018) VOO contains between 130 and 300 ppm of Vit. E (mg/kg). According to the study by Caporaso et al. (2015), a daily intake of 30 g of olive oil with an average concentration of 200 ppm is able to provide about 50% of the daily needs of vitamin E. Therefore, olive oil, is a source of vitamin E, as mentioned implicitly in the above-referred annex from the EC Regulation.

Olive oil contains a considerable amount of vitamin K, about 60.2 µg/100 g according to USDA database (2018). Consequently, it can help regulating blood clotting and contribute to the maintenance of normal bones. The RDA forvit Kis about 60 µg according to Codex Alimentarius (CAC/GL 2-1985-revision 2015). Once again, as in the case of Vit. E, some divergences about RDA for Vit. K were found. It is noteworthy that EFSA's value for RDA is about 45 µg, but it varies according to age groups. Thus, RDA considering adjusted requirements to each age category, ranges from 10 µg for infants aged 7–11 months to 70 µg for adults, including for pregnant and lactating women (EFSA 2017). The concentrations of phylloquinone normally found in olive oil are significant in relation to RDA.

It is noteworthy to recall that vitamin K represents a family of fat-soluble compounds. Two of these compounds, Phylloquinone (Vit. K1) and Menaquinone (Vit. K2), are naturally present, and the third one (Vit. K3, Menadione) is a synthetic form. Phylloquinone is stable to heat (during food processing and cooking) but sensitive to light. The predominant sources of Vit. K in the human diet are dark green leafy vegetables and certain plant oils, including canola, soybean, and olive oil (Booth and Suttie 1998; Shearer et al. 2012).

In olive oil, K1 form that is present. Phylloquinone (called also phytonadione or phytomenadion) contains a phytyl group. The EFSA panel on Dietetic Products, Nutrition and Allergies (EFSA 2017) noted that vitamin K acts as a cofactor for the enzymatic conversion of vitamin K-dependent proteins (Gla proteins) into their active form. These proteins are implicated in different physiological processes, such as blood clotting, bone mineralization, and possible control of soft tissue calcification (EFSA 2017). To preserve vitamin K1 in olive oil, it must be protected from light and do not undergo high heat exposures, such as frying at high temperature, repeatedly and/or for long periods.

7.3   Daily Intakes of Olive Oil

Mediterranean diet is complex in nature and a multidisciplinary concept, as explained in the introduction section. The associated dietary pattern is rich in fruits, vegetables, and cereals, and the main source of lipids should be olive oil, providing 15–30% of total daily energy (Keys et al. 1986). The fatty acid content, mainly MUFA accounts for 16–29% of energy (Kris-Etherton 1999), Based on the dietary practices of the people who live along the Mediterranean Sea, the daily intake of olive oil is relevant but less than 50 mL/day. Regarding current daily intake quantities of olive oil in EU, according to a report of the IOC published in 2015 the Greeks stand at the top of the world rankings with 16.3 Kg/capita/year, equivalent to an average daily intake of about 46 mL/capita. Spain followed with 10.4 kg/capita/year (around 30 mL/day), Italy with 9.2 kg (around 27 mL/day) and Portugal with 7.1 kg (around 21 mL/day). Cyprus and Luxembourg, consume about 5.5 and 5.3 kg olive oil/capita/year, respectively (around 16 mL/day). France, Malta, Croatia, Ireland and Belgium, are in the list with average olive oil consumption of 1.2–1.7 kg/capita/year (4–5 mL/day).

Countries consuming between 0.5 (1.4 mL/day) and 1 kg (2.8 mL/day) of olive oil per capita per year are, in descending order: Finland, Latvia, Germany, Netherlands, Sweden, Slovenia, Austria, United Kingdom and Denmark. Finally those countries whose average consumption of olive oil does not reach 400 g per year (around or less than 1 mL/day) are Romania, Poland, Hungary, Bulgaria, Slovakia, Estonia and the Czech Republic.

In the non-European countries, the cases of Tunisia, and Syria are noteworthy. Tunisia, although being a major world producer, shows a lower average daily consumption of olive oil than expected, when compared to its average annual production.

In fact, Tunisians consume less than 4 kg/year, which gives an average per capita daily consumption of 12 mL, a four times a smaller amount than the Greeks and half of the quantity consumed by Spanish, Italians and Portuguese.

Per capita olive oil consumption of Moroccans (3.9 kg/year or 11 mL/day) and Jordanians (3.1 kg/year or 9 mL/day) show a similar trend to that of the Tunisians. In those top olive oil producing countries, where the olive tree has a so strong symbology, its best product is reserved for exportation. Apparently, peoples consume less and less olive oil, aggravating the issues of diverging from the Mediterranean dietary pattern, as the rising obesity rates, and non-communicable diseases. On the other hand, Syria's consumption is close to that of Portugal, with 7.0 kg (20 mL/day). The fact that currently (2018) the country is unstable and insecure will certainly affect trades. Thus, a possible explanation for that figure may lay in the need of the communities to make the best use of their own resources.

For different reasons (whether cultural or climacteric constraints to the olive tree) olive oil is consumed in low quantities in Albania and Lebanon (4.7 and 4.5 kg/year respectively, which is approx. equivalent to 14 mL/day), Palestine 3.2 kg (around 9 mL/day), Israel and Libya (2.4 kg/year; 7 mL/day), Turkey (2.0 kg/year; 6 mL/day). These countries will most probably reveal low adherence scores to the MD dietary pattern, of which olive oil is one of the pillars.

In the range of average consumptions of olive oil, ranging from 1.1 to 1.7 kg per capita and per year, are: Switzerland (1.7 kg; around 5 mL/day), Australia (1.6 kg; 5 mL/day), Algeria (1.5 kg; 4.3 mL/day), and Canada (1.1 kg; 3.3 mL/day).

In Saudi Arabia and Norway, the olive oil consumption is 0.7 kg/year/per capita (equivalent to 2 mL/day). Argentina, Egypt, Iran and Iraq have per capita consumption levels close to those of Hungary, Poland and Romania (0.1 kg/year which is the equivalent of 0.3 mL/day).

Which is surprising in this report (IOC 2013) is the remarkable change in consumption in the United States (non-IOC countries) over the last 25 years, even though consumption per capita is still around 0.9 kg in 2013 (2.7 mL/day), but follows a rising trend.

Major importers like the USA, Brazil, and Japan seem to have a large margin to growth, although per capita olive oil consumption in Japan, Brazil, Taiwan, China's, Russia are still below 0.5 Kg per capita and per year (around 1.2 mL/day). Da Silva et al. (2009) verified that most Mediterranean countries have been drifting away from the MD, while countries in Northern Europe and others from around the world are taking on a Mediterranean-like dietary pattern.

References

Abuznait, A. H., Qosa, H., Busnena, B., El Sayed, K. A., & Kaddoumi, A. (2013). Olive oil-derived oleocanthal enhances β-amyloid clearance as a potential neuroprotective mechanism against Alzheimer's disease: In-vitro and in-vivo studies. *ACS Chemical Neuroscience, 4*(6), 973–982.

Abuznait, A., Qosa, H., Mohamed, L. A., Batarshe, Y. S., & Kaddoumi, A. (2017). Neuroprotective effects of olive oil components in Alzheimer disease. In T. Farooqui & A. A. Farooqui (Eds.), *Neuroprotective effects of phytochemicals in neurological disorders* (p. 299). Hoboken: Wiley Blackwell.

Alloggio, V., Caponio, F., & De Leonardis, T. (1996). Influenza delle tecniche di preparazione della pasta di olive sulla qualità dell'olio. Nota I. Profilo quali-quantitativo delle sostanze fenoliche, mediante HPLC, in olio d'oliva vergine della cv Ogliarola Salentina. *Rivista Italiana Delle Sostanze Grasse, 73*, 355–360.

Amiot, M. T., Fleuriet, A., & Macheix, J. T. (1986). Importance and evolution of phenolic compounds in olive during growth and maturation. *Journal of Agricultural and Food Chemistry, 34*, 823–826.

Ammar, S., Zribi, A., Mansour, A. B., Ayadi, M., Abdelhedi, R., & Bouaziz, M. (2014). Effect of processing systems on the quality and stability of Chemlaly olive oils. *Journal of Oleo Science, 63*, 311–323. https://doi.org/10.5650/jos.ess13180.

Angerosa, F., & Di Giacinto, L. (1995). Crushing influence on the quality characteristics of virgin olive oil. Note II. *Rivista Italiana delle Sostanze Grasse, 72*, 1–4.

Arvaniti, F., Priftis, K. N., Papadimitriou, A., Papadopoulos, M., Roma, E., Kapsokefalou, M., Anthracopoulos, M. B., & Panagiotakos, D. B. (2011). Adherence to the Mediterranean type of diet is associated with lower prevalence of asthma symptoms, among 10–12 years old children: The PANACEA study. *Pediatric Allergy and Immunology, 22*, 283–289. https://doi.org/10.1111/j.13993038.2010.01113.x.

Bach-Faig, A., Berry, E. M., Lairon, D., Reguant, J., Trichopoulou, A., Dernini, S., Medina, F. X., Battino, M., Belahsen, R., Miranda, G., & Serra-Majem, L. (2011). Mediterranean diet pyramid today. Science and cultural updates. *Public Health Nutrition, 14*, 2274–2284. https://doi.org/10.1017/S1368980011002515.

Balatsouras, G. (1997). To Elaiolado. (Olive oil), Vol. II, Athens, Greece (in Greek).

Batarseh, Y. S., Mohamed, L. A., Al Rihani, S. B., Mousa, Y. M., Siddique, A. B., El Sayed, K. A., & Kaddoumi, A. (2017). Oleocanthal ameliorates amyloid-β oligomers toxicity on astrocytes and neuronal cells: In-vitro studies. *Neuroscience, 352*, 204–215.

Beauchamp, G. K., Keast, R. S., Morel, D., Lin, J., Pika, J., Han, Q., Lee, C. H., Smith, A. B., & Breslin, P. A. (2005). Phytochemistry: Ibuprofen-like activity in extra-virgin olive oil. *Nature, 437*, 45–46. https://doi.org/10.1038/437045a.

Blekas, G., Tsimidou, M., & Boskou, D. (1995). Contribution of a-tocopherol to olive oil stability. *Food Chemistry, 52*, 289–294.

Bonaccio, M., Di Castelnuovo, A., Bonanni, A., Costanzo, S., De Lucia, F., Pounis, G., Zito, F., Donati, M. B., de Gaetano, G., & Iacoviello, L. (2013). Adherence to a Mediterranean diet is associated with a better health-related quality of life: A possible role of high dietary antioxidant content. *BMJ, 3*(8), e003003. https://doi.org/10.1136/bmjopen-2013-003003.

Bonaccio, M., Di Castelnuovo, A., De Curtis, A., Costanzo, S., Persichillo, M., Donati, M. B., Cerletti, C., Iacoviello, L., & de Gaetano, G. (2014). Adherence to the Mediterranean diet is associated with lower platelet and leukocyte counts: Results from the Moli-sani study. *Blood, 123*, 3037–3044. https://doi.org/10.1182/blood-2013-12-541672.

Bonaccio, M., Di Castelnuovo, A., Costanzo, S., Gialluisi, A., Persichillo, M., CerlettiC, D. M. B., de Gaetano, G., & Iacoviello, L. (2018). Mediterranean diet and mortality in the elderly: A prospective cohort study and a meta-analysis. *The British Journal of Nutrition, 120*, 841–854. https://doi.org/10.1017/S0007114518002179.

Bonvino, N. P., Liang, J., McCord, E. D., Zafiris, E., Benetti, N., Ray, N. B., Hung, A., Boskou, D., & Karagiannis, T. C. (2018). OliveNet™: A comprehensive library of compounds from *Olea europaea*. *Database*. (2018, *2018*, bay016. https://doi.org/10.1093/database/bay016.

Booth, S. L., & Suttie, J. W. (1998). Dietary intake and adequacy of vitamin K. *Journal of Nutrition, 128*(5), 785–788. https://doi.org/10.1093/jn/128.5.785.

Boskou, D., Blekas, G., & Tsimidou, M. (2005). Phenolic compounds in olive oil and olives. *Current Topics in Nutraceutical Research, 3*(2), 125–136.

CAC. (2013). Codex Alimentarius Commission. Codex Alimentarius.Internacional Food Standards. FAO/WHO. Guidelines for use of nutrition and health claims CAC/GL 23–1997 adopted in 1997. Last amended in 2013. Available: http://www.fao.org/ag/humannutrition/32444-09f5545b8abe9a0c3baf01a4502ac36e4.pdf. Accessed 14 Oct 2018.

CAC. (2016). Commission du Codex Alimentarius. Codex. Alimentarius. Programme mixte FAO\OMS sur les normes alimentaires comité du codex sur la nutrition et les aliments et diététiques ou de régime – projet de VNR-B pour la vitamine E, Trente-huitième session Hambourg, Allemagne 5–9 décembre 2016. Available at: http://www.fao.org/fao-who-codexalimentarius/shproxy/ru/?lnk=1&url=https%253A%252F%252Fworkspace.fao.org%252Fsites%252Fcodex%252FMeetings%252FCX-720-38%252Fnf38_04f.pdf. Accessed 1 Sept 2018.

Caporaso, N., Savarese, M., Paduano, A., Guidone, G., De Marco, E., & Sacchi, R. (2015). Nutritional quality assessment of extra virgin olive oil from the Italian retail market: Do natural antioxidants satisfy EFSA health claims? *Journal of Food Composition and Analysis, 40*, 154–162. https://doi.org/10.1016/j.jfca.2014.12.012.

Cert, A., Alba, J., León-Camacho, M., Moreda, W., & Pérez-Camino, M. C. (1996). Effects of talc addition and operating mode on the quality and oxidative stability of virgin olive oils obtained by centrifugation. *Journal of Agricultural and Food Chemistry, 44*(12), 3930–3934. https://doi.org/10.1021/jf9603386.

CIHEAM/FAO. (2015). Mediterranean food consumption patterns: diet, environment, society, economy and health. A White Paper Priority 5 of Feeding Knowledge Programme, Expo Milan 2015. CIHEAM-IAMB, Bari/FAO, Rome.

CIHEAM-IAMB, Bari/FAO, Rome.Ciqual (2018). French food composition table. ANSES. Available at: https://ciqual.anses.fr/. Accessed on 1 Sept 2018.

Da Silva, R., Bach-Faig, A., Raidó Quintana, B., Buckland, G., Vaz de Almeida, M. D., & Serra-Majem, L. (2009). Worldwide variation of adherence to the Mediterranean diet, in1961–1965 and 2000–2003. *Public Health Nutrition, 12*, 1676–1684. https://doi.org/10.1017/S1368980009990541.

Dabbou, S., Issaoui, M., Esposto, S., Sifi, S., Taticchi, A., Servili, M., Montedoro, G. F., & Hammami, M. (2009). Cultivar and growing area effects on minor compounds of olive oil from autochthonous and European introduced cultivars in Tunisia. *International Journal of Food Science and Technology, 89*, 1314–1325. https://doi.org/10.1002/jsfa.3588.

De la Torre, K., Jauregui, O., Gimeno, E., Castellote, A. I., Lamuela- Raventoa, R. M., & Loapez-Sabater, M. C. (2005). Characterization and quantification of phenolic compounds in olive oils by solid-phase extraction, HPLC-DAD, and HPLC-MS/MS. *Journal of Agricultural and Food Chemistry, 53*, 4331–4340.

Delgado, A. M., Vaz-Almeida, M. D., & Parisi, S. (2017). Olive oil and table olives. In *Chemistry of the Mediterranean diet*. Cham: Springer.

Di Giovacchino, L., Costantini, N., Serraiocco, A., Giulio, S., & Basti, C. (2001). Natural antioxidants and volatile compounds of virgin olive oils obtained by two or three-phases centrifugal decanters. *European Journal of Lipid Science and Technology, 103*(5), 279–285. https://doi.org/10.1002/1438-9312(200105)103.

Di Giovacchino, L., Sestili, S., & Di Vincenzo, D. (2002). Influence of olive processing on virgin olive oil quality. *European Journal of Lipid Science and Technology, 104*, 587–601.

Di Giovacchino, L., Solinas, M., & Miccoli, M. (1994). Effect of extraction systems on the quality of virgin olive oil. *Journal of the American Oil Chemists' Society, 71*(11), 1189–1194. https://doi.org/10.1007/BF02540535.

EC. (2018a). European Commission, Agriculture and Rural Development, Agriculture and food, DOOR; Available at http://ec.europa.eu/agriculture/quality/door/list.html?locale=en. Accessed on 10 Oct 2018.

EC. (2018b). European commission, agriculture and rural development. Committee for the common organisation of the agricultural markets–arable crops and olive oil. Presentation by Lucie Zolichová, AGRI G.4, 28 June 2018.

EFSA. (2012). European Food Safety Authority. *EFSA Journal, 10*(8), 2848. Available at https://efsa.onlinelibrary.wiley.com/doi/epdf/10.2903/j.efsa.2012.2848. Accessed 15 Sept 2018.

EFSA. (2018). European Food Safety Authority. Nutrition. Available at https://www.efsa.europa. eu/en/topics/topic/nutrition-and-health-claims. Accessed 14 Oct 2018

EFSA European Food Safety Authority. (2017). Panel on dietetic products, nutrition and allergies. *EFSA Journal.* 2017, *15*(5), 4780. https://doi.org/10.2903/j.efsa.2017.4780.

FAO. (2001). Human energy requirements. Report of a Joint FAO/WHO/UNU Expert Consultation. Food and nutrition technical report series 1, Rome, 17–24 Oct 2001. FAO, Rome. Available athttp://www.fao.org/3/a-y5686e.pdf. Accessed 10 Oct 2018.

FDA. (2018a). US Food and Drug Administration. Food. Label and Nutrition. Label claims. Available at https://www.fda.gov/Food/LabelingNutrition/ucm2006873.htm. Accessed 14 Oct 2018.

FDA. (2018b). US Food and Drug Administration. Food, labelling and nutrition. Summary of Qualified Health Claims Subject to Enforcement Discretion. Available at: https://www.fda.gov/ food/labelingnutrition/ucm073992.htm#cardio. Accessed on 14 Oct 2018.

Fedeli, E. (1977). Lipids of olives. *Progress in the Chemistry of Fats and Other Lipids, 15*, 57–74.

García-González, D. L., Tena, N., Romero, I., Aparicio-Ruiz, R., Morales, M. T., & Aparicio, R. (2017). A study of the differences between trade standards inside and outside Europe. *Grasas y Aceites, 68*(3), e210. https://doi.org/10.3989/gya.0446171.

Gomez Candela, C., Bermejo Lopez, L. M., & Loria Kohen, V. (2011). Importance of balanced omega 6/omega 3 ratio for the maintenance of health. Nutritional recommendations. *Nutricion Hospitalaria, 26*, 323–329. https://doi.org/10.3305/nh.2011.26.2.5117.

Huijbregts, P., Feskens, E., Rasanen, L., Fidanza, F., Nissinen, A., Menotti, A., & Kromhout, D. (1997). Dietary pattern and 20 year mortality in elderly men in Finland, Italy, and The Netherlands: Longitudinal cohort study. *BMJ, 315*, 13–17. https://doi.org/10.1136/ bmj.315.7099.13.

IOC. (2013). International olive oil council. Market Newsletter No. 78 – December 2013. Available at: http://www.olive-info.eu/media/Bibliotheque/COI/Newsletter-Decembre 2013. Accessed on 14 Oct 2018.

IOC. (2015). International Olive Oil Council. Market Newsletter No.93 April 2015.

IOC. (2018). International Olive Council. Market Newsletter No. 123 – January 2018. Available at: http://www.internationaloliveoil.org/news/view/698-year-2018-news/973-market-newsletter-january-2018. Accessed 13 Oct 2018.

IOC. (2018a). Available from: http://www.internationaloliveoil.org/estaticos/view/78-tips. [Last accessed on 2018 Feb 07].

ISOFAR. (2016). International Society of Organic Agriculture Research. Country Report: Tunisia 2016. Available at: http://isofar.org/isofar/index.php/2-uncategorised/119-country-report-tunisia-2016. Accessed on 20 Sept 2018.

Issaoui, M., Dabbou, S., Brahmi, F., Ben Hassine, K., Hajaïj Ellouze, M., & Hammami, M. (2009). Effect of extraction systems and cultivar on the quality of virgin olive oils. *International Journal Food Science and Technology, 44*(9), 1713–1720. https://doi. org/10.1111/j.1365-2621.2009.01985.x.

Issaoui, M., Gharbi, I., Flamini, G., Cioni, P. L., Bendini, A., Gallina Toschi, T., & Hammami, M. (2015). Aroma compounds and sensory characteristics as biomarkers of quality of differently processed Tunisian virgin olive oils. *International Journal of Food Science and Technology, 50*, 1764–1770. https://doi.org/10.1111/ijfs.12830.

Kalogeropoulos, N., & Tsimidou, M. Z. (2014). Antioxidants in Greek virgin olive oils. *Antioxidants, 3*, 387–413. https://doi.org/10.3390/antiox3020387.

Kastorini, C. M., Milionis, H. J., Esposito, K., Giugliano, D., Goudevenos, J. A., & Panagiotakos, D. B. (2011). The effect of Mediterranean diet on metabolic syndrome and its components: A meta-analysis of 50 studies and 534,906 individuals. *Journal of American College of Cardiology, 57*(11), 1299–1313. https://doi.org/10.1016/j.jacc.2010.09.073.

Keys, A. (1995). Mediterranean diet and public health: Personal reflections. *The American Journal of Clinical Nutrition, 61*(6 suppl), 1321s–1333s.

Keys, A., & Fidanza, F. (1960). Serum cholesterol and relative body weight of coronary patients in different populations. *Circulation, 22*, 1091–1106.

Keys, A. B., & Keys, M. (1975). *How to eat well and stay well the Mediterranean way*. Garden City: Doubleday.

Keys, A., Aravanis, C., Blackburn, I., Buzina, R., Djordjevic, B. S., Dontas, A. S., Fidanza, F., Karvonen, M. J., Kimura, N., Menotti, A., Muhacek, I., Nedeljkovic, S., Puddu, V., Punsar, S., Taylor, H. L., & van Buchem, F. S. P. (1980). *Seven countries-multivariate analysis of death and coronary heart disease. A commonwealth fund book*. Cambridge, MA: Harvard University Press.

Keys, A., Menotti, A., Karvonen, M. J., Aravanis, C., Blackburn, H., Buzina, R., et al. (1986). The diet and 15-year death rate in the seven countries study. *American Journal of Epidemiology, 124*, 903–915.

Kiritsakis, P., Kiritsakis, A. K., Lenart, E. B., Willet, W. C., & Hernandez, R. J. (1998). *Olive oil: From the tree to the table* (2nd ed.). USA: Wiley-Blackwell. isbn:978-0917678424, 348p.

Kris-Etherton, P. M. (1999). Monounsaturated fatty acids and risk of cardiovascular disease. *Circulation, 100*, 1253–1258.

Lassale, C., Batty, G. D., Baghdadli, A., Jacka, F., Sánchez-Villegas, A., Kivimäki, M., & Akbaraly, T. (2018). Healthy dietary indices and risk of depressive outcomes: A systematic review and meta-analysis of observational studies. *Molecular Psychiatry*. https://doi.org/10.1038/s41380-018-0237-8. [Epub ahead of print].

Mannino, S., Cosio, M. S., & Bertuccioli, M. (1993). High performence liquid chromatography of phenolic compounds in virgin olive oils using amperometric detection. *Italian Journal Food Science, 5*, 363–370.

Montedoro, G. F. (1972). I costituenti fenolici presenti negli oli vergini di oliva. *Scienze e Tecnologie Alimentari, 2*, 177–186.

Montedoro, G., Bertuccioli, M., & Anichini, F. (1978). Aroma analysis of virgin olive oil by head space volatiles and extraction techniques. In G. Charalambous & G. E. Inglett (Eds.), *Flavor of foods and beverages, chemistry and technology* (pp. 247–281). New York, NY: Academic.

Mori, T. A. (2014). Dietary n-3 PUFA and CVD: A review of the evidence. *The Proceedings of the Nutrition Society, 73*, 57e64. https://doi.org/10.1017/S0029665113003583.

Mozaffarian, D., & Wu, J. H. Y. (2011). Omega-3 fatty acids and cardiovascular disease: Effects on risk factors, molecular pathways, and clinical events. *Journal of the American College of Cardiology, 58*, 2047e67. https://doi.org/10.1016/j.jacc.2011.06.063.

Owen, R. W., Giacosa, A., Hull, W. E., Haubner, R., Würtele, G., Spiegelhalder, B., & Bartsch, H. (2000). Olive-oil consumption and health: The possible role of antioxidants. *The Lancet Oncology, 1*, 107–112.

Perez-Jimenez, J., Fezeu, L., Touvier, M., Arnault, N., Manach, C., Hercberg, S., Galan, P., & Scalbert, A. (2011). Dietary intake of 337 polyphenols in French adults. *The American Journal of Clinical Nutrition, 93*, 1220–1228. https://doi.org/10.3945/ajcn.110.007096.

Perez-Jimenez, J., Neveu, V., Vos, F., & Scalbert, A. (2010). Identification of the 100 richest dietary sources of polyphenols: An application of the phenol-explorer database. *European Journal of Clinical Nutrition, 64*, S112–S120. https://doi.org/10.1038/ejcn.2010.221.

Psaltopoulou, T., Naska, A., Orfanos, P., Trichopoulos, D., Mountokalakis, T., & Trichopoulou, A. (2004). Olive oil, the Mediterranean diet, and arterial blood pressure: The Greek European prospective investigation into Cancer and nutrition (EPIC) study. *The American Journal of Clinical Nutrition, 80*(4), 1012–1018. https://doi.org/10.1093/ajcn/80.4.1012.

Qosa, H., Batarseh, Y. S., Mohyeldin, M. M., El Sayed, K. A., Keller, J. N., & Kaddoumi, A. (2015). Oleocanthal enhances amyloid-β clearance from the brains of TgSwDI mice and in vitro across a human blood-brain barrier model. *ACS Chemical Neuroscience, 6*(11), 1849–1859.

Ranalli, A., Marchegiani, D., Contento, S., Girardi, F., Nicolosi, M. P., & Brullo, M. D. (2009). Variations of iridoid oleuropein in italian olive varieties during growth and maturation. *European Journal Lipid Science and Technology, 111*, 678–687.

Rigacci, S., & Stefani, M. (2016). Nutraceutical properties of olive oil polyphenols. An itinerary from cultured cells through animal models to humans. *International Journal of Molecule Science, 17*(6), E843. https://doi.org/10.3390/ijms17060843.

Saillard, M. (2014). Les effets « santé » de l'huile d'olive. OCL oil seeds and fats crops and lipids, 21(5) D510. https://doi.org/10.1051/ocl/2014028.

Salazar-Ordóñez, M., Rodríguez-Entrena, M., Cabrera, E. R., & Henseler, J. (2018). Survey data on consumer behaviour in olive oil markets: The role of product knowledge and brand credence. Data in Brief, 18, 1750–1757. https://doi.org/10.1016/j.foodqual.2018.02.010.

Salvador, M. D., Aranda, F., Gomez-Alonso, S., & Fregapane, G. (2003). Influence of extraction system, production year and area on Cornicabra virgin olive oil: A study of five crop seasons. Food Chemistry, 80, 359–366.

Šarolić, M. (2014). Karakterizacija ulja dalmatinskih sorti maslina. Doktorski rad, Prehrambeno-tehnološki fakultet, Osijek.

Sarolić, M., Gugić, M., Tuberoso, C. I., Jerković, I., Suste, M., Marijanović, Z., & Kuś, P. M. (2014). Volatile profile, phytochemicals and antioxidant activity of virgin olive oils from Croatian autochthonous varieties Mašnjača and Krvavica in comparison with italian variety Leccino. Molecules, 19(1), 881–895. https://doi.org/10.3390/molecules19010881.

Shearer, M. J., Fu, X., & Booth, S. L. (2012). Vitamin K nutrition, metabolism, and requirements: Current concepts and future research. Advances in Nutrition, 3(2), 182–195. https://doi.org/10.3945/an.111.001800.

Simopoulos, A. P. (1991). Omega-3 fatty acids in health and disease and in growth and development. The American Journal of Clinical Nutrition, 54, 438–463.

Simopoulos, A. P. (2006). Evolutionary aspects of diet, the omega-6/omega-3 ratio and genetic variation: Nutritional implications for chronic diseases. Biomedicine & Pharmacotherapy, 60, 502e7. https://doi.org/10.1016/j.biopha.2006.07.080.

Stefanoudaki, V., Koutsaftakis, A., & Harwood, J. L. (2011). Influence of malaxation conditions on characteristic qualities of olive oil. Food Chemistry, 127(4), 1481–1486. https://doi.org/10.1016/j.foodchem.2011.01.120.

Tsimidou, M., Lytridou, M., Boskou, D., Pappa Louisi, A., Kotsifaki, F., & Petrakis, C. (1996). On the determination of minor phenolic acids of virgin olive oil by RP-HPLC. Grasas y Aceites, 47(3), 151–157.

Trichopoulou, A., & Lagiou, P. (1997). Healthy traditional Mediterranean diet: An expression of culture, history, and lifestyle. Nutrition Reviews, 55, 383–389. https://doi.org/10.1111/j.1753-4887.1997.tb01578.x.

UN. (2018). United Nations. Sustainable development goals. Available at: https://www.un.org/sustainabledevelopment/biodiversity/. Accessed on 13 Oct 2018.

USDA. (2018). United States Department of Agriculture Agricultural Research Service. National Nutrient Database for Standard Reference Legacy Release. Basic report: 04053, Oil, olive, salad or cooking. Available at https://ndb.nal.usda.gov/ndb/foods/show/04053?fgcd=&manu=&format=&count=&max=25&offset=&sort=default&order=asc&qlookup=virgin+olive+oil&ds=SR&qt=&qp=&qa=&qn=&q=&ing=. Accessed 1 Sept 2018.

Valls-Pedret, C., Sala-Vila, A., Serra Mir, M., Corella, D., et al. (2015). Mediterranean diet and age-related cognitive decline: A randomized clinical trial. JAMA Internal Medicine, 175(7), 1094–1103. https://doi.org/10.1001/jamainternmed.2015.1668.

Vareiro, D., Faig, A. B., Quintana, B., Bertomeu, I., Buckland, G., Almeida, M. D. V., & Majem, L. S. (2009). Availability of Mediterranean and non-Mediterranean foods during the last four decades: comparison of several geographical areas. Public Health Nutrition, 12, 936–941.

Vaz-Almeida, M. D., Parisi, S., & Delgado, A. M. (2017a). Adherence to the Mediterranean diet. In Chemistry of the Mediterranean diet. Cham: Springer.

Vaz-Almeida, M. D., Parisi, S., & Delgado, A. M. (2017b). The Mediterranean diet: What is it? In Chemistry of the Mediterranean diet. Cham: Springer.

Vaz-Almeida, M. D., Parisi, S., & Delgado, A. M. (2017c). Food and nutrient features of the mediterranean diet. In Chemistry of the Mediterranean diet. Cham: Springer.

Vaz-Almeida, M. D., Parisi, S., & Delgado, A. M. (2017d). Concluding remarks. In Chemistry of the Mediterranean diet. Cham: Springer.

Visioli, F., Franco, M., Toledo, E., Luchsinger, Willett, W. C., Hu, F. B., & Martinez-Gonzalez, M. A. (2018). Olive oil and prevention of chronic diseases: Summary of an international conference. *Nutrition Metabolism and Cardiovascular Diseases, 28*(7), 649–656. https://doi.org/10.1016/j.numecd.2018.04.004.

WHO. (2018a). World Health Organization. Non communicable diseases and mental health. Target 7: Halt the rise in obesity. Available at https://www.who.int/nmh/ncd-tools/target7/en/. Accessed 10 Oct 2018.

WHO. (2018b). World Health Organization. Global Strategy on Diet, Physical Activity and Health. Available at http://www.who.int/dietphysicalactivity/goals/en/. Accessed on 10 Oct 2018.

WIPO. (2018). World Intellectual Property Organization, IP services, the lisbon system for the international registration of appellations of origin. Available at http://www.wipo.int/lisbon/en/. Accessed on 10 Oct 2018.

Yang, J., Farioli, A., Korre, M., & Kales, S. N. (2014). Modified Mediterranean diet score and cardiovascular risk in a North American working population. *PLoS One, 9*(2), e87539. https://doi.org/10.1371/journal.pone.0087539.

Part II
Oils From Fruit Nuts

Chapter 5
Virgin Walnut (*Juglans regia* L.) Oil

**Giuseppe Fregapane, Rosa M. Ojeda-Amador,
and Maria Desamparados Salvador**

Abstract The consumption of nuts (e.g. walnuts, pistachios, almonds) has recently received particular attention due to their high content of unsaturated fatty acids, antioxidants and other biologically active compounds and, thus, their potential beneficial health effects. Apart from nuts, a growing interest towards virgin vegetable oils has also increased recently. The main reasons for this trend are marked by consumers that increasingly appreciate the taste and smell of oils related to the raw materials from which they come from as well as by their potential nutritional properties -resulting in novel healthy oils- with added value for the consumers as compared to the common industrial refined vegetable oils. The purpose of the current chapter is to analyze and discuss the current knowledge on the extraction and processing of virgin walnut oil as well as the composition and properties of major (fatty acid profile) and minor components (tocols, phenolics, sterols, carotenoids, flavour and aroma compounds), which are directly related to their organoleptic and functional traits. Virgin walnut oils are characterized by a high content of linoleic (50–65%) and linolenic (up to 20%) acids, phospholipids (16.5 g/kg), sterols (1.5–2.2 g/kg), and tocopherols (220–650 mg/kg). These bioactive phytochemcials have well known health-promoting impacts, as well as a peculiar sensory properties.

Keywords Gourmet oils · Cultivar · Chemical characterization · Bioactive compounds · Health-promoting effect

Abbreviations

FA	Fatty acids
LDL	Low-density lipoprotein
MUFA	Monounsaturated FA
PUFA	Polyunsaturated FA

G. Fregapane (✉) · R. M. Ojeda-Amador · M. D. Salvador
Faculty of Chemistry, Castilla-La Mancha University, Ciudad Real, Spain
e-mail: giuseppe.fregapane@uclm.es; rosamaria.ojeda@alu.uclm.es; amparo.salvador@uclm.es

© Springer Nature Switzerland AG 2019
M. F. Ramadan (ed.), *Fruit Oils: Chemistry and Functionality*,
https://doi.org/10.1007/978-3-030-12473-1_5

SFA   Saturated FA
TAG   Triacylglycerols
TPP   Total polar phenolics
VWO   Virgin walnut oil

1 Nuts and Their Oils

Walnut (*Juglans regia* L.) is a plant of the Juglandaceae family native to Central
Asia, the western Himalayan chain and Kyrgyzstan (Fernández-López et al. 2000;
Salas-Salvadó et al. 2011) that was introduced in Europe before Roman times (1000
B.C.), from where spreads to other regions with a Mediterranean type ecosystem
throughout the world such as USA and North of Africa. Nowadays, it is in continuous expansion thanks to new cultivars that adapt better to different climatic conditions. Currently, China is the major crop producer (1,785,879 tons) followed by the
USA (607,814 tons), Iran (405,281 tons) and Turkey (195,000 tons) (Martínez et al.
2010; FAOSTAT data 2014), although USA and France are the main exporting
countries.

Thanks to their potential to improve lipids profile in blood and to reduce the risk
of coronary heart diseases (CDH), walnuts have been the focus of intense research
studies (Sabaté et al. 2010; Amarowicz et al. 2017). These health benefits could be
attributed to the high amount of unsaturated fatty acids in nuts, as well as to the
dietary fibre and phytosterols that inhibit the absorption of cholesterol in the digestive tract (Bolling et al. 2011). The high unsaturated/saturated fatty acid ratio (10)
found in walnuts (60% of linoleic acid and 15% of oleic and linolenic acid each;
Tables 5.1 and 5.2) was the aim of several investigations, showing a good balance of
n-6 and *n*-3 polyunsaturated fatty acids (PUFA), beneficial in decreasing the incidence of cardiovascular risk and lowering the blood cholesterol (Simopoulos 2002).

Table 5.1 Composition of walnut, its virgin oil and residual cake

Walnut		Virgin Walnut Oil (VWO)		Residual cake	
Moisture (%)	3–5	Fat yield (%)	78–90	Moisture (%)	5.5–6.5
Fat (%)	62–71	TAG (%)	96–98	Fat (%)	6–10
Protein (%)	14–16	SFA (%)	9–11	Protein (%)	26–30
Fibre (%)	6.5–7.5	MUFA (%)	15–24	Fibre (%)	12–15
Phenolics (%)	1.0–1.3	PUFA (%)	66–75	Phenolics (%)	1.6–2.0
		Phospholipids (g/kg)	16.5		
		Sphingolipid (g/kg)	2.9		
		Sterols (g/kg)	1.5–2.2		
		Tocopherols (mg/kg)	220–650		
		Pigments (mg/kg)	7–20		
		Phenolics (mg/kg)	14–26		
		Volatiles (mg/kg)	4–9		

TAG triacylglycerols, *SFA* saturated fatty acids, *MUFA* monounsaturated fatty acids, *PUFA* polyunsaturated fatty acids

Table 5.2 Fatty acid (FA) and triacylglycerols (TAG) composition (%) of VWO

P-C16:0	Po-C16:1	S-C18:0	O-C18:1	L-C18:2	Ln-C18:3	References
Virgin oil						
6.6–7.8	–	1.5–2.2	17–28	50–59	1.9–19	Martínez et al. (2006)
6.3–7.5	0.0–0.4	1.7–1.7	16–24	58–65	9.1–11	Rabrenovic et al. (2011)
6.7–7.3	0.1–0.1	2.4–2.7	15–16	60–62	13–15	Ojeda-Amador et al. (2018b)
6.0	0.1	2.4	14.5	61.5	14.1	Rabadán et al. (2018)
Solvent-extracted oil						
6.3–7.5	0.1	2.2–2.8	14–18	58–63	9.6–13	Amaral et al. (2003)
7.1	0.1	2.3	13.5	63	0.05	Arranz et al. (2008)
5.9	0.0	2.3	18	63	10	Derewiaka et al. (2014)
LLL	**LOL/OLL**	**LLLn**	**OOL**	**Total**		**References**
Virgin oils						
25–29	14–16[a]	15–18	–	–		Bada et al. (2010)
Solvent-extracted oil						
135	111	97	–	–		Holčapek et al. (2005)[b]
–	–	–	–	97		Miraliakbari and Shahidi (2008)
31	29	14	10	–		Tu et al. (2016)

P palmitic, *Po* palmitoleic, *S* stearic, *O* oleic, *L* linoleic, *Ln* Linolenic
[a]OLL+PoLO
[b]Individual triglycerides expressed as mg/kg

Moreover, the PREDIMED study (Prevention with Mediterranean Diet) as well as other clinical trials, have confirmed that a high intake of walnuts can lower the incidence of hypertension, diabetes, cancer and other inflammatory conditions (Salas-Salvadó et al. 2008; Mohammadifard et al. 2015; Amarowicz et al. 2017). The US Food and Drug Administration (FDA) approved in 2003 a health claim regarding the recommended consumption nuts (42 g, 1.5 oz daily) that may reduce the risk of heart disease.

On the other hand, minor compounds (i.e., tocopherols, phenolics and phospholipids) can act as antioxidants, reduce oxidative damage of cellular biomolecules such as lipids, proteins, and nucleic acids and protect against certain types of cancer (Aruoma 1998), apart from improving the oil oxidative stability (Wang et al. 2002; Alasalvar et al. 2003) due to their radical scavenging function. Due to the high energetic value of nuts, studies showed that the intake of nuts not only does not increase the body weight but also promotes satiety (Vadivel et al. 2012; Souza et al. 2017).

Apart from nuts, a growing interest towards their virgin vegetable oils has appeared in recent years, which goes beyond the olive virgin oil. The main reasons for this trend are marked by consumers that increasingly appreciate the taste and smell of the raw materials from which the oil is extracted as well as by their potential nutritional properties, resulting in novel gourmet oils. These possess an added value for the consumers as compared to the common industrial refined vegetable oils (Kamal-Eldin and Moreau 2010), being the denominated "virgin" or "cold-pressed" oils (as defined by FAO-WHO Codex Stan 210 1999) generally recognized as a synonym of quality and enjoyment by consumers (Matthäus 2008). A recent project for amending the standard for named vegetable oils FAO-WHO Codex Stan 210 (CX/FO 15/24/11 2015) is under consideration with the purpose of incorporate composition standards for cold-pressed oils including virgin nut oils (e.g. walnut, pistachio and hazelnut). These virgin oils obtained from nuts are considered as specialty oils due to their flavor and healthy properties and are generally produced in small size mills with a low production and sold in healthy markets mainly employed in gastronomy (Kamal-Eldin and Moreau 2010). Virgin walnut oil (VWO) is used for edible purposes, mainly as salad dressing, and in the cosmetic industry as a component of dry skin creams, antiwrinkle and anti-aging products (Martínez et al. 2010).

More attention has also been focused on the valorization of food processing by-products (Nyam et al. 2009). In this sense, the pressing process of walnut oil extraction produces a residual partially defatted cake, which could result as a disposal problem for the industry. However, this by-product retains nutrients and bioactive compounds (Table 5.1) that may be used as a natural source of phytochemicals and antioxidants for dietary supplements as well as ingredients for functional foods.

2 Extraction and Processing of Nut Oils

The production and consumption of edible virgin oils date from 8000 years ago. More than one century ago, the development of a production technology by means of organic solvents extraction led to an important increase in the yield of this industry, although implies the requirement of a refining of the raw extracted oil, that led to the disappearance of the virgin oils from market (except the virgin olive oil). However, a recent interest for consuming less processed foods appeared among consumers, which demand the production of virgin oils produced in small amounts with a simple production technology (Matthäus 2008). Nevertheless, the industrial production of VWO is still searching for appropriate procedures to rich higher performances and to prevent oxidative degradation during oil extraction due to the high unsaturated fatty acid content (Martínez et al. 2010).

Nut oils can be extracted using pressure systems (expelling by screw press or pressing by hydraulic presses), obtaining virgin oils with a high nutritional value, or employing solvents that require a later refining process. Mechanical systems are considered as cold extraction methods since only pressure is applied, although the screw press system requires the heating of the barrel to obtain satisfactory extraction yields like in the case of virgin pistachio oils (Sena-Moreno et al. 2015; Ojeda-Amador et al. 2018a) or pumpkin seed oils (Rezig et al. 2018). Such heating produces pleasant sensory characteristics with great acceptance by consumers (Álvarez-Ortí et al. 2012). According to Martínez and Maestri (2008), the nut moisture may be more important than heating in terms of oil recovery employing a screw press. The pre-treatment of seed and nuts before extraction is one of the most relevant parameters that influence the process yield. VWO is often produced in small-scale artisanal mills and sold in gourmet and health markets promoted by their sensory characteristics (color, smell and taste) as well as bioactive compounds with functional properties (Ramadan and Elbanna 2017; Ojeda-Amador et al. 2018a).

The oil extraction process generates a partially defatted by-product (denominated residual cake) which retains nutrients and bioactive compounds present in their kernels, being of great interest their valorization as potential functional ingredients (Santos et al. 2013).

The industrial edible vegetable oils are mainly manufactured by solvent extraction, which requires a refining process that removes a great amount of the minor components like antioxidants and aromatics, reducing therefore the quality of this kind of edible oils. In the case of walnut oil, the solvent extraction method seems to be not suitable on an industrial scale due to the great amount of solvent needed to extract the high lipid content of this nut (Martínez et al. 2010); although, this system could be used to deplete the residual cake obtained after pressing.

Finally, supercritical fluid extraction is an alternative extraction method that usually employs compressed carbon dioxide for the extraction of the oil; in the case of walnut oil, the highest oil recovery (12% of residual oil content in the cake) was obtained employing a pressure of 400–500 bar (Salgin and Salgin 2006).

3   Lipid Composition of Walnut Virgin Oil

Walnut is one of the highest oil-rich nut (Table 5.1; Venkatachalam and Sathe 2006). Different studied cultivars showed an oil range between 62% and 71% (Ozkan and Koyuncu 2005; Dogan and Akgul 2005; Gharibzahedi et al. 2014). As concerned the partially defeated cake obtained after screw pressing, its residual oil content was 6–10% (Table 5.1; Ojeda-Amador et al. 2018b).

The fatty acids profile of VWO is shown in Table 5.2, being remarkable the high amount of essential linoleic acid (50–65%; Amaral et al. 2003; Martínez et al. 2006; Rabrenovic et al. 2011; Ojeda-Amador et al. 2018b; Rabadán et al. 2018), which is higher than other seed and nut oils (e.g. 9–12% palm oil; 48–59% soy oil; 16–37% pistachio oil; 15–34%, almond oil; FAO-WHO Codex Stan 210 1999, and its project for amending CX/FO 15/24/11 2015). VWO from Sejnovo variety had the highest and Franquette the lowest linoleate content. Oleic and linolenic acids were the second and third in abundance (14–28% and 9–19%, respectively), showing Franquette the highest proportion in oleic acid, whilst Chandler contributes with a major proportion of linolenic acid. It is important to remark that walnuts are the only tree nut containing a significant amount of α-linolenic acid (C18:3ω-3, Derewiaka et al. 2014).

The major components of tree nut oils are triacylglycerols (TAG, 96–98%; Table 5.1), with smaller amounts of diacylglycerols (DAG), monoacylglycerols, free fatty acids and minor unsaponifiable compounds (Kamal-Eldin and Moreau

2010). In VWO, TAG is constituted mainly by LLL (25–31%; Table 5.2) and LOL/OLL (14–16%) followed by LLLn (15–18%) and OOL (10%) (Bada et al. 2010; Anqi et al. 2016).

On the other hand, 96 isomers of phospholipids (PL) were identified in walnuts with a total content of 16,457 mg/kg nut (Table 5.1). Phosphatidylinositol (PI) is the group with the highest concentration (5335 mg/kg) formed by 18 different species such as C16:0/C18:2 (2164 mg/kg) and C16:0/C18:3 (790 mg/kg) as the main contributors. Phosphatidylethanolamine (PE) represents the second most important component (4309 mg/kg) with 17 species, accounting the species C16:0/C18:2 (1714 mg/kg) and C18:2/C18:2 (1024 mg/kg) the high concentration. Furthermore, phosphatidylcholine (PC) and phosphatidic acid (PA) content 15 (3133 mg/kg) and 9 (2439 mg/kg) species, being for PC the most important one C16:0/C18:2 with 1103 mg/kg and C18:2/C16:0, the principal for PA with 896 mg/kg. Other groups such as phosphatidylglycerol (PG, 670 mg/kg) and phosphatidylserine (PS, 109 mg/kg) were also detected. Walnut oils phospholipids (PL) have been also studied by Miraliakbari and Shahidi (2008) who reported PS as the highest component with 2700 mg/kg oil, followed by PC with very close concentration (2400 mg/kg) and PI (1300 mg/kg).

Sphingolipids where quantified in walnut oils (Miraliakbari and Shahidi 2008; Table 5.1), accounting a total of 2900 mg/kg. Alasalvar and Pelvan (2011) identified 12 ceramides and 8 cerebrosides in this nut oil. Fang et al. (2005) stated a concentration of 25 mg/kg nut ceramides in walnuts. However, effort is needed to establish reliable data for these compound, due to the importance of sphingolipids in metabolic pathways in the human body.

As reported in Table 5.3, VWO showed a sterol content ranging from 1498 to 2187 mg/kg (Gong et al. 2017), which is very close to virgin olive oil (1100–2100 mg/kg; Aparicio and Luna 2002) and several other vegetable oils (Codex Stan 210 1999), whereas, a higher maximum level was apparently observed in solvent-extracted oils (1144–3070 mg/kg (Table 5.3). As expected, β-sitosterol was found as

Table 5.3 Sterol composition (mg/kg) in virgin walnut oil

Chol	Camp	Stigma	β-sito	Total (g/kg)	References
Virgin oil					
–	4.7–5.1	–	87.8–90.0	–	Martínez et al. (2006)[a]
–	63–104	15–26	1034–1515	1498–2187	Gong et al. (2017)
–	5.4	0.3	93.6	1009	Rabadán et al. (2018)
Solvent-extracted oil					
0.8–3.8	61–108	0.0–8.0	1093–1757	1207–2026	Amaral et al. (2003)
0.1–0.1	180–190	330–350	2160–2250	–	Miraliakbari and Shahidi (2007)
	190	350	2250	3070	Alasalvar and Pelvan (2011)
–	101	0	1118	–	Derewiaka et al. (2014)
4.0–6.3	5.0–88.0	1.0–5.8	974–1494	1144–1679	Abdallah et al. (2015)

Chol cholesterol, *Camp* campesterol, *Stigma* stigmasterol, *β-sito* β-sitosterol
[a]Individual sterols expressed as %

the predominant phytosterol with a content of 1034–1515 mg/kg in VWO and 974–2250 mg/kg in solvent-extracted oils; followed by campesterol with 61–190 mg/kg, corresponding to approximately 4–5% (Table 5.3). The walnut variety is a relevant factor both on sterols total content and on their profile. Amaral et al. (2003) performed a study on six different walnut cultivars, showing a range of β-sitosterol from 1093 mg/kg oil (Parisienne variety) up to 1757 mg/kg (Marbot variety), reporting a total sterol concentration of 1207–2026 mg/kg. According to Martínez et al. (2006), little differences were observed between Franquette and Chandler varieties: 90.0% and 87.8% for β-sitosterol and 4.7–5.1% of campesterol, respectively.

Triterpenic and aliphatic alcohols from walnut oils have also been studied accounting a range between 242 mg/kg (Parisienne) and 620 mg/kg (Local). Cycloartenol was the compound with the highest concentration (226–532 mg/kg; Abdallah et al. 2015). For aliphatic alcohols, local variety showed the lowest amount (24.3 mg/kg) and Franquette the highest (51.0 mg/kg), wherein hexacosanol (9.7–28.2 mg/kg) and tetracosanol (4.2–12.7 mg/kg) were the major compounds.

4 Minor Bioactive Components

Tocopherols are a fat-soluble vitamin found in vegetable oils and act as a potent antioxidant reducing the peroxidation of unsaturated lipids via chain-breaking mechanism (Shin et al. 2009). As depicted in Table 5.4, VWO exhibited a high tocopherols content accounting for 223–646 mg/kg (Martínez et al. 2010; Górnas et al. 2014; Gong et al. 2017; Ojeda-Amador et al. 2018b), mainly in the form of its γ-tocopherol isomer, wherein the Hartely variety was the richest (554 mg/kg; quantified as α-tocopherol; Ojeda-Amdor et al. 2018b). This form of vitamin E represents about 85% of the total tocopherols in the VWO; whereas the other main isomers, α- and δ-tocopherols, account for 10–62 mg/kg and 11–38 mg/kg,

Table 5.4 Tocopherols (mg/kg) and pigments (mg/kg) in walnut oil

α-tocoph	γ-tocoph	T tocoph	β-carotene	T carot	T chloro	References
Virgin oil						
63	323	582	2.2	–	4.0	Górnas et al. (2014)
51–68	250–337	337–448	–	–	–	Martínez et al. (2006)
9.6–26	201–235	223–287	–	–	–	Gong et al. (2017)
24–26	517–554	596–647	–	4.2–7.3	3.3–9.9	Ojeda-Amador et al. (2018b)
Solvent-extracted oil						
10–17	208–262	242–297	–	–	–	Amaral et al. (2005)
38	376	402	–	–	–	Miraliakbari and Shahidi (2007)
		249				Arranz et al. (2008)
38	376	437	–	–	–	Alasalvar and Pelvan (2011)
1.9–13	163–359	187–436	0.2–0.6	0.3–0.8	0.2–0.5	Abdallah et al. (2015)

α-tocoph α tocopherol, *γ-tocoph* γ tocopherol, *T tocoph* Total tocopherols, *T carot* Total carotenoids, *T chloro* Total chlorophylls respectively (Table 5.4). Other virgin nut oils such as those obtained from pistachio, almond and hazelnut exhibited similar tocopherols levels, although almond and hazelnut oils show a different isomers distribution, being α-tocopherol the most abundant (Gong et al. 2017). In the solvent-extracted oils, Amaral et al. (2005) found that Lara cultivar had the lowest total tocopherols content (242 mg/kg) and Franquette had the highest (297 mg/kg), as well as the presence of a low quantity of β-tocopherol (0.9–1.8 mg/kg). Other authors reported a total content close to 400 mg/kg for solvent-extracted oils (Table 5.4). Walnut oil is therefore a good source of vitamin E with recognized antioxidant activity, both in vivo and in food stuff (Burton 1994). In this respect, it is worth to remark that γ-tocopherol is recognized as a more efficient food lipid antioxidant than α-tocopherol (Wagner et al. 2004).

Pigments in solvent-extracted walnut oils, have been described in different cultivars (Abdallah et al. 2015). Table 5.4 presents the content of chlorophyll 'a' which is from 0.12 mg/kg in Franquette cultivar to 0.31 mg/kg in Lauzeronne, and a lower content of chlorophyll 'b' was found (0.06–0.16 mg/kg). The carotenoids family (0.33–0.75 mg/kg) in solvent-extracted oils is constituted mainly of β-carotene in a range from 0.22 mg/kg (Local) to 0.62 mg/kg (Hartley), and of lutein (0.01–0.06 mg/kg) and violaxanthin (0.01–0.04 mg/kg). In VWO, a content of 2.16 mg/kg of β-carotene and a total chlorophyll content of 3.97 mg/kg was reported (Górnas et al. 2014). Similar contents were reported by Ojeda-Amador et al. (2018b) in three different VWO varieties: total chlorophylls was 3.3 mg/kg in Hartley and 9.9 mg/kg in Chandler, and for carotenoids 4.2 mg/kg in Hartley and 7.3 mg/kg in Chandler.

Walnut is the nut with a high content in phenolic compounds (Kornsteiner et al. 2006; Yang et al. 2009). Indeed, the total concentration of these compounds measured by Folin-Ciocalteu, ranged from 10,045 to 12,474 mg/kg (Figueroa et al. 2016; Ojeda-Amador et al. 2018b; Table 5.1). The main phenolic families found in walnuts kernel are hydrolyzable tannins, about 70% of total (Slatnar et al. 2015), wherein the most abundant compound was Di-HHDP glucose (Di-hexahydroxydiphenoyl-glucose) with an observed concentration range from 918 mg/kg (Lara) and up to 2226 mg/kg (Hartley; Ojeda-Amador et al. 2018b). Flavanols were the second family in importance, ranging from 26% (796 mg/kg Chandler) to 35% of the total content (2433 mg/kg Hartley; Ojeda-Amador et al. 2018b), wherein procyanidin dimer was the most relevant compound (from 753 to 2366 mg/kg in Chandler and Hartley varieties respectively). Slatnar et al. (2015) found up to 996 mg/kg of flavonols in the Franquette walnut variety.

Nevertheless, VWO contained a low level (Table 5.5) of these compounds, 10–26 mg/kg (Górnas et al. 2014; Ojeda-Amador et al. 2018b), due to their polarity, but similar to other virgin seed oils (e.g. 10–30 mg/kg soybean, sunflower, rapeseed, corn and pumpkin; Siger et al. 2008). Only a few phenolic compounds have been identified in VWO (di-HHDP glucose, glansreginin A, glansreginin B and HHDP galloyl glucose 1; Table 5.5). All identified compounds belonging to the hydrolyzable tannins family, ranging between 0.91 and 2.91 mg/kg accounting Hartley and Chandler with the lowest and highest concentration, respectively. Glansreginin A is

Table 5.5 Polar phenolic compounds (mg/kg) identified in walnut oils

di-HHDP glucose	Glansreginin A	Glansreginin B	HHDP galloyl glucose 1	Total phenolics	References
Virgin oil					
–	–	–	–	9.5	Górnas et al. (2014)
–	0.01–0.24	0.01–0.05	–	19–80	Slatnar et al. (2015)
–	–	–	–	17	Rabadán et al. (2018)
0.11–0.84	0.37–0.85	0.01–0.17	0.21–1.05	14–26	Ojeda-Amador et al. (2018b)

apparently the main compound with a concentration up to 0.85 mg/kg (Slatnar et al. 2015; Ojeda-Amador et al. 2018b).

Hydrocarbons have been studied by Martínez et al. (2006) in three VWO varieties. They identified 15 compounds and the profile of all samples was characterized by the predominance of even n-alkanes from C14 to C20. The composition of the alkane fraction is unique and has been considered as a fingerprint of each crop in virgin olive oil (Osorio-Bueno et al. 2005). These differences were attributed to genotypical and environmental influences in walnut (Martínez et al. 2006). Squalene is a hydrocarbon (30 carbon atoms) that acts as a biosynthetic precursor to all steroids in plants. Moreover, it has been related to protective effect attributed to its possible antioxidant functions decreasing the risk of developing certain cancers and reducing serum cholesterol levels (Smith 2000). Squalene has been detected in a concentration of 9.7 mg/kg in walnut oil (Maguire et al. 2004).

5 Antioxidant and Health-Promoting Effects

A great interest in acquiring more knowledge regarding the health-promoting effects ascribed to nuts exists among the scientific community, since there are increasing evidence from prospective observational studies that nut intake lowers the risk of cardiovascular disease (Souza et al. 2017). Moreover, nuts have a beneficial impact on hypertension, diabetes, inflammation and cancer (Salas-Salvadó et al. 2008; Grosso et al. 2015; Amarowicz et al. 2017), as reported in the PREDIMED study (Prevention with Mediterranean Diet) and in other epidemiological and clinical trials. These benefits are probably due to the content in healthy unsaturated fatty acids and phytochemicals like phenolics, sterols, tocols and other biologically active compounds (Kamal-Eldin and Moreau 2010).

Among nuts, walnuts are known to be a good source of essential linoleic acid, which decreases LDL-cholesterol and increases HDL-cholesterol (Davis et al. 2007). Moreover, walnut oil possesses a 4:1 ratio between *n*-6 and *n*-3 PUFA, which helps to prevent diseases, although makes it more susceptible to oxidation, which is balanced by the high content in natural antioxidants that contributes to health (Gharibzahedi et al. 2014). Furthermore, walnuts also possess phytosterols, dietary fibre, phenolics and tocols that may be beneficial for health (Anderson et al. 2001; Amaral et al. 2005). The consumption of nuts and their oils suppresses appetite, enhance satiety and reduces food intake (Hughes et al. 2008). Phenolic compounds are potentially relevant for human health due to their good antioxidant, antiatherogenic, anti-inflammatory and antimutagenic properties (Anderson et al. 2001; Carvalho et al. 2010). However, in walnut oils its content is low due to their low-fat solubility, and therefore the corresponding antioxidant activity is also low (e.g. 0.10 mg/kg DPPH and 1.0 mg/kg ORAC; Ojeda-Amador et al. 2018b), although similar to other virgin nut oils (Miraliakbari and Shahidi 2008).

6 Contribution to Organoleptic Properties

Volatile compounds in vegetable oils are mainly formed by the oxidative pathways of linoleic acid, due to the activity of the linoleate oxygen oxidoreductase (LOX) which catalyzes the oxidation of PUFA (Robinson et al. 1995; Martínez and Maestri 2008). As depicted in Table 5.6, the profile of the VWO is characterized by aldehydes (22–81%), acids (14–30%) and alcohols (20–25%) with different proportions depending on the variety (Ojeda-Amador et al. 2018b).

The major aldehyde component present in the head-space of virgin walnut oils (mainly hexanal, 2-heptenal, octanal, nonanal, 2-decenal and 2-undecenal) are theoretically related with hydroperoxide precursors (Torres et al. 2005; Choe and Min 2006). Regarding VWO produced from different varieties, Chandler showed the highest aldehydes concentration (3.16 mg/Kg) followed by Lara (2.19 mg/Kg) and Hartley (1.37 mg/Kg), showing significant statistical differences between them (Ojeda-Amador et al. 2018b). Hexanal, related to green descriptor (Morales et al. 2005), is the major contributor within the aldehyde's family accounted for 1.79 mg/kg in Chandler VWO and 0.60 mg/kg in Lara variety (56% and 27%, respectively). On the contrary nonanal,-associated with citrus odour (Kochhar 1993), was the main compound of this family in Hartley variety (0.38 mg/kg, 28%), but also high in Chandler (17%) and Lara VWOs (27%); similarly to what reported by Martínez and Maestri (2008), who measured a range of 16.2–34.7% n-pentano and 15.7–34.1% 2,4-decadienal in VWO.

Table 5.6 Volatile compounds (%) found in virgin walnut oils

Hexanal	2-Decenal	Aldehydes[a]	Alcohols	Acids	Ketones	References
4.9–6.7	3.6–6.3	37–46	0.0–4.3	–	0.0–0.7	Torres et al. (2005)
8.1–21	3.0–6.8	55–81	0.0–1.9	–	–	Martínez et al. (2008)
6.6–32	–	22–42	2.6–12	21–43	1.0–4.8	Bail et al. (2009)
4.7–28	1.6–3.2	22–51	20–25	14–30	1.8–3.6	Ojeda-Amador et al. (2018b)

[a]Total aldehydes

Volatile acids and alcohols showed closed proportions in the headspace composition of VWO (Table 5.6), with a concentration range of 0.87–3.19 mg/kg for acids and 1.25–1.55 mg/kg for alcohols (Ojeda-Amador et al. 2018b). Within the acids family, acetic acid accounted from 5% to 26% of the total (0.25–2.53 mg/kg, depending on the variety), according to studies reported by Bail et al. (2009) and Uriarte et al. (2011), who found acetic acid and hexanal as the main headspace components of the VWO. Although the concentration of acetic acid (expressed as the internal standard 4-methyl-2-pentanol) is apparently high, it is closed to its odor detection threshold, and therefore, it could be not perceived. Regarding alcohols, 1-dodecanol and 1-nonen-3-ol were the most important compounds found at levels of 0.56–0.88 mg/kg and 0.15–0.29 mg/kg, respectively. 1-pentanol (about 5% of this family and 1% of the total volatiles, according to Martínez and Maestri 2008) -related to fruity notes (Morales and Aparicio 1999)- was also present due to the transformation of linoleic acid (Choe and Min 2006).

Other families of volatiles, such as terpenes and ketones were present in much less concentrations, only accounting for 7–13% and 2–3% each one, respectively of the total volatile compounds in VWO (Bail et al. 2009; Ojeda-Amador et al. 2018b). Some commercial VWO have shown to contain pyrazines (up to 13% of totals) and furans (up to 10%), which are formed when a high temperature is used during the extraction or when roasting nuts are used for oil extraction (Bail et al. 2009; Zhou et al. 2016; Ojeda-Amador et al. 2018b).

Regarding the CIE L*a*b* color space in VWO, the 'b*' parameter (yellow-blue component) ranged from 25.5 to 30.9 and from −2.9 to −3.8 for the 'a*' parameter (red-green component) depending on the variety used for oil extraction, resulting in a clear yellow tonality (Ojeda-Amador et al. 2018b), which was in agreement with what observed by Górnas et al. (2014).

References

Abdallah, I. B., Tlilli, N., Martinez-Force, E., Pérez-Rubio, A. G., Perez-Camino, M. C., Albouchi, A., & Boukhchina, S. (2015). Content of carotenoids, tocopherols, sterols, triterpenic and aliphatic alcohols, and volatile compounds in six walnuts (*Juglans regia* L.). *Food Chemistry, 173*, 972–978.

Alasalvar, C., & Pelvan, E. (2011). Fat-soluble bioactives in nuts. *European Journal of Lipid Science and Technology, 113*(8), 943–949.

Alasalvar, C., Shahidi, F., Ohshima, T., Wanasundara, U., Yurtas, H. C., Liyanapathirana, C. M., & Rodrigues, F. B. (2003). Turkish Tombul hazelnut (*Crylus avellane* L.). 2. Lipid characteristics and oxidative stability. *Journal of Agricultural and Food Chemistry, 51*(13), 3797–3805.

Álvarez-Ortí, M., Quintanilla, C., Sena, E., Alvarruiz, A., & Pardo, J. E. (2012). The effects of a pressure extraction system on the quality parameters of different virgin pistachio (*Pistacia vera* L. var. Larnaka) oils. *Grasas y Aceites, 63*(3), 260–266.

Amaral, J. S., Casal, S., Pereira, J. A., Seabra, R. M., & Oliveira, B. P. P. (2003). Determination of sterol and fatty acid composition, oxidative stability, and nutritional value of six walnut (*Juglans regia* L.) cultivars grown in Portugal. *Journal of Agricultural and Food Chemistry, 51*(26), 7698–7702.

Amaral, J. S., Alves, M., Seabra, R., & Oliveira, B. (2005). Vitamin E composition of walnuts (*Juglans regia* L.): A 3-year comparative study of different cultivars. *Journal of Agricultural and Food Chemistry, 53*(13), 5467–5472.

Amarowicz, R., Gong, Y., & Pegg, R. B. (2017). Recent advances in our knowledge of the biological properties of nuts. In I. C. F. R. Ferreira, P. Morales, & L. Barros (Eds.), *Wild plants, mushrooms and nuts: Functional food properties and applications* (1st ed.). Chichester: Wiley.

Anderson, K. J., Teuber, S. S., Gobeille, A., Cremin, P., Waterhouse, A. L., & Steinberg, F. M. (2001). Walnut polyphenolics inhibit in vitro human plasma and LDL oxidation. *The Journal of Nutrition, 131*(11), 2837–2842.

Anqi, T., Zhenxia, D., & Shuping, Q. (2016). Rapid profiling of triacylglycerol for identifying authenticity of edible oils using supercritical fluid chromatography-quadruple time-of-flight mass spectrometry combined with chemometric tools. *Analytical Methods, 8*(21), 4226–4238.

Aparicio, R., & Luna, G. (2002). Characterisation of monovarietal virgin olive oils. *European Journal of Lipid Science and Technology, 104*(9–10), 614–627.

Arranz, S., Cert, R., Pérez-Jiménez, J., Cert, A., & Saura-Calixto, F. (2008). Comparison between free radical scavenging capacity and oxidative stability of nut oils. *Food Chemistry, 110,* 985–990.

Aruoma, O. I. (1998). Free radicals, oxidative stress, and antioxidants in human health and disease. *Journal of the American Oil Chemists' Society, 75*(2), 199–212.

Bada, J. C., León-Camacho, M., Prieto, M., Copovi, P., & Alonso, L. (2010). Characterization of walnut oils (*Juglans regia* L.) from Asturias, Spain. *Journal of the American Oil Chemists' Society, 87*(12), 1469–1474.

Bail, S., Stuebiger, G., Unterweger, H., Buchbauer, G., & Krist, S. (2009). Characterization of volatile compounds and triacylglycerol profiles of nut oils using SPME-GC-MS and MALDI-TOF-MS. *European Journal of Lipid Science and Technology, 111*(2), 170–182.

Bolling, B. W., Oliver-Chen, C. Y., McKay, D. L., & Blumberg, J. B. (2011). Tree nut phytochemicals: Composition, antioxidant capacity, bioactivity, impact factors. A systematic review of almonds, Brazils, cashews, hazelnuts, macadamias, pecans, pine nuts, pistachios and walnuts. *Nutrition Research Reviews, 24*(2), 244–275.

Burton, G. W. (1994). Vitamin E: Molecular and biological function. *The Proceedings of the Nutrition Society, 53*(2), 251–262.

Carvalho, M., Ferreira, P. J., Mendes, V. S., Silva, R., Pereira, J. A., Jerónimo, C., & Silva, B. M. (2010). Human cancer cell antiproliferative and antioxidant activities of *Juglans regia* L. *Food and Chemical Toxicology, 48*(1), 441–447.

Choe, E., & Min, D. B. (2006). Mechanisms and factors for edible oil oxidation. *Comprehensive Reviews in Food Science and Food Safety, 5,* 169–186.

CX/FO 15/24/11. (2015). Codex committee on fats and oils. Discussion paper on cold pressed oils. In 34 session, Malaysia, 9–13 February 2015.

Davis, L., Stonehouse, W., Loots, D. T., Mukuddem-Petersen, J., Van der Westhuizen, F., Hanekom, S. J., & Jerling, J. C. (2007). The effects of high walnut and cashew nut diets on the antioxidant status of subjects with metabolic syndrome. *European Journal of Nutrition, 46*(3), 155–164.

Derewiaka, D., Szwed, E., & Wolosiak, R. (2014). Physicochemical properties and composition of lipid fraction of selected edible nuts. *Pakistan Journal of Botany, 46*(1), 337–343.

Dogan, M., & Akgul, A. (2005). Fatty acid composition of some walnut (*Juglans regia* L.) cultivars from east Anatolia. *Grasas y Aceites, 56*(4), 328–331.

Fang, F., Ho, C., Sang, S., & Rosen, R. T. (2005). Determination of sphingolipids in nuts and seeds by a single quadrupole liquid chromatography-mass spectrometry method. *Journal of Food Lipids, 12*(4), 327–343.

FAOSTAT, Food and Agriculture Organization: FAOSTAT data. (2014). Available at http://faostat. fao.org.

FAO-WHO Codex Stan 210 (1999). Standard for Named Vegetable Oils

Fernández-López, J., Aleta, N., & Alías, R. (2000). *Forest genetic resources conservation of Juglans regia L.* Rome: International Plant Genetic Resources Institute.

Figueroa, F., Marhuenda, J., Zafrilla, P., Martínez-Cachá, A., Mulero, J., & Cerdá, B. (2016). Total phenolics content, bioavailability and antioxidant capacity of 10 different genotypes of walnut (*Juglans regia* L.). *Journal of Food and Nutrition Research, 55*(3), 229–236.

Gharibzahedi, S. M. T., Mousavi, S. M., Hamedi, M., & Khodaiyan, F. (2014). Determination and characterization of kernel biochemical composition and functional compounds of Persian walnut oil. *Journal of Food Technology, 51*(1), 34–42.

Gong, Y., Pegg, R. B., Carr, E. C., Parrish, D. R., Kellett, M. E., & Kerrihard, A. L. (2017). Chemical and nutritive characteristics of tree nut oils available in the U.S. market. *European Journal of Lipid Science and Technology, 119*(8), 1–15.

Górnas, P., Siger, A., Juhņeviča, D., Lācis, G., Šņē, E., & Segliņa, D. (2014). Cold-pressed Japanese quince (*Chaenomeles japónica* (Thunb.) Lindl. ex Spach) seed oil as a rich source of α-tocopherol, carotenoids and phenolics: A comparison of the composition and antioxidant activity with nine other plant oils. *European Journal of Lipid Science and Technology, 116*(5), 563–570.

Grosso, G., Yang, J., Marventano, S., Micek, A., Galvano, F., & Kales, S. N. (2015). Nut consumption on all-cause, cardiovascular, and cancer mortality risk: A systematic review and meta-analysis of epidemiologic studies. *The American Journal of Clinical Nutrition, 101*, 783–793.

Holčapek, M., Lísa, M., Jandera, P., & Kabátová, N. (2005). Quantification of triacylglycerols in plant oils using HPLC with APCI-MS, evaporative light-scattering, and UV detection. *Journal of Separation Science, 28*(12), 1315–1333.

Hughes, G. M., Boyland, E. J., Williams, N. J., Mennen, L., Scott, C., Kirkham, T. C., Harrold, J. A., Keizer, H. G., Keizer, J. C., & Halford, J. C. (2008). The effect of Korean pine nut oil (*Pinno Thin*) on food intake, feeding behaviour and appetite: A double-blind placebo-controlled trial. *Lipids in Health and Disease, 7*, 6.

Kamal-Eldin, A., & Moreau, R. A. (2010). Tree nut oils. In R. A. Moreau & A. Kamal-Eldin (Eds.), *Gourmet and health-promoting specialty oils* (Vol. 3. EE.UU, pp. 127–150). Urbana: Academic Press and AOCS Press.

Kochhar, S. P. (1993). Oxidative pathways to the formation of off-flavours. In M. J. Saxby (Ed.), *Food taints and off-flavours* (pp. 150–201). London: Blackie Academic & Professional.

Kornsteiner, M., Wagner, K. H., & Elmadfa, I. (2006). Tocopherols and total phenolics in 10 different nut types. *Food Chemistry, 98*, 381–387.

Maguire, L. S., O'Sullivan, S. M., Galvin, K., O'Connor, T. P., & O'Brien, N. M. (2004). Fatty acid profile, tocopherol, squalene and phytosterol content of walnuts, almonds, peanuts, hazelnuts and the macadamia nut. *International Journal of Food Sciences and Nutrition, 55*(3), 171–178.

Martínez, M. L., & Maestri, D. M. (2008). Oil chemical variation in walnut (*Juglans regia* L.) genotypes grown in Argentina. *European Journal of Lipid Science and Technology, 110*(12), 1183–1189.

Martínez, M. L., Mattea, M. A., & Maestri, D. M. (2006). Varietal and crop year effects on lipid composition of walnut (*Juglans regia*) genotypes. *Journal of the American Oil Chemists' Society, 83*(9), 791–796.

Martínez, M. L., Mattea, M. A., & Maestri, D. (2008). Pressing and supercritical carbon dioxide extraction of walnut oil. *Journal of Food Engineering, 88*(3), 399–404.

Martínez, M. L., Labuckas, D. O., Lamarque, A. L., & Maestri, D. M. (2010). Walnut (*Juglans regia* L.): Genetic resources, chemistry, by-products. *Journal of the Science of Food and Agriculture, 90*(12), 1959–1969.

Matthäus, B. (2008). Virgin oils –the return of a long known product. *European Journal of Lipid Science and Technology, 110*(7), 595–596.

Miraliakbari, H., & Shahidi, F. (2007). Lipid class compositions, tocopherols and sterols of tree nut oils extracted with different solvents. *Journal of Food Lipids, 15*(1), 81–96.

Miraliakbari, H., & Shahidi, F. (2008). Antioxidant activity of minor components of tree nut oils. *Food Chemistry, 111*(2), 421–427.

Mohammadifard, N., Salehi-Abarghouei, A., Salas-Salvadó, J., Guasch-Ferré, M., Humphries, K., & Sarrafzadegan, N. (2015). The effect of tree nut, peanut, and soy nut consumption on blood pressure: A systematic review and meta-analysis of randomized controlled clinical trials. *The American Journal of Clinical Nutrition, 101*(5), 966–982.

Morales, M. T., & Aparicio, R. (1999). Effect of extraction conditions on sensory quality virgin olive oil. *Journal of the American Oil Chemists' Society, 76,* 295–300.

Morales, M. T., Luna, G., & Aparicio, R. (2005). Comparative study of virgin olive oil sensory defects. *Food Chemistry, 91*(2), 293–301.

Nyam, K. L., Tan, C. P., Lai, O. M., Long, K., & Che Man, Y. B. (2009). Physicochemical properties and bioactive compounds of selected seed oils. *LWT-Food Science and Technology, 42*(8), 1396–1403.

Ojeda-Amador, R. M., Fregapane, G., & Salvador, M. D. (2018a). Composition and properties of virgin pistachio oils and their by-products from different cultivars. *Food Chemistry, 240,* 123–130.

Ojeda-Amador, R. M., Salvador, M. D., Gómez-Alonso, S., & Fregapane, G. (2018b). Characterization of virgin walnut oils and their residual cakes produced from different varieties. *Food Research International, 108,* 396–404.

Osorio-Bueno, E., Sánchez-Casas, J., Montaño-García, A., & Gallardo-González, L. (2005). Discriminating power of the hydrocarbon content from virgin olive oil of Extremadura cultivars. *Journal of the American Oil Chemists' Society, 82*(1), 1–6.

Ozkan, G., & Koyuncu, M. A. (2005). Physical and chemical composition of some walnut (*Juglans regia* L) genotypes grown in Turkey. *Grasas y Aceites, 56*(2), 141–146.

Rabadán, A., Álvarez-Ortiz, M., Pardo, J. E., & Alvarruiz, A. (2018). Storage stability and composition changes of three cold–pressed nut oils under refrigeration and room temperature condition. *Food Chemistry, 259,* 31–35.

Rabrenovic, B., Dimic, E., Maksimovic, M., Sobajic, S., & Gajic-Krstajic, L. (2011). Determination of fatty acid and tocopherol compositions and the oxidative stability of walnut (*Juglans regia* L.) cultivars grown in Serbia. *Czech Journal of Food Sciences, 29*(1), 74–78.

Ramadan, M. F., & Elbanna, K. (2017). The oil of oregano (*Origanum vulgare*). *Inform, 28*(3), 18–20.

Rezig, L., Chouaibi, M., Ojeda-Amador, R., Gómez-Alonso, S., Salvador, M. D., Fregapane, G., & Hamdi, S. (2018). *Cucurbite maxima* Pumpkin Seed Oil: From the chemical properties to the differences extracting techniques. *Notulae Botanicae Horti Agrobotanici, 46*(2), 663–669.

Robinson, D. S., Zecai, W., Claire, D., & Rod, C. (1995). Lipoxygenases and the quality of foods. *Food Chemistry, 54*(1), 33–43.

Sabaté, J., Oda, K., & Ros, E. (2010). Nut consumption and blood lipid levels. A pooled analysis of 25 intervention trials. *Archives of Internal Medicine, 170*(9), 821–827.

Salas-Salvadó, J., Fernández-Ballart, J., Ros, E., Martínez-González, M. A., Fitó, M., Estruch, R., Corella, D., Fiol, M., Gómez-Gracia, E., Arós, F., Flores, G., Lapetra, J., Lamuela-Raventós, R., Ruiz-Gutiérrez, V., Bulló, M., Basora, J., Covas, M. I., & PREDIMED Study Investigators. (2008). Effect of a Mediterranean diet supplemented with nuts on metabolic syndrome status: One-year results of the PREDIMED randomized trial. *Journal of the American Medical Association, 168*(22), 2449–2458.

Salas-Salvadó, J., Casas-Agustench, P., Salas-Huetos, A. (2011). Cultural and historical aspects of Mediterranean nuts with emphasis on their attributed healthy and nutritional properties. Nutrition, *Metabolism and Cardiovascular Diseases, 21,* S1–S6.

Salgin, S., & Salgin, U. (2006). Supercritical fluid extraction of walnut kernel oil. *European Journal of Lipid Science and Technology, 108*(7), 577–582.

Santos, O. V., Corrêa, N. C. F., Carvalho, R. N., Jr., Costa, C. E. F., França, L. F. F., & Lannes, S. C. S. (2013). Comparative parameters of the nutritional contribution and functional claims of Brazil nut kernels, oil and defatted cake. *Food Research International, 51*(2), 841–847.

Sena-Moreno, E., Pardo, J. E., Catalán, L., Gómez, R., Pardo-Giménez, A., & Álvarez-Ortí, M. (2015). Drying temperature and extraction method influence physicochemical and sensory characteristics of pistachio oils. *European Journal of Lipid Science and Technology, 117*(5), 684–691.

Shin, E. C., Huang, Y. Z., Pegg, R. B., Phillips, R. D., & Eitenmiller, R. R. (2009). Commercial runner peanut cultivars in the United States: Tocopherol composition. *Journal of Agricultural and Food Chemistry, 57*(21), 10289–10292.

Siger, A., Nogala-Kalucka, M., & Lampart-Szczapa, E. (2008). The content and antioxidant activity of phenolic compounds in cold-pressed plant oils. *Journal of Food Lipids, 15*(2), 137–149.

Simopoulos, A. P. (2002). The importance of the ratio of omega-6/omega-3 essential fatty acids. *Biomedicine & Pharmacotherapy, 56*(8), 365–379.

Slatnar, A., Mikulic-Petkovsek, M., Stampar, F., Veberic, R., & Solar, A. (2015). Identification and quantification of phenolic compounds in kernels, oil and bagasse pellets of common walnut (*Juglans regia* L.). *Food Research International, 67*, 255–263.

Smith, T. J. (2000). Squalene: Potential chemopreventive agent. *Expert Opinion on Investigational Drugs, 9*(8), 18411848.

Souza, R. G. M., Schincaglia, R. M., Pimentel, G. D., & Mota, J. F. (2017). Nuts and human health outcomes: A systematic review. *Nutrients, 9*, 1311–1334.

Torres, M. M., Martínez, M. L., & Maestri, D. M. (2005). A multivariate study of the relationship between fatty acids and volatile flavor components in olive and walnut oils. *Journal of the American Oil Chemists' Society, 82*(2), 105–110.

Tu, A., Du, Z., Qua, S. (2016). Rapid profiling of triacylglycerols for identifying authenticity of edible oils using supercritical fluid chromatography-quadruple time-of-flight mass spectrometry combined with chemometric tools. *Analitical Methods, 8*, 4226–4238.

Uriarte, P. S., Goicoechea, E., & Guillen, M. D. (2011). Volatile components of several virgin and refined oils differing in their botanical origin. *Journal of the Science of Food and Agriculture, 91*(10), 1871–1884.

Vadivel, V., Kunyanga, C. N., & Biesalski, H. (2012). Health benefits of nut consumption with special reference to body weight control. *Nutrition, 28*(11–12), 1085–1097.

Venkatachalam, M., & Sathe, S. K. (2006). Chemical composition of selected edible nut seeds. *Journal of Agricultural and Food Chemistry, 54*(13), 4705–4714.

Wagner, K. H., Kamal-Eldin, A., & Elmadfa, I. (2004). Gamma-tocopherol an underestimated vitamin? *Annals of Nutrition & Metabolism, 48*(3), 169–188.

Wang, T., Hicks, H. B., & Moreau, R. (2002). Antioxidant activity of phytosterols, oryzanol, and other phytosterol conjugates. *Journal of the American Oil Chemists' Society, 79*(12), 1201–1206.

Yang, J., Liu, R. H., & Halim, L. (2009). Antioxidant and antiproliferative activities of common edible nut seeds. *Food Science and Technology, 42*(1), 1–8.

Zhou, Y., Fan, W., Chu, F., & Pei, D. (2016). Improvement of the flavor and oxidative stability of walnut oil by microwave pretreatment. *Journal of the American Oil Chemists' Society, 93*(11), 1563–1572.

Chapter 6
Almond (*Prunus dulcis*) oil

Slavica Čolić, Gordan Zec, Maja Natić, and Milica Fotirić-Akšić

Abstract The almond tree, *Prunus dulcis*, is a species that belongs to the *Amygdalus* subgenus inside the *Prunus* genus, the *Rosaceae* family and the order *Rosales*. The almond kernel has been used as food for the mankind, due to its oil-rich and high-calorie content. As consumers are more interested in healthy life styles, almonds are emerging as some of the most popular edible nuts. Almonds are a nutrient-dense food, an excellent source of vitamin E, and a good source of manganese, magnesium, copper, phosphorus, fiber and riboflavin. Recent studies have shown that almonds also contain a diverse array of phenolic and polyphenolic compounds. Almond kernel is known as a source of high lipids (44–61% on fresh weight; 20–68% on dry weight). Only 8% of the fatty acids in almond oil are saturated fats, while it is high in monounsaturated fats, which have demonstrated beneficial effects on lipoprotein profiles in the blood and ability lower the risk of cardiovascular diseases. The major fatty acid is oleic acid, representing 50–70% of the total fatty acid content. Other minor components in almond oil include sterols, tocopherols (mainly α-tocopherol) and squalene. The almond oil is used as edible oil, mainly as a salad dressing and in vegetable dips. It is also used in the cosmetic industry, especially in dry skin creams, anti-wrinkle and anti-aging products. Historically, almond oil has been used for its numerous health and beauty benefits in ancient Chinese, Ayurvedic and Greco-Persian schools of medicine. The bitter almond oil contains three basic components, benzaldehyde, amygdalin and hydrogen cyanide that limit its uses to external applications. The sweet almond oil contains large amounts of vitamins E and K that help skin regeneration and maintain elasticity, which is why the oil is used in many cosmetic products. Almond oil is one of the most popular essential oils used in aromatherapy and massage therapy since it is suitable for any skin type.

S. Čolić (✉)
Institute for Science Application in Agriculture, Belgrade, Republic of Serbia

G. Zec · M. Fotirić-Akšić
University of Belgrade, Faculty of Agriculture, Belgrade, Republic of Serbia
e-mail: fotiric@agrif.bg.ac.rs

M. Natić
University of Belgrade, Faculty of Chemistry, Belgrade, Republic of Serbia

© Springer Nature Switzerland AG 2019
M. F. Ramadan (ed.), *Fruit Oils: Chemistry and Functionality*,
https://doi.org/10.1007/978-3-030-12473-1_6

149

Keywords Composition · Oils · *Prunus amygdalus* · Properties · Usability

Abbreviation

BAO Bitter almond oil
CAGR   Compound annual growth rate
DW Dry weight
FA   Fatty acids
FW   Fresh weight
GAE Gallic acid equivalents (mg gallic acid/100 g of FW)
LDL Low-density lipoprotein
MUFAs Monounsaturated fatty acids
PA   Phosphatidic acids
PC   Phosphatidylcholine
PE   Phosphatidylethanolamine
PI   Phosphatidylinositol
PL   Phospholipids
PUFAs Polyunsaturated fatty acids
SAO Sweet almond oil
TG   Triglycerides

1   Introduction

The almond [*Prunus dulcis* (Miller) D.A. Webb] tree, is a species that belongs to the subgenus *Amygdalus* inside the genus *Prunus*, the *Rosaceae* family and the order *Rosales*. Apparently, it is originated from one or more wild species that evolved in the deserts and mountain slopes of central and southwestern Asia (Micke and Kester 1998). Almonds were collected in the wild 10,000 years ago, and they were among the first plants to be domesticated. The earliest archaeological evidence suggests that this occurred in Jordan. Sweet almond is one of the oldest cultivated fruit species and it is appraised for its culinary uses and numerous medicinal properties. The ancient Greek doctors Hippocrates and Galen described almond as 'hot', and noted its 'heating', 'cleansing', 'nourishing' and 'strengthening mental functions' properties (Albala 2009). Almonds are mentioned in the earliest Sumerian culinary texts in a list of banquet menu items (Rosengarten 1984) and they are also among the plants frequently mentioned in the Bible. Almonds are consistently found in archaeological sites including the famous tomb of Tutankhamen in Egypt.

Almond is the most important nut in the world in terms of commercial production. Based on the FAO data (FAOSTAT 2018), almond production is constantly increasing in the world, and it is currently at about 3.0 million tons (average, in the shell) or 1.24 million tons in the kernel (International Nut and Dried Fruit Council

2018). The leaders in production are the USA, Spain and Australia, followed by Iran, Morocco, Italy, Syria, Turkey and Tunisia. Almond is especially well adapted and spread throughout the whole Mediterranean region from which 28% of the world production is obtained. Almond is also grown in California, northwest Mexico, in a small part of Australia and in other areas around the world.

The kernel is the edible part of the almond nut with a high nutritional value (Socias i Company et al. 2008). As consumers are becoming more interested in healthy life styles, almonds are emerging as some of the most popular edible nuts, typically used as snack foods or as ingredients in a variety of processed foods, especially in bakery and confectionery products (Cordeiro and Monteiro 2001). Almond seed is valued for its special physicochemical, sensory and nutritional characteristics, and it has high market potential as an ingredient in many snacks and other processed foods as well as pharmaceutical and cosmetic industries.

Because of high unsaturated fatty acids (e.g., oleic, linoleic, and linolenic acids) content, almond consumption has been associated with a wide range of health benefits, including reduced levels of coronary heart disease as a result of reduction in the low density lipoprotein (LDL) cholesterol (Ahrens et al. 2005; Zacheo et al. 2000), hypertension, type 2 diabetes, obesity and reduction in oxidative stress (Chen et al. 2006). Almonds have high nutritional properties due to their components such as lipids and unique fatty acids (FA), containing mostly unsaturated fat, little saturated fat and no cholesterol, total fiber (containing small amounts of viscous fiber), protein, arginine, α-tocopherol, magnesium, copper, manganese, calcium, phosphorus and potassium (Chen et al. 2006; Yada et al. 2011). The high contents of macronutrients are making almond an important dietary source.

The chemical composition of almonds is affected by many factors. It depends both on genotype and environmental factors such as growing region, cultivation methods, climatic conditions, harvesting years, and kernel ripeness (Sathe et al. 2008; Yada et al. 2011, 2013; Kodad et al. 2014a, b; Muhammad et al. 2015). Consequently, the published results among research groups show great variability. The main components found in almond seeds are lipid fraction, proteins, soluble sugars, minerals and a fibrous fraction. Various phytochemicals such as phenolic acids, flavonoids, stilbenes and ligants appear in low quantities (Manach et al. 2004) but have a large influence on almond quality.

Almonds are a rich source of lipids composed predominantly of monounsaturated fatty acids (MUFAs) and polyunsaturated fatty acids (PUFAs) (Venkatachalam and Sathe 2006). Lipid content is a very important factor in the confectionery industry. Kernels with a high content of lipids can be used to produce nougat or to extract oil, while kernels with a low percentage of lipids are suitable for the production of almond milk and almond flour. The lipid fraction is also a major determinant of kernel flavor particularly following roasting (Socias I Company et al. 2008). Together with moisture level, proteins and oil composition, high lipid content, define the nutritive value of almonds. As reported by Miraliakbari and Shahidi (2008a), oil fraction is composed of various lipid classes, such as triacylglycerols, sterols, sterol esters, phosphotidylserine, phosphotidylinositol, phosphotidylcholine, and sphingolipids. Studies of oil content in commercial or local almond

Table 6.1 Range of variability of almond oil content (Kodad 2017)

Origin	Range of variability (%)	References
Afghanistan	43–66.1[a]	Kaska et al. (2006) and Zamany et al. (2017)
Argentina	63–66[b]	Kodad et al. (2011b)
	48–57[a]	Maestri et al. (2015)
Australia	53.1–63.5[b]	Zhu et al. (2015)
Bulgaria	63.9[b]	Zlatanov et al. (1999)
China	49.26–53.76	Li et al. (2016)
Greece	55.6–61.1	Nanos et al. (2002)
Iran	20.19–62[b]	Mehran and Filsoof (1974), Abaspour et al. (2012), Imani et al. (2012), and Kiani et al. (2015)
Italy	31.16–66.8[b]	Piscopo et al. (2010) and Kodad et al. (2011b)
	42–47[a]	Schirra and Agabbio (1989), Barbera et al. (1994), and Ruggeri et al. (1998)
Morocco	48.6–67.5[b]	Kodad et al. (2013) and Houmy et al. (2016)
Portugal	58.33–63.6[a]	Kodad et al. (2011b)
	44–59[b]	Egea et al. (2009), Cordeiro and Monteiro (2001), and Barreira et al. (2012)
Serbia	36.30–62.86[a]	Čolić et al. (2012, 2013, 2017, 2018)
Spain	40-67[b]	Romojaro et al. (1988), Soler et al. (1988), Garcia-Lopez et al. (1996), García-Pascual et al. (2003), Kodad et al. (2006, 2011b, 2014b), Kodad and Socias i Company (2008), López-Ortiz et al. (2008), Sánchez-Bel et al. (2008), and Rabadan et al. (2017)
Tunisia	56.1–59.8[b]	Ayadi et al. (2006)
Turkey	46.2–63.18[b]	Askin et al. (2007), Çelik and Balta (2011), Özcan et al. (2011), Kirbaşlar et al. (2012), Yildirim et al. (2016), and Matthäus et al. (2018)
USA	35–66[a]	Hall et al. (1958), Abdallah et al. (1998), Ahrens et al. (2005), Venkatachalam and Sathe (2006), López-Ortiz et al. (2008), and Sathe et al. (2008)

[a]Fresh weight basis
[b]Dry weight basis

cultivars/selections have been carried out worldwide (Table 6.1) reporting considerable variability. Oil content is within a range 43–66.1% of kernel fresh weight (FW) and 20–67.5% of kernel dry weight (DW). Conducted studies indicate that oil content depends mainly on the genotype, year as well as growing condition, pollen source (Yada et al. 2013; Kodad et al. 2014b; Alizadeh-Salte et al. 2018), rootstock (Čolić et al. 2018), and harvest time (Matthäus et al. 2018).

Almond oil comes in two variants: bitter and sweet. The bitter almond oil (BAO), which is considered to be an essential oil, is extracted from the bitter almonds (*Prunus amygdalus*, var. *amara*). It contains a glycoside called amygdalin ranging between 33.0 and 53.9 g/kg (Lee et al. 2013) that decomposes into glucose, benzaldehyde and hydrocyanic acid making the fruit non-edible and developing a characteristic cyanide aroma with moisture. Almond trees that grow in the wild produce bitter almonds with varying degrees of this glycoside. BAO does have some medicinal properties and it may be safe for external application. It is traditionally used as a natural alternative for treating parasites, fevers, cough, and congestion, although it doesn't have any prominent uses in skin care. Based on studies of Atapour and Karminia (2011), Abu-Hamdeh and Alnefaie (2015), and Al-Tikrity and Ibraheem (2017) BAO showed potential application as new nonedible feedstock for producing biodiesel.

Sweet almond oil (SAO) is an odorless, pale-yellow liquid with a nutty taste. The fixed SAO comes exclusively from edible almonds (*Prunus amygdalus* var. *dulcis*) selectively cultivated for their sweet taste. Fresh SAO, extracted using supercritical fluids (CO_2) or by pressing, shows low contents in free fatty acids (FFA), peroxides and phosphatides, and therefore, it can be consumed directly and used for culinary purposes without refining, as a 100% virgin product (Roncero et al. 2016). SAO is mostly used as a carrier oil in medicine and cosmetics (oils, gels, lotions, shampoos), and even for making candles. In addition, it can be used in the production of cosmetic detergents, creams and soaps, as well as in perfumery industries. SAO is often used for skin and hair care products because it is rich in FFA, vitamins, and antioxidant.

The demand for almond oil is on the rise globally due to oil's health benefits and high nutritional value. The rise in demand for aromatherapy products and growing use of almond oil in cosmetic formulations, which is due to increase in preferences for natural ingredients, are the primary factors that drive the global almond oil market. Almond oil is widely utilized in the production of face creams, body lotions, hair oils, and other cosmetic products for moisturizing and cleansing. The research study "Global Almond Oil Market Trends, Analysis Future & Forecast Reports by 2025" prepared by Market Research Hub anticipates that the sector for food industry would rise at a positive Compound Annual Growth Rate (CAGR) during the period 2018–2025. However, one downside is that production costs will be augmented because of the relatively higher cost of almond oil (Table 6.2) compared to other vegetable oils on the market (Mateus Martins et al. 2017).

According to a report "Almond Oil Market by Type, Application and Distribution Channel: Global Opportunity Analysis and Industry Forecast, 2017–2023" published by Allied Market Research, the global almond oil market was valued at $1118 million in 2016, and is projected to reach $2680 million by 2023, growing at a

Table 6.2 Average price of almond oils depending on type

Type	Volume (mL)	Price (USA $)
Carrier	500	8
Organic	100	22
Cosmetic grade	1000	9.5
Cooking	250	2.5

https://www.mysupermarket.co.uk/asda-compare-prices/Oils_And_Vinegar/Tesco_Almond_Oil_250ml.html

CAGR of 13.2% from 2017 to 2023 (https://www.prnewswire.com/news-releases/global-almond-oil-market-expected-to-reach-2680-million-by-2023%2D%2D-allied-market-research-679269793.html; accessed 04 May 2018).

In terms of revenue, SAO accounted for more than half of the market, in 2016 as the oil is being widely utilized in pharmaceutical, skin care, and color-cosmetic industries because of its healthy moisturization properties and natural stability. Cosmetic segment holds the largest share, in terms of revenue, and is expected to grow at a CAGR of 13.5%. Use of almond oil is increasing in cosmetic formulations due to rise in preference of consumers toward organic and healthy ingredients.

Top five exporters of almond oil in 2016 by value were India (29%), USA (8%) and Spain (3%), and by quantity were Indonesia (27%), Brazil (22%), India (10%), USA (6%) and China (5%).

1.1 Almond Cake as a Source of Nutrients and Bioactive Compounds

After extracting almond oil by squeezing the kernels, the remaining substance called "flour" or "cake". Due to its high protein content, the almond cake can be used for human or animal feeding. Sarkis et al. (2014) and Karaman et al. (2015) analyzed almond oil byproducts and reported that a high phenolic content remains in the flour making it usable for improvement of nutritional and functional properties of many food products that contain oil and fat. Flour is rich in minerals, fiber and energy. Also, it possesses significant amounts of bioactive compounds including phenolics, flavonoids, and tannins and can be used for pharmaceutical purposes such as, the production of enzymes, antibiotics, bio-pesticides, vitamins and other biochemical products. Rabadan et al. (2017) found significant differences in the content of carbohydrates, proteins and total mineral content and the presence of specific minerals in almond flours obtained from ten Spanish cultivars, thus resulting in flours with different potential uses. Almond cake extracts also showed antibacterial activity depending on their concentration, while almond flour did not show any antibacterial activity concerning the tested microorganisms, except for *Listeria monocytogenes* (Karaman et al. 2015).

Over the last several years, a lot of research has been done on the compositional and chemical characterization of the cultivated sweet almond. A comprehensive review by Yada et al. (2011) compiled the scientific knowledge and available literature on the macronutrients and micronutrients present in cultivated almonds. The later review is especially important as it points to the difference in individual components in relation to the geographical origin of almonds. The existing literature on almond chemical composition and nutritional value is reviewed in a series of book chapters (Alasalvar and Shahidi 2009b). The following sections summarize work that has been published so far and complement the research done in the past few years.

Similarly, to other tree nuts, almonds are a rich source of proteins. The protein content is usually measured by Kjeldahl digestion and expressed as total nitrogen multiplied with a conversion factor of 5.18 (Askin et al. 2007). As summarized by Yada et al. (2011), in various almond cultivars from California, Spain, Italy, India, and Turkey, protein content was similar, and it ranged from 14 to 26 g/100 g for Turkish selections, which is the largest range so far reported. Drogoudi et al. (2013) analyzed variation in the protein content among the 72 different almond genotypes found in Greece, France and Italy, and concluded that protein content depends on genotype rather than origin. For the majority of samples, values were from 15.1 to 25.9 mg/100 g. Kodad et al. (2013) investigated genotypes from four Moroccan regions and reported the protein content in the range 14.1–35.1% of kernel DW (the total N content obtained by the Dumas method and applying the conversion factor of 6.25). Recently, protein content in almond kernels from Xinjiang was studied, and the results were in the range from 27.17 to 31.13 g/100 g edible portion (Li et al. 2016). High protein content in flour was observed by Rabadan et al. (2017) in Spanish cultivars Ayles (581.3 g/kg) and Marcona (559.3 g/kg). This could be interesting due to the positive influence of the protein content on the structure of dough during baking, thus, enabling a higher proportional replacement of wheat flour by almond flour (Pineli et al. 2015).

Considering studies from different regions, almonds are characterized by high amounts of carbohydrates (20–27%), sugars (2–7.5%) and fiber (11–5%) (Yada et al. 2011). Sucrose is the major sugar, followed by raffinose, glucose, fructose, and trace amounts of arabinose and xylose. Sucrose, as predominant sugar, was found in the range from 11.5 to 22.2 g/100 g DW. Other individual sugars were detected in minor concentrations: raffinose (0.71–2.11 g/100 g), glucose (0.42–1.30 g/100 g), maltose (0.29–1.30 g/100 g), and fructose (0.11–0.59 g/100 g) based on DW (Balta et al. 2009). Starch is not a major carbohydrate component in almonds and its content was reported to be up to 1.4 g/100 g (Ruggeri et al. 1998) and in almonds from Serbia as reported by Dodevska et al. (2015). Of all the edible nuts, almonds have the highest total dietary fiber content (Mandalari et al. 2010), consisting of non-starch polysaccharides (cellulose, hemicellulose, pectins, gums, β-glucans), oligosaccharides, resistant starch and lignin associated with the dietary fiber polysaccharides.

Almonds are valued as a good source of minerals and vitamins. Almonds contain vitamin A and different B-complex vitamins, such as thiamine (B1), riboflavin (B2), niacin (B3), pantoneic acid, pyridoxine (B6), and folate. Ascorbic acid (vitamin C) and vitamin K were not reported in almond (Yada et al. 2011). The proximate content of ash (which represents the mineral content) in almonds is 3 g ash/100 g FW (Yada et al. 2011). Of the 14 mineral identified as principal minerals for the human body, up to nine appear in almond flour in relevant concentrations (Rabadan et al. 2017). Potassium and phosphorus are usually predominant components, followed by calcium and magnesium. Also, almonds are recognized as a good source of iron, zinc and manganese.

2   Extraction and Processing of Almond Oil

2.1   Almond Oil Extraction

Extraction procedures and processing method largely influence the flavors and quality of nut oils. Non-volatile almond oil is extracted from ripe dried seed and if it is intended to be used for human consumption, it must be prepared from the sweet varieties. Similar to other nut oils, the process of obtaining almond oil consists of several steps, (i) harvesting, (ii) de-hulling (removal of the mesocarp), (iii) drying, (iv) cracking, and finally (v) oil extraction (Harris 2013). Drying and extraction are critical steps influencing the quality of the final product. Their optimization is often a subject of research. Almond drying can be done in different ways: direct sun exposure, hot air oven, with the use of a fan, or in a hot air dryer (Piscopo et al. 2011). Adequate drying prolongs the shelf life and prevents the growth of fungi, and aflatoxin.

So far, numerous extraction techniques have been studied and employed to obtain lipid fractions (Liu et al. 2017). These procedures could be classified as mechanical methods such as maceration, cold pressing and expeller press, and as chemical methods that include solvent extraction, CO_2 extraction, enzyme-assisted three-phase partitioning (Panadare and Rathod 2017), or salt-aided aqueous method adopted by Liu et al. (2017). Out of them, pressing and solvent extraction are considered the most important methods because of their high oil recovery (Martinez et al. 2013; Balvardi et al. 2015). Recently, the most common methods used for obtaining oil (solvent extraction, extraction with supercritical fluids (CO_2) and pressure systems (hydraulic and screw presses) were reviewed by Roncero et al. (2016).

2.2   Solvent Extraction

Solvent extraction has good industrial performance and gives the highest oil yield, even compared to pressing (Roncero et al. 2016). Solvents are also used to extract the pastes obtained after pressing (Sharma and Gupta 2004; Matos and Acuña 2010). Other factors such as pH of the aqueous phase, temperature, enzymes, ultrasound or high pressure affect the efficiency of solvent extraction (Hou et al. 2013). However, oils obtained using solvents are of the worst quality due to the appearance of undesirable flavors. After refining, which includes several steps (neutralization, bleaching, degumming, and deodorization) oil becomes suitable for use, but cannot be considered as virgin oil as it is obtained by chemical treatments at high temperatures.

Hexane is widely accepted for solvent extraction because of its solubility and price. However, in spite of the effectiveness of solvent extraction procedure, all authors have agreed that more environmental friendly oil extraction needs to be designed to ensure a safe product of high quality, especially as hexane is considered to be hazardous and highly flammable air pollutant (Vidhate and Singhal 2013; Tan et al. 2016). Ethanol and terpenes have been recommended as an alternative to hexane (Sahad et al. 2014).

Other alternative procedures for the almond oil extraction are mostly convenient on a laboratory scale. Three-phase partitioning has been evaluated for extraction of oil from different plant sources (Sharma and Gupta 2004), and it was recently considered as an easy, simple, efficient and economic method for oil extraction with industrial potential (Panadare and Rathod 2017). The efficiency of aqueous extraction is low, but is another safe alternative oil extraction process that should be more explored (Hanmoungjui et al. 2000). To overcome water dispersibility problems and to improve oil yield, some researchers have used enzymes, but the high cost of the proposed procedure and instability of enzymes are main limitations (Amarasinghe et al. 2009; Li et al. 2013). Salt-aided aqueous extraction (SAAE) method was recently implemented for the BAO extraction (Liu et al. 2017), with an oil yield of 90.9%.

2.3   Supercritical Fluid Extraction

When the nuts are extracted by supercritical fluid extraction (SFE), compressed CO_2 is used. This procedure is considered to be a good alternative to conventional solvent extraction giving oils of high quality, in lower yields, with lower oxidative stability, and in a higher price, when compared to solvent extraction (Martinez et al. 2008; Kamal-Eldin and Moreau 2009). The final product is appropriate for human consumption without further processing and refining and it is characterized as virgin oil with pleasant sensorial characteristics. Different extraction conditions were proposed to improve the effectiveness of the extraction. Oil yield was observed to vary largely with changes in pressure, temperature, and solvent rates (Femenia et al. 2001; Leo et al. 2005). The highest impact was observed for the temperature increase, which is the most probably affecting the oil solubility in CO_2 resulting in four times higher oil yield (Leo et al. 2005).

2.4   Pressing

To release the oil from seeds and nuts, mechanical extraction methods are the best alternative to solvent extraction. By pressing processes, high-quality oils are produced as the oils maintain their physico-chemical and sensorial properties. Pressing could be done by hydraulic and screw presses. The almond oil extracted using hydraulic press is prized for its light color and sweet flavor. Stability to oxidation processes of almond oil extracted using hydraulic press was studied and results were contradictory as reviewed by Roncero et al. (2016). Extraction using a screw press involves the slow breaking of almond nuts under mechanical pressure and requires preheating to achieve more efficient separation. Such a procedure gives the final oil product of lower quality with variable yield (Álvarez-Ortí et al. 2012; Martínez et al. 2013).

Expeller press is a screw-type machine that produces oil at higher temperatures resulting in lower quality than cold-pressed oils. In addition, expeller press could be used on low temperatures. The product of gentle and cold pressing bellow 30 °C is cold pressed oil, while the "extra-virgin oil" refers to the first-pressed oil.

3 Fatty Acids Composition and Acyl Lipids Profile of Almond Fixed Oil

3.1 Fatty Acids (FA)

The FA composition of almond oil has dietary importance. Although almonds contain high amounts of fat, the lipid fraction does not contribute to cholesterol formation in humans, due to a high level of unsaturated (MUFAs and PUFAs) fatty acids (Askin et al. 2007; Beyhan et al. 2011). Intensive research has been developed regarding the FA profile of almond oils (Table 6.3).

The highest number of 26 FA in almond reported in the literature was identified by Beyhan et al. (2011) who studied four commercial and five other almond genotypes from Turkey. In a study of 20 local almond selections originating from North Serbia and cultivars 'Marcona', 'Texas' and 'Troito', Čolić et al. (2017) quantified 16 FA, so did Barreira et al. (2012) for four PDO Portuguese and five commercial cultivars, while Zamany et al. (2017) found only 12 FA in native almond cultivars from Afghanistan. Many of FA such as C14:0, C15:0, C17:0, C18:3, C20:1, C20:3, C21:0, C22:0, C22:2 and C24:0 in those studies were found only in trace amounts (<0.1%). Total number and quantity of FA depend on analyzed genotypes (García-Lopez et al. 1996; Kornsteiner et al. 2006; Askin et al. 2007; Kodad et al. 2014b), quantification method, pollen source (Alizadeh-Salte et al. 2018), ecological conditions, orchard management and harvest time (Kodad and Socias i Company 2008; Sathe et al. 2008), and rootstock (Čolić et al. 2018). In the majority of almond samples, oleic, linoleic, palmitic and stearic acids (in decreasing order) represent over 95% of the total FA content, while other FA represent 5%. Oleic acid and linoleic acid are the most important unsaturated FA found in almond oil (about 90%), while saturated fatty acids, especially palmitic, palmitoleic and stearic, are very low in content. Large variability has been reported for the concentration of oleic and linoleic acids (Askin et al. 2007; Socias et al. 2008).

Loss of nut quality during storage or transport is primarily attributed to kernel FA oxidation into peroxides, which is significantly affected by the percentage of unsaturated fatty acids. Consequently, the fatty acid profile is considered as a quality parameter of the highest interest (Kodad 2017). Oleic acid/linoleic acid (O/L) ratio is a good index of resistance to oil rancidity, with high ratios being preferable (Kester et al. 1993). Kodad and Socias i Company (2008) showed that the ratio could be used to differentiate genotypes, because it does not change over the years. Oil stability and fatty acid composition, essentially the O/L ratio, are considered an important criterion to evaluate kernel quality (Socias i Company et al. 2008).

Table 6.3 Fatty acid profile and range of variability of almond oil (Kodad 2017)

Origin	Range of variability (% of total fatty acids)						References
	Oleic 18:1	Linoleic 18:2	Palmitic 16:0	Stearic 18:0	Palmitoleic 16:1	Myristic 14:0	
Afghanistan	62.54–81.57	8.17–26.53	5.44–8.41	1.16–2.57	0.37–1.5		Zamany et al. (2017)
Argentina	68–77.5	12–22.5	6.01–7.26	1.06–1.77			Kodad et al. (2011b) and Maestri et al. (2015)
Australia	58.5–71.3	15.7–29.9	5.9–7.5	1.0–2.4	0.2–0.62	0.02–0.05	Zhu et al. (2015, 2017)
Brazil	50.4–81.2	6.21–37.1	4.7–15.8	0.3–2.5			Fernandes et al. (2017)
Iran	11.9–79.14	11.9–24.4	5.71–8.1	1–1.89			Mehran and Filsoof (1974), Abaspour et al. (2012), Kiani et al. (2015). and Alizadeh-Salte et al. (2018)
Italy	52.7–79.7	13.9–31.1	4.11–9.57	1.3–5.84	0.21–1.43	0.01–0.03	Amorello et al. (2015)
Morocco	68–70.7	19.2–22.4	6.3–7.5	1.9–2			Kodad and Socias i Company (2008) and Kodad et al. (2013)
Portugal	65–74	17.6–25	6.0–7.0	1.5–2.3	0.38–0.59		Barreira et al. (2012)
Serbia	63.14–77.37	15.57–8.69	4.68–6.48	1.45–2.56	0.24–0.56	0–0.22	Ćolić et al. (2017, 2018)
Spain	65–77	11–23.54	5.4–7.64	1.54–3.21	0.54–0.86		Soler et al. (1988), García-López et al. (1996), Prats-Moya et al. (1999), Kodad et al. (2011b, 2014b), and Rabadan et al. (2017)
Turkey	55.14–88.26	9.27–30.01	4.34–10.11	0.26–4.71	0.21–2.52	0–0.59	Askin et al. (2007), Sabudak (2007), Beyhan et al. (2011), Çelik and Balta (2011), Özcan et al. (2011), Kirbaşlar et al. (2012), Karatay et al. (2014), and Yıldırım et al. (2016)
USA	57.4–77.3	15.4–35.1	5.6–7.8	1.1–2.6			Abdallah et al. (1998), Kodad et al. (2014b), and Sathe et al. (2008)

Low content of linoleic acid is correlated to high oil stability (Zacheo et al. 2000), whereas the high content of oleic acid is considered a positive trait from the nutritional point of view (Socias i Company et al. 2008). Therefore, selection of parents for low linoleic acid and high oil content might be undertaken in a breeding program for increased kernel quality (Kodad et al. 2014b). Results of Zhu et al. (2017) who evaluated FA accumulation during kernel development, suggest that the key timing of almond lipid accumulation is between 95 and 115 days post-anthesis. This is a crucial period to apply orchard management techniques, such as increased water and fertilization, to enhance the lipids, as well as to influence oleic acid and linoleic acid maturation to control the O/L ratio for the long shelf life of kernels. Karatay et al. (2014) found that majority of studied almond genotypes from Turkey had a higher value of O/L compared to Marcona, which is used as a reference for oil stability. The highest O/L of 9.98 was recorded by Zamany et al. (2017) in almonds of the 'Khairodini' cultivar. However, a higher O/L ratio means a lower proportion of linoleic acid, the healthiest fatty acid present in almond lipids (Zhu et al. 2015).

3.2 Triglycerides (TG)

Data available on triglycerides (TG) composition of almond oil is limited. Miraliakbari and Shahidi (2008a, b) examined the content of various lipid classes in almond and found the highest content of triacylglycerol's (98.0 g/100 g oil). Nine triglycerides (LLL, OLL, PLL, OLO, PLO, PLP, OOO, POO, and SOO) in the almond kernel of 19 different cultivars were determined by high-performance liquid chromatography (Martin-Carratala et al. 1999). Nine TG were also identified and determined in almond oil by Prats Moya et al. (1999). In decreasing order of importance these were OOO, OLO, POO, OLL, PLO, SOO, LLL, PLL, and PLP. Cherif et al. (2004) analyzed three almond cultivars in Tunisia and identified ten TG including LLL, LLO, LnOO, LOO, LOP, PLP, OOO, POO, POP, and SOO (O represents oleic acid; L linoleic acid; Ln, linolenic acid; P, palmitic acid; and S, stearic acid). The oil of Mazetto cultivar kernels exhibited a TG composition comparable to that of olive oil. They concluded that the difference in the TG profile could be useful for distinguishing various cultivars. The highest number of TG was identified by Holcapek et al. (2005) who detected 24 triglycerides in almond oil. Houmy et al. (2016) carried out an analysis of triglycerides showing that analyzed almond kernel oils are characterized by the dominance of OOO ranging from 31.48% to 43.82%.

3.3 Polar lipids

Interest in food functional lipids with nutritional and technological impacts has been growing in the last years. Nutritional effect is related to the FA composition of neutral lipids (triglycerides), and to the principal species of unsaponifiable matter (e.g. sterols, and tocochromanols) and polar lipids (phospholipids and glycolipids) which are also responsible for the technological behavior of fats and fatty food during processing (Pasini et al. 2013).

Phospholipids (PLs) are a class of lipids that are a major component of all cell membranes as they can form lipid bilayers. Information about PLs and their FA composition is important for utilization of the nuts and their by-products in the food industry and for the determination of their food value. It is well known that PLs may act synergistically with tocopherols to delay the onset of lipid oxidation (Arranz et al. 2008). Only a few reports concerning the PLs composition of the almond were found. Zlatanov et al. (1999) found that PC, PI and PE are the major components with more than 70% of total PLs in the Bulgarian almonds (38%, 28% and 8% respectively), and no differences in qualitative composition between glyceride oils and PLs. The predominant FA were oleic, linoleic, palmitic, and stearic acid, while the unsaturated: saturated acid ratio was determined to change in the direction PI > PA > PE > PC in almond PLs. Boukhchina et al. (2004) reported a similar trend with the same relative abundance for PE and PC (45%), followed by PI (8%). Miraliakbari and Shahidi (2008a, b) also reported high levels of PS, ranging from 21% to 39% of the total PLs content. According to Malisiova et al. (2004) who analyzed almonds of Crete island, PE showed as the predominant PL in immature seeds (37% of PLs), decreasing in mature seeds (14%) where the predominant PL was phytosphingosine (34%). Also, the same authors found that sphingolipids (~22%) and PLs (~78%) were principal polar lipid classes. To the best of our knowledge, there are no available references on glycolipids in almonds.

4 Minor Bioactive Lipids in Almond Fixed Oil

Unsaponifiable fraction and complex lipids in almond oil consist of tocopherols, tocotrienols, PLs, sterols, phytosterols, phytostanols, sphingolipids, squalene and terpenoids that are present in small quantity but have an importance from the biological and nutritional point of view (Roncero et al. 2016).

4.1 Tocopherols and Tocotrienols

Four tocopherol isomers were reported in almonds at various levels (Table 6.4), with predominant α-tocopherol homologue, which is the most biologically active form of vitamin E. Almonds are considered one of the richest food sources of α-tocopherol (Maguire et al. 2004; Chen et al. 2006; Kornsteiner et al. 2006; Matthäus and Özcan 2009; Yada et al. 2011; Kodad et al. 2011a, 2014a; Barreira et al. 2012; Madawala et al. 2012; Zhu et al. 2015).

Many authors reported a large variability in the content of total and individual tocopherols due to genetic variation and specific environmental factors in different

Table 6.4 Tocopherol homologue concentration (mg/kg oil) in almond kernel oil (Kodad et al. 2018)

Tocopherol homologue	Range of variability (mg/kg oil)	Origin	References
α-Tocopherol	335.3–551.7	Spain	Kodad et al. (2014b)
	309–656.7	Morocco	Kodad et al. (2013)
	350–471	Italy	Rizzolo et al. (1991)
	370–571	Argentina	Maestri et al. (2015)
	21.3–31.71	Australia	Zhu (2014) and Zhu et al. (2017)
	553.4	Greece	Kodad et al. (2011b)
	312.29–467.31	Turkey	Yildirim et al. (2016)
	139.1–355.0	Afghanistan	Zamany et al. (2017)
	233.4–439.5	Brazil	Yang (2009) and Fernandes et al. (2017)
β-Tocopherol	5–8	Italy	Zacheo et al. (2000)
	10.53-	Turkey	Yildirim et al. (2016)
	8.45	Brazil	Fernandes et al. (2017)
γ-Tocopherol	6.1–50.2	Spain	Kodad et al. (2006, 2014b) and Socias i Company et al. (2014)
	75	Italy	Senesi et al. (1996)
	2.4–13.5	Morocco	Kodad et al. (2013)
	57.24–77.87	Turkey	Yildirim et al. (2016)
	10.10–12.5	Brazil	Yang (2009) and Fernandes et al. (2017)
δ-Tocopherol	0.2–22	Spain	Kodad et al. (2006, 2014b) and Socias i Company et al. (2014)
	0.1–0.3	Morocco	Kodad et al. (2013)
	0.87	Brazil	Fernandes et al. (2017)
	0.10–2.86	Turkey	Yildirim et al. (2016)

harvest years (Kodad et al. 2011a, 2014a; Socias I Company et al. 2014; Yildirim et al. 2016). Growing conditions and durability of storage were also recognized to have a strong influence on the content of tocopherols (Nanos et al. 2002; García-Pascual et al. 2003; Kodad et al. 2006; Korekar et al. 2011). A literature reviewed in Yada et al. (2011) pointed to significant variation in the content of α-tocopherols in almond cultivars of California, Italy, and Spain. In the most recent study, Kodad et al. (2018) summarized the existing knowledge of the concentration of the main tocopherol homologues detected in different almond cultivars and genotypes from different countries. The authors indicated the range of variability for the α-tocopherol to be between 21.3 and 656.7 mg/kg oil, for γ-tocopherol between 2.4 and 50.2 mg/kg, and for β-tocopherol between 0.1 and 22.0 mg/kg oil. Limited literature available on tocotrienols (α-, β-, γ-, δ- tocotrienol) indicates that they are also present in almond oil, but to much less extent (Piironen et al. 1986; Robbins et al. 2011). Piironen et al. (1986) reported 0.24 mg α-tocotrienol and 0.02 mg γ -tocotrienol per 100 g almonds.

4.2 Phytosterols

These members of the triterpene family have been studied for their hypocholesterolemic, anticarcinogenic, and other health effects (Ostlund 2002). Ranged between 2178 and 4497 mg/kg oil, they make up the greatest portion of the unsaponifiable fraction of lipids (Miraliakbari and Shahidi 2008a; Fernández-Cuesta et al. 2012). This group of compounds has been included inside the phytochemical concept by several authors (Alasalvar and Bolling 2015).

Study of 23 local almond seedling populations from Morocco (Kodad et al. 2015) showed that phytosterols fraction consisted mainly of β-sitosterol (78.90–87.26%) and Δ^5-avenasterol (6.26–13.04%), which together accounted for more than 90% of phytosterols, similar to other reports on almond (Dulf et al. 2010; Fernandes et al. 2017). Other phytosterols detected in minor amounts were campesterol (2.25–3.20%), $\Delta^{5,24}$-stigmastadienol (0.83–1.76%), Δ^7-avenasterol (0.64–2.09%), cholesterol (0.68–1.23%), A7-stigmastenol (0.27–1.65%), Δ^7-campesterol (0.02–2.04%) and stigmasterol (0.04–0.36%). Similar values for minor phytosterols were obtained by Fernandes et al. (2017) who describes for the first time the presence of terpenic and aliphatic alcohols in almond oil. Major phytosterol components, β-sitosterol, Δ^5-avenasterol, and campesterol were the most variable almond kernel phytosterols among genotypes (Kodad et al. 2015).

4.3 Phytochemicals Originating from Almonds

Almond, and especially almond skins are a rich source of phenolic acids and flavonoids (Milbury et al. 2006; Monagas et al. 2007; Urpi-Sarda et al. 2009; Mandalari et al. 2010). Total phenols values in almond oil reported by Bolling (2017) were in the range 12.4–16.8 mg gallic acid equivalents (GAE).

Among phytochemicals, polyphenols are considered to possess potential bioactivity and health-promoting properties. They act as free radical scavengers and have a crucial role in preventing the auto-oxidation of unsaturated fatty acids, and thereby increasing oil's shelf life. Sakar et al. (2017) found that content of polyphenols varies significantly among cultivars and ranged from 0.56 mg (Marcona) to 0.76 GAE/g oil (Fournat de Breznaud). Investigations done by Kornsteiner et al. (2006) showed that almonds with skin have a higher content of polyphenols than almonds without skin (130–456 GAE vs. 4–49 GAE/100 g FW). The results confirm the earlier observation that almond skin contains high amounts of phenolics. Sang et al. (2002a, b) determined nine phenolic compounds from the almond skin.

Vast varieties of phenolic acids and flavonoids have been reported in almonds. Protocatechuic acid was predominant in the majority of published papers (Senter et al. 1983; Sang et al. 2002a; Milbury et al. 2006; Wijeratne et al. 2006a) together with *p*-hydroxybenzoic acid and vanillic acid. Other phenolic acids included caffeic, *p*-coumaric, ferulic, sinapic, syringic, gallic, and ellagic acid. Flavonoids are mainly present as glycosylated forms as glucosides, rhamnosides, rutinosides, and galactosides. The presence of aglycones isorhamnetin, kaempferol, catechin, quercetin, and epicatechin was reported by Chen et al. (2005) and Milbury et al. (2006). Wijeratne et al. (2006b) reported the presence of flavonol morin and flavonol glycosides, quercitrin, astragalin, kaempferol 3-O-rutinoside, and isorhamnetin 3-O-glucoside. Isorhamnetin rutinoside was the most abundant flavonol glycoside reported in the study by Frison-Norrie and Sporns (2002).

Monagas et al. (2007) identified 33 compounds corresponding to flavanols, flavonols, dihydroflavonols and flavanones, and other non-flavonoid compounds in the almond skin. Further, Bolling et al. (2009) detected catechins, as well as flavonoids such as naringenin, quercetin and kaempferol, predominantly as glucosides or rutinosides. Stilbenes, piceid (a derivative of resveratrol) was also reported in the almond skin (Xie and Bolling 2014).

Significant differences were observed among the studied cultivars, although sample freshness, particle size, and solvent type used for the extraction also have an influence on phenolic profile in almonds. Yang (2009) studied Brazilian almonds and reported soluble free phenolics of 83.0 mg/100 g, and bound phenolics of 129.9 mg/100 g, while the content of flavonoids was 39.8 mg/100 g and 53.7 mg/100 g for free and bound form, respectively. According to Kornsteiner et al. (2006), the content of total phenolics in almonds was 239 mg GAE/100 g. Amarowicz et al. (2005) reported values for the crude extract of almond containing 16.1 mg (+)-catechin equivalents (CE)/g of crude extract (mg CE/g), and for low molecular weight fraction (7.14 mg CE/g) and high molecular weight fraction (80.4 mg CE/g). Vanillic acid, caffeic acid, p-coumaric acid, ferulic acid, quercetin, kaempferol, isorhamnetin, delphinidin and procyanidins B2 and B3 were reported in the extract (Amarowicz et al. 2005).

Yildirim et al. (2016) studied commercial almond cultivars grown in Turkey and reported the presence of catechin (27.35–39.87 mg/kg). Other phenolics reported were gallic acid, caffeic acid, chlorogenic acid, epicatechin, ferulic acid, kaempferol, naringenin, and p-coumaric acid. Recently, Čolić et al. (2017) reported high diversity and unique phenolic profile both for 20 almond samples originating from Serbia and for cultivars Marcona, Troito and Texas. Catechin was predominant, followed by chlorogenic acid, naringenin, rutin, apigenin and astragalin. Unique phenolic profiles corresponded with observed isoenzyme polymorphism (Čolić et al. 2010) that qualified phenolic profile as a valuable criterion for the almond characterization.

Total phenolics, flavonoids, and antioxidant activity were most recently found in traditional and commercial Portuguese almond cultivars (Oliveira et al. 2018) and these values were higher than the results reported by Milbury et al. (2006) and Kornsteiner et al. (2006). Ahrens et al. (2005) analyzed Carmel, Mission, and Nonpareil for the content of tannins and reported a range between 0.12% and 0.18% (expressed as catechin equivalents). Hydrolysable tannins, ellagitannins and gallotannins have been identified in almond after hydrolysis, with ranges of 53–57 mg/100 g and 20–34 mg/100 g, respectively (Xie et al. 2012). Using Sephadex LH-20 column chromatography with ethanol and 50% acetone as the mobile phase for the separation of phenolics, Amarowicz et al. (2005) found that content of tannins, in fraction II (296) (expressed as absorbance value at 500 nm/g) was ten times higher than that in the crude extract (28.3). Other phytochemicals, such as diterpene glycoside amygdaloside have also been reported in almonds (Sang et al. 2003). Total amounts of the carotenoids including α and β-carotene, zeaxanthin, lutein, cryptoxanthin and lycopene in almond oil were marginal according to Kornsteiner et al. (2006) and Stuetz et al. (2017).

5 Composition of the Almond Essential Oil

The bitter almond oil (BAO) which is considered an essential oil is extracted from the bitter almonds (*Prunus amygdalus* var. *amara*). Almond oils extracted from sweet and bitter almonds showed similar initial composition, with identical fatty acid compositions and total tocopherol contents (Salvo et al. 1986). Extracted BAO has more active chemical components and does not have the same therapeutic properties as SAO. The main difference is that bitter almond kernels contain significant concentrations of amygdalin ranging between 33.00 and 53.99 mg/g (Lee et al. 2013). Amygdalin present in bitter almonds is poisonous and turns into benzaldehyde and toxic hydrocyanic (prussic) acid on processing. That hinders the use of BAO in human consumption, but makes BAO appropriate for medical purposes. When the hydrogen cyanide (product of amygdalin degradation) is removed in the refining step, the oil is not toxic and could be used as a flavoring agent.

Characterization of almond oil made by Aziz et al. (2013) showed that SAO extract contains 72.81% oleic acid and 27.19% linoleic acid, while bitter oil extract contains oleic, linoleic and α-linoleic acids with the value of 59.4%, 22.5% and 17.9%, respectively. Amount of unsaturated fatty acids (oleic and linoleic) were higher in SAO, while BAO extracts contained α-linolenic acid that not found in SAO.

6 Volatile Compounds in Almond Oil

Off-flavor development in almonds is primarily caused by lipid oxidation of unsaturated fatty acids and the presence of riboflavin, which acts as a photosensitizer in photo-oxidation (Mexis et al. 2009). The process of oxidation of edible oils depends on the oxidative conditions, mostly temperature. Thus, different compounds will be produced if the oxidation is done at room temperature or at higher temperatures (Guillén et al. 2008; Sanchez-Prado et al. 2009). Almonds are sensitive to lipid oxidation because of 48–67% of the almond oil is composed of 63–78% oleic acid (Mandalari and Faulks 2008) which is a precursor to many aldehydes and alcohols (Perez et al. 1999).

The first study related to volatiles in almond oil was done by Pićurić-Jovanović and Milovanović (1993) who identified 20 compounds including alkylfuranones, n-alkanes, cyclopentadiene and aromatic compounds such as benzaldehyde, methylphenol, benzyl alcohol and some alkylbenzenes. The main compounds with high concentrations were methyl benzene, benzaldehyde and benzyl alcohol. Generally, benzaldehyde (a breakdown product of amygdalin) was the predominant volatile and it was associated to the almond-like flavor in the volatile detection in oils. Benzaldehyde was determined as the main volatile component in almond oil in a study by Caja et al. (2000). They also determined some levels of hexanal, nonanal, decadienal and (E,E)-2,4-decadienal. Those aldehydes have been reported in the previous study by Vázquez Araújo et al. (2008) and they play an important role in the aroma of almond oils. Hexanal (formed by autoxidation of linoleic acid, as well as the thermal oxidation of linoleates) has a low odor threshold and is considered an indicator of oil quality (Mexis et al. 2009).

Beltrán Sanahuja et al. (2011) identified 22 compounds in the oil from different almond cultivars (Marcona, Guara and Butte) including aldehydes, alkanes, alcohols and aromatic hydrocarbons. The presence of octanal, nonanal, decanal, dodecanal and tetradecanal, and the quantity of octanal and nonanal (decomposition product of oleic acid) were lower in samples from Butte cultivar compared with Spanish Guara and Marcona cultivar. This was connected to the fact that the Butte cultivar has the lowest oleic and stearic acids concentrations and the highest linoleic acid content (Beltrán Sanahuja et al. 2009). Finally, Beltrán Sanahuja et al. (2011) determined 22 main compounds in a Marcona oil including alkanals, 2-alkenals, 2,4-alkadienals, ketones, acids, esters, alcohols as well as aliphatic and aromatic hydrocarbons. One of the most important volatiles that formed during oxidation was unsaturated aldehydes (derivate in the decomposition of linoleic acid).

7 Health-Promoting Traits of Almond Oil and Oil Constituents

Almond oil has long been used in complementary medicine for its numerous health benefits (Ahmad 2010). The Greek physician Hippocrates (ca. 460–370 b.c.e.) records that 'Almonds are burning but nutritious' (burning because they are oily and nutritious because they are fleshy) (Jones 1967). Ancient Greeks categorized almonds as a hot and dry food that stimulated choler in the body, healing colds and other phlegmatic disorders. According to the ancient documents, almonds have the ability to prevent intoxication and inebriation, if consumed before drinking. Raw or toasted almonds were consumed extensively in Rome. Pliny described that bitter almonds provoke sleep and sharpen the appetite, act as a diuretic and emmenagogue. They are also used for a headache, namely in cases that involve fever. Almonds are also useful for lethargy and epilepsy, and the head is anointed with them for the cure of epinyctis (Gradziel 2011). Almond oil has been used in ancient Chinese and Ayurvedic practices to treat skin conditions like eczema and psoriasis, and to support the brain and nervous system. The Greco-Persian Unani-Tibb medical system considers almond oil as a natural aphrodisiac for massage and internal consumption (Zohary and Hopf 2000; Cantor et al. 2006).

Nowadays, almonds are recognized for their health-promoting qualities, particularly for their role in reducing the risk of cardio-vascular diseases (including stroke and other chronic ailments), which is probably due to the favorable lipid profile and low-glycaemic nature (Abbey et al. 1994; Fraser 1999; Li et al. 2011). Also, almonds and their oil have numerous properties including anti-inflammatory, immunity-boosting, and anti-hepatotoxicity effects (Hyson et al. 2002; Sultana et al. 2007). Almond oil is a great source of the potent antioxidant vitamin E. As a matter of fact almond provides up to 72.7% of the daily 15 mg recommended dose for vitamin E for adults (Alasalvar and Shahidi 2009a). The lipid fraction of the almond seed ranges from 50% to 60% and, within this fraction, 80–90% are MUFAs or PUFAs (Nanos et al. 2002; Kazantzis et al. 2003). Both almonds and almond oil are rich in MUFAs and PUFAs that are shown to lower levels of "bad" LDL cholesterol and total cholesterol (Wien et al. 2010). In that way, according to Hyson et al. (2002), a diet rich in almond oil significantly lowered both LDL and total cholesterol levels, while it increased "good" cholesterol (high-density lipoprotein) by 6%. Mori et al. (2011) proved that participants who consumed a breakfast with added almond oil had lower blood sugar, both after the meal and throughout the day, compared to participants who did not eat almond oil. Further, almond oil has a beneficial effect in the management of irritable bowel syndrome, and reduces the incidence of colonic cancer (Davis and Iwahashi 2001). Besides that it is often used together with phenol and injected into hemorrhoids (Lenhard 1990). Almond oil is commonly used in otolaryngology for conditions affecting the outer ear such as otosclerosis due to its anti-sclerosant properties (Davis and Iwahashi 2001).

8 Food and Non-food Applications of Almond Oil

8.1 Food Applications

Since almond oil is a vegetable oil, there are many uses in the food preparation. Both cold-pressed and refined almond oil can be used for cooking, but the ways for cooking with almond oil depends on which oil type is used. Refined oil has strong taste with nutty flavor while cold-pressed oil has mild flavor. Almond oil is rich source of essential fatty acids, which are necessary for the body's normal functioning. Using one tablespoon of almond oil for cooking or serving food is yielding 120 calories. It can be used in baking, frying goodies and salad dressing. Almond oil improves the functional characteristics of mayonnaise, whipped cream, cake fillings and toppings. Cold-pressed almond oil should be restricted to cold dishes and drizzling over salads. Exposure to high heat reduces its nutty flavor, and nutrition.

8.2   Non-food and Cosmetic Applications

The interest in natural products that have a hydrating and restorative skin activity has increased in recent years. Almond oil is well known to possess these properties and it is widely used in the cosmetics industry (Ahmad 2010; Özcan et al. 2011) because of its moisturizing, softening, anti-inflammatory and anti-aging properties. SAO is light, clear pale yellow with medium viscosity and a faintly sweet and nutty aroma, which is easily absorbed by the skin and is very nutritious. It is recommended for all skin types. Because of its numerous health benefits, it is now often included in soaps, lotions, creams and moisturizers.

Almond oil is naturally endowed with highly potent antioxidant vitamin E. In a study of Sultana et al. (2007), almond oil was shown to prevent the structural damage caused by UV radiation, having natural 'sun protection factor' (SPF) 5. Regular application of the oil can protect skin from oxidative stress keeping it soft and supple. Almond oil is light in texture, and can easily penetrate deep into the skin, softening and dislodging the dirt and debris accumulated in the skin pores and hair follicles. This prevents blackheads and acne. Thanks to the vitamin A content in the oil, it may even help reduce acne flare-ups. It is demonstrated that massage with BAO may be effective in reducing the visibility of current striae gravidarum, the prevention of new striae or striae itching (Taşhan and Kafkasli 2012; Hajhashemi et al. 2017; Malakouti et al. 2017). It is also a gentle and moisturizing conditioner for hair, repairing damaged and dry hair and restoring it back to its original luster and shine.

Holistic beauty experts have often praised the soothing effects of almond oil on the skin. This is also known as a healing touch of massage therapy (Soden et al. 2004; Chang 2008). Due to its rich concentration of oleic and linoleic acids (Koriyama et al. 2005) almond oil is one of the most popular oils used in aromatherapy and massage therapy.

9   Adulteration and Authenticity of Almond Oil

Food authentication is a rapidly growing field due to increasing public awareness concerning food quality and safety (Danezis et al. 2016). Authenticity has been a major concern of consumers, producers and regulators. Authenticity testing is a quality criterion for food and food ingredients, increasingly a result of legislative protection of regional foods.

Almond oil is used in the cosmetic, pharmaceutical and food industries. Since expensive, it is often substituted by or adulterated with oils extracted from kernels of fruits of the same family (*Prunus spp*). One of the common adulterants is apricot oil, which rather difficult to detect. In a mixture with almond oil, additions of at least 30% apricot oil may be detected by the Bieber test. For detection of smaller amounts of the added apricot oil (5%), Gutfinger and Letan (1973) described procedure based on the determination of tocopherols.

According to German food guidelines, raw pastes that are declared as "marzipan" may only contain almonds (*Prunus amygdalus* var. *dulcis*) as an oilseed ingredient (Brüening et al. 2011). If other seeds are used, declaration of the respective ingredient is required. Due to high almond market values, dilutions of marzipan with cheaper seeds may occur. Most of the raw paste producers is small- and medium-sized companies producing both marzipan and persipan raw paste, in most cases on the same production lines. Especially in the case of almond substitution with apricot, the close relationship of both species presents a challenge for the analysis.

Several approaches were developed in order to characterize the origin of BAO, which is derived from bitter almonds (*Prunus amygdalus* var. *amara*) from BAO derived from apricots, peaches, plums, and cherries. Different methods were developed to detect adulteration of almond with apricot oil, such as SNIF-NMR method (Remaud et al. 1997), Polymerase Chain Reaction based methods (Brüening et al. 2011; Haase et al. 2013), and recently fatty acid fingerprinting (Esteki et al. 2017).

The use of natural flavors is often seen as a strong marketing advantage, but since they are more expensive than artificial/synthetic flavors, are prone to adulteration (Culp and Noakes 1992). Controls of the origin of flavor compounds are therefore currently routinely performed by the food and flavor industries. There is increasing demand for the development of fast, easy, and low-cost analytical methods for adulteration testing. Considering both fundamental economic implications of any fraud and food authenticity as one of the most crucial issues in food control and safety, it is necessary to develop a chemical methodology that can confirm food labels identifying the geographic origin. Most of these studies characterized the almond based on analysis of essential oils (Tiên et al. 2015), fatty acid (Maestri et al. 2015) or phytosterols (Zhu et al. 2015).

References

Abaspour, M., Imani, A., & Hassanlo, T. (2012). Effects of almond genotype and growing location on oil percentage and fatty acid composition of its seeds. *Journal of Nuts, 3*(3), 5–12.

Abbey, M., Noakes, M., Belling, G. B., & Nestel, P. J. (1994). Partial replacement of saturated fatty acids with almonds or walnuts lowers total plasma cholesterol and low-density-lipoprotein cholesterol. *The American Journal of Clinical Nutrition, 59*(5), 995–999. https://doi.org/10.1093/ajcn/59.5.995.

Abdallah, A., Ahumada, M. H., & Gradziel, T. M. (1998). Oil content and fatty acid composition of almond kernels from different genotypes and California production regions. *The Journal of the American Society for Horticultural Science, 123*(6), 1029–1033.

Abu-Hamdeh, N. H., & Alnefaie, K. A. (2015). A comparative study of almond and palm oils as two bio-diesel fuels for diesel engine in terms of emissions and performance. *Fuel, 150*, 318–324. https://doi.org/10.1016/j.fuel.2015.02.040.

Ahmad, Z. (2010). The uses and properties of almond oil. *Complementary Therapies in Clinical Practice, 16*(1), 10–12. https://doi.org/10.1016/j.ctcp.2009.06.015.

Ahrens, S., Venkatachalam, M., Mistry, A. M., Lapsley, K., & Sathe, S. K. (2005). Almond (*Prunus dulcis* L.) protein quality. *Plant Foods for Human Nutrition, 60*(3), 123–128. https://doi.org/10.1007/s11130-005-6840-2.

Alasalvar, C., & Bolling, B. W. (2015). Review of nut phytochemicals, fat-soluble bioactives, anti-oxidant components and health effects. *The British Journal of Nutrition, 113*(S2), S68–S78. https://doi.org/10.1017/S0007114514003729.

Alasalvar, C., & Shahidi, F. (2009a). Natural antioxidants in tree nuts. *European Journal of Lipid Science and Technology, 111*(11), 1056–1062. https://doi.org/10.1002/ejlt.200900098.

Alasalvar, C., & Shahidi, F. (2009b). *Tree nuts: Composition, phytochemicals, and health effects.* Boca Raton: CRC Press Taylor & Francis Group.

Albala, K. (2009). Almonds along the silk road: The exchange and adaptation of ideas from West to East. *Petits Propos Culinaires, 88,* 17–32. http://works.bepress.com/ken-albala/9/.

Alizadeh-Salte, S., Farhadi, N., Arzani, K., & Khoshghalb, H. (2018). Almond oil quality as related to the type of pollen source in Iranian self incompatible cultivars. *International Journal of Fruit Science, 18,* 29–36. https://doi.org/10.1080/15538362.2017.1367983.

Al-Tikrity, E. T. B., & Ibraheem, A. F. A. F. K. K. (2017). Biodiesel production from bitter almond oil as new non-edible oil feedstock. *Energy Sources, Part A Recovery Utilization and Environmental Effects, 39*(7), 1–8. https://doi.org/10.1080/15567036.2016.1243172.

Álvarez-Ortí, M., Quintanilla, C., Sena, E., Alvarruiz, A., & Pardo, J. E. (2012). The effects of a pressure extraction system on the quality parameters of different virgin pistachio (*Pistacia vera* L. var. Larnaka) oils. *Grasas y Aceites, 63*(3), 260–266. https://doi.org/10.3989/gya.117511.

Amarasinghe, B. M. W. P. K., Kumarasiri, M. P. M., & Gangodavilage, N. C. (2009). Effect of method of stabilization on aqueous extraction of rice bran oil. *Food and Bioproducts Processing, 87*(2), 108–114. https://doi.org/10.1016/j.fbp.2008.08.002.

Amarowicz, R., Troszynska, A., & Shahidi, F. (2005). Antioxidant activity of almond seed extract and its fractions. *Journal of Food Lipids, 12*(4), 344–358. https://doi.org/10.1111/j.1745-4522.2005.00029.x.

Amorello, D., Orecchio, S., Pace, A., & Barreca, S. (2015). Discrimination of almonds (*Prunus dulcis*) geographical origin by minerals and fatty acids profiling. *Natural Product Research, 30*(18), 1–4. https://doi.org/10.1080/14786419.2015.1107559.

Arranz, S., Cert, R., Pérez-Jiménez, J., Cert, A., & Saura-Calixto, F. (2008). Comparison between free radical scavenging capacity and oxidative stability of nut oils. *Food Chemistry, 110*(4), 985–990. https://doi.org/10.1016/j.foodchem.2008.03.021.

Askin, M. A., Balta, M. F., Tekintas, F. E., Kazankaya, A., & Balta, F. (2007). Fatty acid composition affected by kernel weight in almond [*Prunus dulcis* (Mill.) D.A. Webb.] genetic resources. *Journal of Food Composition and Analysis, 20*(1), 7–12. https://doi.org/10.1016/j.jfca.2006.06.005.

Atapour, M., & Kariminia, H. R. (2011). Characterization and transesterification of Iranian bitter almond oil for biodiesel production. *Applied Energy, 88*(7), 2377–2381. https://doi.org/10.1016/j.apenergy.2011.01.014.

Ayadi, M., Ghrab, M., Gargouri, K., Elloumi, O., Zribi, F., Ben Mimoun, M., Boulares, C., & Guedri, W. (2006). Kernel characteristics of almond cultivars under rainfed conditions. *Acta Horticulturae, 726,* 377–381. https://www.actahort.org/books/726/726_61.htm.

Aziz, H. M., Ahmed, R. M., Muhammed, B. J. (2013). Characterization of bioactive compounds by HPLC from sweet and bitter almond fruits. Kurd Acad J (KAJ) – A- , Special Issue: 1st international conference of agricultural sciences held by Faculty of Agricultural Sciences, University of Sulaimani, and Kurdistan Academics Association, 20–21 Nov 2013, pp. 151–157.

Balta, F., Battal, P., Balta, M. F., & Yoruk, H. I. (2009). Free sugar compositions based on kernel taste in almond genotypes Prunis dulcis from eastern Turkey. *Chemistry of Natural Compounds, 45*(2), 221–224. https://doi.org/10.1007/s10600-009-9296-z.

Balvardi, M., Rezaei, K., Mendiola, J. A., & Ibanez, E. (2015). Optimization of the aqueous enzymatic extraction of oil from Iranian wild almond. *Journal of the American Oil Chemists' Society, 92*(7), 985–992. https://doi.org/10.1007/s11746-015-2671-y.

Barbera, G., La Mantia, T., Monastra, F., De Palma, L., & Schirra, M. (1994). Response of Ferragnes and Tuono almond cultivars to different environmental conditions in southern Italy. *Acta Horticulturae, 373,* 125–128. https://www.actahort.org/books/373/373_16.htm.

Barreira, J. C. M., Casal, S., Ferreira, I. C. F. R., Peres, A. M., Pereira, J. A., & Oliveira, M. B. P. P. (2012). Supervised chemical pattern recognition in almond (*Prunus dulcis*) Portuguese PDO cultivars: PCA – and LDA-based triennial study. *The Journal of Agricultural and Food Chemistry, 60*(38), 9697–9704. https://doi.org/10.1021/jf301402t.

Beltran Sanahuja, A., Prats Moya, M. S., Maestre Perez, S. E., GraneTeruel, N., & Martin Carratala, M. L. (2009). Classification of four almond cultivars using oil degradation parameters based on FTIR and GC data. *The Journal of the American Oil Chemists' Society, 86*(1), 51–58. https://doi.org/10.1007/s11746-008-1323-x.

Beltran Sanahuja, A. B., Santonja, M. R., Grané Teruel, N., Carratala, M. L. M., & Garrigos Selva, M. C. (2011). Classification of almond cultivars using oil volatile compound determination by HS-SPME–GC–MS. *Journal of the American Oil Chemists' Society, 88*(3), 329–336. https://doi.org/10.1007/s11746-010-1685-8.

Beyhan, Ö., Aktas, M., Yilmaz, N., Simsek, N., & Gerçekçioğlu, R. (2011). Determination of fatty acid compositions of some important almond (*Prunus amygdalus* L.) varieties selected from Tokat province and Eagean region of Turkey. *The Journal of Medicinal Plants Research, 5*(19), 4907–4911. http://www.academicjournals.org/journal/JMPR/article-full-text-pdf/C90FEF626367.

Bolling, B. W. (2017). Almond polyphenols: Methods of analysis, contribution to food quality, and health promotion. *Comprehensive Reviews in Food Science and Food Safety, 16*, 346–368. https://doi.org/10.1111/1541-4337.12260.

Bolling, B. W., Dolnikowski, G., & Blumberg, J. B. (2009). Quantification of almond skin polyphenols by liquid chromatography-mass spectrometry. *Journal of Food Science, 74*(4), C326–C332. https://doi.org/10.1111/j.1750-3841.2009.01133.x.

Boukhchina, S., Sebai, K., Cherif, A., Kallel, H., & Mayer, P. M. (2004). Identification of glycerophospholipids in rapeseed, olive, almond, and sunflower oils by LC–MS and LC–MS–MS. *Canadian Journal of Chemistry, 82*(7), 1210–1215. https://doi.org/10.1139/v04-094.

Brüening, P., Haase, I., Matissek, R., & Fischer, M. (2011). Marzipan: Polymerase chain reaction-driven methods for authenticity control. *Journal of Agricultural and Food Chemistry, 59*(22), 11910–11917. https://doi.org/10.1021/jf202484a.

Caja, M. M., del Castillo, M. R., Alvarez, R. M., Herraiz, M., & Blanch, G. P. (2000). Analysis of volatile compounds in edible oils using simultaneous distillation-solvent extraction and direct coupling of liquid chromatography with gas chromatography. *European Food Research and Technology, 211*(1), 45–51. https://doi.org/10.1007/s002170050587.

Cantor, D., Fleischer, J., Green, J., & Israel, D. L. (2006). The fruit of the matter. *Mental Floss, 5*(4), 12.

Çelik, F., & Balta, M. F. (2011). Kernel fatty acid composition of Turkish almond (*Prunus dulcis*) genotypes: A regional comparison. *Journal of Food, Agriculture and Environment, 9*(1), 171–174.

Chang, S. Y. (2008). Effects of aroma hand massage on pain, state anxiety and depression in hospice patients with terminal cancer. *Journal of Korean Academy of Nursing, 38*(4), 493–502. https://doi.org/10.4040/jkan.2008.38.4.493.

Chen, C. Y., Milbury, P. E., Lapsley, K., & Blumberg, J. B. (2005). Flavonoids from almond skins are bioavailable and act synergistically with vitamins C and E to enhance hamster and human LDL resistance to oxidation. *The Journal of Nutrition, 135*(6), 1366–1373. https://doi.org/10.1093/jn/135.6.1366.

Chen, C. Y., Lapsley, K., & Blumberg, J. (2006). A nutrition and health perspective on almonds. *Journal of the Science of Food and Agriculture, 86*(14), 2245–2250. https://doi.org/10.1002/jsfa.2659.

Cherif, A., Khaled, S., Boukhchina, S., Belkacemi, K., & Arul, J. (2004). Kernel fatty acid and triacylglycerol composition for three almond cultivars during maturation. *The Journal of the American Oil Chemists' Society, 81*(10), 901–905. https://doi.org/10.1007/s11746-004-0999-z.

Čolić, S., Milatović, D., Nikolić, D., & Zec, G. (2010). Isoenzyme polymorphism of almond genotypes selected in the region of northern Serbia. *Horticultural Science (Prague), 37*(2), 56–61. https://www.agriculturejournals.cz/publicFiles/19956.pdf.

Čolić, S., Rakonjac, V., Zec, G., Nikolić, D., & Akšić Fotirić, M. (2012). Morphological and biochemical evaluation of selected almond [*Prunus dulcis* (Mill.) D.A.Webb] genotypes in northern Serbia. *Turkish Journal of Agriculture and Forestry, 36*(4), 429–438. http://journals.tubitak.gov.tr/agriculture/issues/tar-12-36-4/tar-36-4-5-1103-50.pdf.

Čolić, S., Rahović, D., Bakić, I., Zec, G., & Janković, Z. (2013). Kernel characteristics of the almond genotypes selected in Northern Serbia. *Acta Horticulturae, 981*, 123–126. https://www.actahort.org/books/981/981_14.htm.

Čolić, S., Fotirić Akšić, M., Lazarević, K., Zec, G., Gašić, U., Dabić Zagorac, D., & Natić, M. (2017). Fatty acid and phenolic profile of almond grown in Serbia. *Food Chemistry, 234*, 455–463. https://doi.org/10.1016/j.foodchem.2017.05.006.

Čolić, S., Zec, G., Bakić, I., Janković, Z., Rahović, D., Fotirić Akšić, M. (2018). Rootstock effect on some quality characteristics of almond cultivars Troito, Marcona and Texas. *Acta Horticulturae, 1219*, 19–24. https://doi.org/10.17660/ActaHortic.2018.1219.4

Cordeiro, V., & Monteiro, A. (2001). Almond growing in Trás-os-Montes region (Portugal). *Acta Horticulturae, 591*, 161–165. https://www.actahort.org/books/591/591_22.htm.

Culp, R. A., & Noakes, J. E. (1992). Determination of synthetic components in flavors by deuterium/hydrogen isotopic ratios. *Journal of Agricultural and Food Chemistry, 40*(10), 1892–1897. https://doi.org/10.1021/jf00022a033.

Danezis, G. P., Tsagkaris, A. S., Camin, F., Brusic, V., & Georgiou, C. A. (2016). Food authentication: Techniques, trends & emerging approaches. *TrAC, 85*(Part A), 123–132. https://doi.org/10.1016/j.trac.2016.02.026.

Davis, P. A., & Iwahashi, C. K. (2001). Whole almonds and almond fractions reduce aberrant crypt foci in a rat model of colon carcinogenesis. *Cancer Letters, 165*(1), 27–33. https://doi.org/10.1016/S0304-3835(01)00425-6.

Dodevska, M., Šobajić, S., & Djordjević, B. (2015). Fibre and polyphenols of selected fruits, nuts and green leafy vegetables used in Serbian diet. *Journal of the Serbian Chemical Society, 80*(1), 21–33. http://www.doiserbia.nb.rs/img/doi/0352-5139/2015/0352-51391400062D.pdf.

Drogoudi, P. D., Pantelidis, G., Bacchetta, L., De Giorgio, D., Duval, H., Metzidakis, I., & Spera, D. (2013). Protein and mineral nutrient contents in kernels from 72 sweet almond cultivars and accessions grown in France, Greece and Italy. *International Journal of Food Sciences and Nutrition, 64*(2), 202–209. https://doi.org/10.3109/09637486.2012.728202.

Dulf, F. V., Unguresan, M. L., Vodnar, D. C., & Socaciu, C. (2010). Free and esterified sterol distribution in four Romanian vegetable oils. *Notulae Botanicae Horti Agrobotanici Cluj-Napoca, 38*(2), 91–97. http://www.notulaebotanicae.ro/index.php/nbha/article/viewFile/4753/4517.

Egea, G., González-Real, M. M., Baille, A., Nortes, P. A., Sánchez-Bel, P., & Domingo, R. (2009). The effects of contrasted deficit irrigation strategies on the fruit growth and kernel quality of mature almond trees. *Agricultural Water Management, 96*(11), 1605–1614. https://doi.org/10.1016/j.agwat.2009.06.017.

Esteki, M., Farajmand, B., Kolahderazi, Y., & Gandara, J. S. (2017). Chromatographic fingerprinting with multivariate data analysis for detection and quantification of apricot kernel in almond powder. *Food Analytical Methods, 10*(10), 3312–3320. https://doi.org/10.1007/s12161-017-0903-5.

FAOSTAT. (2018). *FAO Statistical data base*. http://www.fao.org/faostat/en/#data/QC. Accessed 5 May 2018.

Femenia, A., Garcia, M., Simal, S., Rossello, C., & Blasco, M. (2001). Effects of supercritical carbon dioxide (SC-CO2) oil extraction on the cell wall composition of almond fruits. *Journal of Agricultural and Food Chemistry, 49*(12), 5828–5834. https://doi.org/10.1021/jf010532e.

Fernandes, G. D., Gómez-Coca, R. B., Pérez-Camino, M. C., Moreda, W., & Barrera-Arellano, D. (2017). Chemical characterization of major and minor compounds of nut oils: Almond, hazelnut, and pecan nut. *Journal of Chemistry*. Article ID 2609549, 11 pages. https://doi.org/10.1155/2017/2609549.

Fernández-Cuesta, A., Kodad, O., Socias I Company, R., & Velasco, L. (2012). Phytosterol variability in almond germplasm. *Journal of the American Society for Horticultural Science, 137*(5), 343–348. http://journal.ashspublications.org/content/137/5/343.full.pdf+html.

Fraser, G. E. (1999). Nut consumption, lipids, and risk of a coronary event. *Clinical Cardiology, 22*(S3), 11–15. https://doi.org/10.1002/clc.4960221504.

Frison-Norrie, S., & Sporns, P. (2002). Identification and quantification of flavonol glycosides in almond seedcoats using MALDI-TOF MS. *Journal of Agricultural and Food Chemistry, 50*(10), 2782–2787. https://doi.org/10.1021/jf0115894.

García-López, C., Grané-Teruel, N., Berenguer-Navarro, V., García-García, J. E., & Martín-Carratalá, M. L. (1996). Major fatty acid composition of 19 almond cultivars of different origins. A chemometric approach. *Journal of Agricultural and Food Chemistry, 44*(7), 1751–1756. https://doi.org/10.1021/jf950505m.

García-Pascual, P., Mateos, M., Carbonell, V., & Salazar, D. M. (2003). Influence of storage conditions on the quality of shelled and roasted almonds. *Biosystems Engineering, 84*(2), 201–209. https://doi.org/10.1016/S1537-5110(02)00262-3.

Gradziel, T. M. (2011). Origin and dissemination of almond. *Horticultural Reviews, 38*, 23–82. https://doi.org/10.1002/9780470872376.ch2.

Guillén, M. D., Goicoechea, E., Palencia, G., & Cosmes, N. (2008). Evidence of the formation of light polycyclic aromatic hydrocarbons during the oxidation of edible oils in closed containers at room temperature. *Journal of Agricultural and Food Chemistry, 56*(6), 2028–2033. https://doi.org/10.1021/jf072974h.

Gutfinger, T., & Letan, A. (1973). Detection of adulteration of almond oil with apricot oil through determination of tocopherols. *Journal of Agricultural and Food Chemistry, 21*(6), 1120–1123. https://doi.org/10.1021/jf60190a039.

Haase, I., Brüning, P., Matissek, R., & Fischer, M. (2013). Real-time PCR assays for the quantitation of rDNA from apricot and other plant species in marzipan. *Journal of Agricultural and Food Chemistry, 61*(14), 3414–3418. https://doi.org/10.1021/jf3052175.

Hajhashemi, M., Rafieian, M., Rouhi Boroujeni, H. A., Miraj, S., Memarian, S., Keivani, A., & Haghollahi, F. (2017). The effect of *Aloe vera* gel and sweet almond oil on striae gravidarum in nulliparous women. *The Journal of Maternal-Fetal & Neonatal Medicine, 31*(13), 1703–1708. https://doi.org/10.1080/14767058.2017.1325865.

Hall, A. P., Moore, J. G., Gunning, B., & Cook, B. B. (1958). The nutritive value of fresh and roasted, California-grown Nonpareil almonds. *Journal of Agricultural and Food Chemistry, 6*(5), 377–382. https://doi.org/10.1021/jf60087a008.

Hanmoungjui, P., Pyle, D. L., & Niranjan, K. (2000). Extraction of rice bran oil using aqueous media. *Journal of Chemical Technology and Biotechnology, 75*(5), 348–352. https://doi.org/10.1002/(SICI)1097-4660(200005)75:5<348::AID-JCTB233>3.0.CO;2-P.

Harris, L. J. (2013). *Improving the safety and quality of nuts*. Cambridge, UK: Woodhead Publishing Limited.

Holcapek, M., Lisa, M., Jandera, P., & Kabátová, N. (2005). Quantitation of triacylglycerols in plant oils using HPLC with APCI-MS, evaporative light-scattering, and UV detection. *Journal of Separation Science, 28*(12), 1315–1333. https://doi.org/10.1002/jssc.200500088.

Hou, L. X., Shang, X. L., Wang, X., & Liu, J. (2013). Application of enzyme in aqueous extraction of sesame oil. *European Food Research and Technology, 236*(6), 1027–1030. https://doi.org/10.1007/s00217-013-1955-4.

Houmy, N., Mansouri, F., Benmoumen, A., Elmouden, S., Boujnah, M., Sindic, M., Fauconnier, M. L., Serghini-Caid, H., & Elamrani, A. (2016). Characterization of almond kernel oils of five almonds varieties cultivated in Eastern Morocco. *Options Méditerranéennes: Série A. Séminaires Méditerranéens, 119*, 317–321. http://om.ciheam.org/om/pdf/a119/00007414.pdf.

Hyson, D. A., Schneeman, B. O., & Davis, P. A. (2002). Almonds and almond oil have similar effects on plasma lipids and ldl oxidation in healthy men and women. *The Journal of Nutrition, 132*(4), 703–707. https://doi.org/10.1093/jn/132.4.703.

Imani, A., Hadadi, A., Amini, S. H., Vaeizi, M., & Jolfaei, B. (2012). The effect of genotype and year on the average percentage of oil seed content of almond. *International Journal of Nuts and Related Sciences, 3*(1), 37–40. http://ijnrs.damghaniau.ac.ir/article_515729_47cd7997e366922671f6e383700b73ee.pdf.

International Nut and Dried Fruit Council. (2018). https://www.nutfruit.org/what-we-do/publications/technical-resources.

Jones, W. H. S. (1967). *Hippocrates* (Vol. IV). Cambridge, MA: Harvard University Press.

Kamal-Eldin, A., & Moreau, R. A. (2009). Tree nut oils. In R. A. Moreau & A. K. E. Urbana (Eds.), *Gourmet and health-promoting specialty oils* (pp. 126–149). Urbana: AOCS Press.

Karaman, S., Karasu, S., Tornuk, F., Toker, O., Geçgel, U., & Sagdic, O. (2015). Recovery potential of cold press byproducts obtained from the edible oil industry: Physicochemical, bioactive, and antimicrobial properties. *Journal of Agricultural and Food Chemistry, 63*(8), 2305–2313. https://doi.org/10.1021/jf504390t.

Karatay, H., Şahin, A., Yılmaz, Ö., & Aslan, A. (2014). Major fatty acids composition of 32 almond (*Prunus dulcis* (Mill.) D.A. Webb) genotypes distributed in east and southeast of Anatolia. *Turkish Journal of Biochemistry, 39*(3), 307–316. http://www.turkjbiochem.com/2014/307-316.pdf.

Kaska, N., Kafkas, S., Padulosi, S., Wassimi, N., & Ak, B. E. (2006). Characterization of nut species of Afghanistan: I–Almond. *Acta Horticulturae, 726*, 147–155. https://www.actahort.org/books/726/726_23.htm.

Kazantzis, I., Nanos, G. D., & Stavroulakis, G. G. (2003). Effect of harvest time and storage conditions on almond kernel oil and sugar composition. *Journal of Science and Food Agriculture, 83*(4), 354–359. https://doi.org/10.1002/jsfa.1312.

Kester, D. E., Cunningham, S., & Kader, A. A. (1993). Almonds. In R. Macrae, R. K. Robinson, & M. J. Sadler (Eds.), *Encyclopedia of food science, food technology and nutrition* (pp. 121–126). London: Academic.

Kiani, S., Rajabpoor, S., Sorkheh, K., & Ercisli, S. (2015). Evaluation of seed quality and oil parameters in native Iranian almond (*Prunus* L. spp.) species. *Journal of Forest Research, 26*(1), 115–122. https://doi.org/10.1007/s11676-014-0009-5.

Kirbaşlar, F. G., Türker, G., Özsoy-Güneş, Z., Ünal, M., Dülger, B., Ertaş, E., & Kizilkaya, B. (2012). Evaluation of fatty acid composition, antioxidant and antimicrobial activity, mineral composition and calorie values of some nuts and seeds from Turkey. *Records of Natural Products, 6*(4), 339–349. http://www.acgpubs.org/RNP/2012/Volume%206/Issue%201/48-RNP-1103-529.pdf.

Kodad, O. (2017). Chemical composition of almond nuts. In R. Socias I Company & T. Gradziel (Eds.), *Almonds: Botany, production and uses* (pp. 428–448). Wallingford: CABI.

Kodad, O., & Socias i Company, R. (2008). Variability of oil content and of major fatty acid composition in almond (*Prunus amygdalus* Batsch) and its relationship with kernel quality. *Journal of Agricultural and Food Chemistry, 56*(11), 4096–4101. https://doi.org/10.1021/jf8001679.

Kodad, O., Socias I Company, R., Prats, M. S., & Lopez Ortiz, M. C. (2006). Variability in tocopherol concentrations in almond oil and its use as a selection criterion in almond breeding. *The Journal of Horticultural Science and Biotechnology, 81*(3), 501–507. https://doi.org/10.1080/14620316.2006.11512094.

Kodad, O., Estopanan, G., Juan, T., Mamouni, A., & Socias i Company, R. (2011a). Tocopherol concentration in almond oil: Genetic variation and environmental effects under warm conditions. *Journal of Agricultural and Food Chemistry, 59*(11), 6137–6141. https://doi.org/10.1021/jf200323c.

Kodad, O., Alonso, J. M., Espiau, M. T., Estopanan, G., Juan, T., & Socias I Company, R. (2011b). Chemometric characterization of almond germplasm: Compositional aspects involved in quality and breeding. *Journal of the American Society for Horticultural Science, 136*(4), 273–281. http://journal.ashspublications.org/content/136/4/273.full.pdf+html.

Kodad, O., Estopañán, G., Juan, T., & Socias I Company, R. (2013). Protein content and oil composition of almond from Moroccan seedlings: Genetic diversity, oil quality and geographical origin. *Journal of the American Oil Chemists' Society, 90*(2), 243–252. https://doi.org/10.1007/s11746-012-2166-z.

Kodad, O., Estopañán, G., Juan, T., & Socias i Company, R. (2014a). Tocopherol concentration in almond oil from Moroccan seedlings: Geographical origin and post-harvest implications. *Journal of Food Composition and Analysis, 33*(2), 161–165. https://doi.org/10.1016/j.jfca.2013.12.010.

Kodad, O., Estopanán, G., Juan, T., Alonso, J. M., Espiau, M. T., & Socias i Company, R. (2014b). Oil content, fatty acid composition and tocopherol concentration in the Spanish almond genebank collection. *Scientia Horticulturae, 177*, 99–107 https://doi.org/10.1016/j.scienta.2014.07.045.

Kodad, O., Fernández-Cuesta, A., Karima, B., Velasco, L., Ercişli, S., & Socias I Company, R. (2015). Natural variability in phytosterols in almond (*Prunus amygdalus*) trees growing under a southern Mediterranean climate. *The Journal of Horticultural Science and Biotechnology, 90*(5), 543–549. https://doi.org/10.1080/14620316.2015.11668712.

Kodad, O., Socias i Company, R., & Alonso, J. M. (2018). Genotypic and environmental effects on tocopherol content in almond. *Antioxidants, 7*(1), 6. https://doi.org/10.3390/antiox7010006.

Korekar, G., Stobdan, T., & Arora, R. (2011). Antioxidant capacity and phenolics content of apricot (*Prunus armeniaca* L.) kernel as a function of genotype. *Plant Food for Human Nutrition, 66*(4), 376–383. https://doi.org/10.1007/s11130-011-0246-0.

Koriyama, H., Watanabe, S., Nakaya, T., Shigemori, I., Kita, M., Yoshida, N., Masaki, D., Tadai, T., Ozasa, K., Fukui, K., & Imanishi, J. (2005). Immunological and psychological benefits of aromatherapy massage. *Evidence-Based Complementary and Alternative Medicine, 2*(2), 179–184. https://doi.org/10.1093/ecam/neh087.

Kornsteiner, M., Wagner, K. H., & Elmadfa, I. (2006). Tocopherols and total phenolics in 10 different nut types. *Food Chemistry, 98*(2), 381–387. https://doi.org/10.1016/j.foodchem.2005.07.033.

Lee, J., Zhang, G., Wood, E., Castillo, C. R., & Mitchell, A. E. (2013). Quantification of amygdalin in nonbitter, semibitter, and bitter almonds (*Prunus dulcis*) by UHPLC-(ESI)QqQ MS/MS. *Journal of Agricultural and Food Chemistry, 61*(32), 7754–7759. https://doi.org/10.1021/jf402295u.

Lenhard, B. H. (1990). Phenol almond oil for sclerosing of hemorrhoids. *Hautarzt, 41*(12), 699. [in German].

Leo, L., Rescio, L., Ciurlia, L., & Zacheo, G. (2005). Supercritical carbon dioxide extraction of oil and α -tocopherol from almond seeds. *Journal of the Science of Food and Agriculture, 85*(13), 2167–2174. https://doi.org/10.1002/jsfa.2244.

Li, S. C., Liu, Y. H., Liu, J. F., Chang, W. H., Chen, C. M., & Chen, C. Y. (2011). Almond consumption improved glycemic control and lipid profiles in patients with type 2 diabetes mellitus. *Metabolism, Clinical and Experimental, 60*(4), 474–479. https://doi.org/10.1016/j.metabol.2010.04.009.

Li, Y., Zhang, Y., Sui, X., Zhang, Y., Feng, H., & Jiang, L. (2013). Ultrasound-assisted aqueous enzymatic extraction of oil from perilla (*Perilla frutescens* L.) seeds. *CyTA – Journal of Food, 12*(1), 16–21. https://doi.org/10.1080/19476337.2013.782070.

Li, S., Geng, F., Wang, P., Lu, J., & Ma, M. (2016). Proteome analysis of the almond kernel (*Prunus dulcis*). *Journal of the Science of Food and Agriculture, 96*(10), 3351–3357. https://doi.org/10.1002/jsfa.7514.

Liu, L., Yu, X., Zhao, Z., Xua, L., & Zhanga, R. (2017). Efficient salt-aided aqueous extraction of bitter almond oil. *Journal of the Science of Food and Agriculture, 97*(11), 3814–3821. https://doi.org/10.1002/jsfa.8245.

López-Ortiz, C. M., Prats-Moya, S., Beltrán Sanahuja, A., Maestre-Pérez, S. E., Grané - Teruel, N., & Martín-Carratalá, M. L. (2008). Comparative study of tocopherol homologue content in four almond oil cultivars during two consecutive years. *Journal of Food Composition and Analysis, 21*(2), 144–151. https://doi.org/10.1016/j.jfca.2007.09.004.

Madawala, S. R. P., Kochhar, S. P., & Dutta, P. C. (2012). Lipid components and oxidative status of selected specialty oils. *Grasas y Aceites, 63*(2), 143–151. https://doi.org/10.3989/gya.083811.

Maestri, D., Martínez, M., Bodoira, R., Rossi, Y., Oviedo, A., Pierantozzi, P., & Torres, M. (2015). Variability in almond oil chemical traits from traditional cultivars and native genetic resources from Argentina. *Food Chemistry, 170*, 55–61. https://doi.org/10.1016/j.foodchem.2014.08.073.

Maguire, L. S., O'Sullivan, S. M., Galvin, K., O'Connor, T. P., & O'Brien, N. M. (2004). Fatty acid profile, tocopherol, squalene and phytosterol content of walnuts, almonds, peanuts, hazelnuts and the macadamia nut. *International Journal of Food Sciences and Nutrition, 55*(3), 171–178. https://doi.org/10.1080/09637480410001725175.

Malakouti, J., Khalili, A. F., & Kamrani, A. (2017). Sesame, sweet almond & sesame and sweet almond oil for the prevention of striae in primiparous females: A triple-blind randomized controlled trial. *Iranian Red Crescent Medical Journal, 19*(6), e33672. https://doi.org/10.5812/ircmj.33672.

Malisiova, F., Hatziantoniou, S., Dimas, K., Kletstas, D., & Demetzos, C. (2004). Liposomal formulations from phospholipids of Greek almond oil. Properties and biological activity. *Zeitschrift für Naturforschung. Section C, 59*(5–6), 330–334.

Manach, C., Scalbert, A., Morand, C., Remesy, C., & Jimenez, L. (2004). Polyphenols: Food sources and bioavailability. *The American Journal of Clinical Nutrition, 79*(5), 727–747. https://doi.org/10.1093/ajcn/79.5.727.

Mandalari, G., & Faulks, R. (2008). Release of protein, lipid, and vitamin E from almond seeds during digestion. *Journal of Agricultural and Food Chemistry, 56*(9), 3409–3416. https://doi.org/10.1021/jf073393v.

Mandalari, G., Tomaino, A., Arcoraci, T., Martorana, M., LoTurco, V., Cacciola, F., Rich, G. T., Bisignano, C., Saija, A., Dugo, P., Cross, K. L., Parker, M. L., Waldron, K. W., & Wickham, M. S. J. (2010). Characterization of polyphenols, lipids and dietary fibre from almond skins (*Amygdalus communis* L.). *Journal of Food Composition and Analysis, 23*(2), 166–174. https://doi.org/10.1016/j.jfca.2009.08.015.

Martín-Carratalá, M. L., Llorens-Jordá, C., Berenguer-Navarro, V., & Grané-Teruel, N. (1999). Comparative study on the triglyceride composition of almond kernel oil. a new basis for cultivar chemometric characterization. *Journal of Agricultural and Food Chemistry, 47*(9), 3688–3692. https://doi.org/10.1021/jf981220n.

Martínez, M. L., Mattea, M. A., & Maestri, D. M. (2008). Pressing and supercritical carbon dioxide extraction of walnut oil. *Journal of Food Engineering, 88*(3), 399–404. https://doi.org/10.1016/j.jfoodeng.2008.02.026.

Martínez, M. L., Penci, M. C., Marin, M. A., Pablo, D., Ribotta, P. D., & Maestri, D. M. (2013). Screw press extraction of almond (*Prunus dulcis* (Miller) D.A. Webb): Oil recovery and oxidative stability. *Journal of Food Engineering, 119*(1), 40–45. https://doi.org/10.1016/j.jfoodeng.2013.05.010.

Mateus Martins, I., Chen, Q., Chen, C. Y. O. (2017). Emerging functional foods derived from almonds. In: I. Ferreira, L. Barros, P. Morales (Eds.), *Wild plants, mushrooms, and nuts: Functional food properties and applications* (pp. 445–469). Wiley. Chichester, West Sussex, PO19 8SQ, United Kingdom.

Matos, A., & Acuña, J. (2010). Influencia del tiempo, tamaño de partícula y proporción sólido-líquido en la extracción de aceite crudo de la almendra de durazno (*Prunus persica*). *Review Investigational Ciência e Tecnologia de Alimentos, 1*, 1–6.

Matthäus, B., & Özcan, M. M. (2009). Fatty acids and tocopherol contents of some *Prunus* spp. kernel oils. *Journal of Food Lipids, 16*(2), 187–199. https://doi.org/10.1111/j.1745-4522.2009.01140.x.

Matthäus, B., Özcan, M. M., Juhaimi, F. A., Adiamo, O. Q., Alsawmahi, O. N., Ghafoor, K., & Babiker, E. E. (2018). Effect of the harvest time on oil yield, fatty acid, tocopherol and sterol contents of developing almond and walnut kernels. *Journal of Oleo Science, 67*(1), 39–45. https://doi.org/10.5650/jos.ess17162.

Mehran, M., & Filsoof, M. (1974). Characteristics of Iranian almond nuts and oils. *Journal of the American Oil Chemists' Society, 51*(10), 433–434. https://doi.org/10.1007/BF02635147.

Mexis, S. F., Badeka, A. V., Chouliara, E., Riganakos, K. A., & Kontominas, M. G. (2009). Effect of γ-irradiation on the physicochemical and sensory properties of raw unpeeled almond kernels (*Prunus dulcis*). *Innovative Food Science and Emerging Technologies, 10*(1), 87–92. https://doi.org/10.1016/j.ifset.2008.09.001.

Micke, W. C., & Kester, D. E. (1998). Almond growing in California. *Acta Horticulturae, 470*, 21–28. https://doi.org/10.17660/ActaHortic.1998.470.1.

Milbury, P. E., Chen, C. Y., Dolnikowski, G. G., & Blumberg, J. B. (2006). Determination of flavonoids and phenolics and their distribution in almonds. *Journal of Agricultural and Food Chemistry, 54*(14), 5027–5033. https://doi.org/10.1021/jf0603937.

Miraliakbari, H., & Shahidi, F. (2008a). Lipid class compositions, tocopherols and sterols of tree nut oils extracted with different solvents. *Journal of Food Lipids, 15*(1), 81–96. https://doi. org/10.1111/j.1745-4522.2007.00104.x.

Miraliakbari, H., & Shahidi, F. (2008b). Antioxidant activity of minor components of tree nut oils. *Food Chemistry, 111*(2), 421–427. https://doi.org/10.1016/j.foodchem.2008.04.008.

Monagas, M., Garrido, I., Lebrón-Aguilar, R., Bartolome, B., & Gómez-Cordovés, C. (2007). Almond (*Prunus dulcis* (Mill.) D.A. Webb) skins as a potential source of bioactive polyphenols. *Journal of Agricultural and Food Chemistry, 55*(21), 8498–8507. https://doi.org/10.1021/jf071780z.

Mori, A. M., Considine, R. V., & Mattes, R. D. (2011). Acute and second-meal effects of almond form in impaired glucose tolerant adults: A randomized crossover trial. *Nutrition and Metabolism, 8*(1), 6. https://doi.org/10.1186/1743-7075-8-6.

Muhammad, S., Sanden, B. L., Lampinen, B. D., Saa, S., Siddiqui, M. I., & Smart, D. R. (2015). Seasonal changes in nutrient content and concentrations in a mature deciduous tree species: Studies in almond (*Prunus dulcis* (Mill.) DA Webb). *European Journal of Agronomy, 65*, 52–68. https://doi.org/10.1016/j.eja.2015.01.004.

Nanos, G. D., Kazantzis, I., Kefalas, P., Petrakis, C., & Stavroulakis, G. G. (2002). Irrigation and harvest time affect almond kernel quality and composition. *Scientia Horticulturae, 96*(1–4), 249–256. https://doi.org/10.1016/S0304-4238(02)00078-X.

Oliveira, I., Meyer, A., Afonso, S., Ribeiro, C., & Goncalves, B. (2018). Morphological, mechanical and antioxidant properties of Portuguese almond cultivars. *Journal of Food Science and Technology, 55*(2), 467–478. https://doi.org/10.1007/s13197-017-2955-3.

Ostlund, R. E., Jr. (2002). Phytosterols in human nutrition. *Annual Review of Nutrition, 22*, 533–549. https://doi.org/10.1146/annurev.nutr.22.020702.075220.

Özcan, M. M., Ünver, A., Erkan, E., & Arslan, D. (2011). Characteristics of some almond kernel and oils. *Scientia Horticulturae, 127*(3), 330–333. https://doi.org/10.1016/j.scienta.2010.10.027.

Panadare, D. C., & Rathod, V. K. (2017). Three phase partitioning for extraction of oil: A review. *Trends in Food Science and Technology, 68*, 145–151. https://doi.org/10.1016/j.tifs.2017.08.004.

Pasini, F., Riciputi, Y., Verardo, V., & Caboni, M. F. (2013). Phospholipids in cereals, nuts and some selected oilseeds. *Recent Research Development Lipids, 9*, 139–201. https://pdfs.semanticscholar.org/2fe3/a31213da45eeed3fc782f71b8c710c394610. pdf?_ga=2.129723084.1563405966.1527686296-2059612530.1527588704.

Perez, A. G., Sanz, C., Olias, R., & Olias, J. M. (1999). Lipoxygenase and hydroperoxide lyase activities in ripening strawberry fruits. *Journal of Agricultural and Food Chemistry, 47*(1), 249–253. https://doi.org/10.1021/jf9807519.

Pićurić-Jovanović, K., & Milovanović, M. (1993). Analysis of volatile compounds in almond and plum kernel oils. *Journal of the American Oil Chemists' Society, 70*(11), 1101–1104. https://doi.org/10.1007/BF02632149.

Piironen, V., Syväoja, E. L., Varo, P., Salminen, K., & Koivistoinen, P. (1986). Tocopherols and tocotrienols in Finnish foods: Vegetables, fruits, and berries. *Journal of Agricultural and Food Chemistry, 34*(4), 742–746. https://doi.org/10.1021/jf00070a038.

Pineli, L. L. O., Carvalho, M. V., Aguiar, L. A., Oliveira, G. T., Celestino, S. M. C., Botelho, R. B. A., & Chiarello, M. D. (2015). Use of baru (Brazilian almond) waste from physical extraction of oil to produce flour and cookies. *LWT - Food Science and Technology, 60*(1), 50–55. https://doi.org/10.1016/j.lwt.2014.09.035.

Piscopo, A., Romeo, F., Petrovicova, B., & Poiana, M. (2010). Effect of the harvest time on kernel quality of several almond varieties (*Prunus dulcis* (Mill.) D.A. Webb). *Scientia Horticulturae, 125*(1), 41–46. https://doi.org/10.1016/j.scienta.2010.02.015.

Piscopo, A., Romeo, F. V., & Poiana, M. (2011). Effect of drying process on almond (*Prunus dulcis* (Mill.) D.A. Webb) kernel composition. *Rivista Italiana Delle Sostanze Grasse, 88*(3), 153–160.

Prats-Moya, S., Grané-Teruel, N., Berenguer-Navarro, V., & Martín-Carratalá, M. L. (1999). A chemometric study of genotypic variation in triacylglycerol composition among selected almond cultivars. *Journal of the American Oil Chemists' Society, 76*(2), 267–272. https://doi. org/10.1007/s11746-999-0229-6.

Rabadán, A., Álvarez-Ortí, M., Gómez, R., Pardo-Giménez, A., & Pardo, J. E. (2017). Suitability of Spanish almond cultivars for the industrial production of almond oil and defatted flour. *Scientia Horticulturae, 225*, 539–546. https://doi.org/10.1016/j.scienta.2017.07.051.

Remaud, G., Debon, A. A., Martin Martin, Y., Martin, G. G., & Martin, G. J. (1997). Authentication of bitter almond oil and cinnamon oil: Application of the SNIF-NMR method to benzaldehyde. *Journal of Agricultural and Food Chemistry, 45*(10), 4042–4048. https://doi.org/10.1021/jf970143d.

Rizzolo, A., Baldo, C., & Polesello, A. (1991). Application of high-performance liquid chromatography to the analysis of niacin and biotin in Italian almond cultivars. *Journal of Chromatography, 553*, 187–192. https://doi.org/10.1016/S0021-9673(01)88487-9.

Robbins, K. S., Shin, E. C., Shewfelt, R. L., Eitenmiller, R. R., & Pegg, R. B. (2011). Update on the healthful lipid constituents of commercially important tree nuts. *Journal of Agricultural and Food Chemistry, 59*(22), 12083–12092. https://doi.org/10.1021/jf203187v.

Romojaro, F., Riquelme, F., Giménez, J. L., & Llorente, S. (1988). Study on carbohydrate fractions in some almonds cultivars of the Spanish south-east. *Fruit Science Reports, 15*, 1–6.

Roncero, J. M., Álvarez-Ortí, M., Pardo-Giménez, A., Gómez, R., Rabadán, A., & Pardo, J. E. (2016). Virgin almond oil: Extraction methods and composition. *Grasas y Aceites, 67*(3), e143. https://doi.org/10.3989/gya.0993152.

Rosengarten, F. J. (1984). *The book of edible nuts.* New York: Walker and Co.

Ruggeri, S., Cappelloni, M., Gambelli, L., Nicoli, S., & Carnovale, E. (1998). Chemical composition and nutritive value of nuts grown in Italy. *Italian Journal of Food Science, 10*(3), 243–251.

Sabudak, T. (2007). Fatty acid composition of seed and leaf oils of pumpkin, walnut, almond, maize, sunflower and melon. *Chemistry of Natural Compounds, 43*(4), 465–467. https://doi. org/10.1007/s10600-007-0163-5.

Sahad, N., Md Som, A., & Sulaiman, A. (2014). Review of green solvents for oil extraction from natural products using different extraction methods. *Applied Mechanics and Materials, 661*, 58–62. https://doi.org/10.4028/www.scientific.net/AMM.661.58.

Sakar, E. H., El Yamani, M., & Rharrabti, Y. (2017). Variability of oil content and its physico-chemical traits from five almond (*Prunis dulcis*) cultivars grown in Northern Morocco. *Journal of Materials and Environmental Science, 8*(8), 2679–2686. https://www.jmaterenvironsci.com/Document/vol8/vol8_N8/287-JMES-Sakar.pdf.

Salvo, F., Alfa, M., & Dugo, G. (1986). Variation de l'indice de peroxyde, des indices spectrométriques, de la composition en acides gras et stérols. *Rivista Italiana Delle Sostanze Grasse, 63*, 37–40.

Sánchez-Bel, P., Egea, I., Martínez-Madrid, M. C., Flores, B., & Romojaro, F. (2008). Influence of irrigation and organic/inorganic fertilization on chemical quality of almond (*Prunus amygdalus* cv. Guara). *Journal of Agricultural and Food Chemistry, 56*(21), 10056–10062. https://doi.org/10.1021/jf8012212.

Sanchez-Prado, L., Risticevic, S., Pawliszyn, J., & Psillakis, E. (2009). Low temperature SPME device: A convenient and effective tool for investigating photodegradation of volatile analytes. *Journal of Photochemistry and Photobiology A, 206*(2–3), 227–230. https://doi.org/10.1016/j.jphotochem.2009.07.009.

Sang, S., Lapsley, K., Li, G., Jeong, W. S., Lachance, P. A., Ho, C. T., & Rosen, R. T. (2002a). Antioxidative phenolic compounds isolated from almond skins (*Prunus amygdalus* Batsch). *Journal of Agricultural and Food Chemistry, 50*(8), 2459–2463. https://doi.org/10.1021/jf011533+.

Sang, S., Kikuzaki, H., Lapsley, K., Rosen, R. T., Nakatani, N., & Ho, C. T. (2002b). Sphingolipid and other constituents from almond nuts (*Prunus anygdalus* Batsch). *Journal of Agricultural and Food Chemistry, 50*(16), 4709–4712. https://doi.org/10.1021/jf020262f.

Sang, S., Li, G., Tian, S., Lapsley, K., Stark, R. E., Pandey, R. K., Rosen, R. T., & Ho, C. T. (2003). An unusual diterpene glycoside from the nuts of almond (*Prunus amygdalus* Batsch). *Tetrahedron Letters, 44*(6), 1199–1202. https://doi.org/10.1016/S0040-4039(02)02794-6.

Sarkis, J. R., Correa, A. P., Michel, I., Brandeli, A. C. I., & Tessaro, I. C. (2014). Evaluation of the phenolic content and antioxidant activity of different seed and nut cakes from the edible oil industry. *Journal of the American Oil Chemists' Society, 91*(10), 1773–1782. https://doi. org/10.1007/s11746-014-2514-2.

Sathe, S. K., Seeram, H. H., Kshirsagar, D., & Lapsley, K. A. (2008). Fatty acid composition of California grown almonds. *Journal of Food Science, 73*(9), 607–614. https://doi. org/10.1111/j.1750-3841.2008.00936.x.

Schirra, M., & Agabbio, M. (1989). Influence of irrigation on keeping quality of almond kernels. *Journal of Food Science, 54*(6), 1642–1645. https://doi.org/10.1111/j.1365-2621.1989. tb05178.x.

Senesi, E., Rizzolo, A., Colombo, C., & Testoni, A. (1996). Influence of pre-processing storage conditions on peeled almond quality. *Italian Journal of Food Science, 2*, 115–125.

Senter, S. D., Horvat, R. J., & Forbus, W. R. (1983). Comparative GLC-MS analysis of phenolic acids of selected tree nuts. *Journal of Food Science, 48*(3), 798–799. https://doi. org/10.1111/j.1365-2621.1983.tb14902.x.

Sharma, A., & Gupta, M. N. (2004). Oil extraction from almond, apricot and rice bran by three-phase partitioning after ultrasonication. *European Journal of Lipid Science and Technology, 106*(3), 183–186. https://doi.org/10.1002/ejlt.200300897.

Socias i Company, R., Kodad, O., Alonso, J. M., & Gradziel, T. M. (2008). Almond quality: A breeding perspective. In J. Janick (Ed.), *Horticultural reviews* (pp. 197–238). Hoboken: Wiley. https://doi.org/10.1002/9780470380147.ch3.

Socias I Company, R., Alonso, J. M., Kodad, O., Espada, J. L., & Andreu, J. (2014). Kernel quality of local Spanish almond cultivars: Provenance variability and end uses. *Nucis, 16*, 16–19.

Soden, K., Vincent, K., Craske, S., Lucas, C., & Ashley, S. (2004). A randomized controlled trial of aromatherapy massage in a hospice setting. *Palliative Medicine, 18*(2), 87–92. https://doi. org/10.1191/0269216304pm874oa.

Soler, L., Canellas, J., & Saura-Calixto, F. (1988). Oil content and fatty acid composition of developing almond seeds. *Journal of Agricultural and Food Chemistry, 36*(4), 695–697. https://doi. org/10.1021/jf00082a007.

Stuetz, W., Schlörmann, W., & Glei, M. (2017). B-vitamins, carotenoids and α-/γ-tocopherol in raw and roasted nuts. *Food Chemistry, 221*, 222–227. https://doi.org/10.1016/j.foodchcm.2016.10.065.

Sultana, Y., Kohli, K., Athar, M., Khar, R. K., & Aqil, M. (2007). Effect of pre-treatment of almond oil on ultraviolet B-induced cutaneous photoaging in mice. *Journal of Cosmetic Dermatology, 6*(1), 14–19. https://doi.org/10.1111/j.1473-2165.2007.00293.x.

Tan, Z., Yang, Z., Yi, Y., Wang, H., Zhou, W., Li, F., et al. (2016). Extraction of oil from flaxseed (*Linum usitatissimum* L.) using enzyme-assisted three-phase partitioning. *Applied Biochemistry and Biotechnology, 179*(8), 1325–1335. https://doi.org/10.1007/s12010-016-2068-x.

Taşhan, T. S., & Kafkasli, A. (2012). The effect of bitter almond oil and massaging on striae gravidarum in primiparous women. *Journal of Clinical Nursing, 21*(11–12), 1570–1576. https://doi. org/10.1111/j.1365-2702.2012.04087.x.

Tiên, D. T. K., Hadji-Minaglou, F., Antoniotti, S., & Fernandez, X. (2015). Authenticity of essential oils. *Trends in Analytical Chemistry, 66*, 146–157. https://doi.org/10.1016/j. trac.2014.10.007.

Urpi-Sarda, M., Garrido, I., Monagas, M., Gómez - Cordovés, C., Medina-Remón, A., Andres-Lacueva, C., & Bartolomé, B. (2009). Profile of plasma and urine metabolites after the intake of almond (*Prunus dulcis* (Mill.) D.A. Webb) polyphenols in humans. *Journal of Agricultural and Food Chemistry, 57*(21), 10134–10142. https://doi.org/10.1021/jf901450z.

Vazquez Araujo, L., Enguix, L., Verdú, A., García García, E., & Carbonell Barrachina, A. (2008). Investigation of aromatic compounds in toasted almonds used for the manufacture of turrón. *European Food Research and Technology, 227*(1), 243–254. https://doi.org/10.1007/ s00217-007-0717-6.

Venkatachalam, M., & Sathe, S. K. (2006). Chemical composition of selected edible nut seeds. *Journal of Agricultural and Food Chemistry, 54*(13), 4705–4714. https://doi.org/10.1021/ jf0606959.

Vidhate, G. S., & Singhal, R. S. (2013). Extraction of cocoa butter alternative from kokum (*Garcinia indica*) kernel by three phase partitioning. *Journal of Food Engineering, 117*(4), 464–466. https://doi.org/10.1016/j.jfoodeng.2012.10.051.

Wien, M., Bleich, D., Raghuwanshi, M., Gould-Forgerite, S., Gomes, J., Monahan-Couch, L., & Oda, K. (2010). Almond consumption and cardiovascular risk factors in adults with prediabetes. *Journal of the American College of Nutrition, 29*(3), 189–197. https://pdfs.semanticscholar.org/23ff/7cb5723ad8485a507c63c2d3f5fe62ec6f4a.pdf?_ga=2.103272256.1164948374.1527717399-2132208873.1525105899.

Wijeratne, S. S. K., Amarowicz, R., & Shahidi, F. (2006a). Antioxidant activity of almonds and their by-products in food model systems. *Journal of the American Oil Chemists' Society, 83*, 223–230. https://doi.org/10.1007/s11746-006-1197-8.

Wijeratne, S. S. K., Abou-Zaid, M. M., & Shahidi, F. (2006b). Antioxidant polyphenols in almond and its coproducts. *Journal of Agricultural and Food Chemistry, 54*(2), 312–318. https://doi.org/10.1021/jf051692j.

Xie, L., & Bolling, B. W. (2014). Characterisation of stilbenes in California almonds (*Prunus dulcis*) by UHPLC-MS. *Food Chemistry, 148*, 300–306. https://doi.org/10.1016/j.foodchem.2013.10.057.

Xie, L., Roto, A. V., & Bolling, B. W. (2012). Characterization of ellagitannins, gallotannins, and bound proanthocyanidins from California almond (*Prunus dulcis*) varieties. *Journal of Agricultural and Food Chemistry, 60*(49), 12151–12156. https://doi.org/10.1021/jf303673r.

Yada, S., Lapsley, K., & Huang, G. (2011). A review of composition studies of cultivated almonds: Macronutrients and micronutrients. *Journal of Food Composition and Analysis, 24*(4–5), 469–480. https://doi.org/10.1016/j.jfca.2011.01.007.

Yada, S., Huang, G., & Lapsley, K. (2013). Natural variability in the nutrient composition of California-grown almonds. *Journal of Food Composition and Analysis, 30*(2), 80–85. https://doi.org/10.1016/j.jfca.2013.01.008.

Yang, J. (2009). Brazil nuts and associated health benefits: A review. *LWT - Food Science and Technology, 42*(10), 1573–1580. https://doi.org/10.1016/j.lwt.2009.05.019.

Yildirim, A. N., Yildirim, F., Şan, B., Polat, M., & Sesli, Y. (2016). Variability of phenolic composition and tocopherol content of some commercial almond cultivars. *Journal of Applied Botany and Food Quality, 89*, 163–170. https://doi.org/10.5073/JABFQ.2016.089.020.

Zacheo, G., Cappello, M. S., Gallo, A., Santino, A., & Cappello, A. R. (2000). Changes associated with post-harvest ageing in almond seeds. *Lebensmittel-Wissenschaft und -Technologie - Food Science and Technology, 33*(6), 415–423. https://doi.org/10.1006/fstl.2000.0679.

Zamany, A. J., Samadi, G. R., Kim, D. H., Keum, Y. S., & Saini, R. K. (2017). Comparative study of tocopherol contents and fatty acids composition in twenty almond cultivars of Afghanistan. *Journal of the American Oil Chemists' Society, 94*(6), 805–817. https://doi.org/10.1007/s11746-017-2989-8.

Zhu, Y. (2014). Almond (*Prunus dulcis* (Mill.) D.A. Webb) fatty acids and tocopherols under different conditions. Ph.D. Thesis, University of Adelaide, Adelaide, Australia.

Zhu, Y., Wilkinson, K. L., & Wirthensohn, M. G. (2015). Lipophilic antioxidant content of almonds (*Prunus dulcis*): A regional and varietal study. *Journal of Food Composition and Analysis, 39*, 120–127. https://doi.org/10.1016/j.jfca.2014.12.003.

Zhu, Y., Wilkinson, K., & Wirthensohn, M. (2017). Changes in fatty acid and tocopherol content during almond (*Prunus dulcis* cv. Nonpareil) kernel development. *Scientia Horticulturae, 225*, 150–155. https://doi.org/10.1016/j.scienta.2017.07.008.

Zlatanov, M., Ivanov, S., & Aitzetmueller, K. (1999). Phospholipid and fatty acid composition of Bulgarian nut oils. *European Journal of Lipid Science and Technology, 101*(11), 437–439. https://doi.org/10.1002/(SICI)1521-4133(199911)101:11<437::AID-LIPI437>3.0.CO;2-T.

Zohary, D., & Hopf, M. (2000). *Domestication of plants in the old world* (3rd ed., p. 186). London: Oxford University Press.

Chapter 7
Virgin Pistachio (*Pistachia vera* L.) Oil

Maria Desamparados Salvador, Rosa M. Ojeda-Amador,
and Giuseppe Fregapane

Abstract Dietary consumption of nuts (e.g. walnuts, pistachios, and almonds) has recently received particular attention due to their high content of unsaturated fatty acids, antioxidants and other biologically active compounds and, thus, their potential beneficial health effects. Apart from nuts themselves, a growing interest towards virgin vegetable oils has also appeared in recent years, which goes beyond the virgin oil for excellence, the one produced from olives. The main reasons for this trend are marked by consumers that increasingly appreciate the taste and smell of oils related to the raw materials from which they come from as well as by their potential nutritional properties -resulting in novel gourmet and healthy oils- with added value for the consumers as compared to the most common industrial refined vegetable oils. The purpose of the current chapter is the analysis and discussion of the current knowledge on the extraction and processing of virgin pistachio (*Pistachia vera* L.) oil and particularly on the composition and properties of major (mainly the fatty acid) and minor components (tocols, phenolics, sterols, carotenoids, flavour and aroma compounds), which are directly related to their organoleptic and functional properties. Virgin pistachio oils are characterized by a high content in oleic acid (51–81%) -close to virgin olive oil-, linoleic acid (13–31%), phospholipids (8 g/kg), sterols (2.5–7.6 g/kg) and tocopherols (300–900 mg/kg). All those bioactive components possess known health-promoting effects, as well as a peculiar and appreciated sensory properties.

Keywords Cultivars · Chemical characterization · Bioactive compounds · Health-promoting effects · Organoleptic properties

M. D. Salvador · R. M. Ojeda-Amador · G. Fregapane (✉)
Faculty of Chemistry, Castilla-La Mancha University, Ciudad Real, Spain
e-mail: amparo.salvador@uclm.es; rosamaria.ojeda@alu.uclm.es;
giuseppe.fregapane@uclm.es

© Springer Nature Switzerland AG 2019
M. F. Ramadan (ed.), *Fruit Oils: Chemistry and Functionality*,
https://doi.org/10.1007/978-3-030-12473-1_7

Abbreviations

FA   Fatty acids
LDL   Low-density lipoprotein
MUFA   Monounsaturated FA
PUFA   Polyunsaturated FA
SFA   Saturated FA
TAG   Triacylglycerols
TPP   Total polar phenolics
VPO   Virgin pistachio oil

1   Nuts and their Oils

Pistacia spp. is a plant of the Anacardiaceae family, native in central and western Asia and estimated to be ~80 million years old. It was introduced in Europe before Roman times and later spread in several regions around the world with a Mediterranean ecosystem, like the USA and North Africa. Nowadays, *Pistacia vera* L. is cultivated as an agricultural crop in the Middle East, California and Mediterranean Europe, being in continuous expansion thanks to the presence of large geographical areas with favorable climatic conditions and to the existence of new varieties (Couceiro et al. 2017). Iran has been a major pistachio producer for 3000–4000 years and currently is one of the main growers (315,151 tons/year), alongside USA (406,646 tons/years) and Turkey (170,000 tons; FAOSTAT data 2016). In central Spain, this crop is quickly growing reaching a production of 2418 tons per year (10th world position).

Table 7.1 Composition of pistachio kernel, its virgin oil and residual cake

Pistachio kernel		Virgin pistachio oil (VPO)		Residual cake	
Moisture (%)	3–5	Lipids yield (%)	53–73	Moisture (%)	4–6
Lipids (%)	48–63	TAG (%)	92–98	Lipids (%)	20–24
Protein (%)	18–22	SFA (%)	10–16	Protein (%)	27–31
Fibre (%)	8–12	MUFA (%)	56–77	Fibre (%)	11–19
Phenolics (%)	0.6–1.0	PUFA (%)	10–31	Phenolics (%)	0.9–1.5
		Phospholipids (g/kg)	18		
		Sphingolipid (g/kg)	8.3		
		Sterols (g/kg)	2.1–7.6		
		Tocopherols (mg/kg)	300–900		
		Pigments (mg/kg)	65–110		
		Phenolics (mg/kg)	16–60		
		Volatiles (mg/kg)	16–45		

TAG triacylglycerols, *SFA* saturated fatty acids, *MUFA* monounsaturated fatty acids, *PUFA* polyunsaturated fatty acids

Nuts are considered a great source of biologically active compounds, due to their high content of essential unsaturated fatty acids and phenolic compounds (Alasalvar and Shadihi 2008; Salas-Salvadó et al. 2008; Amarowicz R et al. 2017). In July 2003, the United States Food and Drug Administration (FDA) approved the first health claim specific to nuts and the reduction of the risk of heart disease: "*scientific evidence suggests but does not prove that eating 1.5 oz (42.5 g) per day of most nuts, such as pistachios, as part of a diet low in saturated fat and cholesterol may reduce the risk of heart disease*".

Pistachios are rich in protein (18–22%), dietary fibre (8–12%) and lipids (48–63%; Table 7.1), with a balanced content of mono- (56–77%) and polyunsaturated (10–31%; Table 7.1) fatty acids, which could help the reduction of LDL-cholesterol levels and therefore the risk of coronary heart disease (Kris-Etherton et al. 2001; Amarowicz et al. 2017). Moreover, they present a high content of bioactive compounds, such as tocopherols, phytosterols and phenolic compounds (Bulló et al. 2015), being among the top 50 foods with a high antioxidant potential (Halvorsen et al. 2006).

Currently, the main uses of pistachio are as snack followed by the pastry and ice-cream industries with high requirements in terms of nut quality. However, other alternatives are being explored for pistachios that do not satisfy these high-quality specifications; and among them the production of its virgin nut oil with outstanding sensory qualities should be highlighted.

Nut oils can be extracted using mechanical systems, obtaining oils with a great organoleptic and nutritional value, or employing organic solvents that requires a downstream refining process to make them edible. However, mechanical oil extraction gives low yields resulting in an economic loss for this industry. For this reason, more than one century ago, an appropriated production technology was set up to enhance oil yields by means of solvent extraction; but as a result, virgin oils disappeared from the market, with the exception of virgin olive oil (Matthäus 2008). It was only 25 years ago when people began to be greatly concerned about consuming more health-promoting and less-processed food products, that should keep the natural composition and properties of the fresh raw materials.

With the increasing demand for novel edible oils, virgin oils obtained from nuts are receiving particular attention due to their potential nutritional properties –generally denominated as *healthy oils*– and their attractive and peculiar sensory characteristics -*gourmet oils*-, which provide added value to the consumer as compared to traditional refined vegetable oils (Kamal-Eldin and Moreau 2010). Cold-pressed oils are defined by the FAO-WHO Codex Stan 210 as natural products obtained without altering the nature of the oil, by mechanical procedures only e.g. expelling or pressing without the application of heat; they can be purified by washing with water, settling, filtering and centrifugation only. Today they are recognized as the highest quality ones that are naturally free from *trans*-fat and full of natural antioxidants; and are often produced in small-scale artisanal mills and sold in gourmet and health markets.

The virgin pistachio oil (VPO) pleasant sensory odor and taste characteristics that remember the raw nut, even with a light roasted flavor, and its green color made it very adequate for "*haute cuisine*" culinary purposes and salad dressing, as well as for bakery, sweets and confectionary (Kamal-Eldin and Moreau 2010; Rabadán et al. 2017a). Furthermore, it is also a valuable product for cosmetic care formulations, such as lotions, soaps, skin creams, lip balms, shampoos and hair conditioner (Hannon 1997; Bail et al. 2009), and also for therapeutic products with anti-inflammatory properties (Zhang et al. 2010).

Pistachio oil is not yet described by the current Codex Alimentarius on Fats and Oils (FAO-WHO), however a recent project for amending the standard for named vegetable oils Codex Stan 210 (CX/FO 15/24/11 2015) is under study and consideration with the purpose of incorporate composition standards for cold pressed oils including virgin nut oils (e.g. walnut, pistachio, hazelnut as well as avocado fruit oil). Commercial pistachio oil products, prized as a specialty oil owing to its beneficial effects on human health, are sold in several Middle Eastern and European countries, but not being yet very extended due to its high market prices.

On the other hand, it is very relevant to remark that increased attention has also been focused on the valorization of food processing by-products (Nyam et al. 2009). In this sense, the pressing process of pistachio oil extraction produces a residual partially defatted cake, leading to a disposal problem for the industry. Nevertheless, this by-product retains nutrients and bioactive compounds (Table 7.1) of great interest, which may be used as a natural source of phytochemicals and antioxidants as ingredients for functional foods as well as dietary supplements.

2 Extraction and Processing of Nut Oils

Cold pressed oils are the oldest types of edible oil consumed by the humankind. Indeed, cold pressing is the traditional and natural way to produce oil, which is squeezed out of the nut at temperatures below 50 °C that ensures that the full essence and character of the oil is preserved. No external heat is applied, wherein the heat generated only by the pressure applied and the rotational friction.

Nowadays, edible oils from nuts can be manufactured using three different systems: pressure, solvent and supercritical fluids extraction. Systems employing pressure are still the most common to obtain commercial pistachio oils, due to their easy use and the positive quality parameters of the resulted virgin oils, which are regarded as "virgin" since only mechanical procedures are used as already mentioned (FAO-WHO Codex Stan 210 1999). Despite the lower performance yield compared with solvent extraction (Catalán et al. 2017), this method requires less expensive installations and involves lower environmental risk. Two different kinds of pressure systems, expeller and hydraulic presses, are mostly used. They are both generally referred as cold extraction methods, although the screw press requires the heating of the barrel (50–60 °C) to obtain higher extraction yields (Sena-Moreno et al. 2015; Ojeda-Amador et al. 2018). Moreover, the use of heating during the extraction process plays an important role in the sensory characteristics of the VPO, highlighting their sensory properties such as color, smell and taste (Álvarez-Ortí et al. 2012; Rabadan et al. 2017a). The performance of the process is slightly higher in the screw press (~40%) than in the hydraulic press (~30%; Oseni et al. 2002; Sena-Moreno et al. 2015). Therefore, the screw press system recognized as most efficient and economic than the hydraulic one.

Solvent extraction (using mainly *n*-hexane) have been used for a long time for producing seed oils. With the purpose of enhancing oil yields, several processing parameters such as the relation between solid/solvent, granulometry and temperature have been studied and optimized (Patricelli et al. 1979). However, as already mentioned this technique requires a refining step to make edible the raw extracted oil that greatly reduces its content of valuable minor components, like antioxidants and volatiles. In the case of pistachio oils, solvent extraction has not been used for commercial purposes probably due to the great negative reduction in its aroma compounds that drastically reduces the oil quality.

Supercritical fluids extraction emerged in the mid-1980s as an alternative method to replace the conventional solvent systems and in particular to reduce the use of organic solvents (green chemistry). This method employs 'clean' solvents for the oil extraction, such as carbon dioxide, ethanol and water among others, being carbon dioxide the most commonly used because of its low cost and safety, allowing the absence of traces of organic solvent in the final product (Herrero et al. 2006). The oily seed or nut is exposed to the solvent under supercritical conditions (i.e. 31 °C and 74 bar), that dissolves the solute of interest in the supercritical fluid, which is later separated by decreasing the pressure (Nielsen 1998). Carbon dioxide in the supercritical state increases the capacity to solubilize polar compounds also due to the better penetration into the solids and improve the extraction of thermo-sensible compounds and bioactive molecules (Dunford et al. 2003). A higher yield is obtained when mechanical and supercritical techniques are combined leading to a lower residual oil content in the by-products (Martínez and Vance 2008). On the other hand, the use of ethanol (10%) combined with CO_2 at 60 °C allows the reduction of the pressure applied and the operating cost (Palazoğlu and Balaban 1998) in pistachio oils extraction.

3 Lipid Composition of Pistachio Virgin Oil

The average oil content found in Turkish and Iranian pistachio cultivars is about 55%, ranging from 48% in Avdat variety to 63% in Kastel (Agar et al. 1998) giving an extraction yield using a screw-press from 53% (in Avdat), to 66–68% (Larnaka, Kerman and Mateur) and up to 72–73% (Kastel and Aegina; Ojeda-Amador et al. 2018).

Several studies have reported the fatty acid (FA) composition of pistachio oils, as depicted in Table 7.2, which shows a high content of oleic acid from 51% to 81%, close to olive oil (FAO-WHO Codex Stan 33-1981) and higher than many other seed oils (FAO-WHO Codex Stan 210-1999). Kerman cultivar presented the lowest amount in this healthy fatty acid, highlighting, on the other hand, Sridique cultivar

Table 7.2 Fatty acid and triacylglycerides composition (%) of pistachio oil

P-C16:0	Po-C16:1	S-C18:0	O-C18:1	L-C18:2	Ln-C18:3	Reference
Virgin oil						
13	2.0	2.8	51	30	0.6	Givianrad et al. (2013)
9.7–10	0.8–0.9	2.7–3.0	75–77	9.6–11	0.3	Sena-Moreno et al. (2015)
12	1.3	1.1	55	30	0.5	Martínez et al. (2016)
11	1.1	1.1	55	31	0.3	Ling et al. (2016)
9.2–12	0.8–1.2	1.0–2.1	55–74	13–30	0.6–0.8	Ojeda-Amador et al. (2018)
Solvent-extracted oil						
8.5–12	1.3–1.5	2.5–3.5	65–70	15–19	–	Yildiz et al. (1998)
8.4–9.9	0.5–0.8	2.3–3.9	72–76	10–14	0.2–0.3	Seferoglu et al. (2006)
8.5–10	0.9–1.2	0.9–2.1	52–68	12–27	0.3–0.5	Tsantili et al. (2010)
12	1.1	1.5	57	29	0.3	Robbins et al. (2011)
OLO	**OLL**	**LOP**	**OOO**	**Total**		**References**
Virgin oils						
–	7[a]	22[b]	14	–		Saber-Tehrani et al. (2012)
Solvent-extracted oil						
20–25	17–25	12	9–13	–		Holcapek et al. (2003)
20	17	–	13	–		Ballistreri et al. (2010)
–	–	–	–	95.8		Miraliakbari and Shahidi (2007)

P palmitic, *Po* palmitoleic, *S* stearic, *O* oleic, *L* linoleic, *Ln* Linolenic
[a]SLL+POL
[b]OLL+POL

as the one with the highest oleic acid content. Intermedia content (62–75%) was found in the remaining varieties such as Larnaka, Mateur and Sirora (Tsantili et al. 2010; Catalán et al. 2017; Ojeda-Amador et al. 2018). Linoleic acid is the second most abundant fatty acid (13–31%) found in VPO, with the greatest proportion (~30%) found in Kerman, as expected by its lowest oleate value. Palmitic acid (10%) takes up the third position in the FA profile of VPO; appearing palmitoleic, stearic and linolenic acids in percentages of about 1–4% (Givianrad et al. 2013; Sena-Moreno et al. 2015; Catalán et al. 2017). This wide variation in the profile of FA allows the phenotype selection to produce oils with differentiated nutritious value.

Glycerolipids in edible oils are generally constituted mainly by triacylglycerols (TAG; up to 99%) and in much lower amounts or even traces by diacylglycerols (DAG) and monoacylglycerols (MAG). A study in pistachio oils showed that TAG accounts for 91–99% of total lipid, showing DAG only a 0.5–3.5% (Chahed et al. 2008). The different isomers of TAG, according to their esterified fatty acids, are distributed between OLO (20–25%) and OLL (17–25%) as the major glycerolipids compounds, followed by LOP (12%) and OOO (9–13%; Holcapek et al. 2003; Ballistreri et al. 2010) as reported in Table 7.2 with a total content of 958 g/kg (Miraliakbari and Shahidi 2007). A somewhat different distribution was also described, SLL+PLO (22%), SOL+POO (17%), OOLn+PLL (16%) and OOO

(14%; Saber-Tehrani et al. 2012; Givianrad et al. 2013). Regarding DAG, only LL, OL, LP, OO and OP have been identified, although not quantified, in pistachio oils (Holcapek et al. 2003).

On the other hand, phospholipids (PL) possess higher nutritional interest as compared to glycerolipids, however, they have been rarely studied in VPO. Nevertheless, pistachio oils are recognized as one with highest PL content among nut oils, showing phosphatidylcholine (PC) the greatest content (3000 mg/kg oil), followed by phosphatidylserine (PS; 2800 mg/kg) and phosphatidylinositol (PI; 1200 mg/kg; Miraliakbari and Shahidi 2007). A high content of PL (18,824 mg/kg) is observed in the pistachio kernel, being phosphatidic acid (PA), PI and PC its main components with 5289 mg/kg, 4909 mg/kg and 3637 mg/kg, respectively (Song et al. 2018). Little is known about sphingolipids and glycolipids in pistachio nut or its oil. Miraliakbari and Shahidi (2007) quantified in 8300 mg/kg the sphingolipids in pistachio oils.

VPO possess a high sterol content (2100–7600 mg/kg), as shown in Table 7.3, as compared to virgin olive oil (1100–2100 mg/kg; Aparicio and Luna 2002) and several other vegetable oils (FAO-WHO Codex Stan 210). A remarkable difference in the total sterols was observed among varieties: Sirora showing the lowest amount (3600 mg/kg), while Aegina, Kastel and Kerman a similar intermediate content (3900–4110 mg/kg) and Avdat, Larnaka, Mateur and Napoletana possessing the highest levels (7300–7600 mg/kg). As observed in other vegetable oils, β-sitosterol is the predominant phytosterol also in this kind of nut oils as depicted in Table 7.3. In VPO, β-sitosterol apparent ranged from 93% to 96% with 50–88% β-sitosterol and 2.2–5.1% campesterol (Saber-Tehrani et al. 2012; Sena-Moreno et al. 2016; Gong et al. 2017; Ojeda-Amador et al. 2018). On the other hand, solvent-extracted pistachio oils have reported a much lower content of total sterols (1500–1700 mg/kg, Miraliakbari and Shahidi 2007; 1840 mg/kg, Alasalvar and Pelvan 2011). As shown in Table 7.3, campesterol found in solvent-extracted oil apparently was much higher (12%), than in virgin ones (2.2–5.1).

Table 7.3 Sterol composition (%) in pistachio oil

Chol	Camp	Stigma	β-sito	β-sito ap.	Total (g/kg)	References
Virgin oil						
0.4	4.4	1.0	88	–	2.1	Saber-Tehrani et al. (2012)
0.1	4.0–4.1	0.6–0.7	–	93–94	4.2–4.5	Sena-Moreno et al. (2016)
–	4.7–5.1	1.4–1.8	50–53	–	2.6–3.5	Gong et al. (2017)
0.1–0.6	2.2–4.9	0.5–1.3	–	93–96	3.6–7.6	Ojeda-Amador et al. (2018)
Solvent-extracted oil						
–	0.3–0.9	5.5–7.4	85–89	–	–	Yildiz et al. (1998)
2.0–2.3	12–13	6.5–6.6	70–75	–	1.5–1.7	Miraliakbari and Shahidi (2007)
–	11	6.0	65	–	1.8	Alasalvar and Pelvan (2011)

Chol cholesterol, *Camp* campesterol, *Stigma* stigmasterol, *β-sito* β-sitosterol, *β-sito ap.* β-sitosterol apparent

Table 7.4 Tocopherols (mg/kg) and pigments (mg/kg) in pistachio oil

α-tocoph	γ-tocoph	T tocoph	Lutein	T Carot	T Chloro	References
Virgin oil						
510	105	–				Gentile et al. (2007)
70	162	–				Ballitreri et al. (2009)
328	–	–				Alasalvar and Pelvan (2011)
379	21	410	5.2	15	17	Saber-Tehrani et al. (2012)
33	804	898	–	48	41	Martínez et al. (2016)
34	309	367	–	–	12	Ling et al. (2016)
40–46	278–325	322–378	–	–	–	Gong et al. (2017)
–	–	–	22	29	29	D'Evoli et al. (2017)
6.5–33	548–719	562–758	–	36–61	28–50	Ojeda-Amador et al. (2018)
Solvent-extracted oil						
–	100–434	–	–	–	–	Kornsteiner et al. (2006)
–	–	–	29	47	128	Giuffrida et al. (2006)
–	–	–	26	–	25–200	Bellomo et al. (2009)
287–328	31–48	334–398	–	–	–	Miraliakbari and Shahidi (2007)
328	48	398	–	–	–	Alasalvar and Pelvan (2011)

α-tocoph α-tocopherol, *γ-tocoph* γ-tocopherol, *T tocoph* Total tocopherols, *T carot* Total carotenoids, *T chloro* Total chlorophylls

4 Minor Bioactive Components

A high content of tocopherols is observed in VPO (322–898 mg/kg), as depicted in Table 7.4. Relevant differences among the proportion of α- and γ- isomers have been observed. Several authors found γ-tocopherol as the main tocopherols component (162–804 mg/kg; Ballistreri et al. 2009; Ling et al. 2016; Martinez et al. 2016; Gong et al. 2017; Ojeda-Amador et al. 2018) in VPO and in solvent-extracted oils (up to 434 mg/kg; Kornsteiner et al. 2006). Among cultivars, VPO from the Kerman variety showed the highest content (719 mg/kg) compared with the rest of the varieties (~600 mg/kg, Ojeda-Amador et al. 2018). γ-Tocopherol is considered an important functional compound and demonstrates a similar bioavailability to α-tocopherol (Jiang et al. 2001), acting both in vivo and in vitro as an antioxidant and being even more efficient antioxidant in food lipid matrices (Wagner et al. 2004). On the contrary, according to the results of other authors, α-tocopherol would be the main isomer (287–510 mg/kg) in pistachio oil as stated in Table 7.4. Furthermore, β-tocopherol was not detected in this kind of nut oils, whereas small amounts of δ-tocopherol (22 mg/kg; Alasalvar and Pelvan 2011) and of two tocotrienols (δ-tocotrienol and γ-tocotrienol, 1–2% of the total; Ojeda-Amador et al. 2018) were described.

The content of pigments, mainly carotenoids and chlorophylls, in pistachio oils depend on variety, the degree of ripeness, environmental conditions and geographical origin (Giuffrida et al. 2006). The total content in carotenoids was reported between 15 and 61 mg/kg, as depicted in Table 7.4, being their main constituents lutein (22–29 mg/kg), β-carotene (4–7 mg/kg), neoxanthin (0.2–5 mg/kg), luteo-xanthin (3–10 mg/kg) and violaxanthin (0.3–3 mg/kg). The total chlorophylls content was much higher in solvent-extracted pistachio oils (128–200 mg/kg) as compared to VPO (12–50 mg/kg), made up by 39–59 mg/kg of chlorophyll a, 32–38 mg/kg of chlorophyll b, 30 mg/kg of pheophytin a and much lower amounts of and pheophytin b (0.7 mg/kg; Giuffrida et al. 2006; Bellomo et al. 2009). On the contrary, in cold-pressed oils, it was found that luteoxanthin was apparently the main specie (10.4 mg/kg), followed by lutein (5.2 mg/kg) and neoxanthin (0.15 mg/kg; Saber-Teherani et al. 2012).

Pistachio kernels serve as a very good source of total polar phenolic compounds (TPP; 6023–9550 mg/kg; Ojeda-Amador et al. 2018) as reported in Table 7.1, with a corresponding high antioxidant potential (9.7–28.4 mmol/kg DPPH; 61–282 mmol/kg ORAC; Ojeda-Amador et al. 2018). On the contrary, as expected due to their physicochemical properties (high water solubility), only low TPP concentrations are found in VPO (16–58 mg/kg) according to the literature (Table 7.5), but similar to other virgin seed oils, such as soybean, sunflower, rapeseed and corn (10–40 mg/kg; Siger et al. 2008). Ojeda-Amador et al. (2018) described TPP contents in cold-pressed VPO between 16 and 23 mg/kg, studying eight different pistachio cultivars; whereas higher values (25–54 mg/kg) were reported by other authors (Givianrad et al. 2013; Sena-Moreno et al. 2015; Ling et al. 2016) as well as in commercial VPO (Ojeda-Amador et al. 2018). These observed differences in TPP concentrations could be explained by the different extraction conditions used. Indeed, solvent-extracted pistachio oil apparently contains a much higher amount of TPP (380 mg/kg; Miraliakbari and Shahidi 2008).

Table 7.5 Polar phenolic compounds (mg/kg) identified in pistachio oil

E-7-glc	Catechin	Caffeic acid	Gallic acid	Protocat acid	*p*-Coumaric acid	TPP	References
Virgin oil							
–	–	1.96	–	–	0.36	58	Saber-Tehrani et al. (2012)
–	–	–	–	–	–	25–54	Sena-Moreno et al. (2015)
–	–	–	–	–	–	39–47	Ling et al. (2016)
–	–	–	–	–	–	16–23	Ojeda-Amador et al. (2018)
2.26	1.33	0.08	0.11	1.15	0.63	–	Sonmezdag et al. (2018)
Solvent-extracted oil							
–	–	–	–	–	–	379	Miraliakbari and Shahidi (2008)
–	0.8–1.1	–	3.1–3.2	4.4–4.6	0.23–0.25	–	Saitta et al. (2014)

E-7-glc Eriodictyol-7-O-glucoside, *protocat* protocatechuic, *TPP* total polar phenolics

The profile of individual phenolics in VPO has been little studied probably due to its lower content. Recently, Sonmezdag et al. (2018) identified 12 phenolic compounds in the Turkish Uzun cultivar, including 6 phenolic acids, 5 flavonols, and 1 flavan-3-ol phenolic compounds. Among them, eriodictyol-7-O-glucoside was the compound with the highest content (2.26 mg/kg), followed by catechin (1.33 mg/kg) and protocatechuic acid (1.15 mg/kg). The antioxidant capacity of catechin has been established as the one of the most powerful among the flavanols (Saitta et al. 2014). Other phenolics detected in lower amounts were ferulic acid with 0.86 mg/kg, p-coumaric acid with 0.63 mg/kg and eriodictyol with 0.54 mg/kg. Luteolin and rutin accounted 0.27 mg/kg and 0.21 mg/kg respectively, showing gallic acid (0.11 mg/kg) and caffeic acid (0.08 mg/kg) smaller amount. Some of these phenolic compounds have been also identified in other nuts such as almonds and peanuts (Talcott et al. 2005; Amarowicz et al. 2005), as well as in seed and hull of pistachio (Tomaino et al. 2010). A different profile was described in cold-pressed Iranian pistachio oils by Saber-Teherani et al. (2012), wherein 1.96 mg/kg of caffeic acid, 0.67 mg/kg cinnamic acid and 0.64 mg/kg pinoresinol, being the described this kind of nut oils.

Saitta et al. (2014) studied the phenolic profile of solvent-extracted pistachio oils from Italy and Turkey, appearing protocatechuic acid and gallic acid as the compounds with the highest amount (4.5 mg/kg and 3.2 mg/kg), followed by 4-hydroxybenzoic acid (1.4 mg/kg). Compounds such as salicylic acid (0.2 mg/kg), pyrogallol (0.1 mg/kg), tyrosol (0.4 mg/kg) and epicatechin (0.15 mg/) where reported for the first time.

A few works detected the presence of squalene in the oil of pistachio nut, wherein 200 mg/kg were measured in Sicilian Bronte PDO pistachios (Protected Denomination of Origin; Salvo et al. 2017), while a smaller amount (82 mg/kg) was quantified in pistachio oil from Iran (Derewiaka et al. 2014).

5 Antioxidant and Health-Promoting Effects

As already mentioned, nuts are considered a fundamental component of a healthy diet. They are rich in protein and lipids, with a balanced content of mono- and poly-unsaturated fatty acids (Table 7.1), and contain several bioactive compounds, such as antioxidants that can beneficially impact health outcomes (Kamal-Eldin and Moreau 2010; Salas-Salvadó et al. 2011). Indeed, tree nuts have been referred to as a natural functional food due to the synergistic interactions amongst their many bioactive constituents, which may favorably influence human physiology (Amarowicz et al. 2017).

Evidence suggests that nuts and nut oils can lower LDL-cholesterol levels and hence reduce the risk of coronary heart disease due to lipid composition and secondary metabolites, denominated as phytochemicals, with diverse bioactivities (Kris-Etherton et al. 2001; Atanasov et al. 2018). This has been confirmed by the PREDIMED study (Prevention with Mediterranean Diet) and other epidemiological or clinical trials, which have also indicated that a high intake of nuts (approx. 40 g daily) can lower the incidence of hypertension, metabolic syndrome, diabetes, cancer, other inflammatory conditions and total mortality (Salas-Salvadó et al. 2008; Mohammadifard et al. 2015).

Among nuts, pistachios (*Pistacia* spp.) exhibit interesting nutritional properties because they contain cardioprotective constituents, such as a high oleic acid content, phytosterols, phenolics and tocopherols, leading to a potential high antioxidant and anti-inflammatory food product (Yildiz et al. 1998; Bulló et al. 2015). The intake of oleic acid showed a beneficial effect in the cholesterol reduction (Fonolla-Joya et al. 2016) and exerts anti-inflammatory cellular effect (Perdomo et al. 2015). Furthermore, pistachio is one of the nut with the highest phytosterol content (Phillips et al. 2005), which are able to reduce blood cholesterol, as well as to decrease the risk of certain types of cancer and enhance immune function (Moreau et al. 2002; Awad and Fink 2000; Bouic 2001; Amarowicz et al. 2017).

As discussed, pistachio oil is rich in different forms of tocopherols and tocotrienols, all together referred as vitamin E, possessing a powerful lipid-soluble antioxidant activity, which protects cell membrane lipids from oxidation (Yokota et al. 2001). Furthermore, α-tocopherol resulted to be a potent modulator of gene expression and γ-tocopherol appears to be highly effective in preventing cancer-related processes (Brigelius-Flohé 2006). These dietary components may contribute to the antioxidant defense and counteract oxidative damage and oxidative stress by scavenging and neutralizing free radicals or by providing compounds that can induce the gene expression of the endogenous antioxidants (Blomhoff 2005; Moskaug et al. 2005).

6   Contribution to Organoleptic Properties

The color of the kernel is an important quality characteristic of pistachio, preferring the food industry an intense green. In VPO, the CIE L*a*b* color parameters showed a 'b*' value ranging from 24 to 87 and 'a*' from −3.4 to −8.3 (Ling et al. 2016; Rabadán et al. 2017b; Ojeda-Amador et al. 2018), corresponding to a high green tonality oil.

On the other hand, the aroma of any food commodity is attributed to be a complex mixture of hundreds and heterogeneous volatile chemicals (alcohols, aldehydes, esters, ketones, terpenes, pyrazines, and acids) each of them with a different

Table 7.6   Volatile compounds (mg/kg) found in virgin pistachio oil

α-Pinene	Limonene	Terpenes[a]	Alcohols	Acids	Aldehydes	References
11	33	45	1.6	12	2.4	Ling et al. (2016)
15–37	0.1–3.5	16–41	0.0–1.1	0.0–0.9	0.0–0.7	Ojeda-Amador et al. (2018)
1.5	1.1	6.0	2.2	2.1	1.1	Sonmezdag et al. (2018)

[a]Total terpenes

contribution to the whole aroma perception (Belitz et al. 2004). Flavour and aroma have a great influence on the acceptance of the food product by consumers, being the odour of roasted nuts stronger than that of raw ones due to some volatile compounds formed during the roasting (Aceña et al. 2010; Rodríguez-Bencomo et al. 2015; Carbonell-Barrachina et al. 2015; Hojjati et al. 2015). Furthermore, the knowledge on the volatile profile constitutes an important source of information about the characterization of a food product and for the definition of the optimal parameters of its manufacture.

The volatile compounds retained in the VPO during the extraction process of the oil are responsible for their typical and peculiar aroma very appreciated by consumers. Aroma main compounds and families in VPO are reported in Table 7.6. Terpenes, which are associated with appreciated sensory descriptors, such as pine (α- and β-pinene), citrus (limonene and 3-carene), floral (myrcene) and berry (α-terpinene), are the major compounds (97%) of total volatiles (Ling et al. 2016; Ojeda-Amador et al. 2018). Their contents widely ranged from 16 to 45 mg/kg, depending on the variety, being Kastel and Mateur the varieties with the lowest and highest terpene concentrations, respectively, with intermediate values for Avdat, Larnaka and Sirora (Ojeda-Amador et al. 2018). Within the terpene family, α-pinene and limonene have been reported as the main components (Kendirci and Onoğur 2011; Hojjati et al. 2013; Rodríguez-Bencomo et al. 2015; Ling et al. 2016; Ojeda-Amador et al. 2018) as well as in the essential oil from hull and leaf of *Pistachia vera* (Tsokou et al. 2007). Other compounds of this family, found in lower concentration, are β-pinene, myrcene, α-terpinolene and 3-carene (Ling et al. 2016; Sonmezdag et al. 2018).

Alcohols contribute with floral (hexanol) and green (nonanol) flavors. Moreover, aldehydes such as furfural and nonanal contribute with almond and green sensory notes. These aroma components, including alcohols, aldehydes, acids and hydrocarbons represent only ~3% of the total volatile in VPO (Ojeda-Amador et al. 2018). Alcohols are formed by compounds such as hexanol (produced from LOX), 1-octen-3-ol and nonanol, contributing hexanal, nonanal, furfural and benzaldehyde to the aldehydes one (Ling et al. 2016; Sonmezdag et al. 2018).

Apparently, commercial VPO contained a lower terpene concentration (60% of the total volatiles; Ojeda-Amador et al. 2018) compared to the virgin oils from raw pistachios (88–97%), due to the presence of pyrazines, furans and pyrroles, which appear when high temperatures are implemented during the extraction process or when roasted pistachios are used, which contribute to roasted nutty notes.

Ojeda-Amador et al. (2018) reported the sensory profile of VPO, which were described as having an intense green appearance, marked persistence of pistachio aroma and flavor, quite sweet, neither bitter nor pungent, with an oily in mouth sensation and without typical oil defects, such as rancid. Furthermore, the Mateur and Larnaka varieties presented a marked green appearance, higher intensities of the roasted nuts attribute and are appreciated for their flavor intensity and the persistence of the pistachio aroma in the mouth.

Sena-Moreno et al. (2015) and Rabadán et al. (2017a) studied consumer's acceptance of VPO extracted by different extraction systems (expeller and hydraulic press) under different conditions. Consumers preferred pistachio oils extracted with the screw press than with hydraulic one, probably due to the higher temperature required for the first process system that originates greener and taster oils characterized by the presence of roasting aromas.

References

Aceña, L., Vera, L., Guasch, J., Busto, O., & Mestres, M. (2010). Comparative study of two extraction techniques to obtain representative aroma extracts for being analysed by gas chromatography-olfactometry: Application to roasted pistachio aroma. *Journal of Chromatography. A, 1217*(49), 7781–7787.

Agar, I. T., Kafkas, S., & Kaska, N. (1998). Lipid characteristics of Turkish and Iranian pistachio kernels. In L. Ferguson & D. Kester (Eds.), *Second international symposium on pistachios and almonds* (Vol. 1, pp. 378–384). Leuven: International Society Horticultural Science.

Alasalvar, C., & Pelvan, E. (2011). Fat-soluble bioactives in nuts. *European Journal of Lipid Science and Technology, 113*(8), 943–949.

Alasalvar, C., & Shadihi, F. (2008). Tree nuts: Composition, phytochemicals and health effects: An overview. In C. Alasalvar & F. Shahidi (Eds.), *Tree nuts*. Boca Raton: CRC Press.

Álvarez-Ortí, M., Quintanilla, C., Sena, E., Alvarruiz, A., & Pardo, J. E. (2012). The effects of a pressure extraction system on the quality parameters of different virgin pistachio (*Pistacia vera* L. var. Larnaka) oils. *Grasas y Aceites, 63*(3), 260–266.

Amarowicz, R., Troszyńska, A., & Shahidi, F. (2005). Antioxidant activity of almond seed extract and its fractions. *Journal of Food Lipids, 12*(4), 344–358.

Amarowicz, R., Gong, Y., & Pegg, R. B. (2017). Recent advances in our knowledge of the biological properties of nuts. In I. C. F. R. Ferreira, P. Morales, & L. Barros (Eds.), *Wild plants, mushrooms and nuts: Functional food properties and applications* (1st ed.). Chichester: Wiley.

Aparicio, R., & Luna, G. (2002). Characterisation of monovarietal virgin olive oils. *European Journal of Lipid Science and Technology, 104*(9–10), 614–627.

Atanasov, A. G., Sabharanjak, S. M., Zengin, G., Mollica, A., Szostak, A., Simirgiotis, M., Huminiecki, Ł., Horbanczuk, O. K., Mohammad, S., & Mocan, A. (2018). Pecan nuts: A review of reported bioactivities and health effects. *Trends in Food Science and Technology, 71*, 246–257.

Awad, A. B., & Fink, C. S. (2000). Phytosterols as anticancer dietary components: Evidence and mechanism of action. *The Journal of Nutrition, 130*(9), 2127–2130.

Bail, S., Stuebiger, G., Unterweger, H., Buchbauer, G., & Krist, S. (2009). Characterization of volatile compounds and triacylglycerol profiles of nut oils using SPME-GC-MS and MALDI-TOF-MS. *European Journal of Lipid Science and Technology, 111*(2), 170–182.

Ballistreri, G., Arena, E., & Fallico, B. (2009). Influence of ripeness and drying process on the polyphenols and tocopherols of *Pistachia vera* L. *Molecules, 14*, 4358–4369.

Ballistreri, G., Arena, E., & Fallico, B. (2010). Characterization of triacylglycerols in *Pistacia vera* L. oils from different geographic origins. *Italian Journal of Food Science, 22*(1), 69–75.

Belitz, H. D., Grosch, W., Schieberle, P. (2004). Aroma compounds. In *Food chemistry* (pp. 342–408), 3rd revised ed. Berlin: Springer.

Bellomo, M. G., Fallico, B., & Muratore, G. (2009). Stability of pigments and oil in pistachio kernels during storage. *International Journal of Food Science and Technology, 44*(12), 2358–2364.

Blomhoff, R. (2005). Dietary antioxidants and cardiovascular disease. *Current Opinion in Lipidology, 16*(1), 47–54.

Bouic, P. J. (2001). The role of phytosterols and phytosterolins in immune modulation: A review of the past 10 years. *Current Opinion in Clinical Nutrition and Metabolic Care, 4*(6), 471–475.

Brigelius-Flohé, R. (2006). Bioactivity of vitamin E. *Nutrition Research Reviews, 19*(2), 174–186.

Bulló, M., Juanola-Falgarona, M., Hernández-Alonso, P., & Salas-Salvadó, J. (2015). Nutrition attributes and health effects of pistachio nuts. *The British Journal of Nutrition, 113*(2), S79–S93.

Carbonell-Barrachina, A., Memmi, H., Noguera-Artiaga, L., Gijon-Lopez, M. C., & Perez-Lopez, D. (2015). Quality attributes of pistachio nuts as affected by rootstock and their irrigation. *International Journal of the Science Food and Agriculture, 95*, 2866–2873.

Catalán, L., Alvarez-Ortí, M., Pardo-Giménez, A., Gómez, R., Rabadán, A., & Pardo, J. E. (2017). Pistachio oil: A review on its chemical composition, extraction systems, and uses. *European Journal of Lipid Science and Technology, 119*(5), 1–8.

Chahed, T., Bellila, A., Dhifi, W., Hamrouni, I., M'hamdi, B., Kchouk, M. E., & Marzouk, B. (2008). Pistachio (*Pistacia vera*) seed oil composition: Geographic situation and variety effects. *Grasas y Aceites, 59*(1), 51–56.

Couceiro, J. F., Guerrero, J., Gijón, M. C., Moriana, A., Pérez, D., & Rodríguez de Francisco, M. (2017). *El cultivo del pistacho* (2nd ed.). Spain: Mundi-Prensa.

CX/FO 15/24/11 (2015). Codex committee on fats and oils -Discussion paper on cold pressed oils. In 34 session, Malaysia (9–13 February 2015).

D'Evoli, L., Lucarini, M., Gabrielli, P., Aguzzi, A., & Lombardi-Boccia, G. (2017). Carotenoids and chlorophylls in Bronte's pistachio (*Pistacia vera* L.) and pistachio processed products. *International Journal of Food Nutrition and Sciences, 6*(3), 41–47.

Derewiaka, D., Szwed, E., & Wolosiak, R. (2014). Physicochemical properties and composition of lipid fraction of selected edible nuts. *Pakistan Journal of Botany, 46*(1), 337–343.

Dunford, N. T., Teel, J. A., & King, J. W. (2003). A continuous counter current supercritical fluid deacidification process for phytosterol ester fortification in rice bran oil. *Food Research International, 36*(2), 175–181.

FAOSTAT (2016). Food and Agriculture Organization: FAOSTAT data. Available at http://faostat. fao.org

FAO-WHO Codex Stan 210 (1999). Standard for named vegetable oils. Revision 2009 and Amendment 2015.

FAO-WHO Codex Stan 33 (1981). Standard for olive oils and olive pomace oils. Revision 2015 and Amendment 2013.

Fonolla-Joya, J., Reyes-García, R., García-Martín, A., López-Huertas, E., & Muñoz-Torres, M. (2016). Daily intake of milk enriched with n-3 fatty acids, oleic acid, and calcium improves metabolic and bone biomarkers in postmenopausal women. *Journal of the American College of Nutrition, 35*(6), 529–536.

Gentile, C., Tesoriere, L., Butera, D., & Fazzari, M. (2007). Antioxidant activity of Sicilian pistachio (*Pistachio vera* L. var Bronte) nut extract and its bioactive components. *Journal of Agricultural and Food Chemistry, 55*, 643–648.

Giuffrida, D., Saitta, M., La Torre, L., Bombaci, L., & Dugo, G. (2006). Carotenoid, chlorophyll and chlorophyll-derived compounds in pistachio kernels (*Pistacia vera* L.) from Sicily. *Italian Journal of Food Science, 18*(3), 313–320.

Givianrad, M. H., Saber-Tehrani, M., & Jafari Mohammadi, S. A. (2013). Chemical composition of oils from wild almond (*Prunus scoparia*) and wild pistachio (*Pistacia atlantica*). *Grasas y Aceites, 64*(1), 77–84.

Gong, Y., Pegg, R. B., Carr, E. C., Parrish, D. R., Kellett, M. E., & Kerrihard, A. L. (2017). Chemical and nutritive characteristics of tree nut oils available in the U.S. market. *European Journal of Lipid Science and Technology, 119*(8), 1–15.

Halvorsen, B. L., Carlsen, M. H., Phillips, K. M., Bøhn, S. K., Holte, K., Jacobs, D. R., & Blomhoff, R. (2006). Content of redox-active compounds (i.e. antioxidants) in foods consumed in the United States. *The American Journal of Clinical Nutrition, 84*(1), 95–135.

Hannon, J. (1997). Pistachio nut oil: A natural emollient for the cosmetic formulator. *Drug and Cosmetic Industry, 160*(2), 30–34. 36, 80.

Herrero, M., Cifuentes, A., & Ibañez, E. (2006). Sub and supercritical fluid extraction of functional ingredients from different natural sources: Plants, food by-products, algae and microalgae: A review. *Food Chemistry, 98*(1), 136–148.

Hojjati, M., Calín-Sánchez, A., Razavi, S. H., & Carbonell-Barrachina, A. A. (2013). Effect of roasting on colour and volatile composition of pistachios (*Pistacia vera* L.). *International Journal of Food Science and Technology, 48*(2), 437–443.

Hojjati, M., Noguera-Artiaga, L., Wojdylo, A., & Carbonell-Barrachina, A. (2015). Effects of microwave roasting on physicochemical properties of pistachios (*Pistacia vera* L). *Food Science and Biotechnology, 24*(6), 1995–2001.

Holčapek, M., Jandera, P., Zderadička, P., & Hrubá, L. (2003). Characterization of triacylglycerol and diacylglycerol composition of plant oils using high-performance liquid chromatography-atmospheric pressure chemical ionization mass spectrometry. *Journal of Chromatography A, 1010*(2), 195–215.

Jiang, Q., Christen, S., Shigenaga, M. K., & Ames, B. N. (2001). γ-Tocopherol, the major form of vitamin E in the US diet, deserves more attention. *The American Journal of Clinical Nutrition, 74*(6), 714–722.

Kamal-Eldin, A., & Moreau, R.A. (2010). Tree nut oils. In R. A. Moreau & A. Kamal-Eldin (Eds.), Gourmet and health-promoting specialty oils (vol. 3, pp. 127–150). EE.UU.: Academic Press/ AOCS Press.

Kendirci, P., & Onoğur, T. A. (2011). Investigation of volatile compounds and characterization of flavor profiles of fresh pistachio nuts (*Pistacia vera* L.). *International Journal of Food Properties, 14*(2), 319–330.

Kornsteiner, M., Wagner, K., & Elmadfa, I. (2006). Tocopherols and total phenolics in 10 different nut types. *Food Chemistry, 98*(2), 381–387.

Kris-Etherton, P. M., Zhao, G. X., Binkoski, A. E., Coval, S. M., & Etherton, T. D. (2001). The effects of nuts on coronary heart disease risk. *Nutrition Reviews, 59*(4), 103–111.

Ling, B., Xuanmin, Y., Li, R., & Wang, S. (2016). Physicochemical properties, volatile compounds, and oxidative stability of cold pressed kernel oils from raw and roasted pistachio (*Pistacia vera* L. Var Kerman). *European Journal of Lipid Science and Technology, 118*(9), 1368–1379.

Martinez, J. L., & Vance, S. W. (2008). Supercritical extraction plants: Equipment, process and costs. In J. L. Martinez (Ed.), *Supercritical fluid extraction of nutraceuticals and bioactive compounds* (pp. 25–49). Boca Raton: CRC Press, Taylor and Francis Group.

Martinez, M. L., Fabani, M. P., Baroni, M. V., Huaman, R. N. M., Ighani, M., Maestri, D. M., Wunderlin, D., Tapia, A., & Feresin, G. E. (2016). Argentinian pistachio oil and flour: A potential novel approach of pistachio nut utilization. *Journal of Food Science and Technology, 53*(5), 2260–2269.

Matthäus, B. (2008). Virgin oils – the return of a long known product. *European Journal of Lipid Science and Technology, 110*, 595–596.

Miraliakbari, H., & Shahidi, F. (2007). Lipid class compositions, tocopherols and sterols of tree nut oils extracted with different solvents. *Journal of Food Lipids, 15*(1), 81–96.

Miraliakbari, H., & Shahidi, F. (2008). Antioxidant activity of minor components of tree nut oils. *Food Chemistry, 111*(2), 421–427.

Mohammadifard, N., Salehi-Abargouei, A., Salas-Salvadó, J., Guasch-Ferré, M., Humphries, K., & Sarrafzadegan, N. (2015). The effect of tree nut, peanut, and soy nut consumption on blood pressure: A systematic review and meta-analysis of randomized controlled clinical trials. *The American Journal of Clinical Nutrition, 101*(5), 966–982.

Moreau, R. A., Whitaker, B. D., & Hicks, K. B. (2002). Phytosterols, phytostanols, and their conjugates in foods: Structural diversity, quantitative analysis, and health-promoting uses. *Progress in Lipid Research, 41*(6), 457–500.

Moskaug, J. O., Carlsen, H., Myhrstad, M. C., & Blomhoff, R. (2005). Polyphenols and glutathione synthesis regulation. *The American Journal of Clinical Nutrition, 81*(1), 277–283.

Nielsen, S. S. (1998). Food analysis. In *Crude fats analysis* (pp. 203–214). West Lafayette: Purdue University.

Nyam, K. L., Tan, C. P., Lai, O. M., Long, K., & Man, Y. B. C. (2009). Physicochemical properties and bioactive compounds of selected seed oils. *LWT- Food Science and Technology, 42*(8), 1396–1403.

Ojeda-Amador, R. M., Fregapane, G., & Salvador, M. D. (2018). Composition and properties of virgin pistachio oils and their by-products from different cultivars. *Food Chemistry, 240*, 123–130.

Oseni, K., Owolarafe, M., Faborode, O., & Obafemi, O. A. (2002). Comparative evaluation of the digester–screw press and a hand-operated hydraulic press for palm fruit processing. *Journal of Food Engineering, 52*(3), 249–255.

Palazoğlu, T. K., & Balaban, M. O. (1998). Supercritical CO2 extractions of lipids from roasted pistachio nuts. *Journal of American Society of Agricultural Engineering, 41*(3), 679–684.

Patricelli, A., Assogna, A., Emmi, E., & Sodini, G. (1979). Fattori che influenzano lèstrazione del lipidi da semi decorticati di girasole. *Rivista Italiana Sostanze Grasse, 61*, 136–142.

Perdomo, L., Beneit, N., Otero, Y. F., Escribano, O., Díaz-Castroverde, S., & Gómez-Hernández, A. (2015). Protective role of oleic acid against cardiovascular insulin resistance and in the early and late cellular atherosclerotic process. *Cardiovascular Diabetology, 14*(75), 1–12.

Phillips, K. M., Ruggio, D. M., & Ashraf-Khorassani, M. A. (2005). Phytosterol composition of nuts and seeds commonly consumed in the United States. *Journal of Agricultural and Food Chemistry, 53*(24), 9436–9445.

Rabadán, A., Álvarez-Ortí, M., Gómez, R., Alvarruiz, A., & Pardo, J. E. (2017a). Optimization of pistachio oil extraction regarding processing parameters of screw and hydraulic presses. *LWT-Food Science and Technology, 83*, 79–85.

Rabadán, A., Pardo, J. E., Gómez, R., Alvarruiz, A., & Álvarez-Ortí, M. (2017b). Usefulness of physical parameters for pistachio cultivar differentiation. *Scientia Horticulturae, 222*, 7–11.

Rodríguez-Bencomo, J. J., Kelebek, H., Sonmezdag, A. S., Rodríguez-Alcalá, L. M., Fontecha, J., & Selli, S. (2015). Characterization of the aroma-active, phenolic, and lipid profiles of the pistachio (*Pistacia vera* L.) nut as affected by the single and double roasting process. *Journal of Agricultural and Food Chemistry, 63*(35), 7830–7839.

Robbins, K. S., Shin, E. C., Shewfelt, R. L., Eitenmiller, R. R., & Pegg, R. B. (2011). Update on the Healthful Lipid Constituents of Commercially. Important Tree Nuts. *Journal of Agricultural and Food Chemistry, 59*, 12083–12092.

Saber-Tehrani, M., Givianrad, M. H., Aberoomand-Azar, P., Waqif-Husain, S., & Jafari Mohammadi, S. A. (2012). Chemical composition of Iran's *pistacia atlantica* cold-pressed oil. *Journal of Chemistry, 2013*, 1–6.

Saitta, M., La Torre, G. L., Potorti, A. G., Di Bella, G., & Dugo, G. (2014). Polyphenols of pistachio (*Pistacia vera* L.) oil samples and geographical differentiation by principal component analysis. *Journal of the American Oil Chemists' Society, 91*(9), 1595–1603.

Salas-Salvadó, J., Fernández-Ballart, J., Ros, E., Martínez-González, M. A., Fitó, M., Estruch, R., Corella, D., Fiol, M., Gómez-García, E., Arós, F., Flores, G., Lapetra, J., Lamuela-Raventós, R., Ruiz-Gutiérrez, V., Bulló, M., Basora, J., & Covas, M. I. (2008). Effect of a Mediterranean diet supplemented with nuts on metabolic syndrome status: One-year results of the PREDIMED randomized trial. *Journal of the American Medical Association, 168*(22), 2449–2458.

Salas-Salvadó, J., Casas-Agustench, P., & Salas-Huetos, A. (2011). Cultural and historical aspects of Mediterranean nuts with emphasis on their attributed healthy and nutritional properties. *Nutrition, Metabolism, and Cardiovascular Diseases, 21*(1), S1–S6.

Salvo, A., La Torre, G., Di Stefano, V., Capocchiano, V., Mangano, V., Saija, E., Pellizzeri, V., Casale, K. E., & Dugo, G. (2017). Fast UPLC/PDA determination of squalene in Sicilian P.D.O. pistachio from Bronte: Optimization of oil extraction method and analytical characterization. *Food Chemistry, 221*, 1631–1636.

Sena-Moreno, E., Pardo, J. E., Catalán, L., Gómez, R., Pardo-Giménez, A., & Álvarez-Ortí, M. (2015). Drying temperature and extraction method influence physicochemical and sensory characteristics of pistachio oils. *European Journal of Lipid Science and Technology, 117*(5), 684–691.

Sena-Moreno, E., Pardo, J. E., Pardo-Giménez, A., Gómez, R., & Álvarez-Ortí, M. (2016). Differences in oils from nuts extracted by means of two pressure systems. *International Journal of Food Properties, 19*(12), 2750–2760.

Seferoglu, S., Seferoglu, H.G., Tekintas, F.E., Balta, F. (2006) Biochemical composition influenced by different locations in Uzun pistachio cv. (Pistacia vera L.) grown in Turkey. *Journal of Food Composition and Analysis, 19*, 461–465,

Siger, A., Nogala-Kalucka, M., & Lampart-Szczapa, E. (2008). The content and antioxidant activity of phenolic compounds in cold-pressed plant oils. *Journal of Food Lipids, 15*(2), 137–149.

Song, S., Cheong, L., Wang, H., Man, Q., Pang, S., Li, Y., Ren, B., Sonmezdag, A. S., Kelebek, H., & Selli, S. (2018). Pistachio oil (*Pistacia vera* L. cv. Uzun): Characterization of key odorants in a representative aromatic extract by GC-MS-olfactometry and phenolic profile by LC-ESI-MS/MS. *Food Chemistry, 240*, 24–31.

Sonmezdag, A. S., Kelebek, h., & Selli, S. (2018). Pistachio oil (*pistachio vera* L. cv Uzun): Characterization of key odorants in a representative aromatic extract by GC-MS-olfatometry and phenolic profile LC-ESI-MS/MS. *Food Chemistry, 240*, 24–31.

Talcott, S. T., Passeretti, S., Duncan, C. E., & Gorbet, D. W. (2005). Polyphenolic content and sensory properties of normal and high oleic acid peanuts. *Food Chemistry, 90*(3), 379–388.

Tomaino, A., Martorana, M., Arcoraci, T., Monteleone, D., Giovinazzo, C., & Saija, A. (2010). Antioxidant activity and phenolic profile of pistachio (*Pistacia vera* L., variety Bronte) seeds and skins. *Biochimie, 92*(9), 1115–1122.

Tsantili, E., Takidelli, C., Christopoulos, M. V., Lambrinea, E., Rouskas, D., & Roussos, P. A. (2010). Physical, compositional and sensory differences in nuts among pistachio (*Pistachia vera* L.) varieties. *Scientia Horticulturae, 125*(4), 562–568.

Tsokou, A., Georgopoulou, K., Melliou, E., Magiatis, P., & Tsitsa, E. (2007). Composition and enantiomeric analysis of the essential oil of the fruits and the leaves of *Pistacia vera* from Greece. *Molecules, 12*(6), 1233–1239.

Wagner, K., Kamal-Eldin, A., & Elmadfa, I. (2004). Gamma-tocopherol –an underestimated vitamin? *Annals of Nutrition & Metabolism, 48*(3), 169–188.

Yildiz, M., Gürcan, Ş. T., & Özdemir, M. (1998). Oil composition of pistachio nuts (*Pistacia vera* L.) from Turkey. *Fett-Lipid, 100*(3), 84–86.

Yokota, T., Igarashi, K., Uchihara, T., Jishage, K., Tomita, H., Inaba, A., Li, Y., Arita, M., Mizusawa, H., & Arai, H. (2001). Delayed-onset ataxia in mice lacking α-tocopherol transfer protein: Model for neuronal degeneration caused by chronic oxidative stress. *Proceedings of the National Academy of Sciences, 98*(26), 15185–15190.

Zhang, J., Kris-Etherton, P. M., Thompson, J. T., & Vanden Heuvel, J. P. (2010). Effect of pistachio oil on gene expression of IFN-induced protein with tetratricopeptide repeats 2: A biomarker of inflammatory response. *Molecular Nutrition & Food Research, 54*(1), S83–S92.

Chapter 8
Chestnut (*Castanea sativa*) Oil

Mustafa Kiralan, S. Sezer Kiralan, Gülcan Özkan, and Erkan Karacabey

Abstract Chestnut (*Castanea sativa*), also known as sweet chestnut or European chestnut, belongs to the botanical family of Fagaceae. The high contribution to the world production of this fruit is mainly originated from China, Bolivia and Turkey. Chestnut promises health and nutritional benefits for consumers due to its rich nutrients including dietary fibers, minerals, essential fatty acids, vitamins, essential amino acids, antioxidants and other important bioactive components. Leaves, bark, twigs and nuts are the available parts for usage. Chestnut is commonly consumed as a raw and/ or in a roasted form. Additionally, chestnut flour takes place in food markets as another popular and alternative product. According to the literature, low-fat content was reported for chestnuts. However, the valuable nutritional composition of chestnut oil leads interests to these nuts as a potential oil source. Chestnut oil is rich in omega fatty acids such as linoleic and oleic acids, in tocopherols such as γ-tocopherol, which contribute to human health. The current chapter serves as a guide which presents the history of chestnuts oil from past to today to shape the future studies.

Keywords Chestnut · Fatty acids · Tocopherols · Functional properties

1 Introduction

Chestnut (*Castanea sativa*) is one of the oldest edible fruits in the World. According to FAOSTAT (FAO Statistics Division) data, world annual chestnut production reached to 2.261.589 tones over an area of 602.718 ha in 2016. China is the major chestnut producer with 1.879.031 tones, followed by Bolivia with 1.879.031 tones

M. Kiralan (✉) · S. S. Kiralan
Faculty of Engineering, Department of Food Engineering, Balıkesir University, Balıkesir, Turkey

G. Özkan · E. Karacabey
Faculty of Engineering, Department of Food Engineering, Suleyman Demirel University, Isparta, Turkey
e-mail: gulcanozkan@sdu.edu.tr; erkankaracabey@sdu.edu.tr

© Springer Nature Switzerland AG 2019
M. F. Ramadan (ed.), *Fruit Oils: Chemistry and Functionality*,
https://doi.org/10.1007/978-3-030-12473-1_8

and Turkey with 64.750 tones (FAO 2018). The chestnut tree is a member of Fagacease family and chestnut belongs the genus Castanea. There are four important economic species of chestnut: *C. dentata* (North American), *C. mollissima* (Chinese), *C. sativa* (European), and *C. crenata* (Japanese) (Barakat et al. 2009). One of them, *C. sativa* is named as sweet chestnut and is the only native species of the genus in Europe (Conedera et al. 2004). This chestnut species is distributed from North-Western Africa (Morocco) to North-Western Europe (southern England, Belgium) and from south-western Asia (Turkey) to Eastern Europe (Romania), the Caucasus (Georgia, Armenia) and the Caspian Sea along the Mediterranean basin with the favorable climatic conditions (Conedera et al. 2016).

Chestnut is generally consumed as a dried and rarely as a fresh item. In order to get favorable taste, drying is applied at home or industrial scale. With the help of drying process, organoleptic properties and digestibility of the chestnut fruit improved. In industrial scale, different types of processed chestnut products could be seen in markets such as frozen, sterilized in aluminum bags, tinned, stored in flasks and dried. During the production of these forms, high-quality chestnuts are preferred. The poor quality chestnuts (lower caliber, polyspermic and broken) are also economically utilized as purées, chestnut creams, flours, soups or yogurts (De Vasconcelos et al. 2010). Chestnut floor, being one of the favorable products for the food industry, contains many healthy nutrients such as protein (6.92%), dietary fiber (4.19%), and also a low amount of fat (2.05%) (Sacchetti et al. 2004). Besides to these nutritional compounds, its flour is rich in lignans (the amount of total lignans is 980.03 μg/100 g d.w.) which contributes to antioxidant activity (Durazzo et al. 2013).

Although chestnut fruits also have oil content, classified as "good fat", but at a small amount, it could be alternative for the food industry as new fruit oil. This fruit oil is important for human diet due to health-promoting fatty acid distribution. The main fatty acid in the profile is linoleic acid which could reduce blood cholesterol level and thereby lower the risk of the development of atherosclerosis (Jandacek 2017).

2   Extraction and Processing of Fruit Oil

Plant-based oils are generally obtained by solvent extraction under laboratory conditions. Different solvents have been reported including dichloromethane:methanol (Ferreira-Cardoso et al. 1998), *n*-hexane (Zlatanov et al. 2013), ether (Ertürk et al. 2006) for the extraction of oils from *C. sativa*. Dichloromethane:methanol was used as a solvent in oil extraction from seven different chestnut varieties (Bebim, Benfeit, Lada, Longal, Negral, Aveleira and Boaventura) in Portugal, and crude fat content in samples were ranged between 1.02% (cv. Lada) and 1.76% (cv. Negral) (Ferreira-Cardoso et al. 1998). Oil extraction was carried for chestnut fruits in Bulgaria via *n*-hexane and yield was measured as 20 g oil/kg (Zlatanov et al. 2013). In another study, chestnut cultivars and genotypes belonging to the species *C.sativa* in Turkey were studied and the oil content was ranged from 0.49 to 2.01 g/100 g (Ertürk et al. 2006).

3 Fatty Acids Composition and Acyl Lipids Profile of Fruit Fixed Oil

Vegetable oils are mainly constituted by triacylglycerols (95–98%) and complex mixtures of minor compounds (2–5%) of a wide range of chemical nature (Cert et al. 2000). Triglyceride composition has been used as a measurement of the quality and purity of vegetable oils (Aparicio and Aparicio-Ruíz 2000). Vegetable oils are characterized by their effective carbon number, equal to the mean carbon number of their triglycerides, so triglycerides could be used to confirm authenticity (Sovova et al. 2001; Buchgraber et al. 2004). Besides, triglyceride profiles could be used for a chemometric classification in accordance with the origin of vegetable oils (Ulberth and Buchgraber 2000) and represents the physical properties of an oil (Man et al. 1999).

Barreira et al. (2012b) reported triacylglycerol profile of four Portuguese *Castanea sativa* Miller cultivars (Aveleira, Boaventura, Judia and Longal) for a 3 years period. Thirteen compounds were determined in chestnut samples including LLnLn, LLLn, LLL, OLLn, PLLn, LLO, PLL, OLO, PLO, PLP, OOO, POO and PPO (L, linoleoyl; Ln, linolenoyl; P, palmitoyl; O, oleoyl). 1-oleoyl-2-lin-oleoyl-3-linoleoyl-sn-glycerol, 1-linoleoyl-2-linoleoyl-3-palmitoyl-sn-glycerol, 1-oleoyl-2-linoleoyl-3-oleoyl-sn-glycerol and 1-linoleoyl-2-oleoyl-3-palmitoyl-sn-glycerol were the prevalent triacylglycerols in chestnut samples. As can be seen in Table 8.1, the major compound in chestnut cultivars was LLO comprised of 23–26% of

Table 8.1 Triacylglycerol profile of chestnuts	**Triglyceride**	Portuguese chestnuts (g/100 g TAG)[a]	Portuguese chestnuts (%)[b]
	LLnLn	–	0.20–0.26
	LLLn	2–4	0.69–3.7
	LLL	4–10	3.8–9.8
	OLLn	2–4	1.6–3.9
	PLLn	1.8–3	1.3–2.7
	LLO	23–26	–
	OLL	–	18.9–24.5
	PLL	14–17	–
	LLP	–	14.4–17.4
	OLO	14–16	13.2–18.4
	PLO	12–17	–
	LOP	–	11.5–20.5
	PLP	1.2–2.0	0.79–1.8
	OOO	5–11	4.54–10.9
	POO	2.9–6	–
	OOP	–	3.0–8.1
	POP	–	0.08–0.36

[a]Barreira et al. (2012)
[b]Barreira et al. (2009)

triacylglycerols. The other identified major triacylglycerols were PLL, OLO and PLO representing 14–17%, 14–16% and 12–17% of the total, respectively.

Barreira et al. (2009) used triacylglycerol profile to distinguish chestnuts cultivars. Thirteen triacylglycerols were identified in chestnuts samples (Aveleira, Boa Ventura, Judia and Longal) including LLnLn, LLLn, LLL, OLLn, PLLn, OLL, LLP, OLO, LOP, PLP, OOO, OOP and POP (L, linoleoyl; Ln, linolenoyl, P, palmitoyl; O, oleoyl). OLL was the major triacylglycerol in chestnuts from cv. Boa Ventura (23.5%), cv. Judia (24.2%) and cv. Longal (24.5%). In cv. Aveleira, the main TAG was determined as LOP (20.5%).

Polyunsaturated fatty acids (PUFA) are important compounds for human health, wherein omega-3 and omega-6 fatty acids are essential for humans, since not be synthesized in the human body. Importance of omega-3 fatty acids have been reported and emphasized on its potential role to prevent or to ameliorate some certain diseases such as hypertension, diabetes, Crohn disease, rheumatoid arthritis, other inflammatory and autoimmune disorders, and cancer (Simopoulos 1999; Connor 2000).

Although the oil amount is low, the fatty acid composition of the chestnut oil is important for human nutrition. Chestnuts are rich in unsaturated fatty acid (USFA). The majority of USFA are PUFA. For 17 Portuguese chestnut cultivars, the oil samples were analyzed and results revealed that percent of saturated fatty acids (SFA) and USFA were 17% and 83%, respectively. Additionally, USFA was examined separately as monounsaturated fatty acid (MUFA) and PUFA. MUFA and PUFA were found to comprise 31% and 52% of total fatty acids, respectively. Palmitic acid was the most important one in the group of SFA and ranged from 12.54% (cv. Aveleira) to 16.80% (cv. Carreiro). As a MUFA, oleic acid was the main fatty acid of Portuguese chestnut oils varying from 20.66% cv. Cota to 37.60% for cv. Trigueira. Linoleic acid (MUFA) was the major compound of fatty acid profiles of oil samples, being in different levels from 37.57% in cv. Trigueira to 50.93% in cv. Lada (Borges et al. 2007).

Barreira et al. (2009) studied the fatty acid composition of four Portuguese chestnut cultivars (Aveleira, Boa Ventura, Judia and Longal) from the "Castanha da Terra Fria" protected designation of origin. As given in Table 8.2, oil samples have 14 common fatty acids including C14:0, C15:0, C16:0, C:17:0, C18:0, C20:0, C22:0, C24:0 (SFA), C16:1, C17:1, C18:1, C20:1 (MUFA), C18:2, C18:3 (PUFA). SFA content of samples changed in the range of 16.2–19.4%. According to study results, main fatty acid groups were MUFA and PUFA ranged between 30.9–38.7% and 42.0–51.9%, respectively. Barreira et al. (2009) also reported linoleic (37.9–45.5%), oleic (29.6–37.4%) and palmitic (14.2–17.3%) acids as main fatty acids of chestnut cultivars.

Similar fatty acid compositions for chestnut cultivars originated from Turkey and Greece have been reported and according to these studies, abundant fatty acids were C16:0 for SFA, C18:1 for MUFA, C18:2 for PUFA (Barreira et al. 2012a; Kalogeropoulos et al. 2013). In another study, Ferreira-Cardoso et al. (1998) also studied the fatty acid composition of chestnut oils, but the analysis was focused on two fractions, non-polar and polar lipids. In a non-polar fraction, lauric acid was found to be major saturated fatty acid, in the range of 17.73–64.16%, as oleic acid (8.59–27.44%) and linoleic acid (15.57–33.17%) were the main fatty acid members of MUFA and PUFA, respectively. In a polar fraction, palmitic acid was identified

Table 8.2 Fatty acid composition of chestnut oil

	Portuguese chestnut cultivars (%)[a]	Portuguese chestnut cultivars (%)[b]	Turkish chestnut (%)[c]	Greece chestnut (%)[d]
Saturated fatty acids				
C14:0	0.11–0.31	0.11–0.16	0.14	0.14
C15:0	0.08–0.23	0.09–0.13	–	0.19
C16:0	12.54–16.80	14.2–17.3	14.0	20.61
C17:0	0.10–0.20	0.13–0.33	0.11	0.30
C18:0	0.70–1.12	0.86–0.95	0.82	1.16
C20:0	0.20–0.34	0.31–0.40	0.36	0.84
C22:0	0.12–0.42	0.23–0.33	0.26	0.23
C24.0	0.05–0.19	0.07–0.12	0.14	–
∑SFA	14.08–18-64	16.2–19.4	15.6	23.54
Unsaturated fatty acids				
C16:1	0.62–1.38	0.28–0.34	0.37	0.75
C17:1	0.06–0.12	0.12–0.17	–	–
C18:1	20.66–37.60	29.6–37.4	32.00	27.01
C20:1	0.44–0.81	0.73–0.83	0.65	0.47
∑MUFA	22.46–39.29	30.9–38.7	33.00	28.44
C18:2	37.57–50.93	37.9–45.5	45.00	40.11
C18:3	4.40–10.02	4.0–6.4	6.00	6.01
∑PUFA	41.98–60.11	42.0–51.9	51.00	46.12

[a]Borges et al. (2007)
[b]Barreira et al. (2009)
[c]Barreira et al. (2012)
[d]Kalogeropoulos et al. (2013)

as a predominant SFA, varying from 20.89% to 30.59%. Oleic acid (8.11–30.81%) was detected as a major MUFA. Linoleic acid (PUFA) was also detected in high levels within the range of 43.71–55.75%.

Antioxidant activity is another significant functional property of foods, and it is evaluated for edible oils, since including constituents with high antioxidant activity. Phospholipids content is in focus of researches from this point of view and considered as important for vegetable oil due to their antioxidant potentials. In sardine oil, phospholipids exhibited strong antioxidant activity (Saito and Ishihara 1997). These constituents could play as synergists with α-tocopherol (Bandarra et al. 1999) as well. Besides, phospholipids have been reported as a potential metal scavenger (Pokorný and Korczak 2001). As a result, phospholipids content is one of the important topics for chestnut oil. However, limited experimental studies were reported in the literature to this extent. One of these exceptional studies has been conducted by Zlatanov et al. (2013), in which the phospholipid composition of sweet chestnut oil obtained from the region of southern Bulgaria was determined and reported amount was around 49 g/kg oil. Phosphatidylcholine, phosphatidylethanolamine, phosphatidic acids and phosphatidylinositol were predominant phospholipids with the amounts of 277, 167, 155, and 112 g/kg, respectively (Table 8.3).

Table 8.3 Phospholipid
content of chestnut oil

Phospholipid[a]	Content (mg/kg)
Phosphatidylcholine	277
Phosphatidylinositol	167
Phosphatidylethanolamine	112
Phosphatidic acids	155
Monophosphatidylglycerol	12
Diphosphatidylglycerol	64
Phosphatidylserine	39
Lysophosphatidylcholine	60
Lysophosphatidylethanolamine	69
Sphingomyelin	45

[a]Zlatanov et al. (2013)

Monophosphatidylglycerol, phosphatidylglycerol, phosphatidylserine, lysophos-phatidylcholine, lysophosphatidylethanolamine and sphingomyelin were also iden-tified in chestnut oil, but their levels were less than 100 g/kg (Zlatanov et al. 2013).

4 Minor Bioactive Lipids in Fruit Fixed Oil

Although minor bioactive compounds of vegetable oils cover a wide range of con-stituents with strong functional activities, there is limited information about minor compounds of chestnut oil. Zlatanov et al. (2013) reported one of the examples in which minor components of sweet chestnut oil from Bulgaria were examined. In that study, the amount of unsaponifiable compounds in the oil sample was around 32 g/kg oil. Sterols, phospholipids and tocopherols of oil were investigated and the amounts of these compounds were 8 g/kg oil, 49 g/kg oil, and 1920 mg/kg oil, respectively.

Sterols synthesized by animals and plants play a key role in functional activities. Plant sterols are naturally found in vegetable products, especially oils. Although there are many kinds of sterols identified, β-sitosterol, campesterol, and stigmas-terol are the most abundant ones in the plant kingdom (Piironen et al. 2000, 2003). Phytosterols exhibited some beneficial health effects including the cholesterol-lowering effect. Moreover, sterols have been recently shown to be related to the prevention of age-related diseases. Extensive discussion about effects of sterols on human body has been also handled by Law (2000), Quilez et al. (2003), and Rudkowska (2010). In this sense, sterol content of chestnut oil is of great interest. Total amount of sterol in chestnut oil was reported to be as 8 g/kg oil. There was apparent difference in between the free (812 g/kg) and esterified sterol forms (188 g/kg) in chestnut oil. The individual sterols in free and esterified form of chestnut oil were tabulated in Table 8.4. Sterol profile of chestnut oil sample was characterized by a relatively high content of β-sitosterol (569 g/kg in free sterols and 736 g/kg in esterified sterols), followed by stigmasterol (301 g/kg in free sterols and 73 g/kg in esterified sterols) and campesterol (117 g/kg in free sterols and 155 g/kg in esteri-fied sterols) (Zlatanov et al. 2013). Chestnut oil sample was examined for its sterol

Table 8.4 Composition of free and esterified sterols in sterol fraction (g/kg)

Sterol	Free sterols	Esterified sterols
Cholesterol	3	20
Brassicasterol	1	4
Campesterol	117	155
Stigmasterol	301	73
β-Sitosterol	569	736
Δ^5-Avenasterol	4	7
Δ^7-Stigmasterol	3	2
Δ-Avenasterol	2	3

Zlatanov et al. (2013)

profile by Kalogeropoulos et al. (2013), and the level of total sterols was found to be 41.25 mg/100 g fresh weight in Greece chestnut samples. Among the identified sterols, β-sitosterol was the major one (20.84 mg/100 g fresh weight), followed by stigmasterol (11.81 mg/100 g fresh weight) and campesterol (4.98 mg/100 g fresh weight) (Kalogeropoulos et al. 2013).

Another minor group affecting the functional power of vegetable oils are tocopherols and tocotrienols being strong antioxidants (Lee and Wan 2000). Antioxidant potential changes depending on isomer forms of these compounds. Tocotrienols are introduced as powerful antioxidants. Moreover, their significant biological activities such as hypocholesterolemic, anti-thrombotic and anti-tumor effects have been emphasized, since having potential against cardiovascular disease and cancer. Health-promoting effects of tocotrienols were discussed in detail by Theriault et al. (1999).

Tocopherols and tocotrienols composition of chestnut fruits and its oil sample were presented in Table 8.5. γ-tocopherol was the major compound being in the range of 3770–4784 ng/g fresh fruit in Portuguese chestnut varieties (Aveleira, Boaventura, Judia, and Longal). In the same work, the abundant tocotrienol isomer was γ- tocotrienol in chestnut oils and its amount changed between 137 and 420 ng/g fresh fruit (Barreira et al. 2009). Bulgarian sweet chestnut oil was found to be a poor source of γ-tocopherol, δ- tocopherol and γ- tocotrienol compared to Portuguese chestnut oils. Moreover, δ- tocotrienol was not identified in Bulgarian sweet chestnut oil sample (Zlatanov et al. 2013). In another study, the tocopherol profile of chestnut samples from Turkey was determined and results showed that γ- tocopherol was major isomer with the amount of 1.0 mg/100 g dw (Barreira et al. 2012). Kalogeropoulos et al. (2013), in their studies about Greece chestnut samples, reported that sum of β- and γ-tocopherol isomers were higher than the other tocopherol isomers (α- and δ- tocopherol).

5   Health-Promoting Traits of Fruit Oil and Oil Constituents

Due to its potential health beneficial effects, chestnuts and chestnut oil popularity have been increased and taken part in the human diet. Literature indicated chestnuts as a good source of essential fatty acids. PUFA and MUFA levels of chestnut oil are

Table 8.5 Tocopherol and tocotrienol composition of chestnut and chestnut oils

	Portuguese chestnut (ng/g fresh fruit)[a]	Bulgarian sweet chestnut oil (g/kg oil)[b]	Turkish chestnut (mg/100 g dw)[c]	Greece chestnut (mg/100 g fresh weight)[d]
Tocopherol				
α-tocopherol	22–100	23	0.0019	1.51
β-tocopherol	–	Trace	–	7.12[e]
γ-tocopherol	3770–4784	927	1.0	
δ- tocopherol	195–332	43	0.07	0.31
Tocotrienol				
α-tocotrienol	–	–	–	
β- tocotrienol	–	–	–	
γ- tocotrienol	137–420	7	–	
δ- tocotrienol	64–83	–	–	

[a]Barreira et al. (2009)
[b]Zlatanov et al. (2013)
[c]Barreira et al. (2012)
[d]Kalogeropoulos et al. (2013)
[e]This value is given sum of β- and γ-tocopherol isomers

at satisfied levels which are thought to be useful in order to protect consumers from cardiovascular diseases and some cancer types (Whelan and Rust 2006). Reports emphasized that the increasing linoleic acid intake from diet could decrease plasma cholesterol (Horrobin and Huang 1987).

Sterols are significant minor constituents of chestnut oil and show reducing the effect on cholesterol absorption, as a result of reducing the serum total and LDL cholesterol levels in a human. Health experts recommend sterols intake from daily diet to reduce the risk of coronary heart disease (Wester 2000; AbuMweis and Jones 2008). Moreover, phytosterols could be anti-atherogenic effects, especially, β-sitosterol could have immune stimulating and anti-inflammatory activities. Plant sterols also could play an important role in different types of cancers such as colorectal, breast and prostate cancers (Trautwein and Demonty 2007).

Chestnut oil is also rich in γ-tocopherol and γ- tocotrienol. In recent years, γ-tocopherol has taken more attention in the studies about human health due to its potential anti-inflammatory and antioxidant properties (Dietrich et al. 2006). γ- tocotrienol has also some health properties such as hypocholesterolemic, anti-cancer and neuroprotective effects.

References

AbuMweis, S. S., & Jones, P. J. (2008). Cholesterol-lowering effect of plant sterols. *Current Atherosclerosis Reports, 10*(6), 467.
Aparicio, R., & Aparicio-Ruíz, R. (2000). Authentication of vegetable oils by chromatographic techniques. *Journal of Chromatography A, 881*(1–2), 93–104.

Bandarra, N. M., Campos, R. M., Batista, I., Nunes, M. L., & Empis, J. M. (1999). Antioxidant synergy of α-tocopherol and phospholipids. *Journal of the American Oil Chemists' Society, 76*(8), 905–913.

Barakat, A., DiLoreto, D. S., Zhang, Y., Smith, C., Baier, K., Powell, W. A., Wheeler, N., Sederoff, R., & Carlson, J. E. (2009). Comparison of the transcriptomes of American chestnut (*Castanea dentata*) and Chinese chestnut (*Castanea mollissima*) in response to the chestnut blight infection. *BMC Plant Biology, 9*(1), 51.

Barreira, J. C., Alves, R. C., Casal, S., Ferreira, I. C., Oliveira, M. B. P., & Pereira, J. A. (2009). Vitamin E profile as a reliable authenticity discrimination factor between chestnut (*Castanea sativa* mill.) cultivars. *Journal of Agricultural and Food Chemistry, 57*(12), 5524–5528.

Barreira, J. C., Antonio, A. L., Günaydi, T., Alkan, H., Bento, A., Botelho, M. L., & Ferreira, I. C. (2012a). Chemometric characterization of gamma irradiated chestnuts from Turkey. *Radiation Physics and Chemistry, 81*(9), 1520–1524.

Barreira, J. C., Casal, S., Ferreira, I. C., Peres, A. M., Pereira, J. A., & Oliveira, M. B. P. (2012b). Chemical characterization of chestnut cultivars from three consecutive years: Chemometrics and contribution for authentication. *Food and Chemical Toxicology, 50*(7), 2311–2317.

Borges, O. P., Carvalho, J. S., Correia, P. R., & Silva, A. P. (2007). Lipid and fatty acid profiles of *Castanea sativa* Mill. Chestnuts of 17 native Portuguese cultivars. *Journal of Food Composition and Analysis, 20*(2), 80–89.

Buchgraber, M., Ulberth, F., Emons, H., & Anklam, E. (2004). Triacylglycerol profiling by using chromatographic techniques. *European Journal of Lipid Science and Technology, 106*(9), 621–648.

Cert, A., Moreda, W., & Pérez-Camino, M. (2000). Chromatographic analysis of minor constituents in vegetable oils. *Journal of Chromatography A, 881*(1–2), 131–148.

Conedera, M., Manetti, M. C., Giudici, F., & Amorini, E. (2004). Distribution and economic potential of the sweet chestnut (*Castanea sativa* mill.) in Europe. *Ecologia Mediterranea, 30*(2), 179–193.

Conedera, M., Tinner, W., Krebs, P., De Rigo, D. & Caudullo, G. (2016) Castanea sativa in Europe: distribution, habitat, usage and threats. European Atlas of Forest Tree Species (ed. by J San-Miguel-Ayanz, D De Rigo, G Caudullo, T Durrant & A Mauri), pp. 78–79. Publications Office of the European Union, Luxembourg.

Connor, W. E. (2000). Importance of n-3 fatty acids in health and disease. *The American Journal of Clinical Nutrition, 71*(1), 171S–175S.

De Vasconcelos, M. C., Bennett, R. N., Rosa, E. A., & Ferreira-Cardoso, J. V. (2010). Composition of European chestnut (*Castanea sativa* mill.) and association with health effects: Fresh and processed products. *Journal of the Science of Food and Agriculture, 90*(10), 1578–1589.

Dietrich, M., Traber, M. G., Jacques, P. F., Cross, C. E., Hu, Y., & Block, G. (2006). Does γ-tocopherol play a role in the primary prevention of heart disease and cancer? A review. *Journal of the American College of Nutrition, 25*(4), 292–299.

Durazzo, A., Turfani, V., Azzini, E., Maiani, G., & Carcea, M. (2013). Phenols, lignans and antioxidant properties of legume and sweet chestnut flours. *Food Chemistry, 140*(4), 666–671.

Ertürk, Ü., Mert, C., & Soylu, A. (2006). Chemical composition of fruits of some important chestnut cultivars. *Brazilian Archives of Biology and Technology, 49*(2), 183–188.

FAO F. (2018). Food and Agriculture Organization of the United Nation. *The Statistics Division*. http://www.fao.org.

Ferreira-Cardoso J., Sequeira, C.A., Torres-Pereira, J.M.G., Rodrigues, L., & Gomes, E.F. (1998) *Lipid composition of Castanea sativa Mill fruits of some native Portuguese cultivars* (pp 133–138). Bordeaux: Second international symposium on chestnut.

Horrobin, D., & Huang, Y.-S. (1987). The role of linoleic acid and its metabolites in the lowering of plasma cholesterol and the prevention of cardiovascular disease. *International Journal of Cardiology, 17*(3), 241–255.

Jandacek, R.J. (2017) Linoleic acid: A nutritional quandary. In: Healthcare. vol 2. Multidisciplinary Digital Publishing Institute, Basel (p 25).

Kalogeropoulos, N., Chiou, A., Ioannou, M. S., & Karathanos, V. T. (2013). Nutritional evaluation and health promoting activities of nuts and seeds cultivated in Greece. *International Journal of Food Sciences and Nutrition, 64*(6), 757–767.

Law, M. (2000). Plant sterol and stanol margarines and health. *BMJ [British Medical Journal], 320*(7238), 861.

Lee, C.-Y. J., & Wan, F. (2000). Vitamin E supplementation improves cell-mediated immunity and oxidative stress of Asian men and women. *The Journal of Nutrition, 130*(12), 2932–2937.

Man, Y. C., Haryati, T., Ghazali, H., & Asbi, B. (1999). Composition and thermal profile of crude palm oil and its products. *Journal of the American Oil Chemists' Society, 76*(2), 237–242.

Piironen, V., Toivo, J., & Lampi, A.-M. (2000). Natural sources of dietary plant sterols. *Journal of Food Composition and Analysis, 13*(4), 619–624.

Piironen, V., Toivo, J., Puupponen-Pimiä, R., & Lampi, A. M. (2003). Plant sterols in vegetables, fruits and berries. *Journal of the Science of Food and Agriculture, 83*(4), 330–337.

Pokorny, J., & Korczak, J. (2001). Preparation of natural antioxidants. In J. Pokorny, N. Yanishlieva, & M. Gordon (Eds.), *Antioxidants in Food. Practical Applications* (pp. 311–330). Boca Raton: CRC Press.

Quilez, J., Garcia-Lorda, P., & Salas-Salvado, J. (2003). Potential uses and benefits of phytosterols in diet: Present situation and future directions. *Clinical Nutrition, 22*(4), 343–351.

Rudkowska, I. (2010). Plant sterols and stanols for healthy aging. *Maturitas, 66*(2), 158–162.

Sacchetti, G., Pinnavaia, G., Guidolin, E., & Dalla Rosa, M. (2004). Effects of extrusion temperature and feed composition on the functional, physical and sensory properties of chestnut and rice flour-based snack-like products. *Food Research International, 37*(5), 527–534.

Saito, H., & Ishihara, K. (1997). Antioxidant activity and active sites of phospholipids as antioxidants. *Journal of the American Oil Chemists' Society, 74*(12), 1531–1536.

Simopoulos, A. P. (1999). Essential fatty acids in health and chronic disease. *The American Journal of Clinical Nutrition, 70*(3), 560s–569s.

Sovova, H., Zarevucka, M., Vacek, M., & Stránský, K. (2001). Solubility of two vegetable oils in supercritical CO_2. *Journal of Supercritical Fluids, 20*(1), 15–28.

Theriault, A., Chao, J.-T., Wang, Q., Gapor, A., & Adeli, K. (1999). Tocotrienol: A review of its therapeutic potential. *Clinical Biochemistry, 32*(5), 309–319.

Trautwein, E. A., & Demonty, I. (2007). Phytosterols: Natural compounds with established and emerging health benefits. *Oléagineux, Corps Gras, Lipides, 14*(5), 259–266.

Ulberth, F., & Buchgraber, M. (2000). Authenticity of fats and oils. *European Journal of Lipid Science and Technology, 102*(11), 687–694.

Wester, I. (2000). Cholesterol-lowering effect of plant sterols. *European Journal of Lipid Science and Technology, 102*(1), 37–44.

Whelan, J., & Rust, C. (2006). Innovative dietary sources of n-3 fatty acids. *Annual Review of Nutrition, 26*, 75–103.

Zlatanov, M. D., Antova, G. A., Angelova-Romova, M. J., & Teneva, O. T. (2013). Lipid composition of *Castanea sativa* mill. and *Aesculus hippocastanum* fruit oils. *Journal of the Science of Food andAagriculture, 93*(3), 661–666.

Chapter 9
Coconut (*Cocos nucifera*) Oil

Ali Osman

Abstract The coconut (*Cocos nucifera* L., family Arecaceae) is an important fruit tree in the world. Copra, the dried kernel, is mainly used for oil extraction. Coconut oil is one of the most important edible oils for domestic use. The oil is rich in medium chain fatty acids (MCFA) and exhibits good digestibility. Different coconut oils are produced from different parts of coconut by different means. Copra as well as refined, bleached and deodorized (RBD) oils are produced from dried coconut kernel, with a difference that RBD oil undergoes chemical refinement and bleaching. The brown testa of the coconut is used for the preparation of coconut testa oil (CTO), which is actually a byproduct of coconut oil preparation. Compared to corpa oil (CO) and RBD oils, virgin coconut oil (VCO) is extracted depend on a "wet method" using fresh coconut milk. As there is no specific method of preparation of VCO has been established, all types of preparations that do not involve refinement and alterations in the oil are considered as a virgin. This chapter will cover the preparation of different oils from coconut, chemistry of coconut oils, blending of coconut oil with other edible oils and biological activities of coconut oils.

Keywords Phenolic compounds · Antimicrobial activity · Anti-inflammatory · Extraction

1 Introduction

The coconut is an important fruit tree in the world providing food for millions of people, especially in the tropical and subtropical regoins and with its many uses it is often called the tree of life. Copra, the dried kernel which is mainly used for oil extraction, contains about 65–75% oil (DebMandal and Mandal 2011). Coconut oil is one of the most important edible oils for domestic use. The oil is rich in medium chain fatty acids (MCFA) and exhibits good digestibility (Shahidi 2006). Coconut

A. Osman (✉)
Agricultural Biochemistry Department, Faculty of Agriculture, Zagazig University, Zagazig, Egypt

© Springer Nature Switzerland AG 2019
M. F. Ramadan (ed.), *Fruit Oils: Chemistry and Functionality*,
https://doi.org/10.1007/978-3-030-12473-1_9

oil is produced by crushing copra, the dried kernel, which contains about 60–65% oil. The oil has the natural sweet taste of coconut and contains 92% of saturated fatty acids (in the form of triglycerides), most of them (about 70%) are lower chain saturated fatty acids known as MCFA. MCFA are not common to different vegetable oils with lauric acid (45–56%) as a main MCFA in coconut oil. Various fractions of coconut oil have medium chain triglycerides which considered as an excellent solvent for flavors, essences, and emulsifiers. These fatty acids are used in the preparation of emulsifiers, as drugs and also in cosmetics (Krishna et al. 2010). Various methods have been developed to extract coconut oil, either through dry or wet processing. Dry processing is the most widely used form of extraction. Clean, ground and steamed copra are pressed by wedge press, screw press or hydraulic press to obtain coconut oil, which then goes through the refining, bleaching, and deodorizing processes (O'Brien and Timms 2004). This chapter will cover the preparation of different oils from coconut, chemistry of coconut oils, blending of coconut oil with other vegetable oils and biological activities of coconut oils.

2 Preparation of Different Oils from Coconut

Different coconut oils are produced from different parts of coconut by different methods. Copra as well as refined, bleached and deodorized (RBD) oils are produced from dried coconut kernel, with a difference that RBD oil undergoes chemical refinement and bleaching. The brown testa of the coconut is used for the preparation of coconut testa oil (CTO), which is actually a byproduct of coconut oil preparation. Compared to corpa oil (CO) and RBD oils, virgin coconut oil (VCO) is extracted depend on a "wet method" using fresh coconut milk. As there is no specific method of preparation of VCO has been established, all types of preparations that do not involve refinement and alterations in the oil are considered as a virgin (Narayanankutty et al. 2018).

2.1 Corpa Oil

Copra is the dried coconut kernel. The fresh coconut kernel is dried in the oven or sunlight and the oil is collected by mechanical milling. The oil is collected and sundried to remove the moisture content (Narayanankutty et al. 2018).

2.2 Testa Oil

Coconut testa oil (CTO) is the emerging form, which can be extracted using isopropyl alcohol from the coconut testa (Zhang et al. 2016). CTO is best obtained at a temperature of 60 °C for a period of 3 h with the substrate to solvent ratio of 1:4 and having a yield up to 63–76%. Since the extraction involves chemical solvents, CTO has not yet been widely used for edible purposes.

2.3 Virgin Coconut Oil

Virgin coconut oil (VCO) is extracted naturally from the fresh coconut kernel without the application of high temperature or chemical treatment. Based on the mode of preparation, several types of VCO are available (Narayanankutty et al. 2018).

2.3.1 Cold Extraction (C-VCO)

Cold processing is the method of VCO extraction without the aid of heat. In this method, the coconut milk is subjected to chilling (2–8 °C) overnight and the separated oil is collected by centrifugation, filtered and stored. This is a simpler and cheapest method available (Narayanankutty et al. 2018).

2.3.2 Hot Extraction (H-VCO)

Hot extraction is traditionally used in Southern India for VCO preparation. In this method, the coconut milk is subjected to a moderate temperature of up to 100 °C. The processing lasts for 60 min or until the oil get completely separated from the milk then the oil is collected by filtration. This heating process helps to increase the release of bound phenolic acids into the oil and also yield is much higher. The oil prepared in this way is being used conventionally in the Ayurvedic system of medicine for skin ailments, especially for children (Narayanankutty et al. 2018).

2.3.3 Fermentation Technique (F-VCO)

The fermentation method uses bacterial activity to generate VCO has also been proposed. It is mainly of two types-natural fermentation as well as induced fermentation. In the natural fermentation method, the freshly grated coconut kernel is extracted with its water to collect the coconut milk. It is then kept for 24–48 h under room temperature (or up to a temperature of 45 °C) to allow fermentation and separation of oil layer, which is scooped out, filtered and stored (Nevin and Rajamohan 2010). Masyithah (2017) prepared VCO by induced fermentation technique, where they used *Saccharomyces cerevisiae* and *Lactobacillus plantarum* (strain 1041 IAM) for the extraction of VCO from coconut milk. *L. plantarum*, and *L. delbrueckii* are also used in the fermentation process. However, studies using induced fermentation are quite rare and VCO produced by natural fermentation method is often regarded as F-VCO.

2.3.4 Enzymatic Extraction Technique

Extraction of oil can be carried out using the enzymes in the aqueous extraction process. This is due to the fact that plant cell walls consist of complex carbohydrate molecules such as cellulose, hemicellulose, mannans, galactomannans, arabinogalactans, pectin substances and protein (Shah et al. 2005). Coconut meat contains about 10% of carbohydrate, in which 50% is cellulose and 75% of the cellulose is made up with α-cellulose (Rosenthal et al. 1996). Oil can be found inside plant cells, linked with proteins and a wide range of carbohydrate such as starch, cellulose, hemicellulose, and pectins. Cell-wall degrading enzymes can be used to extract oil by solubilizing the structural cell wall components of the oilseed. Man et al. (1996) successfully extracted coconut oil with 1% enzyme mixture of cellulase, α-amylase, polygalacturonase, and protease with an oil yield of 74%. The polygalacturonase hydrolyses α-linkages of polygalacturonic acid of the polymer randomly from the ends. An α-amylase randomly hydrolyzed α-linkages to liquefy starch and produced maltose, whereas bacterial proteases were used to hydrolyze the plant protein. The study showed that different enzymes were required to degrade components of the structural cell wall including mannan, galactomannan, arabinoxylogactan and cellulose.

2.3.5 Wet Extraction

Wet processing or aqueous processing is the term used for the extraction of coconut oil directly from coconut milk. This method eliminates the use of a solvent, which reportedly may lower the investment cost and energy requirements. Furthermore, it eliminates the RBD process (Villarino et al. 2007). Even though the concept appears potentially attractive, however, the method yields comparatively low content of oil, which has discouraged its commercial application (Rosenthal et al. 1996). The wet processing can only be carried out by means of coconut milk by breaking the emulsion. This is rather difficult due to the high stability emulsion of coconut milk. Destabilization can be done through three mechanisms. The first stage is creaming by the action of gravitational force resulting in two phases, with the higher specific gravity takes place at the top phase and the lower specific gravity phase moves downward. The second stage is flocculation or clustering in which the oil phase moves as a group, which does not involve the rupture of the interfacial film that normally surrounds each globule and therefore does not change the original globule. The last stage, coalescence is the most critical phase in destabilization. During this stage, the interfacial area is ruptured; the globules joined together and reduced the interfacial area (Onsaard et al. 2005). The wet process appears more desirable due to the free usage of chemical solvents, thus more environmental friendly than the solvent extraction. The method is also much simpler, which can be carried out at home by anyone who is interested in producing natural oil.

2.3.6   Chilling, Freezing and Thawing Techniques

Attempts have been made to break the protein stabilized oil-in-water emulsion including heating and centrifugation, freezing and thawing, chilling and thawing wherein the coconut cream obtained after centrifugation (Seow and Gwee 1997). The emulsion was centrifuged before chilling and thawing to allow better packing of the coconut oil globules. The temperature used were 10 °C and −4 °C for chilling and freezing process, respectively while the thawing process was carried out in a water bath at 40 °C until the coconut cream reached room temperature (25 °C). In addition, this action also helps in removing undissolved solids after extraction. The removal of solids present in high percentages in the dispersion of oilseed was important for efficient recovery of oil by centrifugation (Rosenthal et al. 1996). The centrifugation step was followed to enable the packing of cream oil globule to crystallize on lowering the temperature. Centrifugation process was carried out from 2000 to 5000 g up to 6 min. During thawing, the oil coalesced due to loss of spherical shape and formed large droplets of varying sizes. Robledano-Luzuriage and Krauss-Maffei were two processes known to apply freeze and thaw operation in the extraction of coconut oil (Marina et al. 2009). In the Robledano-Luzuriage process, fresh coconut kernel was comminuted and pressed to obtain approximately equal amounts of emulsion and residue. The residue was pressed again to obtain more emulsion and residue. The emulsion was centrifuged to obtain a cream, skim milk and some solids or protein. The cream was subjected to enzymatic action under closely controlled temperature and pH conditions. After the freeze-thaw operation, the cream was centrifuged again to obtain the oil. The protein in the skim milk was coagulated by heating, subsequently filtered and dried to produce protein concentrate. The oil recovery reported in the Krauss-Maffei was 89%, which was less than the conventional expeller process (95%). In this technique, husked coconuts were autoclaved, shelled, and the coconut meat (the white solid endosperm inside the coconut fruit) first sent through cutter and subsequently through a roller mill. Then it was pressed in a hydraulic press and the emulsion was centrifuged to obtain the cream and skim milk. The cream was heated to 92 °C and filtered to obtain high-quality oil. The skim milk is heated to 98 °C in a flow heater to coagulate protein, which was separated by centrifuging then drying. The residue from the hydraulic press was dried and ground to obtain edible coconut flour. The study showed that a high recovery of oil was obtained but the temperature employed was slightly high which might destroy some of its minor components such as phenolic compounds.

3   Chemical Composition of Coconut Oil

3.1   Triacylglycerol and Fatty Acid Composition

Coconut oil contains a high proportion of glycerides of lower chain fatty acids. The oil is highly stable towards atmospheric oxidation. The oil is characterized by a low iodine value, high saponification value, high saturated fatty acids content and the oil is liquid at 27 °C. Medium chain triglycerides (MCT) are a class of lipids in which three saturated fats are bound to a glycerol backbone. What distinguishes MCT from other triglycerides is the fact that each fat molecule is between 6 and 12 carbons in length (Babayan 1988). MCT are a component of many foods, with coconut and palm oils being the dietary sources with the highest concentration of MCT. MCT are also available as a dietary supplement (Heydinger and Nakhasi 1996). MCT have a different pattern of absorption and utilization than long-chain triglycerides (LCT) that makeup 97% of dietary fats. For absorption of LCT to occur, the fatty acid chains must be separated from the glycerol backbone by the lipase enzyme. These fatty acids form micelles, are then absorbed and reattached to glycerol, and the resultant triglycerides travel through the lymphatics *en route* to the bloodstream. Up to 30% of MCT are absorbed intact across the intestinal barrier and directly enter the portal vein. This allows for much quicker absorption and utilization of MCT compared to LCT. MCT are transported into the mitochondria independent of the carnitine shuttle, which is necessary for LCT mitochondrial absorption. Oxidation of MCT provides 8.3 calories per gram, while LCT provides 9.2 calories per gram (Hoagland and Snider 1943). All fats and oils are composed of triglyceride molecules, which are triesters of glycerol and fatty acids. The fats upon hydrolysis yield fatty acids and glycerol. There are two methods of classifying fatty acids, monounsaturated fatty acids (MUFA), and polyunsaturated fatty acids (PUFA). The second method of classification is based on molecular size or length of the carbon chain in the fatty acid. The vast majority of the fats and oils whether they are saturated or unsaturated or from an animal or a plant, are composed of LCT. All fats we eat consist of LCT while, coconut oil is unique because it is composed predominantly of MCT. The size of the fatty acid is extremely important because physiological effects of medium-chain fatty acids in coconut oil are distinctly different from the long-chain fatty acids more commonly found in our diet (Furman et al. 1965). It is the MCT in coconut oil that makes it different from all other fats and for the most part gives it its unique character and healing properties. Almost all of the medium-chain triglycerides used in research, medicine, and food products come from coconut oil. MCT are easily digested, absorbed, and put to use nourishing the body. Unlike other fats, they put little strain on the digestive system and provide a quick source of energy necessary to promote healing. This is important for patients who are using every ounce of strength they have to overcome serious illness or injury. It's no wonder why MCT are added to infant formulas. MCT is not only found in coconut oil but also are natural and vital components of human breast milk. MCT are considered essential nutrients for infants as well as for people with serious digestive problems like cystic fibrosis (St-Onge et al. 2003). Like other essential nutrients, one must get them directly from the diet. Bhatnagar et al. (2009) recorded that the coconut (*Cocos nucifera*) contains 55–65% oil, having C12:0 as the major fatty acid. Coconut oil has 90% saturates and is deficient in monounsaturates (6%), polyunsaturates (1%), and total tocopherols (29 mg/kg). However, coconut oil contains medium chain fatty acids (58%), which are easily absorbed into the body.

Fatty acid profiles (Figs. 9.1 and 9.2) are found to be similar in all the different varieties of coconut oils such as CO, VCO or RBD oil. In CTO, there is comparatively

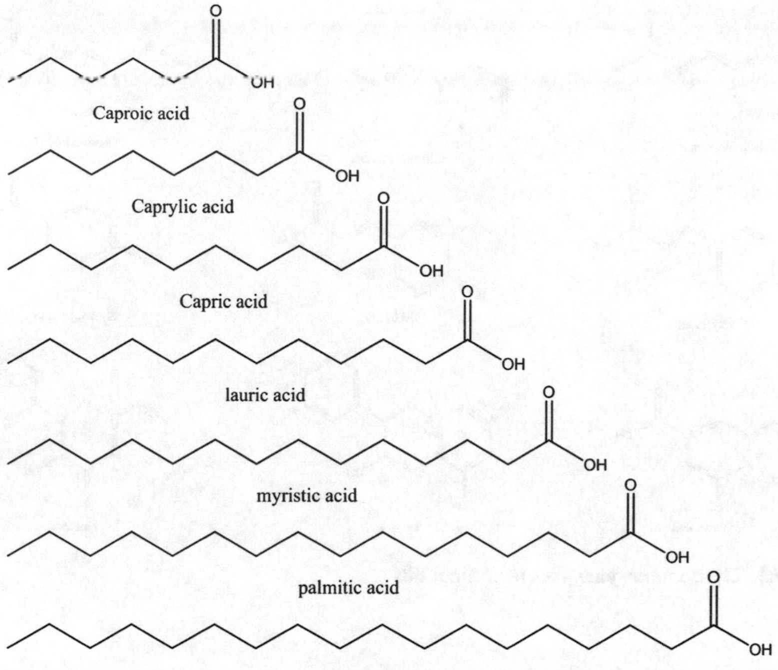

Fig. 9.1 Saturated fatty acids profiles in the different varieties of coconut oils

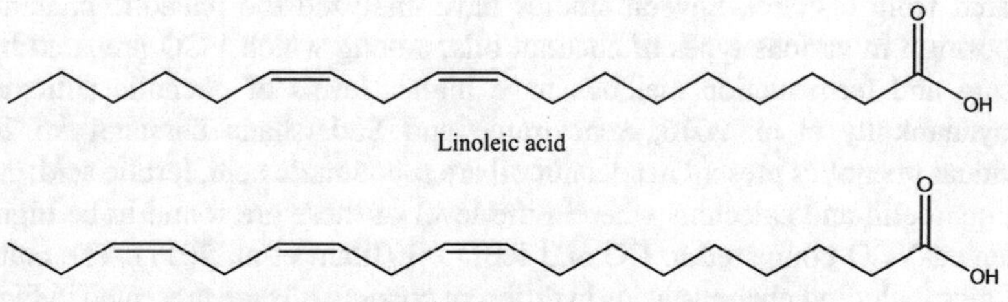

Fig. 9.2 Unsaturated fatty acid profiles in the different varieties of coconut oils

a higher level of unsaturated fatty acids than CO and VCO, with a concomitant reduction in the medium-chain saturated fatty acid (MCFA) (Appaiah et al. 2014). The other oils contain high amounts of MCFA. Among these, lauric acid (C12:0) forms the predominant fatty acid (45–52%), followed by myristic acid (15–19%) and palmitic acid (10–11%). As indicated by the triglyceride composition of these oils, the major triacylglycerol is formed by tri-lauryl glycerols (22.2–23.9%). On the other hand, CTO contains higher levels of Capryl-lauryl-lauric glycerol (18.7%), followed by tri-lauryl glycerol (14.3%) and lauryl-lauryl-oleic glycerol (13.4%) (Appaiah et al. 2014).

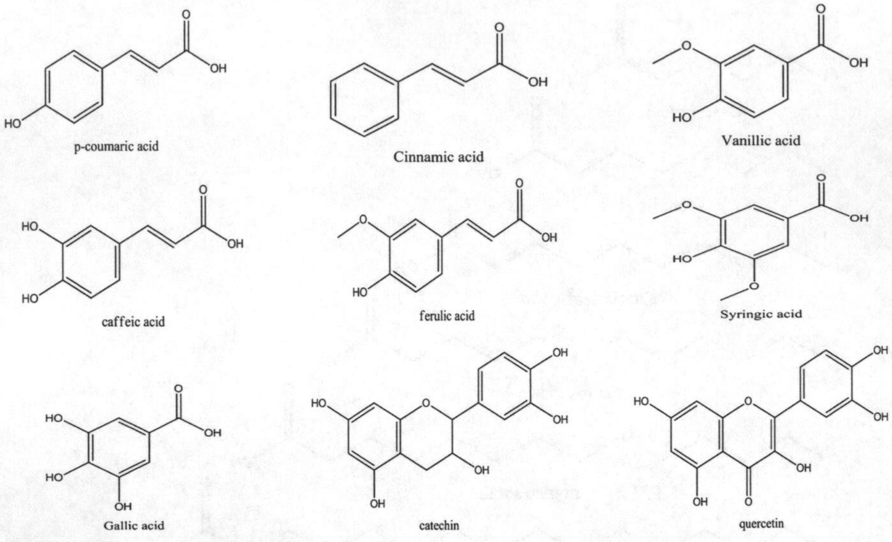

Fig. 9.3 The common phenolics in coconut oils

3.2 Phenolic Compounds Composition

Phenolics are another main group of bioactive compounds present in edible oils prepared from coconut. Several studies have analyzed the phenolic content and composition in various types of coconut oils, among which VCO prepared by hot pressing and fermentation methods have higher levels of phenolic antioxidants (Narayanankutty et al. 2016; Seneviratne and Sudarshana Dissanayake 2008). Individual phenolics present in coconut oil are *p*-coumaric acid, ferulic acid, caffeic acid, quercetin and catechins wherein the level of these are found to be higher in fermented-VCO compared to CO and RBD oil (Illam et al. 2017). The common phenolics including phenolic acids in different coconut oils are presented in Fig. 9.3.

4 Blending of Coconut Oil with Other Vegetable Oils

Coconut oil, because of its long shelf life and melting point of 24.4 °C, was frequently used in the baking industry. However, a negative campaign against saturated fats and tropical oils resulted in the replacement of coconut oil with hydrogenated fats (Enig et al. 1990). Coconut oil is rich in saturated fatty acids (SFA) (93%). However, coconut oil also contains MCFA (60%), especially C12:0 (50%). MCFA (C6:0, C8:0, C10:0, C12:0) are smaller than the standard storage unit of fat (C14), and hence are burnt for energy rather than stored in the body (Kiyasu et al. 1952). Consumption of a diet rich in MCT improves upper body adiposity in overweight men, hence MCT may be considered as a potential tool in the prevention of weight gain and obesity (St-Onge et al. 2003).

C12:0 (50%) and C10:0 (6%), which are found in coconut oil in high amounts, are known for their unique antiviral, antibacterial, and antiprotozoal properties. Lauric acid in coconut oil becomes 2-mono-laurin in the gut and dissolves the lipid envelope that protects most pathogenic bacteria and viruses. It also attacks pathogenic yeast and parasites (Enig 1998). Research has also shown that natural coconut oil in the diet leads to a normalization of body lipids, protects against alcohol damage to the liver, and improves the immune system's anti-inflammatory response (Enig 1993).

Addition of coconut oil to other vegetable oils improves their oxidative stability indicating that coconut oil can be used as a natural antioxidant through the blending process. Coconut oil is very stable against oxidation and hence not prone to peroxide formation. Therefore, the incorporation of coconut oil to other oxidation susceptible vegetable oils increases the stability of the blends. The prolonged use of coconut oil for cooking by coconut oil consumers make their diets deficient in PUFA, MUFA, and natural antioxidants which are present in other vegetable oils. Bhatnagar et al. (2009) prepared blends of coconut oil (20–80% incorporation of coconut oil) with other vegetable oils (i.e. palm, rice bran, sesame, mustard, sunflower, groundnut, safflower, and soybean). Seven blends prepared for coconut oil consumers contained higher amounts of monounsaturates (8–36%), polyunsaturates (4–35%), total tocopherols (111–582 mg/kg), and increased up to 5–33% of DPPH (2,2-diphenyl1-picrylhydrazyl) scavenging activity. In addition, seven blends prepared for non-coconut oil consumers contained 11–13% of medium chain fatty acids. Coconut oil:sunflower oil and coconut oil:rice bran oil blends also exhibited 36.7–89.7% and 66.4–80.5% reductions in peroxide formation in comparison to the individual sunflower oil and rice bran oil, respectively. It was concluded that blending coconut oil with other vegetable oils provides MCFA and oxidative stability to the blends, while coconut oil will be enriched with polyunsaturates, monounsaturates, and natural antioxidants.

5   Biological Effects of Coconut Oil

5.1   Anti-inflammatory

Literature suggested that inflammatory mediators such as IL-6, tumor necrosis factor-α (TNF-α) and expression of nuclear factor-kappa B (NF-κB) may be involved in MTX hepatorenal damage (El-Sheikh et al. 2015). Anti-inflammatory activity of VCO is proven several years back and thereafter, various studies have proposed that the anti-inflammatory activity of VCO and its mechanism of action in various experimental models. Studies conducted by Intahphuak et al. (2010) showed the protective effect of F-VCO in granuloma formation in chemically induced ear and paw oedema models. F-VCO also reduced adjuvant-induced arthritis in rats, by downregulating the expressions of cyclooxygenase, inducible nitric oxide synthase and TNF-α (Vysakh et al. 2014). A study conducted by Zakaria et al. (2011) observed that F-VCO efficiently reduce acute inflammation, however in chronic models, it was found to be less effective. Together with anti-inflammatory activities, anti-nociceptive and analgesic activities are also reported for F-VCO (Intahphuak et al. 2010; Zakaria et al. 2011). Other VCO preparations are not yet evaluated for their anti-inflammatory activity.

Varma et al. (2019) studied the anti-inflammatory in vitro and skin protective properties of VCO. VCO has been traditionally used as a moisturizer since centuries by people in the tropical region. Clinical studies have revealed that VCO improved the symptoms of skin disorders by moisturizing and soothing the skin. However, the mechanistic action of VCO and its benefits on skin has not been elucidated in vitro. The cytotoxicity (CTC50) of VCO was 706.53 and 787.15 mg/mL in THP-1 (Human monocytes) and HaCaT (Human keratinocytes) cells, respectively. VCO inhibited TNF-a (62.34%), IFN-g (42.66%), IL-6 (52.07%), IL-8 (53.98%) and IL-5 (51.57%), respectively in THP-1 cells. Involucrin (INV) and filaggrin (FLG) content increased by 47.53% and 40.45%, respectively in HaCaT cells. VCO increased the expression of Aquaporin-3 (AQP3), involucrin (INV) and filaggrin (FLG) and showed moderate UV protection in HaCaT cells. In vitro skin irritation studies in the Reconstructed human epidermis (RHE) and NIH3T3 cells showed that VCO is a non-skin irritant (IC_{50} >1000 mg/mL) and non-phototoxic (PIF <2). It was suggested that the anti-inflammatory activity of VCO could be due to suppressing inflammatory markers and protecting the skin by enhancing skin barrier function. Overall, the results warrant the use of VCO in skin care formulations.

5.2   Antimicrobial Activity

VCO is traditionally used as an antibacterial agent. It has been shown that F-VCO possesses antibacterial activities against a variety of strains including Candida (Ogbolu et al. 2007) and *Staphylococcus* (Tangwatcharin and Khopaibool 2012). VCO also reduces plaque-related gingivitis, which is a bacterial infection induced oral disease (Peedikayil et al. 2015). Possible antimicrobial activities of VCO may be attributed to lauric acid, which forms the main fatty acids of VCO (Nakatsuji et al. 2009). Monolaurin compounds are the major metabolite which is responsible for its activity (Manohar et al. 2013).

Peng et al. (2016) studied the effect of enhanced VCO, containing about 57.4% triglycerides, 26.8% diglycerides, 1.51% monoglycerides and 14.1% free fatty acids, against clinical mastitis pathogens [*S. aureus* (ATCC 31885), *S. agalactiae* (ATCC 12927) and *S. dysagalactiae* (ATCC 27957)]. The enhanced VCO had shown its antimicrobial efficacy against three potent mastitis causal agents. The amount of enhanced VCO used to treat mastitis-induced cows and the number of treatments applied needed to be reduced to avoid the milk clotting and udders swelling as a result of the acidic characteristics of enhanced VCO.

Pritha and Karpagam (2018) extracted coconut shell oil from raw coconut shells, collected from Thiruvallur District using various solvents (i.e., ethanol, chloroform, acetone, petroleum ether and water). The antibacterial activity against the growth of *Staphylococcus aureus, Enterococcus faecalis, Escherichia coli, Klebsiella pneumonia, Pseudomonas aeruginosa* and *Salmonella typhi* was performed. The antifungal activity against the growth of *Epidermophyton floccosum, Aspergillus niger, Penicillium, Microsporum canis, Candida albicans* and *Aspergillus flavus* were tested. The extracts were compared with standards like novobiocin, amoxillin and ketoconazole for antibacterial and antifungal activity, respectively. The results indicated that the zone of inhibition increased when increasing the concentration of the extracts. Among the five extracts of coconut shell oil, the ethanol extract exhibited maximum antibacterial activity against bacterial strains. Ethanol and petroleum ether extract showed maximum inhibitory activity against *Epidermophyton* and *Candida albicans*.

References

Appaiah, P., Sunil, L., Kumar, P. P., & Krishna, A. G. (2014). Composition of coconut testa, coconut kernel and its oil. *Journal of the American Oil Chemists' Society, 91*, 917–924.

Babayan, V. (1988). Medium chain triglycerides. In *Dietary fat requirements in health and development* (pp. 73–86). Champaign: American Oil Chemists Society.

Bhatnagar, A., Prasanth Kumar, P., Hemavathy, J., & Gopala Krishna, A. (2009). Fatty acid composition, oxidative stability, and radical scavenging activity of vegetable oil blends with coconut oil. *Journal of the American Oil Chemists' Society, 86*, 991–999.

DebMandal, M., & Mandal, S. (2011). Coconut (*Cocos nucifera* L.: Arecaceae): in health promotion and disease prevention. *Asian Pacific Journal of Tropical Medicine, 4*, 241–247.

El-Sheikh, A. A., Morsy, M. A., Abdalla, A. M., Hamouda, A. H., & Alhaider, I. A. (2015). Mechanisms of thymoquinone hepatorenal protection in methotrexate-induced toxicity in rats. *Mediators of Inflammation, 2015*, 859383.

Enig, M. (1993). *Diet, serum cholesterol and coronary heart disease. Coronary heart disease: the dietary sense and nonsense* (pp. 36–60). London: Janus Publishing.

Enig, M. G. (1998). Lauric oils as antimicrobial agents: Theory of effect, scientific rationale. *Nutrients and Foods in AIDS, 17*, 21–100.

Enig, M. G., Atal, S., Keeney, M., & Sampugna, J. (1990). Isomeric trans fatty acids in the US diet. *Journal of the American College of Nutrition, 9*, 471–486.

Furman, R. H., Howard, R. P., Brusco, O. J., & Alaupovic, P. (1965). Effects of medium chain length triglyceride (MCT) on serum lipids and lipoproteins in familial hyperchylomicronemia (dietary fat-induced lipemia) and dietary carbohydrate-accentuated lipemia. *The Journal of Laboratory and Clinical Medicine, 66*, 912–926.

Heydinger, J. A., & Nakhasi, D. K. (1996). Medium chain triacylglycerols. *Journal of Food Lipids, 3*, 251–257.

Hoagland, R., & Snider, G. G. (1943). Digestibility of certain higher saturated fatty acids and triglycerides. *The Journal of Nutrition, 26*, 219–225.

Illam, S. P., Narayanankutty, A., & Raghavamenon, A. C. (2017). Polyphenols of virgin coconut oil prevent pro-oxidant mediated cell death. *Toxicology Mechanisms and Methods, 27*, 442–450.

Intahphuak, S., Khonsung, P., & Panthong, A. (2010). Anti-inflammatory, analgesic, and anti-pyretic activities of virgin coconut oil. *Pharmaceutical Biology, 48*, 151–157.

Kiyasu, J., Bloom, B., & Chaikoff, I. (1952). The portal transport of absorbed fatty acids. *The Journal of Biological Chemistry, 199*, 415–419.

Krishna, A. G., Gaurav, R., Singh, B. A., Kumar, P. P., & Preeti, C. (2010). Coconut oil: Chemistry, production and its applications-a review. *Indian Coconut Journal, 53*, 15–27.

Man, Y. C., Asbi, A., Azudin, M., & Wei, L. (1996). Aqueous enzymatic extraction of coconut oil. *Journal of the American Oil Chemists' Society, 73*, 683–686.

Manohar, V., Echard, B., Perricone, N., Ingram, C., Enig, M., Bagchi, D., & Preuss, H. G. (2013). In vitro and in vivo effects of two coconut oils in comparison to monolaurin on *Staphylococcus aureus*: Rodent studies. *Journal of Medicinal Food, 16*, 499–503.

Marina, A., Che Man, Y., Nazimah, S., & Amin, I. (2009). Monitoring the adulteration of virgin coconut oil by selected vegetable oils using differential scanning calorimetry. *Journal of Food Lipids, 16*, 50–61.

Masyithah, Z. (2017). Parametric study in production of virgin coconut oil by fermentation method. *Oriental Journal of Chemistry, 33*, 3069–3076.

Nakatsuji, T., Kao, M. C., Fang, J.-Y., Zouboulis, C. C., Zhang, L., Gallo, R. L., & Huang, C.-M. (2009). Antimicrobial property of lauric acid against propionibacterium acnes: Its therapeutic potential for inflammatory acne vulgaris. *Journal of Investigative Dermatology, 129*, 2480–2488.

Narayanankutty, A., Illam, S. P., & Raghavamenon, A. C. (2018). Health impacts of different edible oils prepared from coconut (*Cocos nucifera*): A comprehensive review. *Trends in Food Science & Technology, 80*, 1–7.

Narayanankutty, A., Mukesh, R. K., Ayoob, S. K., Ramavarma, S. K., Suseela, I. M., Manalil, J. J., Kuzhivelil, B. T., & Raghavamenon, A. C. (2016). Virgin coconut oil maintains redox status and improves glycemic conditions in high fructose fed rats. *Journal of Food Science and Technology, 53*, 895–901.

Nevin, K., & Rajamohan, T. (2010). Effect of topical application of virgin coconut oil on skin components and antioxidant status during dermal wound healing in young rats. *Skin Pharmacology and Physiology, 23*, 290–297.

O'Brien, R., & Timms, R. (2004). Fats and oils-formulating and processing for applications. *European Journal of Lipid Science and Technology, 106*, 451–451.

Ogbolu, D. O., Oni, A. A., Daini, O. A., & Oloko, A. (2007). In vitro antimicrobial properties of coconut oil on Candida species in Ibadan, Nigeria. *Journal of Medicinal Food, 10*, 384–387.

Onsaard, E., Vittayanont, M., Srigam, S., & McClements, D. J. (2005). Properties and stability of oil-in-water emulsions stabilized by coconut skim milk proteins. *Journal of Agricultural and Food Chemistry, 53*, 5747–5753.

Peedikayil, F. C., Sreenivasan, P., & Narayanan, A. (2015). Effect of coconut oil in plaque related gingivitis-a preliminary report. *Nigerian Medical Journal: Journal of the Nigeria Medical Association, 56*, 143.

Peng, K. S., Harun, D., Amin, M. M., & Long, K. (2016). Enhanced Virgin Coconut Oil (EVCO) as natural postmilking teat germicide to control environmental mastitis pathogens. *International Journal of Biotechnology for Wellness Industries, 5*, 128–134.

Pritha, S. D. S. J., & Karpagam, S. (2018). Antimicrobial activity of coconut shell oil. *International Journal of Pharmaceutical Sciences and Research, 9*(4), 1628–1631. https://doi.org/10.13040/IJPSR.0975-8232.9(4).1628-31.

Rosenthal, A., Pyle, D., & Niranjan, K. (1996). Aqueous and enzymatic processes for edible oil extraction. *Enzyme and Microbial Technology, 19*, 402–420.

Seneviratne, K. N., & Sudarshana Dissanayake, D. M. (2008). Variation of phenolic content in coconut oil extracted by two conventional methods. *International Journal of Food Science & Technology, 43*, 597–602.

Seow, C. C., & Gwee, C. N. (1997). Coconut milk: Chemistry and technology. *International Journal of Food Science & Technology, 32*, 189–201.

Shah, S., Sharma, A., & Gupta, M. (2005). Extraction of oil from *Jatropha curcas* L. seed kernels by combination of ultrasonication and aqueous enzymatic oil extraction. *Bioresource Technology, 96*, 121–123.

Shahidi, F. (2006). *Nutraceutical and specialty lipids, nutraceutical and specialty lipids and their co-products* (pp. 11–35). Boca Raton: CRC Press.

St-Onge, M. P., Ross, R., Parsons, W. D., & Jones, P. J. (2003). Medium-chain triglycerides increase energy expenditure and decrease adiposity in overweight men. *Obesity Research, 11*, 395–402.

Tangwatcharin, P., & Khopaibool, P. (2012). Activity of virgin coconut oil, lauric acid or mono-laurin in combination with lactic acid against *Staphylococcus aureus*. *The Southeast Asian Journal of Tropical Medicine and Public Health, 43*, 969–985.

Varma, S. R., Sivaprakasam, T. O., Arumugam, I., Dilip, N., Raghuraman, M., Pavan, K., Rafiq, M., & Paramesh, R. (2019). In vitro anti-inflammatory and skin protective properties of virgin coconut oil. *Journal of Traditional and Complementary Medicine, 9*, 5–14.

Villarino, B. J., Dy, L. M., & Lizada, M. C. C. (2007). Descriptive sensory evaluation of virgin coconut oil and refined, bleached and deodorized coconut oil. *LWT- Food Science and Technology, 40*, 193–199.

Vysakh, A., Ratheesh, M., Rajmohanan, T., Pramod, C., Premlal, S., & Sibi, P. (2014). Polyphenolics isolated from virgin coconut oil inhibits adjuvant induced arthritis in rats through antioxidant and anti-inflammatory action. *International Immunopharmacology, 20*, 124–130.

Zakaria, Z., Somchit, M., Jais, A. M., Teh, L., Salleh, M., & Long, K. (2011). In vivo antinoci-ceptive and anti-inflammatory activities of dried and fermented processed virgin coconut oil. *Medical Principles and Practice, 20*, 231–236.

Zhang, Y., Zheng, Y., Duan, K., & Gui, Q. (2016). Preparation, antioxidant activity and protective effect of coconut testa oil extraction on oxidative damage to human serum albumin. *International Journal of Food Science & Technology, 51*, 946–953.

Chapter 10
Hazelnut (*Corylus avellana*) Oil

Mustafa Topkafa, Hamide Filiz Ayyildiz, and Huseyin Kara

Abstract Hazelnut (*Corylus avellana*), which is present since pre-agricultural times, belongs to the Betulaceae family and cultivated in many countries, such as Turkey, USA, Azerbaijan, Georgia, Spain, Portugal and France as well as in regions with temperate climate of the northern hemisphere. Hazelnut, which is annually produced approximately 850,000 tons worldwide, is used in chocolate, bakery products, snacks, and edible oil industry. The most important nutrients in the hazelnut are lipids, protein, carbohydrate, phytosterols, vitamins and minerals. The most concentrated content of these components is the oil (50–70%). Hazelnut oil is used in food, cake-biscuit, paint, cosmetics, and soap industry. In this chapter, potential and economic value of hazelnut, production methods of oil, chemical and nutritional properties, nutrient value and chemical composition of hazelnut oil are discussed. In addition, triglyceride, fatty acid, tocopherols, sterols and volatile compounds composition of hazelnut oil were highlighted. The health effects of bioactive lipids such as the proper functioning of the human body, growth and physiological functions are reported in the main headings. The chemical profile of hazelnut oil is very similar to olive oil, wherein olive oil is adulterated with hazelnut oil. For this reason, adulteration of olive oil with hazelnut oil is evaluated in last main heading.

Keywords Betulaceae · Fatty acid · Tocopherols · Sterols · Volatile compounds

M. Topkafa (✉)
Vocational School of Technical Sciences, Department of Chemistry and Chemical Technologies, Konya Technical University, Konya, Turkey

H. F. Ayyildiz
Faculty of Pharmacy, Department of Basic Pharmaceutical Sciences, Selcuk University, Konya, Turkey

H. Kara
Faculty of Science, Department of Chemistry, Selcuk University, Konya, Turkey

© Springer Nature Switzerland AG 2019
M. F. Ramadan (ed.), *Fruit Oils: Chemistry and Functionality*,
https://doi.org/10.1007/978-3-030-12473-1_10

1 Introduction

Hazelnut, belongs to *Corylus avellana* L. family, is a popular nut in the world and is grown in Turkey's Black Sea coast, in southern Europe (Italy, Spain, Portugal and France), in some parts of US (Oregon and Washington) and in regions with temperate climate of the northern hemisphere. More than a hundred varieties of hazelnuts are cultivated in the world. Table 10.1 presents the world hazelnut production (FAO 2018). Turkey is the largest hazelnut producer in the world (FAO 2018) with approximately 56% and exporter with 82% (Fig. 10.1). Hazelnut, which is annually produced approximately 850,000 tons worldwide, is used in many industrial areas such as chocolate, biscuits, confectionery, sweet, cake and ice cream as a chopped, sliced, ground and whole. The most important nutrients of the hazelnut are lipids, protein, carbohydrate, vitamins, minerals, sugars and sterols, wherein their levels are vary according to the type of the hazelnuts. Table 10.2 shows the chemical composition of hazelnut cultivars grown in Portugal (Amaral et al. 2006). The most concentrated content of these components is the oil with about 50–70%.

Because hazelnut is used as a snack, chopped, flour and puree, the production of hazelnut edible oil is limited. However, the oil production increases when hazelnut production is abundant (Atalayoğlu and Çakmak 2010; Doğan and Bircan 2010; Özçakmak and Çetinkaya 2016). Hazelnut kernel is annually used approximately 100,000–150,000 tones for crude hazelnut oil production in Turkey (Tasan et al. 2008). Hazelnuts to be used as edible oil must be stored under suitable conditions. While the crude hazelnut oils are generally sold in the domestic vegetable oil industry, the price is quite high compared to other ordinary edible oils such as soybean, canola, sunflower and corn oils. Hazelnut oils are used as an edible oil after refining.

Hazelnut oil meal is a by-product of crude oil production remaining after ordinary, expeller and *n*-hexane extraction, wherein the meal is used as an animal feed (Özçakmak and Çetinkaya 2016). Because hazelnut oil meal has a high content of protein, carbohydrates, vitamins and poor in cellulose (Table 10.3), research is continuing on the effects of feeding on the hazelnut meal on the growth, and weight gain in dairy cattle, fish feed, broiler chickens, rams and small animals (Atalayoğlu and Çakmak 2010; Doğan and Bircan 2010; Özçakmak and Çetinkaya 2016; Ozen and Erener 1992; Saricicek 2000). When the amino acid components of hazelnut seeds are analyzed (Bilgin et al. 2007), it appears that they are rich in arginine (4.53%), leucine (2.77%), isoleucine (1.80%), phenylalanine (1.21%), valine (1.58%), but poor in lysine (0.99%) and methionine (0.15%).

2 Extraction and Processing of Hazelnut Oil

There are different techniques for obtaining vegetable edible oils such as mechanical screw pressing (expeller press), cold pressing, supercritical fluid extraction (supercritical carbon dioxide), microwave and organic solvent extraction (Ayyildiz

Table 10.1 World hazelnut production by years (FAO 2018)

	2006	2007	2008	2009	2010	2011	2012	2013	2014	2015	2016
Georgia	23,500	21,200	18,700	21,800	28,800	31,100	24,700	39,700	33,800	35,300	29,500
Iran	14,544	18,987	18,607	17,492	18,443	18,758	19,532	20,655	10,098	12,723	16,327
Italy	142,109	128,231	111,841	106,600	90,270	128,940	85,232	112,650	75,456	101,643	120,572
Spain	24,810	16,134	24,330	10,290	15,086	17,590	14,406	15,300	13,542	11,423	15,306
Turkey	661,000	530,000	800,791	500,000	600,000	430,000	660,000	549,000	450,000	646,000	420,000
USA	37,195	33,568	29,030	42,638	25,401	34,927	35,500	40,823	32,659	28,123	34,473
World	961,668	811,773	1,068,329	773,097	854,593	744,725	923,994	869,072	707,894	932,718	743,455

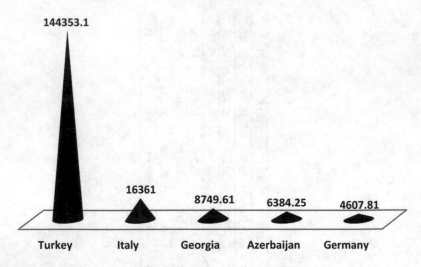

Fig. 10.1 Top 5 exporters of hazelnuts/average 1993–2013 (FAO 2018)

Table 10.2 Chemical composition (g/100 g fresh weight) of 19 hazelnut cultivars grown in Portugal (Amaral et al. 2006)

Cultivar	Moisture	Protein	Lipids	Ash	Carbohydrates	Energy (kcal)
Butler	4.36 ± 0.04	11.27 ± 0.08	61.79 ± 1.90	2.72 ± 0.02	19.86 ± 1.54	681
Campanica	4.68 ± 0.09	10.80 ± 0.20	64.95 ± 0.25	2.74 ± 0.02	16.83 ± 0.37	695
Cosford	4.52 ± 0.23	10.20 ± 0.01	61.96 ± 0.07	2.59 ± 0.02	20.73 ± 0.22	681
Couplat	3.75 ± 0.09	9.90 ± 0.07	67.00 ± 0.22	2.44 ± 0.03	16.91 ± 0.38	710
Daviana	5.35 ± 0.06	11.37 ± 0.06	60.92 ± 0.07	2.90 ± 0.00	19.47 ± 0.07	672
Ennis	6.43 ± 0.04	9.48 ± 0.07	60.19 ± 0.08	2.80 ± 0.09	21.11 ± 0.07	664
F. Coutard	6.29 ± 0.16	10.89 ± 0.09	62.67 ± 0.23	2.69 ± 0.08	17.46 ± 0.37	677
Grossal	5.20 ± 0.17	9.57 ± 0.03	65.29 ± 0.17	2.53 ± 0.08	17.41 ± 0.39	696
Gunslbert	5.85 ± 0.19	12.62 ± 0.16	60.16 ± 0.51	3.39 ± 0.06	17.99 ± 0.82	664
Lansing	4.89 ± 0.05	10.09 ± 0.10	63.05 ± 0.02	2.82 ± 0.01	19.15 ± 0.10	684
L. Espanha	3.79 ± 0.08	10.27 ± 0.00	65.39 ± 0.14	2.62 ± 0.01	17.93 ± 0.06	701
M. Bollwiller	4.61 ± 0.10	12.69 ± 0.06	59.25 ± 0.16	3.06 ± 0.05	20.39 ± 0.02	666
Morell	4.12 ± 0.21	10.83 ± 0.01	64.77 ± 0.09	2.42 ± 0.06	17.87 ± 0.23	698
Negreta	3.46 ± 0.20	10.58 ± 0.03	69.03 ± 0.19	2.43 ± 0.01	14.50 ± 0.28	722
Pauetet	5.62 ± 0.27	10.72 ± 0.08	65.65 ± 0.13	2.57 ± 0.00	15.45 ± 0.09	695
R. Piemont	4.19 ± 0.04	10.93 ± 0.06	66.69 ± 0.01	2.40 ± 0.02	15.79 ± 0.01	707
Segorbe	4.85 ± 0.19	12.38 ± 0.05	63.02 ± 0.32	2.57 ± 0.01	17.19 ± 0.31	685
St. M. Jesus	4.39 ± 0.09	12.40 ± 0.04	68.30 ± 1.48	2.81 ± 0.01	12.09 ± 1.41	713
T. Giffoni	6.30 ± 0.04	9.34 ± 0.17	65.30 ± 0.05	2.63 ± 0.05	16.44 ± 0.18	691
Mean	4.87	10.86	63.97	2.69	17.61	689.57
Range	3.5–6.4	9.3–12.7	59.2–69.0	2.4–3.4	12.1–21.1	664–722

Table 10.3 Nutrient content (%) of hazelnut meal

Dry matter	Protein	Oil	Cellulose	Ash	Nitrogen-free substances	Reference
90.80	39.52	7.62	8.10	6.20	29.36	Şehu et al. (1996)
91.27	43	3.01	7.84	5.98	31.44	Bilgin et al. (2007)

Table 10.4 Analysis of crude hazelnut oil

	Karabulut et al. (2005)	Altuntas et al. (2018)	Fernandes et al. (2017)
FFA (%)	2.096	17.26	–
Total tocopherols (mg kg^{-1})	518.9	580.82	345.22
Fatty acid (%)			
C16:0	5.23	5.12	5.57
C16:1	0.20	0.15	0.23
C17:0	ND	0.04	0.05
C17:1	ND	0.08	0.08
C18:0	2.53	2.28	2.48
C18:1	81.06	80.95	82.29
C18:2	10.65	11.28	8.82
C18:3	0.09	0.10	0.11
Color analysis			
L	40.54	81.29	ND
a	2.01	−0.12	ND
b	23.71	95	ND
Total sterol (mg kg^{-1})	1680.4	–	1787.82
Squalene (mg kg^{-1})	ND	–	340.05
Total terpenic alcohols (mg kg^{-1})	ND	–	26.54

et al. 2015; Bernardo-Gil et al. 2002; de Sousa Ferreira Soares et al. 2012; Dong et al. 2014; Özkal et al. 2005a;Topkafa 2016; Topkafa et al. 2015; Tunç et al. 2014). The oil amounts, desirable and undesirable compound contents obtained by using these methods are different. The mechanical screw pressing method is used in hazelnut oil industry. In this method, the seed is subjected to seed cleaning, breaking of kernels, removal of husk, flaking, roasting and pressing operations. The greatest advantage of this method is the low cost, but oil production efficiency is low and direct consumption is not possible because of undesirable odor, color and contaminants. For this reason, the obtained oil is called crude hazelnut oil that include several impurities such as free fatty acid, mono- and di-acylglycerols, phosphatides, pigments, hydrocarbons, trace metals, waxes, pesticides, aflatoxin, glycerols and other contaminants which considered as impurities depending on the growing conditions, soil structure, climatic conditions, storage conditions, and oil applied operations. Besides these impurities, crude hazelnut oil also contains bioactive components such as sterols, tocols (tocopherols and tocotrienols), squalene and fat-soluble vitamins which are very useful in terms of shelf life of the oil and human health as shown in Table 10.4 (Altuntas et al. 2018; Ayyildiz et al. 2015; Fernandes et al.

2017; Karabulut et al. 2005). These impurities must be removed from the crude oil by refining. Refining is the removal of impurities in crude oil without altering the natural properties of the oil and without altering its composition. The refining process is the most important factor in determining the quality and economic performance of oils. Although the process parameters are correctly determined at refining, it was observed a continuous decrease in the content of bioactive components in the hazelnut oil during refining (Karabulut et al. 2005).

Recently, the consumers like the typical taste, specific aroma and particular color of cold-pressed hazelnut oil and prefer to use it in salads or cold dishes. The cold pressing technique is less efficient compared with other techniques. However, cold-pressed seed oils have high nutritional value and important chemical properties because the raw material is not exposed to a heat treatment, which causes loss of nutrients. The natural beneficial phytochemicals such as tocols, sterols, carotenoids, antioxidants and phenolic compounds as well as essential fatty acids were protected at higher levels in cold-pressed hazelnut oils as given in Table 10.5

Table 10.5 Analysis of cold-pressed hazelnut oil

	Ciemniewska-Żytkiewicz et al. (2014)	Ciemniewska-Żytkiewicz et al. (2015)	Slatnar et al. (2014)	Castelo-Branco et al. (2016)	Damirchi et al. (2005)
Free fatty acids (FFA, %)	0.08–0.09	0.09	ND	3.26	ND
Peroxide value (meq O_2/kg)	0.28–1.16	0.79	ND	8.33	ND
Main fatty acids (%)					
C16:0	6.1–7.1	4.7	ND	6.61	ND
C16:1	0.2–0.3	0.2	ND		ND
C18:0	2.9–3.1	1.5	ND	1.84	ND
C18:1	81.0–82.1	80.4	ND	79.4	ND
C18:2	7.9–8.2	13.2	ND	11.3	ND
C18:3	0.1	0.1	ND	0.08	ND
Total antioxidant capacity (mmol TE/kg)	ND	ND	ND	0.67	ND
Total phenolic content (mg GAE g^{-1} oil)	ND	ND	0.15–0.26	0.09	ND
DPPH (%)	ND	ND	17.7–29.9	ND	ND
Total tocopherols (mg/kg)	ND	342.6	ND	414	ND
Total 4-desmethylsterols (µg/g)	ND	ND	ND	ND	593.2–761.6
Total 4-monomethylsterols (µg/g)	ND	ND	ND	ND	28.5–40.5

(Castelo-Branco et al. 2016; Ciemniewska-Żytkiewicz et al. 2014, 2015; Damirchi et al. 2005; de Sousa Ferreira Soares et al. 2012; Parker et al. 2003; Parry et al. 2006; Ramadan 2013; Slatnar et al. 2014; Topkafa 2016).

The supercritical fluid extraction (SFE) process has begun to attract attention as an alternative to common extraction methods. In this method, extraction yield close to conventional extraction methods (expeller and cold pressing) depending on particle size, critical pressure and temperature. The CO_2 used in this method has the most important advantages of being non-toxic, non-flammable and non-corrosive, and can be removed from oil because of its low critical temperature and pressure (Bernardo-Gil et al. 2002; Özkal et al. 2005a, b). Different solvents such as water, ethanol and chloroform were also added to the CO_2 to increase the extraction efficiency (Brignole 1986). Although industrial applications of SFE on the process are very limited, laboratory applications are quite common for hazelnut oil. The studies carried out with this method in laboratory applications aimed to increase the oil efficiency and achieve high yields of beneficial phytochemicals in the edible oil.

In a research by Bernardo-Gil and coworkers (2002), they compared the SFE and *n*-hexane Soxhlet extraction in terms of efficiency of methods, free fatty acids, sterols, triacylglycerols and tocopherols composition. Results have shown that the acidity index of hazelnut oil obtained from SFE was lower than obtained from *n*-hexane solvent-extracted oil, while fatty acid compositions were similar in both hazelnut oils. Although the triglyceride species of the hazelnut oils obtained from both methods were the same, the values of species were different from each other. The total tocopherol and sterol contents of the hazelnut oil obtained with SFE were higher than with *n*-hexane solvent-extracted oil. The oxidative stability of SFE-extracted hazelnut oil was more protected against oxidation than *n*-hexane solvent-extracted hazelnut oil (Bernardo-Gil et al. 2002).

3   Fatty Acids Composition and Acyl lipids of Fixed Oil

Edible oils are common food ingredients and contain essential fatty acids such as linoleic acid C18:2 (a precursor of the family of *omega*-6) and α-linolenic acid C18:3 (a precursor of *omega*-3). Among edible oils, hazelnut oil plays a major role in human nutrition and health and shows a high degree of unsaturation because of its special fatty acid composition which includes oleic (C18:1) and linoleic acids (Alasalvar et al. 2003). Depending on the geographical origin, climatic conditions and cultivar, there is the difference in the fatty acid, triglyceride, tocopherol, sterol and phospholipid compositions of hazelnut oils. For this reason, a codex about hazelnut oil in Table 10.6 was published by the European Union, Codex Committee on Fats and Oils (CCFO) at 2017 (EU 2017). As seen in Table 10.6, oleic acid, which is a monounsaturated fatty acid (MUFA), is a major fatty acid in hazelnut oil and is in the range of 66.0–85.0%. Hazelnut oil also contains linoleic acid (5.7–25.0%) and linolenic acid (0–0.2%), which are poly-unsaturated fatty acid (PUFA). Palmitic acid (16:0) with 4.0–9.0% is the major

Table 10.6 Levels of fatty acids, tocols, desmethylsterols and physicochemical characteristics of hazelnut oil (EU 2017)

Fatty acid	(%)	Relative density (x° C/water at 20 °C)	0.908–0.915
		Apparent density (g/mL)	25 °C/water 25 °C
C6:0	ND	Refractive index (ND 40 °C)	1.456–1.474
C8:0	ND	Saponification value (mg KOH/g oil)	188–197
C10:0	ND	Iodine value	80–100
C12:0	ND	Unsaponifiable matter (g/kg)	≤10
C14:0	0.0–0.1		
C16:0	4.0–9.0	Cholesterol	0.0–0.6
C16:1	0.1–0.3	Campesterol	4.0–7.0
C17:0	ND	Stigmasterol	0.0–3.0
C17:1	ND	Beta-sitosterol	75.0–96.0
C18:0	1.0–4.0	Delta-5-avenasterol	1.0–7.0
C18:1	66.0–85.0	Delta-7-stigmastenol	ND
C18:2	5.7–25.0	Delta-7-avenasterol	≤1
C18:3	0–0.2	Others	ND
C20:0	0.0–0.3	Total sterols (mg/kg)	1200–2200
C20:1	0.1–0.3		
C20:2	ND	Alpha-tocopherol	100–460
C22:0	0.0–0.1	Beta-tocopherol	ND-12
C22:1	ND	Gamma-tocopherol	18–194
C22:2	ND	Delta-tocopherol	ND –10
C24:0	ND	Alpha-tocotrienol	ND
C24:1	ND	Gamma-tocotrienol	ND
C18:1 *t*	≤0.1	Delta-tocotrienol	ND
C18:2 *t* + C18:3	≤0.1	Total (mg/kg)	200–600

saturated fatty acid (SFA), followed by stearic acid with 1.0–4.0%. Other SFA, such as C14:0, C20:0 and C22:0 (0.0–0.1%, 0.0–0.3% and 0.0–0.1%, respectively) are minor fatty acids. Although natural hazelnut oil does not contain *trans* fatty acids, refined hazelnut oil may contain small amounts of *trans* fatty acids depending on the conditions of the refining process.

Triglycerides (TAG) are the main components of vegetable oils. They represent the most concentrated source of energy in the diet and act as carriers for fat-soluble vitamins such as vitamins A, D, E, K (Topkafa 2016). Hazelnut oil has a characteristic pattern of TAG consisting of different fatty acids. The main TAG fragments in hazelnut oil are OOO, OLO, POO, SOL and SOO containing oleic acid (Bada et al. 2004; Parcerisa et al. 1999). In a research by Bada and coworkers (2004), they examined the TAG composition of fifteen different hazelnut oils in Asturias (Spain) and defined 12 different TAG species. The major TAG were triolein which found in a range of 50.97–60.16%, followed by OLO + POO (12.57–19.05%) as given in Table 10.7. Besides, different hazelnut oils contained significant amounts of PLL, SOL and POO containing saturated fatty acids such as stearic and palmitic acids (Bada et al. 2004).

Table 10.7 Triglyceride composition of hazelnut oil (Bada et al. 2004)

Sample	LLL	OLLn	PLL+ LnOO	POLn	OLO+ POO	PLO	OOO	SOL+POO	PLS+ POP	GOO	AOL+ SOO	POS
1	0.96	0.10	4.86	0.75	18.59	3.59	50.97	13.29	0.67	0.19	5.45	0.53
2	0.84	0.11	4.02	0.68	17.81	3.46	53.60	13.17	0.51	0.22	5.07	0.48
3	0.71	0.10	3.79	0.68	16.34	3.15	54.77	13.26	0.77	0.35	5.50	0.53
4	0.65	0.10	3.81	0.69	16.22	3.18	54.19	13.19	0.66	0.40	6.01	0.65
5	1.03	0.12	4.03	0.74	15.66	3.23	54.68	13.09	0.86	0.12	5.94	0.44
6	0.89	0.13	4.19	0.70	18.04	3.34	52.68	12.43	0.52	0.28	6.11	0.69
7	0.80	0.12	3.77	0.55	15.86	1.82	58.66	12.56	0.51	0.39	4.42	0.46
8	0.76	0.16	3.75	0.56	15.95	2.68	57.09	13.46	0.46	0.23	4.42	0.38
9	0.72	0.10	3.25	0.63	14.52	2.66	58.79	13.30	0.81	0.20	4.61	0.36
10	0.77	0.12	4.13	0.86	16.64	3.01	56.39	12.05	0.47	0.43	4.60	0.49
11	0.73	0.11	3.88	0.58	16.55	3.02	56.01	13.36	0.49	0.34	4.48	0.39
12	0.69	0.10	3.34	0.56	15.00	3.28	58.27	12.90	0.51	0.20	4.64	0.45
13	0.80	0.11	3.81	0.70	16.33	2.98	56.63	12.85	0.54	0.23	4.58	0.51
14	1.67	0.23	5.58	0.82	19.05	3.97	51.70	11.18	0.65	0.30	4.51	0.27
15	0.60	0.07	2.73	0.40	12.56	2.40	60.16	13.56	0.60	0.25	6.03	0.56

*The letter abbreviations of fatty acids in the triglyceride molecule are as follows: P = Palmitic Acid; S = Stearic Acid; O = Oleic Acid; L = Linoleic Acid; Ln = Linolenic Acid; A = Arachidic Acid; G = Gondoic Acid

Table 10.8 Fatty acid composition of hazelnut oil and phospholipids (Zlatanov et al. 1999)

	C12:0	C13:0	C14:0	C16:0	C16:1	C18:0	C18:1	C18:2	C18:3	Unsaturated./ saturated fatty acid
Oil	0.4	–	tr.	6.6	tr.	2.5	80.8	8.7	1.0	90.5: 9.5
PC	2.1	13.4	1.0	16.2	1.9	6.2	59.2	tr.	tr.	61.1:38.9
PI	11.3	–	1.4	13.8	1.6	1.3	44.5	26.1	tr.	72.2:27.8
PE	tr.	5.5	tr.	28.7	tr.	11.5	34.8	19.5	tr.	54.3:45.7
PA	1.9	tr.	8.9	69.5	0.4	tr.	17.4	1.9	tr.	19.7:80.3

Glycolipids and phospholipids are polar lipids that present in hazelnut oil at about 2–4% (Parcerisa et al. 1997; Zlatanov et al. 1999). The major polar compound was monogalactosyl diacylglycerol. Phospholipids are a class of lipids that contain one or two fatty acid and phosphate groups. Major individual phospholipids are phosphatidylcholine (PC), phosphatidylethanolamine (PE), phosphatidylinositol (PI), and phosphatidic acids (PA) (Parcerisa et al. 1999; Zlatanov et al. 1999). In a research by Zlatanov and coworkers, total phospholipid content of hazelnut oil was found to be 2.8%. Although phosphatidylcholine (50%), phosphatidylinositol (18%) and phosphatidylethanolamine (16%) were major phospholipids, phosphatidic acids were minor phospholipids with 4%. In the same study, it was also found that there was no difference between fatty acid compositions of glycerides and phospholipids as shown in Table 10.8 (Zlatanov et al. 1999).

4   Minor Bioactive Lipids in the Fixed Oil

Sterols, which are steroid alcohol, are found in animal and vegetable oils. Vegetable sterols, termed phytosterols, are found in the form of free and fatty acid esters in oils which include generally cholesterol, campesterol, campestanol, stigmasterol, cholesterol, β-sitosterol, β-sitostanol Δ^5-avenasterol, Δ^7-stigmasterol, and Δ^7-avenasterol. Hazelnut oils contain sterols with changing concentrations from 1200 to 3469 mg/kg oil, while β-sitosterol was the major sterol constituting 75–87% of total sterols (Alasalvar et al. 2009; Amaral et al. 2006; Bada et al. 2004; Bernardo-Gil et al. 2002; EU 2017). The other major sterols were campesterol and Δ^5-avenasterol, each constituting between 3% and 7% of total sterols. Cholesterol, campestanol, stigmasterol, cholesterol, β-sitostanol, Δ^7-stigmasterol, and Δ^7-avenasterol were minor sterols and contributed to less than 6% to the total sterols. Studies conducted by different research groups have found that there was no significant difference in the composition of sterols in hazelnut oils obtained from different varieties (Alasalvar et al. 2009; Amaral et al. 2006; Bada et al. 2004). Fatty acid and TAG compositions of olive oil and hazelnut oil are very similar, so hazelnut oil is added to olive oil for adulteration. For this reason, it is very difficult to distinguish these oils from each other. However, sterol compositions of olive oil and hazelnut oil are useful to detect adulteration, because they contain qualitative and quantitative differences in terms of sterols. Tocopherols and tocotrienols known as tocols are important bioactive substances that show antioxidant properties and, resistant to oxidation by increasing the shelf life of edible oils. Eight naturally occurring vitamin E isomers include α-, β-, γ -, and δ-tocopherols as well as α-, β-, γ-, and δ-tocotrienols. Although many ordinary oils such as sunflower and corn oils do not include tocotrienols, hazelnut oil is an excellent source of both tocopherol and tocotrienols. Several researchers have reported that different type of hazelnut oils such as crude, refined and pressed oils include tocols with different concentrations from 211 to 900 mg/kg oil (Alasalvar et al. 2009; Ayyildiz et al. 2015; Bada et al. 2004; Castelo-Branco et al. 2016; Fernandes et al. 2017). In all types of hazelnut oils, α-tocopherol was the most abundant (70–90% of the total of tocols). Hazelnut oil also contain a significant amount of γ-tocopherol and β-tocopherol, while has a low quantity of δ-tocopherol. Besides, total tocotrienols are found in small quantities ranging from 0.6% to 1.8% of the total tocols (Alasalvar et al. 2009). When European Union, Codex Committee on Fats and Oils (CCFO) (Table 10.6) and researches compared the amounts of tocols, it was found that the codex values were significantly lower than those reported in the literature (Alasalvar et al. 2009; Ayyildiz et al. 2015; Bada et al. 2004; EU 2017; Fernandes et al. 2017) (Table 10.9).

Oxidation is one of the important markers showing deterioration of products and causes the destruction of the important bioactive substances and nutrients in the edible oils. Antioxidants, which are defined as substances that delay or inhibit oxidation, prevent the incomplete reduction of oxygen and the formation of various free radicals such as hydrogen peroxide and hydroxyl radical, which are formed in small quantities during normal oxygen metabolism. Natural antioxidants are mainly vitamins (C, E and A vitamins), sterols, flavonoids, carotenoids and phenolics. Although the antioxidant traits of these bioactive molecules have been studied in recent years, the individual effect of each molecule has not been fully described. Therefore, the effect of these molecules is evaluated as the total antioxidant capacity (TAC) and total phenolic compounds. Many methods such as 2,2-diphenyl-1-picrylhydrazyl radical (DPPH·), ferric reducing antioxidant power (FRAP), cupric reducing antioxidant capacity (CUPRAC), trolox equivalent antioxidant capacity (TEAC), and oxygen radical antioxidant capacity (ORAC) have been developed (Apak et al. 2004; Castelo-Branco et al. 2016; Espín et al. 2000; Miraliakbari and Shahidi 2008; Singleton et al. 1999; Szydłowska-Czerniak et al. 2008; Tuberoso et al. 2007). Contents of the total phenolic compounds are expressed in mg of gallic acid equivalents (GAE)/100 g of oil. Several researchers have reported that total phenolic compounds vary from 0.15 to 338 mg GAE kg^{-1} although trolox equivalent antioxidant capacity (TEAC) vary from 0.67 to 412 µM trolox/g oil in hazelnut oils (Castelo-Branco et al. 2016; Miraliakbari and Shahidi 2008; Slatnar et al. 2014). The TAC could be different according to applied methods, extraction procedures and cultivar of hazelnut oil.

Oxidative stability is a significant parameter for evaluating the shelf life of oils and fats. It is affected by the fatty acids composition and level of minor components such as tocols, sterols, flavonoids, carotenoids, phenolics, free fatty acids, peroxide

Table 10.9 Volatile compounds in hazelnut oil (Mildner-Szkudlarz and Jeleń 2008)

	Peak area (TIC*10^6)
1. Butanal	19.28[a]
2. Acetic acid[A]	442.64[a]
3. 3-Methyl-1-butanal	18.33[b]
4. 2-Methyl-1-butanal	88.46[a]
5. Pentanal[A]	37.34[a]
6. 1-Octene	6.92[b]
7. Hexanal[A]	122.92[b]
8. 2-Octene	25.42[b]
9. Unidentified	43.14[a]
10. Furfural	149.50[a]
11. 2-Furanmethanol	107.04[a]
12. 2-Heptanone	6.44[b]
13. 1-Heptanal[A]	12.07[b]
14. Unidentified	21.69[a]
15. Pyrazine-2,6-dimethyl	22.59[a]
16. Unidentified	23.24[d]
17. E-2-heptenal[A]	10.85[b]
18. 2-Furancarboxaldehyde-5-methyl	25.33[a]
19. 1-Heptanol	2.14[b]
20. Hexanoic acid	5.65[b]
21. Octanal[A]	20.62[b]
22. E-2-octenal	2.76[a]
23. 1-Octanol	4.19[b]
24. 2-Nonanone	5.08[b]
25. Nonanal[A]	12.95[b]
26. E-2-nonenal	0.72[d]
27. Octanoic acid	1.41[e]
28. 2-Decanone	1.00[b]
29. Decanal	0.45[a]
30. E-2-decenal[A]	3.95[b]
31. E,E-2,4-decadienal	0.80[c]
32. E-2-undecenal	0.54[c]

[a](RSD) <5%; [b]5 < (RSD) < 15%; [c]15 < (RSD) < 25%; [d]25 < (RSD) < 50%; [e](RSD) > 50%, n =3. [A] Identity of compounds by comparison of their mass spectra with standards, rest of compounds identified tentatively using NBS 75 K mass spectra library search

value and iodine value. The rancimat method has been used to test the oxidative stability of oils. The rancimat method monitors the conductivity of a sample heated at higher temperatures, through which streams of air are passed. Several articles have shown that cold-pressed and supercritical fluid extracted hazelnut oils had the highest induction periods with 8.88 h, while refined hazelnut oils had lower induction period with 3.16 h (Ayyildiz et al. 2015; Bernardo-Gil et al. 2002; Castelo-Branco et al. 2016). In some studies, it was found that induction time varies from 8.9 to 19 h depending on the hazelnut cultivar (Amaral et al. 2006; Ciemniewska-Żytkiewicz et al. 2014).

The volatile compounds in edible oils are important markers of the characteristic flavor and aroma as well as the quality of oils. However, these volatile species can be observed either in cold-pressed or in unrefined oils. Especially in the neutralization, bleaching and deodorizing stages of refining, these compounds are almost completely removed from the oil (Karabulut et al. 2005). Hazelnut oil contains nearly 32 volatile compounds. These volatile compounds are ketones, aldehydes, alcohols, aromatic hydrocarbons, and furans. Furfural, 2-furanmethanol, 2,6-dimethyl pyrazine and 5-methyl-2-furancarboxaldehyde, E-5-methyl-hept-2-en-4-one, phenylacetaldehyde, sabinene, octanol, decanal, 2-acetyl pyrrole and terpineol are also volatile species specific to hazelnut oil (Caja et al. 2000; Mildner-Szkudlarz and Jeleń 2008).

5 The Contribution of Bioactive Compounds in Oil to Organoleptic Properties

Edible oils contain many important bioactive compounds such as TAG composed of saturated and unsaturated fatty acids, glycolipids, phospholipids, sterols, tocols, carotenoids, phenolic compounds, hydrocarbons, flavor and aroma compounds. Although saturated and unsaturated fatty acids, glycolipids, phospholipids, sterols, tocols, carotenoids and phenolic compounds contribute to the functioning of oils and functional foods, the hydrocarbons, flavor and aroma compounds are important for organoleptic properties. Flavor and aroma compounds such furfural, 2-furanmethanol, 2,6-dimethyl pyrazine and 5-methyl-2-furancarboxaldehyde, E-5-methyl-hept-2-en-4-one, phenylacetaldehyde, sabinene, octanol, decanal, 2-acetyl pyrrole and terpineol cause a different taste and aroma in cold-pressed hazelnut oil, when compared to other common edible because of they are specific volatile compounds in hazelnut oil (Caja et al. 2000; Mildner-Szkudlarz and Jeleń 2008). For this reason, consumers like the characteristic taste, aroma and intensive color of cold-pressed hazelnut oils as well as their high nutritional value and health beneficial factors, and prefer to use it in salads or cold dishes. However, these special flavor and aroma compounds are almost completely removed from the crude hazelnut oil in the bleaching and deodorizing stages of refining at high temperature and vacuum.

Fatty acids, glycolipids, phospholipids, sterols, tocols, squalene, carotenoids, phenolic compounds are very important factors to evaluate oxidative stability such as prevention of oxidation of hazelnut oils, protection of chemical composition and oil quality. Sterols, tocols, carotenoids, squalene, phenolic compounds may exhibit antiradical and antioxidant activities (Miraliakbari and Shahidi 2008). These bioactive components can increase the oxidative stability of hazelnut oil to be up to 19 h (Amaral et al. 2006; Ciemniewska-Żytkiewicz et al. 2014). Moreover, these components prevent the formation of deterioration products such as diacylglycerols, mono-acylglycerols, free glycerol, free fatty acids, peroxides, aldehydes and ketones in hazelnut oil.

Hazelnut oil is also used as a frying oil, salad oil, flavoring ingredient and a component of skin moisturizers and cosmetic components as well as cooking oil because of including a high amount of glycolipids, phospholipids, sterols and tocols (Alasalvar and Shahidi 2008; Ciemniewska-Żytkiewicz et al. 2014).

6 Health-Promoting Traits of Oil and Oil Constituents

In recent years, scientific studies have focused on the effect of dietary oil on the development, prevention, and treatment of diseases/disorders. Fatty acid composition of the edible oils used in diet is one of the most important factors for metabolic syndrome, cardiovascular disease, low-density lipoprotein (LDL), aortic cholesterol accumulation and lipid peroxides (Gillingham et al. 2011; Hargrove et al. 2001; Hatipoglu et al. 2004). According to the results, MUFA are recommended in the diet because of the low prevalence of chronic diseases in populations consuming MUFA rich diets. Dietary MUFA consumption prevents or ameliorates metabolic syndrome and cardiovascular disease risk, improves blood lipid composition, blood pressure and glucose levels, increases insulin sensitivity and decreases obesity risk (Gillingham et al. 2011). Research by Hatipoglu and coworkers (2004) showed that hazelnut oil that is rich MUFA, caused a decrease in glutathione levels, glutathione peroxidase and glutathione transferase activities, reduced plasma, liver and aortic lipid peroxide levels and aortic cholesterol levels as well as decreased oxidative stress and cholesterol accumulation in rabbit aorta. The same research group indicated that hazelnut oil may have an anti-atherogenic potential that reduces oxidative stress and LDL oxidation (Hatipoglu et al. 2004). In a study on the effect of hazelnut oil on lipids and lipid peroxidation in plasma and apo B 100-containing lipoproteins (VLDL+LDL), erythrocytes and hematological traits in rabbits, it has been found that hazelnut oil reduced lipid peroxide levels in plasma and apolipoprotein B 100-containing lipoproteins as well as aortic atherosclerotic lesions in rabbits, and reduced the hemolytic anemia together with significant decreases in DC and H_2O_2-induced MDA levels (Balkan et al. 2003).

In a research by Sánchez and Lutz (1998), they were found that palmitoleic and oleic acids in phosphatidylcholine and phosphatidylethanolamine, and palmitoleic and vaccenic acids in phosphatidylserine and phosphatidylinositol were found to be increased in the rats fed with hazelnut oil (Sánchez and Lutz 1998).

Tocols which are antioxidants showing vitamin E activity protected vegetable oils against oxidative damage. Each tocol have different biological activity, although α-tocopherol has biologically more active form of vitamin E (Saremi and Arora 2010). It is abundant in nuts such as hazelnut, almond and peanut oils compared to other oils (Ayyildiz et al. 2015; Saremi and Arora 2010). Tocols found in high amounts in hazelnut oil and have different antioxidant and biological activities such as prevention of Parkinson's disease, ataxia various cancers, decreasing LDL oxidation, inhibiting lipid peroxidation in biological membranes, breaking free radical chain reactions and providing natural oxidative protection to the oil (Amaral et al. 2005; Burton 1994; Burton and Traber 1990; Saldeen et al. 1999; Sanagi et al. 2005; Schwartz et al. 2008).

Phytosterols, present in high amounts in hazelnut oil, have been used as blood cholesterol-lowering agents, wherein the intake of 2 g/day of stanols or sterols reduces low-density lipoprotein (LDL) by 10%. Sterol and stanol are recommended for lowering LDL cholesterol levels in people at risk for coronary heart disease (Katan et al. 2003; Kritchevsky and Chen 2005). Besides, some studies have shown that phytosterols may have a protective effect to delay some tumor formations such as colon, breast and prostate cancer (Awad and Fink 2000; Raicht et al. 1980).

7   Adulteration

Due to the high price, increasing request, unique taste characteristics and nutritional benefits of olive oils, they are often subjected to adulterate with oils such as hazelnut, peanut and cotton oils having of similar fatty acid and TAG compositions. Hazelnut oil is one of the most preferred oils in adulteration. In this way, researches on olive oil adulteration developed new methods and techniques for the classification and quantification of the adulteration with hazelnut oil. These techniques included liquid and gas chromatography (HPLC and GC), differential thermal analysis (DTA), Fourier transform infrared spectroscopy (FT-IR), near-infrared spectroscopy (NIR), Fourier transform Raman (FT-Raman) and ^{13}C- and ^{1}H-nuclear magnetic resonance (Baeten et al. 2005; Cercaci et al. 2003; Christy et al. 2004). Although studies have shown that olive oil is not fully distinguishable from adulterated with hazelnut oil, they can detect hazelnut oil up to 10% (Caja et al. 2000; Cercaci et al. 2003; Damirchi et al. 2005; Mildner-Szkudlarz and Jeleń 2008). The most effective methods were tocopherols and volatile compound analyses because the content of tocopherol in hazelnut oil is higher than olive oil and hazelnut oil has specific volatile species such as furfural, 2-furanmethanol, 2,6-dimethyl pyrazine and 5-methyl-2-furancarboxaldehyde, E-5-methyl-hept-2-en-4-one, phenylacetaldehyde, sabinene, octanol, decanal, 2-acetyl pyrrole and terpineol (Benitez-Sánchez et al. 2003; Mildner-Szkudlarz and Jeleń 2008).

References

Alasalvar, C., & Shahidi, F. (2008). *Tree nuts: Composition, phytochemicals, and health effects.* Boca Raton: CRC Press.
Alasalvar, C., Shahidi, F., Ohshima, T., Wanasundara, U., Yurttas, H. C., Liyanapathirana, C. M., & Rodrigues, F. B. (2003). Turkish Tombul hazelnut (*Corylus avellana* L.). 2. Lipid characteristics and oxidative stability. *Journal of Agricultural and Food Chemistry, 51*(13), 3797–3805.

Alasalvar, C., Amaral, J. S., Satır, G., & Shahidi, F. (2009). Lipid characteristics and essential minerals of native Turkish hazelnut varieties (*Corylus avellana* L.). *Food Chemistry, 113*(4), 919–925.

Altuntas, A. H., Ketenoglu, O., Cetinbas, S., Erdogdu, F., & Tekin, A. (2018). Deacidification of Crude Hazelnut Oil Using Molecular Distillation-Multiobjective Optimization for Free Fatty Acids and Tocopherol. *European Journal of Lipid Science and Technology, 113*, 637–643.

Amaral, J. S., Casal, S., Torres, D., Seabra, R. M., & Oliveira, B. P. (2005). Simultaneous determination of tocopherols and tocotrienols in hazelnuts by a normal phase liquid chromatographic method. *Analytical Sciences, 21*(12), 1545–1548.

Amaral, J. S., Casal, S., Citová, I., Santos, A., Seabra, R. M., & Oliveira, B. P. P. (2006). Characterization of several hazelnut (*Corylus avellana* L.) cultivars based in chemical, fatty acid and sterol composition. *European Food Research and Technology, 222*(3), 274–280.

Apak, R., Güçlü, K., Özyürek, M., & Karademir, S. E. (2004). Novel total antioxidant capacity index for dietary polyphenols and vitamins C and E, using their cupric ion reducing capability in the presence of neocuproine: CUPRAC method. *Journal of Agricultural and Food Chemistry, 52*(26), 7970–7981.

Atalayoğlu, G., & Çakmak, M. N. (2010). Pullu Sazan (*Cyprinus carpio* L. 1843) Yemlerinde Fındık Küspesinin Kullanılma Olanaklarının Araştırılması. *Firat University Journal of Science, 22*(2), 71.

Awad, A. B., & Fink, C. S. (2000). Phytosterols as anticancer dietary components: Evidence and mechanism of action. *The Journal of Nutrition, 130*(9), 2127–2130.

Ayyildiz, H. F. T., Mustafa, Kara, H., & Sherazi, S. T. H. (2015). Evaluation of fatty acid composition, tocols profile, and oxidative stability of some fully refined edible oils. *International Journal of Food Properties, 18*(9), 2064–2076.

Bada, J. C., León-Camacho, M., Prieto, M., & Alonso, L. (2004). Characterization of oils of hazelnuts from Asturias, Spain. *European Journal of Lipid Science and Technology, 106*(5), 294–300.

Baeten, V., Fernández Pierna, J. A., Dardenne, P., Meurens, M., García-González, D. L., & Aparicio-Ruiz, R. (2005). Detection of the presence of hazelnut oil in olive oil by FT-Raman and FT-MIR spectroscopy. *Journal of Agricultural and Food Chemistry, 53*(16), 6201–6206.

Balkan, J., Hatipoğlu, A., Aykaç-Toker, G., & Uysal, M. (2003). Influence on hazelnut oil administration on peroxidation status of erythrocytes and apolipoprotein B 100-containing lipoproteins in rabbits fed on a high cholesterol diet. *Journal of Agricultural and Food Chemistry, 51*(13), 3905–3909.

Benitez-Sánchez, P. L., León-Camacho, M., & Aparicio, R. (2003). A comprehensive study of hazelnut oil composition with comparisons to other vegetable oils, particularly olive oil. *European Food Research and Technology, 218*(1), 13–19.

Bernardo-Gil, M. G. G., João, Santos, J., & Cardoso, P. (2002). Supercritical fluid extraction and characterisation of oil from hazelnut. *European Journal of Lipid Science and Technology, 104*(7), 402–409.

Bilgin, Ö., Ali, T., & Tekinay, A. A. (2007). The use of hazelnut meal as a substitute for soybean meal in the diets of rainbow trout (*Oncorhynchus mykiss*). *Turkish Journal of Veterinary and Animal Sciences, 31*(3), 145–151.

Brignole, E. A. (1986). Supercritical fluid extraction. *Fluid Phase Equilibria, 29*, 133–144.

Burton, G. W. (1994). Vitamin E: Molecular and biological function. *Proceedings of the Nutrition Society, 53*(02), 251–262.

Burton, G. W., & Traber, M. G. (1990). Vitamin E: Antioxidant activity, biokinetics, and bioavailability. *Annual Review of Nutrition, 10*(1), 357–382.

Caja, M. d. M., Del Castillo, M. R., Alvarez, R. M., Herraiz, M., & Blanch, G. P. (2000). Analysis of volatile compounds in edible oils using simultaneous distillation-solvent extraction and direct coupling of liquid chromatography with gas chromatography. *European Food Research and Technology, 211*(1), 45–51.

Castelo-Branco, V. N., Santana, I., Di-Sarli, V. O., Freitas, S. P., & Torres, A. G. (2016). Antioxidant capacity is a surrogate measure of the quality and stability of vegetable oils. *European Journal of Lipid Science and Technology, 118*(2), 224–235.

Cercaci, L., Rodriguez-Estrada, M. T., & Lercker, G. (2003). Solid-phase extraction-thin-layer chromatography-gas chromatography method for the detection of hazelnut oil in olive oils by determination of esterified sterols. *Journal of Chromatography A, 985*(1–2), 211–220.

Christy, A. A., Kasemsumran, S., Du, Y., & OZAKI, Y. (2004). The detection and quantification of adulteration in olive oil by near-infrared spectroscopy and chemometrics. *Analytical Sciences, 20*(6), 935–940.

Ciemniewska-Żytkiewicz, H., Ratusz, K., Bryś, J., Reder, M., & Koczoń, P. (2014). Determination of the oxidative stability of hazelnut oils by PDSC and Rancimat methods. *Journal of Thermal Analysis and Calorimetry, 118*(2), 875–881.

Ciemniewska-Żytkiewicz, H., Bryś, J., Sujka, K., & Koczoń, P. (2015). Assessment of the hazelnuts roasting process by pressure differential scanning calorimetry and MID-FT-IR spectroscopy. *Food Analytical Methods, 8*(10), 2465–2473.

Damirchi, S. A., Savage, G. P., & Dutta, P. C. (2005). Sterol fractions in hazelnut and virgin olive oils and 4, 4′-dimethylsterols as possible markers for detection of adulteration of virgin olive oil. *Journal of the American Oil Chemists' Society, 82*(10), 717–725.

Doğan, G., & Bircan, R. (2010). Balık yemlerinde alternatif bitkisel protein kaynağı olarak fındık küspesi kullanımı. *Mehmet Akif Ersoy Üniversitesi Fen Bilimleri Enstitüsü Dergisi, 1*(2), 49–57.

Dong, Z., Zhang, J.-G., Tian, K.-W., Pan, W.-J., & Wei, Z.-J. (2014). The fatty oil from okra seed: Supercritical carbon dioxide extraction, composition and antioxidant activity. *Current Topics in Nutraceutical Research, 12*(3), 75–84.

Espín, J. C., Soler-Rivas, C., & Wichers, H. J. (2000). Characterization of the total free radical scavenger capacity of vegetable oils and oil fractions using 2, 2-diphenyl-1-picrylhydrazyl radical. *Journal of Agricultural and Food Chemistry, 48*(3), 648–656.

EU. (2017). *Codex Alimentarius Commission, Agenda Item 11, CODEX STAN 210-1999* (Vol. 2018). Malaysia: European Union.

FAO. (2018). *Agricultural statistical database* (Vol. 2013). Rome: Food and Agriculture Organization of the United Nations.

Fernandes, G. D., Gómez-Coca, R. B., Pérez-Camino, M. D. C., Moreda, W., Barrera-Arellano, D. (2017). Chemical characterization of major and minor compounds of nut oils: Almond, hazelnut, and pecan nut. *Journal of Chemistry, 2017*, 1–11.

Gillingham, L. G., Harris-Janz, S., & Jones, P. J. (2011). Dietary monounsaturated fatty acids are protective against metabolic syndrome and cardiovascular disease risk factors. *Lipids, 46*(3), 209–228.

Hargrove, R. L., Etherton, T. D., Pearson, T. A., Harrison, E. H., & Kris-Etherton, P. M. (2001). Low fat and high monounsaturated fat diets decrease human low density lipoprotein oxidative susceptibility in vitro. *The Journal of Nutrition, 131*(6), 1758–1763.

Hatipoglu, A., Kanbaglı, Ö., Balkan, J., Küçük, M., Cevikbas, U., Aykaç-Toker, G., Berkkan, H., & Uysal, M. (2004). Hazelnut oil administration reduces aortic cholesterol accumulation and lipid peroxides in the plasma, liver, and aorta of rabbits fed a high-cholesterol diet. *Bioscience, Biotechnology, and Biochemistry, 68*(10), 2050–2057.

Karabulut, I., Topcu, A., Yorulmaz, A., Tekin, A., & Ozay, D. S. (2005). Effects of the industrial refining process on some properties of hazelnut oil. *European Journal of Lipid Science and Technology, 107*(7–8), 476–480.

Katan, M. B., Grundy, S. M., Jones, P., Law, M., Miettinen, T., & Paoletti, R. (2003). Efficacy and safety of plant stanols and sterols in the management of blood cholesterol levels. *Mayo Clinic Proceedings, 78*, 965–978. Elsevier.

Kritchevsky, D., & Chen, S. C. (2005). Phytosterols-health benefits and potential concerns: A review. *Nutrition Research, 25*(5), 413–428.

Mildner-Szkudlarz, S., & Jeleń, H. H. (2008). The potential of different techniques for volatile compounds analysis coupled with PCA for the detection of the adulteration of olive oil with hazelnut oil. *Food Chemistry, 110*(3), 751–761.

Miraliakbari, H., & Shahidi, F. (2008). Antioxidant activity of minor components of tree nut oils. *Food Chemistry, 111*(2), 421–427.

Tasan. M., Gecgel, U., & Daglioglu, O. (2008). Hazelnut oil production in Turkey. In *6th Euro fed lipid congress, oils, fats and lipids in the 3rd millennium*. Athens: Euro Fed Lipid.

Özçakmak, S., & Çetinkaya, A. (2016). HACCP Sistemi Uygulaması ile Fındık Küspesindeki Aflatoksinin Kontrolü. *Balıkesir Üniversitesi Fen Bilimleri Enstitüsü Dergisi, 17*(2), 1–14.

Ozen, N., & Erener, G. (1992). Research note: Utilizing hazelnut kernel oil meal in layer diets. *Poultry Science, 71*(3), 570–573.

Özkal, S., Salgın, U., & Yener, M. (2005a). Supercritical carbon dioxide extraction of hazelnut oil. *Journal of Food Engineering, 69*(2), 217–223.

Özkal, S., Yener, M., Salgın, U., & Mehmetoğlu, Ü. (2005b). Response surfaces of hazelnut oil yield in supercritical carbon dioxide. *European Food Research and Technology, 220*(1), 74–78.

Parcerisa, J., Richardson, D. G., Rafecas, M., Codony, R., & Boatella, J. (1997). Fatty acid distribution in polar and nonpolar lipid classes of hazelnut oil (*Corylus avellana* L.). *Journal of Agricultural and Food Chemistry, 45*(10), 3887–3890.

Parcerisa, J., Codony, R., Boatella, J., & Rafecas, M. (1999). Triacylglycerol and phospholipid composition of hazelnut (*Corylus avellana* L.) lipid fraction during fruit development. *Journal of Agricultural and Food Chemistry, 47*(4), 1410–1415.

Parker, T. D., Adams, D., Zhou, K., Harris, M., & Yu, L. (2003). Fatty acid composition and oxidative stability of cold-pressed edible seed oils. *Journal of Food Science, 68*(4), 1240–1243.

Parry, J., Hao, Z., Luther, M., Su, L., Zhou, K., & Yu, L. L. (2006). Characterization of cold-pressed onion, parsley, cardamom, mullein, roasted pumpkin, and milk thistle seed oils. *Journal of the American Oil Chemists' Society, 83*(10), 847–854.

Raicht, R. F., Cohen, B. I., Fazzini, E. P., Sarwal, A. N., & Takahashi, M. (1980). Protective effect of plant sterols against chemically induced colon tumors in rats. *Cancer Research, 40*(2), 403–405.

Ramadan, M. F. (2013). Healthy blends of high linoleic sunflower oil with selected cold pressed oils: Functionality, stability and antioxidative characteristics. *Industrial Crops and Products, 43*, 65–72.

Saldeen, T., Li, D., & Mehta, J. L. (1999). Differential effects of α-and γ-tocopherol on low-density lipoprotein oxidation, superoxide activity, platelet aggregation and arterial thrombogenesis. *Journal of the American College of Cardiology, 34*(4), 1208–1215.

Sanagi, M. M., See, H., Ibrahim, W. A. W., & Naim, A. A. (2005). Determination of carotene, tocopherols and tocotrienols in residue oil from palm pressed fiber using pressurized liquid extraction-normal phase liquid chromatography. *Analytica Chimica Acta, 538*(1), 71–76.

Sánchez, V., & Lutz, M. (1998). Fatty acid composition of microsomal phospholipids in rats fed different oils and antioxidant vitamins supplement 1. *The Journal of Nutritional Biochemistry, 9*(3), 155–163.

Saremi, A., & Arora, R. (2010). Vitamin E and cardiovascular disease. *American Journal of Therapeutics, 17*(3), e56–e65.

Saricicek, B. (2000). Protected (bypass) protein and feed value of hazelnut kernel oil meal. *Asian-Australasian Journal of Animal Sciences, 13*(3), 317–322.

Schwartz, H., Ollilainen, V., Piironen, V., & Lampi, A.-M. (2008). Tocopherol, tocotrienol and plant sterol contents of vegetable oils and industrial fats. *Journal of Food Composition and Analysis, 21*(2), 152–161.

Şehu, A. Y., Sakine, & Kaya, İ. (1996). Bıldırcın rasyonlarına katılan fındık küspesinin büyüme ve karkas randımanı üzerine etkisi. *Ankara Universitesi Veteriner Fakultesi Dergisi, 43*, 163–168.

Singleton, V. L., Orthofer, R., & Lamuela-Raventós, R. M. (1999). [14] Analysis of total phenols and other oxidation substrates and antioxidants by means of folin-ciocalteu reagent. *Methods in Enzymology, 299*, 152–178. Elsevier.

Slatnar, A., Mikulic-Petkovsek, M., Stampar, F., Veberic, R., & Solar, A. (2014). HPLC-MSn identification and quantification of phenolic compounds in hazelnut kernels, oil and bagasse pellets. *Food Research International, 64*, 783–789.

de Sousa Ferreira Soares, G., Gomes, V. D. M., dos Reis Albuquerque, A., Barbosa Dantas, M., Rosenhain, R., Souza, A. G. D., Persunh, D. C., Gadelha, C. A. D. A., Costa, M. J. D. C., &

Gadelha, T. S. (2012). Spectroscopic and thermooxidative analysis of organic okra oil and seeds from *Abelmoschus esculentus*. *The Scientific World Journal, 2012*, 1–6.

Szydłowska-Czerniak, A., Dianoczki, C., Recseg, K., Karlovits, G., & Szłyk, E. (2008). Determination of antioxidant capacities of vegetable oils by ferric-ion spectrophotometric methods. *Talanta, 76*(4), 899–905.

Topkafa, M. (2016). Evaluation of chemical properties of cold pressed onion, okra, rosehip, safflower and carrot seed oils: Triglyceride, fatty acid and tocol compositions. *Analytical Methods, 8*(21), 4220–4225.

Topkafa, M., Kara, H., & Sherazi, S. T. H. (2015). Evaluation of the triglyceride composition of pomegranate seed oil by RP-HPLC followed by GC-MS. *Journal of the American Oil Chemists' Society, 92*(6), 791–800.

Tuberoso, C. I., Kowalczyk, A., Sarritzu, E., & Cabras, P. (2007). Determination of antioxidant compounds and antioxidant activity in commercial oilseeds for food use. *Food Chemistry, 103*(4), 1494–1501.

Tunç, İ., Çalışkan, F., Özkan, G., & Karacabey, E. (2014). Mikrodalga destekli soxhlet cihazı ile fındık yağı ekstraksiyonunun yanıt yüzey yöntemi ile optimizasyonu. *Akademik Gıda, 12*(1), 20–28.

Zlatanov, M., Ivanov, S., & Aitzetmüller, K. (1999). Phospholipid and fatty acid composition of Bulgarian nut oils. *Lipid/Fett, 101*(11), 437–439.

Chapter 11
Tiger Nut (*Cyperus esculentus* L.) Oil

Adel Abdel Razek Abdel Azim Mohdaly

Abstract Tiger nut tuber of the *Cyperus esculentus* L. plant is an unusual storage system. Recent investigations clearly show that tiger nut is a valuable source of vegetable oils, rich in monounsaturated fatty acids, tocopherols, and phytosterols as well as high-added value compounds (proteins, carbohydrates, vitamins, minerals and bioactive compounds). Several conventional (Soxhlet) and alternative innovative (SC-CO$_2$, enzyme, and high pressure) extraction methods have been developed for the efficient recovery of tiger nut oil (TNO) and high-added value compounds. Moreover, it has shown the potential of tiger nuts by-products in the development of new healthier and functional products such as fiber-rich and oxidative stable foods. This chapter provides an overview of these investigations and tries to expose potential avenues for future research in the commercial exploitation of tiger nuts as well as its oil and by-products as a source of ingredients to be incorporated in new food matrices to improve their technological and functional aspects.

Keywords Extraction methods · Fatty acids · Phytosterols · Residual meals · Tiger nut oil · Tocopherols

Abbreviations

AEE	Aqueous enzymatic extraction
AIDS	Acquired immunodeficiency syndrome
FA	Fatty acid
FAO	Food and Agriculture Organization
GAME	Gas assisted mechanical expression
HDL-C	High-density lipoprotein-cholesterol
HIV	Human immunodeficiency virus

A. A. R. A. A. Mohdaly (✉)
Food Science and Technology Department, Faculty of Agriculture, Fayoum University, Fayoum, Egypt
e-mail: aam01@fayoum.edu.eg

© Springer Nature Switzerland AG 2019
M. F. Ramadan (ed.), *Fruit Oils: Chemistry and Functionality*,
https://doi.org/10.1007/978-3-030-12473-1_11

LDL-C   Low-density lipoprotein-cholesterol
ME   Mechanical expression
SC-CO$_2$ Supercritical carbon dioxide
SEM   Scanning electron microscopy
TNO   Tiger nut oil

1   Introduction

Cyperaceae or sedges are a family of monocotyledonous angiosperms found world-wide both in tropical and temperate regions. They make up the seventh largest angiosperm and third largest monocotyledonous family. Only 10% are used by humans and mostly in the tropics such as in Thailand and Southern India where they are cultivated for matting and for basketry (Simpson et al. 2011). To the untrained eye, they appear similar to grasses (Gramineae family). Some characteristics are given in Table 11.1 to distinguish between grasses and sedges.

Cyperus L., a large genus belonging to the Cyperaceae family comprises more than 500 species and *Cyperus esculentus* is one of these. It is an underutilized crop, which produces rhizomes from the base of the tuber that is somewhat spherical (Devries and Feuke 1999). Pollination is by wind. Young tubers are white, while older tubers are covered by a yellow outer membrane; they are usually found within 6 in. of the ground surface. Vegetative colonies of its plants are often produced from the tubers and their rhizomes. It is a tuber that grow freely and is consumed widely in Egypt, Nigeria, West Africa, East Africa, North and South America, Australia, parts of Europe particularly Spain as well as in the Arabian Peninsula (Abaejoh et al. 2006). It is commonly known as tiger nut, zulu nuts, yellow nut grass, ground almond, rush nuts, earth almond or chufa. The ancient Egyptians recognized the importance of the crop, cultivating and using it for culinary and medicinal purposes. Defelice (2002) reports that the tubers, which date back to the fifth millennium BC, are thought to be the third most ancient domesticated foodstuff of ancient Egypt after emmer wheat and barley. Many thousand years ago, tiger nut, in Spanish called chufa, was cultivated in the region of chufa between Sudan and Egypt on the borders of the Nile River. There are documents that certify this product over 400 years ago. Proof of this is that on many occasion archeologists found earthen jars containing tiger nut in graves of pharaohs (Obadina et al. 2008). It was cultivated in the ancient Mesopotamia between the rivers Tigris and Euphrates. At the same time, historical

Table 11.1 Differences between sedges and grasses (Lowe and Stanfield 1974)

	Sedges	Grasses
Stems	Usually solid, triangular; some hollow, tubular	Hollow, tubular (solid at nodes); solid
Leaves	In 3 (or more) rows, except Coleochloa in 2 rows; sometimes apparently leafless, with bladeless sheaths	In 2 rows; never leafless
Leaf sheath	Entire, i.e. a closed tube	A tube split down 1 side or margins connate
Ligule	Usually absent	Usually present

Persian and Arab documents mentioned the nutritive, and digestive value of tiger nut. During the era, the tiger nut milk was classified as medicinal drink due to it's been highly energetic and diuretic, rich in mineral, predominantly phosphorus and potassium and also vitamins C and E. In the eighth century, the Arab traders introduced the cultivation of tiger nut in the Mediterranean region of Valencia (Spain) for elaboration of tiger nut milk (leche de chufa). It has been reported that grainy sandy group and mild temperatures are special for the cultivation growth of earth tuber (Abaejoh et al. 2006). Tiger nuts tubers appear somewhat long or round in shape with a dimension of 8–16 mm, smaller size however, are not used for human consumption. When hydrated, it is slightly harder (nut texture), but with a rather more intense and concentrated taste. The cultivation time is April to November. Being cultivated through continuance irrigation, tiger nut has to be dried before storage. The drying process is completely natural (i.e., sun drying) and the process can take up to 1 month. The dehydrating process ensures longer shelf life, preventing rot or any other bacterial infection securing their quality and nutritional level. Unfortunately, the dehydration process makes the tiger nut skin wrinkled, a situation that limits its acceptability to some people (Belewu and Abodunrin 2006).

Tiger nut is a very fast growing plant and often confused with a weed in some areas. Tiger nut is preferably grown in well-drained sandy or loamy soils with a pH range of 5.0–7.5 and grows best when temperatures are above 20 °C, more specifically when temperatures alternate between day and night temperatures of ca. 30 °C and 20 °C respectively, taking only 90–110 days to mature (Dyer 2006). If grown in a rich soil or soil with an average fertility rate, it often does not require additional fertilizer (Pascual et al. 2000), and hence decreasing costs of raw materials needed for its cultivation. However, a recent study by Dyer (2006), suggested that the addition of Schultz plant food (a common brand of plant food) can lead to an increase in tuber production and tuber mass. In the same study, tiger nut's production was found to be highly sensitive to time of sowing and temperature. Although it was concluded that tiger nut was not affected by soil type, the effect of competition with wild nutsedge was prominent when sandy soil was used. When other types of soil were used, competition did not pose a threat to tuber production. Sandy soils have been reported to support the growth of the *Cyperus esculentus* in general and the wild nutsedge being a more vigorous variety may have reduced the production of cultivated tiger nut. Plant size was also correlated to tuber production with larger plants having more tubers. The weight of fresh tuber ranges from 70 mg to about 900 mg while the weight of dried tuber ranges from 30 to 350 mg. A dried tuber nut can absorb up to three times its own weight of water. On average, a tiger nut plant can give rise to up to 50 tubers, and have tuber yields of 17,000 kg per hectare (Pascual et al. 2000).

2   Varieties of Tiger Nut Tuber

There are two approaches in describing the varieties of the tiger nut tuber. One refers to its botanical classification, while the other is based on its color. Botanically, amongst the eight varieties, only four wild varieties *(leptostachyus, esculentus, hermanii,* and *macrostachyu*s) and one cultivated variety (*sativus*) are acknowledged (Pascual et al. 2000). The cultivated variety was found to have a higher lipid and sugar content (Defelice 2002). In the color based classification, four varieties have been reported to exist; red, brown, black, and yellow. There is a tendency that the brown and yellow refer to the same variety as the tubers appear brown when dried and unwashed. Once they have been soaked and cleaned, they become lighter in color and appear to take on a yellow hue. They are found in Africa and Spain. The red variety has only been analyzed in Cameroon, while the black has been found in Cameroon and Ghana (Ejoh and Ndjouenkeu 2006; Abano and Amoah 2011). According to Ejoh and Ndjouenkeu (2006), no significant variation exists between the lipid content of the black and brown varieties, but more between the tubers obtained from different areas in Cameroon. The yellow variety is often preferred over other varieties due to its attractive color, bigger size and fleshier body. The yellow variety also yields more milk, contains lower lipids and higher protein and less anti-nutritional factors especially phenolics (Abano and Amoah 2011).

3 Uses and Products

Cultivated tiger nuts are used for food purposes mainly in raw consumption, and making ice-cream, as well as the production of gluten-free flour, milk-type extract, and edible oil. It also finds uses as a flavoring agent for ice cream and biscuits, as well as in making soap, starch and flour. Tiger nut has a unique sweetness that is ideal for use in the baking industry. It can be used to make delicious cakes and biscuits and as a component of fruit flavors. There is a strong belief in the benefits of flour for health reasons as it has been found to be an alternative for dietetics. It is a good alternative to wheat flour, as it is gluten-free and good for people who cannot take gluten in their diet. It is considered a good flour or additive for the baking industry, as its natural sugar (good option for diabetics). Tiger nut flour does not lose any of its nutritional properties in the milling process (Salau et al. 2012). The range of uses has grown increasingly due to an increased interest in the plant. The tuber has been described as having a characteristic taste almost resembling a hazelnut. About 19% of tiger nut is dietary fiber and this has been investigated as a potential source of fiber in food. In Egypt, tiger nut can be eaten as a snack, which can be prepared by soaking in water for a few minutes. It can also be eaten roasted, fried or baked. It tastes best when dried. It is generally dried out in the sun, a process that takes one or more months with occasional turning over. Popular in Europe as carp bait, the applications of tiger nuts are varied, from attracting game to its use in the cosmetic industry. In Spain, particularly in the Valencian region, tiger nuts are extensively used to prepare a cold beverage, known as "horchata de chufa" or tiger nut beverage. In North Africa, the tubers are consumed in their natural form or after being soaked in water for some hours. In the United Kingdom, tiger nut oil (TNO) is mostly used in the fishing industry. It is known to be effective especially among anglers. It is not commonly used as a food ingredient, which may be due to some factors. These may include insufficient information on the nutritional profile of the tuber, oil extraction techniques, usability of the oil and possibly simply an awareness of the potential uses of the tuber.

It has though been identified to be suitable as frying oil (Ali Rehab and El Anany 2012). It can thus be used in the same way as other cooking oils. The oil can also be incorporated in food formulations, and used as a flavoring agent. Historically, it was reported to be used in soap manufacturing, and lubrication of fine apparatuses. In the United States, the primary use of tiger nut as a crop is to attract and feed game, particularly wild turkeys. The caramel from malted tubers of *Cyperus esculentus* may be used to add body, flavor, or color to certain baked products, non-alcoholic malt beverages and dark beers, and in the production of condiments. The starches obtained from tiger nut and rice showed similar properties; the solutions of the starch exhibited a good paste stability, clarity, and adhesive strength. The starch can be used in many starch-based foods as well as in the cosmetic industry, and for laundry, glazing and stiffening. The waste residue after oil extraction could be further modified to produce syrups, flours, or livestock feeds because of the high protein content that most oilseeds possess (Ali Rehab and El Anany 2012; Ezeh et al. 2014).

4   Economics and Nutritional Benefits of Tiger Nut

Tiger nuts have long been recognized for their health benefits as they have a high content of soluble glucose and oleic acid, along with high-energy content (starch, lipids, sugars and proteins). Tiger nuts are rich in minerals such as phosphorous, potassium, calcium, magnesium and iron necessary for bones, tissue repair, muscles, the bloodstream and for body development. Nuts also rich in vitamins E and C. Sugar-free tiger nut milk is suitable for diabetic people and helps in weight control (Martinez 2003), due to its content of carbohydrates with a base of sucrose and starch (without glucose), and its high content of arginine, which liberates the hormone that produces insulin (Bamishaiye and Bamishaiye 2011). It is recommended for those who suffer from indigestion, flatulence and diarrhea because it provides digestive enzymes like the catalase, lipase and amylase. The high content of oleic acid has a positive effect on cholesterol, thereby preventing heart attacks, thrombosis and activates blood content of soluble glucose. Tiger nut reduces the risk of colon cancer and prevents constipation. Tiger nut contains a good quantity of vitamin B1, which assists in balancing the central nervous system and helps to encourage the body to adapt to stress (David 2005). The milk supplies the body with enough quantity of vitamin E, essential for fertility in both men and women. Vitamin E delays cell aging, improves the elasticity of skin and helps to clear the appearance of wrinkles, acne and other skin alterations.

In China, tiger nut milk is used as a liver tonic, heart stimulant, drank to heal serious stomach pain, to promote normal menstruation, to heal mouth and gum ulcers, used in Ayurvedic medicines and is a powerful aphrodisiac (sexual stimulant).

The black species of the tiger nut is an excellent medicine for breast lumps and cancer. The tubers have a relatively high antioxidant capacity, because they contain considerable amounts of water-soluble flavonoid glycosides. Consumption of antioxidants could protect the immune system of malnourished populations. The intake of antioxidant-containing foods may delay the progression of HIV (Human immunodeficiency virus) infection to AIDS (Acquired immunodeficiency syndrome) (Bamishaiye and Bamishaiye 2011).

For many years, the tiger nut tubers have been considered to have adequate properties to fight respiratory infections, and some stomach illnesses. To this date, the *Horchata de chufa* is considered an effective remedy for diarrhea, according to popular tradition in Valencia, Spain. It promotes the production of urine and this is why it is a preventive measure for cyst, prostrate, hernia, rectum deformation and prolapsed (anal feature-small painful flesh at the tip of the anus) and to prevent endometriosis or fibrosis as well as blockage of the tip of the fallopian tube. Tiger nut exhibit anti-inflammatory properties upon inflammation and immune-stimulatory effects (Soha et al. 2017). Agbai and Nwanegwo (2013) reported that oral administration of tiger nut improves reproductive functions in adult male albino rats by altering the plasma levels of gonadotropins, testosterone and sperm functions in a dose-dependent manner.

The TNO reduces low-density lipoprotein-cholesterol (LDL-C), increases high-density lipoprotein cholesterol (HDL-C), reduces the levels of triglycerides (TAG) in the blood and the risk of forming blood clots, thereby preventing arteriosclerosis. It also stimulates the absorption of calcium in bones and the production of new bony material, due to short and medium chain fatty acids, oleic acid and essential fatty acids (Rita 2009). It is recommended for infants and the elderly because of its high content in vitamin E and its antioxidant benefits in the cell membrane (David 2005). TNO shares a similar fatty acid (FA) profile with olive oil, having oleic acid as its most abundant fatty acid. In recent years, the significance of vegetable oils for health has been recognized especially for their effects on heart health. The World Health Organisation recommends oleic acid to make up daily total fat intake after adequate polyunsaturated fatty acid (PUFA) intake. New and emerging research has identified dietary unsaturated fatty acids to play a role in affecting an individual's risk of developing diseases including diabetes, asthma and cancer (Lunn and Theobald 2006). TNO contains these beneficial fatty acids as well as vitamin E and phenolics that contribute to its stability. Countries in Africa could benefit as major growers and exporters of the tuber with myriad uses of the oil from salad oil to biodiesel (Xueshe et al. 2008) and its ease of growth being important factors. TNO could replace the expensive imported olive oil or be utilized to improve the diets of consumers in impoverished areas. The productivity of the tuber was deemed high with a satisfactory coefficient of economic efficiency, i.e. not less than 50%. The potential increase in interest in the tuber was identified as far back as 1964, where the storage characteristics and stability of the oil under various storage conditions of two varieties from Ghana and Nigeria were studied.

The nutritional value of tiger nut derivatives, like oil and milk, arises from the very composition of the tiger nut. The level of anti-nutrients such as tannins, alkaloids and polyphenols is drastically reduced by soaking in water for 6 h, thereby making it free of unwanted elements especially in making the milk.

5 Tiger Nut Tuber Composition

The available data (Table 11.2) reveals that tubers are rich in essential dietary constituents such as proteins (3.28–9.70%), lipids (22.14–44.92%), fibers (5.50–15.47%) and ashes (1.60–4.25%) (Adel et al. 2015; Codina-Torrella et al. 2015; Arafat et al. 2009; Oladele and Aina 2007; Ejoh and Ndjouenkeu 2006). Among these constituents, the industrially relevant compounds that could be recovered are starch, soluble carbohydrates (mainly in the form of horchata), lipids and fibers. The main carbohydrate in tiger nut beverage is sucrose (Roselló-Soto et al. 2018). Tiger nut tubers contain almost twice the quantity of starch as potato or sweet potato tubers (Kuner et al. 2002). Regarding total sugar content, reducing sugar and sucrose, in general tubers have high contents of sugar. When the sugar contents of tiger nut tubers were compared with those of other tubers and nuts, the sugar level of tiger nut was relatively low (Shaker et al. 2009). However, the taste of tiger nut depends on the sugar content to give a very characteristic flavor. Because of its pleasant nutty flavor, tiger nut is consumed as a kind of snack food and could be useful in food technology. The nut is also rich in mineral contents such as sodium, calcium, phosphorus, potassium, magnesium, manganese, zinc, iron and traces of copper (Oladele and Aina 2007). Therefore, tiger nut could be used as supplementation for food industrial to improve their content of minerals.

Table 11.2 Composition (% of dry matter) of tiger nut

Origin	Moisture	Fat	Protein	Fiber	Ash	References
Valencia	8.66	35.21	8.45	9.31	1.97	Codina-Torrella et al. (2015)
Burkina Faso	7.75	25.77	7.32	9.07	1.90	
Burkina Faso	6.38	25.35	5.62	8.42	1.95	
Niger	7.45	28.19	3.28	8.75	1.60	
Egypt	7.30	22.14	4.33	15.47	2.60	Adel et al. (2015)
Egypt	5.77	25.70	7.00	5.50	1.86	Arafat et al. (2009)
Nigeria (Yellow variety)	3.50	32.13	7.15	6.26	3.97	Oladele and Aina (2007)
Nigeria (brown variety)	3.78	35.43	9.70	5.62	4.25	
Cameroon	ND	35.32	8.08	10.31	2.39	Ejoh and Ndjouenkeu (2006)
Cameroon	ND	44.92	7.50	14.49	2.61	
Cameroon	ND	26.88	6.99	8.26	2.28	
Cameroon	ND	43.50	8.30	14.14	2.60	
Cameroon	ND	35.63	6.57	13.70	2.58	
Cameroon	ND	31.66	7.54	12.51	2.32	

ND not detected

The amino acids profile of tiger nut was dominated by aspartic acid, which resulted from the conversion of asparagine (Borges et al. 2008). Other important amino acids were glutamic acid, which resulted from glutamine, followed by leucine, alanine and arginine (Oladele and Aina 2007). In general, tiger nut tubers are a good source of these compounds; however, amino acids profiles are not well-balanced, with certain essential amino acids occurring in limiting concentration when compared to Food and Agriculture Organization (FAO) recommended levels. Tiger nut milk without sugar can be drunk for diabetes for its content in carbohydrate and due to its content of arginine, which liberates the hormone insulin.

Chukwuma et al. (2010) analyzed tiger nut tuber for the presence of phytochemicals, wherein alkaloids, cyanogenic glycosides, resins, tannins, sterols and saponins were present in the raw tuber, however only alkaloids, sterols and resins were present in the roasted tiger nut. Analysis of the anti-nutrient composition yielded oxalates (0.25 g/100 g), tannins (9.50 mg/100 g), phytates (1.97 mg/100 g), saponins (0.88 g/100 g), and cyanogenic glycosides (1.80 mg/100 g). It was also observed that roasting numerically decreased the levels of the anti-nutritive factors.

6 Tiger Nut Oil

The edible oil obtained from the tuber is golden brown in color and has a rich nutty taste. The oil remains in a uniform liquid form at refrigeration temperature. This makes the oil suitable for salad making. TNO has a composition similar to olives and a rich mineral content, especially phosphorus and potassium. Tiger nuts oil is also cholesterol free and has very low sodium content. It has a high oleic acid and low PUFA (linoleic and linolenic acids), enough to cover daily minimum needs for an adult (around 10 g) and low acidity, and so is excellent for the skin. It also has higher oxidative stability than other oils, due to the presence of γ-tocopherol. It is regarded as high-quality oil due to its extraction without external heat (cold pressed oil), and is highly recommended for cooking over other oils because it is more resistant to chemical decomposition at high temperatures (Shaker et al. 2009). Furthermore, less fat is absorbed into the food as it creates a crust on the surface during cooking, preventing the oil itself being absorbed into the product. In the textile industry, the oil is used to waterproof textile fibers. The oil compares well with corn, soybean, olive and cottonseed oil and can thus serve as a substitute for these oils. In addition, the oil is a potential source of biodiesel (Bamishaiye and Bamishaiye 2011).

6.1 Extraction Techniques

6.1.1 Conventional Extraction Methods

Cold pressing is a widely used technique for industrial production of TNO, so, commercially TNO is sold as cold pressed oil (Koubaa et al. 2015). Several researches have shown that tiger nut lipids can be extracted using organic solvents such as

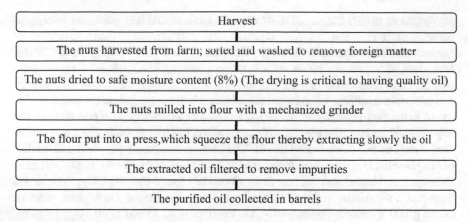

Fig. 11.1 Flow chart of the cold-press extraction process of tiger nut oil (Bamishaiye and Bamishaiye 2011)

n-hexane, petroleum ether, and ethanol, with a substantial yield, more than 95% of oil can be extracted (Ekpe et al. 2016). In such cases, samples are initially homogenized and stirred with a suitable solvent and the lipid portion is recovered. The production process for TNO is shown in Fig. 11.1. Soxhlet extraction can also be used. For instance, a study showed that up to 15.9% (w/w) oil yield can be obtained by using *n*-hexane:2-propanol (3:1, v/v) solvent mixture in 6 h using a Soxhlet apparatus (Yeboah et al. 2012). With hexane as an oil extraction medium, reported yields are assumed to reflect oil content of the material. Reported oil yields of TNO extraction using hexane vary widely from 15.9% to 41.2%. The variation in reported oil yields of TNO may be due to differences in origin of the tubers, genetic history, the age of the tissue or varietal differences. Recently, Ekpe et al. (2016) showed that *n*-hexane offers superior extraction efficiency compared to petroleum ether. However, the use of *n*-hexane has been questioned due to environmental problems and occupational safety as well as their increasing prices (Roselló-Soto et al. 2018).

6.1.2   Non-conventional Extraction Methods

Nowadays, industries are much concerned towards replacing the use of organic solvents for extraction purposes as it is linked with health, safety, environmental and regulatory issues. Therefore, some innovative alternative techniques have been established for the clean recovery of lipids and other added-value compounds form tiger nuts and its by-products.

Compressed Fluids

Supercritical fluid extraction is a promising technology that has been successfully employed in the extraction of diverse compounds from plant materials and other bio-resources (Poojary et al. 2016; Roselló-Soto et al. 2016). This technique minimizes or avoids the use of toxic organic solvents and it is preferred to organic solvents as it is non-toxic, non-explosive and cost-effective, thus considered as 'green' and environmental friendly (Koubaa et al. 2015; Poojary et al. 2016). Moreover, this process offers relatively pure and residual solvent-free extracts, which in turn reduces downstream processing. In addition, it frequently involves the use of certain modifiers (also called co-solvents) to improve efficiency and selectivity of extraction (Poojary et al. 2016; Koubaa et al. 2015). The potential of the supercritical carbon dioxide (SC-CO$_2$) extraction to recover fresh TNO was evaluated (Lasekan and Abdulkarim 2012). The authors compared the yield and quality of oils obtained from tiger nuts through SC-CO$_2$ (20–40 MPa, 40–80 °C, 60–360 min) extraction and traditional Soxhlet extraction. They found that SC-CO$_2$ pressure and time had a significant influence on oil extraction, obtaining the highest yield of 26.28 g/100 g sample through SC-CO$_2$ treatment (30.25 MPa, 60 °C, 210 min). Moreover, they found that fatty acid composition differed based on the processing condition. To avoid long extraction times, the use of the mechanical expression (ME) assisted by supercritical fluids could be used as an alternative to improve the recovery of oil, bioactive compounds, and nutrients(Koubaa et al. 2015). One of such process is referred to as gas-assisted mechanical expression (GAME) and this technique has been applied to extract oil from different matrices of vegetable origin. In case of tiger nuts, Koubaa et al. (2015) evaluated the efficacy of GAME process consisting of both SC-CO$_2$ and ME for the extraction of oil and phenolics. The pressure levels of 20 MPa and 30 MPa were optimal for SC-CO$_2$ and ME, respectively. Additionally, GAME process allowed faster oil extraction from nuts, reaching 50% yield oil recovery after 10 min extraction, compared to only 10% and 20% when using SC-CO$_2$ and ME separately at 20 MPa and 30 MPa pressures, respectively. Furthermore, GAME allowed the recovery of the highest amount of phenolic compounds compared to SC-CO$_2$ and ME processes when applied independently. Phenolics profiling by HPLC-MS revealed that GAME offers a greater diversity of phenolic compounds (57 compounds) in the extracted oils when compared to SC-CO$_2$ (48 compounds), and ME (27 compounds). Scanning electron microscopy (SEM) images revealed that GAME results in a better cell damage. It was concluded that GAME represents a great opportunity to replace conventional methods for oil extraction from plant matrices.

Enzyme and High Pressure-Assisted Extraction

Enzymes have become a popular tool in extraction technologies as they significantly improve the oil extraction yield by hydrolyzing the cell wall components and, consequently, aiding the mass transfer of analytes (Roselló-Soto et al. 2016). Several commercially available cellulolytic or proteolytic enzymes could be used for this purpose based on the type of matrix (Poojary et al. 2017). In the case of tiger nuts, the impact of enzyme-assisted extraction on the recovery TNO rich in bioactive compounds was evaluated (Ezeh et al. 2016). The authors performed aqueous enzymatic extraction using cellulolytic enzymes such as alcalase, α-amylase Viscozyme L and Celluclast, while in another set of experiments was combined with high-pressure processing. Results revealed that aqueous enzyme-assisted extraction yielded comparable amounts of fatty acids, phenolics and tocopherols to that obtained through control pressing, while high-pressure processing treatment prior to enzymatic extraction enhanced tocopherols and total polyphenolic content in oils significantly. In addition, the authors showed that the by-products of oil extraction process are a source of sugars and the methanol extracts of residual meals found to contain glucose, sucrose and certain oligosaccharides. In a similar study, the authors found up to 90% oil recovery when they used enzymatic treatment prior to mechanical extraction (Ezeh et al. 2016).

Aqueous Enzymatic Extraction

An environmentally friendly alternative to solvent extraction is the aqueous extraction in which water is used as the oil extraction medium. Aqueous solvents have proven to be as effective as *n*-hexane with the added advantage of requiring less solvent. The drawback to the use of aqueous solvents is an increase in energy required to remove water from the oil. Aqueous extraction gives relatively poor oil recoveries but offers a safer process with higher quality oil as compared to solvent extraction. Researchers have sought ways to improve this process by studying pretreatments such as extrusion and enzymatic treatment (Lamsal and Johnson 2007). Hydrolytic enzymes such as cellulose are able to enhance oil extraction yields as they degrade the structure of cell walls, which may inhibit the release of oil. Some materials that studied using aqueous enzymatic extraction (AEE) include soybeans, horseradish seeds (*Moringa oleifera*) and corn germ (Mat Yusoff et al. 2015). In AEE, the enzyme chosen is of utmost importance. The cell structure of the oilseed needs to be investigated to ensure the right enzyme or combination of enzymes is used. Other factors affecting oil yields include particle size, enzyme concentration, agitation speed, pH and temperature. Depending on the composition of the oilseed, these factors may affect the oil extraction process in different ways, and so it is important before carrying out an optimization study, to explore how these factors work independently. In addition to the already mentioned factors, other treatments exist which can be used alongside AEE, high-pressure processing being one. Although its conventional use in industry is for food preservation, its capacity to destabilize cell membranes and strengthen cell walls, thereby increasing enhancing cell rupture could allow to extend its applications and investigate its efficacy in enhancing oil extraction. A schematic representing the sequence followed for aqueous enzymatic extraction is given in Fig. 11.2.

The extraction of edible oil from tiger nut by an aqueous enzymatic procedure was investigated by Ezeh et al. (2014), who found that the use of α-amylase, alcalase, and celluclast independently improved extraction oil yields compared to oil extraction without enzymes by 34.5%, 23.4%, and 14.7%, respectively. Combining α-amylase with other enzymes have been reported to augment oil yields, and even more when cell wall degrading enzymes are used (Mat Yusoff et al. 2015). An enzyme

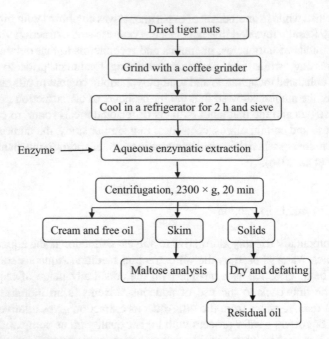

Fig. 11.2 Sequence of aqueous enzymatic extraction of tiger nut oil

mixture of alcalase, α-amylase, and celluclast was used to increase oil yields. Higher yields were obtained when mixed enzyme concentrations were increased from 0.5% to 1%. Agitation during incubation led to gravity separation, decreasing oil yields, and a static incubation unexpectedly gave the highest oil yield. A hydrolysis time of 6 h and the solid/liquid ratio of 1:6 also gave higher oil yields. Swelling which increases the size of the tubers increases the number of cells broken during grinding. A higher number of broken cells mean more passage ways exist for oil to be released. Palonen (2004) observed from several studies that drying lignocellulosic substrates, and the consequential decrease in pore size, decreased the effectiveness of enzymatic hydrolysis. When high-pressure processing is used as a pre-treatment prior to the extraction process, the pressure of 300 MPa employed in increased oil yields by 24.1% in relation to a similar process without pressure pre-treatment. Higher pressures exceeding 300 MPa are likely to encourage lipid-amylose complex formation but this requires a more sensitive analytical method for confirmation.

6.2 Oil Composition

6.2.1 Fatty Acid (FA) Composition

As mentioned earlier, TNO has a similar fatty acid (FA) composition to olive oil (Adel et al. 2015). This means that it predominantly consists of oleic acid, the major unsaturated FA, with values ranging from 59% to 76% compared to values for olive oil from 56% to 85%. Other major fatty acids in TNO are palmitic (10–20%), linoleic (8–16%) and stearic (0.3–5.3%) acids as given in Table 11.3. Traces of myristic, linolenic, and arachidic acids were also detected.

The color, variety and geographical location in which the tubers are grown, and the harvest season have an impact on the relative proportion of FA. The variation in FA composition is distinct in that some FA such as a linolenic acid is present in TNO from certain regions but were not detected in samples from other regions like Ghana. This may be in part due to climatic or environmental factors as well as the cultivated variety, which is not usually specified. In addition, the methods of analysis employed may vary from author to author. GC-MS has the advantage of identifying compounds using both retention time and mass spectrum so it offers more accuracy. The total saturated FA content is low with a minimum of 15% in TNO from Turkey to a maximum of 25% from South Africa. As the FA composition is a determinant of the quality of edible oils, the high concentration of monounsaturated fatty acids (MUFA) makes it desirable due to its good shelf life and potential health benefits. The carbon double bonds in FA are prone to oxidation, producing aldehydes, ketones and hydrocarbons that cause odors and flavors linked with rancidity. Hence, oxidative stability increases with decreased levels of unsaturated FA, most especially PUFA. The low concentrations of PUFA in TNO, therefore, are of course favorable.

6.2.2   Minor Bioactive Lipids

Tocopherol Content

Vitamin E homologs (tocopherols) naturally present in TNO contribute to increasing the oil's shelf life due to their antioxidant properties (Ali Rehab and El Anany 2012). The tocopherol content consists mainly of α-tocopherol (86.7 mg/g) and β-tocopherol (33.4 mg/g) making up a total of 120.1 mg/g (Yeboah et al. 2012). Tocopherol content is affected by mode of oil extraction. Organic solvents are able to penetrate the cells of the oil-containing plant material dissolving more non-polar compounds. Jing et al. (2012) claimed that in vitro studies on the antioxidant capacity of TNO indicated that its radical scavenging ability was equal to that of vitamin C.

Oderinde and Tairu (1992) reported on phospholipid composition of TNO as 5.4% with ethanolamine, inositol, choline and serine glycerophospholipid as the major classes. Choline glycerol phospholipid was the most abundant at 33%. The phospholipid classes were noted to be in agreement with other data for vegetable oils phospholipids. In another study, the phospholipids percentage was 3.1% (Kim et al. 2007). Both reported values of phospholipid are high. High amounts of phospholipid can lead to dark colored oils. Like in soybean oil processing, degumming can be carried out to remove the phospholipids in order to improve its physical stability. Soybean phospholipids are commonly used as emulsifiers in industries such as food, pharmaceutical and cosmetic industry (Ezeh et al. 2014). In order to utilize TNO phospholipid as an emulsifier, its surface tension properties need to be investigated and compared with that of soybean oil. The unsaponifiable matter of TNO was identified as consisting of hydrocarbons, waxes, triterpene alcohols,

Table 11.3 Fatty acid composition (g/100 g of fatty acids) of tiger nut oil

Origin	C14:0	C16:0	C16:1	C18:0	C18:1	C18:2	C18:3	C20:0	SFA	MUFA	PUFA	References
Spain		15.76	0.27	2.57	72.39	8.22	0.18	0.28	19.01	72.61	8.39	Lopéz-Cortés et al. (2013)
Egypt		15.15	0.26	3.35	71.69	8.47	0.04	0.37	19.26	71.98	8.52	Roselló-Soto et al. (2018)
Nigeria		17.87	0.21	3.79	59.44	8.77	0.12	0.39	22.46	59.71	8.88	
South Africa		20.38	0.33	4.05	62.95	10.91	0.11	0.31	25.20	63.28	11.02	Adel et al. (2015)
Egypt		15.19	0.29	5.07	69.25	8.37	0.19	ND	20.59	70.66	8.75	Yeboah et al. (2012)
Ghana	ND	16.32	ND	0.10	65.55	12.13	ND	0.68	22.33	65.55	12.13	Muhammad et al. (2011)
Nigeria	1.7	10.4	ND	0.07	76.10	11.8	0.6	6.1	18.50	76.10	12.40	Arafat et al. (2009)
Egypt	0.80	14.50	1.50	0.16	69.50	8.80	0.40	0.20	18.90	71.00	9.20	Zhang et al. (1996)
China	ND	14.99	ND	0.20	69.32	13.11	0.0	ND	17.55	69.32	13.11	Ezeh et al. (2014)
South Korea	NR	15.4	0.2	0.12	65.5	16.2	0.5	NR	17.6	65.7	16.7	
Turkey	ND	13.34	ND	2.58	71.1	12.71	0.42	ND	15.92	71.1	13.13	

ND not detected, *SFA* saturated fatty acids, *MUFA* monounsaturated fatty acids, *PUFA* polyunsaturated fatty acids, *NR* not reported

sterols esters, higher alcoholic esters and sterols (Oderinde and Tairu 1992). The total unsaponifiables was 3% of the oil. Hydrocarbons constituted the highest of these, followed by sterols.

Sterols

In addition to tocols, other classes of compounds that comprise the unsaponifiable fraction in TNO include phytosterols, phenolics and possibly carotenoids. The most abundant phytosterol class (about 90%) in TNO was identified to be 4-desmethyl sterol and its dominant component was β-sitosterol (43–61 mg/100 g). Other 4-desmethyl sterols present were stigmasterol (17–21 mg/100 g), campesterol (11–17 mg/100 g), and δ^5-avenasterol (Lopéz-Cortés et al. 2013; Yeboah et al. 2012) (Table 11.4). Trace concentrations of 4,4-dimethyl sterols were also found (5.6%). Oderinde and Tairu (1992) analysis on TNO sterols, confirmed β-sitosterol as the major sterol but with a much lower quantity of total sterols (100.02 µg/g oil) with only 0.1% cholesterol. This variation might be a result of the variety of the tuber. Temperature and planting location have been found to have an effect on sterol and tocopherol contents where an increase in growth temperature leads to elevated phytosterol levels.

Phenolic Compounds

The concentration of phenol compounds in TNO was quantified and given as 16.5 mg Gallic Acid Equivalent (GAE) per 100 g oil (Ali Rehab and El Anany 2012). Pellegrini et al. (2001) reported on the phenolics content in refined, virgin and extra virgin oils were 0.4, 1.4–2.4 and 7.3–26.5 mg GAE/100 g oil, respectively. Soybean, sunflower and corn oils have been found to contain 6–8, 0.3–0.4 and less than 0.1 mg/100 g oil, respectively. Compared with these values, TNO (16.5 mg/100 g oil) has similar phenolics content to extra virgin olive oil and much higher than soybean and virgin olive oil. The phenolics content of oils is important in assessing its antioxidant activity. These bioactive compounds play a protective role in the degradation of tocopherols during cooking processes and storage. Parker et al. (2000) used several concentrations of sodium hydroxide to extract esterified phenolic acids of cell wall

Table 11.4 Phytosterols (mg/100 g) and α-tocopherol (µg/g) levels of tiger nut oil

Origin	β-Sitosterol	Campesterol	Stigmasterol	α-tocopherol	References
Spain	60.18	14.01	19.50	4.6	Lopéz-Cortés et al. (2013)
Egypt	43.34	11.44	17.85	28.7	Roselló-Soto et al. (2018)
Nigeria	60.82	16.63	18.76	ND	
South Africa	61.15	16.32	17.35	ND	
Ghana	51.70	15.30	20.6	86.73	Yeboah et al. (2012)

Table 11.5 Total esterified phenolic acids (mg/g) of cell wall material from the skin and peeled tubers of tiger nut (Parker et al. 2000)

Phenolic acid	0.1 M NaOH		1 M NaOH		2 M NaOH	
	Skin	Tuber	Skin	Tuber	Skin	Tuber
Monomeric component						
p-Hydroxybenzoic acid	6.0	0.0	6.3	0.0	6.7	7.0
Vanillic acid	25.3	8.0	17.7	5.8	18.6	3.0
p-Hydroxybenzaldehyde	134.0	15.9	39.3	8.2	26.9	4.0
Vanillin	68.7	34.0	62.5	24.2	48.5	15.5
p-trans-coumaric acid	3239.0	635.0	6801.0	1416.0	4228.0	479.0
Trans-ferulic acid	2025.0	2284.0	1095.0	536.0	404.0	106.0
p-cis-coumaric acid	85.8	0.0	244.0	37.8	121.0	0.0
cis-ferulic acid	142.7	218.0	53.8	27.0	18.0	3.5
Dimeric component						
8-8′ AT diferulic acid	0.0	127.0	0.0	240.0	0.0	0.0
8-5′ diferulic acid	22.3	99.0	0.0	0.0	33.3	0.0
5-5′ diferulic acid	180.4	262.0	49.6	32.1	0.0	5.0
8-O-4′ diferulic acid	384.1	507.0	94.5	41.1	27.8	6.8
8-5′-(B) diferulic acid	178.2	199.0	0.0	0.0	0.0	0.0

material from the skin and peeled tiger nut tubers. The 13 compounds detected are listed in Table 11.5. The most abundant acid in the peeled tubers was ferulic acid released using 0.1 M NaOH, while p-coumaric acid dominated in the skin and required a stronger alkali (2 M NaOH) to be released. It should be expected that the concentration of these compounds will be reduced to a certain degree during oil extraction, depending on the production process employed and the materials used. Pumpkin seed oil was found, for example, to produce oils with a higher antioxidant capacity when a polar solvent was used during its extraction. Using enzymes as a pre-treatment increased the total concentration of some phenolic compounds, most especially for trans-ferulic acid, which had the most concentration of the phenolic compounds. The hemicellulase mixture, Viscozyme is most likely responsible for the increase in trans-ferulic acid as it exists as an esterified component of tiger nut cell walls (Parker et al. 2000). The impact of high pressure appears to have greatly enhanced the release of phenolic compounds into the aqueous enzymatic extracted oil. On the other hand, the use of enzyme as a pre-treatment decreased the total phenolics content in the pressed oil.

The concentration of anti-nutrients (tannins, saponins, phytate, oxalates and cyanogenic glycosides) in raw tiger nut tubers was found to be very low compared to that in nuts such as peanuts. Hence, their concentrations in the oil would be expected to be insignificant. Furthermore, when analyzing the color of a TNO sample in a frying study using a Lovibond tintometer, it was observed to be darker with a value of 4 compared to sunflower oil with 2.50. The darker color was thought to be formed by the degradation of henolic compounds and carotenoids extracted into the oil.

6.3   Physico-Chemical Properties of Tiger Nut Oil

The average values for the refractive index and density of TNO at 25 °C have been reported as 1.518 and 0.915 g/cm^3, respectively (Arafat et al. 2009). Iodine value is a measure of the degree of unsaturation of fats or oils. A higher iodine value simply indicates a higher degree of unsaturation. For TNO, a wide range of iodine value (76.60–142.3) has been reported. Ejoh and Ndjouenkeu (2006) discovered no significant variation between the iodine values of the black and brown varieties obtained from the same area in Cameroon. Variation did exist when compared to the tubers from other locations within the same country. Therefore, it can be considered a nondrying oil or semi-drying oil but most authors describe it as a nondrying oil (Ali Rehab and El Anany 2012). Its iodine value also indicates that it is liquid at room temperature, thus making it suitable for soap manufacture and vegetable-based ice cream.

The concentration of free fatty acids is quantified by the acid value given in mg KOH/g oil. A high acid value indicates oil with a reduced quality and thus acid value is considered an important indicator of the quality of vegetable oils (Kardash and Tur'yan 2005). For TNO, most of its reported acid values are quite low; 0.03–1.38 mg KOH/g oil (Muhammad et al. 2011; Yeboah et al. 2012). Tiger nuts grown in the USA, were reported to have a high acid value of 15.7, but this indicates poor handling and processing of the nuts. Typically, accepted acid values should be less than 4 mg KOH/ g oil. This unusually high value might be a result of a very active fat-splitting enzyme in the tuber. The enzyme was observed to be present in small amounts as it slowly hydrolyzed amygdalin. Amygdalin, a naturally occurring glycoside in food plants forms an ester bond with fatty acids. Thus, the presence of the enzyme may have slowly led to hydrolytic rancidity due to free fatty acids released. Other high values reported were from studies done in Egypt, comparing the composition of germinated and ungerminated tubers. Oil from germinated and ungerminated tubers had acid values of 5.97 and 9.07, respectively. In Nigeria, TNO was exposed to sunlight over a 9 weeks period and its hydrolytic stability was monitored via its acid value (Ezebor et al. 2006). The change in acid value was 6.06 and its initial value was quite high at 8.48. There was no record as to how old the tiger nuts were before the study was conducted, it may mean that hydrolytic rancidity had already begun to take place in the tuber before the study commenced as indicated by the high acid value. The presence of amygdalin also indicates the presence of cyanogenic compounds, which is in agreement with the finding that the raw tubers contain cyanogenic glycosides. Preliminary examinations on the acute toxicity of tiger nut extract revealed that it was not toxic to mice at the administered concentrations (Sánchez-Zapata et al. 2012). Bamishaiye et al. (2010) confirmed this in a study involving rats where their hematological parameters were monitored. It was concluded that dietary treatments did not affect the serum components significantly and hence was free of toxicity. The study also confirmed the anti-infection properties of TNO.

Peroxide value acts as an indicator of an oil's freshness and quality. Peroxides are primary reaction products formed when unsaturated FA undergo oxidation.

Following from that, the higher the peroxide value, the more oxidized the oil has become, and this can be manifested by oxidative rancidity. Typical values of TNO's peroxide value range from 0.30 to 6.9 meq/kg (Arafat et al. 2009), which fall well below the acceptable value of 10 meq/kg for fresh oils. These indicate that TNO has good keeping capacity being able to withstand long time storage without undergoing oxidative peroxidation (Muhammad et al. 2011).

p-Anisidine value is used to detect secondary oxidation products and it provides useful information on non-volatile carbonyl compounds formed during processing. It has been recommended that good quality oils should have a p-anisidine value of less than two. The p-anisidine value for TNO was reported to be negative but negative p-anisidine value has been suggested to be an artifact arising from the presence of water in samples and reagents thus leading to interference in the test. The authors did not explain this but negative values are simply an unreal situation. An in-depth study covering this would be required to determine the p-anisidine value of TNO. Acid and p-anisidine values are commonly used as quality indices for oils (Ezeh et al. 2014).

Changes in acid, peroxide and iodine values were better controlled and significantly decreased when tiger nut/sunflower oil mixtures varied between 20:80 and 50:50 in a deep-frying study. TNO had the lowest change in iodine value of 5.6%. In the study, TNO increased the stability of sunflower oil during frying which was attributed to the FA composition and natural antioxidants in TNO (Ali Rehab and El Anany 2012).

TNO has a viscosity of 46.10 mPa s at 30 °C and 39.8 mPa s at 40 °C. This low value makes it suitable for use in the coating industry and diesel fuel (Pascual et al. 2000).

6.4   Comparison of Tiger Nut Oil Composition with Other Vegetable Oils

Linssen et al. (1988) studied a chemical analysis of tiger nut and olive oils obtained *via* mechanical pressing and discovered that their FA composition and distribution were nearly the same. The taste and color, on the other hand, differed as TNO is neutral in taste and has a golden-yellow color, while olive oil has a characteristic taste and is green in color (Ezeh et al. 2014). The FA composition of TNO is compared with other vegetable oils in Table 11.6.

A high similarity exists between TNO and olive oil as mentioned, although the discrepancy between the ranges of linoleic to linolenic acids is larger. Other similar FA profiles are those of hazelnut, macadamia nut and avocado oil. A lower percentage of PUFA in oils contributes to their increased stability, as observed for olive oil. As such, the lower level of PUFA in TNO gives it the same advantage. TNO's fatty acid profile meets the requirements of oils that are nutritionally beneficial and offer stability for high-temperature applications (high oleic FA and low linolenic FA) without requiring hydrogenation such as high oleic sunflower and soybean oil.

Table 11.6 Fatty acid profiles of tiger nut oils and common vegetable oils

Product	% SFA	% MUFA	% PUFA	%Total n-6	%Total n-3	References
Tiger nut *Cyperus esculentus*	11.6–22.3	65.6–76.1	9.2–13.6	8.8–13.4	0–0.6	Muhammad et al. (2011)
Olive *Olea europaea*	13.2–16.6	73.0–80.9	5.1–10.7	4.5–10.1	0.5–1.0	Stefanoudaki et al. (1999)
Hazelnut *Corylus avellana*	7.15–16.7	74.5–83.2	8.27–16.7	8.17–17.8	0.2–0.5	Madawala et al. (2012)
Macadamia nut	13.2–17.8	78.8–82.4	2.2–4.7	2.0–4.5	0.06–0.4	
Avocado *Persea americana*	13.1–23.7	58.4–76.5	10.3–16.5	4.9–16.1	0.2–1.44	
Rapeseed *Brasica napus*	4.34–6.9	58.6–66.2	27.5–34.6	18.4–22.3	8.2–12.4	
Peanut *Arachis hypogea*	14.8–17.6	38.6–48.3	32.9–46.0	35.5–44.6	0.1–1.37	Maguire et al. (2004)
Pistachio *Pistacia atlantica*	9.3–28.1	55.0–60.1	14.8–33.4	14.4–33.0	0.1–0.4	Ryan et al. (2006)
Argan *Argania spinosa*	15.1–18.5	44.2–45.4	35.7–40.7	35.6–36.9	0.1–3.8	Onyinye Ezeh et al. (2014)
Walnut *Juglans regia*	9.5–12.3	15.7–25.4	62.1–70.5	50.1–60.1	10.4–12.0	
Sunflower *Helianthus annuus*	8.9–11.7	16.9–33.8	56.5–73.7	56.5–73.3	0–0.1	
Corn *Zea mays*	13.4–14.1	25.2–38.7	46.1–60.5	44.7–60.6	0.5–5.8	
Chia *Salvia hispanica*	9.3–14.0	6.0–9.3	76.0–82.9	17.0–22.5	56.9–64.8	
Soybean *Glycine max*	12.2–14.9	22.3–38.1	49.7–61.8	40.7–55.5	6.3–9.0	

Oils like these are commonly found in industry and used especially in frying because of their high stability. Depending on the cultivar of tiger nut tubers, its oil can join the plethora of commercially available oils used for frying applications in processed foods, snack products and spray oils. The use of TNO in these industrial applications would make for an interesting study. The oxidative stabilities of TNO and olive oil have already been assessed by Arafat et al. (2009) using the Rancimat test, expressed in oxidation induction time. The oxidation of lipids initially takes place slowly before a sudden rise in oxidation rate occurs. The sudden rise marks the endpoint of the induction period (Velasco and Dobarganes 2002). Both oils had similar values with 28 h for TNO and 30 h for olive oil, thus indicating a similar shelf life. Longer induction times indicate higher stability and antioxidant activity. The FA of TNO are the only aspect of the oil that can be fairly compared in relation to other oils. This is because of the sparse information on the chemical constituents

Table 11.7 Phytosterols concentration (mg/100 g) in tiger nut oil and common vegetable oils

Oil	β-Sitosterol	Campesterol	Stigmasterol	δ^5-Avenasterol	Total phytosterol	References
Tiger nut	43.3–61.18	11.44–16.63	17.35–20.6	3.8	98.7	Muhammad et al. (2011)
Olive	70.3–201	1.1–4.5	0.9–1.6	6.6–21.8	129–238	Stefanoudaki et al. (1999)
Hazelnut	68.1–219	4.4–16.4	1.2–3.8	6.3–13.9	88.9–263	Madawala et al. (2012)
Coconut	37.9–48.6	5.4–7.8	10.6–12.5	10.9–14.7	68.9–70.0	
Avocado	302.3–320.1	7.5–18.7	0.5–2.5	7.7–21.5	353.0–434.2	
Rapeseed	148.9–419.8	114.5–293.0	0.9–3.5	10.1–38.2	326.7–823	
Peanut	99.0–169.0	15.3–23.7	12–21.9	12.9–17.0	167–228.7	Maguire et al. (2004)
Sesame	331.2–335.4	74.5–91.6	33.0–36.7	41.3–53.4	492.0–620.4	Madawala et al. (2012)
Sunflower	194.2–265.3	27.1–54.5	17.7–33.7	18.7–55.5	263.0–382.2	Ezeh et al. (2014)
Soybean	139.2–183.7	43.3–71.0	48.8–63.9	4.2–13.8	203.0–301.7	

nd not detected, *nr* not reported

of the oil, including research on its unsaponifiable matter, and phospholipids. Despite this, an attempt has been made while being cautious of this limitation. In olive, soybean, cottonseed and coconut oils, the unsaponifiable (0.8–1.5%, 1.5–1.7%, 0.5–1.2% and 0.5%, respectively) were lower than the unsaponifiables in TNO at 3% (Ezeh et al. 2014).

Table 11.7 shows a list of sterols in different vegetable oils. The total sterol content (98.7 mg/100 g) in TNO falls short of the range of sterols in olive oil (129–238 mg/100 g). It is higher than the range given for coconut oil (68.9–70.0 mg/100 g). The value for the predominant -sterol; β-sitosterol was lower in TNO (43.3–61.18 mg/100 g) than in all of the other vegetable oils with the exception of coconut oil. Campesterol and stigmasterol levels are higher in TNO (11.44–16.63 mg/100 g; 17.35–20.6 mg/100 g) than in olive (1.1–4.5 mg/100 g; 0.9–1.6 mg/100 g), coconut (5.4–7.8 mg/100 g; 10.6–12.5 mg/100 g), and hazelnut (4.4–16.4 mg/100 g; 1.2–3.8 mg/100 g) oils. It has been shown that campesterol is absorbed more efficiently than sitosterol with a ratio of 2.29 after consumption (Kritchevsky and Chen 2005). However, much higher levels of sitosterol in other vegetable oils still ensure that it is absorbed when included in the diet.

The tocopherol content of TNO consists mainly of α-tocopherol (86.7 mg/g) and β-tocopherol (33.4 mg/g) making up a total of 120.1 mg/g oil. This falls within the range for olive oil as shown in Table 11.8. The high oxidative stability of olive oil is related to its higher α-tocopherol content which is more stable than β-tocopherol. It would be premature to assume that TNO is less stable due to lower amounts of α-tocopherol. With the exception of coconut oil, the vitamin E content of TNO is much lower than that in sunflower, peanut oils and most notably soybean oil (Table 11.8), but this is fully consistent with the known occurrence of high concentrations of vitamin E in tissues rich in PUFA (Ezeh et al. 2014).

Table 11.8 Tocols content (mg/g oil) of tiger nut oil and common vegetable oils

Oil	Tocopherols				Tocotrienols		References
	α	β	γ	δ	α	β	
Tiger nut	86.73	33.37	–	–	–	–	Muhammad et al. (2011)
Olive	96–240	0–6.0	7–14	0	–	–	Stefanoudaki et. (1999)
Coconut	2–5	–	0–1.2	0–6	5–30	1.0–1.7	Madawala et al. (2012)
Rapeseed	153–256	–	340–510	9.8–19	0.4	–	
Peanut	130–141	0–4	131–214	9–21	–	–	Maguire et al. (2004)
Sunflower	487–590	23–24	4–51	0–8	0–1.1	–	Ezeh et al. (2014)
Soybean	75–11	15.0–17.0	578–797	263–266	0–2	0–1	

nd not detected, *nr* not reported

The value of phenolics content in TNO (16 mg GAE/100 g) was noted to be higher than the phenolic content of sunflower oil (5 mg GAE/100 g). It is also higher than that of other common vegetable oils such as soybean (6–8 mg GAE/100 g), sunflower (0.3–0.4 mg GAE/100 g), and corn oil with less than 0.1 mg GAE/100 g (Marfil et al. 2011), but much lower than 84 mg GAE/100 g of virgin coconut oil.

The phospholipid composition is higher in TNO (3.15%) than in soybean oil (1.1–1.9%), 1.2% in crude sunflower oil, and 2% in crude canola oil. The major phospholipids present in TNO are similar to those in sunflower and soybean oils. In a deep frying study using sunflower oil and TNO, the viscosity of TNO was reported to be lower than sunflower oil's viscosity after 30 h of frying. When the quantity of TNO in the sunflower/TNO blend was increased, the change in viscosity decreased (Ezeh et al. 2014). While simulating a frying process using virgin olive oil, Benedito et al. (2002) recorded the change in viscosity over a 16 h period. Over 10 h, the change in olive oil was much higher at 22.6 mPa s than the change using TNO at 3.9 mPa s. An increase in viscosity during frying is an expected phenomenon due to the formation of high molecular weight polymers. The polymers that form are rich in oxygen and accelerate the degradation of oil further, thus increasing the viscosity even more, although under genuine industrial frying conditions, the oxygen content of oil is low and the polymers formed are not rich in oxygen. Deterioration proceeds rapidly once the antioxidants are consumed. As a result, more viscous oil indicate oils with a higher degree of deterioration. On comparing deep frying studies, the acid value change for TNO (0.42 units) was noted to be lower than that of sunflower, cottonseed, and olive oils with 0.56, 0.88 and 0.50 units, respectively despite being used for 30 h for frying compared to only 2.5 h for cottonseed and olive oils (Ezeh et al. 2014).

7 Tiger Nut Wastes and By-Products

In 2007–2008, 4.2 tons of tiger nuts wastes and by-products were produced in the Valencian region, Spain. These by-products could be a valuable source of oil and high-added value compounds including nutrients (carbohydrates, proteins, vitamins, and minerals) and bioactive compounds (i.e., polyphenols). The nutrients can be used as an energy source, while bioactive compounds have prospective application in nutraceutical industries (Gil-Chávez et al. 2013). Moreover, these by-products also have a high fiber content, which can be used in the development of new products, thus responding to society's demand for potential functional foods (Granato et al. 2017). Moreover, in recent years, stringent rules have been implemented in several countries for disposal of agricultural wastes considering their associated health and environmental implications (Galanakis 2016). The developed countries are promoting prospective utilization of agricultural wastes as a source of energy, nutrient and medicine. For instance, Spanish and European legislative frameworks for waste management prioritizes the recycling and recovery of by-products from their disposal in order to preserve natural resources, protect the environment and consumer health. Substantial research works have evidenced that tiger nut is a good source of oil and its by-products are rich in various nutrients and bioactive compounds. Several reports have revealed that the extracts or by-products of tiger nuts exhibit antioxidant activity (Badejo et al. 2014), which is ascribed to its phenolics content.

The use of tiger nut by-products obtained during the production of horchata has attracted the interest of both the food industry and the scientific community for the development of healthier products mainly due to their high dietary fiber content (59.71 g/100 g, 99.8% dietary fiber) (Sánchez-Zapata et al. 2009). They can also be used in the production of bio-based composites or as a carbon source for probiotic bacterial growth (Sánchez-Zapata et al. 2013). However, it is necessary to use a preservation method prior to their addition to food products due to their high microbial load. Moreover, tiger nut by-products have extraordinary water- and oil-holding capacities, high emulsifying ability as well as low water absorption compared to other dietary fiber sources (Sánchez-Zapata et al. 2009), which make them ideal for improving the functional properties of the developed products. Aguilar et al. (2015) studied the impact of the use of tiger nut beverage by-products as an alternative to soy, which is a known allergen, for the production of gluten-free bread, obtaining bread with similar composition and consumer's acceptance to those obtained when soya flours were used. In another study, the influence of the addition of two different tiger nut beverage by-products flours at different percentages (5%, 10% and 20% w/w dry basis) on wheat flour was evaluated (Verdú et al. 2017). The authors observed a slight increase in mass loss, although non-significant, in the baking process when the by-product flours were added compared to control samples. Other authors evaluated the potential of tiger nut by-products, particularly tiger nut fibers, to be used in the development of healthier fiber-rich meat products (Sánchez-Zapata et al. 2010). Tiger nut fibers were used as a source of unsaturated fatty acids to improve the nutritional quality of a traditional dry-cured fermented sausage Longaniza de Pascua (Sánchez-Zapata et al. 2013). Together with walnut oil, the addition of tiger nut fibers enhanced moisture (increased weight), water activity, redness and blueness of the samples, while the lightness value remained unchanged but the pH values was reduced. Moreover, lipid oxidation was prevented, likely due to the high phenolics content of tiger nut by-products.

Furthermore, the tiger nuts liquid by-product exhibited the potential to control lipid oxidation in pork burger, possibly due to its high levels of antioxidant compounds (Sánchez-Zapata et al. 2012). Moreover, it enhanced cooking propertics by reducing the shrinkage of burger and conveyed better consumer acceptability as they were perceived as less fatty (Sánchez-Zapata et al. 2012).

8   Conclusions

The anomaly that tiger nut is an oil containing tuber makes it an interesting crop to study. This, coupled with the fact that it is commonly grown in several countries, which stand to benefit from research into its use, makes it even more fascinating. In addition, development of new foods, natural and healthy ingredients as well as management of wastes derived from food processing have become the major concerns for food industries. Though horchata production provides major marketing and revenue for tiger nut processing industries, the production and use of TNO or valorization of byproducts are almost neglected in such industries.

This chapter confirms that tiger nut is a valuable source of diverse nutrients and bioactive compounds such as vitamin E homologs, phytosterols, and phenolics. Moreover, it is a good source of oil that is rich in MUFA and tocopherols combined with the low level of PUFA that can be used in cooking. Quality indices parameters such as its acid and peroxide values give an indication of the oil's freshness and these have been found to be low, thus concluding that it has a good keeping capacity.

Several investigations revealed that oil could be efficiently extracted using conventional organic solvent extraction, however, recent investigations highlight the advantages of non-conventional extraction techniques, particularly supercritical fluid extraction, in terms of yield and safety. TNO can be extracted using aqueous enzymatic processing methods, with the aid of enzymes and high-pressure processing as pre-treatments respectively. Most oils was obtained when enzyme assisted pressing was utilized. The oilseed meal, a by-product is commonly used as animal feed because it can be rich in protein. There have been suggested alternative uses of oilseed meals including the production of bioethanol and valuable food ingredients such as oligosaccharides.

References

Abaejoh, R., Djomdi, I., & Ndojouenkeu, R. (2006). Characteristics of tiger nut (*Cyperus esculentus*) tubers and their performance in the production of a milky drink. *Journal of Food Processing & Preservation, 30*, 145–163.

Abano, E., & Amoah, K. (2011). Effect of moisture content on the physical properties of tiger nut (*Cyperus esculentus*). *Asian Journal of Agricultural Research, 5*, 56–66.

Adel, A. A. M., Awad, A. M., Mohamed, H. H., & Iryna, S. (2015). Chemical composition, physicochemical properties and fatty acid profile of tiger nut (*Cyperus esculentus* L) seed oil as affected by different preparation methods. *International Food Research Journal, 22*(5), 1931–1938.

Agbai, E., & Nwanegwo, C. (2013). Effect of methanolic extract of *Cyperus esculentus* L. (Tiger nut) on luteinizing hormone, follicle stimulating hormone, testosterone, sperm count and motility in male albino wis tar rat. *Journal of Applied Biosciences, 5*(2), 52–61.

Aguilar, N., Albanell, E., Miñarro, B., Guamis, B., & Capellas. (2015). Effect of tiger nutderived products in gluten-free batter and bread. *Food Science and Technology International, 21*(5), 323–331.

Ali Rehab, F. M., & El Anany, A. M. (2012). Physicochemical studies on sunflower oil blended with cold pressed tiger nut oil during deep frying process. *Grasas y Aceites, 63*, 455–465.

Arafat, S. M., Gaafar, A. M., Basuny, A. M., & Nassef, S. L. (2009). Chufa tubers (*Cyperus esculentus* L.): As a new source of food. *World Applied Sciences Journal, 7*(2), 151–156.

Badejo, A. A., Damilare, A., & Ojuade, T. D. (2014). Processing effects on the antioxidant activities of beverage blends developed from *Cyperus esculentus*, Hibiscus sabdariffa, and Moringa oleifera extracts. *Preventive Nutrition and Food Science, 19*(3), 227–233.

Bamishaiye, E. I., & Bamishaiye, O. M. (2011). Tiger nut: As a plant, its derivatives and benefits. *AJFAND African Journal of Food, Agriculture, Nutrition and Development, 11*, 5157–5170.

Bamishaiye, E., Muhammad, N., & Bamishaiye, O. (2010). Haematological parameters of albino rats fed on tiger nuts (*Cyperus Esculentus*) tuber oil meal-based diet. *Internet J Nutr Wellness, 10*(1), 1–5.

Belewu, M. A., & Abodunrin, O. A. (2006). Preparation of Kunnu from unexploited rich food source: Tiger nut (*Cyperus esculentus*). *World Journal of Dairy & Food Sciences, 1*, 19–21.

Benedito, J., Mulet, A., Velasco, J., & Dobarganes, M. C. (2002). Ultrasonic assessment of oil quality during frying. *Journal of Agricultural and Food Chemistry, 50*, 4531–4536.

Borges, O., Goncalves, B., Sgeoeiro, L., Correia, P., & Silva, A. (2008). Nutritional quality of chestnut cultivars from Portugal. *Food Chemistry, 106*, 976–984.

Chukwuma, E. R., Obiama, N., & Christopher, O. I. (2010). The phytochemical composition and some biochemical effect of Nigerian Tiger nut (*Cyperus esculentus*. L) tuber. *Pakistan Journal of Nutrition, 9*(7), 709–715.

Codina-Torrella, I., Guamis, B., & Trujillo, A. J. (2015). Characterization and comparison of tiger nuts (*Cyperus esculentus* L.) from different geographical origin: Physico-chemical characteristics and protein fractionation. *Industrial Crops and Products, 65*, 406–414.

David, A.B. (2005). *Tiger nut. A dictionary of food and nutrition*. Encyclopedia.com: http://www.encyclopedia.com/doc/1O39-tigernut.html.

Defelice, M. S. (2002). Yellow nutsedge *Cyperus esculentus* L.: Snack food of the gods. *Weed Technology, 16*, 901–907.

Devries, F., & Feuke, T. (1999). Chufa (*Cyperus esculentus*) a weedy cultivar or cultivated weed. *Economic Botany, 45*, 27–37.

Dyer, A. R. (2006). *The ecology of chufa (Cyperus esculentus sativus)*. South Carolina: University of South Carolina.

Ejoh, R. A., & Ndjouenkeu, D. R. (2006). Characteristics of tigernut (*Cyperus Esculentus*) tubers and their performance in the production of a milky drink. *Journal of Food Processing & Preservation, 30*, 145–163.

Ekpe, O. O., Igile, G. O., Williams, I. O., & Eworo, P. (2016). Quality mapping of tiger nut oil and the extraction efficiency between n-hexane and petroleum ether solvents. *Food Science and Quality Management, 50*, 39–48.

Ezebor, F., Igwe, C., Owolabi, F., & Okoh, S. (2006). Comparison of the physico-chemical characteristics, oxidative and hydrolytic stabilities of oil and fat of *Cyperus esculentus* L.(yellow nutsedge) and *Butyrospermum parkii* (shea nut) from Middle-Belt States of Nigeria. *Nigerian Food Journal, 23*, 33–39.

Ezeh, O., Michael, H. G., & Keshavan, N. (2014). Tiger nut oil (*Cyperus esculentus* L.): A review of its composition and physico-chemical properties. *European Journal of Lipid Science and Technology, 116*, 783–794.

Ezeh, O., Gordon, M. H., & Niranjan, K. (2016). Enhancing the recovery of tiger nut (*Cyperus esculentus*) oil by mechanical pressing: Moisture content, particle size, high pressure and enzymatic pre-treatment effects. *Food Chemistry, 194*, 354–361.

Galanakis, C. M. (2016). Innovation strategies in the food industry: Tools for implementation. In C. M. Galanakis (Ed.), *Innovation strategies in the food industry: Tools for implementation* (pp. 1–313). Oxford: Elsevier. https://doi.org/10.1016/C2015-0-00303-3.

Gil-Chávez, G. J., Villa, J. A., Ayala-Zavala, J. F., Heredia, J. B., Sepulveda, D., Yahia, E. M., & González-Aguilar, G. A. (2013). Technologies for extraction and production of bioactive compounds to be used as nutraceuticals and food ingredients: An overview. *Comprehensive Reviews in Food Science and Food Safety, 12*(1), 5–23.

Granato, D., Nunes, D. S., & Barba, F. J. (2017). An integrated strategy between food chemistry, biology, nutrition, pharmacology, and statistics in the development of functional foods: A proposal. *Trends in Food Science and Technology, 62*, 13–22.

Jing, S., Ouyang, W., Ren, Z., Xiang, H., & Ma, Z. (2012). The in vitro and in vivo antioxidant properties of *Cyperus esculentus* oil from Xinjiang, China. *Journal of the Science of Food and Agriculture, 93*, 1505–1509.

Kardash, E., & Tur'yan, Y. I. (2005). Acid value determination in vegetable oils by indirect titration in aqueous-alcohol media. *Croatica Chemica Acta, 78*, 99–103.

Kim, M., No, S., & Yoon, S. (2007). Stereospecific analysis of fatty acid composition of chufa (*Cyperus esculentus* L) tuber oil. *Journal of the American Oil Chemists' Society, 84*, 1079–1080.

Koubaa, M., Barba, F. J., Mhemdi, H., Grimi, N., Koubaa, W., & Vorobiev, E. (2015). Gas Assisted Mechanical Expression (GAME) as a promising technology for oil and phenolic compound recovery from tiger nuts. *Innovative Food Science and Emerging Technologies, 32*, 172–180.

Kritchevsky, D., & Chen, S. C. (2005). Phytosterols-health benefits and potential concerns: A review. *Nutrition Research, 25*, 413–428.

Kuner, Y., Ercan, R., Karababa, E., & Nazlıcan, A. N. (2002). Physical and chemical properties of chufa (*Cyperus esculentus* L) tubers grown in the kurova region of Turkey. *Journal of the Science of Food and Agriculture, 82*, 625–631.

Lamsal, B. P., & Johnson, L. A. (2007). Separating oil from aqueous extraction fractions of soybean. *Journal of the American Oil Chemists' Society, 84*, 785–792.

Lasekan, O., & Abdulkarim, S. M. (2012). Extraction of oil from tiger nut (*Cyperus esculentus* L.) with supercritical carbon dioxide (SC-CO$_2$). *LWT- Food Science and Technology, 47*(2), 287–292.

Linssen, J. P. H., Kielman, G. M., Cozijnsen, J. L., & Pilnik, W. (1988). Comparison of chufa and olive oils. *Food Chemistry, 28*, 279–285.

Lopéz-Cortés, I., Salazar-García, D. C., Malheiro, R., Guardiola, V., & Pereira, J. A. (2013). Chemometrics as a tool to discriminate geographical origin of *Cyperus esculentus* L. based on chemical composition. *Industrial Crops and Products, 51*, 19–25.

Lowe, J., & Stanfield, D. (1974). *Sedges (family Cyperaceae)*. Ibadan: Ibadan University Press.

Lunn, J., & Theobald, H. (2006). The health effects of dietary unsaturated fatty acids. *Nutrition Bulletin, 31*, 178–224.

Madawala, S. R., Kochhar, S. P., & Dutta, P. C. (2012). Lipid components and oxidative status of selected specialty oils. *Grasas y Aceites, 63*, 143–151.

Maguire, L. S., O'Sullivan, S. M., Galvin, K., O'Connor, T. P., & O'Brien, N. M. (2004). Fatty acid profile, tocopherol, squalene and phytosterol content of walnuts, almonds, peanuts, hazelnuts and the macadamia nut. *International Journal of Food Sciences and Nutrition, 55*, 171–178.

Marfil, R., Giménez, R., Martínez, O., Bouzas, P. R., et al. (2011). Determination of polyphenols, tocopherols, and antioxidant capacity in virgin argan oil (*Argania spinosa*, Skeels). *European Journal of Lipid Science and Technology, 113*, 886–893.

Martinez, V. (2003). Scientific analysis of effects of tiger nut on heart diseases and related aspects. *Tiger Nut and Health*, 1–2.

Mat Yusoff, M., Gordon, M. H., & Niranjan, K. (2015). Aqueous enzyme assisted oil extraction from oilseeds and emulsion de-emulsifying methods: A review. *Trends in Food Science and Technology, 41*, 60–82.

Muhammad, N., Bamishaiye, E., Bamishaiye, O., Usman, L., Salawu, M. O., Nafiu, M. O., & Oloyede, O. (2011). Physicochemical properties and fatty acid composition of *Cyperus esculentus* (Tiger Nut) tuber oil. *Biores Bull, 5*, 51–54.

Obadina, A. O., Oyawole, O. B., & Ayoola, A. A. (2008). Quality assessment of Gari produced using rotary drier. In V. C. Bellinghouse (Ed.), *Food processing, methods, techniques and trends*. New York: Nova Science Publishers.

Oderinde, R., & Tairu, A. (1992). Determination of the triglyceride, phospholipid and unsaponifiable fractions of yellow nutsedge tuber oil. *Food Chemistry, 45*, 279–282.

Oladele, A. K., & Aina, J. O. (2007). Chemical composition and functional properties of flour from two varieties of tigernut. *African Journal of Biotechnology, 6*, 2473–2476.

Palonen, H. (2004). *Role of lignin in the enzymatic hydrolysis of lignocellulose*. VTT Technic Res Centre Finland. VTT publications, 520, Aalto University, ISBN: 951-38-6272-0. ISSN: 1455–0849

Parker, M. L., Ng, A., Smith, A. C., & Waldron, K. W. (2000). Esterified phenolics of the cell walls of chufa (*Cyperus esculentus* L.) tubers and their role in texture. *Journal of Agricultural and Food Chemistry, 48*, 6284–6291.

Pascual, B., Maroto, J. V., López-Galarza, S., Sanbautista, A., & Alagarda, J. (2000). Chufa (*Cyperus esculentus* L. var. sativus Boeck.): An unconventional crop. Studies related to applications and cultivation. *Economic Botany, 54*, 439–448.

Pellegrini, N., Visioli, F., Buratti, S., & Brighenti, F. (2001). Direct analysis of total antioxidant activity of olive oil and studies on the influence of heating. *Journal of Agricultural and Food Chemistry, 49*, 2532–2538.

Poojary, M. M., Barba, F., Aliakbarian, B., Donsì, F., Pataro, G., Dias, D., & Juliano, P. (2016). Innovative alternative technologies to extract carotenoids from microalgae and seaweeds. *Marine Drugs, 14*(11), 214. https://doi.org/10.3390/md14110214.

Poojary, M. M., Orlien, V., Passamonti, P., & Olsen, K. (2017). Enzyme-assisted extraction enhancing the umami taste amino acids recovery from several cultivated mushrooms. *Food Chemistry, 234*, 236–244.

Rita, E. S. (2009). The use of tiger-nut (*Cyperus esculentus*), cow milk and their composite as substrates for yoghurt production. *Pakistan Journal of Nutrition, 6*, 755–758.

Roselló-Soto, E., Parniakov, O., Deng, Q., Patras, A., Koubaa, M., Grimi, N., & Barba, F. J. (2016). Application of non-conventional extraction methods: Toward a sustainable and green production of valuable compounds from mushrooms. *Food Engineering Reviews, 8*(2), 214–234.

Roselló-Soto, E., Mahesha, M. P., Francisco, J. B.; Jose, M. L., Jordi, M., & Juan Carlos, M. (2018). Tiger nut and its by-products valorization: From extraction of oil and valuable compounds to development of new healthy products. *Innovative Food Science and Emerging Technologies, 45*, 306–312.

Ryan, E., Galvin, K., O'Connor, T., Maguire, A., & O'Brien, N. (2006). Fatty acid profile, tocopherol, squalene and phytosterol content of brazil, pecan, pine, pistachio and cashew nuts. *International Journal of Food Sciences and Nutrition, 57*, 219–228.

Salau, R. B., Ndamitso, M. M., Paiko, Y. B., Jacob, J. O., Jolayemi, O. O., & Mustapha, S. (2012). Assessment of the proximate composition, food functionality and oil characterization of mixed varieties of *Cyperus esculentus* (tiger nut) rhizome flour. *Continental Journal of Food Science and Technology, 6*(2), 13–19.

Sánchez-Zapata, E., Fuentes-Zaragoza, E., Fernández-López, J., Sendra, E., Sayas, E., Navarro, C., & Pérez-Álvarez, J. A. (2009). Preparation of dietary fiber powder from tiger nut (*Cyperus esculentus*) milk ("Horchata") byproducts and its physicochemical properties. *Journal of Agricultural and Food Chemistry, 57*(17), 7719–7725.

Sánchez-Zapata, E., Muñoz, C. M., Fuentes, E., Fernández-López, J., Sendra, E., Sayas, E., & Pérez-Alvarez, J. A. (2010). Effect of tiger nut fiber on quality characteristics of pork burger. *Meat Science, 85*(1), 70–76.

Sánchez-Zapata, E., Fernández-López, J., & Pérez-Alvarez, J. A. (2012). Tiger nut (*Cyperus esculentus*) commercialization: Health aspects, composition, properties, and food applications. *Comprehensive Reviews in Food Science and Food Safety, 11*(4), 366–377.

Sánchez-Zapata, E., Díaz-Vela, J., Pérez-Chabela, M. L., Pérez-Alvarez, J. A., & Fernández-López, J. (2013). Evaluation of the effect of tiger nut fiber as a carrier of unsaturated fatty acids rich oil on the quality of dry-cured sausages. *Food and Bioprocess Technology, 6*(5), 1181–1190.

Shaker, M. A., Ahmed, M. G., Amany, M. B., & Shereen, L. N. (2009). Chufa tubers (*Cyperus esculentus* L.): As a new source of food. *World Applied Sciences Journal, 7*(2), 151–156.

Simpson, D., Yesson, C., Couch, C., & Muasya, A. (2011). Climate change and Cyperaceae. In T. Hodkinson, M. Jones, S. Waldren, & J. Parnell (Eds.), *Climate change, ecology and systematics* (1st ed.). Cambridge, UK: Cambridge University Press.

Soha, M. H., Amany, M. S., Abdel Karim, M. A., Ayman, M. A., & Alshimaa, M. A. (2017). Protective effect of Hesperidin and Tiger nut against Acrylamide toxicity in female rats. *Experimental and Toxicologic Pathology, 69*, 580–588.

Stefanoudaki, E., Kotsifaki, F., & Koutsaftakis, A. (1999). Classification of virgin olive oils of the two major cretan cultivars based on their fatty acid composition. *Journal of the American Oil Chemists' Society, 76*, 623–626.

Velasco, J., & Dobarganes, C. (2002). Oxidative stability of virgin olive oil. *European Journal of Lipid Science and Technology, 104*, 661–676.

Verdú, S., Barat, J. M., Alava, C., & Grau, R. (2017). Effect of tiger-nut (*Cyperus esculentus*) milk co-product on the surface and diffusional properties of a wheat-based matrix. *Food Chemistry, 224*, 8–15.

Xueshe, Q., Weiming, Z., Gongping, G., & Guanglun, Z. (2008). Comprehensive utilization and cultivation of fuel oil plant *Cyperus esculentus L. Chinese Wild Plant Resource*, 3-004.

Yeboah, S. O., Mitei, Y. C., Ngila, J. C., Wessjohann, L., & Schmidt, J. (2012). Compositional and structural studies of the oils from two edible seeds: Tiger nut, *Cyperus esculentum* and asiato Pachira insignis, from Ghana. *Food Research International, 47*(2), 259–266.

Zhang, H. Y., Hanna, M. A., Ali, Y., & Nan, L. (1996). Yellow nut-sedge (*Cyperus esculentus* L.) tuber oil as a fuel. *Industrial Crops and Products, 5*(3), 177–181.

Part III
Fats From Fruit Seeds and Pulp

Chapter 12
Rambutan (*Nephelium lappaceum*) Fats

Mohamed Elwathig Saeed Mirghani

Abstract Rambutan (*Nephelium lappaceum*) fruit is rich in carbohydrates, lipids, phosphorus, vitamin C, niacin, iron, calcium, copper, protein and fiber. Rambutan seed kernel fat (RSKF) can be a promising alternative edible fat that has the potential to be used in the food industry, especially to replace hydrogenated fat. The main fatty acids in RSKF are arachidic acid (38.3%) and oleic acid (37.1%). These two fatty acids covered 75.7% of the total fatty acids. RSKF exhibited several nutritional, biological and health promoting effects. This chapter reports on the chemical composition and health promoting impacts of RSKF.

Keywords Arachidic acid · Sapindaceae · Kernel fat · Hydrogenated fat

1 Introduction

1.1 The Origin of the Tree, Habitat, Production and Shape

Rambutan (*Nephelium lappaceum* L., family Sapindaceae), is indigenous to Southeast Asia, especially Indonesia, Malaysia, Thailand and Southern China as well. Nowadays it is had been developed in the plantation for commercial agricultural production in many Asian countries including Thailand, Seri Lanka, Philippines, Brunei, Indonesia, Malaysia and Vietnam, in addition to Australia, Hawaii and Central America (Turner et al. 2011). Name "rambutan" originates from the Malayan word "rambut" which means "hair", due to numerous hair-like spikes on the surface of the fruit. The rambutan tree is respected in landscaping since its evergreen, bushy and growing up to 20 m height. Leaves are compound of three to eight leaflets, which are elliptic, 7–20 cm long and 3–8 cm in width (Fig. 12.1). It has small, aromatic light green flowers,

M. E. S. Mirghani (✉)
Department of Biotechnology Engineering, Kulliyyah of Engineering,
International Islamic University Malaysia (IIUM), Kuala Lumpur, Malaysia
e-mail: elwathig@iium.edu.my

© Springer Nature Switzerland AG 2019
M. F. Ramadan (ed.), *Fruit Oils: Chemistry and Functionality*,
https://doi.org/10.1007/978-3-030-12473-1_12

Fig. 12.1 Rambutan tree with oval-shaped fruits covered with dense, bristly hairs or soft spines from out part

without petals and borne on axillary panicles (Augustin and Chua 1988). The oval-shaped fruits are reddish, orange-yellow, about 5 cm long, covered with dense, bristly hairs or soft spines from out part. The pulp inside the fruit is very sweet, glowing white, juicy, edible, white, and misty. Some rambutan trees generate two crops a year, a smaller crop produced in mid-spring along with a bigger one at the end of fall. Several trees are hermaphrodites and most valued because they are producing both female and male flowers; however, some trees are female. Rambutan tree produces about 5000–6000 fruit per season that is considered to be astringent, stomachic, vermifuge, and febrifuge. The fruit seeds are somewhat bitter and of sedative properties (Harahap et al. 2012). There are more than 200 varieties of rambutan today, but only a few are cultivated and used in the human diet. Thailand is the greatest manufacturer of rambutan in the world. The rambutan industry is deep-rooted in Thailand and in Malaysia where both are producing canned rambutan in syrup (Augustin and Chua 1988; Kheiri and bin Mohammad Som 1979).

1.2 Nutritional Value of Rambutan

The rambutan is usually consumed as dessert, just served fresh and uncooked. Rambutan fruit is rich in carbohydrates, fats, phosphorus, vitamin C, niacin (vitamin B3), iron, calcium, copper, protein and fiber (Table 12.1). Research conducted recently learned that consuming nine to ten vegetables and fruit from the rambutan family each day were good at reducing blood pressure.

Table 12.1 Mineral and vitamin contents including nutrition charts of the Rambutan fruits and seed kernels

No	Nutritive value per 100 g of rambutan	
	Principle	**Nutritive value**
1	Protein	1.0 g
2	Cholesterol	0.0 mg
3	Thiamin	0.01 mg
4	Vitamin C	7.4 mg
5	Sodium	16.5 mg
6	Potassium	63.0 mg
7	Calcium	33.0 mg
8	Vitamin A	4.5 IU
9	Iron	0.5 mg
The composition of the seed kernel (on dry weight bases)		
1	Seed	5.6–7.4% of the fruit
2	Protein	11.9–14.1%
3	Crude lipids	37.1–38.9%
4	Crude fiber	2.8–6.6%
5	Ash	2.6–2.9
6	Moisture	34.1–34.6%

Source: Augustin and Chua (1988)

1.3 Rambutan Seeds (Kernels)

Rambutan seeds are within the edible part that is occasionally roasted and also taken as food or mixed with other foods. The seeds of the fruit contain many nutrients including carbohydrates, proteins, fats and minerals and other desirable food products such as antioxidants. The rambutan seeds are about 4–9% of fruit weight. They are removed from the fruits during processing as a waste by-product (Solís-Fuentes et al. 2010). The seed fat was extracted using *n*-hexane in a Soxhlet apparatus as reported by Sirisompong et al. (2011) and Mahisanunt et al. (2017) who reported on solvent fractionation of the lipids extracted from rambutan seed kernel. The results showed that the fractionated rambutan fat might lead to its potential use in specific food products.

Many studies reported that the rambutan seed kernel extract is edible and is not toxic (Eiamwat et al. 2014; Harahap et al. 2012; Manaf et al. 2013). Therefore, rambutan seed kernel fat (RSKF) can be a promising alternative edible fat that has the potential to be used in the food industry, especially to replace hydrogenated fat. Table 12.2 shows the fatty acid composition of RSKF, which show that the main fatty acids are arachidic acid (38.3%) and oleic acid (37.1%). These two fatty acids covered 75.7% of the total fatty acids, which is in constant with previous studies (Harahap et al. 2012; Manaf et al. 2013; Sirisompong et al. 2011; Solís-Fuentes et al. 2010; Sonwai and Ponprachanuvut 2012). The fatty acid profile showed that

Table 12.2 Fatty acid composition of rambutan seed kernel fat

No.	Fatty acid	Rambutan seed kernel fat	
		Structure	% fatty acid per total fatty acids
1	Myristic acid	C14:0	00.02 ± 0.00
2	Palmitic acid	C16:0	04.81 ± 4.69
3	Palmitoleic acid	C16:1	00.46 ± 0.45
4	Stearic acid	C18:0	07.55 ± 0.04
5	Oleic acid	C18:1	37.14 ± 0.20
6	Linoleic acid	C18:2	00.93 ± 0.01
7	Linolenic acid	C18:3	00.08 ± 0.00
8	Arachidic acid	C20:0	38.36 ± 0.30
9	Eicosenoic acid	C20:1	06.59 ± 0.03
10	Behenic acid	C22:0	03.18 ± 0.06
11	Erucic acid	C22:1	00.53 ± 0.09
12	Lignoceric acid	C24:0	00.35 ± 0.02
13	Saturated fatty acids	–	54.27 ± 0.26
14	Unsaturated fatty acids	–	45.73 ± 0.26
15	Ratio (sat. : unsat.)	–	(54%: 46%)

Source: Mahisanunt et al. (2017)

the RSKF can easily be fractionated into a high-melting fraction and low-melting fraction (acetone and ethanol fractions) which were significantly different from the original RSKF (Mahisanunt et al. 2017). That will definitely increase the potential uses of the fats extracted from rambutan seed kernels. Properties of Rambutan seed oil and fatty acid composition in of the oil was also studied by Issara et al. (2014), while the physicochemical characterization of rambutan kernel oil was studied by Kheiri and Som (1979).

2 Health Benefits of Rambutan Fruits and Seed Kernel Fats

2.1 Food

Rambutan fruit is a delicious having a very good moth feel when consumed fresh in the season. Rambutan fruits contain plenteously amount of carbohydrate, protein, and has got full of water content. Therefore, the fruit can be used to get back energy and also drop thirst. As already shown in Table 12.1 (Augustin and Chua 1988), rambutan fruit contains a high amount of vitamin C which makes it a healthy fruit as vitamin C works an antioxidant that helps in reducing toxic free radicals in the cell. Vitamin C also helps to enable the assimilation of the heme ions (iron). The fruits are also rich in potassium, calcium, iron and vitamin A as well as other important nutrients.

Rambutan seed is suitable to be consumed raw, crushed or even combined and shared with various other types of food. However, the seeds are reported somewhat bitter and of seductiveness (Harahap et al. 2012). Rambutan seed oil is rich in saturated fatty acids that help to avoid hydrogenation of vegetable oils. By fractionation, the industry is able to produce both healthy vegetable oil and fat. The rambutan seed fat is softer than cocoa butter at low temperatures and has a harder uniformity at higher temperatures. This performance is doubtless due to the composition of the solid fat content difference (Solís-Fuentes and Durán-de-Bazúa 2004). Therefore, rambutan fat would be useful to replace cocoa butter in the chocolate production industry as a friendly softer filling fat (Lannes et al. 2003; Zzaman et al. 2017).

2.2   Immune System

The higher content of vitamin C, vitamin A and niacin (vitamin B3) of rambutan can improve the metabolic process as well as defense mechanisms, which improve the immune system of the human body. This kind of higher metabolism helps save us from several hazardous illnesses. Therefore, rambutan fruit could improve overall human health (Zhuang et al. 2017). Rambutan seed oil contains fat-soluble vitamins (A, D, E and K) which will definitely improve the immune system when consumed as food.

2.3   Strengthen Bones

One of the most important health benefits of eating rambutan is that it can enhance our bones (Sabbe et al. 2009). This advantage of eating rambutan is brought on by higher calcium, phosphorus and irons content in rambutan fruits and seeds. Therefore, consumption of rambutan and its products could improve health having powerful bones. The oil extracted from rambutan seed is rich in fat-soluble vitamins that include vitamins A, D, E and K. It is well known that vitamin D is very essential for bone growth and strengthening the bones.

2.4   Healthy Digestion and Treats Dysentery

Rambutan also offers fiber that will help to prevent bowel problems. Additionally, rambutan may also destroy parasites within the intestines and enables alleviate signs and symptoms of diarrhea. It is reported that the skin of rambutan had been used in folk medicine to treat dysentery (http://www.a2zsolutions.co). "Cut the skin into tiny bits, add three glasses of water and boil them till the water continues to be half. Let the water to get cool, strain and after that take in the liquid twice a day" (http://www. a2zsolutions.co).

2.5   Improve the Sperm Quality

Vitamin C can is very important for sperm development. The possible lack of vitamin C in males could be restricted in getting children. An enhancement to this needs the duration of 1 month simply by growing vitamin C consumption of 500 mg. The quality as well as the quantity of sperm and its actions could be enhanced by growing consumption of vitamin C. Some semen parameters, including sperm motility, sperm count, and sperm morphology, were studied by Akmal et al. (2006) before and after the vitamin C treatment. Results showed that the mean sperm count was increased after 2 months of vitamin C intake. The sperm motility was increased significantly and mean sperms with normal morphology increased significantly to more than 67%. The study showed that vitamin C supplementation in infertile men might improve sperm count, sperm motility, sperm morphology and might have a place as an additional supplement to improve the semen quality towards conception. Here rambutan is an excellent source of the natural high amount of vitamin C for human health benefit.

2.6   Anticancer Traits

Rambutan is rich in antioxidant components. Research carried out by the University of Chiang Mai in Thailand (Hunter 2014) reported that rambutan fruit, seeds and skin are rich in effective antioxidants especially flavonoids. Several kinds of flavonoids are thought to lessen cholesterol levels, anticancer and anti-inflammatory (Zhuang et al. 2017). The presence of vitamin C was stated to offer great defense tool, which assists in avoiding cancers (breast, lung, bladder, colon, and pancreas). Vitamin C might also decrease free-radicals that trigger cancer since this vitamin is an excessive antioxidant agent.

2.7   Blood Formation

Rambutan fruit, seed and seed fats offer small quantities of copper, which is required to produce more white and red blood cells. Furthermore, and as stated in Table 12.1, rambutan contains iron that help avoids anemia (Vaughan and Geissler 2009). Vitamin C, in addition, decreases the chance of cataracts, strengths the blood vessels walls, and reduces the chance of cardiovascular disease. Scientists think that vitamin C could also prevent aging by maintaining the white blood cells (Khalid Iqbal et al. 2004).

2.8   Cosmetics

The extract of rambutan seed showed limit oxidant-induced cell death (DPPH at 50 μM) by apoptosis to some extent similar to that of other seeds such as grape seeds. This extract might be used either alone or in combination with other active components, in cosmetic, nutraceutical and pharmaceutical applications (Palanisamy et al. 2008). The natural antioxidant in lipid-containing product and lipid-based product such as oil, fat, butter can make these products suitable for the production of moisturizing creams and other skin care products (Febrianto et al. 2012).

Rambutan seeds consumed raw, crushed or even combined with various foods are even good at creating the skin and hair much healthier and softer. The skin tone of the face also will become gentle (Hunter 2014). *N. lappaceum* fruits and seed lipids have specific bioactive compounds had to concern much attention due to health benefit effect (Febrianto et al. 2012). Moreover, these compounds are able to protect the oxidative damage in the human body's cell and tissue.

References

Akmal, M., Qadri, J. Q., Al-Waili, N. S., Thangal, S., Haq, A., & Saloom, K. Y. (2006). Improvement in human semen quality after oral supplementation of vitamin C. *Journal of Medicinal Food, 9*(3), 440–442.

Augustin, M., & Chua, B. (1988). Composition of rambutan seeds. *Pertanika, 11*(2), 211–215.

Eiamwat, J., Pengprecha, P., Tiensong, B., Sematong, T., Reungpatthanaphong, S., Natpinit, P., Hankhunthod, N. (2014). Fatty acid composition and acute oral toxicity of rambutan (*Nephelium lappaceum*) seed fat and oil extracted with SC-CO$_2$. http://www.thaiscience.info/Article%20for%20ThaiScience/Article/2/10031548.pdf.

Febrianto, F., Hidayat, W., Bakar, E. S., Kwon, G.-J., Kwon, J.-H., Hong, S.-I., & Kim, N.-H. (2012). Properties of oriented strand board made from Betung bamboo (*Dendrocalamus asper* (Schultes. f) Backer ex Heyne). *Wood Science and Technology, 46*(1–3), 53–62.

Harahap, S. N., Ramli, N., Vafaei, N., & Said, M. (2012). Physicochemical and nutritional composition of rambutan anak sekolah (*Nephelium lappaceum* L.) seed and seed oil. *Pakistan Journal of Nutrition, 11*, 1073. http://www.a2zsolutions.co/PDFs/Health%20and%20Medicinal%20Benefits%20of%20Rambutan.pdf (last updated 2015).

Hunter, J. P. III. (2014). *Health benefits form foods and spices* (p. 171). Library of Congress 2014, TX#Phil. 4-6-77-9 HMC # LLCP. Washington, DC.

Iqbal, K., Khan, A., Muzaffar Ali, M., & Khattak, K. (2004). Biological significance of ascorbic acid (vitamin C) in human health-a review. *Pakistan Journal of Nutrition, 3*(1), 5–13.

Issara, U., Zzaman, W., & Yang, T. (2014). Rambutan seed fat as a potential source of cocoa butter substitute in the confectionary product. *International Food Research Journal, 21*(1), 25–31.

Kheiri, M., & Bin Mohammad Som, M. N. (1979). Physico-chemical characteristics of rambutan [*Nephelium lappaceum*] kernel fat [Peninsular Malaysia]. Agricultural product utilisation report (Malaysia). no. 186.

Kheiri, M., & Som, M. N. M. (1979). *Physico-chemical characteristics of Rambutan kernel fat.* Serdang: MARDI.

Lannes, S. C. S., Medeiros, M. L., & Gioielli, L. A. (2003). Physical interactions between cupuassu and cocoa fats. *Grasas y Aceites, 54*(3), 253–258.

Mahisanunt, B., Jom, K. N., Matsukawa, S., & Klinkesorn, U. (2017). Solvent fractionation of rambutan (*Nephelium lappaceum* L.) kernel fat for production of non-hydrogenated solid fat: Influence of time and solvent type. *Journal of King Saud University- Science, 29*, 32–46.

Manaf, Y. N. A., Marikkar, J. M. N., Long, K., & Ghazali, H. M. (2013). Physico-chemical characterisation of the fat from red-skin rambutan (*Nephellium lappaceum* L.) seed. *Journal of Oleo Science, 62*(6), 335–343.

Palanisamy, U., Cheng, H. M., Masilamani, T., Subramaniam, T., Ling, L. T., & Radhakrishnan, A. K. (2008). Rind of the rambutan, *Nephelium lappaceum*, a potential source of natural antioxidants. *Food Chemistry, 109*(1), 54–63.

Sabbe, S., Verbeke, W., & Van Damme, P. (2009). Perceived motives, barriers and role of labeling information on tropical fruit consumption: Exploratory findings. *Journal of Food Products Marketing, 15*(2), 119–138.

Sirisompong, W., Jirapakkul, W., & Klinkesorn, U. (2011). Response surface optimization and characteristics of rambutan (*Nephelium lappaceum* L.) kernel fat by hexane extraction. *LWT-Food Science and Technology, 44*(9), 1946–1951.

Solís-Fuentes, J. A., & Durán-de-Bazúa, M. C. (2004). Mango seed uses: Thermal behaviour of mango seed almond fat and its mixtures with cocoa butter. *Bioresource Technology, 92*(1), 71–78.

Solís-Fuentes, J. A., Camey-Ortíz, G., del Rosario Hernández-Medel, M., Pérez-Mendoza, F., & Durán-de-Bazúa, C. (2010). Composition, phase behavior and thermal stability of natural edible fat from rambutan (*Nephelium lappaceum* L.) seed. *Bioresource Technology, 101*(2), 799–803.

Sonwai, S., & Ponprachanuvut, P. (2012). Characterization of physicochemical and thermal properties and crystallization behavior of krabok (*Irvingia Malayana*) and rambutan seed fats. *Journal of Oleo Science, 61*, 671–679.

Turner, N. J., Łuczaj, Ł. J., Migliorini, P., Pieroni, A., Dreon, A. L., Sacchetti, L. E., & Paoletti, M. G. (2011). Edible and tended wild plants, traditional ecological knowledge and agroecology. *Critical Reviews in Plant Sciences, 30*(1–2), 198–225.

Vaughan, J., & Geissler, C. (2009). *The new Oxford book of food plants*. Oxford: OUP.

Zhuang, Y., Ma, Q., Guo, Y., & Sun, L. (2017). Protective effects of rambutan (*Nephelium lappaceum*) peel phenolics on H_2O_2-induced oxidative damages in HepG2 cells and d-galactose-induced aging mice. *Food and Chemical Toxicology, 108*, 554–562.

Zzaman, W., Issara, U., Easa, A. M., & Yang, T. A. (2017). Exploration on the thermal behavior, solid fat content and hardness of rambutan fat extracted from rambutan seeds as cocoa butter replacer. *International Food Research Journal, 24*(6), 2408–2413.

Chapter 13
Chyuri (*Diploknema butyracea*) Butter

Hari Prasad Devkota, Anjana Adhikari-Devkota, Tarun Belwal,
and Dhaka Ram Bhandari

Abstract *Diploknema butyracea* (Roxburgh) H. J. Lam, belonging to family
Sapotaceae, is distributed throughout the Himalayan belt including Nepal, India,
and Bhutan at an altitude of about 300–1500 m. The butter extracted from the seeds
known as "Chyuri butter" or "Phulwara butter" is the most widely used product
from the plant. The main chemical constituents of the butter are triglycerides (TG)
and the fatty acids in these triglycerides are mainly composed of palmitic acid, oleic
acid, stearic acid and linoleic acid. Although the butter is used as vegetable fat and
as medicine by ethnic people, there are not many studies on the product development and commercialization of the Chyuri butter. In this chapter, we summarize the
studies regarding Chyuri butter, its chemical composition and the future potential in
pharmaceuticals and cosmetic industries.

Keywords Roxburgh · Sapotaccac · Chiuri butter · Phulwara butter · Cosmetic

1 Introduction

Diploknema butyracea (Roxburgh) H. J. Lam [Syns: *Bassia butyracea* Roxburgh;
Madhuca butyracea (Roxburgh) Macbride, *Aesandra butyracea* (Roxburgh)
Baehni] belongs to family Sapotaceae (Manandhar 2002). It is commonly known as
"Chyuri" or "Chiuri" in Nepali, "Phulwara" or "Chiura" in Hindi and Indian-butter
nut in English (Fig. 13.1).

H. P. Devkota (✉) · A. Adhikari-Devkota
Graduate School of Pharmaceutical Sciences, Kumamoto University, Kumamoto, Japan
e-mail: devkotah@kumamoto-u.ac.jp

T. Belwal
G. B. Pant National Institute of Himalayan Environment and Sustainable Development,
Almora, Uttarakhand, India

D. R. Bhandari
Institute of Inorganic and Analytical Chemistry, Giessen, Germany

© Springer Nature Switzerland AG 2019
M. F. Ramadan (ed.), *Fruit Oils: Chemistry and Functionality*,
https://doi.org/10.1007/978-3-030-12473-1_13

Fig. 13.1 Photographs of Chyuri tree (**a**), flowers (**b**) and fruits (**c**). (Photos by Dr. Khem Raj Joshi (**a**, **b**) and Mr. Kuber Jung Malla (**c**))

Chyuri tree is of social, cultural and economical importance to the ethnic people living in the Himalayan region. Chepangs, an indigenous ethnic group of people living in the dense forests of mid-Southern Chitwan district of Nepal, call Chyuri tree as "Yoshi" and use various parts of Chyuri plant as food and medicine. For example, fruits are used as dietary supplements. Juice of the bark is used to treat indigestion, asthma, rheumatism, boils and parasitic worms. Juicy pulp of the ripe fruit is eaten fresh and the juice of the corolla is boiled to make a syrup which is used by the villagers as syrupy sugar (Manandhar 2002; Adhikari et al. 2007). Bark and oil cake are used as fish poison. Additionally, oil cake is also used as fertilizer to protect crops from harmful insects and worms (Shakya 2000).

The most widely used product of this plant is the butter extracted from the seeds known as "Chyuri butter", "Phulwara butter" (Fig. 13.2), which is used both as cooking oil and medicine (Detailed uses of butter are provided in Sect. 4). The relative composition of fruit pulp, fruit peel, seed coat and seed kernel was reported to be 48%, 26%, 6% and 20%, respectively (Shakya 2000). The seeds contain about 55–65% of fat (Bushell and Hilditch 1938; Shakya 2000). The extraction of fat (butter) from these seeds and its sale is the only source of income for many ethnic people. However, due to limited research on the composition, utilization, and product formulation, the potential of Chyuri butter in the pharmaceutical and cosmetic field is still not harnessed.

Fig. 13.2 Photographs of (**a**) Chyuri seed and (**b**) seed fat (Chyuri butter)

Fig. 13.3 Traditional
method of extraction of
Chyuri butter

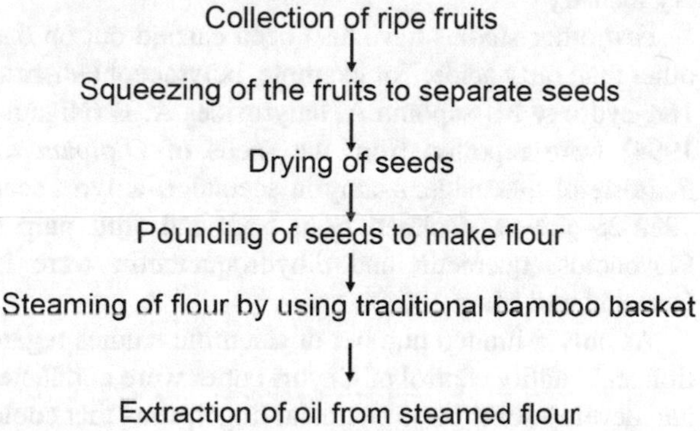

Collection of ripe fruits

↓

Squeezing of the fruits to separate seeds

↓

Drying of seeds

↓

Pounding of seeds to make flour

↓

Steaming of flour by using traditional bamboo basket

↓

Extraction of oil from steamed flour

2   Extraction and Processing of Chyuri Fat (Chyuri Butter)

Most of the local people extract the Chyuri butter using the traditional method as
follows: The ripe Chyuri fruits are collected from the trees and dried. The seeds are
then separated, cleaned, dried and crushed to obtain flour. The seed flour is steamed
in a perforated bamboo basket and placed over a boiling metal pot for steaming. The
steamed flour is squeezed to extract the fat (Fig. 13.3, Shakya 2000). This traditional
method of extraction is very laborious and time-consuming. Development of
advanced methods of extraction of seed fat may increase the yield and purity of the
Chyuri butter.

An average tree yields 200 kg of seeds per season and to obtain 1 L of Chyuri
butter, about 18 kg of seeds are required (Rastogi 2008, http://www.appropedia.org/
Chiuri). The ghee is packed and marketed with a brand name for promotion of
Chyuri products in urban areas through a wider marketing network and advertise-
ment (http://www.appropedia.org/Chiuri).

3 Chemical Composition

The main chemical constituents of Chyuri butter are triglycerides. In 1938, Bushell and Hilditch reported the triglyceride composition of Phulwara butter (Table 13.1). The fatty acids composition of triglycerides included palmitic acid (57%), oleic acid (36%), steric acid (3.6%) and linoleic acid (3.6%). They also reported that the chief components of the glycerides were oleodipalmitins (62%) and palmitodioleins (23%) with small amounts of tripalmitin (8%) and probably oleopalmitostearins (7%). Agarwal et al. (1963) reported similar composition (Table 13.1) for seed fat obtained from Uttar Pradesh, India. Devkota et al. (2012) also reported the similar composition of fatty acids from the Chyuri butter collected from Nepal (Table 13.1, Fig. 13.4). Later in 2017, Bandyopadhyay et al. (2017) reported similar composition in different samples collected from India. These studies suggested that the Chyuri ghee has high potential to be used in the cosmetics, pharmaceutical and confectionery industry.

Few other studies have also been carried out on the fruits and seeds of Chyuri other than fatty acids. For example, butyraceol (Mishra et al. 1991), MI-saponin A, 16α-hydroxy MI-saponin A, butyrosides A, B (Nigam et al. 1992), C, D (Li et al. 1994) were reported from the seeds of *Diploknema butyracea*. α-Spinasterol, β-sitosterol glucoside, α-amyrin acetate, α-amyrin acetate and 3β-palmitoxy-olea-12en-28-ol were isolated from bark and fruit pulp (Awasthi and Mitra 1968). Flavonoids, quercetin and dihydroquercetin were isolated from the nut-shell (Awasthi and Mitra 1962).

As only a limited number of scientific studies regarding the chemical composition and quality control of Chyuri butter were conducted, it prompted us to analyze and develop some basic chemical fingerprints that could be used for quality control of Chyuri butter. Different techniques based on infrared (IR), ultraviolet (UV), nuclear magnetic resonance (NMR) spectroscopic and mass spectrometric (MS) methods were used to generate fingerprints that represent to the chemical profile of the butter. IR and UV spectra are shown in Figs. 13.5 and 13.6, respectively.

Table 13.1 The relative composition of fatty acids in triglycerides present in Chyuri butter

Fatty acid ester	Relative composition (%) of fatty acid in each sample				
Palmitic acid methyl ester	57	55.86	55.6	58.4–66.0	56.8–64.10
Steric acid methyl ester	3.6	1.63	5.2	2.4–2.9	2.4–3.5
Oleic acid methyl ester	36	38.55	35.9	28.9–35.2	28.0–31.3
Linoleic acid methyl ester	3.6	3.69	3.3	2.6–3.7	4.3–5.7
Reference (country of origin of the sample)	Bushell and Hilditch (1938) (India)	Agarwal et al. (1963) (India)	Sengupta and Roy Choudhary (1978) (India)	Devkota et al. (2012) (Nepal)	Bandyopadhyay et al. (2017) (India)

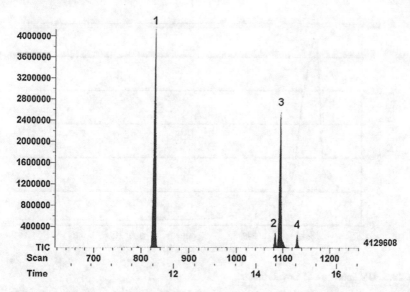

Fig. 13.4 Chromatogram of mixture of fatty acid methyl esters obtained after methanolysis of Chyuri butter (Devkota et al. 2012). The peaks are for palmitic acid methyl ester (**1**), linoleic acid methyl ester (**2**), oleic acid methyl ester (**3**) and stearic acid methyl ester (**4**)

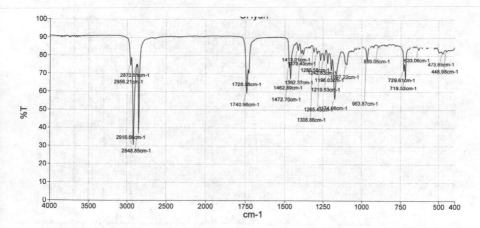

Fig. 13.5 IR spectrum of Chyuri butter

The ^{1}H-NMR data (600 MHz, in CDCl$_3$) and ^{13}C-NMR spectra (150 MHz, in CDCl$_3$) are represented in Figs. 13.7 and 13.8, respectively. Similarly, electron ionization-mass spectrometry (EI-MS) spectrum is represented in Fig. 13.9. Identification of individual triglycerides with these data is challenging as the Chyuri butter is a mixture of different compounds. However, these chemical fingerprint data from the samples can assist in quality control and identification of the butter from different market sources. Moreover, it can be used to trace adulteration for e.g. the commercial ghee (clarified butter from cow and buffalo milk) in Nepal which often adulterated with Chyuri seed oil (Jha 1981) can be detected.

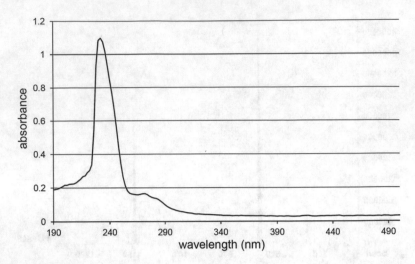

Fig. 13.6 UV spectrum of Chyuri butter

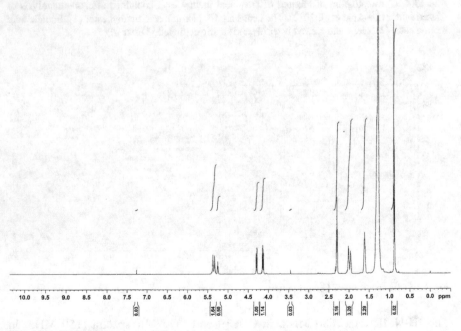

Fig. 13.7 ¹H NMR spectrum of Chyuri butter

4   Medicinal and Other Applications

Various parts of Chyuri plant are used as food and medicine by diverse ethnic groups in Nepal. Seed fat (Chyuri butter) is used for the treatment of headache, rheumatism boils, pimples, burns and used as emollient for chapped hands and feet during winter (Manandhar 2002; Adhikari et al. 2007; Watanabe et al. 2013). Fruits are used as

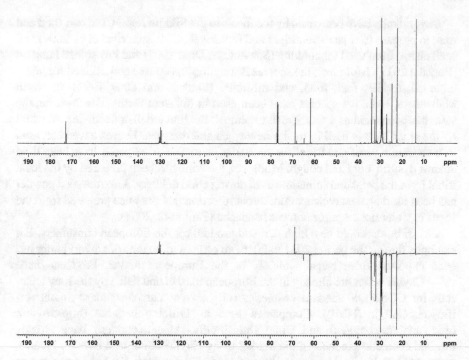

Fig. 13.8 ¹³C NMR spectrum of Chyuri butter [upper: broadband decoupled spectrum, lower: Distortionless Enhancement by Polarization Transfer (DEPT) spectrum]

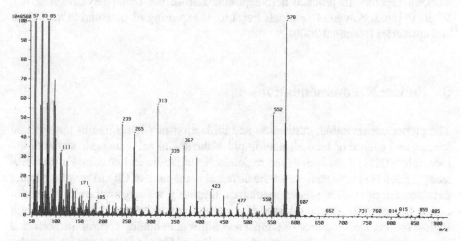

Fig. 13.9 EI-MS spectrum of Chyuri butter

dietary supplement for the Chepang people. The vegetable butter extracted from seed is used for cooking and lighting lamps. Some people use it as hair oil and raw material for soap. Bark and oil cake (after extracting the oil contains saponins) are used as fish poison. Oil cake is also used as fertilizer to protect crops from pest and worms (Shakya 2000).

Chyuri fruits have been used by the locals in the Sikkim region, India as food and used to prepare fruit jam (Sundriyal and Sundriyal 2004; Sundriyal et al. 2004) and seed oil has been used for cooking (Samant and Dhar 1997) and low smoke lamp oil (Punetha 2017). Moreover, the species is a multipurpose tree and utilized for vegetable oil, fodder, fuel, food, and medicine (Samant and Dhar 1997). In North Maharastra, India the species have been used in different forms. For instance, the stem has been used as a toothbrush to remove the foul smell, a decoction prepared by the inner bark is used to treat amenorrhea and diarrhea. Leaves have been used to treat mouth ulcer and to relieve muscular pain. The dried petals have been used against dysuria, cold and cough. Moreover, the wine has been prepared by the local tribal from the petal and sometime used during child delivery. Also, the seed powder has been used against constipation, throat infection and the juice prepared from the seeds in water used against chronic headache (Patil et al. 2004).

Chyuri butter could be a high demand product for the European consumers. For example, during the period 2011 to 2016, an estimated 20 tons of Chyuri butter has been exported from Nepal annually to the European market. Products made using Chyuri butter are already in the European market and listed by the Raw materials for COSMOS-standard cosmetics (http://www.cosmos-standard-rm.org/verifmp.php/?alt_id=94b10). Companies such as Officina Naturae (https://www.officinanaturae.com/en) and Cîme (https://cime-skincare.com/en) have already lunched cosmetic products in the market using Chyuri butter as a base for creams, ointment, and lotion. In Nepal, Alternative Herbal Products (P) Ltd. (http://www.alternative.org.np/), Deuti Herbal Industry and Himalayan Bio-Trade Pvt. Ltd. are working together to produce herbal products from the butter (Rytkönen 2016). While in India, Polygon Chemicals Pvt. Ltd. is exploring its potential in medicines and cosmetics (Nautiyal 2009).

5   Future Recommendations

The proper conservation, cultivation and utilization of Chyuri tree as a medicinal species and source of seed oil may improve the social and economic status of the local inhabitants of the Himalayan region in Nepal, India and Bhutan. Further, bio-assay guided chemical analysis of the different plant parts of Chyuri may help in the development of evidence-based medicines. Similarly, a study on the altitude variation of components in seed oil would be interesting for the selection of proper biological source. Development of simple and affordable quality control methods will also contribute to the further market expansion of Chyuri butter and its formulations. Additionally, physio-chemical studies of the butter such as consistency, pH, spreadability, diffusion, wash ability and drug incorporation and drug release potential should be carried out in the field of herbal formulations (ointment and cream) so that the proper utilization can be achieved in pharmaceutical and cosmetic industries.

References

Adhikari, M. K., Shakya, D. M., Kayastha, M., Baral, S. R., Subedi, M. N. (2007). Bulletin of Department of Plant Resources No. 28. Medicinal Plants of Nepal (revised). Department of Plant Resources, Kathmandu, p. 112.

Agarwal, P. N., Pandey, S. C., Sharma, T. R., & Prakash, O. (1963). The component fatty acids of the seed fat of *Madhuca butyracea. Journal Oil Technologists' Association India, 18,* 8–10.

Awasthi, Y. C., & Mitra, C. R. (1962). Flavonoids of *Madhuca butyracea* nut shell. *The Journal of Organic Chemistry, 27,* 2636–2637.

Awasthi, Y. C., & Mitra, C. R. (1968). *Madhuca butyracea.* Constituents of the fruit pulp and the bark. *Phytochemistry, 7,* 637–640.

Bandyopadhyay, B. B., Joshi, L., & Nautiyal, M. K. (2017). Relationship of the oil content and fatty acid composition with seed characters of *Diploknema butyracea. Academia of Agriculture Journal, 2,* 1–4.

Bushell, W. J., & Hilditch, T. P. (1938). Fat acids and glycerides of solid seed fats. IV. Seed fat of *Madhuca butyracea* (phulwara butter). *Journal of Society of Chemical Industry, 57,* 48–49.

Devkota, H. P., Watanabe, T., Malla, K. J., Nishiba, Y., & Yahara, S. (2012). Studies on medicinal plant resources of the Himalayas: GC-MS analysis of seed fat of Chyuri (*Diploknema butyracea*) from Nepal. *Pharmacognosy Journal, 4,* 42–44.

Li, X., Liu, Y., Wang, D., Yang, C., Nigam, S.K. & Mishra, G. (1994). Triterpenoid saponins from *Madhuca butyracea. Phytochemistry, 37,* 827–829.

Jha, J. S. (1981). Spectrophotometric studies of Cheuri (*Madhuca butyracea*) fat and ghee mixtures: I. *Journal of the American Oil Chemists' Society, 58,* 843–845.

Manandhar, N. P. (2002). *Plants and people of Nepal* (p. 205). Portland: Timber Press. 2002.

Mishra, G., Banerji, R., & Nigam, S. K. (1991). Butyraceol, a triterpenoidal sapogenin from *Madhuca butyracea. Phytochemistry, 30,* 2087–2088.

Nautiyal, S. (2009). Indian butter tree paves way for entrepreneurship despite economic slump. [cited 2018 4/27/2018]; Available from: https://yourstory.com/2009/02/indian-butter-tree-paves-way-for-entrepreneurship-despite-economic-slump-2/.

Nigam, S. K., Li, X., Wang, D., Mishra, G., & Yang, C. (1992). Triterpenoidal saponins from *Madhuca butyracea. Phytochemistry, 31,* 3169–3172.

Patil, D. A., Pawar, S., & Patil, M. V. (2004). Mahuwa tree and the aborigines of North Maharashtra. *Natural Product Radiance, 3,* 356–358.

Punetha, P. (2017). Oil diyas from Chyura seeds in demand this season, in *The times of India, The times of India Dehradun* https://timesofindia.indiatimes.com/city/dehradun/oil-diyas-from-chyura-seeds-in-demand-this-season/articleshow/61136085.cms. Accessed on 30.04.2018.

Rastogi, R. (2008). Let us indentify the useful trees. New Delhi Children Book Trust.

Rytkönen, A. (2016). Forest-based value chains in Nepal. 27–32. https://www.eda.admin.ch/dam/countries/countries-content/nepal/en/Forest_based_Value_Chains-EN.pdf.

Samant, S. S., & Dhar, U. (1997). Diversity, endemism and economic potential of wild edible plants of Indian Himalaya. *International Journal of Sustainable Development and World Ecology, 4,* 179–191.

Sengupta, A., & Roy Choudhary, S. K. (1978). Triglyceride composition of *Madhuca butyraceae* seed fat. *Journal of the American Oil Chemists' Society, 55,* 621–624.

Shakya, M. R. (2000). Chepangs and Chiuri – the use of non timber forest products in Nepal. *Food Chain, 26,* 3–5.

Sundriyal, M., & Sundriyal, R. C. (2004). Wild edible plants of the Sikkim Himalaya: Marketing, value addition and implications for management. *Economic Botany, 58,* 300–315.

Sundriyal, M., Sundriyal, R. C., & Sharma, E. (2004). Dietary use of wild plant resources in the Sikkim Himalaya, India. *Economic Botany, 58,* 626–638.

Watanabe, T., Rajbhandari, K. R., Malla, K. J., Devkota, H. P., & Yahara, S. (2013). In L. E. I. Ayurseed (Ed.), *A handbook of medicinal plants of Nepal supplement I* (pp. 124–125). Kanagawa: Life Environmental Institute.

Chapter 14
Madhuca longifolia Butter

Mohamed Fawzy Ramadan and Jörg-Thomas Mörsel

Abstract Sustainable oil and fat sources are desired to achieve supply chain flexibility and cost-saving opportunities. Non-conventional fruits are considered because of their constituents unique chemical composition that may augment the supply of nutritional and functional products. *Madhuca longifolia* Syn. *M. indica* (family Sapotaceae) is an important economic tree growing throughout the subtropical region of the Indo-Pak. Information concerning the exact composition of mahua (known also as mowrah) butter from fruit-seeds of buttercup or *Madhuca* tree is still few. Few studies investigated mahua butter for its composition, nutritional value, biological activities and antioxidant traits. This chapter summarizes recent knowledge on bioactive compounds, functional traits as well as food and non-food industrial applications of mahua butter.

Keywords Sapotaceae · Mahua butter · Mowrah butter · Buttercup tree · Biofuel

1 Introduction

Oil and fats market is interested in solutions that could address the agricultural supply chain challenges. Few plants produce oils and fats in commercial quantities. Vegetable oils contribute about 85% of the oils or fats available for consumption (Lawson 1995; Yadav et al. 2011a). Search is necessary for new resources of oil-bearing crops, which provide high yield of oils or fats. The genus *Madhuca* (family Sapotaceae) is a multipurpose tree with its species, *M. longifolia*, *M. latifolia* and *M. butyracea* being the most prevalent. Buttercup or mahua, *M. longifolia* (Koenig) (Synonyms, *M. indica* Gmelin), is a large, shady, deciduous tree dotting much of the

M. F. Ramadan (✉)
Agricultural Biochemistry Department, Faculty of Agriculture, Zagazig University, Zagazig, Egypt

J.-T. Mörsel
UBF-Untersuchungs-, Beratungs-, Forschungslaboratorium GmbH, Altlandsberg, Berlin, Germany

© Springer Nature Switzerland AG 2019   291
M. F. Ramadan (ed.), *Fruit Oils: Chemistry and Functionality*,
https://doi.org/10.1007/978-3-030-12473-1_14

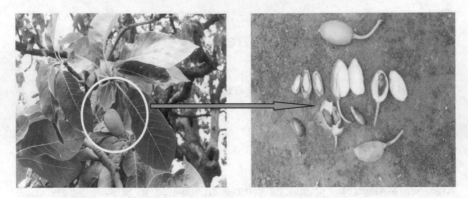

Fig. 14.1 *M. longifolia* tree, a deciduous tree, 10–15 m tall and with a spreading, dense, round, shady canopy. Fruits ovoid, fleshy, 2–4 cm across, greenish, 1–4 seeded. Seeds elongate, 2 cm long, brown, shining (Ramadan et al. 2006)

central Indian landscape, both wild and cultivated. The tree is economically important due to the widespread uses of its fruits, seeds, flowers, and timber (Ramadan et al. 2006). The medicinal traits attributed to this plant are stimulant, demulcent, emollient, heating, and astringent (Ramadan and Moersel 2006; Yadav et al. 2011a, b; Singh et al. 2018).

The buttercup fruit-seeds are generally ellipsoidally shaped (Fig. 14.1). The whole *M. longifolia* seeds contain about 60% lipids, 22% carbohydrates, 16.9% protein, 3.2% fiber, 3.4% ash, 2.5% saponins, and 0.5% tannins. *M. longifolia* fruit seeds contain high quantity of lipids, commercially known as mahua or mowrah butter, and it has many edible, medicinal and non-food applications (Ramadan et al. 2006). This chapter reported on composition, nutritional value, functional properties as well as food and non-food applications of mahua butter.

The deoiled seed cake contains 30% protein, 1% oil, 8.6% fiber, 42.8% carbohydrates, 6% ash, 9.8% saponins, and 1% tannins (Singh and Singh 1991). The levels of saponins could be decreased by treatment with isopropanol. The deoiled *M. longifolia* seed cake showed good oil emulsification properties. The *in vitro* digestibility of isopropanol-treated cake was found to be 81%. Detoxified mahua seed flour seems to be a good source of proteins for food and feed products (Singh and Singh 1991; Ramadan and Moersel 2006). *M. longifolia* cake has been used as a low-grade fertilizer, bio-pesticide, included in animal feed (up to 20%) and in dye removal from wastewaters. Detoxification methods were applied to use of *M. longifolia* cake as an improved animal feed. Gupta et al. (2012) evaluated the use of raw as well as detoxified *M. longifolia* cake (produced from water treatment) for biogas production and mushroom cultivation. Significant enhancement in the biogas (93%) and the mushroom yield (128%) was obtained. Gupta et al. (2013) evaluated the conditions affecting biogas production from raw and detoxified *M. longifolia* seed cake. Detoxified *M. longifolia* seed cake resulted in better properties compared to raw cake. Significant reduction in celluloses (34.4%) and hemicelluloses (29.7%) and an increase in the nutrients of the digested slurry were obtained. Inamdar et al. (2015) studied the impact of deoiled *M. longifolia* seed cake on methane production and nutrient utilization in buffaloes. Defatted *M. longifolia* seed cake reduced total gas production, while the addition of *M. longifolia*, under *in vitro* conditions, decreased methane production.

In recent years, there has been increasing use of oilseed plants and their oils for production of biodiesel. Biodiesel is a fatty acid methyl esters (FAME), which could be derived from any vegetable oil by transesterification. FAME as biodiesel is environmentally safe, non-toxic and biodegradable. Biofuel was derived from edible oil obtained from rapeseed, soybean, palm, sunflower, tree-borne oilseed like jatropha, neem, castor, karanj and kokkam (Carlos et al. 2010; Yadav et al. 2011a). Saponification value, iodine value and cetane number are used to predict the quality of FAME for use as biofuel (Azam et al. 2005). These parameters varied from 198.3 to 202.8, 52.0–68.6 and 58.0–61.6, respectively, in FAME of *M. longifolia* butter (Yadav et al. 2011a, b). It was reported that *M. longifolia* FAME could be used as an alternative for biodiesel in India (Kapilan and Reddy 2008; Ghadge and Raheman 2005; Yadav et al. 2011a). Puhan et al. (2005) tested biofuel from *M. longifolia* butter in a single cylinder, four strokes, direct injection, constant speed, compression ignition diesel engine to evaluate the performance and emissions. The properties of *M. longifolia* FAME were close to those of diesel oil. The fuel properties of *M. longifolia* biodiesel complied the requirements of both the European and American standards for biodiesel (Ghadge and Raheman 2006; Kapilan and Reddy 2008). *M. longifolia* biodiesel gives an equally good performance and lower emissions, which make it a good alternative fuel to operate diesel machines as well as irrigation pumps without engine modifications (Manjunath et al. 2015).

2   Extraction and Processing of *M. longifolia* Fruit Butter

The oil yield is always the key factor to decide plant suitability for industrial purposes from an economic point of view (Lawson 1995; Ramadan and Moersel 2006; Yadav et al. 2011b). The seed of *M. longifolia* is among the under-utilized for oil production. This might be due to the lack of technical information with regard to its properties and potential uses. Unlike several tropical fruit seeds, *M. longifolia* seeds show good commercial potential as a source of oil. The fruit seeds may have 60% oil and hence, a high possibility exists for oil recovery from the seeds. Castor, jatropha and *Simarouba glauca* kernels are the only oilseed sources having oil yield comparable (>50%) to *M. longifolia* while all others yield <50% oil (Manjunath et al. 2015). The high oil content indicates the suitability of *M. longifolia* seeds for industrial purposes as it reduces the production cost. Moreover, mahua butter is suitable for human consumption, in contrast, to other tree borne oilseeds like castor, jatropha, and karanj which have some toxic compounds like ricin in castor, curcurin in jatropha, and pongamin and karakjiin flavonoid in karanj (Yadav et al. 2011a).

The crude fats extracted from the seeds of *M. longifolia* is known in India as mahua butter, which is pale yellow in color and remains as a semi-solid under the tropical temperature conditions. Based on iodine value IV (*ca.* 80), *M. longifolia* butter could be classified as a non-drying oil. The IV of *M. longifolia* butter is above from those reported for Malaysian cocoa butter or Borneo Illipe butter (Marikkar et al. 2010).

Cloud point (CP) and slip melting point (SMP) are important parameters related to the nature of fatty acid and triacylglycerol (TAG) distribution of oils and fats. Marikkar et al. (2010) used acetone to fractionate *M. longifolia* butter into solid and liquid fractions including high-melting fraction (HMF) and low-melting fraction (LMF). The average SMP of *M. longifolia* butter was 35.5 °C, which is comparable to those of Malaysian Cocoa butter and Borneo Illipe butter (Wong 1988). The SMP being below the physiological temperature indicates its suitability for edible applications. Upon fractionation, the SMP of the solid component, HMF, was found to exceed the physiological temperature. However, its value was within the range of the commercially available palm sterine samples. The CP value of LMF is within the range found in most of the commercially available palm olein. Thus, it may have some resistance to clouding effect, particularly, if its intended use is as cooking oil for temperate climatic regions (Marikkar et al. 2010). The solid fat content (SFC) of *M. longifolia* butter at 0 °C was 33%, while that of Malaysian cocoa butter is 94% (Marikkar et al. 2010). This wider difference may support the presumption that the physical nature of *M. longifolia* butter is softer compared to Malaysian cocoa butter (Campos et al. 2002).

3   Fatty Acids and Acyl Lipids of *M. longifolia* Butter

Palmitic acid (P), stearic acid (S), oleic acid (O) and linoleic acid (L) are the main fatty acids in *M. longifolia* butter. *M. longifolia* butter is rich in saturated fatty acids, SFA (39.3–52.7%) and monounsaturated fatty acids, MUFA (32.9–48.6%), while it has low polyunsaturated fatty acids (PUFA) especially linoleic acid (9.36–15.4%). When compared with other oilseed crops (Fig. 14.2), *M. longifolia* fatty acid profile is similar to cocoa butter, palm oil, and shea fat. Therefore, the fatty acid pattern explains the suitability of *M. longifolia* butter to substitute cocoa butter in several pharmaceutical and cosmetic applications (Ramadan et al. 2006; Yadav et al. 2011a, b).

Vegetable oils with high SFA levels are desired in food industry especially to avoid transesterification and hydrogenation processes in the production of solid fat products such as shortening and margarine as well as to avoid the production of *trans* fatty acids. Oleic acid is the main determiner of oil quality and its high amount is favored. The predominance of oleic acid in oils has been favored by nutritionists as it reduces the blood cholesterol, and hereby reducing the incidence of coronary heart disease (CHD). Compared to other fats, *M. longifolia* butter has higher oleic acid than jatropha and neem (45%). Moreover, oleic acid level in *M. longifolia* butter is also higher than palm oil, sal (*Shorea robusta*) fat, and kokum (*Garcinia indica)* fat (Yadav et al. 2011a).

Marikkar et al. (2010) fractionated *M. longifolia* butter into solid and liquid fractions using acetone into HMF and LMF. The fatty acid profile was affected by fractionation, as there were deviations in the fatty acid compositions of the fractions.

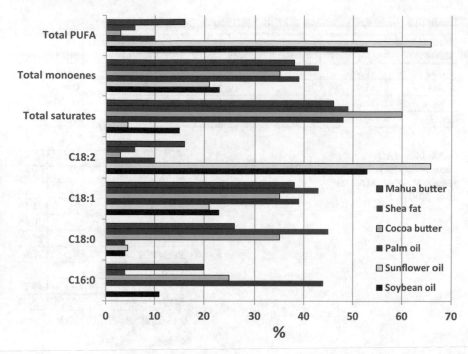

Fig. 14.2 Average amounts of major fatty acids (%) in *M. longifolia* butter compared to common seed oils and natural semi-solid fats

In HMF, the SFA content increased with a decrease in its MUFA and PUFA levels. Its fatty acid composition showed a closer comparison to that of Malaysian cocoa butter, but did not show many comparisons to that of commercial palm stearine (Lipp and Ankalam 1998). The SFA content of LMF fraction decreased with an increase in the MUFA and PUFA content. The liquid fraction rich in oleic acid is beneficial for application in frying medium.

Ramadan et al. (2006) applied different chromatographic methods on silica gel to obtain major lipids classes and subclasses of *M. longifolia* butter. The levels of lipid classes and subclasses presented in mahua butter are presented in Table 14.1. Among total lipids (TL) in the seeds, the level of neutral lipids (NL) was the highest (95.4% of TL), followed by glycolipids GL (0.51% of TL) and phospholipids PL (0.13% of TL), respectively. Subclasses of NL in the *M. longifolia* butter contained triacylglycerol (TAG), free fatty acids (FFA), diacylglycerol (DAG), monoacylglycerol (MAG), esterified sterols (STE) and free sterols (ST). High amount of TAG was measured (95.6% of total NL) followed by FFA (1.3% of total NL), while DAG, MAG and STE were measured in lower amounts. Marikkar et al. (2010) studied the TAG profile of *M. longifolia* butter (Table 14.2). The TAG profile of *M. longifolia* butter showed a closer comparison to palm oil. Among the TAG molecules of *M. longifolia* butter, OOP was the most dominant followed by POS and OOS. A fingerprint comparison with the TAG profile of Malaysian cocoa butter

Table 14.1 Main lipid subclasses of *M. longifolia* butter (Ramadan et al. 2006)

Neutral lipid subclass	mg/g butter	Glycolipid subclass	mg/g butter	Phospholipid subclass	mg/g butter
MAG	9.92	CER	1.433	PS	0.076
DAG	11.9	SG	1.689	PI	0.086
FFA	12.8	MGD	0.076	PC	0.428
TAG	912	ESG	1.843	PE	0.806

Table 14.2 TAG composition (%) of *M. longifolia* butter (Marikkar et al. 2010)

TAG	%
OOL	3.00
PLO	4.26
PPL	1.19
OOO	9.85
OOP	22.9
PPO+POP	11.9
OOS	17.8
POS	19.3
SOS	9.74

O oleic, *P* palmitic, *L* linoleic, *S* steric

showed that *M. longifolia* butter also had a considerable amount of POP, POS, and SOS molecules.

Subclasses of GL in the mahua butter were sulphoquinovosyldiacylglycerol (SQD), digalactosyldiglycerides (DGD), cerebrosides (CER), sterylglycosides (SG), monogalactosyldiglycerides (MGD) and esterified sterylglycosides (ESG). ESG, SG and CER were the prevalent components and made up about 97% of the total GL. PL subclasses in *M. longifolia* butter were separated into major fractions *via* thin layer chromatography. PL fraction revealed that the predominant PL subclasses were PE (57.7%) followed by PC (30.6%), while PI and PS were isolated in lower quantities.

4 Minor Bioactive Lipids in *M. Longifolia* Butter

Another distinct feature of mahua butter is the high unsaponifiables content (8 g/kg lipids). *M. longifolia* butter is characterized by a high amount of sterols (3.9 g/kg lipids). The sterol marker was Δ5-avenasterol which comprised approx. 30.2% of the total lipids (Table 14.3). β-sitosterol and Δ5, 24-stigmastadinol constituted approx. 46% of the total lipids. Sitostanol, campesterol, stigmasterol, lanosterol, Δ7-avenasterol and Δ7-stigmastenol, were detected at lower amounts or traces. β-Sitosterol has been most intensively investigated with respect to its physiological

Table 14.3 Main
tocopherols and sterols in *M.
longifolia* butter (Ramadan
et al. 2006)

Compound	mg/kg
α-Tocopherol	38
β-Tocopherol	189
γ-Tocopherol	1741
Campesterol	37
Lanosterol	34
β-Sitosterol	935
Δ5-Avenasterol	1192
Sitostanol	707
Δ5, 24-Stigmastadinol	904
Δ7-Stigmastenol	31
Δ7-Avenasterol	94

impact in human (Yang et al. 2001). Certain sterols, e.g. β-sitosterol has the beneficial effect of being hypocholesterolemic and Δ5-avenasterol can protect lipids from oxidative polymerization during frying.

The composition of tocopherols in *M. longifolia* butter is given also in Table 14.3. γ-Tocopherol constituted approx. 88.8% of the total identified sterols. The rest were β- tocopherol (approx. 9.6%) and α-tocopherol (approx. 1.9%). α-Tocopherol is the most efficient antioxidant of tocopherols, while β-tocopherol has 25–50% of the antioxidant activity of α-tocopherol, and γ-isomer 10–35%. Despite an agreement that α-tocopherol is the most efficient antioxidant and vitamin E homolog in vivo, however, investiagtions indicated a considerable discrepancy in its absolute and relative antioxidant effectiveness in vitro, especially when compared to γ-tocopherol (Ramadan et al. 2003). Tocols in vegetable oils and fats, are believed to protect PUFA from peroxidation. Levels of tocopherols detected in *M. longifolia* butter may provide to the fat nutritional value and strong stability toward oxidation (Ramadan and Moersel 2006). Moreover, tocols helps to decrease incidences of heart attacks and reduce muscle damage from oxygen free radicals produced during exercise. Epidemiologic studies, suggested that people with lower tocols and other antioxidants intake and plasma levels may be at increased risk for cancer and for atherosclerosis (Kallio et al. 2002).

5 The Contribution of Bioactive Compounds in *M. Longifolia* Butter to Organoleptic Properties

Apart from the stability of edible oils and fats depends on the fatty acid profile, the presence of minor lipid-soluble bioactive and the initial levels of hydroperoxides. The peroxide value (0.24 meq/kg) of *M. longifolia* butter indicated that *M. longifolia* seed may have low levels of oxidative and lipolytic activities or contains high levels of antioxidants (Singh and Singh 1991).

Fig. 14.3 The antiradical action of *M. longifolia* butter and extra virgin olive oil at different incubation times on DPPH radicals

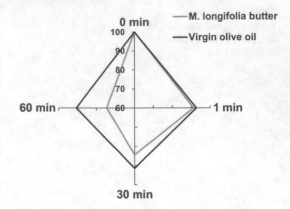

Antiradical properties of the *M. longifolia* butter and extra virgin olive oil were compared using DPPH free radicals. Figure 14.3 shows that *M. longifolia* butter had higher antiradical potential than olive oil. After 60 min of incubation, approx. 25% of DPPH· was quenched by *M. longifolia* butter, while olive oil was able to quench 9.40%. Regarding the composition of *M. longifolia* butter and virgin olive oil, they have a different composition of fatty acids and lipid-soluble bioactives. The antiradical action of oils can be interpreted as the combined action of different endogenous antioxidants (Ramadan and Moersel 2006).

6   Health-Promoting Traits of and Applications of *M. Longifolia* Butter

M. longifolia butter have nutritional and medicinal uses. *M. longifolia* butter has been used as a substitute for cocoa butter. The semisolid mahua fat is used in cooking, adulteration of ghee, and manufacturing chocolates. The seed fat has emuluscent property; it is used for skin disease, rheumatism, headache, laxative, piles and sometimes used as galactagogue (Ramadan and Moersel 2006; Yadav et al. 2011a, 2012; Ramadan et al. 2016; Jha and Mazumder 2018). *M. longifolia* butter can also be utilized in the manufacture of laundry soaps and lubricants (Parrota 2001; Ramadan and Moersel 2006).

Khatoon and Reddy (2005) prepared plastic fats with no *trans* fatty acids for use in bakery and as vanaspati by interesterification of blends of palm hard fraction (PSt) with *M. longifolia* butter. After interesterification for 60 min at 80 °C, blends showed an SFC similar to those of hydrogenated shortenings and vanaspati. Therefore, they could be used instead of hydrogenated fats as they are *trans* free. Jeyarani and Yella Reddy (2010) prepared fats using *M. longifolia* butter and kokum by enzymatic interesterification (IE) using 1,3-specific lipase. There was a significant change in the solids fat content that was attributed to the decrease in monounsaturated and disaturated TAG and the increase in the trisaturated TAG. The melting points of the blend subjected to IE for 60 min resembled that of milk fat and the one interesterified for 6 h showed a wider melting range, was similar to that of hydrogenated fats used in bakery purposes.

References

Azam, M. M., Amtul, W., & Nahar, N. M. (2005). Prospects and potential of fatty acid methyl esters of some non-traditional seed oils for use as biodiesel in India. *Biomass and Bioenergy, 29*, 293–302.

Campos, R., Narine, S. S., & Marangoni, A. G. (2002). Effect of cooling rate on the structure and mechanical properties of milk fat and lard. *Food Research International, 35*, 971–981.

Carlos, M., Moure, A., Giraldo, M., Carrillo, E., Domınguez, H., & Parajo, J. C. (2010). Fractional characterisation of jatropha, neem, moringa, trisperma, castor and candlenut seeds as potential feedstocks for biodiesel production in Cuba. *Biomass and Bioenergy, 34*, 533–538.

Ghadge, S. V., & Raheman, H. (2005). Biodiesel production from mahua (*Madhuca indica*) oil having high free fatty acids. *Biomass and Bioenergy, 28*, 601–605.

Ghadge, S. V., & Raheman, H. (2006). Process optimization for biodiesel production from mahua (*Madhuca indica*) oil using response surface methodology. *Bioresource Technology, 97*, 379–384.

Gupta, A., Chaudhary, R., & Sharma, S. (2012). Potential applications of mahua (*Madhuca indica*) biomass. *Waste and Biomass Valorization, 3*, 175–189.

Gupta, A., Kumar, A., Sharma, S., & Vijay, V. K. (2013). Comparative evaluation of raw and detoxified mahua seed cake for biogas production. *Applied Energy, 102*, 1514–1521.

Inamdar, A. I., Chaudhary, L. C., Agarwal, N., & Kamra, D. N. (2015). Effect of *Madhuca longifolia* and *Terminalia chebula* on methane production and nutrient utilization in buffaloes. *Animal Feed Science and Technology, 201*, 38–45.

Jeyarani, T., & Yella Reddy, S. (2010). Effect of enzymatic interesterification on physicochemical properties of mahua oil and kokum fat blend. *Food Chemistry, 123*, 249–253.

Jha, D., & Mazumder, P. M. (2018). Biological, chemical and pharmacological aspects of *Madhuca longifolia*. *Asian Pacific Journal of Tropical Medicine, 11*(1), 9–14.

Kallio, H., Yang, B., Peippo, P., Tahvonen, R., & Pan, R. (2002). Triacylglycerols, glycerophospholipids, tocopherols and tocotrienols in berries and seeds of two subspecies (ssp. *sinensis* and *Mongolia*) of sea buckthorn (*Hippophaë* rhamnoides). *Journal of Agricultural and Food Chemistry, 50*, 3004–3009.

Kapilan, N., & Reddy, R. P. (2008). Evaluation of methyl esters of mahua oil (*Mahua indica*) as diesel fuel. *JAOCS, 85*, 185–188.

Khatoon, S., & Reddy, S. R. Y. (2005). Plastic fats with zero trans fatty acids by interesterification of mango, mahua and palm oils. *European Journal of Lipid Science and Technology, 107*, 786–791.

Lawson, H. (1995). Sources of oils and fats. In H. Lawson (Ed.), *Food oils and fats, technology, utilization and nutrition* (pp. 39–48). USA: Chapman & Hall.

Lipp, M., & Ankalam, E. (1998). Review of cocoa butter and alternative fats for use in chocolate-Part A. Compositional data. *Food Chemistry, 62*, 73–97.

Manjunath, H., Omprakash Hebbal, B., & Hemachandra Reddy, K. (2015). Process optimization for biodiesel production from Simarouba, Mahua, and waste cooking oils. *International Journal of Green Energy, 12*, 424–430.

Marikkar, J. M. N., Ghazali, H. M., & Long, K. (2010). Composition and thermal characteristics of *Madhuca longifolia* seed fat and its solid and liquid fractions. *Journal of Oleo Science, 59*, 7–14.

Parrota, J. A. (2001). *Healing plants of peninsular India* (pp. 655–657). UK: CABI Publishing, Wallingford.

Puhan, S., Vedaraman, N., Ram, B. V. B., Sankarnarayanan, G., & Jeychandran, K. (2005). Mahua oil (*Madhuca Indica* seed oil) methyl ester as biodiesel-preparation and emission characteristics. *Biomass and Bioenergy, 28*, 87–93.

Ramadan, M. F., & Moersel, J.-T. (2006). Mowrah butter: Nature's novel fat. *Information, 17*, 124–126.

Ramadan, M. F., Kroh, L. W., & Moersel, J. T. (2003). Radical scavenging activity of black cumin (*Nigella sativa* L.), coriander (*Coriandrum sativum* L.) and niger (*Guizotia abyssinica* Cass.) crude seed oils and oil fractions. *Journal of Agricultural and Food Chemistry, 51*, 6961–6969.

Ramadan, M. F., Sharanabasappa, G., Parmjyothi, S., Seshagri, M., & Moersel, J.-T. (2006). Profile and levels of fatty acids and bioactive constituents in mahua butter from fruit seeds of butter cup tree (*Madhca longifolia*). *European Food Research and Technology, 222*, 710–718.

Ramadan, M. F., Mohdaly, A. A. A., Assiri, A. M. A., Tadros, M., & Niemeyer, B. (2016). Functional characteristics, nutritional value and industrial applications of *Madhuca longifolia* seeds: An overview. *Journal of Food Science and Technology, 53*, 2149–2157.

Singh, A., & Singh, I. S. (1991). Chemical evaluation of mahua (*Madhuca indica*) seed. *Food Chemistry, 40*, 221–228.

Singh, D., Singh, M., Yadav, E., Falls, N., Komal, U., Dangi, D. S., Kumar, V., & Verma, A. (2018). Amelioration of diethylnitrosamine (DEN)-induced hepatocellular carcinogenesis in animal models via knockdown oxidative stress and proinflammatory markers by *Madhuca longifolia* embedded silver nanoparticles. *RSC Advances, 8*, 6940.

Wong, S. (1988). *The chocolate fat from the Borneo Illipe trees*. Malaysia: Vivar Printing Sdn Bhd, Pahang.

Yadav, S., Suneja, P., Hussain, Z., Abraham, Z., & Mishra, S. K. (2011a). Prospects and potential of *Madhuca longifolia* (Koenig) J.F. Macbride for nutritional and industrial purpose. *Biomass and Bioenergy, 35*, 1539–1544.

Yadav, S., Suneja, P., Hussain, Z., Abraham, Z., & Mishra, S. K. (2011b). Genetic variability and divergence studies in seed and oil parameters of mahua (*Madhuca longifolia* Koenig) J.F. Macribide accessions. *Biomass and Bioenergy, 35*, 1773–1778.

Yadav, P., Singh, D., Mallik, A., & Nayak, S. (2012). Madhuca lonigfolia (sapotaceae): A review of its traditional uses, phytochemistry and pharmacology. *International Journal of Biomedical Research, 3*, 291–305.

Yang, B., Karlsson, R. M., Oksman, P. H., & Kallio, H. P. (2001). Phytosterols in sea buckthorn (*Hippophaë rhamnoides* L.) berries: Identification and effects of different origins and harvesting times. *Journal of Agricultural and Food Chemistry, 49*, 5620–5629.

Part IV
Oils From Fruit Seeds and Pulp

Chapter 15
Chia (*Salvia hispanica*) Oil

Muhammad Nadeem and Muhammed Imran

Abstract Drying sources of edible oils and prevalence of several metabolic disorders have motivated the researchers to learn the significance of new bases of edible oils. Soybean, canola and sunflower are common versions of industrial edible oils. If soybean with 20% oil content can be used at industrial level then why chia (*Salvia hispanica*) with 40% oil cannot be used at commercial level. Oil industry is not familiar with the massive potential of chia oil. Soybean, sunflower and canola oils have higher concentration of *omega*-6 fatty acids. European population is already suffering from the problem of over intake of *omega*-6 fatty acids. About 65% *omega*-3 fatty acids are present in chia oil. *Omega*-3 fatty acids have cardioprotective effects, antidiabetic, anticancer, antiaging and anti-inflammatory properties. They can help to reduce/minimize the life style related disorders. In industrial processing, edible oils are harshly processed. This harsh processing not only decreases their nutritional value but also lead to the production of polymers, which have been identified as risk factor for cardiovascular disease and cancer causing compounds. Processing of crude oils require massive energy and cost. Results of earlier investigations have shown that crude chia oil can be used in food products without affecting the sensory prospects. Chia oil can be used in bakery products, dairy products, margarine, Vanaspati, poultry, livestock feed and in cosmetics industry. Interesterification, transesterification, partial hydrogenation and fraction can improve the industrial applications of chia oil. This chapter will summarize the composition and functional traits of chia oil.

Keywords *Salvia hispanica* · *Omega*-3 fatty acids · Fractionation · Cosmetics

M. Nadeem (✉)
Faculty of Animal Production and Technology, University of Veterinary and Animal Sciences, Lahore, Pakistan
e-mail: muhammad.nadeem@uvas.edu.pk

M. Imran
Institute of Home and Food Sciences, Faculty of Life Sciences, Government College University Faisalabad, Faisalabad, Pakistan
e-mail: imran@gcuf.edu.pk

© Springer Nature Switzerland AG 2019
M. F. Ramadan (ed.), *Fruit Oils: Chemistry and Functionality*,
https://doi.org/10.1007/978-3-030-12473-1_15

1   Introduction

Fats and oils are important constituent of food, they yield more than two times energy than carbohydrates (Shahidi 2005). In addition to energy, fats and oils also provide essential fatty acids, phytosterols, tocopherols and several other bioactive compounds. Fats and oils are extremely important for food industries, in baking, shortening, margarine, and frying. They are also widely used in cosmetics. In view of the ever increasing population and drying sources of dietary lipids, new types of edible oil should be discovered. Chia botanical name is *Salvia hispanica* L. and belongs to the family *Lamiaceae*. It is an annual plant originated from southern part of Mexico and norther Guatemala. Tropical and subtropical climates are suitable for the growth of chia plant. In view of the massive nutritional potential, it has been regarded as possible ingredient of the novel food industry. For several foods and medicinal applications, chia seeds have been used since thousands of years. Chia seed contains 35% fiber, 32% excellent quality oil, 25% protein, 6% moisture and 2% carbohydrates (Olivos-Lugo et al. 2010). Oil content and physicochemical attributes of chia oil and other commercial oils are given in Tables 15.1 and 15.2. Oil content of the seed is greater than most of the commercial sources of edible oil such as cottonseed oil, soybean oil and sunflower oil. Oil content of cottonseed oil, soybean and sunflower oil are about 18%, 20% and 28%,

Table 15.1 Oil content of chia oil and some vegetable oils

Seed type	Oil content (%)	Reference
Chia seed oil	35–40	Ullah et al. (2016)
Sunflower oil	35–40	Zheljazkov et al. (2008)
Canola oil	42–45	Shahidi (2005)
Soybean oil	18–20	Erickson (1995)
Cottonseed oil	17–20	Nadeem et al. (2014)
Sesame oil	51	Nadeem et al. (2013)
Moringa oleifera oil	37–40	Nadeem et al. (2016a)
Mango kernel oil	12–15	Nadeem et al. (2016b)
Palm fruit	25	Donough et al. (2016)
Palm kernel	47	Siew et al. (1995)

Table 15.2 Physico-chemical characteristics of chia oil and commercial vegetable oils

Parameter	Chia oil	Sunflower oil	Soybean oil
Free fatty acids (%)	0.11	0.55	0.85
Moisture content (%)	0.14	0.17	0.21
Iodine value	192.7	119	130–133
Peroxide value	0.22	0.24	1.29
Saponification value	218	191	196
Unsaponifiable matter	0.85	1.21	1.33
Reference	Nadeem et al. (2017)	Anwar et al. (2010)	Anwar et al. (2007)

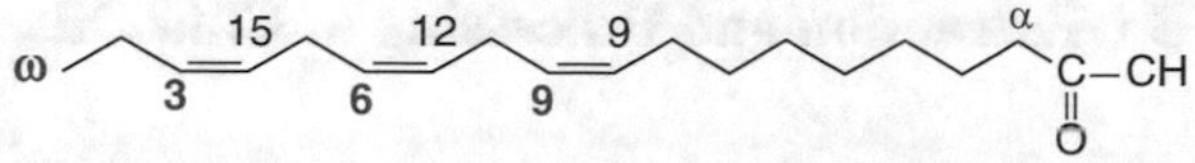

Fig. 15.1 Structural formula of α-linolenic acid (Galli and Risé 2009)

respectively. Further, commercial sources of edible oils contains several impurities such as darker color, higher amount of free fatty acids, and phosphatides. These impurities must be remove for the manufacturing of edible oil worth using in kitchen with reasonable shelf stability. Earlier investigations have shown that oil extracted from chia seeds by mechanical expression can be used in foods without pre-processing (i.e. commercial operations of edible oil refining, bleaching and deodorization). Mechanically expressed chia oil was used in margarine, ice cream and cheddar cheese (Nadeem et al. 2017; Ullah et al. 2017, 2018). Chia oil contains more than 65% *omega*-3 fatty acids, it is regarded as one of the richest source of *omega*-3 from foods of plant origin (Nadeem et al. 2017). Analytical characterization of fish oil revealed that it contains about 14–15% *omega*-3 fatty acids. On an average basis, chia oil contains more than 50% *omega*-3 fatty acids than fish oil (Ullah et al. 2015). *Omega*-3 fatty acids have cardio, hepatic-protective, hypotensive, anti-inflammatory properties, anti-ageing and anti-diabetic properties (Imran et al. 2015). The inflammatory pathways lead to main metabolic disorders such as, diabetes and cardiovascular diseases. Studies on dietary interference have shown that *omega*-3 fatty acids have anti-inflammatory properties. Intake of foods containing *omega*-3 fatty acids may help to avoid metabolic diseases. Clinical evidences and epidemiological studies have established a strong correlation between the consistent intake of *omega*-3 fatty acids and lower occurrences of heart diseases and diabetes (Albert et al. 2005). Scientific trials on laboratory animals revealed that fortification of foods with α-linolenic acid (ALA, Fig. 15.1) raised the cardio-protective properties (Ixtaina et al. 2011).

2   Botanical Description of Chia

Scientific name of Chia is *Salvia hispanica*. Chia belongs to a plant family *Lamieaceae*; Genera *Salvia* Specie; *hispanica* commonly known as chia, Spanish sage, Mexican chia and black chia (Hentry et al. 1990). Chia is herbaceous plant that can be reached about 1 m height. It is squared type stem. Chia leaves color are dark green and its pattern of leaves simple petiolate, oval and serrated on the limb. The leaves size is about 4–8 in. length and 3–5 cm wide. Flowers of Chia have bilabiate cylax. Size of chia is range between 1.5 mm wide × 2 mm long and oval shape. For good production of plant well drained clay and sandy soil is necessary for reasonable acid and salts balance. Chia can be produced 2500 kg/acre for maximum production but on a average 500–600 kg seed/acre has been reported (Cahill 2003).

3 Extraction and Processing of Chia Oil

Chia oil is regarded as mutable chemical composition product. Its production depends upon various factors such as the cultivation environment and the extraction system (Dauksas et al. 2002). Vegetable oils extraction is now coming under the field of green chemistry. It becomes a big challenge for the researchers to get acceptable yield of the desired oils with the use of such type of solvents which have high efficiency and lower harmful effects on the environment. So chia oil comes under the category of food so the solvents used for the extraction of oil from the seeds should be compatible with food industry. In this regard, those solvents which can be used for extraction of oils from the vegetables are highlighted as ethyl acetate (Almeida et al. 2012; Tian et al. 2013), ethanol (Freitas et al. 2008; Dutta et al. 2014) and isopropanol (Seth et al. 2007; Dutta et al. 2014; Ramluckan et al. 2014) and hexane (Ramalho and Suarez 2013). That solvent which have low boiling point are considered perfect for the extraction of oil as they minimize the decomposition of oil. Hexane has the same property and have highest efficiency to extract oil with high yield i.e. 33.55% (Ramalho and Suarez 2013). The solvent on the second number used for extraction are ethyl acetate and isopropanol having yields of 30.23% and 25.56% respectively. There are also new techniques for the extraction of oils from seeds. One of them is the extraction of chia oil by using CO_2 which is regarded as "supercritical extraction with CO_2". Benefit of this method is that the final product have no any residue of solvent. The principle behind this method is the use of pressure and temperature in combination for the isolation of oil from chia seeds. This method provided the best yield (7.18%). The use of pressure technology has more significant effect on the product yield as compared to temperature, which have danger of decomposition (Uribe et al. 2011). Another important technique used nowadays is the use of ultrasonic waves for the extraction of oils from seeds. These waves actually break down the seeds cells by collapsing of cavitation bubbles walls. The penetration of solvent is achieved through ultrasonic jet. The use of ultrasonic technology provided benefit in the sense of high yield and lower extraction time that may be good for commercial oil producers (Rosas-Mendoza et al. 2017).

4 Fatty Acid Profiling and Minor Bioactive Lipids

Fatty acid profiling of chia oil is done by extraction oil from the seeds by using solvent extraction method or by pressing. Then the Fatty acid methyl esters of the sample are made and then they are ready for the profiling by using GC-MS technique. Fatty acid profile of chia oil is provided in Table 15.3. There are also some others oils profiling is also given that are soybean oil, sunflower oil and palm kernel oil. Table 15.3 indicated that appreciable amount of ω-3 alpha-linolenic acid and ω-6 linoleic acids are present in chia oil. The approximately amount of palmitic (6.76%), oleic (8.34%), linoleic (12.1%) and α-linolenic (60.5%) fatty acids are

Table 15.3 Fatty acid profile (%) of chia oil and some common vegetable oils

Fatty acid	Chia oil	Soybean oil	Sunflower oil	Palm kernel oil
C12:0	0.03	–	–	52.4
C14:0	6.79	–		16.2
C16:0	8.16	14.28	6.39	7.52
C18:0	11.88	4.11	1.97	1.88
C18:1	8.33	22.87	42.19	12.5
C18:2	11.65	52.34	46.33	1.69
C18:3	64.19	5.77	0.15	0.12
C20:0	0.03	–	–	0.11
Reference	Ulalh et al. (2016)	Chowdhury et al. (2007)	Chowdhury et al. (2007)	Shahidi (2005)

present chia oil. The maximum amount of fatty acids which are found in the chia oil are in the sequence of palmitic, stearic and oleic. Table 15.4 presents also Triglyceride composition (%) of chia oil.

Phytosterols are actually called sterols and they have same structure as compared with cholesterol but have some modifications. The addition of double bond and methyl or ethyl group make modifications in the structure of sterols. There is also addition of side chain in the sterol. Mainly large amount of phytosterols are present in vegetable oils accompanied with PUFA. The most common sterols are sitosterol, campesterol and stigmasterol. Daily intake quantity of sterols is recommended as sitisterols 65%, campesterols 30% and stigmasterol 3%. Daily intake of sterols imparts many health benefits as they possess such properties which are bactericidal and antifungal effects anticancer effects, anti-inflammatory properties, anti-ulcer effects and antioxidants effects. Chia seed oil contains 1224.3 mg/L of sterols in it. The values of sterol in chia oil and other common oils is given in Table 15.5.

Chia oil contains a notable amount of organic chemical compound that is named as tocopherols. Tocopherols are involved in activity of vitamin E. These compounds can be obtained by solvent extraction and by pressing the seed. The values of tocopherol are given in Table 15.6. Chia seeds oil contains three types of tocopherols which are g-tocopherol (>85%) d-tocopherol and a-tocopherol. Variable concentrations of α-tocopherol are found in chia oil that are (0.4–9.9 mg/kg). Chia oil is deficient in b-tocopherol. In total 238–427 mg/kg of tocopherols are present in chia oil. Chia oil contain higher amount of tocopherol as compared with peanut oil but lower than flaxseed, soybeans and sunflower oils. Maximum yield of tocopherol from oils is also important task so the best way of extraction is through solvents as compared to pressing. Tocopherols are highly associated with long chain fatty acids (Tuberoso et al. 2007).

Plants originated naturally chemicals are called phenolics. They are also called phytochemicals. Table 15.7 presents the total phenolic content of chia oil. The oil contains flavonoids lignans, stilbenes and phenolic acids. Chia oil contains 6×10^{-6} to 2.1×10^{-5} mol/kg of phenolic contents. Chlorogenic and caffeic acids were termed as he major phenolic compounds found in chia oil. There are some other

Table 15.4 Triglyceride composition (%) of chia oil and flax seed oil

Triacylglycerol	Chia oil	Flax seed oil
LnLnLn	21.2–22.5	19–21
LLnLn	21.9–23.7	16–18
OLnLn	11.0–12	18–20
LLLn	7.0–7.8	4–6
LnLnP	7.9–8.3	5–7
LLL	0.1–0.2	0.3–0.46
StLnLn	3.4–4	3–6
OLLn	7.4–7.9	9–11
LnLP	5.8–6.3	2–3
OLnO	2–3.7	5–7
OLL	1.33–1.8	1–4
StLLn	2.6–2.8	1–3
PLnP	0.9–1.0	0.5–0.7
OLnP	2.6–2.9	3.2–4.1
StLL	0.3–0.8	0.2–0.4
OLO	0.6–0.9	1.5–2
LOP	0.8–1.0	0.9–2
OOO	0.12–0.19	0.6–1
StLO	0.3–0.8	0.6–2.0
OPO	0.1–0.7	0.5–1
StOO	0.2–0.4	0.3–7
Others	1.1–1.8	2.9–4
UUU	74–76	77.9–79
TPUFA	51.2–53.7	40.7–43
SUU	23–25	18.8–20.0
Reference	Ciftci et al. (2012)	Ciftci et al. (2012)

Detail of symbols: *ECN* equivalent carbon number, *Ln* linolenic, *L* linoleic, *O* oleic, *P* palmitic, *St* stearic; symbols StLO, PLO, and others indicate fatty acid present in triacyglycerols however do not represent it specific configuration. Triglycerides: UUU, triunsaturated; TPUFA, containing only PUFA; SUU, one saturated fatty acid

Table 15.5 Sterol composition in flax and chia seed lipids (mg/kg lipid)

Sterol	Chia seed oil	Flax seed oil	Sunflower oil
Campesterol	470–475	583–588	6.9–13.8
Stigmasterol	1249–1252	239–245	6.8–12.9
β-sitosterol	2055–2060	1452–1460	52–71
Delta-5-avenasterol	350–360	242–245	6.89
Delta-7-stigmasterol	Not detected	180–189	7–25
Cycloartenol	Not detected	990–998	Not detected
Avenasterol	Not detected	37–40	Not detected
2, 4-methylenecycloartanol	Not detected	340 345	Not detected
Reference	Ciftci et al. (2012)	Ciftci et al. (2012)	Codex Alimentarius (2011)

Table 15.6 Composition of tocopherols and plastochromanol-8 in flax and chia seed lipids (mg/kg)

Tocochromanol	Chia oil	Flax seed oil	Sunflower oil
α-Tocopherol	8–12	7–9	487–669
γ-Tocopherol	590–595	541–545	20–53
δ-Tocopherol	4–6	4–6	2.99–20
Plastochromanol-8	2–5	190–195	Not detected
Reference	Dąbrowski et al. (2017)	Ciftci et al. (2012)	Ciftci et al. (2012)

Table 15.7 Induction period and total phenolic content of chia oil

Oil type	Induction period	Total phenolic contents	Reference
Chia oil	10.26	14.56	Nadeem et al. (2017)
Mango kernel oil	35.8	9.88	Arif et al. (2016)
Moringa oleifera oil	13.5	7.1	Nadeem et al. (2016a)

phenolic compounds found in chia oil are kaempferol, quercetin and myricetin (Reyes-Caudillo et al. 2008; Taga et al. 1984). The magical properties of chia oil shows that those polyphenols present in it are not present in other oil seeds. The higher yield of polyphenols can be achieved by using pressing extraction method (Tuberoso et al. 2007). Vegetable oil color is an important factor for consumer acceptability. So it totally depends upon the presence of color pigment i.e. carotenoids and chlorophyll. Chia oil contains a range of carotenoids i.e. 0.53–1.21 mg/kg. Chia oil is deficient with chlorophyll pigment. Mainly β-carotene is detected in the seed oil while α-carotene is not present in it. Chia oil contains more carotenoids as compared to soybean (0.3 mg/kg), maize (0.9 mg/kg), flaxseed (0.7 mg/kg) and sunflower (0.1 mg/kg) but lower as compared with rapeseed (1.7 mg/kg) (Tuberoso et al. 2007).

5 Health Benefits of *Omega* Fatty Acids

The health benefits of *omega* fatty acids which are long chain polyunsaturated fatty acids (PUFA) are countless. These are considered as dietary essential fats and can be substituted with others. *Omega* fatty acids are supplemented in the human body as they perform unique processes which are anti-inflammatory processes. The proper consumption of *omega*-3 fatty acids by the pregnant females resulted complete fetal development. Eicosapentaenoic acid (EPA) and docosahexaenoic acid (DHA) are two most significant long-chain fatty acids. The proper fetal and brain development and formation of tissue layer is highly associated with the presence of DHA. Health aging phenomena and fetal development is linked with the presence of EPA and DHA. DHA is considered as building block of cell membranes. Retina and brain cells are also rich in DHA. The prevention and treatment of many diseases are linked with the potent lipid mediators which are actually precursors of EPA and DHA. General cardiovascular health, Alzheimer's disease, poor fetal development as well as inflammatory processes are highly in threat due to less supply of EPA and DHA. The consumption of essential *omega*-3 fatty acid has benefited the human health. Clopidogrel hypo-responsiveness, anti-inflammatory properties, improved antiplatelet effects, reduced major coronary events and improved cardiovascular functions are highly linked with the intake of high amount of *omega* fatty acids. There is clear sign of DHA deficiency in the patients of Alzheimer's disease. This deficiency can be cured by providing enough amount of EPA+DHA to the patient. EPA and DHA are proven as safe and inexpensive way of attaining a healthier life with the prevention of pediatric allergies and other diseases.

6   Antioxidant Characteristics of Chia Oil

Chia (*Salvia hispanica* L.) seed oil is considered as fascinating source of polyunsaturated fatty acids (PUFA). It contains higher magnitudes of ALA as compared with flaxseed oil (up to 57%) (Khattab and Zeitoun 2013). The low ratio of n-6/n-3 fatty acid is considered to minimize the risk cardiovascular disease (Simopoulos 2002). Normal growth and development of human body is associated with such types of essential fatty acids, which are present in chia oil. Another important role is the prevention of human body from diabetes, coronary artery disease, hypertension, arthritis, other inflammatory and autoimmune disorders and cancer. Little amount of saturated fatty acids are also present in chia seed oil. Greater amount of *omega* acids are found in chia seed oil as compared to other food sources. For that reason, it is assumed as powerhouse of *omega*-3 fatty acids. Scientific literature have revealed that *omega*-3 fatty acids have health promoting properties including hypotensive effects, anti-inflammatory and cardiao-protective abilities. Chia oil is a potential source of bioactive compounds. These compounds includes carotenoids, tocopherols, polyphenols (Ixtaina et al. 2011), which are proven as having low peroxide values and exhibit a very good oxidative stability. Table 15.7 presents the induction period of the oil which reflect the oil oxidative stability. 3,4-DHPEA-EDA, quercetin, kaempferol, myricetin and chlorogenic acid are major polyphenols present in chia oil. These phenols provide functional components in the dietetic food items at the industrial level due to exhibiting more antioxidant capacity and low level of lipid autoxidation. The presence of secondary auto oxidation products make the oil more stable instead of the presence of high quantities of PUFA. These products are kaempferol, chlorogenic acid, myricetin and quercetin (Ixtaina et al. 2011). Chia oil is magical oil, which contains high quantities of phenolic compounds as compared to other oils, which are deficient to these compounds (Tuberoso et al. 2007). It could be beneficial for humans and in improving their health to consume chia oil.

7 Pharmacological Traits of Chia Oil

Salvia hispanica L. seed oil is taken into account as a powerhouse of natural anti-oxidants and essential fatty acids. The worth of chia oil is due to the presence of bioactive components (Ixtaina et al. 2011; Reyes et al. 2008). The intake of long chain fatty acids is mainly concerned with the intake of α-linolenic acid (ALA). The consumption of ALA is highly linked with a low chances of cardiovascular diseases (Calder 2004). The patients of bronchial asthma and rheumatoid arthritis are offered to use high amount of PUFA. The public hygenic studies have shown that the ability of childhood study and behavior are increased by the intake of such fatty acids having high amount of ALA (Richardson 2004). ALA-rich fatty acids also proven helpful in decreasing the psychiatric illnesses in elders (Freeman et al. 2006). Chia seeds are good source of PUFA along with ALA. This edible oil is also good source of phenolic compounds, minerals, protein, fiber as well as PUFA (Reyes et al. 2008). Edible oil (25–38%) which is carrier of (60%) ALA can be extracted from chia seeds (Ixtaina et al. 2011). After extraction of chia oil the remaining residue contain phenolic compounds and fiber. These phenolic compounds exhibit high level of antioxidant properties (Reyes et al. 2008). Many essential food items are cheered with ALA due to its health benefitting properties. It is also used in cosmetics as it have good effect when used on skin. Literature shows that the regular intake of PUFA is secret of providing good health. The PUFA supplements helps in reduction the chances of hypertension and inflammatory diseases and cardiac problems. The most essential PUFA that is broken down into EPA and DHA with Δ6 and Δ5 by the action of desaturase enzymes. The worldwide increase in the awareness about public health has necessitated the search for functional food with multiple health benefits. Chia seed oil plays an important role in reactive protein interactions, which are highly associated with atherothrombotic disease. It is currently consumed for various health benefits. Owing to the abundance of PUFA, chia seed oil possess anti-hyperlipidemia, anti-hypercholesterolemia, anti-diabetic, anti-cancer property and anti-inflammatory properties. There are many medicinal uses of chia oil, which were reported. The effective dose differ from person to person but it can be standardized by having the requirement of specific person. There are many pharmacological properties associated with chia seed and its oil, however the biological activity of chia oil can be understood by studying the chemistry of bioactive components and fatty acid present in it.

8 Commercial Prospects of Chia Oil

Tables 15.8 and 15.9 present some food and non-food applications of chia oil. Among the commercial sources of edible oils, soybean, sunflower and canola are largely used 'by the edible oil processors to manufacture cooking oils, partially hydrogenated fats for the purpose of cooking and baking. Sources of edible oils

Table 15.8 Food applications of chia oil

Food application	Result	References
Low melting point fraction of chia oil was used in cream	Enrichment of ice cream with olein fraction of chia oil significantly raised the amount of omega-3 fatty acids and antioxidant characteristics of ice cream with no effect on sensory characteristics	Ullah et al. (2017)
Effect of chia oil addition on lipolysis and antioxidant characteristics of cheddar cheese was studied	Addition of chia oil in cheddar cheese considerably improved the antioxidant profile with improved oxidative stability	Ullah et al. (2018)
Chia oil was used to raise the concentration of omega-3 fatty acids in *trans* free margarine	Supplementation of margarine with chia oil increased the concentration of beneficial omega-3 fatty acids. Chia oil supplemented margarine had higher total antioxidant capacity and antioxidant activity in linoleic acid. Oxidative stability of chia oil supplemented margarine was not different from the margarine formulated from partially hydrogenated fats	Nadeem et al. (2017)
Chemical characteristics of chia oil were studied	A comprehensive study was performed on the characterization of chia oil. Physical and chemical assays indicated that oil is edible with a huge potential for commercial applications	Imran et al. (2015)

are not unlimited; to fulfill the nutritional requirements of human population, non-traditional sources of edible oils should be introduced at commercial level. The nature has very kindly blessed us with more than 600 oil producing plants, but it the failure of edible oil industry and research organization to bring them to the consumers. For the estimation of commercial prospects of an edible oil, seed production/acre, oil content, oil quality such as free fatty acids, color, moisture content, phosphatides and oxidative stability are taken into consideration. Chia oil fulfills all these critical points; for example it produces about 2500 kg seed from one acre as compared to soybean which produces about 1200 kg seed/acre. Further, oil content of chia seed ranges from 35% to 40% as compared to soybean that produces about 20% oil. Production of oil from chia seed is about 50% greater than soybean oil; most commonly used commercial vegetable oil. Level of free fatty acids in edible oils should be as low as possible. At industrial level, free fatty acids are usually removed by alkali refining or physical refining both forms require massive electric and steam inputs to decrease the free fatty acids. Higher free fatty acids lead to the excessive soap stock production and huge amount of spray water is required to remove soap from refined oil. In view of the drying sources of energy, oils with lower free fatty acids should be selected. Studies on the characterization of chia oil showed that it had lower concentration of free fatty acids (0.1–0.12% oleic acid) as compared to soybean oil (0.8–0.9% oleic acid) (Ullah et al. 2016, 2018; Nadeem et al. 2017). Soybean, sunflower and canola oils require deodorization to

Table 15.9 Suggested applications of chia oil for food/feed uses

Food product	Concentration/application	Uses of end product
Ice cream	10% replacement	Ready to eat ice cream
	15% replacement	
	20% replacement	
Cheddar cheese	10% replacement	Pizza
	15% replacement	Sandwiches
	20% replacement	Salad dressing
Omega-3 nutraceutical (more than 75% *omega*-3 fatty acids)	10% replacement	Yoghurt
	20% replacement	Value added bakery products
	30% replacement	Ice cream
Interesterified/partially hydrogenated chia oil	Bakery shortening	Vanaspati
	Cooking	Bakery shortening
	Frying fat	
Omega-3 enriched biscuits for school children	Olein fraction of chia oil	Biscuits
	Chia oil	Cakes
	Stearin fraction of chia oil	
Cooking oil	100% chia oil or blend with soybean oil, sunflower oil, canola oil and cottonseed oil in different proportions	For cooking
		For frying
Omega-3 supplements	5–20% depending upon the nature on supplement	For curing skin diseases
		Antidiabetic agent
		Antiaging agent
		Renal dysfunction
Poultry feed	It may be used in poultry feed from 1% to 2% to produce *omega*-3 enriched eggs	Health conscious communities
Livestock feed	To increase the concentration of *omega*-3 fatty acids in milk and value added dairy products. It may be used from 5% to 10% concentration in the form of bypass fat or rumen protected fat	Functional milk and milk products with increased health benefits
Cosmetics	5–20%	Sunscreen

remove the indigenous or processing induced off-flavors. For example, alkali refining induces typical soapy flavor and bleaching earth (acid activated clays) induce musty/earthy odor to the treated oils. All these off-flavors are not desirable for the oil processing industry. These flavors are usually removed in the batch/continuous deodorizers, both require massive installation and operation cost. The cost of steam for the operation of vacuum installations of the deodorization facility is very high. According to the economics, processing of one metric ton vegetable oil require two metric ton saturated steam (pressure 15 kg/cm^2). Chia oil can be incorporated in foods without any pre-processing. This is confirmed by the investigations of Nadeem et al. (2017) that addition of chia oil in *trans* free margarine had no effect on flavor profile of margarine.

9 Fractionation of Chia Oil

The low melting point fractions of vegetable oils had higher concentration of unsaturated fatty acids. To increase the concentration of *omega*-3 and *omega*-6 fatty acids, fractionation of chia oil was performed at −30 °C. For dry fractionation, samples of chia oil were heated to 63 °C and cooled down to −30 °C and held at this temperature for 3 h. Fractionations were obtained by filtration. Fatty acid profile and phenolic compounds were analyzed in olein and stearin fraction of chia oil and compared with parent chia oil. Results showed that concentration of *omega*-3 fatty acids in olein fraction of chia oil was more than 80%. Analysis of phenolic compounds using HPLC indicated that caffeic acid, chlorogenic acid, quercetin and phenolic glycoside were the main phenolic compounds in olein and stearin fraction of chia oil (Ullah et al. 2016). Olein fraction had a higher amount of phenolic compounds than stearin fraction and native chia oil (Ullah et al. 2017).

References

Albert, C. M., Oh, K., Whang, W., Manson, J. E., Chae, C. U., & Tampfer, M. J. S. (2005). Dietary alpha-linolenic acid intake and risk of sudden cardiac death and coronary heart disease. *Circulation, 112*, 3232–3238.

Almeida, P. P., Mezzomo, N., & Ferreira, S. R. S. (2012). Extraction of *Mentha spicata* L. volatile compounds: Evaluation of process parameters and extract composition. *Food and Bioprocess Technology, 5*, 548–559.

Anwar, F., Hussain, A. I., Iqbal, S., & Bhanger, M. I. (2007). Enhancement of the oxidative stability of some vegetable oils by blending with *Moringa oleifera* oil. *Food Chemistry, 103*, 1181–1191.

Anwar, F., Qayyum, H. M. A., Hussain, A. I., & Iqbal, S. (2010). Antioxidant activity 0f 100% and 80% methanol extraction from barley seed (*Hordeum valgave* L.) stabilization of sun flower oil. *Grasasy Aceites, 6*, 237–243.

Arif, A. M., Javed, I., Abdullah, M., Imran, M., Mahmud, A., Nadeem, M., and Ayaz, M. (2016). Chemical Characteristics of Mango (Mangifera indica l.) Kernel Oil and Palm Oil Blends for Probable Use as Vanaspati. *Journal of Oil Palm Research, 28*, 344–352.

Cahill, J. (2003). Ethnobotany of chia, *Salvia hispanica* L. (Lamiaceae). *Economic Botany, 57*, 604–618.

Calder, P. C. (2004). Omega-3 (ω-3) fatty acids and cardiovascular disease: Evidence explained and mechanisms explored. *Clinical Science, 107*, 1–11.

Chowdhury, K., Banu, L. A., Khan, S., & Latif, A. (2007). Studies on the fatty acid composition of edible oil. *Bangladesh Journal of Scientific and Industrial Research, 42*(3), 311–316.

Ciftci, O. N., Przybylski, R., & Rudzińska, M. (2012). Lipid components of flax, perilla, and chia seeds. *European Journal of Lipid Science and Technology, 114*, 794–800.

Codex Alimentarius. (2011). Stan 210, Norma del CODEX para Aceites Vegetales Especificados.

Dąbrowski, G., KonopkaI Czaplicki, S., & Tanska, M. (2017). Composition and oxidative stability of oil from *Salvia hispanica* L. seeds in relation to extraction method. *European Journal of Lipid Science and Technology, 119*, 1600209.

Dauksas, E., Venskutonis, P. R., Sivik, B., & Nilson, T. (2002). Effect of fast CO_2 pressure changes on the yield of lovage (*Levisticum officinale* Koch.) and celery (*Apium graveolens* L.) extracts. *Journal of Supercritical Fluids, 22*, 201–210.

Donough, C. R., Cahyo, A., Wandri, R., Fisher, M., & Oberthur, T. (2016). Plant nutrients in palm oil. *Better Crops Plant Food, 100*(2), 19–22.

Dutta, R., Sarkar, U., & Mukherjee, A. (2014). Extraction of oil from *Crotalaria Juncea* seeds in a modified Soxhlet apparatus: Physical and chemical characterization of a prospective bio-fuel. *Fuel, 116*, 794–802.

Erickson, D. R. (1995). *Practical handbook of soybean processing and utilization.* Champaign: AOCS Press.

Freeman, M. P., Hibbeln, J. R., Wisner, K. L., Davis, J. M., Mischoulon, D., & Peet, M. (2006). Omega-3 fatty acids: Evidence basis for treatment and future research in psychiatry. *The Journal of Clinical Psychiatry, 67*, 1954–1967.

Freitas, L. S., Jacques, R. A., Richter, M. F., Silva, A. L., & Camarão, E. B. (2008). Pressurized liquid extraction of vitamin E from Brazilian grape seed oil. *Journal of Chromatography. A, 1200*, 80–83.

Galli, C., & Risé, P. (2009). Fish consumption, omega 3 fatty acids and cardiovascular disease. The science and the clinical trials. *Nutrition and Health, 20*, 11–20.

Hentry, H. S., Mittleman, M., & McCrohan, P. R. (1990). Introduccion de la chia y la goma de tragacanto en los Estados Unidos. In O. J. Janick & J. E. Simon (Eds.), *Avances en Cosechas Nuevas* (pp. 252–256). Portland: Prensa de la Madera.

Imran, M., Anjum, F. M., Nadeem, M., Ahmad, N., Khan, M. K., Mushtaq, Z., & Hussain, S. (2015). Production of bio-omega-3 eggs through the supplementation of extruded flaxseed meal in hen diet. *Lipids in Health and Disease, 14*, 126. https://doi.org/10.1186/s12944-015-0127-x.

Ixtaina, V. Y., Martinez, M. L., Spotorno, V., Mateo, C. M., Maestri, D. M., & Diehl, B. W. K. (2011). Characterization of chia seed oils obtained by pressing and solvent extraction. *Journal of Food Composition and Analysis, 24*, 166–174.

Khattab, R. Y., & Zeitoun, M. A. (2013). Quality of flaxseed oil obtained by different extraction techniques. *LWT – Food Science and Technology, 53*, 338–345. https://doi.org/10.1016/j.lwt.2013.01.004.

Nadeem, M., Abdullah, M., Hussain, I., Javid, A., & Zahoor, Y. (2013). Antioxidant potential of *Moringa oleifera* leaf extract for the stabilization of butter at refrigeration temperature. *Czech Journal of Food Sciences, 31*, 332–339.

Nadeem, M., Mahud, A., Imran, M., & Khalique, A. (2014). Enhancement of the oxidative stability of whey butter through almond (*prunus dulcis*) peel extract. *Journal of Food Processing and Preservation, 39*(6), 591–598. ISSN 1745-4549. https://doi.org/10.1111/jfpp.12265.

Nadeem, M., Ullah, R., & Ullah, A. (2016a). Improvement of the physical and oxidative stability characteristics of ice cream through interesterified *Moringa oleifera* oil. *Pakistan Journal of Scientific and Industrial Research, 59*(1), 38–43.

Nadeem, M., Imran, M., Iqbal, Z., Abbas, N., & Mahmud, N. (2016b). Enhancement of the oxidative stability of butter oil by blending with mango (*Mangifera indica* L.) kernel oil in ambient and accelerated oxidation. *Journal of Food Processing & Preservation.* https://doi.org/10.1111/jfpp.12957.

Nadeem, M., Imran, M., Taj, I., Ajmal, M., & Junaid, M. (2017). Omega-3 fatty acids, phenolic compounds and antioxidant characteristics of chia oil supplemented margarine. *Lipids in Health and Disease, 16*, 102. https://doi.org/10.1186/s12944-017-0490-x.

Olivos-Lugo, B. L., Valdivia-López, M. Á., & Tecante, A. (2010). Thermal and physicochemical properties and nutritional value of the protein fraction of Mexican chia seed (*Salvia hispanica* L.). *Food Science and Technology International, 16*(1), 89–96.

Ramalho, H. F., & Suarez, P. A. Z. (2013). The chemistry of oils and fats and their extraction processes and refining. *Virtual Journal of Chemistry, 5*, 2–15.

Ramluckan, K., Moodley, K. G., & Bux, F. (2014). An evaluation of the efficacy of using selected solvents for the extraction of lipids from algal biomass by the soxhlet extraction method. *Fuel, 116*, 103–108.

Reyes-Caudillo, E., Tecante, A., & Valdivia-Lopez, M. A. (2008). Dietary fiber content and antioxidant activity of phenolic compounds present in Mexican chia (*Salvia hispanica* L.) seeds. *Food Chemistry, 107*, 656–663.

Richardson, A. J. (2004). Clinical trials of fatty acid treatment in ADHD, dyslexia, dyspraxia and the autistic spectrum. *Prostaglandins Leukotrienes Essential Fatty Acids, 70*, 383–390.

Rosas-Mendoza, M. E., Coria-Hernández, J., Meléndez-Pérez, R., & Arjona-Román, J. L. (2017). Characteristics of chia (*Salvia hispanica* L.) seed oil extracted by ultrasound assistance. *Journal of the Mexican Chemical Society, 61*(4), 326–335.

Seth, S., Agrawala, Y. C., Ghoshb, P. K., Jayasb, D. S., & Singh, B. P. N. (2007). Oil extraction rates of soya bean using isopropyl alcohol as solvent. *Biosystems Engineering, 97*, 209–217.

Shahidi, F. (2005). *Baileys' industrial edible oil and fat products* (6th ed.). New York: Wiley.

Siew, W. L., Chong, C. L., & Tan, Y. A. (1995). Composition of the oil in palm kernel from *Elaeis guineensis*. *Journal of the American Oil Chemists' Society, 72*, 1587–1589. https://doi.org/10.1007/BF02577859.

Simopoulos, A. P. (2002). Omega-3 fatty acids in inflammation and autoimmune diseases. *Journal of the American College of Nutrition, 21*(6), 495–505.

Taga, M. S., Miller, E. E., & Pratt, D. E. (1984). Chia seeds as a source of natural lipid antioxidants. *Journal of the American Oil Chemists' Society, 61*, 928–931.

Tian, Y., Xu, Z., Zheng, B., & Lo, Y. M. (2013). Optimization of ultrasonic-assisted extraction of pomegranate (*Punica granatum* L.) seed oil. *Ultrasonics Sonochemistry, 20*, 202–208.

Tuberoso, C., Kowalczyk, A., Sarritzu, E., & Cabras, P. (2007). Determination of antioxidant compounds and antioxidant activity in commercial oilseeds for food use. *Food Chemistry, 103*, 1494–1501.

Ullah, R., Nadeem, M., Imran, M., Tayyab, M., & Sajid, R. (2015). Antioxidant characteristics of ice cream supplemented with sugarcane (*Saccharum officinarum* L.) juice. *Food Science and Biotechnology, 24*(4), 1227–1232. (2015). https://doi.org/10.1007/s10068-015-0157-1.

Ullah, R., Nadeem, M., Ayaz, M., Imran, M., & Tayyab, M. (2016). Fractionation of chia oil for enrichment of omega 3 and 6 fatty acids and oxidative stability of fractions. *Food Science and Biotechnology, 25*(1), 41–47. https://doi.org/10.1007/s10068-016-0006-x.

Ullah, R., Nadeem, M., Imran, M., & Taj, I. (2017). Omega fatty acids and oxidative stability of ice cream supplemented with olein fraction of chia oil. *Lipids in Health and Disease*. https://doi.org/10.1186/s12944-017-0420-y.

Ullah, R., Nadeem, M., Mahmud, A., Khan, I. T., Shahbaz, M., & Tayyab, M. (2018). Omega fatty acids, phenolic compounds and lipolysis of cheddar cheese supplemented with chia (*Salvia hispanica* L.) oil. *Journal of Food Processing & Preservation*. https://doi.org/10.1111/jfpp.13566.

Uribe, J. A. R., Perez, J. I. N., Kauil, H. C., Rubio, G. R., & Alcocer, C. G. (2011). Extraction of oil from chia seeds with supercritical CO_2. *Journal of Supercritical Fluids, 56*(2), 174–178.

Zheljazkov, V. D., Callahan, A., & Cantrell, C. L. (2008). Yield and oil composition of thirty-eight basil (*Ocimum basilicum* L.) accessions grown in Mississippi. *Journal of Agricultural and Food Chemistry, 56*, 241–245.

Chapter 16
Argan [*Argania spinosa* (L.) Skeels] Oil

Dominique Guillaume, Daniel Pioch, and Zoubida Charrouf

Abstract Argan oil is extracted from the kernels of *Argania spinosa* (L.) Skeels, a tree that almost exclusively grows endemically in southern Morocco. If argan oil was initially only known around its traditional production area, major efforts combining chemical, agronomic and human sciences have led to its international recognition and marketing. In addition, to ensure the sustainable production of a sufficient quantity of argan kernels, a vast and unprecedented program that led to the reforestation of large areas of drylands has been developed in Morocco. Therefore, argan oil production is considered as an economic and ecologic success.

Edible argan oil is prepared by cold-pressing roasted argan kernels. Unroasted kernels afford an oil of cosmetic grade, showing a bitter taste. Both oils, which are not refined and are virgin oils, share a similar fatty acid content that includes oleic and linoleic acids as major components. Additionally, argan oil is rich in antioxidants. Together, these components likely contribute to the oil pharmacological properties that, in humans, traditionally included cardiovascular disease and skin protection. Recent scientific studies have greatly expanded the scope of these pharmacological activities.

Argan oil is now rewarded with a "Geographic Indication" that certifies its exclusive and authentic Moroccan origin and the compliance with strict production rules. In addition, the quality of argan oil can nowadays be ascertained by using an array of physicochemical methods.

By-products, generated in large quantity during argan oil production, are also finding promising development routes.

D. Guillaume (✉)
Faculté de Médecine/Pharmacie, URCA, Reims, France
e-mail: dominique.guillaume@univ-reims.fr

D. Pioch
CIRAD, UR BioWooEB Biorefinery Team, Montpellier, France
e-mail: daniel.pioch@cirad.fr

Z. Charrouf
Université Mohammed V-Agdal, Laboratoire de Chimie des Plantes, Synthèse Organique et Bioorganique, Faculté des Sciences, Rabat, Morocco

© Springer Nature Switzerland AG 2019
M. F. Ramadan (ed.), *Fruit Oils: Chemistry and Functionality*,
https://doi.org/10.1007/978-3-030-12473-1_16

317

Keywords Edible oil · Cosmetic oil · Fatty acid · Tocopherol · Sustainable development · Morocco

1   Introduction

The argan tree [*Argania spinosa* (L.) Skeels] is a spontaneous, thermophilic and xerophytic tree that belongs to the Sapotaceae family. Before the Pleistocene (Quaternary) glaciations, the argan forest likely extensively covered very large areas in Northern Africa. However, the series of glacial events that characterized the Pleistocene glaciations induced the nearly total extinction of the argan tree from the Maghreb and Sahelian band. Providentially, in the Souss Valley (Morocco) a microclimate simultaneously resulting from temperate oceanic conditions and steepness-related gelid wind protection allowed optimum conditions for the argan tree survival. This environment prevented the argan tree complete extinction (Kenny and De Zborowski 2007). Hence, today the argan tree is an 80-million-year-old relic and is the only species of the family Sapotaceae remaining in the subtropical zone.

The argan tree almost exclusively grows in Morocco on an 870,000–1000,000 ha area referred to as "the argan forest" (Charrouf and Guillaume 2008a). Unfortunately, vast parts of the argan forest present a long history of degraded vegetation (McGregor et al. 2009). This decrepitude is the result of a slowly evolving phenomenon that is now properly addressed but that Moroccan authorities and local people took unfortunately a long time to realize.

The argan forest is a unique flat, hilly, or steep landscape, which constitutes a fragile biotope dominantly populated by an Amazigh agrarian population. Imazighen (singular Amazigh) constitute an ethnic group that brings together the descendants of Northern Africa's original inhabitants before the Romans and the Arab Muslim conquests in the seventh century. In the Souss Valley, as well as other remote areas in Morocco such as the central Rif, Amazigh (sometimes mistakenly called Berber) population makes a socially active group mainly characterized by matriarchal traditions. In the inland, the argan tree must co-exist as harmoniously as possible with ancestral practices allowing domestic young goat, and in its southernmost part stray camels, grazing (Chatibi et al. 2016). On the Atlantic coast, the population is more mixed and urban but the economic activity principally based on tourism is another type of threat for the argan tree. The argan forest was designated by the "UNESCO's Man and Biosphere program" as a Biosphere in 1998.

Hence, argan oil is the finest natural production of a unique type of tree and is itself unique in its composition and properties. Multivariate discriminant analysis of its fatty acid content has shown that it resembles to sesame oil and presents some similarities with high-oleic sunflower oil (Rueda et al. 2014).

Argan oil is extracted from argan kernels. In the South-Moroccan Amazigh culture, argan oil has for centuries been used for its nutritional as well as dermatologic properties. In this latter field, argan oil is particularly recommended to cure skin pimples, juvenile acne, and chicken pox pustules (Charrouf and Guillaume 1999). As edible oil, argan oil is the main part of the source of the necessary daily lipid fraction of the rural population of the Souss Valley, thus being the basic ingredient of the Amazigh diet (Charrouf and Guillaume 2010).

For cultural and ancestral reasons, women exclusively achieve the preparation of argan oil. Until the 1980s, argan oil preparation was performed following an atavistic process at the family-scale. Most of argan oil properties were unknown or limited to the traditional knowledge of rural women. A vast scientific and multidisciplinary program, both nationally and internationally funded, and named "the Argan Oil Project" came out in the mid-1980s and lasted for almost 40 years (Charrouf and Guillaume 2011, 2018; Charrouf et al. 2011). The Argan Oil Project cumulatively led to the complete reconstruction of the argan oil production, marketing, and delivery chains (Turner 2016), the improvement and optimization of the oil preparative process, the design of distinct cosmetic and edible argan oil grades, the determination of argan oil detailed composition, the design of scientific methods ascertaining argan oil quality, the building of production units based on a woman-only cooperative system, the establishment of an official quality norm, the recognition of a Geographic Indication along with regulatory certification allowing the marketing of argan oil as "endemic argan oil of Morocco", the sustainable management of the argan forest, and finally it allowed to confirm some of the alleged pharmacological benefits of the oil (Charrouf and Guillaume 2014). In short, the Argan Oil Project revolutionized the fate of argan oil by igniting and perpetuating its worldwide popularization. Accordingly, whereas argan oil export was estimated to be 1 ton in 1996, it was 40 tons in 2003, reaching 400 tons in 2009, to boom to 1000 tons in 2014 (Roumane 2017). Recent estimates predict argan oil market to reach almost 20,000 tons by 2022 (Khallouki et al. 2017a). Today, argan oil is found on the North-American, European and Japanese markets and its largest importers are France and Germany with 650 and 100 tons, respectively, in 2014 (Roumane 2017). In parallel, the price of 1 l of argan oil has jumped from a few dirhams in the 1980s (Moroccan market) to 150 euro in 2017 (European market).

In terms of research and development, the Argan Oil Project is often shown as an unprecedented wonderful success story. However, there were some troubles, at least at the start, and sharp questions have been regularly raised regarding some aspects of the project throughout its establishment (Lybbert et al. 2002, 2011; Simenel et al. 2009; Le Polain de Waroux 2013). Presently, some cooperatives must still strive to reach sustainable growth and special attention should be paid so that Amazigh families do not be deprived of argan oil for their own use (Huang 2017). In addition, the wealth produced by the argan oil growing market has not affected identically all argan forest dwellers (Le Polain de Waroux and Chiche 2013). Anyway, the argan oil project frequently illustrates the work to be done to enhance the value of a traditional production and bring it from a local product to a major actor of the global economy. This chapter is aimed at giving a whole picture of the odyssey of argan oil.

2   The Argan Tree

The argan tree is a slow-growing and thorny evergreen tree, gnarly with aging, particularly well adapted to survive extended periods of drought. It can grow as a bush or, if isolated and under sufficiently fertile and favorable conditions, it can reach 7–10 m high. Indeed, the argan tree growth can be strongly modified by its environment. For example, browsing and human use can reduce tree size and leaf production (Ain-Lahout et al. 2013). Natural conditions also affect the tree growth, the proximity of the ocean and its associated high salinity brings some influence, the coastal population of *A. spinosa* being possibly subjected to higher stress than the inland trees. Aridity and anthropogenic environmental changes also influence argan tree growth (Alados and El Aich 2008; Zunzunegui et al. 2017). Accordingly, it has been proposed that observed oil content variations in *A. spinosa* kernels could result from combined genetic and environmental factors (Aabd et al. 2014). This should be taken into account for selecting elite genotypes to be used for reforestation purposes.

Argan tree can easily live 150 or 200 years. In the Souss Valley, the argan tree has always played an important ecological function (Morton and Voss 1987; Ruas et al. 2016) and its participation in biodiversity preservation is essential. Indeed, if the argan forest experiences generally arid conditions, violent rains can sometimes occur and very strong winds can blow over the forest. Because of it's widely spread root-system, the argan tree, which in some places represents the only permanent vegetation, protects the soil against erosion and is a natural dam to slow the desert progression (Le Polain de Waroux and Lambin 2012). The argan tree shades all kind of domestic cultures, as barley or other types of easy-to-grow cereals, and brings moisture to the soil (Morton and Voss 1987). Doing so, its guarantees and perpetuates the soil natural fertility. The rural Amazigh population in the Souss Valley also uses Argan leaves as hanging forage for cattle.

However, due to the combination of several factors including a prolonged over use of argan trees during the nineteenth century and up to the first half of the twentieth century, several recent consecutive arid years, demographic pressure, and ill-management imposed by the very touristy nearby coastal zone, the argan tree has, once again, been threatened with slow extinction (Mellado 1989). Immediate consequences could have been disastrous with an irremediable advance of the desert without the Argan Oil Project that took steps to raise awareness of desertification in South Morocco and, subsequently, the Moroccan government encouraging a vast reforestation program (Charrouf et al. 2008). The argan oil project, based on a sustainable development approach (Charrouf and Guillaume 2009), started progressively and during the year 2005, the infant year of the agroforestry part of the project, only 500 ha were reforested, causing people to doubt about the feasibility of the project (Fazoui 2015). Then, a cumulative area of 100,000 ha was reforested between 2012 and 2017 according to the Moroccan authorities. In addition, and as a proof of awareness by rural people, during the last two decades, the areas dedicated to organic farming significantly increased. In most other areas, trees are labeled as "Wild collection" meaning that inhabitants respect their environment (Azim 2017).

In natural conditions, multiplication of the argan tree is limited because of its slow pace of growth. Propagation from seeds is difficult and argan tree natural renewal is restricted (Lopez Saez and Alba Sanchez 2009). Indeed, germination is influenced by genotypic parameters and environmental conditions, such as light, temperature or soil moisture, this latter being a frequent limiting factor in the argan forest (Alouani and Bani-Aameur 2004). Nonetheless, recent progresses in horticulture science, as well as specific care brought to young plants during the 1st years following plantation, have allowed to get a satisfactory argan tree planting success rate, and as good or better results than those obtained by using complex and alternative multiplication methods such as vegetative propagation or in vitro culture (Bousselmame et al. 2001; Nouaim et al. 2002; Justamante et al. 2017). As a consequence of these horticultural improvements, the argan oil has become the main national organic product, reaching 72% of the Moroccan organic production (Azim 2017).

Argan tree trunk grows gnarly with age. Argan tree leaves are 2 cm long and 0.5 cm wide, on average and generally permanent unless facing strong climatic stress. The argan tree blooms in August, spring or autumn, depending on climatic conditions. Its hermaphrodite flowers are small and yellowish. Argan fruit appears generally after the autumn rains. It ripens in spring and from April to June and argan trees get covered by the fruit that slowly to turn bright yellow between June and September. The fruit is a drupe weighting from 5 to 20 g and consisting of a pericarp and an endocarp (Sandret 1957). The fruit shape may vary depending on each individual tree: oval, spherical, elongated (fusiform), or pointed (apiculate). The pericarp is made of an epicarp (skin) covered by wax and a mesocarp (pulp). The pericarp, traditionally used as feed for goats and sheep, contains soluble sugars, hemicellulose and cellulose, latex, lipids and polyphenols (Kenny and De Zborowski 2007; Chernane et al. 1999; Pioch et al. 2011, 2015a, b). The endocarp contains a nut very rich in lipids whose extraction provides the argan oil.

3   Extraction and Processing of Argan Oil

In the Souss Valley, women have extracted argan oil from argan kernels for centuries following a hand-based method passed down to each generation from mother to daughter. Based on this age-old know-how, in the argan oil project context, a semi-mechanized method was designed 40 years ago to increase the efficiency of the argan oil extraction process and reduce its hardness. Special care was taken to avoid these introduced mechanical improvements altering the fragile equilibrium of the Amazigh society and to fully respect and preserves its traditions. To fulfill these goals the semi-mechanized method was exclusively implanted in woman cooperatives (Charrouf et al. 2011), to continue to acknowledge the ancestral work of women and their specific skill on this subject. This is particularly the case for the breaking of the nutshell without damaging the contained kernels given the shell hardness and variation in shape and size.

3.1 The Argan Fruit Harvest

The argan forest occupies a special place in the Amazigh culture (Simenel 2011) and the property of the fruit is regulated by ancestral and complex rules known by the Amazigh family. In general, fallen fruits belong to everyone (Charrouf et al. 2011). Schematically, traditional argan oil production by Amazigh women culminates between June and September, concomitantly or shortly after the beginning of the argan fruit pick up the season (Charrouf and Guillaume 2011). During this period of time, women (and sometimes kids), daily stride through the argan forest with large wicker baskets and collect ripe fruit that have fallen on the ground, underneath the trees.

3.2 Argan Fruit Processing Prior Oil Extraction

Argan fruit is air-dried for a few days in a ventilated and sunny place. Air drying makes make the initially latex-rich and sticky peel and flesh easier to be discarded, and so the fruit is more prompt to deliver the nuts. Indeed, once the peel is dry, it gets brown and brittle and it can be removed by rubbing it with stones. Nowadays, this step has been mechanized and is efficiently performed by scratching-machines. Need to mention that goat-peeled nuts -whole fruits eaten by goats, then regurgitated without peel and pulp- were sometimes formerly used for argan oil production. This method has been vastly popularized by tourist magazines. However, it is now prohibited in women cooperatives because of the associated bacteriological concerns and quality deterioration of the resulting oil. In the next step, argan nuts have to be broken to free the kernels from their shell. For this, women hold the nuts between thumb and forefinger and violently hit the nut apex with a stone (Charrouf and Guillaume 2011). Nut-breaking is a painstaking process, inexperienced women frequently crushing their fingers with the heavy stone. Nut-breaking is also a thankless task because 40 kg of nuts must be processed to obtain only about 2 kg of kernels.

Collected kernels are then differently processed depending on whether edible or cosmetic argan oil is to be produced. To prepare cosmetic oil, crude kernels are directly processed. To prepare edible oil, kernels have to be roasted for a few minutes in order to temper natural bitterness and provide argan oil with its typical hazelnut flavor. Traditional open-fire roasting was replaced in woman cooperatives with gas- or electric-burners for allowing a more accurate temperature adjustment. Such technology allows the simultaneous roasting of large quantities of kernels and the preparation of an oil of reproducible chemical and organoleptic quality. To prepare cosmetic oil, crude kernels are processed without roasting.

3.3 Argan Oil Extraction

Kernels (roasted or not) need then to be crushed to deliver argan oil. Traditionally, a millstone made up of a bedrock and a rotating cone-shaped grinding stone was used. Such method allowed the formation of an oily dough whose lipid content was extracted after prolonged hand-kneading. Water was added to the dough to facilitate malaxing and the resulting emulsion was decanted to finally deliver argan oil. Using this method, up to half the initial oil content could not be recovered, trapped in the dough. Nowadays, kernel extraction is performed by use of electric endless screw mechanical expellers (Charrouf and Guillaume 2008a). With such optimized instrumentation (Mountasser and El Hadek 1999), the residual amount of oil in the oil cake is very low and the pressing time is reduced by a factor of 3.5. In woman cooperatives where above described mechanized production methods are applied, four liters of argan oil can be obtained from 100 kg of dry fruit, requiring 16–20 h for one operator (Charrouf et al. 2002). Then, argan oil is only filtered after pressing, for clarification. It is not refined thus keeping highly active minor components. Above described improvement of most steps, opened the way to the large-scale production of argan oil, making this technical progress, compared to the traditional method, an important milestone in the Argan Oil Project.

A critical step in the validation of the large scale and semi-mechanized production of argan oil was the certification that press extraction was fully preserving the nutritional and dietary properties traditionally attributed to hand-extracted argan oil. This was achieved in 2005 by analyzing the composition and physico-chemical properties of 21 randomly selected samples of semi-mechanically or traditionally prepared (including goat-peeling) argan oil through the argan forest (Hilali et al. 2005). This study clearly established the similarity of the composition, in terms of fatty acids and minor components, between hand-extracted and semi-mechanically prepared and the superior quality the latter over the former. Consequently, cold-pressed argan oil, that had already been shown to present quality characteristics similar or better than non-linolenate oils (Yaghmur et al. 2000), was found perfectly suitable for shallow or pan-frying as long as it is not repeatedly used (Gharby et al. 2013b). Moreover, these studies showed the enhanced stability of argan oil during deep fat frying, compared to other oils, significantly contributing to improving the shelf life of culinary fried food products (Yaghmur et al. 2000).

4 Fatty Acid Composition and Acyl Lipid Profile of Argan Oil

Argan oil chemical composition has been investigated. Minute variations could exist depending on the tree genotype (El Adib et al. 2015). Early and pioneering chemical work on argan oil was done by Maurin (1992). Acylglycerols (glycerides), 95% of which being triglycerides, constitute 99% of argan oil. Unsaturated fatty

Table 16.1 Distribution of triglycerides in argan oil (%)

Major oleic acid-containing triglycerides		Minor oleic acid-containing triglycerides		Oleic acid-deprived triglycerides	
O, O, L	19.5	S, O, O	3.4	L, L, L	7.4
O, L, L	13.6	P, P, O	3.2	P, L, L	6.3
P, O, L	13.6	S, O, L	3.0	S, L, L	1.8
O, O, O	12.8	P, S, O	1.8	P, P, L	1.6
P, O, O	11.5	S, O, O	3.4	P, S, L	1.6

acids (UFAs) make 80% of the fatty acid fraction. Oleic (monounsaturated) and linoleic (diunsaturated) acids (46–48%, and 31–35%, respectively) are the two main UFAs. Linolenic acid, a triunsatutrated fatty acid, is only present as traces. Saturated fatty acids found in argan oil are palmitic (11–14%) and stearic acid (5–6%). Other fatty acids frequently found in the composition of most seed oils, although in low amounts like myristic, palmitoleic, arachidic, gadoleic, or behenic acids are here only as trace amount in argan oil.

Regarding the influence of the fruit shape, contrasting results were reported. Oil prepared from spherical and elongated fruits would present higher levels of linoleic or oleic acids, respectively than the average value (Belcadi-Haloui et al. 2008; Gharby et al. 2013a). However, because several factors can act on the fatty acid profile of seed oils, the statistical value of these results over several years and places must be confirmed.

The most abundant triglycerides encountered in argan oil constitutes 19.5% of all triglycerides (Table 16.1). It is composed of one linoleic acid and two oleic acid residues (O, O, L). The four others most abundant triglycerides of argan oil all include at least one oleic acid residue in their composition (Maurin et al. 1992). They are composed of three oleic acid residues (O, O, O), two linoleic and one oleic (O, L, L), one palmitic, one oleic, and one linoleic (P, O, L), two oleic and one palmitic (O, O, P). Minor triglycerides also incorporating an oleic acid residue include a saturated acid (palmitic or stearic) located at the sn1,3 positions of glycerol. In triglycerides incorporating a palmitic or stearic acid residue, these later generally esterify the glycerol extremities (sn1,3 positions) (Khallouki et al. 2008). Therefore, unsaturated fatty acids are likely to be the ones which are available for biosynthesis, being the major acids in the sn2 position (oleic, 67% and linoleic, 29%), which is well suitable in the diet.

5 Minor Bioactive Components in Argan Oil

In addition to glycerides, argan oil is composed of 1% of molecules grouped under the generic name of unsaponifiable-matters (Huyghebaert and Hendrickx 1974). This category includes carotenes (37%), triterpenic alcohols (20%), sterols (20%), phenols [polyphenols, tocopherols, coenzyme Q_{10}, and melatonin] (8%), xanthophylls (5%), wax, and traces of aroma compounds (Farines et al. 1984). Despite the low content of unsaponifiable-matters in argan oil, the presence of highly pharma-cologically active compounds like tocopherols or CoQ_{10} (Venegas et al. 2011) has been frequently used to justify, however without real demonstration, its nutritional value and health properties (Charrouf and Guillaume 1999; Khallouki et al. 2005, 2017b; Cabrera-Vique et al. 2012; Lopez et al. 2013).

5.1   Tocopherols

Tocopherols are of particular importance in argan oil owing to their vitamin E activ-ity. Together with polyphenols and CoQ_{10}, tocopherols act as free radical scaven-gers, and contribute to the total strong antioxidant capacity of argan oil (Marfil et al. 2011). The good preservation properties of argan oil are also likely to result from the presence of these molecules that reduce argan oil oxidative degradation during storage (Chimi et al. 1994). Compared to most edible vegetable oils, argan oil is particularly rich in tocopherols (Madawala et al. 2012). Such specificity is incorpo-rated into the Moroccan official norm, which specifies that argan oil must contain between 60 and 90 mg of tocopherols per 100 g (Snima 2003). In two independent studies, argan oil produced from kernels originating from pointed fruits was found particularly rich in γ-tocopherol (Belcadi-Haloui et al. 2008; Gharby et al. 2013a). If no significant influence of the extraction method on the tocopherol content has ever been observed (Marfil et al. 2011), storage conditions influence the tocopherol content and prolonged storage at 60 °C halves the total tocopherol content. γ-Tocopherol has been proposed as the most pharmacologically active molecule of the tocopherol group (Christen et al. 1997). Interestingly, γ-tocopherol is in the greatest proportion (above 70% of tocols) in argan oil whereas α-, β- and δ-tocopherols represent 10% or less (Cayuela et al. 2008). Comparing argan oil tocopherol distribution with that of eight other vegetable oils using chemometric methods, it was shown that argan oil is close to walnut oil, followed by sesame and linseed oils (Rueda et al. 2016).

5.2   Phenolics

Phenolic content in argan oil is low, in the range of 56 ppm (Charrouf and Guillaume 2007). Major identified phenolic compounds are oleuropein and caffeic, vanillic, ferulic, and syringic acids (Khallouki et al. 2003; Rojas et al. 2005). Identified poly-phenols are ubiquitous molecules like resorcinol, epicatechin, and catechin (Charrouf and Guillaume 2007). Phenolic composition of argan kernels has also been investigated. Unroasted contained (−)-epicatechin and (+)-catechin as major polyphenols (0.6 and 0.4 mg/100 g, respectively). Roasted kernels contain a lower amount of polyphenols than unroasted kernels. This reduction was attributed to a temperature-assisted polyphenol oxidation during the heating step (El Monfalouti et al. 2012).

5.3 Phospholipids

Phospholipids are another class of molecules found as minor constituents in argan oil as a consequence of the absence of refining step. Their level is much higher in food grade oil (0.27 mg/100 mg oil) than in cosmetic oil (0.022 mg/100 mg oil). Because the former presents a better profile in terms of preservation capacity, phospholipids have been suggested to be important compounds for this superior oxidative stability, likely acting indirectly, by extending in time the full antioxidant activity of the tocopherol fraction (Gharby et al. 2012a, b).

5.4 Sterols

As phospholipids, sterols are present in very minute quantity in argan oil. Identified sterols are campesterol, stigmasterol, sitosterol, stigmast-7-en-3β-ol, schottenol, and spinasterol (Farines et al. 1981; Madawala et al. 2012). The implication of the sterol fraction in the physiological properties of argan oil is often suggested (Khallouki et al. 2003). However, the ubiquity of the sterols found in argan oil within the vegetal kingdom is not a favorable factor to support a real physiological interest in argan oil unless it is the consequence of a synergistic effect that still needs to be identified.

5.5 Wax

The wax content -a mixture of fatty acids and alcohols- in argan oil ranges between 7 and 95 mg/kg. The average value is 26.4 mg/kg (Cabrera-Vique et al. 2012). Wax does not alter argan oil nutritional properties but they are disfavorably perceived by consumers. Because argan oil is generally filtered after being pressed, wax content fluctuates as a function of the filtration efficiency; it could also be influenced by the extraction method (Cabrera-Vique et al. 2012). Wax formation generally increases above 20 °C, so for a better preservation, storage of argan oil should be performed in the dark, at temperatures below 20 °C and in an oxygen-poor environment (Harhar et al. 2010a). It should be noted that argan oil filtration could lead to a decrease of its oxidative stability (Kartah et al. 2015). Consequently, exessive filtration is not recommended.

5.6 Miscellaneous Organic Compounds

Large variations in CoQ_{10} and melatonin contents have been observed for traditionally prepared argan oil, whereas press-extracted argan oil presents a more stable content. The concentration of CoQ_{10} ranges between 10 and 30 mg/kg oil. The average content of melatonin is 60.5 ng/kg of oil (Venegas et al. 2011).

5.7   Mineral Content

Mineral content in plants or plant-extracted derivatives depends on a large number of factors among which are the growing site conditions, the soil chemical characteristics, anthropogenic pollution induced by industries or the use of fertilizers, and the genotype of the species (Cataldo and Wildung 1978). Because the argan tree is spontaneously growing and no industry is present in the argan forest, the influence of anthropogenic factors on the element content of argan oil can be discarded. Therefore, the element content in argan oil is considered primarily strictly dependent on the argan tree genotype and its growing area. Consequently, argan oil elements content has recently received a lot of attention. This sudden interest is stimulated by the possible implication of some elements in deleterious oxidative reactions as well as their possible use to detect adulteration by precisely identifying the geographical origin of the analyzed samples. Using oil samples prepared in four of the main area, of production of argan oil (Agadir, Ait Baha, Essaouira, and Tarroudant) little correlation has been found between soil element composition and argan oil element content (Mohammed et al. 2013a) suggesting that the argan forest is globally a homogenous area in terms of soil chemical characteristics influencing the metal content. Metal content in argan oil could also be influenced by its preparative process. Variation in element content in press-extracted or traditionally prepared argan oil has also been shown to not be significantly different (Ennoukh et al. 2017) or eventually semi-mechanized argan oil presented a slightly lower element content (Marfil et al. 2008). Element content in argan oil has been found to be remarkably stable over time suggesting its controlled by the tree genotype. Consequently, precise quantitive profile determination of eight elements (Cd, Cr, Cu, Zn, Fe, K, Mg, and Ca) has been proposed as sufficient to ascertain argan oil authenticity (Mohammed et al. 2013a, b).

6   Composition of the Argan Fruit Essential Oil

Seven volatile substances were primarily isolated from the pulp of the argan fruit. Resorcinol was found to represent 73.5% of the isolated products followed by *E*- and *Z*-but-2-enol (12.5% and 6%, respectively). Minor products include 3-methylbutyric acid, and n-octan (Tahrouch et al. 1998). The composition of the essential oil of *A. spinosa* fruit fresh and dried pulp has been investigated. Extraction of essential oils was performed by hydro-/steam-distillation and microwave-assisted extraction. In both cases, camphor was found to be the major (i.e.: 34%) component of the essential oil. In fresh pulp, 1,8-cineole (eucalyptol), endo-borneol, and cyclohex-3-en-ol were found to be in high concentration (16%, 11.8%, and 11.1%, respectively). In the dried pulp, cyclohex-2-ene-1-one was found to be the second most abundant compound (12.6%), and 1,8-cineole (7.8%) the third. Essential oil of argan dried pulp contains more oxidized compounds (Harhar et al. 2010a). In addition, essential oil of *A. spinosa* leaf extracted by steam distillation has evidenced that essential oil is mainly composed of 1,10 di-epi-cubenol (El Kabouss et al. 2002). A recent study in which essential oil was obtained by microwaved or supercritical fluid extraction indicated that cubenol derivatives are the principal components of argan leaf essential oil (El Amrani et al. 2015).

7 Contribution of Bioactive Compounds in Argan Oil to Organoleptic Properties

Food-grade argan oil is a tasty vegetable oil that is not refined. Its organoleptic properties are consequently only set by the quality of the raw material and the production process (Matthäus et al. 2010). Both parameters have undergone intensive studies that have ultimately led to drafting mandatory to follow specifications included in the Geographical Indication in 2010.

Kernel roasting not only removes argan oil bitterness, but it also induces the formation of odorants since cosmetic and dietary oil present a different content. Therefore, edible argan oil contains in very minute quantity some roasting-induced odorants that are essential for the specific flavor of edible argan oil. Identified odorants belong to various chemical groups as alcohols, aldehyde/ketones, ester/lactones, pyridines, pyrazines, sulfurs, terpenes, and ethers (Charrouf et al. 2006). Odorant composition in high-quality argan oil and that prepared from goat-regurgigated kernels presents some differences, confirming the traditional claim that regurgigation gives a special flavor to the resulting oil (Charrouf et al. 2006). A kernel roasting time of 25 min at 110 °C has been found to be optimum for aroma compound formation in dietary oil (El Monfalouti et al. 2013b).

8 Preservation of Argan Oil

As indicated for its organoleptic properties, argan oil shelf life and stability upon storage and use also depend on that of the kernels and on the production process. The quality of argan kernels can be assessed by Vis/Nir measurements (Guinda et al. 2015). However, implementation of such a method is today impossible in woman cooperatives. Fortunately, a combination of studies has allowed the establishment of semi-empiric rules.

For example, harvest date of argan fruit is of little influences on the fatty acid composition of argan oil. However, harvest date influences the minor component content of argan oil, especially its tocopherol and phospholipid content, which is known to be positively linked to the oil oxidative stability. Tocopherol and phospholipid are at a maximal concentration in argan oil produced from kernels collected from fully ripe fruit, generally harvested in July (Harhar et al. 2014). This confirms the ancestral practice according to which fruit must be collected when fallen on the ground, and thus supposedly fully ripe.

Next, fruit is sun-dried to facilitate its peeling, it is subjected to a daily cycle alternating high and low temperatures and exposure to sun light. Therefore, oxidative events that could occur during this period of time must be minimized. A detailed analysis of several oil physicochemical properties has shown that a sun-drying period of 10–14 days is optimum for preparing high-quality oil. Such a time is sufficiently long to allow an easy depulping but short enough to prevent oxidative degradation of argan oil constituents (Harhar et al. 2010b).

Depending on the women cooperative practices, fruit may be stored as raw and depulped later when necessary, or it may be depulped immediately after sun-drying thus enabling to store only clean kernels. Under carefully controlled conditions preventing mold formation, both methods lead to an oil of similar quality (Harhar et al. 2015).

After nut breaking large quantities of argan, kernels are obtained. In some places, those kernels are sometimes stored prior to extraction, for example for extending the cooperative activity out of the harvest time, or for shipping oversea when the oil is to be extracted in a foreign country. Consequently, the question of whether it was better to store argan nuts or argan kernels has also been addressed. If kernel storage is performed at 4 °C, kernel quality is preserved for 1 year, but at room temperature, storage should not be longer than 10 months (Harhar et al. 2010a).

Roasting is the next step of the preparative process. Significant reduction of tocopherols during seed roasting has been previously reported (Gemrot et al. 2006). Therefore, care must be taken to avoid tocopherol degradation during argan kernel roasting. Using a large array of quality attributes like taste, residual moisture, color, peroxide value, peroxidative value, fatty acid composition, tocopherol content, benzopyrene content, and oxidative stability, it was shown that a roasting time of 25–30 min at 110 °C is suitable for optimal taste and preservation (Harhar et al. 2011; El Monfalouti et al. 2013b). The influence of much longer and warmer roasting time has also been studied. Such conditions favor negative attributes as dark coloration (Demnati et al. 2018).

For many years, argan oil preservation capacity has been considered to be very low. This was a consequence of its traditional preparative process that required the use of water, often of poor bacteriological quality. In addition, unsatisfactory storage conditions favored accelerated fatty acid oxidation. However, argan oil prepared by cold pressing kernels suitably stored according to the Geographic Indication recommendations, displays an excellent preservation capacity that is even better than that of ordinary olive oil (Gharby et al. 2012a, b). Interestingly, edible and cosmetic argan oils present different preservation properties, even though product specifications of the cosmetic or food industry are different, and measured through different laboratory standard.

Cosmetic argan oil is more sensitive to oxidation than edible argan oil (Zaanoun et al. 2014). Nevertheless, to appreciate the oxidative degree, metrics used in the cosmetic industry are acid and peroxide values and the oil specific absorbance. Stored at 25 °C and protected from sunlight, cosmetic argan oil quality is still satisfactory after 12 months according to the official Moroccan norm, but storage should not exceed 6 months to fulfill industrial standards (Gharby et al. 2014).

Regarding edible argan oil, extensive studies have led to identifying odor and taste (Matthäus 2013) as the most important quality parameters to satisfy consumers. Therefore, quality argan oil must be free from foreign and rancid odor and taste. Sensory attributes typical for high-quality argan oil are "nutty" and "roasty". These attributes result from the fruit itself from the roasting process, respectively. Negative attributes are rancid, Roquefort-cheese, bitter, wood-like, burnt, musty, yeast-like, fusty. Rancid can come from improper storage. Burnt comes from an excess of roasting. Roquefort-cheese and fusty come principally from goat-peeled fruit. Negative attributes are very rarely mentioned for cold pressed argan oil. However, they appear more rapidly for traditionally prepared oil confirming its low preservation capacity. Except for the burning taste that results from a roasting performed at an elevated temperature, or prolonged for a too long time, most of the negative attributes of argan oil come from an excessive and uncontrolled oxidative process. Therefore, tocopherols, polyphenols and molecules presenting an anti-oxidative activity actively participate to the taste of edible argan oil by slowing the formation of products bringing the rancid attribute (Matthäus et al. 2010).

Processing also influences the preservation properties of edible argan oil. Mild roasting enhances the preservation properties of argan oil since food grade argan oil is significantly less susceptible to oxidation and to develop negative attributes during storage than cosmetic argan oil (Matthäus et al. 2010). Furthermore, in cold-pressed extracted edible oil, no negative organoleptic attributes were detected even after 20 weeks of storage whereas for traditionally extracted edible oil, negative attributes were reported after 12 weeks of storage at 20 °C. In cosmetic oil the development of the unpleasant fusty and Roquefort cheese attributes was quick, covering the perception of the nutty attribute that was only slightly perceivable at the beginning of the experiment (Matthäus et al. 2010).

Argan oil preservation properties are satisfactory as long as the oil is not stored at a temperature above 64 °C (Alaoui et al. 2001). But argan oil is also sensitive to UV irradiation and the combined effect of elevated temperature and irradiation results in a shorter shelf life (Kondratowicz-Pietruszka and Ostasz 2017). However, after 2 years of storage at a maximum temperature of 25 °C, cold pressed edible argan oil protected from sunlight still displays physicochemical and preservation properties similar to those of freshly prepared argan oil. The shelf life of edible argan oil has consequently been set at 2 years when protected from sunlight (Gharby et al. 2011). Because the exact date of preparation of argan oil might be difficult to know, a storage time of 1 year at home is recommended (Matthäus and Brühl 2015).

9   Health Promoting Traits of Argan Oil

Edible argan oil worldwide success is undoubtedly due to its special taste but its pharmacological properties are also essential to explain its appeal to consumers. In parallel, cosmetic argan oil presents important to skin and hair improvement properties. Traditionally claimed properties of edible and cosmetic argan oil and several reviews have been devoted to this topic (Cherki et al. 2006; Charrouf and Guillaume 2008b, 2010, 2012; Rammal et al. 2009; El Monfalouti et al. 2010; Guillaume and Charrouf 2011a, b, 2013; Cabrera-Vique et al. 2012; El Abbassi et al. 2014).

In addition to the traditionally claimed properties of argan oil, several pharmacological properties are attributed to argan oil without bringing a proof of activity. Alleged properties simply come from the presence of above-detailed components found in its chemical composition (Khallouki et al. 2003). This is particularly regretful since the presence of a bioactive compound within a natural product should not be considered as a proof of its pharmacological properties. Such affirmation could result in undue expectations from argan oil consumers and lead to disappointment. Indeed, criteria for absorption, distribution, and toxicity must be considered prior to evoking a pharmacological activity.

Using an in vitro digestion method, the phenolic fraction of argan oil, which is often presented as the probable promotor of most of argan oil properties, has recently been shown to be strongly affected by the digestive process (Rueda et al. 2017). Even though, polyphenols present in argan oil would possess enough bioavailability to induce a high antioxidant potential and health-giving in vivo properties (Seiquer et al. 2015), a direct correlation between the presence of some amount of phenols and a pharmacological activity is impossible to draw, as is a correlation between the quantity of ingested phenols and the intensity of the pharmacological activity.

In addition, some pharmacological properties attributed to this oil have been evidenced only on animals and their relevance in human has never been clearly demonstrated, yet. Consequently, in this chapter, only well-established effects observed on humans or human cells are detailed.

9.1 Hypolipemiant and Antioxidant Effects and the Cardiovascular Risk

Hypolipemiant properties of argan oil were evidenced in a study including a group composed of 96 healthy subjects, 62 of which being regular consumers of argan oil on a daily basis of 15 g (Drissi et al. 2003). Careful analysis of fasting plasma lipids sampled after overnight dietary restriction excepting water, antioxidant vitamins and LDL oxidation susceptibility indicated that subjects consuming argan oil on a regular basis had significantly reduced levels of plasma LDL cholesterol (12.7%) and lipoprotein-a (25.3%), compared with the non-consumer group. Argan oil consumers also presented a significantly lower level of plasma lipoperoxides (58.3%). This important study brought the first evidence that regular consumption of argan oil induces a lowering of LDL cholesterol and has potent antioxidant properties in human. Hence, it scientifically confirmed the traditionally claimed preventive properties of argan oil on the cardiovascular risk. Another nutritional study performed on

60 volunteers reinforced these results and additionally demonstrated that consuming 25 g/day of argan oil for 3 weeks leads to a lower plasma triglyceride level in men (Derouiche et al. 2005). Additional cohort studies carried out on human consuming either 15 g/day of argan oil for 4 weeks (Sour et al. 2012), 22.5 g/day for 3 weeks (Haimeur et al. 2013), or 27 g/day for 4 weeks, confirmed the above findings (Eljaoudi et al. 2015). So, it is now accepted that argan oil supplementation reduces total cholesterol, low-density lipoprotein cholesterol, and triglycerides and increases high-density lipoprotein cholesterol levels in human (Ursoniu et al. 2017).

Concomitantly to observe the hypolipemiant and antioxidant properties of argan oil, Drissi et al. (2003) noticed a strong positive correlation between increasing phenolic-extract, sterol, and tocopherol concentrations and the LDL-Lag phase. The importance of the phenolic-extract of argan oil on the prevention of cardiovascular diseases was confirmed by observing that this extract inhibits human low-density lipoprotein oxidation and enhances cholesterol efflux from human THP-1 macrophages (Berrougui et al. 2006).

9.2 Antiproliferative Effect

The polyphenol-sterol fraction of argan oil was also shown to induce an antiproliferative effect on human prostate cancer cell lines (Bennani et al. 2007). An interventional study performed by comparing 2 groups of 30 men fed either with 25 mL/day of argan oil or with 25 mL/day of olive oil for 3 weeks. Biochemical measurements suggested that the antiatherogenic effect and antioxidant status of argan oil in human could result from improving the activity of a group of enzymes, namely paraoxonases, involved in anti-inflammatory, anti-oxidative, anti-atherogenic, anti-diabetic properties (Cherki et al. 2005). Modification of the peroxisomal acyl-CoA oxidase type1 (ACOX1) activity has also recently been proposed to be a key protein involved in some of argan therapeutic properties (Vamecq et al. 2018). However, a study with a cohort of 125 healthy elderly subjets (Ostan et al. 2016) has failed to evidence any improvement on inflammation when argan oil (25 mL/day for 8 weeks) was associated with RISTOMED diet, a diet built to reach the recommended daily requirement of nutrients, vitamins and minerals complying with different cultural patterns (Valentini et al. 2014).

The insulin-sensitizing and anti-proliferative effects of argan oil unsaponifiable fraction were reported using the human HT-1080 fibrosarcoma cell line. After confirming that argan oil unsaponifiable matters does not exhibit any cytotoxic activity toward HTC cells, it was shown that, at a concentration as low as 25 µg mL^{-1}, argan oil unsaponifiable reduces specifically the ability of extracellular signal-regulated kinases 1 and 2 (ERK1/2) to respond to increasing doses of insulin. Interestingly, the protein kinase B (Akt) response, and the insulin-induced activation of mitogen-activated protein kinase 1 and 2 (MEK1/2) remained undisturbed. The different action of argan oil unsaponifiable on MEK1/2 and ERK1/2 activity in response to insulin in HTC cells means that its strong anti-proliferative activity could be mediated by the interruption of signaling cascades at the MEK1/2–ERK1/2 interface (Samane et al. 2006).

The antiproliferative effect of argan oil tocopherols was evaluated using the two classical human prostatic cell lines (DU145 and PC3) and the androgen-sensitive LNCaP cell line (Drissi et al. 2006). The best antiproliferative effect of tocopherols was observed with DU145 and LNCaP cell lines with tocopherol concentration inhibiting growth by 50% compared to the controls of 28 µg/mL and 32 µg/mL, respectively. Due to the promising therapeutic activity evidenced for argan oil, its use to prepare nanoemulsions as vehicles possessing anticancer activity has been investigated. If the concept has been validated, its application in human medicine still needs to be established (Jordan et al. 2012).

9.3   Hypoglycemiant Effect

Not surprisingly, considering the above-mentioned findings, it has also been shown that consumption of 25 mL/day of argan oil for 3 weeks may have an antiathero-genic effect by improving lipids, and the susceptibility of LDL to oxidation in type 2 diabetes patients with dyslipidemia (Oud Mohamedou et al. 2011). Argan oil consumption is, therefore, recommended in the nutritional management of type 2 diabetes. Several studies performed on rats support this recommendation.

9.4   Endocrinal Effect

Some studies on human strongly support the idea that, in addition to cardiovascular protection, argan oil possesses endocrinological properties. Hence, a study performed on 60 young men, one half of each receiving an argan oil diet, and the other an olive oil diet, has shown that argan oil consumption at a 25 mL/day dose for 3 weeks induces a positive action on the androgen hormonal profile of men. The effect, which is similar to that of olive oil, could result from an activation of the hypothalamo-pituitary-testicular axis and/or an induction of steroidogenic proteins by argan oil tocopherols (Derouiche et al. 2013). An independent study performed on 151 postmenopausal women has evidenced that an argan oil supplement at a daily dose of 25 mL for 8 weeks reduces postmenopausal symptoms and significantly increases vitamin E concentration (El Monfalouti et al. 2013a).

9.5   Pain Relief and Burn Healing Effects

A randomized controlled clinical trial performed on 100 patients presenting knee osteoarthritis has permitted to evidence the benefits of argan oil consumption at a dose of 30 mL/day for 8 weeks on the relieve of knee osteoarthritis associated pain

(Essouiri et al. 2017). Pain reduction was evaluated by use of the visual analog scale for pain, the determination of the walking perimeter, and the WOMAC index (Bellamy and Buchanan 1986) and Lequesne indexes (Lequesne et al. 1987). Once again, tocopherols have been supposed to actively participate in this improvement (Essouiri et al. 2017). Topical application of argan oil is traditionally recommended to cure skin burns (Charrouf and Guillaume 1999). Evidence that argan oil would accelerate healing of burn injuries have only been observed in rats, so far (Avsar et al. 2016).

9.6 Animal Health

Out of the human therapeutic domain, argan oil can also be used in veterinary medicine. Antibacterial activity against fish pathologic bacteria has been investigated. Argan oil exhibited a MIC value of 62.5 µL/mL against *Yersinia ruckeri,* the causative agent of enteric redmouth disease in fish (Kumar et al. 2015). Against other fish pathologic agents, a MIC value of 125 µL/mL was calculated against *Vibrio anguillarum, Aeromonas hydrophila* and *Citrobacter freundii.* Against *Edwardsiella tarda* and *Lactococcus garvieae*, the MIC was 250 µL/mL (Öntas et al. 2016). Food-grade argan oil, at a concentration of 1% and 2%, increases the survival rate of Nile tilapia (*Oreochromis niloticus* L.) against *Lactococcus garvieae*. No associated negative effect on fish growth or feed efficiency was found (Baba et al. 2017). Cosmetic argan oil can be used to cure lice infestation in cats, when mixed with apple cider has been shown by Hassan et al. (2017).

9.7 Dermocosmetic Properties of Argan Oil

The role of argan oil in dermocosmetology is also well documented (Guillaume and Charrouf 2011b; Suggs et al. 2014; Vaughn et al. 2018; Lin et al. 2018). In just a few years, cosmetic argan oil has managed to occupy an important place in the list of raw materials used in cosmetics and its rapid rise has even made the argan tree often nicknamed "*Argania cosmetosa*" (Guillaume and Charrouf 2011b). Cosmetic grade argan oil (INCI name: *Argania spinosa* kernel oil, CAS: 223747-87-3) currently prepared by cold-pressing unroasted kernels may also be obtained using supercritical CO_2 (45 °C, the pressure of 400 bar) as a green solvent, without alteration of its quality parameters (Taribak et al. 2013). Cosmetic argan oil can be found in the composition of commercial shampoos, hair conditioners, hand-wash lotion, or repair serum. Cosmetic argan oil safety has been assessed (Charrouf and Guillaume 2018) and validated by the Cosmetic Ingredient Review Expert Panel (Burnett et al. 2017). But a few cases of contact allergy have nevertheless been reported (Astier et al. 2010; Foti et al. 2014; Veraldi et al. 2016; Barrientos et al. 2014; Lauriola and Corazza 2016). It should also be mentioned that one case of hypersensitivity pneumonitis has been reported in cosmetic-industry workers exposed to non-sterile argan cake powder (Paris et al. 2015).

Traditionally, the dermatological use of argan oil was mainly for curing skin pimples, juvenile acne, and chicken pox pustules, for reducing dry skin matters and slow down the appearance of wrinkles, brittle fingernails, but also more specifically in the biomedical field for treating psoriasis, eczema, joint pain, skin inflammation, and scabies, and for healing burns and wounds (Charrouf and Guillaume 1999; Lin et al. 2018). It was also said to have a favorable impact on hair loss and dry hair (Karabacak and Dogan 2014). Recent scientific findings confirm some traditional uses. Skin-protecting properties of argan likely come from its main component palmitic and linoleic acids that have emollient and hydration properties for skin (Chelaru et al. 2016). However, recent studies that have confirmed the activity in the dermocosmetic field have focused on the oil antioxidant, hydration, antiaging, and protection properties on the skin.

The efficacy to reduce the greasiness of oily facial skin has been demonstrated with 20 healthy volunteers who applied argan oil on their forehead and cheeks twice daily for a period of 4 weeks (Dobrev 2007). Skin elasticity resulting from the hydration properties of dietary argan oil has also been shown to occur after repeated applications of cosmetic argan oil in the left volar forearm during 60 days (Boucetta et al. 2014). After 2 months of once a day topical application, argan oil also improved skin moisture (Boucetta et al. 2013). On a panel of 60 postmenopausal women, it was shown that argan oil consumption (25 mL/day for 8 weeks) improve skin hydration due to a reduction of transepidermal water loss and an increase water content of the epidermis (Boucetta et al. 2014).

These hydration properties are also expressed by nanostructured lipid carriers (NLCs) that use argan oil as the liquid lipid phase. Such nanostructures present the advantage to synergistically combine the NLC occlusion and argan oil hydration properties (Tichota et al. 2014). The skin-favorable properties in terms of dermophilicity and moisturizing power have made of argan oil a key component of some liposomes in facilitating the drug accumulation in the dermis (Manca et al. 2016). Accordingly, argan oil-nanoparticles designed for local and cosmetic application containing the anti-inflammatory drugs indomethacin (Badri et al. 2015) or diclofenac (Lococco et al. 2012) have been successfully prepared. Microbiological spoilage of argan oil-containing nanostructures can be performed with 10% w/w propylene glycol or 5% pentylene glycol, these two chemicals not altering nanoemulsion stability (particle size or in Zeta potential) for 120 days (Hommoss 2011).

Cosmetic argan oil is a known depigmenting agent for murine cutaneous melanoma cells that could be used against hyperpigmentation troubles. Indeed, it possesses a skin depigmenting effect through the inhibition of melanin biosynthesis, likely via tocopherols or a synergistic effect involving several components. Argan oil causes melanogenesis associated transcription factor (MITF) phosphorylation which subsequently inhibited the transcription of tyrosinase and DOPAchrome tautomerase, two melanogenic enzymes (Villareal et al. 2013). The regulation of melanogenesis by argan oil in uveal melanoma cells would follow a slightly different pathway, likely acting on the ERK1/2 and Akt pathways (Caporarello et al. 2017).

The effect of argan oil on hair is still poorly documented; however, argan oil is frequently used as a reference for testing hair breaking (Del Campo et al. 2017).

10 Miscellaneous Properties of Argan Oils

Observed anti-corrosion properties of argan oil have been attributed to its high antioxidant content (Afia et al. 2011, 2014). However, DFT calculations have demonstrated that under chosenstandard testing conditions, linoleic acid is more stable than oleic acid and hence is likely to be responsible for the corrosion inhibition efficiency in acidic medium (Gece 2017). The continuously increasing demand for energy and the decreasing petroleum resources has led to the search for renewable and sustainable fuels. However, because of its high market price, argan oil is unlikely to be developed as anti-corrosive agent or as fuel (Belgharza et al. 2014). By-products generated during argan oil preparation are much more actively investigated for industrial purposes.

It has also been shown that addition of argan oil to conventional extenders for cryopreservation as tris egg yolk and skim milk may improve the quality of ram semen during liquid storage at different temperatures. This effect is likely due to the argan oil high antioxidant content (Allai et al. 2015). Finally, edible argan oil supplementation with molasses has been shown to enhance fermentative performance and antioxidant defenses of active dry wine yeast possibly by prevention of membrane damage (Gemero-Sandemetrio et al. 2015).

11 Adulteration and Authenticity

Due to its high market price, adulteration of argan oil has become an important issue that needs to be properly considered (Seidemann 1998). Indeed, cases of adulterated argan oil are frequently reported (Momchilova et al. 2014).

Adulteration can result from at least three fraudulent processes. Indeed, adulteration can result from the marketing of a product coming from a country devoid of traditional culture towards this product. Adulteration can also be the blending of something impure with something authentic. It can also result from the mixing of an inferior article with a superior one of the same kind. Both aspects have to be considered in the case of argan oil.

Geographic Indication that certifies the marketing of "endemic argan oil of Morocco", is aimed at protecting argan oil authenticity by preventing the marketing of argan oil prepared from the fruit of trees grown out of Morocco. This protection is particularly important because several countries have considered, or are currently considering, the culture of argan trees to produce argan oil (Nerd et al. 1994, 1998; Kechairi et al. 2018; Falasca et al. 2018). Identifying the geographical origin is essential for consumers since, for example, argan oils from Morocco and from

Israel present a different content in mono-, di-glycerides, and phospholipids (Yaghmur et al. 1999).

Adulteration of argan oil can also mean marketing a cheap seed oil -like sunflower oil- or a mixture of a cheap seed oil and argan oil while presenting it as high-quality argan oil, or can also mean marketing low-quality argan oil while pretending it is high-quality argan oil. This second aspect is much more complex to tackle particularly considering that, traditional argan oil cannot be removed from the local Moroccan market. To detect adulterations, the Moroccan authorities have edited a norm based on physicochemical parameters (Snima 2003). Selected values do not lead to the market exclusion of traditional argan oil as long as its presents satisfactory quality attributes. In other words, the official norm only certifies that the oil is argan oil, but not cold press-extracted argan oil. Only the respect of Geographic Indication rules certifies the compliance to the strict semi-mechanized preparative process.

Adulteration of argan oil by the addition of another oil can be detected by looking for compounds present in very minute amount of argan oil whereas these compounds are present in high quantity in the contaminating product. The official norm uses campesterol as a chemical marker, a sterol found in very minute quantity in argan oil although relatively abundant in most vegetable oils. Indeed, argan oil contains five sterols: campesterol (0.4%), spinasterol (34–42%), stigma-8,22-dien-3-ol (4–7%), schottenol (42–49%), and stigmasta-7, 24-dien-3-ol (2–7%) (Hilali et al. 2005). By quantitative GC-analysis, it was shown that campesterol quantification enables certifying a 95% purity of argan oil (Guillaume and Charrouf 2007). Other methods have been suggested and other minor compounds have been proposed as markers: stigmastadiene, kaurene and pheophytin-a. A similar detection limit of 5% can be reached (Ourrach et al. 2012).

Triacylglycerol profile analysis, by using HPLC-evaporative light scattering detection, has also been proposed as authentification tool, owing to an improved detection limit (Salghi et al. 2014). Analysis of triacylglycerols by UPHPLC or HPTLC based methods has also been developed for this purpose (Pagliuca et al. 2018).

Certification of the authenticity of edible oils being possible to be based on the singularity of chemical element profile, this method has been successfully applied to ascertain the authenticity of argan oil. The element composition can be determined using ICP-OES (Gonzálvez et al. 2010; Mohammed et al. 2013a). Discriminant analysis can be used to evidence the mixture of argan oil with other vegetable oils (Gonzalvez et al. 2010)

A method, based on the formation of gold nanoparticles and spectrophotometric analysis, has also been proposed to determine total phenolic acids in genuine argan oil samples, and ferulic acid, the main phenolic acid in virgin argan oil, was also used as an adulteration marker (Zougagh et al. 2011). Another proposed method combines midinfrared spectroscopy and chemometrics (Oussama et al. 2012).

Specific methods have also been designed. Electronic nose (e-nose) are systems equipped with an array of chemical sensors. Such instruments can be used to detect oil adulteration (Majchrzak et al. 2018). Accordingly, adulteration of argan oil with sunflower oil by using the combination of a voltammetric e-tongue and an e-nose based has been successfully applied (Bougrini et al. 2014). Adulteration with olive oil can be detected by fluorescence spectroscopy (Addou et al. 2016). Synchronous fluorescence spectroscopy (SFS) has been used to detect adulteration of argan oil by corn oil and then the determination of the corn oil content. Four percent is deemed the lowest concentration of adulteration detected (Stokes et al. 2018).

Even though the argan forest is a homogenous area and Moroccan argan oil is presented as a single product, a method based on FTIR and chemometric tools has been suggested recently to identify the geographic origin of oil samples within the argan forest, enabling to discriminate samples prepared from fruit harvested in Essaouira, Agadir, Tiznit, Taroudant, or Ait Baha areas (Kharbach et al. 2017).

12 Current and Potential Uses of Co-products Generated During Argan Oil Production

The food industry is known to generate a large amount of wastes or co-products (Helkar et al. 2016) which are now being increasingly evaluated as a source of nutraceuticals, in addition to traditional uses. The same applies to the argan oil production chain, which makes available, starting from the upstream, dry-pulp, shell, and press-cake.

12.1 Argan Pulp

The current processing of argan oil generates large quantities of pulp estimated to reach 44,500 tons per year (Guinda et al. 2011), and which is usually used as food for cattle. However, new outputs actively prospect. The first chemical analysis of argan pulp was reported in 1987 by Fellat-Zarrouk et al. (1987).

The pulp is the main constituent of the fresh fruit (48–59%), its content in water and volatiles is close to 75% (Pioch et al. 2011). Apart of hemicellulose and cellulose polysaccharides (Ray et al. 2004; Habibi and Vignon 2005), which are the cell-wall polymeric components, main pulp constituents are: soluble sugars, non polar fractions like lipids and including wax from fruit skin, polyisoprene, and miscellaneous secondary metabolites (Kenny and De Zborowski 2007; Chernane et al. 1999; Pioch et al. 2011, 2015a, b; Charrouf et al. 1991) whose content is influenced by fruit morphotype and growing location (Zhar et al. 2016).

The total water-soluble sugars amount 16–24% (relative to pulp dry weight) depending on fruit shape (Pioch et al. 2011). This content showed a fourfold increase during the last month of ripening, just before fruit fall to the ground. Proteoglycans, whose properties and industrial interest need to be addressed, were also isolated from argan pulp (Aboughe-Angone et al. 2008).

The content of neutral lipids in pulp varies from 4.8% to 5.7% (dry weight), and the most abundant fatty acids identified were linoleic (28–31%), palmitic (24–26%) and linolenic (12–13%) acids, oleic acid amounting only 13–15% (Pioch al. 2011).

Polyisoprene, another pulp component, has been investigated in details (Palu et al. 2011). This elastomer, which amounts up to 3.6% in the dry pulp is a mixture of *trans* and *cis*-1,4 polyisoprene, could find uses as chewing-gum or sealing. Both polyisoprene and lipids were at a maximum concentration just before fruit fall to the ground with a total of 11.9% and 7.4%, respectively. This suggests that to value these compounds, fruit should be harvested on the tree, which is not the current practice.

A total polyphenolic content of 1.5% (dry weight) in pulp has been measured (Khallouki et al. 2015). Its composition includes, in addition to catechin, epicatechin, and rutin that were identified early (Chernane et al. 1999), 13 other phenolics that were more recently identified by LC-tandem MS/MS analysis (Charrouf et al. 2010). In a quantitative study, isoquercitrin, hyperoside, and rutin were identified as the major phenolic compounds with a content of 28.4, 21.1, and 9.8 mg/100 g; respectively (El Monfalouti et al. 2012). Using a HPLC-ESI based method, 32 compounds including catechins (39%), flavonoids (28%), procyanidins (26%), and phenolic acids (7%) were identified in argan fruit pulp, together with aminophenols named arganimide A and argaminolics A–C (Klika et al. 2014, 2015; Khallouki et al. 2015).

A large number of parameters including growing conditions, geographical origin, ripening process, ripening level, storage conditions, and tree genotype can influence phenolic content in argan tree parts (Chernane et al. 1999; El Monfalouti et al. 2012; Khallouki et al. 2017b). Antioxidant properties of these phenolics found in argan pulp could be useful in the colorant industry (Chemchame et al. 2015).

Ripening stage and environmental conditions influence the profile and concentration of pentacyclic triterpenes, another class of compounds identified inargan fruit pulp (Guinda et al. 2011). High levels of oleanolic acid were found in unripe fruit pulp originating from the coastal region whereas those from the inland presented high levels of betulinic acid, in addition to ursolic acid present in both cases.

Other compounds of potentially high economic value were found in solvent extracts of argan fruit pulp (Pioch et al. 2015a, b). In addition to pyrocathecol and amyrin, the CH_2Cl_2 extract from the defatted pulp of unripe fruits was found rich in lupeol (48%).

Hence, argan fruit pulp could be better valorized and bring a substantial contribution to farmer's income. Trials at pilot scale have shown that key steps (drying, mechanical depulping and fractionation) can be carried out efficiently with simple hardware (Pioch et al. 2011, 2015a, b). Interestingly, after extraction of the marketable products, the bagasse, rich in cellulose, sugars, and proteins, would remain available as feed for cattle, hence preserving the traditional pulp use.

The argan nutshell is the next by-product generated during argan oil preparation. It is estimated that 60,000 of tons of argan nutshell are annually discarded in Morocco (Tatane et al. 2018). Traditionally this wood-like material has been burnt, including for roasting kernel. Nowadays, several outputs are identified to use argan shell as a cheap, low-density, and eco-friendly material, either simply under a powder form or after pyrolysis. In the dermocosmetic domain, after being finely crushed, argan shell powder can be introduced in peel-off mask formulations and used as exfoliating cleanser owing to its gentle abrasive properties. Argan shell has also been evaluated as a source of saponins, a group of natural products presenting a large variety of pharmacological properties (Güçlü-Üstündag and Mazza 2007) but its low content (0.01%) doe not pay in favor of uses in the food, dermocosmetic, pharmaceutical industry (Alaoui et al. 2002).

In the agricultural domain argan shell powder can also be used to prepare valuable biochar because of is content in major nutrients (Bouqbis et al. 2016, 2017). The use of argan shell to prepare ultra microporous active carbons useful for CO_2 capture is presently actively investigated (Boujibar et al. 2018); but it could simply be used for any other purposes (i) on the market activated carbon (Ennaciri et al. 2014), and (ii) as natural bio-filler in high-density polyethylene composites of which it decreases the decomposition temperature (Essabir et al. 2015, 2016). Carbonization of argan shell can also yield hard carbon that is an attractive material for negative electrodes in sodium-ion batteries (Dahbi et al. 2017). Improvement of earthen construction materials can also be successfully realized using argan nutshell powder whose addition to compressed earth blocks considerably leads to an increase of the resulting blocks physical and thermal properties, and an increase of their mechanical strength (Tatane et al. 2018).

The residue remaining after argan oil extraction is called the "press-cake". Traditionally-obtained, press cake was used as cattle feed (Rojas et al. 2005). However, the shift to screw-press instead of traditional extraction, resulted in huge amounts of press-cake, being now an abundant biomass whose optimized use is currently investigated.

Chemical composition analysis of argan press cake has confirmed the validity of its traditional use as animal food since it possesses a high-energy content due to the presence of glucides and proteins (Rojas et al. 2005). The bitterness of argan press cake results from the presence of large amounts of saponins, whose main component structure has been investigated and elucidated (Charrouf et al. 1992; Chafchaouni-Moussaoui et al. 2013; Henry et al. 2013). Interestingly, some isolated saponins exhibit cytotoxic activity and inhibit proliferation of LNCaP, DU145, and PC3 human prostatic cell lines. This antiproliferative effect on PC3, the most sensitive cell line, is dose-dependent (Drissi et al. 2006).

Additional chemical investigations of argan press cake have evidence that it is also rich in phenolic compounds (El Monfalouti et al. 2012). Press cake from roasted kernels presents even a sevenfold higher phenolic content than press cake from unroasted kernels (173.4 vs 25.4 mg of gallic acid/kg). Sixteen phenols have been identified in press cake from roasted kernel, epicatechin being the most abundant (110 mg/kg) (Rojas et al. 2005). Phenols are well-known antioxidant molecules. Therefore, the high phenolic content of argan press cake has generated several studies aimed at evaluating its possible use as green steel anticorrosive agent. The results look promising (Afia et al. 2012, 2013).

Argan press cake could also be a cheap source of cellulose for the pharmaceutical and and food domain (Hu al 2016), and even magnetic nanocellulose for the food industry (Benmassaoud et al. 2017).

13   Conclusion

A. spinosa, an endangered species endemic to South-West Morocco, makes a specific and complex agro-sylvo-pastoral system, from financial, environmental and social standpoints. The Argan Oil Project, especially the marked and long-lasting research efforts from national and international teams, has brought a tremendous amount of knowledge on all aspects of argan related topics. This ultimately facilitated the commercialization of argan oil now produced in a sustainable way, on various markets with certified quality and origin.

The cosmetic industry has made argan oil a major ingredient in the composition a range of products, and it famous worldwide as a food ingredient whose use is not restricted to Moroccan cuisine. This quick and rather exceptional development resulted in strong changes in the agricultural sector and within the rural population, argan oil now being devoted to exportation owing to its high price on international markets.

Interestingly this oil extracted from the kernels, and making less than 4% of fresh fruit, provides several by-products now available in large amounts, opening a range of potential uses in agriculture, food and non-food industry, even in human health. Taking advantage of these by-products under the frame of a multiple-product biorefinery would further increase the value of the argan production chain, and reduce its dependency on argan oil market in case of breakdown. Avoiding further degradation of the argan forest and protecting this endangered species when climate evolution is uncertain while meeting the increasing demand of argan products is a challenge that already begins to be addressed (Moukrim et al. 2018).

References

Aabd, N. A., Msanda, F., & Mousadik, A. (2014). Evaluation of variability in argan oil content through different environments and preselection of elite genotypes. *Euphytica, 195*, 157–167.

Aboughe-Angone, S., Nguema-Ona, E., Ghosh, P., et al. (2008). Cell wall carbohydrates from fruit pulp of *Argania spinosa*: Structural analysis of pectin and xyloglucan polysaccharides. *Carbohydrate Research, 343*, 67–72.

Addou, S., Fethi, F., Chikri, M., et al. (2016). Detection of argan oil adulteration with olive oil using fluorescence spectroscopy and chemometrics tools. *Journal of Materials and Environmental Science, 7*, 2689–2698.

Afia, L., Salghi, R., Bammou, L., et al. (2011). Testing natural compounds: *Argania spinosa* kernels extract and cosmetic oil as ecofriendly inhibitors for steel corrosion in 1 M HCl. *International Journal of Electrochemical Science, 6*, 5918–5939.

Afia, L., Salghi, R., Zarrouk, A., et al. (2012). Inhibitive action of argan press cake extract on the corrosion of steel in acidic media. *Portugaliae Electrochimica Acta, 30*, 267–279.

Afia, L., Salghi, R., Zarrouk, A., et al. (2013). Comparative study of corrosion inhibition on mild steel in HCl medium by three green compounds: *Argania spinosa* press cake, kernels and hulls extracts. *Transactions of the Indian Institute of Metals, 66*, 43–49.

Afia, L., Salghi, R., Bammou, L., et al. (2014). Anti-corrosive properties of argan oil on C38 steel in molar HCl solution. *Journal of Saudi Chemical Society, 18*, 19–25.

Ain-Lahout, F., Zunzunegui, M., Barradas, M. C. D., et al. (2013). Climatic conditions and herbovory effects on morphological plasticity of *Argania spinosa*. *Natural Product Communications, 8*, 5–9.

Alados, C. L., & El Aich, A. (2008). Stress assessment of argan (*Argania spinosa* (L.) Skeels) in response to land uses across an aridity gradient: Translational asymmetry and branch fractal dimension. *Journal of Arid Environments, 72*, 338–349.

Alaoui, B. F., Zeriouh, A., & Belkbir, L. (2001). Conservation study of the argan oil by thermogravimetry. *Asian Journal of Chemistry, 13*, 144–150.

Alaoui, A., Charrouf, Z., Soufiaoui, M., et al. (2002). Triterpenoid saponins from the shells of *Argania spinosa* seeds. *Journal of Agricultural and Food Chemistry, 50*, 4600–4603.

Allai, L., Druart, X., Contell, J., et al. (2015). Effect of argan oil on liquid storage of ram semen in Tris or skim milk based extenders. *Animal Reproduction Science, 160*, 57–67.

Alouani, M., & Bani-Aameur, F. (2004). Argan (*Argania spinosa* (L.) Skeels) seed germination under nursery conditions: Effect of cold storage, gibberellic acid and mother-tree genotype. *Annals of Forest Science, 61*, 191–194.

Astier, C., El Alaoui Benchad, Y., Moneret-vautrin, D. A., et al. (2010). Anaphylaxis to argan oil. *Allergy, 65*, 662–663.

Avsar, U., Halici, Z., Akpinar, E., et al. (2016). The effects of argan oil in second-degree burn wound healing in rats. *Ostomy Wound Management, 62*, 26–34.

Azim, K. (2017). Country report: Organic agriculture development in Morocco. Retrieved from http://isofar.org/isofar/index.php/2-uncategorised/88-country-report-organic-agriculture-development-in-morocco.

Baba, E., Acar, Ü., Yilmaz, S., et al. (2017). Pre-challenge and post-challenge haemato-immunological changes in *Oreochromis niloticus* (Linnaeus, 1758) fed argan oil against *Lactococcus garvieae*. *Aquaculture Research, 48*, 4563–4572.

Badri, W., Miladi, K., Eddabra, R., et al. (2015). Elaboration of nanoparticles containing indomethacin: Argan oil for transdermal local and cosmetic application. *Journal of Nanomaterials, 2015*, Article ID 935439.

Barrientos, N., Moreno de Vega, M., & Dominguez, J. (2014). Allergic contact dermatitis caused by argan oil in an infant. *Contact Dermatitis, 71*, 316–317.

Belcadi-Haloui, R., Zekhnini, A., & Hatimi, A. (2008). Comparative study on fatty acid and tocopherol composition in argan oils extracted from fruits of different forms. *Acta Botanica Gallica, 155*, 301–305.

Belgharza, M., El Azzouzi, E. H., Hassanain, I., et al. (2014). Viscosity of rapeseed and argan vegetable oils, and their comparison with the mineral oil. *Advances in Environmental Biology, 8*, 225–227.

Bellamy, N., & Buchanan, W. W. (1986). A preliminary evaluation of the dimensionality and clinical importance of pain and disability in osteoarthritis of the hip and knee. *Clinical Rheumatology, 5*, 231–241.

Benmassaoud, Y., Villasenor, M. J., Salghi, R., et al. (2017). Magnetic/non-magnetic argan press cake nanocellulose for the selective extraction of sudan dyes in food samples prior to the determination by capillary liquid chromatography. *Talanta, 166*, 63–69.

Bennani, H., Drissi, A., Giton, F., et al. (2007). Antiproliferative effect of polyphenols and sterols of virgin argan oil on human prostate cancer cell lines. *Cancer Detection and Prevention, 31*, 64–69.

Berrougui, H., Cloutier, M., Isabelle, M., et al. (2006). Phenolic-extract from argan oil (*Argania spinosa* L.) inhibits human low-density lipoprotein (LDL) oxidation and enhances cholesterol efflux from human THP-1 macrophages. *Atherosclerosis, 184*, 389–396.

Boucetta, K. Q., Charrouf, Z., Aguenaou, H., et al. (2013). Does Argan oil have a moisturizing effect on the skin of postmenopausal women? *Skin Research and Technology, 19*, 356–357.

Boucetta, K. Q., Charrouf, Z., Derouiche, A., et al. (2014). Skin hydration in postmenopausal women: Argan oil benefit with oral and/or topical use. *Przeglad Menopauzalny, 13*, 280–288.

Bougrini, M., Tahri, K., Haddi, Z., et al. (2014). Detection of adulteration in argan oil by using an electronic nose and a voltammetric electronic tongue. *Journal of Sensors, 2014*, Article ID 245831.

Boujibar, O., Souikny, A., Ghamouss, F., et al. (2018). CO_2 capture using N-containing nanoporous activated carbon obtained from Argan fruit shells. *The Journal of Environmental Chemical Engineering, 6*, 1995–2002.

Bouqbis, L., Daoud, S., Koyro, H. W., et al. (2016). Biochar from argan shells: Production and characterization. *International Journal of Recycling of Organic Waste in Agriculture, 5*, 361–635.

Bouqbis, L., Daoud, S., Koyro, H. W., et al. (2017). Phytotoxic effects of argan shell biochar on salad and barley germination. *Agriculture and Natural Resources, 51*, 247–251.

Bousselmame, F., Kenny, L., & Chlyah, H. (2001). Optimizing cultural conditions for in vitro rooting of Argan (*Argania spinosa* L.). *Comptes Rendus de l'Académie des Sciences Paris, (Life Sciences), 324*, 995–1000.

Burnett, C. L., Fiume, M. M., Bergfeld, W. F., et al. (2017). Safety assessment of plant-derived fatty acid oils. *International Journal of Toxicology, 36*, 51S–129S.

Cabrera-Vique, C., Marfil, R., Gimenez, R., et al. (2012). Bioactive compounds and nutritional significance of virgin argan oil – an edible oil with potential as a functional food. *Nutrition Reviews, 70*, 266–279.

Caporarello, N., Olivieri, M., Cristaldi, M., et al. (2017). Melanogenesis in uveal melanoma cells: Effect of argan oil. *International Journal of Molecular Medicine, 40*, 1277–1284.

Cataldo, D. A., & Wildung, R. E. (1978). Soil nd plant factors influencing the accumulation of heavy metals in plants. *Environmental Health Perspectives, 27*, 149–159.

Cayuela, J. A., Rada, M., Pérez-Camino, M. C., et al. (2008). Characterization of artisanally and semiautomatically extracted argan oils from Morocco. *European Journal of Lipid Science and Technology, 110*, 159–1166.

Chafchaouni-Moussaoui, I., Charrouf, Z., & Guillaume, D. (2013). Triterpenoids from *Argania spinosa*: 20 years of research. *Natural Product Communications, 8*, 43–46.

Charrouf, Z., & Guillaume, D. (1999). Ethnoeconomical, ethnomedical, and phytochemical study of *Argania spinosa* (L.) Skeels. *Journal of Ethnopharmacology, 67*, 7–14.

Charrouf, Z., & Guillaume, D. (2007). Phenols and polyphenols from *Argania spinosa*. *American Journal of Food Technology, 2*, 679–683.

Charrouf, Z., & Guillaume, D. (2008a). Argan oil, functional food, and the sustainable development of the argan forest. *Natural Product Communications, 3*, 283–288.

Charrouf, Z., & Guillaume, D. (2008b). Argan oil: Occurrence, composition and impact on human health. *European Journal of Lipid Science and Technology, 110*, 632–636.

Charrouf, Z., & Guillaume, D. (2009). Sustainable development in Northern Africa: The Argan Forest case. *Sustainability, 1*, 1012–1022.

Charrouf, Z., & Guillaume, D. (2010). Should the Amazigh diet (regular and moderate argan-oil consumption) have a beneficial impact on human health? *Critical Reviews in Food Science and Nutrition, 50*, 473–477.

Charrouf, Z., & Guillaume, D. (2011). The rebirth of the argan tree, or how to stop the desert while giving a future to Amazigh women in Morocco. In R. N. Harpelle & B. Muirhead (Eds.), *Long-term solutions for a short-term world* (pp. 71–86). Waterloo: WLU Press.

Charrouf, Z., & Guillaume, D. (2012). Argan oil: Health-minded food and health benefits. *Food Style21, 16*, 57–60. (in japanese).

Charrouf, Z., & Guillaume, D. (2014). Argan oil, the 35-years-of-research product. *European Journal of Lipid Science and Technology, 116*, 1316–1321.

Charrouf, Z., & Guillaume, D. (2018). The argan oil project: Going from utopia to reality in 20 years. *Oilseeds Fats, Crops Lipids, 25*, D209.

Charrouf, Z., Fkih-Tetouani, S., Charrouf, M., et al. (1991). Triterpenes and sterols extracted from the pulp of *Argania spinosa* (L.) Skeels, Sapotaceae. *Plant Medicine Phytotherapy, 25*, 112–117.

Charrouf, Z., Wieruszeki, J. M., Fkih-Tetouani, S., et al. (1992). Triterpenoid saponins from *Argania spinosa*. *Phytochemistry, 31*, 2079–2086.

Charrouf, Z., Guillaume, D., & Driouich, A. (2002). The argan tree: An asset for Morocco. *Biofutur, 220*, 54–57.

Charrouf, Z., El Hamchi, H., Mallia, S., et al. (2006). Influence of roasting and seed collection on argan oil odorant composition. *Natural Product Communications, 1*, 399–403.

Charrouf, Z., Harhar, H., Gharby, S., et al. (2008). Enhancing the value of argan oil is the best mean to sustain the argan grove economy and biodiversity, so far. *Oléagineux, Corps Gras, Lipides, 15*, 269–271.

Charrouf, Z., Hilali, M., Jauregui, O., et al. (2010). Separation and characterization of phenolic compounds in argan fruit pulp using liquid chromatography–negative electrospray ionization tandem mass spectroscopy. *Food Chemistry, 100*, 1398–1401.

Charrouf, Z., Dubé, S., & Guillaume, D. (2011). *L'arganier et l'huile d'argane*. Paris: Glyphe.

Chatibi, S., Araba, A., & Casabianca, F. (2016). Problématique de la labellisation du chevreau de l'arganeraie. Pertinence de la médiation pour la levée des oppositions. *Options Méditerranéennes, 115*, 355–360.

Chelaru, C., Ignat, M., Albu, M., et al. (2016). Chemical characterization of vegetable oils – lemon, lavender and argan. *Revista De Chimie Bucharest, 67*, 1680–1683.

Chemchame, Y., Errabhi, A., & Makhloufi, A. (2015). Optimization of the dyeing conditions for wool fiber with natural indigo using the argan's pulp. *American Journal of Applied Chemistry, 2*, 70–74.

Cherki, M., Derouiche, A., Drissi, A., et al. (2005). Consumption of argan oil may have an antiatherogenic effect by improving paraoxonase activities and antioxidant status: Intervention study in healthy men. *Nutrition, Metabolism and Cardiovascular Diseases, 15*, 352–360.

Cherki, M., Berrougui, H., Drissi, A., et al. (2006). Argan oil: Which benefits on cardiovascular diseases? *Pharmacological Research, 54*, 1–5.

Chernane, H., Hafidi, A., El Hadrami, I., et al. (1999). Composition phénolique de la pulpe des fruits d'Arganier (*Argania spinosa* L. Skeels) et relation avec leurs caractéristiques morphologiques. *Agrochimica, 43*, 137–150.

Chimi, H., Cillard, J., & Cillard, P. (1994). Autoxidation of argan oil *Argania spinosa* L. from Morocco. *Sciences des Aliments, 14*, 117–124.

Christen, S., Woodall, A. A., Shingenaga, M. K., et al. (1997). γ-Tocopherol traps mutagenic electrophiles such as NOx and complements α-tocopherol: Physiological implications. *Proceedings of the National Academy of Sciences of the United States of America, 94*, 3217–3222.

Dahbi, M., Kiso, M., Kubota, K., et al. (2017). Synthesis of hard carbon from argan shells for Na-ion batteries. *Journal of Materials Chemistry, 5*, 9917–9928.

Del Campo, R., Zhang, Y., & Wakeford, C. (2017). Effect of miracle fruit (*Synsepalum dulcificum*) seed oil (MFSO®) on the measurable improvement of hair breakage in women with damaged hair: A randomized, double-blind, placebo-controlled, eight-month trial. *Journal of Clinical and Aesthetic Dermatology, 10*, 39–48.

Demnati, D., Pacheco, R., Martinez, L., et al. (2018). Effect of roasting temperature and time on the chemical composition and oxidative stability of argan (Argania spinosa L.) oils. *European Journal of Lipid Science and Technology*. https://doi.org/10.1002/ejlt.201700136.

Derouiche, A., Cherki, M., Drissi, A., et al. (2005). Nutritional intervention study with argan oil in man: Effects on lipids and apolipoproteins. *Annals of Nutrition & Metabolism, 49*, 196–201.

Derouiche, A., Jafri, A., Driouch, I., et al. (2013). Effets of argan oil and olive oil consumption on the hormonal profile of androgen among healthy adult Moroccan men. *Natural Product Communications, 8*, 51–53.

Dobrev, H. (2007). Clinical and instrumental study of the efficacy of a new sebum control cream. *Journal of Cosmetic Dermatology, 6*, 113–118.

Drissi, A., Girona, J., & Cherki, M. (2003). Evidence of hypolipemiant and antioxidant properties of argan oil derived from the argan tree (*Argania spinosa*). *Clinical Nutrition, 23*, 1159–1166.

Drissi, A., Bennani, H., Giton, et al. (2006). Tocopherols and saponins derived from *Argania spinosa* exert, an antiproliferative effect on human prostate cancer. *Cancer Investigation, 24*, 588–592.

El Abbassi, A., Khalid, N., Zbakh, et al. (2014). Physicochemical characteristics, nutritional properties, and health benefits of argan oil: A review. *Critical Reviews in Food Science and Nutrition, 54*, 1401–1414.

El Adib, S., Aissi, O., Charrouf, Z., et al. (2015). *Argania spinosa* var. *mutica* and var. *apiculata*: Variation of fatty-acid composition, phenolic content, and antioxidant and α-amylase-inhibitory activities among varieties, organs, and development stages. *Chemistry & Biodiversity, 12*, 1322–1338.

El Amrani, A., Cayuela, J. A., & Eddine, J. J. (2015). Composition of Moroccan *Argania spinosa* leaf essential oils isolated by supercritical CO_2, microwave and hydrodistillation. *Journal of Essential Oil Bearing Plants, 18*, 1138–1147.

El Kabouss, A., Charrouf, Z., Faid, M., et al. (2002). Chemical composition and antimicrobial activity of the leaf essential oil of *Argania spinosa* L. Skeels. *Journal of Essential Oil Research, 14*, 147–149.

El Monfalouti, H., Guillaume, Denhez, et al. (2010). Therapeutic potential of argan oil: A review. *The Journal of Pharmacy and Pharmacology, 62*, 1669–1675.

El Monfalouti, H., Charrouf, Z., Belviso, S., et al. (2012). Analysis and antioxidant capacity of the phenolic compounds from argan fruit (*Argania spinosa* (L.) Skeels). *European Journal of Lipid Science and Technology, 114*, 446–452.

El Monfalouti, H., Charrouf, Z., El Hamdouchi, A., et al. (2013a). Argan oil postmenopausal Moroccan women: Impact on the vitamin E profile. *Natural Product Communications, 8*, 55–57.

El Monfalouti, H., Charrouf, Z., Giodano, M., et al. (2013b). Volatile compound formation during argan kernel roasting. *Natural Product Communications, 8*, 33–36.

Eljaoudi, R., Elkabbaj, D., Bahadi, A., et al. (2015). Consumption of argan oil improves antioxidant and lipid status in hemodialysis patients. *Phytotherapy Research, 29*, 1595–1599.

Ennaciri, K., Baçaoui, Sergent, M., et al. (2014). Application of fractional factorial and Doehlert designs for optimizing the preparation of activated carbons from argan shells. *Chemometrics and Intelligent Laboratory Systems, 139*, 48–57.

Ennoukh, F. E., Bchitou, R., Mohammed, F. A., et al. (2017). Study of the effects of extraction methods on argan oil quality through its metal content. *Industrial Crops and Products, 109*, 182–184.

Essabir, H., Achaby, E. M., Hilali, E. M., et al. (2015). Morphological, structural, thermal and tensile properties of high density polyethylene composites reinforced with treated argan nut shell particles. *Journal of Bionic Engineering, 12*, 129–141.

Essabir, H., Bensalah, M. O., Rodrigue, D., et al. (2016). Biocomposites based on argan nut shell and a polymer matrix: Effect of filler content and coupling agent. *Carbohydrate Research, 143*, 70–83.

Essouiri, J., Harzy, T., Benaicha, N., et al. (2017). Effectiveness of argan oil consumption on knee osteoarthritis symptoms: A randomized controlled trial. *Current Rheumatology Reviews, 13*, 231–235.

Falasca, S. L., Pitta-Alvarez, S., & Ulberich, A. (2018). The potential growing areas for *Argania spinosa* (L.) Skeels (Sapotaceae) in Argentinean drylands. http://downloads.hindawi.com/journals/ija/aip/9262659.pdf.

Farines, M., Charrouf, M., & Soulier, J. (1981). The sterols of *Argania* spinosa seed oil. *Phytochemistry, 20*, 2038–2039.

Farines, M., Soulier, J., Charrouf, M., et al. (1984). Étude de l'huile des graines d'*Argania spinosa* (L.) Sapotaceae. II. Stérols, alcools triterpéniques et méthylstérols de l'huile d'argan. *Revue Francaise des Corps Gras, 31*, 443–448.

Fazoui, H. (2015). The impact of market evolution of argan oil on Moroccan arganeraie. *Revista Geologica de America Central, 55*, 199–222.

Fellat-Zarrouk, K., Smoughen, S., & Maurin, R. (1987). Etude de la pulpe du fruit de l'arganier (*Argania spinosa*) du Maroc. *Matières Grasses et Latex, Actes Institute Agronomy Veterinary, 7*, 17–22.

Foti, C., Romita, P., Ranieri, L. D., et al. (2014). Allergic contact dermatitis caused by argan oil. *Contact Dermatitis, 71*, 183–184.

Gece, G. (2017). Theoretical basis for the corrosion inhibition feature of argan oil. *Bulgarian Chemical Communications, 49*, 846–851.

Gemero-Sandemetrio, E., Torrellas, M., Rabena, M. T., et al. (2015). Food-grade argan oil supplementation in molasses enhances fermentative performance and antioxidant defenses of active dry wine yeast. *AMB Express, 5*, 75.

Gemrot, F., Barouh, N., Vieu, J.-P., et al. (2006). Effect of roasting on tocopherols of gourd seeds (*Cucurbita pepo*). *Grasas y Aceites, 57*, 409–414.

Gharby, S., Harhar, H., Guillaume, D., et al. (2011). Oxidative stability of edible argan oil: A two-year study. *LWT - Food Science and Technology, 44*, 1–8.

Gharby, S., Harhar, H., El Monfalouti, et al. (2012a). Chemical and oxidative properties of olive and argan oils sold on the Moroccan market. A comparative study. *Medicine Journal of Nutrition and Metabolism, 5*, 31–38.

Gharby, S., Harhar, H., Guillaume, D., et al. (2012b). The origin of virgin argan oil's high oxidative stability unraveled. *Natural Product Communications, 7*, 621–624.

Gharby, S., Harhar, H., Kartah, B., et al. (2013a). Can fruit-form be a marker for argan oil production? *Natural Product Communications, 8*, 25–28.

Gharby, S., Harhar, H., Kartah, B., et al. (2013b). Chemical changes in extra virgin argan oil after thermal treatment. *Natural Product Communications, 8*, 29–31.

Gharby, S., Harhar, H., Kartah, B., et al. (2014). Oxidative stability of cosmetic argan oil: A one-year study. *Journal of Cosmetic Science, 65*, 81–87.

Gonzalvez, A., Armenta, S., & de la Guardia, M. (2010). Adulteration detection of argan oil by inductively coupled plasma optical emission spectrometry. *Food Chemistry, 121*, 878–886.

Gonzálvez, A., Ghanjaoui, M. E., El Rhazi, M., et al. (2010). Inductively coupled plasma optical emission spectroscopy determination of trace element composition of argan oil. *Food Science and Technology International, 16*, 65–71.

Güçlü-Üstündag, Ö., & Mazza, G. (2007). Saponins: Properties, applications and processing. *Critical Reviews in Food Science and Nutrition, 47*, 231–258.

Guillaume, D., & Charrouf, Z. (2007). Detection of argan oil adulteration using quantitative campesterol GC-analysis. *Journal of the American Oil Chemists' Society, 84*, 761–764.

Guillaume, D., & Charrouf, Z. (2011a). Argan oil. *Alternative Medicine Review, 16*, 275–279.

Guillaume, D., & Charrouf, Z. (2011b). Argan oil and other argan products: Use in dermocosmetology. *European Journal of Lipid Science and Technology, 113*, 403–408.

Guillaume, D., & Charrouf, Z. (2013). Argan oil for nutritional and skin care applications. *H&PC Today, 8*, 28–30.

Guinda, A., Rada, M., Delgado, T., et al. (2011). Pentacyclic triterpenic acids from *Argania spinosa*. *European Journal of Lipid Science and Technology, 113*, 231–237.

Guinda, A., Rada, M., Benaissa, M., et al. (2015). Controlling argan seed quality by NIR. *Journal of the American Oil Chemists' Society, 92*, 1143–1151.

Habibi, Y., & Vignon, M. R. (2005). Isolation and characterization of xylans from seed pericarp of *Argania spinosa* fruit. *Carbohydrate Research, 340*, 1431–1436.

Haimeur, A., Messaouri, H., Ulmann, L., et al. (2013). Argan oil prevents prothrombotic complications by lowering lipid levels and platelet aggregation, enhancing oxidative status in dyslipidemic patients from the area of Rabat (Morocco). *Lipids in Health and Disease, 12*, 107.

Harhar, H., Gharby, S., Ghanmi, M., et al. (2010a). Composition of the essential oil of *Argania spinosa* (Sapotaceae) fruit pulp. *Natural Product Communications, 5*, 935–936.

Harhar, H., Gharby, S., Guillaume, D., et al. (2010b). Effect of argan kernel storage conditions on argan oil quality. *European Journal of Lipid Science and Technology, 112*, 915–920.

Harhar, H., Gharby, S., Kartah, B., et al. (2011). Influence of argan kernel roasting-time on virgin argan oil composition and oxidative stability. *Plant Foods for Human Nutrition, 6,* 163–168.

Harhar, H., Gharby, S., Kartah, B., et al. (2014). Effect of harvest date of *Argania spinosa* fruits on argan oil quality. *Industrial Crops and Products, 56,* 156–159.

Harhar, H., Gharby, S., Guillaume, D., et al. (2015). Influence of argan fruit peel on the quality and oxidative stability of argan oil after prolonged storage. *Emirates Journal of Food and Agriculture, 27,* 522–526.

Hassan, A. S. U., Al-Fayyadh, F. F., & Abdul-Jaleel, Y. S. (2017). Employment of combined-extraction from argan oil and diluted apple cider vinegar (ACV-AO) for refuting pediculosis in cats. *Kufa Journal For Veterinary Medical Sciences, 8,* 203–213.

Helkar, P. B., Sahoo, A. K., & Patil, N. J. (2016). Review: Food industry by-products used as funtionnal food ingredients. *International Journal of Waste Resources, 6,* 248. https://doi.org/10.4172/2252-5211.1000248.

Henry, M., Kowalczyk, M., Maldini, M., et al. (2013). Saponin inventory from *Argania spinosa* kernel cakes by liquid chromatography and mass spectrometry. *Phytochemical Analysis, 24,* 616–622.

Hilali, M., Charrouf, Z., Soulhi, A. E. A., et al. (2005). Influence of origin and extraction method on argan oil physico-chemical characteristics and composition. *Journal of Agricultural and Food Chemistry, 53,* 2081–2087.

Hommoss, A. (2011). Preservative system development for argan oil-loaded nanostructured lipid carriers. *Pharmazie, 66,* 187–191.

Hu, Y., Hamed, O., Salghi, R., et al. (2016). Extraction and characterization of cellulose from agricultural waste argan press cake. *Cellulose Chemistry and Technology, 51,* 263–272.

Huang, P. (2017). Liquid gold: Berber women and the argan oil co-operatives in Morocco. *International Journal of Intangible Heritage, 12,* 140–155.

Huyghebaert, A., & Hendrickx, H. (1974). Quelques aspects chimiques, physiques et technologiques de l'huile d'argan. *Oléagineux, 41,* 29–31.

Jordan, M., Nayel, A., Brownlow, B., et al. (2012). Development and evaluation of tocopherol-rich argan oil-based nanoemulsions as vehicles possessing anticancer activity. *Journal of Biomedical Nanotechnology, 8,* 944–956.

Justamante, M. S., Ibanez, S., Villanova, J., et al. (2017). Vegetative propagation of argan tree (*Argania spinosa* (L.) Skeels) using in vitro germinated seeds and stem cuttings. *Scientia Horticulturae, 225,* 81–87.

Karabacak, E., & Dogan, B. (2014). Natural remedies in hair care and treatment. *Türkderm, 48,* 60–63.

Kartah, B. E., El Monfalouti, H., Harhar, H., et al. (2015). Effect of filtration on virgin argan oil: Quality and stability. *Journal of Materials and Environmental Science, 6,* 2871–2877.

Kechairi, R., Ould Safi, M., & Benmahioul, B. (2018). Etude comparative de deux plantations d'*Argania spinosa* (L.) Skeels (Sapotaceae) dans le Sahara occidental algérien (Tindouf et Adrar). *International Journal of Environmental Studies, 75,* 294–308.

Kenny, L., & De Zborowski, I. (2007). Biologie de l'arganier. In *Atlas de l'arganier et de l'arganeraie* (p. 190). Rabat: Hassan II IAV.

Khallouki, F., Younos, C., Soulimani, R., et al. (2003). Consumption of argan oil (Morocco) with its unique profile of fatty acids, tocopherols, squalene, sterols and phenolic compounds should confer valuable cancer chemopreventive effects. *European Journal of Cancer Prevention, 12,* 67–75.

Khallouki, F., Spiegelhalder, B., Bartsch, H., et al. (2005). Secondary metabolites of the argan tree (Morocco) may have disease prevention properties. *African Journal of Biotechnology, 4,* 381–388.

Khallouki, F., Mannina, L., Viel, et al. (2008). Thermal stability and long-chain fatty acid positional distribution on glycerol of argan oil. *Food Chemistry, 110,* 57–61.

Khallouki, F., Haubner, R., Ricarte, I., et al. (2015). Identification of polyphenolic compounds in the flesh of argan (Morocco) fruits. *Food Chemistry, 179,* 191–198.

Khallouki, F., Eddouks, M., Mourad, A., et al. (2017a). Ethnobotanic, ethnopharmacologic aspects and new phytochemical insights into Moroccan argan fruits. *International Journal of Molecular Sciences, 18*, 2277. https://doi.org/10.3390/ijms18112277.

Khallouki, F., Voggel, J., Breuer, A., et al. (2017b). Comparison of the major polyphenols in mature argan fruits from two regions of Morocco. *Food Chemistry, 221*, 1034–1040.

Kharbach, M., Kamal, R., Bousrabat, M., et al. (2017). Characterization and classification of PGI Moroccan argan oils based on their FTIR fingerprints and chemical composition. *Chemometrics and Intelligent Laboratory Systems, 162*, 182–190.

Klika, K. D., Khallouki, F., & Owen, R. W. (2014). Amino phenolics from the fruit of the argan tree *Argania spinosa* (Skeels L.). *Zeitschrift für Naturforschung. Section C, 69*, 363–367.

Klika, K. D., Khallouki, F., & Owen, R. W. (2015). Carboxy methyl and carboxy analogs argaminolics B and C. *Records of Natural Products, 9*, 597–602.

Kondratowicz-Pietruszka, E., & Ostasz, L. (2017). Oxidation changes in argan oil caused by heating and UV irradiation. *Polish Journal of Commodity Science, 1*, 103–115.

Kumar, G., Menanteau-Ledouble, S., Saleh, M., et al. (2015). *Yersinia ruckeri*, the causative agent of enteric redmouth disease in fish. *Veterinary Research, 46*, 103.

Lauriola, M., & Corazza, M. (2016). Allergic contact dermatitis caused by argan oil, neem oil, and *Mimosa tenuiflora*. *Contact Dermatitis, 75*, 388–390.

Le Polain de Waroux, Y. (2013). The social and environmental context of argan oil production. *Natural Product Communications, 8*, 1–4.

Le Polain de Waroux, Y., & Chiche, J. (2013). Market integration, livelihood transitions and environmental change in areas of low agricultural productivity: A case study from Morocco. *Human Ecology, 41*, 535–545.

Le Polain de Waroux, Y., & Lambin, E. F. (2012). Monitoring degradation in arid and semi-arid forests and woodlands: The case of the argan woodlands (Morocco). *Applied Geography, 32*, 777–786.

Lequesne, M. G., Mery, C., Samson, M., et al. (1987). Indexes of severity for osteoarthritis of the hip and knee. Validation-value in comparison with other assessment tests. *Scandinavian Journal of Rheumatology. Supplement, 65*, 85–89.

Lin, T. K., Zhong, L., & Santiago, J. L. (2018). Anti-inflammatory and skin barrier repair effects of topical application of some plant oils. *International Journal of Molecular Sciences, 19*, 70. https://doi.org/10.3390/ijms19010070.

Lococco, D., Mora-Huertas, C. E., Fessi, H., et al. (2012). Argan oil nanoemulsions as new hydrophobic drug-loaded delivery system for transdermal application. *Journal of Biomedical Nanotechnology, 8*, 843–848.

Lopez Saez, J. A., & Alba Sanchez, F. (2009). Ecología, etnobotánica y etnofarmacología del argán (*Argania spinosa*). *Boletín Latinoamericano y del Caribe de Plantas Medicinales y Aromáticas, 8*, 323–341.

Lopez, L. C., Cabrera-Vique, C., Venegas, C., et al. (2013). Argan oil-contained antioxidants for human mitochondria. *Natural Product Communications, 8*, 47–50.

Lybbert, T. J., Barrett, C. B., & Nar, H. (2002). Market-based conservation and local benefits: The case of argan oil in Morocco. *Ecological Economics, 41*, 125–144.

Lybbert, T. J., Aboudrare, A., Chaloud, D., et al. (2011). Booming markets for Moroccan argan oil appear to benefit some rural households while threatening the endemic argan forest. *Proceedings of the National Academy of Sciences of the United States of America, 108*, 13963–13968.

Madawala, S. R. P., Kochhar, S. P., & Dutta, P. C. (2012). Lipid components and oxidative status of selected speciality oils. *Grasas y Aceites, 63*, 143–151.

Majchrzak, T., Wojnowski, W., Dymerski, T., et al. (2018). Electronic noses in classification and quality control of edible oils: A review. *Food Chemistry, 246*, 192–201.

Manca, M. L., Matricardi, P., Cencetti, C., et al. (2016). Combination of argan oil and phospholipids for the development of an effective liposome-like formulation able to improve skin hydration and allantoin dermal delivery. *International Journal of Pharmaceutics, 505*, 204–211.

Marfil, R., Cabrera-Vique, C., Gimenez, R., et al. (2008). Metal content and physicochemical parameters used as quality criteria in virgin argan oil: Influence of the extraction method. *Journal of Agricultural and Food Chemistry, 56*, 7279–7284.

Marfil, R., Gimenez, R., Martinez, et al. (2011). Determination of polyphenols, tocopherols, and antioxidant capacity in virgin argan oil (*Argania spinosa*, Skeels). *European Journal of Lipid Science and Technology, 11*, 886–893.

Matthäus, B. (2013). Quality parameters for cold pressed edible argan oils. *Natural Product Communications, 8*, 37–41.

Matthäus, B., & Brühl, L. (2015). Quality parameters for the evaluation of cold-pressed edible argan oil. *Journal für Verbraucherschutz und Lebelnsmittelsicherheit, 10*, 143–154.

Matthäus, B., Guillaume, D., Gharby, S., et al. (2010). Effect of processing on the quality of edible argan oil. *Food Chemistry, 120*, 426–432.

Maurin, R. (1992). Argan oil *Argania spinosa* (L.) Skeels, Sapotaceae. *Revue Francaise des Corps Gras, 39*, 139–146.

Maurin, R., Fellat-Zarrouck, K., & Ksir, M. (1992). Positional analysis and determination of triacylglycerol structure of *Argania spinosa* seed oil. *Journal of the American Oil Chemists' Society, 69*, 141–145.

McGregor, H. V., Dupont, L., Stuut, J. B., et al. (2009). Vegetation change, goats, and religion: A 2000-year history of land use in southern Morocco. *Quaternary Science Reviews, 28*, 1434–1448.

Mellado, J. (1989). SOS Souss: Argan forest destruction in Morocco. *Oryx, 23*, 87–93.

Mohammed, F. A. E., Bchitou, R., Bouhaouss, A., et al. (2013a). Can the dietary element content of virgin argan oils really be used for adulteration detection? *Food Chemistry, 136*, 105–108.

Mohammed, F. A., Bchitou, R., Boulmane, M., et al. (2013b). Modeling of the distribution of heavy metals and trace elements in argan forest soil and parts of argan tree. *Natural Product Communications, 8*, 21–23.

Momchilova, S. M., Taneva, S. P., Dimitrova, R. D., et al. (2014). Evaluation of authenticity and quality of argan oils sold on the Bulgarian market. *Rivista Italiana Delle Sostanze Grasse, 93*, 95–103.

Morton, J. F., & Voss, G. L. (1987). The argan tree (*Argania sideroxylon*, sapotaceae), a desert source of edible oil. *Economic Botany, 41*, 221–233.

Moukrim, S., Lahssıni, S., Rhazi, M., et al. (2018). Climate change impacts on potential distribution of multipurpose agro-forestry species: *Argania spinosa* (L.) Skeels as case study. *Agroforestry Systems*. https://doi.org/10.1007/s10457-018-0232-8.

Mountasser, A., & El Hadek, M. (1999). Optimisation of factors influencing the extraction of argan oil through pressing. *Oléagineux Corps Gras Lipides, 6*, 273–279.

Nerd, A., Eteshola, E., Borowy, N., et al. (1994). Growth and oil production of argan in the Negev desert of Israel. *Industrial Crops and Products, 2*, 89–95.

Nerd, A., Irijimovich, V., & Mizrahi, Y. (1998). Phenology, breeding system and fruit development of argan (*Argania spinosa*, Sapotaceae) cultivated in Israel. *Economic Botany, 52*, 161–167.

Nouaim, R., Mangin, G., Breuil, et al. (2002). The argan tree (*Argania spinosa*) in Morocco: Propagation by seeds, cuttings and in-vitro techniques. *Agroforestry Systems, 5*, 71–81.

Öntas, C., Baba, E., Kaplaner, E., et al. (2016). Antibacterial activity of *citrus limon* peel essential oil and *Argania spinosa* oil against fish pathogenic bacteria. *Kafkas Üniversitesi Veteriner Fakültesi Dergisi, 22*, 741–749.

Ostan, R., Béné, M. C., Spazzafumo, L., et al. (2016). Impact of diet and nutraceutical supplementation on inflammation in elderly people. Results from the RISTOMED study, an open-label randomized control trial. *Clinical Nutrition, 35*, 812–818.

Oud Mohamedou, M. M., Zouirch, K., El Messal, M., et al. (2011). Argan oil axerts an antiatherogenic effect by improving lipids and susceptibility of LDL to oxidation in type 2 diabetes patients. *International Journal of Endocrinology, 2012*, 747835.

Ourrach, I., Rada, M., Pérez-Camino, M. C., et al. (2012). Detection of argan oil adulterated with vegetable oils: New markers. *Grasas y Aceites, 63*, 355–364.

Oussama, A., Elabadi, F., & Devos, O. (2012). Analysis of argan oil using infrared spectroscopy. *Spectroscopy Letters, 45*, 458–463.

Pagliuca, G., Bozzi, C., Gallo, F. R., et al. (2018). Triacylglycerol "hand-shape profile" of argan oil. Rapid and simple UHPLC-PDA-ESI-TOF/MS and HPTLC methods to detect counterfeit argan oil and argan-oil-based products. *Journal of Pharmaceutical and Biomedical Analysis, 150*, 121–131.

Palu, S., Pioch, D., Suchat, S., et al. (2011). Valorisation of the argan fruit pulp: A source of latex? (pp. 453–461). Proceedings of the 1st international congress on argan tree, Agadir, Morocco.

Paris, C., Herin, F., Reboux, G., et al. (2015). Working with argan cake: A new etiology for hypersensitivity pneumonitis. *BMC Pulmonary Medicine, 15*, 18.

Pioch, D., Buland, F.-N., Pingret de Sousa, D., et al. (2011). Valorization of the pulp of *Argania spinosa* L., processing and products; towards an optimized valorization of the Argan fruit? In *International congress on Argan tree, proceedings of the 1st international congress on argan tree* (pp. 215–222). Agadir.

Pioch, D., Buland, F.-N., Pingret de Sousa, D., et al. (2015a). The pulp of Argan fruit: A valuable but highly variable potential agro-industrial feedstock. In *Proceedings of the 2nd international congress on Argan tree* (pp. 289–293). Agadir.

Pioch, D., Palu, S., Buland, F.-N., et al. (2015b). Towards an optimized valorization of *Argania spinosa* (L.) Skeels fruit pulp; an alternative to the current weak single-product argan kernel oil Production. In *Proceedings of the 2nd international congress on Argan tree* (pp. 311–317). Agadir.

Rammal, H., Bouayed, J., Younos, C., et al. (2009). Notes ethnobotanique et pharmacologique d'*Argania spinosa* L. *Phytothérapie, 7*, 157–160.

Ray, B., Loutelier-Bourhis, C., Lange, C., et al. (2004). Structural investigation of hemicellulosic polysaccharides from *Argania spinosa*: Characterisation of a novel xyloglucan motif. *Carbohydrate Research, 339*, 201–208.

Rojas, L. B., Quideau, S., Pardon, P., et al. (2005). Colorimetric evaluation of phenolic content and GC-MS characterization of phenolic composition of alimentary and cosmetic argan oil and press cake. *Journal of Agricultural and Food Chemistry, 53*, 9122–9127.

Roumane, A. (2017). Cooperatives and the economic value of biodiversity: The case of the argan oil sector in Morocco. *RECMA, 346*, 59–72.

Ruas, M. P., Ros, J., Terral, J. F., et al. (2016). History and archaeology of the emblematic argan tree in the medieval Anti-Atlas Mountains (Morocco). *Quaternary International, 404*(A), 114–136.

Rueda, A., Seiquer, I., Olalla, M., et al. (2014). Characterization of fatty acid profile of argan oil and other edible vegetable oils by gas chromatography and discriminant analysis. *Journal of Chemistry, 2014*, Article ID 843908.

Rueda, A., Samniego-Sanchez, C., Olalla, M., et al. (2016). Combination of analytical and chemometric methods as a useful tool for the characterization of extra virgin argan oil and other edible virgin oils. Role of polyphenols and tocopherols. *Journal of AOAC International, 99*, 489–494.

Rueda, A., Canterero, S., Seiquer, I., et al. (2017). Bioaccessibility of individual phenolic compounds in extra virgin argan oil after simulated gastrointestinal process. *LWT – Food Science and Technology, 75*, 466–472.

Salghi, R., Armbuster, W., & Schwack, W. (2014). Detection of argan oil adulteration with vegetable oils by high-performance liquid chromatography-evaporative light scattering detection. *Food Chemistry, 153*, 387–392.

Samane, S., Noel, J., Charrouf, Z., et al. (2006). Insulin-sensitizing and anti-proliferative effects of *Argania spinosa* seed extracts. *Evidence-based Complementary and Alternative Medicine, 3*, 317–327.

Sandret, F. (1957). La pulpe d'argane, composition chimique et valeur fouragère: variation en cours de maturation. *Annales de la Recherche Fôrestière au Maroc, 1*, 151–157.

Seidemann, J. (1998). Falsification of fatty oil of *Argania spinosa* (L.) Skeel. *Deutsche Lebensmittel-Rundschau, 94*, 26–27.

Seiquer, I., Rueda, A., Olalla, M., et al. (2015). Assessing the bioavailability of polyphenols and antioxidant properties of extra virgin argan oil by simulated digestion and Caco-2 cell assays. Comparative study with extra virgin olive oil. *Food Chemistry, 188*, 496–503.

Simenel, R. (2011). How to domesticate a forest without the men? An ethno-historical ecology of the argan forest in southwest Morocco. *Techniques & Culture, 56*, 224–247.

Simenel, R., Michon, G., Auclair, L., et al. (2009). L'argan: l'huile qui cache la forêt domestique. De la valorisation du produit à la naturalisation de l'écosystème. *Autrepart, 50*, 51–74.

Snima (Service de normalisation industrielle marocaine). (2003). *Huiles d'argane. Spécifications.* Norme marocaine NM 08.5.090. Rabat.

Sour, S., Belarbi, M., Khalid, D., et al. (2012). Argan oil improves surrogate markers of CVD in humans. *British Journal of Nutrition, 107*, 1800–1805.

Stokes, T. D., Foteini, M., Brownfield, B., et al. (2018). Feasibility assessment of synchronous fluorescence spectral fusion by application to argan oil for adulteration analysis. *Applied Spectroscopy, 72*, 432–441.

Suggs, A., Oyetakin-White, P., & Baron, E. D. (2014). Effect of botanicals on inflammation and skin aging: Analyzing the evidence. *Inflammation & Allergy Drug Targets, 13*, 168–176.

Tahrouch, S., Rapior, S., Bessière, J. M., et al. (1998). Les substances volatiles de *Argania spinosa* (Sapotaceae). *Acta Botanica Gallica, 145*, 259–263.

Taribak, C., Casas, L., Mantell, C., et al. (2013). Quality of cosmetic argan oil extracted by super-critical fluid extraction from *Argania spinosa* L. *Journal of Chemistry, 2013*, Article ID 408194.

Tatane, M., Elminor, H., Ayeb, M., et al. (2018). Effect of argan nut shell powder on thermal and mechanical behavior of compressed earth blocks. *International Journal of Applied Engineering Research, 13*, 4740–4745.

Tichota, D. M., Silva, A. C., Sousa, L., et al. (2014). Design, characterization, and clinical evaluation of argan oil nanostructured lipid carriers to improve skin hydration. *International Journal of Nanomedicine, 9*, 3855–3864.

Turner, B. (2016). Supply-chain legal pluralism: Normativity as constitutive of chain infrastructure in the Moroccan argan oil supply chain. *Journal of Legal Pluralism and Unofficial Law, 48*, 378–414.

Ursoniu, S., Sahebkar, A., Serban, M. C., et al. (2017). The impact of argan oil on plasma lipids in humans: Systematic review and meta-analysis of randomized controlled trials. *Phytotherapy Research, 32*, 377–383.

Valentini, L., Pinto, A., Bourdel-Marchasson, I., et al. (2014). Impact of personalized diet and probiotic supplementation on inflammation, nutritional parameters and intestinal microbiota – the "RISTOMED project": Randomized controlled trial in healthy older people. *Clinical Nutrition, 34*, 593–602.

Vamecq, J., Andreoletti, P., El Kebbaj, R., et al. (2018). Peroxisomal acyl-coA oxidase type 1: Anti-inflammatory and anti-aging properties with a special emphasis on studies with LPS and argan oil as a model transposable to aging. *Oxidative Medicine and Cellular Longevity, 2018*, 6986984.

Vaughn, A. R., Clark, A. K., Sivamani, R. K., et al. (2018). Natural oils for skin-barrier repair: Ancient compounds now backed by modern science. *American Journal of Clinical Dermatology, 19*, 103–117.

Venegas, C., Cabrera-Vique, C., Garcia-Corzo, L., et al. (2011). Determination of coenzyme Q_{10}, coenzyme Q_9, and melatonin contents in virgin argan oils: Comparison with other edible vegetable oils. *Journal of Agricultural and Food Chemistry, 59*, 12102–12108.

Veraldi, S., Mascagni, P., Tosi, D., et al. (2016). Allergic contact dermatitis caused by argan oil. *Dermatitis, 27*, 391.

Villareal, M. O., Kume, S., Bourhim, T., et al. (2013). Activation of MITF by argan oil leads to the inhibition of the tyrosinase and dopachrome tautomerase expressions in B16 murine melanoma cells. *Evidence-Based Complementary and Alternative Medicine, 2013*, 340107.

Yaghmur, A., Aserin, A., Mizrahi, A., et al. (1999). Arganoil-in-water emulsions: Preparation and stabilization. *Journal of the American Oil Chemists' Society, 76*, 15–18.

Yaghmur, A., Aserin, A., & Garti, N. (2000). Evaluation of argan oil for deep-fat frying. *Lebensmittel-Wissenschaft und-Technologie, 34*, 124–130.

Zaanoun, I., Gharby, S., Bakass, I., et al. (2014). Kinetic parameter determination of roasted and unroasted argan oil oxidation under Rancimat test conditions. *Grasas y Aceites, 65*, e33.

Zhar, N., Naamani, K., Dihazi, A., et al. (2016). Comparative analysis of some biochemical param-
    eters of argan pulp morphotypes (*Argania spinosa* (L) Skeels) during maturity and according to
    the continentality in Essaouira region (Morocco). *Physiology and Molecular Biology of Plants,
    22,* 361–370.
Zougagh, M., Salghi, R., Dhair, S., et al. (2011). Nanoparticle-based assay for the detection of
    virgin argan oil adulteration and its rapid quality evaluation. *Analytical and Bioanalytical
    Chemistry, 399,* 2395–2405.
Zunzunegui, M., Boutaleb, S., Díaz Barradas, M. C., et al. (2017). Reliance on deep soil water
    in the tree species *Argania spinosa. Tree Physiology.* https://doi.org/10.1093/treephys/tpx152.

Chapter 17
Avocado (*Persea americana* Mill.) Oil

Chin Xuan Tan and Hasanah Mohd Ghazali

Abstract Avocado oil is growing in popularity as a source of specialty oil with health-promoting properties. Unlike typical fruit oils, avocado oil is extracted from the pulp instead of the seed and can be consumed in its crude form without the necessity of refining. The oil can be graded into extra virgin, virgin or pure, depending on the extraction methods and conditions. Over the last decades, several methods have been developed on avocado oil extraction and these methods are summarized and discussed. The intake of avocado oil is recommended in order to gain the full benefit of essential nutrients and health-promoting minor bioactive lipids that they contain, along with their desirable aroma and taste. Lately, much interest in the health benefits of avocado oil has led to numerous animal and human intervention studies. The therapeutic effects and other issues associated with avocado oil such as oxidative stability, authenticity and toxicity, are also compiled and highlighted.

Keywords Avocado oil · *Persea americana* · Functional oil

1 Introduction

Avocado (*Persea americana* Mill.) is an evergreen dicotyledonous plant from the Lauraceae family, which encompasses approximately 45 genera and 2850 species (Christenhusz and Byng 2016). The avocado plant has been cultivated in Central America since the pre-Columbian period. The plant grows in tropical or sub-tropical countries and can reach up to 30 m height when fully grown. The leaves are 15–25 cm long with well-developed petioles that are spirally arranged near the branch ends (Afahkan 2012). More than 500 varieties of avocado have been identified and the fruiting seasons vary according to the variety (Yahia and Woolf 2011). Avocado fruit begins to ripen once it is detached from the tree. The largest avocado

C. X. Tan · H. M. Ghazali (✉)
Department of Food Science, Faculty of Food Science and Technology,
Universiti Putra Malaysia, Serdang, Selangor, Malaysia
e-mail: hasanah@upm.edu.my

© Springer Nature Switzerland AG 2019
M. F. Ramadan (ed.), *Fruit Oils: Chemistry and Functionality*,
https://doi.org/10.1007/978-3-030-12473-1_17

fruit producing country in 2016 was Mexico (1889 thousand tons), followed by Dominican Republic (601 thousand tons) and Peru (455 thousand tons) (FAO 2017).

An avocado fruit can be divided into three anatomical parts, namely, peel, pulp and seed. The edible yellowish pulp constitutes the largest proportion of the fruit (65%) whereas the peel (15%) and the seed (20%) constitute the rest (Costagli and Betti 2015). Unlike typical fruits with acidic and/or sweet taste, avocado pulp has a smooth and butter-like consistency taste. It has been reported that avocado pulp contains abundant amounts of lipids and essential minerals like calcium, phosphorus, magnesium and potassium (USDA 2011). Owing to a high lipid content (15–36%) in the pulp (Maitera et al. 2014; USDA 2011), avocado fruit is also known as 'butter fruit' and 'vegetable butter'.

Plants with high amounts of lipids in either the pulp nor the seed can serve as an important raw material for edible oil extraction. One of the industrial usages of avocado fruit is the production of avocado oil from its fleshy pulp. Unlike other plant oils, avocado oil is extracted from the pulp instead of the seed as the seed contains hepatotoxic agents and a small quantity of oil (<2%) (Qin and Zhong 2016). Recently, avocado oil has been promoted as functional oil because of its high concentrations of oleic monounsaturated fatty acid and minor bioactive lipids, which are associated with human health.

2 Proximate Composition of Avocado Pulp

The proximate composition of avocado pulp has been reported by the United States Department of Agriculture (USDA 2011) as follows: moisture, 72.33%; crude oil, 15.41%; crude protein, 1.96%; dietary fiber, 6.80%; ash, 1.66%; carbohydrate (by difference), 8.64%. The oil content is comparable to the levels reported by Bora et al. (2001) while Maitera et al. (2014) reported a higher oil content (36.40%) in the pulp of avocado fruit grown in the tropical region of Nigeria.

3 Extraction and Processing of Avocado Oil

The initial step in avocado oil production involves peeling and destoning of the ripe fruit. Avocado pulp has a moderately uniform cellular composition, consisting of thin-walled parenchyma cells and thick-walled polyhedral- or round-shaped idioblast cells (Fig. 17.1). The latter are evenly distributed and are surrounded by parenchyma cells in a circular arrangement. The parenchyma cells contain finely dispersed oil emulsion whereas the idioblast cells contain a large droplet of oil sac (Qin and Zhong 2016). Both are the oil-bearing cells in the avocado pulp. Depending on the extraction method used, the avocado pulp may need to be dehydrated first before it is subjected to oil extraction. Avocado oil can be used in the crude form without the

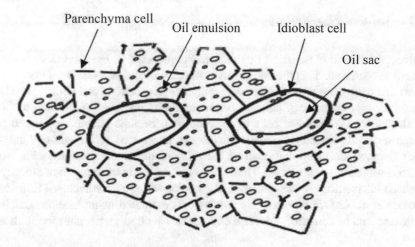

Fig. 17.1 Cellular structure of avocado pulp

Table 17.1 Classifications of avocado oil

	Extra virgin	Virgin	Pure
Fruit properties	Healthy or fruit with minor levels of physiological disorders and rots	Healthy or fruit with some levels of physiological disorders and rots	NR
Extraction conditions	Natural or mechanical approaches at temperatures <50 °C and without undergoing refining	Natural or mechanical approaches at temperatures <50 °C and without undergoing refining	NR
Flavor and odor	>40% grassy and mushroom or butter with some smoky	>20% grassy and mushroom or butter with some smoky	Bland or matches description of flavor
Oil color	Intense green	Green or yellow	Pale yellow
Smoke point (°C)	≥250	≥200	≥250
Acid value (%)	≤1.0	≤2.0	≤0.2
FFA (% oleic acid)	≤0.5	0.8–1.0	≤0.1
PV (meq O_2/kg)	≤4.0	<8.0	<0.5
Stability[a] (months)	24	18	>24

Woolf et al. (2009)
NR Not reported, *FFA* free fatty acids, *PV* peroxide value
[a]Stability: the oil flushed with nitrogen and stored in the dark area at ambient temperature

necessity of refining and can be graded into extra virgin, virgin or pure, depending on the extraction methods and conditions (Table 17.1). A number of studies demonstrating the extraction and processing of avocado oil have been undertaken and these are outlined below.

3.1 Solvent Extraction

Solvent extraction is the most common technique used at the laboratory scale to extract avocado oil. This method involves the use of heat or agitation. Organic solvents that have been used are typically nonpolar (e.g. hexane and petroleum ether) and/or alcohol-based (e.g. ethanol, isopropanol and acetone). According to Ortiz et al. (2004), the idioblast cells of avocado pulp became irregularly shaped and rough-surfaced after oil extraction with hexane. In contrast, deformation of the cellular structure occurred and most of the oil held inside the idioblast cells when acetone was used to extract oil. These observations imply the extraction efficacy of avocado oil is strongly dependent on the choice of organic solvents. Meanwhile, Mostert et al. (2007) reported that avocado oil extracted using hexane contained large amounts of non-triacylglycerol components such as gums, phospholipids and waxes.

Solvent extraction using a Soxhlet extractor, also known as Soxhlet extraction, is a standard technique for analyzing the lipid content in foods and the main reference method for evaluating the performance of new oil extraction alternatives (Wang and Weller 2006). Soxhlet extraction outperforms other conventional extraction techniques except for, in a limited field of applications, the extraction of thermolabile bioactive compounds (Azmir et al. 2013). The Association of Official Analytical Chemists (AOAC) official method for extraction of crude avocado oil has been applied in many studies (Abaide et al. 2017; Mostert et al. 2007; Tan et al. 2018a; Yanty et al. 2011). This standardized method extracts avocado oil using either petroleum ether or hexane in a Soxhlet apparatus for 8–12 h at 60–70 °C before recovering the oil using evaporation. Soxhlet extraction has been used to determine the theoretical maximum overall extraction yield of avocado oil (Corzzini et al. 2017).

Santana et al. (2015) employed an agitation-based solvent method to extract avocado oil. They mixed dried avocado pulp with the solvent [ethanol at 60 °C and pulp-to-solvent ratio of 1:4 (w/w) or petroleum ether at 45 °C and pulp-to-solvent ratio of 1:3 (w/w)] and agitated the mixture using a shaking water bath for an hour, followed by oil recovery using evaporation. Their study showed that the use of petroleum ether was capable of extracting a greater amount of oil yield than ethanol. However, this disagrees with Gatbonton et al. (2013), who reported that the use of alcohol-based organic solvents was better able at extracting a greater oil yield compared to the use of nonpolar organic solvents. The differences might be attributable to the extraction conditions such as extraction temperature and time.

Solvent extraction in oil manufacturing is acknowledged by the US Environmental Protection Agency (US EPA) as being the main contributor to hazardous air pollution (Gunstone 2011). The National Emission Standards for Hazardous Air Pollutants (NESHAP) for oil production has since been established. Moreover, the oil derived from solvent extraction contain residual solvent, which indicates the need of a refining step for further applications such as edible oil, cosmetic and pharmaceutical uses (Yahia and Woolf 2011).

3.2 Expeller Pressing

Expeller pressing is a mechanical approach to extracting oil from plant materials. This method is widely adopted in extracting oil from plant materials with high oil content such as peanut, flax seed, sunflower seed, palm kernel and cottonseed (Gunstone 2011). Traditionally, the high moisture content of avocado pulp is dehydrated first before it is subjected to expeller pressing using a screw press device (Southwell et al. 1990). A screw press device consists of a barrel made of narrowly spaced bars, in which a worm shaft rotates and presses the plant materials (Chapius et al. 2014). This device utilizes the continuous pressure and friction from the screw drives to compress the avocado pulp through a caged barrel-like cavity. Avocado oil is then expelled and seeps through the small openings into a container. Santana et al. (2015) reported the yield of avocado oil obtained from screw pressed was in the range of 55.7–61.2%. Frictional heat is generated when the worm shaft slides against the extracted materials and this might affect the quality of the extracted oil (Chapius et al. 2014).

3.3 Centrifugation

Centrifugation involves the application of centrifugal force to separate a suspension or mixture of liquid and solid particles into the supernatant and pellet through spinning. Extraction of avocado oil using this approach was first reported by Werman and Neeman (1987). They combined raw avocado pulp with water at a ratio of 1:3 (w/w) and agitated the mixture under different conditions [temperatures (25–85 °C), pH (4.5–8) and sodium chloride solution (0–8%)] for 30 min. Then, they centrifuged the mixture to obtain the top oil layer. The highest oil yield was obtained under the optimum conditions of 75 °C, pH 5.5 and the addition of 5% sodium chloride solution.

Based on the centrifugation method described above, Bizimana et al. (1993) examined the interrelationships between factors such as pulp-to-water ratio (1:3–1:5, w/w), temperatures (75–98 °C), pH (4–5.5), types (calcium carbonate, calcium sulfate, calcium chloride and sodium chloride) and concentrations (5–20%) of inorganic salt solutions and centrifugal forces (6000–12,300 g) on avocado oil yield. Their study showed the addition of calcium carbonate, calcium sulfate or sodium chloride solutions at concentrations ranging between 5% and 15% was capable of increasing the oil yield. The researchers advocated the necessity to limit the usage of inorganic salt solutions at 5% level as prolonged use of high concentrations (>5%) of inorganic salt solutions can lead to equipment corrosion. Thus, in this case, optimum conditions with highest oil yield were 1:5 (w/w) of pulp-to-water ratio, 98 °C, pH 5.5, addition of 5% inorganic salt (calcium carbonate, calcium sulfate or sodium chloride) solutions and centrifugal force of 12,300 × g.

In the late 1990s, Alfa Laval cooperated with a food company in New Zealand to develop a new method for the extraction of avocado oil using cold-pressing (Costagli and Betti 2015). The method was based on the modifications of existing cold-pressing olive oil method, whereby the extraction is carried out at low temperatures (<50 °C) and without the use of inorganic salts, enzymes or organic solvents (Wong et al. 2005). A 3-outlet decanter centrifuge is used to separate the malaxed avocado paste into oil, water and defatted solid cake. The recommended malaxing time and temperature for production of cold-pressed avocado oil were less than 50 °C and 90 min, respectively (Costagli and Betti 2015). The time and temperature used in malaxing avocado paste are longer than for olive paste due to the presence of finely dispersed emulsion in the avocado pulp cells.

3.4 Enzymatic Extraction

Enzymes are macromolecular biological catalysts. Addition of enzymes to plant-based oil bearing materials before or during oil extraction facilitates the degradation of cell wall components and hydrolysis of the structural polysaccharides and lipid bodies, thereby enhancing oil recovery. Due to the structural complexity of plant materials, the efficacy of enzymatic oil extraction is affected by numerous factors such as concentration and composition of enzymes, particle size and moisture content of the plant materials, pH, temperature, hydrolysis time and the solid-to-solvent ratio (Azmir et al. 2013).

Previously, enzymatic extraction of avocado oil has been carried out in three different ways, namely, aqueous enzymatic extraction, enzyme-assisted solvent extraction and enzyme-assisted expeller pressing (Buenrostro and López-Munguia 1986; Santana et al. 2015). Aqueous enzymatic extraction of avocado oil using different commercial enzymes (cellulase, α-amylase, pectinase and protease) was reported by Buenrostro and López-Munguia (1986). They added 1% of the enzymes separately to diluted avocado paste and incubated at 40 °C for an hour before recovering the oil by centrifugation. Their study showed the addition of 1% of α-amylase drastically increased the avocado oil yield compared with other enzymes studied and the negative control. Later, Santana et al. (2015) proposed the use of enzyme-assisted solvent extraction and enzyme-assisted expeller pressing of avocado oil. In their study, pectinase (0.05%) was added to the avocado paste before drying (45–60 °C) to constant weight. The researchers reported that the addition of enzyme to the pulp followed by solvent extraction or expeller pressing did not enhance the oil yield.

3.5 Ultrasonic Extraction

Ultrasound is a type of acoustic waves with frequencies greater than 20 kHz, which is above the threshold of human hearing range. It creates compression and expansion cycles when passing through a solid, liquid or gaseous medium. In liquid medium, the expansion cycle generates bubbles and produces intensive localized pressure. The bubbles undergo implosive collapse once they have expanded to a certain degree. This process produces a phenomenon known as cavitation, which is capable of degrading the wall of the oil-bearing cells and the structure of the oil emulsion, thereby releasing their intracellular components into the extracting solvent (Tan et al. 2018a). The efficacy of ultrasonic oil extraction is influenced by numerous factors such as particle size and moisture content of plant materials, sonication conditions (temperatures, time and frequencies) and the choices of extracting solvents (Azmir et al. 2013).

Reddy et al. (2012) reported on the ultrasound-assisted solvent extraction of avocado oil where they sonicated avocado pulp with hexane in an ultrasonic bath at 60 °C for an hour. The oil was then recovered using evaporation. Tan et al. (2018a) introduced ultrasound-assisted aqueous extraction of avocado oil. The researchers added water to the avocado powder at various ratios and sonicated the mixtures in an ultrasonic bath under different conditions [temperature (20–40 °C) and time (10–30 min)], followed by expeller pressing and centrifugation to obtain the top oil layer. The highest oil yield was obtained under the optimum conditions of 6 mL/g of water-to-powder ratio, 35 °C of sonication temperature and 30 min of sonication time. Extraction of avocado oil under the optimum conditions of ultrasound-assisted aqueous extraction yielded 73% of the oil obtained using Soxhlet extraction.

The application of ultrasound in malaxed and non-malaxed avocado paste in avocado oil extraction was investigated by Martínez-Padilla et al. (2018). Their study showed that low (18 + 40 kHz) and high (2 MHz) frequencies ultrasound increased the oil yield obtained after centrifugation.

3.6 Supercritical Fluid Extraction

Supercritical fluid extraction is a unitary mass transfer operation involving the use of a fluid at a temperature and pressure above its critical point. The critical point is the point where pressure and temperature for any fluid become identical or supercritical. A basic supercritical fluid system consists of a tank of the mobile phase, a pump to pressurize the gas, a controller to regulate the pressure inside the system, co-solvent vessel and pump, an oven that contains the extraction vessel and a trapping vessel (Azmir et al. 2013). The efficacy of supercritical fluid oil extraction is affected by several factors like the choices of supercritical fluids, extraction conditions (temperature, pressure and time) and plant materials preparation (Wang and Weller 2006).

Supercritical carbon dioxide (CO_2) extraction of avocado oil at different temperatures (37–81 °C) and pressure (350–532 atm) was examined by Botha and McCrindle (2004). Their study showed the oil yield increased with increasing pressure and temperature. After 120 min of supercritical CO_2 extraction, the highest oil yield was obtained at operating temperature and pressure of 81 °C and 532 atm, respectively. Mostert et al. (2007) also reported on the extraction of avocado oil using supercritical CO_2 at 37 °C and 35 atm for an hour. These conditions yielded 83–90% of the oil obtained with Soxhlet extraction.

A two-step supercritical fluid extraction of avocado oil was reported by Corzzini et al. (2017). This is a sequential oil extraction method and each stage utilizes solvent with different polarity. Utilization of solvents with different polarities during oil extraction exploits the different solubilities of lipid components, thereby increasing the oil yield. The solvents used for the first and the second stages of the two-step supercritical fluid extraction of avocado oil were CO_2 and ethanol-CO_2 mixtures, respectively. The overall extraction time was 4.5 h, where the extraction time for the first stage was 3 h and the second stage was 1.5 h. When conducted at 400 bar and 60 °C or 80 °C, this extraction method yielded 98% of the oil obtained with Soxhlet extraction (Corzzini et al. 2017).

Recently, Tan et al. (2018b) reported on the subcritical CO_2 extraction of avocado oil. Unlike supercritical CO_2 extraction, subcritical CO_2 extraction operates below the critical pressure (72.9 bar) and temperature (31.1 °C) of CO_2. This allows the reduction in overall cost as the requirements of operating pressure and temperature are much lower when compared to supercritical CO_2 extraction. In comparison to Soxhlet extraction, subcritical CO_2 extraction of avocado oil at 27 °C and 68 bar for 7.5 h yielded 82% of the oil.

3.7 Pressurized Fluid Extraction

Pressurized fluid extraction, also known as accelerated fluid extraction, is the application of high pressure to maintain solvents remain beyond their normal boiling point, thereby promoting the process of oil extraction. This method allows rapid oil extraction due to the utilization of high temperatures and pressure, but a small amount of solvent. According to Azmir et al. (2013), the combination of high temperatures and pressure facilitates the mass transfer rate and solubility, while reducing the surface tension and viscosity of solvents, thereby, improving extraction rate.

Extraction of avocado oil using pressurized fluid extraction was conducted by Abaide et al. (2017). In their study, they extracted avocado oil using compressed liquefied petroleum gas for 10 min under various extracting pressure (0.5–2.5 MPa) and temperatures (293–313 K). The highest oil yield was achieved by compressed liquefied petroleum gas at 0.5 MPa and 293 K, which yielded 60% of the oil compared to Soxhlet extraction.

3.8 Refining of Avocado Oil

Depending on the extracting conditions and fruit quality, crude avocado oil may contain high levels of free fatty acids (Santana et al. 2015), which can affect the quality and edibility of the oil. Yahia and Woolf (2011) suggested that crude avocado oil extracted using 'harsh' technologies (e.g. high heat and/or organic solvents) to be further processed into refined avocado oil. The main focuses of avocado oil refining are to maximize the removal of undesirable components (e.g. free fatty

acids and undesirable organoleptic) and to minimize the loss of desirable components (e.g. bioactive components). This process considers other factors, such as
increasing the conversion process, stability characteristics of oil, consumer preferences in taste, flavor and color of oil as well as the end use of oil such as cooking
oil, shortening and margarine (Yahia and Woolf 2011). Refining of avocado oil
involves four main steps, namely, bleaching, deodorizing, winterizing and neutralizing (Finau 2007). The resulting oil is light in yellow color, bland flavor and containing lower levels of health beneficial bioactive components.

4 Fatty Acid Composition and Acyl Lipids

Fatty acid composition affects the nutritional quality and physical properties of
edible oil (Tan and Azlan 2016). The fatty acid profile of avocado oil has been relatively consistent among studies, but the relative concentration of each fatty acid
component was found to vary considerably due to the geographical region
(Ratovohery et al. 1988; Yanty et al. 2011), variety (Yanty et al. 2011), oil extraction
method (Reddy et al. 2012; Werman and Neeman 1987) and harvest time (Ozdemir
and Topuz 2004). The study by Woolf et al. (1999) showed that avocado oil extracted
from fruit that was sun-exposed during growth contained a greater amount of saturated fatty acids (SFA), but a lesser amount of monounsaturated fatty acids (MUFA)
than shaded fruit. Another study conducted by Werman and Neeman (1987) indicated the solvent-extracted avocado oil contained a greater proportion of SFA than
centrifuge-extracted avocado oil.

Yanty et al. (2011) studied the variation in the fatty acid profile of four avocado
varieties collected from different geographical regions. Irrespective of the variety
and/or geographical location, oleic acid (43.65–63.73%), palmitic acid (14.80–
30.37%) and linoleic acid (12.75–17.45) were the most abundant fatty acids in avocado oil (Table 17.2). A major portion of the avocado oil comprises MUFA (58%),
followed by SFA (26%) and polyunsaturated fatty acids (PUFA) (16%). This is
comparable to cashew nut oil (SFA 20%, MUFA 58% and PUFA 21%) (Ryan et al.
2006).

Avocado lipids can be classified into four main categories: (1) neutral lipids (tri-,
di-, and monoacylglycerols); (2) glycolipids; (3) phospholipids and (4) free fatty
acids. The main component of avocado oil is neutral lipids, which constitutes 96%
of the total lipid content (Caballero et al. 2015). Similar to other plant oils, triacylglycerols (89.04–97.81%) comprises the majority of the neutral lipids of avocado
oil, with small amounts of diacylglycerols (2.19–3.51%) and monoacylglycerols
(0–6.39%) (Msika and Legrand 2010). The physical characteristics of oils are influenced by its triacylglycerols (TAG) profile. As can be seen in Table 17.3, the greatest proportions of TAG molecular species in avocado oil were dioleoyl-palmitoyl
glycerol (22.42–27.41%), trioleoyl glycerol (11.42–29.00%), linoleoyl-oleoyl-
palmitoyl glycerol (11.05–19.29%) and linoleoyl-dioleoyl glycerol (7.71–20.22%).
It has been reported the TAG composition of the avocado oil is strongly affected by
the variety and geographical regions (Yanty et al. 2011).

Table 17.2 Fatty acid composition of avocado oil

Fatty acid	Mean (%)	Range (%)
Palmitic acid (C16:0)	24.80	14.80–30.37
Palmitoleic acid (C16:1)	5.48	4.40–7.44
Stearic acid (C18:0)	1.04	0.27–1.56
Oleic acid (C18:1)	52.45	43.65–63.73
Linoleic acid (C18:2)	14.82	12.75–17.45
Linolenic acid (C18:3)	1.43	1.09–2.03
SFA	25.84	15.07–31.66
MUFA	57.93	48.87–68.59
PUFA	16.25	13.95–19.48

Yanty et al. (2011)

SFA total saturated fatty acids, *MUFA* total monounsaturated fatty acids, *PUFA* total polyunsaturated fatty acids

Table 17.3 Triacylglycerol composition of avocado oil

Triacylglycerol	Mean (%)	Range (%)
LLL	0.45	0–0.85
LLO	3.64	2.32–5.04
PLL	2.85	2.17–4.21
LOO	11.67	7.71–20.22
LOP	15.38	11.05–19.29
LPP	4.02	0–9.61
OOO	18.21	11.42–29.00
POO	24.29	22.42–27.41
LLLn	1.31	0.90–1.87
PPO	9.04	2.80–12.43
SOO	0.74	0.41–1.27
SOP	0.43	0–0.75
SPP	0.10	0–0.18
Others	7.86	3.42–11.69

Yanty et al. (2011)

L linoleic, *O* oleic, *S* stearic, *P* palmitic, *Ln* linolenic

5 Minor Bioactive Lipids

5.1 Tocopherols

Tocopherols can act as natural antioxidants to retard the autocatalytic lipid peroxidation process and the development of free radicals. The food industry utilizes tocopherols to enhance the stability and shelf life of food products. Four tocopherol isomers, namely, α-, β-, ϒ- and σ-tocopherols, have been reported in avocado oil. Of these, α-tocopherol, which act as a radical chain-breaking antioxidant in lipoproteins and membranes, was the most abundant (89 µg/g), followed by ϒ-tocopherol (38 µg/g), σ-tocopherol (14 µg/g) and β-tocopherol (8 µg/g) (Madawala et al. 2012). The total tocol content of avocado oil (149 µg/g) was greater than that in hazelnut (115 µg/g) and macadamia nut (54 µg/g) oils (Madawala et al. 2012).

Jorge et al. (2015) reported the total tocopherols of centrifuge-extracted avocado oil (30.47–36.73 mg/kg) contained almost similar levels as the commercial cold-pressed avocado oil (30.87 mg/kg). A recent study demonstrated the total tocopherols of avocado oil extracted *via* a two-step supercritical fluid method (15.4–28.2 mg/100 g) was five to nine times greater than a single step supercritical fluid method (3 mg/100 g) (Corzzini et al. 2017). The great differences of the tocopherols content may be due to the extracting conditions (solvent and time).

5.2 Phytosterols

Phytosterols are cholesterol-like compounds found in plant origin foods and it represents a major portion of the unsaponifiable fraction of plant oils. It plays an important role in inhibiting intestinal cholesterol absorption, inclusive of circulating endogenous biliary cholesterol, a key step in cholesterol elimination (Gupta et al. 2011). Piironen et al. (2003) evaluated the phytosterol content of several local fruits in Finland and results indicated the phytosterol content of avocado pulp (0.75 g/kg fresh fruit weight) was superior to those in the edible parts of grape, banana, apple, plum, orange and kiwi, with contents ranging from 0.12 to 0.23 g/kg fresh fruit weight. As most of the phytosterols are lipophilic in nature, they are most likely to be present in significant amounts in the oil. The most abundant phytosterols in avocado oil were sitosterol (3023 μg/g), followed by Δ-5-avenasterol (215 μg/g), campesterol (187 μg/g) and stigmasterol (7 μg/g) (Madawala et al. 2012). Centrifuge-extracted avocado oil (943.10–999.60 mg/kg) contained greater amounts of phytosterol content than commercial cold-pressed avocado oil (755.60 mg/kg) (Jorge et al. 2015), indicating that the extraction method had an important impact on the phytosterol yield.

Berasategi et al. (2012) compared changes in the phytosterol content of avocado and olive oils under high-temperature heat treatment. Both oils contained sitosterol as the main phytosterol, in which the level was twofold more in avocado oil compared to olive oil. Before heating, the total phytosterols of the avocado oil (339.64 mg/100 g) were greater than olive oil (228.27 mg/100 g). After heating at 180 °C for 9 h, the total phytosterols of avocado and olive oils were reduced by 20.37% and 7.87%, respectively. It has been reported the stability of phytosterols is affected by heating temperatures and time, lipid composition and sterol structure (Igoumenidis et al. 2011).

5.3 Carotenoids

Carotenoids are natural pigments that contribute to the red, orange or yellow colors of plant oils. Other than being vitamin A precursors, carotenoids also play an important role in preventing free radical chain reactions. Avocado oil was reported to contain carotenoids such as lutein (1.6 μg/g), neoxanthin (0.2 μg/g), antheraxanthin (<0.5 μg/g) and violaxanthin (<0.5 μg/g). The total carotenoid content of cold-pressed avocado oil was reported to be 1.9 μg/g (Woolf et al. 2009). This value was far lower than the total carotenoid content of commercial cold-pressed avocado oil as reported by Flores et al. (2014), which was 11.1–46.9 μg/g. Carotenoids analytical techniques, plant cultivar and climatic conditions are the key factors affecting the carotenoids level detected in plant oils (Nehdi et al. 2010).

5.4 Total Phenolics

Phenolic compounds constitute a large portion of the phytochemicals found abundantly in plants (Tan and Azlan 2017). The total phenolic content (TPC) of plant oil is commonly determined using the Folin-Ciocalteu reagent assay. The TPC of selected crude and refined plant oils, expressed as μg caffeic acid equivalents (CAE)/g, were reported by Haiyan et al. (2007). In comparison to crude camellia, pumpkin and sesame oils (11.9–22.7 μg CAE/g), crude avocado oil (11.6 μg CAE/g) contained the least TPC. In contrast, refined avocado oil had the greatest TPC (12.8 μg CAE/g) compared to other three refined plant oils (camellia, pumpkin and soybean), with a range of 3.9–4.2 μg CAE/g. On the other hand, a comparison of changes in TPC of cold-pressed avocado and olive oils across various temperatures (25–290 °C) were performed by Forero-Doria et al. (2017). At 290 °C, the TPC of avocado oil and olive oil were reduced by 50.7% and 82.3%, respectively, when compared with their respective TPC at 25 °C. This shows that the phenolic compounds of the avocado oil are more resistant towards the temperature degradation.

The application of the ultrasound approach in extracting plant oils, which has been receiving considerable attention in the food industry in recent years, showed that this approach is capable of promoting the diffusion of phenolic compounds into the oil phase. As evidenced by Martínez-Padilla et al. (2018), the greatest TPC was detected in the avocado oil extracted from the high frequency (2 MHz) malaxed puree.

6 Volatiles Composition

Volatile compounds do not only affect the flavor of plant oils, but also its perception and acceptability. Table 17.4 summarizes the volatile compounds that have been identified in avocado oil. The volatile compounds of avocado oil extracted using organic solvents, namely, hexane and acetone were evaluated by Moreno et al. (2003). Their study indicated the hexane-extracted avocado oil contained more aromatic hydrocarbons while the acetone-extracted avocado oil was composed mainly of aldehydes, terpenoids and short-chain fatty acids. On the other hand, Haiyan et al. (2007) found that the volatile compounds present in cold-pressed avocado oil

Table 17.4 Volatile compounds of avocado oil

Method	Volatiles	Total[a]	Reference
Hexane (solvent) extraction[b]	1,2–Dimethylbenzene, 1,2-Dimethyl-4-ethylbenzene, 1,2,4-Trimethylbenzene, 1,4–Dimethylbenzene, 2-Proponoic acid, 2-ethylhexyl ester, 2,6-Dimethylundecane, Benzoic acid 2-hydroxymethyl ester, Decane, Dodecane, Pentadecane, Propanoic acid, 2-methyl-1-(1,1-dimethyl)-2-methyl-1,3-propamedyl ester, Propanoic acid, 2-methyl-3-hydroxy-2,4-trimethylpenthl ester, Tetradecane, Tridecane, Undecane	15	Moreno et al. (2003)
Acetone (solvent) extraction[b]	1,2–Dimethylbenzene, (*E, E*)-α-Farnesene, α-Bergamotene, α-Copaene, α-Cubebene, α-Humolene, β-Bisabolene, β-Caryophyllene, Caryophyllene oxide, Germacrene D, *trans, trans*-2,4–Decadienal, *trans, trans*-2,4-Decadienal isomer	12	Moreno et al. (2003)
Cold-pressed	1-Hexanol, 3-Carene, (*E*)-2-Hexenal, α-Piene, β-Piene, Acetic acid, Hexanal, Nonanal, Pentanal	9	Haiyan et al. (2007)
Refined	1-Penten-3-ol, 2-Butenal, Acetic acid, Furfural, Heptanal, Hexanal, Nonanal, Pentanal, Pentenal isomer, Toluene	10	Haiyan et al. (2007)
Cold-pressed	α-Piene, (*E*)-2-Hexanol, (*E*)-2-Hexenal, Acetic acid, Hexanal, Hexanol, Propanol	7	Woolf et al. (2009)

[a]Total identified volatile compounds
[b]Avocado pulp was vacuum oven dried at 70°C before oil extraction

were different from those in refined avocado oil. Woolf et al. (2009) reported that the avocado oil extracted from good quality fruit (healthy or fruit with minor levels of rots) contained higher levels of desirable volatiles like (E)-2-hexanol, (E)-2-hexenal, hexanal and hexanol. Conversely, avocado oil extracted from poor quality fruit (fruit with major levels of rots) contained a high level of undesirable volatile like acetic acid, most likely due to microbial activities.

7 Oxidative Stability

The stability of plant oils against oxidation depends on the degree of unsaturation and the concentration of minor bioactive lipids. By using the Rancimat method, Madawala et al. (2012) found the induction time at 100 °C of plant oils decreased in the order of avocado oil > hazelnut oil > almond oil > grapeseed oil > walnut oil. This indicates avocado oil was more oxidation resistant compared to the rest (hazelnut, almond, grapeseed and walnut oils). Moreover, the oxidative stability of avocado oil under high-temperature heat treatment was compared against olive oil using thiobarbituric acid reactive substances (TBAR) assay (Berasategi et al. 2012). After heating at 180 °C for 9 h, the TBAR value of avocado oil was lower than olive oil, indicating avocado oil was more heat stable than olive oil.

8   Health-Promoting Properties of Avocado Oil

Numerous scientific evidence demonstrates that regular intake of avocado oil lowers
the risk of chronic diseases (Carvajal-Zarrabal et al. 2014; Furlan et al. 2017; Toro-
Equihua et al. 2016; Torre-Carbot et al. 2015). This is due to the synergistic combination of compounds such as tocopherols, unsaturated fatty acids and bioactive
phytochemicals (e.g. phytosterols, carotenoids and phenolics) that are responsible
for the functionality and health-promoting properties of the avocado oil. Table 17.5
summarizes the health benefits of avocado oil. Up to now, health benefits of avocado
oil are associated with lipid-lowering effect, hypotensive effect, cardio-protective
property, diabetes management, anti-arthritic property, periodontal disease management, skin and wound healing and enhancement of nutrients absorption. Details of
these health benefits are discussed below.

8.1   Lipid-Lowering Effect

Lipid-lowering effect of several plant oils (avocado, grape seed, canola, soybean,
safflower and partially hydrogenated plant oils) was evaluated by Torre-Carbot et al.
(2015) using Wistar rats. The rats were initially supplemented with 5.7% plant oils
for 2 weeks before switching to a higher concentration of oils (11%) for another
3 weeks. Results showed that rats supplemented with avocado oil had the lowest
concentration of total cholesterol and low-density lipoprotein (LDL) cholesterol.
The researchers also reported the LDL/high-density lipoprotein (HDL) index of
avocado oil supplemented group was the lowest, indicating the vascular-protective
potential of avocado oil.

8.2   Hypotensive Effect

Study of the blood pressure response to angiotensin II and the fatty acid composition of renal and cardiac membranes after consumption of a diet enriched with 10%
avocado oil was performed by Salazar et al. (2005). After 2 weeks of avocado oil
intake, they found the fatty acid profile of renal and cardiac microsomes had altered
and angiotensin II-induced blood pressure response was greater compared to the
normal control. The researchers postulated the modification of essential fatty acid
levels at the renal and cardiac membranes, as a result of avocado oil consumption,
changed the way that kidney and cardiac respond to the hormone that modulate
blood pressure.

Table 17.5 Health-promoting properties of avocado oil

Health effect	Study design/model	Results	Reference
Lipid-lowering	Healthy rats	Improved blood lipid profile and reduced food intake and weight gain	Torre-Carbot et al. (2015)
Hypotensive	Angiotensin II-induced rats	Modified fatty acid content in renal and cardiac membranes in a tissue-specific manner	Salazar et al. (2005)
Cardioprotective	Sucrose-induced metabolic alteration rats	Improved blood lipid profile and partially reversed the inflammatory process	Carvajal-Zarrabal et al. (2014)
	Cohort overweight human	Alleviated atherosclerosis risk factors and inflammation and potentially endotoxemia improvement	Furlan et al. (2017)
Diabetes management	Streptozotocin-induced diabetic rats	Improved brain mitochondrial function, decreased oxidative stress and modified diabetic dyslipidemia	Ortiz-Avila et al. (2015)
	Sucrose-induced insulin resistance rats	Improved glucose tolerance and insulin resistance while reducing the body weight gain	Toro-Equihua et al. (2016)
Anti-arthritic	Review of randomized, placebo-controlled and double-blind trials	Effective for the symptomatic treatment of osteoarthritis	Ernst (2003)
	Review of in vitro and animal studies	Positively regulated the altered phenotype of osteoarthritis subchondral bone osteoblasts, stimulated the synthesis of proteoglycans in chondrocytes cultures and decreased the synthesis of collagenases by synovial cells	Henrotin (2018)

(continued)

Table 17.5 (continued)

Health effect	Study design/model	Results	Reference
Periodontal disease management	Alveolar bone and periodontal ligament cells	Opposed cytokine effect and reversed the inhibiting effect of interleukin-1 beta (IL-1β)	Andriamanalijaona et al. (2006)
Skin and wound healing	Randomized, prospective clinical trial	No differences in the PASI score and 20 MHz sonography results of vitamin B_{12} ointment containing avocado oil and calcipotriol ointment	Stücker et al. (2001)
	Incisional and excisional cutaneous rats	Reduced inflammatory and improved tensile strength and collagen density	Oliveira et al. (2013)
Enhancement of nutrients absorption	Crossover trial	Improved carotenoids absorption	Unlu et al. (2005)

8.3   Cardioprotective Effect

The cardioprotective potential of avocado oil supplementation in sucrose-induced metabolic alteration rats was evaluated by Carvajal-Zarrabal et al. (2014). Supplementation with 7.5% avocado oil resulted in the reduction of high sensitivity C-reactive protein (hs-CRP), triacylglycerol, LDL cholesterol and very low-density lipoprotein (VLDL) cholesterol levels compared to the untreated rats. The researchers concluded that avocado oil can partially reverse the inflammatory process attributed to the sucrose solution consumption. Another cohort study conducted by Furlan et al. (2017) showed the replacement of butter with avocado oil regulated the negative physiological impact associated with a hypercaloric-hyperlipidic meal. Replacement of 9.3% butter with 9.6% avocado oil in the experimental meal consisted of bacon, eggs, iced sugar, potatoes and wheat bread improved the postprandial levels of interleukin-6, C-reactive protein, total cholesterol, LDL cholesterol, triacylglycerol, insulin and glycemia of overweight subjects. Their study highlighted the importance of correct fat quality choice as a fundamental of cardiometabolic risk modulation.

8.4   Diabetes Management

Ortiz-Avila et al. (2015) evaluated the effect of oral administration of avocado oil at a dose of 1 mL/250 g body weight on the oxidative status and brain mitochondrial function of streptozotocin (STZ)-induced diabetic rats for a period of 90 days. Compared to untreated STZ-rats, although administration of avocado oil to the STZ-rats did not normalize serum glucose level, it was capable of reducing serum cholesterol and triacylglycerol levels. Avocado oil was also able to reduce lipid peroxidation and reactive oxygen species levels while improving the reduced/oxidized glutathione ratio. The impairment of mitochondrial transmembrane potential and mitochondrial respiration were also hindered after administration of avocado oil. The researchers suggested the potential of avocado oil on delaying the onset of diabetic encephalopathy.

In another study, Toro-Equihua et al. (2016) evaluated the effect of avocado oil supplementation on sucrose-induced insulin resistance rats. Addition of 5–20% avocado oil to the diet increased insulin sensitivity and normalized insulin resistance induced by a sucrose-rich diet. Besides that, avocado oil supplementation also reduced body weight gain while improving glucose tolerance.

8.5   Anti-Arthritic Effect

Avocado and soybean unsaponifiables (ASU) mixture is an anti-arthritic agent produced by mixing the unsaponifiable fraction of avocado and soybean oils at a ratio of 1:2, respectively (Henrotin 2018). The ASU mixture is now commercialized as an over-the-counter medicine known as Piascledine® 300 (Henrotin 2018). A systematic review proved the effectiveness of ASU mixture consumption in alleviating the symptoms of osteoarthritis (Ernst 2003). This systematic review includes only the "gold standard" of epidemiologic studies, which is the database of randomized, double-blind and placebo-controlled clinical trials. Most of the literature data in this review showed that consumption of ASU mixture at a dose of 300 mg/day exhibited superior outcomes than placebo on relieving the symptoms of osteoarthritis with no major adverse health effect being documented. The effect of ASU mixture on various animal (rat, mice, dog and sheep) and *in vitro* (chondrocytes culture, osteoblasts culture, chondrocytes/osteoblasts co-culture, synovial cells culture) models were recently reviewed by Henrotin (2018). The author concluded that ASU mixture provides health benefits on the metabolic changes in the tissues (cartilage, synovium and subchondral bone) associated with the pathophysiology of osteoarthritis.

8.6   Periodontal Disease Management

Periodontal disease is a pathological condition that involves inflammation of the tooth-supporting structure. An in vitro study to evaluate the effect of ASU mixture on the expression of bone morphogenetic protein-2 (BMP-2), transforming growth factor (TGF)-β_1 and TGF-β_2 in alveolar bone and periodontal ligament cells when exposed to the destructive interleukin-1 beta (IL-1β) was performed by Andriamanalijaona et al. (2006). They found that the expression of TGF-β_1, TGF-β_2

and BMP-2 of the cells were strongly reduced by IL-1β, the main driver of bone loss and tissue destruction in gum disease. In contrast, ASU mixture at a dose of 10 μg/mL was found to stimulate the production of TGF-β$_1$, TGF-β$_2$ and BMP-2. The researchers concluded that the ASU mixture can exert a preventive action on the erosive damage caused by IL-1β.

8.7   Skin and Wound Healing Property

The therapeutic effect of vitamin B$_{12}$ cream containing avocado oil (VBAO) on the treatment of plaque psoriasis was reported by Stücker et al. (2001). The effectiveness of VBAO was compared against calcipotriol cream, a common treatment for psoriasis. Both creams were applied to psoriatic plaques on the contralateral body sides of each patient twice daily. After 12 weeks of therapy, the results of Psoriasis Area Severity Index (PASI) and 20 MHz sonography showed the VBAO produced similar effects as calcipotriol cream on alleviating plaque psoriasis. The researchers highlighted the potential of VBAO as a well-tolerated, long-term topical therapy of psoriasis.

Another study on the wound healing properties of avocado oil was demonstrated by an in vivo study that used incisional and excisional cutaneous rats (Oliveira et al. 2013). After 14 days of therapy, the percentages of wound contraction and reepithelialization of the rats applied with a semisolid formulation of avocado oil (SFAO) were better than the rats applied with petroleum jelly. Topical application of SFAO has been shown to reduce the number of inflammatory cells in the scar tissue and enhanced collagen synthesis during wound healing process. The researchers recommended SFAO as a new alternative for skin wounds treatment.

8.8   Enhancement of Nutrients Absorption

The health benefit of avocado oil supplementation on the bioavailability of carotenoids was demonstrated in a crossover trial by Unlu et al. (2005). Salad, which was prepared from low-fat but high-carotenoid ingredients (carrot, spinach, lettuce, read and salad dressing), served as an experimental diet. The absorption of carotenoids (α-carotene, β-carotene and lutein) by the human body was low when consumed the control meal (salad) alone. In comparison to the control meal, an addition of 24 g of avocado oil to the salad drastically increased carotenoids absorption (8.9 times greater for α-carotene, 17.4 times greater for β-carotene and 6.7 times greater for lutein) by the human body. The researchers pointed out the significant role of dietary lipids for effective carotenoid absorption and the necessity of taking dietary interactions into account when providing nutritional recommendations. Table 17.5 summarizes the health benefits of avocado oil.

9 Edible and Non-edible Applications of Avocado Oil

Depending on the extraction and processing methods, the flavors of crude avocado oil varies from buttery or grassy to mushroom-like or slightly nutty. These flavors are suitable for the application of uncooked dishes like dressing salads and marinades. Owing to a high smoke point (>200 °C) (Table 17.1), crude avocado oil is also suitable in high-temperature cooking like pan frying, barbeque, baking and roasting (Woolf et al. 2009). The pharmaceutical industry uses crude avocado oil to produce an array of dietary supplement and healthcare products due to its bioactive compounds. Refined avocado oil, which is light yellow color and has a bland flavor, is mostly utilized in cosmetics application rather than pharmaceutical application.

10 Other Issues

10.1 Authenticity and Adulteration

The scientifically proven health benefits of avocado oil, as highlighted in Table 17.5, command it a higher market price than other plant oils. Thus, there is a potential of avocado oil adulteration with other cheaper or inferior quality plant oils. Thermal analysis using a differential scanning calorimetry has been purposed as a rapid and simple method to authenticate the adulteration of plant oils. In a study conducted by Yanty et al. (2017), the addition of 5–9% of palm stearin and cocoa butter to avocado oil shifted the onset of crystallization temperatures and resulted in higher endset melting temperatures. The researchers concluded the occurrence of these phenomena was due to the compositional changes, resulted from the increment of saturated fatty acids and di-/tri-saturated triacylglycerols.

10.2 Allergies and Toxicity

There have not been any reported cases of adverse effects, allergies or toxicity due to avocado oil.

References

Abaide, E. R., Zabot, G. L., Tres, M. V., Martins, R. F., Fagundez, J. L., Nunes, L. F., Druzian, S., Soares, J. F., Dal Prá, V., Silva, J. R. F., Kuhn, R. C., & Mazutti, M. A. (2017). Yield, composition, and antioxidant activity of avocado pulp oil extracted by pressurized fluids. *Food and Bioproducts Processing, 102*, 289–298.

Afahkan, M. E. (2012). *Effects of petroleum ether and methanolic fractions of avocado pear (Persea americana Mill.) seeds on Wistar rats fed a high fat-high cholesterol diet*. Nigeria: Ahmadu Bello University.

Andriamanalijaona, R., Benateau, H., Barre, P. E., Boumediene, K., Labbe, D., Compere, J. F., & Pujol, J. P. (2006). Effect of interleukin-1β on transforming growth factor-beta and bone morphogenetic protein-2 expression in human periodontal ligament and alveolar bone cells in culture: Modulation by avocado and soybean unsaponifiables. *Journal of Periodontology, 77*(7), 1156–1166.

Azmir, J., Zaidul, I. S. M., Rahman, M. M., Sharif, K. M., Mohamed, A., Sahena, F., Jahurul, M. H. A., Ghafoor, K., Norulaini, N. A. N., & Omar, A. K. M. (2013). Techniques for extraction of bioactive compounds from plant materials: A review. *Journal of Food Engineering, 117*(4), 426–436.

Berasategi, I., Barriuso, B., Ansorena, D., & Astiasarán, I. (2012). Stability of avocado oil during heating: Comparative study to olive oil. *Food Chemistry, 132*(1), 439–446.

Bizimana, V., Breene, W. M., & Csallany, A. S. (1993). Avocado oil extraction with appropriate technology for developing countries. *Journal of the American Oil Chemists' Society, 70*(8), 821–822.

Bora, P. S., Narain, N., Rocha, R. V., & Paulo, M. Q. (2001). Characterization of the oils from the pulp and seeds of avocado (Cultivar: Fuerte) fruits. *Grasas y Aceites, 52*(3–4), 171–174.

Botha, B., & McCrindle, R. (2004). Supercritical fluid extraction of avocado oil. *South African Avocado Growers' Association Yearbook, 27*, 24–27.

Buenrostro, M., & López-Munguia, A. (1986). Enzymatic extraction of avocado oil. *Biotechnology Letters, 8*(7), 505–506.

Caballero, B., Finglas, P., & Toldrá, F. (2015). *Encyclopedia of food and health*. Cambridge, MA: Academic.

Carvajal-Zarrabal, O., Nolasco-Hipolito, C., Aguilar-Uscanga, M., Melo-Santiesteban, G., Hayward-Jones, P., Barradas-Dermitz, D. (2014) Avocado oil supplementation modifies cardiovascular risk profile markers in a rat model of sucrose-induced metabolic changes. *Disease Markers, 2014*, 1–8.

Chapius, A., Blin, J., Carre, P., & Lecomte, D. (2014). Separation efficiency and energy consumption of oil expression using a screw-press: The case of *Jatropha curcas* L. seeds. *Industrial Crops and Products, 52*, 752–761.

Christenhusz, M., & Byng, J. (2016). The number of known plants species in the world and its annual increase. *Phytotaxa, 261*(2), 201–217.

Corzzini, S. C. S., Barros, H. D. F. Q., Grimaldi, R., & Cabral, F. A. (2017). Extraction of edible avocado oil using supercritical CO_2 and a CO_2/ethanol mixture as solvents. *Journal of Food Engineering, 194*, 40–45.

Costagli, G., & Betti, M. (2015). Avocado oil extraction processes: Method for cold-pressed high-quality edible oil production versus traditional production. *Journal of Agricultural Engineering, 46*(3), 115.

Torre-Carbot, K. D. L., Chávez-Servín, J., Reyes, P., Ferriz, R., Gutiérrez, E., Escobar, K., Aguilera, A., Anaya, M. A., García-Gasca, T., García, O. P., & Rosado, J. L. (2015). Changes in lipid profile of Wistar rats after sustained consumption of different types of commercial vegetable oil: A preliminary study. *Universal Journal of Food and Nutrition Science, 3*(1), 10–18.

Del Toro-Equihua, M., Velasco-Rodríguez, R., López-Ascencio, R., & Vásquez, C. (2016). Effect of an avocado oil-enhanced diet (*Persea americana*) on sucrose-induced insulin resistance in Wistar rats. *Journal of Food and Drug Analysis, 24*(2), 350–357.

Ernst, E. (2003). Avocado-soybean unsaponifiables (ASU) for osteoarthritis – a systematic review. *Clinical Rheumatology, 22*(4–5), 285–288.

FAO (2017). *Commodities by country*. Available at: http://www.fao.org/faostat/en/#rankings/commodities_by_country.

Finau, K. A. (2007). *Literature review on avocado oil for SROS technological purposes*. Samoa: Nafanua.

Flores, M. A., Perez-camino, M. D. C., & Troca, J. (2014). Preliminary studies on composition, quality and oxidative stability of commercial avocado oil produced in Chile. *Journal of Food Science and Engineering, 4*, 21–26.

Forero-Doria, O., García, M. F., Vergara, C. E., & Guzman, L. (2017). Thermal analysis and anti-oxidant activity of oil extracted from pulp of ripe avocados. *Journal of Thermal Analysis and Calorimetry, 130*(2), 959–966.

Furlan, C. P. B., Valle, S. C., Östman, E., Maróstica, M. R., & Tovar, J. (2017). Inclusion of Hass avocado-oil improves postprandial metabolic responses to a hypercaloric-hyperlipidic meal in overweight subjects. *Journal of Functional Foods, 38*, 349–354.

Gatbonton, G., De Jesus, A., & Lorenzo, K. (2013). Soxhlet extraction of Philippine avocado fruit pulp variety 240. *Chemical Engineering, 1*(1), 1–8.

Gunstone, F. D. (2011). *Vegetable oils in food technology: Composition, properties and uses.* New Jersey: Wiley.

Gupta, R., Sharma, A., & Dobhal, M. (2011). Antidiabetic and antioxidant potential of β-sitosterol in streptozotocin-induced experimental hyperglycemia. *Journal of Diabetes, 3*(1), 29–37.

Haiyan, Z., Bedgood, D. R., Bishop, A. G., Prenzler, P. D., & Robards, K. (2007). Endogenous biophenol, fatty acid and volatile profiles of selected oils. *Food Chemistry, 100*(4), 1544–1551.

Henrotin, Y. E. (2018). Avocado/soybean unsaponifiables (Piacledine®300) show beneficial effect on the metabolism of osteoarthritic cartilage, synovium and subchondral bone: An overview of the mechanisms. *AIMS Medical Science, 5*(1), 33–52.

Igoumenidis, P., Konstanta, M., Salta, F., & Karathanos, V. (2011). Phytosterols in frying oils: Evaluation of their absorption in pre-fried potatoes and determination of their destruction kinetics after repeated deep and pan frying. *Procedia Food Science, 1*, 608–615.

Jorge, T., Polachini, T., Dias, L., & Jorge, N. (2015). Physicochemical and rheological character-ization of avocado oils. *Ciência e Agrotecnologica, 39*(4), 390–400.

Madawala, S. R. P., Kochhar, S. P., & Dutta, P. C. (2012). Lipid components and oxidative status of selected specialty oils. *Grasas y Aceites, 63*(2), 143–151.

Maitera, O., Osemeahon, S., & Barnabas, H. (2014). Proximate and elemental analysis of avocado fruit obtained from Taraba State, Nigeria. *Indian Journal of Science and Technology, 2*(2), 67–73.

Martínez-Padilla, L. P., Franke, L., Xu, X. Q., & Juliano, P. (2018). Improved extraction of avo-cado oil by application of sono-physical processes. *Ultrasonics Sonochemistry, 40*, 720–726.

Moreno, A. O., Dorantes, L., Galíndez, J., & Guzmán, R. I. (2003). Effect of different extraction methods on fatty acids, volatile compounds, and physical and chemical properties of avocado (*Persea americana* Mill.) oil. *Journal of Agricultural and Food Chemistry, 51*(8), 2216–2221.

Mostert, M. E., Botha, B. M., Du Plessis, L. M., & Duodu, K. G. (2007). Effect of fruit ripeness and method of fruit drying on the extractability of avocado oil with hexane and supercritical carbon dioxide. *Journal of the Science of Food and Agriculture, 87*(15), 2880–2885.

Msika, P., & Legrand, L. (2010). Process for producing refined avocado oil rich in triglycerides, and oil obtainable by said process. *U.S.Patent and Trademark Office.*

Nehdi, I., Omri, S., Khalil, M. I., & Al-Resayes, S. I. (2010). Characteristics and chemical com-position of date palm (*Phoenix canariensis*) seeds and seed oil. *Industrial Crops and Products, 32*(3), 360–365.

Oliveira, A. P. D., Franco, E. D. S., Barreto, R. R., Cordeiro, D. P., Melo, R. G. D., Aquino, C. M. F. D., Silva, A. A. R., Medeiros, P. L. D., Silva, T. G. D., Goes, A. J. D. S., & Maia, M. B. D. S. (2013). Effect of semisolid formulation of *Persea americana* Mill. (Avocado) oil on wound healing in rats. *Evidence-Based Complementary and Alternative Medicine, 2013*, 1–8.

Ortiz, M. A., Dorantes, A. L., Gallndez, M. J., & Cardenas, S. E. (2004). Effect of a novel oil extraction method on avocado (*Persea americana* Mill.) pulp microstructure. *Plant Foods for Human Nutrition, 59*(1), 11–14.

Ortiz-Avila, O., Esquivel-Martínez, M., Olmos-Orizaba, B. E., Saavedra-Molina, A., Rodriguez-Orozco, A. R., & Cortés-Rojo, C. (2015). Avocado oil improves mitochondrial function and decreases oxidative stress in brain of diabetic rats. *Journal Diabetes Research, 2*(2), 23–30.

Ozdemir, F., & Topuz, A. (2004). Changes in dry matter, oil content and fatty acids composition of avocado during harvesting time and post-harvesting ripening period. *Food Chemistry, 86*(1), 79–83.

Piironen, V., Toivo, J., Puupponen-Pimiä, R., & Lampi, A.-M. (2003). Plant sterols in vegetables, fruits and berries. *Journal of the Science of Food and Agriculture, 83*(4), 330–337.

Qin, X., & Zhong, J. (2016). A review of extraction techniques for avocado oil. *Journal of Oleo Science, 65*(11), 1–8.

Ratovohery, J., Lozano, Y., & Gaydou, E. (1988). Fruit development effect on fatty acid composition of *Persea americana* fruit Mesocarp. *Journal Agricultural of Food Chemistry, 36*(2), 287–293.

Reddy, M., Moodley, R., & Jonnalagadda, S. B. (2012). Fatty acid profile and elemental content of avocado (*Persea americana* Mill.) oil-effect of extraction methods. *Journal of Environmental Science and Health. Part. B, Pesticides, Food Contaminants, and Agricultural Wastes, 47*(6), 529–537.

Ryan, E., Galvin, K., O'Connor, T. P., Maguire, A. R., & O'Brien, N. M. (2006). Fatty acid profile, tocopherol, squalene and phytosterol content of Brazil, pecan, pine, pistachio and cashew nuts. *International Journal of Food Sciences and Nutrition, 57*(3–4), 219–228.

Salazar, M. J., Hafidi, M. E., Pastelin, G., Ramírez-Ortega, M. C., & Sánchez-Mendoza, M. A. (2005). Effect of an avocado oil-rich diet over an angiotensin II-induced blood pressure response. *Journal of Ethnopharmacology, 98*(3), 335–358.

Santana, I., dos Reis, L. M. F. F., Torres, A. G., Cabral, L. M. C. C., & Freitas, S. P. (2015). Avocado (*Persea americana* Mill.) oil produced by microwave drying and expeller pressing exhibits low acidity and high oxidative stability. *European Journal of Lipid Science and Technology, 117*(7), 999–1007.

Southwell, K., Harris, R., & Swetman, A. (1990). Extraction and refining of oil obtained from dried avocado fruit using a small expeller. *Tropical Science, 30*(2), 121–131.

Stücker, M., Memmel, U., Hoffmann, M., & Hartung, J. (2001). Vitamin B12 cream containing avocado oil in the therapy of plaque psoriasis. *Dermatology, 203*(2), 141–147.

Tan, C. X., & Azlan, A. (2016). Nutritional, phytochemical and pharmacological properties of Canarium odontophyllum Miq. (Dabai) fruit. *Pertanika Journal of Scholar Research, 2*(1), 80–94.

Tan, C. X., & Azlan, A. (2017). Dietary fiber and total phenolic content of selected raw and cooked beans and its combinations. *International Food Research Journal, 24*(5), 1863–1868.

Tan, C. X., Chong, G. H., Hamzah, H., & Ghazali, H. M. (2018a). Optimization of ultrasound-assisted aqueous extraction to produce virgin avocado oil with low free fatty acids. *Journal of Food Process Engineering, 41*, e12656.

Tan, C. X., Chong, G. H., Hamzah, H., & Ghazali, H. M. (2018b). Comparison of subcritical CO_2 and ultrasound-assisted aqueous methods with the conventional solvent method in the extraction of avocado oil. *The Journal of Supercritical Fluids, 135*, 45–51.

Unlu, N. Z., Bohn, T., Clinton, S. K., & Schwartz, S. J. (2005). Carotenoid absorption from salad and salsa by humans is enhanced by the addition of avocado or avocado oil. *Human Nutrition and Metabolism, 135*(3), 431–436.

USDA. (2011). *Avocado, almond, pistachio and walnut composition. Nutrient Data Laboratory. USDA National Nutrient Database for Standard Reference.* Washington, DC: USDA.

Wang, L., & Weller, C. L. (2006). Recent advances in extraction of nutraceuticals from plants. *Trends in Food Science & Technology, 17*(6), 300–312.

Werman, M. J., & Neeman, I. (1987). Avocado oil production and chemical characteristics. *Journal of the American Oil Chemists' Society, 64*(2), 229–232.

Wong, R. C., McGhie, T., Wang, Y., Eyres, L., & Woolf, A. (2005). Recent research on the health components in cold pressed avocado oil. Wellington, New Zealand. Available at: https://www.slideserve.com/jorn/recent-research-on-the-health-components-in-cold-pressed-avocado-oil.

Woolf, A. B., Ferguson, I. B., Requejo-Tapia, L. C., Boyd, L., Laing, W. A., & White, A. (1999). Impact of sun exposure on harvest quality of 'Hass' avocado fruit. *Revista Chapingo Serie Horticultura, 5*, 353–358.

Woolf, W. M., Eyres, L., McGhie, T., Lund, C., Olsson, S., Wang, Y., Buliey, C., Wang, M., Friel, E., & Requejo-Jackman, C. (2009). *Avocado oil. Gourmet and health-promoting specialty oils* (pp. 73–125). USA: AOCS Press.

Yahia, E., & Woolf, A. (2011). *Avocado (Persea americana* Mill.*). Postharvest biology and technology of tropical and subtropical fruits: Acai to citrus*. Cambridge, UK: Woodhead Publishing.

Yanty, Marikkar, J. M. N., & Long, K. (2011). Effect of varietal differences on composition and thermal characteristics of avocado oil. *Journal of the American Oil Chemists' Society, 88*(12), 1997–2003.

Yanty, N., Marikkar, N., Mustafa, S., & Mat Sahri, M. (2017). Composition and thermal analysis of ternary mixtures of avocado fat: Palm stearin: Cocoa butter (Avo:PS:CB). *International Journal of Food Properties, 20*(2), 465–474.

Chapter 18
Gac (*Momordica cochinchinensis* (Lour) Spreng.) Oil

Huynh Cang Mai and Frédéric Debaste

Abstract Gac (*Momordica cochinchinensis* (Lour) Spreng.) fruit, originating from South-Eastern Asia, is considered as a superfruit thanks to the unequaled content of lycopene and other carotenoids of its arils. Direct uses of the fruit can be considered in cooking or traditional medicine, yet, most interesting and large-scale applications in food, cosmetic and pharmaceutical require to extract the gac oil with its carotenoids content. Gac oil production is subject to an increase attention by the scientific and engineering domain, but is still in its infancy compared to other oils production. In this chapter, it is proposed to summarize the state of the art of gac oil processing by following the valorization chain. First, the properties of the fruit are presented. The steps used to store the fruit and produce the oil (drying, freezing, and oil extraction) are reported. For each step, the different known options are compared in terms of process conditions and quality of the oil. The properties of the gac oil are reported with an emphasis on the carotenoid content and antioxidant activities. Further processing of the oil (concentration, and carotenoids crystallization) is also addressed. The main existing and foreseen applications of gac oils are finally discussed.

Keywords Fruit oils · Superfruit · Carotenoids crystallization · Nontraditional oils

H. C. Mai (✉)
Department of Chemical Engineering, Nong Lam University, Ho Chi Minh City, Vietnam
e-mail: maihuynhcang@hcmuaf.edu.vn

F. Debaste
Transfers, Interfaces and Processes- Chemical Engineering Unit, Ecole polytechnique de Bruxelles, Université libre de Bruxelles, Brussels, Belgium
e-mail: fdebaste@ulb.ac.be

© Springer Nature Switzerland AG 2019
M. F. Ramadan (ed.), *Fruit Oils: Chemistry and Functionality*,
https://doi.org/10.1007/978-3-030-12473-1_18

1 Gac Fruit and Its Properties

1.1 Origin and Name

Gac (*Momordica cochinchinensis* (Lour) Spreng.) is botanically classified into the family of *Cucurbitaceae*, the genus of *Momordica*, and the species of *Cochinchinensis*. This perennial vine was given the name of *Muricia cochinchinensis* by Lou-reiro, and then *Flora cochinchinensis* by a Portuguese priest in 1790. In 1826, Sprengel concluded that this plant belonged to the genus of Linné *Momordica* and changed the name to *Momordica cochinchinensis* Spreng (Bailey 1937). Gac is known for its carotenoid content, which was identified for the first time in 1941 by Guichard and Bui (Vuong et al. 2006).

Gac is found throughout the Southeast Asian region from South China to Northeastern Australia. It is not only a traditional fruit in Vietnam but also a native fruit of China, Japan, India, Thailand, Laos, Cambodia, the Philippines, Malaysia and Bangladesh. Its common names in different countries are presented in Table 18.1 (Vuong and King 2003). A large, bright-red fruit, gac fruit is known as "sweet gourd" and esteemed as "the fruit from heaven" because of its ability to promote longevity, vitality, and health (Vuong 2000; Kuhnlein 2004).

1.2 Culture and Production

In Vietnam, gac is cultivated in different regions, from the hills and mountains to the delta and coastal areas. In the Mekong Delta regions of Vietnam, gac grows on dioe-cious vines and is usually harvested from climbing hedges or wild plants. The vines are generally found on hedges of houses in the province or gardens (Vuong 2000). Traditionally, in Vietnam gac includes two varieties: gac Nep and gac Te. Gac Nep fruit has a bigger size, thicker and darker red aril than gac Te has. Therefore, gac Nep is cultivated more in Vietnam for its size, color and nutritive composition. Nowadays, in Vietnam, gac Nep is being cultivated on an industrial scale to extract oil from the aril for its colorant and healthy benefits.

Table 18.1 Common names of gac in different countries

Language	Name	Language	Name
Latin	*Momordica cochinchinensis Spreng.*	English	Spiny bitter gourd
	Muricia cochinchinensis Lour.		Sweet gourd
	Muricia mixta Roxb.		Cochinchin gourd
Chinese	Mu Bie Guo	Lao	Mak kao
Malay	Teruah	Thai	Fak kao
Japanese	Kushika, Mokubetsushi	Hindi	Hakur, Kakrol

The plant can be grown either from seeds or from root tubers. Humidity, heat, air circulation and light are required for gac seeds germination, which is very sensitive to the cold and dry condition. Gac seeds are easily germinated in 7–10 days in comfortable conditions. It is cultivated once but harvested for several years. The vines can live up to 15–20 years. According to people in the Mekong delta, a gac vine on a frame of 50 m² can produce from 100 to 200 fruits per year. Gac leaves are alternated and deeply divided from three to five lobes with serrated edges. Gac flowers are pale yellow and solitary in the axils of the leaves (Vuong 2000). The plant begins to bloom about 2 months after tubers have been planted. The flowers are pollinated by insects. Several vines must be grown together in the vicinity to ensure at least one mature male flower for mature female flowers nearby.

Flowering usually occurs in April and continues to August and/or September. It takes about 18–20 days for a mature fruit since the emergence of the female flower bud. A plant produces about 30–60 fruits weighing 1–3 kg each in its season. The ripened fruit is harvested from August to February (Shadeque and Baruah 1984).

1.3 Fruit Structure

The fruits of the gac are round or oblong, mature to a size of about 13 cm in length and 10 cm in diameter, densely aculeate, and green when becoming dark orange or red when ripen. Gac fruits are picked when they are the optimal size, weight and color. However, gac fruit is mainly harvested in developmental stages, while gac fruits are orange/red and the seeds are hardened (Bhumsaidon and Chamchong 2016).

Figure 18.1 shows the morphology of the gac fruit from the outside to the inside including exocarp, mesocarp, aril, and seed. The weight distribution of gac fruit is presented in Table 18.2 (Ishida et al. 2004):

- The exocarp of gac fruit is thorny, firstly green and turns orange or red when ripen. It is hard and covered with small spines 4.5 cm in height (Vuong 2000). In some fruits, the spines are smooth and dense, while the others are hard and widely spaced.
- The mesocarp represents nearly 50% of the weight of a gac fruit, which is spongy, orange and 4 cm of thickness (Vuong 2000). The mesocarp contains aril covering a black or brown seed inside and yellow connective tissue in the middle (Vuong 2000).
- The aril of gac fruit, accounting for 25% of the fruit weight, is red, soft and sticky, 1–3 mm of thickness, and is used for cooking. The aril texture is supple and spongy, similar to raw chicken livers. The mesocarp and the aril of gac fruit have a slight taste of sweet as a cucumber (Vuong et al. 2002). The aril of the gac fruit has high antioxidant activities thanks to its carotenoids content and valuable fatty acids.
- The seeds of gac fruit represent about 25% of the weight of the weight of the fruit. The seeds and arils are prepared with rice for a lustrous appearance and rich in oil, a slight flavor for rice. The seeds of gac fruits are brown and look like small meteorites with gagged edges and black lines running through them.

Fig. 18.1 Morphology of the gac fruit

Table 18.2 Weight distribution of Gac fruit (Ishida et al. 2004)

Fruit part	Fresh weight (FW) (g)	% total FW (%)
Aril	190.0	24.6
Seed	130.0	16.8
Skin	55.0	7.1
Mesocarp	373.7	48.4
Connective tissue	22.6	2.9
Whole fruit	772.0	100

1.4 Fruit Composition

The chemical composition of the fresh gac fruit arils is presented in Table 18.3. The gac fruit aril has a high water content (around 76.8% fresh weight (FW)). The oil content of the aril is about 17.3% in dry weight (DW). Variability of the water and oil content is related to maturity, variety and cultivation conditions of gac fruit (Mai et al. 2013a, b). It has also shown that gac fruit contains a protein that can inhibit the proliferation of cancer cells (Chuyen et al. 2015).

The carotenoids content, especially β-carotene and lycopene, in the gac aril was found to be much higher than that in other common carotenoid-rich fruit (Aoki et al. 2002; Vuong et al. 2006). Different reports of carotenoids analysis in gac fruit are presented in Table 18.4. Gac fruit mesocarp has significantly lower carotenoid contents than gac fruit arils (Aoki et al. 2002; Ishida et al. 2004). The lycopene content of gac fruit arils is greater than that of other fruits considered to be rich in lycopene such as tomato, watermelon, and guava. The total concentration of lycopene in mature gac is about 3053 μg/g FW, compared to 40–50 μg/g FW of tomatoes (Ishida et al. 2004).

All trans-lycopene is the major pigment in ripen gac fruit and has been studied based on its potential health benefits, bioavailability, and changes that occur during fruit ripening and subsequent processing (Kubola et al. 2013; Müller-Maatsch et al. 2017).

The β-carotene content is approximately ten times higher than in western common vegetables rich in β-carotene content, such as carrots (Vuong 2000). The total tocopherol content is around 76 µg/g tocopherol (FW) (Vien 1995; Vuong 2000; Vuong and King 2003; Vuong et al. 2006).

Significant amounts of carotenoids are also present in the fruit mesocarp (Vuong 2000; Aoki et al. 2002; Vuong and King 2003; Vuong et al. 2006) and peel (Kubola and Siriamornpun 2011; Chuyen et al. 2017a). Interestingly, these tissues appear to be richer in carotenoids, particularly lutein, prior to maturity (Kubola and Siriamornpun 2011). The arils and mesocarps of gac fruit contain a significant levels of fatty acids (from 17% to 22%), which is essential for efficient absorption and transport of β-carotene and other soluble vitamins (Vuong et al. 2002; Vuong and King 2003; Kuhnlein 2004). The aril oil composition is detailed in Sect. 4.

Table 18.3 Chemical compositions of Gac arils (Vuong et al. 2002; Mai et al. 2013b)

Composition	Value
Water content (%FW)	76.8 ± 3.3
Oil content (%DW)	17.3 ± 2.6
Total carotenoids content (TCC, mg/g DW)	6.1 ± 0.2
Crude protein (%)	8.2 ± 0.2
Crude fiber (%)	8.7 ± 1.4
Carbohydrate (g/100 g aril)	10.5
Starch (g/100 g)	0.14
Pectin (g/100 g)	1.25
Cellulose (g/100 g)	1.8

Results are expressed as mean values ± S.E.M (standard error of the mean), $N = 3$

Table 18.4 Carotenoids concentration in different reports (µg/g FW)

References	Mesocarp			Aril		
	β-carotene	Lycopene	TCC	β-carotene	Lycopene	TCC
West and Poortvliet (1993)				188,1		891,5
Vien (1995)				458		
Vuong (2000)				355		
Vuong et al. (2002)				175	802	
Aoki et al. (2002)	7–37	0,2–1,6	6–40	60–140	310–460	481
Ishida et al. (2004)	16,3–58,3			636,2–836,3	1546,5–3053,6	2926
Vuong et al. (2006)			283	83,3 ± 40,4	408,4 ± 178,6	497,4 ± 153,7

Extended from Vuong et al. (2006)

TCC total carotenoid content (µg/g FW)

2 Fruit Storage and Preservation Approaches

Fresh ripe gac fruit does only conserve for a few weeks after collection: carotenoid content has been shown to decrease fast after 12 days (Bhumsaidon and Chamchong 2016). Therefore, useful fruit parts conservation is an efficient approach to avoid losses and allow oil extraction or fruit usage during a longer time period (Mai et al. 2013a, b). The two main technics for the fruit conservation are drying and freezing. Freezing of the full fruit as received little attention. Drying of gac aril is the most studied transformation process of gac. Indeed, dried gac fruit aril, plain or in powder form, have their own commercial potential (Tran et al. 2008). Also, drying ease further processing such as like pulp separation from seeds (Mai et al. 2013a, b), and has a huge impact on the yield of oil extraction (Kha et al. 2014).

Most of the developments focus on the aril of gac fruit drying. However, recently, conservation approaches for the further valorization of skin (Chuyen et al. 2017a, b) and mesocarp (Trirattanapikul and Phoungchandang 2016) have emerged.

As drying is a thermal process, potentially in presence of oxygen, the key issue of gac aril drying is the preservation of its carotenoids. As the color of the fruit and its antioxidant activity are mostly controlled by the carotenoids content, these parameters show the same global evolution during drying as the carotenoids content (Mai et al. 2013a, b). However, it should be noted that the total antioxidant activity evolution can show more complex behavior as thermal isomerization between different carotenoids can have an impact on the antioxidant activity for the same total carotenoid content (Phan-Thi and Waché 2014).

Usually, the gac fruit arils are dried to a wet based humidity of 6% at which the water activity is around 0.1 (Tran et al. 2008). Limiting drying to 18% of humidity was shown to be a viable option for shorter-term conservation (Mai et al. 2013a, b). Peels are dried to a wet based humidity between 2% and 6% (Chuyen et al. 2017a). Five main techniques are encountered: oven air drying, vacuum oven drying, heat pump drying spray drying and freeze drying.

2.1 Oven Air-Drying

Gac fruit arils are directly placed in an oven in which hot air flow is achieved. For the carotenoids preservation, optimal drying is achieved around 40 °C or 60 °C (Kha et al. 2011; Mai et al. 2013a, b), as a compromise between drying time and degradation of carotenoids. Still, air drying allows only limited conservation of the carotenoids, around 65% at best both for the arils (Kha et al. 2011; Mai et al. 2013a, b). Pretreatment by soaking with ascorbic acid or bisulfite offer an improvement up to 10% (Kha et al. 2011).

Similar results have been obtained for gac fruit skin drying, with results less dependent on air temperature. An optimal air temperature at 80 °C was identified (Chuyen et al. 2017a) to retain 55% of the carotenoids. Pretreatment with as ascorbic acid allowed to retain 10% more (Chuyen et al. 2017b). Despites a low efficiency in retaining the carotenoids, air drying is widely used due to its simplicity and limited investment costs (Chuyen et al. 2017a).

2.2 Vacuum Drying

Vacuum drying, similar to air-drying but in a reduced total pressure, allows a faster drying and a contact with less oxygen, limiting the carotenoids decompositions. Up to 90% of the carotenoids from arils could be retained at an optimal temperature of 60 °C (Tran et al. 2008; Mai et al. 2013a, b). For the vacuum drying of gac skin, no significant improvement compared to air drying was observed (Chuyen et al. 2017a).

2.3 Heat Pump Drying

Heat pump drying allows efficient lower temperature operation than an air vacuum technics. This approach was only tested for gac fruit skin drying. At 30 °C, the results for carotenoids retention were not significantly different from air drying at 60 °C due to a longer time of operation (Chuyen et al. 2017a). The interest of investing in heat pump drying for gac fruit skin seems limited.

2.4 Spray Drying

Spray drying of gac fruit aril allows the direct production of marketable powder. Direct spray drying of pulp allows only to retain only 6% of the carotenoids (Tran et al. 2008). Optimizing the spray drying, by using an air temperature of 120 °C and adding 10% of maltodextrin as encapsulation agent, allows conserving close to 50% of the carotenoids. (Kha et al. 2010). Further optimization using whey gum as a carrier (Kha et al. 2014c) and using surface response methodology for process optimization lead the conservation of 90% of the carotenoids (Kha et al. 2014b). The achieved powders can be stored during several months with limited loss in their micronutrients content (Kha et al. 2015).

2.5 Freeze-Drying

Freeze-drying is the costliest way to dry gac aril but it offers unequaled carotenoids conservation thanks to its very low temperature and limited contact with air. Studies show a conservation of carotenoids ranging from 82% (Mai et al. 2013a, b), 88%

(Bruno et al. 2018) to 100% (Tran et al. 2008). Yet, as this method has high investment and operational costs compared to the previously presented methods, it is seldom used outside of the laboratory for gac fruit drying.

3 Oil Extraction

Although the other parts of the fruit potentially contain oil, gac oil refers to the oil contained extracted from arils. Gac oil receives a growing attention, as it is the most direct way of accessing the liposoluble carotenoids contained in the arils. Two main approaches can be used to extract oil: mechanical approaches, mostly pressing (Kha et al. 2013), and physicochemical approaches, based on oil solubility in solvents. The two main type of solvent used is organic solvent (Kubola et al. 2013) and supercritical fluids (Tai and Kim 2014). Yet, despites oil not being soluble in water, enzymatically assisted water extraction is also possible (Mai et al. 2013a, b; Thi et al. 2016).

For all the techniques, various pretreatment of the arils can be considered to enhance the extraction efficiency such as drying (Kubola et al. 2013; Kha et al. 2014), microwave heating coupled with steaming (Kha et al. 2013), ohmic heating (Aamir and Jittanit 2017) or enzymatic treatment (Kha et al. 2013; Mai et al. 2013a, b; Thi et al. 2016). This treatment, through the heat of enzymatic activity affect gac aril cell structure, easing oil extraction. The various pretreatment and extraction methods can have a varied impact on the oil content, including on the heat-sensitive carotenoids (Kha et al. 2013).

3.1 Mechanical Extraction

Pressing is the most common extraction method encountered. While low cost and technologically simple, it usually allows extracting only around 70% of the available oil. For gac oil, this yield can be achieved with air-dried arils with 170 kg/m^2. Combining microwave drying (at 630 W for 65 min) and partial re-humidification by steaming, it was possible to reach 93% of yield. This treatment seems to offer a better conservation of carotenoids present in the oil, doubling the amount compared to air-dried aril oil (Kha et al. 2013).

3.2 Organic Solvents Extraction

Various organic solvent have been tested for gac oil extraction including chloroform/methanol mix, petroleum ether (Kubola et al. 2013), and n-hexane (Thuat 2010; Kubola et al. 2013; Aamir and Jittanit 2017). About 95% of the oil contained in air-dried gac aril can be extracted with n-hexane at 50 °C but the process can take up to

18 h (Thuat 2010). Treating the dried arils with commercial enzymatic mixtures Viscozyme L (composed of arabanase, cellulase, hemicellulase, β-glucanase and xylanase) can reduce the extraction time to 2 h. Fresh aril submitted to ohmic heating during the extraction exhibit yields close to 100% (Aamir and Jittanit 2017). On top of the hazardous question raised by an organic solvent, the use of gac oil for food and pharmaceutical applications tend to reduce the interest of organic solvent extraction that has to be totally removed from the final product before consumption.

3.3   Supercritical Fluid Extraction

Supercritical fluid extraction combines the efficiency of the organic solvent extraction with the ease of separation of the supercritical fluid from the extracted oil. On top of that, the extraction is usually faster than with organic solvents. The main opposition to the use of supercritical fluid come from the investment cost and the requirement of a specifically skilled workforce of this advanced technology (Martins et al. 2015). Supercritical carbon dioxide is the most common supercritical fluid and is the only one tested for gac oil extraction as well as for a specific extraction of carotenoids (Kha et al. 2014).

For other extraction methods, the pretreatment applied to the gac aril has a drastic impact on the amount of oil that is effectively extracted. Supercritical carbon dioxide at 200 bar and 50 °C flowing at 70 $kg.h^{-1}.kg_{gac}^{-1}$ allowed to extract 95% of the total oil content of 50 °C air dried arils in 3 h. An additional treatment with 0.1% of pectinase enzymes allowed to achieve the same yield in only 2 h (Kha et al. 2014). However, the most efficient way to accelerate the extraction is to rise the pressure wherein 95% of the oil of air-dried aril can be achieved in 30 min at 400 bar (Tai and Kim 2014).

3.4   Enzymatic Aqueous Extraction

Using enzyme to free the oil contained in gac aril is a low investment alternative cost to solvent extraction. A high yield of oil extraction requires the synergic effect of multiple enzymes. A mix of pectinase, cellulase, protease and α-amylase allowed to extract 82% in slightly more than 2 h (Mai et al. 2013a, b). However, the operational cost of the enzymes acquisition can hinder the profitability of such a process.

4   Oil Properties

The main properties of the extracted oil depend on the oil extraction process and upstream processing as well as on the initial fruit that is used and its storage conditions (Bhumsaidon and Chamchong 2016).

4.1   Fatty Acid Composition

Table 18.5 summarizes the main fatty acid present in gac oil as obtained by different authors with various extraction methods. Variations are observed between the studies. They can be attributed to the difference in fruit initial content, extraction technics as well as quantification methods. However, global trends can be highlighted. The majority (56–75%) of the fatty acid present in gac oil are unsaturated with mainly oleic acid. Most of the rest of the unsaturated fatty acid are polyunsaturated with a dominance of linoleic acid. In this fraction, significant concentration in ω-3 fatty acid is present, mainly in the form of α-linolenic acid. In the saturated fatty acid, palmitic acid is the most present. The oil also contains a significant amount of stearic acid. The mixture of unsaturated, saturated, poly- and mono-unsaturated fatty acids in gac oil improves the absorption and bioavailability of nutrients and carotenoids (Müller-Maatsch et al. 2017).

4.2   Physicochemical Properties

Table 18.6 summarizes typical physicochemical properties of gac oil including non-saponification matter, refraction, melting point, viscosity and density value (Mai et al. 2016). The iodine value of gac aril oil corresponds to a high degree of unsaturated oil (76 g I_2/100 g oil). The high saponification value (715 mg KOH/g) indicates that the triglycerides of gac oil are composed of short fatty acids.

Table 18.7 presents acid value and peroxide values depending on the treatment. The acid value of gac oil is between 0.69 and 3.6 mg KOH/g depending on the extraction method, which is lower than that of some common oils such as soybean oil (about 6 mg KOH/g). The lowest peroxide value measured (around 0.89 meq O_2/kg oil) characterizes the purity and stability of this oil at ambient temperature.

4.3   Carotenoids Content and Other Minor Bioactive Compounds

Fruits and processing variation have a significant impact on the carotenoids content. As carotenoids are a thermosensitive molecule, the processing conditions have an exacerbate impact on the carotenoids content. Studies have shown qualitatively that, for any given treatment, rising the extraction yields also leads to an increase of the retrieved carotenoids concentration (Kha et al. 2013; Mai et al. 2013a, b; Tai and Kim 2014; Thi et al. 2016). This can probably be attributed to the fact that the carotenoids are preferably stored in gac fruit regions that are more difficult to reach.

Table 18.8 summarizes the total carotenoids content, lycopene and β- carotene concentrations obtained by the various author with different processing. The total

Table 18.5 Fatty acid composition of gac oil for different methods of extraction

Study	Vuong et al. (2002)	Thuat (2010)	Kha et al. (2014a)	Thi et al. (2016)	Mai et al. (2013b)	Kha et al. (2014a)	Kha et al. (2014a)	Bruno et al. (2018)
Extraction method	Organic solvent	Organic solvent	Organic solvent	Enzyme and water	Enzyme and water	Press	Microwave and press	Supercritical CO_2
Lauric (C12:0)				0.02	0.02			
Myristic (C14:0)	0.87	0.21	1.09	0.37	0.22	0.63	0.41	0.8
Pentadecanoic (C15:0)				0.1				
Palmitic (C16:0)	22.04	20.27	34.73	24.18	17.31	34.89	24.99	30.1
Palmitoleic (C16:1)	0.26	0.23	0.19	0.16	0.18	0.18	0.4	
Margaric (C17:0)		0.23		0.15	0.14			
Stearic (C18:0)	7.06	5.35	8.45	3.52	7.45	7.78	6.85	5.1
Oleic (C18:1)	35.21	49.57	45.04	48.99	59.5	40.58	48.25	44.5
Linoleic (C18:2)	31.43	23.19	10.14	21.09	13.98	15.6	18.28	19.6
α-linolenic (C18:3)	2.14	0.94	0.37	0.86	0.52	0.34	0.83	
Arachidic (C20:0)	0.39			0.21	0.32			
Eicosa-11-enoic (C20:1)	0.15			0.23	0.17			
Arachinodic (C20:4)	0.1							
Docosanoic (C22:0)	0.19			0.03				
Erucic (C22:1)				0.07	0.1			
Docosahexanoic (C22:6)				0.02				
Tetracosanoic (C24:0)	0.14			0.04				

Table 18.6 Physicochemical properties of the Gac oil (Mai et al. 2016)

Index	Value
Iodine value (g I_2/100 g oil)	76.58 ± 1.9
Saponification value (mg KOH/g)	715.16
Non-saponification matter (%)	0.5
Refraction value ($n^{25}D$)	1.47
Melting point (°C)	12
Viscosity (Pa.s)	0.0466 ± 0.0004
Density (g/mL)	0.955 ± 0.012

Results are expressed as mean values ± S.E.M (standard error of the mean), $N = 3$

Table 18.7 Acidity value and peroxide values depending on the treatment

Authors	Treatment	Acidity value (mg KOH/g)	Peroxide (meq O_2/kg oil)
Thuat (2010)	Hexane extraction	3.58	8.7
Kha et al. (2014a)	Microwave + press	0.69	1.8
Kha et al. (2014a)	Press	1.8	7.7
Kha et al. (2014a)	Solvent extraction	2.19	33.54
Mai et al. (2013b)	Enzymatic extraction	2.55	0.89

carotenoids content in the gac oil can go up to 10 mg/g, in which lycopene and β-carotene are representing a majority, up to 5 mg/g for each.

These amounts in carotenoids are larger than what is reported in usual fruit and vegetables. Common source of lycopene compound includes tomato (31 µg/g), watermelon (41 µg/g), guava (54 µg/g) and pink grapefruit (33.6 µg/g) (Mangels et al. 1993; Aoki et al. 2002).

Carotenoids predominantly occur in their all-*trans* configuration, which is thermos-dynamically the most stable isomer. All-*trans*-lycopene may be converted to *cis* configuration during processing and they possess different biological properties (Lee and Chen 2002). Several reports have demonstrated that the cis-isomers of lycopene are absorbed into the body more easily and play a more important part in biological function than all-*trans*-lycopene (Böhm and Bitsch 1999; Failla et al. 2008).

cis-isomers of lycopene correspond to 2.7–13.2% of the total while *cis*-isomer of β-carotene range between 6.1% and 25.3% of the total. The α-carotene was also found at a lower concentration (1% of the total carotenoids) in gac oil (Ishida et al. 2004; Bruno et al. 2018). Moderate heat treatment (exposure of oil at 80 °C for 4 h) lead to isomerization of lycopene contained in the oil from a fraction of all-*trans*-isomer to different *cis*-isomer, mainly 13-*cis* (22% of the total) and 9-*cis* (16%) isomers (Phan-Thi and Waché 2014).

Carotenoids extracted from a natural source are usually in the free form or as fatty acid esters. Zeaxanthin and β-cryptoxanthin have hydroxyl groups in the six-membered ring, and these hydroxyl groups can bind with fatty acids to form carotenoid esters. Therefore, zeaxanthin and β-cryptoxanthin were found in saponified

Table 18.8 Mass concentration (mg/g oil) of total carotenoids, lycopene and β-carotene using various processes

Authors	Treatment	Total carotenoids (mg/g)	Lycopene (mg/g)	β-carotene (mg/g)
Thuat (2010)	Hexane			3.2
Thi et al. (2016)	Hexane + enzymes	7.7		
Aamir and Jittanit (2017)	Ohmic heating + hexane		1.5	5.8
Kubola et al. (2013)	Chloroform + methanol extraction		0.49	1.18
Mai et al. (2014)	Enzymatic extraction	7.9	3.4	2.69
Kha et al. (2014a)	Press		2.51	0.57
Kha et al. (2014a)	Microwave + press		4.33	1.46
Kha et al. (2014)	Supercritical CO_2		5.08	0.83
Tai and Kim (2014)	Supercritical CO_2	10.9		
Bruno et al. (2018)	Supercritical CO_2	4.63	2.4	1.57

oil samples. The degree of esterification of carotenoids in some fruit increases during ripening (Subagio and Morita 1997). The analysis of the saponified samples detected also a trace of zeaxanthin and β-cryptoxanthin (Aoki et al. 2002).

4.4 Antioxidant Activity

The antioxidant capacity of gac oil strongly depends on the content and bioavailability of carotenoids, especially, β-carotene and lycopene (Mai et al. 2013a, b) like for other oils (Thaipong et al. 2006). Other phenolic compounds, vitamin E and unsaturated fatty acids including oleic and linoleic acid, contribute also on the antioxidant capacity of gac oil.

Lycopene exhibits a high physical quenching rate of singlet oxygen (Di Mascio et al. 1989; Perretti et al. 2013), which is directly related to its antioxidant activity. The antioxidant activity of lycopene in multi-lamellar liposomes is superior to other lipophilic natural antioxidants such as α-tocopherol, α-carotene, β-cryptoxanthin, zeaxanthin, β-carotene and lutein (Stahl and Sies 2007).

While the existence of a significant antioxidant activity of gac oil is generally well accepted, their quantification leads to potentially contradictory results in the few existing studies. Most studies do not directly measure the antioxidant activity of the oil but rather of dry powder (Kha et al. 2010, 2011; Mai et al. 2013a, b). Some hydrophilic compounds also play a role in the antioxidant activity (Mai et al. 2013a, b).

The comparison of the antioxidant properties in various conditions is also hardened by the discrepancies between the methods used to evaluate the antioxidant activity. Moreover, the different process has complex impacts on these activities. Indeed, while treatment induce can induce a loss in carotenoids, moderate heat leads to isomerization that can raise the antioxidant activity (Phan-Thi and Waché 2014).

With the ABTS antioxidant assay (Thaipong et al. 2006), it was shown that the global antioxidant activity is lower with the drying at higher temperatures, in the range of 0.3 mmol Trolox equivalent per gram of powder (TEAC) at 40 °C to 0.2 TEAC at 80 °C (Kha et al. 2011). This seems contradictory, or at least underlying more complex phenomena, with a study showing a doubling of the TEAC with a treatment of oil at 80 °C during 4 h (Phan-Thi and Waché 2014).

Regarding the DPPH· test (Thaipong et al. 2006), one study highlighted similar result as with ABTS, with a reduction of the activity from 0.25 TEAC to 0.2 TEAC (Kha et al. 2011) while another suggests that the activity is best conserved at 60 °C where 0.18 TEAC (80% of the activity observed for the fresh sample) would be conserved. The same study shows results going accordingly with FRAP, while DMPD tests gave uncorrelated results (Mai et al. 2013a, b).

5   Oil Concentration and Carotenoids Extractions

Further transformation of the oil can be considered to enhance its carotenoids content or to valorize pure carotenoids. Few works have dealt with that question. Crossflow filtration of gac oil with a cut-off size of 5 nm was shown to allow a concentration of total carotenoids in the retentate of factor 8.6 while the acid index was divided by a factor 40 (Mai et al. 2014). Crystallization of the carotenoids by mixing the oil with propylene glycol followed by saponification with KOH, allows to achieved crystals containing 94% of carotenoids (Mai et al. 2016). Other processes could be considered for oil concentration. An example of potentially interesting development would be the use of supercritical carbon dioxide, which was proved to efficiently extract the oil to fractionate the oil, like it was done for tomato oil (Perretti et al. 2013).

6   Existing and Foreseen Gac Oil Product

Gac oil is a premier source of carotenoids, especially β-carotene and lycopene. Because of the high content of carotenoids and fatty acid and its high bioactivity, gac fruit can be consumed as a natural treatment for vitamin A deficiency in children in developing countries (Vuong et al. 2002; Vuong and King 2003). Gac oil can be used in many food applications, for example, cooking oil, salad oil, seasoning and food coloring, or in cosmetic applications including soap and skin oils.

6.1 Food Applications

In Vietnam, only the aril of ripening gac fruit is traditionally used as natural colorant and additive for cooking. For example, it is added into sticky rice to produce a brilliant orange rice dish known as "xoi gac". This meal is prepared by mixing araculae of gac fruit with cooked rice to give a red color, a lustrous appearance and a distinct flavor. This food preparation is served as one of the special meals at New Year celebration, wedding and other important celebrations. Consumption of this traditional food could produce a substantial increase in β-carotene intake (Vuong 2000). In addition, women and children in Vietnam readily accepted consumption of gac oil because it can help reducing lard intake (Vuong et al. 2002; Vuong and King 2003).

In Thailand, gac aril is cooked and eaten with chili paste or cooked in a curry (Kubola et al. 2013). The incorporation of 1.0% gac aril powder can, therefore, be used to reduce the amount of nitrite added to Vienna sausage from 125 to 75 ppm, resulting in more red and darker sausage with higher lycopene content (Wimontham and Rojanakorn 2016).

Gac oil is used as an additional nutrition and natural food additives. β-carotene is well known for its pro-vitamin A activity. Lycopene is added to foods to increase the nutritional value of products, in particular for dairy products, energy drinks and fruit juices. β-carotene can also be added to livestock feeds to improve the quality of milk or eggs. The orange to red colors of lycopene and β-carotene are widely used in foods and beverages as natural colors, which are usually considered to be safer than other artificial colors. These natural colors are used to enhance, change or contribute to the color of food products such as fruit juices, sweets, butter, cheeses and sauces. In addition, they can act as antioxidants to extend product shelf life.

6.2 Non-food Applications

Gac oil is also used as a traditional medicine. For example, gac fruit has been applied to treat conditions of the eyes, burns, skin problems and wounds. It was reported that when applied to wounds, skin infections and burns, gac oil stimulates the growth of new skin and the healing of wounds. The fruit is also frequently used as a traditional remedy for arthritis and cardiovascular diseases and degeneration of the macula (Burke et al. 2005).

Because of its high concentration of β-carotene, gac fruit is a valuable aid in preventing or treating vitamin A deficiency. Therefore, the gac aril is used to make a tonic for children and lactating or pregnant women, and to treat xerophthalmia and night blindness (Guichard and Bui 1941). In many developing countries, vitamin A deficiency is epidemic because it can lead to poor night vision, blindness, higher rates of maternal mortality, poor embryonic growth, and reduce the ability to fight infections and lactation. Supplementation with gac fruit extract can alleviate chronic vitamin A deficiency, and help to reduce these health problems (West and Poortvliet 1993; Vuong 2000).

The lycopene and β-carotene in gac fruit enhance skin health by mitigating oxidative damage in tissue. The various antioxidants in gac fruit boost heart health by specifically combating atherosclerosis. Additionally, both lycopene and β-carotene show protective activity against the risk of heart attack.

Gac oil could be used as pharmaceutical ingredients. Pure lycopene and β-carotene are commonly used in pharmaceuticals, such as multiple vitamin tablets for a nutritional supplement or for vitamin A deficiency patients. Several new gac products are currently being developed on the world market. All products obtained from gac have antioxidant characteristics, determined by the bioactive compounds it contains, such as lycopene, β-carotene and vitamin C.

In Vietnam, Vnpofood is the largest manufacturer with a capacity of 3000 tons of gac fruit per year. Their products include gac oil capsules (Vinaga). These gac oil capsules contain pure gac oil. Another brand of gac oil capsules is Garotene, produced by the University of Hanoi Pharmacy. These capsules are enriched with vitamin E. There is also a product called G3, produced by Pharmanex, a company in the United States. It is a combination of *lycium chinensis* fruit phytonutrients, Siberian pineapple and gac fruit.

References

Aamir, M., & Jittanit, W. (2017). Ohmic heating treatment for Gac aril oil extraction: Effects on extraction efficiency, physical properties and some bioactive compounds. *Innovative Food Science & Emerging Technologies*. Elsevier, *41*, 224–234. https://doi.org/10.1016/J.IFSET.2017.03.013.

Aoki, H., Ieu, N. T. M., Kuze, N., Tomisaka, K., & Van Chuyen, N. (2002). Carotenoid pigments in gac fruit (*Momordica cochinchinensis* Spreng). *Bioscience, Biotechnology, and Biochemistry*. Japan Society for Bioscience, Biotechnology, and Agrochemistry, *66*(11), 2479–2482. https://doi.org/10.1271/bbb.66.2479.

Bailey, L. H. (1937). *The garden of gourds*. New York: Macmillan.

Bhumsaidon, A., & Chamchong, M. (2016). Variation of lycopene and beta-carotene contents after harvesting of gac fruit and its prediction. *Agriculture and Natural Resources*. Elsevier, *50*(4), 257–263. https://doi.org/10.1016/J.ANRES.2016.04.003.

Böhm, V., & Bitsch, R. (1999). Intestinal absorption of lycopene from different matrices and interactions to other carotenoids, the lipid status, and the antioxidant capacity of human plasma. *European Journal of Nutrition*. Steinkopff-Verlag, *38*(3), 118–125. https://doi.org/10.1007/s003940050052.

Bruno, A., Durante, M., Marrese, P. P., Migoni, D., Laus, M. N., Pace, E., Pastore, D., Mita, G., Piro, G., & Lenucci, M. S. (2018). Shades of red: Comparative study on supercritical CO_2 extraction of lycopene-rich oleoresins from gac, tomato and watermelon fruits and effect of the α-cyclodextrin clathrated extracts on cultured lung adenocarcinoma cells' viability. *Journal of Food Composition and Analysis*. Academic Press, *65*, 23–32. https://doi.org/10.1016/J.JFCA.2017.08.007.

Burke, D. S., Smidt, C. R., & Vuong, L. T. (2005). *Momordica Cochinchinensis*, *Rosa Roxburghii*, wolfberry, and sea buckthorn—highly nutritional fruits supported by tradition and science. *Current Topics in Nutraceutical Research, 3*(4), 259–265.

Chuyen, H. V., Nguyen, M. H., Roach, P. D., Golding, J. B., & Parks, S. E. (2015). Gac fruit (*Momordica cochinchinensis* Spreng.): A rich source of bioactive compounds and its potential health benefits. *International Journal of Food Science and Technology*. Wiley/Blackwell (10.1111), *50*(3), 567–577. https://doi.org/10.1111/ijfs.12721.

Chuyen, H. V., Roach, P. D., Golding, J. B., Parks, S. E., & Nguyen, M. H. (2017a). Effects of four different drying methods on the carotenoid composition and antioxidant capacity of dried Gac peel. *Journal of the Science of Food and Agriculture*. Wiley-Blackwell, *97*(5), 1656–1662. https://doi.org/10.1002/jsfa.7918.

Chuyen, H. V., Roach, P. D., Golding, J. B., Parks, S. E., & Nguyen, M. H. (2017b). Effects of pre-treatments and air drying temperatures on the carotenoid composition and antioxidant capacity of dried gac peel. *Journal of Food Processing & Preservation*. Wiley/Blackwell (10.1111), *41*(6), e13226. https://doi.org/10.1111/jfpp.13226.

Di Mascio, P., Kaiser, S., & Sies, H. (1989). Lycopene as the most efficient biological carotenoid singlet oxygen quencher. *Archives of Biochemistry and Biophysics*. Academic Press, *274*(2), 532–538. https://doi.org/10.1016/0003-9861(89)90467-0.

Failla, M. L., Chitchumroonchokchai, C., & Ishida, B. K. (2008). In vitro micellarization and intestinal cell uptake of cis isomers of lycopene exceed those of all-trans lycopene. *The Journal of Nutrition*. Oxford University Press, *138*(3), 482–486. https://doi.org/10.1093/jn/138.3.482.

Guichard, F., & Bui, D. S. (1941). La matiere colorante du fruit du Momordica cochinchinnensis Spr. *Annales de l'ecole Superieure de Medecine et de Pharmacie de l'Indochine, 5*, 41–42.

Ishida, B. K., Turner, C., Chapman, M. H., & McKeon, T. A. (2004). Fatty acid and carotenoid composition of gac (*Momordica cochinchinensis* Spreng) fruit. *Journal of Agricultural and Food Chemistry, 52*(2), 274–279. https://doi.org/10.1021/jf030616i.

Kha, T. C., Nguyen, M. H., & Roach, P. D. (2010). Effects of spray drying conditions on the physicochemical and antioxidant properties of the Gac (*Momordica cochinchinensis*) fruit aril powder. *Journal of Food Engineering*. Elsevier, *98*(3), 385–392. https://doi.org/10.1016/J.JFOODENG.2010.01.016.

Kha, T. C., Nguyen, M. H., & Roach, P. D. (2011). Effects of pre-treatments and air drying temperatures on colour and antioxidant properties of Gac fruit powder. *International Journal of Food Engineering, 7*(3). https://doi.org/10.2202/1556-3758.1926.

Kha, T. C., Nguyen, M. H., Roach, P. D., & Stathopoulos, C. E. (2013). Effects of Gac aril microwave processing conditions on oil extraction efficiency, and β-carotene and lycopene contents. *Journal of Food Engineering*. Elsevier, *117*(4), 486–491. https://doi.org/10.1016/J.JFOODENG.2012.10.021.

Kha, T. C., Phan-Tai, H., & Nguyen, M. H. (2014). Effects of pre-treatments on the yield and carotenoid content of Gac oil using supercritical carbon dioxide extraction. *Journal of Food Engineering*. Elsevier, *120*, 44–49. https://doi.org/10.1016/J.JFOODENG.2013.07.018.

Kha, T. C., Nguyen, M. H., Roach, P. D., & Stathopoulos, C. E. (2014a). Effect of drying pre-treatments on the yield and bioactive content of oil extracted from Gac aril. *International Journal of Food Engineering, 10*(1), 103–112. https://doi.org/10.1515/ijfe-2013-0028.

Kha, T. C., Nguyen, M. H., Roach, P. D., & Stathopoulos, C. E. (2014b). Microencapsulation of Gac oil: Optimisation of spray drying conditions using response surface methodology. *Powder Technology*. Elsevier, *264*, 298–309. https://doi.org/10.1016/J.POWTEC.2014.05.053.

Kha, T. C., Nguyen, M. H., Roach, P. D., & Stathopoulos, C. E. (2014c). Microencapsulation of Gac oil by spray drying: Optimization of wall material concentration and oil load using response surface methodology. *Drying Technology, 32*(4), 385–397. https://doi.org/10.1080/07373937.2013.829854.

Kha, T. C., Nguyen, M. H., Roach, P. D., & Stathopoulos, C. E. (2015). A storage study of encapsulated gac (*Momordica cochinchinensis*) oil powder and its fortification into foods. *Food and Bioproducts Processing*. Elsevier, *96*, 113–125. https://doi.org/10.1016/J.FBP.2015.07.009.

Kubola, J., & Siriamornpun, S. (2011). Phytochemicals and antioxidant activity of different fruit fractions (peel, pulp, aril and seed) of Thai gac (*Momordica cochinchinensis* Spreng). *Food Chemistry*. Elsevier, *127*(3), 1138–1145. https://doi.org/10.1016/J.FOODCHEM.2011.01.115.

Kubola, J., Meeso, N., & Siriamornpun, S. (2013). Lycopene and beta carotene concentration in aril oil of gac (*Momordica cochinchinensis* Spreng) as influenced by aril-drying process and solvents extraction. *Foodservice Research International*. Elsevier, *50*(2), 664–669. https://doi. org/10.1016/J.FOODRES.2011.07.004.

Kuhnlein, H. V. (2004). Karat, pulque, and gac: Three shining stars in the traditional food galaxy. *Nutrition Reviews, 62*(11), 439–442.

Lee, M., & Chen, B. (2002). Stability of lycopene during heating and illumination in a model system. *Food Chemistry*. Elsevier, *78*(4), 425–432. https://doi.org/10.1016/S0308-8146(02)00146-2.

Mai, H. C., Truong, V., Haut, B., & Debaste, F. (2013a). Impact of limited drying on *Momordica cochinchinensis* Spreng. Aril carotenoids content and antioxidant activity. *Journal of Food Engineering, 118*(4). https://doi.org/10.1016/j.jfoodeng.2013.04.004.

Mai, H. C., Truong, V., & Debaste, F. (2013b). Optimization of enzyme-aided extraction of oil rich in carotenoids from gac fruit (*Momordica cochinchinensis* Spreng.). *Food Technology and Biotechnology, 51*(4), 488–499.

Mai, H. C., Truong, V., & Debaste, F. (2014). Carotenoids concentration of gac (*Momordica cochinchinensis* Spreng.) fruit oil using cross-flow filtration technology. *Journal of Food Science*. Wiley/Blackwell (10.1111), *79*(11), E2222–E2231. https://doi.org/10.1111/1750-3841.12661.

Mai, H. C., Truong, V., & Debaste, F. (2016). Carotenoids purification from gac (*Momordica cochinchinensis* Spreng.) fruit oil. *Journal of Food Engineering, 172*. https://doi.org/10.1016/j. jfoodeng.2015.09.022..

Mangels, A. R., Holden, J. M., Beecher, G. R., Forman, M. R., & Lanza, E. (1993). Carotenoid content of fruits and vegetables: An evaluation of analytic data. *Journal of the American Dietetic Association*. Elsevier, *93*(3), 284–296. https://doi.org/10.1016/0002-8223(93)91553-3.

Martins, P. F., de Melo, M. M. R., & Silva, C. M. (2015). Gac oil and carotenes production using supercritical CO2: Sensitivity analysis and process optimization through a RSM–COM hybrid approach. *The Journal of Supercritical Fluids*. Elsevier, *100*, 97–104. https://doi.org/10.1016/J. SUPFLU.2015.02.023.

Müller-Maatsch, J., Sprenger, J., Hempel, J., Kreiser, F., Carle, R., & Schweiggert, R. M. (2017). Carotenoids from gac fruit aril (*Momordica cochinchinensis* [Lour.] Spreng.) are more bioaccessible than those from carrot root and tomato fruit. *Food Research International*. Elsevier, *99*, 928–935. https://doi.org/10.1016/J.FOODRES.2016.10.053.

Perretti, G., Troilo, A., Bravi, E., Marconi, O., Galgano, F., & Fantozzi, P. (2013). Production of a lycopene-enriched fraction from tomato pomace using supercritical carbon dioxide. *The Journal of Supercritical Fluids*. Elsevier, *82*, 177–182. https://doi.org/10.1016/J. SUPFLU.2013.07.011.

Phan-Thi, H., & Waché, Y. (2014). Isomerization and increase in the antioxidant properties of lycopene from *Momordica cochinchinensis* (gac) by moderate heat treatment with UV–vis spectra as a marker. *Food Chemistry*. Elsevier, *156*, 58–63. https://doi.org/10.1016/J. FOODCHEM.2014.01.040.

Shadeque, A., & Baruah, G. K. S. (1984). Sweet gourd: A popular vegetable of Assam. *Indian Farming, 34*, 25–35.

Stahl, W., & Sies, H. (2007). Carotenoids and flavonoids contribute to nutritional protection against skin damage from sunlight. *Molecular Biotechnology*. Humana Press Inc, *37*(1), 26–30. https:// doi.org/10.1007/s12033-007-0051-z.

Subagio, A., & Morita, N. (1997). Changes in carotenoids and their fatty acid esters in banana peel during ripening. *Food Science and Technology International, Tokyo, 3*(3), 264–268. https://doi. org/10.3136/fsti9596t9798.3.264.

Tai, H. P., & Kim, K. P. T. (2014). Supercritical carbon dioxide extraction of gac oil. *The Journal of Supercritical Fluids*. Elsevier, *95*, 567–571. https://doi.org/10.1016/J.SUPFLU.2014.09.005.

Thaipong, K., Boonprakob, U., Crosby, K., Cisneros-Zevallos, L., & Hawkins Byrne, D. (2006). Comparison of ABTS, DPPH, FRAP, and ORAC assays for estimating antioxidant activity from guava fruit extracts. *Journal of Food Composition and Analysis, 19*(6–7), 669–675. https://doi.org/10.1016/j.jfca.2006.01.003.

Thi, T., Nhi, Y., & Tuan, D. Q. (2016). Enzyme assisted extraction of gac oil (*Momordica cochinchinensis* Spreng) from dried aril. *Journal of Food and Nutrition Sciences, 4*(1), 1–6. https:// doi.org/10.11648/j.jfns.20160401.11.

Thuat, B. Q. (2010). Research on extraction technology to improve yield and quality of oil from gac aril (*Momordica cochinchinensis* spreng L.). *Vietnam Journal of Science and Technology, 48*(1). https://doi.org/10.15625/0866-708X/48/1/1089.

Tran, T. H., Nguyen, M. H., Zabaras, D., & Vu, L. T. T. (2008). Process development of Gac powder by using different enzymes and drying techniques. *Journal of Food Engineering*. Elsevier, *85*(3), 359–365. https://doi.org/10.1016/J.JFOODENG.2007.07.029.

Trirattanapikul, W., & Phoungchandang, S. (2016). Influence of different drying methods on drying characteristics, carotenoids, chemical and physical properties of Gac fruit pulp (*Momordica cochinchinensis* L.). *International Journal of Food Engineering, 12*(4), 395–409. https://doi.org/10.1515/ijfe-2015-0162.

Vien, D. D. (1995). *Thanh Phan Dinh Duong Thuc An Viet Nam [Food Products in Viet Nam Composition and Nutritive Value]*. Hanoi: Nha Xuat Ban Y Hoc.

Vuong, L. T. (2000). Underutilized β-carotene–rich crops of Vietnam. *Food and Nutrition Bulletin*. SAGE PublicationsSage CA: Los Angeles, CA, *21*(2), 173–181. https://doi.org/10.1177/156482650002100211.

Vuong, L. T., & King, J. C. (2003). A method of preserving and testing the acceptability of Gac fruit oil, a good source of β-carotene and essential fatty acids. *Food and Nutrition Bulletin*. SAGE PublicationsSage CA: Los Angeles, CA, *24*(2), 224–230. https://doi.org/10.1177/156482650302400209.

Vuong, L. T., Dueker, S. R., & Murphy, S. P. (2002). Plasma β-carotene and retinol concentrations of children increase after a 30-d supplementation with the fruit *Momordica cochinchinensis* (gac). *The American Journal of Clinical Nutrition.*. Oxford University Press, *75*(5), 872–879. https://doi.org/10.1093/ajcn/75.5.872.

Vuong, L. T., Franke, A. a., Custer, L. J., & Murphy, S. P. (2006). *Momordica cochinchinensis* Spreng. (gac) fruit carotenoids reevaluated. *Journal of Food Composition and Analysis, 19*(6–7), 664–668. https://doi.org/10.1016/j.jfca.2005.02.001.

West, C. E., & Poortvliet, E. J. (1993). *The carotenoid content of foods with special reference to developing countries*. Arlington: USAID.

Wimontham, T., & Rojanakorn, T. (2016). Effect of incorporation of Gac (*Momordica cochinchinensis*) aril powder on the qualities of reduced-nitrite Vienna sausage. *International Food Research Journal, 23*(3), 1048–1055.

Chapter 19
Goldenberry (*Physalis peruviana*) Oil

Mohamed Fawzy Ramadan and Jörg-Thomas Mörsel

Abstract Non-traditional fruits play an important role in human nutrition as an excellent source of bioactive phytochemicals. Highly valued for its unique flavor, texture and color, recent research had shown *Physalis peruviana* to be rich in lipid active compounds. Total lipids in the whole berry were 2.0%. In *P. peruviana* oil, linoleic and oleic were the main unsaturated fatty acids, while palmitic and stearic acids were the major saturates. Neutral lipids comprised more than 95% of total lipids in whole berry oil. Triacylglycerols were the predominant lipid class and constituted about 81% of total neutral lipids in the whole berry oil. The oil is also rich in phytosterols and tocopherols. Campesterol and β-sitosterol were the main sterols, while β- and γ-tocopherols were the main components in whole *P. peruviana* berry oil. β-Carotene and vitamin K_1 were also measured in high levels in *P. peruviana* pulp/peel oil. This chapter provides a valuable source for current knowledge on bioactive lipids in *P. peruviana*.

Keywords Cape gooseberry · Seed oil · Pulp/peel oil · Phytosterols · Tocopherols · β-carotene · Vitamin K_1

1 Introduction

Berries have been shown to provide health benefits because of their high natural antioxidants, minerals, and vitamins (Zhao 2007; Ramadan 2011; Mokhtar et al. 2018). *Physalis peruviana* (family Solanaceae) plants are annuals or short-lived perennials. The flowers are bell-shaped, but the most distinctive feature is the fruiting calyx, which enlarges to cover the fruit and hangs downwards like a lantern.

M. F. Ramadan (✉)
Agricultural Biochemistry Department, Faculty of Agriculture, Zagazig University, Zagazig, Egypt

J.-T. Mörsel
UBF-Untersuchungs-, Beratungs-, Forschungslaboratorium GmbH, Altlandsberg, Berlin, Germany

© Springer Nature Switzerland AG 2019
M. F. Ramadan (ed.), *Fruit Oils: Chemistry and Functionality*,
https://doi.org/10.1007/978-3-030-12473-1_19

Fig. 19.1 *P. peruviana* in opened calyx. The fruit is a berry, with smooth, waxy, orange-yellow skin and juicy pulp containing small yellowish kernels. The fruit composed of husk (approx. 5%) and berry (approx. 95%). The berries can be further subdivided into seeds (approx. 17%) and pulp/peel fraction (approx. 83%)

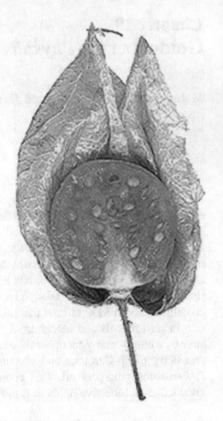

The flowers, produced in winter, are yellow with purple blotches. *P. peruviana* is an herbaceous, semi-shrub, upright, and perennial in subtropical zones plant, it can grow until reach 0.9 m. *P. peruviana* is forming a domed shrub to 1 m (Sharoba and Ramadan 2011).

P. peruviana fruit (Fig. 19.1) with an approximate weight of 4–5 g is protected by an accrescent calyx and covered by a brilliant yellow peel (Mayorga et al. 2001). It is somewhat tomato-like in appearance and flavor, though the taste (sweet and sour) is rich with a hint of tropical luxuriance. *P. peruviana* plant is adaptable to a wide variety of soils and good crops are obtained on the poor sandy ground (Popenoe et al. 1990; Ramadan and Mörsel 2004, 2007). *P. peruviana* plant has been grown in South Africa, India, Egypt, New Zealand, and Australia (Morton 1987; McCain 1993; Rehm and Espig 1991; Ramadan and Mörsel 2003).

International markets exist for non-conventional fruits, wherein the processing of exotic fruits started in several countries (Hassanien and Moersel 2003; Ramadan and Mörsel 2007). *P. peruviana* single plant might yield 300 fruits and carefully tended plants could provide 20–33 tons per hectare. Fruits are long lasting, can be stored for several months and freeze well. World's *P. peruviana* cultivation area is nearly 30,622 ha which yields about 162,386 tons (İzli et al. 2014; Yıldız et al. 2015). *P. peruviana* could become the fruit of particular interest to the food industry.

This strategy established markets for kiwifruits in the 1960s and led to a multimillion-dollar annual crop.

The fruit has been used as a good source of pro-vitamin A, minerals, vitamin C and vitamin B-complex. *P. peruviana* juice plus Adriamycin (ADR) exhibited an enhanced antitumor impact in hepatocellular carcinoma (HCC) and this combination might have an important value in the treatment of HCC. *P. peruviana* juice was more effective than ADR, and it has a remarkable role in the management of hepatic disorders besides its success as a chemo-sensitizer for ADR treatment of hepatocellular carcinoma (Hassan et al. 2017).

P. peruviana fruit contains 15% soluble solids (mainly sugars) and its high level of fructose makes it valuable for diabetics. The phosphorus level is high for a fruit. Its high content of dietary fiber is of importance, wherein fruit pectin acts as an intestinal regulator (McCain 1993; Ramadan and Mörsel 2003). *P. peruviana* waste powder contains 5.87% moisture, 13.72% lipids, 15.89% protein, 3.52% ash, 16.74% fiber and 61% carbohydrates (Mokhtar et al. 2018). In addition, *P. peruviana* by-products powder exhibited techno-functional traits including water absorption index, swelling index, foaming capacity and stability (3.38 g/g, 5.24 mL/g, 4.09 and 72.0%, respectively).

2 Extraction and Processing of *P. peruviana* Oil

Fruit processing produces a large amount of wastes which considered as a rich source of bioactive compounds. *P. peruviana* pomace (seeds and peels) represent a large portion of the by-products generated during *P. peruviana* juice processing (*ca.* 27.4% of fruit weight). *P. peruviana* pomace contains 19.3% lipids, 28.7% fiber, 17.8% protein, 3.10% ash, and 24.5% carbohydrates (Ramadan 2011). Aqueous enzymatic extraction was investigated for oil recovery from *P. peruviana* fruit pomace (Ramadan and Mörsel 2007, 2009). Different extraction methods were checked for the highest pomace oil yield. Enzymatic treatment with Cellulases and Pectinases followed by centrifugation in an aqueous system or followed by solvent extraction was studied for recovery of oil from *P. peruviana* pomace. Enzymatic hydrolysis of pomace followed by solvent extraction reduced the extraction time and increased oil extractability up to 7.6% (Ramadan et al. 2008).

3 Fatty Acids and Acyl Lipid Profiles of *P. peruviana* Oil

Lipids profile of *P. peruviana* Egyptian (Table 19.1) and Colombian (Table 19.2) cultivars was studied (Ramadan and Mörsel 2003, 2007). Whole berries contained *ca.* 2.0% oil, in which seed oil (SO) comprised about 90% (1.8% oil of the berry weight) and pulp/peel oil (PO) constituted about 10% (0.2% oil of the berry weight). In the whole berry oil (WBO), SO and PO linoleic, oleic, palmitic, stearic and

Table 19.1 Levels of sterols, tocopherols, β-carotene and fatty acids in the Egyptian *Physalis peruviana* pulp oil

Compound	g/kg oil	Fatty acid	%
Ergosterol	9.23	C14:0	1.09
Campesterol	12.2	C16:0	19.3
Stigmasterol	6.23	C16:1n-7	7.52
Lanosterol	6.55	C18:0	1.87
β-Sitosterol	5.23	C18:1n-9	22.2
Δ5-Avenasterol	12.5	C18:2n-6	22.7
α-Tocopherol	28.3	C18:3n-6	18.8
β- Tocopherol	15.2	C18:3n-3	0.63
γ- Tocopherol	45.5	C20:3n-6	2.31
δ- Tocopherol	1.50	C24:1n-9	1.12
β-Carotene	4.32	S/U ratio (%)[a]	29.4

Source: Ramadan and Mörsel (2007)
[a]Ratio of saturated fatty acids to unsaturated fatty acids

Table 19.2 Main fatty acids (relative content %) of Colombian *P. peruviana* oils

	Whole berry oil	Pulp/peel oil	Seed oil
C16:0	8.62	9.58	7.29
C18:0	2.57	2.92	2.51
C18:1n-9	13.0	20.1	11.7
C18:2n-6	70.5	44.4	76.1
C18:3n-6	1.79	8.66	0.31
C18:3n-3	0.11	1.09	0.02
S/U ratio (%)[a]	15.1	19.2	12.8

Source: Ramadan and Mörsel (2003)
[a]Ratio of saturated fatty acids to unsaturated fatty acids

γ-linolenic (GLA) were the major fatty acids (Tables 19.1 and 19.2). Pulp/peel oil was rich in saturated fatty acids which comprised more than 16%. The content of PUFA in PO was about 11.7% and the oil was characterized by high level of GLA (8.66%), while α-linolenic acid and DHGLA were estimated in lower levels. WBO and SO contained more neutral lipids (NL, about 95% of total lipids) than polar lipids (PL). Triacylglycerols (TAG) were the predominant NL class and constituted 81.6%, 86.6% and 65.1% of NL in WBO, SO and PO, respectively (Table 19.3). MAG, DAG and FFA were measured in higher amounts in the PO when compared with SO and WBO. Saturated fatty acids were detected in high amounts in all lipid classes especially MAG which characterized by high levels of palmitic acid. The oils contain nine TAG molecular species, but three species namely C54:3, C52:2 and C54:6 accounted for 91% or above (Ramadan and Mörsel 2003).

Recently Mokhtar et al. (2018), reported that fatty acids profile of *P. peruviana* by-products powder contained linoleic acid as the main fatty acid followed by oleic, palmitic and stearic acids. Iodine value (109.5 g/100 g oil), saponification value

Table 19.3 Lipid classes (g/100 g oil) in *P. peruviana* oils

	TAG	MAG	DAG	FFA	STE	PL
Whole berry oil	78.2	1.23	1.65	3.13	0.49	4.15
Seed oil	84.0	1.04	1.36	2.12	0.34	2.99
Pulp/peel oil	60.3	2.76	2.46	5.16	0.65	7.34

Source: Ramadan and Mörsel (2003)
Abbreviations: *MAG* monoacylglycerols, *DAG* diacylglycerols, *TAG* triacylglycerols, *FFA* free fatty acids, *STE* sterol esters, *PL* polar lipids

(183.8 mg KOH/g oil), acid value (2.36 mg KOH/g oil), peroxide value (8.2 meq O_2/kg oil) and refractive index (1.4735) were comparable to those of sunflower and soybean oils. *P. peruviana* by-product oil exhibited absorbance in the UV range at 100–400 nm.

4 Minor Bioactive Lipids in *P. peruviana* Oil

In *P. peruviana* oils, tocopherols amount was high in PO (8.6% of TL), while tocopherols estimated at lower amounts in WBO and SO. β- and γ-Tocopherols were the main tocopherol components in WBO and SO, while γ- and α-tocopherols were the main constituents in PO. Tocopherols are the best-known antioxidants in nature to protect lipids from oxidation (Toyosaki et al. 2008). Phytosterols were determined at high levels in *P. peruviana,* wherein there were no significant differences between WBO and SO in terms of phytosterols content and profile. In *P. peruviana* WBO and SO, campesterol and β-sitosterol were the dominant sterols while the main sterols in PO were Δ5-avenasterol and campesterol. There has been a rise of interest in phytosterols. Most of this interest has focused on cholesterol-lowering properties of 4-desmethyl plant sterols and phytostanols. An evidence of this phenomenon included clinical investigations on several commercial phytosterols products (Costa et al. 2010).

β-Carotene was determined at the highest level in PO followed by WBO then SO, the latter characterized by light yellow hues. Besides their potential health-promoting impacts, carotenoids are used as pigments to color food products. Since the consumer purchase decision of food is, *inter alia*, based on the color, the market position of a product can be improved, in particular by application of natural pigments in contrast to the synthetic colorant, since there is an increased demand for natural mild processed food (Wackerbarth et al. 2009). A positive correlation was observed between ingestion of vegetables and fruits rich in carotenoids and prevention of several chronic-degenerative diseases (Coyne et al. 2005; Fraser et al. 2005). Carotenoids from *physalis* were determined by HPLC-PDA-MS/MS, wherein 22 compounds has been identified. All-*trans*-β-Carotene was the main carotenoid, contributing 76.8% to the total carotenoid content, followed by 9-*cis*-β-carotene and all-*trans*-α-cryptoxanthin, contributing around 3.6% and 3.4%, respectively (De Rosso and Mercadante 2007). The level of carotenoid esters calculated as lutein

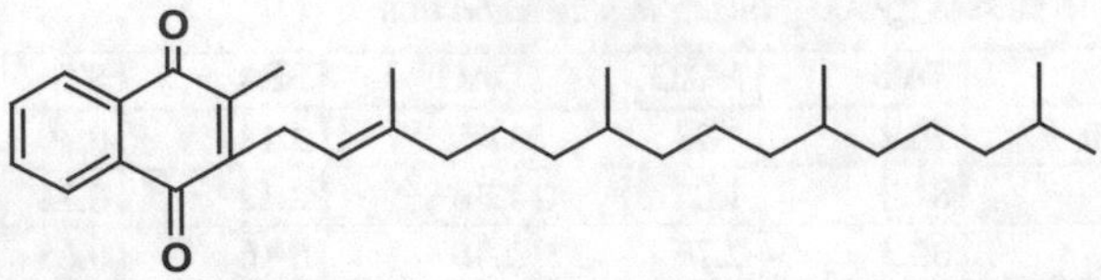

Fig. 19.2 Vitamin K₁ (phylloquinone)

dimyristate equivalents was <0.5 mg/100 g (Breithaupt and Bamedi 2001). In addition, carotenoid profiles of *P. peruviana* fruits differing in ripening states and in different fruit fractions (peel, pulp, and calyx of ripe fruits) were assayed by HPLC-DAD-APCI-MSn. Out of the numerous carotenoids detected, 42 were identified. The carotenoid profile of unripe fruits was dominated by (all-E)-lutein (51%), whereas in ripe fruits, (all-E)-β-carotene (55%) and several carotenoid fatty acid esters, especially lutein esters esterified with myristic and palmitic acid as monoesters or diesters, were detected (Etzbach et al. 2018).

P. peruviana oils were also characterized by a high level of vitamin K₁ (phylloquinone), which was comprised of more than 0.2% of total lipids in PO (Ramadan and Mörsel 2003). Vitamin K₁ (Fig. 19.2) levels are very low in most foods (<10 mg/100 g), wherein the majority of vitamin K₁ is obtained from a few green and leafy vegetables (Jakob and Elmadfa 2000; Piironen et al. 1997).

5   Health-Promoting Traits of *P. peruviana* Oil

P. peruviana oil appears to be nutritionally valuable, as the high content of linoleic acid is associated with the prevention of cardiovascular diseases (CVD) and linoleic acid is known to be the precursor of structural components of plasma membranes and of some metabolic regulatory compounds. The level of tocopherols in *P. peruviana* oil, which is significantly high, identifies the oil as nutritionally valuable. High level of vitamin k₁ is the unique health-promoting characteristic of *P. peruviana* oil. Vitamin K functions as a coenzyme and is involved in the synthesis of a number of proteins participating in blood clotting and bone metabolism (Shearer 1992; Damon et al. 2005). Vitamin K kills cancer cells, reduces the risk of heart disease, and enhances skin health and might have antioxidant traits (Otles and Cagindi 2007). High phylloquinone intakes are markers of a dietary and lifestyle pattern that is associated with lower coronary heart disease (CHD) risk (Erkkilä et al. 2007). The phylloquinone requirement of the adult human is extremely low. Vitamin K₁ is obtained from a greens and leafy vegetables (i.e. spinach and broccoli). Among edible oils, the best sources of phylloquinone were soybean oil (1.30 ug/g), and rapeseed oil (1.5 ug/g) (Piironen et al. 1997).

P. peruviana oil can be a very interesting candidate for the processing of new functional foods and nutraceuticals. The preparation of new α-tocopherol-β-carotene drinks based on *P. peruviana* could greatly extent the distribution and marketing of the fruit. *P. peruviana* pulp, seed and pomace oils might serve as excellent dietary sources for vitamin K_1, α-linolenic acid, essential fatty acids, tocopherols, and carotenoids. On the other side, *P. peruviana* oil is a promising candidate plant for the development of a phytomedicine against many diseases.

References

Breithaupt, D. E., & Bamedi, A. (2001). Carotenoid esters in vegetables and fruits: A screening with emphasis on β-cryptoxanthin esters. *Journal of Agricultural and Food Chemistry, 49*, 2064–2070.

Costa, P. A., Ballus, C. A., Teixeira-Filho, J., & Godoy, H. T. (2010). Phytosterols and tocopherols content of pulps and nuts of Brazilian fruits. *Food Research International, 43*, 1603–1606.

Coyne, T., Ibiebele, T. I., Baadr, P. D., Dobson, A., McClintock, C., Dunn, S., Leonard, D., & Shaw, J. (2005). Diabetes mellitus and serum carotenoids: Findings of a population-based study in Queensland. *American Journal of Clinical Nutrition, 82*, 685–693.

Damon, M., Zhang, N. Z., Haytowitz, D. B., & Booth, S. L. (2005). Phylloquinone (vitamin K_1) content of vegetables. *Journal of Food Composition and Analysis, 18*, 751–758.

De Rosso, V. V., & Mercadante, A. Z. (2007). Identification and quantification of carotenoids, by HPLC-PDA-MS/MS, from Amazonian fruits. *Journal of Agricultural and Food Chemistry, 55*, 5062–5072.

Erkkilä, A. T., Booth, S. L., Hu, F. B., Jacques, P. F., & Lichtenstein, A. H. (2007). Phylloquinone intake and risk of cardiovascular diseases in men. *Nutrition, Metabolism, and Cardiovascular Diseases, 17*, 58–62.

Etzbach, L., Pfeiffer, A., Weber, F., & Schieber, A. (2018). Characterization of carotenoid profiles in goldenberry (*Physalis peruviana* L.) fruits at various ripening stages and in different plant tissues by HPLC-DADAPCI-MS[n]. *Food Chemistry, 245*, 508–517.

Fraser, M. L., Lee, A. H., & Binns, C. W. (2005). Lycopene and prostate cancer: Emerging evidence. *Expert Review of Anticancer Therapy, 5*, 847–854.

Hassan, H. A., Serag, H. M., Qadir, M. S., & Ramadan, M. F. (2017). Cape gooseberry (*Physalis peruviana*) juice as a modulator agent for hepatocellular carcinoma-linked apoptosis and cell cycle arrest. *Biomedicine & Pharmacotherapy, 94*, 1129–1137.

Hassanien, M. F. R., & Moersel, J.-T. (2003). Das Physalisbeerenoel: Eine neuentdeckte Quelle an essentiellen Fettsaeuren, Phytosterolen und antioxidativen Vitaminen. *Fluessiges-Obst, 7*, 398–402.

İzli, N., Yıldız, G., Ünal, H., Isık, E., & Uylaşer, V. (2014). Effect of different drying methods on drying characteristics, colour, total phenolic content and antioxidant capacity of goldenberry (*Physalis peruviana* L.). *International Journal of Food Science and Technology. 2014, 49*, 9–17.

Jakob, E., & Elmadfa, I. (2000). Rapid and simple HPLC analysis of vitamin K in food, tissue and blood. *Food Chemistry, 68*, 219–221.

Mayorga, H., Knapp, H., Winterhalter, P., & Duque, C. (2001). Glycosidically bound flavor compounds of cape gooseberry (*Physalis peruviana* L.). *Journal of Agricultural and Food Chemistry, 49*, 1904–1908.

McCain, R. (1993). Goldenberry, passionfruit and white sapote: Potential fruits for cool subtropical areas. In J. Janick & J. E. Simon (Eds.), *New crops* (pp. 479–486). New York: Wiley.

Mokhtar, S. M., Swailam, H. M., & Embaby, H. E. (2018). Physicochemical properties, nutritional value and techno-functional properties of goldenberry (*Physalis peruviana*) waste powder concise title: Composition of goldenberry juice waste. *Food Chemistry, 248*, 1–7.

Morton, J. F. (1987). Cape Gooseberry. In J. F. Morton (Ed.), *Fruits of warm climates* (pp. 430–434). Winterville: Creative Resource Systems.

Otles, S., & Cagindi, O. (2007). Determination of vitamin K_1 content in olive oil, chard and human plasma by RP-HPLC method with UV-Vis detection. *Food Chemistry, 100*, 1220–1222.

Piironen, V., Koivu, T., Tammisalo, O., & Mattila, P. (1997). Determination of phylloquinone in oils, margarines and butter by high-performance liquid chromatography with electrochemical detection. *Food Chemistry, 59*, 473–480.

Popenoe, H., King, S. R., Leon, J., & Kalinowski, L. S. (1990). Goldenberry (cape gooseberry). In National Research council (Ed.), *Lost crops of the Incas, little-known plants of the Andes with promise for worldwide cultivation* (pp. 241–252). Washington, DC: National Academy Press.

Ramadan, M. F. (2011). Bioactive phytochemicals, nutritional value, and functional properties of cape gooseberry (*Physalis peruviana*): An overview. *Food Research International, 44*, 1830–1836.

Ramadan, M. F., & Mörsel, J.-T. (2003). Oil goldenberry (*Physalis perviana* L.). *Journal of Agricultural and Food Chemistry, 51*, 969–974.

Ramadan, M. F., & Mörsel, J.-T. (2004). Goldenberry: A novel fruit source of fat soluble bioactives. *Inform, 15*, 130–131.

Ramadan, M. F., & Mörsel, J.-T. (2007). Impact of enzymatic treatment on chemical composition, physicochemical properties and radical scavenging activity of goldenberry (*Physalis peruviana* L.) juice. *Journal of the Science of Food and Agriculture, 87*, 452–460.

Ramadan, M. F., & Mörsel, J.-T. (2009). Oil extractability from enzymatically-treated goldenberry (*Physalis peruviana* L.) pomace: Range of operational variables. *International Journal of Food Science and Technology, 44*, 435–444.

Ramadan, M. F., Sitohy, M. Z., & Mörsel, J.-T. (2008). Solvent and enzyme-aided aqueous extraction of goldenberry (*Physalis peruviana* L.) pomace oil: Impact of processing on composition and quality of oil and meal. *European Food Research and Technology, 226*, 1445–1458.

Rehm, S., & Espig, G. (1991). Fruit. In R. Sigmund & E. Gustav (Eds.), *The cultivated plants of the topics and subtropics, cultivation, economic value, utilization* (pp. 169–245). Weikersheim: Verlag Josef Margraf.

Sharoba, A. M., & Ramadan, M. F. (2011). Rheological behavior and physicochemical characteristics of goldenberry (*Physalis peruviana*) juice as affected by enzymatic treatment. *Journal of Food Processing and Preservation, 35*, 201–219.

Shearer, M. J. (1992). Vitamin K metabolism and nutriture. *Blood, 6*, 92–104.

Toyosaki, T., Sakane, Y., & Kasai, M. (2008). Oxidative stability, trans,trans-2,4-decadienals, and tocopherol contents during storage of dough fried in soybean oil with added medium-chain triacylglycerols (MCT). *Food Research International, 41*, 318–324.

Wackerbarth, H., Stoll, T., Gebken, S., Pelters, C., & Bindrich, U. (2009). Carotenoid-protein interaction as an approach for the formulation of functional food emulsions. *Food Research International, 42*, 1254–1258.

Yıldız, G., İzli, N., Ünal, H., & Uylaşer, V. (2015). Physical and chemical characteristics of goldenberry fruit (*Physalis peruviana* L.). *Journal of Food Science and Technology, 52*(4), 2320–2327.

Zhao, Y. (2007). In Y. Zhao (Ed.), *Berry fruit, value-added products for health promotion*. Boca Raton, NW: CRC Press. ISBN:0-8493-5802-7.

Chapter 20
Olive (*Olea europea L.*) Oil

Ines Gharbi and Mohamed Hammami

Abstract Many fruits, seeds, vegetables and plants contain edible oils. Their oils are similar in many respects, but a few minor differences have a significant effect on the characteristics of the oil. The olive tree (*Olea europea* L.) is one of the most important fruit trees in Mediterranean countries. Its oil is an important component of the Mediterranean diet which is consumed worldwide. The salutistic properties of olive oil such as its high nutritional value, excellent digestibility, high oxidative stability even when used for cooking, strong capacity of prevention of heart and vascular troubles explain the reasons for the increased popularity of olive oil. The large increase in demand for high-quality olive oils is related to their peculiar organoleptic characteristics that play an important role in human nutrition. All these properties are intimately linked to its chemical composition, in particular to several minor components, which are strongly affected by the operative conditions of processing and could be considered as analytical markers of its quality. Virgin olive oil is unique because, in contrast to other vegetable oils, it is consumed in its crude state without any further physical-chemical treatments of refining. Thus, natural components having a great biological action are retained, although the potential presence of some contaminants is increased.

Keywords Mediterranean diet · Organoleptic · Phospholipids · Diterpene alcohols · Virgin oil

1 Introduction

Olive oil, the oil of the olive tree (*Olea europaea* L.) is an exceptionally versatile food product. Long recognized by Mediterranean populations as essential for health and a key food element in their diet, it is widely appreciated for its nutritional and

I. Gharbi (✉) · M. Hammami
Laboratory LR12ES05 Lab-NAFS 'Nutrition – Functional Food & Vascular Health',
Faculty of Medicine, University of Monastir, Monastir, Tunisia
e-mail: mohamed.hammami@fmm.rnu.tn

© Springer Nature Switzerland AG 2019
M. F. Ramadan (ed.), *Fruit Oils: Chemistry and Functionality*,
https://doi.org/10.1007/978-3-030-12473-1_20

Fig. 20.1 Average
composition of olive fruit

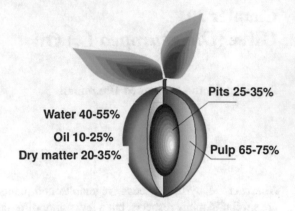

organoleptic properties (Rocha et al. 2016). Properly extracted from fresh, mature fruit of good quality, the oil has a characteristic sensory profile. Its fatty acid composition is characterized by a good balance between saturated, monounsaturated, and polyunsaturated fatty acids. It is also unique among common vegetable oils in that it can be consumed in the crude form, thus conserving vitamin content and phenolic compounds of nutritional importance.

The olive is a member of the botanical family *Oleaceae*, which contains 30 species such as jasmine, ash and privet. The only edible species is *Olea europaea* L., which is cultivated for its large, fleshy, oil-containing fruits (Luchetti 2002). The olive fruit is an ovoid drupe weighting between 2 and 12 g (some varieties may weigh as much 20 g). It has a wax-coated epicarp that turns from green to black as ripening progresses. The mesocarp contains the bitter compound oleuropein, and has high oil content (10–30%) that varies from variety to variety and low sugar content. The endocarp of the olive fruit is made up of tough, fibrous lignin. It is ovoid in shape and grooved to varying degrees, again depending on the variety. The endocarp encloses a kernel (olive seed) accounting for some 3% of fruit weight and containing 2–4% oil (Fig. 20.1).

Oil quantity and quality are highly dependent on the variety. The best oil varieties in the world have developed their reputation over centuries of production for fruit yields, oil content, flavor, keeping quality, maturity date, and ease of harvest. Most olive cultivars range in oil content from 10% to 35% of their fresh weight at full maturity. Growing varieties with an average oil yield of less than 20% ~ 45 gal of oil per fresh ton of fruit are not usually profitable to use for oil. The most prominent oil varieties in the world are Piqual, Empeltre, Arbequina, Frantoio, Coratina, Aglandaou, Picholine, Leccino, Chemlali, and Koroneiki (Ilarioni and Proietti 2014) as given in Table 20.1.

Virgin olive oil production is spread overdeveloped and developing countries. The main producing countries accounted for 96.7% of this production; the European Union, that's the largest producer, alone held a 75.8% share. Accounting for more than 4% of the olive oil produced in the world, Tunisia is holding an important

Table 20.1 Primary world olive oil cultivars. International Olive Oil Council, World Catalogue of Olive Varieties (Barranco et al), Il Germoplasma dell Olivo in Toscana (Cimato et al.), Variedades del Olivo (Tous & Romero) El Cultivo del Olivo (Barranco et al.), Olive Pollination in CA (Griggs et al.)

Cultivar	% Oil	Oil appreciation	Fruit size	Polyphenol content	Primary country
Arbequina	25–27	Very high	Small	Low	Spain
Aglandaou	23–27	High	Medium	Medium	France
Ascolano	15–22	High	Large	Medium	USA
Barouni	13–18	Low	Large	Medium	USA
Bouteillan	20–25	High	Medium	Medium	France
Chemlali	26–28	High	Very small	High	Tunisia
Coratina	23–27	Very high	Medium	Very high	Italy
Cornicabra	23–27	High	Medium	Very high	Spain
Empeltre	18–25	High	Medium	High	Spain
Frantoio	23–26	Very high	Medium	Medium	Italy
Farga	23–27	High	Medium	Medium	Spain
Koroneiki	24–28	Very high	Very small	Very high	Greece
Leccino	22–27	Medium	Medium	Medium	Italy
Lucca	22–24	Medium	Medium	Medium	USA
Manzanillo	15–26	Medium	Large	Very high	USA
Maurino	20–25	Medium	Medium	High	Italy
Mission	19–24	Medium	Medium	Medium	USA
Moraiolo	18–28	Medium	Small	Very high	Italy
Pendolino	20–25	Medium	Medium	Medium	Italy
Picudo	22–24	High	Large	Medium	Spain
Piqual/Nevadillo	24–27	Low	Medium	Very high	Spain
Picholine	22–25	High	Medium	Very high	France
Sevillano	12–17	Low	Very large	Low	USA

position in the olive oil market; it exports about 75% of its production and is ranked second largest exporter after the European Union with an average of 115,000 tons per year over the last years (ONII 2015).

Olive oil is a food product obtained from the decanted and purified juice of the olive tree fruit, the olive, which is part of the traditional Mediterranean diet being regarded as a health food promoter. Several studies showed that olive oil is associated with a low incidence of cardiovascular diseases (Keys 1995; Lipworth et al. 1997) and some types of cancer, probably due to its high content of oleic acid (73–80%, w/w) and the antioxidant properties of phenolics (Wiseman et al. 1996). The total amount of polyphenols in olive oils depends directly on the variety of olive, growth conditions and olive maturation and, consequently, the information on the composition of phenolics in olive oil could have a useful contribution to the biochemical as well as for the control of the production process.

2 Extraction and Processing

2.1 Harvest

The optimal harvesting time is when oil levels are high in the olive fruit. Harvest should begin before natural fruit drop. In normal-ripening varieties, the time to start harvesting can be judged by the color of the fruit skin. When there are no green olives left on the tree, perhaps some fruits at color-change, oil biosynthesis has ceased and harvesting can begin (Tombesi et al. 1996).

Methods used to harvest olives depend on cultural techniques, tree size and shape, and orchard terrain. Most olives are harvested by hand and/or with shakers. Newly-planted orchards are more likely to be mechanically harvested. The high trees of some varieties are harvested with the aid of nets after the natural drop of the fruit. Precautions should be taken to avoid fruit breakage through mechanical damage and fruit contamination by soil. Olive transportation and storage should be considered as critical phases for controlling both mechanical damage and temperature. Improper handling during these phases can result in undesirable enzymatic reactions and the growth of yeasts and molds. The best way to transport the olives is in open-mesh plastic crates that allow air to circulate and prevent the harmful heating caused by the catabolic activity of the fruit (Kiritsakis 1998). When stored before processing, the olives must be spread in shallow layers and kept in well-ventilated, cool, and dry areas. Storing of the olives in jute sacks should be avoided. To ensure that the olives retain the quality characteristics they possessed at the time of harvesting they must be delivered immediately to the extraction plant for processing.

2.2 Washing and Leaf Removal

The purpose of preliminary washing is to remove any foreign materials that could damage machinery or contaminate the oil. Wet fruit is also much more likely to ferment than dry fruit. Small quantities of leaves are not detrimental to the oil and sometimes leaves are added to produce a chlorophyll (green) color and flavor in the oil. Branches and wood material are however very detrimental to olive oil flavor producing a woody taste.

2.3 Milling

The objective of the first true step of olive oil production, crushing the olives, is to produce a paste with easily extracted oil droplets. Two types of machines are used to crush olives: stone mills and stainless steel hammer mills. Each has advantages. A new system just introduced, removes the olive pits prior to crushing.

2.3.1   Stone Mills

The older of the two methods, stone crushers consist of a stone base and upright millstones enclosed in a metal basin, often with scrapers and paddles to guide the fruit under the stones and to circulate and expel the paste. The slow movement of the stone, crushers does not heat the paste and results in less emulsification, so the oil is easier to extract without as much mixing (malaxation). The major disadvantages of this method are the bulky machinery and its slowness, its high cost, and its inability to be continuously operated. The stones are also more difficult to clean, and the slow milling time can increase oxygen exposure and paste fermentation. Because of their inefficiency, stone mills have been replaced by hammer mills in many large operations.

2.3.2   Hammer Mills

Generally consist of a metal body that rotates at high speed, hurling the olives against a metal grate. The major advantage of metal crushers is their speed and continuous operation, which translate into the high output, compact size, and low cost. Their major disadvantage is the type of paste produced. The oil is more emulsified, requiring a longer mixing period to achieve a good oil extraction and the speed of metal crushing can produce elevated temperatures and possible metal contamination. Both factors reduce oil quality. The hammer mill is easier to clean and much faster, allowing for the deployment of a continuous flow system. Oil produced from hammer mills is generally greener since the skins are broken up more. The emulsification problem is overcome by malaxation for a slightly longer period and new stainless steel mills do not impart a metallic flavor into the oil.

2.4   *Mixing of the Olive Paste (Malaxation)*

Malaxation prepares the paste for separation of the oil from the pomace. This step is particularly important if the paste was produced in a hammer mill. The mixing process optimizes the amount of oil extracted through the formation of larger oil droplets and a reduction of the oil-water emulsion. Malaxation usually requires 45 min to 1 h. The longer the contact between the oil and the fruit water, the more the final phenolic content of the oil is reduced. The temperature of the paste during malaxation is very important. It should be warm (80–86 °F, which is still cold to the touch) to improve the viscosity of the oil and improve extractability. Temperatures above 86 °F can cause problems such as loss of fruit flavors, increases in bitterness, and increases in astringency.

2.5 *Oil Extraction from the Paste*

The next step is extracting the oil from the paste and fruit water (water of vegetation). The oil could be extracted by pressing, centrifugation, percolation, or through combinations of the different methods.

2.5.1 Traditional Press

Pressing is the oldest method of oil extraction. The method involves applying pressure to stacked filter mats, smeared with paste, alternate with metal disks; a central spike allows the expressed oil and water (olive juice) to exit. The machinery, however, is cumbersome, the process requires more labor than other extraction methods, the cycle is not continuous, and the filter mats can easily become contaminated.

The olive paste or olive juice containing both water and oil was allowed to sit in containers until the oil, with a lower specific gravity, rose to the top naturally. The oil was decanted away from the remaining water and solid material. This natural separation takes considerable time and the contact of oil with enzymes, breakdown products and fermenting fruit water produced defective oils.

2.5.2 Selective Filtration-Sinolea Process

This process is the opposite of the press system since no pressure is applied to the paste. It operates on the principle that in a paste containing solid particles, water and oil the oil alone will adhere to metal. The machine has stainless steel blades that dip into the paste, the adhering oil then drips off the blades into a separate container while the solids and water are left behind. The lack of pressure produces light oil with unique quality and value. The equipment is complicated and requires frequent cleaning, maintenance of the stainless steel blade mechanisms, and a constant heat source to keep the paste at an even temperature. Extraction is stopped when vegetable water begins to appear in the oil.

2.5.3 Decanters, 3-Phase and 2-Phase

Modern decanters are large horizontal centrifuges that separate the oil from the solids and water in the same process as in a decantation tank, just faster. The savings in time increases the efficiency of the system, but also decreases the time of contact between oil and the fermenting fruit water. The decanters spin at approximately 3000 rpm. Centrifugal force moves the heavier solid materials to the outside; a lighter water layer is formed in the middle with the lightest oil layer on the inside. There is no exact line of separation between the three phases of solid, water, and oil, so the solid phase usually has some water in it, wherein the water has some oil in it and the oil contains some water. In the latter case, which extracts the maximum quantity of oil, an additional vertical centrifugation is needed to remove more of the vegetation water from the oil.

The 3-Phase system decanter separates the paste into a relatively dry solid, fruit-water, and oil. Water is added to this system to get it to flow through the decanter. A minimum quantity of water is added to separate the solid material better and to retain water-soluble phenolics as much as possible. 2-Phase system decanters were introduced in the early 1990s. They function under the same principle as 3-Phase decanters except that the solid and fruit-water exit together. No water needs to be added to the 2-Phase system, which induces a better retention of phenolics.

3 Olive Oil Chemical Composition

The composition of olive oil is primarily triacylglycerols (~99%) and secondarily free fatty acids, mono- and diacylglycerols, and an array of lipids such as hydrocarbons, sterols, aliphatic alcohols, tocopherols, and pigments. A plethora of phenolics and volatile compounds are also present. Some of these compounds contribute to the unique character of the oil.

3.1 Fatty Acids, Triacylglycerols, and Glycerides

Fatty acids present in olive oil are palmitic (C16:0), palmitoleic (C16:1), stearic (C18:0), oleic (C18:1), linoleic (C18:2), and linolenic (C18:3). Myristic (C14:0), heptadecanoic and eicosanoic acids are found in trace amounts. Scano and coworkers (1999), using 13C-Nuclear Magnetic Resonance Spectroscopy, detected traces of 11-*cis*-vaccenic and eicosenoic acids. Fatty acid compositional limits adopted in the most recent editions of Codex Alimentarius and International Olive Oil Council. The fatty acid composition may differ from sample to sample, depending on the zone of production, the latitude, the climate, the variety, and the stage of maturity of the fruit. Greek, Italian, and Spanish olive oils are low in linoleic and palmitic acids and they have a high percentage of oleic acid. Tunisian olive oils are high in linoleic and palmitic acids and lower in oleic acid.

The triacylglycerols found in significant proportions in olive oil are OOO (40–59%), POO (12–20%), OOL (12.5–20%), POL (5.5–7%) and SOO (3–7%) (Boskou 1996). Smaller amounts of POP, POS, OLnL, LOL, OLnO, PLL, PLnO and LLL are also encountered (European Commission Regulation 282, 1998). The presence of partial glycerides in olive oil is due either to incomplete triacylglycerol biosynthesis or hydrolytic reactions. In virgin olive oil, the concentration of diacylglycerols (DG) ranges from 1% to 2.8% (Frega et al. 1993; Kiosseoglou and Kouzounas 1993). In the diacylglycerol fraction, C-34 and C-36 compounds prevail (Leone et al. 1988; Frega et al. 1993). Monoacylglycerols are present in much smaller quantities (less than 0.25%), whereas 1-species are considerably higher than the respective 2-monoglycerides. Their ratio depends on oil acidity (Paganuzzi 1999). Storage conditions affect the distribution of fatty acids. 1,2-Diacylglycerols present in fresh oil tend to isomerize to the more stable 1,3-diacylglycerols. This rearrangement gives information about the age of oil and storage conditions.

3.2   Minor Components

3.2.1   Hydrocarbons

Two hydrocarbons are present in considerable amounts in olive oil, squalene and β- carotene. Squalene (2,6,10,15,19,23-hexamethyl-2,6,10,14,18,22-tetracosahexaene) is the last metabolite preceding sterol ring formation. Its presence is regarded as partially responsible for the beneficial health effects of olive oil and its chemopreventive action against certain cancers (Rao et al. 1998; Smith et al. 1998). It is the major constituent of the unsaponifiable matter and makes up more than 90% of the hydrocarbon fraction (Perrin 1992). Except for squalene, the hydrocarbon fraction of virgin olive oil is composed of diterpene and triterpene hydrocarbons, isoprenoidal polyolefins, and n-paraffins (Lanzón et al. 1994).

3.2.2   Tocopherols

Research on the occurrence and levels of tocopherols in virgin olive oils has shown that from the eight known "E-vitamers" the α-homologue comprises the 90% of the total tocopherol content. α-Tocopherol is found in the free form. The levels reported indicate a wide range of milligrams α-tocopherol per kg oil that depends on the cultivar potential and technological factors. Tocopherol concentration seems to be reduced in the ripe fruits. Data on the influence of the extraction system vary (Psomiadou and Tsimidou 1998; Beltran et al. 2005). Refining or hydrogenation causes loss of tocopherols (Andrikopoulos et al. 1989; Rabascall and Riera 1987).

3.2.3   Pigments

Virgin olive oil color is the result of green and yellow hues due to the presence of chlorophylls and carotenoids. It is influenced by olive cultivar, maturation index, production zone, extraction system, and storage conditions. Therefore, it is considered as a quality index though no standardized method exists for its measurement. Chlorophylls are encountered as pheophytins. Among the latter, pheophytin α (Pheo α) is predominant (Mínguez-Mosquera et al. 1990). Chlorophyll α can only be found in recently obtained oils. Chlorophyll and pheophytin β are also present though in minute amounts. The main carotenoids present in olive oil are lutein and β-carotene (Mínguez-Mosquera et al. 1990). Levels reported are related to the analytical method used. The presence of carotenoids in olive oil is closely related to that of green pigments and is influenced by the same factors.

3.2.4 Aliphatic and Aromatic Alcohols

Aliphatic and aromatic alcohols present in olive oil are found in the free and esterified forms. The most important are fatty alcohols and diterpene alcohols. Alkanols and alkenols with less than ten carbon atoms in their molecule, which are present in the free and esterified form, and some aromatic alcohols (benzyl alcohol and 2-phenylethanol) are constituents of the olive oil volatile fraction. Benzyl esters of hexacosanoic and octacosanoic acid have been also found in olive oil (Reiter and Lorbeer 2001).

3.2.5 Fatty Alcohols

This class of minor constituents consists of linear saturated alcohols with more than 16 carbon atoms, which are present in the free and esterified form. The main fatty alcohols present in olive oil are docosanol, tetracosanol, hexacosanol, and octacosanol (Tiscornia et al. 1982; Frega et al. 1992). Fatty alcohols with odd carbon atoms (tricosanol, pentacosanol, and heptacosanol) may be found in trace amounts (Boskou et al. 1983). Esters of fatty alcohols with fatty acids (waxes) are important minor olive oil constituents because they can be used as a criterion to differentiate various olive oil types (EC Regulation 2568, 1991). The main waxes detected in olive oil are esters of oleic or palmitic acid with 36, 38, 40, 42, 44, and 46 carbon atoms (Reiter et al. 2001).

3.2.6 Diterpene Alcohols

Phytol and geranylgeraniol are two acyclic diterpenoids found in olive oil.

3.2.7 Sterols

Sterols are important lipids related to the quality of the oil and broadly used for checking its enuineness. Four classes of sterols occur in olive oil: common sterols (4- desmethylsterols), α-methylsterols, triterpene alcohols (4, 4-dimethylsterols), and triterpene dialcohols. Olive oil contains common sterols mainly in free and esterified form (Grob et al. 1990). However, these sterols have also been found as sterylglucosides and lipoproteins. The main components of this sterol fraction are β-sitosterol, Δ5-avenasterol, and campesterol. Other sterols present in smaller quantities or in trace amounts are stigmasterol, cholesterol, brassicasterol, cholesterol, ergosterol, sitostanol, campestanol, Δ7-avenasterol, Δ7-cholestenol, Δ7-campestenol, Δ7-stigmastenol, Δ5,23-stigmastadienol, Δ5,24-stigmastadienol, Δ7,22-ergostadienol, Δ7,24-ergostadienol, 24-methylene-cholesterol, and 22,23-dihydrobrassicasterol (Mariani 1998).

3.2.8 Volatile and Aromatic Compounds

Approximately 280 compounds have been identified in the volatile fraction of virgin olive oils. They are hydrocarbons (more than 80 compounds), alcohols (45 compounds), aldehydes (44 compounds), ketones (26 compounds), acids (13 compounds), esters (55 compounds), ethers (5 compounds), furan derivatives (5 compounds), thiophene derivatives (5 compounds), pyranones (1 compound), thiols (1 compound), and pyrazines (1 compound) (Boskou et al. 2006).

Olive oil obtained from healthy olives, harvested at the right ripening stage, and by proper extraction techniques contain mainly volatiles derived from linoleic and linolenic acid decomposition through the lipoxygenase pathway (Olías et al. 1993). The most abundant compounds are hexanal, (E)-2- hexenal, (Z)-3-hexenal, hexan-1-ol, (Z)-3-hexen-1-ol, hexyl acetate, and (Z)-3-hex-enyl acetate. This volatiles are responsible for the green and fruity perception of the unique virgin olive oil aroma. C5 aldehydes, ketones and alcohols, and pentene dimmers also arise from a cleavage reaction of 13-hydroperoxide of linolenic acid (Salch et al. 1995), are usually present at levels lower than their odor threshold. The formation of these C6 and C5 compounds is affected by the cultivar, the degree of fruit ripeness, the storage time of fruits prior to oil extraction, and by the processing (Angerosa et al. 1998). In addition to these volatiles, linolenic, linoleic, and oleic acids autoxidation decomposition products, mainly aldehydes, and also alcohols, and esters deriving from biochemical transformations of amino acids such as isoleucine, leucine, phenylalanine, or valine, have a greater number of volatile compounds and have been found in virgin olive oils of poorer quality. Some of this volatiles give rise to sensory defects when they are present at high levels. They are produced by over-ripening of the fruit, significant attack of the fruits by molds and bacteria, when they are stored for a long period prior to oil extraction, and by advanced autoxidation of the unsaturated fatty acids due to adverse storage conditions. Acids, esters, alcohols, aldehydes and ketones are mainly responsible for the most frequent off-flavors developed in virgin olive oil (Morales et al. 2005).

The presence of terpene alcohols, monoterpene and sesquiterpene hydrocarbons in the virgin olive oil, as well as aromatic hydrocarbons is also interesting. However, it is not clear whether all aromatic hydrocarbons are naturally present compounds or contaminants (Biedermann et al. 1995). Other compounds identified in the volatile fraction are ethyl esters. Oils obtained from altered olive fruits or olive pomace have been found to contain high levels of ethyl palmitate, ethyl oleate and ethyl linoleate, while the levels of these esters in extra virgin olive oils are low (Pérez-Camino et al. 2002).

3.2.9 Phospholipids

Phosphatidylcholine, phosphatidylethanolamine, phosphatidylinositol, and phosphatidylserine were reported to be the main phospholipids present in olive oil (Alter and Gutfinger 1982). The level of phospholipids may be important because these compounds have an antioxidant activity.

3.2.10 Proteins

Mainly in unfiltered oils minute quantities of proteins may be detected.

4 Olive Oil Quality, Adulteration and Authenticity

Quality of olive oil is defined as the combination of its attributes that have significance in determining the degree of its acceptability by the consumer, and may be also defined from commercial, nutritional or organoleptic perspectives. The nutritional value of extra virgin olive oil (EVOO) originates from its high levels of oleic acid content and minor components, such as phenolic compounds that donate the oil its aroma (Romani et al. 1999). Therefore, these quality parameters promote the consumption demands and price of olive oil in comparison with other edible oils ranking it superior among vegetable oils (Visioli et al. 1998). There is a need to develop reliable analytical methods to ensure compliance of olive oil quality with labeling, and to determine the genuineness of the product by the detection of eventual defects during adulterations, processing and storage conditions. Therefore, the International Olive Oil Council (IOOC) and European Communities Legislation (EC) define the identity characteristics of olive oil by specifying analytical methods and standard limit values of the quality parameters such as peroxide value (PV), acidity, Ultra violet (UV) absorbance values (K232 and K270) and organoleptic characteristics (odor, taste and color) for olive oils in order to improve oil quality, expand international trade, and raise its consumption. The chemical tests and the organoleptic properties categorize olive oil into an extra virgin, virgin, and lampant oil indicating its edible quality and marketable values. The EVOO is the highest grade and must contain zero defects and greater than zero positive attributes as evaluated by a certified taste panel. Also, the oil must have a free acidity of less than 0.8%, peroxide value doesn't exceed 20 milliequivalent O_2 kg^{-1} oil and should have clear flavor that reflect the fruit from which it is produced (Wiseman et al. 1996).

References

Alter, M., & Gutfinger, T. (1982). Phospholipids in several vegetable oils. *Rivista Italiana Delle Sostanze Grasse, 59*, 14–18.

Andrikopoulos, N., Hassapidou, M., & Manoukas, A. (1989). The tocopherol content of Greek olive oils. *Journal of the Science of Food and Agriculture, 46*, 503–509.

Angerosa, F., d'Alessandro, N., Basti, C., et al. (1998). Biogeneration of volatile compounds in virgin olive oil: Their evolution in relation to Malaxation time. *Journal of Agricultural and Food Chemistry, 46*, 2940–2944.

Beltran, G., Aguilera, A., del Rio, C., Sanchez, S., & Martinez, L. (2005). Influence of fruit ripening process on the natural antioxidant content of Hojiblanca virgin olive oils. *Food Chemistry, 89*, 207–215.

Biedermann, M., Grob, K., & Morchio, G. (1995). On the origin of benzene, toluene, ethylbenzene and xylene in extra virgin olive oil. *Zeitschrift für Lebensmittel-Untersuchung und -Forschung, 200*, 266–272.

Boskou, D. (1996). In D. Boskou (Ed.), *Olive oil: Chemistry and technology* (pp. 101–120). Champaign: AOCS Press.

Boskou, D., Stefanou, G., & Konstandinidis, M. (1983). Tetracosanol and hexacosanol content of Greek olive oils. *Grasas y Aceites, 34*, 402–404.

Boskou, D., Tsimidou, M., & Blekas, D. (2006). In D. Boskou (Ed.), *Polar phenolic compounds, in olive oil, chemistry and technology* (pp. 73–92). Champaign: AOCS Press.

EC Regulation No 2568/91. (1991). Commission of the European Communities. Official Journal of the European Communities No L 248.

EC Regulation No 282/98. (1998). Commission of the European Communities. Official Journal of European Communities No L 285.

Frega, N., Bocci, F., & Lercker, G. (1992). Direct gas chromatographic analysis of the unsaponifiable fraction of different oils, by using a polar capillary column. *Journal of the American Oil Chemists' Society, 69*, 447–450.

Frega, N., Bocci, F., & Lercker, G. (1993). Free fatty acids and diacylglycerols as quality parameters for extra virgin olive oil. *Rivista Italiana Delle Sostanze Grasse, 70*, 153–156.

Grob, K., Lanfranchi, M., & Mariani, C. (1990). Evaluation of olive oils through the fatty alcohols, the sterols and their esters by coupled LC-GC. *Journal of the American Oil Chemists' Society, 67*, 626–634.

Ilarioni, L., & Proietti, P. (2014). Olive tree cultivars. In C. Peri (Ed.), *The extra-virgin olive oil handbook edition: First chapter: 5*. Chichester: Wiley.

Keys, A. (1995). Mediterranean diet and public health personal reflections. *American Journal of Clinical Nutrition, 61*, 1321S–1323S.

Kiosseoglou, V., & Kouzounas, P. (1993). The role of diglycerides, monoglycerides and free fatty acids in olive oil minor surface-active lipid interaction with proteins at oil-water interface. *Journal of Dispersion Science and Technology, 14*, 527–531.

Kiritsakis, A. (1998). *Olive oil from the tree to the table* (2nd ed.). Trumbull: Food and Nutrition Press.

Lanzón, A., Albi, T., Cert, A., & Gracián, J. (1994). The hydrocarbon fraction of virgin olive oil and changes resulting from refining. *Journal of the American Oil Chemists' Society, 71*, 285–291.

Leone, A., Santoro, M., Liuzzi, V. A., La Notte, E., & Gambacorta, G. (1988). The structure of diglycerides and their occurrence in olive oils as a means to characterize high quality products. *Rivista Italiana Delle Sostanze Grasse, 65*, 613–622.

Lipworth, L., Martinez, M. E., Angell, J., Hsien, C. C., & Trichopoulos, D. (1997). Olive oil and human cancer: An assessment of evidence. *Preventive Medicine, 26*, 81–190.

Luchetti, F. (2002). Importance and future of olive oil in the world market – an introduction to olive oil. *European Journal of Lipid Science and Technology, 104*, 559–563.

Mariani, C. (1998). Ergosterol in olive oils. *Rivista Italiana Delle Sostanze Grasse, 75*, 3–10.

Mínguez-Mosquera, M., Gandul-Rojas, B., Garrido-Fernández, J., & Gallardo-Guerrero, L. (1990). Pigments present in virgin olive oil. *Journal of the American Oil Chemists' Society, 67*, 192–196.

Morales, M., Luna, G., & Aparicio, R. (2005). Comparative study of virgin olive oil sensory defects. *Food Chemistry, 91*, 293–301.

Olías, J., Pérez, A., Ríos, J. J., & Sanz, L. C. (1993). Aroma of virgin olive oil: Biogenesis of the "green" odor notes. *Journal of Agricultural and Food Chemistry, 41*, 2368–2373.

ONH. (2015). *National oil office.* web site www.onh.com.tn.

Paganuzzi, V. (1999). Monoglycerides in vegetable oils. Note IV: Raw oils of law unsaturation. *Rivista Italiana Delle Sostanze Grasse, 76*, 457–471.

Pérez-Camino, C., Moreda, W., Mateos, R., & Cert, A. (2002). Determination of esters of fatty acids with low molecular weight alcohols in olive oils. *Journal of Agricultural and Food Chemistry, 50*, 4721–4725.

Perrin, J. (1992). Minor components and natural antioxidants of olives and olive oils. *Revue Francaise des Corps Gras, 39*, 25–32.

Psomiadou, E., & Tsimidou, M. (1998). Simultaneous HPLC determination of tocopherols, carotenoids, and chlorophylls for monitoring their effect on virgin olive oil oxidation. *Journal of Agricultural and Food Chemistry, 46*, 5132–5138.

Rabascall, N. H., & Riera, J. B. (1987). Variations of the tocopherols and tocotrienols content in the obtention, refining and hydrogenation processes of edible oils. *Gracas y Aceites, 38*, 145–148.

Rao, C., Newmark, H., & Reddy, B. (1998). Chemopreventive effect of squalene on colon cancer. *Carcinogenesis, 19*, 287–290.

Reiter, B., & Lorbeer, E. (2001). Analysis of the wax ester fraction of olive oil and sunflower oil by gas chromatography and gas chromatography-mass spectrometry. *Journal of the American Oil Chemists' Society, 78*, 881–888.

Rocha, J. M., Xavier, F., Malcata, & Balcaol, V. M. (2016). Extra-virgin olive oil: The importance of authentication and quality control. *International Journal Nutritional Science and Food Technology, 2*, 70–72.

Romani, A., Mulinacci, N., Pinelli, P., Vincieri, F. F., & Cimato, A. (1999). Polyphenolic content in five Tuscany cultivars of *Olea europaea* L. *Journal of Agricultural and Food Chemistry, 47*, 964–967.

Salch, Y., Grove, M., Takamura, H., & Gardner, H. W. (1995). Characterization of a C-5,13-cleaving enzyme of 13(S)- hydroperoxide of linolenic acid by soybean seed. *Plant Physiology, 108*, 1211–1218.

Scano, P., Casu, M., Lai, A., Saba, G., Dessi, M. A., Deiana, M., Corongiu, F. P., & Bandino, G. (1999). Recognition and quantitation of cis-vaccenic and eicosenoic fatty acids in olive oils by ^{13}C nuclear magnetic resonance spectroscopy. *Lipids, 34*, 757–759.

Smith, T., Yang, G., Seril, D., Liao, J., & Kim, S. (1998). Inhibition of 4-(methylnitrosamino)-1-(3-pyridyl)-1-butanone induced lung tumorigenesis by dietary olive oil and squalene. *Carcinogenesis, 19*, 703–706. (1998).

Tiscornia, E., Fiorina, N., & Evangelisti, F. (1982). Chemical composition of olive oil and variations induced by refining. *Rivista Italiana Delle Sostanze Grasse, 59*, 519–555.

Tombesi, A., Michelakis, N., & Pastor, M. (1996). Recommendations of the working group on olive farming production techniques and productivity. *Olivae, 63*, 38–51.

Visioli, F., Bellomo, G., & Galli, C. (1998). Free radical-scavenging properties of olive oil polyphenols. *Biochemical and Biophysical Research Communications, 247*, 60–64.

Wiseman, S. A., Mathot, J. N., Fouw, N. J., & Tijburg, L. B. (1996). Dietary non-tocopherol antioxidants present in extra virgin olive oil increase the resistance of low density lipoproteins to oxidation in rabbits. *Atherosclerosis, 120*, 15–23.

Chapter 21
Vaccinium Genus Berry Waxes and Oils

M. Klavins and L. Klavina

Abstract *Vaccinium* is a common and widespread genus of about 450 species of shrubs or dwarf shrubs in the heath family (*Ericaceae*). Most of *Vaccinium* species produce edible berries and many berries are widely cultivated at an industrial scale, for example, bilberries (*Vaccinium myrtillus* L.), highbush blueberries (*Vaccinium corymbosum* L.), lingonberries (*Vaccinium vitis-idaea* L.), and cranberries (*Vaccinium oxycoccos* L.). These berries are important articles in the markets of berries in Northern countries both raw, and processed. *Vaccinium* berries are at first valued considering high vitamin concentrations as well as high concentrations of phenolics. However, also oils and waxes (lipids) of *Vaccinium* berries have high diversity in respect to their composition as well as the high potential of their applications. *Vaccinium* berry oils and waxes can be obtained and produced from berry press residues-berry juice processing wastes. From *Vaccinium* berry seeds oils, a high amounts of triglycerides as well as fatty acids, alkanes, alkanols, terpenoids and other lipids can be isolated. Several sterols also can be found in *Vaccinium* berry oils. From the perspective of practical applications, following groups of substances are of interest: sterols, terpenoids, and polyunsaturated fatty acids.

Keywords Ericaceae · *Vaccinium corymbosum* · *Vaccinium myrtillus* · *Vaccinium corymbosum* · *Vaccinium vitis-idaea* · *Vaccinium oxycoccos*

1 Introduction

To *Vaccinium* genus, belong about 450 species of shrubs or dwarf shrubs in the family *Ericaceae*. Most of *Vaccinium* species produce edible berries, however, under industrial scale, bilberries (*Vaccinium myrtillus* L.), highbush blueberries (*Vaccinium corymbosum* L.), lingonberries (*Vaccinium vitis-idaea* L.), and cranberries (*Vaccinium oxycoccos* L.) are cultivated. These berries are important articles in the

M. Klavins (✉) · L. Klavina
University of Latvia, Riga, Latvia
e-mail: maris.klavins@lu.lv

© Springer Nature Switzerland AG 2019
M. F. Ramadan (ed.), *Fruit Oils: Chemistry and Functionality*,
https://doi.org/10.1007/978-3-030-12473-1_21

markets of berries in Northern countries and they are used in traditional cultures thousands of years ago in Europe, North America and Asia. The traditional fields of usage demonstrate significant potential of the *Vaccinium* berries, however their composition at first was associated with phenolics (Klavins et al. 2018). Studies of *Vaccinium* berry lipids composition are relatively recent, at first related to the development of analytical capacities and much more rare than studies of lipids of American cranberry (Paredes-López et al. 2010).

Lipids in berries can be found as berry waxes (epicuticular and cuticular waxes), cytoplasmatic lipids and seed lipids. Lipids can be analyzed in each of the berry compartment separately as well as in whole berries. *Vaccinium* berry seeds are rich in oil (Johansson et al. 1997) and berry seeds contain high levels of polyunsaturated fatty acids (PUFA) with desirable ratios between *n*-6 and *n*-3 (Johansson et al. 1997). The unique fatty acid composition, often in combination with high contents of lipid-soluble antioxidants, makes the seeds of wild and cultivated *Vaccinium* berries valuable raw materials for nutraceuticals and functional ingredients of foods (Yang et al. 2003).

2　Extraction and Analysis of *Vaccinium* Lipids

Vaccinium berries are relatively rich in lipids as indicated the screening of the seed oil fatty acids of berry species that grow wild in Northern Europe (Johansson et al. 1997) wherein the highest oil content is found in the seeds (*ca*. 30% d.w.). The smaller the seed of a berry species within a genus, the higher was the oil content. Fatty acids of triacylglycerols were analyzed as methyl esters by gas chromatography. Typically the most abundant fatty acids were linoleic, α-linolenic, oleic and palmitic acids The content and composition of the seed oil of various berry species within the same genus, or within the same family, were similar, thus supporting the commonly accepted taxonomic classification.

For analysis of berry lipids, mostly gas chromatography-mass spectrometry (GC-MS) based methods were used (Lowenthal et al. 2013). For qualitative characterization of metabolites found in *Vaccinium* berries, Standard Reference Materials (SRM) can be efficiently used. Definitive identifications of typical lipids in *Vaccinium* species can be achieved using available SRM. Lipids were enriched using an organic liquid/liquid extraction, and derivatized prior to GC-MS analysis. Electron ionization fragmentation spectra can be searched against electron ionization spectra of authentic standards compiled by the National Institute of Standards and Technology's (NIST) mass spectral libraries, as well as spectra selected from the literature. Lipid identifications were further validated using a retention index match along with prior probabilities and compared with results obtained using collision-induced dissociation (Lowenthal et al. 2013).

To extract lipids from fresh berries, the main task is to dehydrate fresh, homogenized berries. To do this, an approach suggested by Bligh-Dyer (Bligh and Dyer 1959) can be used. A mixture of $CHCl_3$ and CH_3OH can be used, ensuring simultaneous dehydration of the berry mass and transfer of lipids to the $CHCl_3$ phase. The yield of lipid fraction using Bligh-Dyer extraction was from 2.70 mg dry residue g^{-1} of cloudberries to 8.43 mg dry residue g^{-1} of blueberries (Klavins et al. 2016). In recent years, berry powders (dried, lyophilized berries) are used more frequently in the food industry (Nile and Park 2014). To estimate the most abundant lipids in berry powders, single-solvent extraction was tested using five different solvents, considering their possible extraction efficiency, perspectives of their application at an industrial scale and environmental aspects of their applications. Single-solvent extraction from berry powder gives much higher yields of dry residue than the use of $CHCl_3$ and CH_3OH mixture. The highest concentration of dry residue was found in the powdered blueberry extracts with diethyl ether, giving 75.81 mg dry residue g^{-1} of berries, while the highest yield of Bligh-Dyer dry residue was 8.62 mg g^{-1} of berries (Table 21.1).

A supercritical CO_2 (SC-CO_2) extraction is a prospective tool for extraction not only from whole berries, but also from a waste product (the solid waste generated in industrial berry juice production residues). The solid residues generated from blueberry press residues after pressing were extracted by conventional solvent extraction or by SC-CO_2 extraction. The composition of *Vaccinium* berry lipids much depends on the used approach. In this respect, supercritical fluid extraction technique have advantages of the high penetrating and solvating power of supercritical fluids for extraction of lipids and other bioactive substances (King and List 1996). SC-CO_2 extraction is the most commonly used process for obtaining the lipophilic extracts free of residues of conventional organic solvents (Lenucci et al. 2010). SC-CO_2 extractions are often carried out at a mild temperature in absence of oxygen; thus, it is possible to avoid thermal and oxidative damages to the bioactive components in the extract. The solid residues generated from blueberry press residues after pressing were extracted by conventional solvent extraction or by SC-CO_2

Table 21.1 Lipid extraction yields from *Vaccinium* berries (Klavins et al. 2016)

Extraction type	Berry	Lipid yield, mg g^{-1} berries (DW)
Bligh-Dyer (Bligh and Dyer 1959) ($CH_3OH/CHCl_3$ 2:1)	Fresh blueberry	8.62
	Fresh bilberry	3.62
	Fresh lingonberry	5.68
	Fresh highbush blueberry (cv. Blue Ray, cv. Chippewa)	2.36
		3.06
Bligh-Dyer, 3 extractions	Fresh blueberry	8.43
Hexane	Dry blueberry	46.22
Chloroform	Dry blueberry	75.81
Petroleum ether	Dry blueberry	40.15
Ethyl acetate	Dry blueberry	52.53
Diethyl ether	Dry blueberry	44.62
Diethyl ether, 3 extractions	Dry blueberry	43.54
Diethyl ether, 3 extractions	Dry lingonberry	32.03

extraction. The effect of particle size and extraction time on the extraction yield produced by conventional solvents was assessed. Such comparison demonstrated the efficiency of SC-CO$_2$ extraction.

In most of the *Vaccinium* berry seed oils analyzed (Yang et al. 2011), the two essential fatty acids, linoleic (18:2n-6) and α-linolenic (18:3n-3) acids, were the major fatty acids. Typically, the two fatty acids together represented 60–80% of the total fatty acids, and the ratio between both fatty acids ranging from 1:1 to 2:1. These oils are optimal as food ingredients or food supplements for increasing the intake of n-3 fatty acids. The content of oleic acid (18:1n-9) was 11–46% of total fatty acids in the most seed oils. Dietary supplementation with γ-linolenic acid (18:3n-6) and stearidonic acid (18:4n-3) may bypass deficiency in Δ6-desaturase activity. Oils from seeds of lingonberry contained exceptionally high levels of γ-tocotrienol (120 mg/100 g oil).

3 Composition of *Vaccinium* Lipids

Vaccinium lipids have been studied not only in Northern countries of Europe but also in species grown in America, and Asia. Blueberries are widely cultivated in North America, China, Korea, but can be cultivated in South America, even though they can potentially be grown in all seasons. The geographical origin does not much influence the lipid composition of berries and the studied species were rich in carotene, tocopherols and other lipophilic substances (Pertuzatti et al. 2014). Composition of berries obtained in a gradient field in Europe (from Poland to Finland) have a significant differences. Lipids extracted from fruits and leaves of lingonberry (*Vaccinium vitis-idaea* L.) collected in Finland and Poland were studied. The main lingonberry triterpenoid profile consisted of α-amyrin, β-amyrin, betulin, campesterol, cycloartanol, erythrodiol, fern-7-en-3β-ol, friedelin, lupeol, sitosterol, stigmasterol, stigmasta-3,5-dien-7-one, swert-9(11)-en-3β-ol, taraxasterol, urs-12-en-29-al, uvaol, oleanolic acid, and ursolic acid. The influence of geographical origin on the level of individual triterpenoid compounds was examined, and considerable variations in triterpenoid profile between berries and leaves obtained from two locations were observed. The most striking difference concerned the occurrence of fernenol and taraxasterol, which were the major triterpenol in lingonberries of Finnish and Polish origin (Szakiel et al. 2012).

The total oil content as well as the composition of fatty acids and phytosterols of five Transylvanian (Romania) pomaces of wild and cultivated blueberries (*Vaccinium myrtillus*), wild cowberry (*Vaccinium vitis-idaea*) and raspberry (*Rubus idaeus*), and cultivated black chokeberry (*Aronia melanocarpa*), were determined by capillary gas chromatography (Dulf et al. 2012). Out of the five pomace oils, the percentages of PUFA ranged from 37% to 69%. The lipid classes (polar lipids, triacylglycerols, sterol esters) were separated and identified using thin-layer chromatography (TLC). Triacylglycerols (TAG) showed the highest PUFA content (ranging from 41.9% to 72.5%) and PUFA/saturated fatty acids ratios were in the range of 5.8–33.1%. In the case of polar lipids and sterol esters fractions, the levels of saturated fatty acids were significantly higher than in TAG. The total amount of sterols was in the range of 101.6–168.2 mg per 100 g of lipids of the analyzed pomaces. The predominant phytosterols were β-sitosterol, stigmastanol + isofucosterol, and campesterol. The results indicated that the investigated pomace oils, due to their good balance between *n*-6 and *n*-3 fatty acids (except for chokeberry) and high β-sitosterol content, could be excellent sources of PUFA and phytosterols, thus suggesting potential value-added utilization of berry waste oils for preparing functional foods or food supplements.

The substances with highest concentrations in the blueberry extracts were C18 unsaturated fatty acids (26.44–102.1 mg 100 g^{-1} berries) (Table 21.2), which were also found in the previous studies (Johansson et al. 1997; Croteau and Fagerson 1969; Dulf et al. 2012).

One of the most important groups of lipids in berries is sterols. Sterols in the seeds of wild Finnish blueberry (*Vaccinium myrtillus*) and lingonberry (*Vaccinium vitis-idaea*) were analyzed as TMS derivatives by GC-MS (Yang et al. 2003). Free and esterified sterols constituted 0.7% and 0.3% of the seed oil of *V. myrtillus*, respectively. In the seed oil of *V. Vitis idaea*, the sterols in the two fractions were equally represented (0.5–0.6%). Sitosterol (85% in *V. myrtillus* and 80% in *V. vitis-idaea*) and campesterol (7% in *V. myrtillus* and 6% in *V. vitis-idaea*) were the dominant free sterols. In comparison to free sterols, steryl esters were found to have considerably lower proportions of sitosterol (40–60%) and campesterol (3–5%), accompanied by higher levels of intermediates in the biosynthetic cascade. Although both species contained the same major sterols, they differed in their relative abundances of individual compounds. Specifically, blueberry contained a higher proportion of sitosterol and campesterol and a lower proportion of isofucosterol and cycloartenol. Comparison of berries collected from northern and southern Finland showed that growth conditions have little effect on the sterol composition in the berries.

Freely available lipids obtained using the Bligh-Dyer extraction method from the 9 chosen types of berries were determined. In total, 111 different substances were identified (Table 21.2) by comparing their mass spectra and retention index with the reference mass spectra and reference retention index. The highest numbers of substances were found in the lingonberry (79) extracts. The lowest numbers of substances were found in the highbush blueberry cv. BlueRay (63) and blueberry (65) extracts. Seventy substances were identified in the highbush blueberry cv. Chippewa extract and 73 and 75 substances in the bilberry and cranberry extracts, respectively. Substances like benzoic acid (0.64–164.40 μg g^{-1} berries), nonanoic acid (0.34–1.43 μg g^{-1}), m-hydroxybenzoic acid (0.16–0.52 μg g^{-1}), squalene (0.37–2.04 μg g^{-1}), α-tocopherol (0.65–3.51 μg g^{-1}) and β-sitosterol (4.23–84.64 μg g^{-1}) were found in all berries at various concentrations. Some of the substances were found in one berry type only-for example, lanosterol in lingonberries (4.92 μg g^{-1}), and chlorogenic acid in both cultivars of highbush blueberry (0.24–1.37 μg g^{-1}) (Table 21.2). The substance with the highest concentration was benzoic acid (164.40 μg g^{-1}) in lingonberries. Also, all of the C18 unsaturated fatty acids were found in high concentrations (up to 102.1 μg g^{-1} blueberries).

Table 21.2 Lipid analysis of *Vaccinium* berries

Compound	Cranberry	Lingonberry	HB cv. Blue Ray	HB cv. Chippewa	Bilberry	Blueberry
Hexanoic acid	0.95	1.51	1.25	0.54	0.83	0.53
Benzyl alcohol	0.69	3.00	ND	ND	ND	ND
Heptanoic acid	0.47	0.58	0.85	0.41	0.57	0.71
Benzoic acid	37.08	164.40	4.40	0.64	0.51	6.13
Octanoic acid	0.87	1.31	1.86	1.00	0.67	1.63
Phenylacetic acid	ND	ND	0.13	0.25	0.27	0.23
Fumaric acid	0.32	ND	ND	ND	0.12	0.08
o-Toluic acid	ND	ND	0.59	0.13	ND	ND
Nonanoic acid	0.70	1.43	1.40	0.44	0.65	0.61
m-Toluic acid	ND	ND	0.40	0.08	ND	0.08
Hydrocinnamic acid	ND	0.24	0.13	0.08	0.13	0.19
2-Deoxytetronic acid	ND	0.41	0.24	0.23	0.46	0.34
Cinnamic acid	0.22	1.42	ND	ND	ND	ND
9-Decenoic acid	0.16	0.37	0.18	0.10	0.17	0.20
Decanoic acid	ND	0.79	2.22	1.51	0.48	1.12
Butanedioic acid	41.68	0.44	ND	ND	3.37	4.61
Terpinol	0.15	ND	ND	ND	ND	ND
p-Anisic acid	ND	0.33	ND	ND	ND	ND
10-Undecenoic acid	ND	ND	0.23	0.10	0.17	0.27
Undecanoic acid	0.20	0.35	1.15	0.13	0.26	0.25
m-Hydroxybenzoic acid	0.26	0.52	0.28	0.16	0.29	0.30
β-Phenyllactic acid	0.14	ND	ND	ND	0.13	ND
Pimelic acid	0.35	0.25	ND	ND	ND	ND
Dodecanoic acid	0.87	1.92	1.72	0.36	1.06	0.64
Octanedioic acid	0.46	0.37	ND	ND	ND	ND
9-Tridecenoic acid	0.36	0.63	ND	ND	0.41	0.21
n-Tridecanoic acid	0.17	ND	ND	ND	ND	0.19
Nonadioic acid	0.65	1.49	ND	ND	0.15	0.17
Tetradecanoic acid	0.73	1.65	5.76	ND	0.41	0.58
Phenyloctanoic acid	0.39	0.96	0.48	0.20	0.35	3.16
p-Coumaric acid	0.33	1.10	ND	ND	ND	0.08
Pentadecanoic acid	0.40	0.62	2.31	0.26	0.27	0.44
Hexadecanoic acid	16.39	23.48	45.80	14.62	21.41	38.46
Heptadecanoic acid	0.46	0.64	1.41	0.43	0.44	0.57
9,12-Octadecadienoic acid	19.81	22.81	1.99	5.01	17.83	26.44
Trans-9-octadecenoic acid	32.16	62.83	20.54	13.82	40.41	102.10
Trans-11-octadecenoic acid	10.88	22.71	17.98	7.03	16.12	51.16
Octadecanoic acid	5.86	6.47	13.97	5.00	8.85	21.31
Linoleic acid	11.17	14.39	0.50	3.44	10.87	0.08

(continued)

Table 21.2 (continued)

Compound	Cranberry	Lingonberry	HB cv. Blue Ray	HB cv. Chippewa	Bilberry	Blueberry
Nonadecanoic acid	ND	ND	0.60	0.35	0.24	0.18
11-eicosenoic acid	0.53	0.40	0.65	0.20	0.19	ND
Eicosanoic acid	2.63	13.31	2.16	4.81	4.21	5.52
Butyl 9,12-octadecadienoate	1.76	4.78	1.44	1.65	3.46	7.96
Butyl 9,12,15-octadecatrienoic acid	3.44	7.98	2.73	2.71	7.25	19.34
Butyl 11-octadecenoic acid	2.24	0.76	0.82	0.64	1.26	5.28
Pentacosane	ND	ND	ND	0.20	0.54	ND
Heneicosanoic acid	0.38	0.24	0.34	0.28	0.29	ND
1-docosanol	0.14	1.13	0.18	0.15	ND	ND
α-Monopalmitin	0.37	0.72	0.27	0.24	0.30	0.33
Tetracosanal	ND	0.46	ND	ND	0.48	ND
Docosanoic acid	2.40	2.72	1.31	1.04	0.85	0.72
Heptacosane	ND	0.31	ND	ND	1.42	0.42
Tricosanoic acid	0.43	0.62	0.26	0.28	0.47	ND
Tetracosan-1-ol	0.33	1.63	0.36	0.26	0.33	0.59
α-Monostearin	0.32	0.70	0.13	0.10	0.32	0.35
Squalene	1.17	2.04	1.16	0.54	0.82	0.82
Tetracosanoic acid	3.35	2.50	0.24	0.72	5.91	2.11
1-Pentacosanol	ND	ND	0.09	0.12	0.21	0.28
Nonacosane	ND	2.48	0.12	0.45	0.22	ND
Pentacosanoic acid	0.34	0.24	ND	0.08	ND	ND
1-Hexacosanol	ND	0.82	0.45	0.58	0.97	4.20
γ-tocopherol	0.13	0.23	0.61	0.45	0.13	0.25
Hexacosanoic acid	2.00	2.92	ND	ND	0.68	ND
Octacosanal	0.66	1.35	0.39	0.82	13.96	2.76
Chlorogenic acid	ND	ND	0.24	1.37	ND	ND
α-tocopherol	1.13	2.22	1.44	1.19	0.78	0.65
1-octacosanol	ND	0.21	0.56	0.72	1.72	0.40
Campesterol	2.98	3.91	0.51	0.36	0.47	12.74
1-Triacontanal	1.30	1.89	0.32	0.50	0.56	0.95
β-Sitosterol	6.48	11.87	8.59	7.25	8.52	84.64
β-Amyrin	0.38	1.31	1.03	0.86	1.14	11.44
α-Amyrin	0.55	1.79	1.31	0.88	0.16	5.30
Cycloartenol	0.23	2.39	ND	3.65	ND	ND
Lanosterol	ND	4.92	ND	ND	ND	ND
Nonacosanoic acid	0.81	ND	ND	ND	ND	ND
Triacontanoic acid	0.81	ND	ND	ND	ND	ND

(continued)

Table 21.2 (continued)

Compound	Cranberry	Lingonberry	HB cv. Blue Ray	HB cv. Chippewa	Bilberry	Blueberry
Lupeol	0.20	0.77	ND	ND	0.33	0.51
Uvaol	ND	0.91	ND	ND	ND	ND
Betulin	ND	0.88	0.42	0.36	0.57	0.53
Oleanolic acid	9.98	6.67	0.52	1.42	0.11	14.85
Ursolic acid	53.18	30.54	1.31	3.94	26.76	20.96

ND-substance not detected; all values are expressed as µg of substance g^{-1} of berries (Klavins et al. 2016)

Single-solvent extractions were done on dry berry powders to find the best solvent for lipid extractions. Five solvents were used: n-hexane, petroleum ether, diethyl ether, ethyl acetate and chloroform (Table 21.3). The largest amount of substances was extracted using diethyl ether (2.9 mg g^{-1} berry powder) and hexane (1.5 mg g^{-1}), while the least amounts were extracted with petroleum ether (0.36 mg g^{-1}). Hexane and diethyl ether extracts contained large amounts of β-sitosterol (341.4 and 334.3 mg g^{-1} berry powder). Diethyl ether extracts contained large amounts of C18 unsaturated fatty acids (101.9–818.5 µg g^{-1}) and malic acid (402.1 µg g^{-1}).

The high content of fatty acids in the studied berries is because berries have a lot of seeds, where the energy is stored in the form of fatty acids. Blueberry, bilberry and cultivars of highbush blueberries are related, which can also be seen in their chemical profiles.

4 Waxes of *Vaccinium* Berries

The outermost layer of *Vaccinium* berry fruits is covered by cuticular wax, which is visible as a bluish-white coating (Gross et al. 2016). It is known that the cuticular wax layer acts as the first protective barrier against biotic and abiotic stresses, and plays a pivotal role in non-stomatal moisture retention, reduction of pathogenic and insect attacks, protection against mechanical damage, and attenuating ultraviolet radiation (Yeats and Rose 2013). Researches have suggested that cuticular wax also plays an important role in fruit quality, postharvest storability, and pathogen susceptibility in postharvest horticultural crops (Lara et al. 2014; Martin and Rose 2014). For example, the cuticular wax is closely related to fruit postharvest water loss. The wax contents have also been reported to be correlated positively with the resistance of grape berries to postharvest infection by *Botrytis cinerea* (Gabler et al. 2003; Laroze et al. 2010). In addition, the cuticular wax may be related to postharvest physiological disorders such as the chilling injury of grapefruit and cuticular cracking of sweet cherry. Therefore, a clear understanding of the components and amounts of fruit wax is important for obtaining better fruit quality, improving

Table 21.3 Effect of single-solvent extraction on the blueberry composition

Substance	Hexane	Petroleum ether	Diethyl ether	Ethyl acetate	Chloroform
Benzoic acid	65.97	22.81	64.74	16.90	49.04
Nonanoic acid	2.75	1.92	2.77	2.65	2.65
Butanedioic acid	0.00	0.00	402.19	0.00	0.00
Dodecanoic acid	2.80	1.78	4.32	1.92	3.35
Citric acid	0.00	0.00	77.33	0.00	0.00
Glucofuranoside	13.85	2.10	10.51	3.18	9.12
Palmitic acid	82.23	16.13	121.34	1.83	118.75
9,12-Octadecadienoic acid	75.15	5.59	320.75	718.16	223.80
9, 12, 15-Octadecatrieonoic acid	298.11	10.94	818.52	105.18	37.24
trans-11-Octadecenoic acid	53.61	2.58	101.98	36.65	65.51
Octadecanoic acid	12.77	2.58	41.65	2.87	2.70
Butyl 9,12-octadecadienoate	87.43	27.08	97.48	309.95	90.98
Butyl 9,12,15-octadecatrienoate	261.20	65.93	296.93	38.41	323.27
Butyl octadecanoate	34.52	11.54	39.22	10.25	30.88
Heptacosane	13.70	8.76	8.31	1.73	6.83
Nonacosane	15.05	2.01	15.14	3.21	15.22
Octacosanal	62.53	18.23	77.55	2.42	77.70
Triacontanal	31.15	3.14	27.83	1.86	36.62
β-Sitosterol	341.31	132.43	334.47	52.08	321.37
β-Amyrin	59.78	14.51	57.93	13.54	57.90
α-Amyrin	16.36	7.13	15.04	26.91	18.22
Betulin	8.58	3.84	15.51	17.91	24.63
Total, $\mu g\ g^{-1}$ berries	1538	361	2951	1367	1515

All values expressed as μg of substance g^{-1} of berry powder (Klavins et al. 2016)

disease resistance and developing postharvest treatment strategies. The outer surface of the *Vaccinium* berries is covered by lipid-soluble epicuticular waxes. Epicuticular waxes can be found in different forms, including amorphous to crystalline deposits. Berry waxes consists of complex mixtures of long-chain aliphatic and cyclic components, including primary alcohols (C_{26}, C_{28}, and C_{30}), hydrocarbons (C_{29}, and C_{31}), esters, fatty acids and triterpenoids (Riederer and Schonherr 1985). Another group of waxes (intracuticular) waxes are embedded in the cutin polymer matrix and consists of polymeric glycolipids (Baker 1982).

Cuticular wax is generally comprised of a complex mixture of very-long-chain (VLC) aliphatic compounds (e.g. fatty acids, alcohols, alkanes, aldehydes, ketones and alkyl esters) and cyclic compounds (e.g. triterpenoids and steroids) (Jetter et al. 2007). Chemical composition and morphology of cuticular wax show great variation between plant species, organ and developmental stage (Lara et al. 2014). Triterpenoid ursolic acid and alkane nonacosane were identified as the predominant

compounds in sweet cherry wax, whereas triterpenoid amyrin and alkane hentriacontane as the main compounds in tomato wax (Kosma et al. 2010).

To extract epicuticular waxes, chemical, mechanical and thermal pre-treatments
can be used to reduce the effect of skin hydrophobicity and promote water transport
during drying of whole berries. Chemical pre-treatment involves immersion of the
product in alkaline or acid solutions of oleate esters prior to drying. Alkaline dip-
ping facilitates drying by forming cracks on the fruit surface (Salunkhe et al. 1991).
However, the high temperature (100 °C) of the chemical solution may cause texture
degradation (Grabowski and Marcotte 2003). Mechanical pre-treatments might
replace or complement chemical pre-treatments (Grabowski et al. 2007) and it consists of peeling, or abrasion of the surface, and puncturing the skin. This kind of
pre-treatment may cause leaking and loss of internal mass during dehydration.
Some other pre-treatments include exposure to sulphur dioxide, and thermal pre-
treatments such as blanching (immersion in hot water) or steaming (Grabowski
et al. 2007) also can be used. However, blanching may cause the loss of soluble
substances like proteins and mineral elements while high temperatures may induce
the loss of heat labile compounds. As an approach to achieve water loss, immersions
in liquid nitrogen can be used (Ketata et al. 2013). Depending on the blueberry species, a reduction of dehydration time from 45% to 65% was obtained for liquid
nitrogen-treated samples when compared to control blueberries. Microscopic observations of the blueberry skin before and after the liquid nitrogen pre-treatment
revealed a decrease of the cuticle thickness and dewaxing of the skin surface. The
dewaxing of the blueberry skin due to liquid nitrogen immersions was confirmed by
wax quantification before and after the cryogenic pre-treatments (Ketata et al.
2013).

Comparison of the *Vaccinium* waxes indicate that the hydrocarbon content of
wax was the lowest in *Vaccinium uliginosum,* but in the other deciduous *Vaccinium*
species (*V. myrtillus) the* hydrocarbon content was the highest (Salaso 1987). The
alkane content of the hydrocarbon fraction varied between 98% and 100%. There is
no evidence for the presence of branched alkanes or unsaturated hydrocarbons
amongst berry waxes. The alkane distribution patterns conformed to the usual criteria, wherein odd-carbon alkanes were higher than their adjacent even-carbon neigh-
bors, and the odd-carbon alkane percentages showed only one peak in each species.
In most species, the major alkane was hentriacontane followed by nonacosane in
Vaccinium vitis-idaea nonacosane assumed the the highest percentage, while in
V. uliginosum the chain lengths were considerably shorter and spread more evenly,
the content of the major alkane, heptacosane, being only 20% (Salaso 1987).

The chemical composition and morphology of cuticular wax in mature fruit of
nine blueberry cultivars were investigated using GC-MS and scanning electron
microscope. Triterpenoids and β-diketones were the most prominent compounds,
accounting for on average 64.2% and 16.4% of the total wax, respectively. Ursolic
or oleanolic acid was identified as the most abundant triterpenoids differing in cultivars. Two β-diketones, hentriacontan-10,12-dione and tritriacontan-12,14-dione,
were detected in cuticular wax of blueberry fruits. Notably, hentriacontan-10,12-
dione and tritriacontan-12,14-dione were only detected in highbush (*V. corymbosum*)

and rabbiteye (*V. ashei*) blueberries, respectively. The results showed that a large amount of tubular wax deposited on the surface of blueberry fruits. There was no apparent difference in wax morphology among the nine cultivars.

5   Potential Applications

Many of the berry lipids have biological activities that might influence the application potential of lipid extracts. Berry lipid components and especially sterols have antioxidant properties and the resulting health benefits (Szajdek and Borowska 2008). In addition to pro-vitamin activity and antioxidant traits, carotenoids, including β-carotene, have several other functions in the body at the molecular level where they act as immunomodulators, inhibit mutagenesis and prevent malignant transformations. The composition of sterols, PUFA and other lipid groups can make *Vaccinium* berry lipids as a unique material for the production of food additives, pharmaceuticals and substances for use in cosmetics (Klavins et al. 2015; Klavins et al. 2017). Considering relatively high yields of lipids from berry press residues, there is a huge potential of *Vaccinium* berry lipid production (Yang et al. 2011). Another group of lipids of interest is *Vaccinium* berry waxes. They have the potential of the applications as surface coatings as well as again in cosmetics. However to achieve large-scale applications much more research on berry lipids should be done.

Acknowledgments The study was supported by the European Regional Development Fund within the project No. 1.1.1.1/16/A/047 "Genus *Vaccinium* berry processing using green technologies and innovative, pharmacologically characterized biopharmaceutical products".

References

Baker, E. A. (1982). Chemistry and morphology of plant epicuticular waxes. In D. F. Cutler, K. L. Alvin, & C. E. Price (Eds.), *The plant cuticle* (pp. 139–166). London: Academic.

Bligh, E. G., & Dyer, W. J. (1959). A rapid method of total lipid extraction and purification. *Canadian Journal of Biochemistry and Physiology, 37*, 911–917.

Croteau, R., & Fagerson, S. (1969). Seed lipids of the American cranberry (vaccinium macrocarpon). *Phytochemistry, 8*(11), 2219–2222.

Dulf, F. W., Andrei, S., Bunea, A., & Socaciu, C. (2012). Fatty acid and phytosterol contents of some Romanian wild and cultivated berry pomaces. *Chemical Papers, 66*, 925–934. https://doi.org/10.2478/s11696-012-0156-0.

Gabler, F. M., Smilanick, J. L., Mansour, M., Ramming, D. W., & Mackey, B. E. (2003). Correlations of morphological, anatomical, and chemical features of grape berries with resistance to *Botrytis cinerea*. *Phytopathology, 93*, 1263–1273.

Grabowski, S., & Marcotte, M. (2003). Pretreatment efficiency in osmotic dehydration of cranberries. In J. Welti-Chanes, F. Velez-Ruiz, & G. V. Barbosa-Cinovas (Eds.), *Transport phenomena in food processing* (pp. 83–94). New York: CRC Press.

Grabowski, S., Marcotte, M., Quan, D., Taherian, A. R., Zareifard, M. R., Poirier, M. R., & Kudra, T. (2007). Kinetics and quality aspects of Canadian blueberries and cranberries dried by osmo-convective method. *Drying Technology, 25*, 367–374.

Gross, K. C., Wang, C. Y., & Saltveit M. (2016). *The commercial storage of fruits, vegetables, and florist and nursery stocks*. US Dept Agriculture Handbook No 66.

Jetter, R., Kunst, L., & Samuels, A. L. (2007). Composition of plant cuticular waxes. In M. Riederer & C. Müller (Eds.), *Biology of the plant cuticle* (pp. 145–181). Oxford: Blackwell Publishing Ltd.

Johansson, A., Laakso, P., & Kallio, H. (1997). Characterization of seed oils of wild, edible Finnish berries. *Zeitschrift für Lebensmittel-Untersuchung und -Forschung, 204*, 300–307.

Ketata, M., Desjardins, Y., & Ratti, C. (2013). Effect of liquid nitrogen pretreatments on osmotic dehydration of blueberries. *Journal of Food Engineering, 116*, 202–212.

King, J. W., & List, C. R. (1996). *Supercritical fluid technology in oil and lipid chemistry*. Champaign: AOCS Press.

Klavins, L., Klavina, L., Huna, A., & Klavins, M. (2015). Polyphenols, carbohydrates and lipids in berries of *Vaccinium* species. *Environmental and Experimental Biology, 13*, 147–158.

Klavins, L., Kviesis, J., Steinberga, I., Klavina, L., & Klavins, M. (2016). Gas chromatography-mass spectrometry study of lipids in northern berries. *Agronomy Research, 14*, 1328–1347.

Klavins, M., Kukela, A., Kviesis, J., & Klavins, L. (2017). Valorisation of berry pomace: From waste to bioactive compounds. In A. Kallel, M. Ksibi, H. B. Dhia, & N. Khelifili (Eds.), *Recent advances in environmental science from the Euro-Mediterranean and surrounding regions* (pp. 1145–1147). Berlin: Springer.

Klavins, L., Kviesis, J., Nakurte, I., & Klavins, M. (2018). Berry press residues as a valuable source of polyphenolics: Extraction optimization and analysis. *LWT- Food Science and Technology, 93*, 583–591. https://doi.org/10.1016/j.lwt.2018.04.021.

Kosma, D. K., Parsons, E. P., Isaacson, T., Lü, S., Rose, J. K. C., Jenks, M. A. (2010). Fruit cuticle lipid composition during development in tomato ripening mutants. *Physiologia Plantarum, 139*(1), 107–117.

Lara, I., Belge, B., & Goulao, L. F. (2014). The fruit cuticle as a modulator of postharvest quality. *Postharvest Biology and Technology, 87*, 103–112.

Laroze, L. E., Dıaz-Reinoso, B., Moure, A., Zuniga, M. E., & Domınguez, H. (2010). Extraction of antioxidants from several berries pressing wastes using conventional and supercritical solvents. *European Food Research and Technology, 231*, 669–677. https://doi.org/10.1007/s00217-010-1320-9.

Lenucci, M. S., Caccioppola, A., Durante, M., Serrone, L., Leonardo, R., Piro, G., et al. (2010). Optimisation of biological and physical parameters for lycopene supercritical CO_2 extraction from ordinary and high-pigment tomato cultivars. *Journal of Science and Food Agriculture, 90*, 1709–1718.

Lowenthal, M. S., Andriamaharavo, N. R., Stein, S. E., & Phinney, K. W. (2013). Characterizing vaccinium berry standard reference materials by GC-MS using NIST spectral libraries. *Analytical and Bioanalytical Chemistry, 405*, 4467–4476. https://doi.org/10.1007/s00216-012-6610-6.

Martin, L. B., & Rose, J. K. (2014). There's more than one way to skin a fruit: Formation and functions of fruit cuticles. *Journal of Experimental Botany, 65*, 4639–4651.

Nile, S. H. & Park, S. W. (2014). Edible berries: bioactive components and their effect on human health. *Nutrition, 30*(2), 134–144.

Paredes-López, O., Cervantes-Ceja, M. L., Vigna-Pérez, M., & Hernández-Pérez, T. (2010). Berries: Improving human health and healthy aging, and promoting quality life-a review. *Plant Foods for Human Nutrition, 65*, 299–308. https://doi.org/10.1007/s11130-010-0177-1.

Pertuzatti, P. B., Barcia, M. T., Rodrigues, D., da Cruz, P. N., Hermoskn-Gutiérrez, I., Smith, R., & Godoy, T. N. (2014). Antioxidant activity of hydrophilic and lipophilic extracts of Brazilian blueberries. *Food Chemistry, 164*, 81–88.

Riederer, M., & Schonherr, J. (1985). Accumulation and transport of (2.4 dichlorophenoxy) acetic acid in plant cuticles. *Ecotoxicology and Environmental Safety, 9*, 196–208.

Salaso, I. (1987). Alkane distribution in epicuticular wax of some heath plants in Norway. *Biochemical Systematics and Ecology, 15*, 663–665.

Salunkhe, D. K., Bolin, H. R., & Reddy, N. R. (1991). *Storage, processing and nutritional quality of fruits and vegetables* (pp. 1–2). Boca Raton: CRC Press.

Szajdek, A., & Borowska, E. J. (2008). Bioactive compounds and health-promoting properties of berry fruits: A review. *Plant Foods for Human Nutrition, 63*, 147–156. https://doi.org/10.1007/s11130-008-0097-5.

Szakiel, A., Paczkowski, C., Koivuniemi, K., & Huttunen, S. (2012). Comparison of the triterpenoid content of berries and leaves of lingonberry *Vaccinium vitis-idaea* from Finland and Poland. *Journal of Agricultural and Food Chemistry, 60*, 4994–5002.

Yang, B., Koponen, J., Tahvonen, R., & Kallio, H. (2003). Plant sterols in seeds of two species of Vaccinium (*V. myrtillus and V. vitis-idaea*) naturally distributed in Finland. *European Food Research and Technology, 216*, 34–38. https://doi.org/10.1007/s00217-002-0611-1.

Yang, B., Ahotupa, M., Määttä, P., & Kallio, H. (2011). Composition and antioxidative activities of supercritical CO_2-extracted oils from seeds and soft parts of northern berries. *Food Research International, 44*, 2009–2017.

Yeats, T. H., & Rose, J. K. C. (2013). The formation and function of plant cuticles. *Plant Physiology, 163*, 5–20.

Chapter 22
Crambe abyssinica Hochst. Oil

Caroline Mariana de Aguiar, Kátia Andressa Santos, Sílvio César Sampaio, and Clayton Antunes Martin

Abstract *Crambe abyssinica* Hochst is an oilseed of the family Brassicaceae, rich in oil with important properties for chemical applications. It is comprised of mono-unsaturated fatty acids (erucic, palmitoleic, oleic, gadoleic and nervonic acids) and antioxidants like phytosterols, tocopherols, carotenoids and chlorophyll. The oil can be extracted by mechanical pressing or using organic solvents. Some studies have used supercritical carbon dioxide and subcritical propane in oil extraction, resulting in higher yields and lower degradation of minor components. The oxidative stability of vegetable oils has been attributed to the content of minor compounds and the structure of fatty acids. Due to the presence of erucic acid, crambe oil is not suitable for human consumption, but it offers great potential for the production of erucic acid, lubricants, detergents, cosmetics, surfactants, pharmaceuticals, corrosion inhibitor, polyethylene films, behenic acid, coatings, nylons, refrigerant fluid, photographic materials, and insulation fluid. In addition, this oil presents great competitiveness and advantages over other vegetable oils in the biodiesel production. Refined oils can be used in fish feed. Cake and meal can be used as protein supplements in ruminant feed, in the removal of metallic ions and pollutants in bioremediation. Therefore, knowledge and studies on the composition of inedible vegetable oils, such as crambe oil, is particularly important for stability and conservation, considering the numerous uses of this oil.

Keywords *Crambe abyssinica* · Fatty acids composition · Antioxidants · Supercritical extraction · Phenolic compounds · Carotenoids · Chlorophyll

C. M. de Aguiar (✉) · C. A. Martin
Federal Technological University of Parana, Toledo, Paraná, Brazil
e-mail: cmaguiar@utfpr.edu.br

K. A. Santos
State University of Western Paraná, Toledo, Paraná, Brazil

S. C. Sampaio
State University of Western Paraná, Cascavel, Paraná, Brazil
e-mail: Silvio.Sampaio@unioeste.br

© Springer Nature Switzerland AG 2019
M. F. Ramadan (ed.), *Fruit Oils: Chemistry and Functionality*,
https://doi.org/10.1007/978-3-030-12473-1_22

1   Introduction

Crambe (*Crambe abyssinica* Hochst) is an oilseed of the family Brassicaceae (Cruciferae), which has around 338 genera and 3709 species (Warwick et al. 2006). Crambe is a genus comprised of 44 species (Prina and Martínez-Laborde 2008) divided into 3 sections: Sarcocrambe, Dendocrambe, and Leptocrambe; the latter contains *Crambe abyssinica* Hochst, which has oilseeds and important properties for chemical applications (Onyilagha et al. 2003). *Crambe abyssinica* Hochst, *Crambe orientalis* and *Crambe tataria* are the most frequently cultivated species (Desai 2004; Tutus et al. 2010).

Studies and commercial production of *Crambe abyssinica* started in the 1980s in the United States, United Kingdom, Italy, France, and Portugal, and then spread in Australia, South Africa, Paraguay, and Brazil (Pitol et al. 2010; Cremonez et al. 2015).

Crambe seeds vary from 0.8 to 2.6 mm (Atabani et al. 2013; Castleman et al. 1999) and can be harvested, transported and stored using machines and structures that are similar to those used with soybean and of corn, requiring only minor adaptations. The low density of seeds, on average 340 kg/m^3, is the main obstacle for crop consolidation (Plein et al. 2010), as it increases transport and storage costs. However, seed husking raises its density, making it similar to the soybean density. Figure 22.1

Fig. 22.1   (**a**) Harvest of *Crambe abyssinica*, (**b**) crambe grains after the harvest, and (**c**) crambe oil

shows *Crambe abyssinica* at the moment of harvest (a), the grains after harvest (b) and the raw crambe oil (c).

Oil is the main product of crambe. The whole grains contain 35.6–42.8% oil (Atabani et al. 2013; Castleman et al. 1999). This oil has significant technological characteristics in relation to edible oils such as sunflower, rapeseed and soybean (Bondioli et al. 1998) and, due to the presence of erucic acid (C22:1), it cannot be used for human consumption (Colodetti et al. 2012; Cremonez et al. 2015).

The main variables related to the production costs of a crambe crop refer to seeds, drying, planting and harvest and transport operations (Bassegio et al. 2016). Jasper et al. (2010) compared the costs of oilseed production and observed that crambe presented lower production costs than other oilseeds (i.e., canola, sunflower and soybean).

Bassegio et al. (2016) report that crambe does not have the potential to significantly impact the supply of raw materials for short-term biodiesel production. However, it is an option for the oil in the future, since the transesterification of crambe oil into biodiesel is viable, with a good percentage of the fatty acids capable of being converted into methyl esters.

According to Onorevoli et al. (2014), crambe oil has fatty acids, mostly erucic acid (22:1n-9) and oleic acid (18:1n-9), whereas the bio-oil from which it is obtained through pyrolysis of residues has a more complex composition, including phenols, acids, ketones, ethers, esters, hydrocarbons, alcohols and nitrogen compounds. The presence of these elements in the bio-oil allows it to be used in chemical and pharmaceutical applications and its potential use as an alternative fuel.

2   Extraction and Processing of Fruit Oil

Traditionally, vegetable oils are extracted from grains by cold pressing or using organic solvents, such as *n*-hexane, providing higher yields (Li et al. 2007). Both methods involve long extraction time followed by additional steps to separate solid residues and the solvent (Santos et al. 2015). The method with organic solvents uses a Soxhlet extractor, a consolidated reference technique for comparison with other extraction methods (Wu et al. 2011).

Pederssetti (2008) argued that some sectors are interested in finding low-cost and safer solvents, especially using supercritical technology. The use of supercritical or subcritical pressure fluids is an option to replace conventional methods. This technology is considered clean because the solvent can be fully removed through system depressurization and recovered (Santos 2014).

Mechanical pressing can be used in the extraction of crambe oil, without any solvent, with 70% efficiency of oil content in the grains; where for every 100 kg of grains used in oil pressing, it is possible to extract, on average, 25 kg of oil. With an extruder and a press, and using a solvent, this percentage increases to almost 100% of available oil content (Pitol et al. 2010). When the mechanical extraction is used, the resulting product is the cake, with residual oil of around 20%; while in extraction with solvents, the extraction is more efficient, with residual oil of about 2% in the meal (Favaro et al. 2010).

Boss (2000) and Freitas (2007) reported some drawbacks related to the quality of the products generated when using solvents, such as hazardous production environment, environmental impact, and processing costs. Vegetable oils are important sources of high value-added products, such as vitamins, pigments and phosphorus lipids, which can be destroyed or not fully used in the conventional process due to the extraction time and high temperature used in the process.

Hexane offers many advantages: it dissolves the oil easily without acting on other components of the grains; it has a homogeneous composition and a small boiling temperature range; it can be mixed with water; it has low latent heat of melting; and it extracts high amounts of oil (Kim et al. 1999; Gomes Jr 2010). On the other hand, it presents some disadvantages, such as high flammability and high cost (Gomes Jr 2010). According to Corso (2008), the extraction of vegetable oils using hexane as the solvent is interesting versus oil pressing, but it has several drawbacks: it is extracted from a non-renewable source of raw material and it is toxic and flammable. Besides, separating the solvent from the extracted oil is a difficult task, involving the risk of having some solvent remaining in the oil (Kim et al. 1999).

The supercritical technology uses specific properties of fluids when they are close to their critical point. In this region, the thermodynamic properties are particularly sensitive to changes in temperature and pressure. Small changes in pressure or temperature cause changes in density, and consequently, and in the solubilization power (Corso 2008).

Jachmanián et al. (2006) stated that although some oils extracted with a supercritical fluid or conventional solvents have similar compositions, some differences were reported in terms of concentration of some minor components, such as tocopherols, sterols, hydrocarbons and pigments. The oxidative stability of oils has been attributed to the content of these compounds, the position of fatty acids in the triacylglycerol molecule, the presence of carotenoids, the grain type, and the processing conditions (Malecka 2002; Jachmanián et al. 2006; Merrill et al. 2008).

Many studies have been conducted with supercritical CO_2 in the extraction of vegetable oils (Li et al. 2007; Corso et al. 2010, Nimet et al. 2011; Da Porto et al. 2012; Solati et al. 2012; Moslavac et al. 2014; Ruttarattanamongkol et al. 2014; Aladic et al. 2015; Danlami et al. 2015; Przygoda and Wejnerowska 2015; Shao et al. 2015). Onorevoli et al. (2014) studied the composition of fatty acids and yield of crambe oil extracted with hexane, subcritical propane, and mechanical press. Santos et al. (2015) analyzed the effect of temperature and pressure on the yield of crambe oil extracted with propane under subcritical conditions and compared the chemical composition and oxidative stability of the oil in relation to the Soxhlet extraction with hexane and dichloromethane. Aguiar (2016) compared the methods of crambe oil extraction with supercritical CO_2 and hexane in terms of minor components (free fatty acids, sterols, tocopherols, total carotenoids, chlorophyll a and phenolic compounds), color, oxidative stability and total antioxidant activity (Table 22.1).

Table 22.1 The composition of crambe oil in terms of color, antioxidant activity and oxidative stability parameters as affected by extraction method

Oil parameter	Soxhlet (*n*-hexane) Mean ±SD	Supercritical (CO_2)[a] Mean ±SD	*p*-value[b]
Total free fatty acids (g 100 g^{-1})	1.68 ± 0.21	3.11 ± 0.22	0.00388*
Total phytosterols (mg 100 g^{-1})	200.44 ± 0.74	874.92 ± 16.38	0.00020*
Total tocopherols (mg 100 g^{-1})	66.024 ± 1.83	277.29 ± 6.73	0.00025*
Total carotenoids (mg Kg^{-1})	196.28 ± 4.44	76.84 ± 4.84	0.00050*
Chlorophyll *a* (mg Kg^{-1})	56.45 ± 4.09	n.d.[c]	0.00174*
Determination of color[d]			
C*	57.34 ± 2.72	41.62 ± 1.95	0.00288*
L*	85.98 ± 3.47	91.55 ± 3.08	0.13938^{ns}
a*	−7.03 ± 1.04	−7.38 ± 0.26	0.60292^{ns}
b*	56.97 ± 2.95	40.96 ± 2.00	0.00339*
Total phenolic compounds[e] (mg GAE Kg^{-1})	103.95 ± 7.67	122.30 ± 5.41	0.01997*
Total antioxidant activity (mmol trolox Kg^{-1})	3.86 ± 0.06	10.23 ± 0.21	0.000005*
Oxidative stability (°C)[f]			
T_{on}	206.89 ± 0.98	193.14 ± 3.53	0.00803*
T_{p1}	273.82 ± 2.82	263.09 ± 1.31	0.01378*
T_{p2}	324.02 ± 0.88	318.15 ± 2.08	0.02774*

*Significant ($p < 0.05$)
ns not significant ($p > 0.05$)
[a]Extraction with SC-CO_2 at 20 MPa and 40 °C
[b]p-value for paired t-test at 5% significance
[c]Not detected; for the statistical test, zero was used as the result for this parameter
[d]C* indicates color intensity (chroma), L* indicates brightness (white-black), a* (red-green), b* (yellow-blue)
[e]Phenolic compounds in gallic acid equivalents (GAE)
[f]T_{on} indicates the oxidation start temperature, T_{p1} indicates the first peak and T_{p2} indicates the second peak

Regarding the antioxidant content, Table 22.1 shows that supercritical CO_2 (SC-CO_2) extracted higher concentrations of phytosterols and tocopherols and lower concentrations of carotenoids ($p < 0.05$) than with *n*-hexane. Aladic et al. (2015) and Nimet et al. (2011) obtained higher concentrations of tocopherols with SC-CO_2 extraction when compared to Soxhlet extraction in *Cannabis sativa* and sunflower oils, attributing this difference to high temperature and long extraction time in Soxhlet extraction. In this study, this relation was approximately four times greater.

Li et al. (2007) observed the same behavior for the content of phytosterols (Table 22.1) in relation to extractions with SC-CO_2 and Soxhlet for sea buckthorn oil, attributing these results to the different solubility of these compounds in the used solvents. Ruttarattanamongkol et al. (2014) also found significantly higher concentrations of phytosterols in the oil of *Moringa oleífera* using the supercritical method (CO_2; 35 °C; 15 MPa) in relation to Soxhlet.

    The behavior for total carotenoid content (Table 22.1) was similar to that obtained by Jachmanián et al. (2006) and Uquiche et al. (2012) for canola and rapeseed oils, respectively. Sahena et al. (2009) reported that hexane extracts both polar and non-polar pigments, justifying the results found in this study.

    Regarding total phenolic compounds, Table 22.1 shows a difference between the extractions ($p < 0.05$), with a higher concentration of these compounds in the oil extracted with SC-CO$_2$. In vegetable oils, phenolic compounds can be polyphenols, simple phenols or acids (Dimitrios 2006). The phenolic compounds present a certain polarity (Angelo and Jorge 2007) due to the presence of hydroxyls in the molecule and, as hexane extracts both polar and non-polar compounds (Sahena et al. 2009), it may have extracted higher concentrations of these compounds, but due to a higher temperature of the extraction (70–80 °C) than the supercritical extraction (40 °C), they may have been degraded because they are thermosensitive (Boutin and Badens 2009; Corso et al. 2010; Moslavac et al. 2014).

    The results of color can be attributed to the presence of carotenoids, which ensures a yellow to red color (Uenojo et al. 2007), in agreement with the study conducted by Tuberoso et al. (2007), in which parameter b* was correlated to carotenoids in several vegetable oils.

    The antioxidant activity of crambe oil is higher in the oil extracted with SC-CO$_2$ ($p < 0.05$). These results can be attributed to higher concentrations of tocopherols, sterols and phenolic compounds in the oil. Table 22.2 shows the results of free fatty acids, sterols and tocopherols obtained by Santos et al. (2015) in oil extracted with hexane and subcritical propane.

    The quantification of free fatty acids (Table 22.2) shows that, with the two extraction methods, the concentrations were lower than 2%, indicating the oil can be stored for long periods without lipid deterioration. In the quantification of phytosterols, a significant difference was observed between the two extraction methods, indicating the temperature and pressure conditions used in the extraction with propane affected the phytosterol content. The extraction using the highest pressure (16 MPa) showed that the factor with the greatest effect on the extraction of these compounds was the vapor pressure of the solute. Regarding tocopherols, a significant difference was observed between the extraction methods. It can be concluded that higher temperatures and lower pressures lead to higher levels of tocopherols in crambe oil.

Table 22.2 The composition of crambe oil extracted with n-hexane and subcritical propane in terms of total free fatty acids, phytosterols and tocopherols

Oil parameter	Soxhlet (n-hexane) Mean ±SD	Supercritical (propane) Mean ±SD
Total free fatty acids (%)	1.19 ± 0.01[b]	1.16[1] ± 0.04[b]
Total phytosterols (mg 100 g^{-1})	180.79 ± 0.4[a]	201.05[2] ± 0.45[b]
Total tocopherols (mg 100 g^{-1})	163.41 ± 1.67[a]	202.18[3] ± 0.11[b]

Different letters on the same line indicate significant difference at 5% confidence level
[1]Extraction with propane 16 MPa at 80 °C
[2]Extraction with propane 16 MPa at 40 °C
[3]Extraction with propane 8 MPa at 80 °C

3 Fatty Acid Composition

Vegetable oils are constituted of triglycerides (up to 95–98%) and a mixture of minor components (2–5%). Chemically, triglycerides are esters of fatty acids with glycerol and the minor components present different qualitative and quantitative compositions (Cert et al. 2000; Rittner 2002). The different fatty acids linked with the glycerol chain define the fatty acid profile. According to Knothe and Dunn (2001), since each fatty acid has peculiar chemical properties, the fatty acid profile is the parameter of greatest influence on the characteristics of lipids from which they originate. The amount of unsaturated fatty acids in oils and fats influences the development of oxidative rancidity, which is favored in the following order: linolenic acid (18:3) > linoleic acid (18:2–6) > oleic acid (Robey and Shermer 1994).

Crambe, canola, peanut and palm oils show higher resistance to oxidation than soybean and sunflower oils due to their different fatty acid composition (Cremonez et al. 2015). The differentiation of crambe oil is its high content of monounsaturated fatty acids (MUFA): erucic acid (53–63%), palmitoleic acid (0.05–2.3%), oleic acid (15–23%), gadoleic acid (1–6%), and nervonic acid (1%), totaling 72.3–88.1% (Knights 2002; Favaro et al. 2010; Onorevoli 2012; Gomes Jr 2010; Lalas et al. 2012; Aguiar et al. 2017). These characteristics ensure higher oxidation stability to the oil when compared, for example, to sunflower, cotton, soybean and babassu oils (Cremonez et al. 2015), since the oxidation rate increases considerably with the double bonds of fatty acids (Kodali 2002).

Erucic acid is a monounsaturated long-chain fatty acid with 22 naturally occurring carbons in the cruciferous family (No et al. 2013). Its molecular formula is $C_{22}H_{42}O_2$, noted as 22:1 ω-9, also known as *cis*-13-docosenoic acid. Table 22.3 shows the fatty acid composition of different crambe oils. Zhu et al. (2016) isolated the oil from crambe seeds and found that each mg of seeds contained approximately 400 mg of triacylglycerol. In the oil composition isolated from crambe seeds, Gurr et al. (1974) reported approximately 88.6% of triglycerides and the remainder consisting of phospholipids including phosphatidylcholine (4.8%), phosphatidylinositol (4.8%) and phosphatidylethanolamine (1.9%).

4 Minor Bioactive Lipids

The main minor compounds present in crambe oil are tocopherols, phytosterols, phenolics, carotenoids and chlorophyll, which can act as antioxidants. Recent studies have been conducted on the characterization of crambe oil, oxidative stability, fatty acid composition and antioxidant concentration including those performed by Lalas et al. (2012), Onorevoli et al. (2014), Santos et al. (2015), and Aguiar et al. (2017). Due to their antioxidant properties, tocopherols inhibit the peroxidation of polyunsaturated fatty acids (PUFA) and other compounds, preventing rancidity during storage, and can be used as an additive for food and cosmetics (Freitas et al. 2008).

Table 22.3 Fatty acid profile (%) of cramble oil samples

Fatty acid	Oil 1[a]	Oil 2[b]	Oil 3[c]	Oil 4[d]	Oil 5[e]	Oil 6[f]	Oil 7[g]
Lauric acid (C12:0)	–	0.05	–	–	–	–	–
Myristic acid (C14:0)	–	0.09	–	–	–	–	–
Palmitic acid (C16:0)	–	1.97	1.96	2	0.88	2.3	2.24
Palmitoleic acid (C16:1)	–	0.14	–	–	–	0.3	0.13
Margaric acid (C17:0)	–	0.05	–	–	–	–	–
Margaroleic acid (C17:1)	–	0.04	–	–	–	–	–
Stearic acid (C18:0)	0.9	1.1	1.24	0.7	0.53	1.16	1.14
Oleic acid (C18:1)	15.9	19.03	23.19	18	15.07	15.87	20.58
Linoleic acid (C18:2)	8.7	7.59	7.03	9.4	13.16	10.1	4.71
Linolenic acid (C18:3)	8.7	4.35	–	6.5	–	4.85	1.07
Arachidic acid (C20:0)	–	1.17	–	0.8	0.63	1.24	1.17
Gadoleic acid (C20:1)	–	5.34	1.12	2	2.4	6.38	4.21
Behenic acid (C22:O)	–	2.14	1.69	–	2.14	1.96	2.2
Erucic acid (C22:1)	56.4	54.77	63.78	55.9	63.77	53.91	61.82
Lignoceric acid (C24:0)	–	0.75	–	–	0.44	0.74	–
Nervonic acid (C24:1)	–	1.42	–	–	0.99	1.2	–
Other	9.4	–	–	–	–	1.01	0.74
Saturated	0.9	7.32	4.89	3.5	4.62	7.39	6.75
Monounsaturated	72.3	80.74	88.09	75.9	82.23	77.66	86.74
Polyunsaturated[h]	17.4	11.94	7.03	15.9	13.16	14.95	5.78

Sources:
[a]Knights (2002)
[b]Favaro et al. (2010)
[c]Onorevoli (2012)
[d]Gomes (2010)
[e]Lalas et al. (2012)
[f]Aguiar et al. (2017)
[g]Tavares et al. (2017)
[h]Fatty acids with two or more double bonds

Tocopherols (α, β, γ and δ), together with the tocotrienols (Fig. 22.2), are generally referred to as vitamin E, differing from each other in the number and location of groups in the chromanol ring (Ramalho and Jorge 2006).

Lalas et al. (2012), Santos et al. (2015), and Aguiar et al. (2017) found α, β, γ, and δ- tocopherols in crambe oil extracted with n-hexane and supercritical CO_2. Lalas et al. (2012) reported that tocopherols present in crambe oil may protect it against oxidation during storage and processing. Phytosterols, mainly β-sitosterol, campesterol and stigmasterol are components of the cell membrane of plants and abundant in vegetable oils and grains (Ryan et al. 2007). Lechner et al. (1999), Lalas et al. (2012), Santos et al. (2015), and Aguiar et al. (2017) quantified β-sitosterol, stigmasterol and campesterol in crambe oil. The literature has studies reporting the antioxidant properties of phytosterols and their action to protect the oil from oxida-

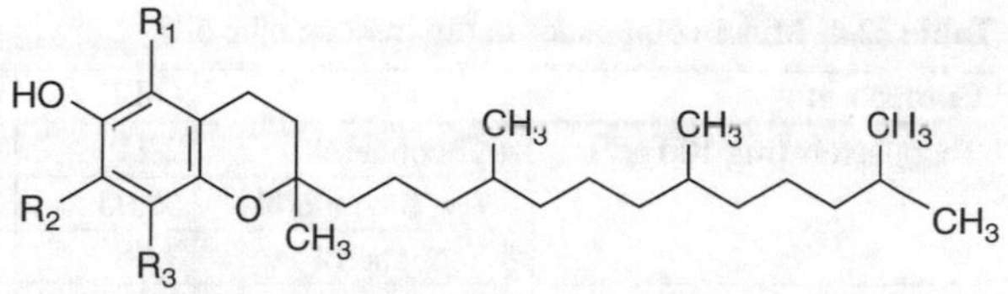

Fig. 22.2 Molecules of tocopherols (Ramalho and Jorge 2006)

α - tocoferol: $R_1 = R_2 = R_3 = CH_3$

β - tocoferol: $R_1 = R_2 = CH_3$; $R_2 = H$

γ - tocoferol: $R_1 = H$; $R_2 = R_3 = CH_3$

δ - tocoferol: $R_1 = R_2 = H$; $R_3 = CH_3$

tion (Ryan et al. 2007; Kmiecik et al. 2009; Tsaknis and Lalas 2002; Winkler and Warner 2008). These compounds can be used as raw material for the synthesis of hormones and drugs, cosmetics, and as additives in thermoplastic resins for the production of rubber materials (Beveridge et al. 2002).

Regarding phenolic compounds, they may be polyphenols, simple phenols or acids, and they can act as antioxidants (Dimitrios 2006), and are pigments or products of secondary metabolism of the plant in reactions to protect against environmental aggression (Silva et al. 2010). They are important because, besides the nutritional issue, they can act as antibiotics, natural pesticides, attractive for pollinators, protection against ultraviolet radiation, maintenance of gas and water impermeable cell walls, and as a structural material to provide stability to plants (Heldt 2005). When evaluating the effect of the extraction method on total phenolic content, Aguiar et al. (2017) reported that the content of these compounds in cambe oil is influenced by the type of extraction, being 17.7% greater for the extraction with SC-CO$_2$ in relation to the extraction with *n*-hexane.

Regarding the total antioxidant activity, Aguiar et al. (2017) observed in their study that it is related to some phenolic compounds that may be present, such as flavonoids and phenolic acids (syringic, ferulic, and synaptic acids) (Tuberoso et al. 2007; Angelo and Jorge 2007), and tocopherols (Szydłowska-Czerniak et al. 2008; Guinazi et al. 2009), more than to phytosterols (Tsaknis et al. 1999) or carotenoids of crambe oil. Tuberoso et al. (2007) determined the concentration of antioxidant compounds and antioxidant activity in several vegetable oils and found the variability of correlations between the antioxidant activity and the composition of oils can be attributed to the concentrations of squalene, chlorophylls, carotenoids, and phenolic compounds of the oils and their mutual interactions.

Chlorophyll is a chromophore that exists in different forms in plants, with the primary forms being chlorophyll *a* and *b* (Bianchi et al. 2015) in the ratio of 1:3, respectively. Chlorophyll *a* is the main pigment of photosynthesis (Heldt 2005), while the other pigments help in light absorption and transfer of energy to the reaction centers, called accessory pigments, for example, chlorophyll *b* and carotenoids (Streit et al. 2005). Carotenoids have a yellow-to-red color (Uenojo et al. 2007) and are associated with higher oxidative stability of vegetable oils (Jachmanián et al. 2006). Aguiar et al. (2017) also determined the presence of chlorophyll *a* and carot-

Table 22.4 Minor compounds in different crambe oils

Component		Oil 1[a]	Oil 2[b]	Oil 4[c]	Oil 5[d]
Tocopherols (mg 100 g^{-1})	α-Tocopherol	2.21	2.63	0.77	–
	β + γ-Tocopherol	59.95	138.59	12.5	–
	δ-Tocopherol	3.86	22.1	0.39	–
Phytosterols (mg 100 g^{-1})	β-Sitosterol	150.17	103.28	–	232.8
	Stigmasterol	34.93	–	–	4.8
	Campesterol	12.74	54.09	–	101.2
	Brassicasterol	–	23.42	–	–
Total phenolic compounds (mg GAE kg^{-1})		–	–	137.73	–
Antioxidant activity (mmol trolox kg^{-1})		–	–	4.59	–
Chlorophyll a (mg kg^{-1})		–	–	41.34	–
Total carotenoids (mg kg^{-1})		–	–	163.94	–

Source:
[a]Aguiar et al. (2017)
[b]Santos et al. (2015)
[c]Lalas et al. (2012)
[d]Lechner et al. (1999)

enoids in crambe oil. Rodriguez-Amaya et al. (2008) reported some studies with plants belonging to the same family as that of crambe oil, in which β-carotene, lutein and zeaxanthin were quantified. Silva et al. (1999) stated that oxidation is a phenomenon that directly affects the commercial value of lipid compounds and all products formulated from them. Therefore, knowledge and studies on the composition of inedible vegetable oils, such as crambe oil, is particularly important for stability and conservation, considering the numerous uses of this oil. Table 22.4 shows the composition of minor compounds in different crambe oils.

It should be noted that the extraction method may influence the concentration of some minor components of oils, such as tocopherols, phytosterols, free fatty acids and pigments. The oxidative stability of oils has been attributed to the content of these compounds as well as the position of fatty acids in the triglyceride molecule, the presence of carotenoids, the type of grain, and the processing conditions (Malecka 2002; Jachmanián et al. 2006; Merrill et al. 2008).

Considering the above, Fig. 22.3 shows the oxidation curves of crambe oil samples extracted through Soxhlet and supercritical CO_2 using the non-isothermal method (Aguiar 2016).

Table 22.1 and Fig. 22.3 show the temperatures for the points observed during the oxidation of crambe oil (T_{on}, T_{p1} and T_{p2}) are statistically higher ($p < 0.05$) for the Soxhlet extraction in relation to the supercritical method, indicating that the oil extracted with SC-CO_2 is less stable to oxidation, as reported by Uquiche et al. (2012), who compared the stability of rapeseed oil extracted with hexane (exhaustive extraction) and SC-CO_2 (40 MPa, 60 °C, 60 min).

Micic et al. (2015) reported that the first two peaks are the main indicators of the oxidation process; the first one (T_{p1}) is considered in the oxidative evaluation of the oil in the non-isothermal method. Antioxidants are consumed during the induction

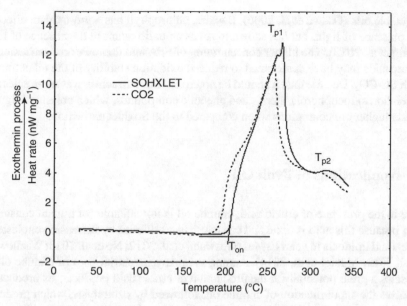

Fig. 22.3 Oxidation curves of crambe oil using the non-isothermal method (DSC) at 10 °C min⁻¹, flow of 50 mL O_2 min⁻¹; T_{on} – start temperature, T_{p1} – first peak, T_{p2} – second peak

period (up to T_{on}), and then hydroperoxides are produced after the oxygen attacks the double bonds of fatty acids (up to T_{p1}), and the second peak is a result of decomposition of these hydroperoxides in other products (Litwinienko et al. 1997; Litwinienko and Kasprzycka-Guttman 1998; Micic et al. 2015).

The values of T_{on}, T_{p1} and T_{p2} were 206.89; 273.82 and 324.02 °C, respectively, for the oil extracted with *n*-hexane, and 193.14; 263.09 and 318.15 °C for the oil extracted with SC-CO_2 (Table 22.1). Oil stability can be influenced by the presence and composition of tocopherols, carotenoids and sterols, type of grain, and processing conditions (Merrill et al. 2008; Jachmanián et al. 2006). Jachmanián et al. (2006) observed that the stability of canola oil extracted with SC-CO_2 was lower than the stability for the oil extracted with *n*-hexane; they attributed this fact to the higher content of carotenoids, in agreement with this study. Uquiche et al. (2012) also observed the same behavior in rapeseed oil and attributed this fact to the higher concentration of antioxidants in the oil extracted with *n*-hexane (tocopherols, phenolic compounds and phospholipids). In addition, other compounds such as phospholipids, not determined in this study, may contribute to the higher oxidative stability of the oil extracted with *n*-hexane, since it can also extract polar compounds. Przybylski et al. (1998) observed that the oxidation stability of canola oil extracted with SC-CO_2 increased when phospholipids were present and the concentration of free fatty acids (FFA) was reduced, concluding that these compounds were the main factors affecting the stability of canola oil. This factor should be taken into account, because the oil extracted with SC-CO_2 presented higher concentration of FFA and the release of FFA contribute to the oxidative process of vegetable oils (Osawa et al. 2006). Besides, chlorophyll has a pro-oxidant effect in the presence of light, but has shown to act as an antioxidant in the absence of light (Yang et al. 2013). The higher concentration of FFA and the lower concentration of carotenoids may have contributed to reduced oxidation stability of the oil extracted with SC-CO$_2$. Despite this, it should be noted that this extraction was more selective in relation to tocopherols, sterols and phenolic compounds, which presented significantly higher concentrations when compared to the Soxhlet extraction.

5   Applications of Fruit Oil

Due to the presence of erucic acid, crambe oil is not suitable for human consumption because this acid is toxic and can cause heart disease by increasing cholesterol levels and lipidosis in heart tissues (Goswami et al. 2012; No et al. 2013; Wazilewski et al. 2013; Maciel et al. 2014). Among the many industrial uses of crambe oil, it presents a great potential for the production of erucic acid products. Its production involves the saponification of crambe oil, followed by ozonolysis, which produces brassidic and pelargonic acids (Lazzeri et al. 1994). In addition, erucamide can be obtained, a substance used in the production of cosmetics and other industrial applications (Falasca et al. 2010). Estimates indicate the annual consumption of 30,000 tons of erucic acid derivatives, such as erucamide, with 3–5% annual growth (Zanetti et al. 2009). The largest producers of erucamide use rapeseed (*Brassica napus*), which presents a percentage of approximately 47% erucic acid. However, the current demand for this acid cannot be fulfilled with this culture only.

Crambe oil appears as the better alternative source of erucic acid, since it presents 8–9% more of this fatty acid, with the advantage of presenting good adaptation to tropical areas -the largest producers are in the United States (Jasper 2009; Zanetti et al. 2009). Although its main application is in the manufacture of erucamide, erucic acid and/or its derivatives also have wide industrial applications. Its long carbon chain allows the oil to withstand high temperatures and remain liquid at low temperatures, making it a high-quality lubricating oil, much more effective and biodegradable than mineral oils (Wang et al. 2000).

Besides the erucic acid applications, crambe oil can be used in the manufacture of lubricants, detergents, cosmetics, surfactants, pharmaceutical products, corrosion inhibitor, polyethylene film preparation, behenic acid (C22:0), coatings, nylons, coolants, photographic materials, and emulsifiers (Lazzeri et al. 1994; Leonard 1994; Mulder and Mastebroek 1996; Temple-Heald 2004; Carlsson 2009; Falasca et al. 2010; Shashidhara and Jayaram 2010). According to Oliveira et al. (2013), vegetable oil-based insulating fluids available on the market are obtained from oilseeds, mainly soybeans, corn and sunflower. However, these plants also provide oil to the food industry. In this context, Gomes Jr. (2010) reported the viability of crambe oil as an electrical insulator for transformers, since this oil was four times more stable than soybean oil.

For the biodiesel market, crambe oil presents great competitiveness and advantages when compared to other vegetable oils. It represents an alternative source in oil supply diversification for biodiesel, as a potential raw material (Favaro et al. 2010; Jasper et al. 2013; Mello et al. 2017). Despite its qualities, crambe oil is still little explored for the production of biodiesel, with few studies found in the literature on this subject (Tavares et al. 2017). One of the main obstacles to its exploration is the presence of phospholipids in its composition. Phospholipids are one of the main impurities in crude oils, especially when they are used for biodiesel production, because they affect the catalytic conversion (Lobo et al. 2009). As mentioned above, crambe oil has up to about 11% phospholipids. Costa et al. (2018) found 86.3 mg L^{-1} of phosphorus (P) in crambe oil and biodiesel produced with this oil had a P concentration above the level allowed in EN 14214 (British Standard 2008), that is, above of 4 mg L^{-1}. Due to the presence of phospholipids, water degumming was not effective to reduce this oil component.

In addition to the uses and applications of crambe oil, by-products of oil extraction including cake and/or meal can be used as protein supplements in ruminant feed (Carlson et al. 1996) and in the removal of metal ions (Gonçalves 2013; Gonçalves et al. 2013; Rubio et al. 2013a, b) and pollutants in bioremediation (Feng et al. 2011).

Meal presents excellent nutritional quality, with values of up to 45% crude protein, with the similar ruminal digestibility of organic matter to soybean meal (Pitol et al. 2010; Mendonça 2012; Herculano 2013). The use of crambe cake in animal feed is an important strategy for the reuse of crambe residues, with the potential to add value to farming applications and promote sustainability of the production chain (Bassegio et al. 2016).

References

Aguiar, C. M. (2016). Efeito da adubação fosfatada sobre características agronômicas e qualidade do óleo da cultura de crambe. Dissertation. Universidade estadual do oeste do paraná.

Aguiar, C. M., Sampaio, S. C., Santos, K. A., Silva, E. A., Piana, P. A., Richart, A., & Reis, R. R. (2017). Total fatty acid content, antioxidant composition, antioxidant activity, and content of oil from crambe seeds cultivated with phosphorus. *European Journal of Lipid Science and Technology, 119*(10). https://doi.org/10.1002/ejlt.201700043.

Aladic, K., Jarni, K., Barbir, T., Vidovic, S., Vladic, J., Bilic, M., & Jokic, S. (2015). Supercritical CO_2 extraction of hemp (*Cannabis sativa* L.) seed oil. *Industrial Crops and Products, 76*, 472–478.

Angelo, P. M., & Jorge, N. (2007). Compostos fenólicos em alimentos – Uma breve revisão. *Revista do Instituto Adolfo Lutz, 66*, 1–9.

Atabani, A. E., Silitongaa, A. S., Onga, H. C., Mahliac, T. M. I., Masjukia, H. H., Badruddina, I. A., & Fayaza, H. (2013). Non-edible vegetable oils: A critical evaluation of oil extraction, fatty acid compositions, biodiesel production, characteristics, engine performance and emissions production. *Renewable and Sustainable Energy Reviews, 18*, 211–245.

Bassegio, D., Zanotto, M. D., Santos, R. F., Werncke, I., Dias, P. P., & Olivo, M. (2016). Oilseed crop crambe as a source of renewable energy in Brazil. *Renewable and Sustainable Energy Reviews, 66*, 311–321.

Beveridge, T. H. J., Li, T. S. C., & Drover, J. C. G. (2002). Phytosterol content in American ginseng seed oil. *Journal of Agricultural and Food Chemistry, 50*, 744–750.

Bianchi, L. M., Duncan, S. E., Webster, J. B., Neilson, A. P., & O'Keefe, S. F. (2015). Contribution of chlorophyll to photooxidation of soybean oil at specific visible wavelengths of light. *Journal of Food Science, 80*, 252–261.

Bondioli, P., Folegatti, L., Lazzeri, L., & Palmieri, S. (1998). Native *Crambe abyssinica* oil and its derivatives as renewable lubricants: An approach to improve its quality by chemical and biotechnological processes. *Industrial Crops and Products, 7*, 231–238.

Boss, E. A. (2000). Analise do desempenho de plantas de extração de óleo convencionais e de processos supercríticos. Dissertation. Universidade Estadual de Campinas. Faculdade de Engenharia Quimica. 105p.

Boutin, O., & Badens, E. (2009). Extraction from oleaginous seeds using supercritical CO_2: Experimental design and products quality. *Journal of Food Engineering, 92*, 396–402.

British Standard EN 14214. (2008). Automotive fuels – fatty acid methyl esters (FAME) for diesel engines – requirements and test methods.

Carlson, K. D., Gardner, J. C., Anderson, V. L., & Hanzel, J. J. (1996). Crambe: New crop success. In J. Janick (Ed.), *Progress in new crops* (pp. 306–322). Alexandria: ASHS Press.

Carlsson, A. S. (2009). Plant oils as feedstock alternatives to petroleum – a short survey of potential oil crop platforms. *Biochimie, 91*, 665–670.

Castleman, G., Pymer, S., & Greenwood, C. (1999). Potential for crambe (*C. abyssinica*) in Mallee/Wimmera of Australia. In *10th international rapessed congress*.

Cert, A., Moreda, W., & Pérez-Camino, M. C. (2000). Chromatographic analysis of minor constituents in vegetable oils. *Journal of Chromatography. A, 881*, 131–148.

Colodetti, T. V., Martins, L. D., Rodrigues, W. M., Brinate, S. V. B., & Tomaz, M. A. (2012). Crambe: aspectos gerais da produção agrícola. *Enciclopédia Biosfera, 8*(14), 258–269.

Corso, M. P. (2008). Estudo da extração de óleo de sementes de gergelim (*Sesamun indicum* L.) empregando os solventes dióxido de carbono supercrítico e n-propano pressurizado. Dissertação (Mestrado). Programa de Pós-Graduação em Engenharia Química da Universidade Estadual do Oeste do Paraná.

Corso, M. P., Fagundes-Klen, M. R., Silva, E. A., Cardozo Filho, L., Santos, J. N., Freitas, L. S., & Dariva, C. (2010). Extraction of sesame seed (*Sesamun indicum* L.) oil using compressed propane and supercritical carbon dioxide. *Journal of Supercritical Fluids, 52*, 56–61.

Costa, E., Almeida, M. F., Ferraz, M. C. A., & Dias, J. M. (2018). Effect of *Crambe abyssinica* oil degumming in phosphorus concentration of refined oil and derived biodiesel. *Renewable Energy, 124*, 27–33. https://doi.org/10.1016/j.renene.2017.08.089.

Cremonez, P. A., Feroldi, M., Nadaleti, W. C., Rossi, E., Feiden, A., Camargo, M. P., Cremonez, F. E., & Klajnb, F. F. (2015). Biodiesel production in Brazil: Current scenario and perspectives. *Renewable and Sustainable Energy Reviews, 42*, 415–428.

Da Porto, C., Decorti, D., & Tubaro, F. (2012). Fatty acid composition and oxidation stability of hemp (*Cannabis sativa* L.) seed oil extracted by supercritical carbon dioxide. *Industrial Crops and Products, 36*, 401–404.

Danlami, J. M., Zaini, M. M. A. A., Arsad, A., & Yunus, M. A. C. (2015). Solubility assessment of castor (*Ricinus communis* L) oil in supercritical CO_2 at different temperatures and pressures under dynamic conditions. *Industrial Crops and Products, 76*, 34–40.

Desai, B. B. (2004). *Seeds handbook: Biology, production processing and storage* (2nd ed., p. 787). New York: Marcel Dekker.

Dimitrios, B. (2006). Sources of natural phenolics antioxidants. *Trends in Food Science and Technology, 17*, 505–512.

Falasca, S. L., Flores, N., Lamas, M. C., Carballo, S. M., & Anschau, A. (2010). *Crambe abyssinica*: An almost unknown crop with a promissory future to produce biodiesel in Argentina. *International Journal of Hydrogen Energy, 35*, 5808–5812.

Favaro, S. P., Roscoe, R., Dalmontes, A. M. A., Mendonça, B. P. C., & Souza, A. D. V. (2010). Produtos e Co-produtos. In *FUNDAÇÃO MS. Tecnologia e produção: crambe* 2010 (pp. 48–59). Maracaju: Fundação MS.

Feng, N., Guoa, X., Lianga, S., Zhub, Y., & Liub, J. (2011). Biosorption of heavy metals from aqueous solutions by chemically modified orange peel. *Journal of Hazardous Materials, 185*, 49–54.

Freitas, L. S. (2007). Desenvolvimento de procedimentos de extração de óleo de semente de uva e caracterização química dos compostos extraídos. Universidade Federal do Rio Grande do Sul 227 p. Thesis.

Freitas, L. S., Jacques, R. A., Richter, M. F., Silva, A. L., & Caramão, E. B. (2008). Pressurized liquid extraction of vitamin E from Brazilian grape seed oil. *Journal of Chromatography A, 1200*, 80–83.

Gomes, Jr. S. B. (2010). Technical and economic evaluation of vegetable oil application crambe as an electrical insulator compared with soybean oil. Dissertation. UTFPR.

Gonçalves, A. C., Jr. (2013). Decontamination and monitoring water and soil in the Amazon region using alternative adsorbent materials, aimed at removing toxic heavy metals and pesticides. *Incorporated Society, 6*, 105–113.

Gonçalves, A. C., Jr., Rubio, F., Meneghel, A. P., Coelho, G. F., Dragunski, D. C., & Strey, L. (2013). Use *Crambe abyssinica* seeds how adsorbent in removing metals. *Water Revista Brasileira de Engenharia Agrícola e Ambiental, 17*, 306–311.

Goswami, D., Basu, J. K., & De, S. (2012). Optimal hydrolysis of mustard oil to erucic acid: A biocatalytic approach. *Chemical Engineering Journal, 182*, 542–548.

Guinazi, M., Milagres, R. C. R. M., Pinheiro-Sant'Ana, H. M., & Chaves, J. B. P. (2009). Tocoferois e Tocotrienois em óleos vegetais e ovos. *Quim Nova, 32*, 2098–2103.

Gurr, M. I., Blades, J., Appleby, R. S., Smith, C. G., Robinson, M. P., & Nichols, B. W. (1974). Studies on seed-oil triglycerides triglyceride biosynthesis and storage in whole seeds and oil bodies of *Crambe abyssinica*. *European Journal of Biochemistry, 43*, 281–290.

Heldt, H. W. (2005). *Plant biochemistry* (3rd ed.). London: Elsevier Academic Press.

Herculano, B. N. (2013). Bran crambe in feeding dairy. Dissertation. UFVJM.

Jachmanián, I., Margenat, L., Torres, A., & Grompone, M. (2006). Estabilidad oxidativa y contenido de tocoferoles en el aceite de canola extraído con CO_2 supercrítico. *Grasas y Aceites, 2*, 155–159.

Jasper, S. P. (2009). Cultura do crambe (*Crambe abyssinica* Hochst): avaliação energética, de custo de produção e produtividade em sistema de plantio direto. Universidade Estadual Paulista, 103 p. Thesis.

Jasper, S. P., Biaggioni, M. A. M., & Silva, P. R. A. (2010). Comparison of crambe production cost (*Crambe abyssinica* hochst) with other oilseed crops in no–till system. *Energy and Agriculture, 25*, 141–153.

Jasper, S. P., Biaggioni, M. A. M., & Silva, P. R. A. (2013). Caracterização físico-química do óleo e do biodiesel de *Crambe abyssinica* Hochst. *Nucleus, 10*, 183–190.

Kim, H., Lee, S., Park, K., & Hong, I. (1999). Characterization of extraction and separation of rice bran oil rich in EFA using SFE process. *Separation and Purification Technology, 15*(1), 1–8. https://doi.org/10.1016/s1383-5866(98)00048-3.

Kmiecik, D., Korczak, J., Rudzinska, M., Michałowska, A. G., & Hes, M. (2009). Stabilization of phytosterols in rapeseed oil by natural antioxidants during heating. *European Journal of Lipid Science and Technology, 111*, 1124–1132.

Knights, E. G. (2002). Crambe: A North Dakota case study (p. 25). A report for the rural industries research and development corporation, RIRDC Publication n. W02/005.

Knothe, G., & Dunn, R. O. (2001). Biofuels derivated from vegetable oils and fats. In *Oleochemical manufacture and applications* (pp. 106–163). Liverpool: Sheffield Academic.

Kodali, D. R. (2002). High performance ester lubricants from natural oils. *Industrial Lubrication and Tribology, 54*, 165–170.

Lalas, S., Gortzi, O., Athanasiadis, V., & Dourtoglou, V. (2012). Full characterization of *Crambe abyssinica* Hochst. *Seed Oil and Journal of the American Oil Chemists, 89*, 2253–2258.

Lazzeri, L., Leoni, O., Conte, L., & Palmieri, S. (1994). Some technological characteristics and potential uses of *Crambe abyssinica* products. *Industrial Crops and Products, 3*, 103–112.

Lechner, M., Reiter, B., & Lorbeer, E. (1999). Determination of free and esterified sterols in potential new oil seed crops by coupled on-line liquid chromatography-gas-chromatography. *European Journal of Lipid Science and Technology, 101*, 171–177.

Leonard, C. (1994). Sources and commercial applications of high erucic vegetable oils. *Lipid Technology, 4*, 79–83.

Li, T. S. C., Beveridge, T. H. J., & Drover, J. C. G. (2007). Phytosterol content of sea buckthorn (*Hippophae rhamnoides* L.) seed oil: Extraction and identification. *Food Chemistry, 101*, 1633–1639.

Litwinienko, G., & Kasprzycka-Guttman, T. (1998). A DSC study on thermoxidation kinetics of mustard oil. *Thermochimica Acta, 319*, 185–191.

Litwinienko, G., Kasprzycka-Guttman, T., & Studzinski, M. (1997). Effects of selected phenol derivatives on the autoxidation of linolenic acid investigated by DSC non-isothermal. *Thermochimica Acta, 307*(1), 97–106.

Lobo, I. P., Ferreira, S. L. C., & da Cruz, R. S. (2009). Biodiesel: Parâmetros de Qualidade e Métodos Analíticos. *Quim Nova, 32*, 1596–1608.

Maciel, A. M., Ming, C. C., Ribeiro, A. P. B., Silva, R. C., Gioielli, L. A., & Gonçalves, L. A. G. (2014). Physicochemical properties of interesterified blends of fully hydrogenated *Crambe abyssinica* oil and soybean oil. *Journal of the American Oil Chemists' Society, 91*, 111–123.

Malecka, M. (2002). Antioxidant properties of theunsaponifiable matter isolated from tomato seeds, oat grains and wheat rem oil. *Food Chemistry, 79*, 327–330.

Mello, B. T. F., Gonçalves, J. E., Rodrigues, G. M., Cardozo-Filho, L., & Silva, C. (2017). Hydroesterification of crambe oil (*Crambe abyssinica* H.) under pressurized conditions. *Industrial Crops and Products, 97*, 110–119.

Mendonça, B. P. (2012). Co product crambe in cattle feed. PhD thesis Viçosa Univ.

Merrill, L. I., Pike, O. A., Ogden, L. V., & Dunn, M. L. (2008). Oxidative stability of conventional and high-oleic vegetable oils with added antioxidants. *Journal of the American Oil Chemists, 85*, 771–776.

Micic, D. M., Ostojic, S. B., Simonovic, M. B., Krstic, G., Pezo, L. L., & Simonovic, B. R. (2015). Kinetics of blackberry and raspberry seed oils oxidation by DSC. *Thermochimica Acta, 601*, 39–44.

Moslavac, T., Jokic, S., Subaric, D., Aladic, K., Vukoja, J., & Prce, N. (2014). Pressing and supercritical CO_2 extraction of *Camelina sativa* oil. *Industrial Crops and Products, 54*, 122–129.

Mulder, J. H., & Mastebroek, H. D. (1996). Variation for agronomic characteristics in *Crambe hispanica*, a wild relative of *Crambe abyssinica*. *Euphytica, 89*, 267–278.

Nimet, G., Silva, E. A., Palú, F., Dariva, C., Freitas, L. S., Neto, A. M., & Cardozo Filho, L. (2011). Extraction of sunflower (*Heliantus annuus* L.) oil with supercritical CO_2 and subcritical propane: Experimental and modeling. *Chemical Engineering Journal, 168*, 262–268.

No, D. S., Zhao, T., Kim, B. H., Choi, H. D., & Kim, I. H. (2013). Enrichment of erucic acid from crambe oil in a recirculated packed bed reactor via lipase-catalyzed ethanolysis. *Journal of Molecular Catalysis Enzymatic, 87*, 6–10.

Oliveira, R. C., Viecelli, C. A., Primieri, C., Barth, E. F., Bleil Junior, H. G., & Sanderson, K. (2013). Crop crambe. *Assoeste, 70*.

Onorevoli, B. (2012). Estudo do *Crambe abyssinica* como fonte de matérias primas oleaginosas: óleo vegetal, ésteres metílicos e bio-óleo. Dissertation. Instituto de Química – Programa de Pós-Graduação em Ciência dos Materiais. Universidade Federal do Rio Grande do Sul.

Onorevoli, B., Machado, M. E., Dariva, C., Franceschi, E., Krause, L. C., Jacques, R. A., & Caramão, E. B. (2014). A one-dimensional and comprehensive two-dimensional gas chromatography study of the oil and the bio-oil of the residual cakes from the seeds of *Crambe abyssinica*. *Industrial Crops and Products, 52*, 8–16.

Onyilagha, J., Bala, A., & Hallett, R. (2003). Leaf flavonoids of the cruciferous species, *Camelina sativa*, *Crambe* spp., *Thlaspi arvense* and several other genera of the family *Brassicaceae*. *Biochemical Systematics and Ecology, 31*, 1309–1322.

Osawa, C. C., Gonçalves, L. A. G., & Ragazzi, S. (2006). Titulação potenciométrica aplicada na determinação de ácidos graxos livres de óleos e gorduras comestíveis. *Quim Nova, 29*, 593–599.

Pederssetti, M. M. (2008). *Analysis of the effect of the temperature and pressure in the super-critical extraction of the canola essential oil with supercritical carbon dioxide and pressurized n-propano*. Dissertation. Universidade Estadual do Oeste do Paraná.

Pitol, C., Broch, D. L., & Roscoe, R. (2010). *Tecnologia e produção: crambe*. Maracaju: Fundacão MS.

Plein, G. S., Favaro, S. P., Souza, A. D. V., Souza, C. F. T., Santos, G. P., Miyahia, M. A. M., & Roscoe, R. (2010). Caracterização da fração lipídica em sementes de crambe armazenadas com e sem casca. In *IV Congresso brasileiro de mamona e I simpósio internacional de oleaginosas energéticas* (Vol. 1, pp. 1812–1816).

Prina, A. O., & Martinez-laborde, J. B. (2008). A taxonomic revision of Crambe section Dendocrambe (*Brassicaceae*). *Botanical Journal of the Linnean, 156*, 291–304.

Przybylski, R., Lee, Y., & Kim, I. (1998). Oxidative stability of canola oils extracted with super-critical carbon dioxide. *Lebensmittel Wissenschaft und Technologie, 31*, 687–693.

Przygoda, K., & Wejnerowska, G. (2015). Extraction of tocopherol-enriched oils from Quinoa seeds by supercritical fluid extraction. *Industrial Crops and Products, 63*, 41–47.

Ramalho, V. C., & Jorge, N. (2006). Antioxidantes utilizados em óleos, gorduras e alimentos gordurosos. *Quim Nova, 29*, 755–760.

Rittner, H. (2002). Tecnologia das matérias graxas: vol. 2 – purificação e refinação. Impressão autorizada. 367 p.

Robey, W., & Shermer, W. (1994). The damaging effects of oxidation. *Feed Mix, 2*, 22–26.

Rodriguez-Amaya, D. B., Kimura, M., Godoy, H. T., & Amaya-Farfan, J. (2008). Updated Brazilian database on food carotenoids: Factors affecting carotenoid composition. *Journal of Food Composition and Analysis, 21*, 445–463.

Rubio, F., Gonçalves Junior, A. C., Meneghel, A. P., Tarley, C. R., Schwantes, D., & Coelho, G. F. (2013a). Removal of cadmium from water using by–product *Crambe abyssinica* Hochst seeds as biosorbent material. *Water Science and Technology, 68*, 227–233.

Rubio, F., Gonçalves Junior, A. C., Dragunski, D. C., Tarleyc, C. R. T., Meneghel, A. P., & Schwantes, D. (2013b). A *Crambe abyssinica* seed by–product as biosorbent for lead (II) removal from. *Desalination and Water Treatment, 51*, 1–10.

Ruttarattanamongkol, K., Siebenhandl-Ehn, S., Schreiner, M., & Petrasch, A. M. (2014). Pilot-scale supercritical carbon dioxide extraction, physico-chemical properties and profile charac-terization of *Moringa oleifera* seed oil in comparison with conventional extraction methods. *Industrial Crops and Products, 58*, 68–77.

Ryan, E., Galvin, K., O'Connor, T. P., Maguire, A. R., & O'Brien, N. M. (2007). Phytosterol, squalene, tocopherol content and fatty acid profile of selected seeds, grains, and legumes. *Plant Foods for Human Nutrition, 62*, 85–91.

Sahena, F., Zaidul, I. S. M., Jinap, S., Karim, A. A., Abbas, K. A., Norulaini, N. A. N., & Omar, A. K. M. (2009). Application of supercritical CO_2 in lipid extraction – a review. *Journal of Food Engineering, 95*, 240–253.

Santos, K. A. (2014). Extraction of crambe (*Crambe abyssinica*) seed oil using propane subcriti-cal: Characterization of oil and bran. Dissertation. Universidade Estadual do Oeste do Parana.

Santos, K. A., Bariccatti, R. A., Cardozo-Filho, L., Schneider, R., Palú, F., Silva, C., & Silva, E. A. (2015). Extraction of crambe seed oil using subcritical propane: Kinetics, characterization and modeling. *Journal of Supercritical Fluids, 104*, 54–61.

Shao, P., Liu, Q., Fang, Z., & Sun, P. (2015). Chemical composition, thermal stability and antioxi-dante properties of tea seed oils obtained by different extraction methods: Supercritical fluid extraction yields the best oil quality. *European Journal of Lipid Science and Technology, 117*, 355–365.

Shashidhara, Y. M., & Jayaram, S. R. (2010). Vegetable oils as a potential cutting fluid – an evolu-tion. *Tribology International, 43*, 1073–1081.

Silva, F. A. M., Borges, M. F. M., & Ferreira, M. A. (1999). Métodos para avaliação do grau de oxidação lipídica e da capacidade antioxidante. *Quim Nova, 22*, 94–103.

Silva, M. L. C., Costa, R. S., Santana, A. S., & Koblitz, M. G. B. (2010). Compostos fenólicos, carotenóides e atividade antioxidante em produtos vegetais. *Semina: Ciências Agrárias, 31*, 669–682.

Solati, Z., Baharin, B. S., & Bagheri, H. (2012). Supercritical carbon dioxide (SC-CO₂) extraction of *Nigella sativa* L. oil using full factorial design. *Industrial Crops and Products, 36*, 519–523.

Streit, N. M., Canterle, L. P., Canto, M. W., & Hecktheuer, L. H. H. (2005). As Clorofilas. *Ciência Rural, 35*, 748–755.

Szydłowska-Czerniak, A., Dianoczki, C., Recseg, K., Karlovits, G., & Szłyka, E. (2008). Determination of antioxidant capacities of vegetable oils by ferric-ion spectrophotometric methods. *Talanta, 76*, 899–905.

Tavares, G. R., Gonçalves, J. E., Santos, W. D., & Silva, C. (2017). Enzymatic interesterification of crambe oil assisted by ultrasound. *Industrial Crops and Products, 97*, 218–223.

Temple-Heald, C. (2004). *Rapeseed and canola oil: Productions, processing, properties and uses* (pp. 111–130). Oxford: Blackwell Publishing.

Tsaknis, J., & Lalas, S. (2002). Stability during frying of *Moringa oleifera* seed oil variety "Periyakulam 1". *Journal of Food Composition and Analysis, 15*, 79–101.

Tsaknis, J., Lalas, S., Gergis, V., Dourtoglou, V., & Spiliotis, V. (1999). Characterization of *Moringa oleifera* variety Mbololo seed oil of Kenya. *Journal of Agricultural and Food Chemistry, 47*, 4495–4499.

Tuberoso, C. I. G., Kowalczyk, A., Sarritzu, E., & Cabras, P. (2007). Determination of antioxidant compounds and antioxidant activity in commercial oilseeds for food use. *Food Chemistry, 103*, 1494–1501.

Tutus, A., Comlekcioglu, N., Karaman, S., & Alma, M. H. (2010). Chemical composition and fiber properties of *Crambe orientalis* and *Crambe tataria*. *International Journal of Agriculture and Biology, 12*, 286–290.

Uenojo, M., Maróstica Junior, M. R., & Pastore, G. M. (2007). Carotenóides: propriedades, aplicações e biotransformação para formação de compostos de aroma. *Quim Nova, 30*, 616–622.

Uquiche, E., Romero, V., Ortíz, J., & Del Valle, J. M. (2012). Extraction of oil and minor lipids from cold-press rapeseed cake with supercritical CO₂. *Brazilian Journal of Chemical Engineering, 29*, 585–597.

Wang, Y. P., Tang, J. S., & Chu, C. Q. (2000). A preliminary study on the introduction and cultivation of *Crambe abyssinica* in China, an oil plant for industrial uses. *Industrial Crops and Products, 12*, 47–52.

Warwick, S. I., Francis, A., & Al-Shehbaz, I. A. (2006). *Brassicaceae*: Species checklist and database on CD-Rom. *Plant Systematics and Evolution, 259*, 249–258.

Wazilewski, W. T., Bariccatti, R. A., Martins, G. I., Secco, D., Souza, S. M. N., Rosa, H. A., & Chaves, L. I. (2013). Study of the methyl crambe (*Crambe abyssinica* Hochst) and soybean biodiesel oxidative stability. *Industrial Crops and Products, 43*, 207–212.

Winkler, J. K., & Warner, K. (2008). The effect of phytosterol concentration on oxidative stability and thermal polymerization of heated oils. *European Journal of Lipid Science and Technology, 110*, 455–464.

Wu, H., Shi, J., Xuea, S., Kakuda, Y., Wanga, D., Jiang, Y., Yee, X., Lif, Y., & Subramanian, J. (2011). Essential oil extracted from peach (*Prunus persica*) kernel and its physicochemical and antioxidant properties. *Food Science and Technology, 44*, 2032–2039.

Yang, M., Zheng, C., Zhou, Q., Huang, F., Liu, C., & Wang, H. (2013). Minor components and oxidative stability of cold-pressed oil from rapeseed cultivars in China. *Journal of Food Composition and Analysis, 29*, 1–9.

Zanetti, F., Vameralib, T., & Mosca, G. (2009). Yield and oil variability in modern varieties of high-erucic winter oilseed rape (*Brassica napus* L. var. oleifera) and Ethiopian mustard (*Brassica carinata* A. Braun) under reduced agricultural inputs. *Industrial Crops and Products, 30*, 265–270.

Zhu, L. H., Krens, F., Smith, M. A., Li, X., Qi, W., Van Loo, E. N., Iven, T., Feussner, I., Nazarenus, T. J., Huai, D., Taylor, D. C., Zhou, X. R., Green, A. G., Shockey, J., Klasson, K. T., Mullen, R. T., Huang, B., Dyer, J. M., & Cahoon, E. B. (2016). Dedicated industrial oilseed crops as metabolic engineering platforms for sustainable industrial feedstock production. *Scientific Reports, 6*, 22181.

Chapter 23
Kenaf (*Hibiscus cannabinus* L.) Seed Oil

Sook Chin Chew and Kar Lin Nyam

Abstract Kenaf (*Hibiscus cannabinus* L.) has received attention worldwide for its commercial value as fiber applications. Kenaf seeds, a by-product from kenaf plant yield kenaf seed oil with no toxicity and primarily contributed by triacylglycerols (99.81%) followed by free fatty acids, diacylglycerols, and monoacylglycerols. Extensive research has related to the processing and applications of kenaf seed oil, which highlighted its potential to use as functional edible oil that advantageous in the food, nutraceutical, and pharmaceutical industry. A chemical refining process with different parameters in each stage has been studied to produce refined kenaf seed oil with removed gums, hydroperoxides, and free fatty acid, as well as no 3-monochloro-1,2-propanediol ester detected. Oleic acid (omega-9) and linoleic acid (omega-6) make up the majority of kenaf seed oil's fatty acid composition, which is associated with cholesterol-lowering ability. Kenaf seed oil possesses significant health benefits and pharmacological activities such as antioxidant activity, anti-hypercholesterolemic, anti-cancer, anti-inflammatory, anti-ulcer, and anti-thrombotic due to the presence of bioactive compounds (tocopherols, tocotrienols, phytosterols, and phenolics). Nanoencapsulation and microencapsulation have been applied to the kenaf seed oil to improve its bioaccessibility and bioavailability in the gastrointestinal tract. Oxidative stability of kenaf seed oil has been extended through microencapsulation techniques (spray drying and co-extrusion) and suitable to apply in the functional product development. The chemistry and functionality of kenaf seed oil are reviewed in this chapter to stimulate future research and impending applications.

Keywords Oleic acid · Linoleic acid · Tocopherol · Tocotrienol · Phytosterol · Phenolic · Antioxidant

S. C. Chew · K. L. Nyam (✉)
Department of Food Science and Nutrition, Faculty of Applied Sciences, UCSI University, Kuala Lumpur, Malaysia
e-mail: nyamkl@ucsiuniversity.edu.my

© Springer Nature Switzerland AG 2019
M. F. Ramadan (ed.), *Fruit Oils: Chemistry and Functionality*,
https://doi.org/10.1007/978-3-030-12473-1_23

Abbreviations

3-MCPD	3-Monochloro-1,2-propanediol
ABTS	2,2′-Azino-bis (3-ethylbenzothiazoline-6-sulphonic acid)
BCB	β-carotene bleaching
DAG	Diacylglycerol
DKSM	Defatted kenaf seed meal
DPPH	2,2- diphenyl-1-picrylhydrazyl
EM	Emulsifier mixtures
FFA	Free fatty acid
IV	Iodine value
KSOM	Kenaf seed oil-in-water macroemulsion
KSON	Kenaf seed oil-in-water nanoemulsion
LDL-C	Low-density-lipoprotein cholesterol
MAG	Monoacylglycerol
MDA	Malondialdehyde
MUFA	Monounsaturated fatty acid
p-AV	p-Anisidine value
PUFA	Polyunsaturated fatty acid
PV	Peroxide value
RNS	Reactive nitrogen species
ROS	Reactive oxygen species
TAG	Triacylglycerol
TOTOX	Total oxidation
β-CD	β-cyclodetxrin

1 Introduction

Kenaf is an annual dicotyledonous herbaceous plant, which classified to the Malvaceae family. This plant is related to okra (*Hibiscus esculentus*), hollyhock (*Althaea rosea*), cotton (*Gossypium hirsutum* L.), and roselle (*Hibiscus sabdariffa*) (Scott and Taylor 1990; Pascoal et al. 2015). According to Dempsey (1975), kenaf is a short day plant, which has a great potential for fiber, energy, and feedstock. Kenaf is cultivated for the soft bast fiber in its stem. Due to its greater adaptability and ease of handling than allied fiber crops, kenaf has received the greatest attention from the world. Kenaf has been planted in Africa for more than 4000 years and can be found in many countries like China, India, United States of America, and Thailand (LeMahieu et al. 2003). Kenaf is cultivated in more than 20 countries and the total production of kenaf and allied fibers is 216,200 tonnes in 2014/2015 (FAO 2017). Nowadays, the main kenaf producers are India and China, which accountable for 46% and 26%, respectively of the total production of kenaf and allied fibers in this world in 2014–2015, and the rest of 28% was contributed by the other countries (FAO 2017). Malaysia government has recognized kenaf as a fiber crop for future development of economic growth in Malaysia and established National Kenaf and Tobacco Board to provide better strategies for the kenaf industry. The kenaf plantation and seed production have been increasing since 2006. The kenaf plantation of 2000 ha in Malaysia had been reported in 2004 and is targeted to extend to 10,000 ha in 2020 (National Kenaf and Tobacco Board 2014).

The fibers of kenaf are normally used in making ropes, sacks, canvases, and carpets (Ayadi et al. 2011). New applications of kenaf have been explored like pulping, paper making, board making, oil absorption and potting media, filtration media, and animal feed. Kenaf is a fast-growing plant that meets the high demand for fiber and forage industries to reduce massive deforestation. U.S. Department of Agriculture is recognized kenaf as the most promising crop for the fiber source in the paper and pulp industries (Ashori et al. 2006; Ramesh 2016). Thus, there has been more attention focused on the fast and ease growing plant fiber such as kenaf. Besides that, less energy and chemical is required in the pulping process of kenaf compared to the standard wood sources (Ayadi et al. 2011). Kenaf has been prescribed as a medicinal plant in Africa long years ago (Ryu et al. 2006). Intensive research is highly demanded in order to maximize the usage of the kenaf plant.

Kenaf plant cultivates in tropical and temperate atmospheres and flourishes with abundant sunlight and high rainfall. It is suitable adapted to the tropics or subtropics where the average daily temperature during the growing season is more than 20 °C. The fleshy kenaf fruits can produce seed capsules with 1 cm long, which containing about 20–26 pieces of kenaf seeds, which are dark brown or black in color, with a wedge or triangular with acute angles shapes inside each capsule. There is approximately 6 mm long and 4 mm wide for the kenaf seeds. Then, there is about 36,000–40,000 seeds/kg can be produced from the kenaf varieties (Chan and Ismail 2009). The seeds are rather hard and contain a yellowish kernel which is strongly attached to the black colored seed coat or episperm (Lewy 1947).

Kenaf seeds contain 9.6% moisture, 21.4% crude protein, 20.4% oil, 12.9% crude fiber, and 6.4% ash (Lewy 1947). Nyam et al. (2009) reported that kenaf seeds contain 9.1% moisture, 21.8% crude protein, 20.8% oil, 28.8% total carbohydrate, 13.6% crude fiber, and 5.9% ash. Kenaf seeds can store at a relative humidity level of 8% at −10, 0, or 10 °C to preserve its viability for 5.5 years, and relative humidity level of 12% at −10 or 0 °C for 5.5 years (Webber and Bledsoe 2002). Kenaf could be considered as a profitable oilseed crop if a persistent seeds yield of 1500 kg/ha could be obtained based on the oil yield of 20% in kenaf seeds (Dempsey 1975; Chan and Ismail 2009).

2 Extraction and Processing

2.1 Oil Extraction

There are four extraction methods to extract kenaf seed oil reported in the literature, which are pressing, solvent extraction, ultrasonic-assisted solvent extraction, and supercritical fluid extraction. An oil yield of 14.0% and 20.0% were obtained from the kenaf seeds by pressing and solvent extraction with petroleum ether, respectively, reported in Lewy (1947). Solvent extraction can use to extract oil from the source by using organic solvents such as petroleum or n-hexane. The oil yields of 21.4–26.4% with an average of 23.7% were obtained from the kenaf seeds using sonification by n-hexane-isopropanol from the study of nine kenaf genotypes (Mohamed et al. 1995). An average of 19.8% of oil yield was obtained from the study of eight kenaf varieties by using solvent extraction with chloroform: methanol at a ratio of 2:1 (Coetzee et al. 2008). An oil yield of 20.8% was obtained from the kenaf seeds using Soxhlet extraction with petroleum ether (solid to the solvent ratio, 1:10) at 40–60 °C for 8 h (Nyam et al. 2009). Approximately of 20% of the oil yield could be obtained from the kenaf seeds, rely on which extraction method utilized. The oil content of kenaf seeds is alike to cottonseed oil (23.2–25.7%) and is more than soybean oil (15.6–23.4%) (Mohamed et al. 1995).

The solvent is needed to be removed after the extraction to recover the kenaf seed oil using solvent evaporator. However, there might be a risk of the existence of residual solvent if the solvent removal process is not complete. Thus, supercritical carbon dioxide extraction stands as an alternative extraction method to prevent solvent contamination. Supercritical fluid extraction utilizes the carbon dioxide fluid to extract oil, which is non-toxic, non-flammable, cost effective, environmentally friendly, and time-saving (Vaquero et al. 2006; Araújo and Sandi 2007; Chan and Ismail 2009). Moreover, this method enables to conduct at low temperature to minimize the thermal degradation of the bioactive compounds presented in the oil (Mariod et al. 2011) and complete removal of the fluid at the end of the extraction (Chan and Ismail 2009).

Chan and Ismail (2009) reported the oil yields of 2.1–20.2% with 150 min of extraction and utilization of 3.75 kg of carbon dioxide using supercritical fluid extraction. The oil yields were obtained in the following sequence by studied the parameters (pressure, bar/temperature, °C): 600/80 > 600/60 > 600/40 > 400/80 > 400/40 > 400/60 > 200/40 > 200/60 > 200/80. A rise in extraction pressure would increase the extraction yield of kenaf seed oil and accelerate the extraction process simultaneously. A rise in the extraction pressure at constant temperature will increase the density of carbon dioxide and increase the solubility of the solute (oil) in the fluid, subsequently increase the oil yield (Foo et al. 2011).

Rapid and conventional Soxhlet extractions were also studied using 140 min and 12 h, respectively, which yield 24.81% and 22.4% of kenaf seed oil, respectively. Conventional soxhlet extraction was showed to give a better extraction yield (24.8%), followed by rapid Soxhlet extraction, ultrasonic, and supercritical fluid extraction. However, conventional Soxhlet extraction used up to 12 h to complete the extraction process, compared to 140 min by rapid Soxhlet extraction, 150 min by supercritical fluid extraction, and 3–4 h by ultrasonic extraction. The oil yields obtained by the supercritical fluid extraction at 600 bars/80 °C (20.2%), was not significantly difference with the rapid Soxhlet extraction (22.4%), and ultrasonic-assisted solvent extraction (21.1%) (Chan and Ismail 2009).

2.2 Oil Refining

Free fatty acids (FFA), phosphates, gums, waxes, colour pigments, pesticides, odoriferous compounds represented as the undesirable components in the crude vegetable oils and fats that may affect the nutritional quality and oxidative stability of the crude oil (Ghazani and Marangoni 2013). Some of the crude oils are advised to undergo a chemical or physical refining process to remove their high contents of undesirable compounds to yield the refined oil with stable and consumer acceptance. Degumming, neutralization, bleaching, and deodorization are the four stages of the chemical refining process. Chew et al. (2016) had conducted a refining process on the crude kenaf seed oil using the described refining methods in Wang and Johnson (2001) and Verleyen et al. (2002), with the conditions given in Table 23.1.

Table 23.2 shows the comparison of the quality of refined kenaf seed oil produced in the Chew et al. (2016) that mentioned above and Chew et al. (2017a, b, c,

Table 23.1 Summary table of conditions used in different stages of the refining process

Stage	Temperature (°C)	Time (min)	Materials and chemicals used
Degumming	70	10	Phosphoric acid of 85% concentration (0.3% w/w)
		30	Ultra-pure water (3% w/w)
Neutralization	65	30	NaOH solution (12° Baume) (3.93 g)
Bleaching	95	30	Acid-activated bleaching earth (1.2% w/w)
Deodorization	200	60	–

Table 23.2 Physico-chemical properties of crude and refined kenaf seed oils from the previous studies

Physico-chemical property	Chew et al. (2016)		Chew et al. (2017e)	
	Crude	Refined	Crude	Refined
Refractive index	1.4657 ± 0.0007	1.4701 ± 0.0002	1.4634 ± 0.0003	1.4693 ± 0.0002
Specific gravity	0.91 ± 0.01	0.92 ± 0.01	0.89 ± 0.01	0.92 ± 0.01
Color				
L*	11.55 ± 0.29	6.57 ± 0.15	12.15 ± 0.06	7.86 ± 0.05
a*	-1.94 ± 0.15	-2.14 ± 0.07	2.93 ± 0.12	-2.28 ± 0.05
b*	14.50 ± 0.54	6.43 ± 0.28	19.35 ± 0.05	3.16 ± 0.05
PV (meq O_2/kg)	2.64 ± 0.26	0.55 ± 0.10	1.80 ± 0.00	0.00 ± 0.00
p-Aniside value	2.41 ± 0.32	3.41 ± 0.34	2.15 ± 0.14	3.61 ± 0.09
TOTOX	7.70 ± 0.84	4.51 ± 0.34	5.75 ± 0.14	3.61 ± 0.09
FFA (%)	1.72 ± 0.04	0.61 ± 0.04	0.71 ± 0.02	0.05 ± 0.00
Phosphorus content (mg/kg)	–	–	62.87 ± 2.89	0.73 ± 0.04
Total phytosterol (mg/100 g oil)	702.96 ± 31.89	544.80 ± 54.67	549.34 ± 7.54	375.87 ± 15.96
Total tocopherol (mg/100 g oil)	64.79 ± 2.54	65.67 ± 5.33	64.11 ± 2.23	53.15 ± 1.28

d, e) using optimized refining process. The refining process in Chew et al. (2016) had totally removed 79.2% of peroxide value (PV), 41.4% of total oxidation (TOTOX) value, and 64.5% of FFA from the crude kenaf seed oil. However, the refined oil shall not contain more than 0.2% of FFA according to the Malaysia Food Regulation 1985, which is closely followed to the CODEX standard (Legal Research Board 2013). Thus, parameters in the refining process should be studied in order to meet the requirements of the standards. Different types of vegetable oils require different parameters in the refining process as different intrinsic components presented in the crude vegetable oils. The parameters should be optimized with the maximize removal of undesirable matter in the crude oil and least possible damage to the bioactive compounds.

2.2.1 Degumming

The first stage takes place in the refining process is degumming. Degumming functions to exclude phospholipids, traces metal ions, and mucilaginous materials that presented in the crude oil. Phospholipids are good as natural antioxidants in the oil but phospholipids may form dark color and precipitate in the oil during the storage time. Phospholipids may carry trace metal ions (pro-oxidants) to reduce the oxidative stability of the oil. The removal of phospholipids in the degumming stage is prior to the other refining stages to avoid refining losses (Chew et al. 2017a).

Hydratable and non-hydratable are the two types of phospholipids that commonly presented in the crude oil. Acid degumming will convert the non-hydratable into hydratable phospholipids. Then, oil-soluble phospholipids (hydratable phospholipids) need to undergo a hydration process (water degumming) to become oil-insoluble, and subsequently removed by centrifugation (Zufarov et al. 2008; Ghazani and Marangoni 2013). Phosphatidyl ethanolamine (12.8%) and phosphatidyl acid (4.9%) are the non-hydratable phospholipids identified in the crude kenaf seed oil (Mohammed et al. 1995), which are required to remove by acid degumming (Zufarov et al. 2008). Phosphoric acid solution (85% concentration) is commonly used in the acid degumming at a dosage of 0.10–0.15% (Wang and Johnson 2001; Ghazani and Marangoni 2013). Phosphatidyl choline (21.9%) is the hydratable phospholipid identified in the crude kenaf seed oil (Mohammed et al. 1995), which is required to remove by a water degumming process (Zufarov et al. 2008). Water is commonly added to a dosage of 1–3% w/w to precipitate the hydratable phospholipids (Farr 2000; Ghazani and Marangoni 2013). However, a higher water dosage (10–20% w/w) was also reported in the study of palm oil, rapeseed oil, and *Moringa oleifera* seed oil (Franke et al. 2009; Sánchez-Machado et al. 2015). The degumming stage using 0.3% w/w of phosphoric acid and 3% w/w of water at 70 °C produced the degummed kenaf seed oil with a phosphorus content of 35.01 mg/kg, which only removes 26.2% of the phosphorus content from crude oil (Chew et al. 2016, 2017a).

Therefore, Chew et al. (2017a) conducted an optimization study on the parameters of temperature, phosphoric acid dosage, and water dosage in the degumming stage of kenaf seed oil. The degumming stage was successfully removed 85.9% of phosphorus content from the crude kenaf seed oil using the optimized parameters (temperature of 40 °C, a phosphoric acid dosage of 0.09% w/w and a water dosage of 22.4%). The degummed kenaf seed oil presented an acceptable standard of phosphorus content (6.70 mg/kg oil) as it is required for the oil with a phosphorus content of lower than 10 mg/kg before carrying to the final refining stage to obtain good quality refined oil (Tyagi et al. 2012).

2.2.2 Neutralization

Neutralization is a stage to neutralize the FFA by addition of a stoichiometric amount of sodium hydroxide (NaOH) with excess percentages, and remove it as insoluble soap by centrifugation (Chew et al. 2017b). FFA is not desired to present in the oil, as it will speed up the decomposition of hydroperoxide leading to further oxidation and formation of offensive odor and flavor (Ghazani and Marangoni 2013). FFA content indicates the degradation of triglycerides in oil caused by the enzyme lipase, light and heat. Higher FFA in the oil will affect the oil quality and not suitable for consumption. Neutralization helps to remove residual phospholipids, pigments, and waxes that might affect the oxidative stability of the oil (Wei et al. 2015).

NaOH is commonly added in a stoichiometric amount based on the FFA content of the oil coupled with an excess of 0.1–0.5% (Wang and Johnson 2001; Lee et al. 2014; Wei et al. 2015). However, an excess of 10% and 25% were also used in the neutralization of hazelnut oil and silkworm oil, respectively in the study of Karabulut et al. (2005) and Ravinder et al. (2015). The neutralized oil is washed with 10–20% of water to ensure no residual soap and NaOH presented in the neutralized oil. Neutralization is commonly conducted at a temperature of 65–90 °C (Wang and Johnson 2001; Karabulut et al. 2005; Lee et al. 2014). Therefore, Chew et al. (2017b) conducted an optimization study on the parameters of temperature, the excess percentage of NaOH, and time in the neutralization of degummed kenaf seed oil. The neutralization of kenaf seed oil was optimized with a utilization of an excess percentage of NaOH of 3.75% at 40 °C for 20 min. The results showed that the neutralization with optimum parameters yields the neutralized kenaf seed oil with a FFA of 0.12% and a PV of 1.57 meq O$_2$/kg, compared with the FFA of 1.94% and PV of 2.90 meq O$_2$/kg in the crude oil.

The results showed that 40–50 °C was more suitable to neutralize the FFA, which supported by the study of Wei et al. (2015) reported the similar neutralization effect on the temperature of 45, 65, and 85 °C for tea seed oil. A lower temperature is recommended to preserve the bioactive compounds in the oil. The increase of PV in the neutralized oil was observed in the previous studies due to the high temperature employed in the neutralization (Zacchi and Eggers 2008; Bachari-Saleh et al. 2013). Soap will bring along with the hydroperoxides and separate in the neutralization stage, thus resulting in the decrease of PV if the optimum temperature is used.

2.2.3   Bleaching

Bleaching functions to exclude colour pigments, residual soaps and phospholipids, metal ions (iron and copper), and hydroperoxides with the utilization of a bleaching agent. Bleaching earth is commonly used as the bleaching agent due to its high adsorption capacity and low cost. Generally, bleaching earth will undergo an acid pre-treatment to expose more specific surface area and porosity to enhance its adsorption capacity. Thus, acid-activated bleaching earth is commonly used in the bleaching compared to the neutral bleaching earth, which only acts as adsorptive agent only. Colour pigments will be excluded through physical adsorption by bleaching earth while the residual impurities will be excluded through covalent or ionic bonds by bleaching earth (Chew et al. 2017c).

Bleaching is commonly conducted at the temperature of 85–110 °C with a reduced pressure or vacuum (Ortega-García et al. 2006; Farhoosh et al. 2009; Lee et al. 2014). However, the lower temperature was reported to be more suitable for bleaching as the adsorption equilibrium shifts towards desorption and adsorbed particles might disintegrate back into the oil at a higher temperature (Chew et al. 2017c). Higher temperature will accelerate the decomposition of hydroperoxides by bleaching earth also. The time required in the bleaching relies on the temperature and the bleaching earth used. Thus, Chew et al. (2017c) studied the parameters of the concentration of bleaching earth, temperature, and time. High dosage of bleaching earth will enhance the removal of impurities in the oil, but will also decompose more hydroperoxides into further oxidation compounds. Thus, a decrease of PV coupled with an increase of p-AV will be observed in the bleaching stage with an increased dosage of bleaching earth. The bleaching stage was optimized at 70 °C for 40 min with the addition of 1.5% w/w of acid-activated bleaching earth.

2.2.4   Deodorization

Deodorization functions to get rid of FFA, odoriferous compounds, moisture, and colour pigments using vacuum-steam distillation conducted at high temperature under a vacuum pressure. FFA is required to remove at this stage to the acceptable standard of edible oil. The majority of FFA is removed in the stages of neutralization and deodorization in the chemical refining process. The stripping steam gas will remove the volatile secondary oxidation products such as aldehydes and ketones in the deodorization stage. Besides that, tocopherols in the oil are easily deteriorated by the high temperature of deodorization as tocopherols are more volatile than the triglycerides (Wang and Johnson 2001). High temperature and sparge steam are the two major elements to deteriorate the tocopherols in the oil by oxidation and polymerization when the oil contact to the air or heat (Ortega-García et al. 2006). Thus, retention of tocopherols is important in the deodorization to prevent loss of nutritional value and oxidative stability of the oil.

Chew et al. (2017d) conducted an optimization study on the temperature (180–260 °C) and time (0.5–2.5 h) used in the deodorization of kenaf seed oil. The tocopherols and tocotrienols of kenaf seed oil decrease with the increases of temperature and time. The deodorization of kenaf seed oil was optimized at the temperature of 220 °C for 1.5 h. Deodorization had no effect on the PV of kenaf seed oil at 200 °C for 1 h (Chew et al. 2016). However, the deodorization had destructed all the hydroperoxides with PV of 0 meq O_2/kg in the kenaf seed oil. This showed that the deodorization parameters are important in the removal efficiency of deodorization. The FFA of the refined oil has met the requirement of the CODEX standard for refined oil, which is suitable for consumption (Legal Research Board 2013). Deodorization parameters were significantly influenced the tocopherol contents in vegetable oil (Wang and Johnson 2001; Ortega-García et al. 2006). However, deodorization stage had no significant effect on the tocopherols and tocotrienols contents of kenaf seed oil, showed in the study of Chew et al. (2017e).

2.2.5   Overall Refining Process

Table 23.3 summarizes the qualities of kenaf seed oil in each stage of the optimized refining process in Chew et al. (2017e). There was a slightly increased in the secondary oxidation status in the final refined oil, as compared to the crude oil.

Table 23.3 Qualities of kenaf seed oil during the refining process

Analysis	Crude	Degummed	Neutralized	Bleached	Deodorized
PV (meq O_2/ kg oil)	1.80 ± 0.00^a	1.57 ± 0.05^b	1.15 ± 0.06^c	0.23 ± 0.03^d	0.00 ± 0.00^e
p-Anisidine value	2.15 ± 0.14^c	2.35 ± 0.15^c	3.52 ± 0.05^b	4.63 ± 0.11^a	3.61 ± 0.09^b
TOTOX	5.75 ± 0.14^{ab}	5.50 ± 0.23^b	5.83 ± 0.13^a	5.08 ± 0.06^c	3.61 ± 0.09^d
FFA (%)	0.71 ± 0.02^a	0.67 ± 0.01^b	0.09 ± 0.00^c	0.09 ± 0.00^c	0.05 ± 0.00^d
Iodine value (g I_2/100 g oil)	89.56 ± 1.53^b	89.50 ± 3.29^b	94.10 ± 1.50^{ab}	95.07 ± 3.11^{ab}	97.22 ± 0.56^a
Phosphorus content (mg/ kg)	62.87 ± 2.89^a	6.70 ± 0.44^b	0.76 ± 0.01^c	0.73 ± 0.02^c	0.73 ± 0.04^c
Total phytosterol (mg/100 g)	549.34 ± 7.54^a	492.27 ± 15.63^b	427.81 ± 11.51^c	410.18 ± 16.90^c	375.87 ± 15.96^d
Total tocotrienol and tocopherol (mg/100 g)	64.11 ± 2.23^a	63.13 ± 0.36^a	57.29 ± 0.38^b	53.89 ± 1.18^c	53.15 ± 1.28^c

Values followed by different superscript letters [abcde] within the same row are significantly different ($p < 0.05$) according to Tukey's test

The optimized refining process was removed 37.2% of TOTOX value, 93% of FFA and 98.8% of phosphorus content from the crude kenaf seed oil. The optimized refining process was totally removed 31.6% of phytosterols content and 17.1% of the tocopherols and tocotrienols contents in kenaf seed oil. The phytosterol content was majorly reduced (13.1%) in the neutralization stage through the removal of soaps. Besides that, phytosterols are degraded through the acid-catalyzed dehydration of bleaching earth and a high temperature of deodorization leads to the dehydration of sterols and formation of steradienes (Karabulut et al. 2005; Ortega-García et al. 2006).

Chew et al. (2016) showed the conventional refining process had reduced 22.5% of the total phytosterol content in kenaf seed oil, which is lesser than the study of Chew et al. (2017e). This might be due to the parameters used in the neutralization stage of the conventional refining process were not efficient to form the soapstock to remove the FFA content. Thus, lesser phytosterol content was removed through the soapstock, as the FFA value still remained high in the refined kenaf seed oil after the conventional refining process (Chew et al. 2016). The reduction of total tocopherols and tocotrienols contents of kenaf seed oil was majorly affected in the neutralization stage as tocopherols are unstable in the alkali condition and adsorbed onto the soaps (Chew et al. 2017e). There was a removal of tocopherol content of 28.5% in safflower oil (Ortega-García et al. 2006) and 29.2% in sunflower oil (Suliman et al. 2013) reported in the literature. Thus, optimal processing parameters are important to preserve the bioactive compound.

3   Physical and Chemical Properties

Table 23.4 shows the physicochemical properties of kenaf seed oil reported from the previous studies (Lewy 1947; Nyam et al. 2009; Chew et al. 2017e). The specific gravity of crude, degummed, neutralized, bleached, and deodorized kenaf seed oil ranged from 0.89 to 0.92 (Chew et al. 2017e). The specific gravity of kenaf seed oil is comparable with other commercial vegetable oils such as sunflower oil (0.916–0.923), soybean oil (0.917–0.924), and linseed oil (0.925–0.932) (Guner et al. 2006). The refractive index of kenaf seed oil reported in Chew et al. (2017e) was

Table 23.4 Physico-chemical properties of crude kenaf seed oil from the previous studies

Physico-chemical property	Lewy (1947)	Nyam et al. (2009)	Chew et al. (2017e)
Specific gravity	0.9175	–	0.89 ± 0.01
Refractive index	1.4657	–	1.4634 ± 0.0003
Acid value (mg KOH/g oil)	4.7	1.6 ± 0.0	–
FFA (% oleic acid)	–	0.8 ± 0.0	0.72 ± 0.02
Saponification value (mg KOH/g oil)	189.8	171.0 ± 6.1	–
Insaponifiable matter (%)	1.7	–	–
Iodine value (g I_2/100 g oil)	99.7	86.3 ± 1.0	89.56 ± 1.53
Peroxide value (meq O_2/kg oil)	–	2.3 ± 0.3	1.80 ± 0.00

ranged from 1.4634 to 1.4693, which is in agreement with the refractive value of kenaf seed oil in the previous study (1.4657) (Lewy 1947). Lewy (1947) reported the kenaf seed oil is clear yellow color, and the hot pressed oil has a mild odor that similar to the cottonseed oil, and the cold pressed oil is odorless. Nyam et al. (2009) reported the kenaf seed oil is yellow color and its contained the highest L* and b* values, which is lighter-colored and more yellow than the bittermelon, Kalahari melon, pumpkin, and roselle seed oils.

The FFA reported in kenaf seed oil is 0.72–0.80%, which is considered low. However, several studies reported the FFA values of kenaf seed oil ranged from 1.72% to 1.94% (Ng et al. 2014; Chew et al. 2016, 2017b). This might due to the longer storage of the seeds that might increase the FFA content caused by the moisture. IV indicates the degree of unsaturation in the oil, which attributed by the MUFA and PUFA of kenaf seed oil in this study. The IV of kenaf seed oil is comparable to other commercial vegetable oils (9.4–145.0 g I_2/100 g) reported in the previous study (Tan et al. 2002). The PV reported in the kenaf seed oil is fulfilled to the CODEX standard, whereby the PV of vegetable oil cannot exceed than 10 meq O_2/kg oil (Codex Alimentarius Commission 1982).

4   Composition of Kenaf Seed Oil

4.1   Phospholipids Composition

Kenaf seed is made up of a different combination of triglycerides, fatty acids, phospholipids, and sterols. The major phospholipids of kenaf seed oil were identified by Mohamed et al. (1995) using thin layer chromatography as phosphatidyl choline (21.9%), phosphatidyl ethanolamine (12.8%), phosphatidyl glycerol (8.9%), lysophosphatidyl choline (5.3%), phosphatidic acid (4.9%), sphingomyelin (4.4%), cardiolipin (3.6%), phosphatidyl serine (2.9%), and phosphatidyl inositol (2.7%). The phospholipid content of kenaf seed oil from nine kenaf genotypes was ranged from 3.9 to 10.3 g phosphatidyl choline/100 g oil, with a mean of 6.0 g phosphatidyl choline/100 g oil. The percentage of phospholipids content is higher than other edible oils such as soybean oil (1.5–3.0 g/100 g) and cottonseed oil (<2.0 g/100 g). Phosphatidyl choline and lyso phosphatidyl choline are function in the synthesis of the lipid bilayer and liposome formation in the cell membrane, as well as can use as an emulsifier in food and pharmaceutical industries.

4.2   Triacylglycerides Composition

Monoacylglycerol (MAG) and diacylglycerol (DAG) are the products from the hydrolysis of triacylglycerol (TAG) in the oil by the enzyme lipase. Hydrolysis of the TAG would lead to lipid oxidation and bring down the nutritional value of the

Table 23.5 Acylglycerol composition of kenaf seed oil during the refining process

Acylglycerol (%)		Crude	Degummed	Neutralized	Bleached	Deodorized
MAG		0.09 ± 0.01	ND	ND	ND	ND
DAG		0.10 ± 0.01^a	0.09 ± 0.01^b	0.04 ± 0.00^c	0.05 ± 0.00^c	0.04 ± 0.00^c
ECN	TAG					
40	LLLn	0.23 ± 0.02^a	0.24 ± 0.02^a	0.23 ± 0.03^a	ND	ND
	LnLnO	0.03 ± 0.00^b	0.04 ± 0.00^{ab}	0.04 ± 0.00^a	ND	ND
42	LLL	0.37 ± 0.03^a	0.39 ± 0.02^a	0.36 ± 0.01^a	0.07 ± 0.00^b	0.04 ± 0.01^b
	OLLn	0.12 ± 0.01^a	0.12 ± 0.01^a	0.12 ± 0.01^a	ND	ND
43	LLH	4.74 ± 0.17^a	4.61 ± 0.13^{ab}	4.44 ± 0.17^{abc}	4.12 ± 0.29^c	4.23 ± 0.14^{bc}
44	OOLn	0.06 ± 0.00^{bc}	0.05 ± 0.01^c	0.07 ± 0.01^b	0.10 ± 0.01^a	0.06 ± 0.01^{bc}
	OLL	13.42 ± 0.41^a	12.86 ± 0.39^{ab}	12.42 ± 0.32^b	12.05 ± 0.30^b	12.46 ± 0.65^b
	LLP	12.75 ± 0.33^a	12.21 ± 0.33^a	12.06 ± 0.30^a	11.85 ± 0.20^a	12.43 ± 0.76^a
46	LLS	14.55 ± 0.25^a	14.42 ± 0.35^a	14.13 ± 0.41^a	14.28 ± 0.25^a	15.09 ± 1.33^a
	POL/OOL	22.97 ± 0.61^{ab}	22.43 ± 0.38^{ab}	22.24 ± 0.39^b	22.46 ± 0.55^{ab}	24.45 ± 1.92^a
	PLP	5.26 ± 0.16^b	5.39 ± 0.12^{ab}	5.58 ± 0.21^{ab}	5.64 ± 0.09^{ab}	5.87 ± 0.46^a
48	OOO/ALL	7.85 ± 0.46^b	8.31 ± 0.23^{ab}	8.70 ± 0.51^a	8.85 ± 0.39^a	6.68 ± 0.29^{ab}
	POO/SOL	12.18 ± 0.78^a	12.65 ± 0.43^a	12.74 ± 0.48^a	12.98 ± 0.27^a	12.59 ± 0.60^a
	PPO	3.94 ± 0.42^c	4.38 ± 0.46^{bc}	4.79 ± 0.36^{ab}	5.22 ± 0.12^a	4.00 ± 0.35^{ab}
50	LLN/LLB	0.91 ± 0.12^b	1.27 ± 0.08^{ab}	1.32 ± 0.11^a	1.48 ± 0.11^a	1.31 ± 0.14^a
	OSS/OLA	0.44 ± 0.04^c	0.55 ± 0.06^{bc}	0.60 ± 0.01^b	0.75 ± 0.00^a	0.63 ± 0.08^a
52	LLLg/ OLN/OLB	ND	ND	0.07 ± 0.00^b	0.05 ± 0.00^c	0.07 ± 0.02^a
54	OLLg/ OON/OOB	ND	ND	0.07 ± 0.00^b	0.06 ± 0.00^c	0.06 ± 0.00^a

Source: Chew et al. (2017e)

Values followed by different superscript letters [abc] within the same row are significantly different ($p < 0.05$) according to Tukey's test

ND not detected, A arachidic acid, B behenic acid, H heptadecenoic acid, L linoleic acid, Lg ligno-ceric acid, Ln linolenic acid, N nervonic, O oleic, P palmitic acid, S stearic acid

oil. Presence of MAG in the oil may generate smoking effect during frying. Thus, it is important to identify the TAG composition in the vegetable oils. Table 23.5 shows the MAG, DAG, and TAG composition of crude, degummed, neutralized, bleached, and deodorized kenaf seed oil. There was 22.97–24.46% of POL and OOL (ECN = 46) contributed to the highest amount of TAG of kenaf seed oil. LLS (ECN = 46), LLP (ECN = 44), OLL (ECN = 44), and POO/SOL (ECN = 48) contributed more than 10%, respectively, which totally dedicated to approximately 53% in the TAG composition of kenaf seed oil. LLH (ECN = 43), PLP (ECN = 46), OOO/ALL (ECN = 46), PPO (ECN = 48), and LLN/LLB (ECN = 50) were totally dedicated to about 22% in the TAG composition of kenaf seed oil. Others TAG, MAG, and DAG were identified in less than 1% respectively in the kenaf seed oil (Chew et al. 2017e).

TAG contributed to 99–99.96% in the acylglycerol composition in kenaf seed oil. The TAG is made up of three fatty acid chains bind to the glycerol with ester bond in plant seeds. Thus, the major TAGs mostly consisted of 44, 46, and 48 acyl carbon atoms due to the high content of linoleic acid, oleic acid, and palmitic acid in the fatty acid composition of kenaf seed oil. The types of TAG presented in the oil depended on its fatty acid composition. There was only 0.2% of MAG and DAG in the crude kenaf seed oil. The degumming stage had removed the MAG through the insoluble gums, while neutralization stage had significantly removed the DAG through the soaps. There was only 0.04% of DAG in the refined kenaf seed oil (Chew et al. 2017e). Refined oil is commonly containing 0–2% of DAG but there was up to 6% of DAG in the refined palm oil. This might be due to the DAG tends to remain in the oil as DAG able to form eutectic mixture with TAG (Gunstone 2004).

4.3 Fatty Acid Composition

The fatty acid composition of kenaf seed oil has been reported in several previous studies in the literature. These reports vary considerably in the fatty acid composition of kenaf seed oil from different varieties and shown in Table 23.6. Coetzee et al. (2008) investigated the fatty acid composition of eight commercial kenaf varieties from various countries. Fatty acids presented in the kenaf seed oil were composed of approximately 21.6–28.2% saturated and 71.8–78.3% unsaturated, in which 28.0–44.5% monounsaturated fatty acids (MUFA) and 28.8–50.3% polyunsaturated fatty acids (PUFA) were identified. There was a variation in the fatty acid composition among different kenaf varieties. From the study of Mohamed et al. (1995) and Coetzee et al. (2008), linoleic acid ($C_{18:2}$) (28.6–50.1%) was the predominant fatty acid, followed by oleic acid ($C_{18:1}$) (24.8–43.4%) and palmitic acid ($C_{16:0}$) (18.2–22.6%).

Palmitoleic acid, arachidic acid, α-linolenic acid, behenic acid, lignoceric acid, and nervonic acid were also found in all kenaf cultivars in minor percentages. However, myristic acid, heptadecenoic acid, heneicosanoic acid, eicosadienoic acid, eicosatrienoic acid, eicosapentaenoic acid were rarely identified (Coetzee et al. 2008). Epoxyoleic acid was found in kenaf seed oil in minor amounts ranged from 0.26% to 1.39% from the study of Mohamed et al. (1995). The minor amount of epoxyoleic acid does not negatively affect the utilization of kenaf seed oil. On the other hand, oleic acid ($C_{18:1}$) (33.0–37.6%) reported as the predominant fatty acid in kenaf seed oil (variety: V36), followed by linoleic acid ($C_{18:2}$) (31.7–36.2%) and palmitic acid ($C_{16:0}$) (19.3–23.3%) (Nyam et al. 2009; Chew et al. 2016, 2017e).

Crude and refined kenaf seed oils presented the oleic acid as the major fatty acid, followed by linoleic and palmitic acids in their fatty acids profiles (Chew et al. 2016, 2017e). The refining process did not make a significant change in the fatty acids composition of kenaf seed oil, as shown in Table 23.7. The unsaturated fatty acids in the crude and refined kenaf seed oil are 75.1% and 75.2%, respectively. Consumption of oil that high in saturated fatty acids will increase the low-density-lipoprotein cholesterol (LDL-C) level in the blood. However, oleic acid prompts in

Table 23.6 Fatty acid composition of kenaf seed oil reported from the previous studies

Fatty acid	Mohamed et al. (1995)	Coetzee et al. (2008)	Nyam et al. (2009)	Razon et al. (2013)	Chew et al. (2016)
C12:0	0.24–0.86	–	–	–	–
C14:0	0.30–0.79	0–0.14, mean 0.06	–	0–0.1	0.15–0.25
C16:0	18.61–21.38	18.22–22.64, mean 20.27	20.3	17.2–19.0	19.42–23.32
C16:1	1.3–2.1	0.35–0.55, mean 0.46	0.4	0.3–0.5	0.52–0.62
C17:1	–	0–0.14, mean 0.05	–	0.2–0.4	0.23–0.25
C18:0	2.03–4.02	2.05–4.30, mean 3.15	3.8	1.9–3.3	4.33–4.57
C18:1	24.8–34.1	26.94–43.42, mean 32.79	37.1	25.4–32.4	33.0–37.62
C18:2	42.0–50.1	28.60–49.75, mean 41.15	36.6	37.4–42.0	31.68–32.32
C18:3n6	0.43–1.14	0.23–0.83, mean 0.39	0.3	–	0.78–0.87
C18:3n3	–			–	0.29–0.30
C20:0	0.26–0.77	0.42–0.75, mean 0.55	0.6	0.4–0.7	0.62–0.70
C20:1	0.10–0.32	–	0.5	–	–
C20:5	–	0–0.28, mean 0.14	–	–	–
C21:0	–		0.5	–	–
C22:0	0.14–0.39	0.28–0.35, mean 0.31	–	0.3–0.4	0.32–0.35
C22:1	0.47–0.84	–	–	–	–
C24:0	0.06–0.18,	0.05–0.14, mean 0.09	0.6	0.1–0.2	0.23–0.37
C24:1	–	0.25–0.79, mean 0.57		–	2.76–3.59
Cis-12,12-epoxyoleic	0.17–1.39	–	–	2.1–6.7	–

decreasing total serum cholesterol and LDL-C. Consumption of oil rich in monounsaturated and polyunsaturated fatty acids confers health benefits on a human. Oleic acid is more stable than linoleic acid to withstand the high heat application. Thus, high level of oleic acid (omega-9) makes the oil desired as cooking oil in the aspects of nutrition and stability (Nyam et al. 2009). There is an increasing demand for the high oleic acid vegetable oils in this competitive market nowadays.

Linoleic acid (omega-6) is an essential fatty acid to increase the nutritional value of the vegetable oil also. Besides that, palmitic acid is a saturated fatty acid that suitable to make margarine, shortening and other fat products (Nyam et al. 2009). A minor amount of stearic acid was identified in the kenaf seed oil, which is a neutral saturated fatty acid that does not increase the LDL-C level in human's blood

Table 23.7 Fatty acid composition of kenaf seed oil during the refining process

Fatty acid (%)	Crude	Degummed	Neutralized	Bleached	Deodorized
C14:0	0.14 ± 0.00^b	0.14 ± 0.00^b	0.14 ± 0.00^b	0.14 ± 0.00^b	0.15 ± 0.00^a
C16:0	19.33 ± 0.12^c	19.48 ± 0.10^c	19.51 ± 0.04^{bc}	19.72 ± 0.08^{ab}	19.76 ± 0.13^a
C16:1	0.53 ± 0.03^a	0.50 ± 0.00^a	0.51 ± 0.00^a	0.51 ± 0.00^a	0.51 ± 0.00^a
C17:1	0.22 ± 0.00^a	0.21 ± 0.00^a	0.22 ± 0.01^a	0.21 ± 0.00^a	0.22 ± 0.00^a
C18:0	4.32 ± 0.05^{ab}	4.35 ± 0.03^a	4.37 ± 0.01^a	4.28 ± 0.01^b	3.84 ± 0.04^c
C18:1n9	36.14 ± 0.12^{bc}	36.13 ± 0.08^{bc}	36.01 ± 0.02^c	36.26 ± 0.04^b	36.53 ± 0.10^a
C18:2n6	36.23 ± 0.05^b	36.23 ± 0.05^b	36.27 ± 0.01^b	36.44 ± 0.03^a	36.52 ± 0.06^a
C18:3n6	0.96 ± 0.07^a	0.85 ± 0.01^b	0.86 ± 0.01^b	0.85 ± 0.00^b	0.88 ± 0.01^b
C18:3n3	0.35 ± 0.02^a	0.32 ± 0.00^b	0.32 ± 0.01^b	0.33 ± 0.00^{ab}	0.33 ± 0.01^{ab}
C20:0	0.65 ± 0.02^a	0.63 ± 0.01^{ab}	0.62 ± 0.01^b	0.63 ± 0.00^{ab}	0.64 ± 0.02^{ab}
C22:0	0.31 ± 0.01^a	0.32 ± 0.01^a	0.31 ± 0.00^a	0.31 ± 0.00^a	0.31 ± 0.01^a
C24:0	0.17 ± 0.01^{ab}	0.18 ± 0.00^a	0.17 ± 0.00^{ab}	0.16 ± 0.00^{bc}	0.15 ± 0.01^c
C24:1	0.65 ± 0.01^b	0.66 ± 0.01^b	0.68 ± 0.00^a	0.15 ± 0.01^c	0.16 ± 0.00^c
SFA	24.93 ± 0.07^c	25.10 ± 0.07^b	25.13 ± 0.03^{ab}	25.25 ± 0.06^a	24.85 ± 0.06^c
MUFA	37.53 ± 0.10^a	37.50 ± 0.07^a	37.42 ± 0.02^a	37.13 ± 0.04^b	37.42 ± 0.10^a
PUFA	37.54 ± 0.11^{bc}	37.40 ± 0.05^d	37.45 ± 0.02^{cd}	37.62 ± 0.03^{ab}	37.73 ± 0.04^a

Source: Chew et al. (2017e)

Values followed by different superscript letters [abcd] within the same row are significantly different ($p < 0.05$) according to Tukey's test

SFA saturated fatty acid, *MUFA* monounsaturated fatty acid, *PUFA* polyunsaturated fatty acid

(Coetzee et al. 2008). α-Linolenic acid was presented in a minor amount (0.3%) in the kenaf seed oil, which is an essential omega-3 fatty acid that will metabolize to eicosapentaenoic acid that will transform to eicosanoids possess with anti-inflammatory and anti-thrombotic functions (Ruiz et al. 2002; Nyam et al. 2009). Thus, the high amount of unsaturated fatty acids makes the kenaf seed oil recommended to use as edible oil for diet enrichment.

4.4 3-Monochloro-1,2-Propanediol Content

3-Monochloro-1,2-propanediol (3-MCPD) content is highly concerned in the vegetable oil as it will bring the carcinogenic effect, infertility, and malfunction of certain organs that shown in the animal studies (Wong et al. 2017). 3-MCPD esters are the precursor of the free 3-MCPD, while 2-MCPD and glycidyl esters are the precursors to form a 3-MCPD ester. 3-MCPD esters were detected in edible oils and fats such as deep fried food, spreads (Hamlet et al. 2011; Ermacora and Hrncirik 2014), infant formula (Zelinková et al. 2009), and human breast milk (Zelinková et al. 2008). There were different concentrations of 3-MCPD esters in different types of oil, including rapeseed oil (1000 µg/kg), palm oil (4400 µg/kg), and peanut oil (440–620 µg/kg) (Franke et al. 2009; Li et al. 2016; Wong et al. 2017). The presence of 3-MCPD esters had called public attention on the safety levels of 3-MCPD, factors and formation mechanism of 3-MCPD esters. A tolerable daily intake (TDI)

Table 23.8 2-, 3-MCPD esters and glycidyl ester contents of kenaf seed oil during the refining process (Chew et al. 2017e)

Oil	2-MCPD ester (µg/kg)	3-MCPD ester (µg/kg)	Glycidyl ester (µg/kg)
Crude	3.5	ND	ND
Degummed	ND	ND	ND
Neutralized	1.0	ND	ND
Bleached	ND	ND	ND
Deodorized	9.0	ND	54.8

ND not detected

of 2 µg/kg body weight for free 3-MCPD set by the European Scientific Committee on Food in 2001 (Li et al. 2016). However, the Panel on Contaminants in the Food Chain set a new TDI of 0.8 µg/kg body weight for free 3-MCPD and 3-MCPD ester in 2016 (EFSA 2016).

Intrinsic precursor presented in the crude oil, preliminary heat treatment on seeds, and refining conditions were the appointed factors to the formation of 3-MCPD esters (Chew et al. 2017e). Deodorization temperature that more than 200 °C used is pointed as the main factor in the formation of 3-MCPD esters in the refined oil (Franke et al. 2009). Chew et al. (2017e) investigated the contents of 3-MCPD ester, 2-MCPD ester, and glycidyl ester in the kenaf seed oil during refining, as shown in Table 23.8. The results showed that the crude and refined kenaf seed oils were not contaminated by 3-MCPD esters and safe for edible. On the other hand, a minor amount of 2-MCPD esters was identified in the crude, neutralized, and deodorized kenaf seed oil. Deodorization has slightly increased the glycidyl esters in kenaf seed oil. The formation of glycidyl esters in the deodorization stage might attribute to the high-temperature heating of DAG and MAG at deodorization stage (Chew et al. 2017e).

Refined kenaf seed oil is free from 3-MCPD esters and contains minor amounts of 2-MCPD esters and glycidyl esters might attribute by the lower composition of MAG and DAG in the crude kenaf seed oil. MAG and DAG act as the precursors of the formation of glycidyl esters (Chew et al. 2017e). Previous studies reported higher amounts of glycidyl esters in palm oil (3000–5000 µg/kg), sunflower oil (269 µg/kg), soybean oil (171 µg/kg), rapeseed oil (166 µg/kg), and peanut oil (148 µg/kg) (ESFA 2016; Wong et al. 2017). This showed that the intrinsic components of the kenaf seed oil and the deodorization conditions did not contribute to the formation of 3-MCPD esters.

5 Minor Bioactive Compounds

5.1 Sterol

Sterols are a group of naturally occurring substances derived from hydroxylated polycyclic isopentenoids with a 1,2-cyclopentanophenanthrene structure (Fig. 23.1). Sterols contain a total of 27–30 carbon atoms in which a side chain with carbon

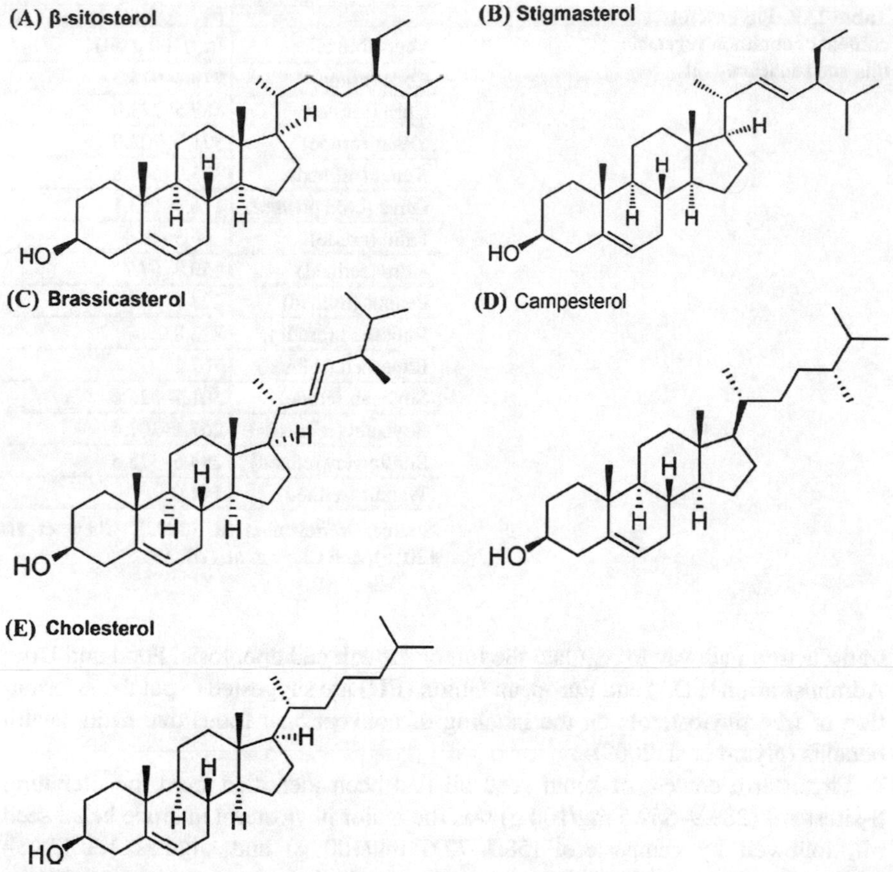

(A) β-sitosterol

(B) Stigmasterol

(C) Brassicasterol

(D) Campesterol

(E) Cholesterol

Fig. 23.1 Structures of (**a**) β-sitosterol, (**b**) stigmasterol, (**c**) brassicasterol, (**d**) campesterol, and (**e**) cholesterol. (Source: Chen et al. 2015)

atoms ≥7 is attached at the position 17 among the carbons (C-17) (Nes et al. 1984). Sterols are unsaponifiable matter in the oils. Sterols present in free or esterified form with fatty acids in vegetable oils. The combined determination of free and esterified phytosterols gives an informative approach to check the authenticity of vegetable oils (Cunha et al. 2006).

Phytosterols help to inhibit the absorption of cholesterol from the small intestine. The previous study reported phytosterols decrease blood LDL-C by intake of 10–15% in a part of a healthy diet (Nyam et al. 2009). Phytosterols had been scientifically proven to reduce 40–50% in the absorption of dietary cholesterol, 6–10% in total serum cholesterol, and 8–14% in serum LDL-C by a daily intake of 2 g of phytosterols (Kritchersky and Chen 2005). In addition, phytosterol possesses anti-bacterial, anti-fungal, anti-inflammatory, anti-tumoral, anti-ulcer, and antioxidant activities. Phytosterols help in the membrane properties and signal

Table 23.9 Phytosterol content of common vegetable oils and kenaf seed oil

Vegetable oil	Phytosterol content (mg/100 g oil)
Corn (crude)	779.6–924.3
Corn (refined)	685.5–773.0
Kenaf (crude)	371.3–703.0
Kenaf (refined)	375.9–544.8
Olive (cold pressed)	176.3–193.1
Palm (crude)	69.3–79.4
Palm (refined)	59.9–67.7
Peanut (refined)	228.7
Rapeseed (crude)	823.8
Rapeseed (refined)	767.1
Soybean (crude)	301.7–326.6
Soybean (refined)	267.1–307.4
Sunflower (refined)	294.6–375.5
Walnut (refined)	144.0

Source: Verleyen et al. (2002), Chew et al. (2016), and Chew et al. (2017e)

transduction pathway to regulate the tumor growth and apoptosis. Food and Drug Administration (FDA) and European Union (EU) are suggested to put the information of free phytosterols on the labeling of conventional foods due to its health benefits (Nyam et al. 2009).

Phytosterol content of kenaf seed oil had been identified from the literature. β-sitosterol (289.9–589.5 mg/100 g) was the major phytosterol in crude kenaf seed oil, followed by campesterol (58.1–72.6 mg/100 g) and stigmasterol (23.3–40.9 mg/100 g), which contributed to the total phytosterol content of 371.3–703.0 mg/100 g oil (Nyam et al. 2009; Chew et al. 2016). On the other hand, refined kenaf seed oil presented a total phytosterol content of 375.9–544.8 mg/100 g, which contributed by β-sitosterol (320.5–463.3 mg/100 g), campesterol (41.2–56.8 mg/100 g) and stigmasterol (14.2–24.7 mg/100 g) (Chew et al. 2016, 2017e). The phytosterol contents of crude and refined kenaf seed oil are higher than the olive oil, palm oil, peanut oil, and walnut oil, as well as comparable to the common vegetable oils, such as corn oil, rapeseed oil, soybean oil, and sunflower oil, as shown in Table 23.9.

5.2 Squalene

Squalene is a highly unsaturated aliphatic hydrocarbon ($C_{30}H_{50}$) that functions in protecting the body lipids and organelle membranes from oxidation. Squalene represents as a precursor of cholesterol. It is recognized in its advantageous to fight against certain types of cancer (Smith 2000). There was in a range of 3.69–14.3 mg/100 g oil was identified in the crude kenaf seed oil (Nyam et al. 2009; Chew et al. 2015).

5.3 Tocopherol

Tocopherols are a group of fat-soluble antioxidants with a chromanol ring and a hydrophobic side chain. Tocopherols are varied in the methylation pattern of the benzopyran ring, so they exist in four different analogs as α-, β-, γ-, and δ- (Schwarz et al. 2008; Boschin and Arnoldi 2011). α-Tocopherol has three methyl groups, β- and γ- tocopherols have two methyl groups, and δ-tocopherol has one methyl group, as shown in Fig. 23.2. Besides tocopherols, tocotrienols are also belonged to the Vitamin E family, with α-, β-, γ-, and δ- analogs.

Tocopherols are presented in a saturated form, while tocotrienols are presented in an unsaturated form and contain an isoprenoid side chain in the vitamin E family (Ahsan et al. 2015). Tocopherols and tocotrienols act as natural antioxidants to interrupt the propagation stage of oxidation reaction by scavenging the free radicals. Tocopherol functions as an antioxidant to regulate the membrane functions by preventing oxidation of body lipids and organelle membranes (Nyam et al. 2009; Ng et al. 2013a). Tocopherol is proven to reduce the risk of getting heart disease, Alzheimer's disease, and cancer. γ-Tocopherol has been reported to be more effective than α-tocopherol to decrease the platelets aggregation, intra-arterial thrombus's aggregation and oxidation of LDL (Li et al. 1999; Saldeen et al. 1999). α-Tocotrienol gives an antioxidant activity about 40–60 times higher than α-tocopherol. Tocotrienol is proven to restrain cholesterol biosynthesis and offer neuroprotective properties (Sen et al. 2007; Khanna et al. 2010).

The tocopherol and tocotrienol contents of crude and refined kenaf seed oil is given in Table 23.10. γ-Tocopherol was reported as the major tocopherol in the kenaf seed oil followed by α-, β-, and δ-tocopherols (Nyam et al. 2009; Chew et al. 2015, 2017e). The total tocopherol content of crude kenaf seed oil was reported in 84.66 mg/100 g oil (Nyam et al. 2009), while the total tocopherol and

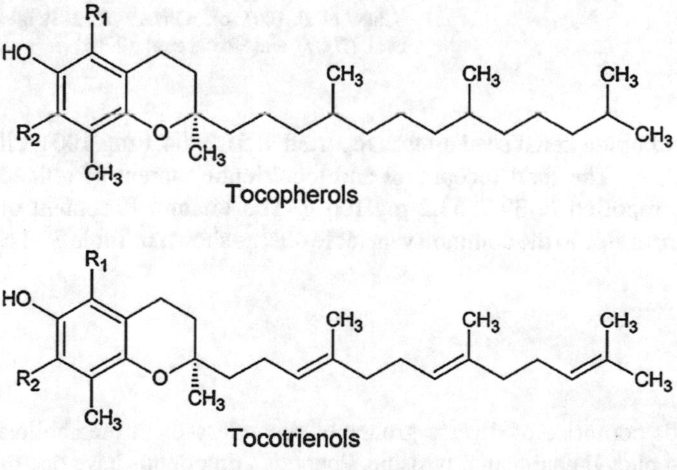

Fig. 23.2 Chemical structures of tocopherols and tocotrienols. (Source: Bartosińska et al. 2016)

Table 23.10 Tocopherol and tocotrienol contents of crude and refined kenaf seed oil

	Crude (Chew et al. 2017d, e)	Refined (Chew et al. 2017d, e)
Tocotrienol (mg/100 g oil)		
α	2.2–3.7	1.2–1.5
γ	2.0–2.4	1.4–1.8
δ	3.0–6.0	2.9–4.4
Tocopherol (mg/100 g oil)		
α	12.2–21.9	8.1–17.0
β	6.4–6.8	5.9–6.1
γ	17.8–27.2	16.0–23.5
δ	1.3–2.5	1.2–1.8
Total (mg/100 g oil)	51.3–64.1	39.7–53.2

Table 23.11 Vitamin E content of kenaf seed oil and common vegetable oils

Vegetable oil	Vitamin E content (mg/100 g oil)
Corn (crude)[a]	100.6
Corn (refined)[a]	70.7
Kenaf (crude)[b]	64.1
Kenaf (refined)[b]	53.2
Olive (extra virgin)[c]	17.7
Palm oil (refined)[d]	50.6–76.6
Rapeseed (crude)[a]	82.3
Rapeseed (refined)[a]	61.7
Soybean (crude)[a]	132.8
Soybean (refined)[a]	72.5
Sunflower (refined)[a]	73.7
Walnut (refined)[a]	63.4

Source: [a]Ergönül and Köseoğlu (2014), [b]Chew et al. (2017e), [c]Gliszczyńska-Świgło et al. (2007), and [d]Rossi et al. (2001)

tocotrienol content of crude kenaf seed oil was reported in 51.3–64.1 mg/100 g oil (Chew et al. 2017d, e). The total tocopherol and tocotrienol content of refined kenaf seed oil was reported in 39.7–53.2 mg/100 g. The vitamin E content of kenaf seed oil is comparable to the common vegetable oils, as shown in Table 23.11.

5.4 Phenolics

Phenolic compounds comprise of diverse groups of secondary plant metabolites that originated from phenylalanine and tyrosine. Phenolic compounds have one or more hydroxyl (OH) groups bind directly to an aromatic ring, varying from

Table 23.12 Phenolic acids of kenaf seed oil

Phenolic acid (mg/100 g oil)	Nyam et al. (2009)	Ng et al. (2013b)
Vanillic acid	0.53 ± 0.01	4.66 ± 0.38
Caffeic acid	0.40 ± 0.01	3.55 ± 0.10
Gallic acid	0.24 ± 0.01	1.38 ± 0.06
p-hydroxybenzoic acid	0.19 ± 0.01	1.04 ± 0.08
Ferulic acid	0.18 ± 0.01	0.91 ± 0.04
p-coumaric acid	0.16 ± 0.01	ND
Benzaldehyde	ND	0.38 ± 0.10
Protocatechuic acid	0.05 ± 0.01	0.31 ± 0.02
Syringic acid	<0.05	ND
Sum	1.75	12.23

ND not detected

common phenolic molecules to polymerized compounds. Flavonoids, phenolic acids, and tannins are the main classes of dietary phenolic compounds (Balasundram et al. 2006). Phenolic acids contain two subgroups, which are hydroxybenzoic acid and hydroxycinnamic acid. Hydroxybenzoic acids have common C_6-C_1 structure and their derivatives include gallic, vanillic, and syringic acids. However, caffeic, ferulic, and p-coumaric acids are the derivatives of hydroxycinnamic acids, each having aromatic compounds with C_6-C_3 side chain (Balasundram et al. 2006). Different types of oil may present different types and content of phenolic compounds. Phenolics will influence the antioxidant activity and flavor of the oils (Nyam et al. 2009).

The content of phenolic acids in the kenaf seed oil determined by Nyam et al. (2009) and Ng et al. (2013a) are given in Table 23.12. The phenolic acids were identified by high-performance liquid chromatography with a reversed phase column. Vanillic acid, caffeic acid, gallic acid, p-hydroxybenzoic acid, ferulic acid, p-coumaric acid, and protocatechuic acid were identified as the main phenolic acids in the kenaf seed oil. Syringic acid was detected only in trace amount in kenaf seed oil (Nyam et al. 2009). Vanillic acid, caffeic acid, gallic acid, benzoic acid, ferulic acid, benzaldehyde, and protocatechuic acid were determined as the main phenolic acids in the kenaf seed oil (Ng et al. 2013a). Phenolic acids enhance the oxidative stability of the vegetable oil. Moreover, plant phenolics can be considered to use as a potential neutral drug in combating cancer or diet enrichment to reduce cancer risk (Wong et al. 2014a).

5.5 Carotenoids

Carotenoids contributed to the vitamin A activity and act as an antioxidant to scavenge free radicals. The simultaneous presence of β-carotene and α-tocopherol in the vegetable oil can offer a synergistic antioxidant effect, while β-carotene can stand as strong oxidation inhibitors (Ghazani and Marangoni 2013). Carotenoids content identified in crude kenaf seed oil was 17.4 mg β-carotene/kg oil and reduced to 6.0 mg β-carotene/kg oil after the refining process (Chew et al. 2016). The carotenoids content of crude kenaf seed oil is considered low. However, carotenoids are easily degraded at high temperature. Compare to other vegetable oils, β-carotene content of crude sunflower oil was 21.2–24.7 mg/kg oil, and reduced to 5.0–5.5 mg/kg oil in refined sunflower oil, while β-carotene content of crude rapeseed oil was 63.6 mg/kg oil, and reduced to 10.2 mg/kg oil in refined rapeseed oil (Kreps et al. 2014).

6   Oxidative Evaluation of Kenaf Seed Oil

Lipid oxidation will affect the nutritional value, functionality, and toxicity of the oil during processing and storage. The natural antioxidants presented in the oil such as phenolics, tocopherols, and phytosterols can help to preserve the oil quality. The intrinsic components, the presence of antioxidant or pro-oxidant, processing and storage conditions will affect the rate of lipid oxidation in the oil. Microwave pre-treatment on kenaf seeds for 2 min able to increase the phenolic content of kenaf seed oil from 49.2 to 75.7 mg GAE/100 g, thus this pre-treatment could slightly improve the oxidative stability of the kenaf seed oil, as reported in the study of Nyam et al. (2015). High temperature favours the liberation of phenolics from bound structures or chemical changes to phenolic compounds (Chew and Nyam 2016a).

The previous study revealed that microencapsulated kenaf seed oil (MKSO) by co-extrusion technology (Chew and Nyam 2016a) able to protect the kenaf seed oil against oxidation during an accelerated storage at 65 °C for 24 days, assessed by PV, p-Anisidine value (p-AV), and TOTOX. Microencapsulation of kenaf seed oil by spray drying at 40% of total solid content of wall materials at an inlet temperature of 160 °C showed a better protection of kenaf seed oil against oxidation compared to the total solid contents of 20% and 30% (Ng et al. 2014) and the processing inlet temperature of 180 and 200 °C (Ng et al. 2013b), assessed by PV, p-AV, and TOTOX. This is because higher microencapsulation efficiency (97.0%) was produced for the MKSO produced at the optimum parameters by spray drying. This indicates the oil is well protected by the shell barrier. However, higher inlet temperature decreases the microencapsulation efficiency of MKSO as the external surface of the particles dried faster than the internal, which resulted in the cracks on the surface that release of more surface oil, and subsequently exposed to lipid oxidation (Ng et al. 2013b).

The spray dried MKSO able to preserve the fatty acid composition well of kenaf seed oil during accelerated storage. There was a decrease of MUFA and PUFA, coupled with an increase of saturated fatty acids, showed in the fatty acid composition of un-encapsulated kenaf seed oil during the storage. The bioactive compounds were maintained in a stable circumstance in the MKSO during the storage. Therefore, microencapsulation was an effective technique to protect the kenaf seed oil against lipid oxidation, as well as preserve the bioactive compounds to maintain its nutritional value (Ng et al. 2013a).

Chew et al. (2017f) conducted an accelerated storage study at 65 °C between crude and refined kenaf seed oils to assess the changes of antioxidant activity and bioactive compounds. The phenolic content and antioxidant activity of refined oil (DPPH and ABTS) were significantly lower than those of crude oil at day 24. This might be due to the majority of the phenolics were removed via the refining process that leads to lower oxidative stability of refined oil than the crude oil. The phenolic extract is proven more effective in protecting the oil stability than α-tocopherol (Gümez-Alonso et al. 2003). It is better to have a higher level of natural antioxidants (phenolics, carotenoids, tocopherols, and phosphatides) to offer the synergistic effects of antioxidants to protect the oil stability.

There was a reduction of 72.5% tocopherol content and 31.1% phytosterol content in the crude oil coupled with a reduction of 67% tocopherol content and 12.1% phytosterol content in the refined oil after the storage. There was a decrease of 81.2% and 82.9% of α-tocopherol in the crude and refined kenaf seed oils, respectively, compared to the decrease of 70% and 62% of γ-tocopherol in the crude and refined kenaf seed oils, respectively upon the storage. α-Tocopherol is less stable than γ-tocopherol when exposed to the heat. Previous studies presented the α-tocopherol was the least stable among the tocopherols isomers during storage, with the order of α->(γ+β)>δ-tocopherol for the degree of degradation of tocopherols (Lampi et al. 1999; Player et al. 2006; Bruscatto et al. 2009). Phytosterol is more stable than tocopherol during the storage, as the results showed that the refined oil could preserve the phytosterol content during the storage (Chew et al. 2017f).

There was no significant difference in tocopherol and phytosterol contents for crude and refined oils after the storage. The results showed that the refined oil was still stable to protect the bioactive compounds than the crude oil, as the rate of degradation of tocopherol and phytosterol contents in refined oil was slower than that in crude oil during the storage. Refining process helps to remove the chlorophyll, FFA, and trace metal ions, which act as the pro-oxidants in the crude oil. Although refining removes the phenolic compound in the oil, refined oil able to protect tocopherol and phytosterol in a better manner compared with crude oil (Chew et al. 2017f).

7 Encapsulation Technologies of Kenaf Seed Oil

Currently, there is a rising demand for the functional food products for the nutritive and healthy purpose in the market. Thus, the research paying attention to the dietary fibers, polyphenols, unsaturated fatty acids, probiotics, and prebiotics has been increased on the delivery the health-promoting traits to human. However, high oxygen environment in the gastric will favor the oxidative deterioration of functional oil before it reaches to the small intestine for absorption. The poor hydrophilic property, physicochemical instability and prone to oxidation of the high unsaturated fatty acids oil limits the application and bioavailability of the kenaf seed oil (Cheong et al. 2016a). Encapsulation of kenaf seed oil by using emulsifier can encounter these issues to improve their functionality.

7.1 Nanoencapsulation

Nanoencapsulation is defined as an approach to encapsulate the core agent in nanoscale size. Nanoemulsion is colloidal dispersions made up by two immiscible fluids, in which one is being scattered in the other, including oil-in-water emulsion and water-in-oil emulsion with a droplet size of 50–1000 nm (Sanguansri and Augustin 2006). Nanoemulsion is in liquid state, which can utilize directly or in a powder form by additional drying step (spray drying or freeze-drying) of the nano-emulsion. Microfluidisation, high-pressure homogenization, and ultrasonication are the three techniques that can produce nanoemulsions. High-pressure homogeniza-tion is applying a high pressure in a range of 100–2000 bar and high shear stress to breakdown the particles of the emulsion into nanoscale size (Ezhilarasi et al. 2013). The nano droplet size will decrease the attractive forces between the droplets, and subsequently increase the kinetic stability of the nanoemulsion against the droplet flocculation and coalescence (Cheong et al. 2016a).

Nanoencapsulation of kenaf seed oil using high-pressure homogenization can produce nanoemulsion to enhance the functionality of kenaf seed oil by improving its water solubility, physiochemical stability, and bioavailability, that beneficial in the functional food development in the food and nutraceutical fields. Previous litera-ture reported kenaf seed oil has been encapsulated in oil-in-water nanoemulsions using high-pressure homogenization by the formulations of sodium caseinate-gum arabic-Tween 20 and sodium caseinate-Tween 20-β-cyclodetxrin (β-CD) under optimum formulations. Both of the nanoemulsions produced particle size of 121.22 nm, polydispersity index (PDI) of 0.16, zeta-potential of −39.63 mV (Cheong et al. 2016a) and particle size of 155.53 nm, PDI of 0.07, zeta potential of −46.67 mV (Cheong et al. 2016b), which indicated a high stability. Figure 23.3 shows a schematic diagram to illustrate a stable nanoemulsion formed by three emulsifier mixtures, which are sodium caseinate, gum arabic, and Tween 20 (Cheong et al. 2016a). Processing pressure of 28,000 psi for 4 cycles was found to be the optimum parameters to produce the kenaf seed oil nanoemulsion with the emulsi-fiers of sodium caseinate-Tween 20-β-CD using high-pressure homogenizer, which produced the particle size of 122.2 nm, PDI of 0.15, zeta potential of −46.6 mV (Cheong and Nyam 2016).

The kenaf seed oil-in-water nanoemulsions have been investigated under *in vitro* gastrointestinal (GI) digestion to assess its bioaccessibility after simulated human digestion. The nanoemulsions showed a good digestibility of kenaf seed oil by releasing 247.7 μmol/mL of FFA (lipolysis rate 84.4%) for the sodium caseinate-gum arabic-Tween 20 (Cheong et al. 2016c), and releasing 250.85 μmol/mL of FFA (lipolysis rate 85.3%) for the sodium caseinate-Tween 20- β-CD, after simulated

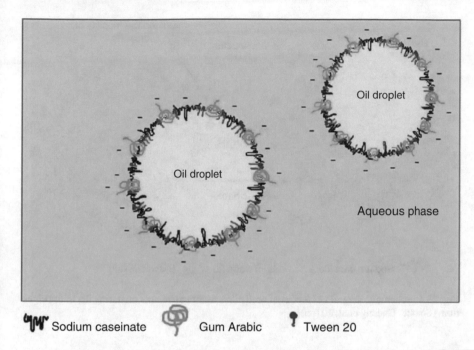

Oil droplet

Oil droplet

Aqueous phase

ᵒᴵᴹᴵᴿ Sodium caseinate   Gum Arabic   Tween 20

Fig. 23.3 Schematic diagram illustrates a stable nanoemulsion stabilised by good synergistic interaction of three-component emulsifiers. (Source: Cheong et al. 2016a)

digestion (Cheong et al. 2016d). The good lipid digestion indicates a prior step for the good release and absorption of the lipophilic key nutrients in the small intestine. Figure 23.4 shows the schematic diagram to show the kenaf seed oil droplets are stabilized by the adsorbed and un-adsorbed β-CD particles during the initial digestion. Most of the sodium caseinate was hydrolyzed and desorbed from the oil droplet interface, but the un-hydrolyzed β-CD particles could still stable the oil droplets to prevent them from coalescence.

Table 23.13 shows the changes in antioxidant activity and bioactive compounds of un-encapsulated kenaf seed oil and kenaf seed oil-in-water nanoemulsion (sodium caseinate-Tween 20- β-CD) before and after *in vitro* digestion. In addition, improved bioaccessibility of bioactives (tocopherols and phenolics) and slower degradation of phytosterols compared to digested un-encapsulated kenaf seed oil were reproted in the studies of Cheong et al. (2016c, d). The emulsifier of β-CD able to form complexation with phenolic compounds in the kenaf seed oil to increase the total phenolic contents in the nanoemulsions, and subsequently enhance the solubility and bioaccessibility of kenaf seed oil under simulated gastrointestinal fluid. Thus, it helped to improve the bioaccessibility of antioxidants, and this is a key step to improve the bioavailability of the lipophilic key nutrients for a good absorption in the small intestine. Thus, kenaf seed oil-in-water nanoemulsions showed a high potential application as functional products such as direct application or supplemen-

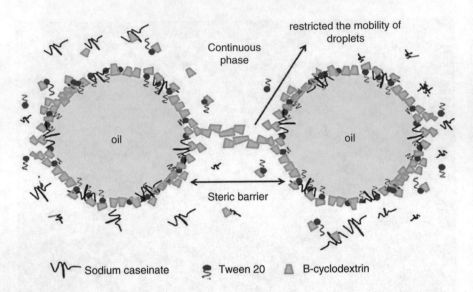

Fig. 23.4 Schematic diagram of the kenaf seed oil-in water nanoemulsion during the initial digestion. (Source: Cheong et al. 2016d)

Table 23.13 Bioactive compounds and antioxidant activity of undigested bulk oil, digested bulk oil, undigested nanoemulsion, and digested nanoemulsion after *in vitro* digestion

Bioactives assay	Undigested bulk oil	Digested bulk oil	Undigested nanoemulsion	Digested nanoemulsion
Total phenolic (mg GAE/100 g oil)	9.6 ± 0.8^a	15.0 ± 0.9^b	6.8 ± 1.1^c	46.7 ± 1.2^d
DPPH (mg Trolox equiv/100 g oil)	35.9 ± 2.1^a	36.7 ± 1.3^a	32.6 ± 0.5^a	9.8 ± 13.2^b
α-Tocopherol	11.8 ± 4.5^a	7.4 ± 0.2^{ab}	5.2 ± 0.2^b	8.0 ± 0.2^{ab}
γ-Tocopherol	42.6 ± 0.8^c	118.9 ± 5.3^a	22.7 ± 2.2^d	91.7 ± 1.7^b
δ-Tocopherol	Tr	27.7 ± 0.5^a	3.5 ± 0.3^c	23.9 ± 0.9^b
Total tocopherol (mg/100 g oil)	54.3 ± 4.9^c	153.9 ± 5.6^a	31.4 ± 2.4^d	123.6 ± 1.4^b
Squalene	83.9 ± 17.1^a	70.6 ± 10.7^a	54.8 ± 9.8^a	14.5 ± 6.2^b
Campesterol	52.2 ± 4.0^a	56.4 ± 2.9^a	28.2 ± 1.2^b	21.5 ± 3.4^b
Stigmasterol	40.1 ± 2.7^a	45.2 ± 0.9^a	14.0 ± 1.3^b	14.4 ± 3.6^b
β-Sitosterol	545.4 ± 52.1^a	574.4 ± 19.9^a	326.5 ± 14.3^b	246.2 ± 26.1^b
Total phytosterols (mg/100 g oil)	721.5 ± 74.9^a	746.7 ± 33.9^a	423.5 ± 16.7^b	296.6 ± 36.8^c

Values followed by different superscript letters [abcd] within the same row are significantly different ($p < 0.05$) according to Tukey's test

tation into beverages to improve the nutritional value that advantageous in the food and nutraceutical industries (Cheong et al. 2016c, d).

In addition, Cheong et al. (2017) investigated three storage temperatures (4, 25, and 40 °C) on the physicochemical properties and oxidative stability of kenaf seed oil-in-water nanoemulsion (sodium caseinate-Tween 20-β-CD). Formation of sediment occurred during the 8 weeks of storage and the nanoemulsions showed the lowest stability with −27.3 mV of zeta potential at 40 °C. The nanoemulsion was more suitable to store at 4 °C as it could maintain the stability with it showed the highest zeta-potential value (−36.6 mV), lowest changes of PDI, pH, and PUFA upon the 12 weeks of storage. The results showed that the nanoemulsions were stable for 8 weeks of storage at 4 and 25 °C. The nanoemulsion stored at 40 °C was stable for 1 week, which is equal to 28 days at room temperature based on the calculation of Arrhenius equation.

7.2 Microencapsulation

Microencapsulation is another approach to encounter the limitations of kenaf seed oil by providing a physical barrier to protect the core agent against the adverse environment. Microencapsulation of kenaf seed oil can help in the development of functional ingredients to deliver lipophilic bioacitves to the human that can be applied to the functional food and beverages. Extrusion, coacervation, freeze-drying, and spray drying are the common technology used to encapsulate the core agent to produce the micro-powder (Chew et al. 2015). Spray drying is the most popular and cost-effective technology used in the food industry to encapsulate the core agent for functional food applications. Spray drying is atomized the oil-in-water emulsions into a drying air at high temperature to evaporate the water very fast and results in the formation of the quick crust and quasi-instantaneous entrapment of the core agent. Microencapsulation by spray drying enables to reduce the oil droplet size into micro-powder form to improve the digestibility of kenaf seed oil. This will result in the increase of the total surface area of MKSO available for the digestion by the enzyme lipase to facilitate the lipid digestion (Chew et al. 2018).

Ng et al. (2013a) worked on the formulation of maltodextrin, sodium caseinate, and soy lecithin to encapsulate kenaf seed oil and achieved a microencapsulation efficiency of 97.02%, which was optimized at total solid contents of 40% and inlet temperature of 160 °C offered better protection against lipid oxidation, compared to the inlet temperature of 180 and 200 °C (Ng et al. 2013b), and 20% and 30% solid content (Ng et al. 2014). In addition, Ng et al. (2013a) investigated the oxidative stability of spray-dried MKSO accelerated storage at 65 °C for 24 days. The unencapsulated kenaf seed oil undergoes a decrease in the MUFA and PUFA during the storage as the double bond of the unsaturated fatty acids were broken down in the lipid oxidation. Besides this, there was a significant degradation of the phenolic acids, tocopherols, and phytosterols of un-encapsulated kenaf seed oil during the storage. The high temperature of the spray drying process will cause a slight deterioration of the bioactive compounds in the kenaf seed oil. However, the physical barrier provided by the spray drying able to preserve the unsaturated fatty acids and bioactive compounds in the kenaf seed oil during the accelerated storage. Thus, Ng et al. (2013a) demonstrated the spray drying offers an effective technique to protect the kenaf seed oil against oxidation.

Chew et al. (2018) studied the potential of gum arabic, sodium caseinate, and β-CD as the wall materials to spray dry the kenaf seed oil and tested in an *in vitro* GI digestion. Spray dried MKSO offered a good protection in the gastric environment by a lower percentage of oil released (1.43–6.44%) after gastric digestion. Then, a high percentage of oil released (81.10–91.19%) was released from the MKSO after gastric and intestinal digestion. The bioaccessibility of phenolics (130.4% increase) and tocopherol and tocotrienol contents (147.7% increase) were increased after the GI digestion, coupled with a slower decrease of phytosterol content (59.4% decrease), compared to the undigested kenaf seed oil. Long chain fatty acids able to form micelles with lipophilic bioactive to improve its solubility in the GI fluid compared to the medium chain fatty acids. Thus, the high contents of long chain fatty acids in the kenaf seed oil such as C16:0 and C18:0 are responsible for the improved solubility capacity for tocopherols and tocotrienols. However, there was only a-tocopherol (decrease 27.7–68.3%) undergo a decrease after the simulated digestion as the a-tocopherol is the most sensitive form, while all the tocopherols and tocotrienols isomers were increased after digestion (Chew et al. 2018).

Complexation of polyphenols with hydroxypropyl-β-CD is the reason that increased the bioaccessibility of the phenolic in the kenaf seed oil. Increased content of phenolics and tocopherols would offer synergism effect in the MKSO models to protect the lipophilic bioactive. Increased phenolic contents helped in the retention of tocopherols and tocotrienols contents and liberate of more free forms of tocopherols and tocotrienols. Phytosterols undergo degradation after the simulated GI digestion as the hydrophobicity of the phytosterol, which is not favorable in the simulated enzyme fluid and reduce its bioaccessibility (Chew et al. 2015, 2018; Cheong et al. 2016d). The detailed results of the antioxidant activity (DPPH and ABTS), phenolic, tocopherol and tocotrienol, as well as phytosterol of the undigested kenaf seed oil, digested kenaf seed oil, and digested MKSO are summarized in Table 23.14 (Chew et al. 2018).

Co-extrusion technology is another approach using the encapsulator instrument with a concentric nozzle which equipped with vibrating nozzle technology to allow its dispersed the droplets into the uniform size of microcapsules. Kenaf seed oil had been encapsulated by using encapsulator equipped with vibrating nozzle technology with a concentric nozzle to produce MKSO microcapsules (Chew and Nyam 2016b) using 1.5% alginate and high methoxyl pectin (HMP)-alginate as the wall materials. Additional chitosan coating had been added to the HMP-alginate microcapsule to strengthen the MKSO by its microencapsulation efficiency increased from 63.7% to 73.3% (Chew et al. 2015) (Fig. 23.5). Chitosan coating has protected the MKSO in the simulated gastric digestion by only 3.84% of oil released. After that, the MKSO with chitosan coating was releasing 83.3% of oil after the intestinal digestion for 4 h. The results showed that the microencapsulation of kenaf seed oil

Table 23.14 Antioxidant activity and bioactive compounds of undigested kenaf seed oil, digested kenaf seed oil, and digested MKSO

Assay	Undigested oil	Digested un-encapsulated oil	Digested MKSO
DPPH (mg Trolox equiv/100 g oil)	29.2 ± 1.5[b]	36.6 ± 1.5[a]	35.1 ± 1.2[a]
ABTS (mg Trolox equiv/100 g oil)	42.9 ± 0.9[a]	43.9 ± 0.7[a]	45.1 ± 3.8[a]
TPC (mg GAE/100 g oil)	3.9 ± 0.1[b]	8.8 ± 1.3[a]	8.9 ± 1.3[a]
δ-Tocotrienol	4.2 ± 0.2[c]	5.1 ± 0.0[b]	6.0 ± 0.2[a]
γ-Tocotrienol	1.8 ± 0.1[b]	2.1 ± 0.1[b]	6.1 ± 0.2[a]
α-Tocotrienol	0.9 ± 0.1[b]	1.0 ± 0.1[b]	5.3 ± 0.2[a]
δ-Tocopherol	1.9 ± 0.0[c]	10.3 ± 1.0[b]	38.3 ± 0.8[a]
(β+γ)-Tocopherol	24.9 ± 0.3[c]	39.3 ± 1.5[b]	57.7 ± 3.0[a]
α-Tocopherol	13.5 ± 0.4[a]	9.8 ± 0.2[b]	5.7 ± 0.2[c]
Total tocopherol and tocotrienol (mg/100 g oil)	48.0 ± 0.8[c]	67.6 ± 2.6[b]	119.0 ± 4.6[a]
Campesterol	46.2 ± 1.3[a]	38.7 ± 4.3[b]	10.9 ± 0.7[c]
Stigmasterol	20.0 ± 2.3[a]	12.6 ± 1.1[b]	7.3 ± .4[c]
β-Sitosterol	374.3 ± 21.1[a]	271.3 ± 32.6[b]	160.8 ± 12.7[c]
Total phytosterol (mg/100 g oil)	440.5 ± 21.1[a]	322.7 ± 35.7[b]	178.9 ± 12.0[c]

Values followed by different superscript letters [abc] within the same row are significantly different (*p* < 0.05) according to Tukey's test

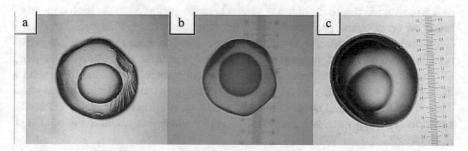

Fig. 23.5 Optical micrographs (Magnification 40×) (**a, b**) chitosan-coated microencapsulated kenaf seed oil (0.1 w/v chitosan) and (**c**) non coated microencapsulated kenaf seed oil. (Source: Chew et al. 2015)

by chitosan-HMP-alginate with co-extrusion offers an effectively controlled release delivery model. This MKSO is potential to be used as a food supplementation for targeted delivery of the key nutrients to specific site of the GI. The *in vitro* study by Chew et al. (2015) showed the MKSO undergoes a decrease in the phenolics and phytosterols contents coupled with an increase in the tocopherols after the simulated GI digestion.

8 Health Promoting Traits of Kenaf Seed Oil

8.1 Antioxidant Activity

Free radicals are originated from reactive oxygen species (ROS) and reactive nitrogen species (RNS) when the body meets the external stimuli or stress. Free radicals will cause the oxidation in the body, and subsequently affect the functions of proteins, lipids, DNA, and other biomolecules, which disrupt the metabolism and cause mutagenesis. Oxidative stress occurred when the body contains higher levels of free radicals and pro-oxidants than the natural antioxidants in the body can scavenge them (Cheng et al. 2016). Kenaf seed oil is suggested to use as functional oil as it contains a high amount of phenolics, flavonoids, tocopherols, and phytosterols.

Different assays can use to assess the antioxidant activity of vegetable oil such as 2,2′-azino-bis (3-ethylbenzothiazoline-6-sulphonic acid) (ABTS), 2,2- diphenyl-1-picrylhydrazyl (DPPH), β-carotene bleaching (BCB), and ferric reducing antioxidant power (Chan and Ismail 2009; Chew et al. 2015, 2017f). DPPH$^{•}$ assay is suitable to assess the hydrophobic system, such as oil affected by lipophilic antioxidants, including carotenoids, chlorophyll, tocopherols, and phytosterols. On the other hand, ABTS$^{•+}$ assay is suitable to assess the hydrophilic and lipophilic systems, including lipophilic bioactives and phenolics, as ABTS$^{•+}$ can dissolve in the aqueous and organic solvents (Chew et al. 2017f).

The antioxidant activity of kenaf seed oil can be attributed to the high content of bioactive compounds presented in the oil. Phenolic compounds consist of their functional groups (hydroxyl, OH−), which are able to scavenge free radicals to disrupt the initiation and propagation phases of the lipid oxidation mechanism. Different types of phenolic compounds can exert their antioxidant activity with a different pathway in different oil systems to decelerate the lipid oxidation process. The oil system, storage temperature and duration, pre-treatment of the oil seeds, extraction technique are the elements to affect the phenolic content in the oil (Chew et al. 2017f)

Tocopherols are the natural compounds, which presented in high content in kenaf seed oil. It can act as an antioxidant to enhance the shelf stability of the vegetable oil. Tocopherol can regulate the cell membrane functions to prevent the lipid oxidation. Tocopherol consists of their hydroxyl groups to scavenge the lipid peroxyl radicals to prevent these radicals to reach with other adjacent fatty acids chains in the cell membrane. α-Tocopherol is donating its hydrogen atom from the OH− group to the peroxyl radical (LOO•), yielding an α-tocopheryl and a hydroperoxide to disrupt the lipid oxidation mechanism. The biological activity of α-tocopherol is double than that of β- and γ- and is 100 times higher than that of δ-tocopherol, so the biological activity of α-tocopherol is the highest among the tocopherols isomers (Chew et al. 2017f).

Chan and Ismail (2009) conducted a comparison study with the kenaf seed oil extracted at different pressure (200, 400, and 600 bar) and temperature (40, 60, and 80 °C) using supercritical fluid extraction with others commercial vegetable oils on the antioxidant activity. The IC_{50} values assessed by DPPH• for kenaf seed oils were ranged in 12.3–39.8 mg/mL, while IC_{50} values for commercial vegetable oils were ranged in 20.6–70.4 mg/mL. The antioxidant activity of oil samples that assessed by DPPH• was arranging in the order as followed: kenaf (200/80) > palm > rice bran > kenaf (200/60) > kenaf (200/40) > kenaf (400/60) > kenaf (400/40) > kenaf (rapid Soxhlet extraction) > soy > canola > sunflower > corn > kenaf (600/60) > kenaf (600/80) > kenaf (ultrasonic extraction) > kenaf (600/40) > kenaf (400/80) > kenaf (conventional Soxhlet extraction) > olive. Kenaf seed oil extracted at the pressure of 200 bar and 80 °C presented the highest antioxidant activity among all the tested samples.

The previous study reported olive oil consists of high content of phenolic (Morello et al. 2005), but it reported in the lowest antioxidant activity in this study. This might due to the olive oil used in the study was refined olive oil, in which the refining process degraded its phenolic compounds, or the DPPH• assay cannot assess the antioxidant activity due to the hydrophilic phenolic well. BCB is suitable to test for lipophilic antioxidants. This assay is assessed by the rate of discoloration of β-carotene, which correlate well with the rate of lipid oxidation and efficacy of antioxidants in the test sample. According to Chan and Ismail (2009), the kenaf seed oil samples presented a higher antioxidant activity, which assessed by the BCB assay, as compared to other commercial vegetable oils (palm, rice bran, canola, corn, sunflower, soy, and olive), except for the kenaf seed oil extracted at 400 bar/80 °C and 600 bar/80 °C. This study showed that kenaf seed oil exhibited comparable antioxidant activity with others commercial vegetable oil and it is high potential to use as high antioxidative edible oil.

8.2   Anti-hypercholesterolemic

Increasing serum total cholesterol, triglyceride, and lipoproteins such as LDL-C, very low-density lipoprotein cholesterol and decrease high-density lipoprotein cholesterol are considered as the symptoms of hyperlipidemia (Tilak et al. 2001). Hyperlipidemia will increase the risk of cardiovascular diseases such as coronary artery disease, atherosclerosis, and cerebral vascular disease (Villanueva et al. 2011). The commercial hypocholesterolemic drug, statin reduces the total cholesterol, LDL-C and reduce the risk of heart disease by 30% effectively (Baigent et al. 2005). However, side effects are accompanied with the statin drug like abdominal pain, headache, fatigue, skin rashes, dizziness, constipation, muscle weakness, blurred version, and/or liver inflammation (Cheong et al. 2018a). Thus, the natural plant is recommended to involve in the drug or food supplementation to avoid the side effects.

Ng et al. (2015) conducted a study on the anti-hypercholesterolemic effect of kenaf seed samples (kenaf seed oil, spray dried MKSO, kenaf seed extract, and defatted kenaf seed meal (DKSM)) and compare with the commercial hypocholesterolemic drug on serum lipids profiles and malondialdehyde (MDA) level in the male Sprague dawley rats. Normal diet or hypercholesterolemic diet (incorporation of cholesterol into the diet) was supplied with or without the treatment of these kenaf samples for 32 days. The hypercholesterolemic diet had resulted in the higher serum total cholesterol, LDL-C and MDA levels in the rats than the normal diet control group. The results showed that the kenaf seed samples offered the anti-hypercholesterolemic effect in the order of kenaf seed extract > kenaf seed oil > MKSO > DKSM, by a significant decrease in the serum triglycerides, total cholesterol and MDA levels. Serum MDA level is an indicator to examine the lipid peroxidation in the body that caused by the reactive oxygen species. The MDA levels of the rats with received treatments were no significant difference with the simvastatin. Thus, kenaf seed oil is suggested to have a comparable cholesterol-lowering property with the commercial hypocholesterolemic drug, simvastatin. Kenaf seed oil can be used in diet enrichment to lowering the serum cholesterol for health effect (Ng et al. 2015).

Cheong et al. (2018a) had conducted a study on the anti-hypercholesterolemic effect of kenaf seed oil, kenaf seed oil-in-water macroemulsion (KSOM), kenaf seed oil-in-water nanoemulsion (KSON), and emulsifier mixtures without the oil (EM) on serum lipid profile, liver oxidative status, and histopathological study in rats. On the first 29 days, EM showed the highest effect in inhibiting the absorption of cholesterol in the hypercholesterolemic diet group, followed by KSON, KSOM, simvastatin, and kenaf seed oil. This is because β-CD used in the emulsifier helped to form inclusion complexes with hydrophobic cholesterol, hence β-CD can inhibit the cholesterol absorption in the small intestine. However, simvastatin showed a lower effect to reduce the cholesterol absorption in the hypercholesterolemic diet group as it is advised that simvastatin drug should be consumed with diet restriction of saturated fat and cholesterol for better drug effect. The cholesterol-lowering effect of simvastatin drug had been showed when the hypercholesterolemic diet group changed back to the normal diet after day 29.

The results showed that all the treatment samples offered cholesterol-lowering properties by lower serum total cholesterol, LDL-C, lipid peroxidation levels in the rats with received treatments, compared to the groups without treatments and normal control group. The histopathological evaluation confirmed the KSON offered the highest cholesterol-lowering activity, weight control, and decreased liver fat. This study showed that the anti-hypercholesterolemic effect was offered by reducing the cholesterol (%) in the order of KSON (−73.4%) > simvastatin (−70.5%) > KSO (−67.2%) > KSOM (−64.6%) > EM (−50.0%). This study showed that the nanoemulsion has helped to improve the functionality of kenaf seed oil.

Linoleic acid, sterol, tocopherol, and phenolics in the kenaf seed oil are believed to play an important role in the cholesterol-lowering activity. Phenolics offer an antioxidant activity to scavenge the body free radical and reduce the oxidative stress, which decreases the risk of getting hypercholesterolemia. This study showed the kenaf seed extract, which rich in phenolics offered the highest anti-hypercholesterolemic effect. Thus, it demonstrated the strong relationship between oxidative stress and hypercholesterolemia. Besides that, phytosterol helps to decrease the body triglycerides, increase high-density lipoprotein, and prohibit the absorption of dietary cholesterol (Ng et al. 2015). α-Tocopherol helps in lowering the serum MDA level and the risk of getting the cardiovascular disease (Burton 1994).

8.3   Anti-cancer

Previous research has been conducted to evaluate the cytotoxic activity of kenaf seed oil against different human cancer cell lines such as human cervical cancer, ovarian cancer, breast cancer, colon cancer, lung cancer and leukemia cancer cell lines. All the findings reported are encouraging the potential use of kenaf seed oil as a natural cytotoxic agent. IC_{50} is an indicator of the cytotoxic activity of the sample tested. The sample with the IC_{50} value of 125–5000 µg/ml is recommended to use as a potential source of natural anti-cancer agents (Foo et al. 2011; Yazan et al. 2011a, b; Ghafar et al. 2013; Wong et al. 2014a). Wong et al. (2014a) examined the cyto-toxic activity of kenaf seed extract and kenaf seed oil on the cell lines of human cervical cancer (Hela CCL-2), human breast cancer (MCF-7), human colon cancer (HCT-116), and human lung cancer (SK-LU1). The findings reported the kenaf seed oil exert cytotoxic activity against all the cell lines, by kenaf seed oil showed a a strong cytotoxic effect against the cell lines of human colon cancer, followed by human breast cancer, human lung cancer, and human cervical cancer.

The kenaf seed oil extracted by supercritical fluid extraction at different extrac-tion pressure (200, 400, and 600 bar) and temperature (40, 60, and 80 °C) were showed cytotoxic effect against human colorectal (HT29) cell line with the IC_{50} values varying from 200 to 3750 µg/mL (Ghafar et al. 2013), and more cytotoxic towards CaOV3 (ovarian cancer) than HT29 (colon cancer) with the IC_{50} values ranged from 100 to 4000 µg/mL (Yazan et al. 2011a). Besides that, Yazan et al. (2011b) reported the IC_{50} values of kenaf seed oil ranged from 125 to 5000 µg/mL towards MOLT-4 (human leukemia) and MDA-MB-231 (human breast cancer) cells, which showed the cytotoxic effects. Foo et al. (2012) reported the treatment of kenaf seed oil had reduced the severity of leukemia in WEHI-3B cells *in vivo* by increased the growth of cytotoxic T cells to kill the leukemia cells, and decreased the growth of immature monocytes and granulocytes in the mice study. There was a decrease in the weights of the spleen and liver of the mice after received the dosage of kenaf seed oil. Table 23.15 shows the IC50 values of the kenaf seed oil (variety: V36) extracted by supercritical fluid extraction towards various cancer cell lines.

Supercritical fluid extraction is more effective to extract the bioactive compounds in the kenaf seed oil. The kenaf seed oil extracted by supercritical fluid extraction exerts higher cytotoxic activity than the Soxhlet extracted kenaf seed oil against all the cell line of leukemia cancer. The conditions of supercritical fluid extraction of 600 bar and 40 °C was more suitable to extract kenaf seed oil as this conditions produced the kenaf seed oil with more cytotoxic effect against the cell lines of leu-kemia cancer (HL-60, WEHI-3B, and K562) with the lowest IC_{50} values of 178.8, 189.4, and 213.3 µg/mL, respectively (Foo et al. 2011). Increase in the extraction

Table 23.15 IC$_{50}$ values of kenaf seed oil extracted by supercritical fluid extraction towards various cell lines

Cell line	IC$_{50}$ (μg/mL)		
	SFE 600/40	SFE 600/60	SFE 600/80
MCF-7 (breast cancer)	>5000	>5000	>5000
MDA-MB-231 (breast cancer)	483.4 ± 32.0	>5000	>5000
4 T1 (breast cancer)	>5000	>5000	>5000
HeLa (cervical cancer)	>5000	>5000	>5000
A549 (lung cancer)	>5000	>5000	>5000
MOLT-4 (leukemia)	153.3 ± 25.4	1657.4 ± 72.8	>5000
HL-60 (leukemia)	178.8 ± 10.5	320.5 ± 11.4	>800
WEHI-3B (leukemia)	189.4 ± 11.6	380.3 ± 15.2	>800
K562 (leukemia)	213.3 ± 15.5	472.3 ± 13.1	>800
CaOV3 (ovarian cancer)	211.7 ± 3.8	187.0 ± 5.2	188.3 ± 10.1

Source: Foo et al. (2011) and Yazan et al. (2011a, b)

temperature of supercritical fluid extraction would deteriorate some bioactive compounds, resulting in the increase of IC$_{50}$ values of kenaf seed oil with lower cytotoxic activity. The kenaf seed oil extracted at 600 bars and 40 °C was showed to have the highest cytotoxic effect as this condition can maximize the extraction of bioactive compounds in the kenaf seed oil (Foo et al. 2011; Yazan et al. 2011a; Ghafar et al. 2013).

The cell viability of the cell lines of leukemia cells (Foo et al. 2011), leukemia (MOLT-4) and breast cancer (MDA-MB-231) (Yazan et al. 2011b) had been reduced with the increase in the concentration of treatment. The cells with kenaf seed oil treatment were showed the symptoms of apoptosis such as cellular shrinkage, membrane blebbing, nuclear compaction, and apoptotic body formation under microscope examination (Foo et al. 2011; Yazan et al. 2011a, b; Wong et al. 2014a). Apoptosis is the features of cell death, and agents with the ability to induce apoptosis in tumors are recommended to use as an anti-cancer agent.

According to Yazan et al. (2011b), α-linolenic acid and phytosterol are the two potential key agents to exert the cytotoxicity towards the cancer cell lines. α-Linoleic acid has been reported to inhibit leukemia (MOLT-4) cell growth *in vitro* (Phoon et al. 2001). Phytosterols have been showed to inhibit the growth and reduce metastatic ability of human breast, colon, leukemia, and prostate cancers cells. Besides that, phytosterols are important in regulating the membrane functions and signal transduction pathways that regulate the growth of a tumour and apoptosis (Awad et al. 2000; Awad and Fink 2000). Moreover, linoleic acid has been showed to inhibit the growth of human breast cancer, colon cancer, skin cancer, stomach cancer, and leukemia *in vitro* and *in vivo* (Hubbard et al. 2000; Kritchevsky 2000; MacDonald 2000; Phoon et al. 2001). Ghafar et al. (2013) pointed the tocopherols content presented in the kenaf seed oil is known to induce apoptosis in the human colorectal cancer cell and inhibit human prostate cancer cell growth. Wong et al. (2014a) pointed the high phenolic contents presented in the kenaf seed extract and oil responsible for the apoptosis of the cancer cells. Thus, high phytosterol, linoleic acid, and phenolic contents in the kenaf seed oil are responsible for the cytotoxic activity of kenaf seed oil. Thus, the findings were supported the kenaf seed oil as a potential source of natural anti-cancer agent.

8.4   Anti-ulcer

Gastric ulcer is the lesions of the gastric mucosa. Imbalance in the secretion of hydrochloric acid and pepsin, blood flow, generation of prostaglandins, imbalance secretion of mucus and bicarbonate are the etiology of gastric ulcer. Regular intake of non-steroidal anti-inflammatory drugs, heavy alcohol intake, stress, and smoking are the factors to cause gastric ulcer formation. There are many synthetic anti-ulcer drugs in the market such as antacids, histamine H2-receptor antagonists, misoprostol, proton-pump inhibitors, and sucralfate, but minor side effects are accompanied with these synthetic drugs (Cheong et al. 2018b). Therefore, natural plant sources are encouraged to use in the prevention and treatment of the ulcers to avoid the side effects.

Nyam et al. (2016) investigated the anti-ulcer activity of kenaf and rossele seed oils in ulcer-induced rats. Treatment of kenaf seed oil (500 mg/kg) showed the protective effect of 54.5% in ethanol-induced ulcer rats, 75.0% in indomethacin-induced ulcer rats, and 60.4% in cold restraint stress-induced ulcer rats, compared with the omeprazole, which offered 35.9, 48.0, and 95.3% in the three induced ulcer models mentioned above. In this study, ethanol able to induce gastric ulcer by disrupts the protective barrier of the gastric mucosa and stimulating microvascular changes after applying for few minutes. Mucus secretion in the gastric mucosa is very important to protect against gastric lesions. Flavonoids in the kenaf seed oil are known to increase the secretion of mucus, bicarbonate, and prostaglandin to strengthen the gastric mucosa and scavenge the free radical to prevent gastric ulcer and lesions (Sachin and Archana 2009).

On the other hand, Cheong et al. (2018b) investigated the gastroprotective and anti-ulcer effect of KSON, KSOM, kenaf seed oil, and EM by measuring ulcer index, stomach tissue oxidative status, and histopathological changes in indomathacin-induced and ethanol-induced ulcer rats. Omeprazole offered the highest protection (97.5%) in the indomethacin-induced ulcer rats, followed by KSON (94.3%), KSOM (81.6%), EM (76.3%), and kenaf seed oil (3.4%). Moreover, KSON offered the highest protection (95.0%), followed by KSOM (84.8%), omeprazole (80.2%), kenaf seed oil (49.7%), and EM (40.5%) in the ethanol-induced ulcer rats. Nanoemulsion had improved the functionality of kenaf seed oil as an anti-ulcer agent. High antioxidant activity in the EM such as sodium caseinate and β-CD worked synergistically with kenaf seed oil in nanoemulsion form to enhance the anti-ulcer functionality. The studies by Nyam et al. (2016) and Cheong et al. (2018b) showed the potential of kenaf seed oil and kenaf seed oil-in-water nanoemulsion as an anti-ulcer agent to scavenge free radicals in the body and reduce the risk of getting peptic ulcer disease.

8.5 Anti-inflammation

Inflammation is a physiological body response that results in the local accumulation of plasmic fluid and blood cells when the body meets the stimuli like infections and injury. Nyam et al. (2015) investigated the anti-inflammatory activity of kenaf seed oil, kenaf seed extract, roselle seed oil, and roselle seed extract using the methods of histamine-induced paw edema, carrageenan-induced paw edema, and arachidonic acid-induced paw edema in the male Sprague dawley rats. Treatment of kenaf seed oil (500 mg/kg) showed 19.9–36.4% of inhibition in the histamine-induced paw edema within the 5 h after injection, 4.1–57.5% of inhibition in the carrageenan-induced paw edema within the 5 h after injection, and 3.5–32.5% of inhibition in the arachidonic acid-induced paw edema within the 5 h after injection. On the other hand, the treatment of indomethacin (5 mg/kg), which is an anti-inflammatory drug showed the inhibition percentage of 4.0–56.7% with an average of 36.7% in three of the induced paw edema models within the 5 h after injection. Thus, kenaf seed oil has shown an inhibitory effect in the inflammation reaction in the induced paw edema in the rats.

Kenaf seed oil is composed of various bioactive compounds such as phenolics (108.5 mg GAE/100 g), flavonoids (52.9 mg catechin/100 g), saponins (68.1 mg saponin/100 g), terpenoids (148.8 mg linalool/100 g), alkaloids (17.4%) (Nyam et al. 2015), tocopherols and phytosterols. The bioactive photochemical are believed to contribute to the anti-inflammatory effect. Phenolic compounds such as flavonoids, cinnamic acid, ferulic acid, caffeic acid, chlorogenic acid and steroids possess anti-inflammatory property, according to the previous study (Borrelli et al. 2002). Phenolic acids are responsible for reducing the pro-inflammatory cytokines level that responsible for the inflammation reaction. Besides this, phenolic compounds will also influence the metabolism of arachidonic acid, alter signal transduction route, and control the pro-inflammatory gene expression that gives anti-inflammatory activity (Miguel 2010). Phenolic compounds showed to inhibit the activity of nitric oxide that stimulates the physiological and pathological response as chronic inflammation (Joseph et al. 2009; Kim et al. 2014).

Saponin is high potential to prevent the acute inflammation, as shown by the study of carrageenan-induced paw edema in the rats (Gepdiremen et al. 2005). Tocopherols (α- and γ-) are showed to take part in the anti-inflammatory activity *in vitro* and *in vivo*, by inhibiting the activity of the kinase, which accounted in the inflammation response. α-Tocopherol will inhibit the activity of protein kinase C in vascular smooth muscle cells to prevent the muscle inflammation (Reiter et al. 2007).

8.6 Anti-thrombotic

Thrombosis is a condition of formation of blood clotting in the blood circulatory system. Anti-platelet and anti-coagulant are the two main drugs available in the market to treat thrombosis. The drugs are used to prevent clotting and platelet clumping. It will lead to fatal in the serious case for vascular blockage and acute coronary disorders. The unsaturated fatty acids and bioactive compounds presented in the kenaf seed oil are suggested to exert the anti-thrombotic effect. PUFA, tocopherols, and phenolic compounds are helped to modulate the platelet aggregation (Natella et al. 2005).

Phenolic compounds offer the anti-thrombotic activity by restraining platelet aggregation and interaction between platelet and leukocyte. Flavonoids can reduce the thromboxane A2 level in the blood and inhibit the activity of cyclooxygenase and lipoxygenase, which will reduce the platelet aggregation. Nutrients supplement with anti-hypercholesterolemic, anti-inflammatory, antioxidants and blood thinning effects can help to reduce the risk of thrombosis (Cheng et al. 2016). Hence, kenaf seed oil possesses with these advantageous is suggested as a potential supplement to prevent thrombosis.

9   Other Issues

9.1   Acute Toxicity of Kenaf Seed Oil

Previous literature reported kenaf seed oil extracted from the kenaf seeds is edible with no toxicity. Foo et al. (2012) conducted an acute toxicity study of kenaf seed oil on the mice. Intake of kenaf seed oil emulsion (1.0, 2.0, 3.0, 4.0 and 5.0 g/kg body weight) did not result in any mortality or clinical symptoms of toxicity at all the concentrations in the mice. The weight loss percentage (relative to the initial starting weight) was calculated and a decrease of weight of more than 15% is considered toxic. The results revealed that no weight loss greater than 15% was showed in all the mice groups, even at the highest concentration (5.0 g/kg). Thus, kenaf seed oil is shown to safe and no toxicity (Foo et al. 2012).

9.2   Co-product Uses

Kenaf seed extract and defatted kenaf seed meal showed an anti-hypercholesterolemic effect in the rat study (Ng et al. 2015). High phenolic acids (5880.6 mg/100 g) was identified in the kenaf seed extract, in which tannic acid, sinapic acid, catechin, gallic acid, 4-hydroxybenzaldehyde, 4-hydroxybenzoic acid, ferulic acid, syringic acid, naringin, and protocatechuic acid were identified (Wong et al. 2014b). Thus, kenaf seed extract had studied for its biological activities and functionality due to its high phenolic content that can be used as therapeutic properties. Kenaf seed extract showed anti-inflammatory (Nyam et al. 2015), anti-ulcer activity (Nyam et al. 2016), and cytotoxic activity towards human cervical cancer, breast cancer, colon cancer, and lung cancer cell lines (Wong et al. 2014a). Kenaf seed extract can also be added as a natural antioxidant in the vegetable oil to improve the oxidative stability of the oil during storage (Nyam et al. 2013).

Massive extraction of kenaf seed oil will produce large amounts of defatted kenaf seed meal, which has been identified with high protein (26.2 g/100 g DKSM) and carbohydrate (57.1 g/100 g DKSM) contents, as well as high phenolic and flavonoid content in DKSM. Gallic acid, catechin, 4-hydroxybenzoic acid, vanillic acid, and syringic acid were identified as the phenolic acids in DKSM. DKSM has been recommended to use as edible four with promising nutritional value and antioxidant activity (Chan et al. 2013). Besides that, DKSM can be used as a source of livestock and animal feed due to its high protein content (Mariod et al. 2010). Besides cooking flour, kenaf seeds can also use for lubrication, making soap, linoleum, paints, and varnishes (LeMahieu et al. 2003). The methyl esters of kenaf seed oil are met to the biodiesel standards for fuel properties, so the kenaf oil methyl esters are potential to be used as biodiesel (Knothe et al. 2013).

References

Ahsan, H., Ahad, A., & Siddiqui, W. A. (2015). A review of characterization of tocotrienols from plant oils and foods. *Journal of Chemical Biology, 8*, 45–49.

Araújo, J. M. A., & Sandi, D. (2007). Extraction of coffee diterpenes and coffee oil using supercritical carbon dioxide. *Food Chemistry, 101*, 1087–1094.

Ashori, A., Harun, J., Raverty, W., & Yusoff, M. (2006). Chemical and morphological characteristics of Malaysian cultivated kenaf (*Hibiscus cannabinus*) fibre. *Polymer-Plastics Technology and Engineering, 45*(1), 131–134.

Awad, A. B., & Fink, C. S. (2000). Phytosterols as anticancer dietary components: Evidence and mechanism of action. *Journal of Nutrition, 130*, 2127–2130.

Awad, A. B., Chan, K. C., Downie, A. C., & Fink, C. S. (2000). Peanuts as a source of beta-sitosterol, a sterol with anticancer properties. *Nutrition Cancer, 36*, 238–241.

Ayadi, R., Hamrouni, L., Hanana, M., Bouzid, S., Trifi, M., & Khouja, M. L. (2011). In vitro propagation and regeneration of an industrial plant kenaf (*Hibiscus cannabinus* L.). *Industrial Crops and Products, 33*, 474–480.

Bachari-Saleh, Z., Ezzatpanah, H., Aminafshar, M., & Safafar, H. (2013). The effect of refining process on the conjugated dienes in soybean oil. *Journal of Agricultural Science and Technology, 15*, 1185–1193.

Baigent, C., Keech, A., Kearney, P. M., & Blackwell, L. (2005). Efficacy and safety of cholesterol lowering treatment: Prospective meta-analysis of data from 90,056 participants in 14 randomised trials of statins. *The Lancet, 366*, 1267–1278.

Balasundram, N., Sundram, K., & Samman, S. (2006). Phenolic compounds in plants and agri-industrial by-products: Antioxidant activity, occurrence, and potential uses. *Food Chemistry, 99*, 191–203.

Bartosińska, E., Buszewska-Forajta, M., & Siluk, D. (2016). GC-MS and LC-MS approaches for determination of tocopherols and tocotrienols in biological and food matrices. *Journal of Pharmaceutical and Biomedical Analysis, 127*, 156–169.

Borrelli, F., Mafia, P., Pinto, L., Lanaro, A., Russo, A., Capasso, F., & Ialenti, A. (2002). Phytochemical compounds involved in the anti-inflammatory effect of propolis extract. *Fitoterapia, 73*(1), S53–S63.

Boschin, G., & Arnoldi, A. (2011). Legumes are valuable sources of tocopherols. *Food Chemistry, 127*, 1199–1203.

Bruscatto, M. H., Zambiazi, R. C., Sganzerla, M., Pestana, V. R., Otero, D., Lima, R., & Paiva, F. (2009). Degradation of tocopherols in rice bran oil submitted to heating at different temperatures. *Journal of Chromatographic Science, 47*, 762–765.

Burton, G. W. (1994). Vitamin E: Molecular and biological function. *The Proceedings of the Nutrition Society, 53*, 251–262.

Chan, K. W., & Ismail, M. (2009). Supercritical carbon dioxide fluid extraction of *Hibiscus cannabinus* L. seed oil: A potential solvent-free and high antioxidative edible oil. *Food Chemistry, 114*, 970–975.

Chan, K. W., Khong, N. M. H., Iqbal, S., Mansor, S. M., & Ismail, M. (2013). Defatted kenaf seed meal (DKSM): Prospective edible flour from agricultural waste with high antioxidant activity. *LWT-Food Science and Technology, 53*, 306–313.

Chen, Y. Z., Kao, S. Y., Jian, H. C., Yu, Y. M., Li, J. Y., Wang, W. H., & Tsai, C. W. (2015). Determination of cholesterol and four phytosterols in foods without derivatization by gas chromatography-tandem mass spectrometry. *Journal of Food and Drug Analysis, 23*, 636–644.

Cheng, W. Y., Akanda, J. M. H., & Nyam, K. L. (2016). Kenaf seed oil: A potential new source of edible oil. *Trends in Food Science and Technology, 52*, 57–65.

Cheong, A. M., & Nyam, K. L. (2016). Improvement of physical stability of kenaf seed oil-in-water nanoemulsions by addition of β-cyclodextrin to primary emulsion containing sodium caseinate and Tween 20. *Journal of Food Engineering, 183*, 24–31.

Cheong, A. M., Tan, K. W., Tan, C. P., & Nyam, K. L. (2016a). Improvement of physical stability properties of kenaf (*Hibiscus cannabinus* L.) seed oil-in-water nanoemulsions. *Industrial Crops and Products, 80*, 77–85.

Cheong, A. M., Tan, K. W., Tan, C. P., & Nyam, K. L. (2016b). Kenaf (*Hibiscus cannabinus* L.) seed oil-in-water Pickering nanoemulsions stabilised by mixture of sodium caseinate, Tween 20 and β-cyclodextrin. *Food Hydrocolloids, 52*, 934–941.

Cheong, A. M., Tan, C. P., & Nyam, K. L. (2016c). *In-vitro* gastrointestinal digestion of kenaf seed oil-in-water nanoemulsions. *Industrial Crops and Products, 87*, 1–8.

Cheong, A. M., Tan, C. P., & Nyam, K. L. (2016d). *In vitro* evaluation of the structural and bioaccessibility of kenaf seed oil nanoemulsions stabilised by binary emulsifiers and β-cyclodextrin complexes. *Journal of Food Engineering, 189*, 90–98.

Cheong, A. M., Tan, C. P., & Nyam, K. L. (2017). Physicochemical, oxidative and anti-oxidant stabilities of kenaf seed oil-in0water nanoemulsions under different storage temperatures. *Industrial Crops and Products, 95*, 374–382.

Cheong, A. M., Koh, J. X. J., Patrick, N. O., Tan, C. P., & Nyam, K. L. (2018a). Hypocholesterolemic effects of kenaf seed oil, macroemulsion, and nanoemulsion in high-cholesterol diet induced rats. *Journal of Food Science.* https://doi.org/10.1111/1750-3841.14038. (In press).

Cheong, A. M., Tan, Z. W., Patrick, N. O., Tan, C. P., Lim, Y. M., & Nyam, K. L. (2018b). Improvement of gastroprotective and anti-ulcer effect of kenaf seed oil-in-water nanoemulsions in rats. *Food Science and Biotechnology.* https://doi.org/10.1007/s10068-018-0342-0. (In press).

Chew, S. C., & Nyam, K. L. (2016a). Oxidative stability of microencapsulated kenaf seed oil using co-extrusion technology. *Journal of American Oil Chemist's Society, 93*(4), 607–615.

Chew, S. C., & Nyam, K. L. (2016b). Microencapsulation of kenaf seed oil by co-extrusion technology. *Journal of Food Engineering, 175*, 43–50.

Chew, S. C., Tan, C. P., Long, K., & Nyam, K. L. (2015). *In-vitro* evaluation of kenaf seed oil in chitosan coated-high methoxyl pectin-alginate microcapsules. *Industrial Crops and Products, 76*, 230–236.

Chew, S. C., Tan, C. P., Long, K., & Nyam, K. L. (2016). Effect of chemical refining on the quality of kenaf (*Hibiscus cannabinus*) seed oil. *Industrial Crops and Products, 89*, 59–65.

Chew, S. C., Tan, C. P., & Nyam, K. L. (2017a). Optimization of degumming parameters in chemical refining process to reduce phosphorus contents in kenaf seed oil. *Separation and Purification Technology, 188*, 379–385.

Chew, S. C., Tan, C. P., & Nyam, K. L. (2017b). Optimization of neutralization parameters in chemical refining of kenaf seed oil by response surface methodology. *Industrial Crops and Products, 95*, 742–750.

Chew, S. C., Tan, C. P., & Nyam, K. L. (2017c). Optimization of bleaching parameters in refining process of kenaf seed oil with a central composite design model. *Journal of Food Science, 82*, 1622–1630.

Chew, S. C., Tan, C. P., & Nyam, K. L. (2017d). Application of response surface methodology for optimizing the deodorization parameters in chemical refining of kenaf seed oil. *Separation and Purification Technology, 184*, 144–151.

Chew, S. C., Tan, C. P., Lai, O. M., & Nyam, K. L. (2017e). Changes in 3-MCPD esters, glycidyl esters, bioactive compounds and oxidation indexes during kenaf seed oil refining. *Food Science and Biotechnology*. https://doi.org/10.1007/s10068-017-0295-8. (In press).

Chew, S. C., Tan, C. P., & Nyam, K. L. (2017f). Comparative storage of crude and refined kenaf (*Hibiscus cannabinus* L.) seed oil during accelerated storage. *Food Science and Biotechnology, 26*, 63–69.

Chew, S. C., Tan, C. P., & Nyam, K. L. (2018). *In-vitro* digestion of refined kenaf seed oil micro-encapsulated in β-cyclodextrin/gum arabic/sodium caseinate by spray drying. *Journal of Food Engineering, 225*, 34–41.

Codex Alimentarius Commission. (1982). *Recommended internal standards edible fats and oils*. Rome: FAO/WHO.

Coetzee, R., Labuschagne, M. T., & Hugo, A. (2008). Fatty acid and oil variation in seed from kenaf (*Hibiscus cannabinus* L.). *Industrial Crops and Products, 27*, 104–109.

Cunha, S. S., Fernandes, J. O., & Oliveira, M. B. (2006). Quantification of free and esterified sterols in Portuguese olive oils by solid-phase extraction and gas chromatography–mass spectrometry. *Journal of Chromatography A, 1128*, 220–227.

Dempsey, J. M. (1975). *Fiber crops*. Gainesville: The University Presses of Florida.

EFSA. (2016). Risks for human health related to the presence of 3- and 2-monochloropropanediol (MCPD), and their fatty acid esters, and glycidyl fatty acid esters in food. *EFSA Journal, 14*(5), 4426.

Ergönül, P. G., & Köseoğlu, O. (2014). Changes in α-, β-, γ- and δ-tocopherol contents of mostly consumed vegetable oils during refining process. *CyTA- Journal of Food, 12*, 199–202.

Ermacora, A., & Hrnčiřík, K. (2014). Development of an analytical method for the simultaneous analysis of MCPD esters and glycidyl esters in oil-based foodstuffs. *Food Additves and Contaminants: Part A, 31*, 985–994.

Ezhilarasi, P. N., Karthik, P., Chhanwal, N., & Anandharamakrishnan, C. (2013). Nanoencapsulation techniques for food bioactive components: A review. *Food and Bioprocess Technology, 6*, 628–647.

FAO. (2017). *Jute, Kenaf, sisal, abaca, coir and allied Fibres. Statistical bulletin 2016* (p. 6). Rome: Food and Agricultural Organization of the United Nations.

Farhoosh, R., Einafshar, S., & Sharayei, P. (2009). The effect of commercial refining steps on the rancidity measures of soybean and canola oils. *Food Chemistry, 115*, 933–938.

Farr, W. E. (2000). Refining of fats and oils. In R. D. O'Brien, W. E. Farr, & P. I. Wan (Eds.), *Introduction to fats and oils technology* (pp. 136–157). Urbana: AOCS Press.

Foo, J. B., Yazan, L. S., Chan, K. W., Md Tahir, P., & Ismail, M. (2011). Kenaf seed oil from supercritical carbon dioxide fluid extraction induced g1 phase cell cycle arrest and apoptosis in leukemia cells. *African Journal of Biotechnology, 10*(27), 5389–5397.

Foo, J. B., Yazan, L. S., Mansor, S. M., Ismail, N., Md Tahir, P., & Ismail, M. (2012). Kenaf seed oil from supercritical carbon dioxide fluid extraction inhibits the proliferation of WEHI-3B leukemia cells in vivo. *Journal of Medicinal Plants Research, 6*, 1429–1436.

Franke, K., Strijowski, U., Fleck, G., & Pudel, F. (2009). Influence of chemical refining process and oil type on bound 3-chloro-1, 2-propanediol contents in palm oil and rapeseed oil. *LWT-Food Science and Technology, 42*, 1751–1754.

Gepdiremen, A., Mshvildadze, V., Süleyman, H., & Elias, R. (2005). Acute anti-inflammatory activity of four saponins isolated from ivy: Alpha-hederin, hederasaponin-C, hederacolchiside-E and hederacolchiside-F in carrageenan-induced rat paw edema. *Phytomedicine, 12*, 440–444.

Ghafar, S. A. A., Ismail, M., Yazan, L. S., Fakurzi, S., Ismail, N., Chan, K. W., & Tahir, P. M. (2013). Cytotoxic activity of kenaf seed oils from supercritical carbon dioxide fluid extraction towards human colorectal cancer (HT29) cell lines. *Evidence-Based Complementary and Alternative Medicine, 2013*, 1–8.

Ghazani, S. M., & Marangoni, A. G. (2013). Minor components in canola oil and effects of refining on their constituents: A review. *Journal of American Oil Chemist's Society, 90*, 923–932.

Gliszczyńska-Świglo, A., Sikorska, E., Khmelinskii, I., & Sikorski, M. (2007). Tocopherol content in edible plant oils. *Polish Journal of Food and Nutrition Sciences, 57*, 157–161.

Gümez-Alonso, S., Fregapane, G., Salvador, M. D., & Gordon, M. H. (2003). Changes in phenolic composition and antioxidant activity of virgin olive oil during frying. *Journal of Agricultural and Food Chemistry, 51*, 667–672.

Guner, F. S., Yusuf, Y., & Erciyes, A. T. (2006). Polymeres from triglyceride oils. *Progress in Organic Coatings, 31*, 633–670.

Gunstone, F. D. (2004). *The chemistry of oils and fat: Sources, composition, properties and uses.* Boca Raton: CRC Press LLC.

Hamlet, C. G., Asuncion, L., Velíšek, J., Doleţal, M., Zelinková, Z., & Crews, C. (2011). Formation and occurrence of esters of 3-chloropropane-1,2-diol (3-CPD) in foods: What we know and what we assume. *European Journal of Food Science and Technology, 113*, 279–303.

Hubbard, N. E., Lim, D., Summers, L., & Erickson, K. L. (2000). Reduction of murine mammary tumor metastasis by conjugated linoleic acid. *Cancer Letters, 150*, 93–100.

Joseph, S., Sabulal, B. V., George, V., Smina, T. P., & Janardhanan, K. K. (2009). Antioxidative and anti-inflammatory activites of the chloroform extract of *Ganoderma lucidum* found in South India. *Scientia Pharmaceutica, 77*, 111–121.

Karabulut, I., Topcu, A., Yorulmaz, A., Tekin, A., & Ozay, D. S. (2005). Effect of the industrial refining process on some properties of hazelnut oil. *European Journal of Lipid Science and Technology, 107*, 476–480.

Khanna, S., Parinandi, N. L., Kotha, S. R., Roy, S., Rink, C., Bibus, D., & Sen, C. K. (2010). Nanomolar vitamin E α-tocotrienol inhibits glutamate induced activation of phospholipase A2 and causes neuroprotection. *Journal of Neurochemistry, 112*, 1249–1260.

Kim, C. H., Park, M. K., Kim, S. K., & Cho, Y. H. (2014). Antioxidant capacity and anti-inflammatory activity of lycopene in watermelon. *International Journal of Food Science and Technology, 49*, 2083–2091.

Knothe, G., Razon, L. F., & Bacani, F. T. (2013). Kenaf oil methyl esters. *Industrial Crops and Products, 49*, 568–572.

Kreps, F., Vrbiková, L., & Schmidt, Š. (2014). Influence of industrial physical refining on tocopherol, chlorophyll and beta-carotene content in sunflower and rapeseed oil. *European Journal of Lipid Science and Technology, 116*, 1572–1582.

Kritchevsky, D. (2000). Antimutagenic and some other effects of conjugated linoleic acid. *British Journal of Nutrition, 83*, 459–465.

Kritchevsky, D., & Chen, S. C. (2005). Phytosterols-health benefits and potential concerns-a review. *Nutrition Research, 25*, 413–428.

Lampi, A. M., Kataja, L., Eldin, A. K., & Vieno, P. (1999). Antioxidant activities of α- and γ-tocopherols in the oxidation of rapeseed oil triacylglycerols. *Journal of American Oil Chemist's Society, 76*, 749–755.

Lee, S. Y., Jung, M. Y., & Yoon, S. H. (2014). Optimization of the refining process of camellia seed oil for edible purposes. *Food Science and Biotechnology, 23*, 65–73.

Legal Research Board. (2013). *Food Act 1983 (Act 281) & Regulations.* Malaysia: International Law Book Services.

LeMahieu, P. J., Oplinger, E. S., & Putnam, D. H. (2003). Kenaf. Alternative field crops manual [Online]. Available from: http://www.corn.agronomy.wisc.edu/FISC/Alternatives/Kenaf.htm. Accessed 10 Nov 2015.

Lewy, M. (1947). Kenaf seed oil. *Journal of the American Oil Chemists' Society, 24*, 3–5.

Li, D., Saldeen, T., Romeo, F., & Mehta, J. L. (1999). Relative effects of alpha- and gamma-tocopherol on low-density lipoprotein oxidation and superoxide dismutase and nitric oxide synthase activity and protein expression in rats. *Journal of Cardiovascular Pharmacology and Therapeutics, 4*, 219–226.

Li, C., Li, L., Jia, H., Wang, Y., Shen, M., Nie, S., & Xie, M. (2016). Formation and reduction of 3-monochloropropane-1,2-diol esters in peanut oil during physical refining. *Food Chemistry, 199*, 605–611.

MacDonald, H. B. (2000). Conjugated linoleic acid and disease prevention: A review of current knowledge. *Journal of the American College of Nutrition, 19*, 111–118.

Mariod, A. A., Fathy, S. F., & Ismail, M. (2010). Preparation and characterisation of protein concentrates from defatted kenaf seed. *Food Chemistry, 123*, 747–752.

Mariod, A. A., Matthäus, B., & Ismail, M. (2011). Comparison of supercritical fluid and hexane extraction methods in extracting kenaf (*Hibiscus cannabinus*) seed oil lipids. *Journal of Americal Oil Chemist's Society, 88*, 931–935.

Miguel, M. G. (2010). Antioxidant and anti-inflammatory activities of essential oils: A short review. *Molecules, 15*, 9252–9287.

Mohamed, A., Bhardwaj, H., Hamama, A., & Webber, C. (1995). Chemical composition of kenaf (*Hibiscus cannabinus* L.) seed oil. *Industrial Crops and Products, 4*, 157–165.

Morello, J. R., Vuorela, S., Romero, M. P., Motilva, M. J., & Heinonen, M. (2005). Antioxidant activity of olive pulp and olive oil phenolic compounds of the arbequina cultivar. *Journal of Agricultural and Food Chemistry, 53*(6), 2002–2008.

Natella, F., Nardini, M., Virgili, F., & Scaccini, C. (2005). Role of dietary polyphenols in the platelet aggregation network- a review of the *in vitro* studies. *Current Topics in Nutraceutical Research, 4*(1), 2–21.

National Kenaf and Tobacco Board. (2014). Profile NKTB [Online]. Available from: http://www.lktn.gov.my/index.php/en/about-us/nktb-history. Accessed 2 Dec 2017.

Nes, W. D., Fuller, G., & Tsai, L. S. (1984). *Isopentenoids in plants: Biochemistry and function* (p. 325). New York: Marcel Dekker.

Ng, S. K., Lau, J. L. Y., Tan, C. P., Long, K., & Nyam, K. L. (2013a). Effect of accelerated storage on microencapsulated kenaf seed oil. *Journal of American Oil Chemist's Society, 90*, 1023–1029.

Ng, S. K., Wong, P. Y., Tan, C. P., Long, K., & Nyam, K. L. (2013b). Influence of the inlet air temperature on the microencapsulation of kenaf (*Hibiscus cannabinus* L.) seed oil. *European Journal of Lipid Science and Technology, 115*, 1309–1318.

Ng, S. K., Choong, Y. H., Tan, C. P., Long, K., & Nyam, K. L. (2014). Effect of total solids content in feed emulsion on the physical properties and oxidative stability of microencapsulated kenaf seed oil. *LWT-Food Science and Technology, 58*, 627–632.

Ng, S. K., Tee, A. N., Lai, E. C. L., Tan, C. P., Long, K., & Nyam, K. L. (2015). Anti-hypercholesterolemic effect of kenaf (*Hibiscus cannabinus* L.) seed on high-fat diet Sprague dawley rats. *Asian Pacific Journal of Tropical Medicine, 8*, 6–13.

Nyam, K. L., Tan, C. P., Lai, O. M., Long, K., & Yaakob, C. M. (2009). Physicochemical properties and bioactive compounds of selected seed oils. *LWT-Food Science and Technology, 42*, 1396–1403.

Nyam, K. L., Wong, M. M., Long, K., & Tan, C. P. (2013). Oxidative stability of sunflower oils supplemented with kenaf seeds extract, roselle seeds extract and roselle extract, respectively under accelerated storage. *International Food Research Journal, 20*(2), 695–701.

Nyam, K. L., Sin, L. N., & Long, K. (2015). Phytochemical analysis and anti-inflammatory effect of kenaf and roselle seeds. *Malaysian Journal of Nutrition, 22*(2), 245–254.

Nyam, K. L., Tang, J. L. K., & Long, K. (2016). Anti-ulcer activity of *Hibiscus cannabinus* and *Hibiscus sabdariffa* seeds in ulcer-induced rats. *International Food Research Journal, 23*(3), 1164–1172.

Ortega-García, J., Gámez-Meza, N., Noriega-Rodriguez, J. A., Dennis-Quiñonez, O., García-Galindo, H. S., Angulo-Guerrero, J. O., & Medina-Juárez, L. A. (2006). Refining of high oleic safflower oil: Effect on the sterols and tocopherols content. *European Food Research and Technology, 223*, 775–779.

Pascoal, A., Quirantes-Piné, R., Fernando, A. L., Alexopoulou, E., & Segura-Carretero, A. (2015). Phenolic composition and antioxidant activity of kenaf leaves. *Industrial Crops and Products, 78*, 116–123.

Phoon, M. C., Desbordes, C., Howe, J., & Chow, V. T. K. (2001). Linoleic and linolelaidic acids differentially influence proliferation and apoptosis of MOLT-4 leukaemia cells. *Cell Biology International, 25*, 777–784.

Player, M. E., Kim, H. J., Lee, H. O., & Min, D. B. (2006). Stability of α-, γ- or δ-tocopherol during soybean oil oxidation. *Journal of Food Science, 71*, 456–460.

Ramesh, M. (2016). Kenaf (*Hibiscus cannabinus* L.) fibre based bio-materials: A review on processing and properties. *Progress in Materials Science, 78–79*, 1–92.

Ravinder, T., Kaki, S. S., Kanjilal, S., Rao, B. V. S. K., Swain, S. K., & Prasad, R. B. N. (2015). Refining of castor and tapioca leaf fed eri silkworm oils. *International Journal of Chemical Science and Technology, 5*(2), 32–37.

Razon, L. F., Bacani, F. T., Evangelista, R. L., & Knothe, G. (2013). Fatty acid profile of kenaf seed oil. *Journal of Americal Oil Chemist's Society, 90*, 835–840.

Reiter, E., Jiang, Q., & Christen, S. (2007). Anti-inflammatory properties of α- and γ-tocopherol. *Molecular Aspects of Medicine, 28*, 668–691.

Rossi, M., Gianazza, M., Alamprese, C., & Stanga, F. (2001). The effect of bleaching and physical refining on colour and minor components of palm oil. *Journal of the American Oil Chemists' Society, 78*, 1051–1055.

Ruiz, M. L., Castillo, D., Dobson, D., Brennan, R., & Gordon, S. (2002). Genotypic variation in fatty acid content of blackcurrant seed. *Journal of Agricultural and Food Chemistry, 50*, 332–335.

Ryu, S. W., Jin, C. W., Lee, H. S., Lee, J. Y., Sapkota, K., Lee, B. G., Yu, C. Y., Lee, M. K., Kim, M. J., & Cho, D. H. (2006). Changes in total polyphenol: Total flavonoid contents andantioxidant activities of *Hibiscus cannabinus* L. *Korean Journal of Medicinal Crop Science, 14*, 307–310.

Sachin, S. S., & Archana, R. J. (2009). Antiulcer activity of methanol extract of *Erythrina indica* Lam. Leaves in experimental animals. *Pharmacognosy Research, 1*, 396–401.

Saldeen, T., Li, D., & Mehta, J. L. (1999). Differential effects of alpha- and gamma- tocopherol on low-density lipoprotein oxidation, superoxide activity, platelet aggregation and arterial thrombogenesis. *Journal of the American College of Cardiology, 34*, 1208–1215.

Sánchez-Machado, D. I., López-Cervantes, J., Núñez-Gastélum, J. A., Mora- López, G. S., López-Hernández, J., & Paseiro-Losada, P. (2015). Effect of the refining process on *Moringa oleifera* seed oil quality. *Food Chemistry, 187*, 53–57.

Sanguansri, P., & Augustin, M. A. (2006). Nanoscale materials development – a food industry perspective. *Trends in Food Science and Technology, 17*(10), 547–556.

Schwarz, H., Ollilainen, V., Piironen, V., & Lampi, A.-M. (2008). Tocopherol, tocotrienol and plant sterol contents of vegetable oils and industrial fats. *Journal of Food Composition and Analysis, 21*(2), 152–161.

Scott, A. W., Jr., & Taylor, C. S. (1990). Economics of kenaf production in the lower Rio Grande Valley of Texas. In J. Janick & J. E. Simon (Eds.), *Advances in new crops* (pp. 292–297). Portland: Timber Press.

Sen, C. K., Khanna, S., & Roy, S. (2007). Tocotrienols in health and disease: The other half of the natural vitamin E family. *Molecular Aspects of Medicine, 28*, 692–728.

Smith, T. J. (2000). Squalene: Potential chemopreventive agent. *Expert Opinion on Investigational Drugs, 9*, 1841–1848.

Suliman, T. E. M. A., Jiang, J., & Liu, Y. F. (2013). Chemical refining of sunflower oil: Effect on oil stability, total tocopherol, free fatty acids and colour. *International Journal of Engineering Science and Technology, 5*(2), 449–454.

Tan, C. P., Che Man, Y. B., Selamat, J., & Yusoff, M. S. A. (2002). Comparative studies of oxida-tive stability of edible oils by differential scanning calorimetry and oxidative stability index methods. *Food Chemistry, 76*, 385–389.

Tilak, K. S., Veeraiah, K., & Koteswara Rao, D. K. (2001). Restoration on tissue antioxidants by fenugreek seeds (*Trigonella foenum* Graecum) in alloxan-diabetic rats. *Indian Journal Physiology Pharmacology, 45*, 408–420.

Tyagi, K., Ansari, M. A., Tyagi, S., & Tyagi, A. (2012). A novel process for physically refining rice bran oil through degumming. *Advances in Applied Science Research, 3*, 1435–1439.

Vaquero, E. M., Beltrán, S., & Sanz, M. T. (2006). Extraction of fat from pigskin with supercritical carbon dioxide. *The Journal of Supercritical Fluids, 37*, 142–150.

Verleyen, T., Sosinska, U., Ioannidou, S., Verhé, R., Dewettinck, K., Huyghebaert, A., & De Greyt, W. (2002). Influence of the vegetable oil refining process on free and esterified sterols. *Journal of American Oil Chemist's Society, 79*, 947–953.

Villanueva, M. J., Yokoyama, W. H., Hong, Y. J., Barttley, G. E., & Rupérez, P. (2011). Effect of high-fat diets supplemented with okara soybean by-product on lipid profiles of plasma, liver and faeces in Syrian hamsters. *Food Chemistry, 124*, 72–79.

Wang, T., & Johnson, L. A. (2001). Refining high-free fatty acid wheat germ oil. *Journal of American Oil Chemist's Society, 78*(1), 71–76.

Webber, C. L., III, & Bledsoe, V. K. (2002). Kenaf yield components and plant components. In J. Janick & A. Whipkey (Eds.), *Trends in new crops and new use* (pp. 348–357). Alexandria: ASHS Press.

Wei, J., Chen, L., Qiu, X. Y., Hu, W. J., Sun, H., Chen, X. L., Bai, Y. Q., Gu, X. Y., Wang, C. L., Chen, H., Hu, R. B., Zhang, H., & Shen, G. X. (2015). Optimizing refining temperature to reduce the loss of essential fatty acids and bioactive compounds in tea seed oil. *Food and Bioproducts Processing, 94*, 136–146.

Wong, Y. H., Tan, W. Y., Tan, C. P., Long, K., & Nyam, K. L. (2014a). Cytotoxic activity of kenaf (*Hibiscus cannabinus* L.) seed extract and oil against human cancer cell lines. *Asian Pacific Journal of Tropical Biomedicine, 4*(Supp 1), S510–S515.

Wong, Y. H., Lau, H. W., Tan, C. P., Long, K., & Nyam, K. L. (2014b). Binary solvent extraction system and extraction time effects on phenolic antioxidants from kenaf seeds (*Hibiscus can-nabinus* L.) extracted by a pulsed ultrasonic-assisted extraction. *The Scientific World Journal, 2014*, 1–7.

Wong, Y. H., Muhamad, H., Abas, F., Lai, O. M., Nyam, K. L., & Tan, C. P. (2017). Effects of temperature and NaCl on the formation of 3-MCPD esters and glycidyl esters in refined, bleached and deodorized palm olein during deep-fat frying of potato chips. *Food Chemistry, 219*, 126–130.

Yazan, L. S., Foo, J. B., Ghafar, S. A. A., Chan, K. W., Md Tahir, P., & Ismail, M. (2011a). Effect of kenaf seed oil from different ways of extraction towards ovarian cancer cells. *Food and Bioproducts Processing, 89*, 328–332.

Yazan, L. S., Foo, J. B., Chan, K. W., Md Tahir, P., & Ismail, M. (2011b). Kenaf seed oil from supercritical carbon dioxide fluid extraction shows cytotoxic effects towards various cancer cell lines. *African Journal of Biotechnology, 10*, 5381–5388.

Zacchi, P., & Eggers, R. (2008). High-temperature pre-conditioning of rapeseed: A polyphenol-enriched oil and the effect of refining. *European Journal of Lipid Science and Technology, 110*, 111–119.

Zelinková, Z., Novotný, O., Schůrek, J., Velíšek, J., Hajslová, J., & Doleţal. (2008). Occurrence of 3-MCPD fatty acid esters in human breast milk. *Food Additives and Contaminants: Part A, 25*(6), 669–676.

Zelinková, Z., Doleţal, M., & Velíšek, J. (2009). Occurrence of 3-chloropropane-1,2-diol fatty acid esters in infant and baby foods. *European Food Research and Technology, 228*(4), 571–578.

Zufarov, O., Schmidt, S., & Sekretár, S. (2008). Degumming of rapeseed and sunflower oils. *Acta Chimica Slovenica, 1*, 321–328.

Chapter 24
Apple (*Malus pumila*) Seed Oil

Ali Abbas, Farooq Anwar, and Naveed Ahmad

Abstract Apple (*Malus pumila*) is one of the most common sweet fruits consumed by humans either directly in fresh or in processed form. Apple has an impressive range of medicinal benefits due to its rich level of high-value components and nutrients. Apple fruit contains a considerable amount of seeds, which are considered as agro-waste. The seeds contain appreciable content of oil, which is a good source of antioxidants (tocopherols) and other bioactive components. The apple seed oil is characterized by a high level of linoleic acid and oleic acid. The seed oil also contains different phytosterols with β-sitosterol as the most prevalent component. Because of acceptable physicochemical properties and nutrients composition, the apple seed oil has significant potential food and nutra-pharmaceutical for applications. This chapter describes the overall quality and biochemical composition and food applications of apple seed oil.

Keywords Rosaceae · Fruit oil · Linoleic acid · β-sitosterol · Tocopherols

1 Introduction

Apple is widely used as common table fruit due to its nutritional and health benefits. It is included in the list of the most widely consumed fruits all over the world. A member of Rosaceae family and Maloideae sub-family, apple has around 28 genera, and 1100 species worldwide (Arain et al. 2012). The apple fruit is commonly cultivated in North America and Europe, but is also distributed in several other parts of the world (Rohrer et al. 1994). The total production of apple all over the world in

A. Abbas
Department of Chemistry, Government Postgraduate Taleem-ul-Islam College,
Chenab Nagar, Chiniot, Pakistan

F. Anwar (✉)
Department of Chemistry, University of Sargodha, Sargodha, Pakistan
e-mail: farooq.anwar@uos.edu.pk

N. Ahmad
Department of Chemistry, University of Education, Faisalabad, Pakistan

© Springer Nature Switzerland AG 2019
M. F. Ramadan (ed.), *Fruit Oils: Chemistry and Functionality*,
https://doi.org/10.1007/978-3-030-12473-1_24

2016 was estimated to be 89,329,179 mt. Apple is also widely cultivated in Northern and Western regions of Pakistan, especially 1000 m above the sea level in the colder areas of the country such as Ziarat, Pishin, Mustang, Quetta, Kashmir, Kalat, Hunza, Swat, and Chitral (Tareen et al. 2003).

Apple is recognized as an important food commodity with high nutritional value. The fruit contains a high range of high-value nutrients including carbohydrates, minerals, vitamins and proteins in a balanced quantity. A wide range of food products can be prepared using apple fruit such as jellies, jams, marmalades, salads, pickles, pudding, slice and sauces. Apple cider, a fermented apple juice, can be prepared using the sour varieties of apple (Hulme 1970). As a natural functional food, apple fruit possess multiple biological and nutraceutical properties and their consumption is linked with the reduced prevalence of different chronic diseases, which can be prevented by the direct use of apples and its juice (Boyer and Liu 2004; Hamauzu et al. 2005; Pearson et al. 1999).

The bioactive phytochemicals such as phenolics in apple have been demonstrated to show antioxidant, antimicrobial and anti-proliferative effects against human cancer cells (Manzoor et al. 2012; Wolfe et al. 2003; Oleszek et al. 1988; He and Liu 2008; Liu et al. 2005).

2 Apple Seeds and Seed Oil

During the production of apple cider, two important byproducts are obtained (i.e. seeds and skins). In the cider industry, the seeds produced are treated as a waste material and are not used for the production of oil or value-addition (Bada et al. 2014). The apple seeds (Figs. 24.1 and 24.2) contain a considerable amount of edible fixed oil that can be used in the cosmetics, oleo-chemicals, and food industries (Stone and Kushner 2000; Kris-Etherton and Etherton 2003). A number of

Fig. 24.1 Apple fruit with seeds inside

Fig. 24.2 Seeds of apple
fruit

researchers have published data on the physicochemical properties of apple seed oil
(Lei-Tian et al. 2010; Yu et al. 2007; Yukui et al. 2009). Different variables such as
agro-climatic and varietal/genetic factors are reported to affect the contents and
chemical characteristics of apple seed oil (Marjan et al. 2007; Yu et al. 2007; Yukui
et al. 2009). On the other hand, volatile components can be obtained from seeds
using head space volatile trapping or solvent extraction (Nieuwenhuizen et al.
2013). Apple fruits are reported to have a list of VOCs (volatile organic compounds)
which include ketones, esters, alcohols and aldehydes (Dixon and Hewett 2000;
Dimick and Hoskin 1983). Although terpenes contribute a little portion of total
VOCs, however different terpenes are identified from apple (Rowan et al. 2009;
Hern and Dorn 2003; Fuhrmann and Grosch 2002). There are various factors that
can change the composition of VOCs including fruit ripening stages, cultivar, environmental conditions and maturity (Rapparini et al. 2001; Dixon and Hewett 2000;
Vallat et al. 2005).

3   Extraction and Fatty Acid Composition of Apple Seed Oil

The vegetable oils from oilseeds and/or fruit seeds can be obtained using a simple
technique such as Soxhlet extraction using *n*-hexane or petroleum ether as extraction solvent (Arain et al. 2012; Matthäus and Özcan 2015). Similarly, the fixed oil
can be obtained from apple seed using the classical methods of extraction such as
Soxhlet extraction using different solvents. Arain et al. (2012) and Adebayo et al.
(2012) used *n*-hexane, while Matthäus and Özcan (2015) used petroleum ether for
extraction of apple seed oil. Gao et al. (2007) used ultrasound-associated extraction
for the isolation of oil from an apple seed. The percentage yield of oil reported by
Adebayo et al. (2012) was 10.6–10.8%, while Arain et al. (2012) reported the oil
content as high as 26.8–28.9%. In another study, Matthäus and Özcan (2015)

evaluated the seed oil content of two apple varieties, golden and starking and accounted for 21.9% and 25.6%, respectively. The oil content reported by Lei-Tian et al. (2010) from apple seeds ranged from 20.6% to 24.3%, while Yukui et al. (2009) reported 29.0%.

Linoleic acid and oleic acid are the principal fatty acids detected in the apple seed oil (Marjan et al. 2007; Yu et al. 2007; Yukui et al. 2009). The composition of fatty acids from different varieties of apple seed oil is given in Table 24.1. Arain et al. (2012) found linoleic acid as a major fatty acid in three different varieties including Royal Gala, Red Delicious and Pyrus Malus (45.1%, 47.8%, and 49.6%, respectively), while the same fatty acid was found in less percentage (40.5%) in Golden Delicious variety. Golden Delicious variety contain oleic acid (45.5%) as a major component. A relatively less percentage (39.3%) of oleic acid was obtained from Red Delicious as compared to other varieties. Pyrus Malus contained a lower percentage of palmitic acid (6.1%), stearic acid (3.1%) and ecosenoic acid (1.0%). Some other fatty acids were also noted in trace amount (<1%) including linolenic, docosanoic, heptadecanoic, palmitoleic and 11-ecosenoic acids.

Matthäus and Özcan (2015) studied the fatty acid composition of selected fruit seeds including two varieties of apple seeds (i.e. Golden and Starking). They found the major fatty acid as a linoleic acid with the contribution of 51.7% and 48.1% followed by oleic acid with 35.7% and 40.4%, respectively. The other fatty acids were palmitic acid, stearic acid and gondoic acid. In another study on apple seed oil (from the origin of Asturias and Spain) conducted by Bada et al. (2014), the main component was linoleic acid in the Limon Montes species (60.7%) and Riega species (60.0%). Other researchers noted a relatively smaller amount of linoleic acid (Lei-Tian et al. 2010; Yukui et al. 2009). The concentration of oleic acid and stearic acid detected in apple seed oil of Collaos species was 36.5% and 2.30%, respectively. This data agreed with those investigated by Yu et al. (2007) for the characterization of apple seed oil. Górnaś et al. (2014) found that apple seed oil mainly contained oleic acid (20.6–29.0%), palmitic acid (5.78–8.33%) and linoleic acid (59.3–67.9%). Such a difference in the composition of a fatty acid of apple seed oil may be due to the genetic makeup as well as the agro-climatic features of fruit (Minnocci et al. 2010). These data revealed that the apple seed oils are one of the potential sources of linoleic and oleic acids. Lipids, including linoleic acid and oleic acids, have a health-promoting effect (Finley and Shahidi 2001). It could be concluded that due to the presence of favorable fatty acid profile, the apple seed oil has an edible potential (Yukui et al. 2009).

A balanced diet contains oils and fats as one of the major ingredients because they provide physical energy and essential fatty acids. The vegetable sources of lipids contribute approximately three-fifths of the total world's consumption of fats and oils and the remaining from marine and terrestrial animal (Adebayo et al. 2012). Plant-based foods such as plant seeds are a healthy source of fat for animals and humans because they contain mainly essential fatty acids (*omega* fatty acids). Vegetable oils are mainly produced from some conventional oilseed crops including palm, rapeseed, cotton, soybean, sesame, sunflower and flaxseed although there are numerous non-conventional sources of these oils (Adebayo et al. 2012).

Table 24.1 Fatty acid composition of different apple seed oils

Sr. #	Variety/cultivars	Palmitic (C16:0)	Palmitoleic (C16:1)	Heptadecanoic (C17:0)	Stearic (C18:0)	Oleic (C18:1)	Linoleic (C18:2)	Linolenic (C18:3)	Eicosanoic (C20:0)	Reference
1	Royal Gala	7.4	0.1	0.1	2.5	41.7	45.1	0.3	1.7	Arain et al. (2012)
	Red Delicious	6.7	0.1	0.1	2.3	39.3	47.8	0.3	2.0	
	Pyrus Malus	6.1	0.2	0.0	2.0	38.7	49.6	0.4	0.9	
	Golden Delicious	7.1	0.1	0.1	3.1	45.5	40.5	0.3	2.0	
2	Golden Apple	7.0	0.1	–	1.9	35.7	51.7	0.6	–	Matthäus and Özcan (2015)
	Starking Apple	6.8	0.1	–	2.0	40.4	48.1	0.3	–	
3	Blanquina	8.5	0.1	–	1.9	32.7	54.0	0.3	1.5	Bada et al. (2014)
	Raxao	8.7	0.1	–	2.0	30.5	56.3	0.2	1.3	
	Collaos	8.1	0.1	–	2.3	36.6	50.3	0.2	1.5	
	Durona	8.3	0.1	–	1.9	34.3	53.0	0.3	1.3	
	Riega	9.2	0.1	–	1.8	27.0	60.0	0.3	1.1	
	Solarina	9.0	0.1	–	2.0	31.8	54.5	0.4	1.3	
	Limon Montes	8.9	0.1	–	1.8	27.0	60.0	0.3	1.1	
4	Fuji	6.5	–	–	1.8	37.5	50.7	–	1.5	Lei Tian et al. (2010)
	New Red Star	6.6	–	–	2.0	38.5	51.4	–	1.5	

4 Bioactive Components in Apple Seed Oil

Two types of lipid components are reported in the apple seed oil, including major component (triglycerides) and minor bioactive components (tocopherols and phytosterols). Apple seed oil, likewise other vegetable oils, is mainly consisted of triglycerides. The composition of triglycerides is made up of LLP in the range of 41.1–39.3% followed by LLL in the range of 27.1–17.8%. The low concentration of triglycerides was SOL in the range of 0.81–0.50% and PPO in the range of 0.31–0.17%. Overall, the triglycerides of unsaturated fatty acids are predominant as compared to triglycerides of saturated fatty acids (Bada et al. 2014).

The minor bioactive components of the seed oil contain tocopherols and phytosterols as important constituents, which not only enhance the oil stability at high temperatures but also act as a polymerization reaction inhibitors (Velasco and Dobarganes 2002). The phytosterols, with β-sitosterol as the dominating component, are linked with lowering the cholesterol and thus contribute towards reducing the incidence of heart diseases (Vivacons and Moreno 2005).

Earlier studies investigated the profile of minor components of apple seed oil. Arain et al. (2012) reported that β-sitosterol was predominant in all varieties of apple seed oil in contrast to other phytosterols. The highest level of β-sitosterol was found in the seed oil of Royal Gala species (16.1%) followed by Red Delicious seed oil (14.7%), Pyrus Malus seed oil (13.6%) and Golden Delicious seed oil (15.9%). The seed oil of Golden Delicious contained a maximum percentage of 9,19-cyclolanost-24-en-3-ol (4.8%) as compared to other studied varieties. This component acts as an intermediate in the production of plant sterols biosynthetically (Kamisak et al. 1987). Stigmast-4-en-3-one was found in higher amount in the seed oil of Pyrus Malus (4.6%) relative to other varieties and this was reported to have hypoglycemic effects (Alexander-Lindo et al. 2004). A minor amount (<1%) of avenasterol and campesterol were recorded in all the studied varieties of apple seed oils. In another study, Bada et al. (2014) reported the sterols content in the apple seed oils from different origins and found that β-sitosterol was the major sterol in all the selected varieties of an apple seed. Another variety, Solarina apple seed oil contained the maximum amount of β-sitosterol at levels as high as 558.52 mg 100 g^{-1} oil. However, campesterol had the highest content in Limon montes apple seed oil at a level of 49.25 mg 100 g^{-1}. The other major components identified were stigmasterol and Δ-5 avenasterol.

Tocopherols, as natural antioxidant components, are recognized for their effectiveness to increase the oil stability along with other health-promoting effects (Traber and Atkinson 2007; Herrera and Barbas 2001). Four different derivatives of tocopherols and tocotrienols (α-, β-, γ-, and δ-), collectively known as tocols, are reported in different vegetable oils. The main function of tocols includes the defense of polyunsaturated fatty acids (PUFA) against peroxidation (Kamal-Eldin and Andersson 1997; Beringer and Dompert 1976). Biological effects of these compounds are mainly because of vitamin-E-active constituents. They protect against the lipid peroxidation; a serious problem in the living system, which relates to different health disorders such as aging, inflammation and certain cancers (Stephens et al. 1996). The biological properties such as the antioxidant activity of α-tocopherol have been well established, the other isomers, especially γ-tocopherol is also important due to the fact that this compound exhibits additional biological effects (Jiang et al. 2000, 2002). It has been revealed that γ-tocopherol possess relatively a stronger anticancer activity than α-tocopherol (Hensley et al. 2004). Arain et al. (2012) reported the highest amount of α-tocopherol (6.4%) in Royal Gala, while β-tocopherol (1.8%) was major in Golden Delicious variety. Bada et al. (2014) found that the major component in apple seed oil was β-tocopherol (125.2 mg kg^{-1} oil) followed by α-tocopherol (84.68 mg kg^{-1} oil).

Besides phytosterols and tocopherols, in the study of the unsaponifiable lipid fraction of apple seed oil, ethyl oleate was found to be an important constituent along with octadecanoic acid, hexadecanoic acid, squalene, 9-hexadecenal and hydrocarbons such as 1-hexacosene, 3-eicosene, 1-docasene, docosane and octacosane (Arain et al. 2012). The unsaponifiable lipid fraction of vegetable oil contains squalene as a major hydrocarbon, which is about 90% of total hydrocarbon (Lanzón et al. 1994). Squalene prevents the lipid peroxidation of human skin and reduces the amount of triglyceride levels in hypercholesterolemia and low-density lipoproteins (Kohno et al. 1995; Kelly 1999).

Phytol, a minor component of the human diet, is found in almost all plants and act as a major component to produce vitamins E and K$_1$. The cellular and biological attributes of phytols have been studied (Christiane et al. 1986; Hibasami et al. 2002). Pyrus Malus seed oil contained the lowest amount of phytol (0.6%) as compared to other varieties. The results showed that the seed oil of different apple varieties contains a different concentration of minor components (Arain et al. 2012).

5 Applications

Plant seed oils can be used in different ways such as cooking as well as for oleochemicals preparations. They are employed in the food industry for manufacturing of fats and oil for cooking purposes, in the dressing of ice-cream and salads (Elaine 1975). There are a variety of plant seeds and nuts that are in use for the extraction of vegetable oils to be used for food, cosmetics, nutraceuticals, industrial lubricants and biofuels preparations along with aromatherapy applications. Vegetable oils have also been used for illumination and lubricating purpose and to produce detergents, cosmetics and for coatings and paint for many centuries (Ibemesi 1992). Apple seed oil, which is rich in linoleic acid and also contained an appreciable amount of antioxidant nutrients have significant potential for the edible oil industry. Besides, apple seed oil was reported to possess interesting in vitro biological activities, emerging from its chemical composition. Apple seed oil has been noted to possess antibacterial activity. Yeasts were more sensitive to the apple seed oil than mildews with MIC value between 0.3 and 0.6 mg/mL. The studied antimicrobial and antioxidant activities support the potential uses of apple seed oil for nutra- pharmaceutical industry (Lei-Tian et al. 2010).

References

Adebayo, S. E., Orhevba, B. A., Adeoye, P. A., Musa, J. J., & Fase, O. J. (2012). Solvent extraction and characterization of oil from African star apple (*Chrysophyllum albidum*) seeds. *Academic Research International, 3*(2), 178–183.

Alexander-Lindo, R. L., Morrison, E. Y., & Nair, M. G. (2004). Hypoglycaemic effect of stigmast-4-en-and its corresponding alcohol from the of *Anacardium occidentale* (cashew). *Phytotherapy Research, 18*, 403–407.

Arain, S., Sherazi, S. T. H., Bhanger, M. I., Memon, N., Mahesar, S. A., & Rajput, M. T. (2012). Prospects of fatty acid profile and bioactive composition from lipid seeds for the discrimination of apple varieties with the application of chemometrics. *Grasas y Aceites, 63*(2), 175–183.

Bada, J. C., Leon-Camacho, M., Copovi, P., & Alonso, L. (2014). Characterization of apple seed oil with denomination of origin from Asturias, Spain. *Grasas y Aceites, 65*(2), e027. https://doi.org/10.3989/gya.109813.

Beringer, H., & Dompert, W. U. (1976). Fatty acid and tocopherol pattern in oil seeds. *Fette Seifen Anstrichm, 78*, 228–231.

Boyer, J., & Liu, R. H. (2004). Apple phytochemicals and their health benefits. *Nutrition Journal, 3*(5), 1–15.

Christiane, V. B., Joseph, V., Ingrid, W., & Frank, R. (1986). Phytol and peroxisome proliferation. *Pediatric Research, 20*, 411–415.

Dimick, P. S., & Hoskin, J. C. (1983). Review of apple flavor – state of the art. *Critical Reviews in Food Science and Nutrition, 18*, 387–409.

Dixon, J., & Hewett, E. W. (2000). Factors affecting apple aroma/flavour volatile concentration: A review. *New Zealand Journal of Crop and Horticultural Science, 28*, 155–173.

Elaine, M. A. (1975). *A Unilever educational booklet*. Information division, Unilever Limited. London, UK.

Finley, J. W., & Shahidi, F. (2001). The chemistry, processing and health benefits of highly unsaturated fatty acids: an overview, Omega-3 fatty acids, chemistry, nutrition and health effects. *American Chemical Society Washington, 1*, 258–279.

Fuhrmann, E., & Grosch, W. (2002). Character impact odorants of the apple cultivars Elstar and Cox Orange. *Nahrung, 46*, 187–193.

Gao, X., Qiu, N. X., Pang, F. K., & Liu, X. Y. (2007). Ultrasound-associated extraction of apple seed oil. *Chinese Journal of Oil Crop Sciences, 29*(1), 78.

Górnaś, P., Rudzińska, M., & Segliņa, D. (2014). Lipophilic composition of eleven apple seed oils: A promising source of unconventional oil from industry by-products. *Industrial Crops and Products, 60*, 86–91.

Hamauzu, Y., Yasui, H., Inno, T., Kume, C., & Omanyuda, M. (2005). Phenolic profile, antioxidant property, and antiinfluenza viral activity of Chinese quince (*Pseudocydonia sinensis* Schneid.), quince (*Cydonia oblonga* Mill.), and apple (*Malus domestica* Mill.) fruits. *Journal of Agriculture and Food Chemistry, 53*, 928–934.

He, X., & Liu, R. H. (2008). Phytochemicals of apple peels: Isolation, structure elucidation, and their antiproliferative and antioxidant activities. *Journal of Agricultural and Food Chemistry, 56*(21), 9905–9910.

Hensley, K., Benaksas, E. J., Bolli, R., Comp, P., Grammas, P., Hamdheydari, L., Mou, S., Pye, Q. N., Stoddard, M. F., Wallis, G., & Williamson, K. S. (2004). New perspectives on vitamin E: γ-tocopherol and carboxyethylhydroxychroman metabolites in biology and medicine. *Free Radical Biology and Medicine, 36*(1), 1–15.

Hern, A., & Dorn, S. (2003). Monitoring seasonal variation in apple fruit volatile emissions in situ using solid- phase microextraction. *Phytochemical Analysis, 14*, 232–240.

Herrera, E., & Barbas, C. (2001). Vitamin E: Action, metabolism and perspectives. *Journal of Physiology and Biochemistry, 57*, 43–56.

Hibasami, H., Kyohkon, M., Ohwaki, S., Katsuzaki, H., Imai, K., Ohnishi, K., Ina, K., & Komiya, T. (2002). Diol- and triol-types of phytol induce apoptosis in lymphoid leukemia Molt 4B cells. *International Journal of Molecular Medicine, 10*, 555–559.

Hulme, A. C. (1970). The biochemistry of fruits and their products. *Academic Press London and New York, 1*, 376–377.

Ibemesi, J. A. (1992). Vegetable oils as industrial raw materials. *The Nigerian Perspective* (A Monograph). Enugu: SNAAP Press.

Jiang, Q., Elson-Schwab, I., Courtemanche, C., & Ames, B. N. (2000). γ-Tocopherol and its major metabolite, in contrast to α-tocopherol, inhibit cyclooxygenase activity in macrophages and epithelial cells. *Proceedings of the National Academy of Sciences, 97*, 11494–11499.

Jiang, Q., Lykkesfeldt, J., Shigenaga, M. K., Shigeno, E. T., Christen, S., & Ames, B. N. (2002). γ-Tocopherol supplementation inhibits protein nitration and ascorbate oxidation in rats with inflammation. *Free Radical Biology and Medicine, 33*, 1534–1542.

Kamal-Eldin, A., & Andersson, R. A. (1997). A multivariate study of the correlation between tocopherol content and fatty acid composition in vegetable oils. *Journal of the American Oil Chemists' Society, 74*, 375–380.

Kamisak, W., Honda, C., Suwa, K., & Isoi, K. (1987). Studies of 13C NMR spectra of 13C-enriched cycloartenol biosynthesized from [1-13C]-,[2-13C]-and [1, 2-13C2]-acetate. Revised 13C NMR spectral assignments of cycloartenol and cycloartanol and 13C NMR spectral support for the generally accepted skeleton formation mechanism of cycloartenol. *Magnetic Resonance in Chemistry, 25*(8), 683–687.

Kelly, G. S. (1999). Squalene and its potential clinical uses. *Alternative Medicine Review: A Journal of Clinical Therapeutic, 4*, 29–36.

Kohno, Y., Egawa, Y., Itoh, S., Nagaoka, S., Takahashi, M., & Mukai, K. (1995). Kinetic study of quenching reaction of singlet oxygen and scavenging reaction of free radical by squalene in *n*-butanol. *Biochimica et Biophysica Acta (BBA)-Lipids and Lipid Metabolism, 1256*, 52–56.

Kris-Etherton, P. M., & Etherton, T. D. (2003). The impact of the changing fatty acid profile on fats on the diet assessment and health. *Journal of Food Composition and Analysis, 16*, 373–378.

Lanzón, A., Albi, T., Cert, A., & Gracian, J. (1994). The hydrocarbon fraction of virgin olive oil and changes resulting from refining. *Journal of the American Oil Chemists' Society, 71*, 285–291.

Lei-Tian, H., Zhan, P., & Li, K. X. (2010). Analysis of components and study on antioxidant and antimicrobial activities of oil in apple seeds. *International Journal of Food Science and Nutrition, 61*, 395–403.

Liu, R. H., Liu, J., & Chen, B. (2005). Apples prevent mammary tumors in rats. *Journal of Agriculture and Food Chemistry, 53*, 2341–2343.

Manzoor, M., Anwar, F., Saari, N., & Ashraf, M. (2012). Variations of antioxidant characteristics and mineral contents in pulp and peel of different apple (*Malus domestica* Borkh.) cultivars. *Molecules, 17*, 390–407.

Marjan, S., Melita, K., Janez, H., & Rajko, V. (2007). Influence of cultivar and storage time on the content of higher fatty acids. *Vegetable Crops Research Bulletin, 66*, 197–203.

Matthäus, B., & Özcan, M. (2015). Oil content, fatty acid composition and distributions of vitamin-E-active compounds of some fruit seed oils. *Antioxidants, 4*(1), 124–133.

Minnocci, A., Iacopini, P., Martinelli, F., & Sebastiani, L. (2010). Micromorphological, bio-chemical, and genetic characterization of two ancient, late-bearing apple varieties. *Journal of Horticulture Science, 75*, 1–7.

Nieuwenhuizen, N. J., Green, S. A., Chen, X., Bailleul, E. J., Matich, A. J., Wang, M. Y., & Atkinson, R. G. (2013). Functional genomics reveals that a compact terpene synthase gene family can account for terpene volatile production in apple. *Plant Physiology, 161*(2), 787–804.

Oleszek, W., Lee, C. Y., Jaworski, A. W., & Price, K. R. (1988). Identification of some phenolic compounds in apples. *Journal of Agricultural and Food Chemistry, 36*(3), 430–432.

Pearson, D. A., Tan, C. H., German, J. B., Davis, P. A., & Gershwin, M. E. (1999). Apple juice inhibits human low density lipoprotein oxidation. *Life Sciences, 64*, 1913–1920.

Rapparini, F., Baraldi, R., & Facini, O. (2001). Seasonal variation of monoterpene emission from *Malus domestica* and *Prunus avium*. *Phytochemistry, 57*(5), 681–687.

Rohrer, J. R., Robertson, K. R., & Phipps, J. B. (1994). Floral morphology of Maloideae (Rosaceae) and its systematic relevance. *American Journal of Botany, 81*(5), 574–581.

Rowan, D. D., Hunt, M. B., Alspach, P. A., Whitworth, C. J., & Oraguzie, N. C. (2009). Heritability and genetic and phenotypic correlations of apple (Malus x domestica) fruit volatiles in a genetically diverse breeding population. *Journal of Agricultural and Food Chemistry, 57*(17), 7944–7952.

Stephens, N. G., Parsons, A., Brown, M. J., Schofield, P. M., Kelly, F., Cheeseman, K., & Mitchinson, M. J. (1996). Randomised controlled trial of vitamin E in patients with coronary disease: Cambridge Heart Antioxidant Study (CHAOS). *The Lancet, 347*(9004), 781–786.

Stone, N. J., & Kushner, R. (2000). Effects of dietary modification and treatment of obesity: Emphasis on improving vascular outcomes. *Medical Clinics of North America, 84*(1), 95–122.

Tareen, M. J., Tareen, A. Q., Kamal, J. A. Sıddıquın, B. N. (2003). Influence of MM-106 and M-9 root stocks on Starking delicious apple. *International Journal of Agriculture & Biology, 5*(3), 339–340.

Traber, M. G., & Atkinson, J. (2007). Vitamin E, antioxidant and nothing more. *Free Radical Biology and Medicine, 43*(1), 4–15.

Vallat, A., Gu, H., & Dorn, S. (2005). How rainfall, relative humidity and temperature influence volatile emissions from apple trees in situ. *Phytochemistry, 66*(13), 1540–1550.

Velasco, J., & Dobarganes, C. (2002). Oxidative stability of virgin olive oil. *European Journal of Lipid Science and Technology, 104*(9–10), 661–676.

Vivancos, M., & Moreno, J. J. (2005). β-Sitosterol modulates antioxidant enzyme response in RAW 264.7 macrophages. *Free Radical Biology and Medicine, 39*(1), 91–97.

Wolfe, K., Wu, X., & Liu, R. H. (2003). Antioxidant activity of apple peels. *Journal of Agricultural and Food Chemistry, 51*(3), 609–614.

Yu, X., van de Voort, F. R., Li, Z., & Yue, T. (2007). Proximate composition of the apple seed and characterization of its oil. *International Journal of Food Engineering, 3*(5), 1–8.

Yukui, R., Wenya, W., Rashid, F., & Qing, L. (2009). Fatty acids composition of apple and pear seed oils. *International Journal of Food Properties, 12*(4), 774–779.

Chapter 25
Apricot (*Prunus armeniaca* L.) Oil

Mustafa Kiralan, Gülcan Özkan, Erdogan Kucukoner,
and M. Mustafa Ozcelik

Abstract The apricot (*Prunus armeniaca* L.) is an important agricultural crop that widely cultivated in most of the Mediterranean and Central Asian countries. As known, the fruit of apricot has an important place in human nutrition, and can be consumed as fresh or processed. World apricot production is about 2.5 million tonnes. However, apricot kernels are produced as byproducts and often considered a waste product of fruits processing industry. They have potential to be economically-valuable resource, since they are a rich source of dietary protein as well as fiber. In addition, the kernels are considered as potential sources of oils. Apricot kernels have a high oil yield, which is comparable to the commonly used oils of oilseed crops such as soybean, canola and sunflower. Oil from these kernels can be obtained by solvent extraction or cold pressing method. The oil contains a high percentage of unsaturated fatty acids and is a rich source of minor compounds such as sterols, tocochromanols and squalene, hence attracting interest for the utilization in food and pharmaceutical industry. Due to its nutritional chemical composition and functional properties, apricot kernel oil can be used as edible oil and in many applications like food products formulation, cosmetics as well as functional and medicinal supplements. In this chapter, particular attention has also been given to the composition and applications of kernel oil.

Keywords Kernel oil · Bioactive compounds · Functional properties

M. Kiralan (✉)
Faculty of Engineering, Department of Food Engineering, Balıkesir University,
Balıkesir, Turkey

G. Özkan · E. Kucukoner · M. M. Ozcelik
Faculty of Engineering, Department of Food Engineering, Suleyman Demirel University,
Isparta, Turkey
e-mail: gulcanozkan@sdu.edu.tr; erdogankucukoner@sdu.edu.tr

© Springer Nature Switzerland AG 2019
M. F. Ramadan (ed.), *Fruit Oils: Chemistry and Functionality*,
https://doi.org/10.1007/978-3-030-12473-1_25

1 Introduction

Apricot (*Prunus armeniaca* L.) is classified under the Prunus species of Prunodae sub family of the Rosaceae family of the Rosales group and is widely cultivated in the Mediterranean and Central Asia. The plant is well known for its medicinal and economical importance. The high production of apricot fruits generates a huge quantity of apricot seeds (Femenia et al. 1995; Yıldız 1994; Hummer and Janick 2009; Manzoor et al. 2012). The largest producer of Apricots is Turkey (730.000 tonnes annually). World apricot production is about 3.8 million tonnes/year. The top ten apricot producers of the world are Turkey, Uzbekistan, Iran, Algeria, Italy, Pakistan, Spain, France, Egypt and Japan respectively. First five countries Algeria, Iran, Italy, Pakistan, Turkey and Uzbekistan produced more than 170.000 tonnes as an annual average (FAOSTAT 2018). The production data for last 5 years were given below (Fig. 25.1).

As known, the fruit of apricot is not only consumed fresh but also used to produce dried apricot, frozen apricot, jam, jelly, marmalade, pulp, juice, nectar, and extrusion products (Yıldız 1994). Moreover, apricot kernels are used in the production of oils, benzaldehyde, cosmetics, active carbon, and aroma perfume (Yıldız 1994).

Apricot is rich in minerals such as potassium and vitamins such as β-carotene. β-carotene, which is the pioneer substance of vitamin A, is necessary for epithelia tissues covering our bodies and organs, eye-health, bone and teeth development and

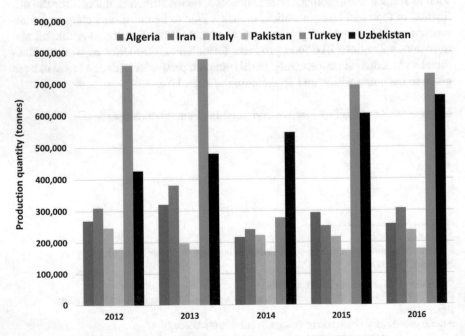

Fig. 25.1 Production quantity of apricots according to countries (FAOSTAT 2018)

working of endocrine glades. Moreover, vitamin A plays important role in reproduction and growing functions of our bodies, in increasing body resistance against infections (Hacıseferoğulları et al. 2007). Physicochemical properties of apricot could be changed according to variety, location, and harvest time. Dry matter content of different apricots varieties (Hacıhaliloğlu, Hasanbey, Soğancı, Kabaaşı, Çöloğlu, Çataloğlu, Hacıkız, Tokaloğlu and Alyanak apricot varieties) from Turkey are ranged from 11.8% to 25.8%. Ash contents of the apricots varied from 0.50% to 0.89%. Apricot fruits contain more minerals including macro (K > Mg > Ca > P > Na) and micro (Fe > Zn > Mn > Ni > Se) elements. Sucrose is a major sugar and its content of apricot varieties changed from 22.96 to 56.83 mg/100 g of dry weight. Apricot fruits are also rich in functional components such as phenolics, carotenoids and organic acids. Total phenolics and total carotenoids content changed between 4233.70 and 8180.49 mg of gallic acid equivalents (GAE) per 100 g of dry weight and 14.83–91.89 mg of β-carotene equivalents per 100 g of dry weight, respectively. Malic acid was determined as the predominant organic acid in all apricot varieties, ranging from 973.4 to 2341.1 mg/100 g of dry weight. Otherwise, the varieties had considerable amount of citric acid (7697.3 and 9997.1 mg/100 g of dry weight) (Akin et al. 2008).

Apricot (*Prunus armeniaca* L.) kernels are by-products and a large number of fruit seeds are usually discarded by the food processing industry (El-Adawy and Taha 2001; Schieber et al. 2001). The kernel is added to bakery products as the whole kernel or grounded and also consumed as appetizers. Apricot kernel is an important source of oil (49.8–56.1%) as well as dietary protein (22.4–29.3%) and fiber. The kernel contain essential amino acids and high concentrations of potassium and magnesium minerals, B group vitamins and oil. The oil is high in unsaturated fatty acids, especially oleic acid (31–80%), linoleic acid (6.3–51%), and γ-tocopherol (Femenia et al. 1995). Therefore, both the oil and kernel from the fruit seeds could be utilized.

In recent years, apricot kernels research receives growing interest due to their high amount of oil (>50%) on a dry basis (Slover et al. 1983; Femenia et al. 1995; Uluata 2016) and has a high concentration of hydrophilic and lipophilic bioactive components (Schieber et al. 2001). Apricot kernel oils are edible and can be used for salads and frying oil compared to the oil yield of some commercial seed oils such as soybean (15–20%), cottonseed (30–36%), sunflower (36–55%) (Shariatifar et al. 2017). The oil is quite similar to almond oil and its applications are also similar. This oil has a deep, nutty flavor so this makes it an excellent ingredient for desserts and other recipes (CBI 2014). Apricot oil has been used in Germany and the United States in preparing fixed oil and macaroon paste (Femenia et al. 1995).

Apricot kernel oil has been also enjoyed throughout history due to their unique nutrient unsaturated fatty acid profile and bioactive components as natural antioxidant are proving to play a beneficial role in many areas of health such as cardiovascular diseases, cancer, tumors and ulcers (Mensink and Katan 1987; Timmermann 1990; Sies and Murphy 1991; De Jong et al. 2003; Quezada 2003; Hensley et al. 2004; Alpaslan and Hayta 2006; Bendini et al. 2007; Ramadan and Moersel 2006; Chen et al. 2008; Warleta et al. 2010; Nikokavouraa et al. 2011).

2 Extraction and Processing of Fruit Oil

The oil from apricot kernels is generally extracted by traditional methods such as cold pressing and solvent extraction, or combination of these methods (Uluata 2016). In addition, different modern extraction techniques such as ultrasound-assisted extraction (Gayas et al. 2017; Dolatowski et al. 2007), supercritical fluid extraction (Özkal et al. 2005) and aqueous enzymatic oil extraction (Sharma and Gupta 2006) can be used to get high-quality oil.

Cold-pressing technique is simple and does not require much energy. In addition, heating and chemical treatments are not used in the technique that can provide advantages (Siger et al. 2007). While extraction method had no effect on the color of oil and fatty acid composition, the oil yield of this technique is lower than hot pressing and solvent extraction. On the other hand, cold-pressed oils contain more biologically active substances (natural antioxidants) such as polar phenolics compounds and apolar tocopherols (Siger et al. 2007; Bhatnagar and Gopala Krishna 2014; Prescha et al. 2014; Uluata 2016). Most apricot kernel oils are sold cold-pressed and you can also find refined oil. Cold-pressed apricot kernel oil is mid-yellow colored with a characteristic smell and taste (apricot), whilst the refined product is pale to mid-yellow, with a milder taste and smell (CBI 2014).

Ultrasound treatment is one of the modern solvent extraction methods, developed as an alternate to the Soxhlet. Ultrasound-assisted extraction is a non-thermal and novel processing technology to extract biologically active components and is considered one of the most used methods to enhance mass transfer phenomena (Corrales et al. 2008). Gayas et al. (2017) suggested that ultrasonic-assisted extraction could be a good option for the extraction of oil from apricot kernels and they found the optimum conditions of temperature, time and a solvent to sample ratio were 51.7 °C, 43.95 min and 19.8:01, respectively. Apperantly, ultrasound offers advantages over traditional analytical technique because measurements are rapid, precise and fully automated (Zbigniew et al. 2007). The method is called green extraction technique because of reducing the use of solvents as well as shortens the time of extraction.

Supercritical fluid extraction has emerged as a striking separation technique because it does not introduce any residual organic chemicals (Patil et al. 2013). Supercritical CO_2 is the most commonly used supercritical fluid (Matricardi et al. 2002), another fluid is food grade butane (Sanders 2001). Effects of process parameters on the extraction of apricot kernel oil with supercritical carbon dioxide (SC-CO_2) were investigated by Özkal et al. (2005). In the study, the parameters included particle size (mean particle diameter <0.425–1.5 mm), solvent flow rate (1–5 g/min), pressure (300–600 bar), temperature (40–70 °C) and co-solvent concentration (up to 3.0 wt. % ethanol). About 99% apricot kernel oil recovery was possible and mass transfer coefficient in the fluid phase increased with a decrease in particle size and pressure, and with an increase in solvent flow rate, temperature and ethanol concentration. Supercritical CO_2 extraction is the excellent technique for nonpolar analytes and CO_2 is readily available at low cost and has low toxicity

(Patil et al. 2013). Even though SC-CO$_2$ has a poor solubility for polar compounds, especially phenolics compounds as natural antioxidants (Prescha et al. 2014; Uluata 2016). Modification such as adding a small amount of ethanol and methanol enable it to extract biologically active polar compounds and shelf life of apricot kernel oil can more extend.

Aqueous enzymatic oil extraction is also one of the ecofriendly modern extraction methods and based on simultaneous isolation of oil and protein from oilseed by dispersing finely ground seed in water and separating the dispersion by centrifugation into oil, solid, and aqueous phases. The presence of certain enzymes during extraction enhances oil recovery by breaking cell walls and oil bodies (Rosendahl et al. 1996; Zhang et al. 2011). Sharma and Gupta (2006) were used a combined method of aqueous enzymatic oil extraction with ultrasonic pre-irradiation to extract the oil from almond and apricot seeds. They found that the enzymatic method of a mixture of three proteases with ultrasonic pre-irradiation to extract oil from apricot kernel reduced the extraction time (about 6 h) and marginally increased the oil yield (77% w/w).

3 Fatty Acid Composition and Acyl Lipids

Apricot oil contains mainly triacylglycerols (98%), followed by phospholipids (1.1%) and free fatty acids (0.2%), while unsaponifiable matters are 0.7% of the oil weight (Zlatanov and Janakieva 1998). Triglyceride profiles of apricot oil are shown in Table 25.1. There are differences in triglyceride composition due to incompletely separation of some triglicerides. Hassanein (1999) demonstrated that triglyceride profile showed the presence of OOO (35–42%), LOO (22–28%), LLO (7–16%), LOP and LLS (6–7%), OOP (6–10.4%). The other studies in Turkish cultivars; Turan et al. (2007) reported that major triglycerides were mixture of OOO+POO (42.38–57.23%), followed by OOL+POL (28.47–34.91%), while Uluata (2016) reported OOO (39.8–39.9%), OOL (24.2–24.4%), OLL (10.2–10.4%), POS (9.7%), and POO (9.1–9.2%).

The fatty acid composition of oil and distribution of major fatty acids according to cultivars and locations are exhibited Table 25.2. Oleic acid and linoleic acid, in that order, consist of more than 80% of all fatty acids in oils of apricot seeds. The major fatty acid is oleic acid in the oils from different cultivars ranging between 62.34% and 80.97%. The other important fatty acid, linoleic acid varied from 13.33% to 30.33%. The highest and the lowest oleic acid and linoleic acid content were determined in the oils from Pakistan cultivars. The most abundant of the saturated fatty acids are palmitic acid (3.35–5.93%).

Beside to fatty acids in oils, the fatty acid composition of fractionated glycerides from oil is examined. Triglyceride and diacylglycerol fractions contain a higher percentage of oleic acid which accounted for 72.1% and 62.7%, respectively, whereas these fractions were rich in terms of linoleic acid (22.2% and 17.8%, respectively) (Hassanein 1999).

Table 25.1 Triglyceride composition of apricot oil

Triglycerides	1	2
LLL	1.1	1.80–3.49
LLO	7.6	–
OLL	–	10.19–17.67
LLP	1.0	–
PLL	–	0.06–0.17
LOO	22.0	–
OOL+POL	–	28.47–34.91
OOO+POO	–	42.38–57.23
LLS+LOP	6.9	–
LPP	0.2	–
SOO	–	1.39–2.23
OOO	43.8	39.8–39.9
LOS	2.2	–
OOP	10.4	–
POP	0.2	–
OOS	4.6	–

1 Hassanein (1999)
2 Turan et al. (2007) and Uluata (2016)

Table 25.2 Distribution of major fatty acids (%) from apricot oils

Fatty acid	Oils from Turkey[a]	Oil from New Zeland[b]	Oils from Pakistan[c]	Oil from China[d]	Oil from Macedonia[e]
C16:0	4.50–6.4	4.4	3.35–5.93	4.57–4.87	4.25
C16:1	0.50–0.74	0.3	0.32–0.71	0.62–0.71	–
C18:0	1.03–1.33	0.2	1.10–1.68	0.85–0.91	1.36
C18:1	66.53–72.51	69.0	62.34–80.97	70.29–71.25	70.90
C18:2	17.17–24.83	26.0	13.33–30.33	22.31–23.00	20.93
C18:3	0.07–0.14	0.1	0–1.03	0.15–0.18	0.74

[a]Data from Turan et al. (2007) and Uluata (2016)
[b]Data from Beyer and Melton (1990)
[c]Data from Manzoor et al. (2012)
[d]Data from Zhou et al. (2016)
[e] Kostadinović Veličkovska et al. (2015)

The phospholipids (PL) in apricot oil comprised of phosphatidylcholine (PC) (43.0%), phosphatidylinositol (PI) (31.6%), phosphatidylethanolamine (PE) (15.2%), phosphatidic acid (PA) (3.1%). The percentage of other identified phospholipids is lower than 1.5%. Apart from PA, oleic acid was mainly present in the form of PC, PI and PE ranging between 43.5% and 51.6%, whereas palmitic acid was the major fatty acid in PA (Zlatanov and Janakieva 1998).

4 Minor Bioactive Lipids

Apricot kernel oil is a good source of phytosterols, wherein the total phytosterol content was determined as 1762.1 µg/g of oil (Hassanien et al. 2014). Rudzińska et al. (2017) have described apricot oil from 15 different cultivars with β-sitosterol in the range of 184.6–821.5 mg/100 g oil. The percentage of β-sitosterol was 76–86% of the total sterols. Besides, campesterol, Δ5-avenasterol, cholesterol, 24-methylene–cycloartanol, gramisterol, Δ7-stigmasterol, Δ7-avenasterol and citrostadienol were recorded at moderate and low concentrations. In the oils from different varieties of Turkey, β-sitosterol was the major phytosterol with amounts 251.40–294.22 mg/100 g. Δ5-avenasterol was found to be the second most abundant sterol in apricot oils with an abundance of 21.92–56.83 mg/100 g (Turan et al. 2007). Campesterol was present at a moderate level with 130 mg/kg in cold-pressed apricot oil. Stigmasterol, Δ5,24- stigmastanol and Δ7-stigmastanol were also identified and their concentrations were lower than 35 mg/kg (Ramadan et al. 2011).

Apricot seed oils are a good source of tocochromanols. Tocopherols in kernels from 16 different apricot varieties were examined by RP-HPLC/FLD (reversed-phase HPLC with fluorescence detection) method. The major tocochromanol isomer, γ-tocopherol was found between 42.48 and 73.27 mg/100 g, followed by α-tocopherol (1.38–4.41 mg/100 g) and δ-tocopherol (0.77–2.09 mg/100 g) (Górnaś et al. 2015). The number of total tocochromanols in apricot oil was 0.59 mg/g (Hassanien et al. 2014). In addition, Uluata (2016) compared the tocopherols (total, α-, β-, γ- and δ) of apricot kernel oil, extracted using both cold-pressing and solvent extraction methods, and found that total tocopherol content of cold-pressed oil was higher than that of solvent-extracted one. Similar results were reported by Turan et al. (2007) and Alpaslan and Hayta (2006).

Total tocochromanols in oils from 15 apricot genotypes were between 78.8 and 258.5 mg/100 g. Individual tocochromanol isomers changed according to cultivars, locations and these variations are exhibited in Table 25.3. In all apricot cultivars, predominant tocochromanol isomer was γ-tocopherol, and its concentration ranged

Table 25.3 Tocochromanol content of apricot oils

Tocochromanols	1	2	3	4
α-Tocopherol	1.5–11.6	27.4–39.6	14.8–40.4	15.53–26.87
β-Tocopherol	tr-0.1	9.2–11.3	–	0.19–0.71
γ-Tocopherol	73.7–237.2	318.9–498.5	330.8–520.8	346.53–563.40
δ-Tocopherol	3.4–8.4	15.1–17.2	28.5–60.2	8.56–18.94
α-Tocotrienol	0.1–1.6	–	–	–
β-Tocotrienol	tr-0.1	–	–	–

1 Data from Górnaś et al. (2017), Results are given as mg/100 g, tr: trace amount (below 0.05 mg/100 g oil)
2 Data from Uluata (2016), Results are given as mg/kg
3 Data from Manzoor et al. (2012), Results are given as mg/kg
4 Data from Turan et al. (2007), Results are given as mg/kg

from 73.7 to 237.2 mg/100 g in oils from Latvian genotypes (Górnaś et al. 2017), between 346.53–563.40 and 346.53–563.40 mg/kg in oils from Turkish cultivars (Turan et al. 2007; Uluata 2016) and 330.8–520.8 mg/kg in oils from Pakistan varieties (Manzoor et al. 2012). α- and δ-Tocopherol isomers were found in higher concentrations in apricot oils. Beside to tocopherols, tocotrienols were identified in one work by Górnaś et al. (2017) who found α- and β- tocotrienol isomers as a minor part of tocochromanols.

The total carotenoids concentration in apricot kernel oils from 15 genotypes was found in the range between 0.15 and 0.53 mg/100 g oil. Lutein, zeaxanthin, β-cryptoxanthin and β-carotene constituted 76–94% of the total carotenoids, and one of them found as dominant carotenoid compound for each specific genotype (Górnaś et al. (2017).

The total content of phenolic compounds and flavonoids were determined in apricot oil with a level of 37.95 mg gallic acid/kg oil and 19.17 mg luteolin/kg oil, respectively (Kostadinović Veličkovska et al. 2015). The content of squalene, as a precursor of steroid hormones, vitamin D and cholesterol, ranged from 12.6 to 43.9 mg/100 g oil in 15 apricot varieties (Warleta et al. 2010; Rudzińska et al. 2017).

Volatile compounds of apricot oils were identified using SPME/GC/MS techniques in oils from Longwangmo apricot, which is China's main and most well-known variety. Nine volatile compounds (benzaldehyde, 2-methyl-propanal, 2-methyl-butyl aldehyde, furfural, nonanal, methylpyrazine, 2,5-dimethyl-pyrazine, methoxy pyrazine, and 3-ethyl-2,5-dimethyl-pyrazine) are found in different amounts in all apricot oils. Aldehydes were a major chemical group in the volatiles of apricot oils. The predominant compound, benzaldehyde is found in the range of 2462.09–4596.18 μg/ kg oil. Especially, aldehydes contribute to the aroma of aldehydes (Zhou et al. 2016).

5   Composition of Fruit Essential Oil

Greger and Schieberle (2007) studied most odor-active compounds in an extract isolated from fresh apricots by using aroma extract dilution analysis. γ-decalactone, (E)-β-damascenone, linalool, δ-decalactone followed by γ-octalactone and (Z)-1,5-octadien-3-one were identified compounds with the highest flavor dilution (FD) factors. (Z)-3-hexenal, 1-octen-3-one, butanoic acid, and 2-methyl-butanoic acid, 3-methyl-butanoic acid and 2-methoxyphenol were also identified as odor-active constituents in apricots, but with somewhat lower FD factors.

Thirty-four volatile compounds included 12 esters, eight terpenes, nine alcohols, one aldehyde, two ketones and two lactones were detected in the fruit tissues of the 12 apricot cultivars. Apricot cultivars could be divided into three groups based on total concentrations of volatiles, Group 1 (Ninfa, Mandorlon, Ouardy and Fracasso cultivars) included the highest values ranging from 6 to >9 μg/g, group 2 (Goldrich, Orange Red and Palummella cultivars) included intermediate values ranging from

2.5 to 5 μg/g, and group 3 (Silvercot, Pinkot, Pellecchiella, Bulida and Alba cultivars) that included compounds ranging from 0.8 to 2 μg/g. (Z)-3-Hexenol, (Z)-2-hexenol (green leaf odor) and linalol (floral odor) were identified volatiles in all cultivars (Lo Bianco et al. 2010).

6 Contribution of Bioactive Compounds in Fruit Oil to Organoleptic Properties and Functional Foods

Apricot kernel has a high amount of oil, which is rich in unsaturated fatty acids (Alpaslan and Hayta 2006; Orhan et al. 2008). The presence of monounsaturated fatty acids (MUFA) and polyunsaturated fatty acids (PUFA), appears particularly desirable. However, fats with a high content of unsaturated fatty acids are labile food ingredients, which may undergo chemical changes during storage and thermal treatment. The lipid oxidation is a major problem for the food industry, because of its negative effects on the nutritional quality and shelf-life of foods containing lipid. During this process, the amount of unsaturated fatty acids decreases, simultaneously a number of off-flavor and toxic compounds arise. This leads to deterioration of both nutritional value and sensory attributes of oils and food products containing them. Synthetic and natural antioxidants are used to prevent lipid oxidation by scavenging and reducing the activity of free radicals (Decker 2002).

Apricot kernel oil has not only used as a source of unsaturated fatty acids but also as a source of phenolic compounds, tocopherols, carotenoids and these components are important natural antioxidants that prevent lipid oxidation (Bozan and Temelli 2008; Waraho et al. 2011). Especially phenolic compounds have a great impact on the oil stability as an antioxidant, sensory and nutritional characteristics (Decker 2002). In addition, four different derivatives of tocopherols (α, β, γ, and δ) are found as antioxidants against oxidative deterioration of the oil in nature. Their antioxidant activity increases in the order of α to δ, whereas their biological activity is the opposite of antioxidant activity (Timmermann 1990; Sies and Murphy 1991; Matthaus and Ozcan 2009).

Uluata (2016) reported that phenolic compounds contribute to the antioxidant capacity of oils. Total phenolic content oil was found between 24.9 and 26.9 μg gallic acid/g oil and antioxidant activity using DPPH, ABTS and ORAC assays of the oils were 137.5–238.9 μg Trolox/g oil, 151.2–168.8 μg Trolox/g oil and 110.6–162.6 μmol/100 g oil, respectively (Uluata 2016). Lipid oxidation was monitored in these oils wherein the lag phase of the lipid hydroperoxides formation was 22 and 30 days with lipid hydroperoxide method, while the induction period of the oils were 15.1–20.1 h by Rancimat apparatus. Apricot oil generally showed high oxidative stability. Induction period was found higher in apricot oil (35.9 h) compared to the other oils such as tomato seed oil (25.4 h), black cumin oil (16.9 h) and grapeseed oil (8.0 h) (Hassanien et al. 2014). Antiradical activity of methanolic extract from oil determined by Trolox equivalent antioxidant capacity (TEAC) assay and apricot oil indicated antioxidant activity equal to 44.55 mg of Trolox/kg of oil (Kostadinović Veličkovska et al. 2015). The oxidative stability of apricot oil against thermal and photo-oxidation conditions was reported by Kiralan et al. (2018). They were reported that peroxide value (PV) and diene values (K_{232}) of apricot samples increased slowly during storage at conditions mentioned above. After 6 days storage at 60 °C, PV exceeded the value of 20 meq O_2/kg. In the same conditions, K_{232} value (approximately 3.5) exhibited similar behavior up to 10 days storage. Similar results for PV and K232 were observed in photo-oxidation conditions.

7   Health-Promoting Traits of Fruit Oil and Oil Constituents

A number of studies were reported that apricot kernel oil is a good source of MUFA (most important oleic acid, ω-9) and PUFA (most important linoleic acid, ω-6) (Alpaslan and Hayta 2006; Mensink and Katan 1987). Both unsaturated fatty acids are important nutrients for human health because they have potent lipid and cardio-protective effects, cholesterol-lowering effects and, decrease the risk of heart diseases as well as cancer (Esterbauer et al. 1991; Hicks and Moreau 2001; Kutlu et al. 2009). The effect of apricot oil supplementation on the cholesterol, MDA levels, glutathione peroxidase (GPx) and CAT activity was studied in hypercholesteremic rats (Kutlu et al. 2009; Ramadan et al. 2011). The authors concluded that apricot kernel oil causes improvement in the liver antioxidant condition of rats in consequence of a decline in CAT and GPx enzymes.

Apricot kernel oil contains more biologically active substances such as phenolics, carotenoids, tocopherols, phytosterols and squalene as minor bioactive compounds (Siger et al. 2007; Warleta et al. 2010; Bhatnagar and Gopala Krishna 2014; Prescha et al. 2014; Kostadinović Veličkovska et al. 2015; Uluata 2016; Górnaś et al. 2017). Those compounds are positively attributed to antioxidant activity and, are of consumers' awareness due to their proven beneficial influence on human health (Bendini et al. 2007; Ramadan and Moersel 2006; Nikokavouraa et al. 2011). α-tocopherol is the antioxidants against oxidative deterioration and is vitamin E in human nutrition (Timmermann 1990; Sies and Murphy 1991). Recent research has shown that γ-tocopherol is a better negative risk factor for certain types of cancer and myocardial infections than α-tocopherol (Hensley et al. 2004).

Phytosterols (β-sitosterol, campesterol, and stigmasterol) are membrane constituents of plants that effectively reduce serum LDL cholesterol and atherosclerotic risk (De Jong et al. 2003). In addition, existing differences between sterol compositions make them the most suitable for determining the botanical origin of oils and, hence, detecting adulteration among vegetable oils (Aparicio and Apariqcio-Ruiz 2000). Plant stanols and sterols are steroid compounds that belong to the cholesterol analogs group with carbon side chains with/without a double bond. The most common forms of sterols in the plant world are β-sitosterol, campesterol and stigmasterol. Plant sterols, despite the similarity to cholesterol, have reported ability for reducing the levels of cholesterol in blood serum (Chen et al. 2008).

The minor components of squalene in apricot kernel oil have some important functions such as the precursor of steroid hormones, vitamin D and cholesterol (Warleta et al. 2010). Squalene (2,6,10,15,19,23-hexamethyl-2,6,10,14,18,20-tetracosahexane) is a triterpenic hydrocarbon, the precursor of steroid hormones, vitamin D and cholesterol, present at high concentrations in shark liver and in some oils such as virgin olive oil. Squalene reduces the levels of reactive oxygen species in vitro and protects against oxidative DNA damage in human mammary epithelial cells. Hence, virgin olive oil rich in squalene may be an important food-stuff, which has been stated to play a significant role in the prevention of breast cancer (Warleta et al. 2010).

In addition to above discussion, apricot kernel oil has been mentioned in TCM (Traditional Chinese medicine) where it is used for treating tumors and ulcers (Quezada 2003; Lewis and Elvin-Lewis 2003). The kernel oil has been also used in the formulation of many cosmetic products as a carrier oil and as a pharmaceutical agent (laxative and expectorant) (Chevallier 1996; Quezada 2003).

8   Edible Applications of Fruit Oil

In the last decades, there has been a growing interest in exploring new sources of edible oils with high nutritional, industrial and pharmaceutical values. Apricot kernel can be an important new source of oil for potential nutraceutical, functional and oleochemical uses (Manzoor et al. 2012).

When compared to other nut and vegetable oils, including almond (60.93 g/100 g of lipid), cashew nut (61.15 g/100 g of lipid), macadamia nut (58.51 g/100 g of lipid), pecan (66.66 g/100 g of lipid), pistachio (50.29 g/100 g of lipid) (Venkatachalam and Sathe 2006), berry seed (ranges from 12.4 to 22.9 g/100 g of oil) (Parry et al. 2005), coconut (7%), palm (43%), rapeseed (54%), and soybean (25%) (Harwood and Yaqoob 2002), apricot kernel oil contains the highest proportion of oleic acid (Alpaslan and Hayta 2006). Also apricot kernel oils contain a considerable amount of tocopherols, consisting mainly of γ-tocopherol compared to many vegetable and nut oils. Tocopherols are the best known and most widely used antioxidants (Clifford 2001; Kamal-Eldin and Andersson 1997). Although it is generally accepted that α-tocopherol has higher hydrogen donation ability than γ-tocopherol, the latter was often found to be a better antioxidant in some cases (Huang et al. 1994). Consequently, apricot kernel oils can be used for edible purposes alone as salad and frying oils or mixed with the other vegetable oils to enhance their oxidative stability. Moreover, the oils are used externally as cosmetic skincare products (creams, soaps and skin lotions), massage oil, sunbathing oil and a base with other aromatherapy oils (CBI 2014).

9 Other Issues

Previous studies on some Rosaceae kernel oils were reported by some authors (Hassanein 1999; Farine et al. 1986). Kernel oils of peach and apricot were used as adulterants or substitutes for some expensive oils, particularly sweet almond oil (Egan et al. 1981). Adulteration of almond oil with apricot oil can be detected by determination of tocopherol (Gurfinger and Letan 1973). Although, the kernel oil market volume is low, organic standards are particularly important in Germany. The price sharply differentiates between organic and conventional apricot kernel oil (CBI 2014).

10 Conclusion

The apricot oil, extracted from the kernel of *P. armeniaca*, has been traditionally used as edible oil for cooking, food and cosmetic products formulation, massaging, and functional and medicinal supplements. The oil contains a number of main secondary metabolites such as unsaturated fatty acids, phenolics, carotenoids, volatiles, sterols, tocochromanols and squalenes. Several studies revealed that the apricot oil could play a beneficial role in the prevention of cardiovascular, cancer and ulcer diseases due to its high contents of specific antioxidants and mono- and polyunsaturated fatty acids.

References

Akin, E. B., Karabulut, I., & Topcu, A. (2008). Some compositional properties of main Malatya apricot (*Prunus armeniaca* L.) varieties. *Food Chemistry, 107*(2), 939–948.

Alpaslan, M., & Hayta, M. (2006). Apricot kernel: Physical and chemical properties. *Journal of the American Oil Chemists' Society, 83*(5), 469–471.

Aparicio, R., & Apariqcio-Ruiz, R. (2000). Authentication of vegetable oils by chromatographic techniques. *Journal of Chromatography, 881*, 93–104.

Bendini, A., Cerretani, L., Carrasco-Pancorbo, A., Gómez-Caravaca, A. M., Segura-Carretero, A., Fernández-Gutiérrez, A., & Lercker, G. (2007). Phenolic molecules in virgin olive oils: A survey of their sensory properties, health effects, antioxidant activity and analytical methods. An overview of the last decade Alessandra. *Molecules, 12*, 1679–1719. https://doi.org/10.3390/12081679.

Beyer, R., & Melton, L. D. (1990). Composition of New Zealand apricot kernels. *New Zealand Journal of Crop and Horticultural Science, 18*(1), 39–42.

Bhatnagar, A. S., & Gopala Krishna, A. G. (2014). Lipid classes and subclasses of cold-pressed and solvent extracted oils from commercial Indian Niger (*Guizotia abyssinica* L.f. Cass.) seed. *Journal of the American Oil Chemists' Society, 91*(7), 1205–1216.

Bozan, B., & Temelli, F. (2008). Chemical composition and oxidative stability of flax, safflower and poppy seed and seed oils. *Bioresource Technology, 99*(14), 6354–6359.

CBI Ministry of Foreign Affairs (2014). CBI product fact sheet: Apricot Kernel Oil in Germany. https://www.cbi.eu/sites/default/files/market_information/researches/product-factsheet-apricot-kernel-oil-germany-vegetable-oils-oilseeds-2014.pdf.

Chen, Z. Y., Jiao, R., & Ma, K. Y. (2008). Cholesterol-lowering nutraceuticals and functional foods. *Journal of Agricultural and Food Chemistry, 56*, 8761–8773.

Chevallier, A. (1996). *The encyclopedia of medicinal plants*. New York: DK Publishing.

Clifford, H. (2001). III. Sources of natural antioxidants: Oilseeds, nuts, cereals, legumes, animal products and microbial sources. In J. Pokorny, N. Yanishlieva, & M. Gordon (Eds.), *Antioxidants in food practical applications*. Cambridge, UK: Woodhead Publishing Limited.

Corrales, M., Toepfl, S., Butz, P., Knorr, D., & Tauscher, B. (2008). Extraction of anthocyanins from grape by-products assisted by ultrasonics, high hydrostatic pressure or pulsed electric fields: A comparison. *Innovative Food Science and Emerging Technologies, 9*, 85–91.

De Jong, N., Plat, J., & Mensink, R. P. (2003). Metabolic effects of plant sterols and stanols. *The Journal of Nutritional Biochemistry, 4*, 362–369.

Decker, E. A. (2002). Antioxidant mechanism. In C. C. Akoh & D. B. Min (Eds.), *Food lipids: Chemistry, nutrition and biotechnology*. New York: Marcel Dekker. 475–492p.

Dolatowski, Z. J., Stadnik, J., & Stasiak, D. (2007). Application of ultrasound in food technology. *Acta Scientiarum Polonorum Technologia Alimentaria, 6*(3), 89–99.

Egan, H., Ronald, K. S., & Ronald, S. (1981). *Pearson's chemical analysis of foods* (8th ed., pp. 507–547). Edinburgh, London, Melbourne and New York: Churchill Livingstone.

El-Adawy, T. A., & Taha, K. M. (2001). Characterization and composition of different seed oils and flours. *Food Chemistry, 74*, 47–54.

Esterbauer, H., Dieber-Rotheneder, M., Striegl, G., & Waeg, G. (1991). Role of vitamin E in preventing the oxidation of low-density lipoprotein. *The American Journal of Clinical Nutrition, 53*, 314–321.

FAOSTAT. (2018). http://www.fao.org/faostat/en/#data/QV.

Farine, M., Soulier, J., & Comes, F. (1986). Etude de la fraction glyceridique des huiles degraines de quelques *Rosaceae prunoides*. *Reviev des Frances Corps Gras, 33*(83), 115–117.

Femenia, A., Rossello, C., Mulet, A., & Canellas, J. (1995). Chemical composition of bitter and sweet apricot kernels. *Journal of Agricultural and Food Chemistry, 43*(2), 356–361.

Gayas, B., Kaur, G., & Gul, K. (2017). Ultrasound-assisted extraction of apricot kernel oil: Effects on functional and rheological properties. *Journal of Food Process Engineering, 40*(3), e12439.

Górnaś, P., Mišina, I., Grāvīte, I., Soliven, A., Kaufmane, E., & Segliņa, D. (2015). Tocochromanols composition in kernels recovered from different apricot varieties: RP-HPLC/FLD and RP-UPLC-ESI/MSn study. *Natural Product Research, 29*(13), 1222–1227.

Górnaś, P., Radziejewska-Kubzdela, E., Mišina, I., Biegańska-Marecik, R., Grygier, A., & Rudzińska, M. (2017). Tocopherols, tocotrienols and carotenoids in kernel oils recovered from 15 apricot (*Prunus armeniaca* L.) genotypes. *Journal of the American Oil Chemists' Society, 94*(5), 693–699.

Greger, V., & Schieberle, P. (2007). Characterization of the key aroma compounds in apricots (*Prunus armeniaca*) by application of the molecular sensory science concept. *Journal of Agricultural and Food Chemistry, 55*(13), 5221–5228.

Gurfinger, T., & Letan, A. (1973). Detection of adulteration of almond oil with apricot oil through determination of tocopherols. *Journal of Agricultural and Food Chemistry, 21*, 1120.

Hacıseferoğulları, H., Gezer, I., Özcan, M. M., & Murat Asma, B. (2007). Post-harvest chemical and physical-mechanical properties of some apricot varieties cultivated in Turkey. *Journal of Food Engineering, 79*(1), 364–373.

Harwood, J. L., & Yaqoob, P. (2002). Nutritional and health aspects of olive oil. *European Journal of Lipid Science and Technology, 104*, 685–697.

Hassanein, M. M. (1999). Studies on non-traditional oils: I. Detailed studies on different lipid profiles of some Rosaceae kernel oils. *Grasas y Aceites, 50*(85), 379–384.

Hassanien, M. M., Abdel-Razek, A. G., Rudzińska, M., Siger, A., Ratusz, K., & Przybylski, R. (2014). Phytochemical contents and oxidative stability of oils from non-traditional sources. *European Journal of Lipid Science and Technology, 116*(11), 1563–1571.

Hensley, K., Benaksas, E. J., Boli, R., Comp, P., Grammas, P., Hamdheydari, L., Mou, S., Pye, Q. N., Stoddard, M. F., Wallis, G., Williamson, K. S., West, M., Wechter, W. J., & Floyd, R. A. (2004). New perspectives on vitamin E: Gamma tocopherol and carboxyethyl hydroxyl chroman metabolites in biology and medicine. *Free Radical Biology & Medicine, 36*, 1–15.

Hicks, K. B., & Moreau, R. A. (2001). Phytosterols and phytostanols: Functional food cholesterol busters. *Food Technology, 55*, 63–67.

Huang, S. W., Frankel, E. N., & German, B. (1994). Antioxidant activity of α- and γ-tocopherols in oil-in-water emulsions. *Journal of Agricultural and Food Chemistry, 42*, 2108–2114.

Hummer, K. E., & Janick, J. (2009). Rosaceae: Taxonomy, economic importance, genomics. In *Genetics and genomics of Rosaceae* (pp. 1–17). New York: Springer.

Kamal-Eldin, A., & Andersson, R. (1997). A multivariate study of the correlation between tocopherol content and fatty acid composition in different vegetable oils. *Journal of the American Oil Chemists' Society, 74*, 375–380.

Kiralan, M., Kayahan, M., Kiralan, S. S., & Ramadan, M. F. (2018). Effect of thermal and photo oxidation on the stability of cold-pressed plum and apricot kernel oils. *European Food Research and Technology, 244*(1), 31–42.

Kostadinović Veličkovska, S., Brühl, L., Mitrev, S., Mirhosseini, H., & Matthäus, B. (2015). Quality evaluation of cold-pressed edible oils from Macedonia. *European Journal of Lipid Science and Technology, 117*(12), 2023–2035.

Kutlu, T., Durmaz, G., Ateş, B., & Erdoğan, A. (2009). Protective effect of dietary apricot kernel oil supplementation on cholesterol evels and antioxidant status of liver in hypercholesteremic rats. *Journal of Food, Agriculture and Environment, 7*(3–4), 61–65.

Lewis, W. H., & Elvin-Lewis, M. P. F. (2003). *Medicinal botany: Plants affecting human health* (p. 214). Hoboken: Wiley.

Lo Bianco, R., Farina, V., Indelicato, S. G., Filizzola, F., & Agozzino, P. (2010). Fruit physical, chemical and aromatic attributes of early, intermediate and late apricot cultivars. *Journal of the Science of Food and Agriculture, 90*(6), 1008–1019.

Manzoor, M., Anwar, F., Ashraf, M., & Alkharfy, K. M. (2012). Physico-chemical characteristics of seeds oils extracted from different apricot (*Prunus armeniaca* L.) varieties from Pakistan. *Grasas y Aceites, 63*, 193–201.

Matricardi, M., Hesketh, R., & Farrell, S. (2002). *Technical Note-20. Supercritical fluid*. Newark: Technologies.

Matthaus, B., & Ozcan, M. M. (2009). Fatty acids and tocopherol contents of some Prunus spp. Kernel oil. *Journal of Food Lipids, 16*, 187–199.

Mensink, R. P., & Katan, M. B. (1987). Effect of monounsaturated fatty acids versus complex carbohydrates on high-density lipoproteins in healthy men and women. *Lancet, 329*, 122–125.

Nikokavouraa, A., Christodouleas, D., Yannakopouloua, E., Papadopoulos, K., & Calokerinos, A. C. (2011). Evaluation of antioxidant activity of hydrophilic and lipophilic compounds in edible oils by a novel fluorimetric method. *Talanta, 84*, 874–880.

Orhan, I., Koca, U., Aslan, S., Kartal, M., & Kusmenoglu, S. (2008). Fatty acid analysis of some Turkish apricot seed oils by GC and GC-MS techniques. *Turkish Journal of Pharmaceutical Sciences, 5*(1), 29–34.

Özkal, S. G., Yener, M. E., & Bayındırlı, L. (2005). Mass transfer modeling of apricot kernel oil extraction with supercritical carbon dioxide. *The Journal of Supercritical Fluids, 35*(2), 119–127.

Parry, J., Su, L., Luther, M., Zhou, K., Yurawecz, M. P., Whittaker, P., & Yu, L. (2005). Fatty acid composition and antioxidant properties of cold-pressed marionberry, boysenberry, red raspberry, and blueberry seed oils. *Journal of Agricultural and Food Chemistry, 53*, 566–573.

Patil Sachin, B. S., Wakte, P. S., & Shinde, D. B. (2013). Optimization of supercritical fluid extraction and HPLC identification of wedelolactone from Wedelia calendulacea by orthogonal array design. *Journal of Advanced Research, 5*, 629–635.

Prescha, A., Grajzer, M., Dedyk, M., & Grajeta, H. (2014). The antioxidant activity and oxidative stability of cold-pressed oils. *Journal of the American Oil Chemists' Society, 91*, 1291–1301.

Quezada, R. S. (2003). U.S. Patent No. 6,582,736. Washington, D.C.: U.S. Patent and Trademark Office.

Ramadan, M. F., & Moersel, J. T. (2006). Screening of the antiradical action of vegetable oils. *Journal of Food Composition and Analysis, 19*, 838–842.

Ramadan, M. F., Zayed, R., Abozid, M., & Asker, M. M. S. (2011). Apricot and pumpkin oils reduce plasma cholesterol and triacylglycerol concentrations in rats fed a high-fat diet. *Grasas y Aceites, 62*(4), 443–452.

Rosendahl, A., Pyle, D. L., & Niranjan, K. (1996). Aqueous and enzymatic processes for edible oil extraction. *Enzyme and Microbial Technology, 19*, 402–420.

Rudzińska, M., Górnaś, P., Raczyk, M., & Soliven, A. (2017). Sterols and squalene in apricot (*Prunus armeniaca* L.) kernel oils: The variety as a key factor. *Natural Product Research, 31*(1), 84–88.

Sanders, T. H. (2001). Individual oils: Peanut oil. In R. F. Wilson (Ed.), *Proceedings of the world conference on oilseed processing and utilization* (pp. 141–144). Champaign: American Oil Chemist's Society Press.

Schieber, A., Stintzing, F. C., & Carle, R. (2001). By-products of plant food processing as a source of functional compounds-recent developments. *Trends in Food Science and Technology, 12*(11), 401–413.

Shariatifar, N., Pourfard, I. M., Khanıkı, G. J., Nabızadeh, R., Akbarzadeh, A., & Nejad, A. S. M. (2017). Mineral composition, physico-chemical properties and fatty acids profile of *Prunus armeniaca* apricot seed oil. *Asian Journal of Chemistry, 29*(9), 2011–2015.

Sharma, A., & Gupta, M. N. (2006). Ultrasonic pre-irradiation effect upon aqueous enzymatic oil extraction from almond and apricot seeds. *Ultrasonics Sonochemistry, 13*(6), 529–534.

Sies, H., & Murphy, M. E. (1991). Role of tocopherols in the protection of biological systems against oxidative damage. *Journal of Photochemistry and Photobiology B: Biology, 8*, 211–224.

Siger, A., Nogala-Kalucka, M., & Lampart-Szczapae, E. (2007). The content and antioxidant a ctivity of phenolic compounds in cold-pressed plant oils. *Journal of Lipids, 15*, 137–149.

Slover, H. T., Jr., Thompson, H. R., & Merola, G. V. (1983). Determination of tocopherols and sterols by capillary gas chromatography. *Journal of the American Oil Chemists' Society, 60*, 1524–1528.

Timmermann, F. (1990). Tocopherole – Antioxidative wirkung bei fetten und ölen. *Fat Science Technology, 92*, 201–206.

Turan, S., Topcu, A., Karabulut, I., Vural, H., & Hayaloglu, A. A. (2007). Fatty acid, triacylg-lycerol, phytosterol, and tocopherol variations in kernel oil of Malatya apricots from Turkey. *Journal of Agricultural and Food Chemistry, 55*, 10787–10794.

Uluata, S. (2016). Effect of extraction method on biochemical properties and oxidative stability of apricot seed oil. *Academic Food Journal, 14*(4), 333–340.

Venkatachalam, M., & Sathe, S. K. (2006). Chemical composition of selected edible nut seeds. *Journal of Agricultural and Food Chemistry, 54*, 4705–4714.

Waraho, T., McClements, D. J., & Decker, E. A. (2011). Mechanisms of lipid oxidation in food dispersions. *Trends in Food Science and Technology, 22*(1), 3–13.

Warleta, F., Campos, M., Allouche, Y., Sánchez-Quesada, C., Ruiz-Mora, J., Beltrán, G., & Gaforio, J. J. (2010). Squalene protects against oxidative DNA damage in MCF10A human mammary epithelial cells but not in MCF7 and MDA-MB-231 human breast cancer cells. *Food and Chemical Toxicology, 48*(4), 1092–1100.

Yıldız, F. (1994). New technologies in apricot processing. *Journal of Standard, Apricot Special Issue, Ankara*, 67–69.

Zbigniew, J., Dolatowski, J. S., & Dariusz, S. (2007). Application of ultrasound in food technology. *Acta Scientrum Polonorum Technologia Alimentaria, 6*(3), 89–99.

Zhang, S. B., Lu, Q. Y., Yang, H., & Li Yu Wang, S. (2011). Aqueous enzymatic extraction of oil and protein hydrolysates from roasted peanut seeds. *Journal of the American Oil Chemists' Society, 88*, 727–732.

Zhou, B., Wang, Y., Kang, J., Zhong, H., & Prenzler, P. D. (2016). The quality and volatile-profile changes of Longwangmo apricot (*Prunus armeniaca* L.) kernel oil prepared by different oil-producing processes. *European Journal of Lipid Science and Technology, 118*(2), 236–243.

Zlatanov, M., & Janakieva, I. (1998). Phospholipid composition of some fruit-stone oils of Rosaceae species. *European Journal of Lipid Science and Technology, 100*(7), 312–315.

Chapter 26
Citrus Oils

Ali Osman

Abstract Citrus essential oils (EOs) has been identified in different parts of fruits as well as in leaves (particularly present in fruit flavedo), showing that limonene, β-myrcene, α-pinene, p-cymene, β-pinene, terpinolene, and other elements are the major aromatic compounds of many citrus species. EOs from citrus fruits are a rich source of bio-functional compounds with various health properties, including anti-oxidant, antimicrobial, anti-inflammatory, and cytoprotective activities. They are mixtures of more than 200 components that can be grouped into 2 fractions which contain monoterpenes and sesquiterpenes hydrocarbons as well as their oxygenated derivatives along with aliphatic aldehydes, alcohols and esters. Today citrus fruit, juice or peel oils are currently used in everything from food or food preparations, soft drinks, ice cream, candy, pharmaceutical preparations, air fresheners, cleaning products, solvents as well as colognes and fine perfumes. This chapter will cover the production, chemistry of citrus essential oils and its applications.

Keywords Essential oil · Limonene · Antioxidant · Antimicrobial · Cytoprotective activity

1 Introduction

Plants that produce essential oils (EOs) represent a large part of the natural flora and an important resource in various fields such as pharmaceutical, food and cosmetic industries, due to their flavor, fragrance and biological activity (Swamy et al. 2016). EOs play a pivotal role in the growth and colonization of plants, giving color and scent to reproductive organs, attracting pollinators, favoring seed dispersion (Sharifi-Rad et al. 2017). EOs are volatile, natural, complex compounds characterized by a strong odor and formed by aromatic plants as secondary metabolites. They are usually obtained by steam or hydro-distillation first developed in the middle ages by

A. Osman (✉)
Agricultural Biochemistry Department, Faculty of Agriculture, Zagazig University, Zagazig, Egypt

© Springer Nature Switzerland AG 2019 521
M. F. Ramadan (ed.), *Fruit Oils: Chemistry and Functionality*,
https://doi.org/10.1007/978-3-030-12473-1_26

Arabs. Known for their antiseptic, i.e. bactericidal, virucidal and fungicidal, and medicinal properties and their fragrance, they are used in embalmment, preservation of foods and as antimicrobial, analgesic, sedative, anti-inflammatory, spasmolytic and locally anesthetic remedies. Up to the present day, these characteristics have not changed much except that more is known about some of their mechanisms of action, particularly at the antimicrobial level. The EOs from citrus fruits are a rich source of bio-functional compounds with various health properties, including antioxidant, antimicrobial, anti-inflammatory, and cytoprotective activities (Amorim et al. 2016; Cirmi et al. 2016; Swamy et al. 2016; Tundis et al. 2012). Citrus oils constitute the largest sector of the world production of EOs. Citrus EOs obtained from the peel of fruit is used in the food and perfume industries. They are mixtures of more than 200 components that can be grouped into 2 fractions which contain monoterpenes and sesquiterpenes hydrocarbons as well as their oxygenated derivatives along with aliphatic aldehydes, alcohols and esters (Shaw 1979). Citrus oils are largely employed as aromatizes in the food and pharmaceutical industries. In particular, lemon and bergamot EOs are used in the cosmetic industry, for the production of perfumes, detergents and body-care products (Dugo and Di Giacomo 2002). Today citrus fruit, juice or peel oils are currently used in several products including food or food preparations, soft drinks, ice cream, candy, pharmaceutical preparations, air fresheners, cleaning products, solvents as well as colognes and fine perfumes (Berger et al. 2010; Ferhat et al. 2007). This chapter will cover the production, chemistry of citrus EOs and its applications.

2   Production

The quality of citrus EOs obviously depends to large extent on factors (provenance, type of soil, climate, citrus variety), but the processing of the fruit also has a significant effect. Essential oil (EO) is a complex mixture of volatile substances generally present at low concentrations. Before such substances can be analyzed, they have to be extracted from the matrix. The extraction method used has an effect on the physical properties of citrus oils. The qualitative characteristics of an EOs are always closely related to the yield obtained. An exhaustive extraction procedure produces a larger quantity of high-boiling components, with high molecular weights. As a result, the oil has high specific gravity, non-volatile residue and refractive index values, while the optical rotation value is lower because of the lower relative percentage of d-limonene.

Existing industrial methods include hydro-distillation (Conde-Hernández et al. 2017), solvent extraction (Zagklis and Paraskeva 2015), supercritical carbon dioxide fractionation (Rubio-Rodríguez et al. 2008), molecular distillation (Xiong et al. 2013), and aqueous sodium hydroxide extraction (Jia et al. 2016); these can assist in extracting seed oil, essential oil, phenolics, aroma compounds, protein, and polysaccharides. The factors critically affecting the yield are the matrix properties of the sample, solvent type, temperature, pressure, and processing duration (Azmir et al. 2013).

In addition, non-conventional methods have been developed for clean processing in recent decades. Ultra-sonication (Ma et al. 2015), microwave (Dahmoune et al. 2015), as well as high intensity pulsed electric field (HIPEF) (Vorobiev and Lebovka 2010) or moderate electric field (MEF) (Sensoy and Sastry 2004) have been used at the lab-scale to enhance extraction processes. Alternating electric fields, which function via electro permeabilization effect on the cell membrane, are commonly applied for plant material processing and pretreatment, but necessitate that electrodes are inserted into the reaction media (Salengke et al. 2012). The use of metal electrodes leads to adverse electrochemical reactions and heavy metal ion leakage which contaminate the products (Jaeger et al. 2016). Extraction process typically needs a batch stirred tank at different locations. The disadvantages of this approach are low mass transfer and low heat transfer. The conventional methods for the extraction of citrus EOs have some disadvantages. When using cold pressing (CP), citrus EO is agitated vigorously with water and a gradual diminution in citral and terpene alcohols contents will be observed. Furthermore, during agitation, air is thrashed into the liquid, thereby creating conditions favorable for hydrolysis, oxidation and resinification. For steam distillation and hydrodistillation (HD), the elevated temperatures and prolonged extraction time can cause chemical modifications of the oil components and often a loss of the most volatile molecules (De Castro et al. 1999; Pollien et al. 1998).

3 Chemical Composition

Citrus EOs have been identified in different parts of fruits as well as in leaves (particularly present in fruit flavedo), showing that limonene, β-myrcene, α-pinene, p-cymene, β-pinene, and terpinolene are the major aromatic compounds of many citrus species (Buettner et al. 2003; Hérent et al. 2007; Sharma and Tripathi 2006).

Because of their high economic importance, numerous studies have investigated the chemical composition of the peel, leaf, and flower EOs of different citrus species. There is a great variation in the chemical composition of citrus oils due to differences in origin, genetic background, season, climate, age, ripening stage, and method of extraction (da Silva et al. 2017; De Pasquale et al. 2006; Dosoky et al. 2016).

The structure of the major aromatic compounds of many citrus species are presented in Fig. 26.1. Sweet orange, bitter orange, mandarin, and grapefruit EOs are rich in monoterpenes with the major component being d-limonene (65.3–95.9%) (Tisserand and Young 2013). The main components in the essential oil of bitter orange leaf are linalyl acetate and linalool (De Pasquale et al. 2006), while the flower EO contained linalool as the major component, followed by d-limonene and linalyl acetate (Ammar et al. 2012).

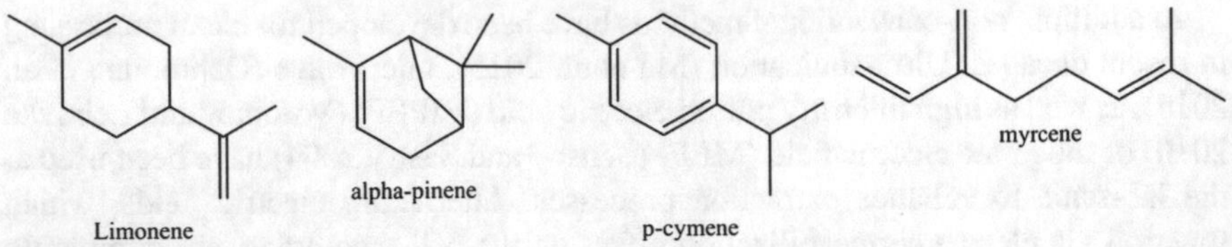

Limonene   alpha-pinene   p-cymene   myrcene

Fig. 26.1 Structure of the major aromatic compounds in citrus species

3.1   Essential Oil from Sweet Orange (Citrus sinensis)

The sweet orange, *Citrus sinensis* (L.) Osbeck, is the most popularly cultivated citrus species in the citricultural regions of the world (Hodgson 1967). Sweet oranges, by virtue of their recognized nutritional value and desirable flavor, have found wide acceptance by consumers. They are mainly utilized as fresh fruit, juice, salads, desserts and preserves (jam, jelly, and marmalades). In addition, they provide valuable EOs of wide commercial applications in food processing, pharmaceutical preparations, perfumery and cosmetics. Commercial production of sweet oranges in many countries has continued to increase, and the current world annual production of 66.2 million tons is the highest of any fruit. The main producing countries are Brazil, USA, Mexico, China, Spain, India, Egypt and Morocco. Volatile chemical compounds of citrus EOs are among the most distinctive components for identification and evaluation of varieties. The occurrence of the components varies among EOs of fruits from different environmental origins, varieties and preparation methods. Several studies on the volatile components of peel EOs of sweet oranges from different origins have been reported (Njoroge et al. 2009; Qiao et al. 2008; Sawamura et al. 2005). Limonene at varying levels has been detected as the most dominant monoterpene hydrocarbon in all the reported sweet orange oils. The other monoterpene hydrocarbons at relatively prominent levels were myrcene, β-pinene and sabinene. γ-Terpinene, a popular component of sour citrus peel oils, has not been reported in most sweet orange oils (Sawamura et al. 2005). Sesquiterpene hydrocarbons have been reported at minor concentrations in most sweet orange peel oils (Njoroge et al. 2005, 2009; Sawamura et al. 2005). The main components are usually β -caryophyllene, (E,E)-α-farnesene, (E)-β -farnesene, β-elemene, γ-cadinene and germacrene D (Sawamura et al. 2005). The oxygenated compounds, octanal, nonanal, decanal, α-sinensal, β-sinensal, linalool, α-terpineol, terpinen-4-ol, carvone, nootkatone, carveol and citral have been found in most of the oils (Sawamura et al. 2005).

Sweet orange (*Citrus sinensis*) is a plant member of the citrus family and mainly cultivated in subtropical regions. Citrus EOs contain 85–99% volatile and 1–15% non-volatile components. The volatile constituents are a mixture of monoterpene (limonene) and sesquiterpene, hydrocarbons and their oxygenated derivatives (Smith et al. 2001).

Njoroge et al. (2005) reported that the volatile components of peel EOs of Salustiana (*Citrus sinensis* Osbeck *forma Salustiana*), Valencia (*C. sinensis* Osbeck *forma Valencia*) and Washington navel (*C. sinensis* Osbeck *forma Washington navel*) sweet oranges grown in Kenya were isolated by cold-pressing and determined by GC and GC-MS. A total of 56 components were identified in Salustiana, 73 in Valencia and 72 in Washington navel peel oils. The identified components amounted to 98.7%, 97.8% and 97.4% of the total volatiles of each oil, respectively. Monoterpene hydrocarbons largely dominated in the volatile fraction of the Salustiana (96.9%), Valencia (94.5%) and Washington navel (92.7%) oils. In each oil, limonene, α-pinene, sabinene and α-terpinene were the major compounds. Sesquiterpene hydrocarbons amounted to 0.1% of the total volatiles of Valencia and Washington navel oils, wherein (*E, E*)-α-farnesene was the main compound. The total oxygenated compounds amounted to 1.7%, 3.4% and 4.5% of the Salustiana, Valencia and Washington navel volatiles. Linalool, decanal, (Z)-carvone, (Z)-carveol, (E)-carveol, nootkatone and sabina ketone were the main components. The three sweet orange varieties could be differentiated by the unique presence of α-phellandrene and γ-terpinene in Salustiana, β-phellandrene, (Z)-nerolidol, aromadendrene in Valencia and *p*-cymene, β-sinensal and dodecanoic acid in Washington navel oils.

Ademosun et al. (2016) reported that the EOs were obtained simultaneously by hydrodistillation and the yields based on the dry weight were 1.54% and 0.98% for the sweet orange peels and seeds, respectively. The chemical composition of the EOs was investigated using GC-FID techniques. The percentage and the retention indices of the identifiable component in the EOs were estimated. Forty compounds were identified in the peel Eos, while 44 compounds were identified in the EOs from the seeds, accounting for 99.80% and 98.78%, respectively. These compounds were divided into four classes, namely monoterpenes, hydrocarbons, oxygenated monoterpenes and sesquiterpene hydrocarbons. Monoterpenes hydrocarbon (96.42%) dominated the peel Eos, while oxygenated hydrocarbon (34.96%) dominated the EOs from seeds. The EOs were characterized by very high percentage of limonene with 92.14% in peel EOs and 13.73% in seed EOs, which constitutes predominantly the majority of *Citrus* spp. EOs (Fisher and Phillips 2006). Limonene, β-myrcen, α-thujene and γ-terpinene were found to be the major components among the monoterpene hydrocarbon. Seed EO was found to be rich in sesquiterpene hydrocarbon (10.58%) and oxygenated hydrocarbon than peel EO [sesquiterpene hydrocarbon [0.78%], and oxygenated hydrocarbon (2.38%)]. The other class includes volatile fatty acids, which found only in seed EOs, whereas decanal, an aldehyde compound, was present in both peel (0.22%) and seed (0.40%) EOs.

Xiao et al. (2016) reported that volatiles of six sweet orange EOs from different origins were characterized by descriptive sensory analysis, gas chromatography-mass spectrometry (GC-MS) and gas chromatography-olfactometry (GC-O). Six attributes (green, fruity, peely, fatty, floral and wood) were selected to assess sweet orange EOs, in which 64 volatile compounds were detected by GC-MS. Monoterpenes constituted the largest chemical group among the volatiles of the EOs. Thirty-one aroma compounds having more than a 50% detection frequency and large span were selected as major odor active compounds correlated with sensory evaluation assessed by partial least squares regression (PLSR). The correlation result showed that α-pinene, sabinene, limonene, δ-terpinolene, hexanal, octanal, decanal and dodecanal were typical aroma compounds, which co-varied with a characteristic aroma of the sweet orange EOs.

3.2 Essential Oil from Bitter Orange (Citrus aurantium)

Bitter orange (*Citrus aurantium* L.) vary from sweet orange by the acidity of its pulp and the bitterness of its rind. Due to these characteristics bitter orange is not consumed as fresh fruit, but its peel is used in the food and drink sector to aromatize liquors, soft drinks and baked goods (Kirbaslar and Kirbaslar 2003) as well as in perfumery and for the aromatization of several drugs. Production of bitter orange peel oils is 25–30 tons/year. Several studies reported the composition of the volatile part of the bitter orange oil. The percentage content of monoterpene hydrocarbons varies from 97.3% to 97.8%, while sesquiterpene hydrocarbons have a percentage of about 0.2%. Oxygenated compounds have a percentage between 1.8% and 2.2%. Esters are the main oxygenated class. Their content ranges from 0.8% to 1.4%, carbonyl compounds from 0.35% to 0.63% and alcohols from 0.33% to 0.46%. The main component of bitter orange EO is limonene, its content being more than 93%. Among the minor monoterpene hydrocarbons, only myrcene always exceeds 1%. Its content ranges from 1.7% to 1.8%, and pinene has a percentage of about 1%. Linalyl acetate is the main oxygenated compound and it represents the major part of the esters. Linalool is the main alcohol, while octanal and decanal the main carbonyl compounds (Dugo et al. 2011; Minh Tu et al. 2002; Tranchida et al. 2012).

Information on all the sesquiterpene hydrocarbons discovered by high-performance liquid chromatography and high-resolution gas chromatography in bitter orange oil were given (Mondello et al. 1995). Nine sesquiterpenes were identified in the sesquiterpene fraction of bitter orange oil. The specific components represent 93.2% of the fraction, with germacrene D and β-caryophyllene as principal components (Mondello et al. 1995). Bitter orange oil has a wide range of variability that can be due to several factors such as geographical location, (Boussaada et al. 2007) season (Dugo et al. 2011) and variety (Lota et al. 2001). Kirbaslar and Kirbaslar (2003) compared Turkish, Spanish and Italian oil compositions and found the main components were limonene (93.68–94.32%), myrcene (1.73–1.86%), linalyl acetate (1.17–1.86%), linalool (0.33–0.46%), β-pinene (0.40–0.57%) and α-pinene (0.39–0.45%). Boussaada et al. (2007) showed that ripe bitter orange peels contained higher concentrations of aliphatic aldehydes and oxygen-containing monoterpenes and sesquiterpenes than did unripe peels.

Lota et al. (2001) studied 39 volatile molecules of 26 cultivars from *Citrus aurantium* L. peel oils. Authors suggested that the limonene/β-pinene composition might help to discriminate the bigarade group. The relative terpene hydrocarbon composition, such as the ratio of α-pinene to sabinene, has been used as a typical marker of sour orange authenticity (Berger 2007).

3.3 Essential Oil from Lemon (Citrus limon L.)

Lemon is a flowering medicinal plant belongs to the family *Rutaceae*. It is a small evergreen tree native to Asia (Ahmad et al. 2006). Several varieties of lemon are available with ellipsoidal yellow fruits (AL-Jabri and Hossain 2014). The lemon is considered a major citrus fruit after oranges and mandarins (Gmitter and Hu 1990). The EO isolated from lemon consisted of a mixture of terpenes (78.9%), alcohols, acids, aldehydes and ester compounds. In these terpenes, limonene (48%) was the main ingredient, followed by β-terpinene (17%) (Sun et al. 2018).

Alfonzo et al. (2017) mentioned that a total of 37 volatile compounds were identified and quantified by GC/MS analysis. Phytochemical groups including the monoterpene hydrocarbons, oxygenated monoterpenes and sesquiterpene hydrocarbons were recognized. The most quantitatively relevant monoterpene hydrocarbons were D-limonene, β-pinene, γ-terpinene and *p*-cymene. The α-terpineol, α-citral, β-citral, neryl acetate and 4-terpineol accounted for the highest concentration within a group of oxygenated monoterpenes; the sesquiterpene hydrocarbons were also detected but at low level.

AL-Jabri and Hossain (2018) reported that the hydro distillation method was used to isolate and identify EOs from locally available Omani lemon fruit samples (sweet and sour). The obtained EOs were analyzed and identified GC–MS. Both EOs contained 22 active compounds with a variation of percentage identified based on GC retention. The main bioactive compounds with high content in Omani sweet lemon EO were limonene (84.73%), α-pinene (1.06%), α-terpineol (2.80%), β-myrcene (2.16%), β-pinene (3.36%), terpinen-4-ol (1.16%) and α-terpinolene (2.33%) and several other minor compounds. Similarly, Omani sour lemon EO was composed of limonene (53.57%), α-terpineol (14.69%), β-pinene (8.23%), α-pinene (1.84%), β-myrcene (1.51%), α-terpinolene (4.33%), terpinen-4-ol (3.38%), cymene (1.80%), β-bisabolene (1.43%), β-linalool (0.85%) and E-citral (1.08%) with several other minor chemical compounds.

3.4 Essential Oil from Grapefruits (Citrus paradisi)

Grapefruit (*Citrus paradisi* Macf.) was discovered in Barbados in the Caribbean Island at the beginning of the nineteenth century and introduced in Texas in 1821 by Don Philippe (Swingle 1967). In the literature, the chemical compositions of *C. paradisi* leaf EO are not homogeneous (Dugo et al. 2011; Dugo and Di Giacomo 2002). Indeed, sabinene is the major compound of grapefruit oils extracted from USA (Duncan and Marsh; 42–59%) (Attaway et al. 1967), from Taiwan (61.91%)

(Cheng and Lee 1981) and from China (Duncan and Marsh; 50.38–50.57%) (Huang and Leonas 2000). Ocimenes (7.18–13%), linalool (3.33–24%) and terpinen-4-ol (1.4–14%), followed by β-pinene (0.0–4.5%), limonene (1.6–4.38%) and myrcene (2.7–4.3%) were also identified. However, in some grapefruit cultivars, sabinene was present in smaller amounts (Duncan and Marsh; 2.49–18.5%) (Ekundayo et al. 1991) or absent (cultivar not specified) (Gurib-Fakim and Demarne 1995). In these cultivars, linalool, terpinen-4-ol, limonene or p-cymene were the major components.

3.5   Essential Oil from Bergamot (Citrus bergamia)

Bergamot (*Citrus bergamia*), which belongs to the Citreae tribe in the Aurantioideae subfamily of the Rutaceae plant family, originates from the Mediterranean ecoregion, particularly from southern Italy and Greece. Its volatile oil, which is produced from the exocarp by means of cold pressing (Setzer 2009), is in high demand for a wide range of perfumes, cosmetics, and especially for aromatherapy (Rotiroti and Bagetta 2009; Saiyudthong and Marsden 2011).

Leggio et al. (2017) mentioned that the distilled bergamot EO used was preliminarily analyzed to define its composition. The individual analytes present in the oil were identified by GC-MS by comparing the corresponding retention times and mass spectra with those of authentic sample. High contents of limonene (30.2%), linalool (21.82%), linalyl acetate (16.21%), and α-terpinene (0.16%) are observed in analogy with the data reported in the literature (Belsito et al. 2007; Fantin et al. 2010; Snow and Slack 2002).

3.6   Essential Oil from Mandarin (Citrus reshni)

Citrus reshni also known as Cleopatra mandarin is a citrus tree that is commonly used in agriculture as a rootstock of different cultivated species of citrus, mostly orange, grapefruit, tangerine and lemon. It originated in India and later was introduced to Florida from Jamaica in the mid-nineteenth century.

Nagy et al. (2018) studied the EOs of the fresh leaves, unripe and ripe fruit peels of *C. reshni* extracted by hydro distillation method. The obtained oils were yellow in color with an intense aromatic mandarin like odor. The percentage yields were 0.4, 1.8, 1.9% v/w calculated on basis of fresh weights and refractive indices were 1.4694, 1.4716, 1.4716 for leaves, unripe and ripe peel oils, respectively. Oil samples were analyzed by GC-MS and 28 components were detected in the leaves representing 98.05% of the total oil composition with sabinene being the major constituent (40.52%) followed by β-linalool (23.25%) and terpinen-4-ol (8.33%) and monoterpene hydrocarbons (63.48%) (Fig. 26.2) as the main class of compounds followed by oxygenated monoterpenes (33.7%). These results are in agreement

Fig. 26.2 Structure of the major compounds in *C. reshni* leaves oil

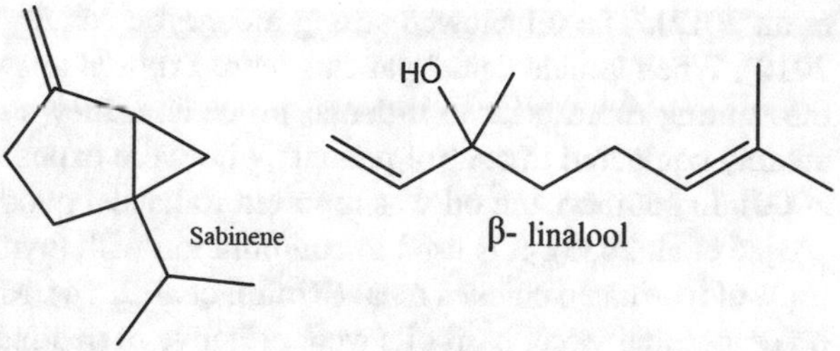

Sabinene   β- linalool

with those reported by Lota et al. (2001). Unlike the results reported by Hamdan et al. (2013), who stated that linalool was the main constituent followed by sabinene in oil of *C. reshni* leaves.

4 Biological Activities of Citrus EOs

Citrus peels and seeds are by-products of the citrus juice industry and are mainly considered as waste products in many countries. El-aal and Halaweish (2010) reported that these wastes have been shown to be excellent sources of bioactive compounds such as phenolics with medicinal properties. The EO is the most vital by-product of citrus processing. Citrus EOs are broadly used as natural food additives in several food and beverage products (Ferhat et al. 2006) because they have been classified as generally recognized as safe (Tisserand and Young 2013). Furthermore, citrus EOs are used as natural preservatives due to their broad spectrum of biological activities including antimicrobial and antioxidant effects (Mitropoulou et al. 2017). The presence of terpenes, flavonoids, carotenes, and coumarins, is thought to be responsible for the strong anti-oxidative and antimicrobial activities (Ali et al. 2017; Viuda-Martos et al. 2008a, b; Yu et al. 2017). Due to their pleasant refreshing smell and rich aroma, citrus EOs are also used in air-fresheners, household cleaning products, perfumes, cosmetics, and medicines.

4.1 Sweet Orange EO

Sweet orange EO showed anticarcinogenic potential via inducing apoptosis in human leukemia (HL-60) cells (Hata et al. 2003) and human colon cancer cells (Murthy et al. 2012), as well as inhibiting angiogenesis and metastasis (Murthy et al. 2012). Olfactory stimulation using orange EO induced physiological and psychological relaxation. Inhalation of orange EO for 90 s caused a significant decrease in oxyhemoglobin concentration in the right prefrontal cortex of the brain which increases comfortable, relaxed, and natural feelings (Igarashi et al. 2014). The odor of sweet orange decreased the symptoms of anxiety and improved the mood (Goes et al. 2012). The oil showed strong anxiolytic activity in Wistar rats (Faturi et al. 2010). When female dental patients were exposed to sweet orange odor diffused in the waiting room prior to a dental procedure, they showed lower levels of state-anxiety compared to control patients who were exposed to air only (Lehrner et al. 2000). In addition, the oil was reported to have a good radical-scavenging activity (Asjad et al. 2013). It is used in combination with thyme oil to improve the quality traits of marinated chicken meat (Rimini et al. 2014). Moreover, formulations based on orange and sweet basil oils were effective in treating acne (Matiz et al. 2012).

4.2   Bitter Orange EO

Bitter orange EO is used as a mild sedative and hypnotic for its soothing, calming, and motor relaxant effects (Dosoky et al. 2016). It also enhances sleeping time and used to treat insomnia (Carvalho-Freitas and Costa 2002). Bitter orange odor decreases the symptoms of anxiety, improves mood, and creates a sense of well-being (Dosoky et al. 2016). It showed strong anxiolytic activity in rodents without any motor impairment, even after 15 consecutive days of treatment (Carvalho-Freitas and Costa 2002). It increased social interactions for rats (time spent in active social interaction), and increased exploration time in the open arms of the elevated plus-maze (de Moraes Pultrini et al. 2006). It was also effective in treating the symptoms of anxiety in patients with chronic myeloid leukemia prior to the collection of medullary material (Pimenta et al. 2016). It exerted its antianxiety effects by regulating serotonin (5-HT) receptors in rats (Costa et al. 2013) and its antidepressant effects through the monoaminergic system in mice (Yi et al. 2011). Bitter orange EO was effective in reducing the severity of first-stage labor pain and anxiety in primiparous women (Namazi et al. 2014), as well as in alleviating moderate and severe knee pain (Yip and Tam 2008). Bitter orange EO is used as a natural antiseizure and anticonvulsant agent. It has been used in treating epilepsy and seizures (Carvalho-Freitas and Costa 2002). It has been reported to have an antispasmodic effect and to enhance sexual desire (Namazi et al. 2014). Bitter orange EO showed good radical-scavenging activity (Dosoky et al. 2016), largely due to the high d-limonene content (Yu et al. 2017) and its microencapsulated form, which was effective in reducing oxidative stress in acute otitis media rats (Lv et al. 2012). Due to its free radical-scavenging properties, bitter orange extract showed nephron protective effects against gentamicin-induced renal damage (Ullah et al. 2014). The antibacterial activity of bitter orange EO was manifested by inhibiting the growth of *Listeria innocua, Salmonella enterica, Escherichia coli, Pseudomonas fluorescens,* and *Aeromonas hydrophila* (Friedman et al. 2004; Iturriaga et al. 2012). It was also effective in controlling multi-species biofilms (Oliveira et al. 2014). Due to its antimicrobial effects, bitter orange EO is used for treating colds, dull skin, flu, gums and mouth, and chronic bronchitis, as well as a food preservative (Dosoky and Setzer 2018). The diluted oil is used to treat pimples and acne (Dosoky and Setzer 2018). In addition, bitter orange EO inhibits the growth of *Penicillium digitatum*, and *P.*

italicum (Dosoky and Setzer 2018). The oil was mentioned as a topical treatment for skin fungal infections like ringworm, jock itch, and athlete's foot (Ramadan et al. 1996). Furthermore, bitter orange EO showed potent fumigant and anti-cholinesterase activities against the silver leaf whitefly, *Bemisia tabaci* (Zarrad et al. 2015).

4.3 Mandarin EO

Mandarin EO showed an antiproliferative effect against human embryonic lung fibroblasts (HELFs) and showed protective effects against bleomycin (BLM)-induced pulmonary fibrosis in rats. The mechanism is thought to be through adjusting the unbalance of oxidation and antioxidation, down-regulating the expressions of connective tissue growth factor (CTGF) and mRNA in lung tissues, and reducing collagen deposition and fibrosis (Zhou et al. 2012). *C. reticulata* EO showed a moderate radical scavenging activity (Yi et al. 2018) mainly due to the high d-limonene content (Yu et al. 2017). Mandarin oil is well known for its broad-spectrum antibacterial and antifungal actions. It inhibits the growth of several bacteria including *Escherichia coli, Bacillus subtilis, Pseudomonas aeruginosa,* and *Staphylococcus aureus* (Yi et al. 2018), as well as several fungi including *Penicillium italicum, P. digitatum, P. chrysogenum, Aspergillus niger, A. flavus, Alternaria alternata, Rhizoctonia solani, Curvularia lunata, Fusarium oxysporum,* and *Helminthosporium oryzae* (Chutia et al. 2009; Matan and Matan 2008; Tao et al. 2014; Wu et al. 2014).

4.4 Grapefruit EO

Because of its anti-obesity effects, grapefruit EO is called the "dieter's friend" (Stiles 2017). The fragrance of grapefruit EO causes activation of the sympathetic nerve activity innervating the WAT, which facilitates lipolysis, then results in a suppression of body weight gain (Nagai et al. 2014; Niijima and Nagai 2003). It efficiently inhibits adipogenesis via inhibiting the accumulation of triglycerides (Dosoky and Setzer 2018). When mixed with patchouli oil, grapefruit EO is known to lower cravings and hunger, which makes it a great tool to lose weight in a healthy way (Stiles 2017). The bright, refreshing scent of grapefruit EO energizes and uplifts the senses. Grapefruit EO promotes body cleansing and removal of toxins and excess fluids (Stiles 2017). Grapefruit EO was cytotoxic against human prostate and lung cancer cells (Zu et al. 2010). Moreover, it showed a strong antibacterial activity against *Bacillus cereus, Enterococcus faecalis, Escherichia coli, Klebsiella pneumonia, Pseudococcus sp., Salmonella thyphimurium, Shigella flexneri,* and *Staphylococcus aureus* and a strong antifungal activity against *Aspergillus niger, Candida albicans, Cladosporium cucumerinum, Penicillium digitatum, P. italicum,* and *P. chrysogenum* (Castro-Luna and Garcia-de-la-Guarda 2016; Okunowo et al. 2013).

4.5   Bergamot EO

Bergamot EO is widely used in the perfumery, pharmaceutical, cosmetic, and food industries (Russo et al. 2014). It is used in suntan preparations due to the presence of bergapten, which is the active melanogenic component (Moysan et al. 1993). Bergamot EO is used in complementary medicine to treat chronic nociceptive and neuropathic pain via modulating the sensitive perception of pain (Lauro et al. 2016; Rombolà et al. 2016). Bergamot EO was reported to be cytotoxic against SH-SY5Y human neuroblastoma cells, suppressing their growth rate through a mechanism related to both apoptotic and necrotic cell death (Berliocchi et al. 2011; Navarra et al. 2015). Bergamot EO inhibited tumor formation by the carcinogen NDMA in vitro by more than 70% (Sawamura 2011). Bergamot EO showed a good radical scavenging activity evaluated by β-carotene bleaching test (IC50 = 42.6 μg/mL) (Tundis et al. 2012) due to the high d-limonene content (Yu et al. 2017). Bergamot EO inhibits the growth of several bacteria including *Escherichia coli, Staphylococcus aureus, Bacillus cereus, Salmonella enterica, S. typhimurium, Pseudomonas putida, Arcobacter butzleri, Enterococcus faecium, E. faecalis,* and *Listeria monocytogenes* (Kirbaşlar et al. 2009).

4.6   Lemon EO

Lemon EO is a natural stress reliever. Inhaling lemon EO causes anti-stress effects through modulating the 5-HT and dopamine (DA) activities in mice (Ogeturk et al. 2010). Lemon EO showed cytotoxic effects against human prostate, lung, and breast cancer cells (Zu et al. 2010). It also induced apoptosis in HL-60 cells due to the presence of citral, decanal, and octanal (Hata et al. 2003). Lemon EO causes activation of the sympathetic nerve activity innervating the white adipose tissue (WAT), which increases lipolysis and results in the suppression of body weight gain (Niijima and Nagai 2003). Lemon EO significantly reduces lipid peroxidation levels and nitrile content, but increases reduced glutathione (GSH) levels, as well as superoxide dismutase, catalase, and glutathione peroxidase activities in mouse hippocampus (Lopes Campêlo et al. 2011). The neuroprotective effect of lemon EO is attributed to its remarkable radical-scavenging activity (Choi et al. 2000). Lemon EO improves creativity and mood, and is thought to affect the heart rhythm (Ceccarelli et al. 2002). The analgesic effect of lemon EO is induced by dopamine-related activation of the anterior cingulate cortex (ACC) and the descending pain inhibitory system (Ikeda et al. 2014). Inhalation of lemon EO reduces the intensity of nausea and vomiting of pregnancy by 33% (Safajou et al. 2014). It also showed anti-spasmodic activity (Ogeturk et al. 2010). Lemon EO significantly enhanced attention level, concentration, cognitive performance, mood, and memory of students during the learning process (Akpinar 2005). In addition, lemon EO is a potent antibacterial against *Bacillus cereus, Mycobacterium smegmatis, Listeria monocytogenes,*

Lactobacillus curvatus, L. sakei, Micrococcus luteus, Escherichia coli, Klebsiella pneumoniae, Pseudococcus aeruginosa, Proteus vulgaris, Enterobacter gergoviae, E. ammnigenus, Staphylococcus aureus, S. carnosus, and *S. xylosus* (Viuda-Martos et al. 2008a, b, 2011), and a strong antifungal against *Aspergillus niger, A. flavus, Penicillium verrucosum, P. chrysogenum, Kluyveromyces fragilis, Rhodotorula rubra, Candida albicans, Hanseniaspora guilliermondii,* and *Debaryomyces hansenii* (Viuda-Martos et al. 2008a, b). Lemon EO has insect repellent effects against the *malaria vector* and *Anopheles stephensi* (Oshaghi et al. 2003).

References

Ademosun, A. O., Oboh, G., Olupona, A. J., Oyeleye, S. I., Adewuni, T. M., & Nwanna, E. E. (2016). Comparative study of chemical composition, in vitro inhibition of cholinergic and monoaminergic enzymes, and antioxidant potentials of essential oil from peels and seeds of sweet orange (*Citrus sinensis* [L.] Osbeck) fruits. *Journal of Food Biochemistry, 40*, 53–60.

Ahmad, M. M., Iqbal, Z., Anjum, F., & Sultan, J. (2006). Genetic variability to essential oil composition in four citrus fruit species. *Pakistan Journal of Botany, 38*, 319.

Akpinar, B. (2005). The effects of *olfactory stimuli* on scholastic performance. *The Irish Journal of Education/Iris Eireannach an Oideachais, 36*, 86–90.

Alfonzo, A., Martorana, A., Guarrasi, V., Barbera, M., Gaglio, R., Santulli, A., Settanni, L., Galati, A., Moschetti, G., & Francesca, N. (2017). Effect of the lemon essential oils on the safety and sensory quality of salted sardines (*Sardina pilchardus* Walbaum 1792). *Food Control, 73*, 1265–1274.

Ali, N. A. A., Chhetri, B. K., Dosoky, N. S., Shari, K., Al-Fahad, A. J., Wessjohann, L., & Setzer, W. N. (2017). Antimicrobial, antioxidant, and cytotoxic activities of *Ocimum forskolei* and *Teucrium yemense* (Lamiaceae) essential oils. *Medicines, 4*, 17.

AL-Jabri, N. N., & Hossain, M. A. (2014). Comparative chemical composition and antimicrobial activity study of essential oils from two imported lemon fruits samples against pathogenic bacteria. *Beni-Suef University Journal of Basic and Applied Sciences, 3*, 247–253.

AL-Jabri, N. N., & Hossain, M. A. (2018). Chemical composition and antimicrobial potency of locally grown lemon essential oil against selected bacterial strains. *Journal of King Saud University-Science, 30*, 14–20.

Ammar, A. H., Bouajila, J., Lebrihi, A., Mathieu, F., Romdhane, M., & Zagrouba, F. (2012). Chemical composition and in vitro antimicrobial and antioxidant activities of *Citrus aurantium* L. flowers essential oil (Neroli oil). *Pakistan Journal of Biological Sciences, 15*, 1034–1040.

Amorim, J. L., Simas, D. L. R., Pinheiro, M. M. G., Moreno, D. S. A., Alviano, C. S., da Silva, A. J. R., & Fernandes, P. D. (2016). Anti-inflammatory properties and chemical characterization of the essential oils of four citrus species. *PLoS One, 11*, e0153643.

Asjad, H. M. M., Akhtar, M. S., Bashir, S., Din, B., Gulzar, F., Khalid, R., & Asad, M. (2013). Phenol, flavonoid contents and antioxidant activity of six common citrus plants in Pakistan. *Journal of Pharmaceutical and Cosmetic Sciences, 1*, 1–5.

Attaway, J. A., Pieringer, A. P., & Barabas, L. J. (1967). The origin of citrus flavor components— III.: A study of the percentage variations in peel and leaf oil terpenes during one season. *Phytochemistry, 6*, 25–32.

Azmir, J., Zaidul, I., Rahman, M., Sharif, K., Mohamed, A., Sahena, F., Jahurul, M., Ghafoor, K., Norulaini, N., & Omar, A. (2013). Techniques for extraction of bioactive compounds from plant materials: A review. *Journal of Food Engineering, 117*, 426–436.

Belsito, E. L., Carbone, C., Di Gioia, M. L., Leggio, A., Liguori, A., Perri, F., Siciliano, C., & Viscomi, M. C. (2007). Comparison of the volatile constituents in cold-pressed bergamot oil and a volatile oil isolated by *vacuum distillation*. *Journal of Agricultural and Food Chemistry,* *55*, 7847–7851.

Berger, R. G. (2007). *Flavours and fragrances: Chemistry, bioprocessing and sustainability.* Berlin: Springer Science & Business Media.

Berger, R. G., Krings, U., & Zorn, H. (2010). *Biotechnological flavour generation, Food flavour technology* (pp. 89–126). Chichester: Wiley-Blackwell.

Berliocchi, L., Ciociaro, A., Russo, R., Cassiano, M. G. V., Blandini, F., Rotiroti, D., Morrone, L. A., & Corasaniti, M. T. (2011). Toxic profile of bergamot essential oil on survival and proliferation of SH-SY5Y neuroblastoma cells. *Food and Chemical Toxicology, 49*, 2780–2792.

Boussaada, O., Skoula, M., Kokkalou, E., & Chemli, R. (2007). Chemical variability of flowers, leaves, and peels oils of four sour orange provenances. *Journal of Essential Oil Bearing Plants, 10*, 453–464.

Buettner, A., Mestres, M., Fischer, A., Guasch, J., & Schieberle, P. (2003). Evaluation of the most odour-active compounds in the peel oil of clementines (*Citrus reticulata* Blanco cv. clementine). *European Food Research and Technology, 216*, 11–14.

Carvalho-Freitas, M. I. R., & Costa, M. (2002). Anxiolytic and sedative effects of extracts and essential oil from *Citrus aurantium* L. *Biological and Pharmaceutical Bulletin, 25*, 1629–1633.

Castro-Luna, A., & Garcia-de-la-Guarda, R. (2016). Antifungal effect of Citrus paradisi "grapefruit" on strains of *Candida albicans* isolated from patients with denture stomatitis Diana Eugenia Churata-Oroya1, Donald Ramos-Perfecto2, Hilda Moromi-Nakata2, Elba Martínez-Cadillo2. *Revista Estomatológica Herediana, 26*, 78–84.

Ceccarelli, I., Masi, F., Fiorenzani, P., & Aloisi, A. M. (2002). Sex differences in the citrus lemon essential oil-induced increase of hippocampal acetylcholine release in rats exposed to a persistent painful stimulation. *Neuroscience Letters, 330*, 25–28.

Cheng, Y.-S., & Lee, C.-S. (1981). Composition of leaf essential oils from ten Citrus species. Proceedings of the National Science Council.

Choi, H.-S., Song, H. S., Ukeda, H., & Sawamura, M. (2000). Radical-scavenging activities of citrus essential oils and their components: Detection using 1, 1-diphenyl-2-picrylhydrazyl. *Journal of Agricultural and Food Chemistry, 48*, 4156–4161.

Chutia, M., Bhuyan, P. D., Pathak, M., Sarma, T., & Boruah, P. (2009). Antifungal activity and chemical composition of *Citrus reticulata* Blanco essential oil against phytopathogens from North East India. *LWT-Food Science and Technology, 42*, 777–780.

Cirmi, S., Bisignano, C., Mandalari, G., & Navarra, M. (2016). Anti-infective potential of *Citrus bergamia* Risso et Poiteau (bergamot) derivatives: A systematic review. *Phytotherapy Research, 30*, 1404–1411.

Conde-Hernández, L. A., Espinosa-Victoria, J. R., Trejo, A., & Guerrero-Beltrán, J. Á. (2017). CO2-supercritical extraction, hydrodistillation and steam distillation of essential oil of rosemary (*Rosmarinus officinalis*). *Journal of Food Engineering, 200*, 81–86.

Costa, C. A., Cury, T. C., Cassettari, B. O., Takahira, R. K., Flório, J. C., & Costa, M. (2013). *Citrus aurantium* L. essential oil exhibits anxiolytic-like activity mediated by 5-HT 1A-receptors and reduces cholesterol after repeated oral treatment. *BMC Complementary and Alternative Medicine, 13*, 42.

da Silva, J. K., da Trindade, R., Moreira, E. C., Maia, J. G. S., Dosoky, N. S., Miller, R. S., Cseke, L. J., & Setzer, W. N. (2017). Chemical diversity, biological activity, and genetic aspects of three Ocotea species from the Amazon. *International Journal of Molecular Sciences, 18*, 1081.

Dahmoune, F., Nayak, B., Moussi, K., Remini, H., & Madani, K. (2015). Optimization of microwave-assisted extraction of polyphenols from *Myrtus communis* L. leaves. *Food Chemistry, 166*, 585–595.

De Castro, M. L., Jimenez-Carmona, M., & Fernandez-Perez, V. (1999). Towards more rational techniques for the isolation of valuable essential oils from plants. *TrAC Trends in Analytical Chemistry, 18*, 708–716.

de Moraes Pultrini, A., Galindo, L. A., & Costa, M. (2006). Effects of the essential oil from *Citrus aurantium* L. in experimental anxiety models in mice. *Life Sciences, 78*, 1720–1725.

De Pasquale, F., Siragusa, M., Abbate, L., Tusa, N., De Pasquale, C., & Alonzo, G. (2006). Characterization of five sour orange clones through molecular markers and leaf essential oils analysis. *Scientia Horticulturae, 109,* 54–59.

Dosoky, N., & Setzer, W. (2018). Biological activities and safety of *Citrus* spp. essential oils. *International Journal of Molecular Sciences, 19,* 1966.

Dosoky, N. S., Moriarity, D. M., & Setzer, W. N. (2016). Phytochemical and biological investigations of *Conradina canescens*. *Natural Product Communications, 11,* 25–28.

Dugo, G., & Di Giacomo, A. (2002). *The genus Citrus*. London: Taylor & Francis Book Ltd.

Dugo, G., Bonaccorsi, I., Sciarrone, D., Costa, R., Dugo, P., Mondello, L., Santi, L., & Fakhry, H. A. (2011). Characterization of oils from the fruits, leaves and flowers of the bitter orange tree. *Journal of Essential Oil Research, 23,* 45–59.

Ekundayo, O., Bakare, O., Adesomoju, A., & Stahl-Biskup, E. (1991). Composition of the leaf oil of grapefruit (*Citrus paradisi* MACF.). *Journal of Essential Oil Research, 3,* 55–56.

El-aal, H. A., & Halaweish, F. (2010). Food preservative activity of phenolic compounds in orange peel extracts (*Citrus sinensis* L.). *Lucrări Ştiinţifice, 53,* 233–240.

Fantin, G., Fogagnolo, M., Maietti, S., & Rossetti, S. (2010). Selective removal of monoterpenes from bergamot oil by inclusion in deoxycholic acid. *Journal of Agricultural and Food Chemistry, 58,* 5438–5443.

Faturi, C. B., Leite, J. R., Alves, P. B., Canton, A. C., & Teixeira-Silva, F. (2010). Anxiolytic-like effect of sweet orange aroma in Wistar rats. *Progress in Neuro-Psychopharmacology and Biological Psychiatry, 34,* 605–609.

Ferhat, M. A., Meklati, B. Y., Smadja, J., & Chemat, F. (2006). An improved microwave Clevenger apparatus for distillation of essential oils from orange peel. *Journal of Chromatography A, 1112,* 121–126.

Ferhat, M. A., Meklati, B. Y., & Chemat, F. (2007). Comparison of different isolation methods of essential oil from Citrus fruits: cold pressing, hydrodistillation and microwave 'dry'distillation. *Flavour and Fragrance Journal, 22*(6), 494–504.

Fisher, K., & Phillips, C. A. (2006). The effect of lemon, orange and bergamot essential oils and their components on the survival of *Campylobacter jejuni, Escherichia coli* O157, *Listeria monocytogenes, Bacillus cereus* and *Staphylococcus aureus* in vitro and in food systems. *Journal of Applied Microbiology, 101,* 1232–1240.

Friedman, M., Henika, P. R., Levin, C. E., & Mandrell, R. E. (2004). Antibacterial activities of plant essential oils and their components against *Escherichia coli* O157: H7 and *Salmonella enterica* in apple juice. *Journal of Agricultural and Food Chemistry, 52,* 6042–6048.

Gmitter, F. G., & Hu, X. (1990). The possible role of Yunnan, China, in the origin of contemporary Citrus species (Rutaceae). *Economic Botany, 44,* 267–277.

Gocs, T. C., Antunes, F. D., Alves, P. B., & Teixeira-Silva, F. (2012). Effect of sweet orange aroma on experimental anxiety in humans. *The Journal of Alternative and Complementary Medicine, 18,* 798–804.

Gurib-Fakim, A., & Demarne, F. (1995). Aromatic plants of Mauritius: Volatile constituents of the leaf oils of *Citrus aurantium* L., *Citrus paradisi* Macfad and *Citrus sinensis* (L.) Osbeck. *Journal of Essential Oil Research, 7,* 105–109.

Hamdan, D. I., Abdulla, R. H., Mohamed, M. E., & El-Shazly, A. M. (2013). Chemical composition and biological activity of essential oils of *Cleopatra mandarin* (Citrus reshni) cultivated in Egypt. *Journal of Pharmacognosy and Phytotherapy, 5,* 83–90.

Hata, T., Sakaguchi, I., Mori, M., Ikeda, N., Kato, Y., Minamino, M., & Watabe, K. (2003). Induction of apoptosis by Citrus paradisi essential oil in human leukemic (HL-60) cells. *In Vivo (Athens, Greece), 17,* 553–559.

Hérent, M.-F., De Bie, V., & Tilquin, B. (2007). Determination of new retention indices for quick identification of essential oils compounds. *Journal of Pharmaceutical and Biomedical Analysis, 43,* 886–892.

Hodgson, R. W. (1967). Horticultural Varieties of Citrus. In W. Reuther, H. J. Webber & L. D. Batchelor (Eds.), *The Citrus Industry*, University of California, Berkeley, 431–459.

Huang, W., & Leonas, K. K. (2000). Evaluating a one-bath process for imparting antimicrobial activity and repellency to nonwoven surgical gown fabrics. *Textile Research Journal, 70*, 774–782.

Igarashi, M., Ikei, H., Song, C., & Miyazaki, Y. (2014). Effects of olfactory stimulation with rose and orange oil on prefrontal cortex activity. *Complementary Therapies in Medicine, 22*, 1027–1031.

Ikeda, H., Takasu, S., & Murase, K. (2014). Contribution of anterior cingulate cortex and descending pain inhibitory system to analgesic effect of lemon odor in mice. *Molecular Pain, 10*, 14.

Iturriaga, L., Olabarrieta, I., & de Marañón, I. M. (2012). Antimicrobial assays of natural extracts and their inhibitory effect against *Listeria innocua* and fish spoilage bacteria, after incorporation into biopolymer edible films. *International Journal of Food Microbiology, 158*, 58–64.

Jaeger, H., Roth, A., Toepfl, S., Holzhauser, T., Engel, K.-H., Knorr, D., Vogel, R. F., Bandick, N., Kulling, S., & Heinz, V. (2016). Opinion on the use of ohmic heating for the treatment of foods. *Trends in Food Science and Technology, 55*, 84–97.

Jia, H., Shao, T., Zhong, C., Li, H., Jiang, M., Zhou, H., & Wei, P. (2016). Evaluation of xylitol production using corncob hemicellulosic hydrolysate by combining tetrabutylammonium hydroxide extraction with dilute acid hydrolysis. *Carbohydrate Polymers, 151*, 676–683.

Kirbaslar, F. G., & Kirbaslar, S. I. (2003). Composition of cold-pressed bitter orange peel oil from Turkey. *Journal of Essential Oil Research, 15*, 6–9.

Kirbaşlar, F. G., Tavman, A., Dülger, B., & Türker, G. (2009). Antimicrobial activity of Turkish citrus peel oils. *Pakistan Journal of Botany, 41*, 3207–3212.

Lauro, F., Ilari, S., Giancotti, L. A., Morabito, C., Malafoglia, V., Gliozzi, M., Palma, E., Salvemini, D., & Muscoli, C. (2016). The protective role of bergamot polyphenolic fraction on several animal models of pain. *PharmaNutrition, 4*, S35–S40.

Leggio, A., Leotta, V., Belsito, E. L., Di Gioia, M. L., Romio, E., Santoro, I., Taverna, D., Sindona, G., & Liguori, A. (2017). Aromatherapy: Composition of the gaseous phase at equilibrium with liquid bergamot essential oil. *Chemistry Central Journal, 11*, 111.

Lehrner, J., Eckersberger, C., Walla, P., Pötsch, G., & Deecke, L. (2000). Ambient odor of orange in a dental office reduces anxiety and improves mood in female patients. *Physiology & Behavior, 71*, 83–86.

Lopes Campêlo, L. M., Moura Gonçalves, F. C., Feitosa, C. M., & de Freitas, R. M. (2011). Antioxidant activity of *Citrus limon* essential oil in mouse hippocampus. *Pharmaceutical Biology, 49*, 709–715.

Lota, M. L., de Rocca Serra, D., Jacquemond, C., Tomi, F., & Casanova, J. (2001). Chemical variability of peel and leaf essential oils of sour orange. *Flavour and Fragrance Journal, 16*, 89–96.

Lv, Y.-X., Zhao, S.-P., Zhang, J.-Y., Zhang, H., Xie, Z.-H., Cai, G.-M., & Jiang, W.-H. (2012). Effect of orange peel essential oil on oxidative stress in AOM animals. *International Journal of Biological Macromolecules, 50*, 1144–1150.

Ma, Q., Fan, X.-D., Liu, X.-C., Qiu, T.-Q., & Jiang, J.-G. (2015). Ultrasound-enhanced subcritical water extraction of essential oils from *Kaempferia galangal* L. and their comparative antioxidant activities. *Separation and Purification Technology, 150*, 73–79.

Matan, N., & Matan, N. (2008). Antifungal activities of anise oil, lime oil, and tangerine oil against molds on rubberwood (*Hevea brasiliensis*). *International Biodeterioration & Biodegradation, 62*, 75–78.

Matiz, G., Osorio, M. R., Camacho, F., Atencia, M., & Herazo, J. (2012). Effectiveness of antimicrobial formulations for acne based on orange (*Citrus sinensis*) and sweet basil (*Ocimum basilicum* L) essential oils. *Biomédica, 32*, 125–133.

Minh Tu, N., Thanh, L., Une, A., Ukeda, H., & Sawamura, M. (2002). Volatile constituents of Vietnamese pummelo, orange, tangerine and lime peel oils. *Flavour and Fragrance Journal, 17*, 169–174.

Mitropoulou, G., Fitsiou, E., Spyridopoulou, K., Tiptiri-Kourpeti, A., Bardouki, H., Vamvakias, M., Panas, P., Chlichlia, K., Pappa, A., & Kourkoutas, Y. (2017). *Citrus medica* essential oil exhibits significant antimicrobial and antiproliferative activity. *LWT-Food Science and Technology, 84,* 344–352.

Mondello, L., Dugo, P., Bartle, K. D., Dugo, G., & Cotroneo, A. (1995). Automated HPLC-HRGC: A powerful method for essential oils analysis. Part V. Identification of terpene hydrocarbons of bergamot, lemon, mandarin, sweet orange, bitter orange, grapefruit, clementine and Mexican lime oils by coupled HPLC-HRGC-MS (ITD). *Flavour and Fragrance Journal, 10,* 33–42.

Moysan, A., Morlière, P., Averbeck, D., & Dubertret, L. (1993). Evaluation of phototoxic and photogenotoxic risk associated with the use of photosensitizers in suntan preparations: Application to tanning preparations containing bergamot oil. *Skin Pharmacology and Physiology, 6,* 282–291.

Murthy, K. N. C., Jayaprakasha, G. K., & Patil, B. S. (2012). D-limonene rich volatile oil from blood oranges inhibits angiogenesis, metastasis and cell death in human colon cancer cells. *Life Sciences, 91,* 429–439.

Nagai, K., Niijima, A., Horii, Y., Shen, J., & Tanida, M. (2014). Olfactory stimulatory with grapefruit and lavender oils change autonomic nerve activity and physiological function. *Autonomic Neuroscience, 185,* 29–35.

Nagy, M. M., Al-Mahdy, D. A., Abd El Aziz, O. M., Kandil, A. M., Tantawy, M. A., & El Alfy, T. S. (2018). Chemical composition and antiviral activity of essential oils from Citrus reshni hort. ex Tanaka (*Cleopatra mandarin*) cultivated in Egypt. *Journal of Essential Oil Bearing Plants, 21,* 264–272.

Namazi, M., Akbari, S. A. A., Mojab, F., Talebi, A., Majd, H. A., & Jannesari, S. (2014). Effects of *citrus aurantium* (bitter orange) on the severity of first-stage labor pain. *Iranian journal of Pharmaceutical Research: IJPR, 13,* 1011.

Navarra, M., Ferlazzo, N., Cirmi, S., Trapasso, E., Bramanti, P., Lombardo, G. E., Minciullo, P. L., Calapai, G., & Gangemi, S. (2015). Effects of bergamot essential oil and its extractive fractions on SH-SY5Y human neuroblastoma cell growth. *Journal of Pharmacy and Pharmacology, 67,* 1042–1053.

Niijima, A., & Nagai, K. (2003). Effect of olfactory stimulation with flavor of grapefruit oil and lemon oil on the activity of sympathetic branch in the white adipose tissue of the epididymis. *Experimental Biology and Medicine, 228,* 1190–1192.

Njoroge, S. M., Koaze, H., Karanja, P. N., & Sawamura, M. (2005). Essential oil constituents of three varieties of Kenyan sweet oranges (*Citrus sinensis*). *Flavour and Fragrance Journal, 20,* 80–85.

Njoroge, S. M., Phi, N. T. L., & Sawamura, M. (2009). Chemical composition of peel essential oils of sweet oranges (*Citrus sinensis*) from Uganda and Rwanda. *Journal of Essential Oil Bearing Plants, 12,* 26–33.

Ogeturk, M., Kose, E., Sarsilmaz, M., Akpinar, B., Kus, I., & Meydan, S. (2010). Effects of lemon essential oil aroma on the learning behaviors of rats. *Neurosciences, 15,* 292–293.

Okunowo, W. O., Oyedeji, O., Afolabi, L. O., & Matanmi, E. (2013). Essential oil of grape fruit (*Citrus paradisi*) peels and its antimicrobial activities. *American Journal of Plant Sciences, 4,* 1.

Oliveira, S. A. C., Zambrana, J. R. M., Di Iorio, F. B. R., Pereira, C. A., & Jorge, A. O. C. (2014). The antimicrobial effects of *Citrus limonum* and *Citrus aurantium* essential oils on multispecies biofilms. *Brazilian Oral Research, 28,* 22–27.

Oshaghi, M., Ghalandari, R., Vatandoost, H., Shayeghi, M., Kamali-Nejad, M., Tourabi-Khaledi, H., Abolhassani, M., & Hashemzadeh, M. (2003). Repellent effect of extracts and essential oils of *Citrus limon* (Rutaceae) and *Melissa officinalis* (Labiatae) against main malaria vector, Anopheles stephensi (Diptera: Culicidae). *Iranian Journal of Public Health, 32,* 47–52.

Pimenta, F. C. F., Alves, M. F., Pimenta, M. B. F., Melo, S. A. L., Almeida, A. A. F. D., Leite, J. R., Pordeus, L. C. D. M., Diniz, M. D. F. F. M., & Almeida, R. N. D. (2016). Anxiolytic effect of *Citrus aurantium* L. on patients with chronic myeloid leukemia. *Phytotherapy Research, 30,* 613–617.

Pollien, P., Ott, A., Fay, L., Maignial, L., & Chaintreau, A. (1998). Simultaneous distillation-extraction: Preparative recovery of volatiles under mild conditions in batch or continuous operations. *Flavour and Fragrance Journal, 13*, 413–423.

Qiao, Y., Xie, B. J., Zhang, Y., Zhang, Y., Fan, G., Yao, X. L., & Pan, S. Y. (2008). Characterization of aroma active compounds in fruit juice and peel oil of Jinchen sweet orange fruit (*Citrus sinensis* (L.) Osbeck) by GC-MS and GC-O. *Molecules, 13*, 1333–1344.

Ramadan, W., Mourad, B., Ibrahim, S., & Sonbol, F. (1996). Oil of bitter orange: New topical antifungal agent. *International Journal of Dermatology, 35*, 448–449.

Rimini, S., Petracci, M., & Smith, D. P. (2014). The use of thyme and orange essential oils blend to improve quality traits of marinated chicken meat. *Poultry Science, 93*, 2096–2102.

Rombolà, L., Amantea, D., Russo, R., Adornetto, A., Berliocchi, L., Tridico, L., Corasaniti, M., Sakurada, S., Sakurada, T., & Bagetta, G. (2016). Rational basis for the use of bergamot essential oil in complementary medicine to treat chronic pain. *Mini Reviews in Medicinal Chemistry, 16*, 721–728.

Rotiroti, D., & Bagetta, G. (2009). Effects of systemic administration of the essential oil of bergamot (BEO) on gross behaviour and EEG power spectra recorded from the rat hippocampus and cerebral cortex. *Functional Neurology, 24*, 107.

Rubio-Rodríguez, N., Sara, M., Beltrán, S., Jaime, I., Sanz, M. T., & Rovira, J. (2008). Supercritical fluid extraction of the omega-3 rich oil contained in hake (*Merluccius capensis–Merluccius paradoxus*) by-products: Study of the influence of process parameters on the extraction yield and oil quality. *The Journal of Supercritical Fluids, 47*, 215–226.

Russo, R., Cassiano, M. G. V., Ciociaro, A., Adornetto, A., Varano, G. P., Chiappini, C., Berliocchi, L., Tassorelli, C., Bagetta, G., & Corasaniti, M. T. (2014). Role of D-limonene in autophagy induced by bergamot essential oil in SH-SY5Y neuroblastoma cells. *PLoS One, 9*, e113682.

Safajou, F., Shahnazi, M., & Nazemiyeh, H. (2014). The effect of lemon inhalation aromatherapy on nausea and vomiting of pregnancy: A double-blinded, randomized, controlled clinical trial. *Iranian Red Crescent Medical Journal, 16*, e14360.

Saiyudthong, S., & Marsden, C. A. (2011). Acute effects of bergamot oil on anxiety-related behaviour and corticosterone level in rats. *Phytotherapy Research, 25*, 858–862.

Salengke, S., Sastry, S., & Zhang, H. (2012). Pulsed electric field technology: Modeling of electric field and temperature distributions within continuous flow PEF treatment chamber. *International Food Research Journal, 19*, 1255.

Sawamura, M. (2011). *Citrus essential oils: Flavor and fragrance*. Somerset: Wiley.

Sawamura, M., Tu, N. T. M., Yu, X., & Xu, B. (2005). Volatile constituents of the peel oils of several sweet oranges in China. *Journal of Essential Oil Research, 17*, 2–6.

Sensoy, I., & Sastry, S. (2004). Extraction using moderate electric fields. *Journal of Food Science, 69*, FEP7–FEP13.

Setzer, W. N. (2009). Essential oils and anxiolytic aromatherapy. *Natural Product Communications, 4*, 1305–1316.

Sharifi-Rad, J., Sureda, A., Tenore, G. C., Daglia, M., Sharifi-Rad, M., Valussi, M., Tundis, R., Sharifi-Rad, M., Loizzo, M. R., & Ademiluyi, A. O. (2017). Biological activities of essential oils: From plant chemoecology to traditional healing systems. *Molecules, 22*, 70.

Sharma, N., & Tripathi, A. (2006). Fungitoxicity of the essential oil of *Citrus sinensis* on post-harvest pathogens. *World Journal of Microbiology and Biotechnology, 22*, 587–593.

Shaw, P. E. (1979). Review of quantitative analyses of citrus essential oils. *Journal of Agricultural and Food Chemistry, 27*, 246–257.

Smith, D. C., Forland, S., Bachanos, E., Matejka, M., & Barrett, V. (2001). Qualitative analysis of citrus fruit extracts by GC/MS: An undergraduate experiment. *The Chemical Educator, 6*, 28–31.

Snow, N. H., & Slack, G. C. (2002). Head-space analysis in modern gas chromatography. *TrAC Trends in Analytical Chemistry, 21*, 608–617.

Stiles, K. (2017). *The essential oils complete reference guide: Over 250 recipes for natural wholesome aromatherapy*. Salem, MA: Page Street Publishing.

Sun, Y., Chen, S., Zhang, C., Liu, Y., Ma, L., & Zhang, X. (2018). Effects of sub-minimum inhibitory concentrations of lemon essential oil on the acid tolerance and biofilm formation of Streptococcus mutans. *Archives of Oral Biology, 87*, 235–241.

Swamy, M. K., Akhtar, M. S., & Sinniah, U. R. (2016). Antimicrobial properties of plant essential oils against human pathogens and their mode of action: An updated review. *Evidence-Based Complementary and Alternative Medicine, 2016*, 3012462.

Swingle, W. T. (1967). The botany of citrus and its wild relatives. In W. Reuther, H. J. Webber & L. D. Batchelor (Eds.), *The Citrus Industry*, University of California, Berkeley, pp. 190–430.

Tao, N., Jia, L., & Zhou, H. (2014). Anti-fungal activity of *Citrus reticulata* Blanco essential oil against *Penicillium italicum* and *Penicillium digitatum*. *Food Chemistry, 153*, 265–271.

Tisserand, R., & Young, R. (2013). *Essential oil safety-e-book: A guide for health care professionals*. Elsevier Health Sciences, Churchill Livingstone (pp. 784).

Tranchida, P. Q., Bonaccorsi, I., Dugo, P., Mondello, L., & Dugo, G. (2012). Analysis of Citrus essential oils: State of the art and future perspectives. A review. *Flavour and Fragrance Journal, 27*, 98–123.

Tundis, R., Loizzo, M. R., Bonesi, M., Menichini, F., Mastellone, V., Colica, C., & Menichini, F. (2012). Comparative study on the antioxidant capacity and cholinesterase inhibitory activity of *Citrus aurantifolia* Swingle, *C. aurantium* L., and *C. bergamia* Risso and Poit. peel essential oils. *Journal of Food Science, 77*, H40–H46.

Ullah, N., Khan, M. A., Khan, T., & Ahmad, W. (2014). Nephroprotective potentials of *Citrus Aurantium*: A prospective pharmacological study on experimental models. *Pakistan Journal of Pharmaceutical Sciences, 27*, 505–510.

Viuda-Martos, M., Ruiz-Navajas, Y., Fernández-López, J., & Pérez-Álvarez, J. (2008a). Antifungal activity of lemon (*Citrus lemon* L.), mandarin (*Citrus reticulata* L.), grapefruit (*Citrus paradisi* L.) and orange (*Citrus sinensis* L.) essential oils. *Food Control, 19*, 1130–1138.

Viuda-Martos, M., Ruiz-Navajas, Y., Fernández-López, J., & Perez-Álvarez, J. (2008b). Antibacterial activity of lemon (*Citrus lemon* L.), mandarin (*Citrus reticulata* L.), grapefruit (*Citrus paradisi* L.) and orange (*Citrus sinensis* L.) essential oils. *Journal of Food Safety, 28*, 567–576.

Viuda-Martos, M., Mohamady, M., Fernández-López, J., ElRazik, K. A., Omer, E., Pérez-Alvarez, J., & Sendra, E. (2011). In vitro antioxidant and antibacterial activities of essentials oils obtained from Egyptian aromatic plants. *Food Control, 22*, 1715–1722.

Vorobiev, E., & Lebovka, N. (2010). Enhanced extraction from solid foods and biosuspensions by pulsed electrical energy. *Food Engineering Reviews, 2*, 95–108.

Wu, T., Cheng, D., He, M., Pan, S., Yao, X., & Xu, X. (2014). Antifungal action and inhibitory mechanism of polymethoxylated flavones from *Citrus reticulata* Blanco peel against *Aspergillus niger*. *Food Control, 35*, 354–359.

Xiao, Z., Ma, S., Niu, Y., Chen, F., & Yu, D. (2016). Characterization of odour-active compounds of sweet orange essential oils of different regions by gas chromatography-mass spectrometry, gas chromatography-olfactometry and their correlation with sensory attributes. *Flavour and Fragrance Journal, 31*, 41–50.

Xiong, Y., Zhao, Z., Zhu, L., Chen, Y., Ji, H., & Yang, D. (2013). Removal of three kinds of phthalates from sweet orange oil by molecular distillation. *LWT-Food Science and Technology, 53*, 487–491.

Yi, L.-T., Xu, H.-L., Feng, J., Zhan, X., Zhou, L.-P., & Cui, C.-C. (2011). Involvement of monoaminergic systems in the antidepressant-like effect of nobiletin. *Physiology & Behavior, 102*, 1–6.

Yi, F., Jin, R., Sun, J., Ma, B., & Bao, X. (2018). Evaluation of mechanical-pressed essential oil from Nanfeng mandarin (*Citrus reticulata* Blanco cv. Kinokuni) as a food preservative based on antimicrobial and antioxidant activities. *LWT-Food Science and Technology, 95*, 346–353.

Yip, Y. B., & Tam, A. C. Y. (2008). An experimental study on the effectiveness of massage with aromatic ginger and orange essential oil for moderate-to-severe knee pain among the elderly in Hong Kong. *Complementary Therapies in Medicine, 16*, 131–138.

Yu, L., Yan, J., & Sun, Z. (2017). D-limonene exhibits anti-inflammatory and antioxidant properties in an ulcerative colitis rat model via regulation of iNOS, COX-2, PGE2 and ERK signaling pathways. *Molecular Medicine Reports, 15*, 2339–2346.

Zagklis, D. P., & Paraskeva, C. A. (2015). Purification of grape marc phenolic compounds through solvent extraction, membrane filtration and resin adsorption/desorption. *Separation and Purification Technology, 156*, 328–335.

Zarrad, K., Hamouda, A. B., Chaieb, I., Laarif, A., & Jemâa, J. M.-B. (2015). Chemical composition, fumigant and anti-acetylcholinesterase activity of the Tunisian *Citrus aurantium* L. essential oils. *Industrial Crops and Products, 76*, 121–127.

Zhou, X., Zhao, Y., He, C. C., & Li, J. (2012). Preventive effects of *Citrus reticulata* essential oil on bleomycin-induced pulmonary fibrosis in rats and the mechanism. *Zhong Xi Yi Jie He Xue Bao. Journal of Chinese Integrative Medicine, 10*, 200–209.

Zu, Y., Yu, H., Liang, L., Fu, Y., Efferth, T., Liu, X., & Wu, N. (2010). Activities of ten essential oils towards *Propionibacterium acnes* and PC-3, A-549 and MCF-7 cancer cells. *Molecules, 15*, 3200–3210.

Chapter 27
Guava (*Psidium guajava*) Oil

Syed Tufail Hussain Sherazi, Sarfaraz Ahmed Mahesar, Anam Arain, and Sirajuddin

Abstract The composition and functionality of guava (*Psidium guajava*) seed oil are reported in this chapter. Guava seed oil was extracted using a Soxhlet apparatus to determine the fatty acid composition of the oil. The oil content of seed on the dry weight basis was 11.1%. The iodine value, acid value, free fatty acid, peroxide value and saponification value were 120.55 g of I_2/100 g oil, 3.74 g/100 g oil, 1.86 g/100 g oil, 4.13 meq/kg oil, and 190.74 mg/100 g of oil, respectively. The fatty acid composition was analyzed using Fourier Transform-Infrared (FT-IR) spectroscopy and gas chromatography-mass spectrometry (GC-MS). The FT-IR spectrum indicates the presence of functional groups related to saturated and unsaturated fatty acids. The results from GC-MS revealed the presence of total 18 fatty acids including 12 saturated and 6 unsaturated fatty acids. The linoleic (60.0%), palmitic (14.8%), oleic (12.5%), stearic (9.08%), and arachidic (1.31%) were the major fatty acids. GC-MS results indicated that guava seed oil is a good source of essential fatty acids. The chemical composition of essential oil was also determined in fruits, seeds and leaves of *Psidium guajava* by hydrodistillation method using GC-MS. The identified essential oils components are bioactive and have many biological potential applications.

Keywords *Psidium guajava* · Guava seed oil · FT-IR · GC-MS · Fatty acid

Abbreviation

EO   Essential oils
FA   Fatty acids
FAME   Fatty acid methyl ester
FT-IR   Fourier transform infrared spectroscopy
GC   Gas chromatography

S. T. H. Sherazi (✉) · S. A. Mahesar · A. Arain · Sirajuddin
National Centre of Excellence in Analytical Chemistry, University of Sindh, Jamshoro, Pakistan

© Springer Nature Switzerland AG 2019
M. F. Ramadan (ed.), *Fruit Oils: Chemistry and Functionality*,
https://doi.org/10.1007/978-3-030-12473-1_27

541

MUFA   Mono saturated fatty acids
RT   Retention time
SD   Standard deviation
SFA   Saturated fatty acids
UFA   Unsaturated fatty acids

1 Introduction

Psidium guajava L. (genus: *Psidium*; family: Myrtaceae) is a perennial tree, food crop and growing in tropical and subtropical areas. It is included in medicinal plants and widely used in folk medicine. Taxonomic classification of the plant is represented in Table 27.1 (Flores et al. 2015; Moussa and Almaghrabi 2016; Qin et al. 2017). *Psidium guajava* is a large dicotyledonous shrub small tree which is 10 m high, many branches. The stems are crooked and the bark is light to reddish brown, thin, smooth and continuously flaking. The root system is generally superficial and very extensive, frequently extending well beyond the canopy. The leaves are opposite and simple; stipules are absent, petiole short, 3–10 mm long; blade oblong to elliptic, veins 5–15 cm long and prominent, gland-dotted (Gutiérrez et al. 2008). The flowers are white, up to 2 cm long, incurved petals two or three in the leaf axils; they are fragrant, with four to six petals and yellow anthers (Dakappa et al. 2013). The fruit, is small pear-shaped and fleshy yellow about 5 cm in diameter, 3–6 cm long, containing numerous small hard white seeds, 2–3 mm in diameter, and length embedded in the white or pink edible flesh (Opute 1978). The fruits and leaves of the plant are popular for food and medicinal values (Kumar et al. 2012). There has been a tremendous interest in this plant as evidenced by Gutiérrez et al. (2008).

Table 27.1 Taxonomic classification of plant

Botanical name	*Psidium guajava*
Kingdom	Planate
Order	Myrtales
Division	Magnoliophyta
Class	Magnoliopsida
Family	Myrtaceae
Subfamily	Myrtoideae
Genus	*Psidium*
Species	*Guajava*
English name	Guava

2 World Production of Guava

The guava fruit is popularly consumed worldwide for its palatable taste and diverse health benefits (Qin et al. 2017). *Psidium guajava* trees are growing in India, Pakistan, Brazil, Mexico, Bangladesh, USA, Thailand and number of other countries. According to a record of 2017 and 2018, the leading world producers of guava included India, China, Thailand, Pakistan, Mexico, Indonesia, Brazil, Bangladesh, Philippines, and Nigeria (www.thedailyrecords.com). Among all, Pakistan is the fourth most abundantly guava fruit producing (0.55 million tons/annual) country (Usman et al. 2013). Guava fruit is extremely nutritious, delicious in taste, high nutritional value, low price and trees produce fruit all year round (Mehmood et al. 2014).

3 Uses of *Psidium guajava* in Traditional Medicine

The ethno-pharmacological survey of *Psidium guajava*, shows that different parts of the plant used to treat a number of diseases including diabetes, caries, wounds, anti-inflammatory, reducing fever, pain relief, and hypertension (Heryanto et al. 2017). *Psidium guajava* leaves are easily available and used as eco-friendly biosorbent, due to its excellent capability to remove fluoride from water. Some of the countries with a long history of traditional medicinal uses of guava are summarized in Table 27.2 (Gutiérrez et al. 2008; Joseph and Priya 2011).

4 Commercial Value of *Psidium guajava*

In addition to the medicinal uses, *some parts of Psidium guajava* are employed as food, in dying, in carpentry, and removal of fluoride from aqueous solution using Psidium guajava leaves as shown in Table 27.3 (Gutiérrez et al. 2008; Shukla et al. 2017; Qin et al. 2017).

5 *Psidium guajava* Fruit

Fruits are an important part of a healthy diet and become increasingly important in human diet due to their nutrient composition and health-promoting effects. Guava fruit is used as an important ingredient for the preparation of beverage, juice, and wine. Representative figure of *Psidium guajava* fruit is shown in Fig. 27.1 and some important ingredients reported in the literature are shown in Table 27.4 (Medina and Pagano 2003; Conway 2001; Fujita et al. 1985; Iwu 1993; Jain et al. 2003; Jordán et al. 2003; Nadkarni 1996; Joseph and Priya 2011; Uchôa-Thomaz et al. 2014).

Table 27.2 Ethno-medical uses of *Psidium guajava*

Country	Used part(s)	Ethno-medical uses	Preparation(s)	References
Colombia, Mexico	Leaves	Gastroenteritis, diarrhea, dysentery, rheumatic pain, wounds, ulcers, and toothache	Decoction and poultice	Gutiérrez et al. (2008), Joseph and Priya (2011), Kumar (2012) and Dakappa et al. (2013)
Indigenous Maya, Nahuatl, Zapotec and Popoluca of the region Tuxtlas, Veracruz, Mexico	Leaves	A cough, diarrhea	Decoction or infusion	Gutiérrez et al. (2008), Joseph and Priya (2011), Kumar (2012) and Dakappa et al. (2013)
Latin America, Mozambique	Leaves	Diarrhea, stomach ache	Infusion or decoction	Gutiérrez et al. (2008)
Mexico	Leaves and fruits	Febrifuge, expel the placenta after childbirth, cold, cough hypoglycaemic, affections of the skin, caries, vaginal hemorrhage, wounds, fever, dehydration, respiratory disturbances	Decoction, poultice	Gutiérrez et al. (2008), Kumar (2012) and Dakappa et al. (2013)
Panama, Cuba, Costa Rica, México, Nicaragua, Panamá, Perú, Venezuela, Mozambique, Guatemala, Argentina	Leaves	Anti-inflammatory, for cold, dysentery	Externally applied hot on inflammations	Gutiérrez et al. (2008), Joseph and Priya (2011), Kumar (2012) and Dakappa et al. (2013)
South Africa	Leaves	Diabetes mellitus, hypertension	Infusion or decoction	Gutiérrez et al. (2008)
Caribbean	Leaves	Diabetes mellitus	Infusion or decoction	Gutiérrez et al. (2008)
China	Leaves	Diarrhoea, antiseptic, Diabetes mellitus	Infusion or decoction	Gutiérrez et al. (2008)
Philippines	Leaf and fruit	Astringent, ulcers, wounds, diarrhea	Decoction and poultice	Gutiérrez et al. (2008), Joseph and Priya (2011), Kumar (2012) and Dakappa et al. (2013)

(continued)

Table 27.2   (continued)

Country	Used part(s)	Ethno-medical uses	Preparation(s)	References
India	Leaves, shoots	Febrifuge, antispasmodic, rheumatism, convulsions, astringent	Decoction or infusion	Gutiérrez et al. (2008), Joseph and Priya (2011), Kumar (2012) and Dakappa et al. (2013)
Ghana				Gutiérrez et al. (2008), Joseph and Priya (2011), Kumar (2012) and Dakappa et al. (2013)
Peru	Leaves	Constipation, cough, diarrhea, digestive problems, dysentery, gastroenteritis, vomiting	Infusion or decoction	Gutiérrez et al. (2008), Joseph and Priya (2011), Kumar (2012), and Dakappa et al. (2013)
Fiji	Leaves and fruit	Diarrhea, coughs, stomach-ache, dysentery, toothaches, indigestion, constipation	Juice, the leaves are pounded, squeezed in salt water	Gutiérrez et al. (2008)
New Guinea, Samoa, Tonga, Niue, Futuna, Tahiti	Leaves	Itchy rashes caused by scabies	Boiled preparation	Gutiérrez et al. (2008)
Cook Islands	Leaves	Sores, boils, cuts, sprains	Infusion or decoction	Gutiérrez et al. (2008)
Trinidad	Leaves	Bacterial infections, blood cleansing, diarrhea, dysentery	Infusion or decoction	Gutiérrez et al. (2008), Joseph and Priya (2011), Kumar (2012), and Dakappa et al. (2013)
Latin America, Central and West Africa, and Southeast Asia	Leaves	Gargle for sore throats, laryngitis and swelling of the mouth, and it is used externally for skin ulcers, vaginal irritation and discharge	Decoction	Gutiérrez et al. (2008)
Panama, Bolivia and Venezuela	Leaves	Dysentery, astringent, used as a bath to treat skin ailments	Decoction	Gutiérrez et al. (2008)

(continued)

Table 27.2 (continued)

Country	Used part(s)	Ethno-medical uses	Preparation(s)	References
Brazil	Ripe fruit, and leaves	Anorexia, cholera, diarrhea, digestive problems, dysentery, gastric insufficiency, inflamed mucous membranes, laryngitis, mouth (swelling), skin problems, sore throat, ulcers, vaginal discharge	Mashed, Decoction	Gutiérrez et al. (2008), Joseph and Priya (2011), Kumar (2012), and Dakappa et al. (2013)
USA	Leaf	Antibiotic and diarrhea	Decoction	Gutiérrez et al. (2008)

Table 27.3 The commercial value of *Psidium guajava*

Fruit	Food: concentrated, confectionery, gelatines, jelly nectar, pastes, stuffed with candies, tinned products, juice, etc.	Most of the countries	Gutiérrez et al. (2008) and Qin et al. (2017)
Leaves	Employed to give a black color to cotton	Southeast Asia	Gutiérrez et al. (2008)
Leaves	Serve to dye matting	Indonesia	Gutiérrez et al. (2008)
Leaves	Carpentry and turnery use the leaves to make a black dye for silk	Malaya	Gutiérrez et al. (2008)
Leaves	Removal of fluoride from aqueous solution using Psidium guajava leaves	Published work from India	Shukla et al. (2017).

Fig. 27.1 *Psidium guajava* fruit

Table 27.4 Ingredients of *Psidium guajava* fruit

Ingredient	Concentration	References
Fat	8.5–14 g/100 g seed	Opute (1978), Habib (1986), and Uchôa Thomaz et al. (2014)
Moisture	77–86 g/100 g	Gutiérrez et al. (2008)
Crude fiber	2.8–5.5 g/100 g	Gutiérrez et al. (2008)
Ash	0.43–0.7 g/100 g	Gutiérrez et al. (2008)
Calcium	9.1–17 mg/100 g	Gutiérrez et al. (2008), Joseph and Priya (2011), Kumar (2012), and Dakappa et al. (2013)
Phosphorus	0.30–0.70 mg/100 g	Joseph and Priya (2011), Kumar (2012), Dakappa et al. (2013), and Uchôa Thomaz et al. (2014)
Iron	0.30–13.8 mg	Gutiérrez et al. (2008) and Uchôa Thomaz et al. (2014)
Thiamine	0.03–0.04 mg	Joseph and Priya (2011), Kumar (2012), and Dakappa et al. (2013)
Riboflavin	0.6–1.068 mg	Joseph and Priya (2011), Kumar (2012), and Dakappa et al. (2013)
Niacin	40 I.U.	Joseph and Priya (2011), Kumar (2012), and Dakappa et al. (2013)
Ascorbic acid	100 mg	Gutiérrez et al. (2008)
Vitamin A	200–400 I.U.	Gutiérrez et al. (2008)
Vitamin B3	35 I.U.	Kumar (2012)
Calories	36–50 kcal/100 g	Gutiérrez et al. (2008)

6 Constituents of *Psidium guajava* Fruit

Caryophyllene oxide, 3-phenylpropyl acetate, 3-caryophyllene and nerolidol were present in essential oil of guava fruit, whereas phenol, (*E*)-2-hexenal, (*E,E*)-2,4-hexadienal, (*Z*)-3-hexenal, γ-butyrolactone, (*Z*)-2-hexenal, (*Z*)-3-hexenyl acetate, and hexanal in fresh white-flesh guava fruit oil was also reported (Paniandy et al. 2000). Protein, glycogen and saccharose in guava fruit were reported in the literature (Dweck 2001; Hwang et al. 2002). The major constituent of fruit skin is ascorbic acid. Guava fruit has important medicinal properties such as antipyretic, spasmolitic, anti-bacterial, anti-inflammatory and analgesic activities (Charles et al. 2006).

Researchers have reported the composition of *Psidium guajava* seeds. The main reported components were oil (9.1–14%), starch (12–13%), and proteins (9.73–15%) (Burkill 1985; Michael et al. 2002). The fatty acid profile of the guava seed powder included lauric, myristic, palmitic, heptadecanoic, stearic, oleic, linoleic, arachidic, gondoic, linolenic, and behenic acids (Uchôa-Thomaz et al. 2014).

7   *Psidium guajava* Seed Oil

Plant seeds are important sources of oils for nutritional, industrial and pharmaceutical applications. The suitability of oil for a particular purpose, however, is determined by its characteristics and fatty acid composition. No oil from any single source has been found to be suitable for all purposes as oils from different sources generally differ in their fatty acid composition (Li et al. 1999). Seeds are often rich sources of oils and other interesting minor compounds. An increasing interest in oils from unconventional seeds has been noted and the by-products generated by fruit and vegetable industry may contribute as important sources for such oils. Unsaturated acids based on C16, C18 are widespread in both leaf, and seed oils and a number of rarer fatty acids are found as lipid components. On the other hand, seed oils with a substantial amount of very long chain of FA have attracted attention because of their value for industrial purposes. Furthermore these compounds can be of chemotaxonomic significance (Bağci and Şahin 2004). Guava oil contains fatty acid such as stearic, oleic, linoleic, linolenic, and palmitic acids (Bontempo et al. 2012; Chandrika et al. 2009; Norshazila et al. 2010; Pelegrini et al. 2008).

Extraction of oil from the seed of guava was performed by the Soxhlet method and the oil content was determined according to the reported method (Tlili et al. 2011). Total oil content of *P. guajava* seed was found to be 11.1% on dry bases, wherein the moisture content was 8.33%.

8   Chemical Composition and Characteristics of Guava Seed Oil

To check the quality of fats and oils, various chemical tests were performed such as value (IV), acid value (AV), free fatty acids value (FFA), peroxide values (PV) and saponification value (SV). The IV is a measurement of the degree of unsaturation and reflects the susceptibility of oil to oxidation. Guava oil contains high IV (120.5 g I_2/100 g oil), indicating that oil is composed of mainly unsaturated fatty acids (Table 27.5). The AV was found to be (3.74 g/100 g oil) indicating the content of

Table 27.5 Chemical parameters of *Psidium guajava* seed oil

Chemical parameter	Guava seed oil
Iodine value (g of I_2/100 g oil)	120.5 ± 0.05
Acid value (g/100 g oil)	3.74 ± 0.07
Free fatty acid value (g/100 g oil)	1.86 ± 0.03
Peroxide value (meq O_2/Kg)	4.13 ± 0.02
Saponification value (mg KOH/100 g oil)	190.7 ± 0.04

Present work, M. Phil of Thesis of Anam Arain 2017, University of Sindh, Jamshoro, Pakistan

FFA in oil. FFA value was found to be 1.86 g/100 g oil. Generally, it is considered that lower the value of FFA, the better the oil quality. PV is the measure of rancidity index of oil, and it was observed that the PV guava seed oil was 4.13 (meq O_2/kg). A higher level of PV is the indication of a rancid oil/fat, but in guava seed oil, PV is relatively low which indicates that there is less chance of rancidity. The SV of guava seed oil was found to be 190.74 (mg/100 g oil), which indicated that oil contains high-molecular weight fatty acids. The obtained SV of guava seed oil is almost comparable with edible vegetable oil.

FT-IR spectral analysis of FAME of guava seed oil confirmed the presence of C=O and C-O characteristic bands at 1742 cm^{-1} and 1169 cm^{-1}. In addition, bands due to unsaturated fatty acids were observed at 3009 cm^{-1} and 1664 cm^{-1}.

8.1 Fatty Acid Composition of Guava Oil

The fatty acid composition of guava oil was determined by chromatographic and spectroscopic methods. The result revealed that 18 fatty acids methyl esters (FAME) were separated identified and quantified in guava oil. These fatty acids include myristic, pentadecanoic, palmitoleic, palmitic, margaric, oleic, stearic, linoleic, nonadecanoic, methyl ricinoleate, 10,13-eicosadienoic, 11-eicosenoic, arachidic, heneicosanoic, behenic, tricosanoic, lignoceric, and cerotic acids on the bases of their elution order. The FAME profile showed that a total of 12 saturated and 6 unsaturated fatty acids, representing 26.45% and 73.55%, respectively, were identified. The predominant unsaturated fatty acids were linoleic acid (60.0%), and oleic acid (12.5%), whereas palmitic acid (14.81%), stearic acid (9.08%), and arachidic acid (1.31%) were found as a predominant saturated fatty acid. General classification of guava seed oil with regard to fatty acids is discussed in Table 27.6, wherein the sequence of MUFA (13.40%) < SFA (26.45%) < PUFA (60.15%) is present. The minimum ratio of PUFA/SFA set by the Department of Health UK is 0.45.

A previous study (Uchôa-Thomaz et al. 2014) on guava seed oil, revealed that lauric, myristic, palmitic, heptadecanoic, stearic, oleic (*n*-9), linoleic (*n*-6), arachidic, gondoic, linolenic, behenic acids were identified. High levels of unsaturated fatty acids (87.0%), mainly linoleic acid (77.3%) and oleic acid (9.42%) were identified. Guava seed oil showed some new fatty acids including palmitoleic, nonadecanoic, methyl ricinoleate, 10,13-eicosadienoic, 11-eicosenoic, heneicosanoic, tricosanoic, lignoceric, and cerotic acids in minor concentration. The major unsaturated fatty acids including linoleic and oleic acids were found to be common, but their concentrations were different from previously reported studies (Habib 1986; Opute 1978; Uchôa-thomaz et al. 2014; Castro-Vargas et al. 2011). These differences may be due to either geographical location origin, time of harvest, climate, drying processes, variety, and extraction technique employed (Opute 1978; Heryanto et al. 2017).

Comparative results of the classes of fatty acids in Table 27.6 indicated that the highest concentration of saturated fatty acids (SFA) was observed by Castro-Vargas

Table 27.6 Fatty acid composition of guava seed oil

Fatty acid	Carbon number	Present work (Area % ±SD)	Uchôa-Thomaz et al. (2014)	Opute (1978)	Habib (1986)
Myristic acid	C14:0	0.21 ± 0.02	0.10	–	0.89
Pentadecanoic acid	C15:0	0.03 ± 0.01	–	–	1.20
Palmitoleic acid	C16:1	0.13 ± 0.02	–	–	–
Palmitic acid	C16:0	14.81 ± 0.02	8.00	9.7	13.3
Margaric acid	C17:0	0.19 ± 0.04	0.07	–	–
Oleic acid	C18:1	12.57 ± 0.09	9.42	7.8	14.0
Stearic acid	C18:0	9.08 ± 0.06	4.48	3.4	11.1
Linoleic acid	C18:2	60.03 ± 0.04	77.35	79.1	52.1
Nonadecanoic acid	C19:0	0.04 ± 0.02	–	–	–
Methyl ricinoleate	C18:1	0.21 ± 0.03	–	–	–
10,13-Eicosadienoic acid	C21:2	0.12 ± 0.01	–	–	–
11-Eicosenoic acid	C21:1	0.49 ± 0.01	–	–	–
Arachidic acid	C20:0	1.31 ± 0.08	0.12	–	–
Heneicosanoic acid	C21:0	0.07 ± 0.01	0.14	–	–
Behenic acid	C22:0	0.33 ± 0.02	0.10	–	–
Tricosanoic acid	C23:0	0.08 ± 0.01	–	–	–
Lignoceric acid	C24:0	0.23 ± 0.01	–	–	–
Cerotic acid	C26:0	0.07 ± 0.06	–	–	–
ΣSFA		26.45	13.01	13.1	26.49
ΣMUFA		13.40	9.42	7.8	14.0
ΣPUFA		60.15	77.35	79.1	52.1
ΣUSFA		73.55	86.77	86.9	66.1
Oil content (%)		11.12 ± 0.03	13.93	9.4	8.9

SD standard deviation, *SFA* saturated fatty acids, *MUFA* monounsaturated fatty acids, *PUFA* poly-unsaturated fatty acids, *USFA* unsaturated fatty acids

et al. (2011) (34.2%), followed by Habib (1986) (26.7%), our result (26.4%), Opute (1978) (13.1%) and Uchôa-Thomaz et al. (2014) (12.6%). On the other hand, the highest concentration of unsaturated fatty acids (USFA) was recorded by Opute (1978) (79.1%), followed by Uchôa-Thomaz et al. (2014) (87.7%), our result (73.5%), Habib (1986) (52.2%), and Castro-Vargas et al. (2011) (35.5%).

Comparative results of the individual major fatty acids presented in Table 27.6 indicated that the highest concentration of linoleic acid was observed by Opute (1978) (79.1%), followed by Uchôa-Thomaz et al. (2014) (77.3%), our results (70.3%), Habib (1986) (52.1%), and Castro-Vargas et al. (2011) (17.7%). Stearic acid recorded the highest level by Castro-Vargas et al. (2011) (18.3%), followed by Habib (1986) (11.1%), our results (8.40%), Uchôa-Thomaz et al. (2014) (4.48%) and Opute (1978) (3.40%).

Uses of fatty acids are increased in soap, synthetic organic detergents, rubber, medicine, plastics, cosmetics, and food products. Biological activities of fats and oils make up the greatest proportion of raw materials in the chemical industry commercially produced fatty acids. The majority of this raw material founds in nature as complex mixtures of triglycerides, others esters and alcohols (Minzangi et al. 2011).

9 Health-Promoting Traits of Guava Seed Oil and Its Constituents

The suitability of oil for a particular purpose, however, is determined by its characteristics and fatty acid composition (Qin et al. 2017). The *omega*-6 and *omega*-3 fatty acids (FA) are commonly in the form of linoleic and α-linolenic acids, which are essential FA for humans and must be obtained in food or dietary supplements. These FA cannot be synthesized by the human body independently. The *omega*-6 and *omega*-3 FA have important biological activities, and their ratio of consumption is associated with human health. A lack of *omega*-3 FA in the dietary intake, has been linked to blood lipids, cardiovascular, autoimmune and inflammatory diseases (Yu et al. 2016).

The chemical characteristics of oils and fats are influenced by the kind and proportion of the FA in the triacylglycerol. The predominant FA present in vegetable oils and fats are saturated and unsaturated compounds with straight aliphatic chains. In general, a higher degree of unsaturation of FA in vegetable oils, the more susceptible they are to oxidative deterioration. Therefore, it is essential to know the composition of FA in oils or fats, to identify their characteristics and to determine the possible adulteration, as well as to know the stability and physicochemical properties of these products (Minzangi et al. 2011).

10 Edible and Non-edible Applications of Guava Seed Oil

The FA composition is very important to determine the nutritional value of edible oil. The oil contains a high level of USFA, therefore, the oil has a significant role in the treatment of atherosclerosis and in reducing blood cholesterol levels. *Omega*-6 (linoleic acid) is a major FA identified in guava seed oil, which is essential FA for humans. It has great importance in the human diet and also it has biological and industrial application. Linoleic acid provides oleochemical uses in paints, surfactants, lubricants, varnishes, cosmetics and pharmaceuticals. The oil also contains a greater amount of palmitic acid and stearic acid which are used in the manufacturing of soaps, cosmetics, lubricants and softening (Uchôa-thomaz et al. 2014).

11   Guava Essential Oils

At present, approximately 3000 essential oils (EO) are known, 300 of which are commercially important especially in the food, cosmetic, pharmaceutical and perfume industry. Various potent biological activities including antimicrobial, antioxidant, anti-inflammatory, and anticancer are attributed to EO. As EO represent a source of antimicrobial, antioxidants and anticancer components, they are currently attracting increasing interest in the scientific community and there is much research being performed on their pharmacological activities (Mothana et al. 2013).

EO can be extracted from different parts of plants such as leaves, stem, flower, seeds, bark and roots. Plant EO gained momentum because of there insecticide and fumigant. The EO is a potential candidate for weed control, and is being tried as pest and disease management. EO is easily extractable, and eco-friendly as they are biodegradable and easily catabolized (breakdown of complex molecules) in environment, do not persevere in water and soil, have little or no toxicity against vertebrates, mammals, birds and fishes (Chalannavar et al. 2012).

The EO of *P. guajava* is yellowish with a special aroma. The yield of EO extracted by hydrodistillation method was found to be 0.4% in the fruit, and 0.6% in leaves on (w/w) dry weight. With regards with previous reports (Adam et al. 2011; Khadhri et al. 2014; Ogunwande et al. 2003; Satyal et al. 2016), the yield was comparable but little differences may be related to a seasonal variation or growing habitats could found (Dhouioui et al. 2016).

11.1   Essential Oil Components from P. guajava Fruit

The results obtained from GC-MS of EO components from *P. guajava* fruit are presented in Table 27.7, based on their eluting order on the HP-5 column. A total of 38 components identified in the fruits corresponding to 97.96% of the total oil. The major constituents of the oil were (-)-globulol (18.55%), caryophyllene (17.36%), tau-cadinol (6.61%), nerolidol 2 (4.76%), copaene (4.48%), α-cadinol (3.77%), aromandendrene (3.55%), ledol (2.69%), humulene (2.68%), D-limonene (2.62%), ±γ-δ-cadinene (2.60%), (Z,Z)-2,6-farnesol (2.20%), cis-α-bisabolene (2.12%), β-bisabolene (2.04%), epiglobulol (2.01%), α-bisabolol (1.51%), and α-selinene (1.20%). Considering the main chemical classes, the major class of oil was terpenoids. Along with terpenoid, hydrocarbons were also identified from EO. The total content of hydrocarbons was 5.35%. The major constituent from hydrocarbon was 3-cyclohexen-1-carboxaldehyde, 3,4-dimethyl (1.00%). Previous study (El-Ahmady et al. 2013), on guava fruit EO showed that the major components of EO were β-caryophyllene, (17.6%), limonene (11.0%), β-caryophyllene-oxide (6.7%), α-selinene (6.6%), β-selinene (6.4%), δ-cadinene (4.9%), daucol (4.6%), β-eudesmol (3.7%), τ-cadinol (2.5%), aromadendrene (2.5%), viridiflorola (2.2%), α-copaenea (2.1%), β-copaene (2.1%), spathulenola (2.1%). The major constituents identified

Table 27.7 Components of *P. guajava* fruit essential oil

S. No.	Compound	RT	Relative (% ±SD)	El-Ahmady et al. (2013)	Chen et al. (2016)
1.	1R- α-Pinene	3.47	0.20 ± 0.09	0.4	10.27
2.	Decane	4.38	0.64 ± 0.02	–	–
3.	D-Limonene	5.00	2.62 ± 0.07	11.0	–
4.	Eucalyptol	5.06	1.99 ± 0.05	–	–
5.	Undecane	6.39	0.31 ± 0.06	–	–
6.	Estragole	8.80	0.27 ± 0.01	–	–
7.	Geranyl acetate	12.93	0.21 ± 0.04	–	–
8.	Copaene	13.30	4.48 ± 0.03	2.1	–
9.	Caryophyllene	14.46	17.36 ± 0.09	17.6	0.31
10.	Bergamotene	14.76	0.57 ± 0.02	–	–
11.	Aromandendrene	14.89	3.55 ± 0.01	2.5	–
12.	Humulene	15.24	2.68 ± 0.09	1.0	0.29
13.	Alloaromadendrene	15.41	0.74 ± 0.06	0.8	–
14.	α-Muurolene	15.76	0.56 ± 0.02	–	–
15.	*cis*-α-Bisabolene	16.36	2.12 ± 0.23	–	–
16.	β-Bisabolene	16.51	2.04 ± 0.09	–	–
17.	γ-Maaliene	16.68	0.90 ± 0.02	–	–
18.	δ-Cadinene	16.89	2.60 ± 0.06	4.9	–
19.	Cubenene	17.10	0.87 ± 0.03	–	–
20.	Epiglobulol	17.77	2.01 ± 0.01	–	–
21.	Nerolidol 2	17.83	4.76 ± 0.06	–	0.26
22.	α-Selinene	17.94	1.20 ± 0.02	6.6	–
23.	(-)-Globulol	18.39	18.55 ± 0.05	–	–
24.	Ledol	18.78	2.69 ± 0.08	–	–
25.	3-Cyclohexen-1-carboxaldehyde, 3,4-dimethyl-	18.92	1.00 ± 0.03	–	–
26.	Caryophylladienol II	19.57	7.05 ± 0.06	–	–
27.	*tau*-Cadinol	19.65	6.61 ± 0.01	2.5	–
28.	β-Eudesmol	19.86	0.47 ± 0.05	3.7	–
29.	α-Cadinol	19.94	3.77 ± 0.07	–	–
30.	α-Bisabolol	20.58	1.51 ± 0.03	–	–
31.	(Z,Z)-2,6-Farnesol	21.33	2.20 ± 0.02	–	–
32.	Octadecane	22.91	0.29 ± 0.07	–	–
33.	Nonadecane	24.93	0.31 ± 0.06	–	–
34.	Eicosane	26.93	0.28 ± 0.01	–	–
35.	Heptadecane	35.70	0.32 ± 0.04	–	–
36.	Tetracosane	36.97	0.40 ± 0.07	–	–
37.	Hexacosane	38.56	0.40 ± 0.05	–	–
38.	2,6,10,14,18,22-Tetracosahexaene, 2,6,10,15,19,23-hexamethyl-, (all-E)-	40.47	1.14 ± 0.04	–	–

by Chen et al. (2016), were β-phellandrene (26.74%), 1,8-cineole (25.85%), β-pinene (15.77%), methyl cinnamate (10.48%), α-pinene (10.27%), p-cymene (2.69%), β-Myrcene (1.96%), and α-terpineol (1.46%).

Several reports regarding the volatile components of *P. guajava* fruit have been published. Most recorded components belonged to the chemical group of sesquiterpenes. The major constituent of our study was (-)-globulol (18.55%), while the major constituents of reported studies were β-caryophyllene (17.6%), by El-Ahmady et al. (2013) and β-phellandrene (26.74%), by Chen et al. (2016). D-limonene found in higher concentration (11.0%), as reported by El-Ahmady et al. (2013), followed by our result (2.62%).

11.2   *Essential Oil Components from* P. guajava *Leaves*

A total of 56 components have been identified in the leaves corresponding to 97.9% of the total oil (Table 27.8). The major constituents of the oil examined in our study were caryophyllene (20.3%), (-)-global (8.20%), nerolidol 2 (7.72%), (+) aromadendrene (4.34%) *cis-α*-bisabolene (3.82%), tetracosane (3.68%), octadecane (3.66%), Z,Z,Z-1,5,9,9-tetramethyl-1,4,7-cycloundecatriene (3.44%), *β*-bisabolene (3.41%), limonene (3.09%), octacosane (2.88%), *δ*-cadinene (2.52%), and 1,4-cadadiene (2.04%). The major constituent of our study was β-caryophyllene (20.3%), while the major constituents in the reported studies were Veridiflorol (36.4%), as reported by Khadhri et al. (2014), β-Caryophyllene (16.9%), as reported by El-Ahmady et al. (2013), and β-phellandrene (26.74%) as reported by Chen et al. (2016).

Several reports regarding the volatile components of *P. guajava* leaves have been published. Most recorded components belonged to the chemical group of sesquiterpenes. The chemical compositions revealed that this leaves had to some extent similar compositions to those of other *P. guajava* EO analyzed. The qualitative and quantitative variations between the results may be attributed to genetic variability and/or to environmental conditions. Differences in analytical methods can also be responsible.

Caryophyllene is the main constituent of the EO obtained from guava fruit. It has been reported that caryophyllene exhibited great medicinal values and, approved by the FDA in many food products. This compound is able to modulate inflammatory processes in humans through the endocannabinoid system. *β*-caryophyllene does not bind the central cannabinoid receptor expressed type 1 (CB1), and therefore do not exert psychoactive effects (Corey et al. 1964). Several biological activities are attributed to *β*-caryophyllene, such as local anesthetic, anticarcinogenic, antioxidant, antibiotic, and anti-inflammatory (Legault and Pichette 2007), antinociceptive (Katsuyama et al. 2013), neuroprotective anxiolytic, antidepressant and anti-alcoholism (Bahi et al. 2014). The results reported in the literature suggest that *β*-caryophyllene has been shown to promote the absorption of 5-fluorouracil across

Table 27.8 Components from *P. guajava* leave essential oil

S. No.	Compound	RT	Relative (% ±SD)	El-Ahmady et al. (2013)	Khadhri et al. (2014)	Chen et al. (2016)
1.	1R-α-Pinene	3.47	0.33 ± 0.06	0.3	–	20.16
2.	D-Limonene	5.00	3.09 ± 0.05	0.2	–	–
3.	Eucalyptol	5.06	2.24 ± 0.07	–	–	–
4.	Caryophyllene	14.46	20.34 ± 0.04	16.9	–	0.65
5.	Alloaromadendrene	15.41	1.35 ± 0.08	2.8	–	–
6.	*cis*-α-Bisabolene	16.36	3.82 ± 0.03	–	–	–
7.	β-Bisabolene	16.51	3.41 ± 0.02	–	–	–
8.	γ-Maaliene	16.68	0.17 ± 0.08	–	–	–
9.	δ-Cadinene	16.89	2.52 ± 0.08	5.3	–	–
10.	Nerolidol 2	17.83	7.72 ± 0.03	–	–	–
11.	(-)-Globulol	18.39	8.20 ± 0.02	–	–	–
12.	Ledol	18.78	1.88 ± 0.08	–	–	–
13.	Caryophylladienol II	19.57	1.30 ± 0.02	–	–	–
14.	α-Cadinol	19.94	1.36 ± 0.06	–	–	–
15.	α-Bisabolol	20.58	0.77 ± 0.05	–	–	–
16.	Octadecane	22.91	3.66 ± 0.04	–	–	–
17.	Eicosane	26.93	0.83 ± 0.02	–	–	–
18.	Heptadecane	35.70	1.56 ± 0.06	–	–	–
19.	Tetracosane	36.97	3.68 ± 0.02	–	–	–
20.	Octacosane	41.5892	2.88 ± 0.09	–	–	–
21.	β-Myrcene	4.29	0.05 ± 0.02	–	–	1.3
22.	β-*cis*-Ocimene	5.37	0.05 ± 0.04	0.7	–	–
23.	Linalool	6.48	0.30 ± 0.09	–	–	–
24.	Terpinen-4-ol	8.38	0.07 ± 0.06	–	–	–
25.	*p*-Allylanisole	8.87	0.71 ± 0.03	–	–	–
26.	Carvacrol	11.18	0.17 ± 0.02	0.2	–	–
27.	α-Cubebene	13.38	1.93 ± 0.01	–	–	–
28.	α-Gurjunene	14.23	0.27 ± 0.01	0.4	–	–
29.	γ-Maaliene	14.70	0.17 ± 0.08	–	–	–
30.	β-Gurjunene	14.80	0.11 ± 0.07	–	–	–
31.	(+)-Aromadendrene	14.99	4.34 ± 0.02	2.8	2.2	–
32.	Z,Z,Z-1,5,9, 9-Tetramethyl-1,4,7-cycloundecatriene	15.33	3.44 ± 0.01	–	–	–
33.	γ-Cadinene	15.76	0.21 ± 0.04	–	–	–
34.	γ-Muurolene	15.84	0.76 ± 0.01	–	–	–
35.	β-Selinene	16.10	0.42 ± 0.02	6.3	–	–
36.	β-Guaiene	16.16	0.23 ± 0.09	–	–	–
37.	β-Bisabolene	16.62	3.41 ± 0.02	–	–	–
38.	δ-Cadinene	16.98	2.52 ± 0.08	5.3	<0.05	–
39.	Isoledene	17.03	0.16 ± 0.06	–	–	–

(continued)

Table 27.8 (continued)

S. No.	Compound	RT	Relative (% ±SD)	El-Ahmady et al. (2013)	Khadhri et al. (2014)	Chen et al. (2016)
40.	γ-Selinene	18.03	0.52 ± 0.04	–	–	–
41.	Heptadecane	20.87	1.56 ± 0.06	–	–	–
42.	Pristane	20.99	0.70 ± 0.03	–	–	–
43.	2-Methyl heptadecane	22.21	0.57 ± 0.09	–	–	–
44.	Phytane	23.20	2.15 ± 0.02	–	–	–
45.	4-Methyl octadecane	24.17	0.71 ± 0.06	–	–	–
46.	2-Methyl octadecane	24.28	1.57 ± 0.02	–	–	–
47.	3-Methyl octadecane	24.44	1.13 ± 0.03	–	–	–
48.	1-Nonadecene	27.20	0.93 ± 0.03	–	–	–
49.	2-Methyl eicosane	28.52	0.86 ± 0.06	–	–	–
50.	Heneicosane	34.64	0.35 ± 0.05	–	–	–
51.	1-Tricosene	34.77	0.57 ± 0.02	–	–	–
52.	Tricosane	35.50	0.75 ± 0.03	–	–	–
53.	Tetracosane	37.04	3.68 ± 0.02	–	–	–
54.	1-Hexacosene	37.50	0.80 ± 0.04	–	–	–
55.	Octacosane	37.90	2.88 ± 0.09	–	–	–
56.	Nonacosane	41.19	2.02 ± 0.03	–	–	–

human skin, suggesting that this compound could increase intracellular accumulation of anticancer agents, thereby, potentiating their cytotoxicity. β-caryophyllene could increase the membrane permeation and consequently the biological activity of drugs. β-caryophyllene facilitates also the passage of paclitaxel through the membrane and thus potentiates its anticancer activity (Legault and Pichette 2007).

References

Adam, F., Vahirua-Lechat, I., Deslandes, E., & Menut, C. (2011). Aromatic plants of French Polynesia. V. Chemical composition of essential oils of leaves of *Psidium guajava* L. and *Psidium cattleyanum* Sabine. *Journal of Essential Oil Research, 23*, 98–101.

Bağci, E., & Şahin, A. (2004). Fatty acid patterns of the seed oils of some *Lathyrus* species L. (Papilionideae) from Turkey, a chemotaxonomic approach. *Pakistan Journal of Botany, 36*(2), 403–413.

Bahi, A., Al Mansouri, S., Al Memari, E., Al Ameri, M., Nurulain, S. M., & Ojha, S. (2014). β-Caryophyllene, a CB2 receptor agonist produces multiple behavioral changes relevant to anxiety and depression in mice. *Physiology & Behavior, 135*, 119–124.

Bontempo, P., Doto, A., Miceli, M., Mita, L., Benedetti, R., Nebbioso, A., & Sica, V. (2012). *Psidium guajava* L. antineoplastic effects: Induction of apoptosis and cell differentiation. *Cell Proliferation, 45*(1), 22–31.

Burkill, H. M. (1985). *The useful plants of west tropical Africa* (Vol. 1: families AD) (2nd ed.). Kew: Royal Botanic Gardens.

Castro-Vargas, H. I., Rodríguez-Varela, L. I., & Parada-Alfonso, F. (2011). Guava (*Psidium guajava* L.) seed oil obtained with a homemade supercritical fluid extraction system using supercritical CO_2 and co-solvent. *Journal of Supercritical Fluids, 56*(3), 238–242.

Chalannavar, R. K., Narayanaswamy, V. K., Baijnath, H., & Odhav, B. (2012). Chemical composition of essential oil of *Psidium cattleianum* var. lucidum (Myrtaceae). *African Journal of Biotechnology, 11*(33), 8341–8347.

Chandrika, U., Fernando, K., & Ranaweera, K. (2009). Carotenoid content and in vitro bioaccessibility of lycopene from guava (*Psidium guajava*) and watermelon (*Citrullus lanatus*) by high-performance liquid chromatography diode array detection. *International Journal of Food Sciences and Nutrition, 60*(7), 558–566.

Charles, W., Philip, E., & Carl, W. (2006). Determination of organic acids and sugars in *guajava* L. cultivars by high-performance liquid chromatography. *Food and Agriculture, 33*, 777–780.

Chen, Z., He, B., Zhou, J., He, D., Deng, J., & Zeng, R. (2016). Chemical compositions and antibacterial activities of essential oils extracted from Alpinia guilinensis against selected foodborne pathogens. *Industrial Crops and Products, 83*, 607–613.

Conway, P. (2001). *Tree medicine: a comprehensive guide to the healing power of over 170 trees*. London: Judy Piatkus (Publishers) Limited.

Corey, E. J., Mitra, R. B., & Uda, H. (1964). Total synthesis of d, l-Caryophyllene and d, l-Isocaryophyllene. *Journal of American Chemical Society, 86*, 485–492.

Dakappa, S. S., Adhikari, R., Timilsina, S. S., & Sajjekhan, S. (2013). A review on the medicinal plant *Psidium guajava* Linn. (Myrtaceae). *Journal of Drug Delivery and Therapeutics, 3*(2), 162–168.

Dhouioui, M., Boulila, A., Chaabane, H., Zina, M. S., & Casabianca, H. (2016). Seasonal changes in essential oil composition of *Aristolochia longa* L. ssp. paucinervis Batt. (Aristolochiaceae) roots and its antimicrobial activity. *Industrial Crops and Products, 83*, 301–306.

Dweck, A. (2001). A review of *Psidium guajava*. *Malayan Journal of Medical Science, 8*, 27–30.

El-Ahmady, S. H., Ashour, M. L., & Wink, M. (2013). Chemical composition and anti-inflammatory activity of the essential oils of *Psidium guajava* fruits and leaves. *Journal of Essential Oil Research, 25*(6), 475–481.

Flores, G., Wu, S. B., Negrin, A., & Kennelly, E. J. (2015). Chemical composition and antioxidant activity of seven cultivars of guava (*Psidium guajava*) fruits. *Food Chemistry, 170*, 327–335.

Fujita, T., Kamei, M., Kanbe, T., Sasaki, K., Yamaguchi, K., & Oshiba, K. (1985). Nutrient contents in fruits and leaves of guava, and in leaves of Japanese Persimmon. *Seikatsu Eisei Journal of Urban Health, 29*(4), 206–209.

Gutiérrez, R. M. P., Mitchell, S., & Solis, R. V. (2008). *Psidium guajava*: A review of its traditional uses, phytochemistry and pharmacology. *Journal of Ethnopharmacology, 117*(1), 1–27.

Habib, M. (1986). Studies on the lipid and protein composition of guava seeds (*Psidium guajava*). *Food Chemistry, 22*(1), 7–16.

Heryanto, R., Permana, D., Tedjo, A., Rohaeti, E., Rafi, M., & Darusman, L. K. (2017). A simple photometer and chemometrics analysis for quality control of sambilotu (*Andrographis Paniculata*) raw material. *Indonesian Journal of Pure and Applied Chemistry, 6*(3), 238–245.

Hwang, J., Yen, Y., Chang, M., & Liu, C. (2002). Extraction and identification of volatile components of guava fruits and their attraction to oriental fruit fly, *Bactrocera dorsalis* (Hendel). *Plant Protection Bulletin (Taipei), 44*(4), 279–302.

Iwu, M. (1993). *Handbook of African medicinal plants, pharmacognostical profile of selected medicinal plants*. Boca Raton, Florida: CRC-Press. https://doi.org/10.1201/b16292

Jain, N., Dhawan, K., Malhotra, S., & Singh, R. (2003). Biochemistry of fruit ripening of guava (*Psidium guajava* L.): Compositional and enzymatic changes. *Plant Foods for Human Nutrition, 58*(4), 309–315.

Jordán, M. J., Margaría, C. A., Shaw, P. E., & Goodner, K. L. (2003). Volatile components and aroma active compounds in aqueous essence and fresh pink guava fruit puree (*Psidium guajava* L.) by GC-MS and multidimensional GC/GC-O. *Journal of Agricultural and Food Chemistry, 51*(5), 1421–1426.

Joseph, B., & Priya, M. (2011). Review on nutritional, medicinal and pharmacological properties of guava (*Psidium guajava* Linn.). *International Journal of Pharma and Bio Sciences, 2*(1), 53–69.

Katsuyama, S., Mizoguchi, H., Kuwahata, H., Komatsu, T., Nagaoka, K., Nakamura, H., Bagetta, G., Sakurada, T., & Sakurada, S. (2013). Involvement of peripheral cannabinoid and opioid receptors in β-caryophyllene-induced antinociception. *European Journal of Pain, 17*, 664–675.

Khadhri, A., El Mokni, R., Almeida, C., Nogueira, J. M. F., & Araújo, M. E. M. (2014). Chemical composition of essential oil of *Psidium guajava* L. growing in Tunisia. *Industrial Crops and Products, 52*, 29–31.

Kumar, A. (2012). Importance for life '*Psidium guava*'. *InternationalJournal of Research in Pharmaceutical and Biomedical Sciences, 3*(1), 137–143.

Legault, J., & Pichette, A. (2007). Potentiating effect of β-caryophyllene on anticancer activity of α-humulene, isocaryophyllene and paclitaxel. *Journal of Pharmacy and Pharmacology, 59*, 1643–1647.

Li, J., Chen, F., & Luo, J. (1999). GC-MS analysis of essential oil from the leaves of *Psidium guajava*. *Zhong Yao Cai= Zhongyaocai= Journal Chinese Medicinal Materials, 22*(2), 78–80.

Ogunwande, I. A., Olawore, N. O., Adeleke, K. A., Ekundayo, O., & Koenig, W. A. (2003). Chemical composition of the leaf volatile oil of *Psidium guajava* L. growing in Nigeria. *Flavour and Fragrance Journal, 18*, 136–138.

Medina, M., & Pagano, F. (2003). Caracterización de la pulpa de guayaba (*Psidium guajava* L.) tipo "Criolla Roja". *Revista de la Facultad de Agronomia Luz, 20*(1), 72–86.

Mehmood, A., Jaskani, M. J., Khan, I. A., Ahmad, S., Ahmad, R., Luo, S., & Ahmad, N. M. (2014). Genetic diversity of Pakistani guava (*Psidium guajava* L.) germplasm and its implications for conservation and breeding. *Scientia Horticulturae, 172*, 221–232.

Michael, H., Salib, J., & Ishak, M. (2002). Acylated flavonol glycoside from *Psidium gauijava* L. seeds. *Pharmazie, 57*(12), 859–860.

Minzangi, K., Kaaya, A., Kansiime, F., Tabuti, J., Samvura, B., & Grahl-Nielsen, O. (2011). Fatty acid composition of seed oils from selected wild plants of Kahuzi-Biega National Park and surroundings, Democratic Republic of Congo. *African Journal of Food Science, 5*(4), 219–226.

Mothana, R. A., Al-Said, M. S., Al-Yahya, M. A., Al-Rehaily, A. J., & Khaled, J. M. (2013). GC and GC/MS analysis of essential oil composition of the endemic Soqotraen *Leucas virgata* Balf. f. and its antimicrobial and antioxidant activities. *International Journal of Molecular Sciences, 14*(11), 23129–23139.

Moussa, T. A., & Almaghrabi, O. A. (2016). Fatty acid constituents of *Peganum harmala* plant using Gas Chromatography–Mass Spectroscopy. *Saudi Journal of Biological Sciences, 23*(3), 397–403.

Nadkarni, A. K. (1996). *[Indian materia medica]; Dr. KM Nadkarni's Indian materia medica: With Ayurvedic, Unani-Tibbi, Siddha, allopathic, homeopathic, naturopathic & home remedies, appendices & indexes*. 1 (Vol. 1). Mumbai: Popular Prakashan. https://trove.nla.gov.au/version/21887031

Norshazila, S., Syed Zahir, I., Mustapha Suleiman, K., Aisyah, M., & Kamarul Rahim, K. (2010). Antioxidant levels and activities of selected seeds of Malaysian tropical fruits. *Malaysian Journal of Nutrition, 16*(1), 149–159.

Opute, F. I. (1978). The component fatty acids of *Psidium guajava* seed fats. *Journal of the Science of Food and Agriculture, 29*(8), 737–738.

Paniandy, J. C., Chane-Ming, J., & Pieribattesti, J. C. (2000). Chemical composition of the essential oil and headspace solid-phase microextraction of the guava fruit (*Psidium guajava* L.). *Journal of Essential Oil Research, 12*(2), 153–158.

Pelegrini, P. B., Murad, A. M., Silva, L. P., dos Santos, R. C., Costa, F. T., Tagliari, P. D., & Franco, O. L. (2008). Identification of a novel storage glycine-rich peptide from guava (*Psidium guajava*) seeds with activity against Gram-negative bacteria. *Peptides, 29*(8), 1271–1279.

Qin, X. J., Yu, Q., Yan, H., Khan, A., Feng, M. Y., Li, P. P., & Liu, H. Y. (2017). Meroterpenoids with antitumor activities from guava (*Psidium guajava*). *Journal of Agricultural and Food Chemistry, 65*(24), 4993–4999.

Satyal, P., Paudel, P., Lamichhane, B., & Setzer, W. N. (2016). Leaf essential oil composition and bioactivity of *Psidium guajava* from Kathmandu, Nepal. *American Journal of Essential Oils and Natural Products, 3*, 11–14.

Shuklaa, S. P., Tiwaria, S., Tiwaria, M., Mohanb, D., & Pandeyc, G. (2017). Removal of fluoride from aqueous solution using *Psidium guajava* leaves. *Desalination and Water Treatment, 62*, 418–425.

Tlili, N., El Guizani, T., Nasri, N., Khaldi, A., & Triki, S. (2011). Protein, lipid, aliphatic and triterpenic alcohol content of caper seeds "*Capparis spinosa*". *Journal of the American Oil Chemists' Society, 88*(2), 265–270.

Uchôa-thomaz, A. M. A., Sousa, E. C., Carioca, J. O. B., Morais, S. M. D., Lima, A. D., Martins, C. G., & Rodrigues, S. P. (2014). Chemical composition, fatty acid profile and bioactive compounds of guava seeds (*Psidium guajava* L.). *Journal of Food Science and Technology, 34*(3), 485–492.

Usman, M., Samad, W. A., Fatima, B., & Shah, M. H. (2013). Pollen parent enhances fruit size and quality in intervarietal crosses in guava (*Psidium guajava*). *International Journal of Agriculture and Biology, 15*(1), 125–129. www.thedailyrecords.com/2018.../world.../world/...guava-producing...world.../6566/.

Yu, S., Du, S., Yuan, J., & Hu, Y. (2016). Fatty acid profile in the seeds and seed tissues of *Paeonia* L. species as new oil plant resources. *Scientific Reports, 6*, 26944.

Chapter 28
Mango (*Mangifera indica* L.) Seed Oil

Bushra Sultana and Rizwan Ashraf

Abstract Mango fruit has its prime importance in the market due to its healthy nutrients and improving well-being by preventing different disease. During processing its seeds are considered as wastes that are mostly dumped due uncommon in use. Recent studies showed that its waste can be employed for food security having latent medicinal traits. The presence of phytochemicals, bioactives and different fatty oils is further diverting the attention of food and nutritional scientist. Introduction of different extraction methods by analytical scientist makes a pathway to access these valuable natural constituents and can be used for different pathological complications. In this chapter, different aspects of mango oil have been described including their extraction and use in food processing as well as medicinal applications.

Keywords Agricultural by-products · Phytochemicals · Fatty oils · Extraction methods · Antioxidants

1 Introduction

The mango (*Mangifera indica* L.) is an important tropical fruit in the world that maintains its demand in the market due to its pleasant taste, rich in nutritional contents and pleasing aroma (Ibarra-Garza & Ingrid 2015). Based on chemical composition mango is supposed as king of fruits and secure a distinction in the market as second most traded fruit in the world. Its world production is estimated to more than 42 million tons per year with India as largest mango producer followed by China, Thailand, Indonesia, Pakistan and Mexico (Faostat 2015).

According to an estimate, during a survey, 35–60% of mango fruit is considered as waste after processing (O'Shea et al. 2012) in particular seeds that are more than million tons annually. After processing, this waste byproduct is considered a threat to contribute to environmental pollution, also its nutrients and phytochemicals are

B. Sultana (✉) · R. Ashraf
Department of Chemistry, University of Agriculture Faisalabad, Faisalabad, Pakistan

© Springer Nature Switzerland AG 2019
M. F. Ramadan (ed.), *Fruit Oils: Chemistry and Functionality*,
https://doi.org/10.1007/978-3-030-12473-1_28

wasted (Ayala-Zavala et al. 2011). This waste can be processed for further commercial utilization. Therefore, mango seeds are getting special attention due to the presence of some bioactive constituents. These worth bioactive are phenolic, carotenoids, vitamins, and dietary fibers that could be employed for treatment of human health issues (Jahurul et al. 2015). It is full of fibers, sugars, vitamins and minerals (Tharanathan et al. 2006). It is also enriched with carbohydrates (58–80%), protein (6–13%) as well as lipids and amino acids (6–16%) that make it convenient for nutritional products (Diarra 2014).

Mango kernel, the inner part of mango seed shell, is obtained from mango stone mechanically. Rukmini and Vijayaraghavan (1984) reported that mango seed kernel is a promising and safe source of fat and edible oil. Mango kernel oil is one of the byproducts of the mango fruit and mostly used in cosmetic and soap industry. It is very secure even could be substituted for any solid fat without adverse effects. Moreover, its physicochemical properties are very similar to those of a commercially employed cocoa butter. Based on origin and variety of mango, a wide variation was observed in the composition of fatty acids. It can be a good medicinally due to rich contents of phytosterols, phenolics, and tocopherols.

It is obvious from above facts that proper utilization of mango waste for a food additive or raw material could produce economic gains for an industry that will contribute to augment nutritional values, promote health, and ultimately reduce environmental complications of this generated waste. This chapter will cover up all those aspects that will transform this waste product to a valuable source of nutrients and food additives viable commercially and good raw material for industry (Ediriweera et al. 2017).

2   Mango Seed Constituents

The mango seed can be mono-embryonic or poly-embryonic depending upon the variety and origin of plants in which mono-embryonic is mostly present in India and Pakistan while poly-embryonic is abundant in Thailand, Philippines and Indonesia (Tharanathan et al. 2006). Depending on the variety, seed represents an average 10–25% of the total weight of fruit, and the kernel is 45–85% of the seed part. Mango seed contains 21% of starch that has similar characteristics with tapioca starch, which is highly indigestible. This limited digestibility could be associated with high tannin content (0.19–0.44%). Some other bioactives having anti-nutritional factors such as cyanogenic glycosides (64 mg kg^{-1}), tannins (56.5 g kg^{-1}), oxalates (42 mg kg^{-1}), and trypsin inhibitory activity (20 TIU g^{-1}) are also present. Being a rich source of nutritional and bioactive components, mango seeds can be a good source of antioxidants. In recent studies, phytochemicals of mango seeds have been the object of intense research owing to their high antioxidant potential (Solís-Fuentes and del Carmen Durán-de-Bazúa 2011).

3   Extraction and Processing of Mango Oil

Mango seed oil can be extracted through a number of techniques commercially including solvent extraction and mechanical shaker. In solvent extraction, a non-polar solvent *n*-hexane, is mostly employed as an extraction medium. Before extraction process, collected mango stones firstly are washed to remove contaminants and sun-dried to reduce the moisture content to 12–15%. These dried seed stone are roasted in a drum roaster and the hull is removed mechanically as shown in Fig. 28.1. The separated kernels are firstly crushed for chop into small pieces. The mango kernel pieces are converted to powder form and employed for solvent extraction. Absolute methanol was also employed for extraction of mango kernel oil (Abdalla et al. 2007b; Masibo and He 2009).

Different solvents including *n*-hexane, petroleum ether and ethanol were applied for extraction of mango seed oil and their results also assure that *n*-hexane is the more appropriate choice for optimum yield (Kittiphoom and Sutasinee 2013). Microwave pretreatment before solvent extraction was very effective to improve the oil yield and reduced the extraction time. The most appropriate pretreatment conditions were microwave application for the 60 s at 110 W that could reduce the time to five times that normally required for conventional extraction (Kittiphoom and Sutasinee 2015).

Hydraulic pressing is used to extract mango kernel oil, which is useful on small scale due to lower initial and operating cost. Moreover, it gives uncontaminated oil

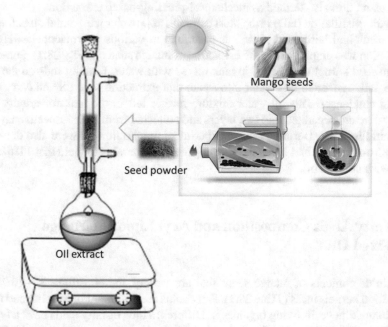

Mango seeds

Seed powder

Oil extract

Fig. 28.1   Solvent-based extraction of mango seed oil

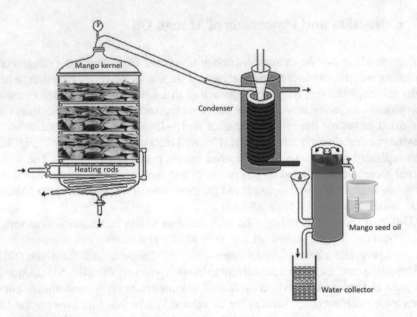

Fig. 28.2 Hydro-distillation extraction for mango seed oil

and pure cake residue as compared to screw press or solvent extraction method (Babaria 2011). In this type of extraction, different process parameters are responsible to customize the quality and yield efficiency of mango kernels oil. Hydraulic press extraction also termed as mechanical-pressing-based extraction.

Hydro-distillation (HD) is the most common approach for essential oil extraction from medicinal herbs and plants, as described in various pharmacopeias. Hydro-distillation was employed for the extraction of mango seed oil (Fig. 28.2). Sun-dried mango seed kernel was placed in tank mixed with water. Heating rods on the base of the tank converted this water into steam that extraction seed oil and carry them toward condenser. This oil-water mixture further left in the tank for gravity base separation and separated into two layers and collected from their respective layer as given in Fig. 28.2. During a typical experiment mango flowers were also deployed for hydro-distilled for 4 h using a Clevenger-type apparatus to yield that 11% of the yellowish colored oil.

4 Fatty Acids Composition and Acyl Lipids of Mango Fixed Oil

The lipids contents of mango seeds that are variety and constitute 3.7–13.7% of total dried kernels mass (Table 28.1). Fatty acids are important metabolite and nutritional constituents in living organisms. Different kinds of fatty acids have a typical role in the regulation of biological and physiological functions (Ajila et al. 2007).

Table 28.1 Comparison of different varieties for their nutritional values (Jin et al. 2018)

Food constituent	Kenya	MKF China	Thailand	Tanzania	India	Malaysia	CB	Palm oil
Fat in kernel (%)	8.5–10.4	5.7–11.1		9.2–10.5	3.7–12.6	6.4–13.7	–	–
IV (g/100 g)	51.1–56.8	42.2–60.7	40.9	41.7–43.5	–	42.9–52.7	34.2–40.7	50.6–55.1
SMP (°C)	25.0–33.0	25.5–32.1	35.7	29.1–30.5	23.5–30.0	35.8–39.1	30.0–35.0	30.8–37.6
Fat composition (%)								
Triacylglycerol	–	85.1–92.7	–	84.9–91.8	–	–	>90.0	>85.0
Diacylglycerol	–	2.5–5.8	–	2.0–4.2	–	–	1.1–2.8	6.6–6.7
Monoacylglycerol	–	tr-0.3	–	0.1–0.8	–	–	–	0.2–0.3
Free fatty acid	2.3–3.8	2.5–6.6	–	3.1–7.5	1.1–4.4	1.6–2.6	1.1–2.3	<5.0
Sterol	–	0.38–0.71	–	0.42–0.84	1.03	–	–	0.04–0.06
Tocopherol	–	0.01–0.09	–	0.10–0.13	–	–	0.01–0.03	0.06–0.010
Squalene	–	0.02–0.9	–	0.07–0.20	–	–	–	0.02–0.05
Triacylglycerol (%)								
PPP	–	tr[d]	tr	tr	–	–	tr	6.9–7.2
POP	–	1.2–2.8	8.9	1.0–1.7	1.1	6.9	13.6–15.5	27.2–30.6
POSt	–	13.1–19.8	5.7	13.2–16.2	17.3	14.8	33.7–40.5	3.7–4.2
POO	–	1.8–7.8	10.8	3.7–8.7	–	–	1.5–6.2	20.7–23.1
StOSt	–	30.0–55.4	29.4	28.9–45.7	38.1	39.3	23.8–31.2	tr-0.2
StOO	–	11.2–23.3	14.6	20.2–32.2	–	–	2.7–9.5	1.6–2.0
OOO	–	3.2–8.6	2.5	4.6–9.6	–	–	tr-1.0	5.1–5.4
StLSt	–	tr	14.6	tr	–	–	1.4–2.0	tr
StOA	–	0.9–4.1	tr	1.0–1.9	–	–	0.8–1.3	tr
Sn-2 fatty acid composition (%)								
P	–	1.1–2.0	–	1.0–2.1	–	–	4.7	–
St	–	1.3–6.9	–	1.3–5.2	–	–	4.7	–
O	–	78.3–84.4	–	80.8–85.7		–	82.8	>58.3
L	–	8.3–14.8	–	7.8–13.0	–	–	6.6	>18.4
A	–	1.1–1.7	–	tr	–	–	tr	–

tr represent the term traces
d represent traces <0.05%

The major contribution among fatty acids in mango kernel oil is oleic acid (45%) and stearic acid (38%) (Shah et al. 2010). Oleic acid is unsaturated fatty acid has 18-carbon in chain length and considered as essential for human nutrition that helps to reduce LDL-cholesterol, triacylglycerols (TAGs), glycemic index and total cholesterol. When incorporated or part of vegetable oil, it increases the stability over-oxidation of the vegetable oil. While stearic acid is also a long chain C18, but saturated fatty acid is responsible to bind and plasticize composites (Netravali 2003), α-helical sites in bio-molecules and human serum albumin (Kittiphoom and Sutasinee 2013).

The other fatty acid profile revealed that TAGs which secure 84.9–92.7% of total fatty acids, followed by free fatty acids (FFAs) contained 1.1–7.5%, diacylglycerols (DAGs, 2.0–5.8%) and mono-acylglycerols with a quantity of 0.8%. The composition of fatty acids is very depending on variety as well as the origin of mango variety that is given in Table 28.1 that depicts the major differences in their TAGs and micronutrients. Lipids extracted from wild mango (Irvingiagabonensis) in Congo have surprisingly different TAGs composition compared to those presented in Table 28.1. It contained lower amount of stearic acid (0.9%) and oleic acid (4.3%), but significant amount of lauric acid (37.1%) and myristic acid (49.8%) (Yamoneka et al. 2015).

When the presence of positional isomers and TAG composition was analyzed, It was observed that oleic acid is the main contributor in sn-2 fatty acids that varies from 78.3% to 89.9% based on variety which resembles canola beads having 82.3% of oleic acid. The StOSt content in these lipids was as high as 65.0–66.3%, indicating they are ideal ingredients for producing the natural CBIs without further modification (Akhter et al. 2016).

The composition of the TAG is major contributor to improve the quality, whereas mineral contents alter the shelf life of lipid-containing food. It is worth noting that most of the mango varieties contain high contents of FFAs that accelerate the oxidation process during storage and shorter the shelf life. These high contents of FFAs also produce smoke on frying or cooking that deterrent its use in food items. A report on the composition of Bangladeshi mango seed oil exposed the presence of high FFAs contents up to 30–37%. Therefore, refining is necessary for processing to reduce high contents of FFAs. There are increasing scientific interests in preparing similar TAG species by exploring new sources and developing modification techniques like fractionation and interesterification. Jin and Jun (2017) selectively fractionated ordinary lipids from mango seeds to produce higher StOSt content from 44.0% to 69.2% as CBI using 2-methylpentane.

Lipids have high nutritional values and highly desired in the food industry due to their effects on the functional properties of food. Mango seed lipids have enticed interest due to cheap but comparable quality source of fats viable for cooking and food usage. It can be observed from Table 28.2 that presence of high-fat content in the seeds of mangos (8.15–13.16%) are responsible for typical characteristics of a vegetable butter (Muchiri et al. 2012). Table 28.1 shows the major fatty acids present in mango seed oil in which stearic and oleic acids are predominant depends upon the variety and country of origin. A Nigerian variety showed a high level of

Table 28.2 Proximate composition of mango seed of different mango varieties (Ribeiro and Schieber 2010)

Variety	Moisture (%)	Lipids (%)	Carbohydrate (%)	Protein (%)	Crude fiber (%)	Ash (%)
Nigeria	44.4	128	32.8	6	2	2
Nigeria	94	13.6	–	6.1	4.6	2.2
Kibangou	45.2	13.3	32.2	6.36	2.02	3.2
Falta	–	11	–	6	2	2
(Egypt) zebda, balady, succary	–	123	–	6.7	2.7	2.5
Zebda (Egypt)	–	8.15	–	7.76	0.26	1.46

oleic acid that is almost thrice than other known varieties. The same variety also reported containing a higher content of unsaturated fatty acids that are approximated to twice that of saturated fatty acids. Similar results were observed by Abdalla et al. (2007a) for Egyptian variety having stearic acid and oleic acid approximately 84.4% and 86.2% of total fatty acids, respectively. These attributes proved that mango oil is suitable for mixing together with other vegetable oils for proper use in the confectionery industry. This stance was strengthened by Abdalla et al. (2007b) who reported that combined use of mango seed extracts and commercial oil has higher antioxidant capacity than the single one. Antioxidants are important ingredients that protect the quality of oils and fats by retarding oxidation. Lipid oxidation is a major constraint in food quality because oxidized products have a foul aroma and bad impact on taste as well as nutritional values. Other traits of mango seed lipids that it lacks *trans* fatty acids. Presence of *trans* fatty acids are directly responsible for various diseases and have adverse effects on human health.

Presence of FFAs and peroxide contents measure the quality of oil. The low value of FFAs in mango seed oil is an indication of high quality. Mango seed oil is almost free from rancidity brought almost by lipases make them able to direct use in industries without neutralization. Iodine value is a measure of unsaturation present in the oil. Saponification value provides the information for average molecular weight (or chain length) of all the fatty acids. Matter of an oily mixture which fails to form soap when blended with NaOH is termed as unsaponifiable matter. The composition of unsaponifiable matter of vegetable oils including sterols, tocopherols and squalene is of great importance for oil characteristics (Ediriweera et al. 2017).

Mango seeds contain on average 6.0–7.76% protein by dry weight (Table 28.2). Even though nuclei have lower protein content but showed a good composition of amino acids. The amino acids were leucine, valine, lysine, phenylalanine, threonine, isoleucine, tyrosine and methionine having 6.9, 5.8, 4.4, 4.3, 3.4, 3.4, 2.7 and 1.2 g of amino acid per 100 g protein, respectively. These amino acids contents in some varieties are more than recommended an essential level of proteins in cooking oil. Some Egyptian mango varieties like Zebda, Balady, and Succary have shown essential amino acids contents are at a higher level that is 70% than the minimum requirement of FDA except threonine, methionine and tyrosine (Abdalla et al. 2007b).

5 Bioactive Compounds in Mango Fixed Oil

The bioactive compounds are natural constituents present in plants/foods with attractive health benefits (Biesalski et al. 2009). Currently, all processed foods have synthetic antioxidants such as butylated hydroxyl anisole (BHA) and butylated hydroxyl toluene (BHT) that are originating from the oil industry (Carocho and Ferreira 2013). The consumption of synthetic antioxidants can cause harmful health effects that can be avoided by using natural antioxidants. In mango seeds, phenolic compounds (mangiferin, isomangiferin, homomangiferin, quercetin, kaempferol and anthocyanins), phenolic acids (gallic, protocatechuic, ferulic, caffeic, coumaric, ellagic and 4-caffeoylquinic acids), and mineral (potassium, copper, zinc, manganese, iron and selenium) are reported (their structures are given in Figs. 28.3 and 28.4) (Ribeiro and Schieber 2010).

The polyphenol mangiferin is extensively studied due to its numerous biological attributes their quantities in different mango varieties are given in Tables 28.3 and 28.4. Other phenolic compounds that are present in kernel oil are fumaric acid, 2-decyl undecyl ester (2.85%), 4,6-di (1,1-dimethylethyl)-2- methyl- (44.6%), apigenin 7- glucoside (1.74%), cis-5-dodecenoic acid, isoheptadecanol (1-hexadecanol, 2-methyl) (2.74%) and (3-cyanopropyl) dimethylsilyl ester (1.47%).

It given in Table 28.5, mango seed has high contents of mineral that varies with variety and origin. Commonly, it has high level of phosphorus, potassium and magnesium (Fowomola 2010). Mineral nutrients are very essential for daily

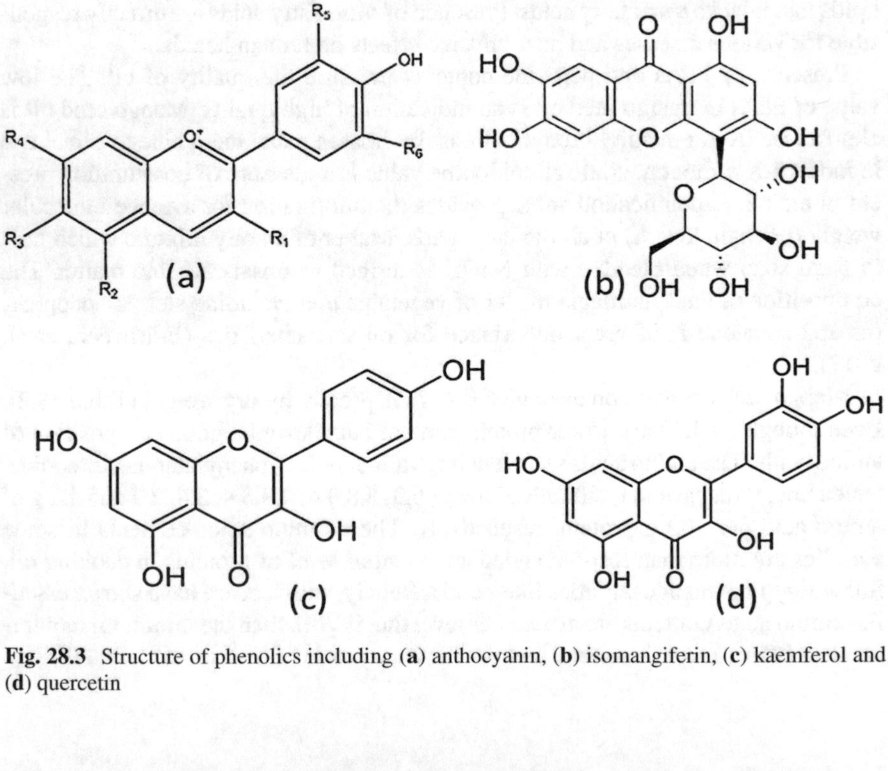

Fig. 28.3 Structure of phenolics including (a) anthocyanin, (b) isomangiferin, (c) kaemferol and (d) quercetin

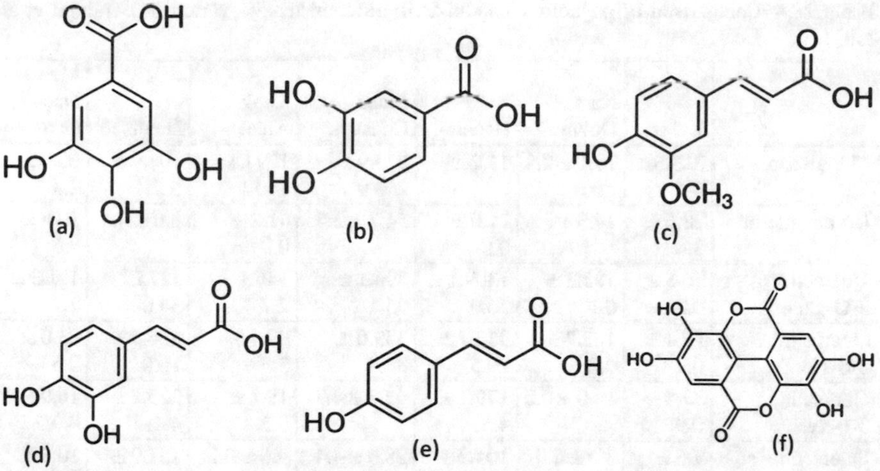

Fig. 28.4 Structure of phenolic acids including (**a**) gallic acid, (**b**) protocatechuic acid, (**c**) ferulic acid, (**d**) caffeic acid, (**e**) coumaric acid and (**f**) ellagic acid

Table 28.3 Comparison of phenolic compounds in different mango varieties (Berardini et al. 2005)

Phytochemical	Tommy Atkins	Manila	Jose	R2E2	Ngowe	Kent	Jose
Mangiferin	1263.2 ± 197.2	43.5 ± 9.8	983.6 ± 50.1	82.9 ± 4.2	775.8 ± 11.7	13.9 ± 1.5	983.6 ± 50.1
Isomangiferin[a]	40.3 ± 0.8	11.5 ± 1.2	45.5 ± 1.9	19.0 ± 0.3	184.7 ± 5.9	4.0 ± 0.3	45.5 ± 1.9
Quercetin 3-O-gal	1217.3 ± 18.0	430.6 ± 18.4	1467.7 ± 42.3	116.5 ± 8.8	797.6 ± 11.8	944.5 ± 38.3	1467.7 ± 42.3
Quercetin 3-0-glc	882.0 ± 4.2	282.5 ± 12.9	1045.3 ± 41.6	124.4 ± 5.9	359.1 ± 6.3	890.0 ± 39.8	1045.3 ± 41.6
Quercetin 3-0-xyl	239.5 ± 3.8	39.2 ± 2.9	278.6 ± 8.4	17.8 ± 0.4	100.9 ± 1.7	150.7 ± 8.2	278.6 ± 8.4
Quercetin 3-0-arap[d]	163.5 ± 2.8	27.6 ± 1.2	191.8 ± 7.5	8.8 ± 0.3	70.7 ± 0.5	91.6 ± 3.4	191.8 ± 7.5
Quercetin 3-0-araf[d]	152.4 ± 2.7	17.9 ± 1.2	119.6 ± 3.5	8.2 ± 0.5	56.8 ± 2.2	84.8 ± 3.8	119.6 ± 3.5
Quercetin 3-0-rha	38.2 ± 1.7	15.6 ± 0.6	116.4 ± 4.3	5.3 ± 0.3	20.6 ± 0.8	58.1 ± 3.5	116.4 ± 4.3
Quercetin 3-0-glc	77.3 ± 5.3	16.8 ± 1.1	171.7 ± 8.8	11.2 ± 1.1	24.1 ± 1.3	30.6 ± 1.8	171.7 ± 8.8
Quercetin 3-0-gal/glc[e]	215.6 ± 4.9	14.6 ± 1.7	374.4 ± 11.1	5.4 ± 0.0	22.6 ± 0.4	70.6 ± 3.4	374.4 ± 11.1

a, Quantified as mangiferin. b, Quantified as mangiferin (including molecular weight correction factor). c, Quantified as Q 3-O-arabinoglucoside. d, Quantified as Q 3-Oxyloside. e, Quantified as isorhamnetin 3-O-glucoside

Table 28.4 Comparison of phenolic compounds in different mango varieties (Berardini et al. 2005)

	Kaew	Nam Dokmai	Haden	Maha Chanock	Chok Anan	Heidi	Mon Duen Gao
Mangiferin	313.6 ± 4.4	78.1 ± 2.8	11.2 ± 0.1	973.9 ± 106.9	1297.1 ± 140.1	108.9 ± 3.5	68.0 ± 2.3
Isomangiferin[a]	28.6 ± 3.6	14.3 ± 1.6	21.0 ± 0.8	54.7 ± 2.5	41.2 ± 0.0	8.0 ± 0.3	10.6 ± 0.6
Quercetin 3-O-gal	76.5 ± 0.2	185.2 ± 0.1	1309.1 ± 26.0	396.1 ± 35.1	146.1 ± 2.7	1275.7 ± 34.0	121.3 ± 2.7
Quercetin 3-0-glc	77.4 ± 0.9	103.7 ± 0.0	912.7 ± 20.5	339.0 ± 38.4	92.4 ± 2.3	814.5 ± 16.0	83.0 ± 2.6
Quercetin 3-0-xyl	77.4 ± 0.9	14.0 ± 0.0	179.1 ± 4.5	42.2 ± 4.7	19.1 ± 0.5	225.7 ± 4.4	16.9 ± 0.0.5
Quercetin 3-0-arap[d]	7.6 ± 0.1	8.3 ± 0.0	104.9 ± 5.1	28.6 ± 4.4	9.6 ± 0.2	131.9 ± 3.4	10.2 ± 0.2
Quercetin 3-0-araf[d]	6.8 ± 0.1	6.1 ± 0.0	70.5 ± 0.8	19.8 ± 2.1	8.2 ± 0.3	123.5 ± 2.6	5.5 ± 0.1
Quercetin 3-0-rha	4.2 ± 0.0	5.9 ± 0.3	52.7 ± 0.6	20.6 ± 3.1	4.2 ± 0.1	41.6 ± 0.5	4.6 ± 0.2
Quercetin 3-0-glc	9.9 ± 0.2	3.6 ± 0.6	43.7 ± 1.1	36.5 ± 4.1	20.9 ± 2.1	73.0 ± 1.2	6.7 ± 0.1

a, Quantified as mangiferin. b, Quantified as mangiferin (including molecular weight correction factor). c, Quantified as Q 3-O-arabinoglucoside. d, Quantified as Q 3-Oxyloside. e, Quantified as isorhamnetin 3-O-glucoside

Table 28.5 Comparison of minerals in different mango varieties

Variety	Potassium (mg 100 g⁻¹)	Magnesium (mg 100 g⁻¹)	Phosphorus (mg 100 g⁻¹)	Calcium (mg 100 g⁻¹)	Sodium (mg 100 g⁻¹)
Nigeria	365	100	140	49	–
Nigeria	60	980	230	450	150
Kibangou (Kongo)	158	22.34	20	10.21	2.7

consumption among which phosphorus and potassium are required in quantities of 10 g/day. The dietary intake of other minerals like calcium, magnesium, zinc and iron are also necessary for good health. The deficiency of these micronutrients are associated with nutritional insecurity and disturb normal metabolic functioning of the human body (Torres-León et al. 2016).

6 Mango Essential Oil

The essential oil of mango seeds contains several components. In an analysis of three variety of mango, almost 32 compounds were identified in which 4-carene was the major components (29.09% of essential oil). Among the monoterpenes, α-terpinolene was the major component of essential oil contributes 44.57% of total monoterpenes followed by cyclohexadecane. In a variety of Falan mango, monoterpenes were the major compounds and accounted for 45.65% of its contribution in essential oil. Meanwhile, the essential oil of Water lily mango was mainly dominated by the presence of sesquiterpenes hydrocarbon (67.11%). The other components were β-caryophyllene (18.91%), *n*-hexadecanoic acid (10.22%). *n*-hexadecanoic acid was supposed as marker phytoconstituents of mango. The stereoisomer of γ-muurolene and *cis*-γ-cadineneare also accounted for 9.94% of total essential oil. Some compounds were observed to present only in selective varieties such as 2, 6-dimethyl-2,4,6-octatriene, α-cubenene, β-cadinene and globulol that found in Water Lily Mango. The unique presence of these compounds in mango varieties could be indirectly responsible for distinctive aroma and flavor.

7   Health-Promoting Traits of Mango Oil Constituents

7.1   Antioxidant Potential

In human body, enzymatic systems in conjunction with oxygen consumption results in the formation of reactive oxygen species (ROS). These ROS have beneficial effects in control amount but its excess could cause severe pathological complications including aging, cancer, neurodegenerative and cardiovascular diseases. To combat this complication, body constantly requires exogenous antioxidants that balance the ROS levels (Atmani et al. 2009; Ma et al. 2011). Among antioxidants, natural origin possesses broad spectral actions showing pharmacological and therapeutic activities against oxidative stress and free radicals (Bagchi et al. 2000). Mango is a good source of antioxidants with important bioactive compounds especially in seed kernel. The use of natural antioxidants in food processing and cosmetic applications has been reported. The limitations to use a plant-based natural antioxidant are cost and availability. Mango is present in bulk quantity and have various phenolic compounds including tocopherols and phytosterols that can be used as good source of natural antioxidants (Puravankara et al. 2000). Ndiaye et al. (2009) indicated that addition of 1% of mango seed kernel oil exhibited antioxidant potency up to a level of BHT. High amounts of antioxidants in the mango kernels propose to extract these antioxidants at larger scale to make it viable commercially.

7.2   Anticancer Properties

The anticancer potential of most of the drugs are directly associated with their antioxidant potential. Given that breast cancer continues to be an incurable disease alternatives such as complementary and alternative medicine (CAM), which uses pharmacological compounds and bioactive compounds such as vitamins, is growing rapidly in the world (Beuzeboc 2015). Abdullah et al. (2014) reported that ethanol extract of mango kernel have anticancer potential against human breast cancer cells with minimum cytotoxicity against normal cells. Other solvent extracts of mango seeds are also known as good alternative of synthetics for cancer treatment and more research on anticancer effect is required to fully explore the potential of crude extracts or pure compounds.

7.3   Antimicrobial Activity

There is a rise in challenges for synthetic antimicrobials to inactivate or inhibit the growth of spoilage and pathogenic microorganisms. Natural antimicrobial agents are gaining increasing attention to control microorganisms. Mango seed oil is a good source of natural antimicrobial due to the presence of potent bioactives. The bioactive phytochemicals in the methanol extract of mango exhibited antimicrobial effects against bacterial pathogens including *Escherichia coli*, *Staphylococcus aureus* and *Vibrio vulnificus*. This antimicrobial potential against Gram-positive and Gram-negative bacteria could be directly linked with its high antioxidant capacity (Khammuang and Sarnthima 2011).

8   Product Design and Use of Seed Kernel

The mango seed has aroused special scientific interest because it has been reported as a bio-waste with high content of bioactive compounds (phenolic compounds, carotenoids, vitamin C, and dietary fiber) that improve human health. The comparison of the composition in fatty acids of mango seed kernel oil with that of vegetable oils indicates that it is rich in stearic and oleic acids. Mango seed kernel extracts enhanced the oxidative stability of ghee and extended its shelf life. This could be attributed to phospholipids and phenolic compounds suggesting a synergistic action of the two types of compounds. Along with phenolics, other compounds including carotenoids and tocopherols may be involved in shelf life extension of ghee products.

Accordingly, mango seed kernel oil is more stable than many other vegetable oils rich in unsaturated fatty acids. Such oils seem to be suitable for blending with vegetable oils, stearin manufacturing and confectionery industry (Jahurul et al. 2015). Mango seed kernel is promising as a source of fat, cocoa butter substitutes, and other food substitutes. Mango flour has been used as main ingredient for infants and adults diet to leverage nutritional requirements. When incorporated in wheat flour, mango flour balanced the requirement of protein, fat and phenolic values and further used as raw material for biscuits manufacturing. Cookies with improved antioxidant properties and good taste were prepared by adding 40% seed powder.

Statistics showed that up to 50% of wheat flour can be substituted with mango seed powder without compromising essential attributes. This attribute makes it more valuable ingredient as a functional additive in bakery products. The proposal made in 1984 reported that mango seed kernel fat can substitute any fat without generating any harmful effects (Rukmini and Vijayaraghavan 1984). When mango seed extract mixed with palm stearin, it optimizes the resultant properties that could be a replacement of cocoa butter, which is used as an additive in the food, pharmaceutical, and cosmetic industries. In addition, mango seed kernel extracts showed the highest degree of tyrosinase-inhibition activities and free radical scavenging when compared with methyl gallate. Phenolic compounds from mango kernel and methyl gallate in the emulsion, cumulatively affect the stability of the cosmetic emulsion. Along with cosmetics it could also be a good source of ingredients that could be incorporated in shampoos, soaps and lotions due to their rich phenolic compounds and microelements like selenium, zinc and copper (Soong and Barlow 2004).

Cocoa butter is one of the most expensive vegetable oils consisting mainly of palmitic acid, stearic acid and oleic acid as well as a trace amount of lauric acid and myristic acid. Therefore, industries have tried to look for cheaper alternative vegetable oils that have chemical and physical properties similar to cocoa butter. When mango kernel oil is mixed with palm oil in different proportions, it could show similar properties to that of cocoa butter. Mango seed kernel oil is a good source of stearic and oleic acids whereas palm oil is a source of palmitic and oleic acids. Kaphueakngam et al. (2009) found that when mango kernel oil and palm oil were mixed in right proportions it attained the same level of fatty acids with similarity in palmitic acid, stearic acid and oleic acid composition that have cocoa butter. The 80/20 (%wt of mango seed kernel oil/palm oil) blend mainly consisted of three fatty acids that were also the main fatty acid components of cocoa butter and resultant properties including melting behavior and slip melting point were closest to that of cocoa butter.

Mango seed powder can be potentially employed for animals feed could have an important alternative to fill the scarcity and competition of feed because mango seeds have been identified as a promising source of nutrients for farm animals having about 80% of carbohydrate contents on dry matter (DM) basis. Mango seed kernels could be a useful energy source and would probably be a good substitute for maize, which is the major source of energy in non-ruminant diets in several tropical countries (de Oliveira Costa et al. 2010).

References

Abdalla, A. E., Darwish, S. M., Ayad, E. H., & El-Hamahmy, R. M. (2007a). Egyptian mango by-product 1. Compositional quality of mango seed kernel. *Food Chemistry, 103*, 1134–1140.
Abdalla, A. E., Darwish, S. M., Ayad, E. H., & El-Hamahmy, R. M. (2007b). Egyptian mango by-product 2: Antioxidant and antimicrobial activities of extract and oil from mango seed kernel. *Food Chemistry, 103*, 1141–1152.

Abdullah, A.-S. H., Mohammed, A. S., Abdullah, R., Mirghani, M. E. S., & Al-Qubaisi, M. (2014). Cytotoxic effects of *Mangifera indica* L. kernel extract on human breast cancer (MCF-7 and MDA-MB-231 cell lines) and bioactive constituents in the crude extract. *BMC Complementary and Alternative Medicine, 14*, 199.

Ajila, C., Naidu, K., Bhat, S., & Rao, U. P. (2007). Bioactive compounds and antioxidant potential of mango peel extract. *Food Chemistry, 105*, 982–988.

Akhter, S., McDonald, M. A., & Marriott, R. (2016). *Mangifera sylvatica* (wild mango): A new cocoa butter alternative. *Scientific Reports, 6*, 32050.

Atmani, D., Chaher, N., Berboucha, M., Ayouni, K., Lounis, H., Boudaoud, H., Debbache, N., & Atmani, D. (2009). Antioxidant capacity and phenol content of selected Algerian medicinal plants. *Food Chemistry, 112*, 303–309.

Ayala-Zavala, J., Vega-Vega, V., Rosas-Domínguez, C., Palafox-Carlos, H., Villa-Rodriguez, J., Siddiqui, M. W., Dávila-Aviña, J., & González-Aguilar, G. (2011). Agro-industrial potential of exotic fruit byproducts as a source of food additives. *Food Research International, 44*, 1866–1874.

Babaria, M. P. M. (2011). *Extraction of oil from mango kernel by hydraulic pressing.* Junagadh: JAU.

Bagchi, D., Bagchi, M., Stohs, S. J., Das, D. K., Ray, S. D., Kuszynski, C. A., Joshi, S. S., & Pruess, H. G. (2000). Free radicals and grape seed proanthocyanidin extract: Importance in human health and disease prevention. *Toxicology, 148*, 187–197.

Berardini, N., Fezer, R., Conrad, J., Beifuss, U., Carle, R., & Schieber, A. (2005). Screening of mango (*Mangifera indica* L.) cultivars for their contents of flavonol O-and xanthone C-glycosides, anthocyanins, and pectin. *Journal of Agricultural and Food Chemistry, 53*, 1563–1570.

Beuzeboc, P. (2015). Cáncer de mama metastásico. *EMC-Ginecología-Obstetricia, 51*, 1–14.

Biesalski, H.-K., Dragsted, L. O., Elmadfa, I., Grossklaus, R., Müller, M., Schrenk, D., Walter, P., & Weber, P. (2009). Bioactive compounds: Definition and assessment of activity. *Nutrition, 25*, 1202–1205.

Carocho, M., & Ferreira, I. C. (2013). A review on antioxidants, prooxidants and related controversy: Natural and synthetic compounds, screening and analysis methodologies and future perspectives. *Food and Chemical Toxicology, 51*, 15–25.

De Oliveira Costa, V. S., Michereff, S. J., Martins, R. B., Gava, C. A. T., Mizubuti, E. S. G., & Câmara, M. P. S. (2010). Species of Botryosphaeriaceae associated on mango in Brazil. *European Journal of Plant Pathology, 127*, 509–519.

Diarra, S. S. (2014). Potential of mango (*Mangifera indica* L.) seed kernel as a feed ingredient for poultry: A review. *World's Poultry Science Journal, 70*, 279–288.

Ediriweera, M. K., Tennekoon, K. H., Samarakoon, S. R. (2017). A review on ethnopharmacological applications, pharmacological activities, and bioactive compounds of *Mangifera indica* (Mango). *Evidence-Based Complementary and Alternative Medicine, 2017*, 1–24.

Faostat, F. (2015). *Agriculture Organization of the United Nations, 2011.* FAO. Retrieved am from http://faostat3.fao.org/faostat-gateway/go/to/download/Q/QC/S. Acceso 20.

Fowomola, M. (2010). Some nutrients and antinutrients contents of mango (*Mangifera indica*) seed. *African Journal of Food Science, 4*, 472–476.

Jahurul, M., Zaidul, I., Ghafoor, K., Al-Juhaimi, F. Y., Nyam, K.-L., Norulaini, N., Sahena, F., & Omar, A. M. (2015). Mango (*Mangifera indica* L.) by-products and their valuable components: A review. *Food Chemistry, 183*, 173–180.

Jin, J. (2017). Production of sn-1, 3-distearoyl-2-oleoyl-glycerol-rich fats from mango kernel fat by selective fractionation using 2-methylpentane based isohexane. *Food Chemistry, 234*, 46–54.

Jin, J., Jin, Q., Akoh, C. C., & Wang, X. (2018). Mango kernel fat fractions as potential healthy food ingredients: A review. *Critical Reviews in Food Science and Nutrition.* https://doi.org/10.1080/10408398.2018.1428527

Ibarra-Garza & Ingrid P. (2015). Effects of postharvest ripening on the nutraceutical and physicochemical properties of mango (*Mangifera indica* L. cv Keitt). *Postharvest Biology and Technology, 103*, 45–54.

Kaphueakngam, P., Flood, A., & Sonwai, S. (2009). Production of cocoa butter equivalent from mango seed almond fat and palm oil mid-fraction. *Asian Journal of Food and Agro-Industry, 2*, 441–447.

Khammuang, S., & Sarnthima, R. (2011). Antioxidant and antibacterial activities of selected varieties of Thai mango seed extract. *Pakistan Journal of Pharmaceutical Sciences, 24*, 37–42.

Kittiphoom, S., & Sutasinee, S. (2013). Mango seed kernel oil and its physicochemical properties. *International Food Research Journal, 20*, 1145–1149.

Kittiphoom, S., & Sutasinee, S. (2015). Effect of microwaves pretreatments on extraction yield and quality of mango seed kernel oil. *International Food Research Journal, 22*, 960–964.

Ma, X., Wu, H., Liu, L., Yao, Q., Wang, S., Zhan, R., Xing, S., & Zhou, Y. (2011). Polyphenolic compounds and antioxidant properties in mango fruits. *Scientia Horticulturae, 129*, 102–107.

Masibo, M., & He, Q. (2009). Mango bioactive compounds and related nutraceutical properties-a review. *Food Reviews International, 25*, 346–370.

Muchiri, D. R., Mahungu, S. M., & Gituanja, S. N. (2012). Studies on mango (*Mangifera indica* L.) kernel fat of some Kenyan varieties in Meru. *Journal of the American Oil Chemists' Society, 89*, 1567–1575.

Ndiaye, C., Xu, S.-Y., & Wang, Z. (2009). Steam blanching effect on polyphenoloxidase, peroxidase and colour of mango (*Mangifera indica* L.) slices. *Food Chemistry, 113*, 92–95.

O'shea, N., Arendt, E. K., & Gallagher, E. (2012). Dietary fibre and phytochemical characteristics of fruit and vegetable by-products and their recent applications as novel ingredients in food products. *Innovative Food Science & Emerging Technologies, 16*, 1–10.

Puravankara, D., Boghra, V., & Sharma, R. S. (2000). Effect of antioxidant principles isolated from mango (*Mangifera indica* L.) seed kernels on oxidative stability of buffalo ghee (butter-fat). *Journal of the Science of Food and Agriculture, 80*, 522–526.

Ribeiro, S. M. R., & Schieber, A. (2010). Bioactive compounds in mango (*Mangifera indica* L.). In *Bioactive foods in promoting health*. Burlington: Elsevier.

Rukmini, C., & Vijayaraghavan, M. (1984). Nutritional and toxicological evaluation of mango kernel oil. *Journal of the American Oil Chemists' Society, 61*, 789–792.

Shah, K., Patel, M., Patel, R., & Parmar, P. (2010). *Mangifera indica* (mango). *Pharmacognosy Reviews, 4*, 42.

Solís-Fuentes, J. A., & Del Carmen Durán-De-Bazúa, M. (2011). Mango (*Mangifera indica* L.) seed and its fats. In *Nuts and seeds in health and disease prevention*. Burlington: Elsevier.

Soong, Y.-Y., & Barlow, P. J. (2004). Antioxidant activity and phenolic content of selected fruit seeds. *Food Chemistry, 88*, 411–417.

Tharanathan, R., Yashoda, H., & Prabha, T. (2006). Mango (*Mangifera indica* L.), "The king of fruits"-an overview. *Food Reviews International, 22*, 95–123.

Torres-León, C., Rojas, R., Contreras-Esquivel, J. C., Serna-Cock, L., Belmares-Cerda, R. E., & Aguilar, C. N. (2016). Mango seed: Functional and nutritional properties. *Trends in Food Science & Technology, 55*, 109–117.

Yamoneka, J., Malumba, P., Blecker, C., Gindo, M., Richard, G., Fauçonnier, M.-L., Lognay, G., & Danthine, S. (2015). Physicochemical properties and thermal behaviour of African wild mango (*Irvingia gubonensis*) seed fat. *LWT- Food Science and Technology, 64*, 989–996.

Chapter 29
Passion Fruit (*Passiflora* spp.) Seed Oil

Massimo Lucarini, Alessandra Durazzo, Antonio Raffo, Annalisa Giovannini, and Johannes Kiefer

Abstract The quality of passion fruit oil is determined by its main components like the lipids but also by minor compounds whose levels are influenced by several factors (i.e. genetic, environmental, etc.). Description of properties of bioactive components in passion fruit seed oil and their potential beneficial role for human health will be a focal point of this chapter. Presence of aroma compounds and their contribution to the organoleptic properties of the fruit and its industrial residues will be discussed. Suitable research strategies as well as the definition of practical possibilities for using passion fruit oils and its constituents represents another focus: conventional procedures and advanced extraction technologies, as well as analytical techniques, with particular attention to green procedures will be taken into account. An integrated and multidisciplinary system of analysis combined with statistical methods is becoming an increasingly valuable tool for analyzing and modeling agro-food systems in their totality. These new directions will be discussed, with particular regards to spectroscopy combined with advanced chemometrics. Such approaches are applied in classification, discrimination and authentication studies. Furthermore, they are used in detection and monitoring of contaminants and adulterants. The use of passion fruit oil in different fields (i.e. pharmaceutical, nutraceutical, cosmetic, bio-based applications) will be discussed.

Keywords Passion fruit seed oil · Genetic variability · Oil composition · Emerging processing techniques · Organoleptic properties · Nutraceuticals · Chemometrics · Spectroscopic technique

M. Lucarini (✉) · A. Durazzo · A. Raffo
CREA Research Centre for Food and Nutrition, Rome, Italy
e-mail: massimo.lucarini@crea.gov.it; alessandra.durazzo@crea.gov.it;
antonio.raffo@crea.gov.it

A. Giovannini
CREA Research Centre for Vegetable and Ornamental Crops, Sanremo, Italy
e-mail: annalisa.giovannini@crea.gov.it

J. Kiefer
Technische Thermodynamik, Universität Bremen, Bremen, Germany
e-mail: jkiefer@uni-bremen.de

© Springer Nature Switzerland AG 2019
M. F. Ramadan (ed.), *Fruit Oils: Chemistry and Functionality*,
https://doi.org/10.1007/978-3-030-12473-1_29

1   Introduction

1.1   Botany (Classification and Morphology)

The plant that is commonly called 'passiflora' belongs to the Plantae Kingdom, Violales order, class Magnoliopsida, family Passifloraceae. The latter is in turn divided into three tribes: Passifloreae, Paropsiae and Abatieae. In the Passifloreae tribe there are 11 genera, some of them with very few species *Ancistrothyrsus* (2 species), *Crossostemma* (1 species), *Deidamia* (5 species), *Dilkea* (3 species), *Efulensia* (2 species), *Hollrungia* (1 species), *Mitostemma* (3 species) and *Schlechterina* (1 species). Two of the genera, *Adenia* and *Basanthe*, have not more than 100 species and one of them, the *Passiflora* genus, has more than 600 species. The *Passiflora* genus is relatively unknown in its complexity and variety of species present in nature. In 1938, the genus has been subdivided by the U.S. botanist Ellsworth Paine Killip in 22 subgenera further subdivided into 'Sections' and 'Series' on the basis of similar morphological characteristics as described in the book 'The American Species of Passifloraceae' (Killip 1938). This classification remained in effect until 2004, when Feuillet and Mac Dougal completely reviewed the genus *Passiflora*. In this classification, also based on morphological characteristics, the subgenera have been reduced to four: Astrophea, Deidamioides, Decaloba and *Passiflora* (Feuillet and Mac Dougal 2004). Studies based on molecular markers supported the monophyletic origin of the genera (Yockteng and Nadot 2004).

The genus *Passiflora* is largely distributed in the American continent, from the South of the United States to the North of Chile and Argentina, in a wide range of habitats, from humid rainforests to semi-arid subtropics. There are few known species of Australian passiflora (*P. aurantia* G. Forst., *P. cinnabarina* Lindl., and *P. herbertiana* Ker Gawl.), 1 in New Zealand (*P. tetranda* Banks ex DC.), and about 20 in the Far East (China, Cambodia, Vietnam, etc.). More than 60 species have edible fruits, some are used in the food industry having nutritional value due to their vitamins and carbohydrates, and for the sweet taste of fruits (*P. edulis f. edulis* Sims or 'maracuja' or 'purple passion fruit' and 'yellow passion fruit' *P. edulis* Sims f. flavicarpa Deg.). Some species have been the basis of many traditional medicines throughout the world for thousands of years and have continued to provide new remedies to mankind. It is one of the richest sources of bioactive compounds. Properties traditionally recognized include anxiolytic, anti-inflammatory, sedative, antioxidant, antispasmodic, and neuroprotective. *P. incarnata* L. 'passion vine' and *P. caerulea* L. 'blue passionflower' are used as sedative and antidepressant (Dhawan et al. 2004). Many other species, since their introduction in the Old World, were grown either outdoors, in a mild climate, or in the glasshouses, for their exotic flowers, and new ornamental hybrids have been obtained (Santos et al. 2012).

The plants are woody or perennial herbaceous creepers, characterized by leaves with different shapes and colors and elegant axillary tendrils, which extend towards the top and then spiral around a suitable support. The stem has a cylindrical or polygonal section on three or more sides. The plants can be hairless, or fully or partly covered by thick hair. Along the stem, at the nodes, there is stipula, which differs according to the taxonomical group the plants belong to. The leaves always alternate and are one of the most important elements of great decorative value. The surface can be velvety glabra or tomentosa. The flower peduncle is born from the leaf axilla and generally carries a single flower, except for rare exceptions, such as *P. racemosa* Brot., in which the flowers are arranged in bunches (Sgaravatti and Zardini 1997). The bracts are present in almost all species and have a lot of different forms. The position of the bracts on the peduncle, their shape and their size are important for classification. The morphology of the flower is rather unique: the calyx can be flat, cup, campanulate or cylindrical. The corolla consists of five sepals, which on the lower page have the same color of the calyx while they present bright colors in the upper one, the margin often has a carenated edge that ends with a spur more or less long and pointed. Even the petals are five, but often they are smaller the sepals and have a lower thickness. The corolla can be arranged perpendicular or parallel to the peduncle, depending on the species. Inside the perianth (set of calyx and corolla) the crown is formed by a series of filaments, from very short to longer than the petals, banned, arranged in circular rings, concentric or fused between them to form a tube, whose shape and color are in relation to the subgenus belonging to the species. This structure is singular in the plant kingdom and makes flowering particularly glaring. The color combinations are very contrasted, with complementary colors placed on the crown and on the corolla, shades of violet and blending blue towards pure white, alternations of pink and violet, of red and white, white and dark violet and petals who dress in a pure white, like a scarlet red. It is definitely one of the most beautiful flowers in the plant world (Vecchia and Giovannini 2011). The reproductive system is located at the top of the long androgynophore and protrudes far above the corolla. The five anthers, often tilting, are supported by five filaments, the ovary is generally round, there are three styles and three stigmas. The plants are pollinated by numerous species of insects. Some *Passiflora* species which are flowering at night, always have pure white color and are pollinated by Chiroptera, nectar and pollen bat eaters. The anthers of *P. mucronata* Lam. are rotated in the same direction in order to better put the pollen on the back of these flying mammals (Ocampo et al. 2010). Some species, with a very long calyx, have established fruitful relationships with sword-billed hummingbirds (Abrahamczyk et al. 2014). Other peculiar morphological structures present in *Passiflora* species are the small glands, called 'nettàri', which secrete nectar to attract insects. The glands can be found on the petiole, in pairs, on the edge of the bracts, on the stipules or on the lower page of the leaf. The different arrangement and shape of the glands allow distinguishing the subgenera and the species. Many species belonging to the subgenus *Decaloba*, have numerous 'nettàri' also on the leaves, aligned along the main ribs in sorted lines. Sometimes they are of a contrasting color, yellow or light green so as to mimic butterfly eggs just laid on the leaf, with a dissuasive function. The ovary after being fertilized produces a globose berry, which contains the seeds immersed in a gelatinous aril, usually edible. The dimensions of egg-shaped and colored fruits could range from a few millimeters in diameter (*P. suberosa* L.) up to 15 cm (*P. macrocarpa* Mast.). The seeds have a flattened shape with one end pointed opposite to another more rounded. They are more or less elliptical, or heart-shaped, and they are covered with a very hard, black, polished tegument, which often has characteristic, transversal cross-linked reliefs, dotted or pitted. The seeds have a good resistance and the period of vitality is always rather long, because they well endure situations of desiccation. It can be said that each species has seeds with a characteristic shape, from which the species of origin can be guessed. Seed dimensions are always small and vary from less than 1 to 5–6 mm in the larger ones (Vecchia and Giovannini 2011).

1.2 History

The discovery of several thousand years old seeds of *Passiflora* from the archaeological sites at Virginia and North America provides strong evidence of the prehistoric use of the fruits by the ancient people. The early European travelers in North America noted that Algonkian Indians in Virginia and Creek people in Florida ate fruits of passiflora from cultivated as well as wild sources. Then the European settlers also consumed the fruit and praised its flavor, thereby, suggesting the prehistoric consumption of *Passiflora* species as a fruit crop (Patel et al. 2011).

Passiflora species have been appreciated by the Spanish in South America. In 1553 Pedro Cieza de Leòn in "Part primera de la cronica del Perù" describes the scented 'granadilla' (small pomegranate), located near the village of Lile in Colombia. In 1554 Nicolas Monardes publishes a manuscript on the medicinal properties of indigo passiflora, with a small section on 'granadilla'. Francisco Hernàndez, personal physician of King Philip of Spain, from 1570 to 1577 travels in the Americas looking for new medicines. His manuscript from 1651 describes a plant of the supersection Cieca, Passiflora sexocellata Schltdl., to which the Aztec name "tzinacanatlapatli" is assigned and reports that the Aztecs used the juice of the leaves to treat eye diseases. Charles Plumier in the 1693 describes and illustrates four species of the supersection Cieca, *P. suberosa* L. and P. *pallida* L. Joseph de Tournefort in 1719 creates two new genera: *Granadilla* with 23 species and *Murucuja* with a species from the filaments of the crown fused together. In 1753, the book Species Plantarum Linneus for the first time describes the genus *Passifora* with 24 species (Vecchia and Giovannini 2011). The first missionaries saw in the morphology of this flower the signs of the passion of Jesus Christ and for this reason it was named from the Spaniards 'La flor del las cinco llagas' (The flower of the five sores). The knight and commander of the order of Malta Giacomo Bosio (1544–1627) describes this symbolism in the book "La trionfante e gloriosa croce Trattato di Iacomo Bosio Lettione varia, e divota; Ad ogni buon Christiano utile, e gioconda", published in Rome in 1610. The characteristic filaments "in the manner of a fringe, of blood color symbolize the lashes used against Jesus Christ". The androgynophore, which supports the ovary becoming the column of the scourging, the three nail-shaped stigmas the nails of the crucifixion, while the five anthers intertwined with each other and recall the crown of thorns. The colored blood specks present in the operculum (calyx) remember the five wounds inflicted on Christ on the cross and the shape of the pointed leaf-like "un ferro di Picca" remembers the spear with which the side of Jesus Christ was stabbed (Bosio 1610). In the book the author draws and describes this marvelous flower: "And not only the cross but many mysteries of the passion of the Lord and Redeemer our Jesus Christ in some flowers are represented. Among these, I do not believe that any flower is neither more marvelous nor stupendous nor has it been seen, nor can we ever see what is born in the Indies of Peru and New Spain; called by the Spaniards The "flor of the las cinco llagas ..." The grain that stands in the middle from which stands the column of nails and the crown as we have said over time it goes swelling and turns into a fruit that from the name of the plant also it is called "granadillo" ... And when it is ripe it shows a yellow color all pinched with some green spots. It is full of a very sweet smell of liquor that almost blows musk and amber with a very delicate taste. By breaking or cutting the bark of the fruit in the tip, it drinks its liquor almost like an egg. Drunk comforts the stomach, helps digestion; and it is of good and healthy nourishment" (author own translation from the original text, kindly made accessible by Dr. Valeria Maria Leonardi, Rome Biblioteca Magistrale of the Sovereign Military Hospital Order of St. John of Jerusalem in Rhodes and Malta).

1.3 Commercial Production of Passion Fruit

Among the numerous edible *Passiflora* species only a few are of any commercial importance. Many are known only in native markets in South and Central America and the West Indies. Commercial production of passion fruit is based on the purple species *Passiflora edulis* Sims and the yellow form *Passiflora edulis* f. *flavicarpa* Degener. Aside from the skin color, the purple and yellow passion fruits differ in horticultural performance and fruit properties. The purple species is more resistant to cold injury, is less acid, and is considered superior in aroma and flavor. The yellow form is faster growing, has a greater resistance to soil fungi, has more vigorous vines, bears crops over longer periods, and has a greater yield of fruit and pulp, larger fruits, and more acid juice. The interest of researchers and producers in these species has been stimulated due to their good nutritional characteristics since it presents a fresh pulp, soft peel, high sugar content and strong exotic flavor. Because passion fruits are often inexpensive and extremely rich in vitamins, their popularity has increased, especially in Europe and in the United States. *Passiflora edulis* Sims. f. *edulis* commonly known as gulupa, chulupa or maracuyá púrpura, is a native species of the southern Andes, growing between 1600 and 2600 m, in climates with average temperatures between 16 °C and 22 °C. The fruits are round-shape, with a diameter between 4 and 6 cm, and green to purple peelings at maturity. Inside, they contain many seeds (as the other Passifloraceae species) surrounded by a gelatinous yellow pulp, that exhibited an intense aroma and sweet-acid taste. These fruits are considered a good source of vitamin A, thiamine, riboflavin, niacin, calcium, phosphorus, and ascorbic acid source. The pulp is used to prepare juices and soft drinks.

P. quadrangularis L. or 'giant granadilla,' widely distributed in the tropics, is the most cultivated species after *P. edulis* in tropical America. It bears the largest fruits, which are more elongated, up to 25 cm long, and fleshy instead of hollow. The seeds are much larger, brownish, and flattened. The rind is not so hard as in the purple and yellow passion fruits. The juice content is much lower and is somewhat inferior in flavor and color. The pulp can be eaten like a melon, with or without the addition of sugar, or cooked with milk. When green, this fruit can be used as a vegetable as green papaya. *P. alata* Curtis or 'fragrant granadilla' is a native species grown in South America, especially in Brazil. It is also found in Peru, Paraguay, and Argentina. This species holds interesting characteristics. Its aroma is quite distinct from other passion fruits and, in addition to being aromatic, its pulp contains higher sugar and lower organic acids concentrations compared to traditional commercial passion fruit species, which makes it sweeter and less sour. For this reason, fragrant granadilla fruit has been highly appreciated for fresh consumption as a dessert, for example, rather than mainly for juice consumption as with other passion fruit species. *P. ligularis* Juss, 'sweet granadilla' or 'water lemon,' is cultivated in the mountains of Mexico and Central America. The fruit is mostly eaten out of hand or used in drinks or ice cream. Its translucent white pulp is almost a liquid, acid with a sweet aroma. The peel is resistant so the fruit can be transported well, without being damaged. Colombia, where this species is cultivated in the Western Cordillera, exports this fruit to Europe. *P. mollissima*, 'banana passion fruit' or 'curuba,' grows widely in the Andes and is distributed from Venezuela and Colombia to Peru and Bolivia. The flavor is more astringent and less acid than *P. edulis*. The sieved pulp is mixed with milk and sugar and served as a drink. It is also used in jams and desserts and for flavoring ice creams. *P. maliformis* L. is known as 'sweet cup' in the West Indies, 'chulupa' in Colombia, and 'granadilla de hueso' in Ecuador. It is a little-known species but may have a good future because of its excellent aroma and flavor.

Tropical fruits are species widely appreciated for their organoleptic properties and they have shown a positive growth rate in the export market since 1995. Their economic value has grown over the last years. For the case of *P. edulis* Sims f. *edulis* or 'purple passion fruit', the main customers are Germany, The Netherlands, the UK and Belgium with sales close to 1,700,000 USD in 2007 and up to 4,100,000 USD in 2008.

2 Extraction and Processing of Fruit Seed Oil (Developments in Extraction and Isolation of High-Value Compounds)

Nowadays Passiflora fruits are normally used to produce juice and are one of the most nutritious fruit juices (Fig. 29.1).

Fruit consumption has increased worldwide for its intense, exotic, tropical scent and attractive taste and also for its disease prevention and health benefits related with the presence of a wide spectrum of nutrients, such as vitamins, fiber and

Fig. 29.1 *Passiflora* fruits: *Passiflora edulis* Sims f. *edulis* 'purple passion fruit' and *P. alata* Curtis 'fragrant granadilla'

minerals, and bioactive components, that make the fruits a healthy addition to diet (Zeraik et al. 2010).

Of the estimated ~600 species of *Passiflora*, only a few have been widely studied and characterized so far, based on popular knowledge and scientific studies. However, the increase of fruit consumption has led to a huge production of waste, formed mainly by seeds and peels. The management of these residues implies operating costs for the industries and may represent an environmental problem. Furthermore, byproducts from different fruits can be important sources of valuable chemicals (Morais et al. 2015). Turning these resources into renewable raw materials for the industry is a major challenge, but at the same time, the extraction of oil from the passion fruit seeds may provide an opportunity for adding value to this agro-industrial waste.

Passion fruit crops are examples of economically important species, whose oils and fats have found various applications in the food, pharmaceutical and cosmetic industries. As the use of Amazonian vegetable oils has increased in recent years, it is important to acquire a better knowledge of their properties in order to optimize their exploitation.

Edible vegetable oils are conventionally extracted by means of mechanical pressing and/or liquid solvent-based extraction processes. The disadvantages of mechanical pressing are mainly related to the possible degradation of bioactive compounds or shelf life reduction (due to exposure to oxygen and light, and low extraction rates). It should be considered that bioactive compounds can be altered during recovery procedures by several factors, such as the extraction process, the solvent and raw material used, the storage conditions, pretreatments, which may trigger chemical reactions that could degrade some of these bioactive compounds.

The use of solvents usually implies high temperatures, long time, substantial energy cost, and can lead to high toxicity, environmental hazard, low selectivity and possible loss of volatile compounds. The most common solvent used for oil extraction is *n*-hexane for its excellent solvent properties in terms of oil recovery and solubility. However, *n*-hexane has been reported as an air pollutant which can react with other pollutants to form ozone and photochemical oxidants (Hanmoungjai et al. 2000).

Nowadays, the development of suitable clean and efficient alternative technologies -extraction, processing, and preservation methodologies- introduced to recover nutraceuticals and to minimize the environmental impact, as well as the use of the rather green solvents (that fitting the green chemistry category) such as acetone and ethanol, have received considerable attention (Viganó et al. 2015; Zuin and Ramin 2018).

As mentioned by Zuin and Ramin (2018), the main emerging green and sustainable separation approaches towards bio-economy and circular economy contexts are the following: solvent processing, microwave, ultrasonication, supercritical fluid, enzymatic processing, alkaline processing, subcritical fluid, ionic liquid, accelerated solvent, microemulsion and others. Among the green techniques, recent examples applied to passion fruit oil are here reported.

For example, cold pressing (Ferreira et al. 2011) and Soxhlet with different solvents (Cardoso de Oliveira et al. 2013) have been used to extract this oil, with similar efficiencies. In particular, Cardoso de Oliveira et al. (2013) studied the extraction of passion fruit oil with green solvents -acetone, ethanol and isopropanol- using ultrasound, shaker and Soxhlet techniques, in comparison with n-hexane; the authors concluded that the ultrasound technique can replace the conventional ones.

Furthermore, Lee et al. (2015) have evaluated the feasibility of green solvent use to passion fruit oil extraction using ultrasound-assisted extraction. They monitored the impact on fatty acids, total phenols and antioxidant properties. Eventually, they showed that, overall, acetone could be a suitable n-hexane replacement for its higher oil recovery, similar fatty acid profile and oil physicochemical properties with higher antioxidant activity. Ethanol-extracted oil showed higher amounts of omega-9 MUFA as well.

Generally, several studies have underlined how supercritical fluid extraction combined with ultrasound could represent a promising process to reduce extraction times as well as to increase the extraction yields with respect to those obtained without ultrasound (Riera et al. 2004; Pasquel-Reátegui 2014). In this regard, Barrales et al. (2015) investigated the effects of pressure, temperature and ultrasonic waves on the supercritical fluid extraction of passion fruit seed oil (*Passiflora edulis* sp.): the application of the ultrasound power of 160 W promoted the oil extraction, and the supercritical fluid extraction global yield achieved 29% (at 40 °C and 16 MPa).

There is a growing interest in recovering phytochemicals from plant source materials in the last decade. In this context, the review of Viganó and Martínez (2015) is worth mentioning. It summarizes updated information on extraction techniques for recovering phytochemicals from passion fruit by-products, by focusing on the application of environmentally friendly extraction techniques, such as supercritical fluid extraction and subcritical fluids. The authors concluded that the passion fruit bagasse (fleshy aryl and seeds) is rich in polar compounds (oil, tocols and carotenoids) that can be recovered stepwise using supercritical fluid extraction. The nonpolar compounds such as polyphenols can be properly obtained by Pressurized Liquid Extraction with ethanol and water as solvents (Viganó and Martínez 2015).

In this regard, the recent work of Pereira et al. (2017), by studying the assessment of subcritical propane, ultrasound-assisted and Soxhlet extraction of oil from sweet passion fruit (*Passiflora alata Curtis*) seeds, showed that subcritical propane higher extraction yield (23.68%) was obtained at 60 °C/2 MPa with an extraction efficiency of 84% compared to Soxhlet extraction using *n*-hexane. Moreover, the authors reported that major values of unsaturated fatty acids were obtained by subcritical propane at 60 °C/2 MPa (86.36%) as well as higher tocopherols contents by Soxhlet extraction using *n*-hexane and subcritical propane at 60 °C/2 MPa (Pereira et al. 2017). Indeed, the authors concluded that the extraction of plant source materials using compressed propane as solvent is a promising technique that allows obtaining free-solvent products. Further work of the same authors (Pereira et al. 2018), on yellow passion fruits concluded that the use of unconventional extraction methods such as compressed propane, a green recovery technology yet still unexplored for seed oil extraction, represents an opportunity to add value for this agroindustrial waste.

Figure 29.2 reported an example of industrial plant scheme of the sustainable process generally applied to oil seeds for isolating bioactive compounds-rich fractions (Romani et al. 2017).

Further studies for the application of sustainable industrial plant based on innovative separation processes performed with physical technologies to passion fruit seeds should be designed.

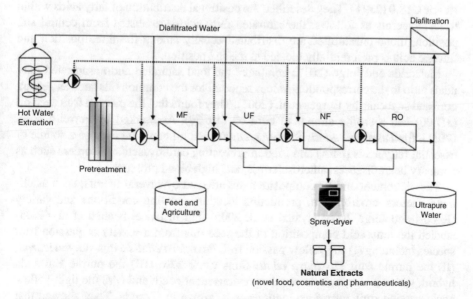

Fig. 29.2 Industrial plant scheme of the sustainable process, for the recovery of bioactive compounds from oilseed using membrane technologies: MF microfiltration, UF ultrafiltration, NF nanofiltration and reverse osmosis (Romani et al. 2017)

3   Composition of Passion Fruit Seed Oil

3.1   Fatty Acids Composition and Acyl Lipids (Neutral Lipids, Glycolipids and Phospholipids) Profile

Passion fruit seeds are a rich source of lipids, proteins, minerals and fiber. In particular, the edible passion fruit seed is known for fiber-rich fractions, including insoluble dietary fiber, alcohol-insoluble solids and water-insoluble solids, which are mainly formed by cellulose, pectic substances and hemicellulose (Chau and Huang 2004).

In recent years, passion fruit seeds have been attracted interest for the production of oil, mainly due to the presence of bioactive compounds (Viganò and Martìnez 2015). Several studies have reported the chemical composition of oils extracted from the seeds of different varieties of passion fruit (Ferrari et al. 2004; Piombo et al. 2006). Investigations have found that passion fruit seeds contain about 30% oil (Malacrida and Jorge 2012), with linoleic, oleic and palmitic acids as the main fatty acids of the oil (Malacrida and Jorge 2012; Barrales et al. 2015; Morais et al. 2017): linoleic acid accounts for about 72–73%, followed by oleic acid (13–16%) and palmitic acid (8–9%) (Liu et al. 2009; Malacrida and Jorge 2012). Passion fruit seed oil has a composition of 85% unsaturated fatty acids, of which 69% are linoleic acid and 18% oleic acid (Ferrari et al. 2004; Ran and Blazquez 2008; Lopes et al. 2010a; de Santana et al. 2015). Piombo et al. (2006) reported a high ratio of linoleic acid C18:3 (73.4%) followed by oleic acid C18:1 (13.7%), palmitic C16:0 (8.8%) and stearic C18:0 (2.4%). They described the positional distribution of fatty acids within triacylglycerols as follows: the saturated fatty acids are absent from central sn2 position, mono-unsaturated are distributed equally among the three position and linoleic acid is preferentially located at the sn2 position.

Malacrida and Jorge (2012) compared the total saturated and unsaturated fatty acids ratio to the corresponding values reported for the common oils such as peanut, corn and soybeans by Borges et al. (2007). They found that the passion fruit seed oil (1/7.06) had a profile similar to corn oil (1/6.70). As marked by Ferreira et al. (2011), the *Passiflora edulis* Sims var. *edulis* seed oil could be used as a source of essential fatty acids (EFA) and used in preventing cardiovascular disorders such as coronary heart diseases, atherosclerosis, and high blood pressure.

The oil content and its composition are affected by several factors, such as climatic factors, environment, production location, growing conditions and variety (Kobori and Jorge 2005; Nyanzi et al. 2005). For instance, Nyanzi et al. (2005) studied the fatty acid composition of the seed oils from a variety of passion fruit species including (I) the yellow passion fruit *Passiflora edulis* Sims var. *flavicarpa*; (II) the purple fruit *Passiflora edulis* Sims var. *edulis*; (III) the purple Kawanda hybrid, which is a cross between the previous mentioned; and (IV) the light-yellow apple passion fruit *Passiflora maliformis* L. grown in Uganda. They showed that linoleic (67.8–74.3%), oleic (13.6–16.9%), palmitic (8.8–11.0%), stearic (2.2–3.1%), and α-linolenic (0.3–0.4%) acids are the most dominant fatty acids. The authors underlined that the fatty acid contents, in addition to the iodine value, in (III) are distinctly closer to the rootstock (I) than the scion (II), indicating that the rootstock influence on the fatty acids content of passion fruit seeds is graft-transmissible.

Regis et al. (2015) studied the oil quality of passion fruit seeds subjected to a pulp-waste purification process. They showed that the passion fruit seeds, which are totally cleaned by the thermal/chemical treatment, lead to a faster drying (less than 50% of the drying time) of the seeds and a bit higher yield of oil extraction (proportionally around 7.7%), without changes in quality of the oil.

The stability to oxidation of passion fruit was determined using the ASTM D7545 method by Machado et al. (2014). The oxidation kinetics was of zero-order, which can be due to mathematical approximations related to first order kinetics. The authors found the lowest stability of this oil may be confirmed by a lower enthalpy of activation and be correlated to a greater amount of unsaturated fatty acids in its composition (Machado et al. 2014).

3.2 Minor Bioactive Compounds (Sterols, Tocols, Carotenoids, Phenolic Compounds, Hydrocarbons)

Besides the unsaturation degree and fatty acid composition, bioactive compounds such as tocopherols, carotenoids, phenolic compounds are characteristic for passion fruit oil. These bioactive species are of crucial importance when it comes to determine the possible technological applications of the oils in food, chemical or pharmaceutical industries.

The quantity and nature of tocopherols naturally present in such unsaturated oils is strictly linked to their oxidative stability. Piombo et al. (2006) showed that passion fruit oil contains both gamma- and delta-tocopherol in comparable amounts (217 and 243 ppm, respectively), whereas the content of alpha-tocopherol was about 5.0 ppm. Concerning the content of phytosterols, the same authors showed that passion fruit oil contains a total phytosterol amount of about of 209 mg/100 g: campesterol (28.2 mg/100 g), equal amounts of beta-sitosterol (87.2 mg/100 g) and stigmasterols (87.1 mg/100 g), and delta-5-avenasterol (6.9 mg/100 g) (Piombo et al. 2006). Generally, for the determination of phytosterol compounds, a lipid extraction step, preceded by alkaline hydrolysis (saponification) is required. Recently, Rotta et al. (2017) investigated the potential use of ultrasonic-assisted saponification in oil samples from different species. Their study considered yellow passion fruit sweet granadilla and sweet passion fruit. Employing gas chromatography with flame ionization detection (GC-FID), they found a content for β-sitosterol ranging 0.90–1.60 mg g^{-1} of oil, followed by stigmasterol (0.70–1.40 mg g^{-1} of oil) and campesterol (0.10–0.30 mg g^{-1} of oil).

Concerning antioxidant properties, Malacrida and Jorge (2012) have reported significant radical scavenging activity of crude oil extracted from yellow passion fruit seed oil (*Passiflora edulis* f. *flavicarpa*).

Ferreira et al. (2011) investigated the fatty acids profile, the total phenolic and carotenoids content, as well as antioxidant activities of *P. edulis* seed oils obtained from different processes: refined oil, cold pressed oil, and Soxhlet extracted oil. The refined oil displayed the highest antioxidant activity among the tested samples (EC50 = 5.74 mg/mL), even if it presented a lower carotenoid content than the cold pressed oil. Moreover, the same authors reported that the Soxhlet extracted seed oil, besides the highest carotenoid content (19.7 mg/g), presented a lower phenol content when compared to its corresponding pressed oil. This is not surprising as it is well established that heat can compromise the phenolic constituents.

Recently, Lee et al. (2015) have evaluated the total phenolic content and antioxidant activity (DPPH value) of passion fruit oil extracted using different solvents. Significant differences in both total polyphenol content and antioxidant activity were found among the oil samples with ethanol-extracted oil containing the highest amount, followed by acetone- and *n*-hexane extraction.

De Santana et al. (2015) have studied and compared the chemical composition and antioxidant capacity of Brazilian passiflora seed oils. All oils possessed similar physicochemical characteristics, except for color, and predominantly contained polyunsaturated fatty acids (PUFA) with a high percentage of linolenic acid (68.75–71.54%). On the other hand, for the total sterol content, the extracted oil from *Passiflora setacea* BRS Pérola do Cerrado seeds had higher quantities (% times higher than the average of all samples) of carotenoids (44%), phenolic compounds (282%), and vitamin E (215%, 56%, 398%, and 100% for the α-tocopherol, β-tocopherol, γ-tocopherol, and δ-tocopherol isomers, respectively (de Santana et al. 2015). The same authors showed that the methanol extracts from *Passiflora setacea* BRS Pérola do Cerrado seed oil exhibit also higher antioxidant activity, which positively correlated with the total phenolics, δ-tocopherol, and vitamin E contents (de Santana et al. 2015).

A comparative study of antioxidant properties of seed was carried out (Lourith and Kanlayavattanakul 2013). Investigating antioxidant activity and phenolics of *Passiflora edulis* seed recovered from juice production residues, they found that ethyl acetate extraction yielded a higher antioxidant activity than aqueous extract. Larger amounts of chlorogenic acid, rosmarinic acid and quercetin were found in the ethyl acetate extract. Also the presence of piceatannol, resveratrol and scirpusin B in the seed and their properties are documented and should be mentioned (Matsui et al. 2010; Sano et al. 2011; Maruki-Uchida et al. 2013, 2018; Matsumoto et al. 2014; Kitada et al. 2017).

3.3 Health-Promoting Traits of Fruit Seed Oil

The recent review of Viganó and Martíncz (2015) summarizes the data on the main components (fatty acids, tocols, carotenoids, phenolic) of passion fruit seeds oil present in the literature by comparing them with other vegetable oils. The fatty acid profile of passion fruit seeds is similar to that of other fruits, such as grape and guava. The authors also reported that the carotenoid concentrations in passion fruit seeds are lower than in pequi and buriti oils (Viganó and Martínez 2015).

Current research is focusing on investigating the health-promoting traits of fruit oil and oil constituents and further effort is needed to obtain a full picture. Some studies reported antioxidant, antitumor and other biological activities of the passion fruit seed oil or from specific compounds present in the oil (Sano et al. 2011; Jorge et al. 2012; Pereira et al. 2018). For instance, Jorge et al. (2012) have studied the synergistic action of *Schinus terebinthifolius* Raddi extract and linoleic acid from *Passiflora edulis* in decreasing melanin synthesis in B16 cells and reconstituted epidermis. The recent work of Pereira et al. (2018) reported high antioxidant performance and antibacterial activity against *Escherichia coli*, *Salmonella enteritidis*, *Staphylococcus aureus* and *Bacillus cereus* of oil from organic yellow passion fruit (*Passiflora edulis* var. *flavicarpa*).

4 Composition of Fruit Essential Oil

The aroma of the yellow passion fruit (*P. edulis* Sims f. *flavicarpa* Deg.) has been supposed to be superior to that of the purple fruit (*P. edulis* f. *edulis* Sims) (Belitz et al. 2009) and most of in-depth investigations dedicated to passion fruit aroma and essential oil composition have focused on the yellow species. The attractive aroma of yellow passion fruit has been described as floral, estery and with an exotic tropical sulfury note (Werkhoff et al. 1998). Formation of volatile compounds responsible for the fruit aroma takes place mainly during the ripening process, when most of them quickly accumulate in the fruit. On the contrary, when harvested at an immature stage, the fruit hardly develops the full aroma characteristic of the fresh ripe fruit (Narain et al. 2010). The volatile profile of passion fruit, as for other tropical fruit, is quite complex and has been extensively investigated in a host of studies leading to the identification of more than 400 volatile compounds. These compounds belong to the chemical classes of esters (143), alcohols (62), sulfur-containing compounds (54), terpenes (49), ketones (42), aldehydes (24), furans (8), lactones (8), acids (7) and other classes (11) (Narain et al. 2010). According to Werkhoff et al. (1998), the attractive tropical flavor note of ripe yellow passion fruit is mainly due to the presence of trace levels of sulfur volatiles in combination with other compounds contributing fruity, estery, floral and green aroma impressions. Unfortunately, characterization of passion fruit aroma compounds is quite a challenging task, mainly because the key odorant sulfur compounds can be hardly detected by the most common GC detectors. This implies that most studies could not detect their presence and report their concentration in the analyzed material. In addition, volatile profiles reported in the literature were strongly influenced by the method applied for volatile isolation (Werkhoff et al. 1998) and by the source material used for aroma isolation (fresh ripe fruit, or unpasteurized fruit juice, or aqueous essence) (Jordán et al. 2002). As mentioned above, sulfur-containing compounds are by far the most interesting components of yellow passion fruit volatile fraction from the sensory point of view. In particular, 3-mercaptohexanol, 3-(methylthio) hexanol as well as the related acetates, butanoates and hexanoates (Fig. 29.3), even if present at trace level, have been recognised as key ingredients of the aroma of yellow passion fruit concentrates, due to their high odour potency and long-lasting character (Engel and Tressl 1991).

Other sulfur compounds recognized as key odorants of yellow passion fruit in a previous study, are the (Z) and (E) isomers of 2-methyl-4-propyl-1,3-oxathiane and 2-methyl-4-propyl-1,3-oxathiane-3-oxide (Winter et al. 1976). Most of sulfur aroma compounds found in yellow passion fruit are chiral, and both their odor quality and odor threshold are tremendously different depending on their configuration. Moreover, it has been shown that sulfur-containing passion fruit volatiles are present at high optical purity and the S configured is the predominant enantiomer (Werkhoff et al. 1998). In particular, it has been demonstrated that the naturally occurring S configured enantiomers of 3-mercaptohexanol and 3-(methylthio)hexanol have the exotic and tropical fruit notes, whereas the R forms were less intense and characterized only by sulfury and herbaceous impressions (Engel and Tressl 1991). The sulfur volatile fraction is certainly quite complex and, besides the above-mentioned compounds, Werkhoff and colleagues (1998), by using a sulfur selective GC detector, could detect more than 100 sulfur volatiles, and identify 47 among them. From a quantitative point of view, the esters are the most important group of the whole volatile profile. In the most representative extract of the aroma of the fresh fruit, obtained by a vacuum headspace sampling technique, Werkhoff and colleagues (1998) found some high boiling components of the ester fraction such as hexyl butanoate, hexyl hexanoate, ethyl 3-hydroxybutyrate, and 3(Z)-hexenyl hexanoate, as its major constituents. On the contrary, a different isolation method (dynamic headspace sampling) preferentially extracted low boiling components, such as ethyl butanoate, butyl acetate, 2- and 3-methyl butyl acetate, ethyl hexanoate, hexyl acetate and 3(Z)-hexenyl acetate (Werkhoff et al. 1998). All of these compounds are expected to contribute to the fruity notes perceived on the fruit. In particular, GC-olfactometric characterization of volatile isolated from an unpasteurized juice and a commercial aqueous essence of yellow passion fruit highlighted an important contribution of 2-methyl butyl hexanoate to the aroma of both products, and of hexyl hexanoate to the aroma of the essence (Jordán et al. 2002). Some unsaturated aliphatic alcohols, such as (E)-3-octen-1-ol, (Z)-4-decenol, 6-methyl-5-hepten-2-ol and (Z)-5-octen-1-ol, are considered important contributors to the green, floral and fruity notes of the overall flavor of yellow passion fruit (Werkhoff et al. 1998). Among other alcohols, a significant contribution of 2-phenyl ethanol and linalool to the aroma of the juice and the essence, respectively, has been

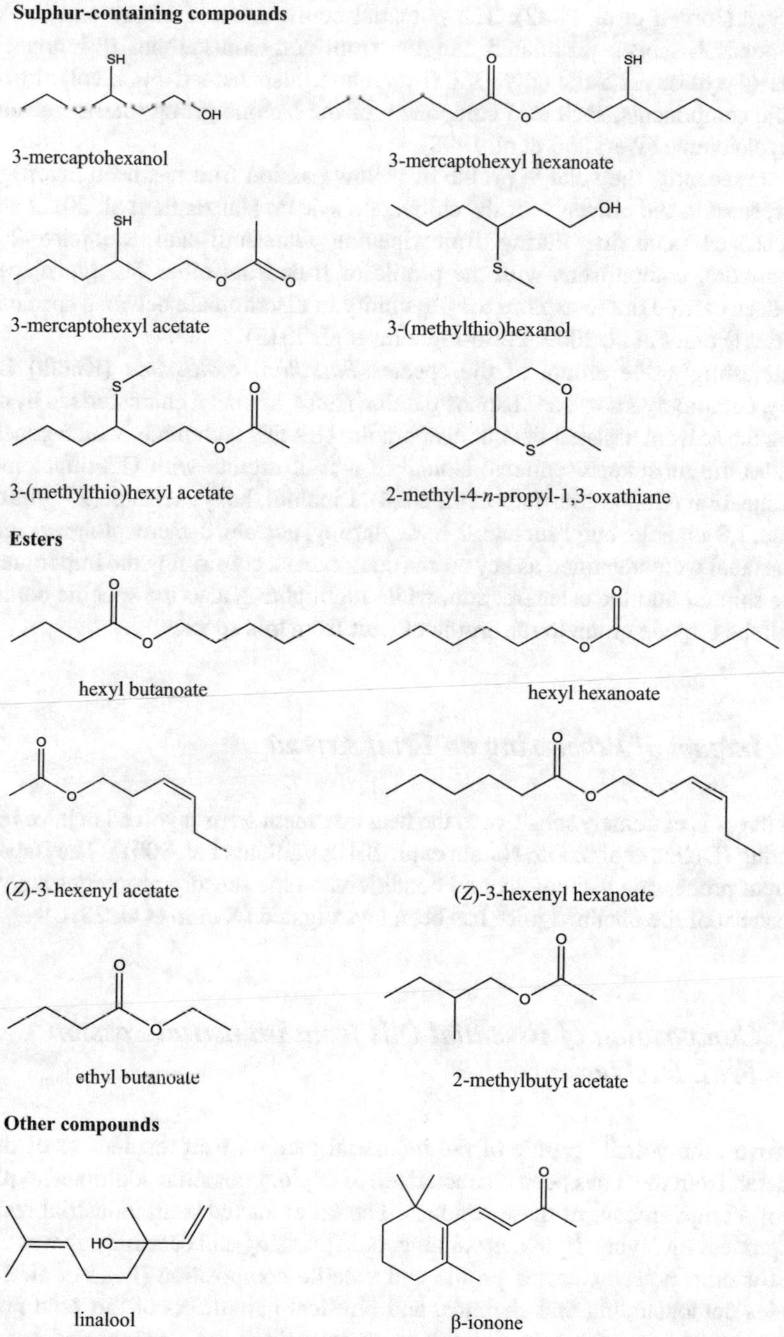

Sulphur-containing compounds

3-mercaptohexanol

3-mercaptohexyl hexanoate

3-mercaptohexyl acetate

3-(methylthio)hexanol

3-(methylthio)hexyl acetate

2-methyl-4-*n*-propyl-1,3-oxathiane

Esters

hexyl butanoate

hexyl hexanoate

(*Z*)-3-hexenyl acetate

(*Z*)-3-hexenyl hexanoate

ethyl butanoate

2-methylbutyl acetate

Other compounds

linalool

β-ionone

Fig. 29.3 Chemical structure of some of the most important odorants of yellow passion fruit

observed (Jordán et al. 2002). The potential contribution of many other volatile compounds has been postulated, ranging from common ketones (β-ionone) and furans (4-hydroxy-2,5-dimethyl-3(2*H*)-furanone, also named Furaneol), to more specific components, such as a component of the lactone group (marmelolactone) and cycloionone (Werkhoff et al. 1998).

More recently, the volatile profile of yellow passion fruit has been investigated with respect to the influence of the cultivation system (Janzantti et al. 2012) and to the changes occurring during fruit ripening (Janzantti and Monteiro 2014). Furthermore, comparisons with the profile of fruit from other *Passiflora* species have been carried out to explore the possibility to discriminate between species and varieties (Pontes et al. 2009; Porto-Figueira et al. 2015).

Interestingly, the aroma of the species *Passiflora mollissima* (Kunth) L. H. Bailey, commonly known as "banana passion fruit", has been characterised by combining the solvent assisted flavour evaporation (SAFE) technique, which generally provides the most representative isolate of a fruit aroma, with GC-olfactometric determination (Conde-Martínez et al. 2014). Linalool, hexyl acetate, (*Z*)-3-hexenyl acetate, 1,8-cineole, butyl acetate, 2-methylpropyl acetate, 3-mercaptohexyl acetate and hexanal were identified as key aroma compounds, confirming the important role of the sulphur and the ester fraction, while highlighting also the specific contribution of the terpene group to the aroma of fruit from this species.

4.1 Impact of Processing on Fruit Aroma

Fruit flavor is extremely sensitive to the heat treatment steps involved in juice manufacturing (Deliza et al. 2005; Narain et al. 2010; Vaillant et al. 2001). The impact of different processing technology and conditions on the sensory properties and volatile pattern of the obtained juice has been investigated (Narain et al. 2010).

4.2 Composition of Essential Oils from Industrial Passion Fruit Residues

Moreover, the volatile profile of the industrial passion fruit residues or of the oil extracted from them has been characterized to explore potential additional exploitation of a large amount of these residues. The oil extracted from industrial residues by a process involving drying, grounding, cold pressing and centrifuging was evaluated for odor potency, aroma profile and volatile composition (Leão et al. 2014). Besides demonstrating that chemical and physical parameters of this cold pressed oil was appropriate for potential use as an ingredient for cleaning and cosmetic products, such as soaps and shampoos, it was also found that it had an odoriferous power between two and three times greater than that of the fresh fruit pulp. The analysis of volatiles showed that the profile of the ester fraction was somewhat similar to that of the fresh fruit, even though its percentage of the total volatile fraction (59%) was lower than that observed in yellow passion fruit juice (91%) by using a similar procedure for volatile isolation (Porto-Figuera et al. 2015). However, it is not straightforward to compare the obtained profile with that of the fresh fruit because important key odorants, such as the sulfur compounds, were not detected and reported at all (Leão et al. 2014), for the above-mentioned reasons. In addition, no compounds related to the impact of the thermal treatment involved in the drying step, was determined, whereas sensory analysis of the oil highlighted the appearance of a cooked aroma that was absent in the fruit juice (Leão et al. 2014). Finally, even though the authors (Leão et al. 2014) claimed that the sensory profile of the oil was similar to that of the fresh fruit, a significant weakening of the fresh, citric and fruity aroma notes was observed. They concluded that the production of an aromatising oil from passion fruit processing residues is feasible and that the volatile profile and aroma quality of the obtained oil provide a basis for its potential use in the manufacture of cosmetics, hygiene products and essential oil or natural flavouring production (Leão et al. 2014). Seeds and shells form the main part of the industrial passion fruit residue and essential oils extracted from both seeds and shells of passion fruit of the species *Passiflora edulis* f. *flavicarpa* (yellow passion fruit) and *Passiflora ligularis* Juss (orange passion fruit) have been characterised in terms of extraction yield and volatile composition (Chez et al. 2015; Chóez-Guaranda et al. 2017). In both cases, the sesquiterpene ionol, ranging from 10% to 19%, was the main compound of the essential oil. Esters of fatty acids (ethyl hexadecanoate), aldehydes (hexadecanal, pentadecanal) and hydrocarbons (hexadecane, heptadecane, squalene) were the other major constituents of these essential oils. The composition of the essential oil extracted from the juice obtained from orange passion fruit also showed ionol as a main component (13%) (Chóez-Guaranda et al. 2017). However, a characterisation of the volatile fraction of the juice from the same species did not confirm the presence of ionol (Porto-Figueira et al. 2015), thus raising doubts about its nature as a genuine component or an extraction artefact in the analysed essential oils. Thus, further investigation seems to be needed to clarify whether the extraction of essential oils from these materials could be exploited to produce useful flavour ingredients for the food industry.

4.3   Composition of Essential Oils from Other Plant Parts

Essential oils extracted from flowers and leaves of three *Passiflora* ornamental hybrids were characterized for their composition to investigate their potential use for production of human wellbeing products (Calevo et al. 2016). The chemical profile of the three hybrids was markedly different, with benzyl alcohol, geraniol, phytol, phenylethanol, eugenol and (Z)-3-hexenal representing the main aroma compounds found in the leaves or the flowers, thus confirming the presence of valuable compounds in the considered hybrids.

5   Edible and Non-edible Applications of Fruit Seed Oil

Besides the several applications of passion fruit waste as a source of low-calorie fiber (Chau and Huang 2004) and protein for the food industry as a dietary supplement, passion fruit seed oil has a promising value in a wide range of uses and applications, both edible and non-edible ones. According to the analyses, it has features similar to conventional edible oils such as soybean oil, and may be a new source for human consumption (Casierra-Posada and Jarma-Orozco 2016).

Passion fruit oil is used in the application of lipids to food production that is dependent on their physical, chemical, and nutritional properties. For instance, Bezerra et al. (2017) studied the technological properties of Amazonian oils and fats and their applications in the food industry. They developed and monitored binary blends with pracaxi and passion fruit oils with palm stearin and cupuassu fat at different ratios. Concerning passiflora oil, the results found for blends PSF (PFS – Passion fruit oil + Palm stearin) 50:50 and 40:60 appeared promising for use in the food industry, for example, as table margarine and functional margarine due to their adequate technological properties.

Barbieri and Leimann (2014) studied the microencapsulation of passion fruit seed oil in poly (ε-caprolactone) in order to avoid the exposure of the oil to atmospheric air that would lead to its oxidation. For this purpose, they evaluated the interaction of the polymeric shell with the encapsulated oil using Fourier Transform Infrared Spectroscopy (FTIR). A recent work of Oliveira et al. (2017) investigated the encapsulation of passion fruit seed oil by means of a supercritical anti-solvent process.

The use of seed byproducts from fruit processing has contributed to increasing the supply of vegetable oils rich in bioactive compounds regarding consolidated cosmetics and functional foods. The passion fruit seeds are used nowadays to produce oil, which is destined to the cosmetic and food industries and may be found commercially. For example, Mattos et al. (2015) emphasized that the passion fruit seed oil has volatile compounds, which can be classified as aromas of industrial interest and have the potential to generate natural essences with high added-value.

Pereira (2011) have characterized a vegetable oil-based nanoemulsion (raspberry: passion fruit: peach oils, 1:1:1). They reported an HLB (hydrophile-lipophile balance) number of 9.0. In another study, the same authors (Pereira et al. 2016) evaluated the influence of lanolin derivatives (ethoxylated and acetylated lanolin) on the physicochemical characteristics and stability of vegetable oil-based nanoemulsions. Eventually, they concluded that nanoemulsions based on vegetable oils and lanolin derivatives are non-toxic since they did not cause any irritation on the skin surface after nanoemulsion application. Therefore, they are recommendable for use as vehicles for pharmaceuticals and cosmetics.

Rocha-Filho et al. (2014) analyzed the influence of lavender essential oil in the formation of passion fruit oil nanoemulsions, water and a blend of nonionic surfactants. Their results showed a reduction in droplet size with the addition of lavender essential oil, due to the possibility of essential oil molecules to penetrate into the interface and change their properties. On the other hand, its seeds have an oil content of 18–30% (Nyanzi et al. 2005; Malacrida and Jorge 2012), which makes them as possible source of base oil for lubricants and promising feedstock for the production of biodiesel. Generally, vegetable oils represent excellent renewable sources for chemical and oleo-chemical industries, considering that they have a lower potential of toxicity, and degrading faster than the traditional mineral oil lubricants as well as they can be functionalized to be utilized in various applications.

As marked by Lopes et al. (2010b), vegetable oils have a great potential to replace petrochemical derivatives and contribute to minimizing environmental impacts, e.g. as base fluids in environment-friendly lubricants. Moreover, polyols obtained from oils open up new possibilities for use as monomers for polyurethanes, which exhibit excellent properties beyond their plain reactivity (Lopes et al. 2010b). The same authors described the synthesis of polymeric materials by thermal polymerization from linseed oil (*Linum usitatissimum* L.) and passion fruit oil (*Passiflora edulis*). They performed a comprehensive characterization by gas chromatographic (GC), Fourier transform infrared (FTIR) spectroscopy, solubility in organic solvents, thermogravimetry (TG), differential scanning calorimetry (DSC) and Raman spectroscopy (Lopes et al. 2009). Indeed, passion fruit oil represents a renewable source of bio-based compounds as also concluded by Rodrigues et al. (2017). In particular, the authors describe a route to produce cost-effective and bio-based polyurethane composites from passion fruit oil methyl esters and coconut husk fibers (Rodrigues et al. 2017).

Although having many environmental benefits and a wide range of applications in the industry, vegetable oils are more susceptible to degradation by oxidation or hydrolytic reactions. As described by Silva et al. (2015), the epoxidation process can improve the oxidative stability and the acid value of both pure vegetable oils. New bio-lubricants with satisfactory properties were obtained successfully using epoxidized passion fruit and moringa oils. Moreover, the authors pointed out that epoxidized oils have a better compatibility with additives, resulting in an increased performance of the fluids (Silva et al. 2015). Farias et al. (2011) evaluated biodiesel blends of passion fruit and castor oil in different proportions and their thermal stability. They showed that the biodiesel blend of passion fruit and castor oil with a 1:1 ratio increased the thermal stability in relation to pure passion fruit derived biodiesel. On the other hand, the biodiesel blend of passion fruit and castor oil with 1:2 presented higher thermal stability, because passion fruit biodiesel has a high content of oleic and linoleic acids (which are more susceptible to oxidation) and castor oil has a high content of ricinoleic acid.

6 Adulteration and Authenticity of Passion Fruit Oil Using Spectroscopy Combined with Chemometrics

In the previous sections, it was shown that passion fruit oils are valuable fluids that can be used in many ways. Therefore, adulteration may be a problem and hence the need for reliable analytical methods that are capable of characterizing an oil.

Spectroscopic methods such as FTIR and UV/Vis absorption spectroscopy are frequently used to identify and analyze oils in the food and cosmetics sectors (Gomez-Caravaca et al. 2016; Rohman 2017). Passion fruit oil is often blended with other types of oils. For example, in cosmetics it may be mixed with pracaxi oil as reported above while in other products blends with olive and sunflower oil are common. Pracaxi oil itself is a high-value product. Therefore, it is unlikely that it is used to dilute or adulterate passion fruit oil. On the other hand, olive oil and sunflower oil can be produced significantly cheaper and therefore may be seen as potential adulterants. Consequently, a suitable analytical method would be capable of distinguishing between these oils and determine the composition of their blends with passion fruit oil.

The number of studies on the application of advanced analytical methods to passion fruit oil is rather limited. Ferreira et al. (2013) used Raman spectroscopy to investigate the carotenoids in a number of Amazonian oils including passion fruit oil. They were able to identify typical carotenoid bands in the oil spectra and to distinguish between the different oil types. Hence, they concluded that Raman spectroscopy in the near infrared spectral range is a suitable tool for analytical quality control of oils. A typification and quality control of Amazonian oils was also possible using matrix-assisted laser desorption/ionization time-of-flight (MALDI-TOF) mass spectrometry as reported by Saraiva et al. (2009). For this purpose, the signatures of triacylglycerol in the mass spectra were taken advantage. De Vasconcelos Vieira Lopes et al. (2010) applied 1H-NMR to passion fruit oil. Their study aimed at analyzing the degree of saturation and effects of desaturation of the unsaturated fatty acids, which it demonstrated successfully.

FTIR spectroscopy represents a possible alternative to the above methods, but has not been applied to passion fruit oil to date, to the best of our knowledge. In comparison with the above methods, FTIR is more straightforward and robust from an experimental point of view. Infrared spectroscopy is an absorption method, in which the molecules are vibrationally excited by taking up a photon. The absorbed photon is removed from the radiation passing the sample, which gives rise to the absorption spectrum. The experimentally simplest and most robust approach to FTIR spectroscopy is attenuated total reflection (ATR) spectroscopy. In an ATR-FTIR experiment, the infrared radiation is propagating in a transmissive high-refractive-index material, often zinc selenide or a diamond crystal. The sample has a lower refractive index and is in contact with the surface of the crystal. The radiation undergoes total internal reflection at this surface, so that the evanescent field can interact with the sample. Consequently, the reflected beam is attenuated and carries the spectroscopic information. Details can be found in the literature (Griffiths and De Haseth 2007; Kiefer et al. 2011). This approach requires virtually no sample preparation. The spectra obtained can be fed into an evaluation algorithm. When the aim of the study is to classify the oil or to identify adulteration, unsupervised chemometric tools that enable classification of spectroscopic data sets are best suited. The most prominent and popular tool for this data analysis is principal component analysis (PCA). PCA is a purely mathematical tool that determines the eigenvectors and eigenvalues of the data set. In other words, it returns the spectral signatures that contribute to the variance of the spectra in descending order. The first principal component represents the features that are characteristic of the largest variance and so on. Plotting the loadings of the principal components (PCs) against each other shows whether or not the individual spectra in the data set group together to form certain clusters. Consequently, if different oils have sufficiently different spectra they will be separated from each other in such a plot. Details about chemometric data analysis and its implementation can be found in references (Wold et al. 1987; Pearson 1901; Hotelling 1933; Jolliffe 2002; Bakeev 2010).

Figure 29.4 shows the FTIR spectra of a commercial passion fruit oil in comparison with sunflower and olive oil. At first glance, all three spectra look very similar. The very weak signature around 3500 cm^{-1} can be attributed to OH stretching vibrations. The low intensity indicates that the amount of OH groups in the oils is small. Between 2800 and 3100 cm^{-1} the CH stretching bands can be observed. Owing to the chemical complexity of the triglyceride molecules in the oils, this band constitutes a multitude of individual sub-peaks originating from different CH groups and their vibrational modes. The strong peak at 1740 cm^{-1} is characteristic of the C=O double bonds in the ester groups. The fingerprint region below 1700 cm^{-1} is very rich in content with overlapping peaks from stretching, rocking, wagging and scissoring modes. In this region of the spectrum, it is difficult to make unambiguous assignments.

The arrows in Fig. 29.4 indicate spectral signatures that exhibit small differences and might allow distinguishing between the different oils. However, a confident discrimination is hardly possible with the naked eye. Therefore, the spectra were processed using principal component analysis. The score plot in Fig. 29.5 shows a clear distinction between the three types of oil. For this purpose, six spectra from each individual oil have been recorded and processed using the PCA algorithm implemented in Matlab. From these results, it could be concluded that the combination of FTIR spectroscopy and chemometric data analysis in terms of PCA is a straightforward and fast method to classify passion fruit oil and distinguish it against potential adulterants.

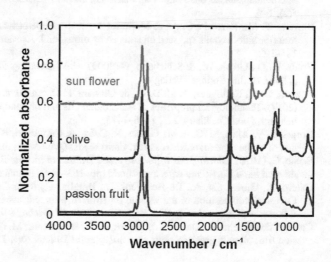

Fig. 29.4 FTIR spectra of sunflower, olive and passion fruit oil. The arrows mark spectral signatures that show small but characteristic differences

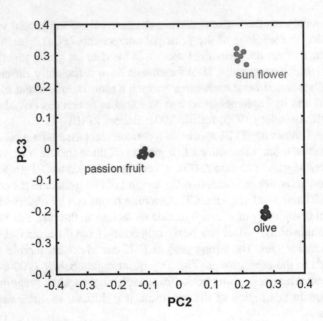

Fig. 29.5 Score plot of a standard PCA evaluation

References

Abrahamczyk, S., Souto-Vilarós, D., & Renner, S. S. (2014). Escape from extreme specialization: Passionflowers, bats and the sword-billed hummingbird. *Proceedings of the Royal Society B, 281*, 20140888. https://doi.org/10.1098/rspb.2014.0888.

Bakeev, K. A. (2010). *Process analytical technology* (2nd ed.). Hoboken: Wiley-Blackwell.

Barbieri, J. C., & Leimann, F. V. (2014). Extração de Óleo da Semente do Maracujá e Microencapsulação em Poli(ε Caprolactona). *Revista Brasileira de Pesquisa em Alimentos, 5*(2), 1–9.

Barrales, F. M., Rezende, C. A., & Martínez, J. (2015). Supercritical CO2 extraction of passion fruit (Passiflora edulis sp.) seed oil assisted by ultrasound. *Journal of Supercritical Fluids, 104*, 183–192.

Belitz, H. D., Grosh, W., & Schieberle, P. (2009). *Food chemistry* (4th revised and extended ed. p. 840). Berlin: Springer-Verlag.

Bezerra, C. V., Rodrigues, A. M. D. C., de Oliveira, P. D., da Silva, D. A., & da Silva, L. H. M. (2017). Technological properties of Amazonian oils and fats and their applications in the food industry. *Food Chemistry, 221*, 1466–1473.

Borges, S. V., Maia, M. C. A. M., Gomes, R. C. M., & Cavalcanti, B. (2007). Chemical composition of umbu (*Spondias tuber*osa Arr. Cam) seeds. *Quím Nova, 30*(1), 49–52.

BoSio, G. (1610). *La trionfante e gloriosa croce Trattato di Iacomo Bosio Lettione varia, e divota; Ad ogni buon Christiano utile, e gioconda* (pp. 163–164). Roma. libro secondo, capitolo sesto.

Calevo, J., Giovannini, A., De Benedetti, L., Braglia, L., della Cuna, F., & Tava, A. (2016). Chemical composition of the volatile oil from flowers and leaves of new Passiflora hybrids. *International Journal of Applied Research in Natural Products, 9*, 21–27.

Cardoso de Oliveira, R., Davantel de Barros, S. T., & Gimenes, M. (2013). The extraction of passion fruit oil with green solvents. *Journal of Food Engineering, 117*, 458–463.

Casierra-Posada, F., & Jarma-Orozco, A. (2016). Nutritional composition of Passiflora species. In M. S. J. Simmonds & V. R. Preedy (Eds.), *Nutritional composition of fruit cultivars* (1st ed., pp. 517–534). London: Academic. https://doi.org/10.1016/B978-0-12-408117-8.00022-2.

Chau, C. F., & Huang, Y. L. (2004). Characterization of passion fruit seed fibres-a potential fibre source. *Food Chemistry, 85*, 189–194.

Chez, I., Herrera, D., Miranda, M., & Manzano, P. (2015). Chemical composition of essential oils of shells, juice and seeds of *Passiflora ligularis* Juss from Ecuador. *Emirates Journal of Food and Agriculture, 27*, 650–653. https://doi.org/10.9755/ejfa.2015.04.039.

Chóez-Guaranda, I., Ortega, A., Miranda, M., & Manzano, P. (2017). Chemical composition of essential oils of *Passiflora edulis* f. Flavicarpa agroindustrial waste. *Emirates Journal of Food Agriculture, 29*, 458–462. https://doi.org/10.9755/ejfa.2016-10-1542.

Conde-Martínez, N., Sinuco, D. C., & Osorio, C. (2014). Chemical studies on curuba (*Passiflora mollissima* (Kunth) L. H. Bailey) fruit flavour. *Food Chemistry, 157*, 356–363. https://doi.org/10.1016/j.foodchem.2014.02.056.

de Santana, F. C., Shinagawa, F. B., Araujo Eda, S., Costa, A. M., & Mancini-Filho, J. (2015). Chemical composition and antioxidant capacity of Brazilian Passiflora seed oils. *Food Science, 80*, C2647–C2654.

de Vasconcelos Viera Lopes, R., Zamian, J. R., Resck, I. S., Araujo Sales, M. J., dos Santos, M. L., & da Cunha, F. R. (2010). Physicochemical and rheological properties of passion fruit oil and its polyol. *European European Journal of Lipid Science and Technology, 112*, 1253–1262.

Deliza, R., MacFie, H. A. L., & Hedderley, D. (2005). The consumer sensory perception of passion-fruit juice using free-choice profiling. *Journal of Sensory Studies, 20*, 17–27.

Dhawan, K., Dhawan, S., & Sharma, A. (2004). Passiflora: A review update. *Journal of Ethnopharmacology, 94*, 1–23.

Engel, K. H., & Tressl, R. (1991). Identification of new sulfur-containing volatiles in yellow passionfruit (*Passiflora edulis* f. flavicarpa). *Journal of Agricultural and Food Chemistry, 39*, 2249–2252.

Farias, R. M. C., Conceição, M. M., Candeia, R. A., Silva, M. C. D., Fernandes, V. J., & Souza, G. A. (2011). Evaluation of the thermal stability of biodiesel blends of castor oil and passion fruit. *Journal of Thermal Analysis and Calorimetry, 106*, 651. https://doi.org/10.1007/s10973-011-1566-x.

Ferrari, R. A., Colussi, F., & Ayub, R. A. (2004). Characterization of by-products of passion fruit industrialization utilization of seeds. *Revista Brasileira de Fruticultura, 26*, 101–102.

Ferreira, B. S., De Almeida, A., Diniz, C. G., & Lúcia, V. (2011). Comparative properties of Amazonian oils obtained by different extraction methods. *Molecules, 16*, 5875–5885.

Ferreira, B. S., de Almeida, C. G., Le Hyaric, M., De Oliveira, V. E., Edwards, H. G. M., & de Oliveira, L. F. C. (2013). Raman spectroscopic investigation of carotenoids in oils from Amazonian products. *Spectroscopy Letters, 46*, 122–127.

Feuillet, C., & Mac Dougal, J. M. (2004). A new infrageneric classification of *Passiflora* L. (Passifloraceae). *Passiflora, 13*, 34–38.

Gomez-Caravaca, A. M., Maggio, R. M., & Cerretani, L. (2016). Chemometric applications to assess quality and critical parameters of virgin and extra-virgin olive oil. A review. *Analytica Chimica Acta, 913*, 1–21.

Griffiths, P. R., & De Haseth, J. A. (2007). *Fourier transform infrared spectrometry* (2nd ed.). Hoboken: Wiley.

Hanmoungjai, P., Pyle, L., & Niranjan, K. (2000). Extraction of rice bran oil using aqueous media. *Journal of Chemical Technology and Biotechnology, 75*, 348–352.

Hotelling, H. (1933). Analysis of a complex of statistical variables into principal components. *Journal of Educational Psychology, 24*, 417.

Janzantti, N. S., & Monteiro, M. (2014). Changes in the aroma of organic passion fruit (*Passiflora edulis* Sims f. flavicarpa Deg.) during ripeness. *LWT- Food Science and Technology, 59*, 612–620. https://doi.org/10.1016/j.lwt.2014.07.044.

Janzantti, N. S., Macoris, M. S., Garruti, D. S., & Monteiro, M. (2012). Influence of the cultivation system in the aroma of the volatile compounds and total antioxidant activity of passion fruit. *LWT- Food Science and Technology, 46*, 511–518. https://doi.org/10.1016/j.lwt.2011.11.016.

Jolliffe, I. (2002). *Principal component analysis*. New York: Wiley.

Jordán, M., Goodner, K. L., & Shaw, P. E. (2002). Characterization of the aromatic profile in aqueous essence and fruit juice of yellow passion fruit (*Passiflora edulis* Sims F. Flavicarpa degner) by GC-MS and GC/O. *Journal of Agricultural and Food Chemistry, 50*, 1523–1528.

Jorge, A. T., Arroteia, K. F., Santos, I. A., Andres, E., Medina, S. P., Ferrari, C. R., Lourenço, C. B., Biaggio, R. M., & Moreira, P. L. (2012). *Schinus terebinthifolius* Raddi extract and linoleic acid from *Passiflora edulis* synergistically decrease melanin synthesis in B16 cells and reconstituted epidermis. *International Journal of Cosmetic Science, 34*(5), 435–440.

Kiefer, J., Frank, K., & Schuchmann, H. P. (2011). Attenuated total reflection infrared (ATR-IR) spectroscopy of a water-in-oil emulsion. *Applied Spectroscopy, 65*, 1024–1028.

Killip, E. P. (1938). The American species of Passifloraceae. *Publication Field Museum of Natural History Botanical, 19*, 1–613.

Kitada, M., Ogura, Y., Maruki-Uchida, H., Sai, M., Suzuki, T., Kanasaki, K., Hara, Y., Seto, H., Kuroshima, Y., Monno, I., & Koya, D. (2017). The effect of piceatannol from passion fruit (*Passiflora edulis*) seeds on metabolic health in humans. *Nutrients, 9*(10), E1142.

Kobori, C. N., & Jorge, N. (2005). Characterization of some seed oils of fruits for utilization of industrial residues. *Ciência e Agrotecnologia, 29*(5), 1008–1014.

Leão, K. M. M., Sampaio, K. L., Pagani, A. A. C., & Da Silva, M. A. A. P. (2014). Odor potency, aroma profile and volatiles composition of cold pressed oil from industrial passion fruit residues. *Industrial Crops and Products, 58*, 280–286. https://doi.org/10.1016/j.indcrop.2014.04.032.

Lee, S. Y., Fu, S. Y., & Chong, G. H. (2015). Ultrasound-assisted extraction kinetics, fatty acid profile, total phenolic content and antioxidant activity of green solvents' extracted passion fruit oil. *International Journal of Food Science and Technology, 50*, 1831–1838.

Liu, S., Yang, F., Zhang, C., Ji, H., Hong, P., & Deng, C. (2009). Optimization of process parameters for supercritical carbon dioxide extraction of Passiflora seed oil by response surface methodology. *Journal of Supercritical Fluids, 48*, 9–14.

Lopes, R. V. V., Loureiro, N. P. D., Zamian, J. R., Fonseca, P. S., Macedo, J. L., dos Santos, M. L., & Sales, M. J. A. (2009). Synthesis and characterization of polymeric materials from vegetable oils. *Macromolecolar Symposia, 286*, 89–94.

Lopes, R. M., Sevilha, A. C., Faleiro, F. G., Silva, D. B., Vieira, R. F., & Agostini-Costa, T. S. (2010a). Estudo comparativo do perfil de ácidos graxos em semente de Passifloras nativas do cerrado brasileiro. *Revista Brasileira de Fruticultura, 32*, 498–506. (in Portuguese). https://doi.org/10.1590/S0100-29452010005000065.

Lopes, R. V. V., Zamian, J. R., Sabioni Resck, I., Araújo Sales, M. J., dos Santos, M. L., & da Cunha, F. R. (2010b). Physicochemical and rheological properties of passion fruit oil and its polyol. *European Journal of Lipid Science and Technology, 12*, 1253–1262.

Lourith, N., & Kanlayavattanakul, M. (2013). Antioxidant activities and phenolics of *Passiflora edulis* seed recovered from juice production residue. *Journal of Oleo Science, 62*, 235–240.

Machado, Y. L., Dantas Neto, A. A., Fonseca, J. L. C., & Dantas, T. N. C. (2014). Antioxidant stability in vegetable oils monitored by the ASTM D7545 method. *Journal of the American Oil Chemists' Society, 91*, 1139–1145.

Malacrida, C. R., & Jorge, N. (2012). Yellow passion fruit seed oil (*Passiflora edulis* f. flavicarpa): Physical and chemical characteristics. *Brazilian Archives of Biology and Technology, 55*, 127–134.

Maruki-Uchida, H., Kurita, I., Sugiyama, K., Sai, M., Maeda, K., & Ito, T. (2013). The protective effects of piceatannol from passion fruit (*Passiflora edulis*) seeds in UVB-irradiated keratinocytes. *Biological & Pharmaceutical Bulletin, 36*(5), 845–849.

Maruki-Uchida, H., Morita, M., Yonei, Y., & Sai, M. (2018). Effect of passion fruit seed extract rich in piceatannol on the skin of women: A randomized, placebo-controlled, double-blind trial. *Journal of Nutritional Science and Vitaminology, 64*(1), 75–80.

Matsui, Y., Sugiyama, K., Kamei, M., Takahashi, T., Suzuki, T., Katagata, Y., & Ito, T. (2010). Extract of passion fruit (*Passiflora edu*lis) seed containing high amounts of piceatannol inhibits melanogenesis and promotes collagen synthesis. *Journal of Agricultural and Food Chemistry, 58*, 11112–11118.

Matsumoto, Y., Gotoh, N., Sano, S., Sugiyama, K., Ito, T., Abe, Y., Katano, Y., & Ishihata, A. (2014). Effects of Scirpusin B, A polyphenol in passion fruit seeds, on the coronary circulation of the isolated perfused rat heart. *International Journal of Medical Research & Health Sciences, 3*, 547–553.

Mattos De Paula, R. C., Gomes Soares, A., & Pereira Freitas, S. (2015). Volatile compounds in passion fruit seed oil (passiflora setacea brs pérola do cerrado and passiflora alata brs doce mel). *Chemical Engineering Transactions, 44*, 103–108. https://doi.org/10.3303/CET1544018.

Morais, D. R., Rotta, E. M., Sargi, S. C., Schimidt, E. M., Bonafé, E. G., Eberlin, M. N., Sawaya, A. C. H. F., & Visentainer, J. V. (2015). Antioxidant activity, phenolics and. UPLC-ESI(-)-MS of extracts from different tropical fruits parts and processed peels. *Food Research International, 77*, 392.

Morais, D. R., Rotta, E. M., Sargi, S. C., Bonafe, E. G., Suzuki, R. M., Souza, N. E., Matsushita, M., & Visentainer, J. V. (2017). Proximate composition, mineral contents and fatty acid composition of the different parts and dried peels of tropical fruits cultivated in Brazil. *Journal of the Brazilian Chemical Society, 28*, 308.

Narain, N., Nigam, N., & De Sousa Galvão, M. (2010). Passion fruit. In Y. H. Hui (Ed.), *Handbook of fruit and vegetable flavors* (pp. 345–389). Hoboken: Wiley.

Nyanzi, S. A., Carstensen, B., & Schwack, W. (2005). A comparative study of acid profiles of Passiflora seed oils from Uganda. *Journal of the American Oil Chemists Society, 82*(1), 41–44.

Ocampo, J., Coppens d'Eeckenbrugge, G., & Jarvis, A. (2010). Distribution of the genus *Passiflora* L. diversity in Colombia and its potential as an indicator for biodiversity management in the coffee growing zone. *Diversity, 2*, 1158–1180.

Oliveira, D. A., Mezzomo, N., Gomes, C., & Ferreira, S. R. S. (2017). Encapsulation of passion fruit seed oil by means of supercritical antisolvent process. *The Journal of Supercritical Fluids, 129*, 96–105.

Pasquel-Reátegui, J. L., Machado, A. P. F., Barbero, G. F., Rezende, C. A., & Martínez, J. (2014). Extraction of antioxidant compounds from blackberry (*Rubus* sp.) bagasse using supercritical CO_2 assisted by ultrasound. *Journal of Supercritical Fluids, 94*, 223–233.

Patel, S. S., Soni, H., Mishra, K., & Singhai, A. K. (2011). Recent updates on the genus Passiflora: A review. *International Journal of Research in Phytochemistry and Pharmacology, 1*(1), 1–16.

Pearson, K. (1901). Principal components analysis. *The London, Edinburgh, and Dublin Philosophical Magazine and Journal of Science, 6*, 559.

Pereira, T. A. (2011). Obtenção e caracterização de nanoemulsões O/A à base de óleo de framboesa, maracujá e pêssego: Avaliação de propriedades cosméticas da formulação. In *Tese de Mestrado – Faculdade de Ciências Farmacêuticas de Ribeirão Preto* (p. 102). Sao Paulo: University of São Paulo.

Pereira, T. A., Guerreiro, C. M., Maruno, M., Ferrari, M., & Rocha-Filho, P. A. (2016). Exotic vegetable oils for cosmetic O/W nanoemulsions: In vivo evaluation. *Molecules, 21*(3), 248.

Pereira, M. G., Hameraski, F., Andrade, E. F., Scheer, A. d P., & Corazza, M. L. (2017). Assessment of subcritical propane, ultrasound-assisted and Soxhlet extraction of oil from sweet passion fruit (*Passiflora alata* Curtis) seeds. *Journal of Supercritical Fluids, 128*, 338–348.

Pereira, M. G., Maciel, G. M., Haminiuk, C. W. I., Bach, F., Hamerski, F., de Paula Scheer, A., & Corazza, M. L. (2018). Effect of extraction process on composition, antioxidant and antibacterial activity of oil from yellow passion fruit (*Passiflora edulis* Var. Flavicarpa) seeds. *Waste and Biomass Valorization*. https://doi.org/10.1007/s12649-018-0269-y.

Piombo, G., Barouh, N., Barea, B., Boulanger, R., Brat, P., Pina, M., & Villeneuve, P. (2006). Characterization of the seed oils from kiwi (*Actinidia chinensis*), passion fruit (*Passiflora eulis*) and guava (*Psidium guajava*). *Oléagineux, Corps Gras, Lipides, 13*(2–3), 195–199.

Pontes, M., Marques, J. C., & Câmara, J. S. (2009). Headspace solid-phase microextraction-gas chromatography-quadrupole mass spectrometric methodology for the establishment of the volatile composition of Passiflora fruit species. *Microchemical Journal, 93*, 1–11. https://doi. org/10.1016/j.microc.2009.03.010.

Porto-Figueira, P., Freitas, A., Cruz, C. J., et al. (2015). Profiling of passion fruit volatiles: An effective tool to discriminate between species and varieties. *Food Research International, 77*, 408–418. https://doi.org/10.1016/j.foodres.2015.09.007.

Ran, V. S., & Blazquez, A. M. (2008). Fatty acid composition of *Passiflora edulis* Sims. Seed oil. *JLST, 40*, 65.

Regis, S. A., de Resende, E. D., & Antoniassi, A. (2015). Oil quality of passion fruit seeds subjected to a pulp-waste purification process. *Ciência Rural, Santa Maria, 45*(6), 977–984.

Riera, E., Golás, Y., Blanco, A., Gallego, A. M., & Mulet, A. (2004). Mass transfer enhancement in supercritical fluids extraction by means of power ultrasound. *Ultrasonics Sonochemistry, 11*, 241–244.

Rocha-Filho, P. A., Camargo, M. F. P., Ferrari, M., & Maruno, M. (2014). Influence of lavander essential oil addition on passion fruit oil nanoemulsions: Stability and in vivo study. *Journal of Nanomedicine and Nanotechnology.* 2014, *5*, 2. https://doi.org/10.4172/2157-7439.1000198.

Rodrigues, J. D. O., Murawski, A., Beckler, B., Lopes, R. V. V., Paterno, L. G., Quirino, R. L., & Sales, M. J. A. (2017). Bio-based polyurethanes and composites from passion fruit oil methyl esters and coconut husk fibers. In A. Shahzad (Ed.), *Biocomposites: Properties, performance and applications* (pp. 125–144). New York: Nova Science Publishers. ISBN: 978-1-53612-120-9.

Rohman, A. (2017). The use of infrared spectroscopy in combination with chemometrics for quality control and authentication of edible fats and oils: A review. *Applied Spectroscopy Reviews, 52*, 589–604.

Romani, A., Scardigli, A., & Pinelli, P. (2017). An environmentally friendly process for the production of extracts rich in phenolic antioxidants from Olea europaea L. and *Cynara scolymus* L. matrices. *European Food Research and Technology, 243*, 1229–1238.

Rotta, E. M., da Silva, M. C., Maldaner, L., & Visentainer, J. V. (2017). Ultrasound-assisted saponification coupled with gas chromatography-flame ionization detection for the determination of phytosterols from passion fruit seed oil. *Journal of the Brazilian Chemical Society, 29*, 1–8.

Sano, S., Sugiyama, K., Ito, T., Katano, Y., & Ishihata, A. (2011). Identification of the strong vasorelaxing substance scirpusin B, a dimer of piceatannol, from passion fruit (*Passiflora edulis*) seeds. *Journal of Agricultural and Food Chemistry, 59*, 6209–6213.

Santos, E. A., Souza, M. M., Abreu, P. P., da Conceicao, L. D. H. C. S., Araujo, I. S., Viana, A. P., de Almeida, A. A. F., & Freitas, J. C. O. (2012). Confirmation and characterization of interspecific hybrids of *Passiflora* L. (Passifloraceae) for ornamental use. *Euphytica, 184*(3), 389–399.

Saraiva, S. A., Cabral, E. C., Eberlin, M. N., & Catharino, R. R. (2009). Amazonian vegetable oils and fats: Fast typification and quality control via triacyglycerol (TAG) profiles from dry matrix-assisted laser desorption/ionization time-of-flight (MALDI-TOF) mass spectrometry fingerprinting. *Journal of Agricultural and Food Chemistry, 57*, 4030–4034.

Sgaravatti, M., & Zardini, P. (1997). In E. Calderini (Ed.), *Passiflore.* Coll. le Gemme Verdi.

Silva, M. S., Arimatéia Júnior, H., Silva, G. F., Dantas Neto, A. A., & Castro Dantas, T. N. (2015). New formulation for hydraulic biolubrificants based on epoxidized vegetable oils: Passion fruit (*Passiflora edulis* Sims f. flavicarpa Degener) and moringa (*Moringa oleifera* Lamarck). *Brazilian Journal of Petroleum and Gas, 9*, 27–36. ISSN 1982-0593.

Vaillant, F., Jeanton, E., Dornier, M., O'Brien, G. M., Reynes, M., & Decloux, M. (2001). Concentration of passion fruit juice on an industrial pilot scale using osmotic evaporation. *Journal of Food Engineering, 47*, 195–202.

Vecchia, M., & Giovannini, A. (2011). Le passiflore: aspetti botanici. *Informatore Botanico Italiano, 43*(1), 47–50.

Viganó, J., & Martínez, J. (2015). Trends for the application of passion fruit industrial by-products: A review on the chemical composition and extraction technique of phytochemicals. *Food Public Health, 5*, 164–173.

Viganó, J., da Fonseca Machado, A. P., & Martínez, J. (2015). Sub- and supercritical fluid technology applied to food waste processing. *Journal of Supercritical Fluids, 96*, 272.

Werkhoff, P., Güntert, M., Krammer, G., et al. (1998). Vacuum headspace method in aroma research: Flavor chemistry of yellow passion fruits. *Journal of Agricultural and Food Chemistry, 46,* 1076–1093.

Winter, M., Furrer, A., Willhalm, B., & Thommen, W. (1976). Identification and synthesis of two new organic sulfur compounds from the yellow passion fruit (*Passiflora edulis* f. flavicarpa). *Helvetica Chimica Acta, 59,* 1613–1620.

Wold, S., Esbensen, K., & Geladi, P. (1987). Principal component analysis. *Chemometrics and Intelligent Laboratory Systems, 2,* 37–52.

Yockteng, R., & Nadot, S. (2004). Phylogenetic relationships among Passiflora species based on the glutamine synthetase nuclear gene expressed in chloroplast (ncpGS). *Molecular Phylogenetics and Evolution, 31,* 379–396.

Zeraik, M. L., Pereira, C. A. M., Zuin, V. G., & Yariwake, J. H. (2010). Maracujá: um alimento funcional? *Revista Brasileira de Farmacognosia, 20,* 459–471.

Zuin, V. G., & Ramin, L. Z. (2018). Green and sustainable separation of natural products from agro-industrial waste: Challenges, potentialities, and perspectives on emerging approach. *Topics in Current Chemistry, 376,* 3.

Chapter 30
Bael (*Aegle marmelos*) Oil

Monika Choudhary and Kiran Grover

Abstract Bael (*Aegle marmelos*) known as temple garden tree has been originated from India and is also grown in most of the countries of Southeast Asia. An extensive literature has also been documented for the medicinal properties of bael. The various components viz. alkaloids, coumarins and steroids isolated from different parts of bael tree are responsible for the pharmacological action of this miraculous tree. The oil extracts and compounds purified from bael have been proven to be biologically active against several major non-communicable diseases like cancer, diabetes and cardiovascular diseases. The essential oil of the leaves contains d-limonene, a-d-phellandrene, cineol, citronellal, citral, pcyrnene, and cumin aldehyde. The limonene-rich oil distilled from the rind of bael fruit is used for scenting the hair oil and used as a yellow dye in calico printing. In addition, the oil extracted from leaves gives relief from frequent cold and respiratory infections. The essential oil obtained from bael is also known to exhibit antifungal properties.

Keywords *Aegle marmelos* · Chemical constituents · Essential oils · Fatty acids

Abbreviations

A. marmelos	*Aegle marmelos*
MUFA	Monounsaturated fatty acid
PUFA	Polyunsaturated fatty acid
SFA	Saturated fatty acid

M. Choudhary (✉)
Punjab Agricultural University, Ludhiana, India

K. Grover
Department of Food and Nutrition, College of Home Science, Punjab Agricultural University, Ludhiana, India
e-mail: kirangrover@pau.edu

© Springer Nature Switzerland AG 2019
M. F. Ramadan (ed.), *Fruit Oils: Chemistry and Functionality*,
https://doi.org/10.1007/978-3-030-12473-1_30

1 Introduction

Worldwide medicinal plants are considered a vast source of several pharmacologically and biologically active principles and compounds, which are commonly used in home remedies against multiple ailments. Bael (*Aegle marmelos*) known as temple garden tree occupies a prestigious reputation in Indian medicinal plants due to its gigantic conventional usage and nutraceutical properties against various diseases (Sharma et al. 2007). Being an inhabitant of India, this tropical fruit plant belongs to Rutaceae family and grown throughout India as well as in Pakistan, Sri Lanka, Bangladesh and most of the Southeast Asian countries. With regard to the morphological characteristics, bael is a moderate-sized, slender, aromatic tree with 6.0–7.5 m height. The bael fruit has a smooth, woody shell with a green or yellow peel. The fibrous yellow pulp of the fruit is very aromatic and tastes like a marmalade (Gupta et al. 2006; Sharma et al. 2011). In last decades, this plant and its parts have extensively been reported for various medicinal properties viz. anticancer, antibacterial, antifungal, antidiabetic, antioxidant activities. All the parts of this religious tree the leaves, bark, roots, fruits and seeds have been documented for its ethno medicinal and other uses in the traditional system of many countries. For example, flowers are used to make sweet-scented water through distillation, marmelle oil is obtained through distillation of rind, the sweet aromatic pulp is used for the preparation of beverages and a yellow colored dye is obtained from the unripe rind. Besides, several bioactive compounds from the different oil extracts of leaves of *A. marmelos* have been identified namely alkaloids, cardiac glycosides, terpenoids, tannins, steroids (Manandhan et al. 1978; Arseculeratne et al. 1981) which are used for various therapeutic purposes such as the treatment of asthma, anaemia, healing of wounds, swollen joints, high blood pressure, jaundice, diarrhoea (Maity et al. 2009; Sharma et al. 2011). However, very few research studies have been conducted which illustrate the food application of *Aegle marmelos* oil.

2 Methods Used for the Extraction of Bael Oil

Steam distillation is one method used for bael oil extraction. This process enables a compound or a mixture of compounds to be distilled at the temperature below the boiling point of that particular compound (Yadav et al. 2013). This method is used for heat sensitive, water-insoluble materials such as oils and hydrocarbons. The oil contains substances with the boiling point up to 200 °C or higher temperatures. However, these substances are volatilized at a temperature close to 100 °C at atmospheric pressure in the presence of steam or boiling water. The raw material is placed in the plant chamber and the steam is allowed to pass through the material under pressure which makes the plant material soft and oil is expelled out of the cells in the form of vapors. The temperature of steam must be high enough to vaporize the oil present. In the condensation chamber, oil vapors are condensed with steam. The oil forms a film on the surface of the water, which is separated, and oil is collected. The method is comparatively faster than others and also does not decompose the components of oil.

There are other methods of extraction from seeds of *A. marmelos* which have been used in various scientific studies (Kulkarni et al. 2012). For instance, bael seeds are collected and cleaned, decorticated, de-shelled, sun-dried and dried at 100–105 °C and are powdered using grinder prior to extraction. The powdered seed kernels are then defatted in a Soxhlet apparatus, using hexane (Boiling point 40–60 °C) and chloroform: methanol (Folch et al. 1957) to yield the oils. The solvent is later removed using rotary evaporator apparatus at 40 °C.

3   Fatty Acid Composition

The total oil content in bael seeds accounts for 33%. The amount of SFAs, MUFAs and PUFAs are 14.8%, 30.0% and 55.2% in *A. marmelos* seed oil (Kulkarni et al. 2012). The palmitic acid (14.4%) and stearic acid (0.4%) make up total SFA content of the oil, whereas MUFA includes oleic acid (18:1, 30.0%) as its major constituent (Table.30.1). The linoleic and linolenic acid constitutes almost an equal proportion (28.1% and 27.1%, respectively) of PUFAs. Similar results have also been reported by other research findings. Bajaniya et al. (2015) have recognized *Aegle marmelos* oil as unsaturated oil rather than saturated oil, which is apparent from the composition of linoleic acid (2452.06 ppm). The amount of oleic and linolenic acids has been recorded as 961.52 ppm and 37.55 ppm, respectively. Moreover, seed oil also contains 12.5% of an unusual fatty acid, ricinoleic acid along with other normal fatty acids (Patkar et al. 2012).

Table 30.1 Fatty acid composition of *A. marmelos* oil

Fatty acid	Kulkarni et al. (2012) Concentration (%)	Bajaniya et al. (2015) Concentration (ppm)
Palmitic acid (C16:0)	14.4	12.08
Stearic acid (C18:0)	0.4	71.54
Oleic acid (C18:1)	30.0	961.52
Linoleic acid (C18:2)	28.1	2452.06
Linolenic acid (C18:3)	27.1	37.55

4   Chemical Constituents of the Essential Oil

4.1   Bael Seed Oil

A total of 23 components have been identified from bale seed oil. The oil is comprised of methyl esters of octanoic acid, dodecanoic acid, methyl tetradecanoate, pentadecanoic acid, 9-hexadecanoic acid, hexadecanoic acid, 9,12- hexadecadienoic acid, heptadecanoic acid, 9-hexadecenoic acid, 9-octadecenoic acid (Z), 9,12-octadecadienoic acid (Bajaniya et al. 2015). The majority of the components are SFA derivatives having pharmacological properties viz. isopropyl myristate is used in cosmetic and topical medicinal preparations and 9,12,15-octadecatrienoic acid (alpha-linolenic acid) is also known as an essential omega-3 fatty acid which has well-demonstrated health-promoting properties.

4.2   Bael Leaf Oil

The analysis of the essential oil (1.5% yield) obtained from the leaves of A. *marmelos* revealed a total of 82 components constituting 99.8% of the essential oil composition. A study conducted on Nepal cultivars has reported limonene (64.1%) as the major constituent of essential oil extracted from the leaf. Other components include (E)-β-ocimene (9.7%), and germacrene B (4.7%), with smaller amounts of (E)-caryophyllene (2.4%), myrcene (2.0%), (Z)-β- ocimene (1.9%), (2Z,6E)-farnesol (1.9%), linalool (1.8%), and γ-curcumene (1.7%) (Satyal et al. 2012). Similarly, Patkar et al. (2012) have also documented limonene (82.4%) as the main constituent from A. *marmelos* leaves. Whereas, α-Phellandrene (56%) has been identified as the common constituent of the essential oil obtained from leaves, twigs and fruits. Other components isolated from leaf oil are p-cymene (17%) and P-Menth-1-en-3, 5-diol. Furthermore, the oil isolated from A. *marmelos* leaves of Cuba sources contains δ-cadinene (12.1%) and β-caryophyllene (10%) as major compounds (Pino et al. 2005). On the other hand, the oil isolated from A. *marmelos* leaves growing in India are composed of α-phellandrene (39.2%) and limonene (26.8%) (Raju et al. 1999). Later studies have reported a different percentage of chemical compounds in the essential oils obtained from the leaves (Table 30.2). For instance, Nabaweya et al. (2015) have reported α-phellandrene (20.97%), α-pinene (17.76%) and δ-carene (16.37%) as the major components, which constitute monoterpenes. Sesquiterpenes hydrocarbons represent 21.03% of which γ-cadinene (8.01%) occupies the major portion. Oxygenated compounds accounts for 11.41% which include monoterpenes alcohols (2.77%), sesquiterpene alcohols (0.43%), aldehyde (0.75%), ester (0.61%) and trans-2-hydroxycinnamic acid (6.85%).

Table 30.2 Differences in chemical compositions of the essential oil from *A. marmelos* leaves grown in different regions

Component (%)	Nabaweya et al. (2015)	Satyal et al. (2012)
α Pinene	17.76	0.2
β-Myrcene	4.32	2.0
α-Phellandrene	20.97	0.0
Limonene	0.0	64.1
*Iso*sylvestrene	1.70	0.0
δ-Carene	16.37	0.0
β-Ocimene	2.51	1.9
trans-2-hydroxycinnmic acid	6.85	0.0
γ-Terpinene	0.66	0.0
Terpenolene	1.02	0.0
Linalool	1.08	1.8
*3-Iso*thujanol	0.45	0.0
4-Terpineol	0.79	0.1
Thuj-3-en-10-al	0.75	0.0
α-Terpineol	0.45	0.1
δ-Elemene	0.80	0.1
α-Cubebene	1.70	0.1
γ-Elemene	3.82	0.1
α -Humulene	0.97	0.8
α-Terpinyl *iso*butyrate	0.61	0.0
γ-Muurolene	1.62	0.0
γ-Curcumene	0.77	1.7
Valencene	0.68	0.0
β-Selinene	0.45	0.0
α-Muurolene	1.32	0.0
β-Bisabolene	0.89	0.0
γ-Cadinene	8.01	0.3
β-Bisabolol	0.43	0.1

5 Biological and Pharmacological Applications of Bael Oil

Many research studies have focussed on medicinal properties of plant fruit of *A. marmelos* (Kuttan and Sabu 2004; Arul et al. 2005; Marzine and Gilbart 2005; Niraj et al. 2017). Leaves of this medicinal plant are equally competent for the treatment of various non-communicable diseases. This is because of the presence of chemical constituents, which forms a major part of essential oils extracted from leaves as discussed previously in the chapter (Bajaniya et al. 2015; Kulkarni et al. 2012). For instance, terpenoid constituents in the essential oil have *in-vivo* antifungal properties (Rana et al. 1997; Balakumar et al. 2011). Tannins, eugenol, and cuminaldehyde exhibit antibacterial activity particularly against *Escherichia coli* (Jyothi and Rao 2010). Besides, skimmianine an active component analyzed from the leaf extract possess anti-cancer properties (Lambertini et al. 2004; Jagetia et al. 2005).

5.1 Antibacterial and Antifungal

Various studies have demonstrated the antifungal and antibacterial activities of the alcoholic extract of *A. marmelos* leaves (Venkatesan et al. 2009; Kothari et al. 2011). The antifungal activity of essential oil isolated from leaves was evaluated using spore germination assay. The oil exhibited variable efficacy against different fungi and the inhibition of spore germination was cent percent at 500 μg/mL (Rana et al. 1997). The leaf extracts significantly inhibits the dermatophytic fungi like *Trichophyton mentagrophytes*, *T. rubrum*, *Microsporum canis*, *M. gypseum* and *Epidermophyton floccosum* (Balakumar et al. 2011; Satyal et al. 2012). Moreover, Nabaweya et al. (2015) documented the antibacterial activity of essential oil against Gram-positive bacteria as *Streptococcus faecalis* with inhibition zone (30 mm) and Gram-negative bacteria as *Pseudomonas aeruginosa* (28 mm). The antifungal activity against *Aspergillus niger* (30 mm) and *Candida albicans* (30 mm) was also reported. The bael essential oil also inhibited the germination of *Aspergillus niger* and *Fusarium oxysporum* spores at different concentrations.

5.2 Hypoglycemic and Hypolipidemic

The components of *A. marmelos* have demonstrated positive effects on heart diseases and diabetes (Kumar et al. 2009). Researchers revealed that the extract of *A. marmelos* significantly reduced the oxidative stress induced by alloxan and produced a reduction in blood sugar (Kuttan and Sabu 2004; Upadhya et al. 2004). Likewise, 67& reduction in blood glucose levels was observed with the dose at 250 mg/kg oral fed for seven consecutive days (Sachdewa et al. 2001). Hema and Lalithakumari (1999) observed that the aqueous and alcoholic extracts of *A. marmelos* at 500 mg/kg dose produced hypoglycemia in normal fasted rabbits. Similar findings were reported by the others (Rao et al. 1995; Seema et al. 1996; Das and Roy 2012). Similarly, the aqueous extracts of bael leaf exhibited cardio protective properties such cardiac stimulant, smooth-muscle relaxant and cardiac depressant (Hema and Lalithakumari 1999). *Aegle marmelos* leaf extracts have potential to decrease the activity of creatine kinase (CK) and lactate dehydrogenase (LDH) in serum and increase them in the heart. In addition, it increase the activity of Na^+/K^+ ATPase while decreases the Ca^{2+}ATPase in the heart and aorta simultaneously. Thereby, bael leaf extract containing essential oils has cardio protective properties (Niraj et al. 2017).

5.3 Anticancer

With regard to anticancer and antineoplastic activities, most of the publications have been reported in the animals which may help in validating the applicability of bael extracts on the human health system (Baliga et al. 2012). Bael extracts and some of its phytochemicals such as marmelin, butyl *p*-tolyl sulfide, 6-methyl-4-chromanone, butylated hydroxyanisole, lupeol, citral, cineole (1, 8 cineole), d-limonene, and eugenol are observed to be effective in selectively inhibiting proliferation of neoplastic cells (Tiku et al. 2004). The preventive effects of bael have been well documented in several preclinical trials. *A. marmelos* has been reported to inhibit the proliferation of transplanted Ehrlich ascites carcinoma in mice with the supplementation of an extract of *A. marmelos* at 400 mg/kg (Jagetia and Venkatesh 2007). Lambertini et al. (2004) also studied the antiproliferative effects of extracts from Bangladeshi medicinal plants on the in-vitro proliferation of human breast cancer cell lines MCF7 and MDA-MB-231.

5.4   Anti-oxidative and Anti-inflammatory

Research studies have proven that the effect of bael leaf extract is as effective as α-tocopherol for the prevention of oxidative stress. Also, the leaves of *A. marmelos* possess the anti-inflammatory, antipyretic and analgesic property which is evident in the clinical trials executed by various researchers (Arul et al. 2005; Jagtap et al. 2004). For instance, *Aegle marmelos* leaf extract at doses of 100 mg/kg and 200 mg/kg body weight for 35 days showed a significant effect on the activities of marker enzymes, lipid peroxides, lipids, lipoproteins and antioxidant enzymes in isoproterenol-treated rats. Thus, the outcomes of such studies may be useful for the clinical applications of bael and its active components in humans but only when the lacunae in the existing knowledge are bridged.

References

Arseculeratne, S. N., Gunatilaka, A. A. L., & Panabokke, R. G. (1981). Studies on medicinal plants of Srilanka: Occurrence of pyrolizidine alkaloids and hepatotoxic properties in some traditional medicinal herbs. *Journal of Ethnopharmacology, 4*, 159–177.

Arul, V., Miyazaki, S., & Dhananjayan, R. (2005). Studies on the anti-inflammatory, antipyretic and analgesic properties of the leaves of *Aegle marmelos* Corr. *Journal of Ethnopharmacology, 96*(4), 159–163.

Bajaniya, V. K., Kandoliya, U. K., Bodar, N. H., Bhadja, N. V., & Golakiya, B. A. (2015). Fatty acid profile and phytochemical characterization of bael seed (*Aegle marmelos* L.) oil. *International Journal of Current Microbiology and Applied Sciences, 4*(2), 97–102.

Balakumar, S., Rajan, S., Thirunalasundari, T., & Jeeva, S. (2011). Antifungal activity of Aegle marmelos (L.) Correa (Rutaceae) leaf extract on dermatophytes. *Asian Pacific Journal of Tropical Biomedicine, 1*(4), 309–312.

Baliga, M. S., Thilakchand, K. R., Rai, M. P., Rao, S., & Venkatesh, P. (2012). Aegle marmelos (L.) Correa (Bael) and its phytochemicals in the treatment and prevention of cancer. *Integrative Cancer Therapies, 12*(3), 187–196.

Das, S. K., & Roy, C. (2012). The protective role of Aegle marmelos on aspirin induced gastroduodenal ulceration in albino rat model: A possible involvement of antioxidants. *Saudi Journal of Gastroenterology, 18*(3), 188–194.

Folch, J., Lees, M., & Stanley, G. H. S. (1957). A simple method for the isolation and purification of total lipides from animal tissues. *The Journal of Biological Chemistry, 226*, 497–509.

Gupta, M., Biswas, T. K., Saha, S., & Debnath, P. K. (2006). Therapeutic utilization of secretory products of some Indian medicinal plants: A review. *Indian Journal of Traditional Knowledge, 5*(4), 569–575.

Hema, C. G., & Lalithakumari, K. (1999). Screening of pharmacological actions of *Aegle marmelos*. *The Indian Journal of Pharmacy, 20*, 80–85.

Jagetia, G. C., & Venkatesh, P. (2007). Inhibition of radiation-induced clastogenicity by *Aegle marmelos* (L.) correa in mice bone marrow exposed to different doses of gamma-radiation. *Human & Experimental Toxicology, 26*, 111–124.

Jagetia, G. C., Venkatesh, P., & Baliga, M. S. (2005). *Aegle marmelos* (L.) correa inhibits the proliferation of transplanted Ehrlich ascites carcinoma in mice. *Biological & Pharmaceutical Bulletin, 28*, 58–64.

Jagtap, A. G., Shirke, S. S., & Phadke, A. S. (2004). Effect of polyherbal formulation on experimental models of inflammatory bowel diseases. *Journal of Ethnopharmacology, 90*(2–3), 195–204.

Jyothi, K. S., & Rao, B. S. (2010). Antibacterial activity of extracts from *Aegle marmelos* against standard pathogenic bacterial strains. *International Journal of PharmTech Research, 2*(3), 1824–1826.

Kothari, S., Mishra, V., Bharat, S., & Tonpay, S. D. (2011). Antimicrobial activity and phytochemical screening of serial extracts from leaves of *Aegle marmelos* (Linn.). *Acta Poloniae Pharmaceutica Drug Research, 68*, 687–692.

Kulkarni, A. S., Khotpal, P. P., Karadbhajane, V. Y., & More, V. I. (2012). Physico-chemical composition and lipid classes of *Aegle marmelos* (Bael) and *citrullus colocylthis (Tumba)* seed oils. *Journal of Chemical and Pharmaceutical Research, 4*(3), 1486–1488.

Kumar, B. D., Mitra, A., & Manjunatha, M. (2009). In vitro and in vivo studies of antidiabetic Indian medicinal plants: A review. *Journal of Herbal Medicine Toxicology, 3*, 9–14.

Kuttan, R., & Sabu, M. C. (2004). Antidiabetic activity of *Aegle marmelos* and its relationship with its antioxidant properties. *Indian Journal of Physiology and Pharmacology, 48*(1), 81–88.

Lambertini E, Piva R, Khan MT, , Lampronti I, Bianchi N, Borgatti M, Gambari R (2004) Effects of extracts from Bangladeshi medicinal plants on in vitro proliferation of human breast cancer cell lines and expression of estrogen receptor alpha gene. International Journal of Oncology 24:419–423.

Maity, P., Hansda, D., Bandyopadhyay, U., & Mishra, D. K. (2009). Biological activities of crude extracts of chemical constituents of Bael, *Aegle marmelos* (L.) Corr. *Indian Journal of Experimental Biology, 47*, 849–861.

Manandhan, M. D., Shoeb, A., Kapil, R. S., & Popli, S. P. (1978). New alkaloids from *Aegle marmelos*. *Phytochemistry, 17*, 1814–1815.

Marzine, P. S., & Gilbart, R. (2005). The effect of an aqueous extract of *A. marmelos* fruits on serum and tissue lipids in experimental diabetes. *Journal of Science and Food Agriculture, 85*(4), 569–573.

Nabaweya, A., Ibrahim, F. S., El-Sakhawy, M. M. D., Mohammed, M. A., Farid, N. A. A., & Abdel-Wahed, D. D. (2015). Chemical composition, antimicrobial and antifungal activities of the essential oil of the leaves of *Aeglemarmelos* (L.) Correa growing in Egypt. *Journal of Applied Pharmaceutical Science, 5*(02), 001–005.

Niraj, Bisht, V., & Johar, V. (2017). Bael (*Aegle marmelos*) extraordinary species of India: A review. *International Journal of Current Microbiology and Applied Sciences, 6*(3), 1870–1887.

Patkar, A. N., Desai, N. V., Ranage, A. A., & Kalekar, K. S. (2012). A review on Aegle marmelos: A potential medicinal tree. *International Research Journal of Pharmacy, 3*(8), 86–91.

Pino, J. A., Marbot, R., & Fuentes, V. (2005). Volatile compounds from leaves of *Aegle marmelos* (L.) Correa grown in Cuba. *Ciencias Químicas, 36*, 71–73.

Raju, P. M., Agarwal, S. S., Ali, M., Velasco-Negueruela, A., & Pérez-Alonso, M. J. (1999). Chemical composition of the leaf oil of *Aegle marmelos* (L.) Correa. *Journal of Essential Oil Research, 11*, 311–313.

Rana, B., Singh, U., & Taneja, V. (1997). Antifungal activity and kinetics of inhibition by essential oil isolated from leaves of *Aegle marmelos*. *Journal of Ethnopharmacology, 57*(1), 29–34.

Rao, V. V., Dwivedi, S. K., Swarup, D., & Sharma, S. R. (1995). Hypoglycaemic and anti-hyperglycaemic effects of Aegle marmelos leaves in rabbits. *Current Science, 69*(1), 932–933.

Sachdewa, A., Raina, D., Srivatsava, A., & Khemani, L. D. (2001). Effect of *Aegle marmelos* and Hibiscus rosa sinensis leaf extract on glucose tolerance in glucose induced hyperglycemic rats. *Journal of Environmental Biology, 22*, 53–57.

Satyal, P., Woods, K. E., Dosoky, N. S., Neupane, S., & Setzer, W. N. (2012). Biological activities and volatile constituents of *Aegle marmelos* (L.) Corrêa from Nepal. *Journal of Medicinally Active Plants, 1*(3), 114–122.

Seema, P. V., Sudha, B., Padayatti, P. S., Abraham, A., Raghu, K. G., Paulose, C. S., & Paulose, C. S. (1996). Kinetic studies of purified malate dehydrogenase in liver of streptozotocin diabetic rats and the effect of leaf extract of *Aegle marmelos* (L.) Correa ex Roxb. *Indian Journal of Experimental Biology, 34*(6), 600–602.

Sharma, P. C., Bhatia, V., Bansal, N., & Sharma, A. (2007). A review on bael tree. *Natural Product Radiance, 6*(2), 171–178.

Sharma, G. N., Dubey, S. K., Sharma, P., & Sati, N. (2011). Medicinal values of bael (*Aegle marmelos*) (L.) Corr.: A review. *International Journal of Current Pharmaceutical Review and Research, 1*(3), 12–22.

Tiku, A. B., Abraham, S. K., & Kale, R. K. (2004). Eugenol as an in vivo radioprotective agent. *Journal of Radiation Research, 45*, 435–440.

Upadhya, S., Shanbhag, K. K., Suneetha, G., Naidu, B. M., & Upadhya, S. (2004). A study of hypoglycemic and antioxidant activity of *Aegle marmelos* in alloxan induced diabetic rats. *Indian Journal of Physiology and Pharmacology, 48*, 476–480.

Venkatesan, D., Karrunakarn, C. M., Kumar, S. S., & Swamy, P. T. P. (2009). Identification of phytochemical constituents of *Aegle marmelos* responsible for antimicrobial activity against selected pathogenic organisms. *Ethnobotanical Leaflets, 13*, 1362–1372.

Yadav, S. D., Muley, R. D., Kulkarni, A. U., & Sarode, B. N. (2013). Study of Bael oil *(Aegle marmelos)* and its properties. Available from https://www.researchgate.net/publication/313722431.

Chapter 31
Papaya (*Carica papaya* L.) Seed Oil

Seok Shin Tan

Abstract Papaya (*Carica papaya* L.) is available in both the tropical and sub-tropical regions around the world. The seeds, size, shape, color and flavor may vary depending on the varieties of papaya. Papaya bears fruits throughout the years and the large amounts of papaya seeds, consisting of about 15–20% in mass, are usually attached to the interior of the fruits in a row. The papaya seeds are edible; however, the majority of the consumers would consider them as wastes. The seeds have the potential to produce 30–34% of oil, especially rich in oleic acid and triacylglycerol; wherein OOO and POO are among the predominant triacylglycerol. The nutritional and functional properties of papaya seed oil are highly similar to olive oil, which in turn makes papaya seed oil a good prospective source of oil. Soxhlet, solvent, aqueous, enzymatic, ultrasound-assisted and supercritical carbon dioxide extractions are among the methods adopted in obtaining the papaya seed oil. The fatty acid compositions of the papaya seed oil yields are within the similar range despite the different extraction methods. Considering the large quantity of discarded papaya seeds globally, oil extraction from the seeds could play an important role in benefiting the food, cosmetic, pharmaceutical and health industries economically.

Keywords *Carica papaya* L. · Papaya seed oil · Extraction

1 Introduction

Papaya (*Carica papaya* L.), under the family of Caricaceae, is available in both the tropical and sub-tropical regions around the world (Yanty et al. 2014). Various species of papaya are available for consumption, which including Sekaki/Hong Kong, Batek Batu, Formosa, Tainoung, Eksotika, Eksotika II, Hawaii, and Chilean. The seeds, size, shape, color and flavor of papaya may vary depending on the varieties and species of the papaya. The papaya fruit flesh is green in color when harvested

S. S. Tan (✉)
Department of Nutrition and Dietetics, School of Health Sciences, International Medical University, Bukit Jalil, Kuala Lumpur, Malaysia

© Springer Nature Switzerland AG 2019
M. F. Ramadan (ed.), *Fruit Oils: Chemistry and Functionality*,
https://doi.org/10.1007/978-3-030-12473-1_31

and gradually becomes yellow to orange or reddish during ripening (Barroso et al. 2016; Yanty et al. 2014).

Papaya bears fruits throughout the year and the large amounts of papaya seeds, consisting about 15–20% in mass, are usually attached to the interior of the fruits in a row (Chielle et al. 2016). Papaya seeds are edible; however, the majority of the consumers would consider them as wastes. The spicy flavor of papaya seeds, attributed by the benzyl isothiocyanate, makes it a substitute for black pepper (Yanty et al. 2014).

The oil obtained from papaya seed can achieve up to 34% (Li et al. 2015; Samaram et al. 2013), with the color ranging from pale to dark yellow, almost odorless and flavorless (Yanty et al. 2014). The nutritional and functional properties of papaya seed oil are highly similar to olive oil, in which the papaya seed oil is especially rich in oleic acid and triacylglycerol; whereby OOO and POO are among the predominant triacylglycerol (Yanty et al. 2014). The edibility of this oil is inconclusive up-to-date and further investigations would be required prior to the commercialization of papaya seed oil for food markets or food industries usage. Nevertheless, the fatty acid composition and triacylglycerol profile of papaya seed oil has in turn, made it a very suitable candidate to be used for cosmetic, pharmaceutical and health industries. This could directly help to reduce, reuse and recycle the high amount of papaya seeds that have been globally labeled as agro waste.

2 Composition of Papaya Seeds

The proximate composition of dry papaya seeds from different cultivars (Sekaki, Batek Batu, Chilean and Formosa) and different drying methods (oven-dried and air-dried) is shown in Table 31.1. The dry papaya seeds contain a high percentage of lipid (28.5%) and protein (27.7%), regardless of the cultivars. Among the different cultivars, papaya seeds from Chilean papaya (*Vasconcellea pubescens*) that grows in colder climates (Briones-Labarca et al. 2015) were significantly low in moisture content but high in fiber, protein and lipid contents. The high lipid content of papaya seeds is especially of economically attractive for the industrial extraction, when compared with other oilseed crops, such as corn (3.1–5.7%) and soybean (18.0–20.0%) (Malacrida et al. 2011).

Table 31.1 Proximate composition (% weight) of dry papaya seed

	Average	Range
Moisture	5.9	3.5–7.2
Lipid	28.5	25.3–30.7
Protein	27.7	24.3–31.8
Ash	5.9	2.4–8.8
Fiber	21.0	17.0–24.4
Carbohydrate	23.1	11.7–32.5

Source: Yanty et al. (2014) and Briones-Labarca et al. (2015)

3 Extraction and Processing of Papaya Seed Oil

Papaya seed oil can be extracted through either the conventional extraction techniques such as solvent extraction, screw press and hydrodistillation, or the non-conventional extraction techniques such as enzyme-assisted, ultrasound-assisted, microwave-assisted, pressurized liquid and supercritical fluid extractions. Selected extraction and processing methods of papaya seed oil are discussed below and the recovery yield of different extraction procedures is shown in Fig. 31.1. The result showed the Soxhlet extraction method recovered the highest yield (30.4%) of papaya seed oil, while screw press extraction method produced the lowest yield (4.2%) as compared to other extraction methods (Puangsri et al. 2005). Besides the extraction methods, the drying process of the papaya seeds prior to the oil extraction does play a role in the levels of oil yielded. The previous study showed the optimum drying temperature that could provide the maximum papaya seed oil yield was at the air temperature of 70 °C and an air velocity of 2.0 m/s (Chielle et al. 2016).

Chemical properties of papaya seed oil obtained from different extraction methods are given in Table 31.2. Iodine value, saponification value, unsaponifiable matter and free fatty acid are among the chemical properties analyzed and discussed. The degree of unsaturation of the oil is determined by the iodine value (Puangsri et al. 2005). Soxhlet and screw press extractions obtained the highest (79.95) and lowest (64.10) iodine value, respectively. As for the saponification value and unsaponifiable matter, screw press and Soxhlet extraction methods recorded the highest

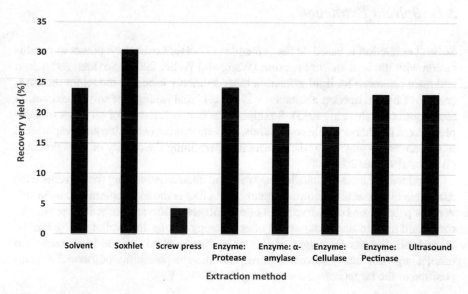

Fig. 31.1 Recovery yield (%) of papaya seed oil obtained by different extraction methods. (Source: Puangsri et al. 2005; Samaram et al. 2013)

Table 31.2 Chemical properties of papaya seed oil obtained from different extraction methods

Extraction method	Iodine value	Saponification value	Unsaponifiable matter (%)	FFA[a] (as % of oleic acid)
Solvent	66.00	154.70	1.39	0.32
Soxhlet	79.95	96.40	1.35	1.27
Screw press	64.10	185.00	4.50	n.d.[b]
Enzyme-assisted				
i. Protease	66.20	154.20	2.15	0.25
ii. α-Amylase	67.60	161.40	2.26	0.23
iii. Cellulase	68.30	158.40	2.58	0.20
iv. Pectinase	69.30	161.70	2.40	0.25
Ultrasound-assisted	71.00	n.d.	1.35	n.d.

Source: Puangsri et al. (2005), Malacrida et al. (2011), Lee et al. (2011), and Samaram et al. (2014)
[a]*FFA* Free fatty acid
[b]*n.d.* Not determined

and lowest readings, respectively. The fatty acid compositions of the papaya seed oil are within the similar range despite the different extraction and processing methods, as given in Table 31.3. Oleic acid is the main fatty acid in papaya seed oil from different extraction methods. Solvent and enzyme (protease) extraction methods have obtained the highest oleic acid (76.8%) among the other extraction methods.

3.1 Solvent Extraction

Solvent extraction is based on the principles of solvent extraction power in combination with the heat and/or agitation (Wang and Weller 2006). Soxhlet, a standard and main reference for lipid extraction that commonly used as a model for the comparison of new extraction alternatives, is the classical example of solvent extraction techniques (Azmir et al. 2013). Soxhlet extraction is suitable for all types of samples except for thermolabile compounds, in view that the extraction and evaporation temperatures do play a significant role in determining the quality of final products (Wang and Weller 2006).

Petroleum ether, *n*-hexane, isopropanol and ethanol are among the solvents used for edible oil extraction. However, petroleum ether is the most commonly used solvent for papaya seed oil extraction. The advantages of solvent extraction are simple, cheap and no filtration requirement after leaching (Wang and Weller 2006). On the other hand, the disadvantages of this extraction method include long extraction time, a large amount of solvent required and also the possibility of thermal decomposition of the target compounds (Wang and Weller 2006).

Table 31.3 Fatty acid composition (%) of papaya seed oil extracted by different methods

Extraction method	Fatty acid (%)									
	Myristic	Palmitic	Palmitoleic	Stearic	Oleic	Linoleic	Linolenic	Arachidic	Eicosenoic
Solvent	0.2	13.9	0.2	4.9	76.8	3.0	0.2	0.4	0.3
Soxhlet	0.2	14.9	0.3	5.2	74.2	3.5	0.2	0.4	0.4
Screw press	0.7	19.7	0.4	6.7	66.7	3.2	0.2	0.4	0.5
Enzyme-assisted									
i. Protease	0.1	12.8	1.8	4.4	76.8	3.2	0.1	0.4	0.3
ii. α-Amylase	0.2	13.3	2.1	4.4	76.0	3.2	0.1	0.4	0.3
iii. Cellulase	0.2	13.4	2.0	4.6	76.5	3.3	0.2	0.4	0.3
iv. Pectinase	0.2	13.6	1.4	4.6	75.9	3.3	0.2	0.4	0.3
Ultrasound-assisted	0.2	15.1	0.3	5.1	74.2	3.5	0.2	0.4	0.4

Source: Puangsri et al. (2005), Samaram et al. (2013), and Lee et al. (2011)

3.2 Enzyme-Assisted Extraction

Enzyme-assisted extraction is based on the addition of specific enzymes during the extraction process to enhance the yield and recovery by breaking the cell wall and hydrolyzing the structural polysaccharides and lipid bodies. Cellulase, α-amylase, protease and pectinase are some examples of enzymes used to support the extraction and yield recovery (Azmir et al. 2013). The advantages of enzyme-assisted extraction include lower extraction temperature, no involvement of explosive solvents and no production of harmful wastes (Puangsri et al. 2005). However, enzyme composition and concentration, the particle size of the sample, solid to water ratio as well as hydrolysis time are among the factors identified which might influence the efficiency of enzyme-assisted extraction (Azmir et al. 2013).

3.3 Ultrasound-Assisted Extraction

Ultrasound, within the range of 20 kHz to 100 MHz, is a special sound wave which beyond human hearing (Azmir et al. 2013). It creates expansion and compression cycles when passes through a medium (Wang and Weller 2006). Ultrasound could improve the mass transfer, induce a greater penetration of the solvent into cellular materials and also facilitate the release of contents through the disruption of biological cell walls (Wang and Weller 2006). This extraction method is recommended for thermolabile compounds that tended to be altered or neglected in the solvent extraction method (Wang and Weller 2006).

The advantages of ultrasound-assisted extraction are simple, inexpensive equipment, reduction of extraction time, temperature, energy and solvent used (Samaram et al. 2013). Nevertheless, there are some factors that might influence the efficiency and effectiveness of ultrasound-assisted extraction, which includes moisture content of the sample, particle size, solvent selection, temperature, pressure, frequency and time of sonication (Azmir et al. 2013).

3.4 Microwave-Assisted Extraction

The utilization of electromagnetic radiations within a frequency from 300 MHz to 300 GHz, to generate heat for the extraction of papaya seed oil, is the principle of microwave-assisted extraction (Wang and Weller 2006). Microwave-assisted extraction is suitable for the extraction of thermosensitive compounds and the advantages of this method include reduction of extraction time and solvent usage while at the same time improved extraction yield (Azmir et al. 2013). Sample particle size, solvent selection and operating conditions are among the factors that influence the efficiency of microwave-assisted extraction (Wang and Weller 2006).

3.5   Supercritical Fluid Extraction

Supercritical, a state that can only be achieved when a substance is exposed to the temperature and pressure that beyond its critical point (Azmir et al. 2013). The supercritical fluid has both the gas-like characteristics of diffusion, viscosity and surface tension as well as the liquid-like characteristics of density and solvation power (Wang and Weller 2006). Carbon dioxide is an ideal solvent for supercritical fluid extraction in view of its critical temperature is close to room temperature (31 °C) and low critical pressure (74 bars) that can easily achieve (Azmir et al. 2013).

Wang and Weller (2006) reported oil extracted with the supercritical fluid method can prevent oxidation of lipids and more protected from the oxidation of unstable polyunsaturated fatty acids (PUFA), as compared to oil extracted with the solvent method. The advantages of supercritical fluid extraction include reduced extraction time, complete extraction, a wider range of solvent selection as well as ideal for thermolabile compounds extraction (Azmir et al. 2013). The choice of supercritical fluids, sample pre-preparation and the extraction conditions are some of the practical issues that will impact the supercritical fluid extraction efficiency (Wang and Weller 2006).

4   Fatty Acid Composition and Acyl Lipids

The fatty acid composition of papaya seed oils extracted from three commercial papaya cultivars (Formosa, Hawaiian and Golden) grown in Brazil is shown in Table 31.4. Regardless of the cultivars, the predominant fatty acids in the papaya seed oil are oleic acid (69.78–72.04%), subsequently followed by palmitic acid (18.20–18.95%), stearic acid (5.07–5.30%) and linolenic acid (3.23–4.84%), respectively. A similar trend is observed by Lee et al. (2011) and Yanty et al. (2014). The oleic acid level of papaya seed oil is comparable with other edible oils, such as olive (71%) and hazelnut (73%) oils (Vingering et al. 2010). It has been reported

Table 31.4 Fatty acid profile (% weight) of papaya seed oil

Fatty acid	Average	Range
Myristic acid	0.21	0.20–0.22
Palmitic acid	18.68	18.20–18.95
Palmitoleic acid	0.28	0.23–0.32
Stearic acid	5.19	5.07–5.30
Oleic acid	70.65	69.78–72.04
Linoleic acid	4.24	3.23–4.84
Arachidic acid	0.38	0.35–0.41
Gadoleic acid	0.34	0.32–0.41

Source: de Melo and de Sousa (2016)

Table 31.5 Triacylglycerol profile (% weight) of papaya seed oil

Triacylglycerol	Average	Range
LOO	3.55	2.54–4.40
LOP	2.27	1.72–2.80
OOO	43.23	41.30–44.60
POO+SOL	30.67	27.70–33.80
OPP	5.81	5.10–6.19
SOO	9.29	8.37–9.80
SOP	3.12	2.41–3.80
Unknown	2.06	0.20–4.80

Source: Samaram et al. (2013)
LOO linoleoyl-dioleoyl glycerol, *LOP* linoleoyl-oleoyl-palmitoyl glycerol, *OOO* trioleoyl glycerol, *POO* palmitoyl-dioleoyl glycerol, *SOL* stearoyl-oleoyl-linoleoyl glycerol, *OPP* oleoyl-dipalmitoyl glycerol, *SOO* stearoyl-dioleoyl glycerol, *SOP* stearoyl-oleoyl-palmitoyl glycerol

that plant oils with a high level of oleic acid have enough oxidative stability in domestic cooking applications like frying (Corbett 2003). Thus, the high-oleic papaya seed oil can potentially be a healthy substitute for partially hydrogenated plant oils.

Triacylglycerols (TAG) represent the major lipid class of papaya seed oil. The TAG composition of papaya seed oil of different cultivars (Sekaki, Batek Batu and Tainoung) and extraction techniques (screw press and solvent extraction) is given in Table 31.5. The predominant TAG molecular species of papaya seed oil are OOO (41.30–44.60%), POO+SOL (27.70–33.80%), SOO (8.37–9.80%) and OPP (5.10–6.89%). The study of Samaram et al. (2013) demonstrated the TAG composition of papaya seed oil significantly affected by extraction techniques, whereby the amounts of OOO, POO+SOL and SOO in ultrasound-assisted solvent-extracted papaya seed oil were significantly lower than solvent-extracted papaya seed oils.

The total phospholipids of papaya seed oil, as measured using the thin layer chromatography silica gel plate, was found to be 0.63%. Three phospholipid components, namely phosphatidylinositol, phosphatidylcholine and phosphatidylethanolamine, have been identified in papaya seed oil. The relative percentages of these components were 34%, 28% and 19%, respectively (Prasad et al. 1987).

5 Minor Bioactive Compounds in Papaya Seed Oil and Their Functions

Minor bioactive compounds of papaya seed oil, including tocopherol, carotenoid, phenolic and flavonoid; are sitting in the unsaponifiable matters upon the oil extraction (Samaram et al. 2014). In addition, sterols, triterpene alcohols, hydrocarbons and the fat-soluble vitamins could be included in the dissolved unsaponifiable

Table 31.6 Minor bioactive compounds compositions of papaya seed oil

Compounds	Value (mg/kg)
Total tocopherols	74.71
α-tocopherol	51.85
β-tocopherol	2.11
γ-tocopherol	1.85
δ-tocopherol	18.89
Total carotenoids	7.05
β-cryptoxanthin	4.29
β-carotene	2.76
Total phenolics[a]	957.60
Total flavonoids[b]	0.60

Source: Malacrida et al. (2011), and Briones-Labarca et al. (2015)
[a]mg of gallic acid equivalent/kg
[b]mg of quercetin equivalent/g

matters as well (Puangsri et al. 2005). Table 31.6 shows the compositions of minor bioactive compounds in the papaya seed oil. Malacrida et al. (2011) reported the low content of tocopherols (74.7 mg/kg) in papaya seed oil as compared to other commercially edible plant oils such as soybean (1797.6 mg/kg), maize (1618.4 mg/kg), and sunflower (634.4 mg/kg) oil. The low content of tocopherols in papaya seed oil might justify the low PUFA content, especially the linoleic and linolenic acids, in the same sample. Both α- and δ-tocopherol are the major tocopherols in the papaya seed oil (Malacrida et al. 2011). The high biological activity and high antioxidant capacity of α- and δ-tocopherol, respectively, are suggested for human consumption (Malacrida et al. 2011).

β-cryptoxanthin is the main carotenoid in papaya seed (Malacrida et al. 2011). The β-carotene content in papaya seed oil is higher than the amount reported for peanut, soybean and corn oils (Malacrida et al. 2011). A similar trend was observed by the same researchers on the total phenolic content of papaya seed oil, whereby the value is much higher than soybean, rice bran, rapeseed, corn and sunflower oils. A study conducted on the effects of the different extraction methods on the yield of total flavonoid content has revealed that high hydrostatic pressure extraction has increased significantly the total flavonoid content as compared to the conventional solvent extraction method (Briones-Labarca et al. 2015).

6 Aroma Profile

The sensory and quality characteristics of edible oils are influenced by their aroma. There have been controversies surrounding the aroma of papaya seed oil. A study by Eckey (1954) reported that papaya seed oil is odorless. However, using an electronic nose (zNose), Yanty et al. (2014) found that papaya seed oil had a distinct aroma profile compared to other seed oils like musk lime, rambutan and honeydew. The author did not report the individual aroma compounds present in papaya seed oil. Further research is needed to better understand the unique aroma properties of papaya seed oil.

7 Oxidative Stability

The oxidative stability index of papaya seed oil, as measured using a Rancimat instrument, was found to be 77.97 h (Malacrida et al. 2011). This value was 6.3–7.8 times longer than the soybean and sunflower oils (Malacrida et al. 2011). The author postulated the high oxidative stability of papaya seed oil is due to its low amounts of PUFA. Another study of de Melo and de Sousa (2016) measured the oxidative stability of papaya seed oil at 65 °C for 25 days. Their study showed papaya seed oil was high in thermo-oxidative stability as the peroxide value recorded at the end of the experiment was lower than value needed for the formation of oxidized compounds.

8 Health-Promoting Traits of Papaya Seed Oil

Limited studies have been conducted on the health-promoting traits of papaya seed oil and its oil constituents to date. This could be due to the lack of studies on the safety assessment and the edibility of this oil. Nevertheless, Castro-Vargas et al. (2016) reported the antibacterial, ovicidal, larvocidal, anti-helminthic, anti-amoebic, anti-inflammatory effects of papaya seed extracts. In addition, Lohiya et al. (2000) found the post-testicular anti-fertility drug potential of papaya seeds in male rabbits. The contraceptive efficacy of papaya seed extracts was also observed in male rats, monkeys and dogs as reported by Castro-Vargas et al. (2016). Benzyl isothiocyanate, a compound that contributed to the spicy flavor of the papaya seeds, has been discovered with cancer preventive property (Yanty et al. 2014). Further investigation on the prospective health benefits of papaya seed oil need to be carried out in view of the high similarity of this oil to olive oil.

9 Edible Applications of Papaya Seed Oil

Papaya seed oil is characterized by high levels of oleic acid (>70%). Plant oils with a high level of oleic acid have sufficient stability to be used in domestic cooking applications like frying (Corbett 2003). The study of Puangsri et al. (2005) suggested the application of papaya seed oil as spray oil for dried fruits, snacks, cereals, crackers and bakery products. This in turn, can enhance the food quality and palatability. Safety assessment of papaya seed oil should be conducted before commercializing for food applications.

10 Other Issues

No publication on the adulteration and authenticity of papaya seed oil is found to date.

References

Azmir, J., Zaidul, I. S. M., Rahman, M. M., et al. (2013). Techniques for extraction of bioactive compounds from plant materials: A review. *Journal of Food Engineering, 117*, 426–436.

Barroso, P. T. W., de Carvalho, P. P., Rocha, T. B., et al. (2016). Evaluation of the composition of *Carica papaya* L. seed oil extracted with supercritical CO_2. *Biotechnology Reports, 11*, 110–116.

Briones-Labarca, V., Plaza-Morales, M., Giovagnoli-Vicuña, C., & Jamett, F. (2015). High hydrostatic pressure and ultrasound extractions of antioxidant compounds, sulforaphane and fatty acids from Chilean papaya (*Vasconcellea pubescens*) seeds: Effects of extraction conditions and methods. *LWT- Food Science and Technology, 60*, 525–534.

Castro-Vargas, H. I., Baumann, W., & Parada-Alfonso, F. (2016). Valorization of agroindustrial wastes: Identification by LC-MS and NMR of benzylglucosinolate from papaya (*Carica papaya* L.) seeds, a protective agent against lipid oxidation in edible oils. *Electrophoresis, 37*, 1930–1938.

Chielle, D. P., Bertuol, D. A., Meili, L., et al. (2016). Convective drying of papaya seeds (*Carica papaya* L.) and optimization of oil extraction. *Industrial Crops and Products, 85*, 221–228.

Corbett, P. (2003). It is time for an oil change! Opportunities for high oleic vegetables oils. *Inform, 14*, 480–481.

Eckey, E. W. (1954). *Vegetable fats and oils*. New York: Reinhold Publishing Corp.

Lee, W. J., Lee, M. H., & Su, N. W. (2011). Characteristics of papaya seed oils obtained by extrusion-expelling processes. *Journal of the Science of Food and Agriculture, 91*, 2348–2354.

Li, Y. M., Su, N., Yang, H. Q., et al. (2015). The extraction and properties of *Carica papaya* seed oil. *Advance Journal of Food Science and Technology, 7*, 773–779.

Lohiya, N. K., Pathak, N., Mishra, P. K., & Manivannan, B. (2000). Contraceptive evaluation and toxicological study of aqueous extract of the seeds of *Carica papaya* in male rabbits. *Journal of Ethnopharmacology, 70*, 17–27.

Malacrida, C. R., Kimura, M., & Jorge, N. (2011). Characterization of a high oleic oil extracted from papaya (*Carica papaya* L.) seeds. *Food Science and Technology Research, 31*, 929–934.

de Melo, M. L. S., & de Sousa, D. P. (2016). Physical and chemical characterization of the seeds and oils of three papaya cultivars (*Carica papaya*). *Journal of Chemical and Pharmaceutical Research, 8*, 870–876.

Prasad, R. B. N., Nagender Rao, Y, & Venkob Rao, S. (1987). Phospholipids of Palash (*Butea monosperma*), papaya (*Carica papaya*), Jancjli Badam (*Sterculia foetida*), coriander (*Cofiandrum safivum*) and carrot (Daucus carota) seeds. *Journal of the American Oil Chemists' Society, 64*, 1424–1427.

Puangsri, T., Abdulkarim, S. M., & Ghazali, H. M. (2005). Properties of *Carica papaya* L.(papaya) seed oil following extractions using solvent and aqueous enzymatic methods. *Journal of Food Lipids, 12*, 62–76.

Samaram, S., Mirhosseini, H., Tan, C. P., & Ghazali, H. M. (2013). Ultrasound-assisted extraction (UAE) and solvent extraction of papaya seed oil: Yield, fatty acid composition and triacylglycerol profile. *Molecules, 18*, 12474–12487.

Samaram, S., Mirhosseini, H., Tan, C. P., & Ghazali, H. M. (2014). Ultrasound-assisted extraction and solvent extraction of papaya seed oil: Crystallization and thermal behavior, saturation degree, color and oxidative stability. *Industrial Crops and Products, 52*, 702–708.

Vingering, N., Oseredczuk, M., Du Chaffaut, L., et al. (2010). Fatty acid composition of commercial vegetable oils from the French market analysed using a long highly polar column. *Oleagineux, Corps Gras, Lipides, 17*, 185–192.

Wang, L., & Weller, C. L. (2006). Recent advances in extraction of nutraceuticals from plants. *Trends in Food Science and Technology, 17*, 300–312.

Yanty, N. A. M., Nazrim Marikkar, J. M., Nusantoro, B. P., et al. (2014). Physico-chemical characteristics of papaya (*Carica papaya* L.) seed oil of the Hong Kong/Sekaki variety. *Journal of Oleo Science, 63*, 885–892.

Chapter 32
Mongongo/Manketti
(*Schinziophyton rautanenii*) Oil

Natascha Cheikhyoussef, Martha Kandawa-Schulz, Ronnie Böck,
and Ahmad Cheikhyoussef

Abstract *Schinziophyton rautanenii* (Schinz) Radcl.-Sm., formerly known as *Ricinodendron rautanenii* Schinz, is a large spreading dioecious tree, typically 15–20 m in height. The tree commonly grows wild in Angola, Botswana, Namibia, South Africa and Zambia and is known to be an important food source to the inhabitants. It is commonly also known as the Mongongo or Manketti tree. The edible yellow oil extracted from the nut of the egg-shaped fruits, contains the unique conjugated fatty acid, α-eleostearic acid among others such as linoleic, oleic and linolenic acids. Findings on the physico-chemical characteristics among different extraction methods of the Manketti oil have shown that its utilization into value-added products in the food, health and the cosmetics sector has great economic potential. This chapter aims at collating the information that underpins the potential of this unique oil to encourage further research and the development of innovative applications of the Manketti oil.

Keywords *Schinziophyton rautanenii* · Manketti oil · α-eleostearic acid ·
Biodiesel · Phytosterols · Southern Africa · Food

Abbreviations

CLN	Conjugated linoleic acids
HPLC	High performance liquid chromatography
ITC	International Trade Centre
NCBI	National Center for Biotechnology Information
α-ESA	Alpha eleostearic acid

N. Cheikhyoussef (✉)
Ministry of Higher Education, Training and Innovation, Windhoek, Namibia
e-mail: natascha.cheikhyoussef@mheti.gov.na

M. Kandawa-Schulz · R. Böck · A. Cheikhyoussef
University of Namibia, Windhoek, Namibia
e-mail: kschulz@unam.na; rbock@unam.na; acheikhyoussef@unam.na

© Springer Nature Switzerland AG 2019
M. F. Ramadan (ed.), *Fruit Oils: Chemistry and Functionality*,
https://doi.org/10.1007/978-3-030-12473-1_32

1   Introduction

The *Schinziophyton rautanenii* (Schinz), commonly known as the Manketti or Mongongo tree, can easily be recognized by a large symmetric rounded crown with its single stem (Hoffmann 2016a) with a height of 15–20 m (Palgrave 1983). This tree is deciduous and dioecious (Vermaak et al. 2011), producing yellowish-white flowers and egg-shaped fruits that are covered with grey-green hairs (Palgrave 1983; Hoffmann 2016a). From April to May, the fruits fall on the ground, at which time a ripening process of the fruit flesh starts and the fruit become a red-brown color (Vermaak et al. 2011). The fruit becomes hard and tough once mature containing nut-like seeds making up about 70% of the fruit (Vermaak et al. 2011; Hoffmann 2016b). The species is formerly known as *Ricinodendron rautanenii* (Schinz) and belongs to the family Euphorbiaceae (Vermaak et al. 2011). The tree is distributed throughout southern Africa, in particular in the countries of Namibia, Angola, Zambia, South Africa, Botswana and Mozambique (Atabani et al. 2014). *S. rautanenii* has been classified as a multipurpose plant species in Southern Africa of which the plant parts are harvested from the wild (Maroyi 2018). The tree is non-domesticated although various attempts and research efforts towards domestication are underway (ITC 2012; Maroyi 2018). The products of the tree are of great economic importance to the rural communities (Hoffmann 2016b), both in terms of food security and income-generation.

The Manketti nut production per annum can range from 250 to 800 kg/ha (Peters 1987) with one tree producing about 45 kg of fruit per annum (Bennet 2006). An estimated 25.6 million nuts providing about 325 million calories of energy are produced in total per year (Gwatidzo et al. 2017a). In Zambia, an estimated production of 3000 mt of Manketti seed could yield about 840 mt of Manketti oil (Juliani et al. 2007). The fruit flesh and the seed or nut can be consumed raw or cooked and added to meat and porridge (Hoffmann 2016b). The edible yellow oil is extracted traditionally and commercially from the nuts of the hard kernel. On the international market, Manketti oil has become a relatively well-known commodity, as part of the products originating from unique natural resources and having entered niche markets. Several international skincare brands (New Agriculturist 2009) now use Manketti oil to develop various cosmetics products. One initiative such as the Kalahari Natural Oils Initiative can produce up to 12 tons of Manketti oil per year for the market (New Agriculturist 2009). The potential trade for ten southern African countries from Manketti derived products was estimated to be at USD 19,677,684 per year (Bennet 2006).

Traditionally, the Manketti oil is used as a body rub to soften and to protect the skin against the sun and in the preparation of food (Juliani et al. 2007; Hoffmann 2016b). The unique characteristics of the Manketti oil provide it with hydrating, regenerating and restructuring properties (Kalahari Biocare 2016) and this gives it the potential to be used in product development for modern cosmetics and nutraceuticals (Juliani et al. 2007; Mitei et al. 2008, 2009; Cheikhyoussef et al. 2018). The Manketti kernel and oil have good shelf life properties. If dried, Manketti fruit or nuts are stored at a constant temperature (10–30 °C) in reasonable air-tight containers, the seeds can remain palatable for about 6 years (Peters 1987).

The suitability of Manketti oil for biodiesel applications has been investigated and has been shown to possess good fuel properties (Rutto and Enweremadu 2011; Atabani et al. 2014; Gandure et al. 2014; Kivevele and Huan 2015). The Manketti biodiesel fuel has a significant calorific value and suitable cold flow properties making it suitable for use as alternative energy sources (Atabani et al. 2014; Gandure et al. 2014). Fuel properties also improve when the Manketti biodiesel fuel is blended with diesel (Atabani et al. 2014).

2   Extraction and Processing of Manketti Oil

Manketti oil is extracted from the nuts found inside the seed kernel of the fruit of the *Schinziophyton rautanenii* (Schinz) tree. Generally, for characterizing the Manketti oil from different regions of the southern part of the African continent, the oil has been extracted using Soxhlet extraction, other organic solvent extraction methods, screw pressing, cold pressing method, mechanical shaking and supercritical fluid extraction. These techniques have resulted in varying oil extraction yields that range between 27% and 59% (Table 32.1).

Generally, these extraction yields are considered high and allow for the Manketti oil to be economically viable for industrial applications (Gwatidzo et al. 2017a). Commercially, the Manketti oil is extracted using cold pressing technologies. Local communities extract the oil using an age-old traditional method that involves boiling the crushed nut with water over a fire for a certain period (Cheikhyoussef et al. 2018). The nut is removed from the fruit with an axe (European Commission 1998) or after being crushed between two rocks (Vermaak et al. 2011). Due to the hardness of the kernel, removing the kernel is difficult and could be disadvantageous for the commercial production of the Manketti oil. Generally, the nuts are removed by the

Table 32.1 Yields of Mongongo/Manketti (*Schinziophyton rautanenii*) oil

Origin	Extraction method	Yield (%)	Reference
Zambia	Organic solvent	42	Chisholm and Hopkins (1966)
Zambia	Traditional hand press	28	Juliani et al. (2007)
Zambia	Hydraulic press	38	Juliani et al. (2007)
Botswana	Soxhlet	41.5	Mitei et al. (2008)
Botswana	Soxhlet	58.6	Gandure et al. (2014)
Namibia	Screw press	39.7	Gwatidzo et al. (2017a, b)
Namibia	Supercritical fluid	44.8	Gwatidzo et al. (2017a, b)
Namibia	Soxhlet	45.3	Gwatidzo et al. (2017a, b)
Namibia	Mechanical shaking	27.3	Gwatidzo et al. (2017a, b)
Botswana	Soxhlet	57.7	Yeboah et al. (2017)
Namibia	Soxhlet	42.6	Cheikhyoussef et al. (2018)

local communities, and the oil is extracted by development initiatives locally, after which the crude oil is exported to international cosmetic brands for further processing and product development. Various small to medium enterprises are producing products from the Manketti oil for sale locally (Cheikhyoussef 2018).

The traditional extraction process of Manketti oil involves a number of steps as has been reported for the Manketti oil extraction in Namibia (Cheikhyoussef et al. 2018; Cheikhyoussef 2018). The seed shell is cracked open to remove the oil-bearing nut from inside. The nuts are slightly roasted by adding hot coals on top of the Manketti nuts, which aims to improve the flavor of the oil. The Manketti nuts are then pounded using mortar (hole in the cemented ground) and pestle (wooden stick). The pounding action produces a sticky nut paste, which is removed from the mortar and put in a bowl. The paste is mixed with some boiled water and the resulting solution is then decanted into a boiling pot. Care is to be taken not to decant the remaining undissolved seed particles. The solution is then boiled for a time period depending on the amount of starting materials. Continuous boiling eventually allows the water to evaporate and the Manketti oil to remain. The oil is poured into glass bottles and stored for future use or meat like chicken is added directly to the pot containing the oil for food preparations. The oil can also be added to spinach and prepared chicken at the final stages of cooking (Personal communications; Cheikhyoussef et al. 2018; Cheikhyoussef 2018).

The oil extraction method used has an effect on the amount and type of phytosterol and tocopherol present in the extracted Manketti oil. Four different extraction methods, Soxhlet extraction, supercritical carbon dioxide, screw pressing and mechanical shaking, resulted in variations of total phytosterol in the Manketti oil, with mechanical shaking obtaining the highest and screw press and Soxhlet extraction, the lowest total sterol content (Gwatidzo et al. 2014). Highest concentrations of the phytosterol, β-sitosterol, Δ^5-avenasterol and campesterol were observed independent of the extraction method used, whilst mechanical shaking and supercritical carbon dioxide extraction resulted in an abundance of stigmasterol and cycloartenol (Gwatidzo et al. 2014). The concentration of total tocopherol in Manketti oil extracted with the traditional process, cold pressing and Soxhlet extraction varied with the highest concentration obtained using Soxhlet and the lowest concentration with the traditional process (Cheikhyoussef et al. 2018). Greatest variation was found with the γ-tocopherol with the highest concentration found using Soxhlet extraction and the lowest concentration with cold pressing extraction (Cheikhyoussef et al. 2018).

The amount of unsaturated and conjugated fatty acids was found to be higher in Manketti oil extracted using screw press and supercritical fluid extraction methods, whilst saturated fatty acids were higher in concentration with Soxhlet extraction and mechanical shaking (Gwatidzo et al. 2017b). Total unsaturated fatty acids in Manketti oil were similar and higher using the traditional extraction process and Soxhlet extraction process as compared to the cold pressing extraction (Cheikhyoussef et al. 2018).

3 Fatty Acid Composition and Acyl Lipid Profile of Manketti Oil

The percentage composition of lipid classes in Manketti oil have been reported by Mitei et al. (2008) to be hydrocarbons (0.49%), triacylglycerols and free fatty acids (79.89%), sterol esters (1.38%), free sterols (3.06%), diacylglycerols (1.18%), monoacylglycerols (0.50%), glycolipids (0.41%) and phospholipids (2.73%). Manketti nut oil contains primarily palmitic acid (7–14%), oleic acid (14–24%), linoleic acid (35–52%), stearic acid (3–16%) and α-eleostearic acid (20–36%) (Table 32.2). Fatty acids such as palmitoleic acid (Gwatidzo et al. 2017b), erucic acid (Chivandi et al. 2008; Gwatidzo et al. 2017b), arachidic acid (Cheikhyoussef et al. 2018), 11-eicosenoic acid (Gwatidzo et al. 2017b; Cheikhyoussef et al. 2018) and linolenic acid (Zimba et al. 2005; Gwatidzo et al. 2017b) have also been reported in some of the characterized Manketti oils (Table 32.2). The Manketti oil has potential applications in the development of paints and varnishes due to its high degree of unsaturation (Booth and Wickens 1988). Triacylglycerol profile of Manketti oil was also determined by reversed-phase high performance liquid chromatography (RP-HPLC) which confirmed the presence of octadecatrienoic (34.6 mol %), linoleic (35.4 mol %), oleic (mol 14.6%), palmitic (7.2 mol %), and stearic acids (Van et al. 2017). Van et al. (2017) further reported that the octadecatrienoic acid is presented mainly by α-eleostearic acid and small impurities of β-eleostearic (2.8 mol %) and jacarandic (0.3 mol %) acids.

One of the unique properties of Manketti oil, is the dominant presence of linoleic acid (Fig. 32.1a) and the α-eleostearic acid (Fig. 32.1b). Alpha eleostearic acid or (9Z,11E,13E)-octadeca-9,11,13-trienoic acid belongs to the group of conjugated linolenic acids (CLN), that is comprised of positional and geometric isomers of octadecatrienoic acids having three conjugated double bonds (Cao et al. 2007). The α-eleostearic acid has been reported to be contained in *Ricinocarpus bowmanii* F Muell (60.9%), *Ricinocarpus tuberculatus* Muell Arg (Euphorbiaceae) (46.8%) (Rao et al. 1991), *Parinari montana* (Chrysobalanaceae) (36%) (Spitzer et al. 1992) and *Prunus mahaleb* L. (white Mahlab) (38.3%) (Sbihi et al. 2014).

Manketti oil has been analyzed with the application of both ^{1}H NMR and ^{13}C NMR (Gwatidzo et al. 2017b; Yeboah et al. 2017; Cheikhyoussef et al. 2018). The presence of the α-eleostearic acid has been confirmed with NMR (Fig 32.2). The characteristic signals for α-eleostearic acid using ^{13}C NMR are in the region of 128.71–130.56 ppm (Gwatidzo et al. 2017b; Yeboah et al. 2017; Cheikhyoussef 2018) whilst, signals for α-eleostearic acid using ^{1}H NMR arising from the protons of α-eleostearic acid are in the region of 5.63–6.37 ppm (Gwatidzo et al. 2017b; Yeboah et al. 2017; Cheikhyoussef et al. 2018).

Table 32.2 Fatty acid compositions (%) of Mongongo/Manketti (*Schinziophyton rautanenii*) oil

Origin	Zambia	Zambia	Zambia	Zimbabwe	Botswana[a]	Kenya[b]	Namibia[a]	Botswana	Namibia
Palmitic acid (16:0)		9.8	8.0	10.8	12.0	10.3	8.74	14.0[c]	10.4–14.3
Palmitoleic acid (16:1 *n*-9)							0.053		
17:0							0.082		
Stearic acid (18:0)		7.7	9.0	3.04	11.8	6.2	6.78	10.3[c]	8.59–16.3
Arachidic acid (20:0)									0.43–0.48
Oleic acid (18:1 *n*-9)		19.2	15.0	15.2	24.4	16.4	17.5	16.5[d]	11.2–13.0
20:0							0.19		
Erucic acid (22:1 *n*-9)				21.5					
11-Eicosenoic acid (11–20:1)							0.35		0.62–0.82
Linoleic acid (18:2 *n*-6)		39.0	37.0	49.5	51.93	46.9	37.8	37.8[d]	31.2–32.2
Linolenic (18:3 *n*-3)		16.7					0.044		
α-eleostearic acid (9c, 11t, 13t-18:3)	23.8		25.0			20.2	26.3	25.3[d]	24.2–35.7
Reference	Chisholm and Hopkins (1966)	Zimba et al. (2005)	Juliani et al. (2007)	Chivandi et al. (2008)	Mitei et al. (2008)	Chander (2010)	Gwatidzo et al. (2017b)	Yeboah et al. (2017)	Cheikhyoussef et al. (2018)

nd not detected, *nr* not reported
[a] Soxhlet (solvent) extraction
[b] Cold pressed
[c] Detection by GC-MS
[d] Detection by ^{13}C NMR

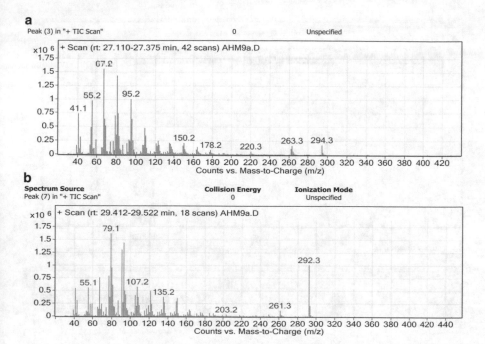

Fig. 32.1 Mass spectra for linoleic acid (**a**) and α-eleostearic acid (**b**) (Cheikhyoussef et al. 2018; Cheikhyoussef 2018)

4   Minor Bioactive Lipids in Manketti Oil

The profiling of the minor bioactive compounds, generally found in the unsaponifiable fraction of oils, is of critical importance as it provides insights into the possible uses of an oil for medicinal and nutritional purposes and also in the detection of adulteration in oils (Mitei et al. 2009). The composition of the tocopherols and phytosterols contained within Manketti oil from different sources are given in Tables 32.3 and 32.4. The concentration of α-tocopherol in the Manketti range between 5.64 and 148 µg/g, whilst that of the γ-tocopherol ranges between 123 and 2233 µg/g. The β-tocopherol is rarely detected and found only in small amounts (2–8 µg/g). Reported literature indicates that only in the Manketti oil from Namibia, the δ-tocopherol was detected (Gwatidzo et al. 2017a; Cheikhyoussef et al. 2018). The α-tocopherol and γ-tocopherol content of Manketti oil from Botswana as obtained by HPLC-FLD was 5.64 µg/g and 2232.99 µg/g, respectively (Mitei et al. 2009). The β-tocopherol, δ-tocopherol and tocotrienol were not detected resulting in a total tocol content of 2238.6 µg/g (Mitei et al. 2009). Tocotrienol presence in Manketti oil reported by Gwatidzo et al. (2017a, b) were also not detected. The tocopherols and tocotrienols, a class of fat-soluble vitamins, known as the vitamin E or E vitamers promote the oxidative stability oils by disrupting peroxidation of unsaturated oils (Boskau 2011; Gwatidzo et al. 2017a).

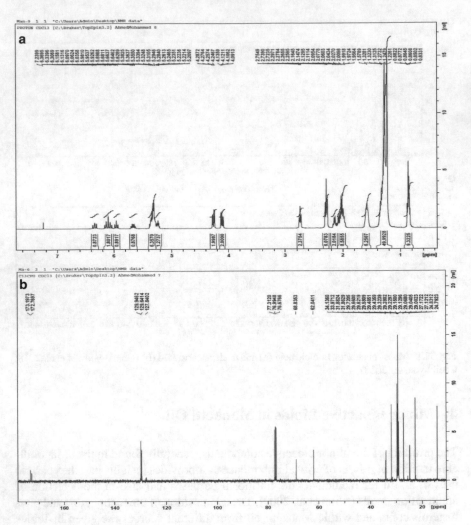

Fig. 32.2 ^{13}C NMR spectrum for Manketti oil (**a**) and ^{1}H NMR spectrum for Manketti oil (**b**) (Cheikhyoussef 2018)

The effect of different extraction methods on the tocopherols and phytosterols content of Manketti oil was investigated by Gwatidzo et al. (2014, 2017a) and reported the presence of lower concentration of tocopherol and Δ^5-avenasterol using methods of Soxhlet extraction and screw press extraction as compared to mechanical shaking and supercritical fluid extraction.

Phytosterols or plant sterols are plant-derived and have important health-promoting properties such as antioxidant and hypocholesterolemic activities (Marineli et al. 2015) and may act in cancer prevention (NCBI 2018). Phytosterols are found in free form (free sterols, FSs), conjugated as steryl esters (SEs), steryl glycosides (SGs) and acyl steryl glycosides (ASGs) (Ferrer et al. 2017). Campesterol,

Table 32.3 Composition of tocopherol and tocotrienol of Mongongo/Manketti (*Schinziophyton rautanenii*) oil

Origin	Botswana	Namibia	Namibia
α-Tocopherol (µg/g)	5.64[a]	36.7[a]	118[a]
		38[b]	148[c]
		39.3[d]	96[e]
		45[f]	
β-Tocopherol (µg/g)	nd	2.31[a]	8[a]
		2.8[b]	nd[c]
		2[d]	7[e]
		2.8[f]	
γ-Tocopherol (µg/g)	2233[a]	234.9[a]	1890[a]
		301.8[b]	1640[c]
		184.6[d]	123[e]
		302.2[f]	
δ-Tocopherol (µg/g)	nd	16.8[a]	445[a]
		12.5[b]	38[c]
		7.1[d]	42[e]
		14.9[f]	
Total tocopherols (µg/g)	2239	290.7[a]	2060[a]
		355.1[b]	1830[c]
		233.1[d]	1370[e]
		366.9[f]	
Tocotrienol (µg/g)	nd	nd	nr
Reference	Mitei et al. (2009)	Gwatidzo et al. (2017a)	Cheikhyoussef et al. (2018)

nd not detected, *nr* not reported
[a]Soxhlet extraction
[b]Supercritical fluid
[c]Traditional
[d]Screw press
[e]Cold press
[f]Mechanical shaking

stigmasterol, β-sitosterol, Δ^5-avenasterol, 22-dihydrospinasterol and Δ^7-avenasterol, lanosterol, $\Delta^{5,23}$-stigmastadienol, Δ^7-campesterol, clerosterol, obtusifoliol, $\Delta^{5,24(25)}$-stigmastadienol, α-amyrin, gramisterol, cycloeucalenol, cycloartenol, stigmasta-8,24-dienol-3-β-ol, 28-methylobtusifoliol, 24-methylenecycloartenol and citrostadienol have been found to be contained in Manketti oil (Gwatidzo et al. 2014). The highest total concentration of phytosterols detected in Manketti oil from Namibia was between 3814 and 22133 mg/100 g depending on the extraction methods used (Gwatidzo et al. 2014). The screw press extraction yielded the lowest concentration of phytosterols, while the mechanical shaking yielded the highest concentration (Gwatidzo et al. 2014). The highest concentration of phytosterols reported was the β-sitosterol (Gwatidzo et al. 2014; Cheikhyoussef et al. 2018). The phytosterol Δ^5-Avenasterol, which has antioxidant activity, has been found to have

Table 32.4 Composition of phytosterols in Mongongo/Manketti (*Schinziophyton rautanenii*) oil

Origin	Botswana[a]	Namibia	Namibia
Campesterol	5.9%	277 mg/100g[a] 1846[b, c]	
Lanosterol		6.0 mg/100g[a] 34.3[d, c]	
Stigmasterol	3.3%	98.3 mg/100g[a] 586.6[b, c]	42.3 mg/100g[a] 44.3 mg/100g[e] 45.3 mg/100g[f]
$\Delta^{5,23}$-Stigmastadienol		15.9 mg/100g[a] 124[b, c]	
Δ^{7}-Campesterol		9.5 mg/100g[a] 113.8[b, c]	
Clerosterol		51.0 mg/100g[a] 231.7[b, c]	
β-Sitosterol	78.5%	1733[a] 13852[b, c]	682 mg/100g[a] 587 mg/100g[e] 668 mg/100g[f]
Δ^{5}-Avenasterol + obtusifoliol		506[a] 2525.9[b, c]	
$\Delta^{5,24(25)}$-stigmastadienol + α-amyrin		60.5[a] 225.1[b, c]	
Gramisterol + cycloeucalenol		20.7[a] 39.5[b, c]	
Cycloartenol + Δ^{7}-sitosterol		135.1[a] 690.5[b, c]	
Δ^{5}-Avenasterol	7.7%		
22-Dihydrospinasterol	2.2%		
Δ^{7}-Avenasterol	1.4%	101.7[a] 493.8[b, c]	
Stigmasta-8,24-dienol-3-β-ol		22.5[a] 97.8[b, c]	
28-Methylobtusifoliol		19.5[a] 85.7[b, c]	
24-Methylenecycloartenol		26.2[a] 135.9[b, c]	
6-Kitositostanol		111.2[a] 312.8[b, c]	
Citrostadienol	1.06%	91.5[a] 393[b, c]	
Reference	Mitei et al. (2009)	Gwatidzo et al. (2014)	Cheikhyoussef et al. (2018)

[a]Soxhlet (solvent) extraction
[b]Mechanical shaking
[c]Highest value
[d]Screw press
[e]Traditional
[f]Cold press

a dominant presence in Manketti oil (Gwatidzo et al. 2014). The presence of Δ^5-avenasterol allows for the oil to be resistant to fast deterioration at higher temperatures (Boskau 2011), which is due to the presence of the ethylene side chain that can retard the oxidative polymerization of oils during heating (Blekas and Boskau 1999). Stigmasterol, found in Manketti oil, has been shown to exhibit anti-osteoarthritic properties for treating osteoarthritis, a disease that originates from an unbalance between cartilage anabolism and catabolism (Gabay et al. 2010). The phytosterols, β-sitosterol, stigmasterol and campesterol have been reported to possess antioxidant activity, scavenge radicals and to stabilize membranes (Yoshida and Niki 2003).

5 Health-Promoting Traits of Manketti Oil and Oil Constituents

Apart from being used in food preparations and cooking, the Manketti oil is a valued emollient (Maroyi 2018) that can be used in skin formulations for cleansing and moisturizing (Kivevele and Huan 2015), as massage oils, soap making and shampoos (ITC 2012). The oil is easily absorbed into the skin allowing it to be developed into various cosmetic formulations (Zimba et al. 2005). The chemical composition in terms of the fatty acid composition of the Manketti oil makes it unique for healing applications such as the treatment of eczema, tissue regeneration, cell repair and treatment of inflammation (Zimba et al. 2005). The α-eleostearic acid has been shown to possess a suppressive effect on tumor growth (Tsuzuki et al. 2004), a property of CLNs that have been reported to have anti-carcinogenic and anti-arteriosclerotic effects (Ha et al. 1987; Lee et al. 1994). A protective film on the skin is provided by the Manketti oil through the presence of α-eleostearic acid, protecting the skin from UV rays (Phytotrade 2012; Kalahari Biocare 2016). According to Phytotrade Africa (2012), α-eleostearic acid is polymerized by UV light producing a film on hair and skin protecting the skin from UV damage.

Three patents have been documented involving the application of Manketti nut oil in different industrial applications. The use of Manketti oil in various formulations, among others, is reported by Lucka and Mullen (2010) in Patent US 2010/0267599 A1 for use as abrasive dispersion agent in developing a microdermabrasion soap bar, its compositions and preparation methods. Another application has been documented by Razzak (2010) in Patent US 2010/0022469 A1 for the development of Anthelminthic (parasitic) formulations for use in livestock and one by Bhagat (2009) in Patent US 2009/0264520 A1 for development and application of nutritional food formulations.

References

Atabani, A. E., Mofijur, M., Masjuki, H. H., et al. (2014). A study of biodiesel production and characterization of Manketti (*Ricinodendron rautonemii*) methyl ester and its blends as a potential biodiesel. *Biofuel Research Journal, 4*, 139–146. https://doi.org/10.18331/BRJ2015.1.4.7.

Bennet, B. (2006). *Natural products: The new engine for African trade growth*. Consultancy to further develop the trade component of the natural resources enterprise programme (NATPRO), Regional Trade Facilitation Programme (RTFP), IUCN The world conservation union and natural resources institute. http://citeseerx.ist.psu.edu/viewdoc/download?doi=10.1.1.610.9509&rep=rep1&type=pdf. Accessed 26 Apr 2018.

Bhagat, U. (2009). Lipid-containing compositions and methods of use thereof. Patent US 2009/0264520 A1.

Blekas, G., & Boskau, D. (1999). Phytosterols and stability of frying oils. In D. Boskau & I. Elmadfa (Eds.), *Frying of food* (pp. 205–222). Lancaster: Technomic Publishing.

Booth, F. E. M., & Wickens, G. E. (1988). *Non-Timber Uses of Selected Arid Zone Trees and Shrubs in Africa*. FAO Conservation Guide 19. Rome: Food and Agriculture Organization of the United Nations, 103-109.

Boskau, D. (2011). Olive oil. In F. Gunstone (Ed.), *Vegetable oils in food technology. Composition, properties and uses* (pp. 244–277). New York: Wiley.

Cao, Y., Yang, L., Gao, H.-L., Chen, J.-N., et al. (2007). Re-characterization of three conjugated linolenic acid isomers by GC-MS and NMR. *Chemistry and Physics of Lipids, 145*, 128–133. https://doi.org/10.1016/j.chemphyslip.2006.11.005.

Chander, A. K. (2010). *Characterization and oxidative stability of speciality plant seed oils*. PhD Thesis, Aston University.

Cheikhyoussef, N. (2018). *Profiling studies of five Namibian indigenous seed oils obtained using three different extraction methods*. PhD Thesis, University of Namibia.

Cheikhyoussef, N., Kandawa-Schulz, M., Böck, R., et al. (2018). Characterization of *Schinziophyton rautanenii* (Manketti) nut oil from Namibia rich in conjugated fatty acids and tocopherol. *Journal of Food Composition and Analysis, 66*, 152–159. https://doi.org/10.1016/j.jfca.2017.12.015.

Chisholm, M. J., & Hopkins, C. Y. (1966). Kamlolenic acid and other conjugated fatty acids in certain seed oils. *Journal of the American Oil Chemists' Society, 43*, 390–391. https://doi.org/10.1007/BF02646796.

Chivandi, E., Davidson, B. C., & Erlwanger, K. H. (2008). A comparison of the lipid and fatty acid profiles from the kernels of the fruit (nuts) of *Ximenia caffra* and *Ricinodendron rautanenii* from Zimbabwe. *Industrial Crops and Products, 27*, 29–32. https://doi.org/10.1016/j.indcrop.2007.06.002.

European Commission. (1998). Data collection and analysis for sustainable forest management in ACP countries-linking national and international efforts. http://www.fao.org/3/a-x6694e/X6694E00.html. Accessed 9 Mar 2017.

Ferrer, A., Altabellaa, T., Arróa, M., & Boronat, A. (2017). Emerging roles for conjugated sterols in plants. *Progress in Lipid Research, 67*, 27–37. https://doi.org/10.1016/j.plipres.2017.06.002.

Gabay, O., Sanchez, C., Salvat, C., et al. (2010). Stigmasterol: A phytosterol with potential anti-osteoarthritic properties. *Osteoarthritis and Cartilage, 18*(1), 106–116. https://doi.org/10.1016/j.joca.2009.08.019.

Gandure, J., Ketlogetswe, C., & Temu, A. (2014). Fuel properties of biodiesel produced from selected plant kernel oils indigenous to Botswana: A comparative analysis. *Renewable Energy, 68*, 414–420. https://doi.org/10.1016/j.renene.2014.02.035.

Gwatidzo, L., Botha, B. M., & McCrindle, R. I. (2014). Extraction and identification of phytosterols in Manketti (*Schinziophyton rautanenii*) nut oil. *Journal of the American Oil Chemists' Society, 91*(5), 783–794. https://doi.org/10.1007/s11746-014-2417-2.

Gwatidzo, L., Botha, B. M., & McCrindle, R. I. (2017a). Influence of extraction method on yield, physicochemical properties and tocopherol content of Manketti (*Schinziophyton rautanenii*) nut oil. *Journal of the American Oil Chemists' Society, 94*(7), 973–980. https://doi.org/10.1007/s11746-017-3004-0.

Gwatidzo, L., Botha, B. M., & McCrindle, R. I. (2017b). Fatty acid profile of Manketti (*Schinziophyton rautanenii*) nut oil: Influence of extraction method and experimental evidence on the existence of α-eleostearic acid. *Journal of Cereals and Oilseeds, 8*(5), 33–44. https://doi.org/10.5897/JCO2017.0175.

Ha, Y. L., Grimm, N. K., & Pariza, M. W. (1987). Anticarcinogens from fried ground beef: Heat-altered derivatives of linoleic acid. *Carcinogenesis, 8*, 1881–1887. https://doi.org/10.1093/carcin/8.12.1881.

Hoffmann, L. (2016a). The Manketti tree (*Schinziophyton rautanenii = Ricinodendron rautanenii*) Part 1. Available via THE NAMIBIAN. https://www.namibian.com.na/148620/archive-read/The-Manketti-tree-(*Schinziophyton-rautanenii-=--Ricinodendron-rautanenii*)-Part-1. Accessed 30 Apr 2018.

Hoffmann, L. (2016b). The Manketti tree (*Schinziophyton rautanenii = Ricinodendron rautanenii*) Part 2. Available via THE NAMIBIAN. https://www.namibian.com.na/148899/archive-read/The-Manketti-tree-(*Schinziophyton-rautanenii-=--Ricinodendron-rautanenii*)-Part-2. Accessed 30 Apr 2018.

International Trade Centre (ITC). (2012). The North American market for natural products prospects for Andean and African products. http://www.intracen.org/The-North-American-Market-for-Natural-Products-Prospects-for-Andean-and-African-Products/. Accessed 17 May 2018.

Juliani, H. R., Koroch, A. R., Simon, J. E., & Wamulwange, C. (2007). Mungongo cold pressed oil (*Schinziophyton rautanenii*): A new natural product with potential cosmetic applications. *Acta Horticulturae (ISHS), 756*, 407–412. https://doi.org/10.17660/ActaHortic.2007.756.43.

Kalahari Biocare. (2016). Mongongo/Manketti. Available via KALAHARI BIOCARE. http://kalaharibiocare.com/devils-claw/mongongo-or-manketti-kernel-oil/. Accessed 25 Apr 2018.

Kivevele, T., & Huan, Z. (2015). Review of the stability of biodiesel produced from less common vegetable oils of African origin. *South African Journal of Science, 11*(9/10), 1–7.

Lee, K. N., Kritchevsky, D., & Pariza, M. W. (1994). Conjugated linoleic acid and atherosclerosis in rabbits. *Atherosclerosis, 108*, 19–25. https://doi.org/10.1016/0021-9150(94)90034-5.

Lucka, L., & Mullen, P. A. (2010). Microdermabrasion soap compositions and methods of preparing same. US Patent US2010/0267599 A1.

Marineli, R. D., Furlan, C. P. B., Marques, A. C., et al. (2015). Phytosterols: Biological effects and mechanisms of hypocholesterolemic action. In V. K. Gupta, M. G. Tuohy, M. Lohani, & A. O'Donovan (Eds.), *Biotechnology of bioactive compounds: Sources and applications*. New York: Wiley.

Maroyi, A. (2018). Contribution of *Schinziophyton rautanenii* to sustainable diets, livelihood needs and environmental sustainability in Southern Africa. *Sustainability, 10*, 581. https://doi.org/10.3390/su10030581.

Mitei, Y. C., Ngila, J. C., Yeboah, S. O., Wessjohan, L., & Schmidt, J. (2008). NMR, GC-MS and ESI-FTICR-MS profiling of fatty acids and triacylglycerols in some Botswana seed oils. *Journal of the American Oil Chemists' Society, 85*(11), 1021–1032. https://doi.org/10.1007/s11746-008-1301-3.

Mitei, Y. C., Ngila, J. C., Yeboah, S. O., Wessjohan, L., & Schmidt, J. (2009). Profiling of phytosterols, tocopherols and tocotrienols in selected seed oils from Botswana by GC-MS and HPLC. *Journal of the American Oil Chemists' Society, 86*, 617–625. https://doi.org/10.1007/s11746-009-1384-5.

NCBI (2018) National Center for Biotechnology Information. U.S. National Library of Medicine, https://pubchem.ncbi.nlm.nih.gov/. Accessed 27 Apr 2018.

New Agriculturist. (2009). Mongongo-a tough nut worth cracking. http://www.new-ag.info/en/focus/focusItem.php?a=794. Accessed 25 Apr 2018.

Palgrave, K. C. (1983). *Trees of Southern Africa*. Cape Town: Struik Publishers. 1868251713.

Peters, C. R. (1987). *Ricinodendron rautanenii* (Euphorbiaceae): Zambezian wild food plant for all seasons. *Economic Botany, 41*(4), 494–502. https://doi.org/10.1007/BF02908143.

Phytotrade. (2012). Mongongo oil. Available via PHYTOTRADE. http://oils.phytotrade.com/portfolio/mongongo-oil/. Accessed 8 Aug 2016.

Rao, K. S., Kaluwin, C., Jones, G. P., et al. (1991). New source of α-eleostearic acid: *Ricinocarpus bowmanii* and *Ricinocarpus tuberculatus* seed oils. *Journal of Science and Food Agriculture, 57*(3), 427–429. https://doi.org/10.1002/jsfa.2740570313.

Razzak, M. (2010). Anthelminitic formulations. US Patent 2010/0022469 A1.

Rutto, H. L., & Enweremadu, C. C. (2011). Optimization of production variables of biodiesel from Manketti using response surface methodology. *International Journal of Green Energy, 8*(7), 767–779.

Sbihi, H. N., Nehdi, I. A., & Al-Resayes, S. I. (2014). Characterization of white mahlab (*Prunus mahaleb* L.) seed oil: A rich source of α-eleostearic acid. *Journal of Food Science, 79*(5), C795–C801. https://doi.org/10.1111/1750-3841.12467.

Spitzer, V., Marx, F., Maia, J. G. S., & Pfeilsticker, K. (1992). Occurrence of alpha eleostearic acid in the seed oil of *Parinari montana* (Chrysobalanaceae). *Fat Science Technology, 94*, 58–60. https://doi.org/10.1002/lipi.19920940206.

Tsuzuki, T., Tokuyama, Y., Igarashi, M., & Miyazawa, T. (2004). Tumour growth suppression by α-eleostearic acid, a linoleic acid isomer with a conjugated triene system, via lipid peroxidation. *Carcinogenesis, 25*(8), 1417–1425. https://doi.org/10.1093/carcin/bgh109.

Van, A. N., Popova, A. A., Deineka, V. I., & Deineka, L. A. (2017). Determination of triacylglycerols of Manketti oil by reversed-phase HPLC. *Journal of Analytical Chemistry, 72*(9), 1007–1012. https://doi.org/10.1134/S1061934817090027.

Vermaak, I., Kamatou, G. P. P., Komane-Mofokeng, B., et al. (2011). African seed oils of commercial importance-cosmetic applications. *South African Journal of Botany, 77*, 920–933. https://doi.org/10.1016/j.sajb.2011.07.003.

Yeboah, E. M. O., Kobue-Lekalake, R. I., Jackson, J. C., et al. (2017). Application of high resolution NMR, FTIR and GC-MS to a comparative study of some indigenous seed oils from Botswana. *Innovative Food Science and Emerging Technologies, 44*, 181–190. https://doi.org/10.1013/j.ifset.2017.05.004.

Yoshida, Y., & Niki, E. (2003). Antioxidant effects of phytosterol and its components. *Journal of Nutritional Science and Vitaminology (Tokyo), 49*(4), 277–280. https://doi.org/10.3177/jnsv.49.277.

Zimba, N., Wren, S., & Stucki, A. (2005). Three major tree nut oils of Southern Central Africa: Their uses and future as commercial base oils. *International Journal of Aromatherapy, 15*, 177–182. https://doi.org/10.1016/j.ijat.2005.10.009.

Chapter 33
Bauhinia purpurea Seed Oil

Mohamed Fawzy Ramadan

Abstract Kachnar (*Bauhinia purpurea*) seed oil is rich in bioactive lipids and could be considered as a raw material for functional and pharmaceutical products. *B. purpurea* oilseeds contain approx. 17.5% as an *n*-hexane extractable lipids. Linoleic, palmitic, oleic and stearic acids were the main fatty acids in the crude *B. purpurea* seed oil. The amounts of neutral lipids in the crude *B. purpurea* seed oil was the highest, followed by glycolipids and phospholipids. *B. purpurea* seed oil is characterized by high levels of sterols, wherein the sterol marker was β-sitosterol. β-Tocopherol was the main identified tocopherol isomer and the rest was δ-tocopherol. This chapter summarizes the chemical composition and functional properties of *B. purpurea* seed oil.

Keywords Kachnar seed oil · Leguminosae · Fatty acids · Sterols · Tocols · Lipid-soluble bioactives

1 Introduction

Interest in new sources of oils and fats has been recently grown. This necessitates the search for new sources of oils. Several plants are grown not only for food and fodder but also for the amazing variety of products with application in industry, including oils and pharmaceuticals. The genus *Bauhinia* consisting of 300 species belongs to the family Leguminosae (Caesalpinioideae). The genus Bauhinia has recently been divided into four sub-genera: Barklya (1 species), Bauhinia (140 species), Elayuna (6 species), and Phanera (150 species). The latter are tendril-bearing species, while the three former taxa comprise tree or shrubby species (Duarte-Almeida et al. 2004).

Bauhinia Purpurea L. is found throughout India and widely grown as an ornamental plant (Soetjipto et al. 2018). It is a small evergreen tree that exhibited

M. F. Ramadan (✉)
Agricultural Biochemistry Department, Faculty of Agriculture, Zagazig University, Zagazig, Egypt
e-mail: mframadan@zu.edu.eg

© Springer Nature Switzerland AG 2019
M. F. Ramadan (ed.), *Fruit Oils: Chemistry and Functionality*,
https://doi.org/10.1007/978-3-030-12473-1_33

Fig. 33.1 *Bauhinia purpurea* pods (**a**) and seeds (**b**)

favorable agro-botanical properties, such as fast vegetative growth and early flowering, along with a high fertility index with a high pod weight (approx. 8.3 g), number of seeds (Fig. 33.1) per pod (eight or nine), weight of seeds per pod (approx. 2.4 g), and seed recovery percentage is 28.9%. The seeds are light in weight and the pods are woody in nature (Rajaram and Janardhanan 1991; Vadivel and Biesalski 2011). The root, stem bark and leaves are being used against many diseases like jaundice, leprosy, cough and used in Ayurvedic medicine formulations (Parrota 2001). The young pods and seeds of *B. purpurea* are known to be cooked and eaten by the tribles like Kathkors and Gondas of India (Rajaram and Janardhanan 1991).

The seed material of *B. purpurea* is a promising under-utilized food that merits wider use in tropical countries. Apart from having excellent nutritive value, *B. purpurea* seed materials also possess various medicinal properties (Vadivel and Biesalski 2011, 2013). *B. purpurea* seeds are a rich source of carbohydrates (51%), lipids (12.3%), protein (25.6%), fiber (5.8%), and minerals (Rajaram and Janardhanan 1991; Vijayakumari et al. 1997; Vadivel and Biesalski 2011, 2013). The fatty acid profile and phenolics@@ profile of the *B. purpurea* seeds has been reported (Ramadan et al. 2006; Ramadan and Moersel 2007).

2 Extraction and Processing of Fruit Oil

For a plant to be suitable for oil production, the oil content must reach the minimum for commercial exploitation and the plant must be suitable for high acreage cultivation (Bockisch 1998). *B. purpurea* seeds, when extracted with *n*-hexane, were found to contain 17.5% crude oil (Ramadan et al. 2006). *B. purpurea* seeds were reported to contain a high level of crude lipids (14.3%) when compared to certain conventional pulses such as *Vigna aconitifolia* (0.69%), *Cicer arietinum* (4.16%), *V. radiate* (0.71%), *V. mungo* (0.45%), and *Phaseolus vulgaris* (0.9%) (Bravo et al. 1999; Vadivel and Biesalski 2011, 2013).

3   Fatty Acids Composition and Acyl Lipids of *B. purpurea* Seed Oil

Soetjipto et al. (2018) extracted the oil from *B. purpurea* seeds with *n*-hexane. The oil yield was 57.3% (w/w) and the chemical characteristic of seed oil included an acid value (13.7 mg KOH/g), and saponification value (153.3 mg KOH/g). *B. purpurea* seed oil included linoleic acid (28.1%), palmitic acid (29.2%), oleic acid (19.8%) and stearic acid (10.7.4%) as the main fatty acids in the crude seed oils. Minor levels of neophytadiena and arachidic acid were also identified in *B. purpurea* seed oil. Table 33.1 presents the physicochemical traits of crude and purified *B. purpurea* seed oil as reported by Soetjipto et al. (2018).

Fatty acid profiles and lipid classes of *B. purpurea* seed oil [neutral lipids (NL), glycolipids (GL), and phospholipids (PL)] are given in Table 33.2 (Ramadan et al. 2006; Ramadan and Moersel 2007). Twelve fatty acids were detected, wherein linoleic followed by palmitic, oleic and stearic acids were the main fatty acids (approx. 97% of total fatty acids). *B. purpurea* seed oil was characterized by the high levels of polyunsaturated fatty acids (*PUFA*). Trienes [(γ-linolenic acid *GLA*, C18:3*n*-6) and (α-linolenic acid *ALA*, C18:3*n*-3)] as well as EPA (C20:5), were estimated in lower amounts. The ratio of unsaturated fatty acids to saturated fatty acids, was higher in neutral lipids than in the corresponding polar lipids. The fatty acid composition of *B. purpurea* seed oil evinces the lipids as a good source of the essential fatty acids. A great deal of interest has been placed in the few oils that contain PUFA especially GLA. The source of natural GLA are few and at present only evening primrose, borage, hemp oils are well known (Kamel and Kakuda 2000). Interest in

Table 33.1 Physical and chemical traits of *B. purpurea* seed oil (Soetjipto et al. 2018)

Parameter	Refined seed oil	Crude seed oil
Color	Yellow	Brownish yellow
Density (g/cm³)	0.82	0.86
Saponification value (mg KOH/g)	187	153
Peroxide value (meq O₂/kg)	15.6	43.5
Acid value (mg KOH/g)	3.87	13.7

Table 33.2 Main fatty acids (%) in *B. purpura* seed oil and lipid classes (Ramadan et al. 2006)

	Crude oil	NL	GL	PL
Fatty acid	Relative content (%)			
C16:0	22.1	21.1	27.3	27.8
C18:0	13.6	13.7	15.4	15.0
C18:1	16.3	16.4	16.6	16.7
C18:2	45.9	46.8	37.7	38.0
C18:3*n*-3	0.29	0.12	0.21	0.19
C18:3*n*-6	0.16	0.17	0.13	0.08
C20:5 EPA	0.30	0.21	0.98	0.59

Table 33.3 Main lipid subclasses (g/kg oil) in *B. purpura* crude seed oil (Ramadan et al. 2006)

NL subclass	g/kg oil	GL subclass	g/kg oil	PL subclass	g/kg oil
MAG	2.97	DGD	0.08	PS	0.14
DAG	6.93	CER	1.19	PI	0.34
FFA	11.8	SG	1.41	PC	1.50
TAG	920	MGD	0.15	PE	0.69
STE	6.63	ESG	1.50		

the PUFA as health-promoting nutrients has expanded dramatically in recent years. A rapidly growing literature illustrates the benefits of PUFA, in alleviating cardiovascular, inflammatory, heart diseases, atherosclerosis, an autoimmune disorder, diabetes and other diseases (Finley and Shahidi 2001; Riemersma 2001; Ramadan et al. 2006; Kiralan and Ramadan 2016; Kiralan et al. 2014, 2017, 2018).

The proportion of lipid classes and subclasses found in *B. purpurea* seed oil is shown in Table 33.3. The amount of NL was the highest (approx. 99%), followed by GL (approx. 0.44%) and PL (approx. 0.27%). NL subclasses included triacylglycerol (TAG), free fatty acids (FFA), diacylglycerol (DAG), esterified sterols (STE), free sterols (ST) and monoacylglycerol (MAG) in decreasing order. GL subclasses of GL included sulphoquinovosyldiacylglycerol (SQD), digalactosyldiglycerides (DGD), cerebrosides (CER), sterylglycosides (SG), monogalactosyldiglycerides (MGD) and esterified sterylglycosides (ESG). ESG, SG and CER were the main components and made up approx. 93% of GL. PL subclasses were phosphatidylcholine (PC) followed by phosphatidylethanolamine (PE), phosphatidylinositol (PI) and phosphatidylserine (PS), respectively. About a 50% of total PL was PC and 25% was PE, while PI and PS were detected in lower amounts. (Ramadan et al. 2006).

4   Minor Bioactive Lipids in *B. purpurea* Fruit Seed Oil

The qualitative and quantitative composition of tocopherols is given in Table 33.4. Two tocopherol isomers were detected in *B. purpurea* seed oil. β-tocopherol constituted approx. 72% of the total identified tocopherols, while δ-tocopherol accounted for approx. 27.8%. Levels of tocopherols detected in *B. purpurea* seed oil may contribute to the oxidative stability of the oil.

B. purpurea seed oil is characterized by high levels of unsaponifiable (approx. 12 g/kg oil). The sterol marker was β-sitosterol, which accounted for approx. 64.5% of the total identified sterols (Table 33.4). The next main component was stigmasterol. Other sterols including campesterol and Δ7-stigmastenol, were detected at equal amounts (approx. 6.0% of the total identified sterols). Lanosterol, brassicasterol, and Δ5, 24-stigmastadinol were not detected in *B. purpurea* seed oil. β-sitosterol has been intensively investigated with respect to its physiological impacts in human (Yang et al. 2001). Plant sterols are of interest due to their antioxidant potential and health-promoting effects. Plant sterols have been added to edible oils as an example of a novel foods (Ramadan 2015).

Table 33.4 Main tocopherols and plant sterols (g/kg oil) in *B. purpura* seed oil (Ramadan et al. 2006)	Compound	g/kg oil
	β-tocopherol	2.57
	δ-tocopherol	0.99
	β-sitosterol	3.83
	Campesterol	0.36
	Stigmasterol	1.22
	Δ7- stigmastenol	0.32
	Δ7-avenasterol	0.17

5 Contribution of Bioactive Compounds in *B. purpurea* Seed Oil to Organoleptic Properties and Functions

The differential scanning calorimetry (DSC) and oxidative stability index (OSI) techniques were applied to test the oxidative stability of *B. purpurea* seed oil (Arain et al. 2009). Both methods confirmed that *B. purpurea* oil is very stable oil when compared to rice bran and cotton seed oil. Due to considerable oxidative stability, *B. purpurea* oil may find some appropriate applications in food and pharmaceutical applications.

6 Health-Promoting Traits of *B. purpurea* Fruit Seed Oil

Vadivel and Biesalski (2013) evaluated the antioxidant and type II diabetes-related enzyme inhibition properties of a phenolic extract from *B. purpurea* seeds. They concluded that *B. purpurea* seeds could be envisaged as a dietary ingredient in the formulation of supplementary foods with therapeutic value to manage type II diabetic patients. *B. purpurea* seeds give a high yield of oil, which is very important for the commercial production of seed oil. Due to the high amounts of essential fatty acids (especially PUFA), sterols and tocols in *B. purpurea* oil, the oil could be used in food and non-food applications.

References

Arain, S., Sherazi, S. T. H., Bhanger, M. I., Talpur, F. N., & Mahesar, S. A. (2009). Oxidative stability assessment of *Bauhinia purpurea* seed oil in comparison to two conventional vegetable oils by differential scanning calorimetry and Rancimat methods. *Thermochimica Acta, 484,* 1–3.

Bockisch, M. (1998). Vegetable fats and oils. In M. Bockisch (Ed.), *Fats and oils handbook* (pp. 174–344). Champaign: AOCS Press.

Bravo, L., Siddhuraju, P., & Saura-Calixto, F. (1999). Composition of underexploited Indian pulses. Comparison with common legumes. *Food Chemistry, 64*, 185–192.

Duarte-Almeida, J. M., Negri, G., & Salatino, A. (2004). Volatile oils in leaves of Bauhinia (*Fabaceae Caesalpinioideae*). *Biochemical Systematics and Ecology, 32*, 747–753.

Finley, J. W., & Shahidi, F. (2001). The chemistry, processing and health benefits of highly unsaturated fatty acids: An overview. In W. J. John & F. Shahidi (Eds.), *Omega-3 fatty acids, chemistry, nutrition and health effects* (pp. 1–13). Washington, DC: American Chemical Society.

Kamel, B. S., & Kakuda, Y. (2000). Fatty acids in fruits and fruit products. In C. K. Chow (Ed.), *Fatty acids in foods and their health implications* (2nd ed., pp. 239–270). New York: Marcel Dekker.

Kiralan, M., & Ramadan, M. F. (2016). Volatile oxidation compounds and stability of safflower, sesame and canola cold-pressed oils as affected by thermal and microwave treatments. *Journal of Oleo Science, 65*, 825–833.

Kiralan, M., Özkan, G., Bayrak, A., & Ramadan, M. F. (2014). Physicochemical properties and stability of black cumin (*Nigella sativa*) seed oil as affected by different extraction methods. *Industrial Crops and Products, 57*, 52–58.

Kiralan, M., Ulaş, M., Özaydin, A. G., Özdemir, N., Özkan, G., Bayrak, A., & Ramadan, M. F. (2017). Blends of cold pressed black cumin oil and sunflower oil with improved stability: A study based on changes in the levels of volatiles, tocopherols and thymoquinone during accelerated oxidation conditions. *Journal of Food Biochemistry, 41*, e12272. https://doi.org/10.1111/jfbc.12272.

Kiralan, M., Çalik, G., Kiralan, S., & Ramadan, M. F. (2018). Monitoring stability and volatile oxidation compounds of coldpressed flax seed, grape seed and black cumin seed oils upon photo-oxidation. *Journal of Food Measurement and Characterization, 12*, 616–621.

Parrota, J. A. (2001). *Healing plants of peninsular India*. Wallingford: CABI Publishing. CABI international walling ford Oxon Ox 10 8DE U.K.

Rajaram, N., & Janardhanan, K. (1991). Chemical composition and nutritional potential of tribal pulses *Bauhinia purpurea*, *B. racemosa* and *B. vahlii*. *Journal of the Science of Food and Agriculture, 55*, 423–431.

Ramadan, M. F. (2015). Oxidation of β-sitosterol and campesterol in sunflower oil upon deep- and pan-frying of French fries. *Journal of Food Science and Technology, 52*(10), 6301–6311.

Ramadan, M. F., & Moersel, J.-T. (2007). Kachnar seed oil. *Information, 18*, 13–15.

Ramadan, M. F., Sharanabasappa, G., Seetharam, Y. N., Seshagiri, M., & Moersel, J.-T. (2006). Characterisation of fatty acids and bioactive compounds of Kachnar (*Bauhinia purpurea* L.) seed oil. *Food Chemistry, 98*(2), 359–365.

Riemersma, R. A. (2001). The demise of the n-6 to n-3 fatty acid ratio? A dossier. *European Journal of Lipid Science and Technology, 103*, 372–373.

Soetjipto, H., Riyanto, C. A., & Victoria, T. (2018) Chemical characteristics and fatty acid profile of butterfly tree seed oil (*Bauhinia purpurea* L). IOP Conf. Series: Materials Science and Engineering 349. (2018) 012024 https://doi.org/10.1088/1757-899X/349/1/012024.

Vadivel V., & Biesalski H. K. (2011) Role of purple camel's foot (*Bauhinia purpurea* L.) seeds in nutrition and Medicine. In V. Preedy, R. Watson, V. Patel (Eds.), *Nuts and seeds in health and disease prevention* (pp. 941–949). ISBN: 9780123756886. Academic Press. doi: https://doi.org/10.1016/B978-0-12-375688-6.10111-2.

Vadivel, V., & Biesalski, H. K. (2013). Antioxidant potential and health relevant functionality of *Bauhinia purpurea* L. seeds. *British Food Journal, 115*, 1025–1037.

Vijayakumari, K., Siddhuraju, P., & Janardhanan, K. (1997). Chemical composition, amino acid content and protein quality of the little-known legume *Bauhinia purpurea* L. *Journal of the Science of Food and Agriculture, 73*, 279–286.

Yang, B., Karlsson, R. M., Oksman, P. H., & Kallio, H. P. (2001). Phytosterols in sea buckthorn (*Hippophaë rhamnoides* L.) berries: Identification and effects of different origins and harvesting times. *Journal of Agricultural and Food Chemistry, 49*, 5620–5629.

Chapter 34
Pongamia pinnata Seed Oil

K. Thirugnanasambandham

Abstract Indian Beech tree (Botanical Name: *Pongamia pinnata* (L.) Pierre) is a deciduous legume that grows up to about 50–80 feet tall and is native to subtropical regions like India. Being a legume, it fixes nitrogen into the soil and is often used as a windbreak between fields on farms. It has a wide spreading canopy making and fragrant flowers making it ideal for ornamental shade applications. The oil obtained from *Pongamia pinnata* seed is non-edible due to bitter tasting flavonoids. The plant has pharmaceutical uses but is not poisonous to the touch like jatropha. It is insect resistant and there is mention of using the press cake as both insecticide and chicken feed. Due to the large availability of *Pongamia pinnata* seed in India, in this present study, an attempt was made to investigate the ultrasound-assisted extraction process to extract the oil from *Pongamia pinnata* seed under various operating conditions such as solvent to sample ratio, sonication time, temperature. Three factors, three-level central composite design (CCD) coupled with desired function methodology was used to optimize and model the extraction process. Optimum extracting conditions for the maximum oil yield were determined using numerical optimization technique. Under these conditions, 72% of oil was extracted. Results confirmed that sonication is efficient method to extract the oil from *Pongamia pinnata* seed.

Keywords *Pongamia pinnata* · Seed oil · Extraction · Ultrasound · Modeling · Optimization

1 Overview of Indian Beech Tree

Pongamia pinnata (L.) Pierre, an arboreal legume, is a member of the subfamily Papilionoideae and family Leguminosae, native to tropics and temperate Asia including part of India, China, Japan, Malaysia, and Australia (Sahu et al. 2017; Muthu et al. 2006). The plant has been synonymously known as *Millettia pinnata*, *Pongamia glabra*, and *Derris indica*; commonly it is referred as karanj, pongam,

K. Thirugnanasambandham (✉)
Department of Chemistry, ECET, Coimbatore, Tamil Nadu, India

© Springer Nature Switzerland AG 2019   647
M. F. Ramadan (ed.), *Fruit Oils: Chemistry and Functionality*,
https://doi.org/10.1007/978-3-030-12473-1_34

and dalkaramch. *Pongamia* is drought resistant, semi-deciduous, nitrogen-fixing leguminous tree. The tree is well suited to intense heat and sunlight and its dense network of lateral roots and thick long tap roots make it drought tolerant. Historically, this plant has been used in India and neighboring regions as a source of traditional medicines, animal fodder, green manure, timber, fish poison and fuel. More importantly, *P. pinnata* has recently been recognized as a viable source of oil for the burgeoning biofuel industry (Prabha et al. 2003). It is one of the widely grown forest trees with 0.11 million tons of seeds collected every year in Andhra Pradesh, Karnataka and Tamil Nadu states of India. Botanical classification of Indian Beech tree is shown in Table 34.1.

The Asian varieties reach an adult height in 4 or 5 years and start bearing seeds at 4–7 years. It can produce 50–100 lbs of seed per tree. Assuming 200 trees per acre, and 25% oil per pound of seed you would get from about 100 to 600 gal of oil per acre. Genetically modified plants produce more. The tree is all over India, used to line roadways and waterways. The seeds can be economically picked up from these plants and the oil expelled in presses in local villages for small quantities of oil that can either be added to diesel used by the village or sold for an income (Ran et al. 2018; Meher et al. 2006). The meal can be used as animal fodder or composted. Nutrient levels in *Pongamia pinnata* leaf and fruit are shown in Table 34.2. Its disadvantage is that the oil may contain high levels of unsaponifiable material. More literally, the oil contains stuff that will not turn into biodiesel. Sometimes a lot of stuff that will not turn into biodiesel.

The tree is known for its multipurpose benefits and as a potential source of biodiesel, besides its application as animal fodder, green manure, timber and fish poison.

The cooked mature seeds are consumed as a food by certain tribal sects, including Lambadi, Uraali and Dravidian in India. The seeds are found to possess 22% of crude protein, 33.4% of lipids, 6.8% of crude fiber, 3.3% of ash and 26% of carbohydrates and all the essential amino acids, except cystein, methionine and tyrosine, in addition to very high protein digestibility (92%). Also, the seeds contain on an average about 28–34% oil with a high percentage of polyunsaturated fatty acids (Table 34.3). *Pongamia pinnata* has received much attention because of its high content of seed oil (~28 to 39%,), which contains high amounts of C18 fatty acids (Ahmad et al. 2004).

Historically, *P. pinnata* is used as a folk medicinal plant, particularly in Ayurveda and Siddha systems of Indian medicine (Chauhan and Chauhan 2002). All parts of the plant are used as a crude drug for the treatment of tumors, piles, skin diseases,

Table 34.1 Botanical classification of Indian Beech tree

Kingdom	Plantae
Division	Magnoliophyta
Class	Magnoliophyta
Order	Fabales
Family	Leguminosae
Genus	Pongamia
Species	Pinnata

Table 34.2 Nutrient levels in *Pongamia pinnata* leaf and fruit

Parameter	Leaf	Fruit
Protein	–	17.40%
Fatty oil	–	27.50%
Ash	–	2.4%
Tannin	–	2.32 g/100 g
K	0.49%	1.30%
N	1.16%	5.10%
Moisture	–	19.10%
Starch	–	5.10%
Mucilage	–	13.50%
Na+	–	0.50%

Table 34.3 Fatty acid composition (%) of *Pongamia pinnata* seed oil

Fatty acid	Relative composition (%)	Carbon value
Palmitic	3.7–7.9%	C16:0
Stearic	2.4–8.9%	C18:0
Arachidic	2.2–4.7%	C20:0
Behenic	4.2–5.3%	C22:0
Lignoceric	1.1–3.5%	C24:0
Oleic	44.5–71.3%	C18:1
Linoleic	10.8–18.3%	C18:2
Eicosenoic	9.5–12.4%	C20:1

itches, abscess, painful rheumatic joints, ulcers, diarrhea, bronchitis, whooping cough and quench dipsia in diabetes. *P. pinnata* is being reported to use by traditional healers of Theni District, Tamilnadu India as an antiseptic, blood purifier and also to treat cuts and wounds. *P. pinnata* has been exploited as a source of biomedicine, specifically as antimicrobial and therapeutic agents. Further, experimental studies also demonstrated its anti-inflammatory, antioxidative, analgesic and antiulcer effects (Azam et al. 2005; Elanchezhiyan et al. 1993).

The effects of ethanolic root extract of *P. pinnata* on the antioxidant status and histopathological changes in acute ischemia-reperfusion injured rat model are investigated. The crude decoction of *P. pinnata* leaves is known to exhibit selective antidiarrheal action with efficacy against cholera and enteroinvasive bacterial strains causing bloody diarrheal episodes. The leaves are scientifically validated for antinociceptive as well as antipyretic activities for the treatment of pain and pyretic disorders (Essa and Subramanian 2006). The flowers of *P. pinnata* had a protective effect against cisplatin and gentamicin-induced renal injury through the antioxidant property. The antihyperglycaemic activity of stem bark extract is proved in alloxan-induced diabetic rats. Pongamol and karanjin isolated from the fruits of *P. pinnata* are reported to produce antihyperglycaemic activity. Further, the seed extract is reported to completely inhibit the growth of herpes simplex viruses (Ballal 2005).

Even though, the nutritional value of seeds and medicinal properties of different parts of *P. pinnata* plant are reported earlier, the information regarding the antioxidant and type II diabetes-related enzyme inhibition properties of seed materials are scarce (Carcache et al. 2003).

Meanwhile, *P. pinnata* seed oil represents a precursor for biodiesel production and has been widely studied as a potential renewable feedstock. The residual seed is suitable for bioethanol production owing to its holocellulose composition. Globally, many countries including Australia (Rural Industries Research and Development Corporation, Australian Government), India (National Oil Seeds and Vegetable Oils Development Board, the Ministry of Agriculture, India) and Hawaii (Source: Biodiesel crop implementation in Hawaii, Hawaii Agriculture Research Centre) have initiated techno-economic modelling and practices for *P. pinnata* plantation programs as a source of renewable feedstock for the biodiesel representing a sustainable energy supply (Brijesh et al. 2006). As a result the annual yield of the seeds reached 200,000 metric tons from India alone. The seeds are composed of 30–35% oil which can be processed to biodiesel; the residue is of current interest in terms of hydrolysis to fermentable sugars for further biofuel production (Bandivdekar and Moodbidri 2002).

The pharmacological activities of *Pongamia pinnata* included anti-plasmodial activity against *Plasmodium falciparum*, and anti-inflammatory activity against different phases (acute, sub acute and chronic) of inflammation (Ahmad et al. 2003). The anti-microbial effect of crude leaf extract of *P. pinnata* evaluates its effect on production and action of enterotoxins. Its extraction has no anti-bacterial, anti-giardial, and anti-rotaviral activities but reduces the production of cholera toxin and bacterial invasion to epithelial cells. *Pongamia pinnata* leaf extract exhibited circulatory lipid peroxidation and antioxidant activity. It has been evaluated in ammonium chloride-induced hyper ammonium rats. The methanol extract of *Pongamia pinnata* roots exhibited protection against aspirin and has a tendency to decrease acetic acid-induced ulcer after 10-days treatment. The oral administration of ethanol extract of *P. pinnata* flower exhibited anti-hyperglycaemic and anti-lipid peroxidative effect and also enhance antioxidant defense system in alloxan-induced diabetic rats (Ahmad et al. 2003).

2 A Statistical Tool for Seed Oil Extraction

The conventional extraction of oil is carried out using a change of one process variable with keeping all other constant. This method consumes more time, chemicals and man power. Nowadays, central composite design (CCD) was used for extraction of oil from the seed. The CCD was employed in the experimental design of extraction of oil. CCD was employed to study and optimize the effect of process variables in the various process. On single factor analysis, process variables and their ranges were selected and independent variables were coded at five levels between −1 and 1. The coding of the variables was done by the following equation

$$x_i = \frac{X_i - X_z}{\Delta X_i} \quad i = 1, 2, 3, \ldots k \qquad (34.1)$$

where x_i, is the dimensionless coded value of an independent variable; X_i, the real value of an independent variable; X_z, the real value of an independent variable at the center point; and ΔX_i, a step change of the real value of the variable i. The total number of experiments (N) was calculated by the following equation (Karmee and Chadha 2005)

$$N = 2^K + 2K + C_p \qquad (34.2)$$

where, K is the number of the process variable, 2^K is the number of factorial points, $2K$ is the number of the axial points (Karoshi and Hegde 2002). In this study, the experimental run was randomized in order to reduce the error arising from the experimental process due to the extraneous factors. A nonlinear regression method was used to fit the second order polynomial to the experimental data and express the mathematical relationship between process variables (Thirugananasambandham and Sivakumar 2015). The generalized form of the second order polynomial equation is shown below in Eq. 34.3.

$$Y = \beta_0 + \sum_{j=1}^{k} \beta_j x_j + \sum_{j=1}^{k} \beta_{jj} x_j^2 + \sum_{i}\sum_{<j=2}^{k} \beta_{ij} x_i x_j + e_i \qquad (34.3)$$

where Y is the response; x_i and x_j are variables (i and j range from 1 to k); β_0 is the model intercept coefficient; β_j, β_{jj} and β_{ij} are interaction coefficients of linear, quadratic and the second-order terms, respectively; k is the number of independent parameters ($k = 4$ in this study); and e_i is the error. The final mathematical second order polynomial model includes four linear terms, six two factor interaction terms, four squared terms and one intercept term.

2.1 Statistical Analysis

Design expert 8.0.7.1 statistical software package (Stat-Ease Inc., USA) was used to analyze the experimental data. Multiple regression analysis and Pareto analysis of variance (ANOVA) were used to evaluate the experimental data and the ANOVA table was generated. Significant terms in the model (linear, interactive and quadratic) for the response were found by analysis of variance (ANOVA) and significance was judged by the F-statistic value calculated from the data. The experimental data was evaluated with various descriptive statistical analysis such as p-value, F value, degrees of freedom (DF), sum of squares (SS), coefficient of variation (CV), determination coefficient (R^2), adjusted determination of coefficient $\left(R_a^2\right)$ and predicted determination of coefficient $\left(R_p^2\right)$ to reflect the statistical significance of the developed quadratic mathematical model. After fitting the data to the models, the model was used for the construction of three-dimensional response surface plots to predict the relationships between independent and dependent variables.

2.2 Determination of Optimal Conditions

After analyzing the polynomial equation depicting the dependent and independent variables, optimization process was carried out by Derringer's desired function methodology. This numerical optimization technique will optimize any combination of one or more goals; these may be either process variables or responses. The possible goals are: maximize, minimize, target, within range, none (for responses only) and set to an exact value (factors only). In this study, goals of the process variables were selected as in a range and the response goals were selected as maximize. A weight factor of 1 was chosen for the response, which can be used to adjust the shape of its particular desirability function. The default value of one creates a linear ramp function between the low value and the goal or the high value and the goal. Increased weight (up to 10) moves the result towards the goal and the reduced weight (down to 0.1) creates the opposite effect. Default importance of three was chosen for the response, which can represent the goals to be equally important.

2.3 Verification of the Predicted Optimized Conditions

After optimization, in order to determine the validity of the optimized conditions, triplicate verification experiments were performed under the optimal conditions as predicted by the model. The average value of the experiments was compared with the predicted values of the developed model in order to find out the accuracy and suitability of the optimized conditions.

3 Sonicated Extraction of Oil

3.1 Raw Sample and Chemicals

Pongamia pinnata seed was collected from Tamil Nadu, India. The seed was cleaned and soaked in petroleum ether for 4 h to separate impurities. The seeds were dried in an oven and crushed into a powder that sieved from three sieves of diverse mesh sizes. The powder from the sieve of 40 μm mesh size was used for oil extraction (Baswa et al. 2001).

3.2 Sonicated Extraction Process

Five grams of *Pongamia pinnata* seed was weighed into a flat-bottomed flask for oil extraction process. The vessels were kept inside the ultrasound device (Fig. 34.1) and sonicated for the defined time in various temperatures with methanol. After sonication, the sample was filtered through the filter paper (Salamatinia et al. 2010). The oil extract was collected and concentrated in the rotary evaporator to acquire *P. pinnata* seed oil. All experiments were performed in triplicate and average value is taken in to account. The extraction yield of the *Pongamia pinnata* seed oil was determined using the following formula (Singh et al. 2005)

$$\text{Oil yield}(\%) = W_1 / W_2 {}^{*}100 \tag{34.4}$$

Here, W_1 is the mass of *P. pinnata* seed oil extracted from the sample (g) and W_2 is the mass of total material processed (g) by standard extraction.

3.3 CCD Design Based Modeling

Sonicated extraction of oil from waste *P. pinnata* seed was examined using RSM coupled with CCD. This method is a particular set of mathematical and statistical methods for designing experiments, building models, evaluating the effects of process parameters, and searching optimum conditions of variables to predict the targeted response (oil yield). Based on preliminary experimental data, a CCD with three factors was examined to investigate and confirm the process parameters affecting the extraction of oil from *P. pinnata* seed. Three independent variables such as temperature, the mass of solvent and sonication time were studied at three levels, and response variable (oil yield (Z_1)) was used to determine the optimum conditions of extracting oil. The range and levels of independent variables are presented in Table 34.4. The statistical software namely Design-Expert 8.0 was used for

Fig. 34.1 Ultrasonic device

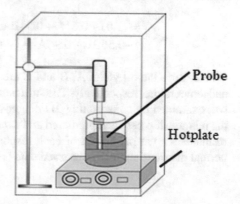

Table 34.4 Process variables and ranges

Process variable	−1	0	1
A	15	45	75
B	5	15	25
C	35	45	55

regression analysis and graphical analysis of the data obtained during the sonication experiment. The statistical significance of the second order polynomial model predicated was evaluated by the analysis of variance (ANOVA). The individual and interactive effects of process parameters on the oil yield was studied using three dimensional (3D) plots.

4 Results and Discussions

4.1 Mathematical Model Development

Three factors with three levels CCD was used to evaluate the effect of process variables on the response namely oil yield from *P. pinnata* seed using sonication. A total number of 20 extraction experiments including five center points were carried out in triplicates using statistically deigned experiments and the results are shown in Table 34.5.

Then the obtained CCD experimental data were analyzed by multi-regression analysis (Table 34.6) and results show that, linear and interactive (2FI) models are exhibited lower F- value and also having high p-values, when compared with the quadratic model (Prakash Maran et al. 2013). Cubic model is found to be aliased. Therefore, the quadratic model is chosen to explain the sonicated extraction of oil from *P. pinnata* seed.

A second-order polynomial model (quadratic) was developed to understand the interactive correlation between the response and the process variables. The final model obtained in terms of coded factors are given below,

$$Z_1 = 72.01 + 0.61A + 0.54B + 0.32C + 0.26AB + 0.41AC$$
$$-0.56BC - 0.96A^2 - 0.76B^2 - 0.36C^2 \tag{34.5}$$

Where, Z_1 is the oil yield. A, B and C are solvent to sample ratio, sonication time and temperature, respectively. The sufficiency of the developed mathematical model was examined by constructing 3D diagnostic plots. These plots help us to find out the relationship between predicted and experimental values, which will indicate the residuals for the prediction of each response is minimum. This also confirms the normal distribution of the observed data (Fig. 34.2).

Table 34.5 CCD design of experiments

S. No	A	B	C	Z1
1	45.0	15.0	45.0	72.0
2	15.0	25.0	35.0	70.0
3	75.0	25.0	35.0	70.9
4	75.0	5.0	35.0	69.1
5	15.0	25.0	55.0	68.0
6	45.0	15.0	28.2	70.4
7	45.0	15.0	45.0	72.0
8	45.0	15.0	45.0	72.0
9	15.0	5.0	35.0	68.3
10	15.0	5.0	55.0	69.5
11	−5.5	15.0	45.0	69.0
12	45.0	15.0	45.0	72.0
13	45.0	−1.8	45.0	68.6
14	75.0	25.0	55.0	71.5
15	45.0	15.0	45.0	72.0
16	45.0	15.0	61.8	72.0
17	45.0	15.0	45.0	72.0
18	45.0	31.8	45.0	71.5
19	95.5	15.0	45.0	70.0
20	75.0	5.0	55.0	71.0

Table 34.6 Multi regression analysis of oil yield

Source	Sum of squares	Df	Mean square	F value	Prob > F	Remarks
Sequential model sum of squares for Z_1						
Mean	88711.20	1.00	88711.20			
Linear	3789.56	3.00	1263.19	3.38	0.0442	
2FI	597.00	3.00	199.00	0.48	0.7010	
Quadratic	5021.14	3.00	1673.71	47.13	<0.0001	Suggested
Cubic	338.44	4.00	84.61	30.46	0.0004	Alised
Residual	16.66	6.00	2.78			
Total	98474.00	20.00	4923.70			

4.2 ANOVA Analysis

Statistical significance of the developed model equation was investigated using Pareto analysis of variance (ANOVA). The higher model F value and lower p-value ($p < 0.0001$) demonstrated that, the developed mathematical model was highly significant (Table 34.7). The goodness of fit of the equation was examined by the determination co-efficient (R^2), adj-R^2 and Pre-R^2. The high R^2 values revealed that, only small variations are not explained by the equation and it demonstrates a significant

Fig. 34.2 Predicted versus actual plot

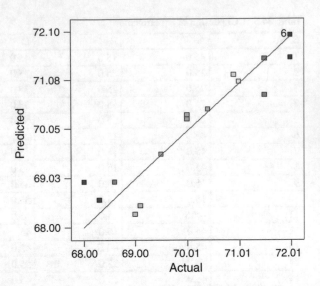

Table 34.7 ANOVA results for oil yield

| Source | Z_1 | | | | | |
	Sum of squares	Df	Mean square	F value	P value	Remarks
Model	35.36348	9	3.929276	12.86381	0.0002	Significant
A	5.144263	1	5.144263	16.84148	0.0021	
B	3.985037	1	3.985037	13.04636	0.0048	
C	1.411725	1	1.411725	4.621758	0.0571	
AB	0.55125	1	0.55125	1.804703	0.2088	
AC	1.36125	1	1.36125	4.456511	0.0609	
BC	2.53125	1	2.53125	8.2869	0.0164	
A^2	13.25602	1	13.25602	43.39806	<0.0001	
B^2	8.425612	1	8.425612	27.58408	0.0004	
C^2	1.847422	1	1.847422	6.048157	0.0337	
R2	0.9636					
Adj-R2	0.9309					
Pre-R2	0.9154					
CV	8.42					
AP	18.54					

and strong correlation between the observed and predicted oil yield. The adequate precision (AP) values of the equation compare the predicted data at the design points with the average prediction error (Srinivasan et al. 2001). In current study, adequate precision values were obtained as >4, indicating an appropriate signal, and recommended that the regression equation could be used to study the design space, forcefully. The coefficient of variance (CV) is the ratio of the standard error of the estimate to the mean value of the observed oil yield was a measure of reproducibility of the equation. CV was considered to be reproducible when it is not greater than 10% (Ansari et al. 2004). Value of CV for Z_1 shows that the response surface model developed was considered to be satisfactory. These ANOVA results confirmed that the second-order regression equation adequately represented the relationship between the oil yield and the variables (A, B and C).

4.3 *Effect of Extraction Variables on Oil Yield*

The surface response and contour plots of the quadratic model with one variable kept at the central level and the other two varying within the experimental ranges are shown in Fig. 34.3.

The solvents used for the extraction of this work is polar (methanol). Polarity is defined as the relative ability of the molecule to engage in the strong interactions with the other polar molecules. It was found that, as the solvent to sample ration increases, the extraction yield of oil increases (Naik et al. 2008) as shown in Fig. 34.3.

Sonication time is the key parameter in the extraction of oil from *Pongamia pinnata* (L.) seed using ultrasound-assisted extraction technique. In order to evaluate the effect of sonication time on the yield of oil, various sonication times were investigated. From the results, it was found that oil yield was increased with increasing sonication time up to 20 min. Where as, there is a drastic decrease in the yield of oil beyond 20 min. Temperature is one of the crucial parameters which influences the extraction of oil from *Pongamia pinnata* (L.) seed, considerably (Meera et al. 2003). In order to investigate the effect of the temperature on the yield of oil, experiments were carried out with the various temperatures and the results were shown in Fig. 34.3. From the results, it was clearly known that yield of oil was increased with increasing temperature up to 35 °C. Thereafter, the yield of oil was decreased. From the results, it was observed that all the individual and combined process variables shows the significant effect on the yield of oil from *Pongamia pinnata* (L.) seed.

4.4 *Optimization and Validation*

According to the numerical optimization step of the program, the desired goal for each independent variable (solvent to sample ratio, sonication time, temperature) and response (oil yield) should be chosen. For the optimization of oil yield from *P. pinnata* seed using ultrasound-assisted extraction technique, a cost-driven approach was preferred. Therefore, in the optimization procedure, the desired goal for the response of oil yield was chosen as "maximum", whereas the independent process variables were selected to be "within range. Optimum extraction conditions for the

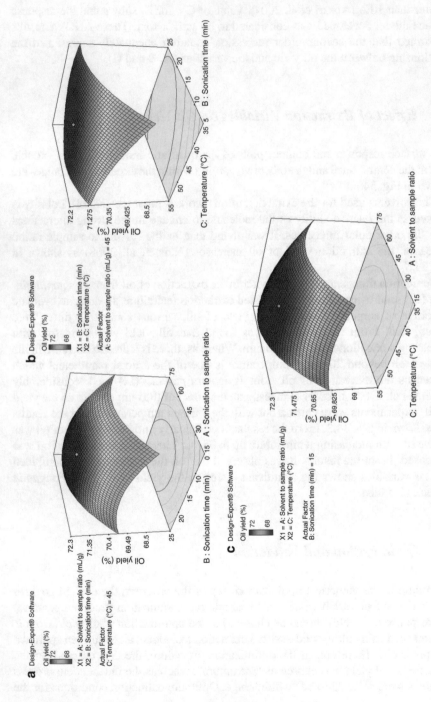

Fig. 34.3 (a) Effects of A and B on oil yield. (b) Effects of B and C on oil yield. (c) Effects of A and C on oil yield

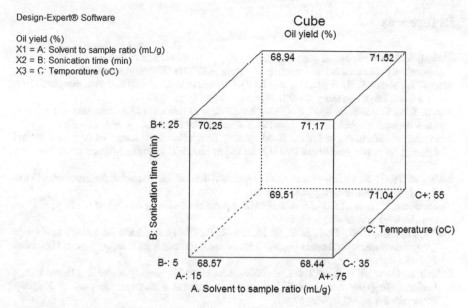

Fig. 34.4 Cube graph for oil yield

maximizing the extraction yield of oil (72%) was found to be: temperature of 35 °C, solvent type of 2 and time of 20 min with a desirability value of 0.996. The suitability of optimum conditions for predicting optimum oil extraction was tested under the same set of optimized conditions. Triplicate experiments were performed under the optimized conditions and the mean values (71.84%) obtained from real experiments, demonstrated the validity of the optimized conditions. The cube graph (Fig. 34.4) of present extraction process also confirms the optimized conditions.

5 Conclusion

Pongamia pinnata tree and its availability in India as well as in the world is discussed. CCD was employed to study and optimize the process variables such as a solvent to sample ratio, sonication time and temperature on the yield of oil from *P. pinnata* seed using ultrasound-assisted extraction process. The second order polynomial model was developed with high R^2 value. Interactive effects of process variables on response were carried out using 3D graphs. The optimum set of the process variables was obtained by Derringer's desired function methodology and it was found to be: solvent to sample ratio of 60, sonication time of 20 min and temperature of 40 °C. Under these optimal conditions, the extraction yield of oil was found to be 72% was closely agreed with the predicted value.

References

Ahmad, S., Ashraf, S. M., & Naqvi, F. (2003). A polyesteramide from *Pongamia glabra* oil for biologically safe anticorrosive coating. *Progress in Organic Coating, 47*, 95–102.

Ahmad, G., Yadav, P. P., & Maurya, R. (2004). Furanoflavonoid glycosides from *Pongamia pinnata* fruits. *Phytochemistry, 65*, 921–924.

Ansari, S. A., Singh, S., & Rani, A. (2004). Inorganic salts influence IAA ionization and adventitious rhizogenesis in *Pongamia pinnata*. *Journal of Plant Physiology, 161*, 117–120.

Azam, M. M., Waris, A., & Nahar, N. M. (2005). Prospects and potential of fatty acid methyl esters of some non-traditional seed oils for use as biodiesel in India. *Biomass and Bioenergy, 29*, 293–302.

Ballal, M. (2005). Screening of medicinal plants used in rural folk medicine for treatment of diarrhea. Internet: http://www.pharmoinfo.net.

Bandivdekar, A. H., & Moodbidri, S. B. (2002). Spermicidal activity of seed oil of *Pongamia glabra*. *Archives of Andrology, 48*, 9–13.

Baswa, M., Rath, C. C., Dash, S. K., & Mishra, R. K. (2001). Antibacterial activity of Karanja (*Pongamia pinnata*) and neem (*Azadirachta indica*) seed oil: A preliminary report. *Microbios, 105*, 183–189.

Brijesh, S., Daswani, P. G., & Tetali, P. (2006). Studies on *Pongamia pinnata* (L.) Pierre leaves: Understanding the mechanism(s) of action in infectious diarrhea. *Journal of Zhejiang University Science B, 7*, 665–674.

Carcache Blanco, E. J., Kang, Y. H., Park, E. J., Su, B. N., Kardono, L. B. S., Riswan, S., Fong, H. H. S., Pezzuto, J. M., & Kinghorn, A. D. (2003). Constituents of the stem bark of *Pongamia pinnata* with the potential to induce quinine reductase. *Journal of Natural Products, 66*, 1197–1202.

Chauhan, D., & Chauhan, J. S. (2002). Flavonoid glycosides from *Pongamia pinnata*. *Pharmaceutical Biology, 40*, 171–174.

Elanchezhiyan, M., Rajarajan, S., & Rajendran, P. (1993). Antiviral properties of the seed extract of an Indian medicinal plant, *Pongamia pinnata* Linn., against herpes simplex viruses: In-vitro studies on Vero cells. *Journal of Medical Microbiology, 38*, 262–264.

Essa, M. M., & Subramanian, P. (2006). *Pongamia pinnata* modulates the oxidant-antioxidant imbalance in ammonium chloride-induced hyperammonemic rats. *Fundamental and Clinical Pharmacology, 20*, 299–303.

Karmee, S. J., & Chadha, A. (2005). Preparation of biodiesel from crude oil of *Pongamia pinnata*. *Bioresource Technology, 96*, 1425–1429.

Karoshi, V. R., & Hegde, G. V. (2002). Vegetative propagation of *Pongamia pinnata* (L.) Pierre: Hitherto a neglected species. *Indian Forester, 128*, 348–350.

Meera, B., Kumar, S., & Kalidhar, S. B. (2003). A review of the chemistry and biological activity of *Pongamia pinnata*. *Journal of Medicinal and Aromatic Plant Science, 25*, 441–465.

Meher, L. C., Vidya, S. D., & Naik, S. N. (2006). Optimization of alkali-catalyzed transesterification of *Pongamia pinnata* oil for production of biodiesel. *Bioresource Technology, 97*, 1392–1397.

Muthu, C., Ayyanar, M., & Raja, N. (2006). Medicinal plants used by traditional healers in Kancheepuram district of Tamil Nadu, India. *Journal of Ethnobiology and Ethnomedicine, 2*, 43–53.

Naik, M., Meher, L. C., Naik, S. N., & Dasa, L. M. (2008). Production of biodiesel from high free fatty acid Karanja (*Pongamia pinnata*) oil. *Biomass and Bioenergy, 32*, 354–357.

Prabha, T., Dora, M., & Priyambada, S. (2003). Evaluation of *Pongamia pinnata* root extract on gastric ulcers and mucosal offensive and defensive factors in rats. *Indian Journal of Experimental Biology, 41*, 304–310.

Prakash Maran, J., Sivakumar, V., Sridhar, R., & Thiruganasambandham, K. (2013). Artificial neural network and response surface methodology modeling in mass transfer parameters predictions during osmotic dehydration of *Carcia papaya* L. *Alexandria Engineering Journal, 52*, 507–516.

Ran, W., Hai, N., Lv, Y., & Jiang, P. (2018). Anti-inflammatory flavone and chalcone derivatives from the roots of *Pongamia pinnata* (L.) Pierre. *Phytochemistry, 149*, 56–63.

Sahu, B., Sahu, A. K., Thomas, V., & Naithani, S. C. (2017). Reactive oxygen species, lipid peroxidation, protein oxidation and antioxidative enzymes in dehydrating Karanj (*Pongamia pinnata*) seeds during storage. *South African Journal of Botany, 112*, 383–390.

Salamatinia, B., Mootabadi, H., Bhatia, S., & Abdullah, A. Z. (2010). Optimization of ultrasonic-assisted heterogeneous biodiesel production from palm oil: A response surface methodology approach. *Fuel Processing Technology, 91*, 441–448.

Singh, K. P., Dhakre, G., & Chauhan, S. V. S. (2005). Effect of mechanical and chemical treatments on seed germination in *Pongamia glabra* L. *Seed Research, 33*, 169–171.

Srinivasan, K., Muruganandan, S., Lal, J., Chandra, S., Tandan, S. K., & Raviprakash, V. (2001). Evaluation of anti-inflammatory activity of *Pongamia pinnata* leaves in rats. *Journal of Ethnopharmacology, 78*, 151–157.

Thirugananasambandham, K., & Sivakumar, V. (2015). Eco-friendly approach of copper (II) ion adsorption on to cotton seed cake and its characterization: Simulation and validation. *Journal of the Taiwan Institute of Chemical Engineers, 50*, 198–204.

Chapter 35
Cranberry Seed Oil

Naveed Ahmad, Farooq Anwar, and Ali Abbas

Abstract Cranberry is a healthy fruit from an evergreen shrub and offers multiple nutritional and nutraceutical benefits. The seeds, separated from the fruit, are a potential source of edible oil, so-called cranberry seed oil. Cranberry seed oil, due to its favorable physicochemical characteristic, good oxidative stability and the pleasant flavor is valued as an edible oil. The cranberry seed oil can be regarded as a good quality functional oil due to the existence of a distinct combination/profile of *omega*-3, *omega*-6 and *omega*-9 fatty acids, as well as tocopherols, tocotrienols and high concentration of antioxidants. Interestingly, cold pressed method is reported to be a better option to extract oil from cranberry seed with higher oxidative stability and lesser nutrient loss. Cranberry seed oil is characterized by higher contents of polyunsaturates such as α-linolenic acid (30–35%) and linoleic acid (35–40%) followed by oleic acid (20–25%), which play vital role in lowering serum cholesterol and thus help to prevent heart related problems, hypertension, certain types of cancer and autoimmune diseases. In addition, cranberry and its seed oil also contain appreciable amounts of tannins, anthocyanins, flavonoids and phenolic acids, which are not only associated to organoleptic properties but also exhibit multiple biological attributes such as antioxidant, anticancer, and anti-inflammatory activities. Based on the profile of high-value compounds and nutrients, cranberry seed oil can be explored as a promising candidate for nutra-pharmaceutical and cosmeceutical applications.

Keywords Ericaceae · Vaccinium · Vaccinium oxycoccos · Anthocyanins · Tannins · Seed oil

N. Ahmad
Department of Chemistry, University of Education, Faisalabad, Pakistan

F. Anwar (✉)
Department of Chemistry, University of Sargodha, Sargodha, Pakistan
e-mail: farooq.anwar@uos.edu.pk

A. Abbas
Department of Chemistry, Government Postgraduate Taleem-ul-Islam College,
Chenab Nagar, Chiniot, Pakistan

© Springer Nature Switzerland AG 2019
M. F. Ramadan (ed.), *Fruit Oils: Chemistry and Functionality*,
https://doi.org/10.1007/978-3-030-12473-1_35

1 Introduction

Cranberry, belonging to the family *Ericaceae,* is an evergreen shrub native to America. It is one of the major agricultural fruit crops cultivated in different states of USA such as Massachusetts, Michigan, New Jersey and Washington (NASS 2001). Cranberry is a short shrub (can rise up to 2–3 m) and mostly grows in sandy, organic, and acidic soils. The flowers of this shrub are of dark pink color and possess distinct petals rendering the flowers liable for pollination by insects (NASS 2001). Cranberry, a fruit derived from cranberry shrub, is popular due to its unique taste/flavor and high nutritional importance. The fruits of cranberry are relatively larger in size compared to the leaves which are white at the initial stage and turn into deep red at the fully ripened stage. Cranberry fruits possess slightly tart and acidic taste, which may overcome its sweetness. Cranberry is valued as a healthy fruit, which contributes nutritional value, flavor, color and functionality to a diverse class of foodstuffs (Thyagarajan 2012). The small red cranberry fruits contain a variety of potent bioactives with the potential to be used as functional ingredients (Porter et al. 2001; Yan et al. 2002; Cunningham et al. 2004). Cranberry has attracted much attention of food chemists and food biotechnologists due to their positive health impacts (Marwan and Nagel 1986; Chen et al. 2001). The cranberry fruits also contain a considerable amount of seeds, which have the potential for oil extraction (Thyagarajan 2012). This chapter is mainly designed to highlight composition and food uses of cranberry seed oil.

2 Extraction and Processing of Cranberry Fruit Seed Oil

Currently, there is great focus on exploration of non-conventional seed oils; especially those derived from different fruit seeds due to their rich profile of minor bioactives and polyunsaturated fatty acids (Liangli et al. 2005; Parry et al. 2005; Anwar et al. 2008, Anwar et al. 2014a, b; Yukui et al. 2009; Manzoor et al. 2012; Ahmad et al. 2014). Due to the limited supply of fruit seeds, such non-conventional oils are supposed to be available in a limited amount; nevertheless, there is prompt need to search potential sources with sustainable and adequate supply of feedstock materials prior to introduce these oils into the market. In order to collect adequate supply for the recovery of oil and other high-value phytochemicals, a large number of fruits have to be processed at centralized facilities where a huge quantity of waste product can be generated/available (Gunstone and Harwood 2007). Due to the richness in important bioactives, fruit seeds oils can be recommended for their potential uses in cosmeceutical and nutraceutical industries or as dietary supplements, if the seeds are treated, handled and preserved under appropriate conditions and the related oils are extracted using green extraction processes as well.

With the increased production and consumption of cranberries, worldwide, coupled with a substantial proportion of the cranberries being processed, a large amount of cranberry seeds is being produced as a by-product, which can be used for oil extraction. Currently, cranberry seed oil is gaining increasing interest of lipid chemists and food industry due to containing a well-balanced fatty acids composition along with a rich profile of high-value phytochemicals and dietary components of medicinal value (Nawar 2004; Liangli et al. 2005). The potential exploitation of cranberry seed oil for commercial applications depends on its availability and quality. In order to recover and extract vegetable oils from plant-based materials (especially the seeds), extraction is one of the key steps (Liu 1999; Shahidi and Wanasundara 2002). In this context a number of extraction techniques such as solvent, ultrasonic-assisted, super-critical fluid, Soxhlet and mechanical pressing are in practice for the extraction of oils (Martinez and Maestri 2008; Sheibani and Ghaziaskar 2008; Lee and Lin 2007). Soxhlet extraction can be employed for obtaining optimum oil yield from seeds, however, high-grade food quality cranberry seed oil is preferably produced by cold pressing (Van Hoed et al. 2009; Thyagarajan 2012).

3 Fatty Acids and Tocols Composition of Cranberry Oil

Table 35.1 depicts the fatty acid and tocols (tocopherols, and tocotrienols) composition of cranberry seed oil. Most of the common vegetable oils contain three fatty acids including palmitic (C16:0, saturated), oleic (C18:1, monounsaturated), and linoleic (C18:2, polyunsaturated) at levels exceeding than 10% on individual basis so that these three acids mostly have a pooled level of 90% or more. This indicated that other acids, especially stearic and linolenic acids are normally present at lower levels. Generally, edible oils can be subdivided into three groups namely high- oleic and high-linoleic oils and those in which both of these acids occur equally at higher levels/extent. Palmitic acid, yet normally present, is rarely the principal component, except palm oil (Gunstone and Harwood 2007). On the other hand, flaxseed oil is exceptionally a rich source of linolenic acid with its content usually greater than 50% (Anwar et al. 2013; Yaqoob et al. 2016).

The specialty oils are sold as high-value oils because of their special/distinct fatty acids and nutrients profile. As far as cranberry seed oil is concerned, it is marketed as a healthy oil for its balanced *omega*-3, *omega*-6 and *omega*-9 fatty acids ratio along with high contents of tocopherols and tocotrienols, together (collectively) known as tocols. A good quality vegetable oil, in its purified form, is reasonably resistant to oxidative stress and effects of light and heat and thus retains its characteristic organoleptic flavor and nutritional value. Depending on the market applications and customer acceptance, natural and/or synthetic preservatives (antioxidant) are added into the oils to protect these from oxidation and thus further enhancing their storage ability/shelf-life (Gunstone and Harwood 2007). It is considered that seed oils with a high amount of antioxidant/natural tocopherols offer

Table 35.1 Phytochemicals and antioxidant potential of cranberry seed oil

Fatty acid composition (g/100 g of fatty acid)	C12:0	ND	ND
	C14:0	ND	ND
	C16:0	7.83 ± 0.06	5–6
	C16:1	ND	ND
	C18:0	1.91 ± 0.01	1–2
	C18:1	22.69 ± 0.02	20–25
	C18:2	44.31 ± 0.03	35–40
	C18:3	22.28 ± 0.01	30–35
	C20:0	ND	0.1
	C20:1	ND	ND
	C20:2	0.98 ± 0.00	ND
	C20:5	ND	1.1
	Sat	9.74 ± 0.06	6.1–8.1
	Mono	22.69 ± 0.02	20–25
	n-6	45.29 ± 0.03	35–40
	n-3	22.28 ± 0.01	31.1–36.1
	n-6/n-3 ratio	ND	1–1.3
	PUFAs	67.57 ± 0.04	66.1–76.1
	Parker et al. (2003)		Liangli et al. (2005)
Tocopherols (g/100 g)	α-tocopherol	0.20	0.048
	β-tocopherol	ND	ND
	γ-tocopherol	0.10	0.091
Tocotrienols (g/100 g)	α-tocotrienol	ND	0.157
	β-tocotrienol	ND	ND
	γ-tocotrienol	1.50	1.235
	Stone and Papas (2003) and Yang (2003)		Van Hoed et al. (2009)
Total Phenolics	TPC (mg GAE/100 g)	161	Yu et al. (2005)
Antioxidant potential	ABTS$^{·+}$ (μmol TE/g)	22.5	
Phytosterols (g/kg)	Campesterol	0.29	Van Hoed et al. (2009)
	Stigmasterol	0.09	
	β-Sitosterol	4.18	
	Δ7-Sitosterol	2.19	
	Δ7-Avenasterol	0.17	

ND Not Determined

greater protective role towards reducing the incidence/occurrence of various oxidative stress-related diseases including cardiovascular, aging, inflammation and certain cancers (Wang and Jiao 2000; Van Hoed et al. 2009). Interestingly, a high oxidative stability and longer shelf life of cranberry seed oil is one of its unique features, as this oil does not deteriorate easily when compared with other vegetable oils such as virgin olive, walnut, and pistachio and hazelnut (Nawar 2004).

It has been noted that cranberry seed oil has a high concentration of linolenic acid (*omega*-3 fatty acid). Linolenic acid has significant potential to serve as an important ingredient in nutraceuticals and is linked with the reduced incidence of coronary heart disease and certain cancer. The ratio of polyunsaturated fatty acids (PUFAs): saturated fatty acids (SFAs) is reported to be very high in cranberry seed oil which is valuable in lowering serum cholesterol and atherosclerosis, and thus helps to prevent heart-related disorders (Heeg et al. 2002). The oils, recovered from almost all types of berries seed, are found to contain a high content of PUFAs. Among PUFAs, linoleic and linolenic acids are recognized as essential fatty acids (EFAs). It is known that these EFAs cannot be produced by any components in an organism by any reported chemical pathway thus must be attained from the diet. The intake of PUFAs including n-6 and n-3 fatty acids have been documented to provide protection against various ailments such as heart diseases, cancer, hypertension, and autoimmune disorders (Connor 2000; Aronson et al. 2001).

According to the literature sources, the oil extracted from cranberry seed contain α-linolenic acid (30–35%), linoleic acid (35–40%) and oleic acid (20–25%) by weight (Liangli et al. 2005), exhibit pleasant flavor, and high oxidative stability. Oils from soybean, fish and canola are almost neutral in flavor and contain a relatively lesser amount of tocotrienols and lack pleasant flavor as compared with cranberry seed oil. Nevertheless, these oils have *omega*-3/n-3 fatty acids; however, the cranberry seed oil has a diverse and unique combination of *omega*-9, *omega*-6 and *omega*-3 fatty acids. The oils, containing an appreciable amount of EFAs, have been documented to play a vital role in the improvement of health and reducing the incidence of certain diseases (Nawar 2004).

The western diet contains a ratio of n-6 and n-3 EFAs lie between 10 and 25 to 1, however, the endorsed ratio is projected to be as 4 to 1 (Parry et al. 2005; Bawa 2008). This imbalance in the ration of n-6 to n-3 PUFAS is attributed to the reduced intake of fish oil and increased utilization of vegetable and seed oil. The oils, derived from diverse berries seeds, exhibited impressive/exciting ratios of n-6/n-3 essential fatty acid in comparison to conventional vegetable/fixed oils (Parker et al. 2003; Parry et al. 2006).

4 Minor Bioactive Components in Cranberry Oil

Due to multiple health benefits of potent components, the seeds and vegetable oils have gained much interest in food chemists/biotechnologists (Shahidi 2000). In this context, different studies have been conducted on the investigation of minor bioactives profiling of various conventional and non-conventional oils (Van Hoed et al. 2009). Cranberries and different products, derived from cranberry, have been reported to contain an appreciable amount of antioxidant nutrients and polyphenolic compounds of medicinal value (Vinson et al. 2008). Cranberry has been noted to contain a wide range of potent phenolic acid derivatives including hydroxycinnamic acid, and hydroxybenzoic acid derivatives (Neto 2007; Ruel and Couillard 2007). Most of these phenolic acids have been found in free forms while others occur in the bound form (Seeram and Heber 2007). Likewise, other fruit seed oils, cranberry seed oil possess a significantly higher amount of tocols, tannins, anthocyanins and flavonoid compounds (Neto 2007; Ruel and Couillard 2007).

Berry seed oils are extensively used as important components/ingredients in cosmeceutical and aromatherapy applications due to their high contents of EFAs and antioxidants (Parry et al. 2005; Bushman et al. 2004; Yu et al. 2005). Berry oils, especially cranberry seed oil (Van Hoed et al. 2009; Oomaha et al. 2000; Parry et al. 2005) has remarkable performance and stability standards in comparison with oils high in EFAs. Berry fruits and fruit seed oils are a rich source of phenolic compounds (such as phenolic acids) as well as anthocyanins, proanthocyanidins, and flavonoids, which display potential health-promoting effects (Hakkinen et al. 1999; Amakura et al. 2000). According to Yu et al. (2005), total phenolic contents in cranberry seed oil were found to be 161 mg GAE/100 g (Table 35.1). The unique bioactives profile and antioxidant properties offered protective effects against multiple diseases (Neto 2007). Phenolic acids and flavonoid compounds in various plant sources and seed oils serve as a viable source of natural antioxidants with the capacity to hinder the lipid oxidation by interrupting free radical chain reactions (Gelmez 2008). There has been evidence supporting positive linkages between consumption/intake of phenolics-rich fruits and prevention of aging-related disorders caused by oxidative stress. Interestingly, fruits that contain physiologically active components (in particular phenolics) have significant efficacy and potential to act as natural antioxidants and therapeutics agent (Vinson et al. 2001; Ahmad et al. 2016). Cranberry fruit and its seed oil has been reported as a valuable source of high-value bioactives with remarkable antioxidant and pharmaceutical potential which is comparable to those of other berries (Amakura et al. 2000).

Studies revealed that different fruit seed oils contain several valuable cosmetically important and effective bioactives including EFAs, natural antioxidants, tocopherols and phytosterols (Parry and Yu 2004). Likewise, berry fruit seed oil is a rich source of valuable nutrients, functional bioactives and cosmo-nutraceuticals. In addition to other bioactives, Van Hoed et al. (2009) reported the presence of campesterol (0.29 g/kg), stigmasterol (0.09 g/kg), β-sitosterol (4.18 g/kg), Δ7-sitosterol (2.19 g/kg) and Δ7-avenasterol (0.17 g/kg) in cranberry seed oil.

An insight into the process of lipid oxidation and study of the influence of different antioxidant enzymes on human health have got considerable attention of researchers because of their critical role in the pathogenesis of different chronic ailments (McCall and Frei 1999) like cancer, arteriosclerosis and allergy (Wittenstein et al. 2002). Antioxidant enzymes including superoxide, dismutases and catalase, and repair enzymes (DNA glycosylases) along with vitamins such as vitamin A (retinol), vitamin E (tocopherol), and vitamin C (ascorbic acid) have been claimed to provide protection to humans and other biological organisms against deteriorative effects of reactive oxygen species (McCall and Frei 1999). Nevertheless, the presence of these important vitamins (in particular vitamin A and E) in the human diet is essential for boosting defense and immune mechanisms. These nutrients play an effective role to scavenge free radicals and thus function as antioxidant to protect against neuromuscular disorders by sustaining proper life time of red blood cells (Morris et al. 1994; Amir et al. 2009). Tocopherols have been documented to provide protection against oxidative stress, caused by free radicals both in biological matrices and lipoproteins (Stone and Papas 2003). Many studies have shown that natural combinations of tocopherols have significant potential for inhibition of aggregation of platelet in humans, slow down the lipid oxidation (Liu et al. 2003).

Vitamin E is a term that is in use for collective tocopherol compounds (Hong et al. 2009). Tocopherols compounds (α, β, γ and δ isomers) have been reported to be naturally found in various vegetable oils, in varying amounts ranging from few mg/kg of oil to 1000 mg/kg of oil, however their actual content is decreased during processing of oils (Rossell 1991; Ruperez et al. 2001). In order to enhance shelf-life, refined, bleached and deodorized vegetable oils are spiked with standardized levels/concentration of antioxidants, especially vitamin E in the form of tocopherol acetate (Rossell 1991). Synthetic antioxidants (i.e. butylated hydroxyanisole, butylated hydroxytoluene and propyl gallate) have been employed by the food and pharma industry over the decades but recently, due to safety concerns of synthetic additives, there is growing interest in the use of natural antioxidants as safer alternatives (Sultana et al. 2007; Sultana et al. 2008; Anwar et al. 2014).

The oil, extracted from cranberry seeds, has been reported to possess a considerable amount of tocopherol and tocotrienol, which is comparable to those of other berries such as blackberry, blueberry, and strawberry, however, it is rather lower than that of red raspberry. It was also identified that cranberry seed oil contains maximum value of γ-tocotrienol as well as α- and γ-tocopherols when compared to those of other berries including blueberry, cranberry and kiwi fruit seed oils (Van Hoed et al. 2009).

5 Functional Food and Nutraceutical Perspectives and Health Benefits of Cranberry Seed Oil

The selected phytochemicals so-called secondary metabolites (especially phenolics and flavonoids) have been reported to be strongly associated with the organoleptic properties of fruits. Moreover, these bioactives exhibit multiple biological attributes and take part in various physiological processes (Robards and Antolovich 1997; Kren and Martinkove 2001). Epidemiological studies demonstrated an inverse correlation between intake of phenol-rich fruits and incidence of certain ailments including cancer, cardiovascular diseases, inflammation and aging-related diseases (Willet 2001).

Cranberries have been known as one of the earliest functional foods that have been used for various food and therapeutic purposes. The cranberry seed oil is known for its unique combination of different *omega* fatty acids and other high-value components such as tocopherols and phenolics, which support potential uses of this oil in nutra-pharmaceutical and cosmeceutical industry. Cranberry and cranberry seed oil has been utilized for the treatment of urinary disorders, wounds, diarrhea, blood poisoning and diabetes. The cranberry seed oil and cranberry fruit have long been used by long ocean voyagers to remain unaffected from scurvy (Henig and Leahy 2000). Cranberries fruit and seed oil have been found to contain biologically active components with potential anticancer, antioxidant and anti-inflammatory activities. The biological attributes of cranberry products can be attributed to the presence of phenolic acids, flavonoid and proanthocyanidin (Shahidi and Weerasinghe 2004).

Cranberry has been found to protect against infectious diseases, including dental decay, stomach ulcer problems and urinary tract infections, caused by pathogenic bacteria and molds (Heinonen 2007). A number of in vitro studies have shown that the phenolic compounds, in berry fruits and fruit seed oil, have significant potential to act as antioxidant, free radical scavenger, metal chelator and also exhibited multiple biological activities including antimicrobial, antiviral, anticarcinogenic, anti-inflammatory and antiproliferative (Nijveldt et al. 2001; Seeram and Heber 2007). This supports that plant-based phenolic compounds are medicinally effective against various disorders such as arthritis, inflammation, allergies and hypertension (Robards and Antolovich 1997; Merken and Beecher 2000).

The presence of ellagic acid and other phenolic acids in cranberry fruit and seed oil has been linked to exhibit different biological properties such as anticancer, anti-mutagenic, anticarcinogenic, antioxidant, and anti-viral (Daniel et al. 1989). Another potent phytochemical, named as resveratrol, which is present in cranberry, has been reported to offer benefits towards maintaining cardiovascular health (Wang et al. 2002). Likewise, *omega*-3 fatty acid and tocotrienols present in cranberry fruit seed oil are reported to exhibit an equivalent beneficial effect as ellagic acid and resveratrol (Shahidi and Weerasinghe 2004).

Cranberry and its derived products have been found to impart various health benefits to humans. Over the decades, cranberry juice extensively employed as a folk remedy for the treatment of different ailments like gastrointestinal disorders and urinary tract infections (Yan et al. 2002). A positive correlation has been established between the consumption of cranberry juices and the reduced incidence of urinary tract infections. Like several other colored fruits, phenolics (phenolic acids, anthocyanins and flavonoids) have also been naturally distributed in cranberry (Vvedenskaya et al. 2004). *p*-Hydroxybenzoic acid, present in cranberry fruit and in the seed oil, has been believed to act as potent bioactive to prevent and cure urinary tract infections (Marwan and Nagel 1986).

Growing evidence has suggested that PUFAs including *omega*-3 (n-3) fatty acids, present in appreciable amount in cranberry seed oil, have potential health benefits to fight against different ailments including hypertension, cardiovascular diseases, cancers and autoimmune disorders (Harel et al. 2002; Tapiero et al. 2002; Villa et al. 2002). The information in this chapter can be valuable for the cranberry growers, consumers, functional food and nutra-pharmaceutical sector.

References

Ahmad, N., Anwar, F., Mahmood, Z., Shahid, A. A., Shakir, I., & Latif, S. (2014). Variation of physico-chemical attributes of seed oil between two developing cultivars of sesame grown under similar agroclimatic conditions in Pakistan. *Asian Journal of Chemistry, 26*(14), 4319–4322.

Ahmad, N., Zuo, Y., Lu, X., Anwar, F., & Hameed, S. (2016). Characterization of free and conjugated phenolic compounds in fruits of selected wild plants. *Food Chemistry, 190*, 80–89.

Amakura, Y., Umino, Y., Tsuji, S., & Tonogai, Y. (2000). Influence of jam processing on the radical scavenging activity and phenolic content in berries. *Journal of Agricultural and Food Chemistry, 48*, 6292–6297.

Amir, W., Rishi, L., Yaqoob, M., & Nabi, A. (2009). Flow- injection determination of retinol and tocopherol in pharmaceuticals with acidic potassium permanganate chemiluminescence. The Japan Society for Analytical Chemistry. *Analytical Sciences, 25*, 407–412.

Anwar, F., Naseer, R., Bhanger, M. I., Ashraf, S., Talpur, F. N., & Aladedunye, F. A. (2008). Physico-chemical characteristics of citrus seeds and seed oil from Pakistan. *Journal of the American Oil Chemists' Society, 85*, 321–333.

Anwar, F., Zreen, Z., Sultana, B., & Jamil, A. (2013). Enzyme-aided cold pressing of flaxseed (*Linum usitatissimum* L.): Enhancement in yield, quality and phenolics of oil. *Grasas Y Aceites, 64*(5), 463–471.

Anwar, F., Mahmood, T., Mehmood, T., & Aladedunye, F. (2014a). Composition of fatty acids and tocopherols in cherry and lychee seed oil. *Journal of Advances in Biology, 5*(1), 586–593.

Anwar, F., Rashid, U., Shahid, S. A., & Nadeem, M. (2014b). Physicochemical and antioxidant characteristics of Kapok (*Ceiba pentandra* Gaertn.) seed oil. *Journal of the American Oil Chemists' Society, 91*(6), 1047–1054.

Aronson, W. J., Glaspy, J. A., Reddy, S. T., Reese, D., Heber, D., & Bagga, D. (2001). Modulation of omega-3/omega-6 polyunsaturated ratios with dietary fish oils in men with prostate cancer. *Urology, 58*, 283–288.

Bawa, S. (2008). The role of omega-3 fatty acids in the prevention and management of depression-Part 1. *Agro Food Industry Hi-Tech, 19*, 70–73.

Bushman, B. S., Phillips, B., Isbell, T., Ou, B., Crane, J. M., & Knapp, S. J. (2004). Chemical composition of cranberry (*Rubus* spp.) seeds and oils and their antioxidant potential. *Journal of Agricultural and Food Chemistry, 52*, 7982–7987.

Chen, H., Zuo, Y., & Deng, Y. (2001). Separation and determination of flavonoids and other phenolic compounds in cranberry juice by high-performance liquid chromatography. *Journal of Chromatography, 913*(1–2), 387–395.

Connor, W. E. (2000). Importance of n-3 fatty acids in health and disease. *The American Journal of Clinical Nutrition, 71*(Suppl. 1), 171–175.

Cunningham, D. G., Vannozzi, S. A., Turk, R., Roderick, R., O'Shea, E., & Brilliant, K. (2004). Cranberry phytochemicals and their health benefits. In F. Shahidi & D. K. Weerasinghe (Eds.), *Utraceutical beverages-chemistry, nutrition and health effects*. Washington, DC: American Chemical Society.

Daniel, E. M., Krupnick, A., Heur, Y. H., Blinzler, J. A., Nims, R. W., & Stoner, G. D. (1989). Extraction, stability and quantification of ellagic acid in various fruits and nuts. *Journal Food Composition and Analysis, 2*, 338–349.

Gelmez, N. (2008). Ultrasound assisted and supercritical carbon dioxide extraction of antioxidants from roasted wheat germ. A M.Sc. thesis. Graduate school of natural and applied sciences. Middle East Technical University, Turkey.

Gunstone, F. D., & Harwood, J. L. (2007). Occurrence and characterization of Oils and Fats. In *The lipid handbook with CD-ROM* (3rd ed.). New York: Taylor and Francis Group, LLC.

Hakkinen, S., Heinonen, M., Karenlampi, S., Mykkanen, H., Ruuskanen, J., & Torronen, R. (1999). Screening of selected flavonoids and phenolic acids in 19 berries. *Food Research International, 32*, 345–353.

Harel, Z., Gascon, G., Riggs, S., Vaz, R., Brown, W., & Exil, G. (2002). Supplementation with omega-3 polyunsaturated fatty acids in the management of recurrent migraines in adolescents. *The Journal of Adolescent Health, 31*(2), 154–161.

Heeg, T., Lager, H., & Bernard, G. (2002). Cranberry seed oil, cranberry seed flour and a method for making. US patent, 6 391 345.

Heinonen, M. (2007). Antioxidant activity and antimicrobial effect of berry phenolics-a Finnish perspective. *Molecular Nutrition & Food Research, 51*, 684–691.

Henig, Y. S., & Leahy, M. M. (2000). Cranberry juice and urinary-tract health: Science supports folklore. *Nutrition, 16*, 684–687.

Hong, J. L., Jihyeung, J., Shiby, P., Jae, Y. S., Andrew, D. C., Amanda, S., Mao, J. L., Chung, S. Y., Harold, L. N., & Nanjoo, S. (2009). Mixed Tocopherols prevent mammary tumorigenesis by inhibiting estrogen action and activating PPAR-γ. *Clinical Cancer Research, 15*, 4242.

Kren, V., & Martinkove, L. (2001). Glycosides in medicine: The role of glycosidic residue in biological activity. *Current Medicinal Chemistry, 8*, 1303–1328.

Lee, M. H., & Lin, C. C. (2007). Comparison of techniques for extraction of isoflavones from the root of Radix Puerariae: Ultrasonic and pressurized solvent extractions. *Food Chemistry, 105*, 223–228.

Liangli, L. Y., Zhou, K. K., & Parry, J. (2005). Antioxidant properties of cold-pressed black caraway, carrot, cranberry and hemp seed oils. *Food Chemistry, 91*, 723–729.

Liu, K. (1999). *Soybean: Chemistry, technology, and utilization*. New York: Aspen Publishers.

Liu, M., Wallmon, A., Olsson-Mortlock, C., Wallin, R., & Saldeen, T. (2003). Mixed tocopherols inhibit platelet aggregation in humans: Potential mechanisms. *The American Journal of Clinical Nutrition, 77*, 700–706.

Manzoor, M., Anwar, F., Ashraf, M., & Alkharfy, K. M. (2012). Physico-chemical characteristics of seed oils extracted from different apricot (*Prunus armeniaca* L.) varieties from Pakistan. *Grasas y Aceites, 63*(2), 193–201.

Martinez, M. L., & Maestri, D. M. (2008). Pressing and supercritical carbon dioxide extraction of walnut oil. *Journal of Food Engineering, 88*(3), 399–404.

Marwan, A. G., & Nagel, C. W. (1986). Characterization of cranberry benzoates and their antimicrobial properties. *Journal of Food Science, 51*, 1069–1070.

McCall, M. R., & Frei, B. (1999). Can antioxidant vitamins materially reduce oxidative damage in humans? *Free Radical Biology & Medicine, 26*, 1034–1053.

Merken, H. M., & Beecher, G. R. (2000). Measurement of food flavonoids by high-performance liquid chromatography. *Journal of Agricultural and Food Chemistry, 48*, 577–599.

Morris, D. L., Kritchevsky, S. B., & Davis, C. E. (1994). Serum carotenoids and coronary heart disease. *Journal of the American Medical Association, 272*, 1439–1441.

National Agricultural Statistics Service (NASS). (2001). *Agricultural statistics board. Cranberries* (Annual reports, Fr Nt 4). Washington, DC: USDA.

Nawar, W. (2004). Cranberry seed oil extract and compositions containing components thereof. US Patent, 0 258 734 A1.

Neto, C. C. (2007). Cranberry and its phytochemicals: A review of in vitro anticancer studies. *The Journal of Nutrition, 137*, 186–193.

Nijveldt, R. J., Van Nood, E., Van Hoorn, D. E. C., Boelens, P. G., Van Norren, K., & Van Leeuwen, P. A. M. (2001). Flavonoids: A review of probable mechanisms of action and potential applications. *The American Journal of Clinical Nutrition, 7*, 418–425.

Oomaha, B. D., Ladetb, S., Godfreya, D. V., Liangc, J., & Girarda, B. (2000). Characteristics of raspberry (*Rubus idaeus* L.) seed oil. *Food Chemistry, 69*, 187–193.

Parker, T. D., Adams, D. A., Zhou, K., Harris, M., & Yu, L. (2003). Fatty acid composition and oxidative stability of cold-pressed edible seed oils. *Journal of Food Science, 68*(4), 1240–1243.

Parry, J., & Yu, L. (2004). Fatty acid content and antioxidant properties of cold-pressed black Raspberry seed oil and meal. *Journal of Food Science, 69*, FCT189–FCT193.

Parry, J., Su, L., Luther, M., Zhou, K., Yurawecz, M. P., Whittaker, P., & Yo, L. (2005). Fatty acid composition and antioxidant properties of cold-pressed marionberry, boysenberry, red raspberry and blueberry seed oils. *Journal of Agricultural and Food Chemistry, 53*, 566–573.

Parry, J., Su, L., Moore, J., Cheng, Z., & Luther, M. (2006). Chemical compositions, antioxidant capacities and antiproliferative activities of selected fruit seed flours. *Journal of Agricultural and Food Chemistry, 54*, 3773–3778.

Porter, M. L., Kruger, C. G., Wiebe, D. A., Cunningham, D. G., & Reed, J. D. (2001). Cranberry proanthocyanidins associate with low-density lipoprotein and inhibit in vitro Cu 2+induced oxidation. *Journal of the Science of Food and Agriculture, 81*, 1306–1313.

Robards, K., & Antolovich, M. (1997). Analytical chemistry of fruit bioflavonoids: A review. *The Analyst, 2*, 11–34.

Rossell, J. B. (1991). Vegetable oil and fats. In J. B. Rossell & J. L. R. Pritchard (Eds.), *Analysis of oilseeds, fats and fatty foods* (pp. 261–328). New York: Elsevier Applied Science.

Ruel, G., & Couillard, C. (2007). Evidences of the cardioprotective potential of fruits: The case of cranberries. *Molecular Nutrition & Food Research, 51,* 692–701.

Ruperez, F. J., Martin, D., Herrera, E., & Barbas, C. (2001). A review of chromatographic analysis of α-tocopherol and related compounds in various matrices. *Journal of Chromatography, 935,* 45–69.

Seeram, N. P., & Heber, D. (2007). Impact of berry phytochemicals on human health: Effects beyond antioxidant. *ACS Symposium Series, 956,* 326–336.

Shahidi, F. (2000). Antioxidant factors in plant foods and selected oilseeds. *BioFactors, 13,* 179–185.

Shahidi, F., & Wanasundara, P. K. J. P. D. (2002). Extraction and analysis of lipids. In C. C. Akoh & D. B. Min (Eds.), *Food lipids: Chemistry, nutrition and biotechnology* (Vol. 2, pp. 133–135). New York: Marcel Dekker.

Shahidi, F., & Weerasinghe, D. K. (2004). *Neutraceutical beverages: Chemistry, nutrition and health effects* (ACS Symposium Series 871). Washington, DC: American Chemical Society.

Sheibani, A., & Ghaziaskar, H. S. (2008). Pressurized fluid extraction of pistachio oil using a modified supercritical fluid extractor and factorial design for optimization. *LWT-Food Science and Technology, 41,* 1472–1477.

Stone, W. L., & Papas, A. (2003). In F. D. Gunstone (Ed.), *Lipids for functional foods and nutraceuticals*. Bridgwater: The Oily Press.

Sultana, B., Anwar, F., & Przybylski, R. (2007). Antioxidant potential of corncob extracts for stabilization of corn oil subjected to microwave heating. *Food Chemistry, 104,* 997–1005.

Sultana, B., Anwar, F., Asi, M. R., & Chatha, S. A. S. (2008). Antioxidant potential of extracts from different agro wastes: Stabilization of corn oil. *Grasas y Aceites, 59*(3), 205–217.

Tapiero, H., Ba, G. N., Couvreur, P., & Tew, K. D. (2002). Polyunsaturated fatty acids (PUFAs) and eicosanoids in human health and pathologies. *Biomedicine & Pharmacotherapy, 56*(5), 215–222.

Thyagarajan, P. (2012). *Evaluation and optimization of cranberry seed oil extraction methods*. M.Sc. Thesis. McGill University, Canada.

Van Hoed, V., Clercq, N. D., Echim, C., Andjelkovic, M., Leber, E., Dewettinick, K., & Verhe, R. (2009). *Berry seeds: A source of specialty oils with high content of bioactives and nutritional value*. Prosser: Department of Organic Chemistry; Department of Food Safety and Food Quality. Faculty of Bioscience Engineering. Ghent University. Apres-Vin Enterprises.

Villa, B., Calbresi, L., Chiesa, G., Rise, P., Galli, C., & Sirtori, C. (2002). Omega-3 fatty acid ethyl esters increase heart rate variability in patients with coronary disease. *Pharmacological Research, 45*(6), 475.

Vinson, A., Xuehui, S., Ligia, Z., & Bose, P. (2001). Phenol antioxidant quantity and quality in foods: Fruits. *Journal of Agricultural and Food Chemistry, 49,* 5315–5321.

Vinson, J. A., Bose, P., Proch, J., Kharrat, H. A., & Samman, N. (2008). Cranberries and cranberry products: Powerful in vitro, ex vivo, and in vivo sources of antioxidants. *Journal of Agricultural and Food Chemistry, 56*(14), 5884–5891.

Vvedenskaya, I. O., Rosen, R. T., Guido, J. E., Russell, D. J., Mills, K. A., & Vorsa, N. (2004). Characterization of flavonols in cranberry (*Vaccinium macrocarpon*) powder. *Journal of Agricultural and Food Chemistry, 52*(2), 188–195.

Wang, S. Y., & Jiao, H. J. (2000). Scavenging capacity of berry crops on superoxide radicals, hydrogen peroxide, hydroxyl radicals and singlet oxygen. *Journal of Agricultural and Food Chemistry, 48,* 5677–5684.

Wang, Y., Catana, F., Yang, Y., Roderick, R., & Van Breemen, R. B. (2002). An LC-MS method for analyzing total resveratrol in grape juice, cranberry juice and in wine. *Journal of Agricultural and Food Chemistry, 50,* 431–435.

Willet, W. C. (2001). *Eat, drink, and be healthy: The Harvard Medical School guide to healthy eating*. New York: Simon and Schuster.

Wittenstein, B., Klein, M., Finckh, B., Ullrich, K., & Kohlschutter, A. (2002). Plasma antioxidants in pediatric patients with glycogen storage disease, diabetes mellitus and hypercholesterolemia. *Free Radical Biology & Medicine, 33*, 103–110.

Yan, X., Murphy, B. T., Hammond, G. B., Vinson, J. A., & Neto, C. (2002). Antioxidant activities and antitumor screening of extracts from cranberry fruit (*Vaccinium macrocarpon*). *Journal of Agricultural and Food Chemistry, 50*, 5844–5849.

Yang, B. (2003). Natural vitamin E: Activities and sources. *Lipid Technology, 15*, 125–130.

Yaqoob, N., Bhatti, I. A., Anwar, F., Mushtaq, M., & Artz, W. E. (2016). Variation in physicochemical/analytical characteristics of oil among different flaxseed (*Linum usittatissimum* L.) cultivars. *Italian Journal of Food Science, 28*(1), 83–89.

Yu, L. L., Zhou, K. K., & Parry, J. (2005). Antioxidant properties of cold-pressed black caraway, carrot, cranberry, and hemp seed oils. *Food Chemistry, 91*, 723–729.

Yukui, R., Wenya, W., Rashid, F., & Qing, L. (2009). Fatty acid composition of apple and pear seed oils. *International Journal of Food Properties, 12*, 774–779.

Chapter 36
Dragon (*Hylocereus megalanthus*) Seed Oil

Sumia Akram and Muhammad Mushtaq

Abstract Dragon fruits or Pitaya are the lithophytes or hemiepiphytes of *Cactaceae* family primarily distributed among tropical regions. This fruit initially attracted the researchers looking for natural food grade dyes for its striking color pigments (betalains), however, investigations established that dragon fruit pulp, peel, and seed contain potentially beneficial carbohydrates, phytoalbumin, biopeptides, vitamins, phenolics, and minerals. The seeds of dragon fruit are exceptionally small, so could not focused for their oleoginous compounds but research up to the date indicates that seeds of this fruit are rich in essential fatty acids principally *omega*-3 and *omega*-6 fatty acids as well as tocopherols. The levels of linoleic acid attributed to dragon seed oil (500 g/Kg of oil) are comparable with flaxseed, canola and sesame oils. The keen survey of literature available regarding the phytochemistry of dragon seed recommends that this oil may become a viable source of good quality oleonutrients for food and cosmetic industries. There is a prompt need to modernize and scale up seed separation and oil extraction methods to revalorize dragon seed phytochemicals.

Keywords Pitaya · Dragon seed oil · Soxhlet extraction · Fatty acid profile · Oleo-nutrients

Abbreviations

DSO   Dragon seed oil
EAE   Enzyme-assisted Extraction
MAE   Microwave-assisted extraction

S. Akram
Department of Chemistry, Minhaj University, Lahore, Pakistan

M. Mushtaq (✉)
Department of Chemistry, Government College University, Lahore, Pakistan

© Springer Nature Switzerland AG 2019   675
M. F. Ramadan (ed.), *Fruit Oils: Chemistry and Functionality*,
https://doi.org/10.1007/978-3-030-12473-1_36

PUFA   Polyunsaturated fatty acids
SE   Soxhlet extraction
SFE   Supercritical fluid extraction

1 Introduction

According to International Organization for Succulent Plant Study, the genus *Hylocereus* falls among the six genera of the *Hylocereeae* tribe. The *Hylocereus spp.* are diploid (2n =22) except *Hylocereus megalanthus* (tetraploid) and commonly known as Pitahaya, Pitaya Blanca, Pitaya roja or dragon fruit. The dragon fruit has also been known under various other traditional and regional names like strawberry pear, night blooming cereus, Belle of the Night, Jesus in the Cradle, American beauty, Cinderella plant, and the interesting one "Queen of night" or "Noble women" assigned for the beauty of its flowers particularly at night. Likewise, all cacti, the origin of *H. undatus* has been assigned to Americas; however there exist many confusions regarding their origin and taxonomic identity (Daubresse Balayer 1999; Mizrahi et al. 1997). The pitaya or dragon fruit comes in a number of varieties including *Hylocereus undatus* (red or purple fruit with white flesh), *Hylocereus polyrhizus* (red colored fruit and flesh), and *Hylocereus megalanthus* (yellow fruit with white flesh) cover a major part of pitaya cultivated for ornamental and fruit purposes (Gunasena et al. 2007).

The dragon fruit plants (Fig. 36.1a) are among the lithophytes or hemiepiphytes distributed in tropical regions. Their aerial roots (Fig. 36.1b) climb on rocks and

Fig. 36.1 The dragon fruit plant (**a**), root (**b**), flower (**c**), fruit (**d**), seed (**e**), and oil products (**f**)

trees and may reach up to a height of 10 m or more. The plants can grow in tropical and subtropical regions of Central and South America, Mexico, Israel, and Southeast Asia with the sunny climate of temperature range 20–30 °C and an average rainfall greater than 1000 mm/year. These plants can withstand at higher of 40 °C and grow lower temperatures up to 10 °C, d but rarely tolerate water logging so they prefer well-drained slightly acidic soils, however, certain cultivars can manage to grow in soils with high salts (Lim 2012).

In general, dragon plant grown from seeds mature within 3 years while those produced from plant cuttings can bloom after 1 year. The flowers generally cover up to 29 cm comprising pure white inner perianth and yellow to green outer perianth (Fig. 36.1c). The well-ripen and mature dragon fruit of *H. polyrhizus* and *H. undatus* varieties (Fig. 36.1e) often measure 10–15 and 15–25 cm in length and 150–300 and 300–800 g mass, respectively. Moreover, the fruits of the first variety carry red, very sweet and slightly acidic flush whereas those of later is white, soft and sweet (Liaotrakoon 2013).

The dragon fruit has become a popular fruit among the people of Thailand, Vietnam, Israel, Sri Lanka, Mexico, Central America, Europe, Southeast Asia, United States, China, Australia, Cyprus, and the Canary islands owing to the presence of iron, calcium, phosphorus, antioxidants, phytonutrients, polyunsaturated fatty acids (PUFA), water-soluble vitamins, carbohydrates, amino acids, and carotenoids. The regular intake of dragon fruit may help to check cholesterol, prevent stomach ailments, control blood sugar, reduce weight, decrease congenital glaucoma, improve hemoglobin, improve vision, strengthen teeth, repair body cells, and boost up immunity. Dragon fruit is enjoyed as fresh fruit for its delicately sweet and refreshing flesh; nevertheless, it can be used in juices, beverages, marmalades, and jellies.

The seeds of dragon fruit occupy only 1–3% by weight of fresh fruit, however the presence of essential fatty acids, tocopherols, flavonoids, terpenes, carbohydrates, protein and other potential bioactives render dragon seed interest worthy and this chapter would cover particular features of dragon seed oil, its photochemistry, organoleptic properties and potential for food, pharmaceutical, and cosmetic industries.

2 Processing of Dragon Seed

The dragon seeds constitute an exceptionally small portion of fruit (i.e. 25–50 g/Kg) on a fresh weight basis. Villalobos-Gutiérrez et al. (2012) separated the highest amount 300 g seed from 7 kg of red pitaya of Central American origin. Although dragon seeds are smaller in size and quantity (i.e. 2–4%) of whole fruit, their phytoconstituents have been ranked with high therapeutic value and increasingly used in food, cosmetic, culinary, and pharmaceutical industries. Besides, the fruit ecotypes and agro-climate conditions, the quality and composition of dragon seed oil also vary with the extraction method applied. The biologists have not tried to develop genetically modified dragon fruit for high oleaginous phytochemicals, however; the ecotypes are there with higher fiber and low calories. In the case of oleaginous attributes, the selection of an appropriate extraction strategy could offer a good quantity of lipids without deteriorating their potential health benefits and antioxidant character.

2.1 Solvent Extraction

The analytical chemists and lipidologists categorize fat extraction methods into conventional and non-conventional methods. The methods of first-generation involve (i) distribution of oleaginous compounds between an organic solvent (solvent extraction) or an aqueous media (distillation) and (ii) mechanical expression (soaking and pressing) while those belonging to non-conventional class work with the assistance of electromagnetic (microwave, ultrasound, and ohmic radiations), enzymes (lipases), detergents and dispersers. The non-conventional methods have often been claimed to be more efficient and green as compared to traditional ones and few of these established to be green like enzyme-assisted extraction and supercritical fluid-based extraction technologies (Mushtaq et al. 2017; Mushtaq et al. 2015). However, the majority of industrialist and food chemists are still following traditional methods for the recovery of fixed and essential oils from agricultural resources.

The oldest and most frequently adopted conventional method (i.e. soxhlet extraction) was initially designed by Soxhlet (1879). The Soxhlet extraction is a semi-continuous method in which non-volatile oil or fat can be separated from food material by recurrent percolation (washing) with an organic solvent. Likewise many other seed oils (canola, flaxseed, cottonseed, and corn), the dragon seed oils are often produced by Soxhlet extractor while applying non-polar solvent like *n*-hexane, petroleum ether or diethyl ether. There is evidence for the use of polar solvents like methanol, ethanol, acetonitrile, and chloroform as extraction solvents.

In the practical method, the dragon fruit flesh is first autoclaved (pressure up to 15 psi) for 40 min, then centrifuged at 8000 g for 20 min to screened out the seeds while destroying mucilage layer embedding seeds into the flesh. The seeds are further dried to particular moisture level, crushed into coarse powder of 40–80 mesh, and loaded into Soxhlet extractor while packing into some cellulosic bag (thimble). The condensed non-polar solvents fall on packed seed powder and recover the lipids present whose composition and degree of unsaturation would vary nature of solvent applied and extraction conditions.

Ariffin et al. (2009) adopted Soxhlet extraction with petroleum ether at 40–60 °C and recovered 29.5% and 32.0% of lipophilic compounds from two varieties of dragon fruit *H. polyrhizus* (red flesh) and *H. undatus* (white flesh), respectively. Similarly, Murugesu et al. (2013), Villalobos-Gutiérrez et al. (2012), Lim et al.

Table 36.1 The summary of attempts made extraction/analysis of dragon seed oil

Variety	Origin	Yield (%)	Extraction/analysis	Major constituents	Biological activities	Reference
H. undatus, H. polyrhizus	Malaysia	29.5–32.0	SE/petroleum ether (40–60 °C)	PUFA-linoleic acid	–	Ariffin et al. (2009)
H. undatus, H. polyrhizus	Malaysia		SE/petroleum ether (40–60 °C)	PUFA-linoleic acid		Lim et al. (2010)
H. polyrhizus	Malaysia	26.9	SE/orbital shaker	–	Antioxidant, antiradical	Adnan et al. (2011)
H. polyrhizus	Malaysia	–	SE/n-hexane	Oleic and linoleic acid	–	Murugesu et al. (2013)
H. polyrhizus	Central America	29.6–30.2	SE/n-hexane (40–60 °C)	PUFA-linoleic acid	–	Villalobos-Gutiérrez et al. (2012)
H. polyrhizus	China	6.13 ± 0.13 SE	SE/n-hexane	PUFA-linoleic acid	–	Rui et al. (2009)
		6.63 ± 0.19 MAE	MAE: variable power			
			EAE: cocktail			
		6.94 ± 0.17 EAE	SFE: CO_2			
		5.54 ± 0.17 SFE				
–	China	31	SC-CO_2 CO_2 20 L/h, 30 MPa, 55 °C	PUFA		Deng et al. (2014)

(2010), and Rui et al. (2009) used Soxhlet-based extraction for the recovery of oleochemicals from dragon seed using n-hexane and petroleum ether as extraction solvents. Table 36.1 provides a comparison of extraction yields and conditions applied against each extraction method. The Soxhlet-based extractions were usually carried out at 70–80 °C for 3–4 h with a solvent to sample ratio of 7:1 (mL:g) and lipids extracted usually exhibited yellowish shade (λ_{max} 453 nm) and a high degree of unsaturation. Although scientific knowledge does not support, still few researchers have also applied orbital shaker type arrangements to extract the lipids from dragon seed (Table 36.1). There is no data available regarding phytochemical profiling of dragon seed oil (DSO) obtained via orbital type extraction.

2.2   Cold Pressing

The second most frequently adopted technique for the recovery of fixed oils, the use of expression machines. These machines mainly of two types: hydraulic presser or screw presser (extruder or expeller) to force the oil out of oleaginous raw materials. The cold pressers have been also tried for the extraction of essential oils. In general, the seeds or fruits are transferred to oil expression machines where these are cut into coarse particles and passed across rasping cylinders that reduce the area available for mass and break pods containing essential oil and release their contents. As this process occurs, water or steam is sprinkled over the mass to collect the expressed mass. The collected mass is filtered and centrifuged which allow the separation of the pure essential oil. Expression machines may provide high-quality essential oils with characteristic fragrances particularly for volatile contents but associated with some recovery limitations. The screw type pressers are continuous and offer slightly higher yields as compared to the hydraulic presser. Although mechanical expression produces good quality oil it is often suitable for small-scale production, special products or a pre-press operation followed by large-scale solvent extraction. Temperature is parameter acting in several ways on the pressing performance. The rise in temperature causes a decrease in oil viscosity favorable to its flowing, but it can also alter the cellular structure and plasticity of the raw material (cooking).

The oils expressed through cold pressers are more resistant towards oxidation and eventually offer extended shelf life but the technique could not be commercialized for dragon seed oil because of near to the ground recovery rates particularly for the seeds containing small quantities of oleaginous compounds. Therefore, the whole oil produced in the United States and many other countries are extracted with n-hexane. The key problems of this solvent are its flammable and explosive nature, non-selective towards fats/oils, costly and potential health-related issues. Moreover, oils produced through n-hexane or ether based solvents may encounter undesirable impurities like pigments, phospholipids, free fatty acids and a reasonable amount of unsaponifiable matter. Consequently, researchers across the globe are under social and economic pressures to discover some alternative, selective and eco-friendly type solvents for extraction of oleochemicals from plants.

Another cite-worthy breakthrough in this context involve the use of carbon dioxide, water and ethanol under the supercritical state for the extraction of biomolecules, especially of non-polar nature. The supercritical fluid extraction (SFE) has become an alternative and green choice and has been commercialized for the decaffeination of coffee. Actually, a supercritical fluid is a dense gas above its critical temperature and critical pressure and its density and ability to solubilize materials changes drastically near critical conditions. Moreover, oils produced by supercritical carbon dioxide (SC-CO_2) contain relatively smaller amounts of pigments, iron, and phosphorus as compared to the oils produced by expellers and Soxhlet type extractors (List et al. 1984). No doubt, SFE is an intelligent and benign choice for the production of good quality/purity fats as SC-CO$_2$ is a non-toxic, non-explosive, cheap, readily available and reusable solvent still limited number of attempt has been taken to extract dragon fruit lipids through this technology. Recently, Deng et al. (2014) optimized SC-CO$_2$ based extraction of pitaya seed oil using an artificial neural network model. The effects of drying time, particle size, and CO$_2$ flux were investigated and optimized by single-factor experiments and the oil was finally extracted from slightly ground seed dried at 80 °C for 1 h. The extraction parameters comprising CO$_2$ flow rate 20 L/h, pressure 30 MPa, temperature 55 °C offered 31% oil during 3 h extraction time. The extracted oil was found to be similar to another kind of vegetable oil with a comparatively high degree of unsaturation and low acid and peroxide values. Moreover, dragon seed oil extracted by SC-CO$_2$ was found to be rich in linoleic and linolenic acid.

The conventional and non-conventional techniques have been adopted for the extraction of lipids from dragon seed. In conventional methods, dragon fruit seed separated following the steps indicated in Fig. 36.2 are either processed under higher pressure (cold pressing) or extracted through Soxhlet extractor using non-polar solvents.

For the non-conventional extraction of dragon seed oil, microwave and ultrasound radiations, supercritical fluids, and enzymes have been often applied. Rui et al. (2009) attempted to compare the efficiency of conventional Soxhlet-based extraction with microwave-assisted extraction (MAE), enzyme-assisted extraction (EAE), microwave-enzyme-assisted extraction (MEAE) and SFE. For the microwave extraction, the powdered dragon seed along with non-polar extraction solvents (*n*-hexane and petroleum ether) were processed under various powers and irradiation intervals that make it known that at elevated microwave power (above 800 w) the lipids liberated more quickly as compared to those which were processed below 500 w.

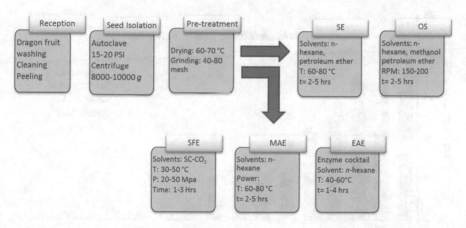

Fig. 36.2 Conventional and non-conventional methods used to extract dragon oils

3 Fatty Acids Composition and Acyl Lipids (Neutral Lipids, Glycolipids, and Phospholipids) Profile of Dragon Fruit Fixed Oil

The lipids not only work as energy reservoirs but also form building blocks of membranes, vesicles and cellular signaling system. The human and likewise other mammals have the ability to synthesize or break down diversity of lipids still certain fatty acids (known as essential fatty acids) such as linoleic acids and linolenic acids (both of these fatty acids are 18-carbon PUFA) cannot be synthesized within the body and should be taken in the diet. It has been established that intake of these PUFA is essential for mental growth in infants and provide first line protection against cardiovascular diseases, depression, cancers and dementia in adults. In this context, a great deal of research has been devoted to exploring indigenous sources of essential fatty acids and nowadays it has become known that vegetable oils such as olive, safflower, corn and sunflower oils are rich in linoleic acid, while linolenic acids are usually found in seeds and nuts.

Triglycerides or triacylglycerols (TAG); an ester of glycerol with three fatty acid constitutes a major portion of animal and plant lipids. The TAG of DSO mainly comprises oleic, linoleic, linolenic and palmitic acids. Initially, the TAG are transformed into fatty acids, monoacyl glycerides, and diacylglycerides out of which fatty acids are oxidized to produce energy or used in the physiology of other biological functions. Therefore, it is crucial to have a fatty acid profile of lipids intended for human consumption. The most widely followed method to profile the fatty acid of lipids involves the conversion of the TAG into their fatty acid methyl esters (FAME) which can be easily determined by gas chromatography coupled with flame ionization detector (GC-FID) or mass spectrometers (GC-MS).

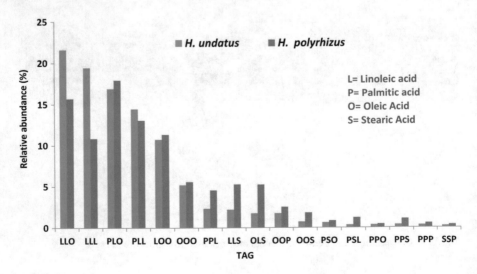

Fig. 36.3 The relative abundance of triacylglycerides (TAG) found in DSO

Figure 36.3 enlightens the TAG composition of dragon seed oil. It is obvious that ester of linoleic (L) and oleic (O) acid dominates in both species of dragon fruit whereas those of palmitic (P) and stearic acid (S) were in traces. In addition to the plant ecotype, soil properties, climate conditions, and postharvest storage or processing, the oil expression technique mainly controls the fatty acid profile of the final product. The general perception in this regards narrates that extraction techniques working at room or below room temperature (cold extraction technique) would offer the oleaginous products rich in unsaturated fatty acids. However, our understanding warns that if the raw materials are not propyl dried/processed, the oil produced may undergo quick auto-oxidation. The data assembled in Table 36.1 indicates that oil extracts via Soxhlet extractors (usually works at 45–60 °C) were also rich in unsaturated fatty acids. A more interesting and informative comparison of the fatty acid profile of DSO has been drawn in Fig. 36.4. The oil produced via supercritical fluid solvents were rich in unsaturated fatty acids up chain length of 18 carbons. It was interesting to note that oils extracted through microwave radiation were rich in a long chain, branched and saturated fatty acids (Fig. 36.4) as compared to conventional Soxhlet extraction with *n*-hexane. Similarly, powdered dragon seed when incubated with enzyme cocktails (pectinase, proteases, and cellulases) and subsequently extracted using non-polar solvents furnished almost same kind of fatty acids as in the oil produced through microwave-assisted extraction.

As stated earlier, the dragon seed comprises less than 5% of fresh fruit and its lipid content vary up to 30% of its dry weight and majority of these lipids are PUFA. However, the dragon fruits are becoming more popular among the researchers because its linoleic acid content is even higher than that of canola, flaxseed, sesame

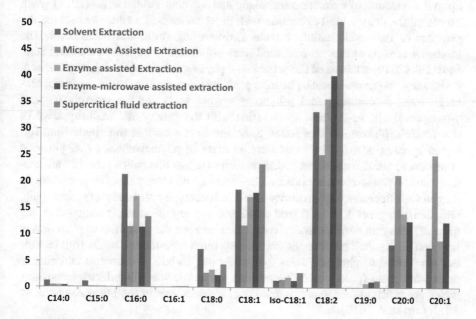

Fig. 36.4 Fatty acid profile of DSO expressed through different techniques (Rui et al. 2009)

and grape seed oils. Ariffin et al. (2009) studied two varieties of dragon fruit (white and red flesh fruits) and observed that both of these ecotypes were rich in linoleic acids. Similarly, Villalobos-Gutiérrez et al. (2012) observed that seeds of red flesh dragon fruit contained 466 g linoleic acid/kg of seed whereas palmitic (182 g), stearic (49 g), and arachidic acid (18 g) were main saturated fatty acids. The mono-unsaturated fatty acids present in dragon seed oil were oleic (239 g), cis-11-vaccenic (45 g), and palmitoleic (3 g) acids. Overall each kilogram of dragon seed furnished 249, 753, 287, and 466 g of saturated, total unsaturated, monounsaturated, and poly-unsaturated (C18:2) fatty acids. A similar sort of results (Table 36.1) had been observed by Lim et al. (2010), who determined saturated, monounsaturated, and polyunsaturated fatty acids 17.99–22.78, 25.20–27.92, 49.30–56.81 g/100 g of oil of *H. polyrhizus* and *H. undatus* seed oils. Murugesu et al. (2013) observed that seed separation methods *i.e.* hot or cold technique applied to get clear mucilage free seed may influence the characteristics of extracted fatty acids. Overall, the free fatty acids and oxidation status of oils extracted from both types of seeds were comparable, however, oleic and linoleic acid contents of oils extracted from seed separated via cold technique were higher.

4 Minor Bioactive Compounds

In addition to fatty acids, the seed of dragon fruit contains a diverse range of functional lipids and organic acids whose recovery and biological potential depends upon the selection of extraction technique and solvents. Adnan et al. (2011) when processed the dragon seed extraction with polar solvents like ethanol observed the presence of flavonoids mainly catechins, quercetins, rutin, and kaempferol. The methanol extracts of dragon fruit seed were also found to be rich in ascorbic acid. Lim et al. (2010) established the presence of phytosterols (cholesterol, stigmasterol, β-sitosterol and campesterol), phenolic acids (caffeic, vanillic, gallic, syringic, protocatechuic, *p*-coumaric, and *p*-hydroxybenzoic acid) and tocopherols (α- and γ-tocopherol), however their level varied with the variety and maturity level of dragon fruit. β-sitosterol was found to be the most abundant functional lipid followed by campesterol, The most familiar class of phytochemicals (i.e. phenolic compounds) are also abundant in dragon fruit to render this fruit a powerful antioxidant and phytotherapeutic choice. The amounts of total phenolic bioactives in dragon fruit fluctuate with maturity level, fruit ecotype and fruit parts (peel, pulp, and seed). In general, dragon fruit contained 1–5 mg of phenolic compounds per gram of fresh fruit with an inverse correlation between the quantities of phenolics in peel and flesh part. The phenolic compounds often reported in dragon fruit include but not limited to phenolic acids (gallic, ferulic, caffeic, *p*-coumaric and sinapic acids), flavonoid (phloretin-2-*O*-glucoside and myricetin-3-*O*-galactopyranoside), which are much higher in concentration in the peel than in the pulp (Adnan et al. 2011; Lim et al. 2010).

The methylated phenols; also known as Vitamin E or tocopherols which carry a pregnancy are an important class of fat-soluble antioxidants. The hydrophobic carbon chain renders these alcohols to penetrate cell membranes and shielded from free oxygen whereas the OH group present on the counter side can deliver an H atom to quench free radicals. The energy transition associated with second transformation (323 kJ/mol) is almost ten times smaller than the O-H bond in other phenols. In consequences, these antioxidants have now become recommended food additives (EU 2010; Food and Administration 2011). The DSO, particularly from *H. polyrhizus* was found to be a rich source of α-tocopherols (Fig. 36.5).

However, during storage at room temperature and under cold conditions the tocopherol contents of DSO continuously declined. The maximum decline has been also reported in α-tocopherol, which might be attributed to its elevated antioxidant character. It was observed that storage at 4 °C for 3 months reduced α-tocopherols from *H. polyrhizus* and *H. undatus* from 477 to 409 and from 293 to 241 mg/kg oil, respectively. The similar storage at 20 °C declined α-tocopherols from *H. polyrhizus* and *H. undatus* up to 343 and 200 mg/kg, respectively (Wijitra et al. 2013). In addition to the seed, the pulp and peel of dragon fruit has been investigated and announced to be good source of high value bioactives like pigments (betacyanins, betaxanthins, and betalains), phenolic acids (gallic acid, protocatechuic, *p*-hydroxybenzoic, vanillic acid, caffeic acid, syringic, and *p*-coumaric acid), flavonoids (phlorizin, quercetin, flavonol glycoside, kaempferol, isorhamnetin, and rutin), catechins, caffeotannic acids, stigmasterol, ascorbic acid, β-carotene, lycopene, and terpenes (Adnan et al. 2011; Suh et al. 2014).

5 Composition of Fruit Essential Oil

The volatiles fraction of oleaginous compounds are often synthesized by plants to protect themselves from infection, pests, predators, bacteria, and fungi, or to attract pollinators. These compounds may belong to alcohols, terpenes, phenols, esters, aldehydes, ketones and oxides and are usually responsible for the aroma and

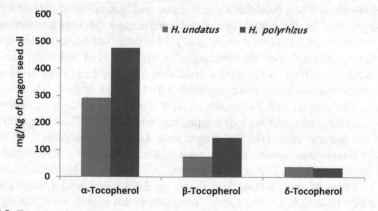

Fig. 36.5 Tocopherol profile of DSO (Wijitra et al. 2013)

therapeutic benefits attributed to essential oils. The exact composition of essential oil may vary with plant ecotype, agro-climate conditions, maturity and ripening stage of fruit, and extraction technique applied. A thorough investigation of literature published regarding dragon seed could not furnish a single report on phytochemistry of dragon seed essential oils. Célis et al. (2012) characterized odor-active compounds of *Hylocereus megalanthus* (yellow pitaya) fruit. The authors blended 200 g of fruit pulp in 500 mL of distilled water containing 0.2 mg of methyl nonanoate (internal standard). The volatiles was collected by simultaneous distillation-extraction (SDE) assembly with redistilled diethyl ether. The aroma compounds of yellow pitaya fruit were analyzed by GC-FID and GC-MS. Overall, 146 volatile bioactives were detected out of which, 121 were positively identified as terpenes (26.0%), esters (18.2%), acids (16.0%), aldehydes (14.3%), alcohols (5.2%), hydrocarbons (4.0%), ketones (8.8%), and others compounds (7.6%). Overall nine boacitives were set responsible for the characteristic odor of the fruit of which the most important were phenylacetaldehyde, 3-methyl butanal, (E)-b-damascenone, hexanal, decanal, octanal, nonanal, 1,8-cinole, and limonene.

6   Contribution of Bioactive Compounds in Fruit Oil to Organoleptic Properties of Dragon Seed Oil

The dragon fruit had striking colors due to the presence of betalains mainly betacyanin (red/purple pigment) and betaxanthin (orange/yellow pigment). Both of these compounds are excellent electron donors so appears to be good free radical scavengers. Betacyanin is actually water soluble pigments extensively studied as the alternative color for foodstuff with potential health benefits against stress-related diseases. The most interesting feature of betalains is their ability to retain color characteristic under acidic conditions (pH 3–7) as compared to well-known phenolic pigments of anthocyanin group (may change color with the change in pH). Previously, betalains extracted from beet sugar have been used to color food products. However, red beet betalains are associated with geosmin and pyrazine derivatives responsible for unfavorable earthy and some nitrates that may lead nitrosamine based carcinogenesis (Esquivel et al. 2007). Moreover, betalains provide first line protection against lipid peroxidation and offer ample level of antioxidant and anti-inflammatory activities. Nowadays, a great deal of interest has been developed to replace anthocyanins based color pigments with betalains in dairy and other food products. The dragon fruit has been ranked at the top for betacyanins prevalence (betanin, hylocerenin, and sephyllocactin) pigments usually 35–50 mg/100 g following the maturity and variety of dragon fruit. Likewise betacyanins, other polyphenolic compounds render dragon fruit characteristic taste, flavor and other organoleptic properties. The color of dragon fruit pulp or oil has been often measured in terms of specification laid down by the International Commission on Illumination under L^*, a^*, b^* system. According to this specification, the lightness of a color is graduated as $L*$ from zero (black) to 100 (Perfect white), its chroma $a*$ from green (negative) to red/magenta (positive) and its hue position ($b*$) from blue (negative) to yellow (positive). It was generally observed that DSO produced via orbital shaker type extractor with petroleum ether as an extraction solvent bears $L*$, $a*$, and $b*$ over than range of 29.33–29.97, 0.80–0.97, and 8.26–9.56, respectively.

7 Health-Promoting Traits of Fruit Oil and Oil Constituents

There is a range of dietary and cosmetic products containing peel, pulp, or seeds of dragon fruits or their extracts. Although a reasonable number of proofs are available regarding their benevolent phytochemistry and potential to protect against various opportunistic pathogens present within the skin, oral cavity and gastrointestinal tract (Pham 2014), none of these products has been evaluated or regulated by the legal agency. The most of scientific studies regarding the use of dragon fruit parts, their extracts or oils directly into food, pharmaceutical, and cosmetic products are under trial and care is due at the consumer end before the use or intake of any such products to treat, cure, or prevent any health-related problem. Nevertheless, the composition of oleaginous constituents of dragon seed percepts their effectiveness against skin dryness, dullness, wrinkles, sagging and signs of aging. There is a wide range of cosmetics continuing DSO that may eliminate eczema and psoriasis and nourish collagen. The fixed and essential oils of dragon seeds contain appreciable amounts of unsaturated fatty acids, phenolics, tocopherols, and terpenes, but scientific literature does not indicate their incorporation into dietary products. Likewise, the flesh and peels of this fruit also contain a good quantity of phenolics and extracts of these parts were able to inhibit various types of food-borne pathogens. The fresh flesh not only offer refreshing sweet flavor but also reduce the risk of obesity, diabetes, colon cancer, coronary heart disease, kidney failure, stroke, and hypertension.

8 Edible Applications of Fruit Oil

The production of DSO could not be scaled on industrial level due to exceptionally small seed to fruit ratio and difficulties involved during the separation of dragon seeds from fruit. Although oil produced from dragon seed was found to be a rich source of fatty acids, vitamins, and lipophilic therapeutics, it has not be used for the edible purpose. The whole of DSO or extracts produced are used in cosmetic industry nevertheless the oil has the potential for edible applications. In order to use DSO in the edible application, there is a prompt need to focus modern extraction techniques for the recovery of oleaginous compounds as well as to investigate what exactly DSO contains and what beneficial and adverse health effect it may impart. There is evidence regarding the use of other fruit parts as food or food products. Besides the presence of distinguished antioxidant pigments, dragon fruit has been also known as an irreplaceable source of sugars like glucose (5 g/100 g), fructose (2 g/100 g), and sorbitol (0.5 g/100 g) with an estimated energy of 285–150 kj/100 g with an energy equal to 150–250 kj/100 g (Luo et al. 2014).

The presence of fiber and prebiotic fractooligosaccharides such as neokestose, nystose, stachyose, bifurcose, kestoses at elevated levels render these fruits to check stomach acidity and accelerated the growth of bifidobacteria and lactobacilli. Moreover, dragon fruit fractooligosaccharides are partially resistant towards salivary α-amylase and gastric acids and eventually improves the digestive functions, glucose and lipid metabolism, and diabetic immunity. Although, an increase in sugars has been linked with the decrease in organic acids (sugar/organic acids is used as fruit ripening characteristic) still, mature/ripen dragon fruits contained reasonable amounts of ascorbic, citramalic, citric and malic acids. Moreover, the presence of carbohydrate hetero polymers like resistant starch, cellulose, hemicelluloses, gums, pectic substances, and inulin render this fruit most suitable for intestinal health disorders.

References

Adnan, L., Osman, A., & Abdul Hamid, A. (2011). Antioxidant activity of different extracts of red pitaya (*Hylocereus polyrhizus*) seed. *International Journal of Food Properties, 14*, 1171–1181.

Ariffin, A. A., Bakar, J., Tan, C. P., Rahman, R. A., Karim, R., & Loi, C. C. (2009). Essential fatty acids of pitaya (dragon fruit) seed oil. *Food Chemistry, 114*, 561–564.

Célis, C. Q., Gil, D. E., & Pino, J. A. (2012). Characterization of odor-active compounds in yellow pitaya (*Hylocereus megalanthus* (Haw.) Britton et Rose).

Daubresse Balayer, M. (1999). Le pitahaya. Fruits oubliés. 1, 15–17.

Deng, C.-J., Liu, S.-C., Tao, L.-Q., & Wu, X.-P. (2014). Optimization of extraction process of pitaya seed oil by supercritical carbon dioxide based on artificial neural network. *Food Research and Development, 10*, 020.

Esquivel, P., Stintzing Florian, C., & Carle, R. (2007). Phenolic compound profiles and their corresponding antioxidant capacity of purple pitaya (*Hylocereus* sp.) genotypes. *Zeitschrift für Naturforschung C, 62*, 636.

EU. (2010). Approved additives and their E Numbers, Food Standards Agency.

Gunasena, H., Pushpakumara, D., & Kariyawasam, M. (2007). *Dragon fruit Hylocereus undatus (Haw.) Britton and Rose. Underutilized fruit trees in Sri Lanka* (pp. 110–142). New Delhi: World Agroforestry Centre.

Liaotrakoon, W. (2013). Characterization of dragon fruit (*Hylocereus* spp.) components with valorization potential. Ghent University.

Lim, T. K. (2012). Hylocereus polyrhizus. In T. K. Lim (Ed.), *Edible medicinal and non-medicinal plants: Volume 1, Fruits* (pp. 643–649). Dordrecht: Springer Netherlands.

Lim, H. K., Tan, C. P., Karim, R., Ariffin, A. A., & Bakar, J. (2010). Chemical composition and DSC thermal properties of two species of *Hylocereus cacti* seed oil: *Hylocereus undatus* and *Hylocereus polyrhizus*. *Food Chemistry, 119*, 1326–1331.

List, G. R., Friedrich, J. P., & Pominski, J. (1984). Characterization and processing of cottonseed oil obtained by extraction with supercritical carbon dioxide. *Journal of the American Oil Chemists' Society, 61*, 1847–1849.

Luo, H., Cai, Y., Peng, Z., Liu, T., & Yang, S. (2014). Chemical composition and in vitro evaluation of the cytotoxic and antioxidant activities of supercritical carbon dioxide extracts of pitaya (dragon fruit) peel. *Chemistry Central Journal, 8*(1), 1.

Mizrahi, Y., Nerd, A., & Nobel, P. S. (1997). Cacti as crops. *Horticultural Reviews, 18*, 291–319.

Murugesu, S., Ariffin, A. A., Ping, T. C., & Chern, B. H. (2013). Physicochemical properties of oil extracted from the hot and cold extracted red pitaya (*hylocereus polyrhizus*) seeds. *Journal of Food Chemistry and Nutrition, 01*(02), 78–83.

Mushtaq, M., Sultana, B., Anwar, F., Adnan, A., & Rizvi, S. S. (2015). Enzyme-assisted supercritical fluid extraction of phenolic antioxidants from pomegranate peel. *The Journal of Supercritical Fluids, 104*, 122–131.

Mushtaq, M., Sultana, B., Akram, S., Anwar, F., Adnan, A., & Rizvi, S. S. H. (2017). Enzyme-assisted supercritical fluid extraction: An alternative and green technology for non-extractable polyphenols. *Analytical and Bioanalytical Chemistry, 409*, 3645–3655.

Pham, T. L. (2014). Antibacterial properties of hylocereus undatus (white dragon fruit) against Opportunistic Pathogens present within the Skin, Oral Cavity and Gastrointestinal Tract. Quinnipiac University.

Rui, H., Zhang, L., Li, Z., & Pan, Y. (2009). Extraction and characteristics of seed kernel oil from white pitaya. *Journal of Food Engineering, 93*, 482–486.

Soxhlet, F. (1879). Die gewichtsaiialytische Bestimmung des Milchfettes; von.

Suh, D. H., Lee, S., Heo, D. Y., Kim, Y.-S., Cho, S. K., Lee, S., & Lee, C. H. (2014). Metabolite profiling of red and white pitayas (*Hylocereus polyrhizus* and *Hylocereus undatus*) for comparing Betalain biosynthesis and antioxidant activity. *Journal of Agricultural and Food Chemistry, 62*, 8764–8771.

US Food and Drug Administration. (2011). Listing of food additives status part II. Retrieved Oct.

Villalobos-Gutiérrez, M. G., Schweiggert, R. M., Carle, R., & Esquivel, P. (2012). Chemical characterization of Central American pitaya (*Hylocereus* sp.) seeds and seed oil. *CyTA – Journal of Food, 10*, 78–83.

Wijitra, L., Nathalie, D. C., Vera, V. H., & Koen, D. (2013). Dragon fruit (*Hylocereus* spp.) seed oils: Their characterization and stability under storage conditions. *Journal of the American Oil Chemists' Society, 90*, 207–215.

Chapter 37
Pomegranate (*Punica granatum*) Seed Oil

Sarfaraz Ahmed Mahesar, Abdul Hameed Kori, Syed Tufail Hussain Sherazi,
Aftab Ahmed Kandhro, and Zahid Husain Laghari

Abstract Pomegranate is an earliest and holy fruit affectionately known as the "jewel of winter" belongs to the Punicaceae family. Throughout the world, ~500 known pomegranate varieties available which reveal different quality characteristics of fruit such as size, shape, color, flavor and taste and seed hardness. The pomegranate seeds contain approximately 3% of total fruit weight, which contains typically oil in the range of 12–20%. Conjugated fatty acids are present in many plant oils with varying concentrations including pomegranate seed oil. Conjugated fatty acids are the geometric and positional isomers of polyunsaturated fatty acids with alternate double bonds. These fatty acids received remarkable interest due to valuable physiological effects on various diseases. The pomegranate seed oil contains higher concentration (>70%) of conjugated fatty acids in the form of punicic acid (9*cis*, 11*trans*, 13*cis*-conjugated linolenic acid). In the present chapter, chemistry and functionality of pomegranate fruit and seed oil especially conjugated fatty acids are reviewed.

Keywords Conjugated fatty acids · Ellagic acid · Tocols · Seed oil · Conjugated fatty acid · Punicic acid

Abbreviations

CLA	Conjugated linoleic acid
CLNA	Conjugated linolenic acid
E. coli	*Escherichia coli*
EA	Ellagic acid
ED	Erectile dysfunction

S. A. Mahesar (✉) · A. H. Kori · S. T. H. Sherazi · Z. H. Laghari
National Centre of Excellence in Analytical Chemistry, University of Sindh,
Jamshoro, Pakistan

A. A. Kandhro
Dr. M.A. Kazi Institute of Chemistry, University of Sindh, Jamshoro, Pakistan

© Springer Nature Switzerland AG 2019
M. F. Ramadan (ed.), *Fruit Oils: Chemistry and Functionality*,
https://doi.org/10.1007/978-3-030-12473-1_37

ETs	Ellagitannins
LDL	Low density lipid
NO	Nitric oxide
PA	Punicic acid
PFLE	Pomegranate flower leaf extract
PJ	Pomegranate juice
PPAR	Peroxisome proliferator-activated receptor
UV	Ultraviolet
α-EA	Ellagic acid

1 Introduction

Human beings use fruits as the main source of food. Researchers have reported that dry and fresh fruits can be used for the medical purpose as well as food (Marwat et al. 2009). Such studies arouse great enthusiasm among researchers and food companies to produce new varietal products and extract bioactive compounds from natural fruits that can have positive effects on human life (Viuda-Martos et al. 2010). Consumers use unprocessed or raw food and fruit juices to obtain rapid energy supply as well as maintain minerals quantity in the body.

Pomegranate (*Punica granatum* L.) is the oldest holy fruit and considered as "jewel of winter", belongs to the Punicaceae family. It follows the Latin name of the fruit Malum granatum, which means "grainy apple (Fig. 37.1). Pomegranate is native to India, Iran and its cultivation stretching all the way to the entire Mediterranean and Southwest American regions since ancient times (Celik et al. 2009; Lansky and Newman 2007). Current world production is estimated around 3.5 million ton per annum (Sinha et al. 2016). The leading producers of pomegranate are India, Iran, China and USA (Holland et al. 2008). In Pakistan pomegranate is harvested in the month of August to October in geographical locations of Gilgit

Fig. 37.1 Pomegranate plant, flower, fruit and oil

Table 37.1 Proximate composition of pomegranate seed

Constituent	Value	Reference
Ash (%)	0.47–1.887	Al-Maiman and Ahmad (2002) and Dadashi et al. (2013)
Oil (%)	13.5–19.3	Dadashi et al. (2013), Fadavi et al. (2006), Habibnia et al. (2012), Melgarejo and Artes (2000), Parashar (2010), Soetjipto et al. (2010), and Laghari et al. (2018)
Moisture (%)	77.72	Al-Maiman and Ahmad (2002)
Protein (%)	4.45–18.34	Al-Maiman and Ahmad (2002), Dadashi et al. (2013), and Laghari et al. (2018)
Crude fiber (%)	13.78–42.4	Dadashi et al. (2013) and Laghari et al. (2018)
Carbohydrates (%)	24.09–35.44	Dadashi et al. (2013) and Laghari et al. (2018)
Energy (Kcal/100 g)	355.52–460.7	Dadashi et al. (2013) and Laghari et al. (2018)

Baltistan, Waziristan, Kurram agency, Dir, Chitral, Hazara, west of Baluchistan and Azad Kashmir. Table 37.1 shows the composition of pomegranate seed.

The pomegranate plant is widely considering as large shrub or small tree (~5 m), mostly grown in hot and dry and in humidity and dry season to get produce high-quality fruit with good yield. The fruit of the pomegranate is considered as a large berry and can be divided into three parts (seed, juice and peel). The pomegranate fruit contains multi-ovule chambers (8–12) which are separated by fleshy mesocarp and membranous walls. The chambers are packed with numerous seeds (arils) and enveloped by a transparent juicy layer. Depending on the variety, the size of arils, the hardness of seed and color of the juicy layer can differ from deep red to white (Holland et al. 2009).

Approximately 3% of the fruit weight contains seed, 30% juice and rest is the peel, including interior membranes (Lansky and Newman 2007). Over hundreds of years, pomegranate has accompanied mankind as a symbol of longevity, life, morality, health, knowledge, and spirituality (Mackler et al. 2013). Table 37.2 shows the main components present in different parts of the pomegranate tree and fruit.

2 Types and Varieties of Pomegranates

Several varieties of pomegranates (~500) with different size, varying shapes, taste and color are cultivated throughout the world. Fruits are round, obvate in shape and vary in diameter from 8 to 12 cm. The rind may be thick or thin and the color ranges from pale yellow to crimson. The pulp in superior types is thick, fleshy and very juicy, while in inferior types it is thin (Fig. 37.1).

The seed coat varies in hardness, some of the softer seeded types known as seedless. There is a number of seedling verities of pomegranate available. Selecting a variety with known qualities is always the better choice. Most horticulturists divide

Table 37.2 Major components of pomegranate tree and fruit

Pomegranate part	Component	References
Juice	Anthocyanins, glucose, organic acid such as ascorbic acid, ellagic acid (EA), ellagitannins (ETs), gallic acid, caffeic acid, catechin, quercetin, rutin, minerals such as iron, sodium, potassium and amono acids	Heber et al. (2007), Ignarro et al. (2006), Jaiswal et al. (2010), Lansky and Newman (2007), Mousavinejad et al. (2009), and Poyrazoğlu et al. (2002)
Seed	Majorly conjugated fatty acids such as punicic acid, eleostearic acid, catalpic acid and other fatty acids linoleic acid, oleic acid, stearic acid, EA, and sterols	El-Nemr et al. (1990), Fadavi et al. (2006), Özgül-Yücel (2005), and Sassano et al. (2009)
Peel and rind	Phenolic punicalagins; gallic acid and other fatty acids; catechin, quercetin, rutin, and other flavonols, flavones, flavonones; anthocyanidins, Luteolin, Kaempferol, EA, punicalagin, punicalin, pedunculagin	Amakura et al. (2000), Seeram et al. (2006), and Van Elswijk et al. (2004)
Flower and leaf extract	Gallic acid, ursolic acid, triterpenoids, others maslinic Asiatic acid, tannins (punicalin and punicafolin) and flavone glycosides, luteolin and apigenin, EA, polyphenols and, punicalagin, punicalin	Aviram et al. (2008), Ercisli et al. (2007), Kaur et al. (2006), and Lan et al. (2009)
Roots and barks	ETs, including punicalin and punicalagin, many piperidine alkaloids	Gil et al. (2000) and Neuhofer et al. (1993)

pomegranate verities into three categories sweet, sweet-tart and sour. Hiwale (2009) divided pomegranate into six groups based on the hardiness of the seed.

1. Soft seeded sweet
2. Soft seeded tart
3. Early variety (mostly sweet)
4. Normal (harder) seeded sweet tart
5. Normal (harder) seeded sweet
6. Sour (nearly always normal seeded)

3 Chemical Composition of Pomegranates

The constituents present in pomegranate fruits vary due to climate, region, cultivation, maturity and environment of storage (Barzegar et al. 2004; Fadavi et al. 2005; Poyrazoğlu et al. 2002). Different researchers have reported variations in fatty and organic acids, sugar, phenolic compounds, minerals, and water-soluble vitamins in pomegranate (Aviram et al. 2000; Çam et al. 2009; Davidson et al. 2009; Mirdehghan and Rahemi 2007; Tezcan et al. 2009). Around 50% weight of pomegranate fruit consist of the peel. It has many significant bioactive compounds including phenolics, ellagitannins (ETs), flavonoids and anthocyanidin (Li et al. 2006), minerals such as nitrogen, phosphorus, calcium, sodium, magnesium and potassium (Mirdehghan and Rahemi 2007) as well as complex polysaccharide (Jahfar et al. 2003). It has been reported that edible parts of fruit mainly consist of 10% seeds and 40% arils. Water is the main part in aril ~85%, total sugars 10% (consists of glucose and fructose), pectin 1.5%, other organic and bioactive compounds like citric acid, ascorbic acids, malic acid, as well as flavonoids and phenols majorly anthocyanins (Aviram et al. 2000; Tezcan et al. 2009).

Pomegranate has strong antioxidant activity due to different compounds of polyphenols mainly ETs, gallotannins, EA acid, and flavonoids such as anthocyanins, quercetin, kaempferol and luteolin glycosides (Tabaraki et al. 2012). Punicalagin, an ETs, is the most abundant polyphenolic compound in pomegranate peel and responsible for biological properties (Bopitiya and Madhujith 2012; Mena et al. 2013). In contrast to pomegranate peels, the seeds are mainly composed by fatty acids and in a lesser extent by antioxidants such as gallic acid, methyl ellagic acid, hydroxycinnamic acids and tocopherols (Lansky and Newman 2007).

The seeds comprise around 3% of total fruit weight with varying chemical compositions (Table 37.3). As far as oil content in seeds is concerned, it has been reported that quantity and quality of oil depends upon maturity and geographical location of cultivated pomegranate fruits.

The pomegranate seed oil consists of >90% polyunsaturated fatty acids (PUFA) such as linoleic, and linolenic acids (Tables 37.4a and 37.4b), as well as other fatty acids such as stearic, oleic, and palmitic acids (Fadavi et al. 2006; Özgül-Yücel 2005). Generally, seed oil of pomegranate contains high proportions of PUFA, especially conjugated fatty acids (Kaufman and Wiesman 2007). The seed also contains fibers, protein, minerals, vitamins, sugars, pectin, polyphenols, the sex steroid, estrone, isoflavones (mainly genistein) and the phytoestrogen coumestrol (El-Nemr et al. 1990; Syed et al. 2007). Tables 37.5a and 37.5b show the tocol and sterol contents of pomegranate seed oil. It is widely accepted that the beneficial health effects of fruits and vegetables in the prevention of disease are due to the bioactive

Table 37.3 Physico-chemical properties of pomegranate seed oil

Parameters	Value	Reference
Specific gravity at 28 °C (g/cm³)	0.9300	Laghari et al. (2018)
Viscosity at 25 °C (*m* Pas.s)	0.037	Laghari et al. (2018)
Free fatty acid (%)	0.96–8.36	Amri et al. (2017a), Dadashi et al. (2013), and Laghari et al. (2018)
Saponification value (mg/KOH)	156–182.5	Dadashi et al. (2013) and Laghari et al. (2018)
Peroxide value (meq O_2/kg)	0.39–3.42	Amri et al. (2017a), Dadashi et al. (2013), and Laghari et al. (2018)
Iodine value (g I_2/100 g)	212–220.34	Dadashi et al. (2013) and Laghari et al. (2018)
Conjugated dienes	4.15	Amri et al. (2017a)
Conjugated trienes	3.95	Amri et al. (2017a)

Table 37.4a Individual fatty acid composition of pomegranate seed varieties available in the world

Fatty acid	Brazil		India		Indonesia		Iran		Italy	Japan	Pakistan	Spain	Tunisia		Turkey		USA
	a	b	c	d	e	f	g	h	i	j	k	l	m	n	o	p	q
C12:0 lauric acid				0.02–0.37				0.01				0.50	0.51	0.02–0.14			
C14:0 myristic acid				0.03–0.04	0.7			0.03–0.04	0.02–0.03				0.36	0.04–0.08			0.35
C14:1 cis-myristoleic acid													0.14	0.04–0.07			
C16:0 palmitic acid	4.04	2.77	2.87	18.16–22.63	5.7	0–9	2.95–3.57	4.04–4.46	2.68–4.28	3.1	2.88	2.58–14.91	22.08	3.77–7.81	2.0	2.45	4.0
C16:1 n7 (trans) trans-9-palmitoleic acid								0.06–0.09					0.40	0.07–0.08			
C16:1 n7 (cis) cis-9-palmitoleic acid				0.22–2.70					0.02–0.04			0.93–2.30	0.27	0.09–0.22			
C 16:1 n9 cis													1.88				
C 17:0 margaric								0.01	0.05–0.08				0.54				
C17:1 w8 cis-9-heptadecyl enic acid/ margaroleic								0.01–0.02					0.83	0.10			
C18:0 stearic acid	2.30	2.42	2.26	8.10–10.42	2.1	9–11	1.99–2.54	2.81–3.0	1.44–1.92	2.0	3.57	1.16–8.98	8.94	2.13–2.26	1.6	1.52	2.92
C18:1 n9 (trans) trans-elaidic acid								0.06–0.07					0.04	3.52–11.10			
C18:1 n9 (cis) cis-oleic acid	5.29	5.74	6.82	24.76–31.26	9.0	19–21	5.71–7.48	8.31–9.77	3.63–7.12	4.5	3.85	3.67–20.25	10.47	0.02–0.82	3.7	4.19	5.68

Fatty acid														
C18:1 n7 (cis) cis-11-octadecanoic acid	7.29	6.05	10.8	20–21						2.12	0.05–22.15			
C18:1 7 trans vaccenic		0.34												
C18:2 (t9,c12) trans-9, cis-12- Octadecadienoic acid						0.36–0.67 cis				0.01	0.04–0.85			
C18:2 (c9,t12) cis-9,trans-12- octadecadienoic acid										0.03	0.34–0.51			
C18:2 w6 (c9, c12) cis-cis-linoleic acid		6.46	31.49–38.61	5.22–7.08	8.11–9.03	4.11–11.32	5.1	2.67	5.19–16.50	28.86	0.11–6.13	3.3	4.49	4.08
C18:3 w6 cis-linolenic acid gamma			0.61–9.94							2.82	0.05–43.13			
C18:3 w3 (cis) cis-linolenic acid alpha	0.39				0.04–0.10	0.05–0.38				1.02	0.06–3.45	0.1		
C18:2(t9,t11) trans-trans octadecadienoic acid/linoelaidic					0.30–0.35	0.02–0.05				0.10	0.16–1.35			

(continued)

Table 37.4a (continued)

Fatty acid	Brazil		India			Indonesia	Iran	Italy		Japan	Pakistan	Spain	Tunisia		Turkey		USA
	a	b	c	d	e	f	g	h	i	j	k	l	m	n	o	p	q
C18:2 (c11,t13) cis-11, trans-13-octadecadienoic acid													0.10	0.04–0.06			
C18:2 (t10,c12) trans-10, cis-12-octadecadienoic acid													0.16	0.35–0.40			
C20:0 arachidic acid	0.50	0.39	0.49	1.06–2.76				0.60–0.64	0.37–0.55		1.22	0.66–2.76	0.91	0.83–1.53	3.0	0.39	0.53
C20:1 w9 eicosenoic acid	0.61	0.39	0.64					0.90–1.08	0.41–1.07				0.43	0.22–1.11	0.1	0.61	
C2O:2 eicosadienoic acid								1.08	0.47–0.73				0.08	0.18–0.43			
C2O:3 w6 dihomo-linolenic acid													0.14	0.04–1.31			
C2O:3 w3 eicosatrienoic acid													0.11	0.45–0.49			
C2O:5 w3 eicosapentaenoic acid									0.47–0.73								
C21:0 heneicosylic acid		2.91															
C22:0 behenic acid				0.23–0.85	0.1				0.1–0.25				1.25	0.19–1.63		0.18	

Fatty acid	a	b	c	d	e	f	g	h	i	j	k	l	m	n	o	p
C22:1 erucic acid	58.14	54.90						0.71–1.54								
C22:2 docosadienoic acid								0.14–0.49								
C18:3 (c9,t11,c13) punicic acid–trichosanic osanic acid		71.76		71.5	9–16	78.25–82.4	72.07–73.31	72.42–84.11	71.7	84.68	43.43–88.22	5.12	2.23–40.10	57.3	74.11	81.22
C18:3 (c8,t10,c12) calendic acid	16.07	4.6										1.41	0.17–0.89			
C18:3 (c9,t11,t13) alpha-eleostearic acid/catalpic	6.71	2.64							2.8	1.13		2.97	0.21–13.79	6.41		
C18:3 (t9,t11,c13) trans-9,trans-11,cis-13-octadecatrienoic/catalpic acid									5.1			3.04	0.31–5.40	7.6	3.48	
C18:3 (t9,t11,t13) trans-beta eta eleostearic acid		0.49							1.6			0.45	0.06–14.57	21.1	1.03	
C24:0 lignoceric acid		0.05	0.14–0.77				0.04–0.10					0.58	0.05–1.08	1.0		
C24:1 nervonic acid	0.24											0.15	0.16–0.36			

a: Melo et al. (2014), b: Melo et al. (2016), c: Sassano et al. (2009), d: Parashar et al. (2009), e: Parashar (2010), f: Soetjipto et al. (2010), g: Habibnia et al. (2012), h: Dadashi et al. (2013), i: Verardo et al. (2014), j: Suzuki et al. (2001), k: Laghari et al. (2018), l: Melgarejo et al. (2000), m: Amri et al. (2017a), o: Ozgul-Yucel. (2005), p: Ozgen et al. (2008), q: Pande (2009)

Table 37.4b Fatty acid composition of phospholipids and glycolipids in pomegranate seed oil

Fatty acid	Glycolipids		Phospholipids		
	Tunisia	Indonesia	Tunisia	Indonesia	Italy
	Amri et al. (2017a)	Soetjipto et al. (2010)	Amri et al. (2017a)	Soetjipto et al. (2010)	Verardo et al. (2014)
C12:0 lauric acid	1.19		0.97		
C14:0 myristic acid	0.42		0.25		
C14:1 cis-myristoleic acid	0.32		0.05		
C16:0 palmitic acid	38.25	0–6	43.00	0–25	9.96–31.33
C16:1 n7 (trans) trans-9-palmitoleic acid	0.37		0.31		1.07–8.19
C16:1 n7 (cis) cis-9-palmitoleic acid	0.45		0.30		
C 16:1 n9 cis	0.30		0.25		
C 17:0 margaric	0.96		0.69		
C17:1 w8 cis-9-heptadecyl enic acid/margaroleic	0.28		0.53		
C18:0 stearic acid	22.40	0–10	24.24	10–12	5.92–21.37
C18:1 n9 (trans) trans-elaidic acid	0.13		0.03		
C18:1 n11 (trans) vccenic acid					
C18:1 n9 (cis) oleic acid	7.74	7–14	8.88	14–34	8.65–32.07
C18:1 n7 (cis) cis-11-octadecanoic acid	1.13		1.36		
C18:2 (t9,c12) trans-9, cis-12-octadecadienoic acid	0.14		0.01		
C18:2 (c9,t12) cis-9,trans-12-octadecadienoic acid	0.13		0.01		
C18:2 w6 (c9, c12) cis-cis-linoleic acid	9.60	6–30	9.98	6–16	6.91–18.80
C18:2(t9,t11)trans-trans octadecadienoic acid/ linoelaidic	0.15		0.21		
C18:2 (c11,t13) cis-11, trans-13-octadecadienoic acid	0.05		0.02		
C18:2 (t10,c12) trans-10, cis-12-octadecadienoic acid	0.12		0.01		
C18:3 w6 cis-linolenic acid	3.64		3.76		
C18:3 w3 (cis) cis-linolenic acid	0.46		0.35		
C18:3 (c9,t11,c13) punicic acid–trichosanic acid	1.42	0–42	0.68	0–22	16.81–62.40

(continued)

Table 37.4b (continued)

	Glycolipids		Phospholipids		
	Tunisia	Indonesia	Tunisia	Indonesia	Italy
Fatty acid	Amri et al. (2017a)	Soetjipto et al. (2010)	Amri et al. (2017a)	Soetjipto et al. (2010)	Verardo et al. (2014)
C18:3 (c8,t10,c12) calendic acid	0.91		0.03		
C18:3 (c9,t11,t13) alpha-eleostearic acid	0.43		0.03		
C18:3 (t9,t11,c13) trans-9,trans-11,cis-13-octadecatrienoic/catalpic acid	0.43		1.20		
C18:3 (t9,t11,t13) trans-beta eleostearic acid	0.25		0.05		
C20:0 arachidic acid	1.28		1.41		
C2O:1 w9 gadoleic acid/eicosenoic	0.06		0.22		
C2O:2 eicosadienoic acid	0.08		0.01		
C2O:3 w6 dihomo-linolenic acid	0.05		0.01		
C2O:3 w3 eicosatrienoic acid	0.13		0.01		
C21:0					
C22:0 behenic acid	1.35		0.90		
C22:1 erucic acid/docosenoic acid					0.49–3.33
C22:2 erucic acid/docosadienoic acid					1.59–16.44
C24:0 lignoceric acid	0.25		0.51		
C24:1 nervonic acid	0.14		0.03		

compounds they contain (Galaverna et al. 2008). Almaiman and Ahmed (2002) reported phenols and ascorbic acid 1.90 and 0.18 mg/100 g, respectively in pomegranate seed. In Tunisian pomegranates, Amri et al. (2017a) has noted following compounds in seed oil like phenols 93.4 mg/kg, flavonoid 59.4 mg/kg, *O*-diphenols 30.1 mg/kg, and pigments (chlorophyll 3.17 mg/kg, and β-carotene 3.17 mg/kg).

4 Pomegranate Seed Oil

Fatty acids are carboxylic acids with a long chain of aliphatic hydrocarbons either saturated or unsaturated. Fatty acids are found in open chain and derived from triglycerides or phospholipids. There are three types of hydrocarbons chain in fatty acids such as small, medium and long chain depends upon a number of carbon atoms present in the chain. The small chain consists of 4–6 carbon atoms, while

Table 37.5a Tocols content in extracted pomegranate seed oil

Country	α-Tocopherol	γ-Tocopherol	β-Tocopherol	δ-Tocopherol	α-Tocotrienol	β-Tocotrienol	Reference
Iran (mg/1000 g)	543.6–1134.6	1856.6–7106	–	–	–	–	Habibnia et al. (2012)
Brazil (mg/100 g)	3.81	153.21	1.03	17.04	–	–	Melo et al. (2016)
Tunisia (mg/100 g)	165.77	107.38	–	27.29	–	–	Elfalleh et al. (2011)
Italy (mg/100 g)	2.54–16.9	61.65–240.08	–	0.78–3.56	0.73–2.76	1.25–5.21	Verardo et al. (2014)

Table 37.5b Content of phytosterols in pomegranate seed oils

| | Country | | | |
| | Iran (g/100 g) | Brazil (mg/100 g) | Tunisia (g/100 g) | Italy (mg/g) |
Components	Habibnia et al. (2012)	Melo et al. (2016)	Amri et al. (2017a)	Verardo et al. (2014)
Cholesterol	0.37–0.40		0–23	
Campesterol	7.56–8.83	49	6–35	0.50–1.23
Stigmasterol	3.14–5.93	12	3–21	0.20–0.57
Beta-sitosterol	85.49–87.7	374	77–94	5.13–11.42
24-Methylene-cholesterol			0–1	
Campestanol			0–8	
Δ5,[23]Stigmastadienol			0–5	
Clerosterol			1–23	
Sitostanol			0–44	
Avenasterol			7–45	
Δ5,[24]-Stigmastadienol			0–93	
Δ7-Stigmastenol			0–27	
Δ5-Avenasterol				0.93–2.42
Δ7-Avenasterol			0–76	
Erythrodiol			0–34	
Uvaol			0.77	
Others		104		
Citrostadienol				0.26–0.82

the medium chain has 8–18 carbons and long chain contains above 18 carbons. Most of the plant seed oils usually contain unsaturated fatty acids in unconjugated form except a few seed oils, which contain conjugated double, triple or tetraenes bonds. Examples of conjugated double and triple bonds are conjugated linoleic acids (CLA) and conjugated linolenic acids (CLNA). Presence of conjugated fatty acids in oil have been an object of studies. Basically, CLNA is a mixture of octa-decatrienoic fatty acid isomers. These isomers include geometrical (cis and/or *trans*) and positional forms of linolenic acid (11,13,15–18:3, 10,12,14–18:3; 9,11,13–18:3; 8,10,12–18:3). A number of plant seeds contain very high concentration (30–70% of lipids) of CLNA isomers as shown in Table 37.6. Among them pomegranate seed oil contains a higher amount of CLNA. The important CLNA present in pomegranate seed oil is PA (Fig. 37.2) (9-*trans*, 11-*cis*, 13-*trans*) which consists of approximately 70–90% of total fatty acids (Abbasi et al. 2008; Tanaka et al. 2011).

Table 37.6 Conjugated linolenic acids contents in some oils (Tanaka et al. 2011)

Seed oil	Type of CLN	Isomers	CLNA (%)
Bitter gourd	α-eleostearic	9c,11t,13t-18:3	>50
	β-eleostearic acid	9t,11t,13t-18:3	
Pot marigold	α -calendic	8t,10t,12c-18:3	>30
	β-calendic	8t,10t,12t-18:3	
Catalpa	Catalpic acid	9t,11t,13c-18:3	>40
Calendula	Calendic acid	8t,10t,12c-18:3	>55
Jacaranda	Jacaric acid	8c,10t,12c-18:3	>35
Pomegranate	Punicic acid	9c,11t,13c-18:3	>70

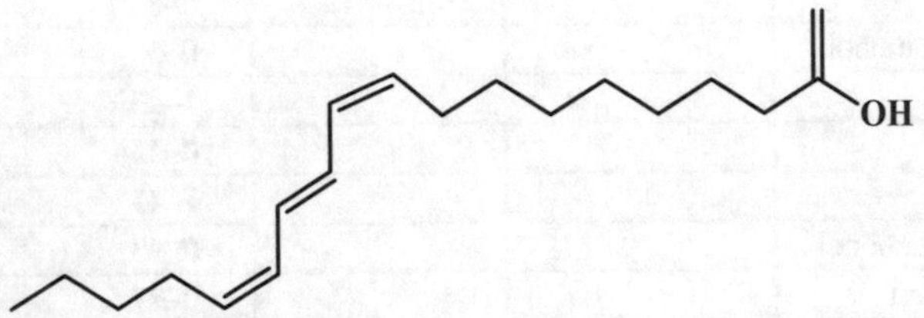

Fig. 37.2 Molecular structure of punicic acid (PA)

4.1   Punic Acid (PA)

It has been reported that PA showed a strong eicosanoid enzyme inhibition properties (Eikani et al. 2012). PA reduced fasting glucose in diabetics II, diet-induced
obesity and insulin resistance, inflammation of colon, bladder, breast and prostate
cancer, nephrotoxic activity, formation of hydroperoxide and improve bone mineral
density (Banihani et al. 2013; Bouroshaki et al. 2010; Boussetta et al. 2009;
Grossmann et al. 2010; Kohno et al. 2004; Lansky et al. 2005; Mukherjee and
Bhattacharyya 2006; Spilmont et al. 2013; Wang and Martins-Green 2014). PA and
α-EA also reduce the activity of sodium arsenite that is responsible for oxidative
stress and deoxy (DNA) damage (Saha and Ghosh 2009).

4.2   FT-IR Spectrum of Pomegranate Seed Oil

The infrared spectrum of pomegranate seed oil is shown in Fig. 37.3. The characteristics functional groups present in pomegranate seed oil are resembled with other
vegetable oils, except in the region of 1050–730 cm^{-1} due to the presence of CLNA.
Prashantha et al. (2009) reported that isomers of eleostearic acid show a strong
spectral band at 993 cm^{-1} corresponding to β-eleostearic acid (*trans*: *trans*: *trans*)
and a doublet with a strong band at 991 cm^{-1} and a weaker band at 963 cm^{-1}

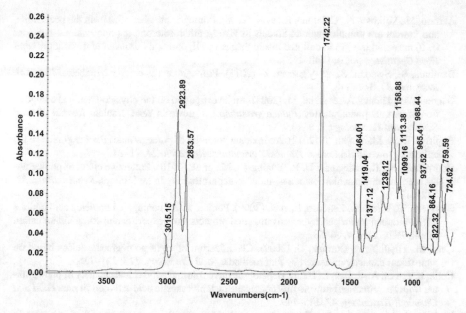

Fig. 37.3 FT-IR spectrum of pomegranate seed oil

corresponding to the α-eleostearic acid (*cis: trans: trans*). The corresponding doublet at 988 and 937 cm⁻¹ in Fig. 37.3, confirmed the presence of CLNA in the pomegranate oil, in general these doublet bands are absent in vegetable oils.

References

Abbasi, H., Rezaei, K., & Rashidi, L. (2008). Extraction of essential oils from the seeds of pomegranate using organic solvents and supercritical CO_2. *Journal of the American Oil Chemists' Society, 85*, 83–89.

Al-Maiman, S. A., & Ahmad, D. (2002). Changes in physical and chemical properties during pomegranate (*Punica granatum* L.) fruit maturation. *Food Chemistry, 76*, 437–441.

Amakura, Y., Okada, M., Tsuji, S., et al. (2000). High-performance liquid chromatographic determination with photodiode array detection of ellagic acid in fresh and processed fruits. *Journal of Chromatography A, 896*, 87–93.

Amri, Z., Lazreg-Aref, H., Mekni, M., et al. (2017a). Oil characterization and lipids class composition of pomegranate seeds. *Biomed Research International*, 2037341, 8 pp. https://doi.org/10.1155/2017/2037341.

Amri, Z., Zaouay, F., Lazreg-Aref, H., et al. (2017b). Phytochemical content, fatty acids composition and antioxidant potential of different pomegranate parts: Comparison between edible and non edible varieties grown in Tunisia. *International Journal of Biological Macromolecules, 104*, 274–280.

Aviram, M., Dornfeld, L., Rosenblat, M., et al. (2000). Pomegranate juice consumption reduces oxidative stress, atherogenic modifications to LDL, and platelet aggregation: Studies in humans and in atherosclerotic apolipoprotein E-deficient mice. *The American Journal of Clinical Nutrition, 71*, 1062–1076.

Aviram, M., Volkova, N., Coleman, R., et al. (2008). Pomegranate phenolics from the peels, arils, and flowers are antiatherogenic: Studies in vivo in atherosclerotic apolipoprotein e-deficient (E 0) mice and in vitro in cultured macrophages and lipoproteins. *Journal of Agricultural and Food Chemistry, 56*(3), 1148–1157.

Banihani, S., Swedan, S., & Alguraan, Z. (2013). Pomegranate and type 2 diabetes. *Nutrition Research, 33*, 341–348.

Barzegar, M., Fadavi, A., & Azizi, M. (2004). An investigation on the physico-chemical composition of various pomegranates (*Punica granatum* L.) grown in Yazd. Iranian. *Journal of Food Science and Technology, 1*, 9–14.

Bopitiya, D., & Madhujith, T. (2012). Antioxidant potential of pomegranate (*Punica granatum* L.) cultivars grown in Sri Lanka. *Tropical Agricultural Research, 24*, 71–81.

Bouroshaki, M. T., Sadeghnia, H. R., Banihasan, M., et al. (2010). Protective effect of pomegranate seed oil on hexachlorobutadiene-induced nephrotoxicity in rat kidneys. *Renal Failure, 32*, 612–617.

Boussetta, T., Raad, H., Lettéron, P., et al. (2009). Punicic acid a conjugated linolenic acid inhibits TNFα-induced neutrophil hyperactivation and protects from experimental colon inflammation in rats. *PLoS One, 4*, e6458.

Çam, M., Hışıl, Y., & Durmaz, G. (2009). Classification of eight pomegranate juices based on antioxidant capacity measured by four methods. *Food Chemistry, 112*, 721–726.

Celik, I., Temur, A., & Isik, I. (2009). Hepatoprotective role and antioxidant capacity of pomegranate (*Punica granatum*) flowers infusion against trichloroacetic acid-exposed in rats. *Food and Chemical Toxicology, 47*, 145–149.

Dadashi, S., Mousazadeh, M., Emam-Djomeh, Z., et al. (2013). Pomegranate (*Punica granatum* L.) seed: A comparative study on biochemical composition and oil physicochemical characteristics. *International Journal of Advanced Biological and Biomedical Research, 1*, 351–363.

Davidson, M. H., Maki, K. C., Dicklin, M. R., et al. (2009). Effects of consumption of pomegranate juice on carotid intima-media thickness in men and women at moderate risk for coronary heart disease. *The American Journal of Cardiology, 104*, 936–942.

Eikani, M. H., Golmohammad, F., & Homami, S. S. (2012). Extraction of pomegranate (*Punica granatum* L.) seed oil using superheated hexane. *Food and Bioproducts Processing, 90*, 32–36.

Elfalleh, W., Tlili, N., Nasri, N., et al. (2011). Antioxidant capacities of phenolic compounds and tocopherols from tunisian pomegranate (*Punica granatum*) fruits. *Journal of Food Science, 76*(5), C707–C713.

El-Nemr, S., Ismail, I., & Ragab, M. (1990). Chemical composition of juice and seeds of pomegranate fruit. *Molecular Nutrition & Food Research, 34*, 601–606.

Ercisli, S., Agar, G., Orhan, E., et al. (2007). Interspecific variability of RAPD and fatty acid composition of some pomegranate cultivars (*Punica granatum* L.) growing in southern Anatolia region in Turkey. *Biochemical Systematics and Ecology, 35*, 764–769.

Fadavi, A., Barzegar, M., Azizi, M., et al. (2005). Note. Physicochemical composition of ten pomegranate cultivars (*Punica granatum* L.) grown in Iran. *Food Science and Technology International, 11*, 113–119.

Fadavi, A., Barzegar, M., & Azizi, M. H. (2006). Determination of fatty acids and total lipid content in oilseed of 25 pomegranates varieties grown in Iran. *Journal of Food Composition and Analysis, 19*, 676–680.

Galaverna, G., Di Silvestro, G., Cassano, A., et al. (2008). A new integrated membrane process for the production of concentrated blood orange juice: Effect on bioactive compounds and antioxidant activity. *Food Chemistry, 106*, 1021–1030.

Gil, M. I., Tomás-Barberán, F. A., Hess-Pierce, B., et al. (2000). Antioxidant activity of pomegranate juice and its relationship with phenolic composition and processing. *Journal of Agricultural and Food Chemistry, 48*, 4581–4589.

Grossmann, M. E., Mizuno, N. K., Schuster, T., et al. (2010). Punicic acid is an ω-5 fatty acid capable of inhibiting breast cancer proliferation. *International Journal of Oncology, 36*, 421–426.

Habibnia, M., Ghavami, M., Ansaripour, M., et al. (2012). Chemical evaluation of oils extracted from five different varieties of Iranian pomegranate seeds. *Journal of Food Biosciences and Technology, 2*, 35–40.

Heber, D., Seeram, N. P., Wyatt, H., et al. (2007). Safety and antioxidant activity of a pomegranate ellagitannin enriched polyphenol dietary supplement in overweight individuals with increased waist size. *Journal of Agricultural and Food Chemistry, 55*(24), 10050–10054.

Hiwale, S. S. (2009). *The pomegranate*. New Delhi: New India Publishing Agency.

Holland, D., Larkov, O., & Bar Ya akov, I. (2008). The pomegranate: New interest in an ancient fruit. *Chronicle Horticulture, 48*, 12–15.

Holland, D., Hatib, K., & Bar-Ya'akov, I. (2009). Pomegranate: Botany, horticulture, breeding. *Horticultural Reviews, 35*, 127–191.

Ignarro, L. J., Byrns, R. E., Sumi, D., et al. (2006). Pomegranate juice protects nitric oxide against oxidative destruction and enhances the biological actions of nitric oxide. *Nitric Oxide, 15*, 93–102.

Jahfar, M., Vijayan, K., & Azadi, P. (2003). Studies on a polysaccharide from the fruit rind of *Punica granatum*. *Research Journal of Chemistry and Environment, 7*, 43–50.

Jaiswal, V., DerMarderosian, A., & Porter, J. R. (2010). Anthocyanins and polyphenol oxidase from dried arils of pomegranate (*Punica granatum* L.). *Food Chemistry, 118*, 11–16.

Kaufman, M., & Wiesman, Z. (2007). Pomegranate oil analysis with emphasis on MALDI-TOF/MS triacylglycerol fingerprinting. *Journal of Agricultural and Food Chemistry, 55*, 10405–10413.

Kaur, G., Jabbar, Z., Athar, M., et al. (2006). *Punica granatum* (pomegranate) flower extract possesses potent antioxidant activity and abrogates Fe-NTA induced hepatotoxicity in mice. *Food and Chemical Toxicology, 44*, 984–993.

Kohno, H., Suzuki, R., Yasui, Y., et al. (2004). Pomegranate seed oil rich in conjugated linolenic acid suppresses chemically induced colon carcinogenesis in rats. *Cancer Science, 95*, 481–486.

Laghari, Z. H., Mahesar, S. A., Sherazi, S. T. H., Memon, S. A., Sirajuddin, Mugheri, G. A., Shah, S. N., Panhwar, T., & Chang, A. S. (2018). Quality evaluation of pomegranate waste and extracted oil. *International Food Research Journal, 25*(3), 1295–1299.

Lan, J. Q., Lei, F., Hua, L., et al. (2009). Transport behavior of ellagic acid of pomegranate leaf tannins and its correlation with total cholesterol alteration in HepG2 cells. *Biomedical Chromatography, 23*, 531–536.

Lansky, E. P., & Newman, R. A. (2007). *Punica granatum* (pomegranate) and its potential for prevention and treatment of inflammation and cancer. *Journal of Ethnopharmacology, 109*, 177–206.

Lansky, E. P., Harrison, G., Froom, P., et al. (2005). Pomegranate (*Punica granatum*) pure chemicals show possible synergistic inhibition of human PC-3 prostate cancer cell invasion across Matrigel™. *Investigational New Drugs, 23*, 121–122.

Li, Y., Guo, C., Yang, J., et al. (2006). Evaluation of antioxidant properties of pomegranate peel extract in comparison with pomegranate pulp extract. *Food Chemistry, 96*, 254–260.

Mackler, A. M., Heber, D., & Cooper, E. L. (2013). Pomegranate: Its health and biomedical potential. *Evidence-Based Complementary and Alternative Medicine*, 903457, 2 pp. https://doi.org/10.1155/2013/903457.

Marwat, S. K., Khan, M. A., Khan, M. A., et al. (2009). Fruit plant species mentioned in the Holy Qura'n and Ahadith and their ethnomedicinal importance. *American-Eurasian Journal of Agricultural & Environmental Sciences, 5*, 284–295.

Melgarejo, P., & Artes, F. (2000). Total lipid content and fatty acid composition of oilseed from lesser known sweet pomegranate clones. *Journal of the Science of Food and Agriculture, 80*, 1452–1454.

Melo, I. L.P., Carvalho, E. B. T., & Filho, J. M. (2014). Pomegranate seed oil (*Punica granatum* L.): A source of punicic acid (conjugated α-linolenic acid). *Journal of Human Nutrition & Food Science, 2*(1), 1-11.

Melo, I. L.P., Carvalho, E. B. T. D., Silva, A. M. O., et al. (2016). Characterization of constituents, quality and stability of pomegranate seed oil (*Punica granatum* L.). *Food Science and Technology, 36*, 132–139.

Mena, P., Vegara, S., Martí, N., et al. (2013). Changes on indigenous microbiota, colour, bioactive compounds and antioxidant activity of pasteurised pomegranate juice. *Food Chemistry, 141*, 2122–2129.

Mirdehghan, S. H., & Rahemi, M. (2007). Seasonal changes of mineral nutrients and phenolics in pomegranate (*Punica granatum* L.) fruit. *Scientia Horticulturae, 111*, 120–127.

Mousavinejad, G., Emam-Djomed, Z., Rezaei, K., et al. (2009). Identification and quantification ofphenolic compounds and their effects on antioxidant activity in pomegranate juices of eight Iranian cultivars. *Food Chemistry, 115*, 1274–1278.

Mukherjee, C., & Bhattacharyya, D. (2006). Oxidative stability of some seed oils containing conjugated octadecatrienoic fatty acids isomers. *Journal of Lipid Science and Technology, 32*, 225–227.

Neuhofer, H., Witte, L., Gorunovic, M., et al. (1993). Alkaloids in the bark of *Punica granatum* L. (pomegranate) from Yugoslavia. *Pharmazie, 48*, 389–391.

Ozgen, M., Durgaç, C., Serçe, S., & Kaya, C. (2008). Chemical and antioxidant properties of pomegranate cultivars grown in the Mediterranean region of Turkey. *Food Chemistry, 111*(3), 703–706.

Özgül-Yücel, S. (2005). Determination of conjugated linolenic acid content of selected oil seeds grown in Turkey. *Journal of the American Oil Chemists' Society, 82*, 893–897.

Pande, G., & Akoh, C. C. (2009). Antioxidant capacity and lipid characterization of six Georgia grown pomegranate cultivers. *Journal of Agricultural and Food Chemistry, 57*, 9427–9436.

Parashar, A. (2010). Lipid content and fatty acid composition of seed oils from six pomegranate cultivars. *International Journal of Fruit Science, 10*(4), 425–430.

Parashar, A., Gupta, C., Gupta, S., et al. (2009). Antimicrobial ellagitannin from pomegranate (*Punica granatum*) fruits. *International Journal of Fruit Science, 9*, 226–231.

Poyrazoğlu, E., Gökmen, V., & Artık, N. (2002). Organic acids and phenolic compounds in pomegranates (*Punica granatum* L.) grown in Turkey. *Journal of Food Composition and Analysis, 15*, 567–575.

Prashantha, M. A. B., Premachandra, J. K., & Amarasinghe, A. D. U. S. (2009). Composition, physical properties and drying characteristics of seed oil of *Momordica charantia* cultivated in Sri Lank. *Journal of the American Oil Chemists' Society, 86*, 27–32.

Saha, S., & Ghosh, M. (2009). Comparative study of antioxidant activity of α-eleostearic acid and punicic acid against oxidative stress generated by sodium arsenite. *Food and Chemical Toxicology, 47*, 2551–2556.

Sassano, G., Sanderson, P., Franx, J., et al. (2009). Analysis of pomegranate seed oil for the presence of jacaric acid. *Journal of the Science of Food and Agriculture, 89*, 1046–1052.

Seeram, N. P., Schulman, R. N., & Heber, D. (2006). *Pomegranates: Ancient roots to modern medicine*. Boca Raton: CRC/Taylor and Francis.

Sinha, S., Thakur, D. S., Mishra, P. K., et al. (2016). D2-analysis suggests wider genetic divergence in pomegranate genotypes. *The Bioscan, 11*(2), 1011–1015.

Soetjipto, H., Pradipta, M., & Timotius, K. (2010). Fatty acids composition of red and purple pomegranate (*Punica granatum* L) seed oil. *Journal of Cancer Chemoprevention, 1*, 74–77.

Spilmont, M., Léotoing, L., Davicco, M.-J., et al. (2013). Pomegranate seed oil prevents bone loss in a mice model of osteoporosis, through osteoblastic stimulation, osteoclastic inhibition and decreased inflammatory status. *The Journal of Nutritional Biochemistry, 24*, 1840–1848.

Suzuki, R., Noguchi, R., Ota, T., Abe, M., Miyashita, K., & Kawada, T. (2001). Cytotoxic effect of conjugated trienoic fatty acids on mouse tumor and human monocytic leukemia cells. *Lipids, 36*(5), 477–482.

Syed, D. N., Afaq, F., & Mukhtar, H. (2007). Pomegranate derived products for cancer chemoprevention. *Seminars in Cancer Biology, 17*(5), 377–385.

Tabaraki, R., Heidarizadi, E., & Benvidi, A. (2012). Optimization of ultrasonic-assisted extraction of pomegranate (*Punica granatum* L.) peel antioxidants by response surface methodology. *Separation and Purification Technology, 98*, 16–23.

Tanaka, T., Hosokawa, M., Yasui, Y., et al. (2011). Cancer chemopreventive ability of conjugated linolenic acids. *International Journal of Molecular Sciences, 12*(11), 7495–7509.

Tezcan, F., Gültekin-Özgüven, M., Diken, T., et al. (2009). Antioxidant activity and total phenolic, organic acid and sugar content in commercial pomegranate juices. *Food Chemistry, 115*, 873–877.

Van Elswijk, D. A., Schobel, U. P., Lansky, et al. (2004). Rapid dereplication of estrogenic compounds in pomegranate (*Punica grantum*) using on-line biochemical detection coupled to mass spectrometry. *Phytochemistry, 65*, 233–241.

Verardo, V., Garcia-Salas, P., Baldi, E., et al. (2014), Pomegranate seeds as a source of nutraceutical oil naturally rich in bioactive lipids. *Food Research International, 65*, 445–452.

Viuda-Martos, M., Fernández-López, J., & Pérez-Álvarez, J. (2010). Pomegranate and its many functional components as related to human health: A review. *Comprehensive Reviews in Food Science and Food Safety, 9*, 635–654.

Wang, L., & Martins-Green, M. (2014). Pomegranate and its components as alternative treatment for prostate cancer. *International Journal of Molecular Sciences, 15*, 14949–14966.

Chapter 38
Sandalwood (*Santalum album*) Oil

Omprakash H. Nautiyal

Abstract Sandalwood (*Santalum album*) oil is an important export commodity in many countries. It is important for the industry to have a capacity for rapid and accurate determination of oil content and quality for commercial samples. The effect of extraction methods (i.e., steam-distillation, hydro-distillation, subcritical CO_2 extraction and solvent extraction) on the oil yield and concentration of major components in a commercial sample was reported. The highest oil yield was obtained from subcritical CO_2 (3.83 g/L) extraction followed by the solvent extraction (2.45–3.7 g/L), hydrodistillations (1.86–2.68 g/L) and steam distillation (1.60 g/L). The highest levels of α- and β-santalol were found in the oils extracted with subcritical CO_2 (83%), ethyl alcohol (84%) and steam distillation (84%). Organoleptic properties were remarkable in case of subcritical CO_2. FTIR analysis has shown the sharp peaks for santalol and santalene in the oil extracted by subcritical CO_2. Three of the four solvent-extracted sandalwood oils were recorded as 'less pleasant' indicating the generally inferior note of oil derived from these methods. Given the highest yield, the highest level of santalols, it could be concluded that subcritical CO_2 is the best technology for sandalwood oil extraction.

Keywords FTIR · Extraction technologies · Carbon dioxide · *Santalum album* · *Santalum yasi* · *Santalum austrocaledonicum*

1 Introduction

Sandalwood is a tree known for its oil and timber, used in the production of perfumes and cosmetics. The golden-brown fruits (nuts), 15–20 mm in diameter, have a thin fleshy layer which can be dried and stored (Fig. 38.1). The sandalwood tree is a small, evergreen tree with a brown-gray trunk and small pink-purple flowers also found in Indonesia, Malaysia and Taiwan. The heartwood and the roots of the tree

O. H. Nautiyal (✉)
Department of Chemistry, Lovely Professional University,
GT Road, Chaheru, Punjab, India

© Springer Nature Switzerland AG 2019 711
M. F. Ramadan (ed.), *Fruit Oils: Chemistry and Functionality*,
https://doi.org/10.1007/978-3-030-12473-1_38

Fig. 38.1 *S. spicatum* nuts
with the husk

employed for distilling essential oil. The tree has to be over 30 years old before it is
ready to produce essential oil is a major factor in the high price of sandalwood
essential oil. Sandalwood fruits (nuts) are a popular food of the emu, which swal-
lows them whole. The nuts are delicious when seasoned and roasted. The nuts also
produce high quality oil, which some Aboriginal people use as liniment on aching
joints (www.bushfoodshop.com.au). Figure 38.2 presents the estimated commercial
production of *S. spicatum* nut.

1.1 Oil Recovery and Content

Sandalwood (*Santalum album*) oil plays an important role as an export commodity
in many countries and its trade depends on quantifying the yield and quality. Steam
distillation, hydro-distillation, solvent extraction, supercritical fluid extraction
($SC-CO_2$) and liquid CO_2 extraction have been used to obtain the volatile oil from
sandalwood (Moretta et al. 1998a, b; Marongiu et al. 2006). Moretta et al. (1998a, b)
found the yield of extractable materials and total volatiles was the highest using
$SC-CO_2$. The levels of five sesquiterpene alcohols were higher in the steam

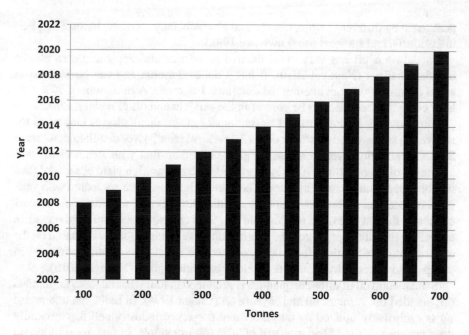

Fig. 38.2 Estimated commercial production of *S. spicatum* nut

distillate. Marongiu et al. (2006) evaluated the yield and composition of heartwood oil extracted from *Santalum album* using SC-CO_2 and hydrodistillation. Higher oil yield (4.11%) was obtained after 1 h with SC-CO_2 compared 1.86% after 30 h hydrodistillation. Hettiarachchi (2008) found that hydrodistillation was a superior method for oil recovery compared with solvent extraction, since it reflects more accurately the results obtained during industrial processing. In this work, the extraction technologies including SCCO$_2$, steam-distillation, hydrodistillation and solvent extraction were compared.

Sesquiterpenic alcohols (90%) are present in sandalwood oil of which 50–60% is the tricyclic α-santalol, while β-santalol comprises 20–25% (Krotz and Helmchen 1994). The composition of the oil depends on the species, region grown, age of the tree, the season of harvest and the extraction method. ISO standards for *S. album* oils are 41–55% α-santalol and 16–24% β-santalol (Lawrence 1991).

SC-CO_2 was employed for the high density of the sandalwood powder, as the density of the fluid is higher compared to supercritical conditions due to SC-CO_2 operation at 28 °C. Thus, the yield obtained of sandalwood oil was higher with SC-CO_2. Supercritical CO_2 extraction was carried out above the critical temperature of CO_2 (31.1 °C) and critical pressure (73.9 bars), while subcritical CO_2 extraction carried below the critical temperature of CO_2 (31.1 °C). The density of subcritical CO_2 is usually higher than supercritical CO_2 extraction and therefore facilitates the yield of extraction higher due to the high mass density of the fluid matching to that of sandalwood oil. More economical and rapid techniques of oil extraction are demanded by different sectors like clinical aromatherapy, perfume industries, pharmaceuticals, and spiritual use (Lawrence 1991).

Sandalwood oil has very good fixative properties and applications in classic blender fixatives (Nautiyal 2010). It has a delicate aroma and can be blended in small quantities without altering the dominant fragrance. A minimum of 90% santalol content is supposed to be present in the sandalwood oil to make it saleable as premium quality in the market. Conventionally steam distillation is employed for recovering sandalwood oil which yields 3.6 wt% oil after 24 h of distillation, whereas SC-CO$_2$ extraction produced much higher yield than that with steam distillation within 1 h of process time. SC-CO$_2$ processed oil contains high yield of santalol than that obtained with steam distillation. These investigations were compared with various conventional techniques. Subcritical fluid extraction is an extraction process utilizing a fluid at an extract temperature below its critical temperature and pressures exceeding its critical pressure. The application of subcritical solvents is based on the experimental observations that many gases exhibit enhanced solvating power when compressed to conditions above and below the critical point (Nautiyal 2010).

Sandalwood oil of different qualities is produced in India that includes unbranded oil, branded oil, Agmark oil and Mysore oil (Chana 1994). In India, the unbranded oil is exclusively utilized by the flavor and fragrance industry and that generally does not comply the Indian standard of 90% total santalols. In fact, most of the oil is used domestically and some of the branded oil is exported. Agmark oil with high content of santalol and an export-quality material complies to be higher than the Indian standard. Mysore and Shimoga distillery produces Mysore oil. Chana (1994) mentioned that the oil content of the heartwood of 10 years old sandalwood trees was 0.9% in comparison with 30 years old trees, which yielded 4.0% oil. It was reported that the total santalols, santalyl acetate and santalenes in the oil from 30 years old heartwood were 89.2%, 3.5% and 2.3%, respectively. Gowda et al. (2006) mentioned that the heartwood from 30-year-old trees possess ~5% oil, while the heartwood of 12 to 15-year-old plantation trees possesses an oil content between 3.5% and 4.0%. The economic feasibility was explored for establishing sandalwood plantations and assumed that after 15 years the economic return would be very lucrative.

East Indian sandalwood oil, the most well-known and the oldest traded type of sandalwood, used over thousands of years. The cultivation center is in India and native to the highlands of southern India and the Malayan Archipelago with the center of production in Mysore. Extending its natural distribution down to Indonesia particularly Timor and later has been introduced into Australia and plantations established in the tropical northwestern areas with an estimated of 8000 ha with annual additions of around 1000 ha. It has also been introduced into a number of the Pacific Islands and plantations have been established in Fiji, Tonga, Vanuatu, and New Caledonia. Australian sandalwood oil (*Santalum spicatum*, syn. *Eucarya spicata*) is native to the desert-like areas of SW Australia, close to Perth. A second sandalwood species, *S. lanceolatum*, is also found in Australia, principally in Queensland, and northwestern part of Western Australia, but commercially is less exploited. *Santalum paniculatum* has its presence in Hawaii and around 7000 ha are reported to be under sustainable management (**Forest Products Commission WA**

Sandalwood Industry Development Plan). Its commercial oil has recently entered the market. *Santalum yasi* cultivar is found in Fiji, Samoa and Tonga. The species has hybridization readily with *S. album* resulting variation in quality of oil depending on the source of the trees. *Santalum austrocaledonicum* is being found in Indonesia, and Papua New Guinea. New *Caledonia* African sandalwood oil, *Osyris lanceolata*, belongs to the same *Santalaceae* family and consumed in the perfumery also known as osyris oil. The tree is found through East and Southern Africa, typically on the dry boundary areas of forests, but rarely in large stands. West Indian Sandalwood *Amyris balsamifera* from the family *Rutaceae*, found in Central American and the Caribbean Island (http://www.fpc.wa.gov.au/).

1.2 Uses of Sandalwood Oil

Sandalwood is extensively used in readymade face packs and as a skin-lightening agent. Powdered sandalwood is available in the market and can be used with a variety of other products for an easy face pack. It can be used either in the form of oil or powder although care must be taken not to use raw sandalwood oil on the skin (Sandalwood essential oil, http://scienceofacne.com/sandalwood-essential-oil/).

Sandalwood or Chandan being beauty ingredient and hence is prized in Ayurveda as reliable and effective. It is derived from the fragrant wood of the Genus Santalum tree and is usually available as a brown-beige looking smooth powder. Various skin ailments are treated with sandalwood essential oil. It is often used as a home remedy for many skin conditions as it has a wide range of medicinal properties. As beneficial to the skin sandalwood oil is imperative in saving from the harmful rays of the sun. It also helps in soothing sunburn with its cooling effect and it has anti-inflammatory proprieties helps in reducing the redness caused due to sunburn. Sandalwood essential oil is very useful in treating insect bites or any other skin wounds. Coagulation of skin proteins, further skin break, abrasions or allergies are being protected. Sandalwood with antiseptic properties prevents pimples, acne and sores from developing. Bacterial growth on the skin is due to exposure to dust and dirt can cause bacterial growth on the skin and can lead further to skin problems. The affected area of the face can be treaedt by sandalwood powder mixed with milk (https://food.ndtv.com/beauty/5-sandalwood-benefits-to-look-out-for-from-tan-removal-to-treating-acne-1745068).

When sandalwood is used as incense or fragrance, it promotes mental clarity and it is frequently used for meditation, prayer or other spiritual rituals. The effect of sandalwood oil on attention and arousal levels was evaluated (https://draxe.com/sandalwood-essential-oil/).

Receiving sandalwood oil and comparing to control subjects extended more mental awareness with behavior demonstration linked to enhanced attention and cognitive clarity. Sandalwood is not only beneficial for mental clarity but also helps to create a feeling of relaxation. Sandalwood is a relaxing agent when smelled or rubbed on the skin. (https://draxe.com/sandalwood-essential-oil/). Studies disclosed that patients receiving palliative care felt much more relaxed and less anxious when they received aromatherapy with sandalwood. It also works as memory booster since one of the sandalwood's benefits is clarity. Its natural aphrodisiac helps increasing libido especially for men and energy provider. It gives great results when mixed to massage oil or topical lotion (http://www.fao.org/docrep/V5350E/V5350e08.htm).

Many aftershaves and facial toners use sandalwood as one of their primary ingredients to help soothe, tighten, and cleanse the skin. The active ingredient (santalol) in sandalwood could decrease inflammation markers in the body called cytokines with belief that santalol acts in a similar manner as NSAID medications without any potential negative side effects. Sandalwood is excellent expectorant that helps to treat coughs, since it has relaxing properties. By adding few drops of sandalwood oil to a tissue or washcloth and on inhalation helps reduce the severity and duration of a cough (https://draxe.com/sandalwood-essential-oil/). Sandalwood oil is very good anti-hypertensive and could decrease blood pressure just by applying the oil directly to the skin.

Sandalwood essential oil may reduce pain and decrease inflammation. Sandalwood oil works prominently as antioxidants that help reducing damage caused by free radicals. Sandalwood oil mixed with unscented lotion was used on the face for anti-aging benefits. It is an anti-spasmodic and can act against spasms of nerves, muscles, and blood vessels due to relaxing benefits. In traditional medicines, sandalwood oil has been used to treat urinary infections, digestive issues, coughs, depression, as well as infections (http://www.fao.org/docrep/V5350E/V5350e08.htm). However, sandalwood oil is not applied directly to the skin, but as practice mixed with a carrier oil or lotion first to dilute it. Almond oil, jojoba oil, or grapeseed oil are included as common carrier oils.

1.3   Economy

1.3.1   Description and Uses

The entire sandnualwood oil traded internationally is known to be East Indian sandalwood oil produced by distillations from the heartwood and roots of *Santalum album*. Figure 38.3 presents the average East Indian sandalwood oil prices from 2011 to 2014. Sandalwood oil from Australian *S. spicatum* and West Indian and African "sandalwood" oils are no longer processed. (https://www.cbi.eu/sites/default/files/market_information/researches/product-factsheet-europe-sandalwood-oil-cosmetics-2015.pdf).

Sandalwood oil with sweet, woody odor is widely used in the fragrance industries, but more specifically in the higher-priced perfumes. It is with excellent blending properties and containing the large proportion of high-boiling constituents in the oil (approx. 90% santalols) also price it valuable as a fixative for other fragrances. In India, it is produced, for the manufacture of traditional attars such as rose attar; the delicate floral oils are distilled directly into sandalwood oil.

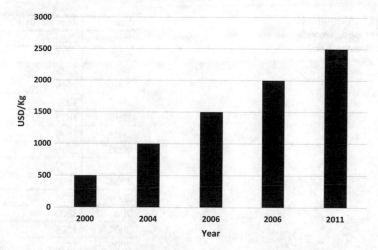

Fig. 38.3 Average East Indian sandalwood oil prices. (Source: Public Ledger)

1.3.2   Demand Trades and World Supply

The main producers and exporters of sandalwood oil are India and Indonesia, but there is no reliable production data available. In India, the domestic consumption is certainly high and thus probably greater than the combined total for the rest of the world. The order of world production/consumption is probably several hundreds of tons annually. The two largest importers of Indian sandalwood oils are the US and France. Imports to the Middle East have been also increased. Indonesian exports are the chief destination to the USA and thus represents the biggest market for sandalwood oil outside India (https://www.cbi.eu/sites/default/files/market_information/researches/product-factsheet-europe-sandalwood-oil-cosmetics-2015.pdf).

Venkatesha Gowda et al. (2006) reported that sandalwood and sandalwood oil is produced in India is approx. 85% of the world demand, while Indonesia (Timor) and other sources (primarily Fiji and New Caledonia) producing of about 10% and 5%, respectively. Tamil Nadu accounts for more than 80% of India's production. Table 38.1 presents the auction sales volume for sandalwood while the handicraft industry utilizing 360–600 tons annually.

1.3.3   Quality and Prices

The level of santalol contributes to the quality of sandalwood oil. The highest quality of the sandalwood oil is native to India. The several Pacific sandalwoods are the same in quality, especially sandalwoods from Fiji (*Santalum yasi*), New Caledonia and Vanuatu (*Santalum austrocaledonicum*). The comparison is in Table 38.2. Generally, heartwood with older sandalwood trees produces higher quality oil, as they are rich in santalol levels. Thirthy years old trees yield high-quality sandalwood oil.

Table 38.1 Auction sales
volume of sandalwood in
India (Lawrence 2009)

Year	Amount (metric tons)
1993	2850
1994	2575
1995	2325
1996	995
1997	1288
1998	1475
1999	1800
2000	1850
2001	1800
2003	1800
2004	1850
2004	1800
2004	1505
2005	1275

Table 38.2 Santalol content (%) in sandalwood species (CBI Product Sheet MoFA)

Trade name	Latin name	Santalol content (%)
Indian sandalwood	*Santalum album*	α-santalol = 41–55
		β-santalol = 16–24
Fiji sandalwood	*Santalum yasi*	α-santalol = 37–39
		β-santalol = 26–28
New Caledonian sandalwood	*Santalum austrocaledonicum*	α-santalol = 48–49
		β-santalol = 20–22
Australian sandalwood	*Santalum spicatum*	α-santalol = 15–25
		β-santalol = 5–20

The heartwood of sandalwood trees normally produced within 10–15 years. Soil texture, rainfall and sun exposure level affects the growth of the heartwood. The development of heartwood becomes rapid with a distinct annual dry period and exposure to full sun and could be harvested after 15–20 years. To obtain the optimum quality sandalwood oil, it must be 100% natural (unmixed with any other oil) (https://www.cbi.eu/sites/default/files/market_information/researches/product-factsheet-europe-sandalwood-oil-cosmetics-2015.pdf).

The stipulation of a minimum free alcohols (santalol) content of 90% is set by an international standard (ISO) for sandalwood oil and there are ranges mentioned for various physicochemical properties. An Essential Oil Association (EOA) specifies the same minimum santalol content In the United States too. The aroma characteristics of sandalwood oil as perfumery oils are very important and are judged by the buyer to be acceptable or not for individual consignments (Krotz and Helmchen 1994).

In the essential oil trade, sandalwood oil is one of the most highly priced items, which reflect the nature of the raw material source. The price was fetched in the late 1980s of almost US$200/kg. London dealers in the year 1992 offered price of Indian origin sandalwood oil of US$140–150/kg as compared to Indonesian oil, which was about US$5 lower. The price of Indian oil rose again to US$180/kg in mid-1993 (McKinnell 2008; Jones 2002; Forest Products Commission 2007; Coakley 2006; Bryne et al. 2003a, b; Broadhurst and Coates 2002; Brand et al. 2007; AVONGRO 2006; Wood 2008).

Location and age of the tree make the yield of the oil to vary. The oil content of the heartwood varies from tree to tree and is higher for older trees. In India, yields of about 0.9% have been reported from the heartwood of 10-year old trees, while mature trees of 30–50 years age have yielded 4% oil. The oil content also varies according to the color of the heartwood. Light-colored wood yields 3–6% oil, while dark brown wood yields about 2.5% oil. The fixed oil is contained by the cotyledons and kernel of sandal seeds and has the drying properties. High protein is present in the oil-free sandal seed meal and could be utilized as an animal feed if available in sufficient quantities (Guenther 1952; Yadav 1993).

Indian sandalwood industry is very large compared to any perfume industry in the world (Gowda 2013). More than 85% of world's production of oil is contributed from India. Tamil Nadu contributes more than 80% of the production. Karnataka was contributing more than 50% of the oil to the market. It is estimated that the present annual production of sandalwood oil production exceeds 3–5 tons. However, these figures do not match with the official production. Domestic industries consume about 80,000 kg of sandalwood oil each year. Sandalwood trade is a very tricky business and one either cannot get the true statistics, on production or on the quantity of oil. Attars production has been occurring in India for centuries with a blend of sandalwood oil and flower oil, such as rose petal, jasmine, and kewda. The concentration of flower effervescence within sandalwood oil excel the quality of attars. Numerous types of attar products are made in India forming an important constituent for the manufacture of incense sticks and Scented tobacco, Pan Masala, Pan Parag, Zarada, and Gutka (Hamilton and Conrad 1990).

Government sales prices in India for *S. album* have increased from about US$9500/t in 1990 to about US$40,000/t in 2005, presumably for de-sapped heartwood that contains the normal level of oil (6%). Lower price fetched for wood with lower oil content (Anon 2006). It is very difficult to estimate potential future nut supplies, as growers do not yet have reliable data on how nut production varies with tree age (McKinnell 1993). For the development of new market other prospects are available ranging from human consumption (nut oil), cosmetics and possibly biodiesel (nut oil), and pharmaceutical products (nut oil or wood oil). Any market will also depend on having a reliable source of supply in sufficient quantity to be commercially attractive. A critical mass of tree farm resource is therefore necessary (Nayar 1988). The following assumptions have been used:

- Only 50% of farm plantings are utilized, as it is assumed the remainder is not set up for nut production or are too remote for economic utilization.
- Farmer plantings continue after 2007 at 1000 ha per year.
- Effective stocking is 300 stems per hectare.
- Production per tree is 0.33 kg oil/year at ages 5–7, 0.50 kg oil/year at ages 8–12 and
- 0.33 kg oil/year at ages 13–25, and nuts are available annually.

Different sandalwood species are in demand in the world market and the estimation of their derivatives is around 4000–7000 tons per year (Sen-Sarma 1982). Sandalwood production official data is not considered on illegal harvesting into account, actual demand and production may be much higher. Both the harvest of sandalwood and the distillation of the essential oil is regulated by the Indian government. Therefore, there is a ban on exports of products. Indian sandalwood official harvest figure in India is around 400–1000 tons per annum. Unofficial harvesting enables sandalwood products smuggling to other countries and presumably adds another 3000–4000 tons per year to the total production. Since trees take at least 30 years prior their harvesting for the essential oil and therefore the interest in cultivating the trees is low. Indonesia is unexpectedly not able to provide significant volumes of sandalwood over the coming 20 years. There is exhaustion approaching of wild-harvested Pacific sandalwoods. However, an establishment of several sandalwood plantations have occurred. Asia is destined for the export of the heartwood from these species.

Around 30 tons of Australian sandalwood oil is produced by Australia per annum (Shankarnarayana and Parthsarathi 1984). Restrictions have been imposed in several countries on the harvesting of sandalwood trees for protecting the trees from extinction (Statham 1990). Australia does both Sandalwood's cultivation and wild harvesting. Global supplies may be affecting the future production of Indian sandalwood in Australia. During 2016–2017, there was an expectation that Australia may begin commercial harvests. Australia may reach the aim of producing 60% of the global supplies of Indian sandalwood by 2029. However, the market is not expected to be saturated with these supplies. For domestic consumption, China has taken up recent plantations for the production of domestic consumption.

Sandalwood oil has the highest potential and best opportunities in fragrances for use in the fragrances market. It is with peculiar and specific scent and often used as a base note. Consumers are tending to pay willingly a higher price in this segment (Table 38.3). The high price of sandalwood oil therefore is less of an issue. It is used in smaller proportions when the oil is used as a fixative in perfumes (http://www.fao.org/docrep/V5350E/V5350e08.htm). Generally, sandalwood is exported for the use of cosmetics in the form of oil. In the country of origin, sandalwood oil can be processed further, by refining the oil, so cosmetics manufacturers can use it directly. Until it is prepared for use in fragrances by the end user, the oil is not processed further.

Table 38.3 Major cosmetic segments and applications of sandalwood oil

Segment	Sub-segment	Benefits of applying sandalwood oil
Fragrances	Perfumes	Fixatives (when used in small amounts) fragrance
Skin care	Facial skin care, body care, moisturizer, anti-aging	Skin condition properties, use on dry cracked and chapped skin
Hair care	Shampoo, conditioner	Fragrance
Toiletries	Soap, body wash, bath salt	Used in small amounts as fixatives fragrance

Indian sandalwood oil and wood powder exhibited skin-conditioning properties. Sandalwood oil is also used as a fragrance in toiletries and hair care products (Suriamihardja 1978). Fairly traded sandalwood oil has a very good market. It is to be noted that fair trade certification is only available for cultivated products, not for wild-harvested sandalwood. Organically certified sandalwood oil has a market. However, it depends primarily on the prospective buyer. Sandalwood oil from India cannot be certified organic due to diversification of production sites. In other countries, it is possible, however, where monitoring of the production sites is more feasible.

2   Extraction and Processing of Sandalwood Oil

Extraction of sandalwood powder (60 µm particle size) was carried out (Nautiyal 2010). The particle size was kept as such so that neither the air channeling would form in case of fine size nor the large particle size would affect the yield. Sandalwood chips were pulverized and sieved through 60 µm to ensure uniformity of the heartwood sample. Subsamples were taken from this homogenized sample and the oil was extracted using one of eight methods including subcritical carbon dioxide (SC-CO$_2$) extraction, solvent extraction (benzene, benzene, diethyl ether, ethyl alcohol, and toluene), hydrodistillation and steam-distillation. Table 38.4 summarizes the methods and batch time used for oil extraction (Nautiyal 2010). The physicochemical properties of the extracted oil were evaluated.

The focus was on micro-distillation methods, which are widely in use and successfully adapted to essential oil research. The major drawbacks are that they require a higher volume of volatile material, whereas young sandalwood trees have <1% essential oil (Jones et al. 2007). Compounds other than essential oil such as waxes, lignans and many other lipophilic compounds are extracted with non-polar solvents.

Table 38.4 Methods and batch time used for oil extraction (Nautiyal 2010)

Extraction process/solvent	Batch time (h)
Liquid CO$_2$ extraction (subcritical condition)	4
Solvent (benzene)	5
Solvent (diethyl ether)	5
Solvent (ethyl alcohol)	5
Solvent (toluene)	12
Hydrodistillation	30
Hydrodistillation (alkaline treated water)	*48
Steamdistillation (pilot plant)	10

To avoid the artifacts formation for long distillation, the yield and quality of the oil under the controlled pH conditions was evaluated
*hydrodistillation taken long hours of processing

Table 38.5 Oil content of four subsamples per heartwood sample extracted by different methods (Hettiarachchi et al. 2010)

Subsample number	Oil content (% w/w) (Modified hydrodistillation)	Oil content (% w/w) (Solvent extraction)	Oil content (% w/w) (Standard hydrodistillation)
UCL 09/01	0.808	2.16	1.45
UCL 09/02	1.371	1.90	1.23
UCL 09/03	0.237	1.80	1.32
UCL 09/04	0.358	2.10	1.45
PT 09/01	–	0.57	0.45
PT 09/02	–	0.55	0.43
PT 09/03	–	0.46	0.38
PT 09/04	–	0.54	0.54

Core samples usually need (1–2 g) to be extracted ensures elevated levels of these compounds. Therefore, calculation of the percentage weight of total lipophilic extract as the volatile composition is inaccurately determined (Hettiarachchi 2008).

Jones et al. (2007) described the use of camphor as an internal standard for inaccurately determined oil content using gas chromatography (GC). This was successfully tested on small wood core samples from young *S. album* trees planted in Kununurra. This method was rapid and reliable in quantifying heartwood oil content and quality in plantation trees. The major drawback is that the volatile material recovered contains additional camphor. This method would be suitable for GC analysis, but the use of these samples for any other form of physical or chemical analysis couldn't be feasible due to the additional camphor. Holding the volatile extracts with camphor as library samples would be unsuitable.

Hettiarachchi (2008) mentioned in his study that hydrodistillation showed 1.36% (w/w) in the wild-harvested sample and 0.43% (w/w) in plantation sample of volatile oil. He noted that the percentages of oil content varying considerably (CV = 74.16%). The results showd difference in all applied methods (Table 38.5). Since these methods lack in the reliability and sensitivity therefore not repeated on smaller trees. It may require several wood core samples for gaining a broader understanding of heartwood oil development across a plantation. Distillation techniques mostly require time and energy as compared to solvent extraction methods. Therefore, solvent extraction of sandalwood oils might be continue for research and development purposes.

3 Developments in the Extraction and Isolation of High-Value Compounds

Chipping and pulverization of heartwood and roots are affected the distillation process. There is an oil yield variation in accordance with the plant part employed, the age of the tree and the agronomical conditions of cultivation. Roots can yield up to 10% oil and heartwood up to 4%. Distillation for 48–72 h is effective to complete the extraction. High-pressure steam distillation provides a higher yield and distillation time was reduced but there were loses of the delicate notes.

The refiners in Kupang extraction of sandalwood oil has long been carried by steam distillation (Ferhat et al. 2006). It generally runs for about 40–70 h. When the distilled oil is found to run out, the distillation process is stopped and is economically unviable. Distillation could be performed at high-pressure to produce higher oil yield in a shorter distillation process. There is raise of temperature because of high-pressure distillation. As a result of rising in temperature, decompose of the oil components make the oil less odoriferous. Kusuma and Mehfud (2017) developed a new method for extracting essential oil by microwave and investigated the use of a new green method for extracting essential oil with low/minimal energy and solvent consideration. They attempted to develop microwave hydrodistillation having combined of hydrodistillation and heating with microwave. The method consisted of three elements: a compressor by injecting air into the distiller containing matrix (parts of the plant to be extracted), microwave, and a condenser as the cooling system. Introducing airflow in the microwave, air hydrodistillation was expected to improve the yield and quality of extracted oil. The addition of airflow was predicted to affect the diffusion through the cell membrane or plant tissue which otherwise is difficult to diffuse. This is because sandalwood powder is with high bulk density containing oil is difficult to extract without an addition of airflow. Figure 38.4 presents the oil content (%) obtained from *Santalum album* powder by different extraction methods.

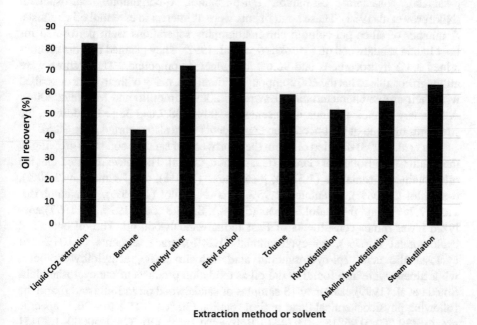

Fig. 38.4 Oil content (%) obtained from *Santalum album* powder by different extraction methods (Nautiyal 2010)

Table 38.6 The yield of sandalwood oil obtained by microwave hydrodistillation and air microwave hydrodistillation (Kusuma and Mahfud 2017)

Extraction method	Air flow rate (L/min)	Sandalwood oil yield (%)
Microwave hydrodistillation	0	1.2184 ± 0.1139
Microwave air hydrodistillation	0.1	1.2248 ± 0.1047
	0.5	1.2985 ± 0.0762
	1.5	1.3170 ± 0.0634
	3.0	1.3170 ± 0.0968
	5.0	1.3170 ± 0.0973

Kusuma and Mahfud (2017) determined the effect of the addition of airflow for extracting sandalwood oil in the microwave air hydrodistillation method. They have performed the extraction using 600 W power; the ratio of 0.05 g mL^{-1} for sandalwood and distilled water. Microwave hydrodistillation yielded 1.218% sandalwood oil. On the other hand, air flow rate of 5.0 L/min yielded 1.317% sandalwood oil by microwave air hydrodistillation (Table 38.6). Ahead of their investigation, the microwave hydrodistillation method consumed 120 min for yielding of 1.218%; on the other hand, microwave air hydrodistillation consumed only 50 min for obtaining the almost same quantity of yield, 1.24% (Chemat et al. 2008).

The revelation of the most intense aroma compounds found in sandalwood oil by employing a GC-sniffing technique were α-santalene α-santalal β-santalal epi-β-santalal, α-santalol, β-santalol, epi-β-santalol, α-bergamotol, spirosantalol (Nikiforov et al. 1988). These components were of interest to essential oil chemists. A number of silica gel column chromatographic separations were performed for East Indian sandalwood oil (Mookherjee et al. 1992). They gleaned that the oil contained 3.4% hydrocarbons and 96.6% oxygenated compounds. The authors were although not able to list the 100 components identified, many of them were described with their aroma characteristics (Lawrence 2009; Jaramillo and Martinez 2004). A summary of their findings is presented in Tables 38.7 and 38.8. In addition, the structures of some of the uncommon constituents were also reported (Fig. 38.5).

Zhu et al. (1993) distilled oil from the heartwood of sandalwood trees growing in the South China Botanical Garden (Guangzhou, China). They have reported that the oil contained α-santalene (1.51%), β-santalene (1.81%), epi-β-santalene (2.30%), α-santalol (22.08%), β-santalol (1.95%), *cis*-α-santalol (11.75%), α-santalol isomers (7.76%), and β-santalol isomers (5.54%). Brunke et al. (1995 and 1997) analyzed fewer minor constituents of East Indian sandalwood oil. The oil contained cyclosantalal (1.67%), epi-cyclosantalal (1.24%), trace amounts (<0.01%) of cyclosantalic acid, epi-cyclosantalic acid, dihydroalbene, acetyldihydroalbene, while albene were also found in the oil as oxidation products of the cyclosantalals. Shukla et al. (1999) examined 13 samples of sandalwood oil in India and found the following physicochemical data: optical rotation (30 °C), −15.1 to −20.4° specific gravity (30 °C), 0.968185–0.974413 Refractive Index (30 °C), 1.50000–1.50451 free OH calculated as total santalols (percent by mass), 88.13–97.01% esters

Table 38.7 Composition (%) of some uncommon components of East Indian and Java sandalwood oils (Lawrence 2009)

Compound	% East Indian oil	% Java oil	Odor characteristic
Santalene oxide	3.1	1.6	Green melon and lavender like
Non-sesquiterpene tricyclic ketone (A)	3.6	–	Diffuse woody and amber like
Trans-α-santalal	12.8	7.7	Pine needle and juniper like
Trans-β-santalal	8.2	4.0	Sweaty and uranicious
(E)-nuciferol	0.7	1.6	Weak odor
Trans-α-photosantalol	4.7	8.4	Weak woody odor
Trans-epi-β-photosantalol	1.4	1.1	Weak woody odor
Trans-β-photosantalol	2.8	6.6	Fatty sandalwood like
Aldehyde 1(B)	0.1	0.1	Sweet melon like
Ketone 1(C)	0.1	0.1	Green woody amber-like
Ketone 2 (D)	0.3	–	A green oily sandal like
Ketone 3 (E)	1.2	1.0	Weak woody melon like
Ketone 4 (F)	0.1	0.1	Strong woody melon like
Aldehyde 2 (G)	1.2	–	Woody ambergris like
Aldehyde 3 (H)	0.2	–	Green oily melon like
Spiroketone (I)	1.7	1.4	Weak uninteresting odor

Table 38.8 Composition (%) of supercritical fluid CO_2 extract of *Santalum album* oil (Lawrence 2009)

Compound	Oil	Extract
E-α-ionone	4.2	0–0.8
epi-β-santalene	1.4	0.2–0.3
α-humulene	1.0	0.3–1.6
β-santalene	1.0	0.2–0.3
cis-artenuic alcohol	5.1	–
cis-α-santalol	35.0	46.1–48.7
β-sinensal	5.7	1.1–6.1
cis-β-santalol	14.0	20.3–21.4

calculated as santalyl acetate (percent by mass), 3.56–6.15% α-santalol, and 46.05–51.32% β-santalol.

Wei et al. (2000) revealed that due to wind damage the *Santalum album* trees probably accelerate the formation of heartwood. The composition of an oil produced from the heartwood from a 25-year-old tree included α-santalene (0.08%), epi-β-santalene (0.12%), β-santalene (0.25%), curcumene (0.05%), α-santalol (43.09%), (Z)-*trans*-α-bergamotol (9.44%), β-santalol (22.5%), nuciferol (9.65%), (Z)-lanceol (0.51%), and epi-β-santalol (3.66%). Chen and Lin (2001) extracted the sandalwood oil with various methods and compared the main composition. α-santalene (0.08–0.65%), epi-β-santalene (0.16–0.89%), β-santalene (0.04–1.42%), curcumene1 (0.09–0.33%), α-santalol2 (30.7–41.5%), α-santalol3 (6.03–8.28%), epi-β-santalol (3.94–6.74%), β-santalol (22.626.8%), nuciferol4 (2.08–2.47%), and lanceol5 (0.46–0.88%) were identified.

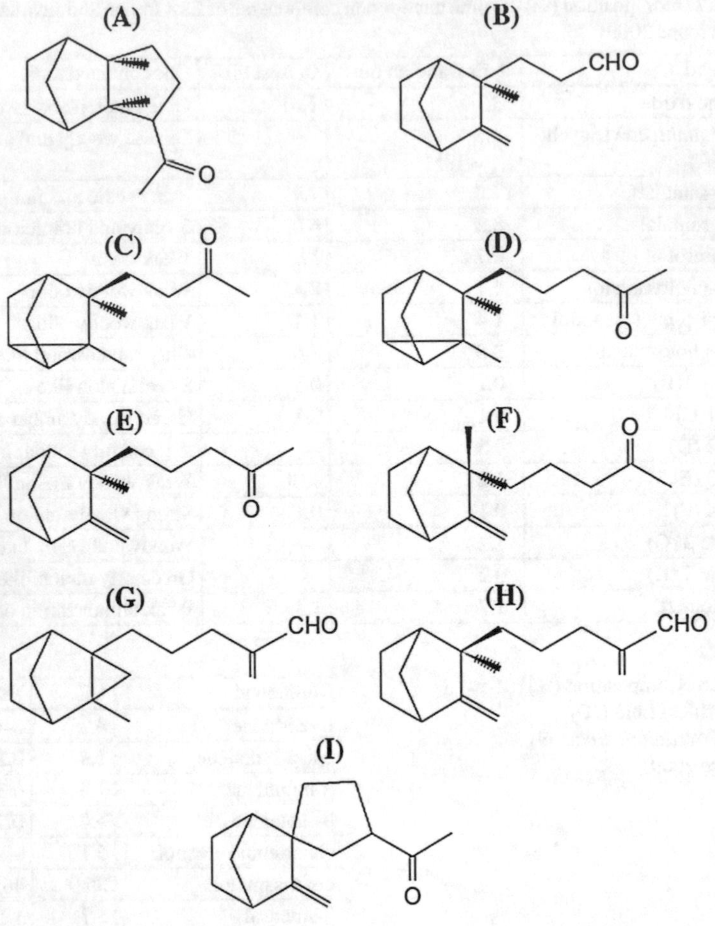

Fig. 38.5 Structures of some uncommon components of sandalwood oil (Lawrence 2009)

Braun et al. (2003) analyzed the oil of East Indian sandalwood obtained commercially in Germany. The composition of the oil was as follow: santene (0.2%), α-santalene (0.7%), α-cedrene (0.1%), *trans*-α-bergamotene (0.2%), epi-β-santalene (0.8%), β-santalene (1.2%), *trans*-β-bergamotene (0.1%), γ-curcumene (0.1%), β-bisabolene (0.1%), β-curcumene (0.2%), ar-curcumene (0.3%), (E)-nerolidol (0.1%), β-bisabolol/epi-β-bisabolol (0.7%), cyclosantalal (0.4%), α-santalal (1.9%), α-bergamotal (0.2%), epi-cyclosantalal (0.3%), α-bisabolol (0.2%), dihydro-α-santalol (0.6%), β-santalal (0.6%), *cis*-α-santalol (41.1%), *trans*-α-bergamotol (6.4%), *trans*-α-santalol (0.4%), *cis*-α-bergamotol (0.2%), epi-β-santalol (3.5%), *cis*-β-santalol (19.8%), fokienol (0.5%), *trans*-β-santalol (1.5%), (Z)-lanceol (1.4%), (Z)-nuciferol (3.4%), spirosantalol (0.9%), and (E)-nuciferol (0.1%). Trace amounts (<0.1%) of episesquithujene, *cis*-α-bergamotene, (E)-β-farnesene, sesquisabinene, β-alaskene, (Z)-γ-bisabolene, α-alaskene, (E)- γ-bisabolene and (E)-α-bisabolene were also found in the oil. Chiral analysis of the β-bisabolols revealed that all four enantomers were present in the oil, although the main stereoisomer was (6R,7R)-β-bisabolol. A water-distilled oil of *S. album* was investigated by Braun et al. (2003) and contained (1R, 4R, 5S)-α-acorenol (0.22%), (1R, 4R, 5R)-β-acorenol (0.11%), (1R, 4S, 5S)-epi-α-acorenol (0.13%), and (1R, 4S, 5R)-epi-β-acorenol (<0.01%).

Caledonium sandalwood (family *Santalum austrocaledonicum* Viell. var. *austro-caledonicum*) oil was reported to be produced from the heartwood. Small quantities of oil were being produced from new *Caledonium* sandalwood (Coppen 1995). Alpha et al. (1996) determined the main constituents of new *Caledonium* sandal-wood oil to be α-santalene, epi-β-santalene, bsantalene, α-santalol, epi-β-santalol and β-santalol. The same authors (Alpha et al. 1997a, b) identified 6, 13-dihydroxy-bisabola-2, 10-diene, 7, 13-dihydroxybisabola-2, 10-diene, (E)-lanceol and cam-pherene-2, 3-diol in the same *Caledonium* sandalwood oil. The main components of this oil were reported (Alpha et al. 1997c) to be α-santalene (1.3%), trans-β-bergamotene (0.2%), epi-β-santalene (1.0%), β-santalene (0.9%), α-santalol (52.9%), (Z)-trans-α-bergamotol (8.1%), epi-β-santalol (3.3%), β-santalol (20.9%), and (Z)-lanceol (4.5%). Braun et al. (2005) analyzed a sample of new *Caledonium* sandalwood oil and reported that it possessed the following composition: santene (0.1%), α-santalene (1.1%), *trans*-α-bergamotene (0.2%), epi-β-santalene (0.9%), β-santalene (0.9%), β-bisabolene (0.3%), β-curcumene (0.2%), α-ekasantalal (0.1%), ar-curcumene (0.3%), dendrolasin (0.1%), α-photosantalol (0.1%), (E)-nerolidol (0.2%), epi-β-photosantalol (0.1%), nor-α-santalenone (0.1%), β-photosantalol (0.1%), nor-β-santalenone (0.1%), β-bisabolol/epi-β-bisabolol (0.8%), cyclosantalal (0.7%), α-santalal, (0.9%), (Z)-trans-α-bergamotal (0.2%), epi-cyclosantalal (0.4%), a-bisabolol (0.7%), *cis*-β-santalal (0.1%), dihydro-α-santalol (0.4%), *trans*-β-santalal (0.2%), campherenol (0.5%), (E, E)-farnesol (1.0%), cis-α-santalol (38.2%), (Z)-*trans*-α-bergamotol (9.9%), *trans*-α-santalol (0.4%), epi-β-santalol (3.8%), *cis*-β-santalol (18.2%), *trans*-γ-curcumen-12-ol (0.6%), (Z)-γ-bisabolen-12-ol (0.6%), *trans*-β-santalol (0.4%), *cis*-β-curcumen-12-ol (1.1%), (Z)-lanceol (9.1%), (Z)-nuciferol (2.1%), spirosantalol (0.8%), and bisabola-2,10-diene-6,13-ol (0.1%). They further found number of trace constitu-ents (<0.01%) such as α-pinene, *p*-cymene, dihydroarbene, furfural, teresantalal, linalool, sesquithujene, epi-α-cedrene, α-cedrene, terpinen-4-ol, a sesquisabinene isomer, γ-acoradiene, citroncllol, sesquiphellandrene, (E)-α-bisabolene, β-ekasantalal, α-acorenol, epi-α-acorenol, β-acorenol, epi-β-acorenol, α-teresantalic acid, 11-epi-6,10-epoxy-bisabol-2-en-12-ol, and cis-12-hydroxysesquicineole in the same oil. The authors determined that the ratio of bisabolol isomers found in the oil was (6R, 7S)-(20%); (6R, 7R)-(63%); (6S, 7R)-(12%); and (6S, 7S)-(5%). The authors had noted that new *Caledonium* sandalwood oil was similar in composition and its odor quality to East Indian sandalwood oil.

Kusuma and Mahfud (2017) carried out the analysis of sandalwood oil recovered by microwave hydrodistillation and microwave air hydrodistillation methods. The investigation revealed that quality of the sandalwood oil complies with the quality standard oil as shown by the composition of the main compounds present in the oil. The compounds detected were α-santalol, β-santalol, α-bergamotol, cis-lanceol, α-santalene, α-santalene, α-bergamotene, and α-curcumene (Table 38.9). The total

Table 38.9 Chemical composition of sandalwood oil obtained by microwave hydrodistillation and microwave air hydrodistillation (Kusuma and Mahfud 2017)

				Area (%)	
Number	Compound	Molecular weight	Molecular formula	MHD	MHAD (5.0 L/min)
1.	N-phenylformamide	$C_7H_7NO_2$	121.14	0.67	0.57
2.	Camphene	$C_{10}H_{16}$	136.23	0.95	0.82
3.	α-pinene	$C_{10}H_{16}$	136.23	0.06	nd
4.	Santolina triene	$C_{10}H_{16}$	136.23	0.07	nd
5.	β-ocimene	$C_{10}H_{16}$	136.23	1.87	0.09
6.	Cis-ocimene	$C_{10}H_{16}$	136.23	0.06	nd
7.	Z-alloocimene	$C_{10}H_{16}$	136.23	1.04	1.11
8.	α-terpinene	$C_{10}H_{16}$	136.23	nd	1.19
9.	1-cyclohexadilene-2-methylpropene	$C_{10}H_{16}$	136.23	0.31	nd
10.	Isoterpenoline	$C_{10}H_{16}$	136.23		
11.	4,6-dimethyl-2-propylpyridene	$C_{10}H_{15}N$	149.23	0.55	0.27
12.	8-methylene-2-exo-norademanatanol	$C_{10}H_{14}O$	150.22	nd	0.04
13.	3a,6-methano-3aH-inden-7(4H)-one-hexahydro-7ad-3aa-6a-7ab	$C_{10}H_{14}O$	150.22	0.72	nd
14.	m-carbomethoxyphenol	$C_8H_8O_3$	152.15	nd	4.05
15.	4-propylresocinol	$C_9H_{12}O_2$	152.19	1.72	nd
16.	Teresantalol	$C_{10}H_{16}O$	152.23	4.58	1.45
17.	Sabinene hydrate	$C_{10}H_{18}O$	154.25	0.10	nd
18.	3-methyl-4-quinazolinone	$C_9H_8N_2O$	160.17	nd	0.37
19.	9-methyl-10-azatricyclo-5,2,2,0-(1,5)-undec-2-ene	$C_{11}H_{17}N$	163.26	nd	0.02
20.	Dispiro-2,1,2,1-octane-1,1,6,6-trimethyl	$C_{12}H_{29}$	169.29	nd	0.02
21.	Cyclododecyne	$C_{12}H_{29}$	169.29	nd	0.31
22.	Vitispirane	$C_{13}H_{20}O$	192.30	nd	0.05
23.	α-curcumene	$C_{15}H_{22}$	202.34	0.19	0.12
24.	α-farnesene	$C_{15}H_{24}$	204.35	0.50	1.88
25.	β-patchoulene	$C_{15}H_{24}$	204.35	0.62	nd
26.	α-santalene	$C_{15}H_{24}$	204.35	nd	0.03
27.	α-cedrene	$C_{15}H_{24}$	204.35	nd	0.05
28.	γ-curcumene	$C_{15}H_{24}$	204.35	nd	0.17
29.	α-guaiene	$C_{15}H_{24}$	204.35	0.66	0.11
30.	Isosativene	$C_{15}H_{24}$	204.35	0.18	nd
31.	β-santalene	$C_{15}H_{24}$	204.35	0.05	0.17
32.	Seychellene	$C_{15}H_{24}$	204.35	0.77	nd
33.	α-patchoulene	$C_{15}H_{24}$	204.35	0.38	nd
34.	Germacrene B	$C_{15}H_{24}$	204.35	5.50	0.10
35.	α-salinene	$C_{15}H_{24}$	204.35	0.06	nd

(continued)

Table 38.9 (continued)

Number	Compound	Molecular weight	Molecular formula	Area (%)	
				MHD	MHAD (5.0 L/min)
36.	δ-guaiene	$C_{15}H_{24}$	204.35	0.48	0.15
37.	γ-elemene	$C_{15}H_{24}$	204.35	nd	3.69
38.	α-bergamotene	$C_{15}H_{24}$	204.35	nd	0.20
39.	Bicyclogermacrene	$C_{15}H_{24}$	204.35	0.64	0.45
40.	2-methyl-6-(4-cyclohex-3-ene-1-ylidene-hept-2-ene)	$C_{15}H_{24}$	204.35	nd	0.71
41.	Acoradiene	$C_{15}H_{24}$	204.35	0.139	0.59
42.	Bicycyloelemene	$C_{15}H_{24}$	204.35	nd	1.36
43.	Viridiflorene (ledene)	$C_{15}H_{24}$	204.35	0.91	nd
44.	Cyclopantanepropionic acid-3-oxo-2-(2-propenyl-methylester trans)	$C_{12}H_{16}O_3$	208.25	nd	0.08
45.	β-costal	$C_{15}H_{22}O$	218.33	0.53	nd
46.	(E)-nuciferol	$C_{15}H_{22}O$	218.33	6.59	7.07
47.	5-hydroxycalamenene	$C_{15}H_{22}O$	218.33	nd	2.15
48.	β-santalol	$C_{15}H_{24}O$	220.35	22.67	24.80
49.	α-santalol	$C_{15}H_{24}O$	220.35	27.81	28.73
50.	β-bergamotol	$C_{15}H_{24}O$	220.35	10.82	10.18
51.	Cis-α-copaene-8-ol	$C_{15}H_{24}O$	220.35	1.86	nd
52.	Cis-lanceol	$C_{15}H_{24}O$	220.35	3.42	2.74
53.	α-cidrol	$C_{15}H_{26}O$	222.35	nd	2.65
54.	Caryophyla-3,8-(15-diene-5-α-ol)	$C_{15}H_{26}O$	222.35	0.47	0.52
55.	Isolongifolol	$C_{15}H_{26}O$	222.35	0.21	0.59
56.	3-phenyl-1,4-(E)-dodecadene	$C_{15}H_{26}O$	242.40	nd	0.21
57.	Dodecenyl succinic anhydride	$C_{16}H_{26}O_3$	266.38	nd	0.19
58.	2-octylcyclopropaneoctanal	$C_{19}H_{36}O$	280.40	nd	0.03
59.	Oleic acid	$C_{18}H_{34}O_2$	282.46	nd	0.07
60.	Monoterpenes	–	–	3.94	3.21

nd not detected

content of santalol (50–70%) meet the standards as set by International Organization for Standardization. Therefore, they have proposed that sandalwood oil obtained by microwave hydrodistillation and microwave air hydrodistillation methods met the quality with commercial sandalwood oil.

From results in Table 38.9, it could be noted that the microwave air hydrodistillation method extracted more constituents than that recovered by microwave hydrodistillation. Microwave air hydrodistillation yielded 37 compounds, while the extraction of sandalwood oil by microwave air hydrodistillation resulted in the identification of 43 compounds. Kusuma and Mehfud (2017) proposed that the addition of airflow in the microwave air hydrodistillation process enhances the extraction of the heavy fraction components in the cell membrane or plant tissue that was otherwise difficult to diffuse. Their data supports extraction of heavier fraction

Table 38.10 Extraction parameters of sandalwood oil by liquid carbon dioxide (Nautiyal 2014)

Time of collection (h)	Yield of oil (g)	Main constituents			
		α-santalene	β-santalene	α-santalol	β-santalol
1	4.11	0.55	1.30	51.30	27.94
2	1.21	0.48	1.08	54.50	28.16
3	0.89	1.00	1.92	50.27	26.18
4	0.30	1.14	2.17	51.99	26.76

components contained in sandalwood oil by microwave air hydrodistillation process in comparison with the oil obtained by microwave hydrodistillation. Sandalwood oil extracted by microwave hydrodistillation contained only two heavy fraction components (MW ≥ 222.37). Whereas, the oil extracted by microwave air hydrodistillation contained seven heavy fraction components (MW ≥ 222.37). These heavy fraction components constituted the most of the components belongs to oxygenated terpenes. Their contribution responsible for aroma and fragrance of the essential oil, compared to other compounds. Therefore, it was concluded that oxygenated terpenes found in sandalwood oil obtained by microwave air hydrodistillation gave better quality of the oil.

GC analysis showed that the peak areas of major constituents, α-santalene and β-santalene extracted in the first hour were 0.55% and 1.30% respectively, whereas peak areas for α-santalol and β-santalol were 51.30% and 27.94% respectively (Nautiyal 2014). In the second hour, the peak areas for α-santalene and β-santalene were 0.48% and 1.08%, and those of α-santalol and β-santalol were 54.50% and 28.16%. In the third hour, the peak areas of α-santalene and β-santalene were 1.00% and 1.92%, and those of α-santalol and β-santalol were 50.27% and 26.18% respectively. Finally, in the fourth hour of extraction, the peak area of α-santalene, β-santalene and α-santalol, β-santalol was 1.14, 2.17, 51.99 and 26.76% respectively. α-santalol and β-santalol contents and the yields of the sandalwood oil were found to maximum in the second hour (Table 38.10). Nautiyal (2014) mentioned that hydrodistillation of sandalwood oil obtained from pre immersed sandalwood in cold water for 72 h was carried out for 36 h. The yield of the oil was found to be 1.71%. The color of the oil was pale yellow with a pleasant odor. In spite of softening the sandalwood chips for a long time, it was difficult for the steam to pierce through medullar ray cell, vessels, wood fibers and wood parenchyma containing oil as it was unpulverized. Gas chromatograph analysis showed the presence of α-santalene and β-santalene in a trace amount, whereas the contents α-santalol and β-santalol were 48.38% and 28.73%, respectively (Table 38.11, Sect. 1). Sandalwood powder (40 mm size) was charged for hydrodistillation, utilizing alkaline water. The extraction was carried out for 48 h and the recovered oil was 2.68%. The yields of α-santalene and β-santalene were 4.25% and 3.01%, and those of α-santalol and β-santalol were 41.90% and 19.89%, respectively. α-Santalene and β-santalene extracted were high. The alkaline medium was used since the pH of water during hydrodistillation plays a major role in the composition of the essential oil. The acidity of water causes transformations of thermo labile monoterpenes. Neutral or alkaline medium minimizes the formation of artifacts during distillation (Table 38.11, Sect. 2).

Table 38.11 Characterization of sandalwood oil extracted by conventional methods (Nautiyal 2014)

Method of extraction	Batch time (h)	Physical/pre-treatment	% concrete Extracted	% oil/absolute Extracted	Oil composition (Major constituents, %)	Color of oil	Odor of oil
Section I							
Hydro distillation	36	Whole sandalwood chips immersed in cold water for 24 h	–	1.71	(1) traces (2) traces (3) 48.38 (4) 28.73	Pale yellow	Pleasant
Solvent extraction (toluene)	5.15	After hydro distillation chips were finely pulverized	5.53 (yellowish red concrete)	0.37	(1) 0.36 (2) 0.83 (3) 39.71 (4) 19.76	Pale yellow	Less pleasant
Hydro distillation of concrete	12	–	–	1.05	(1) 3.98 (2) 4.87 (3) 38.47 (4) 20.42	Pale yellow	Less pleasant
Hydro distillation	30	Pulverized coarse powder	–	1.86	(1) 2.17 (2) 1.26 (3) 40.19 (4) 20.42	Pale yellow	Pleasant
Soxhlet Extraction (toluene)	10	Medium/coarse pulverizing	7.56	2.59	(1) 3.98 (2) 4.80 (3) 29.22 (4) 30.54	Pale yellow	Less pleasant
Hydro distillation	48	0.3%alkaline water, coarse/medium pulverized	–	2.68	(1) 4.25 (2) 3.01 (3) 41.90 (4) 14.89	Pale yellow	Pleasant
Hydro distillation	38	Ungrounded chips immersed in hot water for 24 h	–	1.56	(1) 0.30 (2) 0.91 (3) 56.73 (4) 27.10	Pale yellow	Pleasant
Section II							
Steam distillation	10	Fine pulverized powder	–	1.60	(1) 0.77 (2) 1.80 (3) 54.74 (4) 29.58	Pale yellow	
Soxhlet extraction (benzene)	3	Steam distilled powder	2.07	1.05	(1) 0.85 (2) 1.70 (3) 42.22 (4) 23.26	Pale yellow	Pleasant
Soxhlet extraction (ethyl alcohol)	6	Coarse pulverizing immersed	10.90	3.70	(1) 0.96 (2) 3.28 (3) 50.03 (4) 27.87	Pale yellow	Less pleasant

(continued)

Table 38.11 (continued)

Method of extraction	Batch time (h)	Physical/ pre-treatment	% concrete Extracted	% oil/ absolute Extracted	Oil composition (Major constituents, %)	Color of oil	Odor of oil
Soxhlet extraction (diethyl ether)	5	Coarse pulverizing immersed	5.23	2.58	(1) 0.57 (2) 1.47 (3) 48.82 (4) 14.89	Pale yellow	Less pleasant
Soxhlet extraction (benzene)	5	Previously hydro distilled 30 h coarse powder	4.27	1.25	(1) 3.42 (2) 4.99 (3) 38.21 (4) 23.37	Pale yellow	Pleasant
Soxhlet extraction (toluene)	12	Coarse pulverizing immersed	4.98	2.45	(1) 3.84 (2) 4.03 (3) 37.04 (4) 15.89	Pale yellow	Less pleasant
Soxhlet extraction (benzene)	5	Pulverized fine powder	6.25 (dark red)	3.01	(1) 7.79 (2) 5.12 (3) 30.54 (4) 15.98	Pale yellow	Pleasant

Table 38.12 Comparison of sandalwood oil obtained by different processes (Nautiyal 2014)

Process of extraction	Concrete extracted (Wt %)	Absolute extracted (Wt %)	Composition of oil (%)			
			i	ii	iii	iv
Liquid CO_2 (200 bars, 28 °C, 4 h)	Nil	3.76	0.48	1.08	54.50	28.00
Solvent extraction	Nil	Nil	–	–	–	–
Benzene (5 h)	6.30	3.01	7.86	1.63	30.81	12.18
Diethyl ether (5 h)	5.23	2.58	0.57	1.47	48.82	23.37
Ethyl alcohol (5 h)	10.90	3.70	1.14	0.44	54.55	29.01
Hydro distillation (30 h)	Nil	1.86	2.17	1.26	40.19	12.40
Hydro distillation alkaline treated (45 h)	Nil	2.68	4.25	3.01	41.90	14.89
Steam distillation pilot plant (10 h)	Nil	1.60	0.77	1.80	54.74	29.58

(i) α-Santalene
(ii) β-Santalene
(iii) α-Santalol
(iv) β-Santalol

Nautiyal (2014) further studied the sandalwood oil obtained by different processes (Table 38.12). The major constituents' α-santalol and β-santalol were extracted in a good amount by liquid carbon dioxide as well as by ethyl alcohol. The comparison of extracts of sandalwood oil, obtained by different methods, that has shown that the major constituents α-santalol and β-santalol were extracted in a good amount by liquid carbon dioxide as well as by ethyl alcohol. Some physical and

Table 38.13 Physical and chemical properties of sandalwood oil (Nautiyal 2014)

Method of extraction	Pretreatment	Refractive index	Optical rotation	Acid value
Hydro distillation	Whole chips	1.500	−22.97	Nil
Soxhlet extraction (toluene)	Pulverized coarse size powder	1.500	−19.14	4.67
Hydro distillation of concrete	Nil	1.499	Nil	7.33
Hydro distillation	Pulverized coarse size powder	1.499	−14.46	5.58
Hydro distillation	Pulverized coarse size powder (0.3% alkaline distilled water)	1.502	−19.57	2.66
Steam distillation	Pulverized powder immersed in cold water (48 h)	1.503	−24.67	6.39
Soxhlet extraction (ethanol)	Fine pulverized powder	1.504	−19.56	7.79
Soxhlet extraction (diethyl ether)	Fine pulverized powder	1.503	−14.46	7.79
Soxhlet extraction (benzene)	Fine pulverized powder	1.501	−28.70	6.95
Soxhlet extraction (toluene)	Fine pulverized powder immersed in hot water	1.502	−	6.71
Liquid CO_2	Fine pulverized powder	1.505	−22.97	4.10
Commercial sandalwood oil	Nil	1.504	−19.57	4.15

Required specifications: Refractive index: 1.499–1.506; Acid value: 0.5–8.0; Optical rotation: 15.00–19.20°

Table 38.14 Organoleptic quality oils from sandalwood species

	Santalum album (India)	*Santalum austocaledonicum* (New Caledonia)	*Santalum paniculatum* (Hawai)	*Santalum spicatum* (Australia)
Color	Pale yellow	Colorless	Colorless	Pale, pale yellow
Clarity	Clear	Clear	Clear	Clear
Viscosity	Semi viscous	Semi viscous	Semi viscous	Semi viscous
Intensity	3–4	4	4–5	4–5
Taste	Bitter/butter	Bitter	Bitter	Slightly bitter

Intensity of odor is not how it smells but the strength relative to other odours with 1–10 with 1 = lowest. For example: Bergamot & amber =2, rose geranium = 4–5, tea tree = 6 and massoia = 8

chemical properties of the sandalwood oil, such as refractive index, optical rotation and acid values, are presented in Table 38.13. The optical rotation value obtained by hydrodistillation, steam-distillation, benzene extraction and liquid carbon dioxide extraction were not within the required specification. However, the refractive index and the acid value for all the experiments were within the stipulated values for sandalwood oil (Rozzi and Singh 2002; Skerget and Knez 1997; Moretta et al. 1998a, b). Organoleptic qualities of oil from sandalwood species are represented in Table 38.14.

4 Applications and Health-Promoting Properties of Oil

Sandalwood being an ancient, biblical essential oil holding powerful and precious properties; it has also wide applications in aromatherapy (Arctander 1960a, b). Sandalwood is the oldest materials used in aromatic and perfumery and has been used over 4000 years. Sandalwood was used by the ancient Egyptians to embalm bodies. Sandalwood owes a number of religious connotations and, is mentioned in the bible along with frankincense and myrrh. In order to keep white ants at bay, Indian temples were built up with sandalwood. Sandalwood also finds its uses in Indian meditation ceremonies. In India, sandalwood is combined with rose to produce a scent called Attar. During burial, Muslim countries use sandalwood to ensure a quick ascent of the soul to heaven. Sandalwood was used for respiratory and urinary infections in Ayurvedic medicine and for the revitalization of skin too (Askinson 1915). Chinese Medicine uses sandalwood for skin complaints, stomachache and vomiting. Buddha is honored by the Japanese with the use of sandalwood. Discorides' *De Materia Medica*, mentions sandalwood that is a reference book of many medicinal plants of its time.

Sandalwood oil sedative property makes it useful in meditation, as it induces a feeling of deep peace. Sandalwood is also anti-infectious, a decongestant, anti-depressant, a sexual tonic and an aphrodisiac (Coombs 1995). Its antiseptic properties are useful in the healing process of major burns. Sandalwood oil also inhibits the stimulation of the growth of white blood cells. The chronic bronchial infections and coughs are treated with the sandalwood oil. Sandalwood oil is effective in healing sciatica and lumbago. Soaps, cosmetics and perfumes find the application of sandalwood oil as common fragrance component (Deite 1892).

Sandalwood oil is very useful in treating common colds, bronchitis, fever, dysentery, piles, scabies and infection of the urinary tract, inflammation of the mouth and pharynx, liver and gall-bladder complaints and as an expectorant, stimulant, carminative, digestive and as a muscle relaxant (Burdock and Carabin 2008). In vitro study reported that sandalwood oil is effective on methicillin-resistant *Staphylococcus aureus* (MRSA) and anti mycotic-resistant Candida species (Warnke et al. 2009). A crude extract isolated compounds of sandalwood oil (primarily α- and β-santalol) showed antibacterial activity against *Helicobacter pylori*, a gram-negative bacterium which has a strong link to the development of duodenal, gastric and stomach ulcers (Ochi et al. 2005). Sandalwood oil exhibited virulence against isolates of drug-resistant herpes simplex virus type I (Schnitzler et al. 2007). The oil also showed anti-carcinogenic activity (Burdock and Carabin 2008). Paulpandi et al. (2012) reported that β-santalol exhibited anti-influenza A/HK (H_3N_2) virus activity of 86% without any cytotoxicity at the concentration of 100 μg/mL. On applying sandalwood oil elevates pulse rate, skin conductance level and systolic blood pressure and higher ratings of attentiveness and mood in humans are attained.

Sandalwood oil and its major constituents have been tested for minimal sensitive oral and dermal toxicity in laboratory animals. Sandalwood oil was very potent as an antiviral, anticarcinogenic and bactericidal activity (Arctander 1960a, b).

Australian sandalwood oil was reported to be used both orally and externally for treating gonorrhea, more potent against Candida as compare to tea tree oil. Many skin diseases such as acne and tinea are also treated. The oil is also inhaled for calming or for the respiratory system and utilized as an exciting perfume addition.

5 Adulteration and Authenticity of Sandalwood Oil

The rising demand for sandalwood essential oil resulting in high costs and hence sandalwood oil is one of the most adulterated essential oils (Harman 2015). In the USA, the commercial use of sandalwood oil began in the early 1800s. Its sensory quality, extensive use, and steep hike in the price, often adulterates sandalwood oil with low grade cost effective oils and synthetic or semi-synthetic substitutes such as sandalore (Anonis 1998). British pharmaceutical Codex regulatory agencies face serious problems with adulteration of sandalwood oil as well threat to the health of consumers. Chemical composition and physical properties of the sandalwood oil with substitution and synthetic additives influence and affect the oil quality and the allergic potential. Castor oil, cedar wood oil and low-grade oil from sandalwood species other than *S. album* are reported as the common adulterants (Anonis 1998). There are many methods of adulteration involves in the essential oil industry, by adding a single raw material, cheaper essential oils, cheap synthetics, isolates, and even attempts to pass one essential oil off as another. Rapeseed oil is often used in the EU as a cheap vegetable oil as adding a single raw material to essential oils. (http://sandalwoodoilspecialist.com/sandalwood-oil/adulteration/).

(Burfield, http://www.users.globalnet.co.uk/~nodice/new/magazine/october/october.htm, 2003).

The dilution of sandalwood oil is often done using the essential oils of other species of sandalwood (blended sandalwood essential oil), even to a larger extent with copaiba (*Copaifera*) oil, Atlas cedar (*Cedrus atlantica*) fractions, and amyris (*Amyris balsamifera*) oil. Recreation of sandalwood oil's natural fragrance is brought about with sandalwood fragrance chemicals, synthetic substances, which are not originally from the sandalwood tree. Sandalwood oil adulteration also often created using a technique called "stretching," where odor-free solvents are added to sandalwood oil to increase its quantity.

Chemicals from the actual sandalwood tree used to adulterate sandalwood essential oil involving many synthetic substitutes, which originate from β-santalol as the main constituent of sandalwood oil. Creation of synthetic essential oils is not supported by the experts just by blending the components in sandalwood oil and uneasy to replicate. Many other substances are often searched by synthetic oil manufacturers to find "woody notes" similar to sandalwood oil. It was also noted that the best sandalwood oil substitutes being derivatives of α-campholenic aldehyde yielded from an inexpensive α-pinene, most of it a byproduct of the paper industry. The α-campholenic aldehyde derivatives possess an odor similar to sandalwood.

Since sandalwood oil is in such high demand and often adulterated, therefore it has become utmost important to those purchasing sandalwood essential oil to ensure that the oil is authentic and pure. This issue is important especially for the perfume, pharmaceutical, flavor, and fragrance industries. Sandalwood oil's authenticity can be tested in different ways. Adulteration of sandalwood oil is easy to detect and is conducted by an expert with the right materials requiring advanced technology to analyze. A trained expert can detect adulterated sandalwood oil simply by the aroma.

Kuriakoje et al. (2010) studied the adulteration of sandalwood oil using ear Infrared (NIR) spectroscopy along with multivariate calibration models like principal component regression (PCR) and partial least square regression (PLSR) as rapid analytical techniques. It was concluded that NIR spectroscopy with chemometric techniques could be successfully used as a rapid, simple, instant and non-destructive method for the detection of adulterants, even 1% of the low-grade oils, in the high-quality form of sandalwood oil.

There are various recommendations that the oil from *S. album* should not contain less than 90% w/w of (free) alcohols, calculated as santalols (ISO 1979; Food Chemicals Codex 1981). Assessment of santalol content of sandalwood oil by the acetylation methods (ISO 1976, 2002) described generally lack specificity and accuracy.

There is an increasing demand for the development of a new, rapid, and non-destructive method instead of traditional, time taking and expensive analysis techniques. Nautiyal (2011) examined the quality of extracted sandalwood oil with various techniques. The quality control (QC) of sandalwood essential oils can rely on international norms, (ISO-FDIS 3518:2001) and (ISO 22759:2009), for *S. album* and *S. spicatum*, respectively. Indeed, due to the shortage of genuine East Indian *S. album* essential oil, not everything offered as sandalwood oil is 100% derived from the claimed botanical species. The content of santalol isomers is one of the main criteria for quality assessment. The total alcohol content expressed as santalol should not be <90% in both *S. album* and *S. spicatum* essential oils. In routine QC, gas chromatography is today the method of choice for particularly when dealing with solvent extracts (Nicolas et al. 2011).

References

Alpha, T., Raharivelomanana, P., Bianchini, J. P., Faure, R., Cambon, A., & Joncheray, L. (1996). PF0807_Lawrence_fcx.indd 44 06/2/08 9:38:44 AM 47 Santalenes from Santalum austrocaledonicum. *Phytochemistry, 41*, 829–832.

Alpha, T., Raharivelomanana, P., Bianchini, J. P., Faure, R., & Cambon, A. (1997a). Sesquiterpenoid from Santalum austrocaledonicum. *Phytochemistry, 46*, 1237–1239.

Alpha, T., Raharivelomanana, T. P., Bianchini, J.P., Faure, R., & Cambon, A. (1997b). Identification de deux derives dihydoxyles du bisabolane a partir de Santals oceaniens. *Rivista Ital*. EPPOS, (Numero Speciale), 84–91.

Alpha, T., Raharivelomanana, T. P., Bianchini, J. P., Faure, R., & Cambon, A. (1997c). Identification de deux derives dihydoxyles du bisabolane a partir de Santals oceaniens. *Rivista Ital.* EPPOS, (Numero Speciale), 84–91.

Anon. (2006). *Market overview the Australian Sandalwood Industry.* Australian Agribusiness Group.

Anonis. (1998). Sandalwood and sandalwood compounds. *Perfumer Flavorist, 23*, 19.

Arctander, S. (1960a). Perfume and flavor materials of natural origin.

Arctander, S. (1960b). Sandalwood oil in East India. In *Perfume and flavor materials of natural origin* (pp. 574). Denmark. 12 D. P.

Askinson, G. (1915). *Perfumes and cosmetics.* London: Hodder & Stoughton.

AVONGRO. (2006). Wheatbelt tree cropping incorporated economic development and sustainable production. *Sandalwood economics.*

Brand, J. E., Fox, J. E. D., Pronk, G., & Cornwell, C. (2007). Comparison of oil concentration and oil quality from *Santalum spicatum* and *S. album* plantations, 8-25 years old, with those from mature *S. spicatum* natural stands. *Australian Forestry, 70*(4), 235–241.

Braun, N. A., et al. (2003). Isolation and chiral GC analysis of β-bisabolols-trace constituents from the essential oil of *Santalum album* L. *Journal of Essential Oil Research, 15*(2), 63–65.

Braun, N., Meier, M., & Hammer Schmidt, F. J. (2005). New Caledonium sandalwood oil-a substitute for East Indian sandalwood oil? *Journal of Essential Oil Research, 17*, 477–480. 150.

British Pharmaceutical Codex. (1949) (p. 612). *The British Pharmaceutical Codex 1949.* London: The Pharmaceutical Society.

Broadhurst, L., & Coates, D. (2002). Genetic diversity within and divergence between rare and geographically widespread taxa of the *Acacia acuminata* Benth. (Mimosaceae) complex. *Heredity, 88*, 250–257.

Brunke, E. J., Vollhardt, J., & Schmaus, G. (1995). Cyclosantalal and epicyclosantalal–new sesquiterpene aldehydes from East Indian sandalwood oil. *Flavour and Fragrance Journal, 10*, 211–219.

Brunke, E.J., Falbusch, K.G., Schmaus, G., & Volhardt, J. (1997). *The chemistry of sandalwood fragrance—a review of the last 10 years.* Rivista Ital. EPPOS, (Numero Speciale), 48–83.

Bryne, M., Macdonald, B., Broadhurst, L., & Brand, J. (2003a). Regional genetic differentiation in Western Australian sandalwood (Santalum spicatum) as revealed by nuclear RFLP analysis. *Theoretical and Applied Genetics, 107*, 1208–1214.

Bryne, M., Macdonald, B., & Brand, J. (2003b). Phylogeography and divergence in the chloroplast genome of Western Australian Sandalwood (*Santalum spicatum*). *Heredity, 91*, 389–395.

Burdock, G. A., & Carabin, I. G. (2008). Safety assessment of sandalwood oil (*Santalum album* L.). *Food and Chemical Toxicology, 46*(2), 421–432. https://doi.org/10.1016/j.fct.2007.09.092.

Chana, J. S. (1994). Sandalwood production. *International Journal of Aromatherapy, 6*(4), 11–13.

Chemat, F., Sahraoui, N., Abert-Vian, M., Bornard, I., & Boutekdjiret, C. (2008). Improved microwave steam distillation apparatus for isolation of essential oils comparison with conventional steam distillation. *Journal of Chromatography A, 1210*, 229–233.

Chen, Z. X., & Liu, L. (2001). Influences of various extraction methods on content and chemical components of volatile oil of Santalum album. *Guangzhou Zhongyiyao Daxue Xuebao, 18*(2), 174–177.

Coakley, T. (2006). *Market update by Wescorp International.* Avon Sandalwooder, Autumn 2006.

Coombs, A. J. (1995). *Dictionary of plant names.* Oregon: Timber Press.

Coppen, J. J. W. (1995). Sandalwood oil. In *Non-wood forest products 1. Flavours and fragrances of plant origin* (pp. 53–60). Rome: Food and Agricultural Organization of the United Nations.

Deite, D. C. (1892). *A practical treatise on the manufacture of perfumery.* Philadelphia: Henry Carey Baird & Co..

Ferhat, M. A., Chemat, F., Meklati, B. Y., & Smadja, J. (2006). An improved microwave Clevenger apparatus for distillation of essential oils from orange peel. *Journal of Chromatography A, 1112*, 21–126.

Food Chemicals Codex. (1981). 3rd edn, (p. 268). Washington DC: National Academy Press.

Forest Products Commission. (2007). Annual report 2006/07.

Gowda, V.S.V. (2013). Essential oil production under public sector, private partnership model, Essential oil http://www.ffdcindia.org/essential_oil_24july13.pdf.

Guenther, E. (1952). Oil of sandalwood East Indian. In *The essential oils* (Vol. 5, pp. 173–187). New York: Van Nostrand Co..

http://sandalwoodoilspecialist.com/sandalwood-oil/adulteration/.

http://www.fao.org/docrep/V5350E/V5350e08.htm.

http://www.fpc.wa.gov.au/. Preliminary oil results from a 14-year-old Indian Sandalwood (*Santalum album*) plantation at Kununurra, WA.

https://draxe.com/sandalwood-essential-oil/.

https://food.ndtv.com/beauty/5-sandalwood-benefits-to-look-out-for-from-tan-removal-to-treating-acne-1745068.

https://www.cbi.eu/sites/default/files/market_information/researches/product-factsheet-europe-sandalwood-oil-cosmetics-2015.pdf.

Hamilton, L Conrad, CE (Tech. Coords.) (1990) Proceedings of Symposium on Sandalwood in the Pacific, Honolulu, Hawaii, 9–11 April, 1990. USDA Forest Service, Pacific Southwest Research Station, General Technical Report PSW-122. 84 pp. Berkeley: USDA, PSRS.

Harman, A. (2015). *Harvest to hydrosol*. Fruitland: IAG Botanics.

Hettiarachchi, D. S. (2008). Volatile oil content determination in the Australian sandalwood industry: Towards a standardized method. *Sandalwood Research Newsletter, 23*, 1–4.

Hettiarachchi, D. S., Gamage, M., & Subasinghe, U. (2010). Oil content analysis of sandalwood: A novel approach for core sample analysis. *Sandalwood Research Newsletter, 25*, 1–4.

International Organization for Standardisation (1976) Oil of Sandalwood (Santalum album L.) ISO 3793.

International Organization for Standardisation. (1979). Oil of sandalwood (*Santalum album* Linnaeus) *International Standard ISO 3518-1979 (E)*. 1st edn, 2 pp. International Organization for Standardization.

International Organization for Standardisation (2002) Oil of Sandalwood (Santalum album L). 2nd edn, ISO 3518.

Jaramillo, B. E., & Martinez, J. R. (2004). Comparison of different extraction methods for the analysis of volatile secondary metabolites of *Lippia Alba* (Mill.) N.E. Brown, Grown in Colombia, and evaluation of its in vitro antioxidant activity. *Journal of Chromatography A, 1025*, 93–103.

Jones, P. (2002). *Estimating returns on plantation grown Sandalwood (Santalum spicatum)*. Sandalwood Information Sheet Issue 3. Forest Products Commission.

Jones, C. G., Plummer, J. A., & Barbour, E. L. (2007). Non-destructive sampling of Indian sandalwood (*Santalum album* L.) for oil content and composition. *Journal of Essential Oil Research, 19*(2), 157–164. https://doi.org/10.1080/10412905.2007.9699250.

Krotz, A., & Helmchen, G. (1994). Total syntheses, optical rotations and fragrance properties of sandalwood constituents: (-)-(Z) - and (-)-(E)-β-Santalol and their enantiomers, ent-β-Santalene. *Liebigs Annalen der Chemie, 1994*(6), 601–609. https://doi.org/10.1002/jlac.199419940610.

Kuriakoje, S., Thankappan, X., & Venkatraman, V. (2010). Detection and quantification of adulteration in sandalwood oil through near infrared spectroscopy. *The Analysts Journal, 135*(10), 2676–2681. https://doi.org/10.1039/c0an00261e.

Kusuma, H. S., & Mahfud, M. (2016). Kinetic studies on extraction of essential oil from sandalwood (*Santalum album*) by microwave air-hydro distillation method. *AIP Conference Proceedings, 1755*, 050001-1–050001-6. https://doi.org/10.1063/1.4958484.

Kusuma, H. S., & Mahfud, M. (2017). Kinetic studies on extraction of essential oil from sandalwood (*Santalum album*) by microwave air-hydro distillation method. *Alexandria Engineering Journal*. https://doi.org/10.1016/j.aej.2017.02.007, in press.

Lawrence, B. M. (1991). Sandalwood oil in progress in essential oils. *Perfumer & Flavorist, 1*(1), 5(1976); 1(5), 14(1976); 6(5), 32–34(1981); 16(6), 50–52.

Lawrence, B. M. (2009). A preliminary report on the world production of some selected essential oils and countries. *Perfumer and Flavorist, 34*(1), 38–44. Sandalwood Essential Oil, http://scienceofacne.com/sandalwood-essential-oil/.

Marongiu, B., Piras, A., Porcedda, S., & Tuveri, E. (2006). Extraction of *Santalum album* and *Boswellia carterii* Birdw volatile oil by supercritical carbon dioxide: Influence of some process parameters. *Flavour and Fragance Journal, 21*(4), 718–724.

McKinnell, F. H. (Ed.). (1993). *Proceedings of symposium on sandalwood in the Pacific region held at XVII Pacific science congress, Honolulu, Hawaii, 2 June, 1991* (*ACIAR Proceedings No. 49.* 43 pp). Canberra: ACIAR.

McKinnell, F. H. (2008). *WA sandalwood industry development plan 2008–2020*. Perth: Forest Products Commission.

Mookherjee, B. D., Trenkle, R. W., & Wilson, R. A. (1992). New insights in the three most important natural fragrance products: Wood, amber and musk. In H. Woidich & G. Buchbauer (Eds.), *Proceedings of the 12th International Congress of Flavours, Fragrances and Essential Oils, Oct 4-8 1992* (pp. 234–262). Vienna: Austrian Assoc. Flav. Frag. Industry.

Moretta, P., Ghisalbert, E. L., Piggott, M. J., & Trengove, R. D. (1998a). Extraction of oil from *Santalum spicatum* by supercritical fluid extraction. *ACIAR Proceedings Series, 84*, 83–85.

Moretta, P., Ghisalbert, E. L., Piggott, M. J., & Trengove, R. D. (1998b). *ACIAR Proceedings Series, 84*, 83.

Nautiyal, O. H. (2010). Oil extraction methods in *Santalum album*. *Sandalwood Research Newsletter, 25*, 5–7.

Nautiyal, O. H. (2011). Analytical and Fourier transform infra red spectroscopy evaluation of sandalwood oil extracted with various process techniques. *Journal of Natural Products, 4*(2011), 150–157.

Nautiyal, O. H. (2014). Process optimization of sandalwood (*Santalum album*) oil extraction by subcritical carbon dioxide and conventional technique. *Indian Journal of Chemical Technology, 21*, 290–297.

Nayar, R. (1988). Cultivation, improvement, exploitation and protection of *Santalum album* Linn. *Advances in Forestry Research in India, 2*, 117–151.

Nicolas, B., Céline, D., & Daniel, J. (2011). Phytochemistry of the heartwood from fragrant Santalum species: A review. *Flavour and Fragrance Journal, 26*(1), 7–26.

Nikiforov, A. L. J., Buchbauer, G., & Raverdino, V. (1988). GC-FTIR and GC-MS in odour analysis of essential oils. *Mikrochimica Acta (Vienna), 11*, 193–198.

Ochi, T., Shibata, H., Higuti, T., Kodama, K. H., Kusumi, T., & Takaishi, Y. (2005). Anti-*Helicobacter pylori* compounds from *Santalum album*. *Journal of Natural Products, 68*(6), 819–824. https://doi.org/10.1021/np040188q.

Paulpandi, M., Kannan, S., Thangam, R., Kaver, K., Gunasekaran, P., & Rejeeth, C. (2012). In vitro anti-viral effect of β-santalol against influenza viral replication. *Phytomedicine, 19*(3–4), 231–235. https://doi.org/10.1016/j.phymed.2011.11.006.

Rozzi, N. L., & Singh, R. K. (2002). *Comprehensive Reviews in Food Science and Food Safety, 1*, 33.

Schnitzler, P., Koch, C., & Reichling, J. (2007). Susceptibility of drug-resistant clinical herpes simplex virus type 1 strains to essential oils of ginger, thyme, hyssop, and sandalwood. *Antimicrobial Agents and Chemotherapy, 51*(5), 1859–1862. https://doi.org/10.1128/AAC.00426-06.

Sen-Sarma, P. K. (1982). Sandalwood – its cultivation and utilization. In C. K. Atal & B. M. Kapur (Eds.),. 815 pp *Cultivation and utilization of aromatic plants* (pp. 395–405). Jammu: Regional Research Laboratory, CSIR.

Shankarnarayana, K. H., & Parthsarathi, K. (1984). Compositional differences in sandal oils from young and mature trees and in the sandal oils undergoing colour change on standing. *Indian Perfumer, 28*(3/4), 138–141.

Shukla, B. V., Mohod, R., Shukla, S. V., Lehri, A., & Singh, D. P. (1999). Quality assessment of sandalwood oil using gas chromatograph. *FAFAI Journal, 1*(3), 41–43.

Skerget, M., & Knez, Z. (1997). *Journal of Agricultural and Food Chemistry, 45*, 2066.

Statham, P. (1990). The sandalwood industry in Australia: a history. In *Proceedings of Symposium on Sandalwood in the Pacific, Honolulu, Hawaii, 9–11 April, 1990. USDA Forest Service, Pacific Southwest Research Station, General Technical Report PSW-122* (pp. 26–38). Berkeley: USDA, PSRS.

Suriamihardja, S. (1978). [Problems on sandalwood (*Santalum album* Linn.) silviculture and improving its production]. In Proceedings of the Third Seminar on Volatile Oils, Bogor, Indonesia, July, 1978 (pp. 115–125) Bogor: Balai Penelitian Kimia.

U.S. Dispensatory (1955) *sanadalwood oil is a volatile oil obtained by steam distillation.* Lippincott, Philadephia 25th edn, p. 1836.

Venkatesha Gowda, V. S., Patil, K. B., & Perumal, I. R. (2006). Forest based essential oils viz sandalwood oil production and future scenario. *Indian Perfum, 50,* 45–50; *PAFAI,* 8(1), 63–70.

Verghese, J., Sunny, T. P., & Balkrishanan, K. V. (1990). (+)-α-santalol and (−)-β-santalol (Z) concentration, a new quality determinant of East Indian sandalwood oil. *Flavour and Fragrance Journal, 5,* 223–226.

Warnke, P. H., Becker, S. T., Podschun, R., Sivananthan, S., Springer, I. N., Russo, P. A. J., Wiltfang, J., Fickenscher, H., & Sherry, E. (2009). The battle against multi-resistant strains: Renaissance of antimicrobial essential oils as a promising force to fight hospital-acquired infections. *Journal of Carnio Maxillo Facial Surgery, 37*(7), 392–397. https://doi.org/10.1016/j.jcms.2009.03.017.

Wei, M., Lin, L., Qiu, J. Y., Cai, Y. W., Lu, A., Yuan, L., Liao, H. F., & Xiao, S. G. (2000). Wind damage effects on quality of hardwood of Lignum santal; *albi. Zhongguo Zhongyao Zazhi, 25,* 710–713.

Wood, A. (2008). *Pathogen of the month – January 2008.* South Africa: Australasian Plant Pathology Society (APPS).

Yadav, V. G. (1993). Sandalwood: Its origin, synthetic substitutes and structure-odour relationship. *PAFAI Journal, 15*(4), 21–54.

Zhu, L. F., Li, Y. H., Li, B. L., Lu, B. Y., & Xia, N. H. (1993). *Aromatic plants and essential constituents* (p. 60). Hong Kong: South China Institute of Botany, Chinese Academy of Sciences, Hai Feng Publish.. distributed by Peace Book Co. Ltd..

Chapter 39
Watermelon (*Citrullus lanatus*) Oil

Bushra Sultana and Rizwan Ashraf

Abstract Watermelon (*Citrullus lanatus*), member of the Cucurbitaceae family, has been cultivated for thousands of years as a food source for human consumption. Since developing countries alternately depend on non-conventional sources of protein that augment protein deficiency in diets. Watermelon seeds are reported a worthwhile source of protein due to high protein contents as well as important medicinal constituents. The seeds are rich in oil (37.8–45.4%), and protein (25.2–37%) having in precious amino acids. On the commercial scale, the oil is an important industrial raw material as a drying agent in glass paint and soap industry due to its high iodine value and saponification number. The economic importance, nutritional and medical traits of watermelon seed oil has made it curious to know properly. Therefore, this chapter will cover chemical profiling and functionality of watermelon seed oil along with their medicinal applications.

Keywords Cucurbitaceae · *Citrullus colocynthis* · Carolina cross · Yellow Crimson · Oilseed cake · Fatty acids

1 Introduction

Watermelon (*Citrullus lanatus*) commonly grown in tropical and subtropical regions, having vine-like flowering plant as shown in Fig. 39.1. It has been cultivated mostly for its fruit worldwide and especially in Africa for the last 4000 years. Its genesis is based on two different theories among which one claims that it was derived from perennial relative *Citrullus colocynthis* present in wild archaeological sites, while other stated that it was domesticated from wild forms of *Citrullus lanatus*. From Africa, it was introduced to India at about 800 CE and China at 900 CE that was further extended to Southeast Asia, Japan, Europe and America in the late 1500s (Erhirhie and Ekene 2013). The world's major producers of watermelons are shown in Fig. 39.2.

B. Sultana · R. Ashraf (✉)
Department of Chemistry, University of Agriculture Faisalabad, Faisalabad, Pakistan

© Springer Nature Switzerland AG 2019
M. F. Ramadan (ed.), *Fruit Oils: Chemistry and Functionality*,
https://doi.org/10.1007/978-3-030-12473-1_39

Fig. 39.1 *Citrullus lanatus* flower, seed, fruit, and oil

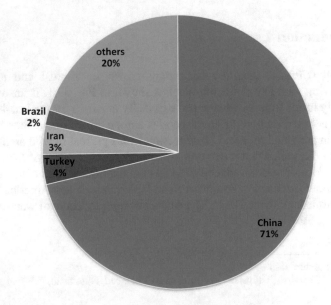

Fig. 39.2 Distribution of watermelon production (%) in the world

This fruit is commonly used owing to its sweet flavor and high nutritional value (Mahla et al. 2018). Watermelon is an annual fruit crop that stands second after tomato in acreage and production. Most of its growth is feasible in open fields usually protected or grown in greenhouses (Erhirhie and Ekene 2013). Its juice and pulp are used for human consumption, while rind and seeds are major solid wastes (Wani et al. 2011). The color of pulp is an important trait of watermelon quality dependent on carotenoid and their accumulations in the chromoplasts during ripening of fruit. Watermelon pulp could have variant colors including red, yellow, salmon or orange (Bang et al. 2007), owing to different carotenoids contents that depend on origin and climate. This scenario predicts that watermelon is a potent source of carotenoids (Yoo et al. 2012). Probably carotenoid composition in red-fleshed cultivars mostly consists of lycopene, about 90% of which as *trans*-lycopene and rest is a *cis*-lycopene isomeric form (Erhirhie and Ekene 2013).

Watermelon fruit is widely consumed in Pakistan yielding a considerable amount of seeds, which are often discarded as wastes. In recent years, the demand for watermelon seeds is increasing day by day due to public health concerns and change in food habits. These seeds have a high nutritional value that is rich in protein 25–37% and oil 37.8–45.4% (Ziyada and Elhussien 2008). Watermelon seeds contain oil, which is known as *Ootanga* and *Kalahari* oil. Conjugated fatty acids (FA) among some Cucurbitaceae oils make them highly useful as drying oils. These nutritional contents support livelihood in the hostile situations where commercial crop diversification is not feasible (Mahla et al. 2018). The seeds are utilized directly for human consumption in various forms such as snacks, additive and stuffing in various products. Melon seeds are used as thickener and emulsifier in soups and stews that are a source of proteins in the diet. The seeds are also reported to possess medicinal traits that help to treat chronic or acute eczema. The seeds are rich in natural antioxidants and phytochemicals such as vitamin C, flavonoids, riboflavin, thiamine and phenolic compounds. Its chemical composition consists of 4.4% carbohydrates, 6.2% ash, 8.2% fiber, 31.9% protein and 57.1% lipids. Among minerals, the seeds contain 7.5 mg iron, 130 mg calcium, 456 mg phosphorus. *Citrullus lanatus* seed extracts possess antioxidant activity wherein the potency of antioxidant activity depends on the type of extract. The watermelon seeds contain about 23% crude protein and 28% crude lipids but corresponding values for the kernel are about 40% and 49%, respectively (Das et al. 2002). The economic importance of oil crops has made it necessary that they should be properly investigated to ascertain their oil quality parameters, since this is an important criterion for marketing and processing of seed oil (Essien and Eduok 2013).

2 Varieties of Watermelon

There are almost 1200 reported varieties of watermelon in various sizes ranging from less than a pound to more than 200 pounds. Watermelon also has a diversity of flesh in colors with red, yellow, orange or white (Erhirhie and Ekene 2013). Some common varities and their appearance are described in Fig. 39.3.

Fig. 39.3 Watermelon
varieties and physical
appearance. (a) Carolina
cross, (b) Yellow crimson,
(c) Orangeglo, (d)
Orangeglo, (e)
Saskatchewan, (f)
Melitopolski, (g) Densuke

(a) **Carolina cross**: this variety produced higher weight melons weighing 262 pounds (119 kg) with green skin and red flesh. It takes about 90 days from planting to harvest (Erhirhie and Ekene 2013).

(b) **Yellow Crimson**: watermelon of this variety has yellow colored flesh. This particular type of watermelon is "sweeter" and honey-flavored than other red flesh watermelon (Ahmad et al. 2017).

(c) **Orangeglo**: this variety has sweet orange pulp with significant weighing 9 kg (20–30 pounds). It has a light green rind with dark greenish stripes. It takes almost 90–100 days from planting to harvest.

(d) **Moon and stars**: this variety is considered as a class of orangeglo and has been found since 1926 (Fig. 39.3d). It has many small yellow circles with purple-black rind having weight 9–23 kg (20–50 pounds) (Janakiraman et al. 2012). It has flesh pink or red color and brown seeds.

(e) **Cream of Saskatchewan**: this variety consists of small round fruits around 25 cm (10 in.) in diameter. It has a quite thin, light green with dark green striped rind, with sweet white flesh and black seeds. It can grow well in cold climates. These watermelons take 80–85 days from planting to harvest.

(f) **Melitopolski**: this variety has small round fruits roughly 28–30 cm (11–12 in.) in diameter. It is an early ripening variety that originated from the Volga River a region of Russia. This variety is seen piled high by vendors in Moscow in summer. This variety takes around 95 days from planting to harvest.

(g) **Densuke**: this variety has round fruit weighed up to 11 kg. It has a black rind that is spot free even have no stripes on the surface. It can only grow on the Island of Hokkaido, Japan, where up to 10,000 watermelons are produced every year. In June 2008, one of the first harvested melons is being sold at an auction for 6300$ making the most expensive watermelon ever sold. The average selling price of this variety is generally around 250$ (Vohra and Kaur 2011).

3 Nutrient Composition

Watermelon (*Citrullus lanatus*) contains about 6% sugar and 92% water by weight. Like other fruits it also contains vitamin C. Watermelon fruit from every aspect has nutritional value, including the rind and the seeds. The flat brown seeds have much higher food value than the flesh. Significant amounts of vitamin C, minerals, lipids, starch and riboflavin have been obtained from the seeds. They can be dried, roasted and eaten as such or ground into flour to make bread. The composition of dried egusi seed without shell per 100 g includes water 5.1 g, protein 28.3 g, lipids 47.4 g, and carbohydrate 15.3 g. Along with other nutritional contents it also contains good mineral profile including calcium 54 mg, phosphorous 755 mg, and iron 7.3 mg. It can cover daily body requirements for these mineral contents. Figure 39.4 shows the percentage daily value (% DV) that can be obtained from watermelon seeds. Watermelon seed has broad vitamin profile that contains thiamin 0.19 mg, riboflavin 0.15 mg, Niacin 3.55 mg and folate 58 µg. The seed is an excellent source of energy and contains no hydrocyanic acid, making it suitable as livestock feed. The seed oil contains glycosides of linoleic, oleic, palmitic and stearic acids. The seed contains a high percentage of oil, which is similar to pumpkin seed oil and can be used in cooking. There are also prospects for use of the seeds in improvement of infant nutrition in view of their high protein and lipids content (Erhirhie and Ekene 2013).

4 Watermelon Seed Oil

The use and dependence on plants as a natural source of medicine has been in existence since ancient time. Vegetable oils account for about 80% of the world's natural oils and fat supply (Ling et al. 2009) with increasing importance in nutrition and commerce owing to their dietary energy, antioxidant, bio-fuels and raw material potentials (Fasina and Colley 2008) and the attendant increasing global demand. The seed kernels from cucurbit fruits and vegetables such as melon, watermelon, cucumber and gourd, traditionally called "Charmagaz" are well-known in the sub-continental region for their therapeutic properties (Mabaleha et al. 2007). According to Pakistan Oil seeds and Products Annual Report, vegetable oils imported to Pakistan have been estimated to be 2.16 million metric tons (MMT) for the marketing year 2011–2012. Nearly 75% of the total domestic consumption (3.5 MMT) of vegetable oil in Pakistan is met through imports (Raziq et al. 2012), while only 25% is derived from the local vegetable oil seed resources (Nehdi 2011).

Watermelon seeds are a rich source of protein 25–37% and oil 37.8–45.4%. The seeds have been utilized for oil production at the subsistence level in different parts of the world including West Africa and the Middle East (Ziyada and Elhussien 2008; Jarret and Levy 2012).

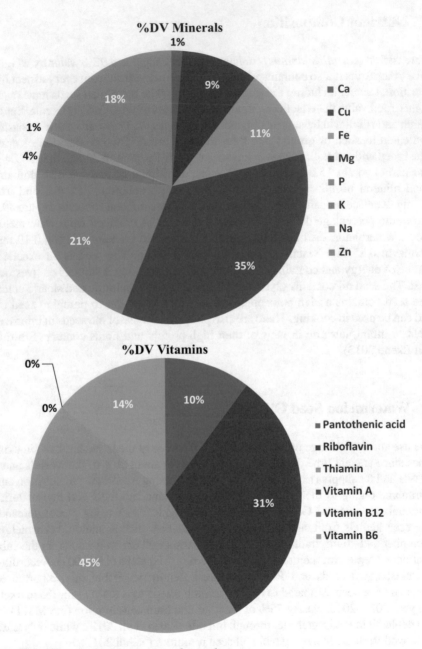

Fig. 39.4 %DV provided from watermelon seeds

Watermelon seed oil characteristics were evaluated to determine whether this oil could be exploited as edible oil and their functional properties are given in Table 39.1. Hexane extraction of watermelon seeds produced yields of 50% (w/w) oil. The refractive index, saponification and iodine value were 1.4712 (at 25 °C), 200 mg

Table 39.1 Chemical properties of watermelon seed oil

Character	Value
Refractive index (40 °C)	1.4630–1.4670
Iodine value	115–125
Saponification value	190–198
Unsaponifiable matter	1.5% max
Moisture	0.5% max
Bulk density (g cm^{-3})	8.33

KOH/g and 156 g I/100 g, respectively. The acid and peroxide values were 2.4 mg KOH/g and 3.24 meq/kg. The induction time of the oil was also 5.14 h at 110 °C (Albishri et al. 2013). Total unsaturation contents of the oil were 81.6%, with linoleic acid (18:2) being the dominant fatty acid (68.3%).

5 Extraction of Oil

To extract oil from watermelon seeds, seeds should be sun-dried to lower the water contents. The common varieties used for extracting oil are *Citrullus lanatus* and *Citrullus vulgaris* but other wild variety like *Citrullus colocynthis* is less in use.

Various types of instruments and techniques have been employed to extract watermelon seeds oil. The most common way of oil extraction is Soxhlet-based extraction having very simple apparatus and most preferably by using different solvents. For the extraction of oil, different solvents were applied, however extractions of oil with *n*-hexane yielded 43.6% (de Conto et al. 2011). In another study by the same author the seed oil was extracted using n-hexane and reported the yield up to 41.3%. In the literature, it was observed that most probable solvent that was employed by researchers for watermelon oil extraction through soxhlet apparatus were non-polar solvents like *n*-hexane and chloroform. Petroleum ether is also reported for the extractions of seed oil form Egusi melon variety and when compared for their chemical properties their values were comparable to other solvents used for extraction (Oluba et al. 2008). The extraction is based on the simple procedure in which boiling solvent in the lower flask evaporates to rise through the vertical tube into the condenser at the top. The condensate liquid drips into thimble, which contains powdered sample wrapped in filter paper subjected to extraction. The extract seeps through the pores of the filter paper and fills the siphon tube, where it flows back down into the round bottom flask. It stayed for sufficient time at least completion of three cycles. After completion of extraction, the solution is subjected to get a ride from solvents by warming through an oven or rotary evaporation and desired pure oil is collected. Then oil is placed in desiccators and weighed again to determine the amount. In addition, the influence of extraction parameters like solvent nature, temperature, time and solvent/kernel ratios are also considerable to maximize the yield. It was observed that oil yield was primarily affected by the solvent/kernel ratio than time and temperature, respectively.

Supercritical fluid based extraction (SFE) is another method for the extraction of melon seed oil. It has provided the necessary impetus and excellent alternative by using of chemical solvents. The extraction by supercritical CO_2 is beneficial in terms of energy consumption as well as more yield could be obtained. In an experiment of extraction for 89 min when the pressure was maintained at 440 bar and temperature raised at 49 °C, the oil yield was 49% which is comparable to conventional methods (Maran and Priya 2015). However, at optimal conditions of pressure 300 bar at 40 °C, the yield can be improved up to 76.3% which is better than Soxhlet extraction that provided 40–46% (Oluba et al. 2008). The supercritical instrumentation is a time economic that required only 2–3 h for proper extraction of melon seed oil. Briefly, seeds or powder was mixed with wool (glass) in a definite ratio most probably 25:1 (w/w) in a sample cartridge. The filled cartridge was inserted in thermal-controlled extraction cell. A flow of Liquified CO_2 introduced into the extraction cartridge with the help of piston pump enwrapped in a cooling jacket that automatically controls the temperature and pressure of the cartridge according to settings. The extract was finally separated from the CO_2 phase and collected in collector at atmospheric pressure and ambient temperature (Rai et al. 2018).

6 Analysis of Oilseed Cake

The oilseed cakes obtained from oil extraction were analyzed for protein, fiber, and ash contents. Watermelon proteins are the major contributors in the deoiled meal but complicated procedures for its extraction hindered its commercial viability. Protein was extracted from watermelon seeds through alkaline water in a water bath (Wani et al. 2008). By optimizing the parameters of extraction, its yield could be improved to 80.7 g/100 g seed meal. In another experiment by Wani et al. (2008), they improved the yield up to 86.06% by optimizing the concentration of alkali and solvent meal ratio. Protein content (percent N × 6.25) could be estimated according to the Association of Official Analytical Chemists AOAC standard method using a Kjeldahl apparatus. The fiber and ash contents were determined according to the ISO method 5983 (ISO 1981) and ISO method 749 (ISO 1977), respectively.

6.1 Fatty Acid Composition

In addition to be a good source of energy, the different chain length fatty acids contribute to the pivotal role in the physiology of body functions. The body functions that are being regulated by fatty acids are intercellular signals, membrane functions and their structures and production of bioactives. It is commonly believed that most of the unsaturated fatty acids provide first line protection especially against cardiovascular diseases and strengthen the human body immune system. The saturated fatty acids (SFA) like stearic acid, palmitic, myristic and lauric acids may raise total

Table 39.2 Fatty acids profile of watermelon seed oil

Fatty acid	Percentage	Property
Myristic acid	0.11	Saturated FA
Stearic acid (18:0)	10.2	Saturated FA
Palmitic acid (16;0)	11.3	Saturated FA
Palmitoleic acid (16:1)	0.29	MUFA
Oleic acid (18:1)	18.1	MUFA
Linoleic acid (18:2)	55.6	PUFA
Alpha linolenic acid (18:3)	0.35	PUFA
Erucic acid (22:1)	–	MUFA
Gondoic acid (20:1)	–	MUFA
Behenic acid (22:0)	–	Saturated FA
Arachidic acid (20:0)	–	Saturated FA

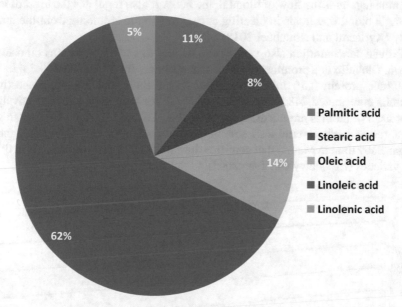

Fig. 39.5 Fatty acid composition (%) of *Citrullus lanatus* seed oil (Essien and Eduok 2013)

and low-density lipid cholesterol to cause blood coagulation, inflammation, coronary heart, cardiovascular diseases and type 2 diabetes (Calder 2015).

The fatty acid profile of melon seed oil was mostly evaluated by GC-MS (gas-liquid chromatography-mass spectrometry). The oil extracted from watermelon seeds often characterized by high contents of linoleic acid up to 60% (Table 39.2, Fig. 39.5). Further analysis revealed the presence of variety of fatty acids including linoleic acid (18:2*n*-6), oleic acid (18:1*n*-9), stearic acid (18:0) and palmitic acid (16:0) as the dominants (Das et al. 2002; Sabahelkhier et al. 2011). Other fatty acids such as behenic acid (22:0), 11-eicosadienoic acid (20:1*n*-9), arachidic (20:0), heptadecanoic acid (17:0), 9-hexadecenoic acid (16:1*n*-7), myristic acid (14:0), and tetracosanoic acid are also present in this oil (Mahla et al. 2018).

6.2   Amino Acids

Watermelon seeds have been reported to contain high level of different types of amino acids. Arginine, glutamic, aspartic and leucine are the predominant amino acids present in watermelon seed. A detailed comparison and concentration of different amino acids are given in Fig. 39.6. Reports are available on biological value, true digestibility, protein efficiency ratio and net protein utilization of watermelon seeds. Watermelon is also a natural source of L-citrulline, a non-essential amino acid precursor of L-arginine which is an essential amino acid (Jayaprakasha and Rao 2011). In a study with 14 watermelon cultivars, the average concentration of citrulline is 2.4 mg g^{-1}. Its administration can effectively increase plasma arginine levels (Mandel et al. 2005; Collins et al. 2007). This arginine regulates the flow of blood in the body. It also regulates the level of nitric oxide in blood that helps in clearing excess metabolic ammonia from the human body (Kyriacou and Rouphael 2018).

Protein concentration also varies from variety to variety as well as climate and origin. Globulin is a prominent protein that accounts for 575.0 and 549.7 g kg^{-1} of the crude protein (CP) in two varieties, Sugar Baby and Matera, respectively. Glutelin contributed 248.0 and 176.4 g kg^{-1} CP for both varieties, respectively. In contrast, Prolamin is present at negligible levels, 4.5 and 1.1 g kg^{-1} CP in both varieties, respectively. Overall total soluble protein content (g kg^{-1} CP) was higher in Sugar Baby than in Matera that confirms the variations in protein contents in different varieties (El-Adawy and Taha 2001).

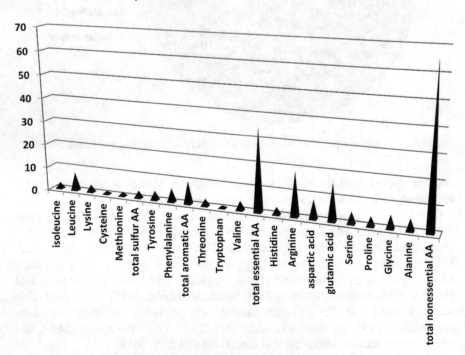

Fig. 39.6 Concentration of amino acids in the watermelon seeds

7   Minor Bioactive Compounds in the Fixed Oil

Watermelon (*Citrullus lanatus* (Thunb.) Mansfeld) is a popular fruit due to its high contents of natural phytochemicals. These constituents are supposed to be a qualitative factor for healthy food. Watermelon is one of the main crop grown and consumed all over the Mediterranean basin. It is much appreciated as refreshing summer fruit. Besides vitamins, mineral salts and specific amino acids, watermelon provides a wide variety of dietary antioxidants such as phenolics (Collins et al. 2007). The major contributors as phenolic acids and flavonoids are given in Table 39.3. In addition to these phenolic compounds some other antioxidants particularly ascorbic acid (AsA), dehydroascorbic acid (DHA) and lycopene along with other flavonoids are also present in seed oil. The phenolic and flavonoids have shown high scavenging potential against free radicals contributing to protective roles in certain disease.

The chromoplasts of watermelon mesocarp cells synthesize and store lycopene as the major carotenoid (70–90%). Lycopene is also responsible for the typical color of the flesh of ripe watermelon (Tadmor et al. 2005). This red pigment in flesh has the highest antioxidant activity among other dietary components present in fruit. Therefore, fresh watermelon constitutes an important source of lycopene for humans. Its bioavailability from fresh watermelon juice is, in fact, similar to that of heat processed tomatoes (Edwards et al. 2003). In addition, watermelon contains moderate but significant amount of phenolic (Brat et al. 2006). secondary metabolites exhibit highly efficient peroxyl-radical scavenging activity and hence pharmacological potential affects. Among phenolics, flavonoids reduce low-density lipoprotein (LDL) oxidation and quench reactive oxygen radicals, decreasing thereby the risk of cardiovascular diseases and cancers (Lila 2004). Watermelon fruits have been identified as a good source of vitamin C, mainly in the reduced form of AsA. In addition to its antioxidant activity against free radicals, AsA has numerous biological functions, which include the synthesis of collagen, steroid and

Table 39.3 Contents of phenolic compounds (mg/100 g) in the seed oil (Mallek-Ayadi et al. 2018)

Phenolic compound	Content	Percentage
Phenolic acids	16.79 ± 0.48	47.78
Gallic acid	4.24 ± 0.03	12.07
Prochatechuic acid	0.93 ± 00	2.65
Chlorogenic acid	1.25 ± 0.01	3.56
4-hydroxybenzoic acid	3.28 ± 0.03	9.33
Vanillic acid	3.87 ± 0.02	11.01
Rosmarinic acid	1.87 ± 0.02	5.32
Phenylacetic acid	1.35 ± 0.02	3.84
Flavonoids	9.54 ± 0.15	27.15
Naringenin-7-*o*-glycoside	4.30 ± 00	12.24
Leuteolin	2.10 ± 0.01	5.98
Apigenin	0.50 ± 0.03	1.42
Flavone	0.86 ± 0.04	2.45
Amentoflavone	1.78 ± 0.24	5.06

peptide hormones and neurotransmitters (Tlili et al. 2011). Melon seed oil is rich in sterols, tocopherols, and phospholipids that enhance the importance with its beneficial effect on the human being. It helps in reducing the risk of cardiovascular diseases and cancers (Parry et al. 2006). It prevents polyunsaturated fatty acids (PUFA) from oxidation inside the body. High concentration of unsaturated lipids makes it good for the heart and cardiovascular system. It protects our lipids as well as the nervous system.

7.1 Watermelon Aroma

The aldehyde and alcohols are the main volatile components of watermelon aroma profile (Saftner et al. 2007). In these functional families, alcohols are more abundant in watermelon and determine key aroma characters. These are also present in some fruits such as fresh melon which contain 3-nonen-1-ol, flower-green (hexanol), herbaceous (1-nonanol) and green melon has 6-nonen-1-ol (Fredes et al. 2016). It case of watermelon, aroma profile considered as an important parameter during fruit ripening that characterizes the evolution of carotenoid pigments and variation in composition (Kyriacou et al. 2018).

8 Uses of Watermelon

8.1 Health-Promoting Traits of Oil Constituents

Watermelon seed oil is light-colored and rich in essential fatty acids. This fatty acid breakdown shows that watermelon seed oil is mostly comprised of PUFA. Besides fatty acids, watermelon seed oil also contains sterols, diglycerides, monoglycerides and phospholipids (Petkova and Antova 2015). The oil is used in cooking, baking and cosmetics. The stable shelf life of watermelon seed oil makes it a suitable emollient. Watermelon seed oil contains high amounts of unsaturated fatty acids with linoleic and oleic acids as the major acids. There are some biological applications that maintained its demand as a remedy for biological complications. The seeds are of great importance in diets because of its nutritional content and diverse medicinal uses. It helps in curing urinary-related problems. The seeds are good vermifuge and showed a strong hypo-tensive action. Research indicates that consumption of watermelon may have antihypertensive effects. See oil as well as aqueous or alcoholic extracts had been reported to paralyze tapeworms and roundworms. The rind of the fruit is prescribed in cases of alcoholic poisoning and diabetes (Duke and Ayensu 1985). *Citrullus lanatus* is used in Northern Sudan for burns, swellings, rheumatism and gout and as a laxative. It is used as a drastic purgative and control diarrhea. Tar extracted from the seeds has been used for the treatment of scabies. The seed oil has an anthelmintic action which is better than that of pumpkin seed oil (Erhirhie and Ekene 2013).

L-citrulline is almost absent from natural foods but interestingly watermelon is a notable source of this component. It is produced in significant amount during drought conditions that help the watermelon to resist drought-mediated oxidative stress in harsh environment. L-citrulline is an amino acid that viewed solely as a metabolic intermediary in the urea cycle. Watermelon juice is an excellent energy drink for athletes to improve their sports performance. Although rich in linoleic acid, watermelon seed oil is preferred as frying and cooking medium in countries in the African and Middle East region due to the unique flavor. It provides various applications as an industrial ingredient in soap production and cosmetics (Wang et al. 2014).

8.2 Anti-aging Effects

Watermelon seed oil is a powerful antioxidant when applied topically on the skin. It provides antioxidants support to the skin cells. Free radical is a key reason that induced damage to the cells to accelerate aging along with UV radiation. In the presence of these nutrients, they are better to deal with free radicals and oxidizing agents in the environment. Watermelon seed oil contains a significant amount of vitamin E. When applied topically or when ingested, vitamin E gets absorbed into our system. This vitamin improves the functioning of the immune system. Application of this oil on the skin improves the ability of our skin to neutralize free radicals before they damage the matrix of collagen (Altaş et al. 2011).

8.3 Hepatoprotective Potential

Watermelon seed oil lead to a reduction in blood ALT and AST levels. Their high levels are actually damaging for the liver. Moreover, watermelon seed oil contains palmitic, stearic, and oleic acids, which considered as the need of body and found in chocolates as well. Literature showed that these types of fatty acids help in lowering LDL cholesterol and increase HDL/good cholesterol. Therefore, watermelon seed oil might be used as a remedy to lower liver damage induced by toxic substances. It also contains vitamins B and arginine which is crucial to release nitric oxide from the body (Altaş et al. 2011).

8.4 Watermelon Seed Oil for Skin and Hair

Watermelon seed oil is good for oily skin. It reduces skin oil (sebum) on the skin by dissolving it away. This fact is helpful for skin especially in conditions that are aggravated by excessive oil production such as acne. It prevents clogging of hair follicles and thus prevents folliculitis like condition. Watermelon seed oil is also helpful for people with dry, irritated skin in conditions like eczema and psoriasis. Watermelon seed oil forms a nice smelling ingredient in hair care formulations. When applied directly to the scalp then washed out, it reduces oiliness on the scalp and hair. Watermelon seed oil reduces frizz in hair and provides it nutrition. It can be used as a hot oil treatment on the hair to achieve smooth, shiny but much less greasy hair (Logaraj 2011).

References

Ahmad, M., Faruk, R., Shagari, K. A., Umar, S. (2017). Analysis of essential oil from watermelon seeds. *SosPoly Journal of Science & Agriculture*. www.uaspolysok.edu.ng/sospolyjsa/view/172201.pdf.

Altaş, S., Kizil, G., Kizil, M., Ketani, A., & Haris, P. I. (2011). Protective effect of Diyarbakır watermelon juice on carbon tetrachloride-induced toxicity in rats. *Food and Chemical Toxicology, 49*, 2433–2438.

Bang, H., Kim, S., Leskovar, D., & King, S. (2007). Development of a codominant CAPS marker for allelic selection between canary yellow and red watermelon based on SNP in lycopene β-cyclase (LCYB) gene. *Molecular Breeding, 20*, 63–72.

Brat, P., Georgé, S., Bellamy, A., Chaffaut, L. D., Scalbert, A., Mennen, L., Arnault, N., & Amiot, M. J. (2006). Daily polyphenol intake in France from fruit and vegetables. *The Journal of Nutrition, 136*, 2368–2373.

Calder, P. C. (2015). Functional roles of fatty acids and their effects on human health. *Journal of Parenteral and Enteral Nutrition, 39*, 18S–32S.

Collins, J. K., Wu, G., Perkins-Veazie, P., Spears, K., Claypool, P. L., Baker, R. A., & Clevidence, B. A. (2007). Watermelon consumption increases plasma arginine concentrations in adults. *Nutrition, 23*, 261–266.

Das, M., Das, S., & Suthar, S. (2002). Composition of seed and characteristics of oil from karingda [*Citrullus lanatus* (Thumb) Mansf]. *International Journal of Food Science & Technology, 37*, 893–896.

De Conto, L. C., Gragnani, M. A. L., Maus, D., Ambiel, H. C. I., Chiu, M. C., Grimaldi, R., & Gonçalves, L. A. G. (2011). Characterization of crude watermelon seed oil by two different extractions methods. *Journal of the American Oil Chemists' Society, 88*, 1709–1714.

Duke, J. A. & Ayensu, E. S. (1985). *Medicinal Plants of China*. Reference Publ., Inc. Algonac, Michigan, 2, 705.

Edwards, A. J., Vinyard, B. T., Wiley, E. R., Brown, E. D., Collins, J. K., Perkins-Veazie, P., Baker, R. A., & Clevidence, B. A. (2003). Consumption of watermelon juice increases plasma concentrations of lycopene and β-carotene in humans. *The Journal of Nutrition, 133*, 1043–1050.

El-Adawy, T. A., & Taha, K. M. (2001). Characteristics and composition of watermelon, pumpkin, and paprika seed oils and flours. *Journal of Agricultural and Food Chemistry, 49*, 1253–1259.

Erhirhie, E., & Ekene, N. (2013). Medicinal values on *Citrullus lanatus* (watermelon): Pharmacological review. *International Journal of Research in Pharmaceutical and Biomedical Sciences, 4*, 1305–1312.

Essien, E. A., & Eduok, U. M. (2013). Chemical analysis of *Citrullus lanatus* seed oil obtained from Southern Nigeria. *Organic Chemistry, 54*, 12700–12703.

Fasina, O., & Colley, Z. (2008). Viscosity and specific heat of vegetable oils as a function of temperature: 35 C to 180 C. *International Journal of Food Properties, 11*, 738–746.

Fredes, A., Sales, C., Barreda, M., Valcárcel, M., Roselló, S., & Beltrán, J. (2016). Quantification of prominent volatile compounds responsible for muskmelon and watermelon aroma by purge and trap extraction followed by gas chromatography–mass spectrometry determination. *Food Chemistry, 190,* 689–700.

ISO749 (1977). Oilseed residues-Determination of total ash.

ISO5983-1 (1981). Determination of nitrogen content and calculation of crude protein content.

Janakiraman, N., Johnson, M., & Sahaya, S. S. (2012). GC–MS analysis of bioactive constituents of *Peristrophe bicalyculata* (Retz.) Nees.(Acanthaceae). *Asian Pacific Journal of Tropical Biomedicine, 2,* S46–S49.

Jarret, R. L., & Levy, I. J. (2012). Oil and fatty acid contents in the seed of *Citrullus lanatus* Schrad. *Journal of Agricultural and Food Chemistry, 60,* 5199–5204.

Jayaprakasha, G., & Rao, L. J. M. (2011). Chemistry, biogenesis, and biological activities of *Cinnamomum zeylanicum. Critical Reviews in Food Science and Nutrition, 51,* 547–562.

Kyriacou, M. C., & Rouphael, Y. (2018). Towards a new definition of quality for fresh fruits and vegetables. *Scientia Horticulturae, 234,* 463–469.

Kyriacou, M. C., Leskovar, D. I., Colla, G., & Rouphael, Y. (2018). Watermelon and melon fruit quality: The genotypic and agro-environmental factors implicated. *Scientia Horticulturae, 234,* 393–408.

Lila, M. A. (2004). Anthocyanins and human health: An in vitro investigative approach. *BioMed Research International, 2004,* 306–313.

Ling, K.-S., Harris, K. R., Meyer, J. D., Levi, A., Guner, N., Wehner, T. C., Bendahmane, A., & Havey, M. J. (2009). Non-synonymous single nucleotide polymorphisms in the watermelon eIF4E gene are closely associated with resistance to Zucchini yellow mosaic virus. *Theoretical and Applied Genetics, 120,* 191–200.

Logaraj, T. V. (2011). Watermelon (*Citrullus lanatus* (Thunb.) Matsumura and Nakai) seed oils and their use in health. In Victor R. Preedy (eds) *Nuts and seeds in health and disease prevention.* London: Academic Press.

Mabaleha, M., Mitei, Y., & Yeboah, S. (2007). A comparative study of the properties of selected melon seed oils as potential candidates for development into commercial edible vegetable oils. *Journal of the American Oil Chemists' Society, 84,* 31–36.

Mahla, H., Rathore, S., Venkatesan, K., & Sharma, R. (2018). Analysis of fatty acid methyl esters and oxidative stability of seed purpose watermelon (*Citrullus lanatus*) genotypes for edible oil. *Journal of Food Science and Technology, 55*(4), 1552–1561.

Mallek-Ayadi, S., Bahloul, N., & Kechaou, N. (2018). Phytochemical profile, nutraceutical potential and functional properties of *Cucumis melo* L. seeds. *Journal of the Science of Food and Agriculture, 99,* 1294–1301.

Mandel, H., Levy, N., Izkovitch, S., & Korman, S. (2005). Elevated plasma citrulline and arginine due to consumption of *Citrullus vulgaris* (watermelon). *Journal of Inherited Metabolic Disease, 28,* 467–472.

Maran, J. P., & Priya, B. (2015). Supercritical fluid extraction of oil from muskmelon (*Cucumis melo*) seeds. *Journal of the Taiwan Institute of Chemical Engineers, 47,* 71–78.

Nehdi, I. (2011). Characteristics, chemical composition and utilisation of *Albizia julibrissin* seed oil. *Industrial Crops and Products, 33,* 30–34.

Oluba, O., Ogunlowo, Y., Ojieh, G., Adebisi, K., Eidangbe, G., & Isiosio, I. (2008). Physicochemical properties and fatty acid composition of *Citrullus lanatus* (Egusi Melon) seed oil. *Journal of Biological Sciences, 8,* 814–817.

Parry, J., Su, L., Moore, J., Cheng, Z., Luther, M., Rao, J. N., Wang, J.-Y., & Yu, L. L. (2006). Chemical compositions, antioxidant capacities, and antiproliferative activities of selected fruit seed flours. *Journal of Agricultural and Food Chemistry, 54,* 3773–3778.

Petkova, Z., & Antova, G. (2015). Proximate composition of seeds and seed oils from melon (*Cucumis melo* L.) cultivated in Bulgaria. *Cogent Food & Agriculture, 1,* 1018779.

Rai, A., Mohanty, B., & Bhargava, R. (2018). Optimization of parameters for supercritical extraction of watermelon seed oil. *Separation Science and Technology, 53,* 671–682.

Raziq, S., Anwar, F., Mahmood, Z., Shahid, S., & Nadeem, R. (2012). Characterization of seed oils from different varieties of watermelon [*Citrullus lanatus* (Thunb.)] from Pakistan. *Grasas y Aceites, 63,* 365–372.

Sabahelkhier, M., Ishag, K., & Sabir Ali, A. (2011). Fatty acid profile, ash composition and oil characteristics of seeds of watermelon grown in Sudan. *British Journal of Science, 1*, 76–80.

Saftner, R., Luo, Y., McEvoy, J., Abbott, J. A., & Vinyard, B. (2007). Quality characteristics of fresh-cut watermelon slices from non-treated and 1-methylcyclopropene-and/or ethylene-treated whole fruit. *Postharvest Biology and Technology, 44*, 71–79.

Tadmor, Y., King, S., Levi, A., Davis, A., Meir, A., Wasserman, B., Hirschberg, J., & Lewinsohn, E. (2005). Comparative fruit colouration in watermelon and tomato. *Food Research International, 38*, 837–841.

Tlili, I., Hdider, C., Lenucci, M. S., Riadh, I., Jebari, H., & Dalessandro, G. (2011). Bioactive compounds and antioxidant activities of different watermelon (*Citrullus lanatus* (Thunb.) Mansfeld) cultivars as affected by fruit sampling area. *Journal of Food Composition and Analysis, 24*, 307–314.

Vohra, A., & Kaur, H. (2011). Chemical investigation of medicinal plant *Ajuga bracteosa*. *Journal of Natural Product Plant Resources, 1*, 37–45.

Wang, Y.-Y., Long, X., Zhong, D.-M., Ge, Y.-B., & Wang, Z.-X. (2014). Anti-tyrosinase and antioxidation activities and contents determining from watermelon peel. *Food Research and Development, 19*, 021.

Wani, A. A., Kaur, D., Ahmed, I., & Sogi, D. (2008). Extraction optimization of watermelon seed protein using response surface methodology. *LWT-Food Science and Technology, 41*, 1514–1520.

Wani, A. A., Sogi, D. S., Singh, P., Wani, I. A., & Shivhare, U. S. (2011). Characterisation and functional properties of watermelon (*Citrullus lanatus*) seed proteins. *Journal of the Science of Food and Agriculture, 91*, 113–121.

Yoo, K. S., Bang, H., Lee, E. J., Crosby, K., & Patil, B. S. (2012). Variation of carotenoid, sugar, and ascorbic acid concentrations in watermelon genotypes and genetic analysis. *Horticulture, Environment, and Biotechnology, 53*, 552–560.

Ziyada, A., & Elhussien, S. (2008). Physical and chemical characteristics of *Citrullus lanatus* var. colocynthoide seed oil. *Journal of Physical Science, 19*, 69–75.

Chapter 40
Semecarpus anacardium Oil

Mohamed Fawzy Ramadan

Abstract Fruits and fruit oil of *Semecarpus anacardium* (family Anacardiaceae) have several applications in the Indian *Ayurvedic* and *Siddha* systems of medicine. Concerning the potential utilization, knowledge of the composition of *Semecarpus anacardium* oil is of major importance. Solvent extractable lipids from *S. anacardium* seeds accounted for approx. 36%, which confirms that *S. anacardium* seeds are a rich source of oil. Linoleic, palmitic and oleic were the major fatty acids. Neutral lipids were the main lipid class followed by glycolipids and phospholipids. β-sitosterol, campesterol and stigmasterol were the main phytosterols detected in the oil, while δ-tocopherol and β-tocopherol were the main tocols. *S. anacardium* seed oil exhibited strong antiradical action toward 1,1-diphenyl-2-picrylhydrazyl (DPPH·) radical and galvinoxyl radical. This chapter reviews the chemical composition and functional traits of *Semecarpus anacardium* seed oil.

Keywords Lipid classes · Lipid-soluble bioactive compounds · Sterols · Tocopherols · Antiradical action

1 Introduction

Vegetable and fruit seeds are important sources of oils of industrial, nutritional and pharmaceutical importance (Ramadan et al. 2007, 2010). *Ayurveda*, literally meaning 'science of life', is an ancient Indian system of medicine that practiced about 3500 years ago. *Ayurveda* system relies strongly on preventive medicine and the promotion of health (Schuppan et al. 1999; Buenz et al. 2004; Dwivedi et al. 2018). Evaluation of the bioactive phytochemicals of *Ayurvedic* plants could be helpful in modern drug development.

Semecarpus anacardium Linn. (family Anacardiaceae), commonly known as 'marking nut', is a moderate-sized deciduous tree, reaching up to a height of

M. F. Ramadan (✉)
Agricultural Biochemistry Department, Faculty of Agriculture, Zagazig University, Zagazig, Egypt
e-mail: mframadan@zu.edu.eg

© Springer Nature Switzerland AG 2019
M. F. Ramadan (ed.), *Fruit Oils: Chemistry and Functionality*,
https://doi.org/10.1007/978-3-030-12473-1_40

Fig. 40.1 *Semecarpus anacardium* tree (**a**) and dried fruits (**b**)

12–15 m, with rough dark brown bark (Fig. 40.1). *S. anacardium* distributed in sub-Himalayan regions of India, and found in Malaysia, Myanmar, Singapore, China, Northern Australia and Africa. The fruit is acrid, anthelmintic, and considered effective in the treatment of acute rheumatism, epilepsy, asthma, neuralgia, and psoriasis. It has been used therapeutically in neurological disorders, leprosy, ulcers, corns, leucoderma and arthritis (Premalatha and Sachdanandam 2000; Parrota 2001; Ramadan et al. 2010).

The nut of *S. anacardium* (Fig. 40.1) and its oil contain several phenolic compounds including biflavonoids (Rastogi and Mehrotra 1991). Trihydroxyflavone, semecarpol, anacardoside and bhilawanols have been identified as the main constituents of *S. anacardium* nut extract (Sujatha and Sachdanandam 2002; Surveswaran et al. 2007). Medicinal traits including acrid, antimicrobial, anti-inflammatory, digestive, antitumor, and antirheumatic have been attributed to the nut extracts (Patwardan et al. 1988; Ramprasath et al. 2006; Nair et al. 2009). The seed inside the *S. anacardium* is known in Hindi as "Godambi" which used as a dry fruit in India. Against AFB_1-induced rat neoplasia in the liver, the nut extract was non-toxic in high doses. The anticancer potential of *S. anacardium* nut extract was also reported (Premalatha et al. 1999; Premalatha and Sachdanandam 2000). Alkaloids isolated from the *S. anacardium* nut oil were effective in eliminating human tumor cells of diverse origin through induction of apoptosis. This is indicative of a potential therapeutic action of those bioactive compounds in human cancer (Chakraborty et al. 2004). In an ethnopharmacological screening, the inhibitory impact on lipid oxidation of *S. anacardium* methanolic extract was studied using bovine brain phospholipid liposomes as model membranes (Sunil Kumar and Muller 1999). Toxicological investigation on *Siddha* preparation from *S. anacardium* nuts revealed that the drug is safe and do not induce any toxic manipulation on the biochemical parameters in animals (Vijayalakshmi et al. 2000; Ramadan et al. 2010).

S. anacardium seed kernel contains reddish-brown oil, which is semi-dry with a pleasant taste. In India, *S. anacardium* seeds are eaten and considered nutritious. Few studies were published on the chemical profile of *S. anacardium* seed oil (Ramadan et al. 2010), while the nut shells had been investigated extensively.

2 Fatty Acids Profile and Acyl Lipids of *S. anacardium* Seed Oil

S. anacardium seeds contain about 36% crude lipids, which confirms that *S. anacardium* seeds are a rich source of oil. Fatty acid composition of crude *S. anacardium* seed oil and lipid classes [neutral lipids (NL), glycolipids (GL), and phospholipids (PL)] are given in Table 40.1. Eight fatty acids were identified in *S. anacardium* seed oil (Ramadan et al. 2010), wherein linoleic followed by palmitic and oleic were the main fatty acids (comprising together approx. 90% of total fatty acids). *S. anacardium* seed oil was characterized by the high level of polyunsaturated fatty acids (PUFA) which accounted for approx. 70% of the total identified fatty acids. Fatty acids in neutral lipids and polar lipids (GL and PL) were not significantly different from each other, whereas linoleic and palmitic acids were the main fatty acids. The ratio of unsaturated to saturated fatty acids, was higher in neutral lipids than in the polar lipids. GL and PL had higher levels of saturates, while saturated fatty acids were detected in fewer amounts in the NL. The fatty acid composition of *S. anacardium* seed oil evinces the oil as a rich source of the essential fatty acids.

The proportions of lipid classes and subclasses found in *S. anacardium* seed oil are given in Table 40.2. The level of NL in *S. anacardium* seed oil is the highest (approx. 99% of oil), followed by GL (approx. 0.55%) and PL (approx. 0.51%). Subclasses of NL in the crude *S. anacardium* seed oil included triacylglycerol (TAG), diacylglycerol (DAG), free fatty acids (FFA), monoacylglycerol (MAG) and esterified sterols (STE) in a decreasing order. The TAG is found in high level

Table 40.1 Main fatty acids in *S. anacardium* seed oil and lipid classes

Fatty acid	Crude oil	Neutral lipids	Glycolipids	Phospholipids
	Relative content (%)			
C16:0	12.6	12.4	12.7	12.8
C18:1n-9	10.5	10.7	10.5	10.5
C18:2n-6	68.8	68.9	68.6	68.3
C18:3n-3	0.96	0.98	0.95	0.93
C20:0	3.42	3.37	3.46	3.53
C22:0	2.83	2.78	2.87	2.95

Table 40.2 Main lipid classes (g/kg oil) in *S. anacardium* seed oil

Neutral lipid class	g/kg oil	Glycolipid class	g/kg oil	Phospholipid class	g/kg oil
MAG	5.67	SQD	0.35	PS	0.25
DAG	7.86	DGD	0.64	PI	0.74
FFA	6.66	CER	1.12	PC	2.55
TAG	885	SG	1.55	PE	1.22
STE	5.43	MGD	0.16		
		ESG	1.32		

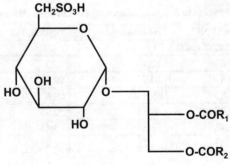

Monogalactosyldiacylglycerol (MGDG)   **Cerebroside (CER)**

Digalactosyldiacylglycerol (DGDG)   R₁= H, **Steryl glucoside (SG)**

R₁= acyl, **Acylated steryl glucoside (ASG)**

Sulfoquinovosyldiacylglycerol (SQD)

Fig. 40.2 Chemical structures of glycolipids found in *S. anacardium* oil

(97.1% of oil) followed by DAG (0.86% of oil), while other subclasses were detected in lower amounts. GL subclasses (Fig. 40.2) were sulfoquinovosyldiac-ylglycerol (SQD), digalactosyldiglycerides (DGD), cerebrosides (CER), sterylg-lycosides (SG), monogalactosyldiglycerides (MGD) and esterified sterylglycosides (ESG) as given in Table 40.2 (Ramadan et al. 2010). SG, ESG and CER were the main measured components. In human, the GL average daily intake was reported to be 90 mg of MGD, 65 mg of SG, 140 mg of ESG, 50 mg of CER, and 220 mg of DGD (Sugawara and Miyazawa 1999; Ramadan and Mörsel 2003). Therefore, *S. anacardium* oil might be applied as a source of GL in the human diet. PL subclasses in *S. anacardium* seed oil were phosphatidylcholine (PC) followed by phosphatidylethanolamine (PE), phosphatidylinositol (PI) and phosphatidylserine (PS).

3 Minor Bioactive Lipids in of *S. anacardium* Seed Oil

In the Indian *S. anacardium* seed oil, β-Sitosterol was the sterol marker and comprised approx. 57% of the total identified sterols (Table 40.3). Campesterol and stigmasterol constituted approx. 35% of the total identified sterols. Δ5-avenasterol, Δ7-Avenasterol, and Δ7-stigmastenol were found at lower amounts. Levels of plant sterols in oils are used for the identification of oils. The profile of plant sterols was reported to be little influenced by environmental factors and/or by cultivation conditions. Plant sterols are of interest due to their antioxidant potential and health-promoting properties (Ramadan et al. 2006).

Profile of tocopherols in the Indian *S. anacardium* seed oil is given in Table 40.3. δ-Tocopherol was the main detected compound, which accounted for approx. 65% of total tocols. β-tocopherol accounted for approx. 17.8%, γ-tocopherol accounted for approx. 10.8% and α-tocopherol accounted for approx. 5.9%. α-Tocopherol is the most efficient antioxidant of tocol isomers, while β-tocopherol has 25–50% of the antioxidant potential of α-tocopherol (Kallio et al. 2002). The nutritionally important phytochemicals such as tocols (vitamin E) improve oil oxidative stability. Tocols in oils and fats are believed to inhibit the oxidation of PUFA. Tocols are the main membrane-localized antioxidants in humans. Deficiency of tocols influences several tissues in bird and mammalian models (Ramadan et al. 2006, 2007; Sokol 1996).

Table 40.3 Main sterols and tocopherols (g/kg oil) in *S. anacardium* oil

Compound	g/kg
Campesterol	1.54
Stigmasterol	1.16
β-Sitosterol	4.65
α-Tocopherol	0.12
β-Tocopherol	0.36
δ-Tocopherol	1.32

4   Contribution of Bioactive Compounds in *S. anacardium* Seed Oil to Organoleptic Properties and Functions

Radical scavenging activity of *S. anacardium* seed oil and extra virgin olive oil were compared using DPPH· and galvinoxyl radicals (Ramadan et al. 2010). *S. anacardium* seed oil characterized by high RSA in comparison with extra virgin olive oil (Fig. 40.3). RSA of oils and fats could be interpreted as the combined action of different endogenous antioxidants. The strong antiradical activity of any oil might be due to (i) the diversity in structural characteristics of potential antioxidants, (ii) the differences in polar lipids and unsaponifiables profiles, (iii) a synergism of polar lipids with other bioactive compounds, and (iv) kinetic behaviors of antioxidants.

Various oils, surfactants and co-surfactants were investigated to prepare a stable self-nano emulsifying drug delivery systems (SNEDDS) of *S. anacardium* oil (Katoch et al. 2016). Stability investigations of formulated SNEDDS were performed at different temperatures (5 °C, 25 °C and 40 °C). The dissolution profile of *S. anacardium* oil loaded SNEDDS in different medium showed that 100% of *S. anacardium* fixed oil released within 10 min irrespective of the pH of the dissolution medium. In addition, the prepared SNEDDS was stable for 3 months.

5   Health-Promoting Traits of *S. anacardium* Seed Oil

Khan et al. (2012) investigated the therapeutic effects of *S. anacardium* nut milk extract (SA) against lipid oxidation and antioxidant status in spleen and thymus of experimental animals induced with 7,12-dimethyl benz[a] anthracene (DMBA). Treatment with SA decreased lipid oxidation and increased antioxidant status in

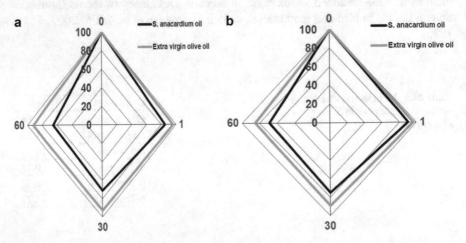

Fig. 40.3 The antiradical action of *S. anacardium* seed oil and virgin olive oil on (**a**) DPPH·radicals and (**b**) galvinoxyl radicals

rats. The results depict the potency of *S. anacardium* antioxidant property by depleting lipid peroxide levels in spleen and thymus and thus its role in enhancing the immune function.

Bioactive phytochemicals responsible for *S.s anacardium* pharmacological traits included bioflavonoid, a phenolic compound, vitamins, minerals and amino acids. The main pharmacological traits *S. anacardium* extracts and ayurvedic formulations were reported (Katoch et al. 2016). Clinical studies showed that *S. anacardium* oil exhibits a wide array of therapeutic traits such as anti-arthritic effect, anti-inflammatory activity, anti-tumor, antineoplastic, cytostatic activity, cytotoxic, hypolipidemic potential, hypocholesterolemic potential, antimicrobial potential, and antistress potential (Gouthaman et al. 2008). Dwivedi et al. (2018) studied the anti-hyperlipidemic activity of *S. anacardium* fixed oil against cholesterol diet induced hyperlipidemia in rats. *S. anacardium* exhibited better impact in lowering serum cholesterol, triglyceride, VLDL in comparison to control group.

It could be said that the *S. anacardium* seeds give a high oil yield of oil wherein the oil could be considered as a good source of essential fatty acids and bioactive lipids. *S. anacardium* seed oil could be nutritionally considered as a new non-conventional supply for edible purposes and pharmaceutical industries.

References

Buenz, E. J., Schnepple, D. J., Bauer, B. A., Elkin, P. L., Riddle, J. M., & Motley, T. J. (2004). Techniques: Bioprospecting historical herbal texts by hunting for new leads in old tomes. *Trends in Pharmacological Sciences, 25*, 494–498.

Chakraborty, S., Roy, M., Taraphdar, A. K., & Bhattacharya, R. K. (2004). Cytotoxic effect of root extract of *Tiliacora racemosa* and oil of *Semecarpus anacardium* nut in human tumour cells. *Phytotherapy Research, 18*, 595–600.

Dwivedi, M. K., Nariya, M., Galib, R., & Prajapati, P. K. (2018). Anti-hyperlipidaemic effects of fresh and cured *Bhallataka Kshaudra* (*Semecarpus anacardium* L.) in animals. *Indian Journal of Natural Products and Resources, 9*, 143–150.

Gouthaman, T., Kavitha, M. S., Ahmed, B. A., Kumar, T. S., & Rao, M. V. (2008). A review on *Semecarpus anacardium* L: An anticancer medicinal plant. In V. K. Singh, N. J. Govil, & R. K. Sharma (Eds.), *Phytopharmacology and therapeutic value* (pp. 193–221). Houston: Studium Press.

Kallio, H., Yang, B., Peippo, P., Tahvonen, R., & Pan, R. (2002). Triacylglycerols, glycerophospholipids, tocopherols and tocotrienols in berries and seeds of two subspecies (ssp. *sinensis* and *mongolica*) of Sea buckthorn (*Hippophaë rhamnoides*). *Journal of Agricultural and Food Chemistry, 50*, 3004–3009.

Katoch, P., Kaur, P., Singh, R., Vyas, M., Singh, S. K., & Gulati, M. (2016). Development and characterization of selfnanoemulsifying drug delivery system loaded with fixed oil of *Semecarpus anacardium* Linn. *Asian Journal of Pharmaceutics, 10*(2), 144–153.

Khan, H. B. H., Vinayagam, K. S., Madan, P., Palanivelu, S., & Panchanatham, S. (2012). Modulatory effect of *Semecarpus anacardium* against oxidative damages in DMBA-induced mammary carcinogenesis rat model. *Comparative Clinical Pathology, 21*, 1275–1284. https://doi.org/10.1007/s00580-011-1278-4.

Nair, P. K. R., Melnick, S. J., Wnuk, S. F., Rapp, M., Escalon, E., & Ramachandran, C. (2009). Isolation and characterization of an anticancer catechol compound from *Semecarpus anacardium*. *Journal of Ethnopharmacology, 122*, 450–456.

Parrota, J. A. (2001). *Healing plants of peninsular India* (pp. 391–392). Wallingford: CABI Publishing, CAB International.

Patwardan, B., Ghoo, R. B., & David, S. B. (1988). A new anaerobic inhibitor of herbal origin. *Indian Journal Pharmacologic Sciences, 50*, 130–132.

Premalatha, B., & Sachdanandam, P. (2000). Potency of *Semecarpus Anacardium* Linn. nut milk extract against aflatoxin B1-induced hepatocarcinogenesis: Reflection on microsomal biotransformation enzymes. *Pharmacological Research, 42*, 161–166.

Premalatha, B., Muthulakshmi, V., & Sachdanandam, P. (1999). Anticancer potency of the milk extract of *Semecarpus anacardium* Linn. nuts against aflatoxin B1 mediated hepatocellular carcinoma bearing Wistar rats with reference to tumour marker enzymes. *Phytotherapy Research, 13*, 183–187.

Ramadan, M. F., & Mörsel, J.-T. (2003). Analysis of glycolipids from black cumin (*Nigella sative* L.), coriander (*Coriandrum sativum* L.) and niger (*Guizotia abyssinica* Cass.) oilseeds. *Food Chemistry, 80*, 197–204.

Ramadan, M. F., Sharanabasappa, G., Seetharam, Y. N., Seshagiri, M., & Moersel, J.-T. (2006). Profile and levels of fatty acids and bioactive constituents in mahua butter from fruit-seeds of Buttercup tree [*Madhuca longifolia* (Koenig)]. *European Food Research and Technology, 222*, 710–718.

Ramadan, M. F., Zayed, R., & El-Shamy, H. (2007). Screening of bioactive lipids and radical scavenging potential of some solanaceae plants. *Food Chemistry, 103*, 885–890.

Ramadan, M. F., Kinni, S. G., Seshagiri, M., & Mörsel, J.-T. (2010). Fat-soluble bioactives, fatty acid profile and radical scavenging activity of *Semecarpus anacardium* seed oil. *Journal of the American Oil Chemists' Society, 87*, 885–894.

Ramprasath, V. R., Shanthi, P., & Sachdanandam, P. (2006). Effect of *Semecarpus anacardium* Linn. nut milk extract on rat neutrophil functions in adjuvant arthritis. *Cell Biochemistry and Function, 24*, 333–340.

Rastogi, R. P., & Mehrotra, B. N. (1991). *Compendium of Indian medicinal plants, Vol 2: Drug research perspective* (p. 369). Lucknow: Central Drug Research Institute.

Schuppan, D., Jia, J.-D., Brinkhaus, B., & Hahn, E. G. (1999). Herbal products for liver diseases: A therapeutic challenge for the new millennium. *Hepatology, 30*, 1099–1104.

Sokol, R. J. (1996). Vitamin E. In E. E. Ziegler & L. J. Filer (Eds.), *Present knowledge in nutrition* (pp. 130–116). Washington, DC: ILSI Press.

Sugawara, T., & Miyazawa, T. (1999). Separation and determination of glycolipids from edible plant by high-performance liquid chromatography and evaporative light-scattering detection. *Lipids, 34*, 1231–1237.

Sujatha, V., & Sachdanandam, P. (2002). Recuperative effect of *Semecarpus anacardium* Linn. nut milk extract on carbohydrate metabolizing enzymes in experimental mammary carcinoma-bearing rats. *Phytotherapy Research, 16*, S14–S18.

Sunil Kumar, K. C., & Muller, K. (1999). Medicinal plants from Nepal; II. Evaluation as inhibitors of lipid peroxidation in biological membranes. *Journal of Ethnopharmacology, 64*, 135–139.

Surveswaran, S., Cai, Y.-Z., Corke, H., & Sun, M. (2007). Systematic evaluation of natural phenolic antioxidants from 133 Indian medicinal plants. *Food Chemistry, 102*, 938–953.

Vijayalakshmi, T., Muthulakshmi, V., & Sachdanandam, P. (2000). Toxic studies on biochemical parameters carried out in rats with Serankottai nei, a siddha drug–milk extract of *Semecarpus anacardium* nut. *Journal of Ethnopharmacology, 69*, 9–15.

Chapter 41
Pumpkin (*Cucurbita pepo* L.) Seed Oil

Hamide Filiz Ayyildiz, Mustafa Topkafa, and Huseyin Kara

Abstract This chapter summarizes the main knowledge of pumpkin seed oil (PSO) such as cultivation conditions, production technology, physicochemical properties, quality control parameters and effects on health. PSO is obtained from pumpkins belong to the Cucurbitaceae family, cultivated in mild and subtropical regions and widely known as palatable and tasty food. It has been produced in Slovenia, Austria, and Hungary and used as cooking oil in many parts in Africa and the Middle East. The oil content is around 50% and it can be obtained from pumpkin seeds with or without husks using solvent extraction, supercritical CO_2 or cold pressed methods. PSO has the dark greenish color as well as very typical strong nutty and roasty flavor. It is a rich natural source of phytosterols, proteins, polyunsaturated fatty acids (PUFA), antioxidant vitamins, carotenoids and tocopherols and various elements and recommended to be in the human diet for health. PSO is highly unsaturated oil containing predominantly oleic and linoleic fatty acids and this feature makes PSO well suited for improving the food nutritive value. PSO also provides many benefits to health such as prevention of the growth and reduction of the size of prostate, mitigation of hypercholesterolemia and arthritis, retardation of the progression of hypertension, alleviation of diabetes by promoting hypoglycemic activity, and lowering levels of various cancer species. These properties are attributed to effective macro- and micro-constituents in the oil. All these features make the PSO a highly nutritious food and a useful source for application in therapeutics and novel foods.

Keywords Cucurbitaceae · Sterols · Fatty acids · Antioxidant vitamins · Carotenoids · Tocopherols

H. F. Ayyildiz (✉)
Faculty of Pharmacy, Department of Basic Pharmaceutical Sciences, Selcuk University, Konya, Turkey

M. Topkafa
Vocational School of Technical Sciences, Department of Chemistry and Chemical Technologies, Konya Technical University, Konya, Turkey

H. Kara
Faculty of Science, Department of Chemistry, Selcuk University, Konya, Turkey

© Springer Nature Switzerland AG 2019
M. F. Ramadan (ed.), *Fruit Oils: Chemistry and Functionality*,
https://doi.org/10.1007/978-3-030-12473-1_41

Abbreviations

DAGs	Diacylglycerols
FFAs	Free fatty acids
MAAEE	Microwave-assisted aqueous enzymatic extraction
NL	Neutral lipids
PC	Phosphatidylcholine
PE	Phosphatidylethanolamine
PI	Phosphatidylinositol
PL	Phospholipid
PL	Polar lipids
PSC	Pumpkin seed cake
PSO	Pumpkin seed oil
SFE-CO$_2$	Supercritical carbon dioxide fluid extraction
TAGs	Triacylglycerols

1 Introduction

Pumpkin is a pepo type fruit, which belongs to the genus Cucurbita and the family Cucurbitaceae. Cucurbitaceae family which is one of the largest families in the plant kingdom, consisting of largest number of edible species, has five major genera: Cucurbitaceae *Citrullus* (watermelons and wild colocynths), Cucurbitaceae *Cucumis* (cucumbers, gherkins and melons), Cucurbitaceae *Lagenaria* (gourds), Cucurbitaceae *Sechium* (chayotte) and Cucurbitaceae *Cucurbita* (Achilonu et al. 2018). The *C. Cucurbita* genus, the most economically important species, includes five species: *C. pepo*, *C. moschata*, *C. maxima*, *C. ficifolia* and *C. turbaniformis* (Gohari Ardabili et al., 2011). *C. pepo* species are usually recognized as the true pumpkin and cultivated for human consumption and for use in traditional medicine (Caili et al. 2006).

The pumpkins believed to have originated in the ancient civilizations of North and Central America before spreading to Europe and other parts of the world during the early sixteenth century. Today, different species of pumpkin are cultivated worldwide in a variety of environmental conditions with high production yield (Phillips et al. 2005). Worldwide production area of pumpkin is more than 260 million tons (FAOSTAT 2016). International statistics for pumpkins are categorized as "pumpkins, squash, and gourds." According to FAOSTAT, in 2016, China (7,838,809 ton) was the largest producer of pumpkins, squash, and gourds followed by India (5,073,678 ton), the Russian Federation (1,224,711 ton), Ukraine (1,209,810 ton), and the United States (1,005,150 ton) (Fig. 41.1). Trade in pumpkins is also active, with the United States being the top importer of pumpkins, followed by France, Japan, Germany, and the United Kingdom (Fig. 41.2). On the flip side, Mexico exports the most pumpkins, followed by Spain, New Zealand, the United States, and Turkey (Fig. 41.3).

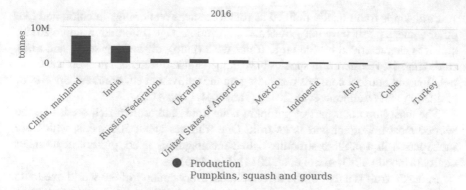

Fig. 41.1 Top countries producing pumpkins, squash and gourds (FAOSTAT 2016)

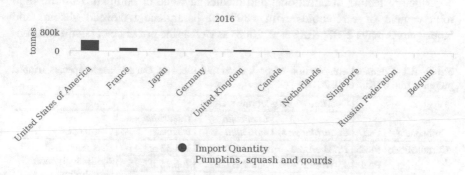

Fig. 41.2 Ten countries importing pumpkins, squash and gourds (FAOSTAT 2016). (Source: FAOSTAT 2018)

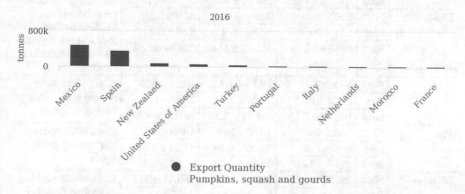

Fig. 41.3 Top countries exporting pumpkins, squash and gourds (FAOSTAT 2016). (Source: FAOSTAT 2018)

Pumpkin is traditionally defined as round in shape and orange in color, and like other winter squash, have long vines and a flowering stage (Filbrandt Katelyn 2012), and it is characterized by the large fruity flesh (pulp), oil-bearing seeds and thick rinds (Fig. 41.4). *Cucurbita* species have considerable differences in chemical composition and nutrient contents depending on the cultivation environment, species or plant/fruit part (Achilonu et al. 2018) (Tables 41.1 and 41.2).

The most important part of pumpkin is its lipid- and protein-rich seeds and the second most important part is its fruit. One fruit has about 500 seeds, which are interspersed in a net-like structure called mucilaginous fibers present at its inner cavity (Mbondo 2013; Adeel et al. 2014) (Fig. 41.5).

Pumpkin fruit is a part of the diet in almost every country of the world due to its good taste and used for making dessert, jam, candy and meal. Its young leaves, tender stem and flowers are also cooked and consumed. Being nutritionally rich, the fruit also possess many medicinal properties (Dhiman et al. 2009). The major contributory factors of nutritional and medicinal value of pumpkin fruit are high total content of carotenoids with >80% of β-carotene (Azevedo-Meleiro and Rodriguez-Amaya 2007; Kurz et al. 2008) as well as the presence of pectin and non-

Table 41.1 Chemical composition of peel, flesh and seeds of Cucurbitaceae species (ripped pumpkin fruits) (Kim et al. 2012)

Component	Part	Concentration (g/kg raw weight)			Comment
		Cucurbita pepo	*Cucurbita moschata*	*Cucurbita maxima*	
Carbohydrate	Flesh	26.23 ± 0.20	43.39 ± 0.84	133.53 ± 1.44	*C. maxima* had significantly more carbohydrates, protein, amino acid, α-tocopherol and β-carotene contents in the flesh and peel than the two species. Methionine was absent in the flesh of other species but in C. *maxima*
	Peel	43.76 ± 0.74	96.29 ± 1.1	206.78 ± 3.25	
	Seed	122.20 ± 7.47	140.19 ± 7.60	129.08 ± 8.25	
Protein	Flesh	2.08 ± 0.1	3.05 ± 0.65	11.31 ± 0.95	
	Peel	9.25 ± 0.12	11.30 ± 0.99	16.54 ± 2.69	
	Seed	308.83 ± 12.06	298.11 ± 14.75	274.85 ± 10.04	
Lipids	Flesh	0.55 ± 0.14	0.89 ± 0.11	4.20 ± 0.23	
	Peel	4.71 ± 0.69	6.59 ± 0.41	8.69 ± 0.99	
	Seed	439.88 ± 2.88	456.76 ± 11.66	524.34 ± 1.32	
Fibre	Flesh	3.72 ± 0.02	7.41 ± 0.07	10.88 ± 0.35	*C. pepo* had more protein in the seeds than other species
	Peel	12.28 ± 0.15	34.28 ± 1.37	22.35 ± 0.01	
	Seed	148.42 ± 0.55	108.51 ± 8.36	161.54 ± 6.79	
Ash	Flesh	3.44 ± 0.04	10.36 ± 0.01	10.53 ± 0.11	*C. maxima* seed had significantly more lipids
	Peel	6.30 ± 0.06	13.96 ± 0.16	11.20 ± 0.64	
	Seed	55.02 ± 1.00	53.15 ± 0.20	44.22 ± 0.36	
Moisture	Flesh	967.70 ± 0.15	942.31 ± 0.08	840.43 ± 0.17	*C. moschata* peel had more fiber than the other species
	Peel	935.98 ± 0.27	871.86 ± 0.09	756.79 ± 0.44	
	Seed	74.06 ± 0.91	51.79 ± 6.04	27.51 ± 0.21	

Values are mean ± SD

Table 41.2 Levels of essential nutrients in peel, flesh and seeds of Cucurbitaceae species (Kim et al. 2012; Zhou et al. 2007)

Compound	Part	Concentration (g/kg raw weight)		
		Cucurbita pepo	*Cucurbita moschata*	*Cucurbita maxima*
α-Tocopherol	Flesh	1.40 ± 0.01	1054 ± 0.99	2.31 ± 0.03
	Peel	4.49 ± 0.72	6.17 ± 2.19	9.62 ± 0.79
	Seed	21.33 ± 3.65	25.74 ± 0.73	20.79 ± 1.33
γ-Tocopherol	Flesh	ND[a]	0.52 ± 0.01	ND
	Peel	0.66 ± 0.09	ND	3.55 ± 0.17
	Seed	61.65 ± 17.66	66.85 ± 4.90	28.70 ± 2.13
β-Tocopherol	Flesh	1.48 ± 0.05	5.70 ± 0.39	17.04 ± 12.18
	Peel	39.48 ± 0.24	68.30 ± 2.02	123.19 ± 30.61
	Seed	17.46 ± 18.29	7.15 ± 1.50	31.40 ± 3.02
β-Tocopherol	Flesh	ND	ND	0.65 ± 0.02
	Peel	0.15 ± 0.02	0.13 ± 0.03	6.57 ± 1.87
	Seed	0.16 ± 0.16	ND	0.21 ± 0.06

Values are mean ± SD
[a]*ND* not detected

Fig. 41.4 Images of Cucurbita plant with flower, leaves and fruits. The fruit can be very variable in size, shape and color

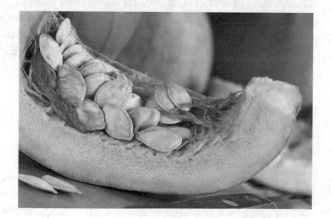

Fig. 41.5 View of a slice of a pumpkin with its seeds. The flesh is orange, and the hollow center contains pulpy loose fibers and flat, oval white seeds

Fig. 41.6 Hulled (a) and hull-less (b) seeds of pumpkin

pectin polysaccharides, minerals (potassium, phosphorus, magnesium, iron, and selenium), vitamins (C, E, K, thiamine (B1) and riboflavin (B2), dietary fiber, phenolic compounds (flavonoids, and phenolic acids) and other active substances beneficial to human health (Nawirska-Olszańska et al. 2014; Zhou et al. 2014).

Pumpkin seeds, known as pepitas, are flat, dark green seeds. They may be encased in a yellow-white husk (hulled or husked seeds), although some pumpkins produce seeds without shells (hull-less or naked seeds) which have only very thin dark-green skin (Abdel-Rahman 2006). These two varieties of pumpkin seeds are shown in Fig. 41.6a, b. Pumpkin seeds have interesting organoleptic properties and are rich in bioactive substances. They are a high-energy source of different biological components such as proteins, oils, carbohydrates, microelements and vitamins and are consumed as an edible pumpkin or as a vegetable oil source in many cultures all over the world (Petkova and Antova 2015). The seeds are uniquely flavored with a nutty taste and are consumed as roasted, salted snack in some parts of Canada, Mexico, USA, Europe and China. Besides, the seeds, rich in oil and zinc, can be eaten raw, roasted, powdered and used as flour, or have the oil extracted and used for cooking or in salads (Patel 2013). They have long been used in indigenous diets as popular medicine thanks to their bioactive and nutraceutical properties (Ozuna and León-Galván 2017; Yadav et al. 2010). The nutritional value of pumpkin seeds is based on high protein content (25–51%) (Bombardelli and Morazzoni 1997) and high percentage of oil that ranging from 40% to 60.8% (Fokou et al. 2009; Achu et al. 2006). The variability in the oil content of pumpkin seed is due to its broad genetic diversity. The nutrient distribution in pumpkin seed is presented in Table 41.3.

2   Extraction and Processing of Oil

Industrially, pumpkin seeds are normally used for the extraction of edible and drying oils which comprise about a half of the seed's weight (Jacks et al. 1972). However, pumpkin seed oil has recently gained much attention not only as edible oil, but also as a potential nutraceutical. It is dichromatic, viscous oil that has been documented for its strong antioxidant activity (Stevenson et al. 2007).

Moreover, pumpkin seed oil (PSO) has the most remarkable health benefits (Nishimura et al. 2014), and thus it requires effective extraction techniques in order to guarantee a high-quality PSO.

Conventional oil extraction is carried out by solvent extraction or pressing. Solvent extraction is usually used for seeds with low lipid content. It is the most efficient method, but its application presents some industrial disadvantages such as security problems and high operation costs. For example, *n*-hexane is a typical solvent that is commercially used in oil extraction from different plant materials. However, environmental safety regulations and increased public health risk are necessitating the industry to consider alternatives to the organic solvents for use in oil extraction (Bhattacharjee et al. 2007). Toxic organic residues were considered a serious problem in the solvent extraction of PSO (Wenli et al. 2004).

The safety and simplicity of the oilseeds mechanical pressing process are more advantageous than efficient solvent extraction equipment. Furthermore, materials pressed out generally have better preserved native properties. However, extraction by just pressing the seeds is relatively inefficient. In this regard, some new extraction methods have been conducted as viable alternatives to organic solvents for the sake of meeting the growing consumer demand and obtaining more natural products. Such alternatives include cold pressing extraction (Neđeral et al. 2014; Rabrenović et al. 2014), supercritical CO_2 extraction (Mitra et al. 2009), aqueous enzymatic extraction (Quan 2012), or microwave-assisted aqueous enzymatic extraction (Jiao et al. 2014). These extraction techniques were subsequently applied to pumpkin seeds obtained from different varieties and originating from different geographical areas all over the world.

PSO is generally extracted by expelling the oil from the seeds mechanically. Both hulled and hull-less pumpkin seeds could be used for oil production. The oil from the hull-less seeds is much easier to extract and more commonly processed. The hull surrounding common pumpkin seeds has been replaced by a membranous seed coat making the oil easier to remove mechanically (Filbrandt Katelyn 2012). The traditional method of expelling oil from pumpkin seeds is accomplished by roasting the seed pulp prior to mechanically expelling the oil (Nakić et al. 2006). During the roasting process of pumpkin seeds a large number of compounds can be identified in the volatile fraction, belonging to various chemical classes such as alcohols, aldehydes, ketones, sulfur compounds, furans and pyrazines (Siegmund and Murkovic 2004). After seeds are removed from the pumpkin, they are dried in order to obtain humidity below 10%. During drying, the rupture of cell walls occurs due to the loss of moisture, which helps in the process of oil extraction by solvent and is of great importance for the yield (Veronezi and Jorge 2012). In practice, the seeds with low moisture content, when stored, maintain most of their chemical and nutritional characteristics unchanged for several months (Schwartzberg 1987). The dried seeds are ground and steeped in salt brine. The increased moisture from the salt brine protects the flavor of the oil during the roasting step. The ground seed pulp is roasted at 212–248 °F for 60 min (Fruhwirth and Hermetter 2008). During the roasting process, proteins are denatured to facilitate the liberation of the oil. Oil is pressed out of the seeds at elevated temperatures (100–130 °C) and between 4351

Table 41.3 Bioactive components and their percentages in pumpkin seed (value per 100 g)

Components	Nutrient value	Percentage of RDA
Energy	559 kcal	28
Carbohydrates	10.71 g	8
Protein	30.23 g	54
Total lipids	49.05 g	164
Cholesterol	0 mg	0
Dietary fiber	6 g	16
Vitamins		
Folate	58 µg	15
Niacin	4.987 mg	31
Pantothenic acid	0.750 mg	15
Pyridoxine	0.143 mg	11
Riboflavin	0.153 mg	12
Thiamine	0.27 mg	23
Vitamin A	16 IU	0.5
Vitamin C	1.9 µg	3
Vitamin E	35.10 mg	237
Electrolytes		
Sodium	7 mg	0.5
Potassium	809 mg	17
Minerals		
Calcium	46 mg	4.5
Copper	1.343 mg	159
Iron	8.82 mg	110
Magnesium	592 mg	148
Manganese	4.543 mg	198
Phosphorus	1.233 mg	176
Selenium	9.4 µg	17
Zink	7.81 mg	71
Phytonutrients		
Carotene-β	9 µg	–
Cryptoxanthin-β	1 µg	–
Lutein-zeaxanthin	74 µg	–

Courtesy: USDA National Database
RDA The Recommended Dietary Allowance

and 8702 psi (Procida et al. 2013). At this point, oil is cooled and packaged for sale. Optimal packaging for PSO is a dark-colored bottle, to retain the beneficial components that can be lost to heat, light, and oxygen (Fruhwirth and Hermetter 2007).

PSO is also produced using a simple cold press without roasting or chemical solvents (Bavec et al. 2007). Cold-pressed PSO is obtained by direct pressing of raw-dried, mostly hull-less pumpkin seeds, with continuous screw presses, with an outlet oil temperature below 50 °C (Dimic 2005). The technological process of cold-pressed PSO includes the following: the seed harvesting in the mid-autumn, immediate washing and drying to a residual water content of around 7%, then storing. Prior to the processing, the seeds pass through the magnetic cleaner, followed by complete removal of organic impurities on the selector. The clean dry seeds are fed to the screw press, which grinds and presses the material, squeezing the oil, which is collected in vessels. Turbid matter resulting from ruptured plant material of the seeds during pressing is removed by sedimentation or filtration at room temperature. Filtered oil is filled into dark glass bottles. Because of this simple process, the oil maintains the natural composition with regard to the constituents and flavor (Vujasinovic et al. 2010).

The PSO produced by cold-press extraction method can be classified as extra virgin oil, because the oil is extracted from the seed once. This method of production preserves bioactive components, such as vitamins, provitamins, phytosterols, phospholipids and squalene, which are, together with some fatty acids, key factors of PSO nutritional value. In addition, it has been proven that these components have a positive effect on human health in different ways; they have anti-inflammatory and diuretic properties, alleviate negative symptoms of benign prostatic hyperplasia, help lower cholesterol levels, and bind free radicals (Caili et al. 2006; Fruhwirth and Hermetter 2007; Sener et al. 2007). However, while cold-press extraction method produces more pristine oil, its extraction efficiency is low and much of the oil remains in the seed pulp, a byproduct of the process (Gorjanović et al. 2011). The solid material remaining after expelling the PSO is pumpkin seed cake (PSC) which is rich in carbohydrates and has a very high content of protein (60–65%, w/w) (Peričin et al. 2008; Popović et al. 2013). However, this nutritive and potentially bioactive byproduct is usually discarded as waste or used as fertilizer or animal feed (Lelley et al. 2009). Pumpkin seed cake has a dark musty green color with a texture typical of almond paste and still contains significant amounts of oil. A second pressing of the seed cake can extract additional oil, but generally results in lower quality salad oils (Fruhwirth and Hermetter 2008).

Another method used in the extraction of PSO is supercritical extraction with carbon dioxide extraction (SFE-CO_2). SFE-CO_2 is getting popular as a cost-effective and environmentally friendly method for extracting useful components such as oils. Supercritical fluids possess both gas-like and liquid-like qualities, and the dual role of such fluids provide ideal conditions for extracting compounds with a high degree of recovery in a short period of time (Reverchon 1997; Wang and Weller 2006). Supercritical carbon dioxide is an odorless, chemically inert, non-toxic, non-flammable, non-corrosive, and low-cost solvent with high purity, and leaves no residue or contamination in the extract. SFE-CO_2 possesses moderately low critical temperature (31.1 °C) and pressure (7380 kPa) and the solvent can be removed by simple depressurization (Venkat and Kothandaraman 1998; Lang and Wai 2001; Wang and Weller 2006). Supercritical carbon dioxide, with its particularly attractive properties, is the preferred solvent for many supercritical fluid extractions. SFE-CO_2 possesses superior mass transfer properties with a higher diffusion coefficient and lower viscosity than liquid solvents. The absence of surface tension allows the rapid penetration of SFE-CO_2 into the pores of heterogeneous matrices and helps to enhance extraction efficiencies (Lang and Wai 2001). Majority of the published work on SFE-CO$_2$, has focused on the oil extraction (Bernardo-Gil and Lopes 2004; Hierro and Santa-Maria 1992; Wenli et al. 2004). The technique is becoming a standard method for oil extraction because of its high extraction efficiency, short extraction time and no residue problems (Bhattacharjee et al. 2007; Lang and Wai 2001; Sovilj 2010).

Aqueous-based solvent systems are among the increasingly preferred ones in the place of traditional organic solvents. Aqueous enzymatic extraction (AEE) technology is environmentally-friendly, cheap and safe. AEE has been applied for the extraction of oils from oilseed crops such as canola (Latif et al. 2008), Kalahari melon (Nyam et al. 2009), grape (Passos et al. 2009), sunflower (Latif and Anwar 2009), flax (Long et al. 2011), soybean (Towa et al. 2010), and pumpkin (Quan 2012; Wang et al. 2011; Libo et al. 2011). Various enzymes were used to hydrolyze structural polysaccharides of cell walls and proteins associated with lipid bodies to enhance oils release. However, one major disadvantage associated with the process is the relatively low efficiency, which can be overcome with the aid of accelerated enzyme-catalyzed reaction technologies. Microwave irradiation, which has proved to be a clean, efficient and convenient energy source, has been widely utilized in accelerated enzyme-catalyzed reactions for natural products and oils extraction (Jiao et al. 2012; Zhang et al. 2012; Ji et al. 2010). Herein, microwave-assisted aqueous enzymatic extraction (MAAEE) is proposed for the extraction of oil from pumpkin seeds.

PSO obtained using the techniques mentioned above is dark green to red, with a specific aroma and taste, and a high content of free fatty acids (FFAs). Due to its color and the foam formation, the oil is not used for cooking (Schuster et al. 1983; Tsaknis et al. 1997). However, PSO is used as salad oil due to high nutrient content in Austria, Slovenia and Hungary, also consumed as an ingredient in soups and minced meat or used as frying oil in some countries. It is also used in diet and medicine (Türkmen et al. 2017).

3   Fatty Acids and Acyl Lipids Profile of Fixed Oil

Lipids are classified into different categories according to their molecular structure and polarity; from non-polar (i.e., hydrocarbons, wax esters, and sterol esters) *via* neutral lipids (NL) (i.e., triglycerides, tocopherols, tocotrienols, FFAs, and sterols) to polar lipids (PL) such as glycolipids and phospholipids. All lipid classes are further divided into molecular species according to the length of their carbon-chains and saturation degree. Incidentally, lipid profiles of edible oils are highly complex, and the content of each lipid class varies with their natural properties (Nordbäck et al. 1998).

As for all oleaginous plants, triacylglycerols (TAGs) are the main component of the pumpkin seeds (92.7–93.4%), whereas phospholipids (PLs) represent only 1.5% of the total lipid content (Yoshida et al. 2004). The content of the other lipid components in PSO, although it varies according to many factors, is as follows: FFAs (2.9–3.5%), sn-1,3-diacylglycerols (1,3-DAGs, 0.4–0.9%), sn-1,2-diacylglycerols (1,2-DAGs, 0.7–0.9%) and steryl esters (SEs, 0.5–1.2%).

PSO is a highly unsaturated oil, with predominantly oleic and linoleic acids present. Very low levels of linolenic acid or other highly unsaturated fatty acids are present, providing PSO with high oxidative stability for storage or industrial purposes and low free radical production in human diets (Stevenson et al. 2007; Jakab et al. 2003). The composition of fatty acids in PSO depends on many factors (such as variety and climatic conditions), but in general, the predominant fatty acids are linoleic (C18:2, 35.6–60.8%), oleic (C18:1, 21.0–46.9%), stearic (C18:0, 3.1–7.4%), and palmitic acid (C16:0, 9.5–14.5%) (Lerma-García et al. 2015). However, some traces of other highly unsaturated fatty acids have been also detected. Its fatty acid profile is similar to that of soybean oil with the exception of linolenic acid, which is very low in PSO (Filbrandt Katelyn 2012; Fruhwirth and Hermetter 2007; Murkovic and Pfannhauser 2000). Studies on PSO triacylglycerols positional isomers found that oleic and linoleic acid distribution patterns are not random (Jakab et al. 2003). The highly unsaturated fatty acid composition of PSO makes it well suited for improving nutritional benefits. PSO has been implicated in providing many health benefits (Caili et al. 2006).

Pumpkin seed, a food byproduct of pumpkin processing, potentially could be used as a source of the fine chemical, simply because of its contents of PLs (El-Adawy and Taha 2001). PLs are natural raw material which widely used as a base component of drug, cosmetic and emulsifier. PLs formulated as anti-hepatotoxic due to its capability to repair of the damaged membrane as result redundant drug usage as well as infection diseases such as hepatitis. PLs could also be constructed as a liposome to be used as drug transporter having low dissolve in water and decreasing a side effect of the drugs (Devlin 2006). Cosmetic industries apply PLs as a base component of lotion as spherical vesicle acts as delivery active compounds (i.e., vitamin E) which mostly insoluble in water (Kanoh et al. 2004). In the food industry, PLs are used as an emulsifier. In chocolate candy, PLs assist in stabilizing a mixture of chocolate with margarine (Schneider 2001). In case of pumpkin seed kernels, many researchers report the present of PLs in the seed (Xanthopoulou et al. 2009; Al-Khalifa 1996). Therefore, it is possible to make pumpkin seed as a new source of PLs. However, detailed information regarding the type of PLs present in the seed as well as the fatty acid composition of the PLs remains little known. This information is essentially needed if the PLs to be applied for certain applications since requirements for PLs as anti-hepatotoxic, emulsifier, liposome needs a different type of PLs and its fatty acid.

The main PLs compound in the kernel of pumpkin seeds is phosphatidylcholine, PC (~55%), followed by phosphatidylinositol, PI (~25%) and phosphatidylethanolamine, PE (~20%) (Yoshida et al. 2005). The fatty acid profile of PLs has a high amount of unsaturated fatty acids, ranging from 68.5% to 70.2% and consisting mainly of oleic and linoleic acids. However, the amount of saturated acids, such as palmitic and stearic acids, is higher in PLs compared with those of TAGs and FFAs (Yoshida et al. 2004).

4 Minor Bioactive Lipids in Fixed Oil

Pumpkin seed oil (about 11–31% of total seed content) has been implicated as a nutraceutical due to its macro- and micro-constituent composition. The seed oil is a rich source of tocopherols, phytosterols, carotenoids, antioxidative phenolic compounds, triterpenes, lignans, and minerals (Applequist et al. 2006; Xanthopoulou et al. 2009; Caili et al. 2006). PSO contains high amounts of vitamin E in the form of α-tocopherols, γ-tocopherol, δ-tocopherol and tocotrienols (Nakić et al. 2006). Tocopherols are non-glycoside compounds of oil, and are natural antioxidants. They are lipophylic compounds, having a chromane head with two rings: a phenolic and a heterocyclic, and a phytyl tail saturated in tocopherols and possessing three double bonds in tocotrienols (Fig. 41.7). It has been documented that the tocopherol content of the PSO ranged from 27 to 75 mg/g of oil for α-tocopherol, from 75 to 493 mg/g for γ-tocopherol, and from 35 to 1110 mg/g for δ-tocopherol (Matus et al. 1993; Stevenson et al. 2007; Glew et al. 2006). While the content of γ-tocopherol is constant during the roasting process, the content of α-tocotrienol is reduced, indicating a low stability of this compound (Siegmund and Murkovic 2004).

Phytosterols are abundant in PSO. Phytosterols are molecules that have a structure similar to cholesterol and can lower cholesterol levels in humans (Fig. 41.8). Sterols are used in functional foods such as spreads to help individuals lower their cholesterol levels. Sterols, up to 0.5% in the PSO (55–60% of the unsaponifiable fraction), are predominantly Δ7 sterols, which are considered to be the key active constituents of pumpkin seed. Much smaller amounts of Δ5- and Δ8-sterols are also present (Bastić et al. 1977; Garg and Nes 1986; Mandl et al. 1999). In contrast, the sterol fractions of most seed oils are predominantly Δ5-sterols including β-sitosterol, campesterol, and stigmasterol (30–60%). Also, sterol content is extremely important in terms of the determination of adulteration in PSO (Wenzl et al. 2002; Dulf et al. 2009).

PSO also had a high concentration of carotenoids. In the human body, carotenoids keep such chemical reactivity as in plants by catching free radicals and active

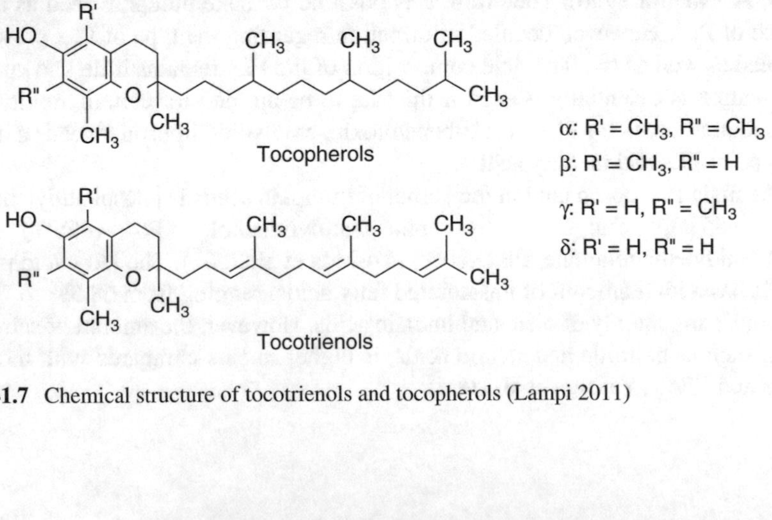

Fig. 41.7 Chemical structure of tocotrienols and tocopherols (Lampi 2011)

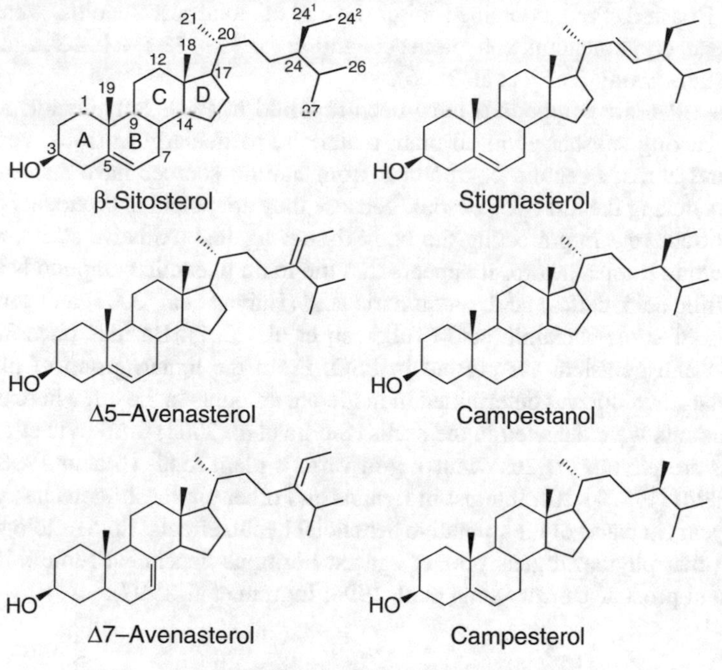

Fig. 41.8 Chemical structure of sterols (Winkler-Moser 2011)

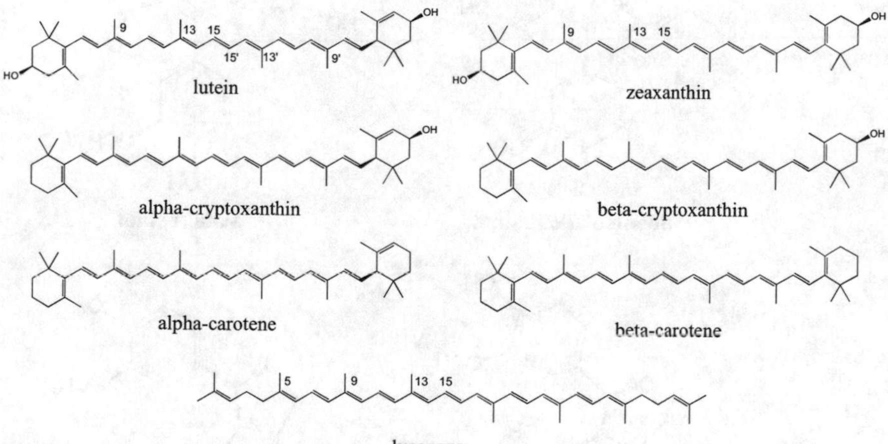

Fig. 41.9 Chemical structure of carotenoids (Lipkie et al. 2015). C atoms are labeled to indicate positions of *cis*- isomers

atomic oxygen (Fig. 41.9). This is particularly important for heavily working peoples, record-seeking sportsmen and people being in the situation of the prolonged stress. The diet rich in carotenoids diminishes coronary disease, tumors of the lungs, urinary diseases and skin problems (Inocent et al. 2007; Konings and Roomans

1997). Roasted PSO contained high levels of total carotenoids, zeaxanthin, β-carotene, cryptoxanthin, and lutein accounting for 71.0, 28.5, 6.0, 4.9, and 0.3 mg/ kg oil, respectively (Parry et al. 2006).

Some phenolic compounds have been reported in PSO. Polyphenols are naturally occurring substances in all plant materials, particularly in fruits, vegetables, seeds and herbs. Phenolic compounds from natural sources have attracted great attention during the last two decades, because they are potent antioxidants that play an important role in protecting the body tissues against oxidative stress, and thus contribute to human health. It appears that the main phenolic compounds are tyrosol, vanillic acid, caffeic acid, *o*-coumaric acid (Haiyan et al. 2007) and some small amounts of *trans*-cinnamic acid (Tuberoso et al. 2007). Besides phenolic acids, daidzein and genistein are present in PSO. From the lignan group of phenolics, secoisolariciresinol was determined in moderate amounts in the oil, whereas significant amounts were detected in the seeds (Sicilia et al. 2003; Murkovic et al. 2004). Lignans are phytoestrogens occurring in various plant foods (Mazur 1998; Mazur et al. 1998) (Fig. 41.10). Interest in lignans and other phytoestrogens has grown in recent years because of their putative beneficial health effects. Epidemiological data suggest that phytoestrogens protect against hormone-dependent tumors including breast and prostate cancer (Sung et al. 1998; Ingram et al. 1997).

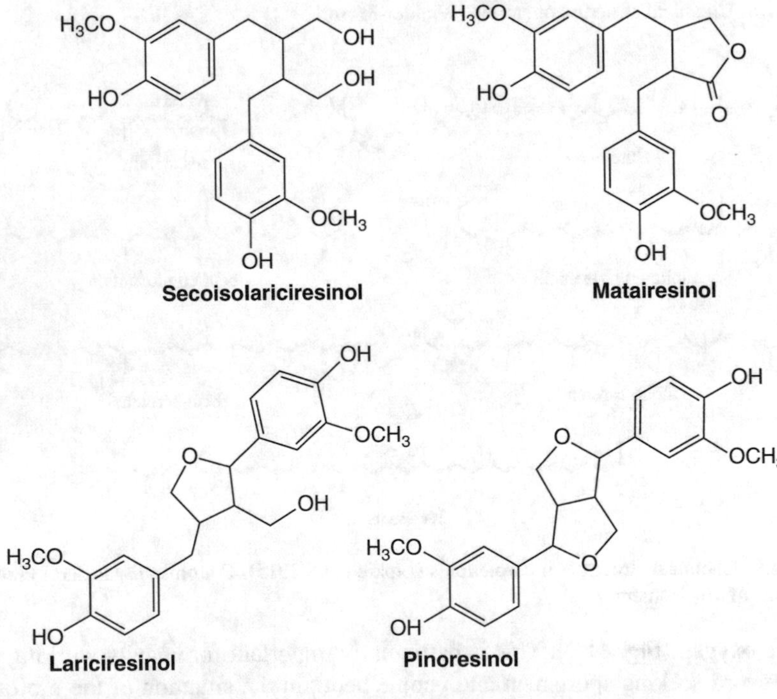

Fig. 41.10 Chemical structure of some dietary lignan precursor (Gang 2010)

PSO also contains high squalene content accounting for 71.6 mg/100 g oil (Kalogeropoulos et al. 2013). Squalene is one of the unsaturated hydrocarbons of triterpene. It plays an important role in the synthesis of cholesterol, steroid hormones and vitamin D and E in the human body (Strandberg et al. 1990). One of the main features of squalene is its antioxidant, anti-bacterial and antifungal traits (Spanova and Daum 2011). Besides, the pumpkin seed is a good source of potassium, phosphorus and magnesium, and also contains moderately high amounts of other trace minerals (calcium, sodium, manganese, iron, zinc, and copper). These elements make pumpkin seed valuable for food supplements (Karanja et al. 2013; Lazos 1986).

5 Contribution of Bioactive Compounds in Pumpkin Seed Oil to Organoleptic Properties

Pumpkin seed and the oil obtained from it have interesting organoleptic properties and are a rich source of bioactive substances. Raw PSO is green, whereas the color of oil obtained from heated pumpkin seed kernel ranges from dark red to dark brown. The dark red to dark brown color of the oil is the result of the higher content of chlorophyll and carotenoids. It has a characteristic odor and taste which are affected by its composition and specific method of production. It is used as edible oil, especially in the preparation of some salads, giving them a very agreeable taste. It is used alone or mixed with other edible oils (Markovic and Bastic 1976). PSO can be also used as an additive to bread, flakes, salads and pastries due to their beneficial phytochemical composition (Xanthopoulou et al. 2009).

6 Health-Promoting Traits of Oil and Oil Constituents

PSO is an extraordinarily rich source of diverse bioactive compounds having functional properties. Hence, it has recently gained much attention not only as edible oil, but also as a potential nutraceutical. It is dichromatic, viscous oil that has been documented for its strong antioxidant activity (Rezig et al. 2018). Researchers have so far focused on the composition and content of fatty acids, tocopherols and sterols in PSO because of their positive health effects (Procida et al. 2013; Rabrenović et al. 2014). The most remarkable health benefits of PSO are prostate disease prevention, retardation of hypertension progression, reduction of the bladder and urethral pressure, urinary disorders prevention, and the alleviation of diabetes by promoting hypoglycemic activity (Nishimura et al. 2014).

Being rich in unsaturated fatty acids especially linoleic and oleic acids and tocopherols and with very high oxidative stability, PSO is suggested to be a healthy addition towards human diet and have potential suitability for food and industrial applications (Stevenson et al. 2007). Essential fatty acids are necessary for human health but the body cannot make them; so they must be taken through food (Eynard and Zinn-Justin 1992). Linoleic acid, a PUFA present in PSO, is known to increase membrane fluidity and allows for osmosis, intracellular and extracellular gaseous exchange (Lovejoy 2002). Also, linolenic and other highly unsaturated fatty acids are present at low levels, providing this oil with high oxidative stability and low free radical production.

Among the biologically active compounds in PSO, tocopherols and tocotrienols, collectively known as tocols, have one of the most important roles in diet and health protection (Shahidi 2000). Diets high in PSO have also been associated with lower level of gastric, breast, lung and colorectal cancer (Srbinoska et al. 2012). Many investigations have shown that vitamin E can ameliorate the risk of developing chronic diseases, particularly heart disease, certain cancers, and Alzheimer's disease (Evans and Bishop 1922; Morris et al. 2002; Wildman 2002). Besides, the consumption of PSO contributes to smoothed skin and increased energy (Filbrandt Katelyn 2012).

Phytosterols present in the PSO are being studied for their role in lowering cholesterol levels. Because of the high content of squalene, γ-tocopherol, carotenoids and $\Delta7$-phytosterols, PSO may be used as a pharmaceutical remedy to treat lipid-associated diseases such as benign prostatic hyperplasia and malignant neoplasms (Vorobyova et al. 2014).

PSO is also rich in carotenoids. In the human body, carotenoids keep such chemical reactivity as in plants by catching free radicals and active atomic oxygen. This is particularly important for heavily working peoples, record-seeking sportsmen and people being in the situation of the prolonged stress. The diet rich in carotenoids diminishes coronary disease, tumors of the lungs, urinary diseases and skin problems (Inocent et al. 2007; Konings and Roomans 1997).

Regular consumption of PSO can promote the health of eyes and boost the immune system remarkably since pumpkin is a rich source of vitamin A. The high vitamin C content in PSO also offers protection against various forms of cancer (Nwokolo and Sim 1987). Vitamin C helps fight free radicals, improves immunity and promotes the production of collagen. PSO is rich in a mixture of minerals such as magnesium, potassium and zinc which is important minerals required for various biological functions. These minerals make PSO a memorable choice for those who want a healthy and glowing skin, also prevent the appearance of wrinkles and to keep your skin hydrated and nourished (Koike et al. 2005; Medzhitov 2008).

7 Edible Applications of Oil

PSO belongs to the group of oils of high nutritive value due to its favorable fatty acid composition and different minor components which have certain beneficial effects on the human organism (Fruhwirth and Hermetter 2008). Besides the positive nutritive and pharmacological properties, PSO is characterized by specific sensory properties, such as color (Kreft and Kreft 2007), odor, taste, and aroma. These traits are significantly different from other kinds of edible oils (Fruhwirth and Hermetter 2008; Vukša et al. 2003; Dimic et al. 2003).

PSO has been well known for a long time and appreciated as salad oil in a number of Southeastern European countries including Austria, Slovenia, Croatia, Hungary, with the consumption expanding in other countries. In this particular market, PSO is known as virgin oil. PSO is also great for refining salads, dips and vegetable dishes. In addition, the oil tastes great in spreads, on sheep's cheese, in soups, on egg dishes and even in cakes (Vujasinovic et al. 2010).

8 Adulteration and Authenticity

Due to the backbreaking cultivation of the plants, the crop of the seeds, which is done manually, and the isolation of the oil, PSO is rather expensive compared to other vegetable oils. Therefore, it is often adulterated by the addition of cheaper oils. Blend with intending to deceive is potentially done by the addition of sunflower or rapeseed oils, which are hardly detectable by the consumer, since PSO is of dark green color with intensive characteristic odor of various pyrazines (Wenzl et al. 2002).

To assess the authenticity of oils, it is fundamental to know, not only the biological origin of seeds, but also the technologies applied, the fat modification techniques used and the chemical composition of the authentic oils and of the potential adulterants (Kamm et al. 2001). There have been several attempts to identify adulteration of edible oils by their fatty acid profile (Wenzel 1987). Identification of adulterations by determination of the fatty acids implies the risk of misinterpretation due to similarities in the fatty acid pattern of some edible oils. A further limitation in the detection of adulteration is provided by variations in the fatty acid profile, which depend on the climate, environmental influences and the area of cultivation. However, some oil blends are identified easily. The admixture of rapeseed oil is detected by the determination of erucic acid, which is normally not present in PSO. More specific is the determination of individual TAGs, whose qualitative profile is not affected by natural quantitative variations in their content of individual fatty acids (Gunstone 1967; Lee et al. 2001).

For plant oils, the use of the secondary metabolites is more convenient as markers of biological or geographical authenticity as well for adulteration and traceability studies, representing the metabolic profile or fingerprint (Socaciu et al. 2009). Specific components in the unsaponifiable fraction of the oil were considered as markers for adulteration. The predominant part of the unsaponifiables includes various hydrocarbons, triterpenoids, carotenoids, tocopherols and sterols. These minor lipid compounds have proven of interest to food analysts and the composition and/or ratios of these trace compounds can provide a "fingerprint" for edible oils (Dulf et al. 2009; Mandl et al. 1999).

For the detection of manipulation of the natural composition of the oil, different techniques were applied. Chromatographic techniques like gas chromatography and high-performance liquid chromatography are dominating, but also spectroscopic methodologies were applied (Davies et al. 2000; El-Hamdy and El-Fizga 1995; Lai et al. 1995).

In case of PSO, adulterated products can be distinguished from pure oils by their sterol composition. In contrast to most edible oils, PSO contains very low amounts of Δ5-sterols like stigmasterol, campesterol or β-sitosterol, while the main phytosterol component consists of Δ7-sterols (Mandl et al. 1999). The approach is based on this analysis concept for the determination of Δ5-sterols in the pumpkin seeds as well as in the refined PSO. The determination of Δ5-sterols, especially β-sitosterol has proven to be a good possibility to detect admixture of cheap vegetable oils to a level below which the economic profit is not significant anymore.

References

Abdel-Rahman, M. K. (2006). Effect of pumpkin seed (*Cucurbita pepo* L.) diets on benign prostatic hyperplasia (BPH): Chemical and morphometric evaluation in rats. *World Journal of Chemistry, 1*(1), 33–40.

Achilonu, M., Nwafor, I., Umesiobi, D., & Sedibe, M. (2018). Biochemical proximates of pumpkin (*Cucurbitaceae* spp.) and their beneficial effects on the general well-being of poultry species. *Journal of Animal Physiology and Animal Nutrition, 102*(1), 5–16.

Achu, M., Fokou, E., Tchiégang, C., Fotso, M., & Tchouanguep, M. (2006). Chemical characteristics and fatty acid composition of Cucurbitaceae oils from Cameroon. *13th World Congress of Food Science & Technology, 2006*, 26–26.

Adeel, A., Sohail, A., & Masud, T. (2014). Characterization and antibacterial study of pumpkin seed oil (*Cucurbita pepo*). *Life Sciences Leaflets, 49*, 53–64.

Al-Khalifa, A. (1996). Physicochemical characteristics, fatty acid composition, and lipoxygenase activity of crude pumpkin and melon seed oils. *Journal of Agricultural and Food Chemistry, 44*(4), 964–966.

Applequist, W. L., Avula, B., Schaneberg, B. T., Wang, Y.-H., & Khan, I. A. (2006). Comparative fatty acid content of seeds of four Cucurbita species grown in a common (shared) garden. *Journal of Food Composition and Analysis, 19*(6–7), 606–611.

Azevedo-Meleiro, C. H., & Rodriguez-Amaya, D. B. (2007). Qualitative and quantitative differences in carotenoid composition among *Cucurbita moschata*, *Cucurbita maxima*, and *Cucurbita pepo*. *Journal of Agricultural and Food Chemistry, 55*(10), 4027–4033.

Bastić, M., Bastić, L., Jovanović, J., & Spiteller, G. (1977). Sterols in pumpkin seed oil. *Journal of the American Oil Chemists' Society, 54*(11), 525–527.

Bavec, F., Grobelnik Mlakar, S., Rozman, Č., & Bavec, M. (2007). *Oil pumpkins: Niche for organic producers*. Alexandria: ASHS Press.

Bernardo-Gil, M. G., & Lopes, L. M. C. (2004). Supercritical fluid extraction of *Cucurbita ficifolia* seed oil. *European Food Research and Technology, 219*(6), 593–597.

Bhattacharjee, P., Singhal, R. S., & Tiwari, S. R. (2007). Supercritical carbon dioxide extraction of cottonseed oil. *Journal of Food Engineering, 79*(3), 892–898.

Bombardelli, E., & Morazzoni, P. (1997). *Cucurbita pepo* L. *Fitoterapia, 68*, 291–302.

Caili, F., Huan, S., & Quanhong, L. (2006). A review on pharmacological activities and utilization technologies of pumpkin. *Plant Foods for Human Nutrition, 61*(2), 70–77.

Davies, A. N., McIntyre, P., & Morgan, E. (2000). Study of the use of molecular spectroscopy for the authentication of extra virgin olive oils. Part I: Fourier transform Raman spectroscopy. *Applied Spectroscopy, 54*(12), 1864–1867.

Devlin, T. M. (2006). *Textbook of biochemistry: With clinical correlations.* Wiley-liss, Hoboken NJ.

Dhiman, A. K., Sharma, K., & Attri, S. (2009). Functional constitutents and processing of pumpkin: A review. *Journal of Food Science and Technology, 46*(5), 411.

Dimic, E. (2005). *Cold-pressed oils.* Novi Sad: The University of Novi Sad, Faculty of Technology.

Dimic, E., Dimic, V., & Romanic, R. (2003). *Process and quality of expeller oil obtained from pumpkin seed.* Hungary: Olaj, szappan, kozmetika.

Dulf, F. V., Bele, C., Unguresan, M., Parlog, R., & Socaciu, C. (2009). Phytosterols as markers in identification of the adulterated pumpkin seed oil with sunflower oil. *Bulletin UASVM Agriculture, 66*(2), 301–306.

El-Adawy, T. A., & Taha, K. M. (2001). Characteristics and composition of watermelon, pumpkin, and paprika seed oils and flours. *Journal of Agricultural and Food Chemistry, 49*(3), 1253–1259.

El-Hamdy, A. H., & El-Fizga, N. K. (1995). Detection of olive oil adulteration by measuring its authenticity factor using reversed-phase high-performance liquid chromatography. *Journal of Chromatography A, 708*(2), 351–355.

Evans, H. M., & Bishop, K. S. (1922). On the existence of a hitherto unrecognized dietary factor essential for reproduction. *Science, 56*(1458), 650–651.

Eynard, B., & Zinn-Justin, J. (1992). The O (n) model on a random surface: Critical points and large-order behaviour. *Nuclear Physics B, 386*(3), 558–591.

Faostat, F. (2016). FAOSTAT statistical database. Publisher: FAO (Food and Agriculture Organization of the United Nations), Rome, Italy.

Filbrandt Katelyn, R. (2012). *Effect of pumpkin seed oil cake on the textural and sensory properties of white wheat.* Menomonie: University of Wisconsin-Stout.

Fokou, E., Achu, M., Kansci, G., Ponka, R., Fotso, M., Tchiegang, C., & Tchouanguep, F. (2009). Chemical properties of some Cucurbitaceae oils from Cameroon. *Pakistan Journal of Nutrition, 8*(9), 1325–1334.

Fruhwirth, G. O., & Hermetter, A. (2007). Seeds and oil of the Styrian oil pumpkin: Components and biological activities. *European Journal of Lipid Science and Technology, 109*(11), 1128–1140.

Fruhwirth, G. O., & Hermetter, A. (2008). Production technology and characteristics of Styrian pumpkin seed oil. *European Journal of Lipid Science and Technology, 110*(7), 637–644.

Gang, D. R. (2010). *The biological activity of phytochemicals.* Publisher: Springer, Dordrecht Heidelberg London New York, Library of Congress Control Number: 2010937099.

Garg, V. K., & Nes, W. R. (1986). Occurrence of Δ5-sterols in plants producing predominantly Δ7-sterols: Studies on the sterol compositions of six Cucurbitaceae seeds. *Phytochemistry, 25*(11), 2591–2597.

Glew, R., Glew, R., Chuang, L.-T., Huang, Y.-S., Millson, M., Constans, D., et al. (2006). Amino acid, mineral and fatty acid content of pumpkin seeds *(Cucurbita* spp) and *Cyperus esculentus* nuts in the Republic of Niger. *Plant Foods for Human Nutrition, 61*(2), 49–54.

Gohari Ardabili, A., Farhoosh, R., & Haddad Khodaparast, M. H. (2011). Chemical composition and physicochemical properties of pumpkin seeds *(Cucurbita pepo* Subsp. pepo Var. Styriaka) grown in Iran. *Journal of Agricultural Science and Technology, 13*, 1053–1063.

Gorjanović, S. Ž., Rabrenović, B. B., Novaković, M. M., Dimić, E. B., Basić, Z. N., & Sužnjević, D. Ž. (2011). Cold-pressed pumpkin seed oil antioxidant activity as determined by a DC polarographic assay based on hydrogen peroxide scavenge. *Journal of the American Oil Chemists' Society, 88*(12), 1875–1882.

Gunstone, F. D. (1967). In F. D. Gunstone (Ed.), *An introduction to the chemistry and biochemistry of fatty acids and their glycerides* (2nd ed.). London: Chapman and Hall Ltd.

Haiyan, Z., Bedgood, D. R., Jr., Bishop, A. G., Prenzler, P. D., & Robards, K. (2007). Endogenous biophenol, fatty acid and volatile profiles of selected oils. *Food Chemistry, 100*(4), 1544–1551.

Hierro, M., & Santa-Maria, G. (1992). Supercritical fluid extraction of vegetable and animal fats with CO_2-a mini review. *Food Chemistry, 45*, 189–192.

Ingram, D., Sanders, K., Kolybaba, M., & Lopez, D. (1997). Case-control study of phyto-oestrogens and breast cancer. *The Lancet, 350*(9083), 990–994.

Inocent, G., Ejoh, R. A., Issa, T. S., Schweigert, F. J., & Tchouanguep, M. (2007). Carotenoids content of some locally consumed fruits and yams in Cameroon. *Pakistan Journal of Nutrition, 6*(5), 497–501.

Jacks, T., Hensarling, T., & Yatsu, L. (1972). Cucurbit seeds: I. Characterizations and uses of oils and proteins. A review. *Economic Botany, 26*(2), 135–141.

Jakab, A., Jablonkai, I., & Forgács, E. (2003). Quantification of the ratio of positional isomer dilinoleoyl-oleoyl glycerols in vegetable oils. *Rapid Communications in Mass Spectrometry, 17*(20), 2295–2302.

Ji, P., Zhou, J., & Liu, X. (2010). Microwave-assisted aqueous extraction of camellia oil. *Modern Food Science and Technology, 26*(5), 486–489.

Jiao, J., Fu, Y.-J., Zu, Y.-G., Luo, M., Wang, W., Zhang, L., et al. (2012). Enzyme-assisted micro-wave hydro-distillation essential oil from *Fructus forsythia*, chemical constituents, and its anti-microbial and antioxidant activities. *Food Chemistry, 134*(1), 235–243.

Jiao, J., Li, Z.-G., Gai, Q.-Y., Li, X.-J., Wei, F.-Y., Fu, Y.-J., et al. (2014). Microwave-assisted aqueous enzymatic extraction of oil from pumpkin seeds and evaluation of its physicochemical properties, fatty acid compositions and antioxidant activities. *Food Chemistry, 147*, 17–24.

Kalogeropoulos, N., Chiou, A., Ioannou, M. S., & Karathanos, V. T. (2013). Nutritional evaluation and health-promoting activities of nuts and seeds cultivated in Greece. *International Journal of Food Sciences and Nutrition, 64*(6), 757–767.

Kamm, W., Dionisi, F., Hischenhuber, C., & Engel, K.-H. (2001). Authenticity assessment of fats and oils. *Food Reviews International, 17*(3), 249–290.

Kanoh, S., Maeyama, K., Tanaka, R., Takahashi, T., Aoyama, M., Watanabe, M., et al. (2004). M. Sakaguchi *(Ed)*, Possible utilization of the pearl oyster phospholipid and glycogen as a cosmetic material. In *Developments in food science* (pp. 179–190). Elsevier, Amsterdam, Netherlands.

Karanja, J., Mugendi, B., Khamis, F., & Muchugi, A. (2013). Nutritional composition of the pump-kin (*Cucurbita* spp.) seed cultivated from selected regions in Kenya. *Journal of Horticulture Letters, 3*(1), 17.

Kim, M. Y., Kim, E. J., Kim, Y.-N., Choi, C., & Lee, B.-H. (2012). Comparison of the chemi-cal compositions and nutritive values of various pumpkin (Cucurbitaceae) species and parts. *Nutrition Research and Practice, 6*(1), 21–27.

Koike, K., Li, W., Liu, L., Hata, E., & Nikaido, T. (2005). New phenolic glycosides from the seeds of *Cucurbita moschata. Chemical and Pharmaceutical Bulletin, 53*(2), 225–228.

Konings, E. J., & Roomans, H. H. (1997). Evaluation and validation of an LC method for the analysis of carotenoids in vegetables and fruit. *Food Chemistry, 59*(4), 599–603.

Kreft, S., & Kreft, M. (2007). Physicochemical and physiological basis of dichromatic colour. *Naturwissenschaften, 94*(11), 935–939.

Kurz, C., Carle, R., & Schieber, A. (2008). HPLC-DAD-MSn characterisation of carotenoids from apricots and pumpkins for the evaluation of fruit product authenticity. *Food Chemistry, 110*(2), 522–530.

Lai, Y., Kemsley, E., & Wilson, R. (1995). Quantitative analysis of potential adulterants of extra virgin olive oil using infrared spectroscopy. *Food Chemistry, 53*(1), 95–98.

Lampi, A. (2011). Analysis of tocopherols and tocotrienols by HPLC. The AOCS lipid library. URL: http://lipidlibrary.aocs.org/topics/tocopherols/index.htm. Updated August 3, 2011.

Lang, Q., & Wai, C. M. (2001). Supercritical fluid extraction in herbal and natural product studies-a practical review. *Talanta, 53*(4), 771–782.

Latif, S., & Anwar, F. (2009). Effect of aqueous enzymatic processes on sunflower oil quality. *Journal of the American Oil Chemists' Society, 86*(4), 393–400.

Latif, S., Diosady, L. L., & Anwar, F. (2008). Enzyme-assisted aqueous extraction of oil and pro-tein from canola (*Brassica napus* L.) seeds. *European Journal of Lipid Science and Technology, 110*(10), 887–892.

Lazos, E. S. (1986). Nutritional, fatty acid, and oil characteristics of pumpkin and melon seeds. *Journal of Food Science, 51*(5), 1382–1383.

Lee, D.-S., Lee, E.-S., Kim, H.-J., Kim, S.-O., & Kim, K. (2001). Reversed-phase liquid chromatographic determination of triacylglycerol composition in sesame oils and the chemometric detection of adulteration. *Analytica Chimica Acta, 429*(2), 321–330.

Lelley, T., Loy, B., & Murkovic, M. (2009). Hull-less oil seed pumpkin. In *Oil crops* (pp. 469–492). New York: Springer.

Lerma-García, M. J., Saucedo-Hernández, Y., Herrero-Martínez, J. M., Ramis-Ramos, G., Jorge-Rodríguez, E., & Simó-Alfonso, E. F. (2015). Statistical classification of pumpkin seed oils by direct infusion mass spectrometry: Correlation with GC-FID profiles. *European Journal of Lipid Science and Technology, 117*(3), 331–337.

Libo, W., Yaqin, X., Yu, Y., & Xin, S. (2011). Aqueous enzymatic extraction of pumpkin seed oil and its physical-chemical properties [J]. *Transactions of the Chinese Society of Agricultural Engineering, 10*, 068.

Lipkie, T. E., Morrow, A. L., Jouni, Z. E., McMahon, R. J., & Ferruzzi, M. G. (2015). Longitudinal survey of carotenoids in human milk from urban cohorts in China, Mexico, and the USA. *PLoS One, 10*(6), e0127729.

Long, J.-J., Fu, Y.-J., Zu, Y.-G., Li, J., Wang, W., Gu, C.-B., et al. (2011). Ultrasound-assisted extraction of flaxseed oil using immobilized enzymes. *Bioresource Technology, 102*(21), 9991–9996.

Lovejoy, J. C. (2002). The influence of dietary fat on insulin resistance. *Current Diabetes Reports, 2*(5), 435–440.

Mandl, A., Reich, G., & Lindner, W. (1999). Detection of adulteration of pumpkin seed oil by analysis of content and composition of specific Δ7-phytosterols. *European Food Research and Technology, 209*(6), 400–406.

Markovic, V., & Bastic, L. (1976). Characteristics of pumpkin seed oil. *Journal of the American Oil Chemists' Society, 53*(1), 42–44.

Matus, Z., Molnár, P., & Szabó, L. G. (1993). Main carotenoids in pressed seeds (*Cucurbitae semen*) of oil pumpkin (*Cucurbita pepo* convar. pepo var. styriaca). *Acta Pharmaceutica Hungarica, 63*(5), 247–256.

Mazur, W. (1998). Phytoestrogen content in foods. *Bailliere's Clinical Endocrinology and Metabolism, 12*(4), 729–742.

Mazur, W., Wähälä, K., Rasku, S., Salakka, A., Hase, T., & Adlercreutz, H. (1998). Lignan and isoflavonoid concentrations in tea and coffee. *British Journal of Nutrition, 79*(1), 37–45.

Mbondo, J. K. (2013). *Formulation and evaluation of pumpkin seed (Cucurbita pepo) tablets.* Nairobi: University of Nairobi.

Medzhitov, R. (2008). Origin and physiological roles of inflammation. *Nature, 454*(7203), 428.

Mitra, P., Ramaswamy, H. S., & Chang, K. S. (2009). Pumpkin (*Cucurbita maxima*) seed oil extraction using supercritical carbon dioxide and physicochemical properties of the oil. *Journal of Food Engineering, 95*(1), 208–213.

Morris, M. C., Evans, D. A., Bienias, J. L., Tangney, C. C., Bennett, D. A., Aggarwal, N., et al. (2002). Dietary intake of antioxidant nutrients and the risk of incident Alzheimer disease in a biracial community study. *JAMA, 287*(24), 3230–3237.

Murkovic, M., & Pfannhauser, W. (2000). Stability of pumpkin seed oil. *European Journal of Lipid Science and Technology, 102*(10), 607–611.

Murkovic, M., Piironen, V., Lampi, A. M., Kraushofer, T., & Sontag, G. (2004). Changes in chemical composition of pumpkin seeds during the roasting process for production of pumpkin seed oil (part 1: Non-volatile compounds). *Food Chemistry, 84*(3), 359–365.

Nakić, S. N., Rade, D., Škevin, D., Štrucelj, D., Mokrovčak, Ž., & Bartolić, M. (2006). Chemical characteristics of oils from naked and husk seeds of *Cucurbita pepo* L. *European Journal of Lipid Science and Technology, 108*(11), 936–943.

Nawirska-Olszańska, A., Biesiada, A., Sokół-Łętowska, A., & Kucharska, A. Z. (2014). Characteristics of organic acids in the fruit of different pumpkin species. *Food Chemistry, 148*, 415–419.

Neđeral, S., Petrović, M., Vincek, D., Pukec, D., Škevin, D., Kraljić, K., et al. (2014). Variance of quality parameters and fatty acid composition in pumpkin seed oil during three crop seasons. *Industrial Crops and Products, 60*, 15–21.

Nishimura, M., Ohkawara, T., Sato, H., Takeda, H., & Nishihira, J. (2014). Pumpkin seed oil extracted from *Cucurbita maxima* improves urinary disorder in human overactive bladder. *Journal of Traditional and Complementary Medicine, 4*(1), 72–74.

Nordbäck, J., Lundberg, E., & Christie, W. W. (1998). Separation of lipid classes from marine particulate material by HPLC on a polyvinyl alcohol-bonded stationary phase using dual-channel evaporative light-scattering detection. *Marine Chemistry, 60*(3–4), 165–175.

Nwokolo, E., & Sim, J. S. (1987). Nutritional assessment of defatted oil meals of melon (*Colocynthis citrullus* L.) and fluted pumpkin (*Telfaria occidentalis* Hook) by chick assay. *Journal of the Science of Food and Agriculture, 38*(3), 237–246.

Nyam, K. L., Tan, C. P., Lai, O. M., Long, K., & Man, Y. B. C. (2009). Enzyme-assisted aqueous extraction of Kalahari melon seed oil: Optimization using response surface methodology. *Journal of the American Oil Chemists' Society, 86*(12), 1235–1240.

Ozuna, C., & León-Galván, M. (2017). Cucurbitaceae seed protein hydrolysates as a potential source of bioactive peptides with functional properties. *BioMed Research International, 2017*(2), 1–16.

Parry, J., Hao, Z., Luther, M., Su, L., Zhou, K., & Yu, L. (2006). Characterization of cold-pressed onion, parsley, cardamom, mullein, roasted pumpkin, and milk thistle seed oils. *Journal of the American Oil Chemists' Society, 83*(10), 847–854.

Passos, C. P., Yilmaz, S., Silva, C. M., & Coimbra, M. A. (2009). Enhancement of grape seed oil extraction using a cell wall degrading enzyme cocktail. *Food Chemistry, 115*(1), 48–53.

Patel, S. (2013). Pumpkin (*Cucurbita* sp.) seeds as nutraceutic: A review on status quo and scopes. *Mediterranean Journal of Nutrition and Metabolism, 6*(3), 183–189.

Peričin, D., Radulović, L., Trivić, S., & Dimić, E. (2008). Evaluation of solubility of pumpkin seed globulins by response surface method. *Journal of Food Engineering, 84*(4), 591–594.

Petkova, Z. Y., & Antova, G. (2015). Changes in the composition of pumpkin seeds (*Cucurbita moschata*) during development and maturation. *Grasas y Aceites, 66*(1), 058.

Phillips, K. M., Ruggio, D. M., & Ashraf-Khorassani, M. (2005). Phytosterol composition of nuts and seeds commonly consumed in the United States. *Journal of Agricultural and Food Chemistry, 53*(24), 9436–9445.

Popović, L., Peričin, D., Vaštag, Ž., Popović, S., Krimer, V., & Torbica, A. (2013). Antioxidative and functional properties of pumpkin oil cake globulin hydrolysates. *Journal of the American Oil Chemists' Society, 90*(8), 1157–1165.

Procida, G., Stancher, B., Cateni, F., & Zacchigna, M. (2013). Chemical composition and functional characterisation of commercial pumpkin seed oil. *Journal of the Science of Food and Agriculture, 93*(5), 1035–1041.

Quan, Q. (2012). Study on extraction of pumpkin seed oil by aqueous enzymatic method. *Journal of Anhui Agricultural Sciences, 12*, 157.

Rabrenović, B. B., Dimić, E. B., Novaković, M. M., Tešević, V. V., & Basić, Z. N. (2014). The most important bioactive components of cold pressed oil from different pumpkin (*Cucurbita pepo* L.) seeds. *LWT-Food Science and Technology, 55*(2), 521–527.

Reverchon, E. (1997). Supercritical fluid extraction and fractionation of essential oils and related products. *The Journal of Supercritical Fluids, 10*(1), 1–37.

Rezig, L., Chouaibi, M., Ojeda-Amador, R. M., Gomez-Alonso, S., Salvador, M. D., Fregapane, G., et al. (2018). *Cucurbita maxima* pumpkin seed oil: From the chemical properties to the different extracting techniques. *Notulae Botanicae Horti Agrobotanici Cluj-Napoca, 46*(2), 663–669.

Schneider, M. (2001). Phospholipids for functional food. *European Journal of Lipid Science and Technology, 103*(2), 98–101.

Schuster, W., Zipse, W., & Marquard, R. (1983). The influence of genotype and growing location on several substances of seeds of the pumpkin (*Cucurbita pepo* L.). *European Journal of Lipid Science and Technology, 85*, 56–64.

Schwartzberg, H. G. (1987). Leaching—organic materials. In R. W. Rousseau (Ed.), Handbook of separation process technology (pp. 540–577). New York: Wiley

Sener, B., Orhan, I., Ozcelik, B., Kartal, M., Aslan, S., & Ozbilen, G. (2007). Antimicrobial and antiviral activities of two seed oil samples of Cucurbita pepo L. and their fatty acid analysis. *Natural Product Communications, 2*(4), 395–398.

Shahidi, F. (2000). Antioxidants in food and food antioxidants. *Food/Nahrung, 44*(3), 158–163.

Sicilia, T., Niemeyer, H. B., Honig, D. M., & Metzler, M. (2003). Identification and stereochemical characterization of lignans in flaxseed and pumpkin seeds. *Journal of Agricultural and Food Chemistry, 51*(5), 1181–1188.

Siegmund, B., & Murkovic, M. (2004). Changes in chemical composition of pumpkin seeds during the roasting process for production of pumpkin seed oil (part 2: Volatile compounds). *Food Chemistry, 84*(3), 367–374.

Socaciu, C., Ranga, F., Fetea, F., Leopold, L., Dulf, F., & Parlog, R. (2009). Complementary advanced techniques applied for plant and food authentication. *Czech Journal of Food Sciences, 27*, S70–S75.

Sovilj, M. N. (2010). Critical review of supercritical carbon dioxide extraction of selected oil seeds. *Acta Periodica Technologica, 41*, 105–120.

Spanova, M., & Daum, G. (2011). Squalene-biochemistry, molecular biology, process biotechnology, and applications. *European Journal of Lipid Science and Technology, 113*(11), 1299–1320.

Srbinoska, M., Hrabovski, N., Rafajlovska, V., & Sinadinović-Fišer, S. (2012). Characterization of the seed and seed extracts of the pumpkins *Cucurbita maxima* D. and *Cucurbita pepo* L. from Macedonia. *Macedonian Journal of Chemistry and Chemical Engineering, 31*(1), 65–78.

Stevenson, D. G., Eller, F. J., Wang, L., Jane, J.-L., Wang, T., & Inglett, G. E. (2007). Oil and tocopherol content and composition of pumpkin seed oil in 12 cultivars. *Journal of Agricultural and Food Chemistry, 55*(10), 4005–4013.

Strandberg, T., Tilvis, R., & Miettinen, T. (1990). Metabolic variables of cholesterol during squalene feeding in humans: Comparison with cholestyramine treatment. *Journal of Lipid Research, 31*(9), 1637–1643.

Sung, M., Lautens, M., & Thompson, L. (1998). Mammalian lignans inhibit the growth of estrogen-independent human colon tumor cells. *Anticancer Research, 18*(3A), 1405–1408.

Towa, L. T., Kapchie, V. N., Hauck, C., & Murphy, P. A. (2010). Enzyme-assisted aqueous extraction of oil from isolated oleosomes of soybean flour. *Journal of the American Oil Chemists' Society, 87*(3), 347–354.

Tsaknis, J., Lalas, S., & Lazos, E. S. (1997). Characterization of crude and purified pumpkin seed oil. *Grasas y Aceites, 48*(5), 267–272.

Tuberoso, C. I., Kowalczyk, A., Sarritzu, E., & Cabras, P. (2007). Determination of antioxidant compounds and antioxidant activity in commercial oilseeds for food use. *Food Chemistry, 103*(4), 1494–1501.

Türkmen, Ö., Özcan, M., Seymen, M., Paksoy, M., Uslu, N., & Fidan, S. (2017). Physico-chemical properties and fatty acid compositions of some edible pumpkin seed genotypes and oils. *Journal of Agroalimentary Processes and Technologies, 23*(4), 229–235.

Venkat, E., & Kothandaraman, S. (1998). Supercritical fluid methods. In *Natural products isolation* (pp. 91–109). Springer.

Veronezi, C. M., & Jorge, N. (2012). Bioactive compounds in lipid fractions of pumpkin (*Cucurbita* sp) seeds for use in food. *Journal of Food Science, 77*(6), C653–C657.

Vorobyova, O., Bolshakova, A., Pegova, R., Kol'chik, O., Klabukova, I., Krasilnikova, E., et al. (2014). Analysis of the components of pumpkin seed oil in suppositories and the possibility of its use in pharmaceuticals. *Journal of Chemical and Pharmaceutical Research, 6*(5), 1106–1116.

Vujasinovic, V., Djilas, S., Dimic, E., Romanic, R., & Takaci, A. (2010). Shelf life of cold-pressed pumpkin (*Cucurbita pepo* L.) seed oil obtained with a screw press. *Journal of the American Oil Chemists' Society, 87*(12), 1497–1505.

Vukša, V., Dimić, E., & Dimić, V. (2003) Characteristics of cold pressed pumpkin seed oil. *9th Symposium: Vitamine und Zusatzstoffe in der Ernährung von Mensch und Tier, Proceedings, Jena/Thüringen*. pp. 493–496.

Wang, L., & Weller, C. L. (2006). Recent advances in extraction of nutraceuticals from plants. *Trends in Food Science & Technology, 17*(6), 300–312.

Wang, Q.-I., Zhang, L., Ji, H., & Yan, H.-Y. (2011). Study on aqueous enzymatic extraction of pumpkin seed oil. *Cereals & Oils, 8*, 006.

Wenli, Y., Yaping, Z., Jingjing, C., & Bo, S. (2004). Comparison of two kinds of pumpkin seed oils obtained by supercritical CO_2 extraction. *European Journal of Lipid Science and Technology, 106*(6), 355–358.

Wenzel, C. (1987). Ernahrung. *Nutrition, 11*, 752–755.

Wenzl, T., Prettner, E., Schweiger, K., & Wagner, F. S. (2002). An improved method to discover adulteration of Styrian pumpkin seed oil. *Journal of Biochemical and Biophysical Methods, 53*(1–3), 193–202.

Wildman, R. E. (2002). *Handbook of nutraceuticals and functional foods*. Boca Raton: CRC press.

Winkler-Moser, J. (2011). Gas chromatographic analysis of plant sterols. The AOCS lipidlibrary, pp. 1–18. The American Oil Chemists' Society, Illinois, USA.

Xanthopoulou, M. N., Nomikos, T., Fragopoulou, E., & Antonopoulou, S. (2009). Antioxidant and lipoxygenase inhibitory activities of pumpkin seed extracts. *Food Research International, 42*(5–6), 641–646.

Yadav, M., Jain, S., Tomar, R., Prasad, G., & Yadav, H. (2010). Medicinal and biological potential of pumpkin: An updated review. *Nutrition Research Reviews, 23*(2), 184–190.

Yoshida, H., Shougaki, Y., Hirakawa, Y., Tomiyama, Y., & Mizushina, Y. (2004). Lipid classes, fatty acid composition and triacylglycerol molecular species in the kernels of pumpkin (*Cucurbita* spp) seeds. *Journal of the Science of Food and Agriculture, 84*(2), 158–163.

Yoshida, H., Tomiyama, Y., Kita, S., & Mizushina, Y. (2005). Roasting effects on fatty acid distribution of triacylglycerols and phospholipids in the kernels of pumpkin (*Cucurbita* spp) seeds. *Journal of the Science of Food and Agriculture, 85*(12), 2061–2066.

Zhang, Y.-I., Li, S., Yin, C.-P., Jiang, D.-H., Yan, F.-F., & Xu, T. (2012). Response surface optimisation of aqueous enzymatic oil extraction from bayberry (*Myrica rubra*) kernels. *Food Chemistry, 135*(1), 304–308.

Zhou, T., Kong, Q., Huang, J., Dai, R., & Li, Q. (2007). Characterization of nutritional components and utilization of pumpkin. *Food, 1*(2), 313–321.

Zhou, C.-L., Liu, W., Zhao, J., Yuan, C., Song, Y., Chen, D., et al. (2014). The effect of high hydrostatic pressure on the microbiological quality and physical-chemical characteristics of Pumpkin (*Cucurbita maxima* Duch.) during refrigerated storage. *Innovative Food Science & Emerging Technologies, 21*, 24–34.

Chapter 42
Palm (*Elaeis guineensis* Jacq.) Oil

Monika Choudhary and Kiran Grover

Abstract Palm oil is the largest produced edible oil across the world especially in the southeast region of Asia. Malaysia and Indonesia are the two largest producers of palm oil accounting for 87% of the world production. Being the richest source of biologically active carotenoids (500–800 ppm), red palm oil has been reported to possess various therapeutic properties such as hypolipidemic, antioxidant and anti-thrombotic. Palm oil inhibits platelets aggregation owing to the presence of vitamin E and helps in either increasing the production of prostacyclin or decreasing the production of thromboxane. Yet, the association between cardiovascular diseases and this oil is controversial. Besides, the therapeutic role of β-carotene-rich red palm oil has also been well demonstrated through the dietary intervention of red palm oil at 0.6 mL for 15 days to the patients of keratomalacia. Palm oil has the cheapest price in the market and has high oxidative stability, and this oil is the most suitable choice for food industry specifically for the preparation of fried products.

Keywords Food application · Olein · Palmitic acid · Palm kernel oil · Stearin · Therapeutic properties · Hypolipidemic · Anti-thrombotic · Thromboxane · Tocotrienols

Abbreviations

CVD   Cardiovascular diseases
FFA   Free fatty acid
HDL   High density lipoprotein
LDL   Low density lipoprotein

M. Choudhary (✉)
Punjab Agricultural University, Ludhiana, India

K. Grover
Department of Food and Nutrition, College of Home Science, Punjab Agricultural University, Ludhiana, India
e-mail: kirangrover@pau.edu

© Springer Nature Switzerland AG 2019
M. F. Ramadan (ed.), *Fruit Oils: Chemistry and Functionality*,
https://doi.org/10.1007/978-3-030-12473-1_42

MUFA   Monounsaturated fatty acid
PKO   Palm kernel oil
PO   Palm oil
PUFA   Polyunsaturated fatty acid
RPO   Red palm oil
SFA   Saturated fatty acid
TG   Triglycerides

1 Introduction

The oil palm is the oldest fruit oil crop, which has evolved as the fastest growing oil economy in the world since last decades. Oil palm (*Elaeis guineensis* Jacq.) is by far the most competent and high yielding oil crop to fulfill growing world demand for vegetable oils that is likely to reach 240 million tons by 2050. Being a member of Arecaceae family, oil palm is grown in humid tropical regions of Africa, Central and South America and Asia (Feintrenie 2012). Per hectare, oil yield of this tropical crop is three to eight times more than any other temperate or tropical oil crop which was witnessed by a gigantic production of this oil i.e., 56.2 million tons from 17.2 million hectares land area in the year 2012. The oil palm fruit can be categorized into three forms with regard to shell thickness. The three major forms are: Dura, the Pisifera and the Tenera. The Dura type yields about 0.5 tons of palm oil per hectare. The Pisifera type yields about an average of 4.5 tons of palm oil per hectare per year owing to genetic improvement for the second cycle Tenera hybrid palms. The Tenera hybrid yields about ten times more oil than soy oil and used chiefly for economic production (Basri et al. 2003). Palm fruit is composed of fresh fruit pulp having 50–60% oil which is 20–22% of bunch weight and fruit kernels with 48–52% oil content which is 2–3% of bunch weight.

Crude palm oil (CPO) is the main product of oil palm, which ensures about one-fourth of world's vegetable oil consumption and represents about 36% of world's plant oil production whose market share is expected to remain increasing in the next decades. Malaysia and Indonesia are the two largest producers of this tropical oil, who together account for generally 85% of the world palm oil production. Palm oil is the main traditional cooking oil in the most of the parts of Africa. It is almost the only oil used for domestic purposes in Southeast Asia, equatorial Africa (Choudhary et al. 2014). The oil palm, thus, plays a major role in the economy of these countries by providing employment and wealth. In 2018, worldwide palm oil production is expected to reach 70.8 million metric tons as compared to 67.9 million metric tons produced in the previous year (Statista 2018). The global consumption of this tropical oil reached 62.92 million metric tons in the year 2017/18, up from 57.52 million metric tons in 2012/13 (Fig. 42.1).

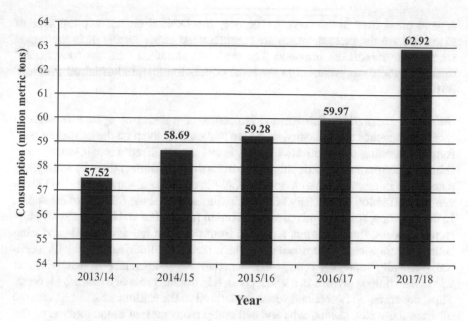

Fig. 42.1 Year wise worldwide consumption of palm oil

2 Extraction and Processing of Fruit Oil

Commercially, there are two types of vegetable oil extracted from the mesocarp of the palm fruit including CPO and palm kernel oil (PKO). Both of these types have a different fatty acid profile, which allows their versatile use in industrial applications. The properties of CPO make it suitable for food products (e.g. margarine, and cooking), while PKO is used for non-food products (e.g. cosmetics). CPO is extracted mechanically by pressing, whereas PKO is extracted chemically by a solvent which involves successive refining operations (Cornet 2001). After harvesting, fruit bunches must be processed within 24 h to prevent oil deterioration by enzymatic hydrolysis. Thereby, only the major producer countries have owned these facilities due to large-scale production.

Conventional Extraction Palm oil is extracted by hand and the pressing is carried through a screw or hydraulic press in the artisanal process. The methods vary greatly, depending on the region, dietary habits and end use of the oil. In the process, the bunches are quartered followed by shedding the fruits immediately or after 2–4 days, heating them in water, crushing and pressing them, collecting the crude oil and clarifying it by prolonged boiling followed by settling. The clarified oil may be boiled again after a second skimming to expel as much water as possible. This process is very laborious and time-consuming; yields poor quality oil as per the international standards. The extraction rate is also lower than in the industrial process. The oil content varies from 10% to 18% by bunch weight depending on the type of press used in the process. The only mechanized elements in this type of extraction are the presses, which are manufactured either locally or in the region whereas the motors are imported. The crude oil extracted from the conventional method is the most preferred to the local population than industrial oil (Verheye 2010).

Industrial Extraction This method is consist of a series of operations such as receiving bunches at the factory, weighing and placing them on the unloading platform and loading the sterilization cages. Fruits are sterilized through steam sterilization to confer an easy stripping of the fruits and to disinfect them by killing any pathogenic microbes and rendering the lipolytic enzymes inactive as well as to prevent the formation of free fatty acids. Stripping separates the fruits from the stalks. In the next step, fruits are mixed and crushed to make a hot and homogeneous pulp. In this process, the pericarp is separated from the palm nut and the oil-producing cells in the mesocarp are crushed well. The mixer is a cylindrical vessel with a central, vertically mounted, rotary shaft wherein the temperature is maintained at 95 °C. The pulp is pressed in a screw press, a hydraulic press or a centrifugal press. Thus, the crude oil is obtained which is clarified in the settling tank to separate the oil from the water, sugars, salts and cell compounds present in the mesocarp. The clarified oil is skimmed off the surface while impurities sink to the bottom. This pure oil is then dehydrated to lower its water content to 0.1–0.2% for better and thus stronger storage (Verheye 2010).

Fractionation Palm oil may be fractionated into two major fractions: palm olein liquid oil (65–70%) and a solid fraction stearin (3035%). The fractionation process consists of crystallization of that portion of the glycerides that solidify at the temperature of operation followed by their separation from the remaining liquid portion. The latter fraction is generally referred to as olein with melting point 1820 °C while the higher melting portion (m.p. 4850 °C) is called stearin. This fractionation increases monounsaturated oleic acid with the simultaneous reduction of palmitic acid, which is the major saturated fatty acid in the palm oil (Koushki et al. 2015).

3 Fatty Acid Composition

The palm oil is made up of major and minor components (Tables 42.1 and 42.2). Triglycerides form the major components of palm oil followed by smaller proportions of diglycerides and monoglycerides. The fatty acids contribute about 95% of the total weight of a triglyceride mole and they are the reactive portion of the molecule. The fatty acids composition (3–5%) of palm oil is made up of a mixture of SFAs, MUFAs and PUFAs (Eqbal et al. 2011). An almost equal amount of saturated and unsaturated fatty acid refers this oil a balanced one. The three main fatty acids of palm oil are palmitic acid (32–47%), oleic acid (40–52%), and linoleic acid (5–7%). The fatty acid composition of palm oil has been presented in Table 42.1.

Table 42.1 Fatty acid composition of palm oil and its fractions

Fatty acid (% by weight)	Palm oil	Palm kernel oil	Palm oil fraction	
			Olein	Stearin
Saturated fatty acids				
Lauric (C12:0)	0.2	47.8	0.1–0.5	0.1–0.6
Myristic (C14:0)	1.1	16.3	0.9–1.4	1.1–1.9
Palmitic (C16:0)	44.0	8.5	37.9–41.7	47.2–73.8
Stearic (C18:0)	4.5	2.4	4.0–4.8	4.4–5.6
Arachidic (C20:0)	0.1	0.1	0.2–0.5	0.1–0.6
Monosaturated fatty acids				
Oleic (C 18:1)	39.2	15.4	40.7–43.9	15.6–37.0
Polyunsaturated fatty acids				
Linoleic (C 18:2)	10.1	2.4	10.4–13.4	3.2–9.8
Linolenic (C 18:3)	0.4	–	0.1–0.6	0.1–0.6

Source: Hilditich and Williams (1964) and Edem (2002)

Table 42.2 Palm oil major and minor components

Component	Percentage
Triglycerides	>90
Diglycerides	2–7
Monoglycerides	<1
Free fatty acids	3–5
Phytonutrients	1
Carotenes	
Phytoene	1
β-Carotene	56
α-Carotene	35
Cis-α-carotene	2
Lycopene	1
Others	4
Sterols	
β-Sitosterol	60
Campesterol	13
Stigmasterol	24
Cholesterol	3
Tocopherols	
α-Tocopherol	21
α -Tocotrienol	23
γ-Tocotrienol	45
δ-Tocotrienol	11

Source: Malaysian palm oil board and International medical university, Malaysia; Jalani and Rajanaidu (2000)

Palm olein, super olein and CPO have the same major fatty acids. However, palm olein and super olein have more oleic and linoleic acids but less palmitic acid than does palm stearin. Both palm olein and super olein are basically more unsaturated fractions of palm oil. Further, palm oil and its fraction contain less than 1.5% of the lauric and myristic acids. In palm oil, 17–23% of the palmitic acid positioned in the sn-2 configuration, while most of the saturated myristic, and stearic acids are in the sn-1,3-configuration which renders a low thrombotic tendency of palm oil (Koushki et al. 2015).

4 Minor Components

The minor components include the free fatty acids and phytonutrients (Table 42.2). The phytonutrients include carotenes, vitamin E (Tocopherol and tocotrienols), phytosterols, phospholipids, squalene, phenolic acids, flavonoids, co-enzymes Q10; chlorophyll, phospholipids, alcohols and sterols (Obahiagbon 2012). Phytonutrients can be obtained as by-products from palm oil mill and refinery and during the process of producing biodiesel from crude palm oil. Additionally, phytonutrients could be obtained as by-products of palm fatty acids distillate. Nutritionally, the most important of the minor components of palm oil are α- and β-carotenes, which are precursors of vitamin A, tocopherols and tocotrienols. Palm oil contains a high concentration of natural carotenoids (500–700 ppm). The main carotenes present in crude palm oil are β-carotene (56%) and α-carotene (35%). The refining process used to obtain palm oil destroys carotenoids present in it but there are some other refining processes available to refine RPO, which does not destroy the carotenoids. Comparatively, palm oil has 15 times more retinol equivalents than carrots and 300 times more than tomatoes. The content of vitamin E is also lost owing to processing methods (Ravigadevi and Yew-Ai 2000). For instance, some studies have reported that palm oil, palm olein and palm stearin retain approximately 69%, 72% and 76% of the original level of vitamin E present in the crude types, respectively. These percentages vary with regard to differences in the plant conditions and the plant design as both factors influence the amount of vitamin E lost during refining. Vitamin E prefers olein fraction during fractionation of palm oil. For example, the concentration of vitamin E in refined, bleached, deodorized palm olein and palm stearin were 104 and 57%, respectively of the original level of vitamin E in the source palm oil (Sundram 2003).

5 Health Benefits

Several studies have demonstrated the nutritional adequacy of palm oil and its products and explained the contributory transitions in the understanding of the nutritional and physiological effects of palm oil, its fatty acids and minor components

(Obahiagbon 2012). Various feeding trials have also threw light on the health-promoting antioxidant properties of RPO due to the presence of phytonutrients including carotenoids, tocopherols, tocotrienols and phospholipids which have anti-oxidant properties (Mancini et al. 2015).

Palm oil helps in either increasing the production of prostacyclin (prevents blood clotting) or decreasing the production of thromboxane (promotes blood-clotting). Holub et al. (1989) have found that the palm oil inhibits platelets aggregation owing to the presence of vitamin E. Other scientific studies have also supported this property of palm oil (Sundram et al. 1990; Ng et al. 1992). The tocotrienols present in palm oil specifically gamma and delta isomers have been found competent in inhibiting cholesterol synthesis in the liver. Studies have proved that the daily consumption of tocotrienols-enriched fraction of palm oil (200 mg) can result in a significant reduction of serum cholesterol, LDL cholesterol, thromboxame, platelet factor 4 and glucose of hyper cholesterolemic subjects within 4 weeks of administration (Song and DeBose-Boyd 2006). Budin et al. (2009) fed streptozotocin-induced diabetic rats with tocotrienol-rich fractions (200 mg/kg body weight) daily for 8 weeks and evaluated their lipid profiles and oxidative stress markers. It was found that tocotrienol-rich fractions group showed significantly lower levels of plasma total cholesterol, LDL cholesterol, and triglyceride, as compared to the untreated group. Hence, the scientific data reveals that the palm oil has cardio-protective properties.

The carotenoids along with other antioxidants synergistically influence the biological antioxidant network by protecting against oxidative damage caused by reactive oxygen species. Being an adequate source of β-carotene, RPO possesses inevitable properties for the treatment of Vitamin A deficiency. β-carotene splits into two vitamin A molecules which further plays a significant role in visual processes, differentiation of cellular epithelium and genetic regulation (Edem 2002). Nursing mothers should take red palm oil as a supplement with their food in order to prevent vitamin A deficiency (Lietz et al. 2000). In India, children (5–10 years old) with keratomalacia were treated twice a day with an emulsion prepared with RPO. Each dose contained 0.6 mL of RPO oil and therapy was continued for 15 days. The RPO treatment compared well with the results obtained by treating another group of keratomalacia patients with cod liver oil containing a similar dose of vitamin A (Obahiagbon 2012).

Phospholipids, the main building blocks of all living cells (including human) are present in RPO and forms an integral part of lipoproteins. The main phospholipids include phosphatidylcholine, phosphatidylinositol, phosphatidylethanolamine and phosphatidyl-glycerol. Several research reports have advocated positive health effects on a human being such as brain development of the growing fetus and lactating mother, improvement in memory and other brain functions, which tend to decline with age, better hormonal status in athletes by reducing the increase of cortisol levels (Starks et al. 2008). Phospholipid fractions and phosphatidylcholine have also been proven to ease digestion, nutrient absorption and lipid transportation (Lochmann and Brown 1997).

The anti-cancer effects of tocopherols and tocotrienols present in palm oil have also been reported pertaining to breast cancer. Results indicated that the tocotrienols fractions were able to induce an inhibitory action on the human breast cancer cells (Nesaretnam et al. 2004). Besides, supplementation of tocols-rich fractions also have anti-diabetic properties which have been achieved through their significant role in recuperating glycemic status, inhibiting oxidative damage; preventing DNA damage and preventing glycosylation of end-products in serum (Budin et al. 2006).

6 Controversies Related to Palm Oil

The principal allegation against palm oil is that it is a highly saturated fat and its consumption allegedly raises the levels of blood cholesterol due to the presence of palmitic acid (Annamaria et al. 2015), thereby increasing the risk of coronary heart disease (Mensah 2008; De Souza et al. 2015). Despite the fact that palm oil is rich in SFAs, the association between CVDs and this oil is controversial. RPO, despite its SFA content (50%), has not appeared to cause atherosclerosis and/or arterial thrombosis in many research studies (Khosla 2006; Ng et al. 1991). A number of human feeding trials reported a 7–38% reduction in blood cholesterol after consuming palm oil diets (Mattson and Grundy 1985; Bonanome and Grundy 1988). Similarly, a study conducted in Australia revealed that young adults fed on refined palm oil (palm olein) had lower values of total blood cholesterol, triglycerides and HDL-cholesterol when compared with the control group consuming an Australian diet (Choudhury et al. 1995). However, other studies reported the undesirable effects of palm oil consumption on levels of apolipoprotein B and LDL-cholesterol in the blood. Many experts consider the best indicator of heart-disease risk- the ratio of total cholesterol to HDL-cholesterol. Palmitic acid increased the total HDL cholesterol ratio more than other SFA such as lauric acid and myristic acid, which are abundant in palm kernel oil (Enig 1993). Palm oil increases the total HDL-cholesterol ratio more than the average US or British dietary fat. As influenced by the recommendations given by the above findings, palm oil is somewhat more harmful than other edible oils.

Secondly, palm oil has also been considered to cause cancer due to the formation of acrylamide at high frying temperatures. Acrylamide is classified as probably carcinogenic to humans by the International Agency for Research on Cancer (IARC) (IARC 1994). Several types of research have reported the harmful effects on human health when exposed to acrylamide (Mottram et al. 2002). When the oil is heated at a temperature above the smoke point (260–290 °C) higher for the oils like palm oil having a higher content of SFAs and lower content of PUFAs, glycerol is degraded to acrolein which is an unpleasant acrid black and irritating smoke (Muchtaridi et al. 2012). Another concern with palm oil consumption is formation of free fatty acid (FFA) content due to lipase activity. Lipase enzyme is present in the mesocarp of the palm fruit which breaks down TGs into FFAs. The acceptable limit of FFA in any healthy oil should be 5% (Corley and Tinker 2003).

7   Edible Application of Palm Oil, Its Properties and Bioactive Compounds Contributing to the Development of Food Formulations and Dietary Supplements

Palm oil has a unique fatty acid composition and antioxidant properties, which makes it suitable for numerous food applications such as frying, baking, and cooking. A competitive price makes palm oil as one of the most utilized oils by food manufacturers. The levels of saturated-unsaturated fatty acids in fats/oils determine the stability of the frying medium against the oxidation. Fats and oils containing a high level of SFAs have more oxidative stability as compare to the fats/oils with high levels of unsaturated fatty acids. Unsaturated fatty acids are prone to oxidation and produce the undesirable smell and toxic compounds due to the breakdown of unsaturated fatty acids into di- and monoacylglycerols, glycerol, and free fatty acids (FFAs) however many health concerns are associated with the consumption of SFAs. Therefore, the oils must possess a high oxidation stability and gumming, high smoke point, the low formation of FFAs, low rate of foaming and darkening and nutritionally good fatty acid composition (Nor Aini and Miskandar 2007; MPOB 2017) to become suitable for frying. Palm oil is the only vegetable oil with the almost equal composition of SFAs and unsaturated fatty acids, thereby it suitable for cooking and frying; especially its liquid fraction palm olein (Matthäus 2007). Products of fried palm oil include potato chips, French fries, doughnuts, ramen noodles, fritters and nuts. The suitability of palm oil and its products for frying has been widely demonstrated in many investigations. Ismail (2005) studied the frying performances of palm oil and palm olein in three major applications including industrial production of potato chips/crisps, industrial production of pre-fried frozen French fries and in fast food outlets. In the first study, about four tons of potato chips were continuously fried 8 h a day and 5 days a week. The palm olein used (with proper management) performed well and was still in excellent condition and usable at the end of the trial. All the quality indicators did not exceed the maximum discard points for frying oils/fats in the three applications, while the fried food product was well accepted by the in-house train sensory panel using a-nine point hedonic score.

In the past decade, there is growing demand to decease SFAs intake and get rid of *trans* fats in our diets owing to their health effects. Several studies have suggested a direct relationship between *trans* fatty acids and increased risk of coronary heart diseases as well as the rise of plasmatic lipid levels. *Trans* fatty acids are developed during hydrogenation of vegetable oils where some *cis* double bonds are rearranged to *trans* bonds to develop a solid fat. Several studies have reported that *trans* fatty acids specifically increase the risk of other chronic health problems such as cancer (Chavarro et al. 2008), diabetes (van-Dam et al. 2002), obesity (Gosline 2006) and infertility (Chavarro et al. 2007). However, palm oil and its hard fractions can serve the purpose thoroughly and may easily replace the hydrogenated fats due to their high solid fat content at ambient temperature that contributes to structural stability, high oxidative stability, easy affordability and availability (Pande and Akoh 2013). Having stable β′ polymorphic form, palm oil has become an excellent choice for the food manufacture involved in the preparation of cooking and baking fats such as margarine and shortenings. The characteristic such as solidification of fats at ambient temperature is the main structuring agent to the texture of numerous food products contributing to hardness, mouth feel and solid-like behavior. Thus, it is impossible to eliminate this solid fat to improve the health aspects of the product, without sacrificing some of the characteristic properties. For example, the triacylglycerol (TAG) profile is responsible for the narrow melting range of chocolate and the broad melting profile of margarine (Rogers 2009). Saturated TAG are used as structural fats to develop a wider range of structuring agent in a high fat food or solid fat formulations such as margarine and shortening (Omonov et al. 2010). That is the reason for an increasing popularity of palm oil as an excellent substitute to the hydrogenated fats in the food industry to produce plastic fats such as palm-based margarine and shortening (Oliveira et al. 2015).

While preparing a good bakery shortening solid fat content (SFC) must range between 15% and 25% at 20 °C and also should have a melting point of >38 °C. Palm oil with SFC of about 22–25% at 20 °C shows a consistency similar to that in a plastic cake shortening. Further, the addition of PKO may improve the performance of palm oil-based shortening. The amount of palm oil is generally 30–40%, but exceptionally up to 80% in shortenings. Blending of palm products with other commercial oils and fats can also produce shortenings that perform well and comparable to the performance of commercial shortenings, which are made of partially hydrogenated liquid oils (Reshma et al. 2008). Excellent performance of palm-based shortening in the development of baked products such as rolled biscuits (Nor Aini and Miskandar 2007), cakes (Ramli et al. 2008), bread (Chin et al. 2010) has also been reported. Besides, PKO is high in lauric acid and has a sharp melting point. Both these factors are desirable for the development of confectionery fats such as chocolate spreads or soft chocolate. Chocolate spread is lipid fractions dispersed with fine solids having 40–44% fat content. The fat content between 28% and 35% is typically found in the low-fat chocolate spread. Furthermore, palm oil, PKO and their fractions are also act as versatile materials with many applications in non-dairy products. Non-dairy products include non-dairy ice-cream, whipped topping, reconstituted filled milk, coffee creamer and cheese. Vegetable oils are commonly used to replace milk fat in dairy products. These products have better keeping properties than those containing milk fat, especially when palm oil is used. Palm oil and PKO are widely used commercially to replace butterfat in ice-cream formulation. The use of palm oil, PKO and their fractions in ice-cream formulation produces ice-cream of good texture, mouth feel and stability especially upon storage. The sharp melting properties of PKO are found to be suitable for ice-cream. Further, cost reduction in ice-cream production can be achieved when palm oil is mixed with 10–15% palm stearin. Palm oil is also an excellent fat to substitute milk fat in the production of cheese without the need of hydrogenation and being free from cholesterol. Palm oil-based yogurt with 5% fat, formulated using a blend of palm oil and PKO received a high sensory score in terms of texture, color, aroma, taste and overall acceptability.

Palm oil and its fraction have also been reported suitable to be used as an animal fat replacer. A study by Hsu and Yu (2002) shows that partial palm oil substitution is suitable for the making of low-fat emulsified meatball (*Kung-was*). Vural and Javidipour (2002) found that interesterified palm oil was able to partially replace beef fat in frankfurters formulation. In addition, Rafidah et al. (2015) have successfully produced vitamin E-enriched palm fat vegetable nuggets. The fraction of palm oil-palm olein has been found to be very suitable for use in infant formulation when blended with other vegetable oils. Palm olein contains 10–15% palmitic acid in the *sn*-2 position of the TAG molecules. This contributes to the high digestibility of the products. Palm oil-based spray cooking oil in a pressurized container offers a non-stick medium for frying. It is specially designed for the preparation of omelet, pancake, spaghetti, bread, macaroni, fried rice and instant fried noodle. It prevents foods from sticking, making it an excellent choice for baking or pan-frying.

The utilisation of the high technology processes such as microencapsulation would further enhance the nutrition and health benefits, and applications of palm oil and its micronutrients in the food industry. Microencapsulation is a method in which tiny particles or droplets of bioactive components are surrounded by a coating or wall material, to form small capsules or microcapsules (Calvo et al. 2011). The microencapsulated palm oil -based oils can be used in many food applications, *e.g.*, in dry food mixes, powdered functional foods or beverages, emulsions, tablets and nutraceutical products. Development of functional food or nutraceutical products from palm micronutrients such as tocotrienol-rich fraction, carotene, squalene and co-enzyme Q10 is also gaining the interest of the manufactures. Dietary supplements are used to ensure that a person gets enough essential nutrients. Palm oil-based dietary supplement such as vitamin E and carotenes have long been commercially available. RPO or palm olein is suitable to be used in many dishes. One tablespoon of red palm olein a day provides enough carotenes and vitamin E for children requirements (Dian et al. 2017). Apart from liquid forms, palm vitamin E and carotenes supplements are also available in the emulsion, soft gel, capsule and tablet forms. Thus, undesirable tropical oil is now emerging as a well-accepted dietary ingredient worldwide.

References

Annamaria, M., Imperlini, N. E., Montagnese, C., Aurora, D., Orrù, S., & Pasqualina, B. (2015). Biological and nutritional properties of palm oil and palmitic acid: Effects on health. *Molecules, 20*, 17339–17361.

Basri, M. W., Maizura, I., Siti Nor Akmar, A., & Norman, K. (2003). Oil palm. In V. L. Chopra & K. V. Peter (Eds.), *Handbook of industrial crops*. New York: The Haworth Press.

Bonanome, A., & Grundy, S. M. (1988). Effect of dietary stearic acid on plasma cholesterol and lipoprotein levels. *The New England Journal of Medicine, 318*, 1244–1248.

Budin, S. B., Rajab, N. F., Osman, K., Top, A. G. M., Mohammed, W. N. W., Baker, M. A., & Mohamed, J. (2006). Effects of pal vitamin E against oxidative damage in Streptozotocin-induced diabetic rats. *Malaysian Journal of Biochemistry and Molecular Biology, 13*, 11–17.

Budin, S. B., Othman, F., Louis, S. R., Bakar, M. A., Das, S., & Mohamed, J. (2009). The effects of palm oil tocotrienol-rich fraction supplementation on biochemical parameters, oxidative stress and the vascular wall of streptozotocin-induced diabetic rats. *Clinics, 64*, 235–244.

Calvo, P., Castano, A. L., Hernandez, M. T., & Gonzalez-Gomez, D. (2011). Effects of microcapsule constitution on the quality of microencapsulated walnut oil. *European Journal of Lipid Science and Technology, 113*, 1273–1280.

Chavarro, J. E., Rich-Edwards, J. W., Rosner, B. A., & Willet, W. C. (2007). Dietary fatty acid intakes and the risk of ovulatory infertility. *The American Journal of Clinical Nutrition, 85*, 231–237.

Chavarro, J. E., Stampfer, M. J., Campos, H., Kurth, T., Willett, W. C., & MA, J. (2008). A prospective study of *trans* fatty acid levels blood and risk of prostate cancer. *Cancer Epidemiology Biomarkers and Preventive, 17*, 95–101.

Chin, N. L., Abdul Rahman, R., Hashim, M. D., & Kowng, S. Y. (2010). Palm oil shortening effects on baking performance of white bread. *Journal of Food Process Engineering, 33*, 413–433.

Choudhary, M., Kaur, J., & Grover, K. (2014). Conventional and no-conventional edible oils: An Indian perspective. *Journal of the American Oil Chemists' Society, 91*(2), 179–206.

Choudhury, N., Tan, L., & Truswell, A. S. (1995). Comparison of palm oil and olive oil effects on plasma lipids and Vit E in young adults. *The American Journal of Clinical Nutrition, 61*, 1043–1061.

Corley, R. H. V., & Tinker, P. B. (2003). *The oil palm* (4th ed.). Oxford: Blackwell Science. Available http://www.blackwellpublishing.com.

Cornet, D. (2001). Oil palm – Elaeis guineensis Jacq. In Raemaekers, R. H. (Ed.), *Crop production in tropical Africa*, Directorate General for International Cooperation (DGIC), Brussels, (pp 769–797).

De Souza, R. J., Mente, A., Maroleanu, A., Cozma, A. I., Ha, V., Kishibe, T., Uleryk, E., Budylowski, P., Schunemann, H., Beyene, J., & Anand, S. S. (2015). Intake of saturated and trans unsaturated fatty acids and risk of all-cause mortality, cardiovascular disease, and type 2 diabetes: Systematic review and meta-analysis of observational studies. *British Medical Journal, 351*, 3978.

Dian, N. L. H. M., Hamid, R. A. B. D., Kanagaratnam, S., Wan Rosnani, A. W. G. I. S. A., Hassim, N., Ismail, N. H., Omar, Z., & Sahri, M. M. (2017). Palm oil and palm kernel oil: Versatile ingredients for food applications. *Journal of Oil Palm Research, 29*(4), 487–511.

Edem, D. O. (2002). Palm oil: Biochemical, physiological, nutritional, hematological, and toxicological aspects: A review. *Plant Foods for Human Nutrition, 57*, 319–341.

Enig, M. G. (1993). Diet, serum cholesterol and coronary heart disease. In G. V. Mann (Ed.), *Coronary heart disease: The dietary sense and nonsense* (pp. 36–60). London: Janus Publishing.

Eqbal, D., Sani, H. A., Abdullah, A., & Kasim, Z. M. (2011). Effect of different vegetable oils (red palm olein, palm olein, corn oil and coconut oil) on lipid profile in rat. *Food and Nutrition Sciences, 2*, 253–258.

Feintrenie, L. (2012). *Transfer of the Asian model of oil palm development: From Indonesia to Cameroon*. Washington, DC: World Bank Conference on Land and Poverty.

Gosline, A. (2006). *Why fast foods are bad, even in moderation*. New Scientist 6:12.

Hilditch, T. P., & Williams, P. N. (1964). *The chemical constitution of natural fats* (4th ed., p. 745). London: Chapman and Hall.

Holub, B. J., Silicilia, F., Mahadevappa, V. G. (1989). Effect tocotrienol derivatives on collagen and ADP-induced human platelet aggregation. In: *Presented at PORIM international palm oil development conference*, Sept 5–6, Kuala Lumpur.

Hsu, S. Y., & Yu, S. H. (2002). Comparisons on 11 plant oil fat substitutes for low-fat kung-wans. *Journal of Food Engineering, 52*, 215–220.

IARC. (1994). *Monographs on the evaluation of carcinogen risk to humans: Some industrial chemicals*. Lyon: International Agency for Research on Cancer.

Ismail, R. (2005). Palm oil and palm olein frying applications. *Asia Pacific Journal of Clinical Nutrition, 14*, 414–419.

Jalani, B. S., & Rajanaidu, N. (2000). Improvement in oil palm: Yield, composition and minor components. *Lipid Technology, 12*, 5–8.

Khosla, P. (2006). Palm oil: A nutritional overview. J agro. *Food Industries, 17*, 21–23.

Koushki, M., Nahidi, M., & Cheraghali, F. (2015). Physico-chemical properties, fatty acid profile and nutrition in palm oil. *Journal Paramedical Sciences, 6*(3), 117–134.

Lietz, G., Henry, C. J. K., Mulokozi, G., Mugyabuso, J., & Balart, A. (2000). Use of red palm oil for promotion of maternal vitamin A status. *Food and Nutrition Bulletin, 21*, 215–218.

Lochmann, R., & Brown, R. (1997). Soyabean-lecithin supplementation of practical diet for juvenile goldfish (Carussius auratus). *Journal of the American Oil Chemists' Society, 74*, 149–152.

Mancini, A., Imperlini, E., Nigro, E., Montagnese, C., Daniele, A., Orrù, S., & Buono, P. (2015). Biological and nutritional properties of palm oil and palmitic acid: Effects on health. *Molecules, 20*, 17339–17361.

Matthäus, B. (2007). Use of palm oil for frying in comparison with other high-stability oils. *European Journal of Lipid Science and Technology, 109*, 400–409.

Mattson, F. H., & Grundy, S. M. (1985). Comparison of effects of dietary saturated, monounsaturated and polyunsaturated fatty acids on plasma lipids and lipoproteins in man. *Journal of Lipid Research, 26*, 194–202.

Mensah, G. A. (2008). Ischaemic heart disease in Africa. *Heart, 94*(7), 836–843.

Mottram, D. S., Wedzicha, B. L., & Dodson, A. T. (2002). Acrylamide is formed in the Maillard reaction. *Nature, 419*, 448–449.

MPOB. (2017). *Pocketbook of palm oil uses* (7th ed.). Bangi: MPOB.

Muchtaridi, M., levita, J., Rahayu, D., & Rahmi, H. (2012). Influence of using coconut, palm, and corn oils as frying medium on concentration of acrylamide in fried Tempe. *Food and Public Health, 2*(2), 16–20.

Nesaretnam, K. R., Ambra, R., Selvaduray, K. R., Radhakrishnan, Reimann, K., Razak, G., & Virgili, F. (2004). Tocotrienol-rich fraction from palm oil affects gene expression in tumors resulting from MCF-7 cell inoculation in mice. *Lipids, 39*, 459–467.

Ng, T. K. W., Hassan, K., Lim, J. B., Lye, M. S., & Ishak, R. (1991). Non hypercholesterolemic effects of a palm-oil diet in Malaysian volunteers. *The American Journal of Clinical Nutrition, 53*, 1015S–1020S.

Ng, T. K. W., Hayes, K. C., de Witt, G. F., Jegathesan, M., Satgunasingham, N., Ong, A. S. H., & Tan, D. T. S. (1992). Palmitic and oleic and exert similar effects on serum lipid profile in the normal-cholesterolemic humans. *Journal of the American College of Nutrition, 11*, 383–390.

Nor Aini, I., & Miskandar, M. S. (2007). Utilization of palm oil and palm oil products in shortenings and margarine. *European Journal of Lipid Science and Technology, 109*, 422–432.

Obahiagbon, F. I. (2012). A review: Aspects of the African oil palm (*Elaeis guineesis* jacq.) and the implications of its bioactives in human health. *American Journal of Biochemistry and Molecular Biology, 2*(3), 106–119.

Oliveira, G. M., Ribeiro, A. P. B., & Kieckbusch, T. G. (2015). Hard fats improve technological properties of palm oil for applications in fat-based products. *LWT-Food Science and Technology, 63*, 1155–1162.

Omonov, T. S., Bouzidi, L., & Narine, S. S. (2010). Quantification of oil binding capacity of structuring fats: A novel method and its application. *Chemistry Physics of Lipids, 163*, 728–740.

Pande, G., & Akoh, C. (2013). Enzymatic modification of lipids for *trans*-free margarine. *Lipid Technology, 25*, 31–33.

Rafidah, A. H., Mohd Burda, F. A., Noor Lida, H. M. D., Shamsudin, S. Y., AWG ISA, W. R., Kanagaratnam, S., & Mat, S. M. (2015). Optimization of vitamin E-enriched palm fat, oat and xanthan gum in a gluten-based nugget formulation. *Journal of Oil Palm Research, 27*, 168–180.

Ramli, N., Syaliza, A. S., & AWG ISA, W. R. (2008). The effect of vegetable fat on the physico-chemical characteristics of dates ice cream. *International Journal of Dairy Technology, 61*, 265–269.

Ravigadevi, S. K. S., & Yew-Ai, T. (2000). Chemistry and biochemistry of palm oil. *Progress in Lipid Research, 39*, 507–558.

Reshma, M. V., Saritha, S. S., Balachandran, C., & Arumughan, C. (2008). Lipase catalyzed inter-esterification of palm stearin and rice bran oil blends for preparation of zero *trans* shortening with bioactive phytochemicals. *Bioresource Technology, 99*, 5011–5019.

Rogers, M. A. (2009). Novel structuring strategies for unsaturated fats – meeting the zero-*trans*, zero saturated fat challenge: A review. *Food Research International, 42*, 747–753.

Song, B. L., & debose-Boyd, R. A. (2006). Insig-dependent ubiquitination and degradation of 3-hydroxy-3-methylglutaryl coenzyme a reductase stimulated by δ and γ tocotrienols. *The Journal of Biological Chemistry, 281*, 25054–25061.

Starks, M. A., Starks, S. L., Kingsley, M., Purpira, M., & Jager, R. (2008). The effects of phospha-tidylserine on endocrine response to moderate intensity exercise. *Journal of the International Society of Sports Nutrition, 5*, 1. https://doi.org/10.1186/1550-2783-5-11.

Statista (2018). Consumption of vegetable oils worldwide from 2013/14 to 2017/2018, by oil type (in million metric tons). Available https://www.statista.com/statistics/263937/vegetable-oils-global-consumption/.

Sundram, K. (2003). Review article palm fruit chemistry and nutrition. *Asia Pacific Journal of Clinical Nutrition, 12*(3), 355–362.

Sundram, K., Khor, H. T., & Ong, A. S. H. (1990). Effect of dietary palm oil and its fractions on rat plasma and high density lipoprotein. *Lipids, 25*, 187–193.

Van Dam, R. M., Stampfer, M., Wille, W. C., II, Hu, F. B., & Rimm, E. B. (2002). Dietary fat and meat intake in relation to risk of type 2 diabetes in men. *Diabetes Care, 25*, 417–424.

Verheye, W. (2010). Growth and production of oil palm. In W. Verheye (Ed.), *Land use, land cover and Soil Sciences* (Encyclopedia of Life Support Systems (EOLSS)). Oxford: UNESCO-EOLSS Publishers. Available http://www.eolss.net.

Vural, H., & Javidipour, I. (2002). Replacement of beef fat in frankfurters by interesterified palm, cottonseed and olive oils. *European Food Research and Technology, 214*, 465–468.

Chapter 43
Rosehip (*Rosa canina* L.) Oil

Mustafa Kiralan and Gurcan Yildirim

Abstract Rosehip (*Rosa canina* L.), is a member of *Rosaceae* family. The seeds of *Rosa canina* contain approximately 15% crude oil. Different extraction techniques are widely used to extract the oils from the seeds. The traditional extraction techniques are classified into two fundamental classes: (I) pressing and (II) solvent extraction method. The modern extraction techniques such as ultrasound, microwave, sub- and supercritical fluid extraction are the other useful methods to extract the oil from rosehip seeds. Rosehip oil is considered a valuable oil because the oil contains essential fatty acids, tocopherols, sterols and phenolics with functional properties. Major essential fatty acids are linoleic, linolenic and oleic acids. Additionally, β-sitosterol is the predominant phytosterol compound. A γ-Tocopherol isomer of tocols is the most abundant in rosehip seed oil. The anti-cancer effect takes the first place among the several health-promoting effects of rosehip oil. Moreover, the rosehip oil is generally preferred to use in cosmetics because of its therapeutic effect on skin disorders.

Keywords Bioactive compounds · Functional properties · Anti-cancer · Cosmetics

1 Introduction (Fruit Oil Recovery and Content, Usages of Oil-Cake, Economy)

The genus Rosa with over 100 species is one of the most widespread members of the Rosaceae family in Europe, Asia, the Middle East and North America. *Rosa canina* L., known as 'Dog Rose' is considered as the valuable sources of polyphenols and vitamin C (Uggla et al. 2003; Tumbas et al. 2012). Besides, the

M. Kiralan (✉)
Faculty of Engineering, Department of Food Engineering, Balıkesir University, Balıkesir, Turkey

G. Yildirim
Department of Mechanical Engineering, Abant Izzet Baysal University, Bolu, Turkey

© Springer Nature Switzerland AG 2019
M. F. Ramadan (ed.), *Fruit Oils: Chemistry and Functionality*,
https://doi.org/10.1007/978-3-030-12473-1_43

tocopherols, essential fatty acids and carotenoids are other important constituents of rosehips (Barros et al. 2011).

Rosehip consists of vitamin C with the level of 262–213 mg/100 g (dry weight basis) depending on the ripening and different part of hips (Barros et al. 2011). Another vital component of the composition in rosehip is the phenolic compounds. The constituent of catechin with the amount of 19.96 µg/g (dry weight) was the major individual phenolic compound. The phenolic acids including the gallic acid, chlorogenic acid, ferulic acid and 2,5-dihydroxy benzoic acid were also found to be considerable amounts (12.67, 12.11, 10.55 and 10.40 µg/g dry weight, respectively) in the rosehip species (Demir et al. 2014). Among the trace elements, the Fe, Cu, Mn and Zn contents in the rosehip seed were recorded to be about 27, 27, 56 and 30 ppm for dry weight (Ercisli 2007). Other constituents present in rosehip include the carbohydrates (93.1 g/100 g of dry weight), proteins (2.72 g/100 g of dry weight), and ash (3.47 g/100 g of dry weight) (Barros et al. 2010).

The functional compounds including vitamin C, phenolics, minerals, essential fatty acids and carotenoids in the seed of rosehip species make rosehips for a valuable ingredient for the preparation of functional foods, pharmaceuticals, and nutraceuticals (Patel 2013). In food technology, the rosehips have been used widely in the herbal teas, jams, marmalades, beverages, jellies, bread for many years (Patel 2017).

The weight of a rosehip ranges from 1.25 to 3.25 g, from which 70% constitutes the pericarp while the seed remains approximately 30%. Generally, the fruit and its inherit pericarp are both commercially available for the industrial applications (Szentmihályi et al. 2002). The products obtained from rosehip fruits can be classified into two main groups: (I) the first parent is derived from the rosehip fruit flesh, whereas (II) the second group of productions is derived and sub-derived from the seed of rosehip fruit. Fresh fruit could be utilized in the production of jellies, jams and infusions while the herbal teas are made from the dried rosehip fruits. It is also possible to encounter the rosehip in the production bakery products as the powder of dried fruit (Mabellini et al. 2011). The usages of seeds can be observed in large amounts as a by-product of the canning factory. In fact, the residuals of rosehip seeds are generally crushed to be used for the animal nutrition or burned (Szentmihályi et al. 2002). In the more recent years, the seeds with functional constituents of essential fatty acids, tocopherols, sterols and phospholipids have been evaluated as an oil source (Zlatanov 1999). The oil of rosehip seeds could be obtained by different extraction techniques including two main groups: (I) conventional extraction and (II) modern extraction. The former extraction techniques (i.e., organic solvents and cold-pressing) have been utilized to extract the oil from the seeds for several years. However, the researchers tend to prefer mostly the modern extraction techniques to obtain higher and higher quality oils. To illustrate in the literature, the extraction method of supercritical carbon dioxide (SC-CO_2) is used to get more quality rosehip oil in modern extraction techniques (Machmudah et al. 2007). Besides, in modern extraction methods the ultrasound, microwave and subcritical fluid extraction routes are applied to obtain higher quality oils. In the modern techniques, the practical consequence is based on the fact that the solvent-free rosehip oils have been used for the medicinal and cosmetic purposes to attract consumers' attention (Del Valle et al. 2000; Szentmihályi et al. 2002).

The oil yield from the rosehip seeds remains at the low level of 6.29 g/100 g DW (Ilyasoglu 2014). Depending on the extraction techniques, the oil yield in the modern extraction methods (microwave extraction, super- and sub-critical fluid extraction) was found to be higher than that in the classical Soxhlet extraction. In case of application of the modern extraction techniques, the oil content was observed to increase up to the 6.68 g/100 g while the oil amount extracted from the conventional Soxhlet method was obtained to remain at the lower level of 4.85 g/100 g (Szentmihályi et al. 2002). Based on the experimental findings, it is natural to confirm that the modern technique with higher oil yield is superior to the conventional one for the oil extraction of seeds.

Rosehip oils are known to be a rich source of unsaturated fatty acids. The levels of linoleic acid (48.6–54.4%), linolenic acid (16.4–18.4%) and oleic acid (14.7–18.4%) are predominant unsaturated fatty acids in the oils gathered from different locations of Turkey (Özcan 2002). It is noteworthy to declare that the rosehip oils rich in unsaturated fatty acids can be used in various applications such as cosmetics and nutraceuticals.

2 Extraction and Processing of Oil (Developments in Extraction and Isolation of High-Value Lipid Compounds, and Phytochemicals)

The valuable components of oil need to be preserved due to the fact the rosehip oils are generally preferred to use in the products for the cosmetic and therapeutic purposes. In this respect, the suitable extraction methods including conventional and particularly modern routes are endeavored to develop day by day. The SC-CO_2 technique, where the various process conditions such as temperature, pressure and CO_2 flow rate enable the researchers to extract sensitively the oils, is one of the alternatives and modern extraction methods. However, here it should strongly be declared that the working conditions provided affect dramatically the individual fatty acid composition of the oil. For example, it was observed that the content of C18:2 significantly increased with the enhancement in the temperature and pressure (up to 300 bar). On the other hand, the condition of CO_2 flow rate seldom affects the content of C18:2 in the oil. As for the other important fatty acid in oil (C18:3), the increment in the process conditions of temperature, pressure and CO_2 flow rate led to the significant increase in the level of C18:3 in the oil (Machmudah et al. 2007). Another study interested in the comparison of traditional and modern methods demonstrated that linoleic and linolenic acids were detected in higher levels in sub- and supercritical fluid extraction (SFE) methods (Szentmihályi et al. 2002). The usages of rosehip oil are favorable in the medicinal (arteriosclerosis, and eczema) and nutritional application areas (Maddocks-Jennings et al. 2005, Chrubasik et al. 2008). Beside to the fatty acids, the fat-soluble compounds such as the carotenes in oil extracted by SFE method including CO_2 and propane were found at a high level

(145.3 µg/g) as compared to those of Soxhlet method (137.5 µg/g) (Szentmihályi et al. 2002). It is to be mentioned that the carotenoids are considered to provide health benefits to prevent some cancer types and eye disease (Johnson 2002).

The cold pressing method is also one of the most widely used techniques to obtain high-quality rosehip oils. Cold pressing hardly damages bioactive compounds such as fatty acid, phenolics, flavonoids and tocols in the oils. However, it was mentioned that the cold pressing provides a low oil yield (Salgın et al. 2016).

3 Fatty Acids Composition and Acyl Lipids (Neutral Lipids, Glycolipids and Phospholipids) Profile of Fixed Oil

To the best of our knowledge, there is only one study on the triglycerides (TAG) of rosehip oils in the literature. In the work conducted by Topkafa (2016), 12 TAG components were identified in the cold pressed rosehip oil where three highest compounds were found to be LLLn with 27.3%, LLnLn with 18.9% and LLL with 15.0%. Besides, the levels of OLL, OLLn, LnLnLn and PLL constituents were observed to be more than 5% of TAG composition. The other minor TAG belonging to the oil extracted were given in Table 43.1.

The fatty acid composition of rosehip seed oil varied according to variety, location and extraction methods. The main fatty acids identified in the oils are unsaturated fatty acids and especially polyunsaturated fatty acids (PUFA) followed by monounsaturated (MUFA) ones. The most abundant PUFA are linoleic acid (C18:2) and linolenic acid (C18:3), while oleic acid (C18:1) is the most commonly identified MUFA.

Three varieties of *Rosa canina* L. (*R. canina* var. *canina*, *R. canina* var. *corymbifera* and *R. canina* var. *dumalis*) oils exhibited different fatty acid compositions.

Table 43.1 Triglyceride composition (%) of rosehip oil (%)

Triglyceride[a]	Content (%)
LnLnLn	5.9
LLnLn	18.9
LLLn	27.3
LLL	15.0
OLLn	8.2
PLLn	2.7
OLL	9.3
PLL	5.5
OOL	3.0
POL + SLL	2.2
OOO	0.9
SOL	0.7

[a]Topkafa (2016)

Table 43.2 Fatty acid composition (relative content %) of *Rosa canina* L. oils

Fatty acid	1	2	3	4	5			6
					R. canina var. *canina*	*R. canina* var. *corymbifera*	*R. canina* var. *dumalis*	
C16:0	3.60–7.87	3.87	5.26	1.71–3.17	3.54	3.23	3.10	17.8
C18:0	2.45–3.27	3.65	3.13	1.69–2.47	2.06	2.05	2.40	8.8
C18:1	16.25–22.11	21.96	22.14	14.71–18.42	17.6	17.17	18.22	52.6
C18:2	35.94–54.75	51.18	48.84	48.64–54.41	55.70	54.55	50.29	2.1
C18:3	20.29–26.48	18.13	20.65	16.42–18.41	18.60	20.24	22.79	1.6
C22:0	–	–	–	0.14–0.64	0.15	0.13	–	8.0

[1]Szentmihályi et al. (2002)
[2]Ercisli et al. (2007)
[3]Kazaz et al. (2009)
[4]Özcan (2002)
[5]Nowak (2005)
[6] Zlatanov (1999), Results are given as g/kg weight

The contents of C18:0, C18:1 and C18:3 were found to be higher whereas the contents of C16:0 and C18:2 were noted to be lower in the oil for *R. canina* var. *dumalis* in comparison with the oils for the other varieties (Nowak, 2005). The differences between the oils gathered from different countries are thoroughly exhibited in Table 43.2. The most abundant fatty acids in rosehip oils belonging to Turkey were linoleic (48.6–54.4%) and oleic acids (14.7–22.1%), followed by the linolenic acid (16.4–20.6%) (Özcan 2002; Ercisli et al. 2007; Kazaz et al. 2009). Interestingly, in the oil provided from the Plovdiv region in South Bulgaria, the major fatty acid was found to be C18:2 with the weight-ratio of 52.6 g/kg, followed by the palmitic acid (C16:0) with the amount of 18.6 g/kg. However, the contents of linolenic and linoleic acids were identified at lower concentration (1.6 and 2.1 g/kg) as compared to those of Turkish rosehip oils. It is another interesting result inferred from this work that the ratios of stearic and behenic acids are obtained to be high level (Zlatanov 1999). Extraction method also is another quantity affecting the fatty acid composition of oils. One can encounter the variation of fatty acid compositions with respect to the extraction method in Table 43.2.

Phospholipids (PL) content of rosehip oil was determined to be 1.4 g/kg weight. The identified individual PL of oils are listed in Table 43.3. According to the table, the major components of phosphatidylcholine (PC), phosphatidylinositol (PI) and phosphatidylethanolamine (PE) are found to be about 46.3, 20.7 and 12.2 g/kg fresh weight, respectively. On the other hand, phosphatidic acid (PA), lysophosphatidylcholine (LPC), lysophosphatidylethanolamine (LPE), phosphatidylserine (PS), mono-phosphatidylglycerol (MPGl) and di-phosphatidylglycerol (DPGL) were identified as the minor PL components of oils. The content of C22:0 was noticed to be higher in the PC with the amount of 26.6 g/kg weight. Furthermore, C18:1 (33.7 g/kg wt), C16:0 (32.7 g/kg wt) and C18:2 (32.1 g/kg wt) were found to predominate in the fatty acids of PE, PI and PA, respectively (Zlatanov 1999).

Table 43.3 Phospholipids
composition of rosehip oil

Phospholipid class[a]	Content (g/kg fresh weight)
Phosphatidylcholine (PC)	46.3
Phosphatidylinositol (PI)	20.7
Phosphatidylethanolamine (PE)	12.2
Phosphatidic acids (PA)	4.3
Lysophosphatidylcholine (LPC)	1.2
Lysophosphatidylethanolamine (LPE)	1.8
Phosphatidylserine (PS)	1.7
Monophosphatidylglycerol (MPGl)	1.2
Diphosphatidylglycerol (DPGL)	0.9
Sphingomieline	0.9

[a]Zlatanov (1999)

4   Minor Bioactive Lipids (Sterols, Tocols, Carotenoids, Phenolic Compounds, Hydrocarbons, Flavor and Aroma Compounds) in the Oil

Besides to acyl lipids, the rosehip oils also contain the minor bioactive lipid components such as the tocols, sterols and carotenoids. Although observed to be the minor compounds in the oils, the levels of components have many functional properties in human health.

Tocopherol content of rosehip oils from Bulgaria was determined as 89.4 mg/kg (Zlatanov 1999). In another study, amount of tocopherol for the cold pressed rosehip oils from Turkey was obtained to be 1124.2 mg/kg (Topkafa 2016). Individual tocopherols of oils are tabulated in Table 43.4. The γ-tocopherol was noted to be the prevailing tocopherol isomer in rosehip oil, showing the contents between 71.0 and 1058 mg/kg weight With the exception of oils from Poland, α-tocopherol with the weight-ratio ranging from 19.0 to 173.4 mg/kg was observed to take the second important place among the tocopherol isomers in the oils. The rosehip oil from Poland was described to be the higher amount for δ-tocopherol (230.4, 259.9 mg/kg) as compared to the oils from the other countries as listed in Table 43.4. It is visible from the table that the Tocotrienol isomers, amount of which were found to vary between the weight-ratios of 0.7 mg/kg and 3.0 mg/kg, were only identified for the oil extracted from Bulgary.

Phytosterols are listed among the important components of unsaponifiable matter of plant oils due to their antioxidant activity and impact on the health (Nogala-Kalucka et al. 2010). The total amounts of sterols are recorded to be 400 mg/kg in Bulgarian rosehip oils (Zlatanov 1999), while the value is determined as 6630 mg/kg in the oils from Turkey (Ilyasoğlu 2014), as against to 5891.6 and 6485.4 mg/kg in the oil from Poland (Grajzer et al. 2015). β-sitosterol predominates in all the sterol fractions, followed by Δ^5-avenasterol. Brassicasterol was only identified at a higher level similar to the value of Δ^5-avenasterol in hexane-extracted oil from

Table 43.4 Tocols composition of rosehip oil (mg/kg)

Tocopherol	Hexane-extracted oil (Bulgary)	Cold-pressed oil (Turkey)	Cold-pressed oil (Poland)	Hexane-extracted oil (Germany)
α-Tocopherol	19.0[a]	57.5[b]	116.6, 147.3[c]	173.4[d]
β-Tocopherol	–	4.6	–	–
γ-Tocopherol	71.0	105.1	630.4, 777.1	895.4
δ-Tocopherol	1.8	4.1	230.4, 259.9	31.2
α-Tocotrienol	3.0	–	–	–
β-Tocotrienol	0.7	–	–	–
γ-Tocotrienol	2.5	–	–	–
δ-Tocotrienol	2.0	–	–	–

[a]Zlatanov (1999)
[b]Topkafa (2016)
[c]Grajzer et al. (2015)
[d]Fromm et al. (2012)

Table 43.5 Sterols composition of rosehip oil

Sterol	Hexane-extracted oil (Bulgary) (mg/kg)[a]		Solvent- extracted oil (Turkey) (mg/100 g)[b]	Cold-pressed oil (Poland) (mg/kg)[c]
	Free	Esterified		
Campesterol	1.8	1.2	23.3	192.3, 205.4
Brassicasterol	5.4	4.6	–	–
Stigmasterol	3.5	1.6	18.9	60.2, 77.9
β-sitosterol	81.5	81.6	544	4753.3, 5297.3
Δ^5-avenasterol	4.6	6.6	31.6	242.4, 379.1
Δ^7-stigmastenol	–	–	41.4	–
$\Delta^{7,25}$-stigmasterol	1.8	tr[d]	–	–
Δ^7-avenasterol	0.9	tr	1.9	37.2, 55.8
Clerosterol	–	–	1.4	–

[a]Zlatanov (1999)
[b]Ilyasoğlu (2014)
[c]Grajzer et al. (2015)
[d]tr: less than 0.1 mg/kg

Bulgary. Unlike the other sterol compounds, Δ^7-stigmastanol and clerosterol are identified in the oil from Turkey (Table 43.5).

Carotenoids, being one of the most important classes of plant pigments, are responsible for the colors of vegetables and fruits and are known to exhibit antioxidant activity (Kaur and Kapoor 2001). The rosehip oils are generally preferred to use in the medicinal applications due to the fact that the antioxidant activity of oils is an important factor to maintain the long-term stability. The total amounts of carotenoid compounds were found to be 36.4–107.7 mg/kg in the oil from Poland (Grajzer et al. 2015) and only 86.3 mg/kg in the oil from Turkey (Ilyasoğlu 2014). It was noted in the literature that the extraction techniques influence the carotenoid content of rosehip oils (Szentmihályi et al. 2002). The SC-CO$_2$ and propane provides more carotenoids (145.3 µg/g) than those extracted from the traditional method of Soxhlet (137.5 µg/g). The oil obtained by Soxhlet extraction contained the carotenoids in higher concentrations than those in the oils obtained by the other methods; for example, ultrasound extraction (59.8 µg/g), microwave extraction (118.7 µg/g) and SFE with CO_2 (45.8 µg/g) (Szentmihályi et al. 2002). In a large-scale study on carotenoids of rosehip oil, the total carotenoid content was 39.15 mg/kg. β-carotene was defined as the major carotenoid (9.28 mg/kg), accounting for approximately 24% of total carotenoids content. All-trans-isomers of lutein, zea-xanthin, and rubixanthin isomers and lycopene were also present in the appreciable amounts (Fromm et al. 2012).

Phenolics exhibiting high antioxidant potential are the important part of minor constituents in edible oils. Total phenolics of rosehip oils are measured to be 215.4 µg/kg in the oil from Turkey, 783.55 and 570.73 µg/kg in two oil samples from Poland (Grajzer et al. 2015). Grajzer et al. (2015) also examined the individual phenolic compounds of rosehip oils. The oils were characterized by high *p*-coumaric acid methyl ester content (391.77–108.32 µg/kg). 4-Hydroxybenzoic acid and methyl ester ferulic acid were only detected in one sample, while the acetovanilon and sinapinic acid were only identified as phenolics in another sample.

Squalene is a triterpene that is a key intermediate in cholesterol biosynthesis and is abundant in the shark liver oil and olive oil (Smith 2000). The content of squalene in rosehip oil was determined as 110.8 and 115.3 mg/kg in the cold-pressed oils from two manufactures in Poland (Grajzer et al. 2015).

5   The Composition of the Essential Oil

It is well-known that the leaves of *Rosa canina* L. are used to get essential oil by means of the different extraction methods. Rosa petals were extracted *via* super-heated water extraction technique at the combination of temperatures of 50 °C, 100 °C, 150 °C and pressures 25, 50 and 75 bar. The experimental results showed that among the applications, the highest extraction yield was observed in the experimental measurement performed at the temperature of 100 °C and pressure of 50 bar, where five major volatile compounds (more than 1 mg/kg sample) including benz-aldehyde, benzyl alcohol, phenyl ethyl alcohol, 2,6,11-trimethyl dodecane and eico-sane were identified (Özel and Clifford 2004). The methods of hydrodistillation and traditional dry distillation were used for the extraction of essential oil from the flowers of rosehips. At the temperature value of 50 °C for hydrodistillation and dry distillation, the main volatile components were 2-phenethyl alcohol and eugenol in the oils. It is to be mentioned here that 2-phenethyl alcohol being from the charac-teristic volatile compounds, contributes the desirable aroma to the rosehip. Moreover, 2-phenethyl alcohol was found at a rather higher concentration level of 58.4% in case of the dry distillation method exerted at the temperature of 50 °C as compared to that (13.6%) extracted from the hydrodistillation technique (Hosni et al. 2010). The development stages of flowers affect directly the essential oil composition, wherein 2-phenethyl alcohol level was noted to increase from 12.1% to 39.3% towards the end of floral development (Hosni et al. 2011).

6 The Contribution of Bioactive Compounds in the Oil to Organoleptic Properties and Functions in the Oil or Functional Foods

The organoleptic properties and health benefits of oils are related to the minor components, hence the interest in the oils has widely increased. The antiradical activity of rosehip oil was measured by the DPPH· method. According to the experimental findings, it was found that the oil presented much stronger antiradical activity with the level of 2.32 mM TEAC/kg than that of other studied cold-pressed oils (Prescha et al. 2014). It was another probable evidence deduced from the work that the rosehip oil containing considerable amounts of carotenoids may lead to exhibit the strong antioxidant activity (Prescha et al. 2014). Beside to the carotenoids, the compounds of tocopherols, squalene, chlorophyll, carotenoids and phenols are abundant in the rosehip oil. This could be attributed to the fact that the high antiradical activity of oil is stemmed from the interactions between the minor compounds. Furthermore, the measurement of antioxidant activity performed by Grajzer et al. (2015) relies on the fact that the rosehip oil exhibited good activity against the oxidation. This could be related to the fact that the rosehip oil exhibits a high antioxidant capacity (Grajzer et al. 2015).

7 Health-Promoting Traits of the Oil and Oil Constituents

In an animal experiment on hamsters, a diet supplemented with 15% rosehip, sunflower, olive, and coconut oils caused to exhibit a hypolipidemic effect in plasma lipids. The oils rich in MUFA and PUFA were found to present similarly to the HDL-cholesterol. The authors suggested that the hypolipidemic effect could be related to the specific action of series n-6 linoleic acid (Gonzalez et al. 1997). In another animal experiment, the rats fed by *ad libitum* diets containing 5% or 15% rosehip oil during 15 or 60 days showed the similar results for the rats fed by triglycerides. However, the plasma total and high-density lipoprotein (HDL) cholesterol levels for the rats fed by the oil including 15% rosehip oil were observed to be higher than the others (Lutz et al. 1993). The different extracts of rosehip (water, methanol, dichloromethane and hexane) were tested *in vitro* conditions. Based on the results, the extracts showed good inhibition of both COX-1 and COX-2 as a consequence of the presence of certain unsaturated fatty acids (Jäger et al. 2007; Wenzig et al. 2008). Moreover, the rosehip seed oil is generally used in the different skin treatments such as eczema, trophic ulcers of the skin, neurodermatitis, and cheilitis (Shabykin and Godorazhi 1967). It was noted that the rosehip oil used together with an oral poly-vitamin preparation of fat-soluble vitamins might exhibit a synergistic effect (Chrubasik et al. 2008).

8 Other Issues (Adulteration and Authenticity)

Rosehip oil is a valuable oil in the pharmaceutical and cosmetic industries. However, the high economic value could be reduced depending on the mixture of other cheaper or lower quality oils. To determine the adulteration of rosehip oils mixed with the other oils, the mid-infrared (MIR) spectroscopy, which is a cheaper method to compared to the conventional methods such as chromatography, mass spectrometry, and nuclear magnetic resonance (NMR), is widely used in the feasible market areas for the universe economy. Additionally, MIR method coupled with a multivariate analysis method (PLS-DA) was found effective to discriminate between the authentic rosehip oil and adulterated rosehip oil containing soybean, sunflower, or corn oil (de Santana et al. 2016).

References

Barros, L., Carvalho, A. M., & Ferreira, I. C. (2011). Exotic fruits as a source of important phytochemicals: Improving the traditional use of *Rosa canina* fruits in Portugal. *Food Research International, 44*(7), 2233–2236.

Barros, L., Carvalho, A. M., Morais, J. S., & Ferreira, I. C. (2010). Strawberry-tree, blackthorn and rose fruits: Detailed characterisation in nutrients and phytochemicals with antioxidant properties. *Food Chemistry, 120*(1), 247–254.

Chrubasik, C., Roufogalis, B. D., Müller-Ladner, U., & Chrubasik, S. (2008). A systematic review on the *Rosa canina* effect and efficacy profiles. *Phytotherapy Research, 22*(6), 725–733.

De Santana, F. B., Gontijo, L. C., Mitsutake, H., Mazivila, S. J., de Souza, L. M., & Neto, W. B. (2016). Non-destructive fraud detection in rosehip oil by MIR spectroscopy and chemometrics. *Food Chemistry, 209*, 228–233.

Del Valle, J. M., Bello, S., Thiel, J., Allen, A., & Chordia, L. (2000). Comparison of conventional and supercritical CO2-extracted rosehip oil. *Brazilian Journal of Chemical Engineering, 17*(3), 335–348.

Demir, N., Yildiz, O., Alpaslan, M., & Hayaloglu, A. A. (2014). Evaluation of volatiles, phenolic compounds and antioxidant activities of rose hip (Rosa L.) fruits in Turkey. *LWT- Food Science and Technology, 57*(1), 126–133.

Ercisli, S. (2007). Chemical composition of fruits in some rose (Rosa spp.) species. *Food Chemistry, 104*(4), 1379–1384.

Ercisli, S., Orhan, E., & Esitken, A. (2007). Fatty acid composition of Rosa species seeds in Turkey. *Chemistry of Natural Compounds, 43*(5), 605–606.

Fromm, M., Bayha, S., Kammerer, D. R., & Carle, R. (2012). Identification and quantitation of carotenoids and tocopherols in seed oils recovered from different Rosaceae species. *Journal of Agricultural and Food Chemistry, 60*(43), 10733–10742.

Gonzalez, I., Escobar, M., & Olivera, P. (1997). Plasma lipids of golden Syrian hamsters fed dietary rose hip, sunflower, olive and coconut oils. *Revista Espanola de Fisiologia, 53*(2), 199–204.

Grajzer, M., Prescha, A., Korzonek, K., Wojakowska, A., Dziadas, M., Kulma, A., & Grajeta, H. (2015). Characteristics of rose hip (*Rosa canina* L.) cold-pressed oil and its oxidative stability studied by the differential scanning calorimetry method. *Food Chemistry, 188*, 459–466.

Hosni, K., Kerkenni, A., Medfei, W., Ben Brahim, N., & Sebei, H. (2010). Volatile oil constituents of *Rosa canina* L.: Quality as affected by the distillation method. *Organic Chemistry International*. vol. 2010, Article ID 621967, 7 pages, 2010. https://doi.org/10.1155/2010/621967.

Hosni, K., Zahed, N., Chrif, R., Brahim, N. B., Kallel, M., & Sebei, H. (2011). Volatile oil constituents of *Rosa canina* L.: Differences related to developmental stages and floral organs. *Plant Biosystems-An International Journal Dealing with all Aspects of Plant Biology, 145*(3), 627–634.

Ilyasoğlu, H. (2014). Characterization of rosehip (*Rosa canina* L.) seed and seed oil. *International Journal of Food Properties, 17*(7), 1591–1598.

Jäger, A. K., Eldeen, I. M., & van Staden, J. (2007). COX-1 and-2 activity of rose hip. *Phytotherapy Research, 21*(12), 1251–1252.

Johnson, E. J. (2002). The role of carotenoids in human health. *Nutrition in Clinical Care, 5*(2), 56–65.

Kaur, C., & Kapoor, H. C. (2001). Antioxidants in fruits and vegetables–the millennium's health. *International Journal of Food Science & Technology, 36*(7), 703–725.

Kazaz, S., Baydar, H., & Erbas, S. (2009). Variations in chemical compositions of *Rosa damascena* Mill. and *Rosa canina* L. fruits. *Czech Journal of Food Sciences, 27*(3), 178–184.

Lutz, M., Torres, M., Carreño, P., & González, I. (1993). Comparative effects of rose hip and corn oils on biliary and plasma lipids in rats. *Archivos Latinoamericanos de Nutrición, 43*(1), 23–27.

Mabellini, A., Ohaco, E., Ochoa, M. R., Kesseler, A. G., Marquez, C. A., & Michelis, A. D. (2011). Chemical and physical characteristics of several wild rose species used as food or food ingredient. *International Journal of Industrial Chemistry, 2*(3), 158–171.

Machmudah, S., Kawahito, Y., Sasaki, M., & Goto, M. (2007). Supercritical CO_2 extraction of rosehip seed oil: Fatty acids composition and process optimization. *The Journal of Supercritical Fluids, 41*(3), 421–428.

Maddocks-Jennings, W., Wilkinson, J. M., & Shillington, D. (2005). Novel approaches to radiotherapy-induced skin reactions: A literature review. *Complementary Therapies in Clinical Practice, 11*(4), 224–231.

Nogala-Kalucka, M., Rudzinska, M., Zadernowski, R., Siger, A., & Krzyzostaniak, I. (2010). Phytochemical content and antioxidant properties of seeds of unconventional oil plants. *Journal of the American Oil Chemists' Society, 87*(12), 1481–1487.

Nowak, R. (2005). Fatty acids composition in fruits of wild rose species. *Acta Societatis Botanicorum Poloniae, 74*(3), 229–235.

Özcan, M. (2002). Nutrient composition of rose (*Rosa canina* L.) seed and oils. *Journal of Medicinal Food, 5*(3), 137–140.

Özel, M., & Clifford, A. A. (2004). Superheated water extraction of fragrance compounds from *Rosa canina*. *Flavour and Fragrance Journal, 19*(4), 354–359.

Patel, S. (2013). Rose hips as complementary and alternative medicine: Overview of the present status and prospects. *Mediterranean Journal of Nutrition and Metabolism, 6*(2), 89–97.

Patel, S. (2017). Rose hip as an underutilized functional food: Evidence-based review. *Trends in Food Science & Technology, 63*, 29–38.

Prescha, A., Grajzer, M., Dedyk, M., & Grajeta, H. (2014). The antioxidant activity and oxidative stability of cold-pressed oils. *Journal of the American Oil Chemists' Society, 91*(8), 1291–1301.

Salgın, U., Salgın, S., Ekici, D. D., & Uludağ, G. (2016). Oil recovery in rosehip seeds from food plant waste products using supercritical CO2 extraction. *The Journal of Supercritical Fluids, 118*, 194–202.

Shabykin, G. P., & Godorazhi, A. I. (1967). A polyvitamin preparation of fat-soluble vitamins (carotolin) and rose hip oil in the treatment of certain dermatoses. *Vestnik Dermatologii i Venerologii, 41*(4), 71–73.

Smith, T. J. (2000). Squalene: potential chemopreventive agent. *Expert Opinion on Investigational Drugs, 9*(8), 1841–1848.

Szentmihályi, K., Vinkler, P., Lakatos, B., Illés, V., & Then, M. (2002). Rose hip (*Rosa canina* L.) oil obtained from waste hip seeds by different extraction methods. *Bioresource Technology, 82*(2), 195–201.

Topkafa, M. (2016). Evaluation of chemical properties of cold pressed onion, okra, rosehip, safflower and carrot seed oils: Triglyceride, fatty acid and tocol compositions. *Analytical Methods 8*, (21), 4220–4225.

Tumbas, V. T., Čanadanović-Brunet, J. M., Četojević-Simin, D. D., Ćetković, G. S., Đilas, S. M., & Gille, L. (2012). Effect of rosehip (*Rosa canina* L.) phytochemicals on stable free radicals and human cancer cells. *Journal of the Science of Food and Agriculture, 92*(6), 1273–1281.

Uggla, M., Gao, X., & Werlemark, G. (2003). Variation among and within dogrose taxa (*Rosa sect. caninae*) in fruit weight, percentages of fruit flesh and dry matter, and vitamin C content. *Acta Agriculturae Scandinavica, Section B-Plant Soil Science, 53*(3), 147–155.

Wenzig, E. M., Widowitz, U., Kunert, O., Chrubasik, S., Bucar, F., Knauder, E., & Bauer, R. (2008). Phytochemical composition and in vitro pharmacological activity of two rose hip (*Rosa canina* L.) preparations. *Phytomedicine, 15*(10), 826–835.

Zlatanov, M. D. (1999). Lipid composition of Bulgarian chokeberry, black currant and rose hip seed oils. *Journal of the Science of Food and Agriculture, 79*(12), 1620–1624.

Chapter 44
Date (*Phoenix dactylifera* L.) Seed Oil

Monia Jemni, Sofien Chniti, and Said Saad Soliman

Abstract The date palm (*Phoenix dactylifera* L.) production is the principal activity and the source of life of peoples of arid and semiarid regions of the world. The production of dates is increasing every season, but losses during harvesting and postharvest handling and marketing are also high due to the incidence of physical and physiological disorders and pathological diseases and to insect infestation. In addition, there is an expansion of exportation of pitted dates. In consequence, a big biomass of date seed is produced. This biomass presents a problem to the station of conditioning of dates. Actually, date seeds are used in animal feeding or in making non-caffeinated coffee. However, date seeds can be used for many others applications such as the production of oil. In fact, date seeds contain 10–12% of the oil. This later could be used in cosmetic, pharmaceutical and food products.

Keywords *Phoenix dactylifera* · Date seed · Lipids · Date palm · Transformation

1 Introduction

The date (*Phoenix dactylifera* L.) fruit has an exceptional nutritional, biochemical and physicochemical characteristic. Dates are an important part of the diet of many countries and are consumed fresh or at various processed forms. Date palm has a genetic diversity in the world. For example, in Tunisia there are more than 250

M. Jemni (✉)
Laboratoire Technologies de dattes, Centre Régional de Recherches en Agriculture Oasienne
Degueche, Tozeur, Tunisie
e-mail: monia.jemni@iresa.agrinet.tn

S. Chniti
Ecole des Métiers de l'Environnement, Campus de Ker Lann, Bruz, France

Université de Rennes 1, ENSCR, CNRS, Rennes, France

S. S. Soliman
Department of Horticultural Crops Technology, National Research Centre,
Cairo, Dokki, Egypt

© Springer Nature Switzerland AG 2019

M. F. Ramadan (ed.), *Fruit Oils: Chemistry and Functionality*,
https://doi.org/10.1007/978-3-030-12473-1_44

varieties of dates, but the variety *Deglet Nour* predominates in a number of trees and production.

The production of dates is increasing every season, but losses during harvesting; postharvest handling and marketing are very high due to the incidence of physical and physiological disorders, pathological diseases and insect infestation. These losses can achieve 30% of the total production in Tunisia (GIFruits 2018). For that, this biomass is a big source of seeds. In addition, the seeds of date palm are a waste product of many industries, after pitted dates and technological and biotechnological transformation of date fruits. Subsequently, a large quantity of date seeds could be easily collected from the date processing industries or from the losses dates (Basuny and Al-Marzooq 2011; Soliman et al. 2015). Until now, date seeds are used for animal feeds or roasted and ground for production of a caffeine-free drink, which can substitute coffee (Soliman et al. 2015). In fact, many studies have been focused on the chemical composition of date seeds and they proved that date seeds contain high levels of valuable bioactive compounds and dietary fiber (Tafti et al. 2017). Also, date seeds contain crude oil. For that, the oil extraction from the date seeds and their use in cosmetics and pharmaceuticals products can be applied (Basuny and Al-Marzooq 2011; Tafti et al. 2017). This work will highlight the methods of oil extraction and the physical-chemical proprieties of date seeds oil.

2 Presentation of Date's Sector

Date palm (*Phoenix dactylifera* L.) of the family Arecaceae is a key plantation crop of many countries in arid regions of West Asia and North Africa. It is thought to have originated in Mesopotamia (what is now Iraq) and its cultivation spread to the Arabian Peninsula, North Africa, and the Middle Eastern Countries in ancient times (about 5000 years ago) (Kader and Hussein 2009) (Fig. 44.1). Almost every part of this tree is used and its food and industrial products play an important role in the rural communities and economies of many developing countries. Dates are produced in the hot arid regions of the world and marketed worldwide as high-value confectionary (Mahmoudi et al. 2008).

The world production of dates was increased from 6.770 million tons in 2006 to 8.460 million tons in 2016 (Fig. 44.2), distributed in 4 Continents with 60.5% for Asia, 39% for Africa, 0.4% for America and 0.2% for Europe (FAOSTAT 2018a, b). The top ten producing countries are Egypt, Iran, Algeria, Saudi Arabia, United Arab Emirates, Iraq, Pakistan, Sudan, Oman, and Tunisia (Table 44.1). The exportation of dates in the world reached 825,271 tons in 2013 (FAOSTAT 2018a, b). There are thousands of date palm cultivars, including those with soft, semi-dry, and dry fruits (depending on their water content at harvest when fully-ripe), grown in these countries (Kader and Hussein 2009).

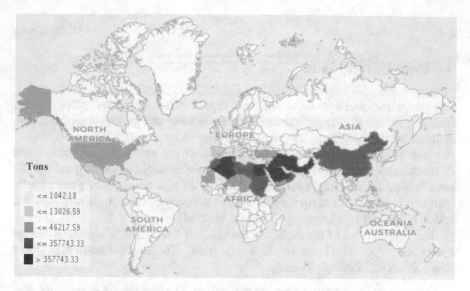

Fig. 44.1 Distribution of *Phoenix* species (2000–2016) (FAOSTAT 2018b)

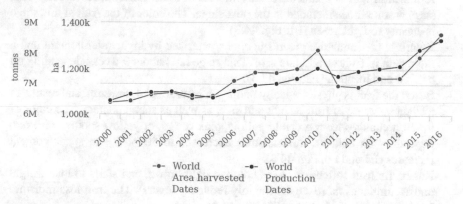

Fig. 44.2 Evolution of world production and area harvested of dates between 2000 and 2016 (FAOSTAT 2018a)

Table 44.1 World production of dates (tons) in 2016 (FAOSTAT 2018a)

Pays	Tons
Egypt	1,694,813
Iran	1,065,704
Algeria	1,029,596
Saudi Arabia	964,536
United Arab Emirates	671,891
Iraq	615,211
Pakistan	494,601
Sudan	439,120
Oman	348,642
Tunisia	241,000
Libya	173,546
China, mainland	159,144
Morocco	125,329
Autres	437,311

3   Fruit Development

The growth and development of date palm fruit involve several physical, physiological, and chemical changes starting with pollination and culminating at harvest. Pollination is one of the most important pre-harvest factors affecting fruit quality of dates. In a commercial plantation, the female trees are artificially pollinated (hand or mechanical pollination) with pollen from male trees. Selection of good a pollinizer is of prime importance in the date palm, as the type of the pollen parent affects fruit size and time of fruit ripening, as well as the chemical composition of the fruit, which is referred to as metaxenia (Lobo et al. 2014).

The development of the fruit is classified into five stages using Arabic terms: *Hababouk*, *Kimri* (also known as *Khimri* or *Jimri*), *Khalal* (also known as *Balah* or *Bisr*), *Rutab* and *Tamr* (or *Tamar*).

- Hababaouk: is the week of female pollination. The Hababaouk stage starts after fertilization and is characterized by the loss of two unfertilized carpels. This stage is sometimes included in the next stage. The color of the fruit at this stage is creamy to light green (F1, Fig. 44.3).
- Kemri: is the immature green stage, characterized by high water content and a rapid gain in fruit weight and size. This stage lasts about 9 weeks depending on cultivar and location (F2, Fig. 44.3).
- Beser: the fruit is physiologically mature; it lasts about 4–5 weeks, and results in a slight decrease in fruit weight and size, as well as starch content. The color of the fruit changes from green to yellow, pink or red, or yellow spotted with red, depending on cultivar, the moisture decreases to 50–60% and sugar content increases (F5 and F6, Fig. 44.3).
- Rutab: the fruit softens, changes color to light brown, and starts to lose weight and accumulates more sugars (mainly reducing sugars). The fruit loss moisture (moisture level of 35–40%) (F7, Fig. 44.3).
- Tamar: The Tamar is the fully-ripe stage of development, loses more moisture and gains more sugars, thus attaining a high sugar-to-water ratio (depending on cultivar). Most dates are harvested at the Tamar stage, when the fruit has about

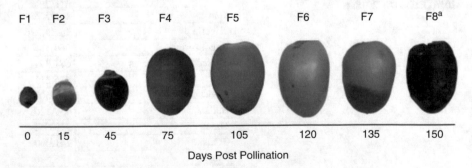

| F1 | F2 | F3 | F4 | F5 | F6 | F7 | F8[a] |

| 0 | 15 | 45 | 75 | 105 | 120 | 135 | 150 |

Days Post Pollination

Fig. 44.3 The ripening stages of a date fruit by days post-pollination (Yin et al. 2012)

60–80% sugar content, depending on location and cultivar; color darkens, which is also marked by low moisture content (10–25%). At this stage, fruit can be harvested soft, semi-dry or dry depending on destination and use. This stage lasts 2–4 weeks and the dates are appropriate for long-term dry storage or processing. Dates can also develop parthenocarpically if not pollinated. However, these fruits will not undergo the five stages described above and will not reach full development (F8, Fig. 44.3) (Al-Khalifah et al. 2014; Lobo et al. 2014).

4 Classification of Dates

At Tamar stage, which is that of optimum harvesting, the varieties of dates can be classified based on several criteria. Indeed, dates are generally elongated oblong or ovoid, however there are some substantially spherical varieties, including *Tinteboucht* of Algeria. Their dimensions vary considerably: from 1.5 to 7 or 8 cm in length and a weight of 2–15 g. Their color ranges from yellowish white to very dark brown, almost black, through amber, red and brown (Lobo et al. 2014).

The varieties of dates are classified and listed, in general according to the sugar composition. This is the main criteria in composition. We distinguish dates with sucrose (sucrose dominant) and dates with reducing sugars (glucose and fructose dominant) (Lobo et al. 2014). According to Coggins et al. (1967), the relative contents of sucrose and reducing sugar, characterizes the variety and influences the texture of the date. A classification was proposed by Hussein et al. (1976) and Barreveld (1993), based on the ratio R = sugar/water, which distinguishes three categories of dates:

- Soft dates R <3.5: generally dates with reducing sugars. These dates do not contain sucrose and having a water content of about 30%.
- Semi-soft dates 2.5 < R <3.5: which have a water content of about 20–30%, with a large concentration of sucrose (in the case of the *Deglet Nour*).
- Dried dates R >3.5: which contain less than 20% water, and which generally have an equivalent amount of reducing sugars and sucrose.

5 Composition of Dates

The fruit of date palm is well known as a stable food. It is composed of a fleshy pericarp and seed as presented in Fig. 44.4 (Basuny and Al-Marzooq 2011). The chemical composition of dates can vary, depending on cultivar, soil conditions, agronomic practices, and the ripening stage. Date fruits pass through several separate stages of maturity, traditionally described by changes in color, texture and taste/flavor. Immature green dates (Kimri stage) are firm in texture, with the highest moisture and tannin contents. At the mature full colored (Besser stage), dates begin

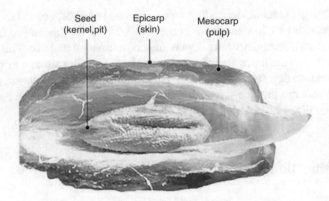

Fig. 44.4 The anatomy of the date fruit at Tamar stage (Ghnimia et al. 2017)

Table 44.2 Composition of the pulp of dates (g/100 g dry weight) (Borchani et al. 2010; Abdul Rahman Assirey 2015)

Component	Dry matter (%)	Total sugars	Ash	Protein	Fat
Alligh	82.94 ± 0.70	84.59 ± 0.18	2.18 ± 0.22	1,22 ± 0.03	0.56 ± 0.19
Deglet Nour	86.42 ± 0.75	88.02 ± 0.60	1.78 ± 0.10	1.71 ± 0.08	0.40 ± 0.11
Bajo	86.88 ± 0.59	79.9 ± 0.31	1.73 ± 0.04	1.28 ± 0.08	0.11 ± 0.04
Boufeggous	88.70 ± 0.68	86.72 ± 0.95	1.58 ± 0.05	1.51 ± 0.16	0.14 ± 0.00
Goundi	90.57 ± 0.37	84.79 ± 0.91	1.85 ± 0.03	2.85 ± 0.20	0.35 ± 0.21
Ikhouat	87.97 ± 0.40	78.86 ± 0.33	2.59 ± 0.52	0.66 ± 0.03	0.07 ± 0.00
Kenta	88.22 ± 0.79	85.11 ± 0.46	1.75 ± 0.02	0.90 ± 0.02	0.06 ± 0.01
Kentichi	87.29 ± 0.18	77.44 ± 0.26	1.74 ± 0.05	0.46 ± 0.01	0.11 ± 0.04
Lagou	73.10 ± 0.60	77.31 ± 0.15	2.08 ± 0.02	1.83 ± 0.05	0.25 ± 0.00
Touzerzaillet	70.66 ± 0.38	78.58 ± 0.77	2.11 ± 0.19	1.49 ± 0.05	0.57 ± 0.04
Tranja	87.85 ± 0.55	83.95 ± 0.35	2.23 ± 0.09	2.42 ± 0.85	0.14 ± 0.07
Ajwa	77.2 ± 0.1	74.3 ± 0.2	3.43 ± 0.01	2.91 ± 0.02	0.47 ± 0.001
Shalaby	84.8 ± 0.2	75.9 ± 0.5	3.39 ± 0.01	4.73 ± 0.01	0.33 ± 0.005
Khodari	80.5 ± 0.1	79.4 ± 0.3	3.42 ± 0.04	3.42 ± 0.03	0.18 ± 0.004
Anabarah	70.5 ± 0.2	78.4 ± 0.2	2.33 ± 0.01	3.49 ± 0.01	0.51 ± 0.004
Sukkari	78.8 ± 0.1	78.5 ± 0.1	2.37 ± 0.05	2.76 ± 0.01	0.52 ± 0.001
Suqaey	85.5 ± 0.1	79.7 ± 0.2	2.29 ± 0.03	2.73 ± 0.04	0.41 ± 0.005
Safawy	76.4 ± 0.3	75.3 ± 0.1	1.68 ± 0.01	2.48 ± 0.02	0.12 ± 0.003
Burni	75.6 ± 0.1	81.4 ± 0.04	2.02 ± 0.01	2.50 ± 0.04	0.67 ± 0.001
Labanah	98.5 ± 0.1	71.2 ± 0.1	3.94 ± 0.02	3.87 ± 0.05	0.72 ± 0.002
Mabroom	78.7 ± 0.1	76.4 ± 0.07	2.79 ± 0.05	1.72 ± 0.05	0.27 ± 0.001

to lose moisture and form considerable quantities of sucrose. In the soft brown (Rutab stage), the loss of moisture is accelerated, the fruits become softer in texture, and sucrose is converted into invert sugars. Dates at the Rutab stage are the most desirable ones since they are at their softest and sweetest. In the final mature Tamar stage, characterized by its good storability, the fruits contain the least amount of moisture and maintain a soft texture and a sweet taste (El Arem et al. 2011). The composition of date's varieties is given in Table 44.2.

According to Borchani et al. (2010), the major sugars found in the date flesh samples were fructose, glucose and sucrose. Sucrose was the major sugar of many varieties such as *Deglet Nour* (51.14 g/100 g dry matter), *Bajo* (38.64 g/100 g dry matter) and *Kentichi* (37.35 g/100 g dry matter), whereas many other cultivars were essentially formed by reducing sugars (glucose and fructose). The richness of these varieties in reducing sugars suggests the existence of a more pronounced invertase activity, which would considerably reduce its content in sucrose.

Dates have a relatively low content of protein. For that, dates were not considered as a good source of protein. Protein content ranged from 0.46 g/100 g to 4.73 g/100 g dry matter (Table 44.2) (Borchani et al. 2010; Abdul Rahman Assirey 2015). Date flesh samples of the different cultivars of dates presented very low-fat contents. It is ranged between 0.06 and 0.72 g/100 g dry matter (Table 44.2). Lipids are mainly concentrated in the skin and have a higher physiological importance in the protection of fruit than contributing to the nutritional value of the date flesh (Borchani et al. 2010; Abdul Rahman Assirey 2015).

6   Chemical Composition of Date Seed

Many researchers have investigated the chemical composition of the different date seeds varieties. According to Besbes et al. (2004a), Al Juhaimi et al. (2012), and Bouhlali et al. (2017), the seeds of some date's varieties are composed by total sugar which ranged between 1.776% and 83.1% (Table 44.3). The protein content of the date seed is relatively high in comparisons with the date flesh. It is ranged between 3.7% and 6.14%. In addition, the same note is observed for the fat content. It is varied between 4.68% for *Khalas* and 12.67% for *Allig*.

Table 44.3 Date seed composition (% dry matter basis) (Besbes et al. 2004; Al Juhaimi et al. 2012; Bouhlali et al. 2017)

	Protein	Lipid	Ash	Moisture	Total sugar
Deglet Nour	5.56 ± 0.02	10.17 ± 0.11	1.15 ± 0.02	9.4 ± 0.18	83.1 ± 0.33
Allig	5.17 ± 0.78	12.67 ± 0.26	1.12 ± 0.05	8.6 ± 0.09	81.0 ± 0.91
Boufagous	4.31 ± 0.19	6.76 ± 0.17	1.30 ± 0.08	7.43 ± 0.34	8.71 ± 0.09
Bousthammi	5.12 ± 0.20	6.97 ± 0.46	1.27 ± 0.04	4.55 ± 0.07	8.70 ± 0.07
Majhoul	6.14 ± 0.07	5.66 ± 0.81	1.09 ± 0.05	8.26 ± 0.27	9.55 ± 0.10
Soughi	4.32 ± 0.00	5.05 ± 0.00	1.17 ± 0.00	9.01 ± 0.00	2.006 ± 0.00
Monaif	5.11 ± 0.00	7.96 ± 0.00	1.05 ± 0.00	7.87 ± 0.00	2.053 ± 0.00
Soulag	3.71 ± 0.00	5.25 ± 0.00	1.15 ± 0.00	8.36 ± 0.00	2.044 ± 0.00
Soukari	5.18 ± 0.00	5.51 ± 0.00	1.26 ± 0.00	8.24 ± 0.00	2.03 ± 0.00
Barhi	5.47 ± 0.00	4.74 ± 0.00	1.13 ± 0.00	7.87 ± 0.00	2.01 ± 0.00
Khalas	4.71 ± 0.00	4.68 ± 0.00	1.07 ± 0.00	8.11 ± 0.00	1.776 ± 0.00
Rozaiz	4.38 ± 0.00	7.30 ± 0.00	1.03 ± 0.00	8.33 ± 0.00	1.933 ± 0.00

Tafti et al. (2017) reported that the composition difference between different cultivars is due to the variety, origin, harvesting time, fertilizer and climatic condition. According to the same authors, the carbohydrate and fat are the main components of date seeds.

7   Oil Extraction

The extraction of oil from date seed is done generally by Soxhlet apparatus after seed grinding. There are many solvents used in extraction. Many researchers have used petroleum ether as a solvent for oil extraction by Soxhlet as reported by Besbes et al. (2004a), Habib et al. (2013), and Herchi et al. (2014). But, there are many others who have used *n*-hexane (Soliman et al. 2015; Nehdi et al. 2018). Ali et al. (2015) have tested four solvents: methanol, 2-propanol, chloroform, *n*-hexane and toluene to determine the best solvent for maximum oil extraction yield. They found that palm seed oil yield is better with non-polar solvents (like chloroform, *n*-hexane and toluene) as compared to those with polar solvents (like methanol and 2-propanol). It was observed that toluene produce a high yield of oil as compared to *n*-hexane solvent because of its high boiling point. Chloroform is not appropriate, because of the low oil yield. Methanol and 2-propanol are not very suitable for the recovery of oil probably because of its polar character. As a result for the work of Ali et al. (2015), the optimal solvent used for oil extraction is *n*-hexane because of it is less costly and its yield is comparable with that of the toluene.

Ali et al. (2015) have also tested the effect of seed size (2 mm, 1 mm and 0.425 mm) and the time of extraction process (1 h, 2 h, 4 h and 6 h) on the yield of oil. They have been noticed that the decrease in particle size leads to an increase of oil yield because of the increased surface area of grounded seed and the increase of the contacted area between seed and solvent. This result was confirmed by Bouallegue et al. (2015). In addition, Ali et al. (2015) have been observed that the difference of oil concentration between the solid phase and solvent phase is greater in the initial extraction process. The oil yields unchanged even after prolongation of the time of extraction process. For that, they concluded that the optimal conditions to obtain the highest oil yield were 120 min, 0.425 mm and *n*-hexane extracted time, the particle size of grounded seed and type of solvent, respectively.

Jamil et al. (2016) have worked also on the effect of time, temperature and ratio solvent: the seed of extraction by Soxhlet on the oil yield. They found that temperature was the most significant parameter on extraction yield. In addition, the combined effects of temperature and solvent to seed ratio and temperature and time were found to be effective for enhancing the yield of oil.

Bouallegue et al. (2015) have tested the effect of texturing pretreatment on the oil yield. They have been observed a positive impact of texturing by instant controlled pressure drop on the oil extraction yields. Hence, the optimal parameters of instant controlled pressure drop texturing were saturated steam pressure P = 0.8 MPa and thermal treatment time t = 70 s.

Ben-Youssef et al. (2017) have tested the effect of the extraction method on the oil yield. They have compared the conventional Soxhlet method with the ultrasound extraction, the microwave oven and the maceration. They found that the extraction of oils from date seed with ultrasound and microwave were better in terms of extraction time comparing to Soxhlet and of extraction yield comparing to maceration.

8 Physicochemical Characteristic of Date Seed Oils

According to Table 44.4, the acid value of seeds oil ranges between 1.06 mg KOH/g for *Deglet Nour* seed oil and 2.10 mg KOH/g for *Allig* seed oil. These results indicate that *Allig* seed oil contains a high amount of free fatty acids than *Deglet Nour*, *Boufgous*, *Bousthammi* and *Majhoul* seeds oil which has the similar free fatty acids content of olive oil that means that it could be edible (Besbes et al. 2004b; Bouhlali et al. 2017).

The peroxide value of date seed oil is ranged between 10.1 and 25 meq O_2/kg (Table 44.4). The highest value was observed for *Allig*. Bouhlali et al. (2017) explained the variation of peroxide value between cultivars to the degree of unsaturation of the fatty acids present in the particular oil, storage, exposure to light, and the content of metals or other compounds that may catalyze the oxidation processes. In addition, they considered the date seeds oil as safe for human consumption, because of its low peroxide value that was less than 30 meq O_2/kg.

The iodine value is ranged between 44.1 and 58.02 g/100 g (Table 44.4). *Bousthammi* seed oil has the highest value. Bouhlali et al. (2017) explained that the higher the unsaturated fatty acids content in the seed oil, the higher the iodine value is. The iodine value gives a measure of the average degree of unsaturation of a lipid. Besbes et al. (2004b) found that the iodine value of date seed oil is lower than those for olive oil, almond oil, sesame oil, soybean oil and sunflower oil and they explained this result by the fact that date seed oil contained less polyunsaturated fatty acids (PUFA).

Table 44.4 Quality indices of date seeds oil (Besbes et al. 2004; Bouhlali et al. 2017)

Quality indices	Deglet Nour	Allig	Boufgous	Bousthammi	Majhoul
Unsaponifiable matter (g/kg)	8.92 ± 0.39	7.76 ± 0.34	1.103 ± 0.062	0.62 ± 0.03	0.827 ± 0.05
Acid value (mg KOH/g oil)	1.06 ± 0.06	2.10 ± 0.14	1.69 ± 0.04	1.083 ± 0.055	1.813 ± 0.035
Peroxide value (meq O_2/100 g)	16.0 ± 0.5	25.0 ± 0.4	10.1 ± 0.3	12.43 ± 0.68	10.43 ± 0.05
Iodine value (g /100 g)	45.5 ± 0.2	44.1 ± 0.3	45.40 ± 1.06	58.02 ± 2.16	50.343 ± 1.066

9 Fatty Acid Composition of Date Seeds Oil

As shown in Tables 44.5, 44.6 and 44.7, the fatty acid content of date seed oil varies slightly with the seeds varieties. According to Habib et al. (2013) and Besbes et al. (2004a), there are 16 fatty acids present in the seed oil of studied varieties. These fatty acids were subdivided into three groups including (i) saturated fatty acids composed of lauric (C12:0), myrsitic (C14:0), pentadeconic (C15:0), palmitic (C16:0), heptadecanoic (C17:0), stearic (C18:0), arachidic (C20:0) and behenic (C22:0), (ii) monounsaturated fatty acid (MUFA) composed of palmitoleic (C16:1), cis-10-heptadeconic (C17:1), oleic (C18:1) and cis-11-eicosenic (C20:1), and (iii) PUFA composed of linoleic (C18:2) and linolenic (C18:3).

Table 44.5 Saturated fatty acid of date seed oil (g/100 g date seed oil) (Besbes et al. 2004; Habib et al. 2013; Bouhlali et al. 2017)

	C10:0	C12:0	C14:0	C15:0	C16:0	C17:0	C18:0	C20:0	C22:0
Deglet Nour	0.80 ± 0.13	17.8 ± 0.60	9.84 ± 0.09	–	10.9 ± 0.17	–	5.67 ± 0.20	–	–
Allig	0.07 ± 0.01	5.81 ± 0.25	3.12 ± 0.06	–	15.00 ± 0.31	–	3.00 ± 0.03	–	–
Boufgous	–	16.74 ± 0.51	10.23 ± 0.23	–	10.91 ± 0.19	0.04 ± 0.01	3.73 ± 0.12	0.46 ± 0.02	0.27 ± 0.01
Bousthammi	–	17.02 ± 0.37	12.28 ± 0.14	–	10.65 ± 0.09	0.01 ± 0.02	2.86 ± 0.06	0.52 ± 0.05	0.21 ± 0.06
Majhoul	–	20.34 ± 0.41	11.85 ± 0.11	–	9.82 ± 0.12	0.02 ± 0.01	2.96 ± 0.09	0.43 ± 0.03	0.25 ± 0.05
Khalas	–	–	13.72 ± 0.02	0.04 ± 0.00	12.52 ± 0.02	0.11 ± 0.00	3.48 ± 0.01	0.44 ± 0.00	nd
Barhe	–	–	14.02 ± 0.01	0.25 ± 0.01	11.72 ± 0.01	0.03 ± 0.00	3.05 ± 0.00	0.31 ± 0.00	0.01 ± 0.00
Lulu	–	–	14.72 ± 0.1	0.17 ± 0.01	11.41 ± 0.02	0.17 ± 0.01	2.80 ± 0.00	0.31 ± 0.00	0.02 ± 0.00
Sokkery	–	–	15.33 ± 0.01	0.67 ± 0.01	12.95 ± 0.06	0.17 ± 0.00	3.31 ± 0.01	0.33 ± 0.00	0.03 ± 0.00
Bomaan	–	–	13.73 ± 0.02	1.22 ± 0.01	13.32 ± 0.01	0.14 ± 0.00	3.83 ± 0.00	0.39 ± 0.00	nd
Sagay	–	–	13.24 ± 0.02	0.27 ± 0.02	12.13 ± 0.02	0.04 ± 0.00	3.08 ± 0.01	0.29 ± 0.00	nd
Maghool	–	–	14.73 ± 0.01	0.73 ± 0.01	12.01 ± 0.01	0.08 ± 0.00	3.52 ± 0.00	0.31 ± 0.00	nd
Maktoomi	–	–	12.73 ± 0.01	0.10 ± 0.00	11.31 ± 0.01	0.09 ± 0.00	2.70 ± 0.00	0.29 ± 0.00	nd
Khodary	–	–	17.92 ± 0.01	0.22 ± 0.00	13.34 ± 0.03	0.08 ± 0.00	2.78 ± 0.01	0.29 ± 0.00	nd
Raziz	–	–	14.72 ± 0.01	0.08 ± 0.01	11.51 ± 0.01	0.10 ± 0.00	2.52 ± 0.00	0.29 ± 0.01	nd

Table 44.6 Monounsaturated fatty acid content of date seed oil (g/100 g date seed oil) (Besbes et al. 2004; Habib et al. 2013; Bouhlali et al. 2017)

	C14:1	C16:1	C17:1	C18:1	C20:1
Deglet Nour	0.09 ± 0.15	0.11 ± 0.19	–	41.3 ± 0.76	–
Allig	0.04 ± 0.03	1.52 ± 0.01	–	47.7 ± 1.11	–
Boufgous	0.01 ± 0.01	0.04 ± 0.01	–	48.38 ± 0.92	0.40 ± 0.04
Bousthammi	0.07 ± 0.01	0.07 ± 0.01	–	46.29 ± 1.17	0.37 ± 0.02
Majhoul	0.03 ± 0.02	0.09 ± 0.03	–	44.92 ± 1.53	0.26 ± 0.04
Khalas	–	0.13 ± 0.01	0.07 ± 0.00	55.10 ± 0.01	0.43 ± 0.00
Barhe	–	0.13 ± 0.02	0.08 ± 0.00	48.01 ± 0.01	0.49 ± 0.00
Lulu	–	0.13 ± 0.00	0.05 ± 0.00	45.60 ± 0.00	0.38 ± 0.00
Sokkery	–	0.13 ± 0.01	0.07 ± 0.00	44.60 ± 0.00	0.56 ± 0.00
Bomaan	–	0.16 ± 0.00	0.07 ± 0.00	45.90 ± 0.01	0.35 ± 0.00
Sagay	–	0.12 ± 0.00	0.06 ± 0.00	42.31 ± 0.01	0.35 ± 0.01
Maghool	–	0.13 ± 0.00	0.06 ± 0.00	47.30 ± 0.00	0.34 ± 0.00
Maktoomi	–	0.15 ± 0.00	0.07 ± 0.00	44.40 ± 0.00	0.37 ± 0.00
Khodary	–	1.14 ± 0.01	0.07 ± 0.00	50.30 ± 0.00	0.41 ± 0.00
Raziz	–	0.11 ± 0.00	0.09 ± 0.00	52.30 ± 0.00	0.41 ± 0.00

Table 44.7 Polyunsaturated fatty acid of date seed oil (g/100 g date seed oil) (Besbes et al. 2004; Habib et al. 2013; Bouhlali et al. 2017)

	C18:2	C18:3
Deglet Nour	12.2 ± 0.5	1.63 ± 0.71
Allig	21.0 ± 0.29	0.81 ± 0.38
Boufgous	8.30 ± 0.08	0.09 ± 0.01
Bousthammi	9.02 ± 0.07	0.15 ± 0.002
Majhoul	8.47 ± 0.10	0.21 ± 0.03
Khalas	8.88 ± 1.73	0.09 ± 0.00
Barhe	10.30 ± 0.00	0.11 ± 0.00
Lulu	9.45 ± 0.00	0.05 ± 0.00
Sokkery	10.20 ± 0.00	0.14 ± 0.0
Bomaan	7.90 ± 0.00	nd
Sagay	13.90 ± 0.01	0.18 ± 0.00
Maghool	11.50 ± 0.00	0.04 ± 0.00
Maktoomi	10.05 ± 0.00	0.03 ± 0.00
Khodary	8.33 ± 0.00	0.10 ± 0.00
Raziz	10.40 ± 0.00	0.09 ± 0.00

Besbes et al. (2004a) reported that the most abundant fatty acids of date seed oil of *Deglet Nour* et *Allig* were oleic (C18:1), linoleic (C18:2), palmitic (C16:0), myristic (C14:0), and lauric (C12:0) which together composed about 92% of the total fatty acids. The major fatty acid found in those cultivars was oleic acid, ranging from 41.3% for *Deglet Nour* seed oil to 47.7% for *Allig* seed oil. *Deglet Nour* seed oil was regarded as an oleic-lauric oil, while *Allig* seed oil was an oleic-linoleic oil.

On the other hand, Bouhlali et al. (2017) reported that among the fatty acids detected in *Boufgous*, *Bousthammi* and *Majhoul* seeds, oleic was the predominant

(44.92–48.38%) followed by lauric (16.74–20.34%), myristic (10.23–12.28%), palmitic (9.82–10.91%), linoleic (8.3–9.02%) and stearic (2.86–3.73%) acids which composed together more than 98% of the total oil. The degree of unsaturation of analyzed date seed oil ranged between 53.98% and 57.23% for *Boufgous* and *Bousthammi* seeds oil, respectively.

Allig and *Bousthammi* seed oil showed a lower amount of saturated fatty acid accounted for 26.3% and 42.79%, respectively, but a higher amount of unsaturated fatty acid accounted for 71% and 57.21% respectively, which makes them more sensitive to oxidation. The highest amount of saturated fatty acid was reported in *Deglet Nour* (44.3%) and *Boufgous* seed oil (46.03%), which contain the lowest amount of unsaturated fatty acid accounted for 45.45% and 53.97%, respectively. The level of unsaturated fatty acid content of date seed oil was lower than other vegetable oils (Besbes et al. 2004a; Bouhlali et al. 2017).

According to Besbes et al. (2004a) and Nehdi et al. (2018), the date seed oil is characterized by the presence of five major fatty acids (C18:1, C18:2, C16:0, C14:0 and C12:0). Nehdi et al. (2018) reported that oleic fatty acid (C18:1) was always the most abundant in date seed oil for that date seed oil can be regarded as monounsaturated.

10 Tocopherol

The different isomers of vitamins E contained in seed oils from different date cultivars are α-tocopherol, β-tocopherol, γ-tocopherol, δ-tocopherol, α-tocotrienol, γ-tocotrienol, and δ-tocotrienol (Nehdi et al. 2018). The average tocol content of the date seed oils was 70.75 mg/100 g, which is higher than that of olive and peanut oils (23.39 and 66.73 mg/100 g, respectively) (Nehdi et al. 2018). According to Biglar et al. (2012) and Habib et al. (2013), α-tocopherol was the predominant component in date seed oil. But, Nehdi et al. (2018), found that γ-tocopherol was the predominant component. In fact, it was observed a positive correlation between tocopherol content in oil and seed cultivars. The content of oil in α-tocopherol was ranged between 10.09 and 33.86 μg vitamin E/g oil for 14 Iranian date seed oils (Biglar et al. 2012). The results reported by Habib et al. (2013) for 18 varieties of date seed oil from the United Arab Emirates were between 1.01 and 1.86 mg/100 g date seed oil. However, Nehdi et al. (2018) found levels higher for α-tocopherol in six Saudi Arabian cultivars ranged between 9.18 and 15.17 mg/100 g oil.

According to Habib et al. (2013), vitamin E is an essential antioxidant, which protects PUFA from free radical damage, thus contributing to the stability of the oil. Moreover, vitamin E deficiency affects nervous system development in children and hemolysis in humans. People with low vitamin E intake may be at increased risk of atherosclerosis.

11   Carotenoids Composition

Habib et al. (2013) have investigated the carotenoid profiles of date seed oil from 18 varieties from the United Arab Emirates and they proved that lutein, cryptoxanthine, echinenone, β -carotene and γ-carotene were found in all date seeds varieties, whereas lycopene and α-carotene were not detected in all date seeds. These authors explained this difference in composition by differences between cultivars, the degree of ripeness, latitude, environmental conditions, processing techniques and storage conditions.

   The composition of the total carotenoid pigment content of the oils is an important quality parameter because they correlate with color, which is a basic attribute for the evaluation of the quality of the oil. Date seed oil has a very intense yellow color in comparison with other vegetable oils. This suggests the presence of a significant quantity of the yellow pigment, carotenoids. This pigment is responsible for the absorption of ultraviolet radiations (Besbes et al. 2004a; AL Juhaimi et al. 2012).

12   Uses of Oil

Monounsaturated oleic acid is the most distributed natural fatty acid in date seed oils and olive oil, wherein high oleic oils normally have positive health aspects because of their low saturated fatty acid levels, minimal trans-isomer levels, potential to reduce LDL cholesterol in the blood and the risk of cardiovascular disease. Liquid oils with high oleic acid contents normally have good flavor and frying stability for that date seed oil can be used as cooking oil and as frying oil. Moreover, the low level of linoleic acid (C18:2) make date oil relatively stable to oxidative deterioration and their presence in oil is indispensable for the healthy growth of human skin (Habib et al. 2013; Nehdi et al. 2018). Furthermore, lauric acid (C12:0) has a more favorable effect on the total cholesterol/HDL cholesterol ratio than any other fatty acid, either saturated or unsaturated. Myristic acid has the greatest effect on serum cholesterol levels (Nehdi et al. 2018).

   Basuny and Al-Marzooq (2011) have used seed oil of khalas variety to replace conventional oil in producing mayonnaise. The data demonstrated that mayonnaise containing date pit oil was superior in sensory characteristics as compared with control manufactured from corn oil.

   Date seed oil can shield against UV-B and UV-A radiation. Thus, date seed oil may be used in the formulation of UV protectors that provide protection against both UV-A and UV-B. The optical transmission of date seed oil was comparable to that of raspberry seed oil, especially in the UV range of 290–400 nm (Besbes et al. 2004a). In addition, it contains natural antioxidants, carotenoids (β-carotene), and some phytochemicals which may consider as an essential oil (Tafti et al. 2017).

   Azeem et al. (2016) and Jamil et al. (2016) have done assays of production of biodiesel from different varieties of dates: *Khadravi*, *Zahidi*, *Basri* and local dates of Muscat. They observed a low chain fatty acids present in date seed oil which make it ideal for the production of biodiesel. For that, date seed oil is a potential source of lipids for the production of the biodiesel, and a diversified source of energy for countries with arid agricultural lands.

References

Abdul Rahman Assirey, E. (2015). Nutritional composition of fruit of 10 date palm (*Phoenix dactylifera* L.) cultivars grown in Saudi Arabia. *Journal of Taibah University for Science, 9*, 75–79.

AL Juhaimi, F., Ghafoor, K., & Özcan, M. M. (2012). Physical and chemical properties, antioxidant activity, total phenol and mineral profile of seeds of seven different date fruit (*Phoenix dactylifera* L.) varieties. *International Journal of Food Sciences and Nutrition, 63*(1), 84–89.

Ali, M. A., Al-Hattab, T. A., & Al-Hydary, I. A. (2015). Extraction of date palm seed oil (*Phoenix Dactylifera*) by soxhlet apparatus. *International Journal of Advances in Engineering & Technology, 8*(3), 261–271.

Al-Khalifah, N.S., Askari, E., & Shanavaskhan, A.E. (2014). *Date palm tissue culture and genetical identification of cultivars grown in Saudi Arabia*. King Abdulaziz City for Science and Technology. King Fahad National Library Cataloging-in-Publication Data, 264p, ISBN: 978-603-8049-45-7.

Azeem, M. W., Hanif, M. A., Al-Sabahi, J. N., Khan, A. A., Naz, S., & Ijaz, A. (2016). Production of biodiesel from low priced, renewable and abundant date seed oil. *Renewable Energy, 86*, 124–132.

Barreveld, W. H. (1993). Date palm product. *FAO Agricultural Bulletin, 101*, 1–216.

Basuny, A. M. M., & Al-Marzooq, M. A. (2011). Production of mayonnaise from date pit oil. *Food and Nutrition Sciences, 2*, 938–943.

Ben-Youssef, S., Fakhfakh, J., Breil, C., Abert-Vian, M., Chemat, F., & Allouche, N. (2017). Green extraction procedures of lipids from Tunisian date palm seeds. *Industrial Crops and Products, 108*, 520–525.

Besbes, S., Blecker, C., Deroanne, C., Lognay, G., Drira, N. E., & Attia, H. (2004a). Quality characteristics and oxidative stability of date seed oil during storage. *Food Science and Technology International, 10*(5), 333–338.

Besbes, S., Blecker, C., Deroanne, C., Drira, N. E., & Attia, H. (2004b). Date seeds: chemical composition and characteristic profiles of the lipid fraction. *Food Chemistry, 84*, 577–584.

Biglar, M., Khanavi, M., Hajimahmoodi, M., Hassani, S., Moghaddam, G., Sadeghi, N., & Oveisi, M. R. (2012). Tocopherol content and fatty acid profile of different Iranian date seed oils. *Iranian Journal of Pharmaceutical Research, 11*(3), 873–878.

Borchani, C., Besbes, S., Blecker, C., Masmoudi, M., Baati, R., & Attia, H. (2010). Chemical properties of 11 date cultivars and their corresponding fiber extracts. *African Journal of Biotechnology, 9*(26), 4096–4105.

Bouallegue, K., Allaf, T., Besombes, C., Ben Younes, R., & Allaf, K. (2015). Phenomenological modeling and intensification of texturing/grinding-assisted solvent oil extraction: case of date seeds (*Phoenix dactylifera* L.). *Arabian Journal of Chemistry*. Article in press.

Bouhlali, E. D. T., Alem, C., Ennassir, J., Benlyas, M., Mbark, A. N., & Zegzouti, Y. F. (2017). Phytochemical compositions and antioxidant capacity of three date (*Phoenix dactylifera* L.) seeds varieties grown in the South East Morocco. *Journal of the Saudi Society of Agricultural Sciences, 16*, 350–357.

Coggins, C. W., Lewis, L. N., & Knapp, J. C. F. (1967). Progress report chemical and histological studies of tough and tender "Deglet-Noor" dates. *Physical, Chemical Date Growers Institute Annals Report, 44*, 15.

El Arem, A., Flamini, G., Saafi, E. B., Issaoui, M., Zayene, N., Ferchichi, A., Hammami, M., Helal, A. N., & Achour, L. (2011). Chemical and aroma volatile compositions of date palm (*Phoenix dactylifera* L.) fruits at three maturation stages. *Food Chemistry, 127*, 1744–1754.

FAOSTAT. (2018a). *Data base.* Available on line: http://www.fao.org/faostat/en/#data/QC/ Accessed May 2018.

FAOSTAT. (2018b). *Distribution of the Phoenix species (2000–2016).* Available on line: http://www.fao.org/faostat/en/#data/QC/visualize/ Accessed May 2018.

Ghnimia, S., Umera, S., Karim, A., & Kamal-Eldin, A. (2017). Date fruit (*Phoenix dactylifera* L.): An underutilized food seeking industrial valorization. *NFS Journal, 6*, 1–10.

Gifruits. (2018). *Groupement interprofessionnel des fruits.* http://www.gifruit.nat.tn/ Accessed May 2018.

Habib, H. M., Kamal, H., Ibrahim, W. H., & Al Dhaheri, A. S. (2013). Carotenoids, fat soluble vitamins and fatty acid profiles of 18 varieties of date seed oil. *Industrial Crops and Products, 42*, 567–572.

Herchi, W., Kallel, H., & Boukhchina, S. (2014). Physicochemical properties and antioxidant activity of Tunisian date palm (*Phoenix dactylifera* L.) oil as affected by different extraction methods. *Food Science and Technology, 34*, 464–470.

Hussein, F., Mostafa, S., El-Samiraea, F., & Al-Zeid, A. (1976). Studies on physical and chemical characteristics of eighteen date cultivars grown in Saudi Arabia. *Indian journal Horticulture, 33*, 107–113.

Jamil, F., Al-Muhtaseb, A. H., Al-Haj, L., Al-Hinai, M. A., Hellier, P., & Rashid, U. (2016). Optimization of oil extraction from waste "Date pits" for biodiesel production. *Energy Conversion and Management, 117*, 264–272.

Kader, A. A., & Hussein, A. (2009). *Harvesting and postharvest handling of dates.* Aleppo: ICARDA. 15 pp. ISBN: 92-9127-213-6.

Lobo, M. G., Yahia, E. M., & Kader, A. A. (2014). Biology and postharvest physiology of date fruit. In M. Siddiq, S. M. Aleid, & A. A. Kader (Eds.), *Dates, postharvest science, processing technology and health benefits* (1st ed., pp. 57–80). Boca Raton: John Wiley & Sons Ltd., Chap 3.

Mahmoudi, H., Hosseininia, G., Azadi, H., & Fatemi, M. (2008). Enhancing date palm processing, marketing and pest control through organic culture. *Journal Organic System, 3*(2), 29 39.

Nehdi, I. A., Sbihi, H. M., Tan, C. P., Rashid, U., & Al-Resayes, S. I. (2018). Chemical composition of date palm (*Phoenix dactylifera* L.) seed oil from six Saudi Arabian cultivars. *Journal of Food Science, 83*(3), 624–630.

Soliman, S. S., Al-Obeed, R. S., & Ahmed, T. A. (2015). Physic-chemical characteristics of oil produced from seeds of some date palm cultivars (*Phoenix dactylifera* L.). *Journal of Environmental Biology, 36*, 455–459.

Tafti, A. G., Dahdivan, N. S., & Ardakani, S. A. Y. (2017). Physicochemical properties and applications of date seed and its oil. *International Food Research Journal, 24*(4), 1399–1406.

Yin, Y., Zhang, X., Fang, Y., Pan, L., Sun, G., Xin, C., Abdullah, M. M., Yu, X., Hu, S., Al-Mssallem, I. S., & Yu, J. (2012). High-throughput sequencing-based gene profiling on multi-staged fruit development of date palm (*Phoenix dactylifera* L.). *Plant Molecular Biology, 78*(6), 617–626.

Chapter 45
Celastrus paniculatus Oil

Mohamed Fawzy Ramadan

Abstract *Celastrus paniculatus* fruit seeds are a rich source of lipids (approx. 45%). Neutral lipids accounted for the highest lipid fraction, followed by glycolipids and phospholipids. Oleic, palmitic and linoleic acids were the predominant fatty acids in *C. paniculatus* fruit seed oil. *Celastrus paniculatus* fruit seed oil is characterized by a high amount of plant sterols, wherein β-sitosterol, campesterol and stigmasterol were the main sterols. γ-Tocopherol was the main tocopherol isomer while the rest was α-tocopherol. *Celastrus paniculatus* fruit seed oil exhibited stronger radical scavenging potential toward the stable DPPH (2,2-diphenyl-1-picrylhydrazyl) radical when compared with and extra virgin olive oil. This chapter reviews the composition, functionality and applications of *Celastrus paniculatus* fruit seed oil.

Keywords Fruit seed oil · Lipid classes · Lipid-soluble bioactives · Antiradical activity

1 Introduction (Fruit Oil Recovery and Content, Uses of Oil Cake, and Economy)

Non-conventional fruit seed oils are used in the healthcare industry, both in the informal and professional versions of health care. An aromatherapy or oil therapy is a natural alternative to the many conventional pharmaceuticals. Ayurveda, a traditional Indian medicinal system, employs several medicinal and aromatic plants in the treatment of numerous diseases (Ramadan et al. 2009).

Celastrus genus includes approx. 100 species of shrubs, distributed in tropical Asia, Japan, China, and North America. Four species confined to India of which *Celastrus paniculatus* Willd. (family Celastraceae) is a well-known Ayurvedic 'Medhya Rasayana' (nervine tonic) (Malik et al. 2017). *C. paniculatus*, a plant

M. F. Ramadan (✉)
Agricultural Biochemistry Department, Faculty of Agriculture, Zagazig University, Zagazig, Egypt
e-mail: mframadan@zu.edu.eg

© Springer Nature Switzerland AG 2019
M. F. Ramadan (ed.), *Fruit Oils: Chemistry and Functionality*,
https://doi.org/10.1007/978-3-030-12473-1_45

831

known as the "Elixir of life", is a climbing shrub, large woody, height up to 18 m and common to all over the hilly parts of India. Fruits are yellow when matured and seeds are reddish brown covered with scarlet aril (Godkar et al. 2004, 2006; Ramadan et al. 2009; Bhagya et al. 2016; Palle et al. 2018).

C. paniculatus plant exhibited therapeutic values in the treatment of cognitive dysfunction, epilepsy, insomnia, rheumatism, gout and dyspepsia (Russo et al. 2001; Godkar et al. 2004). According to Indian Ayurveda, *C. paniculatus* could be employed as a stimulant nervine tonic, rejuvenate, sedative and diuretic (Vaidyaratnam 1997; Kumar and Gupta 2002). *C. paniculatus* methanol extract showed antiradical effects, and was capable of reducing hydrogen peroxide (H_2O_2)-induced cytotoxicity and DNA damage in human non-immortalized fibroblasts (Russo et al. 2001). *C. paniculatus* is commonly utilized as a neuro-protective, and in several central nervous system disorders (Godkar et al. 2006; Malik et al. 2017). In folk medicine, *C. paniculatus* has been used for the prevention and treatment of various diseases and gastrointestinal disturbances, including dyspepsia and stomach ulcers (Palle et al. 2018). In addition, *C. paniculatus* plant was used to treat cognitive deficits in mentally retarded children. *C. paniculatus* fruit seed oil has been exhibited neuroprotective and antioxidant activities (Bhagya et al. 2016; Ramadan et al. 2009).

The fruit seeds extracted with petroleum ether yield dark brown oil known as Malkanguni oil or *Celastrus* oil (Ramadan et al. 2009). Early reports on *C. paniculatus* fruit seed oil stated that oil contains palmitic, stearic, oleic, linoleic and linolenic acids (Sengupta and Bhargava 1970; Sengupta et al. 1987). The oil also contains sesquiterpene alkaloids *viz.,* celapanin, celapanigin and celapagin (CSIR 1999) which exhibited analgesic and anti-inflammatory (Ahmad et al. 1994) and antianxiety (Jadhav and Patwardhan 2003) activities. *C. paniculatus* fruit seed oil enhanced the learning and memory due to decreasing biogenic amines turnover or antioxidant impact. *C. paniculatus* fruit seed oil also acted as a powerful stimulant for the treatment of body and rheumatic pains, scabies, wound eczema and beriberi (Vaidyaratnam 1997; Kapoor 2005). In addition, *C. paniculatus* fruit seed oil showed pharmacological actions such as anti-malarial (Ayudhaya et al. 1987) and anti-spermatogenic (Wangoo and Bidwai 1988), analgesic (Ahmad et al. 1994), and anxiolytic activity (Rajkumar et al. 2007).

Few studies reported the lipids profile of *C. paniculatus* fruit seed oil (Sengupta and Bhargava 1970; Sengupta et al. 1987; Ramadan et al. 2009; Rana and Das 2017). In this chapter, chemistry, functionality and uses of *C. paniculatus* fruit seed oil is highlighted.

2 Fatty Acids Composition and Acyl Lipids of *C. paniculatus* Fruit Seed Oil

The hexane-extractable crude lipids from *C. paniculatus* fruit seeds were reported to be approx. 46%, which confirms *C. paniculatus* fruit seeds are a rich source of lipids (Ramadan et al. 2009). Rana and Das (2017) determined the yield and composition

Table 45.1 Fatty acids, tocopherols and plant sterols profile of *C. paniculatus* fruit seed oil

Fatty acid	Relative content (%)	Tocol	g/kg	Sterol	g/kg
C14:0	0.40	α-Tocopherol	0.52	Campesterol	1.66
C16:0	26.1	γ-Tocopherol	1.04	Stigmasterol	1.44
C18:0	2.75			β-Sitosterol	4.90
C18:1	54.2			Δ5-Avenasterol	0.15
C18:2	11.2			Δ7- Stigmastenol	0.43
C18:3n-3	5.35			Δ7-Avenasterol	0.22

of the *C. paniculatus* fruit seed oil. In the mature seed, the yield of crude lipids was 45.5%. The oil was rich in unsaturated fatty acids (70.1%), while saturated fatty acids accounted for 25.2%. Oleic acid (54.4%), palmitic acid (20.0%), linoleic acid (15.5%), and stearic acid (4.18%) were the main fatty acids in the oil. The study identified novel compounds not previously reported in the literature including 1,4-benzenediol (0.46%), cinnamic acid (0.15%), 2,6-di-tert-butyl-p-benzoquinone (0.03%), butylated hydroxytoluene (0.4%), and eudalene (0.16%).

Six fatty acids were identified in *n*-hexane extract of *C. paniculatus* seeds (Ramadan et al. 2009). Oleic, palmitic and linoleic acids were the main fatty acids, which comprising together more than 90% of total fatty acids (Table 45.1). α-Linolenic acid (ALA, C18:3n-3) was also detected in relatively high amounts. Sengupta and Bhargava (1970) stated that *C. paniculatus* oil mainly contains palmitic (31.2%), stearic (3.5%), oleic (22.5%), linoleic (15.7%), and linolenic acids (22.2%). *C. paniculatus* seed oil was characterized by a high level (54%) of monounsaturated fatty acids (MUFA). MUFA have been shown to lower "bad" LDL cholesterol (low-density lipoproteins) yet retain "good" HDL cholesterol (high-density lipoproteins). The fatty acid composition and high amounts of MUFA make the *C. paniculatus* fruit seed oil a special raw material for functional products applications.

3 Minor Bioactive Lipids of *C. paniculatus* Fruit Seed Oil

C. paniculatus fruit seed oil was characterized by a y high level of unsaponifiable matter (16 g/kg TL), of which approx. 55% plant sterols. Six plant sterols were measured in *C. paniculatus* fruit seed oil (Table 45.1). The sterol marker was β-sitosterol which accounted for about 64% of the total plant sterols content, followed by campesterol and stigmasterol. Δ5-Avenasterol, Δ7-avenasterol and Δ7-stigmastenol were detected in lower amounts (Ramadan et al. 2009). Plant sterols, are of interest due to their antioxidant potential and health-promoting impacts (Yang et al. 2001).

Two of the four tocopherol isomers were identified in *n*-hexane extract of *C. paniculatus* seeds (Ramadan et al. 2009). γ-Tocopherol constituted 66.6% of the total tocopherols, while α-tocopherol accounted for 33.3% (Table 45.1). α- and γ-Tocopherols recognized to be the main tocols in the edible oils. Levels of tocopherols detected in *C. paniculatus* fruit seed oil may contribute to the stability of the oil against oxidation.

4   Composition of *C. paniculatus* Fruit Seed Essential Oil

Arora and Pandey-Rai (2014) studied the chemical composition of the *C. paniculatus* essential oil using chromatography-mass spectrometry. The essential oil yield was 0.09% (v/w), wherein 56 constituents were detected comprising of 99.2% of the total oil. The major component was palmitic acid (38.6%), followed by phytol (11.7%), erucic acid (6.99%), *trans*-beta-copaene (4.78%) and linalool (3.97%).

5   Contribution of Bioactive Compounds in *C. paniculatus* Fruit Seed Oil to Organoleptic Properties and Functions

The antiradical action of *C. paniculatus* fruit seed oil and extra virgin olive oil (EVOO) were compared using stable DPPH·free radicals (Ramadan et al. 2009). *C. paniculatus* fruit seed oil exhibited stronger antiradical action than EVOO. After 60 min of incubation, 24% of DPPH· radicals were quenched by *C. paniculatus* oil, while EVOO quenched 9.40%. The antiradical potential of oils and fats can be interpreted as the combined action of different endogenous antioxidants (Ramadan et al. 2006, 2007). The strong antiradical potential of *C. paniculatus* oil could be due to (i) the content and composition of sterols and tocols (ii) the diversity in structural characteristics of phenolic antioxidants present, and (iv) different kinetic behaviors of antioxidants (Ramadan and Mörsel 2006).

6   Health-Promoting Traits of *C. paniculatus* Fruit Seed Oil and Oil Constituents

Oil extracted from *C. paniculatus* seeds was used to treat acute and chronic immobilization induced experimentally in mice. Levels of antioxidant enzyme regained and markedly increased in the acute and chronic immobilized mice groups. *C. paniculatus* seed oil was highly efficacious in reducing the stress induced by least mobility for hours (Lekha et al. 2010).

Chakrabarty et al. (2012) investigated the effects of *C. paniculatus* seed oil in preventing the onset of chronic aluminum-induced cortico-hippocampal neurodegeneration and oxidative stress. *C. paniculatus* seed oil showed a significant prevention in the onset of aluminum-induced neural insult and systemic oxidative stress, which was corroborated by the enlisted biochemical, neurobehavioral, and histological evidence. It was concluded that *C. paniculatus* seed oil is a putative decelerator of Al-mediated Alzheimer's like pathobiology.

Antioxidant and anti-inflammatory traits of the aqueous, methanol, and chloroform extracts of *C. paniculatus* seeds were evaluated using DPPH (2,2-diphenyl-1-picrylhydrazyl) radical scavenging test, ferric reducing antioxidant power assay

(FRAP), Trolox equivalent antioxidant capacity (TEAC), and lipoxygenase inhibition test. Almost all the assays indicated *C. paniculatus* chloroform extract to have the strongest antioxidant trait and the highest total phenolic content. *C. paniculatus* aqueous extract, however, exhibited the highest anti-inflammatory potential. The main components identified in the *C. paniculatus* essential oil, being strongly antioxidant in nature, reveal the possible cause of *C. paniculatus* being highly efficacious in cognition enhancement. The study also suggests that *C. paniculatus* essential oil could be used as a potential source for new drug development in treating various disorders caused by oxidative stress (Arora and Pandey-Rai 2014).

Bhagya et al. (2016) analyzed the neuroprotective impacts of *C. paniculatus* oil on stress-associated cognitive dysfunctions. The results provided a novel perspective on beneficial effect of *C. paniculatus* oil and herbal therapy on stress-induced cognitive dysfunctions. Malik et al. (2017) evaluated the effect of *C. paniculatus* seeds ethanol extract against 3-nitropropionic acid (3-NP) induced Huntington's disease (HD) like symptoms in Wistar male rats. Ethanol extract of *C. paniculatus* seeds exhibited a protective action against 3-NP induced HD like symptoms due to its high antioxidant impact.

The gastroprotective and antiulcer impacts of *C. paniculatus* seed oil against several gastric ulcer models in rats were studied (Palle et al. 2018). The results demonstrated effective gastro-protection against ethanol- and indomethacin-induced ulcer animal models. In pylorus-ligated rats, *C. paniculatus* seed oil showed gastro-protective traits by decreasing total gastric juice volume and gastric acidity while increasing the gastric pH. The gastro-protection was attributed to effective inhibition of pro-inflammatory cytokines, TNF-α and IL-6, and increase in the levels of IL-10. Treatment with *C. paniculatus* seed oil decreased malondialdehyde levels, which were accompanied by an increase in the activities of superoxide dismutase and catalase. *C. paniculatus* seed oil reduced the rate of gastric emptying but had no effect on gastrointestinal transit. *C. paniculatus* seed oil had a potent gastro-protective effect and supported the folkloric usage of *C. paniculatus* seed oil to treat various gastrointestinal disturbances (Palle et al. 2018) (Fig. 45.1).

Fig. 45.1 *C. paniculatus* fruits and seeds

References

Ahmad, F., Khan, R. A., & Rasheed, S. (1994). Preliminary screening of methanolic extracts of *Celastrus paniculatus* and *Tacomelia undulata* for analgesic and anti-inflammatory activities. *Journal of Ethnopharmacology, 42*, 193–198.

Arora, N., & Pandey-Rai, S. (2014). GC-MS analysis of the essential oil of *Celastrus paniculatus* Willd. seeds and antioxidant, anti-inflammatory study of its various solvent extracts. *Industrial Crops and Products, 61*, 345–351.

Ayudhaya, T. D., Nutakul, W., Khunanck, U., Bhunsith, J., Chawarittumrong, P., Jewawechdumrongkul, Y., Pawarunth, K., Yongwaichjit, K., & Webster, H. K. (1987). Study on the in vitro antimalarial activity of some medicinal plants against *Plarmodium falciparum. Bulletin Department of Medical Science, 29*, 22–38.

Bhagya, V., Christofer, T., & Shankaranarayana Rao, B. S. (2016). Neuroprotective effect of *Celastrus paniculatus* on chronic stress-induced cognitive impairment. *Indian Journal of Pharmacology, 48*(6), 687–693.

Chakrabarty, M., Bhat, P., Kumari, S., D'Souza, A., Bairy, K. L., Chaturvedi, A., Natarajan, A., Rao, M. K., & Kamath, S. (2012). Cortico-hippocampal salvage in chronic aluminium induced neurodegeneration by *Celastrus paniculatus* seed oil: Neurobehavioural, biochemical, histological study. *Journal of Pharmacology and Pharmacotherapeutics, 3*(2), 161–171.

CSIR. (1999). Indian herbal pharmacopoeia council of scientific and industrial research, Jammu Tawi (Vol. 2, pp. 26–34).

Godkar, P. B., Gordon, R. K., Ravindran, A., & Doctor, B. P. (2004). *Celastrus paniculatus* seed water soluble extracts protect against glutamate toxicity in neuronal cultures from rat forebrain. *Journal of Ethnopharmacology, 93*, 213–219.

Godkar, P. B., Gordon, R. K., Ravindran, A., & Doctor, B. P. (2006). *Celastrus paniculatus* seed oil and organic extracts attenuate hydrogen peroxide- and glutamate-induced injury in embryonic rat forebrain neuronal cells. *Phytomedicine, 13*, 29–36.

Jadhav, R. B., & Patwardhan, B. (2003). Anti-anxiety activity of *Celastrus paniculatus* seeds. Indian. *Journal of Natural Products, 19*, 16–19.

Kapoor, L. D. (2005). *Handbook of ayurvedic medicinal plants* (p. 113). New Delhi: Ane Books.

Kumar, M. H. V., & Gupta, Y. K. (2002). Antioxidant property of *Celastrus paniculatus* Willd.: A possible mechanism in enhancing cognition. *Phytomedicine, 9*, 302–311.

Lekha, G., Mohan, K., & Samy, I. A. (2010). Effect of *Celastrus paniculatus* seed oil (Jyothismati oil) on acute and chronic immobilization stress induced in Swiss albino mice. *Pharmacognosy Research, 2*(3), 169–174. https://doi.org/10.4103/0974-8490.65512.

Malik, J., Karan, M., & Dogra, R. (2017). Ameliorating effect of *Celastrus paniculatus* standardized extract and its fractions on 3-nitropropionic acid induced neuronal damage in rats: Possible antioxidant mechanism. *Pharmaceutical Biology, 55*(1), 980–990.

Palle, S., Kanakalatha, A., & Kavitha, C. N. (2018). Gastroprotective and antiulcer effects of *Celastrus paniculatus* seed oil against several gastric ulcer models in rats. *Journal of Dietary Supplements, 15*(4), 373–385. https://doi.org/10.1080/19390211.2017.1349231.

Rajkumar, R., Kumar, E. P., Sudha, S., & Suresh, B. (2007). Evaluation of anxiolytic potential of Celastrus oil in rat models of behaviour. *Fitoterapia, 78*, 120–124.

Ramadan, M. F., & Mörsel, J.-T. (2006). Screening of the antiradical action of vegetable oils. *Journal of Food Composition and Analysis, 19*, 838–842.

Ramadan, M. F., Sharanabasappa, G., Seetharam, Y. N., Seshagiri, M., & Moersel, J.-T. (2006). Profile and levels of fatty acids and bioactive constituents in mahua butter from fruit-seeds of buttercup tree [*Madhuca longifolia* (Koenig)]. *European Food Research and Technology, 222*, 710–718.

Ramadan, M. F., Zayed, R., & El-Shamy, H. (2007). Screening of bioactive lipids and radical scavenging potential of some Solanaceae plants. *Food Chemistry, 103*, 885–890.

Ramadan, M. F., Kinni, S. G., Rajanna, L. N., Seetharam, Y. N., Seshagiri, M., & Mörsel, J.-T. (2009). Fatty acids, bioactive lipids and radical scavenging activity of *Celastrus paniculatus* Willd. seed oil. *Scientia Horticulturae, 123*(1), 104–109.

Rana, V. S., & Das, M. (2017). Fatty acid and non-fatty acid components of the seed oil of *Celastrus paniculatus* willd. *International Journal of Fruit Science, 17*(4), 407–414.

Russo, A., Izzo, A. A., Cardile, V., Borrelli, F., & Vanella, A. (2001). Indian medicinal plants as antiradicals and DNA cleavage protectors. *Phytomedicine, 8*, 125–132.

Sengupta, A., & Bhargava, H. N. (1970). Investigation of the seed fat of *Celastrus Panzculatus*. *Journal of the Science of Food and Agriculture, 21*, 628–631.

Sengupta, A., Sengupta, C., & Masumder, U. K. (1987). Chemical investigations on *Celastrus paniculatus* seed oil. *Fat Science Technology, 3*, 119–123.

Vaidyaratnam, P. S. V. (1997). *Indian medicinal plants. A compendium of 500 species* (Vol. 2, p. 47). Madras: Orient Longman Ltd.

Wangoo, D., & Bidwai, P. P. (1988). Anti-spermatogenic effect of *Celastrus paniculatus* seed extract on the testis of albino rats. *Fitoterapia, 59*, 377–382.

Yang, B., Karlsson, R. M., Oksman, P. H., & Kallio, H. P. (2001). Phytosterols in sea buckthorn (Hippophaë rhamnoides L.) berries: Identification and effects of different origins and harvesting times. *Journal of Agricultural and Food Chemistry, 49*, 5620–5629.

Chapter 46
Black Seed (*Nigella sativa*) Oil

Omprakash H. Nautiyal

Abstract Black cumin (*Nigella Sativa*) seed is native to Asia and is often known as black seed oil. Studies on the black cumin seed oil have shown combating super-bugs like Methicillin-resistant *Staphylococcus aureus* (MRSA) or *Helicobacter pylori*. *Nigella Sativa* seeds are carminative, helping aid in digestion, gas, bloating, and stomach pain. The seed oil contains myristic acid, palmitic acid, stearic acid, palmitoleic acid, oleic acid, linoleic acid, and arachidonic acid. The seeds also contain proteins and vitamins B1, B2, B3, calcium, folate, iron, copper, zinc and phosphorous. Over 600 studies have shown the effect of black cumin seed oil with promising potential of using black cumin seed oil for dealing with autoimmune disease. The most studied active compounds in black seed oil were crystalline nigellone and thymoquinone. Anti-asthmatic effects of black seed oil have been established and found to be more effective than conventional treatments. Similar properties make it beneficial for relieving allergies in human. Skin diseases, like eczema and psoriasis are also treated with black seed oil. It is helpful in soothing inflammation and improving the healing of skin headings. The growth of colon cancer cells without any side effects have also shown to be inhibited by black seed oil. This chapter will provide a scientific review on the composition and functional properties of black cumin seed oil.

Keywords Black caraway · Black cumin · Nigellone · Thymoquinone · Antimicrobial activity

Abbreviations

AED	Anti-epilepsy drugs
AKT	Regulators of cell signaling
DCFH	2,7-dichlorodihydrofluorescein

O. H. Nautiyal (✉)
Department of Chemistry, Lovely Professional University,
GT Road, Chaheru, Punjab, India

DPPH   2,2′-diphenyl-1-picrylhydrazyl
FFA   Free fatty acids
FRAC   Ferrous reducing antioxidant capacity
NSO   Nigella seed oil
RMSEC   Root mean square error of calibration
TFA   Total fatty acids
UFA   Unsaturated fatty acids

1 Introduction

Natural products play a very important role as an alternative to synthetic ingredients in their clinical and nutritional investigations. A wide variety of plants have lipid-lowering activities (Mohammad et al. 2012). *Nigella sativa* (black seed) is investigated among such plants as well as in several studies for its lipid-lowering effects on animals. Many beneficial pharmacologic actions of this plant have been reported, including hypertensive, antitumor, hypoglycemic, anti-inflammatory, antiasthmatic, anti-allergic, and antioxidant properties. The plant has also exhibited some pharmacological effects like antimicrobial, anorexic, lactagogue and diuretic roles. It can be used as a natural food and as an additive.

Nigella sativa is an annual planting with flowers of the family of Ranunculaceae. It is native to south and southwest Asia. The plant grows to 20–30 cm tall having finely divided and linear leaves. Its flowers (Fig. 46.1) are delicate and usually with pale blue color and white. It contains five to ten petals. The fruits are large and inflated having a composition of three to seven united follicles, each contains numerous seeds.

Nigella sativa is known as black caraway, black cumin, fennel flower, Nigella, Nutmeg flower and Roman coriander. In Hindi it is known as *Kalonji*. In India and Middle Eastern cuisines the seeds of *Nigella sativa* are used as a spice. It tastes like combinations of oregano, onions and black pepper. It is pungent and has bitter taste and smell. *Nigella sativa* was a traditional condiment of the ancient world and were extensively used to the flavor of food (Zohary and Hopf 2000). Bible dictionary of Easton's states the Hebrew word *Ketsah* and was referred to *Nigella sativa*. From the second millennium, seeds were found in Hittite flask in Turkey. The ancient Greek author **Theophrastus** has not given any account about *Nigella* in his "Enquiry into Plants" from the fourth century B.C. (Hort 1916). *Nigella sativa* oil constituted of bioactive compounds including linoleic acid (18:2), thymoquinone, nigellone (dithymoquinone), melanthin, nigilline and trans-anethole.

Melanthion is recommended for numbers of medical uses in Hippocrates as one of them to promote pregnancy. *Nigella sativa* is also known as black cumin has been utilized over centuries, specifically in the Middle East and Southeast Asia. *Nigella sativa* seed effects have been studied as antibacterial (Ferdous et al. 1992; Rathee et al. 1982; Hanafy and Hatem 1991) antitumor (Worthen et al. 1998),

Fig. 46.1 *Nigella sativa*
flower

Table 46.1 Chemical composition of *Nigella sativa* seeds

Total lipids	Moisture	Ash	Protein	Carbohydrate	Reference
32.26	6.67	6.82	19.19	35.04	Mohammad et al. (2016)
31.72	4.99	5.29	23.0	34.9	Solati et al. (2014)
40.35	–	4.41	22.6	32.7	Cheikh-Rouhou et al. (2007)
34.8	7.0	3.7	20.8	33.7	Atta (2003)
37.33	5.40	6.72	20.0	30.5	Khoddami et al. (2011)

anti-inflammatory (Houghton et al. 1995), depressant of central nervous system, analgesic (Khanna et al. 1993), and hypoglycemic (Al-Hader et al. 1993) and smooth muscles relaxant (Aqel 1993, 1995; Aqel and Shaheen 1996), cytotoxic and immune-stimulant (Swamy and Tan 2000). *Nigella sativa* oil (NSO) is widely distributed in Arab countries and Mediterranean regions. NSO comprised of a large number of active compounds and has beneficial health effects such as protective and curative activities. NSO is sold 10–15 times more expensive commercially than other edible oils such as corn oil (CO) and soybean oil (SO). This leads to the frequent occurrence of NSO to be blended with cheaper oils in order to get the maximum profits (Nurrulhidayah et al. 2011).

The chemical composition of the *Nigella sativa* seeds is presented in Table 46.1. *Nigella sativa* seeds are widely used in folk medicine in Arabian countries, Pakistan, and in southern Europe (Heiss and Oeggl 2005). *N. sativa* is regarded a panacea (universal remedy) in Islamic folk medicine and used for mostly the same purposes as passed down from Pliny and Dioscorides (Ballero and Fresu 1993; Ali and Blunden 2003). Analyses showed oestrogenic activity (Agradi et al. 2002) and analgesic properties (Bekemeier et al. 1967) of *N. sativa*. Both *N. sativa* and *N. damascena* are used in bread and cheese, while *N. damascena* also serves as a condiment in sweets (D'Antuono et al. 2002).

Nestle in 2010 filed a patent application for thymoquinone, extracted from black seed against food allergy treatment. Their investigation stated that thymoquinone interacts with opioid receptors and helps to inhibit allergic reactions to food. Nestlé attempted to create a *Nigella sativa* monopoly and enable them to sue anyone using it without Nestlé's permission. More than 100 components are present in the black seed oil and few of which are unidentified till today. The oil is rich in polyunsaturated fatty acids (PUFA). Black seeds are rich in Ca, Fe, Na, and K and most required for the body metabolism. These minerals strengthen the immune system, lactation promotion. It is also used as an antihistamine, antipyretic, anti-inflammatory, antioxidant, antitumor and antibacterial. These acids are building blocks of cells that help produce prostaglandin E1. It has excellent healing action on the skin (psoriasis, eczema, dry skin, joints and scalp) and has oral treatment for asthma, arthritis, immune system.

The seeds and oil of *Nigella sativa* are reported to have strong antioxidant properties *in vitro* (Kruk et al. 2000; Badary et al. 2003; Sultan et al. 2009) and *in vivo* (Houghton et al. 1995; Nagi et al. 1999; Mansour et al. 2001; Khan et al. 2003; Abdel-Wahhab and Ali 2005; Kanter et al. 2006; Hosseinzadeh et al. 2007; Gargari et al. 2009; Shawki et al. 2013).

Boskabady and Shirmohammadi (2002) reported antispasmodic properties of *Nigella sativa*, specifically in gastrointestinal and respiratory diseases. There are records in ancient Iranian medical literature that these seeds were potent in digestive and gynecologic disorders including asthma and dyspnea. Traditional Arabian medicines have used the whole black seeds alone or in combination with honey in the treatment of bronchial asthma. The anti-epileptogenic and antioxidant effects of black seed oil have been studied against pentylenetetrazoline-induced kindling in mice wherein black seed oil had anticonvulsant properties (Ilhan et al. 2005).

Rehman et al. (2009) studied the effect of *Nigella sativa* oil in treating the hyperlipidemia. Their clinical trial evaluated 10 patients with age from 50 to 55. They recorded the physical and pathological history of these patients. One gram of *Nigella sativa* powder was given to these patients before breakfast for 2 months. They observed that the total blood cholesterol level from 261.8 mg/dl was reduced to 216.3 mg/dl. LDL-cholesterol lowered from 238 to 188 mg/dl and the serum HDL-cholesterol was found to increase from 43.2 to 54 mg/dl. The authors also observed a significant decrease in triglycerides in blood from 275.9 to 235 mg/dl. The protective effect of honey and *Nigella* grains on the oxidative stress and the cancer was studied by exposing rats to a strong carcinogen. It was found that the rats fed with black seeds have shown 80% protection against oxidative stress and cancer formation. Whereas the rats fed with a daily dose of both honey and black seeds were protected 100% against oxidative stress, inflammatory responses, and cancer formation (Mabrouk et al. 2002).

The *Nigella sativa* volatile oil exhibited a relaxant effect on different smooth muscles including the aorta and jejunum of rabbits, and isolated tracheal muscles of guinea pigs. Mahfouz and El-Dakhakhny (1960); Boskabady and Shirmohammadi (2002) investigated the protection of guinea pigs from histamine-induced bronchospasm with *Nigella sativa* volatile oil, but the authors noticed H1 receptors in isolated tissues remained unaffected. However, post intravenous administration of the *Nigella sativa* volatile oil demonstrated an increasing respiratory rate after intravenous administration.

1.1   Oil Recovery and Content

Properties of *Nigella sativa* oil depend on methods of extraction (Mohammad et al. 2016). Solvent extraction involves extreme heat and deficient in selectivity, which causes the degradation of the desired components. The conventional method for oil extraction is the cold pressing extraction, which involves no heat and/or chemicals treatment. The disadvantage of the method is its low yields and the oil content in the residual meal is about 10–12%. Supercritical fluid extraction (SFE) employed to extract plant oils and extends favorable features over the traditional techniques in the oil industry. SFE in particular recommended method exploited to recover antioxidant compounds from *N. sativa* oil (NSO) especially thymoquinone. Mohammad et al. (2016) identified 20 compounds in the SFE oil whereas the cold-pressed oil contained 19 compounds. The major components of the SFE-extracted oil were caryophyllene (17.4%) followed by thymoquinone (16.8%), 1,4-Cyclohexadiene (7.17%), longifolene (3.50%), and carvacrol (1.82%). The major components for cold-pressed oil were 1,3,8-p-Menthatriene (23.8%) followed by thymoquinone (16.2%), 1,4-cyclohexadiene (7.17%), longifolene (4.49), and carvacrol (3.90%).

Isolation and identification of active compounds have been reported in different varieties of black seeds (Kishore 2013). The major active compounds were thymoquinone (30–48%), thymohydroquinone, dithymoquinone, p-cymene (7–15%), carvacrol (6–12%), 4-terpineol (2–7%), t-anethol (1–4%), sesquiterpene longifolene (1–8%), α-pinene and thymol. Two different types of alkaloids including isoquinoline alkaloids (nigellicimine and nigellicimine-N-oxide), and pyrazole alkaloids or indazole ring bearing alkaloids (nigellidine and nigellicine) determined. α-Hederin, a water-soluble pentacyclic triterpene and saponin, a potential anticancer agent also present in *N. sativa* seeds. Compounds such as carvone, limonene, citronellol were found in trace amounts. Quinine constituents present in *N. sativa* attributes most of the pharmacological properties of which thymoquinone is the most abundant. Thymoquinone yields dithymoquinone and higher oligocondensation products during storage. *N. sativa* seed analyzed to contain protein (26.7%), lipids (28.5%), carbohydrates (24.9%), crude fiber (8.4%) and ash (4.8%). Vitamins and minerals

Table 46.2 Sterol
composition (%) of the
Moroccan Nigella seed oils
(Gharby et al. 2015)

Sterol	Cold-pressed extracted oil	Solvent extracted oil
Cholesterol	0.9 ± 0.1	0.8 ± 0.4
Campesterol	13.1 ± 0.5	12.8 ± 0.5
Stigmasterol	17.8 ± 0.5	18 ± 1
β-Sitosterol	49.4 ± 1.5	51.3 ± 2.5
D5-Avenasterol	12.4 ± 0.5	8 ± 1
D5-Stigmasterol	0.6 ± 0.1	0.7 ± 0.1
D7-Avenasterol	2.1 ± 0.1	1.3 ± 0.5

like Cu, P, Zn and Fe also found in the seeds. Carotene presence is reported in the seeds, which are converted by the liver to vitamin A.

The seeds are also reported to be rich in unsaturated fatty acids, mainly linoleic acid (50–60%), oleic acid (20%), eicodadienoic acid (3%) and dihomolinoleic acid (10%). Palmitic and stearic acids as saturated fatty acids constitute to about 30%. Sitosterol being major sterol accounting for 44% and 54% of the total sterols in Tunisian and Iranian varieties of black seed oils followed by stigmasterol (6.57–20.9% of total sterols).

Gharby et al. (2015) extensively studied the fatty acid composition of *Nigella* seed oil from the various geographical origin. The authors have observed insignificant differences among the sterols compositions of their cold-pressed and solvent extracted oils (Table 46.2). Authors analyzed the different sterols in *Nigella* seed oil extracted by cold press and solvents and the data have shown very less difference in the contents besides few extracted in higher amounts by both of the methods.

The Moroccan *Nigella* seed oil was found identical in their fatty acid composition with the major exceptions of *Nigella* seed oils from Egypt and Tunisia. The composition of the Egypt and Tunisia *Nigella* seed oil contained myristic acid ten times higher than that of Moroccan *Nigella* seed oil. Tunisian *Nigella* seed oil found to contain twice lesser oleic acid as compared to Moroccan oil with higher content of saturated fatty acids (myristic acid: 3.2% vs. average 1%, and stearic acid: 6.3% vs. average 3%). Moroccan *Nigella* seed oil contained linoleic and oleic acids as the main unsaturated fatty acids and accounted for more than 80% of the total fatty acids. The presence behenic or eicosenoic acids in *Nigella* seed oil has been reported in traces as other fatty acids.

Black cumin seed oil inhibited cancer cell activity and even potent to kill some types of cancer cells (Thomas 2018). Black seed oil has shown that it can be as effective as anti-cancer drugs for some types of cancer. Thymoquinone present in black cumin seed oil has powerful benefits for various inflammatory diseases such as liver cancer, melanoma skin cancer, pancreatic cancer, cervical cancer, breast cancer, bone cancer, stomach cancer, lymphoma, prostate cancer, colon cancer, and brain cancer (Woo et al. 2012; Abukhader 2013). Khan et al. (2011) and Randhawa and Alghamadi (2011) reviewed the use of black seed oil as an anticancer agent. It was a safe and effective agent against cancer in the blood system, lungs, kidneys, liver, prostate, breast, cervix, and skin. The molecular mechanisms behind its anticancer role are still not clearly understood. However, thymoquinone in some studies have shown an antioxidant role and improves the body's defense system. Apoptosis is induced by the black seed oil in helping the body eliminate old cells systematically, unneeded cells, and unhealthy cells (such as cancer cells) with no releasing toxins into the body. Black seed oil was potentially helpful to people receiving radiation treatment for cancer (Cikman et al. 2014).

1.2 Seed Cake

Organic *Nigella* seed pressed cake is a by-product of the production of organic black cumin seed oil. Organic *Nigella* seed cake is rich in PUFA and vitamins and is well suited as feed. Leftover cake was analyzed to determine the nutritional value and revealed the proximate composition of oil cakes obtained after cold pressing or solvent extraction is crude protein (23.3–26.5%), moisture (8.1–11.6%), and cellulose (6.8–7.4%).

1.3 Economy

The market research, besides estimating the black cumin seed oil market potential until 2022, analyzes on who can be the market leaders and what partnerships would help them to capture the market share. The black cumin seed oil industry report gives an overview of the dynamics of the market, by discussing various aspects such as drivers, restraints, value chain, customer acceptance and investment scenario (http://www.wsfa.com/story/37774101/global-black-cumin-seed-oil-market-overview-2017-potential-analysis-and-statistics-of-market-size-share-and-gross-margin-by-2022).

2 Extraction and Processing of *Nigella sativa* Oil

Kausar et al. (2018) studied various methods of extraction of *Nigella sativa* oil. The authors have performed the extraction of thymoquinone through different processing methods like maceration, hot solvent extraction by reflux technique, hot solvent extraction by Soxhlet technique and ultrasound-assisted extraction (UAE). They have employed solvents such as methanol, petroleum ether and hexane for ensuring the extraction of thymoquinone in solvents of varying polarity. The thymoquinone contents were estimated in different extracting solvents (Table 46.3). Mohammad et al. (2016) studied *N. sativa* L. seeds cold pressed at room temperature (25 °C) by mechanical pressing without any heating treatment.

Table 46.3 Thymoquinone content in different extracts of *N. sativa* (Kausar et al. 2018)

Extraction technique	Solvent	Extraction yield (% w/w)	Thymoquinone content
Maceration	Methanol	34.8 ± 9.8	4. 27 ± 11.00
	Petroleum ether	15.7 ± 51.00	2.9 ± 0.08
	Hexane	2.32 ± 6.3	3.67 ± 0.988
Reflux	Methanol	47.95 ± 10.5	5.89 ± 1.1
	Petroleum ether	2.33 ± 6.2	2.44 ± 0.84
	Hexane	17.45 ± 5.4	3.45 ± 0.95
Soxhlet	Methanol	40.1 ± 8.5	6 .77 ± 1.2
	Petroleum ether	32.7 ± 5.1	3.66 ± 1.43
	Hexane	32.2 ± 32.2	4.81 ± 1.8
UAE	Methanol	8.0 ± 1.8	14.89 ± 2.6
	Petroleum ether	12.4 ± 2.4	8.270 ± 2.7
	Hexane	9.8 ± 1.67	10.56 ± 3.5

Table 46.4 Physiochemical properties of *N. sativa* extracted using two extraction methods (Mohammad et al. 2016)

Property	Supercritical fluid extraction	Cold-press extraction
FFA (as oleic %)	5.98	6.15
SV (mg of KOH/g of oil)	243.52	238.26
IV (g I_2/100 g oil)	115.1	104.37
PV (meq O_2/kg oil)	3.4	4.1
Refractive index at (25 °C)	1.47	1.47
Viscosity (mPa s)	6.26	6.38

Essential oil from *N. sativa* L. seed was extracted with supercritical fluid extraction (SFE) (Ismail et al. 2010). The physiochemical properties of black cumin seed oils obtained using two extraction methods are presented in Table 46.4. FFA values of NSO extracted by SFE and cold pressing (CP) were significantly different with values of 6.15% and 5.98% (as oleic acid), respectively. The iodine values were 115 g of I_2/100 g of oil for SFE and 104 g of I_2/100 g of oil for CP oil. The SFE and CP extracted oils showed saponification values of 243.52 (mg of KOH/g of oil) and 238.26 (mg of KOH/g of oil), respectively. The refractive index (RI) of NSO extracted by SFE and CP were 1.47813 and 1.47719, respectively. Peroxide values (PV) of oils obtained by SFE and CP extraction were 3.4 and 4.1 meq O_2/kg oil, respectively.

NSO extracted by SFE showed a high concentration of thymoquinone and total phenolic compounds. The antioxidant activities were high with low IC_{50} value as established by DPPH·(2, 2′-diphenyl-1-picrylhydrazyl) and FRAP (The ferric reducing/antioxidant power) assays. The oil extracted by CP and SFE did not show significant differences in their thermal profiles without any weight loss observed before 200 °C and both oils have shown similar functional groups. GC-MS exhibited significant differences in the two oil profiles having directly affected by the extraction methods employed. The SFE technology could be consider the best process for optimizing the extraction of NSO since the method have enhanced the quality and the yield of the black seed oil in comparison with CP.

2.1   Minor Bioactive Lipids of Black Cumin Seed Oil

The chemical properties of oils are considered the most significant feature to determine the quality of oil samples. The properties of *N. sativa L.* oil extracted using SFE and cold press were established. Free fatty acid and peroxide value were the commonly used indicators for monitoring the quality of seed oils. The low FFA value for the oil extracted with the SFE method showed higher stability than that obtained by CP extraction (Solati et al. 2014; Cheikh-Rouhou et al. 2007). The highest amount of thymoquinone (TQ) was observed in the SFE (6.63 mg/mL of oil), while the concentration of TQ was 1.56 mg/mL oil for the cold-pressed oil.

The oil content in most Nigella seeds was higher than 30% (Khan 1999; Matthaus and Ozcan 2011) and possibly up to 40% (Cheikh-Rouhou et al. 2007). However, this yield can reduce to 13–23% in phenological events as water stress or saline conditions, and eventually due to cool temperatures (Al-Kayssi et al. 2011; Bourgou et al. 2010a; Atta 2003). Nigella seeds from Morocco yield 37% and 27% of oil by hexane or cold pressing extraction, respectively. These yields were identical to those reported for Nigella seeds cultivated in countries with weather conditions close to those encountered in Northwestern Morocco. The low oil yield was observed for cold pressed extracted Nigella seed oil, as compared to the solvent extracted oil, which could be due to the high solvent efficiency to extract oleoresins, compared to cold-pressing (Atta 2003).

The cold pressing extraction found to preserve oil integrity, therefore, the high PVs reported for solvent-extracted Nigella seed oil was associated with oxidative processes. Accordingly, the extinction coefficients of solvent-extracted Nigella seed oil were higher than those of cold-pressed oil confirming a higher oxidative state for solvent-extracted Nigella seed oil, and a better quality of the cold press extracted oil.

Kaskoos (2011) identified 26 fatty acids in the Iraqi *Nigella sativa* oil which represented 95% of the total fatty acid composition (Table 46.5). The oil was comprised of 8 saturated fatty acids (15.13%) and 18 unsaturated fatty acids (79.87%). Linoleic acid (42.76%), oleic acid (16.59%), palmitic acid (8.51%), eicosatrienoic acid (4.71%), eicosapentaenoic acid EPA (5.98%) and docosahexaenoic acid DHA (2.97%) were the major components (Nergiz and Otles 1993; Nickavar et al. 2003). The reports (Nergiz and Otles 1993; Nickavar et al. 2003) suggested that the oils of black cumin varieties contained oleic and linoleic acids (18.9–25.0% and 47.5–60.8%, respectively). The ratios of linoleic to oleic acid and unsaturated to saturated fatty acids were 2.57 and 5.27, respectively (Atta 2003; Ramadan and Morsel 2003a, b).

Table 46.5 The fatty acid composition of *Nigella sativa* oil from Iraq (Kaskoos 2011)

Fatty acid	Percentage (%)
Capric acid	0.03
Myristic acid	0.16
Pentadecanoic acid	0.03
Cis-10-pentadecanoic acid	0.08
Palmitic acid	8.51
Palmitoleic acid	0.16
Heptadceneoic acid	0.03
Cis-10-Heptadceneoic acid	0.04
Stearic acid	2.22
Oleic acid	16.59
Linoleladic acid	0.71
Linoleic acid	42.76
γ-Linolenic acid	0.25
Eicosenoic acid	0.16
Heneicosenoic acid	0.26
Cis-11, 14-Eicosadienoic acid	1.94
Cis-8,10,14-Eicosatrienoic acid	0.33
Arachidonic acid	0.03
Tricosanoic acid	0.32
Cis-13,16-Docosadienoic acid	0.05
Lignoceric acid	3.60
Cis-5,8,11,14,17-Eicosapentenoic acid	0.32
Cis-11,14,17-Eicosatrienoic acid	4.71
Cis-5,8,11,14,17-Eicosapantaenoic acid	5.98
Nervonic acid	2.76
Docosahexaenoic acid	2.97
Total saturated fatty acids	15.13
Monoenoic fatty acids	19.82
Bienoic fatty acids	45.46
Trienoic fatty acids	5.29
Tetraenoic fatty acids	0.03
Pentaenoic fatty acids	6.30
Hexaenoic fatty acids	2.97
Total unsaturated fatty acids	79.87
Total fatty acids	95.00

Toma et al. (2013) investigated the solvent-extracted *Nigella sativa* seed oil from Tunisian regions. The recovered oil contained 12 saturated and unsaturated fatty acids (Table 46.6). The analyzed fatty acids of mature seeds were characterized by the predominant presence (63.71%) of the linoleic acid. The authors also found out that the oil extracted from mature seeds using the petroleum ether as a solvent, is richer in linoleic acid (C18:2) than the one obtained by cold pressing. Moreover, in the case of Tunisian seeds, the literature showed similar results (Ramadan and Morsel (2002a, b). The main fatty acid of the Tunisian *Nigella sativa* seeds was

Table 46.6 The fatty acid composition of *Nigella sativa* oil from Tunisia (Toma et al. 2013)

Fatty acid	Percentage (%)
Myristic acid C14:0	0.14
Palmitic acid C16:0	8.92
Palmitolcic acid C16:1	0.18
Stearic acid C18:0	2.44
Oleic acid C18:1	19.42
Linoleic acid C18:2	63.71
Linolenic acid C18:3	0.44
Arachidic acid C20:0	0.13
Eicosenoic acid C20:1	0.27
Eicosedienoic acid C20:2	0.33
Behenic acid C22:0	2.89
Lignoceric acid C24:0	1.04

confirmed to be the linoleic acid, an essential fatty acid, which cannot be biosynthesized by the mammalian's cell. This fatty acid is biosynthesized in the first stage of seeds formation. In the mature stage, C18:2 represent about 63.71% of the TFA of the *Nigella sativa* seeds. This fatty acid has a great importance in human life due to its essential contribution in the biosynthesis of PUFA with long chain (C20:5; C22:5; C22:6) and in the synthesis of prostaglandins (PG1, PG2 and PG3). During their biosynthesis, the fatty acids are stored in the *Nigella sativa* tissues as triglycerides.

Thymoquinone (TQ) produces detrimental effects by overpowering androgen receptor and E2 F-on cell proliferation of several hormone-refractory prostate cancer and linings of cancerous cell 1 (Tariq et al. 2014). The black seed is comprised of fixed oil, which carries functional importance of antioxidant being rich in volatiles (0.40–1.50%) containing 46% monoterpenes and 18.4–24% TQ. Two aliphatic compounds (i.r. 6-nonadecanone-2 and 16-triecosen-7-ol-1) were extracted from black seeds with *n*-hexane. Black seeds have been reported to be a good source of selenium and polyphenols. The ethyl alcohol extract of *Nigella sativa* found to contain saponins. The essential component like vitamin E, vitamin A and selenium, are essential nutritional constituents necessary for conducting a healthy pattern of life. In correspondence, the concentrations of selenium, DL-γ-tocopherol and vitamin A (all-trans-retinol) in black seeds were 9.027, 0.177, 5.427, 0.277 mg/kg seed, respectively. The *Nigella sativa* comprised of fat-soluble vitamins, which accounted for more than 0.2% of total black seed oil components.

Abundant sterols in the black seed enable exhibiting anticancer activity especially β-sitosterol. The amount of β-carotene is about 593 μg/g oil. One gram of black seed fixed oil comprises 40, 284, 48, 225 μg of β-tocopherol, α-tocopherol, δ-tocopherol and γ-tocopherol, respectively. *Nigella sativa* essential oil comprises phenolic compounds (i.e., p-cymene, thymol and carvacrol) which play an important role in the antimicrobial potential of *Nigella sativa* essential oil (Ali et al. 2012).

The main identified compounds in *Nigella sativa* essential oil presented in Table 46.7. The potent inhibitory action and the wide spectrum of antimicrobial

Table 46.7 Main identified compounds in *Nigella sativa* essential oil

Compound	Area percent
α-Thujene	10.03
p-Cymene	36.20
Limonene	1.76
Terpinen-4-ol	2.37
Thymoquinone	11.27
Carvacrol	2.12
Longifolene	6.32

Table 46.8 Antibacterial activity of *Nigella sativa* essential oil

	Concentration (ppm)	Inhibition zone (mm)			
		Bacillus subtillis	*Staphylococcus aureus*	*Bacillus cereus*	*Pseudomonas aerugenosa*
Nigella sativa essential oil	1000	Cl[a]	Cl	70.4	60.7
	2000	Cl	Cl	Cl	Cl
Ampicillin	1000	–[b]	1.91	–	–
	2000	12.2	2.22	1.44	–
Cloxacillin	1000	–	2.28	1.23	–
	2000	–	2.64	1.33	

[a]Complete inhibition
[b]No inhibition

activity of *Nigella sativa* essential oil highlights the interactions between individual components (Singh et al. 2005). Viuda-Marto et al. (2011) investigated the inhibitory action of the Egyptian *Nigella sativa* essentials oil on the growth of some indicators of spoilage bacteria strains (Table 46.8). They selected three bacterial spices (*Listeria, Pseudomonas* and *Serratia*) associated with refrigerated foods, eggs, meat, milk, poultry, seafood and vegetables. *Nigella sativa* oil exhibited high levels of inhibition (95.89%) of DPPH·radicals and high FRAC values (3.33 mmol/L Trolox). *Nigella sativa* essential oil also has shown inhibitory effects on *Listeria innocua*. Rats fed with an aflatoxin-contaminated diet and treated with *Nigella sativa* oil, were found to be protected against aflatoxicosis (Abdel-Wahhab and Ali 2005). There are evidence that the utilization of *N. sativa* and its bioactive components in a daily diet to improve health (Butta and Sultana 2010).

2.2   Food Applications

Food degradation occurs mainly by oxidation or microorganism, during processing, storage and marketing is an important issue in the food industry. The food industry has used synthetic additives, which diminish microbial growth and delay the oxidation of oxidizable materials such as lipids. However, owing to the economic impact

851

of spoiled foods and consumers' growing concerns over the safety of foods containing synthetic antioxidants, much attention has been paid to natural bioactive compounds (Alzoreky and Nakahara 2003; Viuda-Marto et al. 2011).

Mahgoub et al. (2013) studied the impact of adding *Nigella sativa* oil at 0.1 and 0.2% (w/w) to Domiati cheese supplemented with probiotics cultures on the inhibition of foodborne pathogens (*Staphylococcus aureus, Escherichia coli, Listeria monocytogenes* and *Salmonella enteritidis*) inoculating in cheese during storage. *Nigella sativa* oil has shown antimicrobial effect with the concentration of 0.2% oil with the most effective antimicrobial potential on pathogens as compared to the control. Storage of oil-supplemented cheese was shown to extend its shelf life under refrigerated conditions with low microbial loads. Additionally oil-supplemented Domiati cheeses had shown improved physicochemical and sensory properties. *Nigella sativa* essential oil exhibited strong ex vivo antioxidant activity. The oil found to inhibit the growth of A-549 and DLD-1 cancer cell lines with exerting antibacterial activity against *Staphylococcus aureus* and *Escherichia coli* with IC_{50} values of 12.0 and 62.0 µg/mL (Bourgou et al. 2010a, b; Hanafy and Hatem 1991).

Rohman and Ariani (2013) developed and optimized an authentication of *Nigella sativa* oil employing Fourier transform infrared spectroscopy (FTIR) combining with multivariate calibration of partial least square (PLS) for the analysis of *Nigella sativa* seed oil (NSO) in binary and ternary mixtures with corn oil (CO) and soybean oil (SO). PLS modeling was performed for quantitative analysis of NSO in binary mixtures with CO carried out using the second derivative FTIR spectra at combined frequencies of 2977–3028, 1666–1739, and 740–1446 cm^{-1} revealed the highest value of coefficient of determination (R^2, 0.9984) and the lowest value of

Table 46.9 Functional groups and mode of vibration from FTIR spectra of *Nigella sativa* oil (Rohman and Ariani 2013)

Marker	The peak position of FTIR spectra (cm^{-1})	Assignment of bands	Mode of vibration
(a)	3009	C=CH (cis)	Stretching
(b) and (c)	2922 and 2852	–CH (CH$_3$)	Stretching asymmetric
(d)	1742	–C=O (ester)	Stretching
(e)	1658	–C=C (cis)	Stretching
(f)	1461	–C–H (CH$_2$)	Bending (scissoring)
(g)	1378	–C–H (CH$_3$)	Bending symmetric
(h) and (i)	1235 and 1161	C–O (ester)	Stretching
(j) and (k)	1118 and 1098	C–O	Stretching
(l)	964	trans–CH=CH–	Bending out of plane
(m)	914	trans–CH=CH–	Bending out of plane
(n)	871	trans–CH=CH–	Bending out of plane
(o)	844	trans–CH=CH–	Bending out of plane
(p)	721	cis–CH=CH–	Bending out of plane
(q)	1715	–C=O	Stretching

root mean square error of calibration (RMSEC, 1.34% v/v). NSO in binary mixtures with SO was successfully determined at the combined frequencies of 2985–3024 and 752–1755 cm^{-1} using the first derivative FTIR spectra with RMSEC values of 0.9970% and 0.47% v/v, respectively (Table 46.9). The second derivative FTIR spectra at the combined frequencies of 2977–3028 cm^{-1}, 1666–1739 cm^{-1}, and 740–1446 cm^{-1} were selected for quantitative analysis of NSO in ternary mixture with CO and SO with and RMSEC values of 0.9993% and 0.86% v/v, respectively. FTIR spectrophotometer was an accurate technique for the quantitative analysis of NSO in binary and ternary mixtures with CO and SO.

3 Adulteration and Authentication of *Nigella sativa* Oil

Adulterating the foodstuffs including oil is a serious problem because of the dangerous effects that may arise from additional ingredients that mixed into foods, such as the emergence of an allergic reaction (Lai et al. 1995). Besides, the mixing of incompatible materials can associate with restrictions by a certain religion, like adulteration oils with lard. Several methods such as high-performance liquid chromatography (Lee et al. 2001), carbon isotope ratio (Seo et al. 2010) and electronic nose (Hai and Wang 2006) have been developed to detect the adulteration of edible oils. A rapid and simple technique such as FTIR spectroscopy was also used for routine monitoring of oil adulteration.

Ramadan and Morsel (2002a, b) in their analysis of the free sterols have provided rich information about the quality and the identity of the *Nigella* oil. Their investigation in vegetable oils has established that neither cultivation of new breeding lines nor environmental factors were found to alter content and composition of free sterols in contrast to the fatty acid composition, which has been found to change dramatically by breeding programs. The concentration of total sterols ranged from 0.36% of the total hexane extract and 0.33% of the oil extracted with chloroform and methanol. Examination of extracted oil has shown β-sitosterol (32.2–34.1% of total sterol content) represented the main component of the phytosterols followed by Δ5-avenasterol (27.8–27.9% of total sterol content), and Δ7-avenasterol (18.5–22.0% of total sterol content). Stigmasterol, campesterol, and lanosterol were estimated in small amounts, comprising of 17.6–19.5% of the sterol content.

Plant glycolipids are comprised of compounds, which have been divided into distinct groups with respect to their hydrophobic aglycones (Ramadan and Morsel 2003a). In accordance with their quantitative abundance, glycosyl diacylglycerols usually rank first followed by glycosylated ceramides. Glycolipids analysis requires a methodology for isolation, elucidating their chemical structures, release and separation of constituent building blocks and quantification. HPLC is one of the most reliable tools for the analysis of lipid classes with automation to a considerable degree to give much cleaner fractions in micro preparative applications.

Ramadan and Morsel (2004) detected β-carotene in approximately equal amounts in the crude oils. It was anticipated that carotenoids might play a role as primary antioxidants by trapping free radicals or as secondary antioxidants by quenching singlet oxygen. Farther carotenoids and tocopherols in combination act synergistically. Stabilities of seed oils under investigation were found to correlate with sterol levels. Sterols have been documented to have antioxidant activity by interaction with oil surfaces by inhibiting oxidation. Moreover, sterols oxidize at oil surfaces and inhibit propagation by acting as hydrogen donors.

Ramadan (2007) reported on the composition and functionality of black cumin seed oil extracted with *n*-hexane and chloroform: methanol (2:1, v/v). Higher amounts of total lipids were found in the later mixture of solvents. Linoleic acid, palmitic acid, oleic acid and stearic acid were the major fatty acids while neutral lipids accounted for about 97% of total crude oil followed by glycolipids and phospholipids, respectively. Triacylglycerols were the major neutral lipids (83.1–80.8% of the total neutral lipids); while the neutral lipid profile was characterized by high level of free fatty acids (14.3–16.2% of the total neutral lipids).

4   Conclusions

Black cumin (*Nigella sativa*) seed oil is of particular interest due to its utilization for the creating formulations with enrichment of phytochemicals with significant health-promoting potencies. The black cumin essential oil is a precious source of bioactive ingredients comprising of p-cymene, thymoquinone, α-thujene, longifolene, β-pinene, α-pinene and carvacrol. These high levels of bioactive compounds are of utmost importance in nutritional applications. The oil exhibited different biological activities such as antifungal, antibacterial and antioxidant potential. In the present time's black seed oil has become an important subject for investigation and research worldwide, nevertheless more studies must be emphasized to find newer possible activities of this versatile phyto-therapeutic agent in parallel to clinical trials to prove the therapeutic efficiency of the oil.

References

Abdel-Wahhab, M. A., & Ali, S. E. (2005). Antioxidant property of *Nigella sativa* (black cumin) and *Syzygium aromaticum* (clove) in rats during aflatoxicosis. *Journal of Applied Toxicology, 5*, 218–223.

Abukhader, M. M. (2013). Department of pharmacy, Oman Medical College, Muscat, Sultanate of Oman, "Thymoquinone in the clinical treatment of cancer: Fact or fiction?". *Pharmacognosy Reviews, 7*(14), 117–120. PMID: 24347919.

Agradi, E., Fico, G., Cillo, F., Francisci, C., & Tomè, F. (2002). Estrogenic activity of *Nigella damascena* extracts, evaluated using a recombinant yeast screen. *Phytotherapy Research, 16*, 414–416.

Al-Hader, A., Aqel, M., & Hasan, Z. (1993). Hypoglycemic effects of the volatile oil of *Nigella sativa* seeds. *International Journal of Pharmacognosy, 31*(2), 96–100.

Ali, B. H., & Blunden, G. (2003). Pharmacological and toxicological properties of *Nigella sativa*. *Phytotherapy Research, 17*, 299–305.

Ali, M. A., Sayeed, M. A., Alam, M. S., Yeasmin, M. S., Khan, A. M., & Muhamad, I. I. (2012). Characteristics of oils and nutrient contents of *Nigella sativa* Linn. and *Trigonella foenum-graecum* seeds. *Bulletin of the Chemical Society of Ethiopia, 26*, 55–64.

Al-Kayssi, A. W., Shihab, R. M., & Mustafa, S. H. (2011). Impact of soil water stress on nigellone oil content of black cumin seeds grown in calcareous-gypsifereous soils. *Agricultural Water Management, 100*, 46–57.

Alzoreky, N. S., & Nakahara, K. (2003). Antimicrobial activity of extracts from some edible plants commonly consumed in Asia. *International Journal of Food Microbiology, 80*, 223–230.

Aqel, M. B. (1993). Effects of *Nigella sativa* seeds on intestinal smooth muscle. *International Journal of Pharmacognosy, 31*(1), 55–60.

Aqel, M. (1995). The relaxing effects of the volatile oil of *Nigella sativa* seeds on vascular smooth muscles. *Dirasat, 19*, 91–100.

Aqel, & Shaheen, R. (1996). Effects of the volatile oil of *Nigella sativa* seeds on the uterine smooth muscle of rat and guinea pig. *Journal of Ethnopharmacology, 52*(1), 23–26.

Atta, M. B. (2003). Some characteristics of Nigella (*Nigella sativa* L.) seed cultivated in Egypt and its lipid profile. *Food Chemistry, 83*, 63–68.

Badary, O. A., Taha, R. A., Gamal el-Din, A. M., & Abdel-Wahab, M. H. (2003). Thymoquinone is a potent superoxide anion scavenger. *Drug and Chemical Toxicology, 26*, 87–98.

Ballero, M., & Fresu, I. (1993). Le piante di uso officinale nella Barbagia di Seui (Sardegna Centrale). *Fitoterapia, 64*, 141–150.

Bekemeier, H., Leuschner, G., & Schmollack, W. (1967). Antipyretische, antiödematöse und analgetische Wirkung von Damascenin im Vergleich mit Acetylsalicylsäure und Phenylbutazon. *Archives Internationales de Pharmacodynamie et de Thérapie, 168*, 199–211.

Boskabady, M. H., & Shirmohammadi, B. (2002). Effect of *Nigella sativa* on isolated guinea pig trachea. *Archives of Iranian Medicine, 5*(2), 103–107.

Bourgou, S., Pichette, A., Marzouk, B., & Legault, J. (2010a). Bioactivities of black cumin essential oil and its main terpenes from Tunisia. *South African Journal of Botany, 76*, 210–216.

Bourgou, S., Bettaieb, I., Saidani, M., & Marzouk, B. (2010b). Fatty acids, essential oil, and phenolics modifications of black cumin fruit under NaCl stress conditions. *Journal of Agricultural and Food Chemistry, 58*, 12399–12406.

Butta, M. S., & Sultana, M. T. (2010). *Nigella sativa*: Reduces the risk of various maladies. *Critical Reviews in Food Science and Nutrition, 50*, 654–665.

Cheikh-Rouhou, S., Besbes, S., Hentati, B., Blecker, C., Deroanne, C., & Attia, H. (2007). *Nigella sativa* L.: Chemical composition and physicochemical characteristics of lipid fraction. *Food Chemistry, 101*(2), 673–681. https://doi.org/10.1016/j.foodchem.2006.02.022.

Cikman, O., Ozkan, A., Aras, A. B., Soylemez, O., Alkis, H., Taysi, S., & Karaayvaz, M. (2014). Radioprotective effects of *Nigella Sativa* oil against oxidative stress in liver tissue of rats exposed to total head irradiation. *Journal of Investigative Surgery, 27*(5), 262–266. PMID: 24679182.

D'Antuono, L. F., Moretti, A., & Lovato, A. F. S. (2002). Seed yield, yield components, oil content and essential oil content and composition of *Nigella sativa* L. and *Nigella damascena* L. *Industrial Crops and Products, 15*, 59–69.

Ferdous, A. J., Islam, S. N., Ahsan, M., Hasan, C. M., & Ahmed, Z. U. (1992). In vitro antibacterial activity of the volatile oil of *Nigella sativa* seeds against multiple drug-resistant isolates of *Shigella* spp. and isolates of Vibrio cholerae and *Escherichia coli*. *Phytotherapy Research, 6*(3), 137–140.

Gargari, B., Attary, V. E., Rafraf, M., & Gorbani, A. (2009). Effect of dietary supplementation with *Nigella sativa* L. on serum lipid profile, lipid peroxidation and antioxidant defense system in hyperlipidemic rabbits. *Journal of Medicinal Plants Research, 3*, 815–821.

Gharby, S., Harhar, H., Guillaume, D., Raudani, A., Boulbaroud, S., Ibrahimi, M., Ahmed, M., Sultana, S., Hadda, T. B., Moussaoui, I. C., & Charrouf, Z. (2015). Chemical investigation of

Nigella sativa L. seed oil produced in Morocco. *Journal of the Saudi Society of Agricultural Sciences, 14*(2), 172–177.

Hai, Z., Wang, J. (2006). Electronic nose and data analysis for detection of maize oil adulteration in sesame oil. *Sensors and Actuators B 119*(2):449–455.

Hamrouni-Sellami, I., Kchouk, M. E., & Marzouk, B. (2003). Lipid and aroma composition of black cumin (*Nigella sativa* L.) seeds from Tunisia. *Journal of Food Biochemistry, 32*, 335–352.

Hanafy, M. S., & Hatem, M. E. (1991). Studies on the antimicrobial activity of *Nigella sativa* seed (black cumin). *Journal of Ethnopharmacology, 34*, 275–278.

Heiss, A. G., & Oeggl, K. (2005). The oldest evidence of *Nigella damascena* L. (Ranunculaceae and its possible introduction to central Europe). *Vegetation History and Archaeobotany, 14*(4), 562–570.

Hort, A. (Ed.). (1916). *Theophrastos Περίφυτικώνίστοριών [Peri phytikon historion, Enquiry into plants] vol 2 reprinted 1961*. Cambridge, MA: Harvard University Press.

Hosseinzadeh, H., Parvardeh, S., Asl, M. N., Sadeghnia, H. R., & Ziaee, T. (2007). Effect of thymoquinone and *Nigella sativa* seeds oil on lipid peroxidation level during global cerebral ischemiareperfusion injury in rat hippocampus. *Phytomedicine, 14*, 621–627.

Houghton, P. J., Zarka, R., de las Heras, B., & Hoult, J. R. (1995). Fixed oil of *Nigella sativa* and derived thymoquinone inhibit eicosanoid generation in leukocytes and membrane lipid peroxidation. *Planta Medica, 61*, 33–36.

Ilhan, A., Gurel, A., Armutcu, F., Kamisli, S., & Iraz, M. (2005). Antiepileptogenic and antioxidant effects of *Nigella sativa* oil against pentylenetetrazol-induced kindling in mice. *Neuropharmacology, 49*, 456–464.

Ismail, M., Al-Naqeep, G., & Chan, K. W. (2010). *Nigella sativa* thymoquinone-rich fraction greatly improves plasma antioxidant capacity and expression of antioxidant genes in hypercholesterolemic rats. *Free Radical Biology & Medicine, 48*(5), 664–672.

Kanter, M., Coskun, O., & Uysal, H. (2006). The antioxidative and antihistaminic effect of *Nigella sativa* and its major constituent, thymoquinone on ethanol-induced gastric mucosal damage. *Archives of Toxicology, 80*, 217–224.

Kaskoos, R. A. (2011). Fatty acids composition of black cumin oil from Iraq. *Research Journal of Medicinal Plants, 5*(1), 85–89.

Kausar, H., Abidin, L., & Mujeeb, M. (2018). Comparative assessment of extraction methods and quantitative estimation of thymoquinone in the seeds of *Nigella sativa* L by HPLC. *International Journal of Pharmacognosy and Phytochemical Research*. https://doi.org/10.25258/phyto.v9i12.11186.

Khan, M. A. (1999). Chemical composition and medicinal properties of *Nigella sativa* Linn. *Inflammopharmacology, 7*, 15–35.

Khan, N., Sharma, S., & Sultana, S. (2003). *Nigella sativa* (black cumin) ameliorates potassium bromate-induced early events of carcinogenesis: Diminution of oxidative stress. *Human & Experimental Toxicology, 22*, 193–203.

Khan, M. A., Chen, H. C., Tania, M., & Zhang, D. Z. (2011). Anticancer activities of *Nigella sativa* (black cumin). *African Journal of Traditional, Complementary, and Alternative Medicines, 8*(5), 226–232. PMID: 22754079.

Khanna, T., Zaidi, F. A., & Dandiya, P. C. (1993). CNS and analgesic studies on *Nigella sativa*. *Fitoterapia, 64*(5), 407–410.

Khoddami, A., Ghazali, H. M., Yassoralipour, A., Ramakrishnan, Y., & Ganjloo, A. (2011). Physicochemical characteristics of Nigella seed (*Nigella sativa* L.) oil as affected by different extraction methods. *Journal of the American Oil Chemists' Society, 88*(4), 533–540.

Kishore, K. (2013). A review on therapeutic potential of *Nigella sativa*: A miracle herb. *Asian Pacific Journal of Tropical Biomedicine, 3*(5), 337–352.

Kruk, I., Michalska, T., Lichszteld, K., Kladna, A., & Aboul-Enein, H. Y. (2000). The effect of thymol and its derivatives on reactions generating reactive oxygen species. *Chemosphere, 41*, 1059–1064.

Lai, Y. W., Kemsley, E. K., & Wilson, R. H. (1995). Quantitative analysis of potential adulterants of extra virgin olive oil using infrared spectroscopy. *Food Chemistry, 53*(1), 95–98.

Lee, D. S., Lee, E. S., Kim, H. J., Kim, S. O., & Kim, K. (2001). Reversed phase liquid chromatographic determination of triacylglycerol composition in sesame oils and the chemometric detection of adulteration. *Analytica Chimica Acta, 429*(2), 321–330.

Mabrouk, G. M., Moselhy, S. S., Zohny, S. F., Ali, E. M., Helal, T. E., Amin, A. A., & Khalifa, A. A. (2002). Inhibition of methylnitrosourea (MNU) induced oxidative stress and carcinogenesis by orally administered bee honey and Nigella grains in Sprague Dawely rats. *Journal of Experimental & Clinical Cancer Research, 21*(3), 341–346. PMID: 12385575.

Mahfouz M, El-Dakhakhny M. (1960) ome chemical and pharmacological properties of the new antiasthmatic drug, Nigellone. *Egyptian Pharmaceutical Bulletin, 6*, 357–360.

Mahgoub, S., Ramadan, M. F., & El-Zahar, K. (2013). Cold pressed *Nigella sativa* oil inhibits the growth of food-borne pathogens and improves the quality of Domiati cheese. *Journal of Food Safety, 33*, 470–480.

Mansour, M. A., Ginawi, O. T., El-Hadiyah, T., El-Khatib, A. S., Al-Shabanah, O. A., & Al-Sawaf, H. A. (2001). Effects of volatile oil constituents of *Nigella sativa* on carbon tetrachloride-induced hepatotoxicity in mice: Evidence for antioxidant effects of thymoquinone. *Research Communications in Molecular Pathology and Pharmacology, 110*, 239–251.

Matthäus, B., & Özcan, M. M. (2011). Fatty acids, tocopherol and sterol contents of some *Nigella* species seed oil. *Czech Journal of Food Sciences, 29*, 145–150.

Mohammad, S. A., Mehrnoush, D., Nizal, S., Sedigeh, A., & Alireza, G. (2012). Clinical evaluation of *Nigella Sativa* seeds for the treatment of hyperlipidemia: A randomized, placebo controlled clinical trial. *Medical Archives Sarajevo, 66*(3), 198. https://doi.org/10.5455/medarh.2012.66.198-200.

Mohammad, N. K., Manap, M. Y. A., Tan, C. P., Belal, J. M., Alhelli, A. M., & Hussin, S. M. (2016). The effects of different extraction methods on antioxidant properties, chemical composition, and thermal behavior of black seed (*Nigella sativa* L.) oil. *Evidence-Based Complementary and Alternative Medicine, 2016*, 1–12. https://doi.org/10.1155/2016/6273817. PMCID: PMC5015008.

Nagi, M. N., Alam, K., Badary, O. A., al-Shabanah, O. A., al-Sawaf, H. A., & al-Bekairi, A. M. (1999). Thymoquinone protects against carbon tetrachloride hepatotoxicity in mice via an antioxidant mechanism. *Biochemistry and Molecular Biology International, 47*, 153–159.

Nergiz, C., & Otles, S. (1993). Chemical composition of *Nigella sativa* L. seeds. *Food Chemistry, 48*, 259–261.

Nickavar, B. F., Javidnia, M. K., & Amoli, M. A. R. (2003). Chemical composition of the fixed and volatile oils of *Nigella sativa* L. from Iran. *Zeitschrift für Naturforschung. Section C, 58*, 629–631.

Nurrulhidayah, A. F., Che Man, Y. B., Al-Kahtani, H. A., & Rohman, A. (2011). Application of FTIR spectroscopy coupled with chemometrics for authentication of *Nigella sativa* seed oil. *Spectroscopy, 25*(5), 243–250.

Ramadan, M. F. (2007). Nutritional value, functional properties and nutraceuticals applications of black cumin (*Nigella sativa* L.): An overview. *International Journal of Food Science and Technology, 42*, 1208–1218.

Ramadan, M. F., & Morsel, J. T. (2002a). Neutral lipids classes of black cumin (*Nigella sativa* L.) seed oil. *European Food Research and Technology, 214*, 202–206. https://doi.org/10.1007/s00217-001-0423-8.

Ramadan, M. F., & Morsel, J. T. (2002b). Direct isocratic normal-phase assay of fat-soluble vitamins and beta-carotene in oilseeds. *European Food Research and Technology 214*(6), 521–527.

Ramadan, M. F., & Morsel, J. T. (2003a). Analysis of glycolipids from black cumin (*Nigella sative* L.), coriander (*Coriandrum sativum* L.) and Niger (*Guizotia abyssinica* Cass.) oil seeds. *Food Chemistry, 80*, 197–204.

Ramadan, M. F., & Morsel, J. T. (2003b). Determination of lipid classes and fatty acid profile of niger (*Guizotia abyssinica* Cass.) seed oil. *Phytochemical Analysis 14*(6), 366–370.

Ramadan, M. F., & Morsel, J. T. (2004). Oxidative stability of black cumin (*Nigella sativa* L.), coriander (*Coriandrum sativum* L.) and Niger (*Guizotia abyssinica* Cass.) crude seed oils upon stripping. *European Journal of Lipid Science and Technology, 106*, 35–43. https://doi.org/10.1002/ejlt.200300895.

Randhawa, M. A., & Alghamdi, M. S. (2011). Anticancer activity of *Nigella sativa* (black seed)-a review. *The American Journal of Chinese Medicine, 39*(6), 1075–1091. PMID: 22083982.

Rathee, P. S., Mishra, S. H., & Kaushal, R. (1982). Antimicrobial activity of essential oil, fixed oil and unsaponifiable matter of *Nigella sativa* Linn. *Indian Journal of Pharmaceutical Sciences, 44*(1), 8–10.

Rohman, A., & Ariani, R. (2013). Authentication of Nigella sativa seed oil in binary and ternary mixtures with corn oil and soybean oil using FTIR spectroscopy coupled with partial least square. *The Scientific World Journal*, 1–6. doi:https://doi.org/10.1155/2013/740142.

Seo, H.-Y., Ha, J., Shin, D.-B., et al. (2010). Detection of corn oil in adulterated sesame oil by chromatography and carbon isotope analysis. *Journal of the American Oil Chemists' Society, 87*(6), 621–626.

Shawki, M., El-Wakeel, L., Shatla, R., EL-Saeed, G., Ibrahim, S., & Badary, O. (2013). The clinical outcome of adjuvant therapy with black seed oil on intractable pediatric seizures: A pilot study. *Epileptic Disorders, 15*(3), 295–301.

Singh, G., Marimuthu, P., Heluani, C. S., & Catalan, C. (2005). Chemical constituents and antimicrobial and antioxidant potentials of essential oil and acetone extract of *Nigella sativa* seeds. *Journal of Science and Food Agriculture, 85*, 2297–2306.

Solati, Z., Baharin, B. S., & Bagheri, H. (2014). Antioxidant property, thymoquinone content and chemical characteristics of different extracts from *Nigella sativa* L. seeds. *Journal of the American Oil Chemists' Society, 91*(2), 295–300. https://doi.org/10.1007/s11746-013-2362-5.

Sultan, M. T., Butt, M. S., Anjum, F. M., Jamil, A., Akhtar, S., & Nasir, M. (2009). Nutritional profile of indigenous cultivar of black cumin seeds and antioxidant potential of its fixed and essential oil. *Pakistan Journal of Botany, 41*, 1321–1330.

Swamy, S. M. K., & Tan, B. K. H. (2000). Cytotoxic and immunopotentiating effects of ethanolic extract of *Nigella sativa* L. seeds. *Journal of Ethnopharmacology, 70*(1), 1–7.

Tariq, S., Masud, T., Tariq, S., & Sohail, A. (2014). Black seed (*Nigella sativa*) possess bioactive compounds act as anti-helicobacter pylori agent. *World Journal of Pharmaceutical Sciences, 2*(2), 203–209.

Thomas, J. P. (2018). Health impact news http://healthimpactnews.com/2014/black-seed-oil-cures-many-cancers-according-to-numerous-studies/

Toma, C. C., Simu, G. M., Hanganu, D., Neliolah, Vata, G. F. M., Hammami, C., & Hammami, M. (2013). Chemical composition of Tunisian *Nigella sativa* note II profile on fatty oil. *Farmácia, 61*(3), 454–458.

Ur Rehman, F., Khan, M. A., & Marwat, S. K. (2009). Effect of prophetic medicine kalonji (*Nigella sativa L.*) on lipid profile of human beings an in vivo approach. *World Applied Sciences Journal, 6*(8), 1053–1057.

Viuda-Marto, M., Mohamady, M. A., Fernández-López, J., Abd ElRazik, K. A., Omer, E. A., Pérez-Alvarez, J. A., & Sendra, E. (2011). In vitro antioxidant and antibacterial activities of essentials oils obtained from Egyptian aromatic plants. *Food Control, 22*, 1715–1722.

Woo, C. C., Kumar, A. P., Sethi, G., & Tan, K. H. (2012). Thymoquinone: Potential cure for inflammatory disorders and cancer. *Biochemical Pharmacology, 83*, 443–451. PMID: 22005518.

Worthen, D. R., Ghosheh, O. A., & Crooks, P. A. (1998). The in vitro anti-tumor activity of some crude and purified components of black seed, *Nigella sativa* L. *Anticancer Research, 18*(3), 1527–1532.

Zohary, D., & Hopf, M. (2000). *Domestication of plants in the old world* (3rd ed., p. 206). Oxford University Press.

Chapter 47
Pear (*Pyrus communis*) Seed Oil

Muhammad Mushtaq, Sumia Akram, Saira Ishaq, and Ahmad Adnan

Abstract The pear (*Pyrus communis*) fruit offers a wide range of health benefits against macular degeneration, type 2 diabetes, osteoporosis, inflammatory problems and skin infections owing to the presence of phenolics, fibers, vitamins, boron, and other micro-nutrients. The pear fruit like many others of *Rosaceae* family contains an exceptionally small quantity of seeds (about ten tiny seeds per fruit) which can furnish 15–31% oleaginous attributes. The most fascinating oleaginous compounds in pear seed oil include unsaturated fatty acids, tocochromanols, and phytosterols. Out of pear seed oil fatty acids, linoleic acid (C18:2) level was found to be higher than most of the frequently used edible oils. Moreover, the presence of γ-tocopherol at an elevated level recommends that pear seed oil can inhibit human cancer progression and cell proliferation. The aroma compounds in pear seeds or their essential oil have not be explored yet but those compounds may provide phytonutrients of high therapautric and industrial importance. This chapter provides a comprehensive review of the nutritional composition of pear seed oil, the key bioactives responsible for its health benefits and organoleptic properties.

Keywords *Pyrus communis* seed oil · Fatty acids · Tocochromanols · Minor bioactive

1 Introduction

The pear covers an array of trees and shrubs of genus *Pyrus* (family *Rosaceae*) that can withstand a wide temperature range and under favorable conditions, can grow up to 33–56 ft (medium size) forming a tall and narrow crown except for few species which propagate in a shrubby pattern like apples. Pears fall among the fruits

M. Mushtaq (✉) · S. Ishaq · A. Adnan
Department of Chemistry, Government College University, Lahore, Pakistan
e-mail: ahmadadnan@gcu.edu.pk

S. Akram
Department of Chemistry, Minhaj University, Lahore, Pakistan

© Springer Nature Switzerland AG 2019
M. F. Ramadan (ed.), *Fruit Oils: Chemistry and Functionality*,
https://doi.org/10.1007/978-3-030-12473-1_47

859

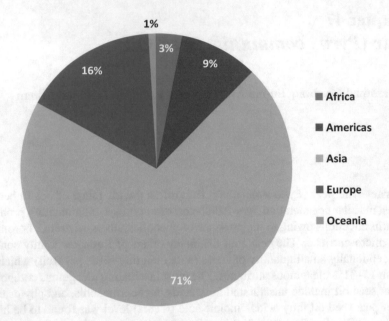

Fig. 47.1 Distribution of pear tree across the planet

of pre-historic times and their current form might have been originated from the foothills of the Tian Shan; a mountain range of Western China in Central Asia. The pre-historic Romans and civilization on the shores Caspian and Atlantic oceans also cultivated and ate this fruit in cooked or raw form. The pears, for their delicious taste and nutritional profile, have been ranked among the utmost popular fruit crops with an annual production higher than 27.6 million tons (Fig. 47.1) whose major portion is produced in Asian regions (Statistics 2016).

According to Pear Bureau Northwest, more than 3000 varieties of 20 recognized pear species are grown worldwide. The pear species *Pyrus communis*, commonly known as a common pear; Occidental pears or the European pear are widely distributed to European regions whereas *Pyrus pyrifolia, Pyrus ussuriensis, Pyrus pashia,* and *Pyrus glabra* are often distributed across Asian regions (Gallardo et al. 2011). Likewise, Bartlett (green pear), Comice (round short juicy, pear in red colors), Forelle (short yellow pear), green or red Anjou (short-necked pears do not undergo color change while ripening), red Bartlett (bright red colored pear), Seckel (the tiniest pear fruits), Beurre d'Amanlis (desert pear), conference (an autumn cultivar of the European pear), Mramornaja (regional variety), Suvenirs, Muizas nr. 4 Petrilas nr. 49, Williams Bon (desert pear), Chretien and Starkrimson (crimson red colored pears), are some other well-known varieties present in scientific literature. Another fruit under the name prickly pears which belongs to a completely different cactus family (Griffith 2004) is erroneously taken as pears or often compared with *Pyrus* family (Hashemi et al. 2018) for their lipids and fatty acid profile.

The pear trees are often propagated by budding, grafting onto a rootstock, or sowing the seeds. The trees grown by the seed may fail to produce fruit or give the

Fig. 47.2 *Pyrus communis* plant (center), flower, seed, fruit, and oil

fruit of different characteristics. In Europe, the main rootstock used is quince (Cydonia oblonga), which produces a dwarfed tree that fruits at an earlier age than most of the trees on pear rootstocks. Pear leaves are arranged 2–12 cm long glossy green with broadly oval to narrow lanceolate appearance and somewhat wedge-shaped at their bases. The flowers of this fruit are usually pink or white (Fig. 47.2) and have five petals and sepals. The pear fruits are moderately elongated starting from the thin circular diameter at the stem and broader one at the opposite end. Likewise, the apples, the pear fruits are generally sweet bearing white, sweet and aromatic pulp which contain a very small amount of tiny brown seed (Fig. 47.2) enclosed inside hard cells (Górnaś et al. 2014a, b).

Pear fruit offers diversity of health benefits when taken as raw or in the form of other products particularly against cancerous diseases, macular degeneration, type 2 diabetes, osteoporosis, inflammatory problems and acne, pimples and skin infections owing to the presence of flavonoids and hydroxycinnamic acids, fibers like pectin, glutathione and quercetin, boron and folate and vitamins (B_{12}, C and E) (Reiland and Slavin 2015). A large quantity of pears has been processed into juices and fresh-cut fruit salads. Consequently, producing a vast amount of byproducts globally, mainly cores, peel and seeds generated during pear processing, though these byproducts are potentially useful. More precisely, the seeds and their oils obtained from fruit byproducts are a valuable source of fatty acids, phytosterols, squalene, and organic acids (Górnaś et al. 2014a, b).

The present chapter will provide a comprehensive report on the nutritional profile of pear seed oil, the key bioactives responsible for its potential health benefits and organoleptic properties. Moreover, the readers would also find profuse information about oil phytochemistry and industrial applications.

2 Extraction and Processing of Fruit Oil

As stated earlier, pear and apple fall among the oldest and most popular fruit. The pears are consumed near and far all over the world for their taste, appearance, nutritional profile and all of above uses in folk remedies. Each part of pear may vary in composition and integrity of bioactive compounds whose recovery or prospective health outcomes are definitively affected by extraction or processing techniques adopted prior to the final product. Therefore, it is necessary for food scientists and engineers to have sound knowledge of pear seed chemistry, understand the working of each oil extraction techniques and their special effects on the nutritional, sensory, and antioxidant character of oleaginous products. The first report in this context dates back to 1650 B.C. when oil from ripened olives was expressed by pressing it between stone mortars or wooden pestles (Kemper 2005). The modern technique often used to isolate oleaginous parts of pear seed should be categorically discussed along with prospective and limitations. However, before opening this discussion, it would be better to have the brief but necessary detail of seed preparation or cleaning steps.

2.1 Seed Selection and Preparation

Pears belongs to the fruits bearing very small (8.4 × 4.8 mm) and reasonably soft seeds which can be easily swollen if the fruit has optimum maturity and taste. Likewise, its relative apple, pear fruit usually own ten seeds distributed among five central cavities. Fromm et al. (2012) observed that certain ecotypes of pear like "Gelbmostler" contain seeds below 1.0% only (insufficient seed to fruit ratio) or even the seeds were not completely developed with an average seed weight below 40.0 mg. Such lower seed yield and seed maturity discourage the processing of pear seeds for oil production. The modernization of extraction techniques, on the other hand, can render the extraction process economical but only if the seeds are properly developed.

A variety of automatic machines are available which can separate out seeds from the fruits while applying gravitational or centrifugal forces. An interesting automatic set up followed for the separation of pear seeds involve the use of a metallic cylinder of specific diameter to cut out the cells which are subsequently centrifuged under mild hammers to separate seed without causing any damage the seeds (Mushtaq 2018). The seeds separated by such machines are usually supplied to agricultural forms. The researchers investigating pear seed phytochemistry have often manually separated seed from fruits.

It has been reasonably understood that the presence of foreign matters in oil-bearing raw materials imposes various sensory, nutritional and shelf life complications. Henceforth, oilseeds before being sent to extraction machines must be cleaned to remove shells, stones, plant stems, leaves (if there) and foreign materials. Similarly, oilseed raw materials when processed without any careful selection can cause wearings of the machine accessories or even may damage cold or screw type pressers. The foreign materials present in the oilseed to be processed into soxhlet type extractors would produce oils of different or even undesirable characteristics. For example, seeds of *Pyrus malus* may contain poisonous cyanoglycosides, so for the safe end, seeds are often spitted out. In the same way, pear seeds like many other oilseeds have strong non-porous coverings (thin layer of endosperm) which must be de-hulled to avoid undesirable color and sensory characteristics of oils produced. Moreover, the presence of hulls would reduce the oil yield by retaining oleaginous constituents in the pressed oil cake or preventing the distribution of extraction solvent for Soxhlet type extractors. Pear seed dehulling prior to screw pressing keeps the feed temperature below 40 °C (less wear friction) during pressing which checks the activity of certain enzymes, auto-oxidation, and transformation of oleaginous constituents into various detrimental products. One more important factor that affects the yield and quality of oleaginous constituents during solvent extraction is moisture level, its presence at elevated levels reduce the distribution of extraction solvent (*n*-hexane or petroleum ether), whereas traces of water render the products more susceptible to oxidation during the extraction (Greenbank and Holm 1924).

After the careful selection and cleaning, seeds are often conditioned to improve the distribution of extraction solvent, meal surface area, and pressing capacity, and avoid any kind of shattering that may take place during the flaking process. The simplest form of conditioning includes moistening and heating of milled or crushed oilseeds and the complex one involve the treatment with some hydrolytic moieties (enzymes or surfactants) to break spherosome membranes. In commercial type extractors, the oilseeds are often crushed prior to hull removal. However, oilseeds bearing sufficiently small size like strawberry, kiwi, and rapeseed need not be crushed and are normally flaked directly. The flaking is often carried out to disrupt cellular structure and increase seed surface area particularly during industrial-scale production of seed oils. For this purpose, oilseed meal is passed through an assembly of two counter-rotating cast-iron rollers bearing smooth surfaces and revolving at a velocity difference of 2–5 RPM. When oilseed meal passes through the rollers the seed coats or cell walls are destroyed and a flake of about 0.20–0.40 mm thickness is formed which offer higher contact area between extraction solvent or presser and meal during both types of extraction methodologies (Lamsal et al. 2006).

Recently, an additional step known as extrusion has been introduced to enhance the oil yield during aqueous solvents based extraction. In this step, oleaginous raw materials are transferred into a rotating type screw fitted inside a closed barrel or same diameter where these undergo a various type of conditioning by heat, pressure, and shear force. The resultant mass is often transformed into cakes of optimal geometry and shape by passing it through various types of dies. The recent developments involve the use of supercritical fluids (under high pressure) which can extrude the mass at a relatively lower temperature and sidestep denaturing of fatty acids, particularly polyunsaturated fatty acids (PUFA) and protein.

2.2 Pear Seed Oil Extraction

A careful review of previously published data regarding the extraction of oleaginous constituents of pear seed indicates three types of approaches. The most frequently adopted one involves the use of non-polar solvents like petroleum ether and *n*-hexane with Soxhlet-type extractors. For Soxhlet extraction, 50–100 g of pre-conditioned seeds are packed in a cellulosic cup (known as thimble) and loaded into an extraction cell of soxhlet type extractor. The vapors of non-polar solvent like *n*-hexane or petroleum ether are used to wash the seeds which are subsequently condensed into siphon top. The extraction solvent comes back to the flask along with oleaginous compounds into extraction once siphon top fills (one cycle). The process continues for many cycles at a temperature between 45 and 60 °C unless the extraction solvent becomes concentrates with oleaginous compounds. The extraction solvents are usually removed under pressure using rotary type evaporators and oil produced is weighed to calculate the percent yield. Likewise apples, pear seeds contain a small amount of cyanide, which will definitely not kill you if you aren't eating stupendous amounts at a time. Pear seeds contain more antioxidants and essential fatty acids than they do cyanide and you could probably eat eight pear fruits, seeds and all, without any effects.

The pear seed lipids are often expressed by simple liquid extraction, where finely ground pear seeds are transferred to flask or centrifuge tube containing sufficient amount of non-polar extraction solvent like *n*-hexane and shaken for a reasonably large interval of time usually overnight. The incorporation of various pre-treatments meant to disrupt vacuoles like surfactants, enzymes or ultrasound radiations can further improve the oil yield. But literature lacks such studies. Although cold pressing machines have been frequently adopted for the expression of lipids from the oils of cotton, rapeseed or soybean, it would be really difficult to apply these machines over a small number of seeds. Therefore, the researchers interested in the phytochemistry of pear seeds have often preferred solvent extraction.

Yukui et al. (2009) observed that seeds of *Pyrus communis* (cv: Dangshau Suli) contain 179 g liposoluble compounds/kg seeds (17.9%). The authors extracted lipids at a relatively higher temperature (100–110 °C) applying petroleum ether as an extraction solvent and observed that oleaginous compounds produced at this temperature mainly composed of saturated, monounsatured and diunsaturated fatty acids. In another study, Górnaś et al. (2016) separated more than 30% (w/w) of oil from petrilas nr. 49 variety of pear fruit. This group followed orbital type extraction mediated with ultrasound radiation. Similarly, Matthäus and Özcan (2015) produced 31.7 g oil from pear seed when extracted *via* Soxhlet extractor using *n*-hexane as an extraction solvent. Recently, Hashemi et al. (2018) isolated lipids equal 22.4 g/100 g of seeds from wild pear (*Pyrus glabra* Boiss) using *n*-hexane as an

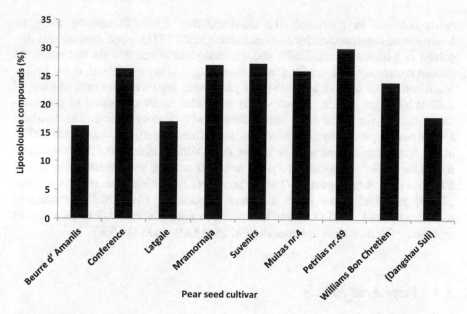

Fig. 47.3 Oleoginous constituents (% w/w) in the seeds of various pear cultivars (Górnaś et al. 2016; Yukui et al. 2009)

extraction solvent. Overall, the literature shows the oleaginous compounds in the seeds of pear fruit vary with agro-climate conditions, and fruit cultivar (Fig. 47.3). It was interesting to note that the reported range of fixed oil in pear seed is comparable with linseed and peanuts, higher than apple and cotton and smaller than apricots.

3 Composition of Pear Seed Oils

The non-volatile fraction of lipids obtained from pear seed mainly contains triglycerides (TAG), phospholipids, steryl glycosides, terpenes, tocochromanols, and hydrocarbons. In many cases, acid or alkaline hydrolysis has been adopted to break fatty acids and steryl esters into free phytosterols and fatty acids. The saponification of fixed oil typically with 1–5 N ethanolic or methanolic KOH at 60–80 °C under continuous stirring in the absence of air may completely saponify TAG into salts of fatty acids which may in turn be converted into fatty acid methyl esters (FAME) in the presence of catalysts (BF_3 or mineral acids), whereas steryl ester are set set free as sterols. The non-saponifiable lipids are often extracted using non-polar solvents like heptane, *n*-hexane, or diethyl ether. Both phases are subsequently separated, evaporated, redissolved in a proper solvent, cleaned up, and finally analyzed by gas chromatography coupled with flame ionization (GC-FID) or mass spectrometery (GC-MS) based detectors. Another alternative for the liquid extraction of unsaponifiable matter was introduced by Nestola and Schmidt (2016), who used aluminum oxide columns to fractionate the unsaponifiable lipids. In another assay by International Organization for Standardization (ISO 1978), plant seed oil was dissolved in n-heptane, mixed with sodium methylate in screw tight test tube, and shaken vigorously for 1–2 min. The resultant aliquot was washed with water, hydrolyzed with mineral acid and non-polar phase was separated. A small amount of sodium hydrogen sulfate (monohydrate) was added to remove traces of polar solvents and the non-polar layer was collected with a sharp needle. The resultant FAME were separated by non-polar capillary column conditions from 150 to 250 °C at 1.5 °C/min using hydrogen as carrier gas. Deineka and Deineka (2004) determined the TAG composition of *Pyrus communis* seed oil and observed that almost 27.6% of this oil belongs to the TAG of linoleic (L) and oleic acids (O) in the forms of L_2O and LnO_2. The other important fraction of TAG (20.3%) containing α-linolenic was L_3+LnLO. It was interesting to note that the TAG of palmitic (P) and stearic (S) acid were in minor except L_2S+LOP+LnOS (11.3%).

3.1 Fatty Acid Profile

In addition to being a viable source of energy, the long and short chain fatty acids participate in the physiology of various body functions. The major functions/body parts regulated by fatty acids include membrane structures and functions, intercellular signals, production of other bioactives and most important one gene expression and transcription. It is generally believed that majority of unsaturated fatty acids provide first line protection against cardiovascular diseases and also helpful for human health, well being and immune system. In contrast, the saturated fatty acids like lauric, myristic, palmitic and stearic acid may raise total and low-density lipid cholesterol to cause blood coagulation, inflammation, coronary heart and cardiovascular diseases and type 2 diabetes (Calder 2015).

The essential fatty acids (EFA) are required for the proper development and working of the brain and nervous system, for the formation of healthy cell membranes and for the manufacture of hormone-like substances called eicosanoids (thromboxanes, leukotrienes, and prostaglandins). These substances regulate several functions in the body including blood pressure, vasoconstriction, blood viscosity, immune and inflammatory responses. Dietary sources of the EFA include some leafy vegetables, seeds nuts, grains, vegetable oils and meats (Davis 2005).

The data plotted in Fig. 47.4 discloses that pear seeds of all cultivars contain polyunsaturated fatty acids (PUFA) in larger quantities as compared to monounsaturated (MUFA) and saturated fatty acid (SFA). The highest fraction (70%) of PUFA was documented by Yukui et al. (2009) in pears (Dangshau Suli) from China. Similarly pear variety "conference" contains the highest percentage of MUFA (38%). The lipids produced from pear seeds of all the investigated cultivars (Fig. 47.4) contained almost the same quantity of saturated fatty acids. Several authors have tried to compare its lipid profile with cactus family (prickly pears) which does not make any sense. However, the PUFA contents of pear seed were

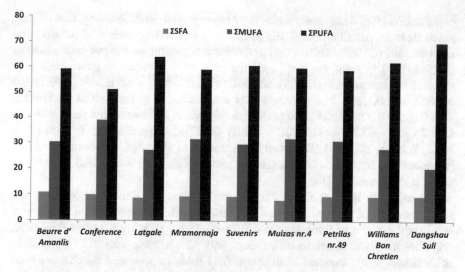

Fig. 47.4 Percent composition of fatty acid present in pear seed oil (Yukui et al. 2009)

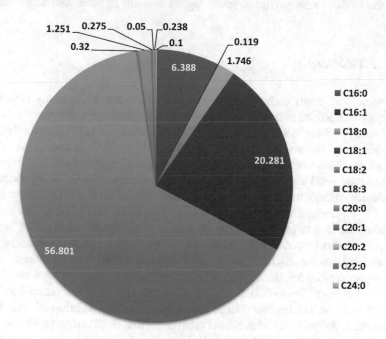

Fig. 47.5 The relative abundance (g/100 g) of various fatty acids in pear seed oil

found to be higher than tap ranked edible oils like olive oil (21%), sesame oil (45%), corn oil (47.7%), soy oil (49.7%) and sunflower oil (49.7%).

The pear seeds, although smaller in quantity, contain numerous saturated and unsaturated fatty acids. Pear seeds oil contains different fatty acids which vary in composition (Fig. 47.5). Four major fatty acids present in different species of pear seeds includes palmitic acid (C16:0), stearic acid (C18:0), oleic acid (C18:1) and linoleic acid (C18:2), while five fatty acids are present in insignificant amounts (0.19–1.08%) in some but not in all the pear seeds, which are palmitoleic acid (C16:1), α-linolenic acid (C18:3), arachidic acid (C20:0), gondoic acid (C20:1) and behenic acid (C22:0). The saturated fatty acids occurred in the highest amount are C16:0 (6.13–8.52%) and C18:0 (1.04–1.31%), while the unsaturated fatty acids are C18:2 (50.73–63.78%) and C18:1 (27.39–38.17%) in pear seeds (Fromm et al. 2012; Yukui et al. 2009). The overall percentage of saturated, monounsatured and polyunsatured fatty acids in pear seed oil was 9.48%, 31.21% and 59.32%, respectively (Górnaś et al. 2016).

According to Benn et al. (1996), plasmatic, oleic and linoleic acids constitute more than 90% of the total fatty acid in pear seed oil. The pear seed oil of contains the largest amount of linoleic acid (50.73–63.78%), an *omega*-6 class fatty acid, while the mostly used canola oil contains only 32% of oleic acid and 15% linoleic acid. However, the amounts of different fatty acids in peer seed oil changes with cultivar, maturation state, and geographical conditions. In general, the oil separated from the seeds of mature fruit contains higher amounts of unsaturated fatty acids.

3.2   Phytosterols

Phytosterols or simply plant sterols are similar to cholesterols in many aspects and necessary to stabilize phospholipid bilayer of cell membranes and many other cell structures. When taken as diet, the absorption of cholesterol in the intestinal mucosa and ultimately can reduce its level of blood by 10–15% (Chen et al., 2008). Phytosterols have been used as a cholesterol-lowering agents for many years. The United States Food and Drug Administration has endorsed the phytosterols "as part of a dietary strategy to reduce the risk of coronary heart diseases". Kritchevsky and Chen (2005) documented the function of phytosterols, according to the author s daily intake of 3 g of phytosterol (or their reduced form stanols) is associated with a consistent and reproducible reduction in LDL cholesterol concentrations up to 10% and reduces the risk of coronary heart disease by 20% over a lifetime.

The pear seed oil has been found to be an outstanding source of phytosterols with a total value from 276 to 600 mg/100 g oil depending upon extraction conditions and fruit cultivar. The highest concentration of sterols was determined from the oil of 'Beurre d' Amanlis' cv whereas oil recovered from cv. 'Petrilas nr.49′ contained the lowest concentration of phytosterols. Moreover, the sterol concentration of pear seed oil was found to be comparable with certain cultivars of its relative "apple seeds". Overall, seven sterols including cholesterol, β-sitosterol, citrostadienol, campesterol, gramisterol, Δ5-avenasterol, and Δ7-stigmasterol were found in pear seed oils. The only β-sitosterol constituted more than 80% (up to 500 mg/100 g oil) of total sterols followed by campesterol (13.9–29.7 mg/100 g oil), Δ5-avenasterol

(9.0–20.6 mg/100 g oil), cholesterol (6.1–26.4 mg/100 g oil) and Δ7-stigmasterol (2.2–15.8 mg/100 g oil), respectively. The gramisterol and citrostadienol were present in traces 2.0–7.8 and 1.1–3.8 mg/100 g oil, respectively. The authors of this investigation, however, observed a negative correlation between the concentration of sterols and oil yield, wherein the cultivars containing higher amounts of oil contained smaller amounts of these bioactives (Górnaś et al. 2016).

3.3 Tocochromanols

Tocochromanols are a class of oleaginous compounds containing polar moiety originated from tyrosine and a polyprenyl side chain (hydrophobic) from the isoprenoid pathway. These compounds can be further with respect to side chain origin, tocopherols (phytyl-derived side chain) and tocotrienols (geranylgeranyl derived side chain) and degree of methylation of the polar moiety (α, β, γ, and δ). Overall these compounds are lipid-soluble and offer a wide range of antioxidants activities towards free radicals and peroxides formed during the storage of oil as well as in body physiology. Moreover, the tocochromanols work as essential dietary nutrients for mammals as tocotrienol and tocopherol homologous (vitamin E) (Eitenmiller and Lee 2004). Recently, the health benefits of tocotrienols gathered increased attention, as less is known about them in contrast to tocopherols. Tocotrienols have great potential against various cancers and many deadly/chronic diseases of the twenty-first century (Aggarwal et al. 2010).

Likewise phytosterols, the pear seed oil also contain noticeable amounts of tocochromanols (120.5–216.1 mg/100 g oil). In this case, γ-tocopherol was exceptionally higher (104–190 mg/100 g oil) than α-, and β-tocopherols which were range 8–19, 0.5–1.1, and 1.8–5.3 mg/100 g –oil, respectively. Moreover, α- and γ- tocopherol contents of pear seed oil vary significantly with fruit ecotype (Fig. 47.6).

The same authors (Górnaś et al. 2015) in another study reported that pear seeds of these cultivars contain α, β, γ, and δ-tocopherols in the concentration range 4.1–6.0, 0.11–0.31, 32–42, and 0.43–1.21 mg/100 g seed. Recently, Hashemi et al. (2018) isolated lipids from the wild pear (*Pyrus glabra* Boiss) seed for the determination of tocochromanols. This group applied Soxhlet extraction with petroleum ether for the extraction of oleaginous compounds. It was interesting to note that total tocochromanols 67.4 mg/100 g oil in wild pear seed oil was much lower than those reported by Górnaś et al. (2016) in *Pyrus communis* seed oils. Similarly, the level of individual tocopherols (α, β, γ, and δ) were also inconsistent, wherein Górnaś et al. (2016) reported γ-tocopherols as outstanding tocochromanol while Hashemi et al. (2018) observed wild pear as the richest source of α-tocopherol (57.60 mg/100 g oil). In the case of tocotrienol, it was observed that pear seed oil of various cultivars contains only traces of these bioactives. Hashemi et al. (2018) found that wild pear seed oil contains the highest amount of α-tocotrienol equal to 1.1 mg/100 g oil. Other tocotrienols (β, γ, and δ) were found in traces.

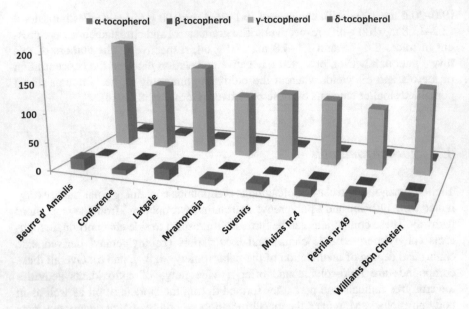

Fig. 47.6 The variation of tocopherols in pear seed oil with fruit ecotype (Górnaś et al. 2016)

4 Minor Bioactive Lipids

Pear seeds in addition to fatty acids and phytosterols, contain various nutritional and non-nutritional phytochemicals of high therapeutic and industrial use. For example, the pear seed falls among seeds which contain squalene (a natural 30-carbon triterpene), which is a precursor for steroids synthesis in plants and animals. These compounds have been frequently used in cosmetic industry for their ability to lubricate and protect skin. In pharmaceutical industries, these phytoconstituents are used as an adjuvant or for the preparation of adjuvant which can stimulate the immune system and improve the efficiency of vaccines. Górnaś et al. (2016) analyzed pear seed oil for squalene content and observed that pear seed oil may contain these phytoconstituents up to 26.4–40.8 mg/100 g oil.

The similarly class of compounds i.e. triterpenoids are famous for their established anticancer properties (Lesellier et al. 2012). In seeds of pear, triterpenoids including betulinic, oleanolic and ursolic acid are present in a small amount almost 736.2 µg/g dry matter. Similarly, carotenoids are natural isoprenoid pigments produced mainly in photosynthetic, but also in some non-photosynthetic organisms such as bacteria and fungi, with more than 700 various known compounds, some of which have pro-vitamin A activity. Low intake of carotenoid can be associated with an increased risk of cataract, cancer, skin damage and cardiovascular diseases (Aust et al. 2001).

Phenolics covers another well known, top-ranked and widely distributed class of bioactives. Phenolics including catechins, flavonoids, and phenolic acids have been often linked with a diversity of health benefits ascribed to fruits and vegetables. Hashemi et al. (2018) examined wild pear seed oil for its phenolic acids and observed 29.8 mg of total phenolic equivalents/100 g oil. Although this quantity is negligible as compared to other fruits, vegetables or their products but the prevalence of phenolic compounds in any natural product depends upon product preparation/extraction methods. The typical solvents applied to the preparation of pear seed oils are usually non-polar, which could hardly isolate the phenolic bioactives from the biological matrix. However, the technical research shows that application of supercritical fluids like supercritical carbon dioxide produces the lipids containing higher levels of phenolic bioactives, which in turn can offer better antioxidant character and an elongated shelf life (Borges et al. 2010).

5   Composition of Fruit Essential Oil

The essential oils include the aromatic compounds and volatile liquid also known as ethereal oils separated out *via* extrusion, expression, extraction, or distillation. The most abundant and eldest method involves the isolation of volatiles using steam or hot water (hydro or steam-distillation). In spite of the being "Gifts of God", pear fruits have not been completely authenticated for their aroma and volatile oils. Although, few researchers have tried to establish the chemical nature of volatile and aroma compounds originating from fruit and peel of pear, there is not a single scientific report on pear seed essential oil chemistry or health benefits. One reason behind this might be the exceptional small seed to fruit ratio but others include its odorless physiology and texture. Nevertheless, other parts of pear tree like flowers, leaves, stem, and fruit have been tested for the presence of volatile compounds and it has been observed that more than 30 volatile derivatives of terpenes (15), alcans (7), aldehydes (2), esters (4) and allyl alkoxy benzene (3) were abundant in pear parts. Individually, pear leaves were found to a rich source of estragol, whereas stem, flower, and fruit contained a good quantity of allyl hexanoate, limonene, and benzyl butanoate, respectively.

The non-technical data indicates that various parts of pear fruits have been used for the preparation of perfumes and fragrances. The ethereal oils and perfumes produced from pear are popular among women, however, the exact sensation, antimicrobial properties and preventive role of essential oils or their products would vary chemical constituents. There is sound evidence to conclude that essential oils, perfumes or cosmetic products produced from pear seed would be rich in ester known as pear ester. For examples, Anapear (methyl (4E)-octa-4,7-dienoate) bearing green pear note, is a major fragrance ingredient of Emporio Armani Elle (Schilling et al. 2010). Similarly, 3(Z)-octenyl propionate, another fine pear odorant, is marketed under the brand name Pearlate® by Bedoukian Research Inc., USA. Overall, the pear seed volatile and order compounds include alcohols, hydrocarbons, aldehydes, esters, ketones and sulfur compounds (Heimeur et al. 2016).

6 Contribution of Bioactive Compounds to Organoleptic Properties of Oil

Organoleptic properties are related to color, texture, flavor and dietary attributes of a food product. The pear fruits are excellent sources of minerals, vitamins, essential and non-essential fatty acids, organic acids, fibers, and phytochemicals of high therapeutic value. The pear seed oil likewise can be an excellent value addition as edible oil or similar products. The seed oil composition and organoleptic properties may vary with fruit ecotype, agro-climate condition, fruit maturity and oil expression methods. Although few researchers have extracted pear seed fixed oil using Soxhlet extractors but none of these have examined physiochemical characteristics of the oil. Similarly, no attempts have been undertaken to investigate the characteristics of compounds responsible for color, odor, and flavor of pear seed oil.

7 Health-Promoting Traits of Pear Seed Oil

To the date, pear seed oil has been explored for the fatty acid profile, tocochromanols (tocopherols and tocotrienols) and selected minor bioactives like phenolics which are powerful antioxidants in different environments. Tocochromanols protect oil and the human body as well against oxidation of lipophilic contents especially unsaturated fatty acids (Neunert et al. 2015; Nogala-Kałucka et al. 2013). The presence of carotenoids renders the food products effective against inflammation, cardiovascular diseases, and age-related macular degeneration (Yahia et al. 2017). Although literature lack any human intervention or *in vivo* application of peer seed oil, but the knowledge of oil phytochemistry suggests that peer seed oil may offer promising health benefits against inflammation, cardiovascular diseases, common cancers, and visual impairment (Olmedilla et al. 2003).

In contrast, other parts of pear plant or their extracts have been evaluated for wound healing, liver-protection, alcohol hangover, weight control, allergic and respiratory problems, cardiovascular diseases, and diabetes (Lairon 2011). The pear juice particularly contains, chlorogenic acid and other potential phenolics, fibre, sorbitol, and fructose which can regular normal bowel function but again there is no report on human intervention.

8 Edible Applications/Prospects of Pear Seed Oil

Pear fruit oil was found to be an outstanding source of α-linoleic acid, an *omega*-6 fatty acid that lowers the risks of various cancerous diseases. The presence of *omega*-6 unsaturated fatty acids suggests that pear seed oil may be beneficial against atherosclerosis, obesity, and diabetes (Simopoulos 2008). It contains vitamin E derivatives used in different cosmetic products. So, it can be used in cosmetic oleic acid effectively used in the prevention of ischemic heart diseases (Massimo et al. 2009). It has been shown that oleic acid inhibits platelet aggregation induced by platelet aggregation factor as well as the secretion of serotonin. Moreover, the presence of γ-tocopherol at elevated level recommends that pear seed oil can inhibit human cancer progression and cell proliferation. The aroma compounds in pear seeds or their essential oil have not be explored yet but they may contain phytochemicals of high therapeutic and industrial importance.

References

Aggarwal, B. B., Sundaram, C., Prasad, S., & Kannappan, R. (2010). Tocotrienols, the vitamin E of the 21st century: Its potential against cancer and other chronic diseases. *Biochemical Pharmacology, 80*, 1613–1631.

Aust, O., Sies, H., Stahl, W., & Polidori, M. C. (2001). Analysis of lipophilic antioxidants in human serum and tissues: Tocopherols and carotenoids. *Journal of Chromatography. A, 936*, 83–93.

Benn, F. W., Dattilo, M., Cornell, W. L. (1996). Flotation of lead sulfides using rapeseed oil. United States patent No US 5,544,760. 1996 Aug 13.

Borges, G., Mullen, W., & Crozier, A. (2010). Comparison of the polyphenolic composition and antioxidant activity of European commercial fruit juices. *Food & Function, 1*, 73–83.

Calder, P. C. (2015). Functional roles of fatty acids and their effects on human health. *Journal of Parenteral and Enteral Nutrition, 39*, 18S–32S.

Chen, Z.-Y., Jiao, R., & Ma, K. Y. (2008). Cholesterol-lowering nutraceuticals and functional foods. *Journal of Agricultural and Food Chemistry, 56*, 8761–8773.

Davis, B. (2005). Essential fatty acids in vegetarian nutrition. Andrews University Nutrition Department. Available at https://www.andrews.edu; Accessed on August. 18, 2005.

Deineka, V. I., & Deineka, L. A. (2004). Type composition of triglycerides from seed oils. II. Triglycerides from certain cultivated plants of the Rosaceae Family. *Chemistry of Natural Compounds, 40*, 293–294.

Eitenmiller, R. R., & Lee, J. (2004). *Vitamin E: Food chemistry, composition, and analysis.* Boca Raton: CRC Press.

Fromm, M., Bayha, S., Kammerer, D. R., & Carle, R. (2012). Identification and quantitation of carotenoids and tocopherols in seed oils recovered from different Rosaceae species. *Journal of Agricultural and Food Chemistry, 60*, 10733–10742.

Gallardo, R. K., Kupferman, E. M., Beaudry, R. M., Blankenship, S. M., Mitcham, E. J., & Watkins, C. B. (2011). Market quality of Pacific Northwest pears. *Journal of Food Distribution Research, 42*, 89–99.

Górnaś, P., Rudzińska, M., & Segliņa, D. (2014a). Lipophilic composition of eleven apple seed oils: A promising source of unconventional oil from industry by-products. *Industrial Crops and Products, 60*, 86–91.

Górnaś, P., Siger, A., Czubinski, J., Dwiecki, K., Segliņa, D., & Nogala-Kalucka, M. (2014b). An alternative RP-HPLC method for the separation and determination of tocopherol and tocotrienol homologues as butter authenticity markers: A comparative study between two European countries. *European Journal of Lipid Science and Technology, 116*, 895–903.

Górnaś, P., Mišina, I., Lāce, B., Lācis, G., & Segliņa, D. (2015). Tocochromanols composition in seeds recovered from different pear cultivars: RP-HPLC/FLD and RP-UPLC-ESI/MSn study. *LWT- Food Science and Technology, 62*, 104–107.

Górnaś, P., Rudzińska, M., Raczyk, M., Mišina, I., Soliven, A., & Segliņa, D. (2016). Chemical composition of seed oils recovered from different pear (*Pyrus communis* L.) cultivars. *Journal of the American Oil Chemists' Society, 93*, 267–274.

Greenbank, G. R., & Holm, G. E. (1924). Some factors concerned in the autoxidation of fats. *Industrial and Engineering Chemistry, 16*, 598–601.

Griffith, M. P. (2004). The origins of an important cactus crop, *Opuntia ficus-indica* (Cactaceae): New molecular evidence. *American Journal of Botany, 91*, 1915–1921.

Hashemi, S. M. B., Khaneghah, A. M., Barba, F. J., Lorenzo, J. M., Rahman, M. S., Amarowicz, R., & Movahed, M. D. (2018). Characteristics of wild pear (*Pyrus glabra* Boiss) seed oil and its oil-in-water emulsions: A novel source of edible oil. *European Journal of Lipid Science and Technology, 120*, 1700284.

Heimeur, N., Idrissi Hassani, L. M., Serghini, M. A., Bessiere, J. M. (2016). Study of volatile compounds of *Pyrus mamorensis* Trab. a characteristic plant of Mamora forest (north-western Morocco). Moroccan Journal of Chemistry, 4(1), 1-4.

ISO. (1978). *Animal and vegetable fats and oils: preparation of methyl esters of fatty acids.* Geneva: International Organization for Standardization.

Kemper, T. G. (2005). Oil extraction. Bailey's industrial oil and fat products. John Wiley & Sons, Inc. pp: 63-68

Kritchevsky, D., & Chen, S. C. (2005). Phytosterols-health benefits and potential concerns: A review. *Nutrition Research, 25*, 413–428.

Lamsal, B. P., Murphy, P. A., & Johnson, L. A. (2006). Flaking and extrusion as mechanical treatments for enzyme-assisted aqueous extraction of oil from soybeans. *Journal of the American Oil Chemists' Society, 83*, 973–979.

Lesellier, E., Destandau, E., Grigoras, C., Fougère, L., & Elfakir, C. (2012). Fast separation of triterpenoids by supercritical fluid chromatography/evaporative light scattering detector. *Journal of Chromatography. A, 1268*, 157–165.

Lairon D. (2011). Nutritional quality and safety of organic food. A review. *Médecine & Nutrition. 47*(1):19–31.

Massimo, C., Lucio, T., Jesus, M., Giovanni, L., & Caramia, G. M. (2009). Extra virgin olive oil and oleic acid. *Nutrición Clínica y Dietetica Hospitalaria., 29*, 12–24.

Matthäus, B., & Musazcan Özcan, M. (2015). Oil content, fatty acid compon and distributions of vitamin-E-active compounds of some fruit seed oils. *Antioxidants, 4*, 124.

Mushtaq, M. (2018). Extraction of fruit juice: An overview A2 – Rajauria, Gaurav. In B. K. Tiwari (Ed.), *Fruit juices* (pp. 131–159). San Diego: Academic.

Nestola, M., & Schmidtositi, T. C. (2016). Fully automated determination of the sterol composition and total content in edible oils and fats by online liquid chromatography-gas chromatography-flame ionization detection. *Journal of Chromatography. A, 1463*, 136–143.

Neunert, G., Górnaś, P., Dwiecki, K., Siger, A., & Polewski, K. (2015). Synergistic and antagonistic effects between alpha-tocopherol and phenolic acids in liposome system: Spectroscopic study. *European Food Research and Technology, 241*, 749–757.

Nogala-Kałucka, M., Dwiecki, K., Siger, A., Górnaś, P., Polewski, K., & Ciosek, S. (2013). Antioxidant synergism and antagonism between tocotrienols, quercetin and rutin in model system. *Acta Alimentaria, 42*, 360–370.

Olmedilla, B., Granado, F., Blanco, I., & Vaquero, M. (2003). Lutein, but not α-tocopherol, supplementation improves visual function in patients with age-related cataracts: A 2-y double-blind, placebo-controlled pilot study. *Nutrition, 19*, 21–24.

Reiland, H., & Slavin, J. (2015). Systematic review of pears and health. *Nutrition Today, 50*, 301.

Simopoulos, A. P. (2008). The importance of the omega-6/omega-3 fatty acid ratio in cardiovascular disease and other chronic diseases. *Experimental Biology and Medicine, 233*, 674–688.

Statistics, F. (2016). Productions, crops. Retrieved 04 Aug 2018.

Yahia, E. M., de Jesús Ornelas-Paz, J., Emanuelli, T., Jacob-Lopes, E., Zepka, L. Q., & Cervantes-Paz, B. (2017). Chemistry, stability, and biological actions of carotenoids. *Fruit and Vegetable Phytochemicals: Chemistry and Human Health, 2*, 285.

Yukui, R., Wenya, W., Rashid, F., & Qing, L. (2009). Fatty acids composition of apple and pear seed oils. *International Journal of Food Properties, 12*, 774–779.

Chapter 48
Amla (*Emblica officinalis* L.) Oil

Monika Choudhary and Kiran Grover

Abstract Amla (*Emblica officinalis* L.), the native of India belongs to Euphorbiaceae family and is widely distributed in other tropical countries of Asia such as China, Indonesia and Malaysia.

Amla is considered as nature's blessing to the mankind due to its nutritional components and various nutraceutical properties which have been well documented in conventional medication system. The amla fruit having greenish-yellow color tastes sour with an intriguing sweet flavor. Owing to its nutritional components such as vitamin C, minerals, polyphenols including ellagitannins (tannins), flavonoids, ellagic acid and other phytochemicals, amla has numerous pharmacological and therapeutic properties viz. Analgesic, adaptogenic, anticancer, cardio protective and gastro protective. Besides, amla seeds are comprised of a significant quantity of essential fatty acid viz. Linolenic acid and linoleic acid. The antibacterial activity of essential oils of amla has also been well documented in the literature and amla oil has been used for centuries in medicine world for its external applications; for instance nourishing hair and scalp.

Keywords Active compounds · Essential oils · Functional properties

Abbreviations

MUFA   Monounsaturated fatty acid
PUFA   Polyunsaturated fatty acid
SFA   Saturated fatty acid

M. Choudhary (✉)
Punjab Agricultural University, Ludhiana, India

K. Grover
Department of Food and Nutrition, College of Home Science, Punjab Agricultural University, Ludhiana, India
e-mail: kirangrover@pau.edu

© Springer Nature Switzerland AG 2019
M. F. Ramadan (ed.), *Fruit Oils: Chemistry and Functionality*,
https://doi.org/10.1007/978-3-030-12473-1_48

1 Introduction

Amla (*Emblica officinalis* L.) also known as Indian gooseberry belongs to Euphorbiaceous family and is widely distributed in tropical and subtropical countries such as India, China, Pakistan, Uzbekistan, Sri Lanka, Indonesia and Malaysia (Liu et al. 2008). The fruit is well documented in the Indian traditional systems of medicine like Ayurveda, Siddha for its enormous health-promoting properties (Vasant et al. 2013). Amla is a small to medium-sized deciduous tree, which grows about 8–18 m height with thin light grey bark having light green subsessile leaves. The most useful part of this miraculous tree- the amla fruit which is fleshy, pale yellow with six obscure vertical furrows and spherical in shape (Indian Medicinal Plants 1997). The amla fruit is well known for its pharmacological and medicinal properties owing to the presence of bioactive compounds, which are used to treat various diseases. The amla fruit has been evaluated by various researchers for its bioactive components and their implications on human health (Arora et al. 2011). Amla fruit can be used by itself or in combination with other plants to treat many ailments. However, a very few studies have been conducted with regard to pharmacological aspects of amla oil except its use for hair growth. Amla fruit seed yields 18% oil, which contains bioactive components and essential fatty acids. The oil is of brownish yellow color and is liquid at room temperature with a strong, musky smell. Several studies have reported the occurrence of chemical constituents in amla fruit; however there are very few reports highlighting the essential oil of the fruits.

2 Processing and Extraction

Preparation of amla oil Traditionally, amla oil is prepared by immersing dried amla fruit in base oils such as mineral oil, coconut oil, or sesame oil (Mithal and Shah 2000; Sanju et al. 2006; Hiremath 2007). After soaking for few days, the fruit is removed and the oil is filtered and purified. In mechanical method, a multi-step process is followed for the extraction of amla oil. Initially, amla fruit is crushed and boiled to expel the natural fruit oils. Then, fruit oils are infused under vacuum pressure and the final hydro-extract is warmed in base oil which may be any of the above-mentioned oils and the remaining moisture is removed. The seeds are expeller pressed separately and the pressed seed oil is collected and added to the base oil extract to produce the final amla oil.

Extraction of essential oils The principle of the method is that raw plant material is kept into distillation apparatus over water. The steam produced from the heated water is passed through the plant material, which ultimately vaporizes the volatile compounds. The vapors flow through a coil where condensation takes place and liquid is then collected in the receiving vessel (Saxena and Patil 2014). One kilogram fruits are used to obtain the essential oil by hydrodistillation using Clevenger apparatus (80–100 °C, 4 h). The aqueous layer is extracted using *n*-hexane. The obtained oil is dried using anhydrous sodium sulfate and stored in dark glass vial in a refrigerator until it is analyzed. The yield of the oil is calculated based on the weight of fresh fruits (Amir et al. 2014).

3   Fatty Acid Composition

The amla oil contains myristic acid (C14:0), palmitic acid (C16:0), stearic acid (C18:0) and oleic acid (C18:1). The amla seed oil contains saturated fatty acids (SFAs) (9.0%) of which myristic, palmitic and stearic acids contribute 3.6%, 2.3% and 3.1%, respectively. The amount of fatty acids varies somewhat depending upon the variety of amla grown in different regions of the world (Pathak et al. 2003; Singh et al. 2005). Oleic acid predominantly constitutes 26.4% of the monounsaturated fatty acids (MUFAs). Also, *Emblica officinalis* fruit is the potential source of linoleic acid (ω-6) which is evident with its share of 51.0% in polyunsaturated fatty acids (PUFAs). Amla seed oil also contains a reasonable amount of omega-3 fatty acid (linolenic acid) i.e., 11.8% (Table 48.1). Several research reports have documented the inevitable role of omega-3 fatty acid in preventing coronary artery diseases (Lee and Lip 2003).

4   Major and Minor Components in Amla Oil and Their Biological Activity

Many researchers who studied benefits of amla have also documented various major and minor biological active components in amla oil (Zhang et al. 2003; Mishra and Mahanta 2014) such as polyphenols, tannins, gallic acid, ellagic acid, emblicanin A

Table 48.1 Fatty acid composition of *Emblica officinalis* oil

Fatty acid	% by weight
Saturated fatty acids	
Lauric (C12:0)	–
Myristic (C14:0)	3.6
Palmitic (C16:0)	2.3
Stearic (C18:0)	3.1
Monosaturated fatty acids	
Oleic (C18:1)	26.4
Polyunsaturated fatty acids	
Linoleic (C18:2)	51.0
Linolenic (C18:3)	11.8

Source: Arora et al. (2011)

& B, phyllembein, quercetin and ascorbic acid. The major components include 9,12,15 octadecatrienoic acid, tetradecanoic acid and linoleates, while benzoic acid, 6 tetradecansulfonic acid, hydroquinone, dodecane 1-fluoro, phthalic acid 2-cyclohexylethyl isobutyl ester are the minor constituents of amla seed oil. These components have also been studied for their biological activities such as tetradecanoic acid for antioxidant activity and 9, 12, 15, octadecatrienoic acid for anti-inflammatory and antiarthritic properties (Lalitharani et al. 2009; Maruthupandian and Mohan 2011). Dodecane and phthalic acid are also present in small quantities, which are reported to have antibacterial properties (Adeleye et al. 2011).

5 Composition of Fruit Essential Oil

Essential oils are natural aromatic compounds, which represent an extremely multifaceted class of natural product chemistry (Saxena and Patil 2014). These are found in the seeds, bark, stems, roots, flowers, and other parts of plants and are being used in many industries such as cosmetics perfumes, beverage, ice creams, confectionary and backed food products (Asghari et al. 2012). In the Egyptian history the extensive use of these aromatic compounds in various sectors such as medical, beauty treatment, food preparation and in religions ceremony have been preserved.

However, very few researchers exhibited their concern for the essential oils present in amla fruit. Researchers investigated the essential oil from the amla fruit grown in Guangdong Province of China using hydrodistillation (HD-EO) and supercritical fluid extraction (SFE-EO) wherein 31 and 26 compounds were found in HD-EO and SFE-EO, respectively (Liu et al. 2009). Decanal, β-caryophyllene, β-bourbonene, camphor, β-elemene, limonene, methyl eugenol, 1-octen-3-ol, borneol, nerol and myrecene were the major constituents of HD-EO. SFE-EO contained high amounts of tetracosane and palmitic acid which were absent in HD-EO. In another study, the essential oils were extracted using steam distillation from amla fruits grown in Sichuan region of China. The analysis identified 43 compounds in which α-furfural, 2-chloro-bicyclooct-5-ene-2- carbonitrile, methyl salicylate, trans-2-decenal, hexahydrofarnesyl acetone were the major constituents indicating that the constituents of the essential oil of *P. emblica* differ according to the habitat from place to another (Wang et al. 2009). Whereas, Amir et al. (2014) analyzed 42 compounds constituting 96.13% of the essential oil and categorized into esters, hydrocarbons, aldehydes, ketenes and alcohols. Esters contributed a major portion (33.2%) of essential oil of which 14.2% was constituted by methyl salicylate. The percentage of hydrocarbons was 30.2 wherein undecane shared the major portion (7.55%). Other major components present in the essential oil extracted from amla oil were aldehydes (20.9%) in which benzaldehyde (11.9%) occupied the topmost position. The percentage of alcohols and ketones were 6.23% and 5.31%, respectively (Table 48.2).

Table 48.2 Components of essential oil extracted from *Emblica officinalis*

Essential oil constituent	Percentage
Aldehydes	
Benzaldehyde	11.98
Cumin aldehyde	4.64
Myristaldehyde	0.23
Sterayl aldehyde	0.12
Vertocitral	0.65
β-cyclocitral	0.37
2-decenal	0.8
γ-terpinen-7-al	2.2
Alcohols	
2,4-hexadienol	0.47
Menth-2-en-1-ol	0.48
Coahuilensol	4.57
Nerolidol-Z-	0.32
Dihydro apofarnesol	0.44
Esters	
Methyl salicylate	14.28
2-methyl butyl acetate	8.6
Isopropyl,2-methyl butyrate	0.03
Ethyl benzoate	8.03
Sabinene hydrate acetate	0.42
Ethyl hexadecanoate	0.33
Ethyl cinnamate-E-	1.57
Hydrocarbons	
n-hexadecane	0.50
n-Eicosane	0.34
n-tricosane	0.73
Tetracosane	0.78
Pentacosane	0.82
Hexacosane	0.64
Heptacosane	0.48
Nonacosane	3.54
Menthane	0.68
Decane	6.82
Undecane	7.55
Butyl cyclohexane	0.31
2,6-dimethyl undecane	0.47
Ketones	
Acetophenone	4.16
2-pentadecanone-6,10,14- trimethyl	0.49
Farnesyl acetone	0.32
β-damascenone-Z-	0.34
Terpenes	
Capillene	1.03
α-farnesene-E,E	2.21
1-octadecene	0.20
Butyl cyclohexene	2.97
2-methyl decalin	0.22

Source: Amir et al. (2014)

6   Health Benefits

Pharmacological research has reported various efficacious properties of amla fruit with regard to prevention and treatment of several non-communicable diseases like cardiovascular diseases, metabolic syndrome, cancer, gastrointestinal disorders, deficiency diseases, and hepatic diseases and disorders (Sharma et al. 2004; Yokozawa et al. 2007; Santoshkumar et al. 2013; Dasaroju and Gottumukkala 2014). However, these properties with reference to amla oil have not been studied and data pertaining to the pharmacological properties of amla oil are scanty. However, amla oil has been used for centuries in Ayurvedic medicine for its wide range of beneficial applications in hair nourishment and healthy scalp (Krishnaveni and Mirunalini 2010). The efficacy of amla oil for nourishing the scalp, conditioning dry and brittle hair and for promoting strong, healthy and shiny hair has been supported by the researchers. Researchers recognized amla oil as a powerful inhibitor of 5α-reductase. The medication finasteride, used to treat male baldness, also works by inhibiting 5α-reductase (Kumar et al. 2012). These exceptional properties have captivated the interest of manufactures to this ancient oil for its effective incorporation into shampoos, scalp treatments and other hair and scalp formulations (Gautam et al. 2012).

Furthermore, amla oil has also been investigated for its antibacterial and antimicrobial potencies (Baratta et al. 1998; Salehi et al. 2005). Antimicrobial resistance is a public health concern at the global level. In developing countries, community-based data has revealed an increase in the burden of antimicrobial resistance (Vyas and Patil 2011). The most common antimicrobial resistance bacteria are *Staphylococcus aureus* found in the environment affecting about 20% of the human populations who become long-term carriers of *S. aureus*. Thereby, powerful antibiotics are needed to destroy these bacteria, which do not cause any life-threatening complications (Sievert et al. 2002). However, a continuous spread of multi-drug resistant pathogens has become a serious threat to public health. Therefore, the discovery of new antimicrobial agents of plant origin which can cure these problems naturally is of utmost importance. Few studies have proven the antibacterial activity of *Emblica officinalis* essential oil, which is used for the treatment of *Staphylococcus aureus* causing diseases. For instance, Saxena and Patil (2014) have revealed that methanol solvent fruit extract of *Emblica officinalis* essential oil exhibits a strong inhibitory effect against *S. aureus* compared to the other solvents extracts of essential oil and positive control Gentamicin. In conclusion, there is an enormous scope for more extensive research to explore and evident the pharmacological aspects of amla oil using modern technologies. In addition, there is need to incorporate the proven medicinal implications of this fruit oil in the food industry as data pertaining to its utilization in various edible formulations is insufficient.

References

Adeleye, I. A., Omadime, M. E., & Daniels, F. V. (2011). Antimicrobial activity of essential oil and extracts of *Gonronema latifolium* decne. on bacterial isolates from blood stream of HIV infected patients. *Journal of Pharmacology and Toxicology, 6*(3), 312–320.

Amir, D. E., AbouZid, S. F., Hetta, M. H., Shahat, A. A., & El-Shanawany, M. A. (2014). Composition of the essential oil of the fruits of *Phyllanthus emblica* cultivated in Egypt. *Journal of Pharmaceutical, Chemical and Biological Sciences, 2*(3), 202–207.

Arora, A., Kumar, I., Sen, R., & Singh, J. (2011). *Emblica officinalis* (amla): Physico-chemical and fatty acid analysis from arid zone of Rajasthan. *International Journal of Basic and Applied Chemical Sciences, 1*(1), 89–92.

Asghari, G., Jalali, M., & Sadoughi, E. (2012). Antimicrobial activity and chemical composition of essential oil from the seeds of *Artemisia aucheri* Boiss. *Journal of Natural Pharmaceutical Products, 6*(2), 11–15.

Baratta, M. T., Dorman, H. J., Deans, S. G., Figueiredo, A. C., Barroso, J. G., & Rubert, G. (1998). Antimicrobial and antioxidant properties of some commercial essential oils. *Flavour and Fragrance Journal, 13*, 235–244.

Dasaroju, S., & Gottumukkala, K. M. (2014). Current trends in the research of *Emblica officinalis (Amla):* A pharmacological perspective. *International Journal of Pharmaceutical Sciences Review and Research, 24*(2), 150–159.

Gautam, S., Dwivedi, S., Dubey, K., & Joshi, H. (2012). Formulation and evaluation of herbal hair oil. *International Journal of Chemical Sciences, 10*(1), 349–353.

Hiremath, S. R. R. (2007). *Textbook of industrial pharmacy* (1st ed., pp. 99–102). Hyderabad: Orient Longaman Pvt. Ltd..

Indian Medicinal Plants (1997) *A compendium of 500 species part 3* (pp. 256–263). New Delhi: Orient Longman Publications.

Krishnaveni, M., & Mirunalini, S. (2010). Therapeutic potential of *Phyllanthus emblica* (amla): The ayurvedic wonder. *Journal of Basic and Clinical Physiology and Pharmacology, 21*, 93–105.

Kumar, N., Rungseevijitprapa, W., Narkkhong, N., Suttajit, M., & Chaiyasuta, C. H. (2012). 5α-reductase inhibition and hair growth promotion of some Thai plants traditionally used for hair treatment. *Journal of Ethnopharmacology, 139*, 765–771.

Lalitharani, S., Mohan, V. R., Regini, G. S., & Kalidass, C. (2009). GC-MS of ethanolic extract of *Pothos scandens* L. leaf. *Journal of Herbal Medicine and Toxicology, 3*, 159–160.

Lee, K. W., & Lip, G. Y. H. (2003). The role of omega-3 fatty acids in the secondary prevention of cardiovascular disease. *The Quarterly Journal of Medicine, 96*, 465–480.

Liu, X., Cui, C., Zhao, M., Wang, J., Luo, W., Yang, B., & Jiang, Y. (2008). Identification of phenolics in the fruit of emblica (*Phyllanthus emblica* L.) and their antioxidant activities. *Food Chemistry, 109*, 909–915.

Liu, X., Zhao, M., Luo, W., Yang, B., & Jiang, Y. (2009). Identification of volatile components in *Phyllanthus emblica* L. and their antimicrobial activity. *Journal of Medicinal Food, 12*(2), 423–428.

Maruthupandian, A., & Mohan, V. R. (2011). GC-MS analysis of ethanol extracts of *Wattakaka volubilis* (L.F) Stapf. leaf. *International Journal of Phytomedicine, 3*(1), 59–62.

Mishra, P., & Mahanta, C. L. (2014). Comparative analysis of functional and nutritive values of amla (*Emblica officinalis*) fruit, seed and seed coat powder. *American Journal of Food Technology, 9*(3), 151–161.

Mithal, B. M., & Shah, R. N. (2000). *A hand book of cosmetics* (1st ed., pp. 141–142). Delhi: Vallabh Prakashan.

Pathak, R. K., Pandey, D., Haseeb, M., & Tandon, D. K. (2003). *The anola*. India: Bulletin CISH Lucknow.

Salehi, P., Sonboli, A., Eftekhar, F., Nejad- Ebrahimi, S., & Yousefzadi, M. (2005). Essential oil composition, antibacterial and antioxidant activity of the oil and various extracts of *Ziziphora clinopodioidies*. *Iranian Biological and Pharmaceutical Bulletin, 28*, 1892–1896.

Sanju, N., Arun, N., & Roop, K. K. (2006). *Cosmetic technology* (1st ed., pp. 379–382). Delhi: Birla Publications Pvt. Ltd.

Santoshkumar, J., Manjunath, S., & Pranavkumar, M. S. (2013). A study of antihyperlipidemia, hypolipedimic and anti-atherogenic activity of fruit of *Emblica officinalis* (amla) in high fat fed Albino rats. *International Journal of Medical Research and Health Sciences, 2*(1), 70–77.

Saxena, R., & Patil, P. (2014). In vitro antibacterial activity of *Emblica officinalis* essential oil against *Staphylococcus aureus*. *International Journal of Theoretical and Applied Sciences, 6*(2), 7–9.

Sharma, S. K., Perianayagam, J. B., Joseph, A., & Christina, A. J. (2004). Evaluation of anti-pyretic and analgesic activity of *Emblica officinalis* Gaertn. *Journal of Ethnopharmacology, 95*, 83–85.

Sievert, D. M., Boulton, M. L., Stoltman, G., Johnson, D., Stobierski, M. G., Downes, F. P., Somsel, P. A., & Rudrik, J. T. (2002). *Staphylococcus aureus* resistant to vancomycin. *US MMWR, 51*, 565–567.

Singh, V., Singh, H. K., & Chopra, C. S. (2005). Studies on processing of aonla (phyllanthus emblica Garten.) fruits. *Beverage and Food World, 32*, 3–54.

Vasant, B. S., Bhaskarrao, D. A., & Bhanudas, S. R. (2013). *Emblica officinalis*- the wonder of ayurvedic medicine. *World Journal of Pharmaceutical Sciences, 3*(1), 285–306.

Vyas, P., & Patil, S. (2011). Antimicrobial activity of essential oils against multidrug resistant enterobacterial pathogens. *Trends in Biosciences, 4*(1), 23–24.

Wang, S.-p., Yuan, M. A., Wang, S.-h., & Chen, F. (2009). Analysis of chemical composition of volatile oil of *Phyllanthus emblica* L. from Sichuan by GC-MS [J]. *West China Journal of Pharmaceutical Sciences, 3*, 1–27.

Yokozawa, T., Kim, H. Y., Kim, H. J., Tanaka, T., Sugino, H., Okubo, T., Chu, D., & Juneja, L. R. (2007). Amla (*Emblica officinalis*Gaertn.) attenuates age-related renal dysfunction by oxidative stress. *Journal of Agricultural and Food Chemistry, 55*, 7744–7752.

Zhang, L. Z., Zhao, W. H., Gua, Y. J., Tu, G. Z., Lin, S., & Xin, L. G. (2003). Studies on chemical constituents in fruits of Tibetan medicine *Phyllanthus emblica*. *Zhongguo Zhong Yao Za Zhi, 28*, 940–943.

Chapter 49
Seje (*Oenocarpus/Jessenia bataua*) Palm Oil

Muhammad Mushtaq, Sumia Akram, and Syeda Mariam Hasany

Abstract The seje palms (*Jessenia/Oenocarpus* complex) are the least-expressed but an equally valuable family of oil-bearing fruits. The plants are native to the swamp and highland forests of Amazon basin in South America. The pulp of these palm fruits; arbitrarily referred as seje, pataua, milpesos, or ungurahuay contains lipids up to 50% (each fruit weight 10–15 g). The major fatty acid constituents are monosaturated acids like oleic and linoleic acids. The lipids recovered from mesocarp of seje fruit were found to be a rich source of sterols, tocochromanols, alcohols, carotenoids, and phenolic compounds. The oil produced from seje fruits was found to be curative for bronchitis, tuberculosis, skin allergies, and hair fall. The researchers believe that the seje palm if cultivated on the commercial scale might be a more economical and viable alternative of olive oil.

Keywords Pataua · Milpesos · Ungurahuay · Fatty acids · Tocochromanols · Alcohols · Carotenoids

1 Introduction

The *Arecaceae* family (commonly known as palms) contains the most extensively cultivated perennial or evergreen plants including climbers, trees, shrubs, and stem fewer plants. This family after grass held high esteem for botanists of intensively cultivated parts of the world. Among the oil-bearing genera of this family, *Oenocarpus* and *Jessenia* can easily adapt plantation cultivation and produce high-quality monounsaturated oils. The genus *Jessenia*; named to honor Dr. Carl Jessen,

M. Mushtaq (✉)
Department of Chemistry, Government College University, Lahore, Pakistan

S. Akram
Department of Chemistry, Minhaj University, Lahore, Pakistan

Department of Chemistry, Kinnaird College for Women University, Lahore, Pakistan

S. M. Hasany
Department of Chemistry, Kinnaird College for Women University, Lahore, Pakistan

© Springer Nature Switzerland AG 2019
M. F. Ramadan (ed.), *Fruit Oils: Chemistry and Functionality*,
https://doi.org/10.1007/978-3-030-12473-1_49

professor of botany at Eldena, Prussia, was recognized by Balick (1980) and consist of single species *Jessenia bataua*. Its members occupy slopes, basins, and above flood level valleys of Amazon, Orinoco, Gulf of Paria in Venezuela, and Andean regions of Colombia. The plants are self-grown or cultivated in areas of high rainfall and well-drained soils usually of low cation exchange capacity. The plants of genus *Jessenia* can reach a height of 15–25 m (Fig. 49.1) and bear smooth trunked, feathery-leaved palms with ringed trunk and a prominent crown shaft below the crown (Collazos and Mejía 1988). The plants grow best in the areas where daytime temperature ranges 20–25 °C but can tolerate a wider range (15–30 °C) and survive a temperature down to about 5 °C. Similarly, these species prefer to grow in areas with an annual rainfall equal to 2000–4000 mm but can tolerate a rainfall from 1500 to 6300 mm.

The word Seje palm has been arbitrarily used for the palm species of two genera *Oenocarpus* and *Jessenia*. The palms are also known with regional common names like Pataua palm, Millepesos palm, Chapil, Mingucha, Chapil, Milpesos, Cola-boca, Cola-pa-chi, Cosa, Kula'po-tci, Shimpi, Shigua, Aricacua, and ungurahuai. Palms of these two genera contribute a major role in the economy and lifestyle of rural areas of South American regions. Taxonomically, these species are poorly

Fig. 49.1 Seje palm plant parts (leaf, stem, blossom, seeds, mesocarp, and oil)

identified or under-represented. Few researchers believe that *Jessenia bataua* is the former name of *Oenocarpus bataua*. Similar confusion exists between their vernacular and botanical information and there is prompt need of field work to understand their biology and taxonomy (Oliveira et al. 1991).

The genera *Oenocarpus* maintained its distinct identity until Burret (1929) found that *Oenocarpus bataua* from Brazil was similar to *Jessenia polycarpa* from Colombia except for flower stamens number. Therefore, Burret redefined *Jessenia bataua* (Mart.) Burrett. These plants are found in a prodigious quantity and some authors had used the term *Jessenia/Oenocarpus* complex for the species of these two genera and same adopted throughout this chapter.

According to Prance and Kallunki (1984), the seje plant bears 1–3 (on average 2.5) panicles (Fig. 49.1). The leaves born in an upward direction and bend downward at the tips, while the petioles form the broad clasping bases. The plant may flower at 3–4 years (domestic palm), and the wild palm blooms at 7–10 years. The blossom is usually 1–2 m long with almost 300 rachillas reaching a length up to 1.3 m. The seje fruits are about 3 cm long and 2 cm wide, purple colored, spherical or ovoid shells which turn further dark when ripen, and encapsulate a 1–5 mm thick juicy mesocarp. The plant reaches maximum production at the age of 40 years, which drops sharply during next years. The fruits of *Jessenia/Oenocarpus bataua* have been used as food by the aboriginal Indians living in the forests. The analysis disclosed that this fruit bears good quality of protein (8–10%) with a bounty of essential amino acids, energy, and oleaginous compounds. Apart from its use in the oil industry, it can also be used to produce high energy milk-like beverages (Bodmer 1991; Collazos and Mejía 1988). It is believed that fruits, irrespective of their plant family, contain palmitic and oleic acids or their esters as key oleaginous compounds followed by linoleic acid. The seje fruit pericarp contains 20–25% oil (a bunch of uncultivated fruit can produce 3–3.5 kg oil). The oil is often produced by local communities and used as an indigenous food source. The seje oil bears a good quantity of unsaturated fatty acids and has been announced virtually similar to olive oil. The seje oil recieved high demand in the local markets of the Amazon region, and current research indicated that the oil bears good quality and long shelf life. The plant seems more adaptable in poor soils and produces a good quantity of fruits. The facts endorse that this oleaginous source should be further explored for industrial-scale production of oil for food, cosmetic, and pharmaceutical benefits. The present chapter will highlight the extraction, composition, and functionality of seje palm (*Jessenia/Oenocarpus bataua*) fruit oil.

2   Extraction and Processing of Seje Fruit Oil

The seje fruit oil has been used as cooking oil by local communities in South America like Brazil, Venezuela, Colombia, Guyana, Brazil, Peru, French Guiana, and Suriname. Recent diversion toward the non-timber forest products (NTFPs) disclosed that seje palm as many other wild plants can offer subsistence and potential value for indigenous peoples. The seje palm fruit, besides eaten singly offers good quality oil rich in monounsaturated fatty acids. A sound perception indicates that olive oil blended with seje palm oil (being similar in colour) has been marketed in Brazil for many years. The historical data further indicates that folk healers often advise seje palm oil as curative medicine for pulmonary problems and hair loss (Miller 2002).

3 Seje Fruit as an Oleaginous Source

The *Jessenia's* palms are not well known like palms of genus *Elaeis*, but their lipids are equally valuable for industrial and health point of view. These palms have been used by local communities of amazon regions for oils, fats, fiber (ropes and mates), fuel, food (solid endosperm is often directly swallowed), beverages, shelter (trunks are used in construction), and cattle field. The seje plant blossom twice in a year and produces about 18.0 kg bunches of nutmeg-sized violet to purple fruit (Fig. 49.1). The fruit pericarp, on the other hand, weighs about 70% of fresh mature fruit and contains 20–24% of oleaginous materials. According to this assumption, 3.0–3.5 kg of oil can be obtained per bunch of seje fruit (Kahn and De Granville 2012). An earlier study by Joyner (1992) assumed the average weight of fruit panicle 10.4 kg and each mature fruit carries about 8.1% oil. The seje fruit could yield approximately 400 kg of oil per hector (an average of 216 trees) which would cost four times cheaper than olive oil. Another report indicated the presence of about 50% oil in fruit mesocarp (Balick and Gershoff 1981).

4 Seed Harvest and Sterilization

The nutritional profile, oil percentage, and moisture level of seje palm fruit change rapidly with maturity and ripening stage; premature fruit contains a very small amount of lipids, which may rise up to 50% of mesocarp under certain circumstance. Likewise, over-ripened fruits may contain a high ratio of free fatty acids and the resultant products become more susceptible to auto-oxidation, rancidity, enzymatic, or microbial degradation. Moreover, the free fatty acid content in damaged or partially injured fruits increases drastically and may initiate the reactions that further increase the level of oxidation prone nutrients. Therefore, care should be taking while harvesting the seje palm fruit for direct consumption or industrial processing. In the case of seje palm, harvesting means cutting of the bunch from the trees, which if fall freely to the ground under gravity would bruise many fruits. Fruits may also be damaged during trimming, transportation, and storage. In any of these cases, care needs to be practiced to check fruit bruising before they enter processing site. The next useful precaution involves the processing of fruit palm within 48 h after harvest, but irony exists which says a few days storage before processing is good. It is known that an increased level of free fatty acids will add 'bite' to the flavor and good laxative effects. However, the free fatty acids are not a quality concern for those who consume the crude oil directly.

The seje palm carries extra-ordinary larger and heavy fruit bunches (1000 or more fruits) than African palm trees. In the case of perennial fruits like seje palm, the concept of harvestable yield is crucial because, under most of the situations, manual harvest mechanism is adopted. According to the above-mentioned philosophy, harvesters need to consider four critical factors *i.e.* ease of harvest, maturity stage of fruit, optimum ripening, and harvest at a minimum loss of fruit. For a long time, the local farmers have been climbing to palm tree *via* certain ropes, belts, or bamboo ladders to cut bunches of fruit which may miss a significant part of the fruit and pose a serious life security threat to the farmer. A useful folk practice involves the harvesting of semi ripens fruit, which is piled under leaves or plastic sheets until they ripe. The modern harvesters use particular types of knives known as sickle, chisel or machetes coupled with an adjustable long arm for manual bunch picking. The use of a particular knife largely depends upon the age of the plant. In modern harvester, similar kind of blades known as mechanically adjustable harvester (MAH) is operated with gasoline based engines and accompanied with particular lifters.

The mature ripen *Oenocarpus/Jessenia* are 10–15 g purplish black elliptic fruits. The literature reports indicated that fruits harvested during December carry higher weight and pulp. In addition to agro-climate conditions, the maturity and ripening stage of fruit would ultimately affect the quantity and quality of lipid compounds separated. After harvest, the fruit or bunches are normally emptied into wooden boxes and transported to processing sites. The palm fruit bunches normally consists of a large number of fruits each may be of different maturity and ripening stage. For manual screening, the fruit-laden spikelets are separated from the bunch stem with an axe, machete or by hand. The manual separation also offers a livelihood to children, women and the elderly living in the vicinity of processing units. The quantity and quality of fruits arriving processing units can be checked manually or by using various sophisticated machines. Although these practices have nothing to do with the initial quality of fruits which had already been decided by plant age, genetics, harvest time, harvest technique, environment or agro-climate conditions, the screening can prevent further deterioration of fruit quality. Moreover, quality standard exercised at receiving unit can evaluate the effectiveness of post-harvesting practices. At large scale production units, a mechanized system consisting of some rotating drums or fixed drum equipped with rotary beating bars detach the fruits from the bunch.

Once threshed, the fruits undergo high-temperature wet heat treatment meant to destroy microorganism, oil-splitting enzymes and control hydrolysis, and autoxidation. This can be accomplished through either sterilization where higher pressure steam is used or cooking in hot water. The cooking often does two jobs, first, it can weaken the fruit stem and expedite its removal from the bunch simply by shaking or tumbling. Second, thermal treatment at this stage coagulates protein and allows the oil-bearing cell flow more easily during subsequent pressing. Cooking also breaks or dissolves the starch, gums, and resins to ease the release of lipids. However, care should be taken regarding sterilization conditions *i.e.* vacuum and temperature; poor vacuum conditions and high temperature can induce oil oxidation. Similarly, boiling at high temperature would deteriorate the nutritional profile of oil cake.

4.1 Digestion

Washing and sterilization at the early stage of process improve the process efficiency and quality of final products. In the industry, oil palm processing units washing plus sterilization is usually accomplished by steam whereas during indigenous traditional processing threshed fruits are cooked in water. These practices result in relatively larger consumption of water that is often reduced by threshing the fruits before washing. On the other high pressure, sterilization allows the fruits to be cleaned before threshing. A major part of bunch waste is used as cooking fuel during local processing; the industrial units, on the other hand, incinerated the waste into fertilizers. The use of high-pressure steam during sterilization causes the moisture and air inside vacuoles to expand which when cooled down releases a major part of oleaginous compounds. Likewise, cooking too ruptures or breaks down the oil-bearing cells. This process is often mediated with particular type rotating shafts equipped with a number of beating arms to accomplish the recovery of oils which has already been started during sterilization. A mechanical improvement, based on the traditional wet method process, is achieved by using a vertical digester with a perforated bottom plate (to discharge the aqueous phase) and a side chute for discharging the solid phase components. Next to fruit health and ripe stage, maceration control the quantity and quality of lipids produced and investigation of non-mechanical maceration techniques like microwave, enzyme, surfactants, and salt-based maceration can offer more economic and efficient digestion (Akram et al. 2018; Mukhtar et al. 2018; Mushtaq et al. 2015).

4.2 Extraction

The separation of lipids from cleaned, sterilized, and macerated or blended fruit mesocarp has been carried out *via* what is known as "dry" and "wet" extraction. In the first case, mashed mesocarp is pressed to squeeze lipids, whereas wet extraction involves the use of various extraction solvents. The dry method; mostly adopted for the recovery of seje palm oils by local villagers or industrialists involve the pressing of digested/mashed fruit mesocarp to squeeze the oil. This can be accomplished *via* manual or some automatic pressers. These pressers are often driven manually or with the help of electric motors. There is a variety of pressers used on the pilot and industrial scales whose configuration and working vary with nature and state of oil-bearing materials. The other chapters of this monograph have comprehensive detail about working and configuration of these pressers. In any of the configuration, the pressure applied, the geometry of plunger, temperature, moisture level, and vacuum affect the yield and quality of oil expressed. In designing equipment for small-scale oil extraction, the key factors to consider are quality consumer demand, minimum free fatty acids (which renders the oil oxidation prone), keep low-level metallic contaminants (iron and other metals work as a catalyst during oxidative deterioration), and keep it adhered to widely acceptable sensory characteristics. For example, the local consumer has nothing to do with its fatty acid profile rather color and flavor would be the only quality factor.

The traditional methods involve the least use of technology (manual extraction) while industrial units follow mechanized and automatic processes. Castro et al. (2013) compared the efficiency of manual household methods with mechanized production of seje oil. During manual production, women or children gather fallen fruits or cut bunches which are threshed manually and charged into hot water (not boiling) tanks for digestion. If the fruits are not properly ripen, they are dumped under leaves until ripening. The soaked fruits are further macerated using handy mortars or crushers. The final mash is transferred to pressers called "tipiti" where oil is expressed by pressing the soft pulp by hand press "sebucan.". In mechanized extraction of seje oil, the fruits were harvested by cutting mature bunches carrying 500–4000 fruits with a specific type of knives. The fruits were further screened for optimum ripeness, a mature seje fruit measures 2.5–3.5 cm length and 2.0–2.5 cm diameter (oblong or ellipsoid) and bears dark purple coloration. Fruits were air cleaned, or disinfected (chlorinated) and pasteurized in a cooker pressure at 130 °C and 25 Psi. The fruit pulp pericarp was now smashed and resultant pulp containing about 40% water was sent to a press filter. The authors claimed that manual method produced 25–40% oil, whereas mechanized extraction furnished 89.5–95.3% oil recoveries.

It is important to note that most of the analysts used wet methods (solvent extraction) of an orbital type shaking or Soxhlet type extractors with non-polar solvents like *n*-hexane, chloroform, dichloromethane, and ethers in the extraction of seje oil in laboratories (Darnet et al. 2011; Santos et al. 2015). No attempt has been undertaken for large-scale processing of these palms till the day. The first technical report on seje palm oleochemistry regarding extraction and composition attributes to Balick (1980) who studied the biology and economics of the palms (Table 49.1). The extraction of lipids from mesocarp was carried out by petroleum ether and lipids separated were subsequently characterized by Gas Chromatography (GC). Balick in his writings continued to recommend that oil of seje palm if produced on a large scale can provide a product of appeal for both developing and industrialized world (Balick 1985; Balick et al. 1996). However, a keen survey of the literature indicated no further research on *Jessenia's* palm for almost two decades.

Montúfar et al. (2010) investigated the lipid profile of wild and cultivated ecotypes of *Oenocarpus bataua* (formerly *Jessenia bataua*). The seeds harvested from Guiana and Peru were dried by silica gel and mesocarp was separated manually. The lipids were extracted using Folch extraction method, which involves the maceration of plant material with chloroform/methanol/methylene chloride using orbital type shaker. The liquid phase was separated after centrifugation and passing through

Table 49.1 Relative abundance of fatty acids (%) in seje palm fruit oil

	Fatty acid	Balick and Gershoff (1981)	Montúfar et al. (2010)	Darnet et al. (2011)	Rodrigues et al. (2010)	Santos et al. (2017)	Castro et al. (2014)	Oliveira et al. (2017)	Carrillo et al. (2018)	Navas Hernández et al. (2009)	Olive oil Carrillo et al. (2018)
14:0	Myristic	–	0.09	–	0.1	–	0.1	0.12	–	0.33	–
15:0	Pentadecanoic	–	0.27	0.3	0.3	–	0.1	0.36	–	–	–
16:0	Palmitic	11–15	15–22	13–14	13.3	28.43	14.6	13.7	9.90	13.2	13.8
16:1-n7	Palmitoleic	0.6–0.8	0.5–1.5	0.5–0.8	0.7	–	0.3	0.59	0.0	0.7	1.4
18:0	Stearic	2.5–4.5	1.0–2.5	3.8–4.5	4.1	1.8	3.3	2.12	3.0	3.2	2.8
18:1-n9	Oleic	74–80	68–75	75–76	76.7	62	79.8	78.3	82.0	78.9	71.6
18:1-n7	cis-vaccenic	–	2.0–3.5	–	–	–	–	–	–	–	–
18:2-n6	Linoleic	1.7–3.5	1.5–2.5	1.5–2.8	3.9	8.0	1.8	4.76	1.6	2.9	9.0
18:3-n3	α-linolenic	0.5–1.0	0.5–1.0	Tr	0.1	–	0.3	–	1.8	0.7	1.0
20:0	Arachidic	–	Tr	0.5–0.8	0.6	–	0.1	–	–	–	–
20:1	Eicosenoic	–	Tr	–	–	–	0.1	–	–	–	–

filter paper. The outcomes indicated that dichloromethane can replace the chloroform, thus avoiding the major and environmental problems. The total lipid recovered (51.6% of dry weight (DW) were significantly higher than other oil-bearing seeds like canola (40–45%), sunflower (35–45%), olive oil (18–35%). There is sound evidence to support the use of steam, ultrasound power, hydrolyzing enzymes, salts, and surfactant to enhance the maceration and subsequent recovery of lipids (Akram et al. 2018; Mukhtar et al. 2018; Teixeira et al. 2013).

4.3 Oil Purification

The manual methods use hot water or steam for the soaking or maceration of fruit mesocarp, which often produces oil with higher moisture and entrained impurities. In addition, the oleaginous fluid may contain fibrous material, cell debris, and suspended solids. First of all, the suspended solids separated from the oil by boiling the press released oil or adding hot water (3:1, v/v) to dissolve suspended solid material. The un-dissolvable solids settle down to the bottom cylinder while lighter oil droplets rise to the top. The oil produced *via* mechanized methods, on the hand, is only boiled to remove the moisture and passed through decanting funnel or filtration screening operating under vacuum. Herbs are often added at this stage to impart characteristics flavor or odor to final oil product. Once cooled, the lighter oil layer formed is at the top of the tank which can be easily skimmed off using a calabash or shallow boils. The clarified oils still contain a reception tank. This clarified oil still contains traces of water and dirt which are poisonous towards oil quality. To check the oil autoxidation (peroxides formation) and auto-hydrolysis (free fatty acids formation), the moisture level best to be zero or at reduced to least 0.15–0.25% only. The residual moisture is removed by re-heating the clarified oil and carefully skimming off the dried oil. In some instances, the traces of free fatty acids and tocopherols, which offer a laxative effect and acidity to the final product, are often desirable.

5 Fatty Acids Composition

Balick and Gershoff (1981) were the first to report the fatty acid profile of seje fruit oil. The oil was extracted using petroleum ether using Soxhlet type extraction assembly. The authors observed that oil mainly consisting of oleic acid, followed palmitic and stearic acids (Table 49.1). The first comprehensive data on the composition of seje palm (*Jessenia bataua*) oil collected from Amazonas State of Venezuela was published by Navas Hernández et al. (2009), who observed that virgin oil contains exceptionally high concentration of oleic acid (78.9% of total fatty acids) with ultimately low amount of palmitic acid (13.2% of total fatty acids). A similar kind of observation was made by Montúfar et al. (2010) who observed that dried mesocarp of seje fruit (*Oenocarpus bataua* Mart) comprises about 50% of lipid including 72.7, 18.1, 2.3, 1.9, 1.7, 0.9 and 0.8% of oleic, palmitic, *cis*-vaccenic, linoleic, stearic, palmitoleic, and α-linolenic, respectively. The pulp of *Oenocarpus bataua* was also investigated by Darnet et al. (2011) for its total fat and fatty acid profile. It was observed that pulp of *Oenocarpus bataua* contained 29.1% oil on a dry mass basis with fatty acid profile similar to olive oil.

Rodrigues et al. (2010) while analyzing amazonian fruits observed that patawa (*Oenocarpus bataua*) carry about 40% oleaginous compounds when calculated on fresh weight bases.

The fatty acid profile of the oil extracted using chloroform and methanol was similar to those observed by other researchers except the oil contained traces of margaric acid (C17:0). Castro et al. (2014) analyzed the quality characteristic and fatty acid profile of seje fruit oil and observed that oleic acid constitute about 80% of seje oil fatty acids. Recently, Santos et al. (2017) analyzed Brazilian palm fruits and fatty acid profile of their mesocarp lipids. In this context, mesocarp of *Oenocarpus bataua* known under vernacular name Bacaba, when extracted with *n*-hexane using.

Soxhlet type extractor produced the lipids containing 28.4, 1.8, 62.0, and 8.0% palmitic, stearic, oleic, and linoleic acids, respectively. According to Oliveira et al. (2017), *Oenocarpus bataua* fruit lipids contain weigh 80% of monoumsaturated fatty acids followed by saturated (16.5%) and polyunsaturated fatty acid (4.76%).

6 Minor Bioactive Lipids

The oil produced from the mesocarp of seje (*Jessenia/ Oenocarpus*) fruit comprises about 1% of minor components including sterols, carotenoids, tocopherols, alcohols, phospholipids, glycolipids, and terpenic and paraffinic hydrocarbons. The oil of seje palm contains triglycerides of oleic acid (O), palmitic (P), and stearic acid (S). Triolein (OOO); a triglyceride of oleic acid covers about 44.2% of oil, followed by palmitodioleine (POO) which weigh up to 27.8%. The others tryglicerides detected in seje palm include 6.7% of SOO and 5.0% of OLL. The plant-based oleaginous compounds can be broadly classified into volatile and non-volatile lipids. If lipids undergo saponification intentionally or unintentionally, it would become rich in free fatty acids, salts of fatty acids, and free sterols. The presence of sterols in the diet can reduce cholesterol level in blood by 10–15% (Chen et al. 2008). According to Kritchevsky and Chen (2005), a daily intake of 3 g of sterols reduces the risk of coronary heart disease by 20% over a lifetime. Besides the fatty acid profile close to olive oil, seje palm contains the lipids that constitute indispensable part of a healthy diet. Navas Hernández et al. (2009) found that seje oil is a rich source of phytosterols with high level (mg/kg of virgin oil) of β-sitosterol (479.2), followed by Δ5-avenosterol (434.7) and stigmasterol (166.1).

Seje fruit oil was found to be rich in carotenoids. Darnet et al. (2011) observed that fruits of bacaba (*Oenocarpus bacaba*), when extracted with petroleum ether for

Table 49.2   Tocochromanols in seje palm fruit oil

Tocochromanol (mg/kg of oil)	Navas Hernández et al. (2009)	Rodrigues et al. (2010)	Darnet et al. (2011)
α-tocopherol	86.6	56.5	78.5
β + γ-tocopherol	–	7.8	10.5
δ-tocopherol	–	7.7	10.7
Total (mg/kg oil)	86.6	59.1	72

6 h offered more than 13 mg carotenoids/kg of oil. The main carotenoids compounds contributing this weight were lutein, *cis*-lutein, β-carotene, and α-carotene (Table 49.2). It was interesting to note that carotenoids contents of seje palm oil, especially β-carotene are higher than that of many vegetable oils like soybean, sunflower, or peanuts (0.1–0.3 mg/kg), while lower than olive oil (6.9 mg/kg) and other oil palm species like *Elaeis oleifera* and *Elaeis guineensis* which contain carotenoids around 800–1500 mg/kg. The presence of carotenoids in seje palm oil makes the oil more resistance towards oxidative deterioration.

Seje palm lipids also contain tocochromanols; another class of lipid-soluble bioactives bearing polar moiety originated from tyrosine and a polyprenyl side chain (hydrophobic) of the isoprenoid pathway. The structural diversity in tocochromanols arises mainly side chain moieties *i.e.*, phytyl-derived side chain (tocopherols), geranyl derived side chain (tocotrienols) and degree of methylation of the polar moiety (α, β, γ, and δ). These compounds offer a wide range of antioxidants activities to oils during storage. Moreover, their presence in the human diet offers a diversity of health benefits (Eitenmiller and Lee 2004). A great deal of research has been diverted to tocotrienols for their preventive effect against various cancers and many deadly/chronic diseases (Aggarwal et al. 2010). Although, plant-derived lipids contain smaller amounts of these biomolecules, their consistency prevalence in oils and fats can easily compensate their deficiency. Seje fruit pulp contains 70–80 g of total tocopherols/kg of oil, whose major part consists of α-tocopherol (Table 49.2). Aliphatic alcohols also contribute a significant fraction of minor bioactive with an average level of about 27.8 mg/kg of oil. The aliphatic alcohols detected in seje fruit oil include octa-, deco-, and hexa-cosanols at 9.8, 5.6, and 3.6 mg/kg of oil.

Navas Hernández et al. (2009) analyzed virgin seje fruit oil collected from different communities of Amazonas State, Venezuela, and observed that phenolic compounds assessed in terms of tyrosol equivalents (µg/kg of oil) vary with cultivation regions. The variation in phenolic acids and flavonoids in the analyzed samples indicates that the level of phenolic bioactives depends on the agroclimate conditions, post-harvest processing, and oil extraction methods. Overall, the presence of phenolic acids likes cinnamic acid, caffeic acid, and ferulic acid and flavonoids like quercetin, vanillin, and naringenin (Table 49.3) indicated that oil might offer elevated resistance to oxidation during storage in the hot and humid conditions.

Table 49.3 Aliphatic alcohols, carotenoids, and phenolics in seje fruit oil

Aliphatic alcohol	Navas Hernández et al. (2009)		Phenolic compounds (tyrosol equivalents (µg/kg oil) Navas Hernández et al. (2009)	
Octacosanol	9.8		Vanillin	348.5
Decosanol	5.6		Naringenin	157.5
Hexacosanol	3.6		Ferulic acid	44.6
Total (mg/kg)	28.7		Quercetin	940.0
Sterol	Navas Hernández et al. (2009)	Montúfar et al. (2010)	Cinamic acid	694.9
Cholesterol	3.0	–	Caffeic acid	39.5
24-metilencholesterol	63.0	–	Methyl-luteolin	76.0
Campesterol	89.1	7.2	Tirosyl acetate	250
Stigmasterol	166.1	19.2		
Clerosterol	9.1	–	**Carotenoids**	Santos et al. (2015)
β-sitosterol	479.2	34.2	β-carotene	3.02
Δ^5-avenosterol	434.7	27.8	Lutein	6.20
Δ^5-24-stigmastodienol	9.0	–	α-carotene	1.05
Total (mg/kg oil)	13.53	368	Total (mg/kg oil)	13.53

7   The Composition of Fruit Essential Oil

In respect to flavor, fragrance, and many other sensory characteristics, the aroma or volatile compounds keep vital importance particularly for virgin oils or oils produced for dietary purposes. Unfortunately, seje palm fruits or seeds have not been evaluated for aroma compounds. The only study undertaken by Navas Hernández et al. (2009) provided a record of volatile compounds along with minor bioactives of virgin seje oil produced by Piaroas communities in the Venezuelan state of Amazonas. In total 16 oil samples, all produced by conventional (non-mechanized) extraction methods were enrich *via* solid phase microextraction, separated in GC column. The eluted volatiles was characterized by their retention times and mass spectrum profile. The analysis revealed the presence of straight chain 6–11 carbon aldehydes (most abundant octanal at 0.78 mg/kg), terpenes and terpenoid compounds (major α- and β-pinenes at 0.36 and 0.47 mg/kg, respectively), tricyclic sesquiterpene (major longifolene 0.29 mg/kg) and camphene (0.08 mg/k). The presence of the volatile compounds like camphene and longifolene may render its oil antimicrobial capacities.

8 The Contribution of Bioactive Compounds in Curing Diseases

The current status of research regarding nutritional profile and oleochemistry broadcast seje palm fruit an exceptional but under-expressed source of valuable dietary and therapeutic bioactives. There is no particular scientific evidence of many biological attributes linked with the intake of seje palm fruits, its lipids or other dietary products. Many of these outcomes have been linked with traditional knowledge while others have been ascribed due to the prevalence of particular class of bioactives. For example, a decoction of seje fruit mesocarp has been used as a folk remedy everywhere in Colombia to treat pulmonary infections, tuberculosis, and bronchitis. The oil of seje fruit has been advised by the folk healers to the peoples suffering tuberculosis and leprosy.

The seje fruit mesocarp has been pronounced to be a rich source of monounsaturated fatty acids whose regular intake often reduces low-density lipoprotein cholesterol and increase high density cholesterol. Moreover, diets rich in oleic acid may reduce the risk of breast cancer and hypertension. Similarly, sterols stabilize phospholipid bilayer of the cell membrane and many other cell structures and their presence in diets compete for the absorption of cholesterol in the intestinal mucosa and can reduce cholesterol level in blood by 10–15% (Chen et al. 2008). The sterols have been used as a cholesterol lowering agents for many years. Kritchevsky and Chen (2005) have observed that daily intake of 3 g of sterols or their reduced form stanols can reduce low density lipoprotein cholesterol up to 10% and reduces the risk of coronary heart disease by 20% over a lifetime. The presence of sterols (13.53 mg/kg of oil) may render this lipid source beneficial against cardiovascular diseases.

The tocochromanols including tocotrienols and tocopheols work as essential dietary nutrients for characteristics health benefits towards longevity, vision, and cardiovascular heath (Eitenmiller and Lee 2004). Moreover, tocotrienols have been found to be more potent against various cancers and many deadly/chronic diseases of the twenty-first century. The seje fruit oil often comprises 50–85 mg of tocopherols/kg oil and α-tocopherol constitutes a major part of these tocopherols. The presence of α-tocopherol at an elevated level makes seje fruit oil a potential candidate for pharmaceutical and cosmetic industries. Likewise, the presence of terpenoids and carotenoids not only improves the storability of oil products but also makes its products good antioxidants. The presence of volatiles bioactives is believed to impart antimicrobial characteristics to the lipids and other dietary products of seje fruit. The antioxidant character of seje fruit oil and its drinks is often believed to be owing to the presence of phenolic acids like caffeic and ferulic acids and flavonoids including quercetin and ambergin. These flavonoids were found capable to neutralize the procarcinogens and protect DNA mutations which may lead to tumor formation.

9   Health-Promoting Traits of Fruit Oil and Oil Constituents

The data collected regarding the seje fruit oleochemistry and nutritional facts is reasonable to declare the plant of high nutritional value. In addition, the fat-protein-carbohydrate composition of *Jessenia bataua* milk is comparable to human milk and more attractive than that supplied by soybean milk. Seje fruit proteins are biologically more attractive, caloric, and safe as compared to cereal grains like rice and corn (Darnet et al. 2011). Epidemiological trials in man and animal indicated that regular intake of essential fatty acids not only strengthen retina, brain, and sperm but also provide energy and form eicosanoids, which in turn play a vital role for endocrine, cardiovascular, pulmonary, and immune systems (Carrillo et al. 2018).

There is paradox about the fatty acid profile and health benefits of seje oil. The nutritionists who believe that health advantages of fats and oils are due to a high ratio of polyunsaturated fatty acids are not satisfied with the composition of seje fruit oil. However, the fatty acid profile vary widely among various ecotypes and agroclimate zone and it might be possible that other closely related species of the seje palm will provide an oil with higher levels of polyunsaturated fatty acids. For example, *Oenocarpus mapora* contained relatively higher amounts of linoleic acid with similar energy, carbohydrates, and proteins profile. Overall, seje plant fall under the example of outcomes of researchers, botanists, agronomists, and nutritionists set out to evaluate little-known foods to combat world hunger.

10   Edible Applications of Fruit Oil

The seje fruit has been used for dietary purposes in a number of ways in different parts of the Amazon region. The fruit bears black purplish color and sweet chocolate flavored pulp (mesocarp) encapsulated under a tough protective epicarp. This pulp is very popular among local habitants and fermented for the preparation of juices, milk, beverage and wines. The fruits often soaked or boiled in water to soften epicarp and mesocarp, which are eaten directly or fried into a delicious food. The blend of pulp with water or alcohol (agua de seche) is considered to be a rich source of protein and lipids. Another blend of fruit pulp with manioc meal produces milk like nutritious drink known as "Seje", "nice milk" or "chichi". In local markets of Belém, chocolate flavored milk of seje fruit mesocarp "wine of patauá" is popular than patauá oil. Due high oleic acid content, seje fruit oil has been used by local habitants as an alternative of olive oil for cooking applications. Finally, the authors feel that seje fruit oil and milk can be an economic, viable, and more sustainable alternative of olive oil.

References

Aggarwal, B. B., Sundaram, C., Prasad, S., & Kannappan, R. (2010). Tocotrienols, the vitamin E of the 21st century: Its potential against cancer and other chronic diseases. *Biochemical Pharmacology, 80*, 1613–1631.

Akram, S., Sultana, B., Asi, M. R., & Mushtaq, M. (2018). Salting-out-assisted liquid-liquid extraction and reverse-phase high-performance liquid chromatographic monitoring of thiacloprid in fruits and vegetables. *Separation Science and Technology, 53*, 1563–1571.

Balick, M. J. (1980). *The biology and economics of the Oenocarpus-Jessenia (Palmae) complex.* Cambridge, MA: Harvard University. 404p.-illus.. En Icones, Maps, Anatomy and morphology, Palynology, Chemotaxonomy, Keys. Thesis: Harvard University: PhD Geog. 4.

Balick, M. J. (1985). Useful plants of Amazonia: A resource of global importance. In G. T. Prance & T. Lovejoy (Eds.), *Amazonia, Key Environments Series.* Pergamon Press, Ltd., pp. 339–368.

Balick, M. J., & Gershoff, S. N. (1981). Nutritional evaluation of the *Jessenia bataua* palm: Source of high quality protein and oil from Tropical America. *Economic Botany, 35*, 261–271.

Balick, M. J., Elisabetsky, E., & Laird, S. A. (1996). *Medicinal resources of the tropical forest: Biodiversity and its importance to human health.* New York: Columbia University Press.

Bodmer, R. E. (1991). Strategies of seed dispersal and seed predation in Amazonian ungulates. *Biotropica*, 255–261.

Burret, M. (1929). Zur Gattung Jessenia Karst. Notizblatt des Botanischen Gartens und Museums zu Berlin-Dahlem. 839–840.

Carrillo, W., Carpio, C., Morales, D., Alvarez, M., Silva, M.. (2018). Fatty *acids content in ungurahua oil (Oenocarpus bataua) from ecuador. Findings on adulteration of ungurahua oil in Ecuador.*

Castro, J., Hernández, M., Gutiérrez, R. (2013). Uses for Amazonian Seje oil (*Oenocarpus bataua*) extracted by mechanical and manual methods. In *III International Conference on Postharvest and Quality Management of Horticultural Products of Interest for Tropical Regions 1047*, (pp. 335–339).

Castro, J. W., Hernández, M. S., & Gutiérrez, R. H. (2014). *Uses for amazonian seje oil (Oenocarpus bataua) extracted by mechanical and manual methods* (pp. 335–339). Leuven: International Society for Horticultural Science (ISHS).

Chen, Z.-Y., Jiao, R., & Ma, K. Y. (2008). Cholesterol-lowering nutraceuticals and functional foods. *Journal of Agricultural and Food Chemistry, 56*, 8761–8773.

Collazos, M. E., & Mejía, M. (1988). Fenología y poscosecha de mil pesos Jessenia bataua (Mart) Burret. *Acta Agronómica, 38*, 53–63.

Darnet, S. H., Silva, L. H. M. D., Rodrigues, A. M. D. C., & Lins, R. T. (2011). Nutritional composition, fatty acid and tocopherol contents of buriti (*Mauritia flexuosa*) and patawa (*Oenocarpus bataua*) fruit pulp from the Amazon region. *Food Science and Technology, 31*, 488–491.

Eitenmiller, R. R., & Lee, J. (2004). *Vitamin E: Food chemistry, composition, and analysis.* Boca Raton: CRC Press.

Joyner, M. (1992). Jessenia bataua: A unique oil-palm with potential for commercial cultivation in the Caribbean. In *28th Annual Meeting, August 9–15, 1992.* Santo Domingo: Caribbean Food Crops Society.

Kahn, F., & De Granville, J.-J. (2012). *Palms in forest ecosystems of Amazonia.* Berlin: Springer Science & Business Media.

Kritchevsky, D., & Chen, S. C. (2005). Phytosterols-health benefits and potential concerns: A review. *Nutrition Research, 25*, 413–428.

Miller, C. (2002). Fruit production of the Ungurahua palm (*Oenocarpus bataua* subsp. bataua, Arecaceae) in an indigenous managed reserve. *Economic Botany, 56*, 165–176.

Montúfar, R., Laffargue, A., Pintaud, J. C., Hamon, S., Avallone, S., & Dussert, S. (2010). *Oenocarpus bataua* Mart. (Arecaceae): Rediscovering a source of high oleic vegetable oil from Amazonia. *Journal of the American Oil Chemists' Society, 87*, 167–172.

Mukhtar, B., Mushtaq, M., Akram, S., & Adnan, A. (2018). Maceration mediated liquid-liquid extraction of conjugated phenolics from spent black tea leaves extraction of non-extractable phenolics. *Analytical Methods, 10*, 4310–4319.

Mushtaq, M., Sultana, B., Anwar, F., Adnan, A., & Rizvi, S. S. H. (2015). Enzyme-assisted supercritical fluid extraction of phenolic antioxidants from pomegranate peel. *The Journal of Supercritical Fluids., 104*, 122–131.

Navas Hernández, P. B., Fregapane, G. M., & Salvador, M. D. (2009). Bioactive compounds, volatiles and antioxidant activity of virgin Seje oils (*Jessenia Bataua*) from the Amazonas. *Journal of Food Lipids, 16*, 629–644.

Oliveira, M., Costa, M., de Andrade, E. (1991). Germplasm conservation of patauá and bacaba (*Oenocarpus/Jessenia* complex). Germplasm conservation of patauá and bacaba (*Oenocarpus/Jessenia* complex).

Oliveira, P. D., Rodrigues, A. M. C., Bezerra, C. V., & Silva, L. H. M. (2017). Chemical interesterification of blends with palm stearin and patawa oil. *Food Chemistry, 215*, 369–376.

Prance, G., & Kallunki, J. (1984). *Ethnobotany of palms in the Neotropics.*

Rodrigues, A. M., Darnet, S., & Silva, L. H. (2010). Fatty acid profiles and tocopherol contents of Buriti (*Mauritia flexuosa*), patawa (*Oenocarpus bataua*), tucuma (*Astrocaryum vulgare*), mari (*Poraqueiba paraensis*) and inaja (*Maximiliana maripa*) fruits. *Journal of the Brazilian Chemical Society, 21*, 2000–2004.

Santos, M., Alves, R., & Roca, M. (2015). Carotenoid composition in oils obtained from palm fruits from the Brazilian Amazon. *Grasas y Aceites, 66*, e086.

Santos, M. D. F. G. D., Alves, R. E., Brito, E. S. D., Silva, S. D. M., & Silveira, M. R. S. D. (2017). Quality characteristics of fruits and oils of palms native to the Brazilian amazon. *Revista Brasileira de Fruticultura, 39*, 1–6.

Teixeira, C. B., Macedo, G. A., Macedo, J. A., da Silva, L. H. M., & Rodrigues, A. M. D. C. (2013). Simultaneous extraction of oil and antioxidant compounds from oil palm fruit (*Elaeis guineensis*) by an aqueous enzymatic process. *Bioresource Technology, 129*, 575–581.

Index

© Springer Nature Switzerland AG 2019
M. F. Ramadan (ed.), *Fruit Oils: Chemistry and Functionality*,
https://doi.org/10.1007/978-3-030-12473-1

Printed in the United States
By Bookmasters